W9-BEZ-356

MCGRAW-HILL
ONLINE RESOURCES

IMPORTANT:

HERE IS YOUR REGISTRATION CODE TO ACCESS

YOUR PREMIUM McGRAW-HILL ONLINE RESOURCES.

For key premium online resources you need THIS CODE to gain access. Once the code is entered, you will be able to use the Web resources for the length of your course.

If your course is using **WebCT** or **Blackboard**, you'll be able to use this code to access the McGraw-Hill content within your instructor's online course.

Access is provided if you have purchased a new book. If the registration code is missing from this book, the registration screen on our Website, and within your WebCT or Blackboard course, will tell you how to obtain your new code.

Registering for McGraw-Hill Online Resources

REGISTRATION CODE

ED2Y-UG2Z-A45A-FWPI-56CD

To gain access to your McGraw-Hill web resources simply follow the steps below:

1. USE YOUR WEB BROWSER TO GO TO: **www.mhhe.com/talaro5**
2. CLICK ON **FIRST TIME USER**.
3. ENTER THE REGISTRATION CODE* PRINTED ON THE TEAR-OFF BOOKMARK ON THE RIGHT.
4. AFTER YOU HAVE ENTERED YOUR REGISTRATION CODE, CLICK **REGISTER**.
5. FOLLOW THE INSTRUCTIONS TO SET-UP YOUR PERSONAL UserID AND PASSWORD.
6. WRITE YOUR UserID AND PASSWORD DOWN FOR FUTURE REFERENCE. KEEP IT IN A SAFE PLACE.

TO GAIN ACCESS to the McGraw-Hill content in your instructor's **WebCT** or **Blackboard** course simply log in to the course with the UserID and Password provided by your instructor. Enter the registration code exactly as it appears in the box to the right when prompted by the system. You will only need to use the code the first time you click on McGraw-Hill content.

Thank you, and welcome to your McGraw-Hill online Resources!

Mc Graw Hill **Higher Education**

0-07-293189-2 T/A **TALARO: FOUNDATIONS IN MICROBIOLOGY, 5/E**

Foundations in
Microbiology

FIFTH EDITION

Basic Principles

Foundations in

Microbiology

FIFTH EDITION

Basic Principles

Kathleen Park Talaro
Pasadena City College

 **Higher Education**

Boston   Burr Ridge, IL   Dubuque, IA   Madison, WI   New York   San Francisco   St. Louis
Bangkok   Bogotá   Caracas   Kuala Lumpur   Lisbon   London   Madrid   Mexico City
Milan   Montreal   New Delhi   Santiago   Seoul   Singapore   Sydney   Taipei   Toronto

The McGraw·Hill Companies

 Higher Education

FOUNDATIONS IN MICROBIOLOGY: BASIC PRINCIPLES, FIFTH EDITION

2 3 4 5 6 7 8 9 0 QPV/QPV 0 9 8 7 6 5 4

ISBN 0–07–255304–9

Publisher: *Martin J. Lange*
Senior sponsoring editor: *Patrick E. Reidy*
Senior developmental editor: *Jean Sims Fornango*
Managing developmental editor: *Patricia Hesse*
Director of development: *Kristine Tibbetts*
Marketing manager: *Tami Petsche*
Senior project manager: *Rose Koos*
Senior production supervisor: *Laura Fuller*
Senior media project manager: *Tammy Juran*
Senior media technology producer: *John J. Theobald*
Design manager: *K. Wayne Harms*
Cover/interior designer: *Lisa Gravunder*
Cover illustration: *"Molly Lenore, Barrett Klein. Courtesy of the American Museum of Natural History, New York."*
Lead photo research coordinator: *Carrie K. Burger*
Photo research: *Karen Pugliano*
Supplement producer: *Brenda A. Ernzen*
Compositor: *The GTS Companies/Los Angeles, CA Campus*
Typeface: *10/12 Times Roman*
Printer: *Quebecor World Versailles Inc.*

Cover image The unusual cover image may give the impression of a fanciful painting, but it is actually a three-dimensional sculpture of a single cell of *Escherichia coli* (0157:H7), displaying its fine fringe of fimbriae and its dangling flagella. This model is from an exhibit at the American Museum of Natural History in Washington, D.C. that surveys the importance of microorganisms as disease agents. The pathogenic *E. coli* attaches to the intestine using its fimbriae, and from there launches a systemic infection that damages the blood cells and kidney. Visit the Talaro Online Learning Center at www.mhhe.com/talaro5 to access an excellent website that allows you to explore this image and many others like it.

The credits section for this book begins on page CR-1 and is considered an extension of the copyright page.

Library of Congress Cataloged the main title as follows:

Talaro, Kathleen P.
  Foundations in microbiology : basic principles / Kathleen Park Talaro. — 5th ed.
    p.   cm.
Includes index.
  ISBN 0–07–255298–0 (hbk. : alk. paper)
  1. Microbiology. 2. Medical microbiology. I. Talaro, Arthur. II. Title.

QR41.2.T35   2005
579—dc22   2003025041
                                              CIP

www.mhhe.com

Subjects as complex as microbiology would be much more daunting to students if not for the contributions of thousands of dedicated college professors from all over the world. These advocates tirelessly and deftly lead you through the many esoteric areas of this subject. Along the way, they bring their own personal insights, tone, and vision of what microbiology means to them and what it can mean to you. They strive with much devotion and hard work to facilitate your learning and understanding, mentor, prepare labs, develop study guides and websites, grade, tutor, and advise. In this spirit, I would like to pay a special tribute to my own first microbiology professors—Dr. Dorothy Faris and Dr. Francis Jarvis—emeritus professors at Idaho State University in Pocatello, Idaho. I was in awe of their impressive command of the subject matter, the scope of their research and contributions to the discipline, and the enthusiasm and love for the subject that just stood out. They helped shape my early science experiences and encouraged me to continue my studies. I can think of no better role models than they were.

Brief Contents

Contents

About the Author

Kathleen Park Talaro

Kathleen Park Talaro is a microbiologist, author, illustrator, photographer, and educator at Pasadena City College. Professor Talaro developed a keen interest in biology and microbiology as a child. From her earliest days on a farm in Blackfoot, Idaho, she was drawn naturally to science and to observing the living world. Her curiosity extended to plants, insects, amphibians, and farm animals, but it was the amazing idea of the tiny creatures swimming around in her father's fishpond that totally captured her attention.

Kathy began her college education at Idaho State University in Pocatello. She found a niche that fit her particular abilities and interests, spending part of her time as a scientific illustrator for one of her professors and part as a biology lab assistant. After graduation with a B.S. in biology, she entered graduate school at Arizona State University, majoring in physiological ecology. During her graduate studies she also participated in two research expeditions to British Columbia with the Scripps Institution of Oceanography. Excited by the diversity and career opportunities available in California, she moved to Pasadena and began working as a microbiologist at Pasadena City College. Kathy continued to expand her background, first finishing a Masters degree at Occidental College, and later taking additional specialized coursework in microbiology at California Institute of Technology and California State University.

If there is one continuing theme reverberating through Kathy's experiences, it is the love of education and teaching. She has been teaching allied health microbiology and major's biology courses for nearly 25 years. "It is so exciting to watch the students develop their early awareness of microorganisms—when they first come face-to-face with the reality of them on their hands, in the air, in their food, and, of course, nearly everywhere. At first, they obsess a little bit about 'germs' and hand washing, but in time, I see the light in their eyes as they begin to understand and apply what they have learned in a sophisticated way. My students come back to visit after going on to professional schools and emphasize the life-changing effect this learning and knowledge of the microscopic realm has had on them."

Professor Talaro is a member of the American Society for Microbiology and the American Association for the Advancement of Science. She keeps active in self-study and research, and continues to attend workshops and conferences to remain current in her field. Her special interests are general medical microbiology and biotechnology. She has been actively involved in science outreach programs for young people by teaching Saturday workshops in microbiology and DNA technology to high school and junior high students.

Preface

A SAMPLING OF CLIPS FROM HEADLINE STORIES AFTER 9/11/01

"The FBI and U.S. Postal Service are offering a $2.5 million reward for information that helps solve the case labeled "Amerithrax." Investigators have interviewed more than 5,000 people and spent more than $13 million on scientific testing."

"In June, the FBI drained and searched a pond in Frederick, Md., but found no evidence. Frederick is the home of the Army Medical Research Institute of Infectious Diseases, one of the nation's main anthrax research centers."

"Five people died, all from inhalation anthrax, between Oct. 5 and Nov. 21, 2001. Seventeen others were sickened by inhalation or cutaneous anthrax. At least three of those are postal workers who continue to recover and have not returned to work."

"The consequences of a successful biological attack, especially if the infection were readily communicable, could far exceed those of a chemical or even a nuclear event," the medical council said. Given the ease of travel and increasing globalization, an outbreak anywhere in the world could be a threat to all nations."

"Smallpox, anthrax, bubonic plague—a smart terrorist could use any of these deadly diseases to wreak havoc on thousands or even millions of Americans. Yet there aren't enough antibiotics and vaccines in stockpile, public-health facilities can't handle a "surge," and most law enforcement agencies have no idea how to cope with a crisis. "I do not believe it is a question of whether a lone terrorist or terrorist group will use infectious disease agents to kill unsuspecting citizens; I'm convinced it's really just a question of when and where."

Microbiology in a New Age

In the fall of 2001, at about the time the previous edition of this textbook was going to press, shattering events occurred that have swept microbiology into the spotlight with a vengeance. The possibility of using microorganisms as weapons was no longer mere speculation. One effect of these revelations has been a heightened awareness of microorganisms unlike any in past memory, especially among the students in my classes. On many occasions, my colleagues and I have had to spend a significant portion of class time in discussions relating to this topic, separating fact from fiction, and expanding on the latest potential "threat" from news reports. Many sessions were spent answering questions such as *"What is anthrax and how is it treated? How could smallpox come back after all these years? Are the vaccines for preventing infections safe? What is the chance I can get SARS, or how did West Nile Valley Fever get to North America?"* Giving voice to these concerns has been a bittersweet educational experience. Of course, it has been gratifying to observe the high level of interest in this subject and to be able to use real life circumstances to teach basic concepts—for example, how to sterilize mail to destroy anthrax spores, the kind of treatments available for biological agents, or exploring the epidemiology of the "pathogen of the week." My other, conflicting, reaction has involved the unfortunate and somewhat overblown fears about microbes these happenings have instilled in many people. More than ever before these small, ubiquitous, mostly beneficial organisms are the subjects of misunderstanding and anxiety.

At such an extraordinary time, there is a serious need for useful, accurate information about microorganisms and their activities, so the public can operate from a basis of knowledge and understanding rather than fear or rumor. Part of this education must help maintain a balanced view of the infectious nature of microbes alongside their undeniable importance in the functioning of the earth. This has been our primary goal for the past four editions, and we have undertaken this fifth revision from that same point of view. Once again, we want our students—all of you—to come away from a course of study prepared for any personal or professional challenges, to develop in your thinking from the unknown to the known, and to be the voices of reason during trying times.

As you will see, microbiology is a complex interdisciplinary subject that covers a broad scope of information, from the basic biology of cells and their function to the roles of microbes in the global ecosystem, in industrial production, and—yes—in disease. It truly does offer something for everyone, whether your purpose is to prepare for a health profession, to gain skills for technical work, to become informed about practical applications in your daily life, or to improve your knowledge of these smallest inhabitants of the biological world.

Emphases and Changes in the Fifth Edition

We have been fortunate to have the assistance of four able specialists for this edition—three microbiologists and one physician—to support what had previously been the responsibility of just the author. These contributors include:

Barry Chess, my fellow instructor at Pasadena City College
Charles Wright of Baltimore County Community College
Marjorie Kelly Cowan of Miami University of Ohio
Steve Hecht of Grand Valley State University

They have shared the daunting task of overseeing many of the miniscule details that all revisions require. I have welcomed their help in writing new sections and feature boxes, suggesting ideas for new and improved figures, editing and updating text, and refreshing chapter overviews, summaries, and questions. Although this textbook has always been a labor of love, I found this collaboration beneficial, and it has released me to concentrate on content and illustrations without having to pore over every word.

As in previous revisions, we have strived to retain basic content that is the foundation of microbiology while introducing pertinent new developments in the many fields that are part of this science. Our aims can be summarized as:

- to keep the book readable and accurate for a wide range of student levels and abilities
- to introduce topics in a logical order so that early concepts and terms can be used as a foundation for later ones
- to use illustrations, maps, flow charts, analogies, and tables to improve assimilation of information by students with different learning styles
- to make difficult concepts accessible and understandable
- to provide a background for practical applications
- to inspire lifelong learning in the subject
- to incite excitement and awe of this dynamic science

IMPROVED ART PROGRAM

A significant element in any science textbook is the visual program. We set our sights anew on fine-tuning the appearance and readability of the artwork and photographs.

- Every figure in the text has been reviewed, and at least 60% of them have been revised or completely redone.
- Consistency in presentation has been imposed so that elements are presented in the same manner from chapter to chapter.
- Whenever feasible, we have moved the legends into the figure itself and placed them close to the step being described.
- We have also simplified certain illustrations that were too crowded and improved the accuracy and flow of others.
- Special attention has been given to the number and accuracy of labels on art and photographs.
- Many hours have been spent searching for top-notch photographs, some of which were supplied by the author.
- Since many users have commented on their appreciation of illustrated tables and flow charts, we have developed several new ones.

STREAMLINING PRESENTATIONS

The breadth of this discipline makes covering it something of a juggling act. Foremost, it is necessary to sufficiently cover the great variety of topics in microbiology. It is also critical to include ample descriptive background to go with the illustrations. On the other hand, we must also acknowledge that this book is generally used for a one-semester course, and that students' time for reading is not unlimited. Given these concerns:

- We have tightened the content by editing, rewording, simplifying, and removing non-essential detail, without cutting important concepts.
- Being mindful to not minimize important concepts some boxes have been removed or re-positioned in the text flow, and some reference material has been moved to the appendix.
- All initial references to figures, tables, and boxes have been boldfaced, and important sequential ideas or concepts have been numbered or bulleted within the text.
- In recognition of the serious nature of bioterrorism, we have included new boxes and sections on this topic as well as on emerging diseases.

Other alterations to streamline the chapters include compressed chapter overviews and more abbreviated summaries. Our extensive review questions have been well received, and so we have retained the three types of questions as in previous editions. Most chapters have additional questions, including several that involve analysis of figures. As in the past, much attention has been paid to vocabulary and pronunciation, and we have added new terms to the glossary.

USING THE INTERNET

In my own Internet searches over the past two years, I have discovered hundreds of excellent websites that cover topics in a pertinent and relevant manner. A computer connected to the World Wide Web can offer dynamic experiences that a textbook cannot. It can display animations, videos, and interactive graphics, quizzes, and case studies that enhance and extend your textbook readings. Since this science advances so rapidly, events that happen tomorrow cannot possibly be found in your textbook. Also, some of these websites were located only after long searches, and I felt that if someone didn't point the readers towards them, they would remain obscure.

- Specific websites that I have found particularly useful and instructive are listed at the chapter level on the Online Learning Center at www.mhhe.com/talaro5.
- Suggestions for exercises you can use with these URL addresses are presented at the end of each chapter.
- Keep in mind that this textbook has its own dedicated website with text-specific activities and study aids that are available at www.mhhe.com/talaro5.

Teaching and Learning Supplements

McGraw-Hill offers various tools and teaching products to support the fifth edition of *Foundations in Microbiology*. Students can order supplemental study materials by contacting your local bookstore. Instructors can obtain teaching aids by calling the Customer Service

Department at 800-338-3987, visiting our Microbiology website at www.mhhe.com, or contacting your local McGraw-Hill sales representative.

Digital Content Manager   This multimedia collection of visual resources allows instructors to utilize artwork from the text in multiple formats to create customized classroom presentations, visually-based tests and quizzes, dynamic course website content, or attractive printed support materials. The digital assets on this cross-platform CD-ROM are grouped by chapter within the following easy-to-use folders.

Active Art Library   Key figures are saved in manipulable layers that can be isolated and customized to meet the needs of the lecture environment.

Animations Library   Numerous full-color animations of key microbiology processes are provided. Harness the visual impact of processes in motion by importing these files into classroom presentations or course websites.

Art Libraries   Full-color digital files of all illustrations in the book, plus the same art saved in unlabeled and gray scale versions, can be readily incorporated into lecture presentations, exams, or custom-made classroom materials. These images are also pre-inserted into blank PowerPoint slides for ease of use.

Photo Libraries   Digital files of instructionally significant photographs from the text can be reproduced for multiple classroom uses.

PowerPoint Lectures   Ready-made presentations that combine art and lecture notes have been specifically written to cover each chapter of the text. Use the PowerPoint lectures as they are, or tailor them to reflect your preferred lecture topics and sequences.

Tables Library   Every table that appears in the text is provided in electronic form. You can quickly preview images and incorporate them into PowerPoint or other presentation programs to create your own multimedia presentations. You can also remove and replace labels to suit your own preferences in terminology or level of detail.

INSTRUCTOR TESTING AND RESOURCE CD-ROM

This cross-platform CD-ROM provides a wealth of resources for the instructor. Supplements featured on this CD-ROM include a computerized test bank utilizing Brownstone Diploma testing software to quickly create customized exams. This user-friendly program allows instructors to search for questions by topic, format, or difficulty level; edit existing questions or add new ones; and scramble questions and answer keys for multiple versions of the same test.

Word files of the test bank are included for those instructors who prefer to work outside of the test generator software.

TRANSPARENCIES

McGraw-Hill provides 450 color *Overhead Transparencies* of text line art and numerous photos.

COURSE DELIVERY SYSTEMS

With help from our partners, **WebCT, Blackboard, TopClass, eCollege,** and other course management systems, professors can take complete control over their course content. These course cartridges also provide online testing and powerful student tracking features.

FOR THE STUDENT

Online Learning Center www.mhhe.com/talaro5
The OLC offers an extensive array of learning and teaching tools. The site includes quizzes for each chapter, links to websites related to each chapter, interactive activities, and case studies.

Student Study Guide by Nancy Boury, Iowa State University
The author provides study objectives, chapter overviews, test-taking strategies, crossword puzzles, multiple-choice questions, critical thinking questions, matching exercises, and pathway mapping problems to reinforce the concepts in each section of the text. Answers to the objective questions are included.

HyperClinic 2 CD-ROM by Lewis Tomalty and Gloria Delisle
Evaluate realistic case studies that include a patient history and description of signs and symptoms. Students can either analyze the results of physician-ordered clinical tests to reach a diagnosis or evaluate a case study scenario and then decide which clinical samples should be taken and which diagnostic tests should be run. More than 200 pathogens are profiled, 105 case studies presented, and 46 diagnostic tests covered. PC compatible.

Microbes in Motion 3 CD-ROM
This interactive, easy-to-use general microbiology CD-ROM helps students actively explore and understand microbial structure and function through audio, video, animations, illustrations, and text. 18 books cover topics from microbial genetics to vaccines. Mac and Windows compatible.

A Message to Students

Most of you are probably taking this course as a prerequisite to a health science program such as nursing, dental hygiene, medicine, pharmacy, or physical therapy. Because you are preparing for a career in a profession that involves interactions with patients, you will be concerned with infection control and precautions, which in turn requires you to think about microbes and how to manage them. This means you must not only be knowledgeable about the characteristics of bacteria, viruses, and other microbes, their metabolism, and primary niches in the world, but you must have a grasp of disease transmission, the infectious process, disinfection procedures, and drug treatments. You will also need to understand how the immune system interacts with microorganisms and the effects of immunization. All of these areas bring their own vocabulary and language—much of it new to you—and mastering it will require time, motivation, and preparation.

TIPS ON LEARNING TO GAIN UNDERSTANDING

The challenge of this type of technical course is, "How can I best learn this information in a way to be successful in the course as well as retain it for the future?"

To begin, probably the most important consideration is how the college and instructor have organized your course. Since there is more information than could be covered in the usual course, your instructor will select what he/she wants to emphasize and construct a reading and problem assignment that corresponds to lectures and discussion sessions. Many instructors have a detailed syllabus or study guide that directs the class to the specific content areas and vocabulary words. Others may have their own website to distribute assignments and even sample exams. Whatever materials are provided, this should be your primary guide in preparing for study.

The next consideration involves your own learning style and what works best for you. To be successful, you must commit essential concepts and terminology to memory. A quick list of how we retain information comes from Edgar Dale, who prioritizes the pyramid of learning as follows: We remember about:

10% of what we read
20% of what we hear
30% of what we see
50% of what we see and hear
70% of what we discuss with others
80% of what we experience personally
95% of what we teach to someone else

It should be obvious from this list that there are many ways to go about assimilating information, but mainly, you need to use all of your senses—to read, to write, draw diagrams, do lab work, take exams, and study with others. This means you must spend the time not just to read the book but also to write down notes as you go along. It means that you attend lecture and discussion sessions to listen to your instructors or teaching assistants explain the material. Notes taken during lecture can be rewritten or outlined to organize the main points. This begins the process of laying down memory. You should discuss concepts with others—perhaps a tutor or study group—and even take on the role of the teacher-presenter part of the time. It is with these kinds of interactions that you will not just rote memorize words but *understand* the ideas and be able to apply them later.

A substantial way to experience the material personally is to give yourself examinations. You may use the exams in the text, study guide, or make up your own. One strategy for self-quizzing is to make a set of flash cards that goes with your notes (which is in itself a great learning exercise). Then you can use the cards to assess what you know one concept or vocabulary word at a time. A beneficial side effect of this technique is that you are getting in the frame of mind for taking an exam, which is how most of your evaluations will be given.

Another big factor in learning is the frequency of studying. It is far more effective to spend an hour or so each day for two weeks than a 14-hour cramming session on one weekend. The brain cannot commit information to memory nonstop—it needs to refresh itself. If you approach the subject in small bites and remain connected with the terminology and topics, over time you will find it all fitting together. Then the day before an exam, all you have to do is review.

FEATURES OF THE TEXT HELP YOU STUDY

This textbook has several features that facilitate learning.

Chapter Overviews are quick statements about chapter content to show you the main concepts in the chapter.

Chapter Capsule with Key Terms contain key words and concepts together in outline form. They also make for a useful guide to review the material.

Chapter Checkpoints serve as a quick "snapshot" review of the main points of a section.

End of chapter questions are varied to give you experience in answering different types of exam questions. At the end of chapter 1, there is a description of how these questions are meant to be answered. Multiple choice questions can be used to quickly test yourself on what you know (answers are in the appendix). They are chosen from random information in the text, and it is likely that if you get them correct, you have adequate preparation. The same is true for the matching questions.

The **concept questions** require more in-depth knowledge and will demand that you write several sentences or even paragraphs to show your mastery of a given idea.

The **critical-thinking questions** are "brain teasers" that give you some problem solving experience after you have learned the basics of a certain concept. Many of them involve case studies, models, interpretations of graphs, comparisons, and lab exercises. They give you experience in applying microbiology to real life situations.

In the final analysis, the process of learning comes down to self-motivation. There is a big difference between having to memorize something to get by and really wanting to know and understand it. We feel strongly about the inherent value and fascination of microbiology. It is up to you to become emotionally involved in it yourself and get caught up in the wonder of learning about this new world. Therein is the key to most success and achievement, no matter what your final goals. And though it is true that mastering the subject matter in this textbook can be challenging, it can certainly be accomplished by any students with a "can do" attitude. Millions of students who have already succeeded will attest to this fact.

IN RECOGNITION AND TRIBUTE

Once again, the team from McGraw-Hill has outdone itself in providing first-rate support throughout this revision. In addition to the valuable services of four new contributors, I am delighted to have Jean Sims Fornango once again as my senior developmental editor. Jean is truly a modern "Renaissance woman," skilled in all areas of publishing, well versed in the arts and sciences, and an all-around inspiration. I can always count on Jean to place things in perspective, to offer great insights, to help make decisions that keep the textbook moving in the right direction. She is the epitome of grace under pressure. Another publishing gem is Rose Koos, reprising her role as the production editor for this project. As involved as the author and developmental editor are in keeping the book project on track, the production editor must have her finger on the pulse of literally "everything." She must tackle the overwhelming task of overseeing the finest details that go into the finished product—from a little typo to a paging problem to an illustration run amok—all the while remaining cool, calm, and collected. I would like to recognize our Senior Sponsoring Editor, Patrick Reidy, who has been a strong supporter of expanding our illustration program and trying new ways of presenting the material and figures.

Other members of the team deserving special mention include Carrie Burger, an experienced and able photo editor; Karen Pugliano, the photo researcher, who negotiated expertly for the very best photographs available. I would like to credit Wayne Harms for using his distinct touch to create an elegant book and cover design. No tribute would be complete without praising the valuable contributions of our publisher, Marty Lange, and our Marketing Manager, Tami Petsche. Many kudos to the folks out in the trenches, those sales representatives who handle the huge responsibility of presenting this textbook to college instructors. You are an integral part of this publishing team and we would be working in a vacuum without you.

I also wish to thank my colleagues in the Natural Science Division at Pasadena City College for their continued encouragement and interest. I want to make special mention of our Dean, Dr. Bruce Carter. He has been instrumental in allowing me to adjust my schedule to work on the text, and has assisted in the upgrading of the microbiology lab to make it a first class place to teach science. I also want to thank Barry Chess, Tom Belzer, and John Stantzos for their friendship and long time support for this project. Last but certainly not least in my heart and mind are my students, who come to the subject eager to learn and prepared to experience the marvelous world of microbiology.

To paraphrase Blanche Dubois, I do "count on the kindness of reviewers" to provide a critical eye and give suggestions for changes and improvements. Thus, I want to salute the new team of reviewers that helped with this edition. I have been a reviewer myself, so I know that it is not an easy job to evaluate another person's work critically yet constructively. I can always rely on them to see things that I have missed or to share their own classroom experiences on how a certain figure or treatment works in the everyday world. This collaboration has been instrumental in shaping the finished product. At this tenth anniversary of *Foundations,* I wish to express my gratefulness for all of the students and instructors who have used the book and written personal comments to me over the years. You have made the many years I have spent in developing this book worthwhile.

This textbook has been thoroughly inspected by the author, editors, reviewers, and contributors, and we have tried to make it as error-free as possible. Since no process is 100% accurate, some errors can still be missed, so we would appreciate any feedback from readers who find typos, missing labels, or omissions, or errors in content. I may be reached by email at ktalaro@aol.com or at www.mcgraw-hill.com.

Reviewers and Focus Group Members for the Fifth Edition:
Benjie Blair, *Jacksonville State University*
Nancy Boury, *Iowa State University*
Barry Chess, *Pasadena City College*
Marjorie Cowan, *Miami University—Ohio*
Kathleen Dannelly, *Indiana State University*
Daniel C. Flynn, *West Virginia University*

Kathy Forman, *Moraine Valley Community College*
Bernard Frye, *University of Texas, Arlington*
Joseph J. Gauthier, *University of Alabama, Birmingham*
Linda Harris-Young, *Motlow State Community College*
Steve Hecht, *Grand Valley State University*
Valeria Howard, *Bismark State College*
Gary E. Kaiser, *Community College of Baltimore County*
Philip Lister, *Creighton University*
Robert Mitchell, *Community College of Philadelphia*
Marcia M. Pierce, *Eastern Kentucky University*
Indiren Pillay, *Southwest Tennessee Community College*
Laraine Powers, *East Tennessee State University*
Karl J. Roberts, *Prince Georges Community College*
Mary Rutter, *Austin Community College*
Patricia Shields, *University of Maryland*
Edward Simon, *Purdue University*
Ken Slater, *Utah Valley State College*
Robert Speed, *George Wallace Community College*
Anthony Streklaukas, *Trident Technical College*
Daniece Williams, *Hinds Community College*
Charles Wright, *Community College of Baltimore County—Essex*

Reviewers of the Fourth Edition:
Kevin Anderson, *Mississippi State University*
Cheryl K. Blake, *Indian Hills Community College*
Bruce Bleakley, *South Dakota State University*
Harold Bounds, *University of Louisiana*
Brenda Breeding, *Oklahoma City Community College*
Karen Buhrer, *Tidewater Community College*
Charles Denny, *University of South Carolina*
Richard Fass, *Ohio State University*
Denise Friedman, *Hudson Valley Community College*
Bernard Frye, *University of Texas*
Louis Giacinti, *Milwaukee Area Technical College*
Ted Gsell, *University of Montana*
Herschel Hanks, *Collin County Community College*
Ann Heise, *Washtenaw Community College*
Valeria Howard, *Bismarck State College*
Harold Kessler, *Lorain County Community College*
George Lukasic, *University of Florida*
Sarah MacIntire, *Texas Women's University*
Lundy Pentz, *Mary Baldwin College*
Hugh Pross, *Queen's University*
Leland Pierson, III, *University of Arizona*
Ken Slater, *Utah Valley State College*
Edward Simon, *Purdue University*
Robert A. Smith, *University of the Sciences, Philadelphia*
Kristine M. Snow, *Fox Valley Technical College*
Cynthia V. Sommer, *University of Wisconsin, Milwaukee*
Linda Harris Young, *Motlow State Community College*
Robert White, *Dalhousie University*

Reviewers of the Second/Third Editions
Rodney P. Anderson, *Ohio Northern University*
Robert W. Bauman, Jr., *Amarillo College*
Leon Benefield, *Abraham Baldwin Agricultural College*
Lois M. Bergquist, *Los Angeles Valley College*

L. I. Best, *Palm Beach Community College—Central Campus*
Bruce Bleakley, *South Dakota State University*
Kathleen A. Bobbitt, *Wagner College*
Jackie Butler, *Grayson County College*
R. David Bynum, *SUNY at Stony Brook*
David Campbell, *St. Louis Community College—Meramec*
Joan S. Carter, *Durham Technical Community College*
Barry Chess, *Pasadena City College*
John C. Clausz, *Carroll College*
Margaret Elaine Cox, *Bossier Parish Community College*
Kimberlee K. Crum, *Mesabi Community College*
Paul A. DeLange, *Kettering College of Medical Arts*
Michael W. Dennis, *Montana State University—Billings*
William G. Dolak, *Rock Valley College*
Robert F. Drake, *State Technical Institute at Memphis*
Mark F. Frana, *Salisbury State University*
Elizabeth B. Gargus, *Jefferson State Community College*
Larry Giullou, *Armstrong State College*
Safawo Gullo, *Abraham Baldwin Agricultural College*
Christine Hagelin, *Los Medanos College*
Geraldine C. Hall, *Elmira College*
Heather L. Hall, *Charles County Community College*
Theresa Hornstein, *Lake Superior College*
Anne C. Jayne, *University of San Francisco*
Patricia Hilliard Johnson, *Palm Beach Community College*
Patricia Klopfenstien, *Edison Community College*
Jacob W. Lam, *University of Massachusetts Lowell*
James W. Lamb, *El Paso Community College*
Hubert Ling, *County College of Morris*
Andrew D. Lloyd, *Delaware State University*
Marlene McCall, *Community College of Allegheny County*
Joan H. McCune, *Idaho State University*
Gordon A. McFeters, *Montana State University*
Karen Mock, *Yavapai College*

Jacquelyn Murray, *Garden City Community College*
Robert A. Pollack, *Nassau Community College*
Judith A. Prask, *Montgomery College*
Leda Raptis, *Queen's University*
Carol Ann Rush, *La Roche College*
Andrew M. Scala, *Dutchess Community College*
Caren Shapiro, *D'Youville College*
Linda M. Sherwood, *Montana State University*
Lisa A. Shimeld, *Crafton Hills College*
Cynthia V. Sommer, *University of Wisconsin—Milwaukee*
Donald P. Stahly, *University of Iowa*
Terrence Trivett, *Pacific Union College*
Garri Tsibel, *Pasadena City College*
Leslie S. Uhazy, *Antelope Valley College*
Valerie Vander Vliet, *Lewis University*
Frank V. Veselovsky, *South Puget Sound Community College*
Katherine Whelchel, *Anoka-Ramsey Community College*
Vernon L. Wranosky, *Colby Community College*
Dorothy M. Wrigley, *Mankato State University*

Reviewers of the First Edition:
Shirley M. Bishel, *Rio Hondo College*
Dale DesLauriers, *Chaffey College*
Warren R. Erhardt, *Daytona Beach Community College*
Louis Giacinti, *Milwaukee Area Technical College*
John Lennox, *Penn State, Altoona Campus*
Glendon R. Miller, *Wichita State University*
Joel Ostroff, *Brevard Community College*
Nancy D. Rapoport, *Springfield Technical Community College*
Mary Lee Richeson, *Indiana University; Purdue University at Fort Wayne*
Donald H. Roush, *University of North Alabama*
Pat Starr, *Mt. Hood Community College*
Pamela Tabery, *Northampton Community College/*

Foundations for Success

Improved Art Program

- 60% of images have been revised
- Increased consistency in figure presentation
- Legends placed within figures
- Improved accuracy

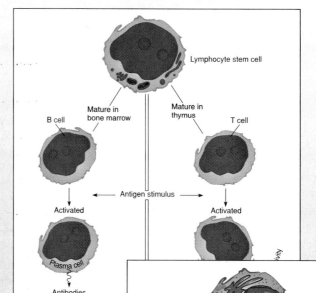

Fourth Edition

Humoral Immunity

FIGURE 14.11

Summary of the general d... lymphocytes, which are t... reactions. B cells and T... diverge into two cell lines. ... cannot differentiate them o... large nucleus—cytoplasm ...

Fifth Edition

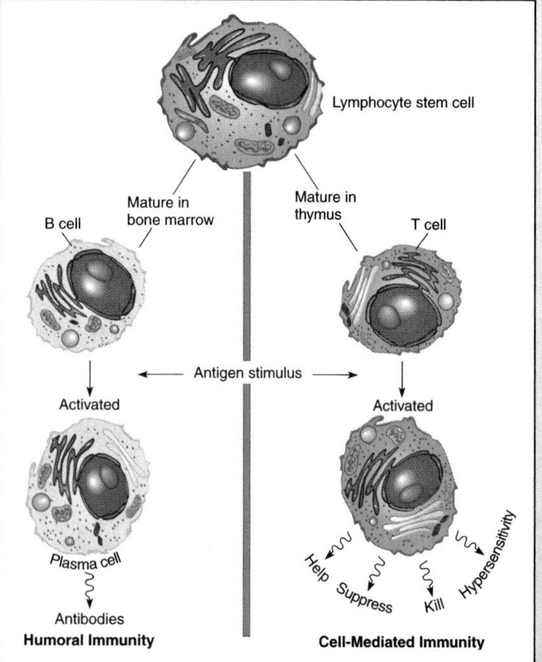

Lymphocyte stem cell

B cell — Mature in bone marrow

Mature in thymus — T cell

Antigen stimulus

Activated

Activated

Plasma cell

Antibodies

Humoral Immunity

Help Suppress Kill Hypersensitivity

Cell-Mediated Immunity

FIGURE 14.10

Summary of the general development and functions of lymphocytes, which are the cornerstone of specific immune reactions. B cells and T cells arise from the same stem cell but later diverge into two cell lines. Their appearances are similar, and one cannot differentiate them on the basis of staining. Note the relatively large nucleus—cytoplasm ratio and the lack of granules.

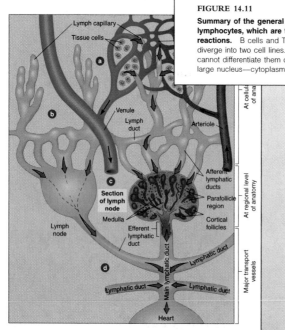

(a) The finest level of lymphatic circulation begins with blind capillaries that pick up fluid, white blood cells, and microbes or other foreign matter from the surrounding tissues and transport this liquid mixture (lymph) away from the extremities via a system of small ducts.

(b) The ducts carry lymph into a circuit of larger ducts that ultimately flow into clusters of specialized filtering organs, the lymph nodes.

(c) The center diagram shows a section through a lymph node to reveal the afferent ducts draining into sinuses that house several types of white blood cells, primarily T lymphocytes, B lymphocytes, macrophages, and dendritic cells. Here, foreign material is filtered out, processed, and becomes the focus of various immune responses.

(d) Lymph continues to trickle from the lymph nodes via efferent ducts into a system of larger drainage vessels, which ultimately connect with large veins near the heart. In this way, cells and products of immunity continually enter the regular circulation.

Lymph capillary
Tissue cells
Venule
Lymph duct
Arteriole
Afferent lymphatic ducts
Section of lymph node
Parafollicle region
Medulla
Cortical follicles
Lymph node
Efferent lymphatic duct
Main lymphatic duct
Lymphatic duct
Lymphatic duct
Lymphatic duct
Lymphatic duct
Heart

At cellular level of anatomy

At regional level of anatomy

Major transport vessels

FIGURE 14.13

Scheme of circulation in the lymphatic vessels and lymph nodes.

Art Program

- Integration of top-notch photographs
- Several new illustrated tables and flow charts have been added throughout the text

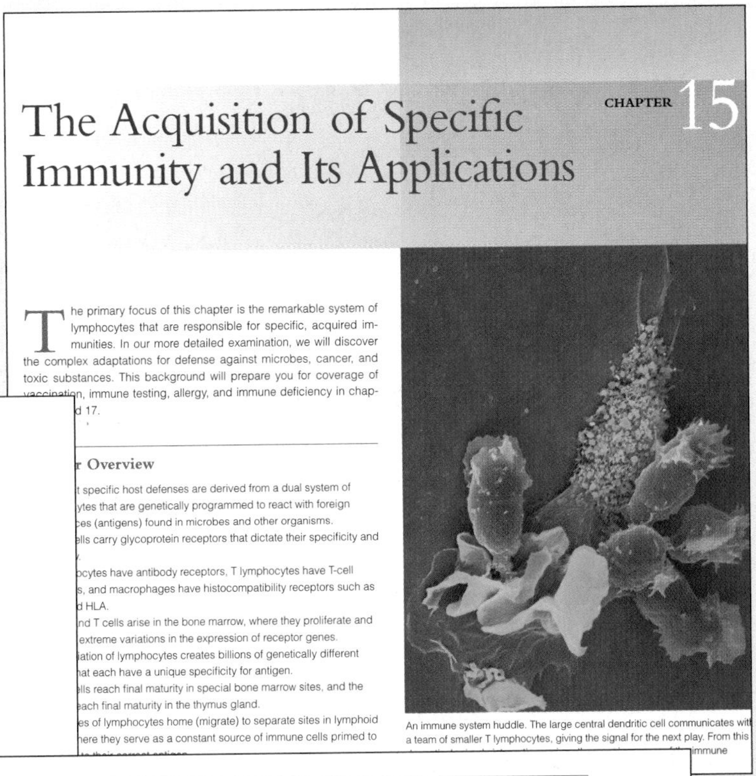

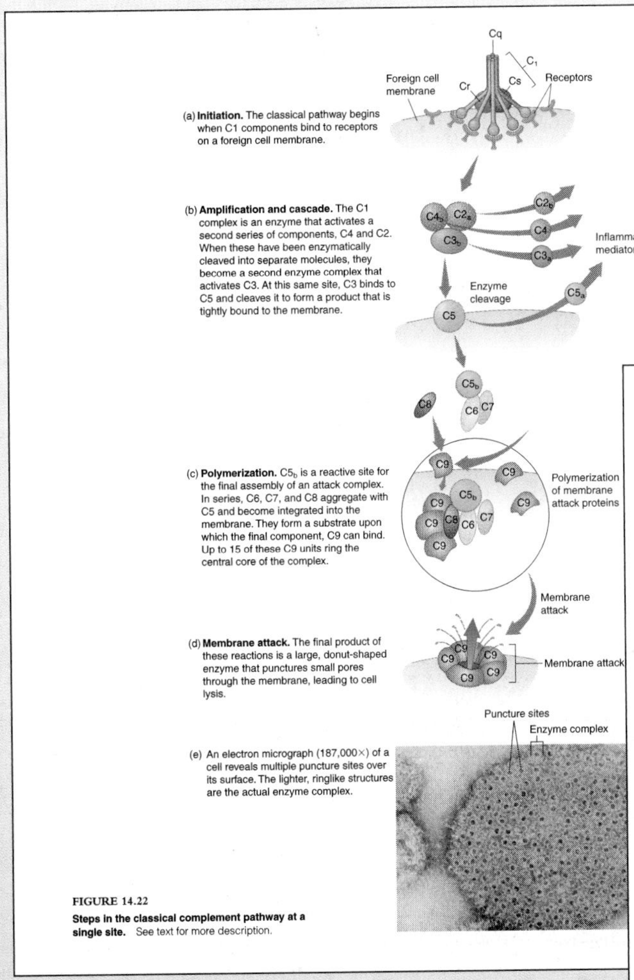

FIGURE 14.22
Steps in the classical complement pathway at a single site. See text for more description.

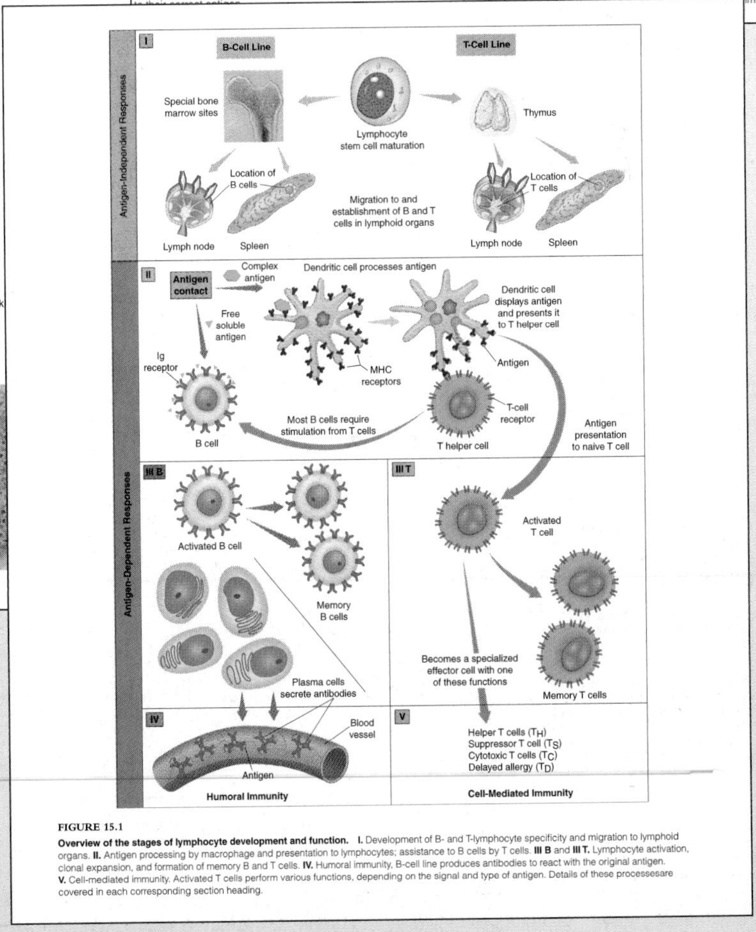

(a) **Initiation.** The classical pathway begins when C1 components bind to receptors on a foreign cell membrane.

(b) **Amplification and cascade.** The C1 complex is an enzyme that activates a second series of components, C4 and C2. When these have been enzymatically cleaved into separate molecules, they become a second enzyme complex that activates C3. At this same site, C3 binds to C5 and cleaves it to form a product that is tightly bound to the membrane.

(c) **Polymerization.** C5b is a reactive site for the final assembly of an attack complex. In series, C6, C7, and C8 aggregate with C5 and become integrated into the membrane. They form a substrate upon which the final component, C9, can bind. Up to 15 of these C9 units ring the central core of the complex.

(d) **Membrane attack.** The final product of these reactions is a large, donut-shaped enzyme that punctures small pores through the membrane, leading to cell lysis.

(e) An electron micrograph (187,000×) of a cell reveals multiple puncture sites over its surface. The lighter, ringlike structures are the actual enzyme complex.

The Acquisition of Specific Immunity and Its Applications

CHAPTER 15

The primary focus of this chapter is the remarkable system of lymphocytes that are responsible for specific, acquired immunities. In our more detailed examination, we will discover the complex adaptations for defense against microbes, cancer, and toxic substances. This background will prepare you for coverage of vaccination, immune testing, allergy, and immune deficiency in chapter 17.

Overview

specific host defenses are derived from a dual system of lymphocytes that are genetically programmed to react with foreign substances (antigens) found in microbes and other organisms. cells carry glycoprotein receptors that dictate their specificity and lymphocytes have antibody receptors, T lymphocytes have T-cell and macrophages have histocompatibility receptors such as and HLA. and T cells arise in the bone marrow, where they proliferate and extreme variations in the expression of receptor genes. of lymphocytes creates billions of genetically different that each have a unique specificity for antigen. cells reach final maturity in special bone marrow sites, and the reach final maturity in the thymus gland. of lymphocytes home (migrate) to separate sites in lymphoid where they serve as a constant source of immune cells primed to

An immune system huddle. The large central dendritic cell communicates with a team of smaller T lymphocytes, giving the signal for the next play. From this immune

FIGURE 15.1
Overview of the stages of lymphocyte development and function. **I.** Development of B- and T-lymphocyte specificity and migration to lymphoid organs. **II.** Antigen processing by macrophage and presentation to lymphocytes; assistance to B cells by T cells. **III B** and **III T.** Lymphocyte activation, clonal expansion, and formation of memory B and T cells. **IV.** Humoral immunity, B-cell line produces antibodies to react with the original antigen. **V.** Cell-mediated immunity. Activated T cells perform various functions, depending on the signal and type of antigen. Details of these processes are covered in each corresponding section heading.

Features that Facilitate Learning

Overviews Chapters open with a vignette that states the relevance of the chapter focus and a bulleted list that outlines the main themes of the chapter.

CHAPTER 11

Physical and Chemical Control of Microbes

The natural condition of humanity is to share surroundings with a large, diverse population of microorganisms. The complete exclusion of microbes from the environment is not only impossible but of questionable value. In many instances, however, our health and comfort can depend on the ability to destroy, inhibit, and remove microbes in the habitats we share. These techniques, also known as antimicrobial control, are very broad in scope. They include routine activities such as cleaning, refrigeration, and cooking. They are also of central importance in the medical, dental, and commercial settings to prevent infection and spoilage, and to ensure the safety of food, water, and other products. Both this chapter and chapter 12 will survey important aspects of microbial control.

Chapter Overview

- The control of microbes in the environment is a constant concern of health care and industry since microbes are the cause of infection and food spoilage, among other undesirable events.
- Antimicrobial control is accomplished using both physical techniques and chemical agents to destroy, remove, or reduce microbes in a given area.
- Many factors must be contemplated when choosing an antimicrobial technique, including the material being treated, the type of microbes involved, the microbial load, and the time available for treatment.
- Antimicrobial agents damage microbes by disrupting the structure of the cell wall or cell membrane, preventing synthesis of nucleic acids (DNA and RNA), or altering the function of cellular proteins.
- Microbicidal agents kill microbes by inflicting nonreversible damage to the cell. Microbistatic agents temporarily inhibit the reproduction of microbes but do not inflict irreversible damage. Mechanical antimicrobial agents physically remove microbes from materials but do not necessarily kill or inhibit them.
- Heat is the most important physical agent in microbial control and can be delivered in both moist (steam sterilization, pasteurization) and dry (incinerators, Bunsen burners) forms.
- Radiation exposes materials to high energy waves that can enter and

A look down the cleaning products aisle of most any store will confirm our preoccupation (some would say obsession) with microbial control.

Controlling Microorganisms

Much of the time in our daily existence, we take for granted tap water that is drinkable, food that is not spoiled, shelves full of products to eradicate "germs," and drugs to treat infections. Controlling our degree of exposure to potentially harmful microbes is a monumen-

---

Chapter Checkpoints serve as quick "snapshot" reviews of the main themes of each major section of a chapter.

346   CHAPTER 11 Physical and Chemical Control of Microbes

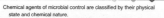

CHAPTER CHECKPOINTS

Chemical agents of microbial control are classified by their physical state and chemical nature.

Chemical agents can be either microbicidal or microbistatic. They are also classified as high-, medium-, or low-level germicides.

Factors that determine the effectiveness of a chemical agent include the type and numbers of microbes involved, the material involved, the strength of the agent, and the exposure time.

Halogens are effective chemical agents at both microbicidal and microbistatic levels. Chlorine compounds disinfect water, food, and industrial equipment. Iodine is used as either free iodine or iodophor to disinfect water and equipment. Iodophors are also used as antiseptic agents.

Phenols are strongly microbicidal agents used in general disinfection. Milder phenol compounds, the bisphenols, are also used as antiseptics.

Alcohols dissolve membrane lipids and destroy cell proteins. Their action depends upon their concentration, but they are generally only microbistatic.

Hydrogen peroxide is a versatile microbicide that can be used as an antiseptic for wounds and a disinfectant for utensils. A high concentration is an effective sporicide.

Surfactants are of two types: detergents and soaps. They reduce cell membrane surface tension, causing membrane rupture. Cationic detergents, or quats, are low-level germicides limited by the amount of organic matter present and the microbial load.

Aldehydes are potent sterilizing agents and high-level disinfectants that irreversibly disrupt microbial enzymes.

Ethylene oxide and chlorine dioxide are gaseous sterilants that work by alkylating protein and DNA.

Chapter Capsule Summaries contain key words and concepts together in outline form. The capsules help students review the most important information in each chapter.

CHAPTER CAPSULE WITH KEY TERMS

I. **Physical methods for controlling microorganisms**
   A. Moist heat denatures proteins and DNA while destroying membranes.
      1. **Sterilization: Autoclaves** utilize steam under pressure to sterilize heat resistant materials while intermittent sterilization can be used to sterilize more delicate items.
      2. **Disinfection:** Pasteurization subjects liquids to temperatures below 100°C and is used to lower the microbial load in liquids. Boiling water can be used to destroy vegetative pathogens in the home.
   B. Dry heat, using higher temperatures than moist heat, can also be used to sterilize.
      1. **Incineration** can be carried out using a Bunsen burner or incinerator. Temperatures range between 600°C and 1800°C.
      2. **Dry ovens** coagulate proteins at temperatures of 15° to 180°C.
   C. Cold temperatures are microbistatic, with refrigeration (0° to 15°C) and freezing (below 0°C) commonly used to preserve food, media and cultures.
   D. Drying and desiccation lead to (often temporary) metabolic inhibition by reducing water in the cell.
   E. Radiation: Energy in the form of radiation is a method of **cold sterilization,** which works by introducing mutations into the DNA of target cells.
      1. **Ionizing radiation,** such as Gamma rays and X-rays, has deep penetrating power and works by causing breaks in the DNA of target organisms.
      2. **Nonionizing radiation** uses **ultraviolet waves** with very little penetrating power and works by creating dimers between adjacent pyrimidines, which interferes with replication.
   F. Filtration involves the physical removal of microbes by passing a gas or liquid though a fine filter and can be used to sterilize air as well as heat sensitive liquids.
II. **Chemical Control of Microorganisms**
   A. Chemicals are divided into **disinfectants, antiseptics, sterilants, sanitizers** and **degermers** based on their level of effectiveness and the surfaces to which they are applied.

   B. Antimicrobic chemicals are found as solids, gases and liquids. Liquids can be either aqueous (water based) or tinctures (alcohol based).
   C. Halogens are chemicals based on elements from group VII of the periodic table.
      1. **Chlorine** is used as chlorine gas, hypochlorites and chloramines. All work by disrupting disulfide bonds and, given adequate time, are sporicidal.
      2. **Iodine** is found both as free iodine (I_2) and iodophors (iodine bound to organic polymers such as soaps). Iodine has a mode of action similar to chlorine and is also sporicidal given enough time.
   D. Phenolics are chemicals based on phenol that work by disrupting cell membranes and precipitating proteins. They are bactericidal, fungicidal and viricidal, but not sporicidal.
      1. Although phenol is now considered too toxic to be used in most circumstances, phenolic compounds (Lysol, Triclosan) are commonly used as, or added to, home and hospital disinfectants.
   E. Chlorhexidine (Hibiclens, Hibitane) is a surfactant and protein denaturant with broad microbicidal properties, although it is not sporicidal. Solutions of chlorhexidine are used as skin degerming agents for preoperative scrubs, skin cleaning and burns.
   F. Ethyl and isopropyl alcohol, in concentrations of 50% to 90%, are useful for microbial control. Alcohols act as **surfactants,** dissolving membrane lipids and coagulating proteins of vegetative bacterial cells and fungi. They are not sporicidal.
   G. Hydrogen peroxide produces highly reactive hydroxyl-free radicals that damage protein and DNA while also decomposing to O_2 gas, which is toxic to anaerobes. Strong solutions of H_2O_2 are sporicidal.
   H. Detergents and soaps
      1. Cationic detergents known as **quaternary ammonium compounds** (quats) act as surfactants that alter the membrane permeability of some bacteria and fungi. They are not sporicidal.
      2. Soaps have little microbicidal activity but rather function by removing grease and soil that contain microbes.

The Learning System

End of Chapter Questions are varied to give students experience in answering different types of questions; Multiple Choice Questions; Concept Questions; and Critical-Thinking Questions.

MULTIPLE-CHOICE QUESTIONS

1. A microbicidal agent has what effect?
   a. sterilizes
   b. inhibits microorganisms
   c. is toxic to human cells
   d. destroys microorganisms

2. Microbial control methods that kill ____ are able to sterilize.
   a. viruses
   b. the tubercle bacillus
   c. endospores
   d. cysts

3. Any process that destroys the non-spore-forming contaminants on inanimate objects is
   a. antisepsis
   b. disinfection
   c. sterilization
   d. degermation

4. Sanitization is a process by which
   a. the microbial load on objects is reduced
   b. objects are made sterile with chemicals
   c. utensils are scrubbed
   d. skin is debrided

5. An example of an agent that lowers the surface tension of cells is
   a. phenol
   b. chlorine
   c. alcohol
   d. formalin

6. High temperatures ____ and low temperatures ____ .
   a. sterilize, disinfect
   b. kill cells, inhibit cell growth
   c. denature proteins, burst cells
   d. speed up metabolism, slow down metabolism

7. The temperature-pressure combination for an autoclave is
   a. 100°C and 4 psi
   b. 121°C and 15 psi
   c. 131°C and 9 psi
   d. 115°C and 3 psi

8. Microbe(s) that is/are the target(s) of pasteurization include:
   a. *Clostridium botulinum*
   b. *Mycobacterium* species
   c. *Salmonella* species
   d. both b and c

9. Ionizing radiation removes ____ from atoms.
   a. protons
   b. waves
   c. electrons
   d. ions

10. The primary mode of action of nonionizing radiation is to
    a. produce superoxide ions
    b. make pyrimidine dimers
    c. denature proteins
    d. break disulfide bonds

11. The most versatile method of sterilizing heat-sensitive liquids is
    a. UV radiation
    b. exposure to ozone
    c. beta propiolactone
    d. filtration

12. ____ is the iodine antiseptic of choice for wound treatment.
    a. Eight percent tincture
    b. Five percent aqueous
    c. Iodophor
    d. Potassium iodide solution

13. A chemical with sporicidal properties is
    a. phenol
    b. alcohol
    c. quaternary ammonium compound
    d. glutaraldehyde

14. Silver nitrate is used
    a. in antisepsis of burns
    b. as a mouthwash
    c. to treat genital gonorrhea
    d. to disinfect water

15. Detergents are
    a. high-level germicides
    b. low-level germicides
    c. excellent antiseptics
    d. used in disinfecting surgical instruments

16. Which of the following is an approved sterilant?
    a. chlorhexidine
    b. betadyne
    c. ethylene oxide
    d. ethyl alcohol

CONCEPT QUESTIONS

1. Compare sterilization with disinfection and sanitization. Describe the relationship of the concepts of sepsis, asepsis, and antisepsis.

2. ... of a tuberculocide.
   ... the effect of a pseudomonicide?
   ... static agent do?

3. ... ow the type of microorganisms present will ... ectiveness of exposure to antimicrobial agents.
   ... numbers of contaminants can influence the ... control them.

4. ... microbial death?
   ... lation of microbes not die instantaneously when ... antimicrobial agent?

5. ... ial processes inhibited in the presence of ... matter?

6. Describe four modes of action of antimicrobial agents, and give a specific example of how each works.

7. a. Summarize the nature, mode of action, and effectiveness of moist and dry heat.
   b. Compare the effects of moist and dry heat on vegetative cells and spores.
   c. Explain the concepts of TDT and TDP, using examples. What are the minimum TDTs for vegetative cells and endospores?

8. How can the temperature of steam be raised above 100°C? Explain the relationship involved.

9. a. What do you see as a basic flaw in tyndallization?
   b. In boiling water devices?
   c. In incineration?
   d. In ultrasonic devices?

---

348 CHAPTER 11 Physical and Chemical Control of Microbes

10. a. What are several microbial targets of pasteurization?
    b. What are the primary purposes of pasteurization?
    c. What is ultrapasteurization?

11. Explain why desiccation and cold are not reliable methods of disinfection.

12. a. What are some advantages of ionizing radiation as a method of control?
    b. Some disadvantages?

13. a. What is the precise mode of action of ultraviolet radiation?
    b. What are some disadvantages to its use?

14. What are the superior characteristics of iodophors over free iodine solutions?

15. a. What does it mean to lower the surface tension?
    b. What cell parts will be most affected by surface active agents?

16. a. Name one chemical for which the general rule that a higher concentration is more effective is *not* true.
    b. What is a sterilant?
    c. Name the principal sporicidal chemical agents.

17. Why is hydrogen peroxide solution so effective against anaerobes?

18. Give the uses and disadvantages of the heavy metal chemical agents, glutaraldehyde, and the sterilizing gases.

19. What does it mean to say that a chemical has an oligodynamic action?

20. a. Define cold sterilization.
    b. Name three totally different methods that qualify for this definition.

CRITICAL-THINKING QUESTIONS

1. What is wrong with this statement: "Prior to vaccination, the patient's skin was sterilized with alcohol"? What would be the more correct wording?

2. For each item on the following list, give a reasonable method of sterilization. You cannot use the same method more than three times; the method must sterilize, not just disinfect; and the method must not destroy the item or render it useless unless there is no other choice. After considering a workable method, think of a method that would not work. Note: Where an object containing something is given, you must sterilize everything (for example, both the jar and the Vaseline in it). Some examples of methods are autoclave, ethylene oxide gas, dry oven, and ionizing radiation.

room air	carcasses of cows with "mad cow" disease
blood in a syringe	inside of a refrigerator
serum	wine
a pot of soil	a jar of Vaseline
plastic Petri plates	fruit in plastic bags
heat-sensitive drugs	talcum powder
cloth dressings	milk
leather shoes from a thrift shop	orchid seeds
a cheese sandwich	metal instruments
human hair (for wigs)	mail contaminated with anthrax
a flask of nutrient agar	spores
an entire room (walls, floor, etc.)	
rubber gloves	
disposable syringes	

3. a. Graph the data on tables 11.3 and 11.4, plotting the time on the Y axis and the temperature on the X axis for three different organisms.
   b. Using pasteurization techniques as a model, compare the TDTs and explain the relationships between temperature and length of exposure.

   c. Is there any difference between the graph for a sporeformer and the graph for a non-sporeformer? Explain.

4. Can you think of situations in which the same microbe would be considered a serious contaminant in one case and completely harmless in another?

5. A supermarket/drugstore assignment: With a partner, look at the labels of 25 different products used to control microbes, and make a list of their active ingredients, their suggested uses, and information on toxicity and precautions.

6. Devise an experiment that will differentiate between bacteriocidal and bacteriostatic effects.

7. There is quite a bit of concern that chlorine used as a water purification chemical presents serious dangers. What alternative methods covered by this chapter could be used to purify water supplies and yet keep them safe from contamination?

8. The shelf-life and keeping qualities of fruit and other perishable foods are greatly enhanced through irradiation, saving industry and consumers billions of dollars. How would you personally feel about eating a piece of irradiated fruit? How about spices that had been sterilized with ETO?

9. The microbial levels in saliva are astronomical ($<10^6$ cells per ml, on average). What effect do you think a mouthwash can have against this high number? Comment on the claims made for such products.

10. Can you think of some innovations the health care community can use to deal with medical waste and its disposal that prevent infection but are ecologically sound?

INTERNET SEARCH TOPICS

1. Look for information on the Internet concerning triclosan-resistant bacteria. Based on what you find, do you think the widespread use of this antimicrobial is a good idea? Why or why not?

2. Find websites that discuss problems with sterilizing reusable medical instruments such as endoscopes and the types of diseases that can be transmitted with them.

3. Handwashing is one of our main protections to ensure good health and hygiene. Visit the student Online Learning Center at www.mhhe.com/talaro5. Go to chapter 11, Internet Search Topics, and log on to the available websites to research information, statistics, and other aspects of handwashing.

Internet Search Topics provide students with additional resources to investigate current topics in microbiology. The URL addresses are listed in each chapter of the Online Learning Center, www.mhhe.com/talaro5

Foundations in

Microbiology

FIFTH EDITION

Basic Principles

CHAPTER 1

The Main Themes of Microbiology

In 1967 the surgeon general of the United States delivered a speech to Congress: "It is time to close the book on infectious diseases," he said. "The war against pestilence is over."

In 1998 Surgeon General David Satcher had a different message. The *Miami Herald* reported his speech with this headline: "Infectious Diseases a Rising Peril; Death Rates in U.S. Up 58% Since 1980."

The middle of the last century was a time of great confidence in science and medicine. With the introduction of antibiotics in the 1940s, and a lengthening list of vaccines that prevented the most frightening diseases, Americans felt that it was only a matter of time before diseases caused by microorganisms (i.e., infectious diseases) would be completely manageable. The nation's attention turned to the so-called chronic diseases, such as heart disease, cancer, and stroke.

So what happened to change the optimism of the 1960s to the warning expressed in the speech from 1998? Dr. Satcher explained it this way: "Organisms changed and people changed." First, we are becoming more susceptible to infectious disease precisely because of advances in medicine. People are living longer. Sicker people are staying alive much longer than in the past. Older and sicker people have heightened susceptibility to what we might call garden-variety microbes. Second, the population has become more mobile. Travelers can crisscross the globe in a matter of hours, taking their microbes with them and introducing them into new "naive" populations. Third, there are growing numbers of microbes that truly are new (or at least, new to us). The conditions they cause are called **emerging diseases.** Changes in agricultural practices and encroachment of humans on wild habitats are just two probable causes of emerging diseases. Fourth, microorganisms have demonstrated their formidable capacity to respond and adapt to our attempts to control them, most spectacularly by becoming resistant to the effects of our miracle drugs.

And there's one more thing: Evidence is mounting that many conditions formerly thought to be caused by genetics or life-style, such as heart disease and cancer, can often be at least partially caused by microorganisms.

Microbes never stop surprising us—in their ability to harm but also to help us. The best way to keep up is to learn as much as you can about them. This book is a good place to start.

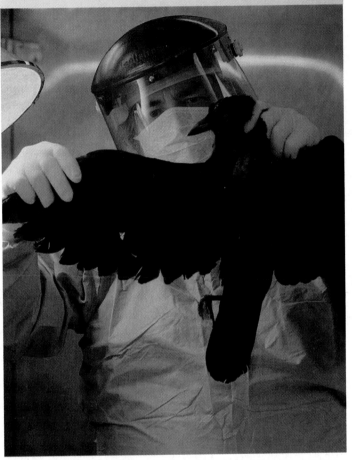

A wildlife pathologist examines a dead crow suspected of dying from infection with West Nile virus. West Nile virus first appeared in the U.S. in 1999.

- In terms of numbers and range of distribution, microbes are the dominant organisms on earth.
- Major groups of microorganisms include bacteria, algae, protozoa, fungi, parasitic worms, and viruses.
- Microbiology involves study in numerous areas involving cell structure, function, genetics, immunology, biochemistry, epidemiology, and ecology.
- Microorganisms are essential to the operation of the earth's ecosystems, as photosynthesizers, decomposers, and recyclers.
- Humans use the versatility of microbes to make improvements in industrial production, agriculture, medicine, and environmental protection.
- The beneficial qualities of microbes are in contrast to the many infectious diseases they cause.
- Microorganisms are the oldest organisms, having evolved over 4 billion years of earth's history to the modern varieties we now observe.

Chapter Overview

- Microorganisms, also called microbes, are organisms that require a microscope to be readily observed.

- Microbiologists use the scientific method to develop theories and explanations for microbial phenomena.
- The history of microbiology is marked by numerous significant discoveries and events in microscopy, culture techniques, and other methods of handling or controlling microbes.
- Microbes are classified into groups according to evolutionary relationships, provided with standard scientific names, and identified by specific characteristics.

The Scope of Microbiology

Microbiology is a specialized area of **biology*** that deals with living things ordinarily too small to be seen without magnification. Such **microscopic*** organisms are collectively referred to as **microorganisms,*** **microbes,*** or several other terms, depending upon the purpose. Some people call them germs or bugs in reference to their role in infection and disease, but those terms have other biological meanings and perhaps place undue emphasis on the disagreeable reputation of microorganisms. There are several major groups of microorganisms that we'll be studying. They are **bacteria, viruses, fungi, protozoa, algae,** and **helminths.** They all have different biological characteristics. A characteristic is a distinguishing trait or feature that can be used to describe or define an organism. The nature of microorganisms makes them both very easy and very difficult to study. Easy, because they reproduce so rapidly and we can quickly grow large populations in the laboratory. Difficult, because we can't see them directly. We rely on a variety of indirect means of analyzing them in addition to using microscopes.

Microbiology is one of the largest and most complex of the biological sciences because it includes many diverse biological disciplines. Microbiologists study every aspect of microbes—their genetics, their physiology, their characteristics that may lead to disease or to benefits, the way they interact with the environment, the way they interact with mammalian hosts, and their uses in industry and agriculture.

Here are some descriptions of different branches of study within microbiology.

Immunology studies the complex web of immune chemicals and cells that are produced in response to infection. It also concerns itself with the study of *hypersensitivity,* inappropriate immune responses that can be harmful to the human host. *Allergy* is one example of hypersensitivity (see chapters 14, 15, 16, and 17).

Public health microbiology and **epidemiology** aim to monitor and control the spread of diseases in communities. The principal U.S. and global institutions involved in this concern are the United States Public Health Service (USPHS) with its main agency, the Centers for Disease Control and Prevention (CDC)

located in Atlanta, Georgia, and the World Health Organization (WHO), the medical limb of the United Nations (see chapter 13). The CDC collects information on disease from around the United States and publishes it in a weekly newsletter called the *Morbidity and Mortality Weekly Report.* The www.cdc.gov site is also one of the most reliable sources of information about diseases you will study in this book. (Visit www.cdc.gov/mmwr/ for the most current report.)

Food microbiology, dairy microbiology, and aquatic microbiology examine the ecological and practical roles of microbes in food and water (see chapter 26).

Agricultural microbiology is concerned with the relationships between microbes and crops, with an emphasis on improving yields and combating plant diseases.

Biotechnology includes any process in which humans use the metabolism of living things to arrive at a desired product, ranging from bread making to gene therapy. It is a tool used in industrial microbiology, which is concerned with the uses of microbes to produce or harvest large quantities of substances such as beer, vitamins, amino acids, drugs, and enzymes (see chapters 7, 10, and 26).

Genetic engineering and recombinant DNA technology involve techniques that deliberately alter the genetic makeup of organisms to mass-produce human hormones and other drugs, create totally novel substances, and develop organisms with unique methods of synthesis and adaptation. This is the most powerful and rapidly growing area in modern microbiology (see chapter 10).

Each of the major disciplines in microbiology contains numerous subdivisions or specialties that in turn deal with a specific subject area or field. In fact, many areas of this science have become so specialized that it is not uncommon for a microbiologist to spend his or her whole life concentrating on a single group or type of microbe, biochemical process, or disease. On the other hand, rarely is one person a single type of microbiologist, and most can be classified in several ways. There are, for instance, bacterial physiologists who study industrial processes, molecular biologists who focus on the genetics of viruses, mycologists doing research on agricultural pests, epidemiologists who are also nurses, and dentists who specialize in the microbiology of gum disease.

Studies in microbiology have led to greater understanding of many general biological principles. For example, the study of microorganisms established universal concepts concerning the chemistry of life (see chapters 2 and 8), systems of inheritance (see chapter 9), and the global cycles of nutrients, minerals, and gases (see chapter 26).

The Impact of Microbes on Earth: Small Organisms with a Giant Effect

The most important knowledge that should emerge from a microbiology course is the profound influence microorganisms have on all aspects of the earth and its residents. For billions of years,

---

*biology Gr. *bios,* life, and *logos,* to study. The study of organisms.

*microscopic (my″-kroh-skaw′-pik) Gr. *mikros,* small, and *scopein,* to see.

*microorganism (my-kroh′-or′-gun-izm)

*microbe (my′-krohb) Gr. *mikros,* small, and *bios,* life.

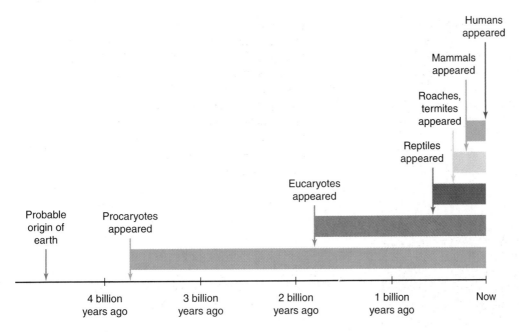

FIGURE 1.1

Evolutionary timeline. The first bacteria appeared approximately 3.5 billion years ago. They were the only form of life for half of the earth's history.

microbes have extensively shaped the development of the earth's habitats and the evolution of other life forms. It is understandable that scientists searching for life on other planets first look for signs of microorganisms.

Bacterial-type organisms have been on this planet for about 3.5 billion years, according to the fossil record. It appears that they were the only living inhabitants on earth for almost 2 billion years. At that time (about 1.8 billion years ago) a more complex type of single-celled organism arose, of a **eucaryotic*** cell type. Eu-cary means *true nucleus,* which gives you a hint that those first inhabitants, the bacteria, had no true nucleus. For that reason they are called **procaryotes*** (pre-nucleus). The early eucaryotes were the precursors of the cell type that eventually formed multicellular animals, including humans. But you can see from **figure 1.1** how long that took! On the scale pictured in the figure, humans seem to have just appeared. The bacteria preceded even the earliest animals by about 3 billion years. This is a good indication that humans are not likely to, nor should we, try to eliminate bacteria from our environment. They've survived and adapted to many catastrophic changes over the course of their geologic history.

Another indication of the huge influence bacteria exert is how **ubiquitous*** they are. Microbes can be found nearly everywhere, from deep in the earth's crust, to the polar ice caps and oceans, to the bodies of plants and animals. Being mostly invisible, the actions of microorganisms are usually not as obvious or familiar as those of larger plants and animals. They make up for their small size by occurring in large numbers and living in places that many other organisms cannot survive. Above all, they play central roles in the earth's landscape that are essential to life.

MICROBIAL INVOLVEMENT IN ENERGY AND NUTRIENT FLOW

Microbes are deeply involved in the flow of energy and food through the earth's ecosystems.[1] Most people are aware that plants carry out **photosynthesis,** which is the light-fueled conversion of carbon dioxide to organic material, accompanied by the formation of oxygen. But microorganisms were photosynthesizing long before the first plants appeared. In fact, they were responsible for changing the atmosphere of the earth from one without oxygen, to one with oxygen. Today photosynthetic microorganisms (including algae) account for more than 50% of the earth's photosynthesis, contributing the majority of the oxygen to the atmosphere (**figure 1.2a**).

Another process that helps keep the earth in balance is the process of biological **decomposition** and nutrient recycling. Decomposition involves the breakdown of dead matter and wastes into simple compounds that can be directed back into the natural cycles of living things (**figure 1.2b**). If it were not for multitudes of bacteria and fungi, many chemical elements would become locked up and unavailable to organisms. In the long-term scheme of things, microorganisms are the main forces that drive the structure and content of the soil, water, and atmosphere. For example:

• The very temperature of the earth is regulated by "greenhouse gases," such as carbon dioxide and methane, that create an insulation layer in the atmosphere and help retain heat. Much of this gas is produced by microbes living in the environment and the digestive tracts of animals.

• Recent estimates propose that, based on weight and numbers, up to 50% of all organisms exist within and beneath the earth's crust in sediments, rocks, and even volcanoes. It is increasingly evident that this enormous

---

*eucaryotic (yoo"-kar-ee-ah'-tik) Gr. *eu,* true or good, and *karyon,* nucleus.

*procaryotic (proh"-kar-ee-ah'-tik) Gr. *pro,* before, and *karyon,* nucleus. Sometimes spelled prokaryotic and eukaryotic.

*ubiquitous (yoo-bĭk'wĭ-təs) being, or seeming to be, everywhere at the same time.

---

1. Ecosystems are any interactions that occur between living organisms and their environment.

(a) Summer pond with a thick mat of algae—a rich photosynthetic community.

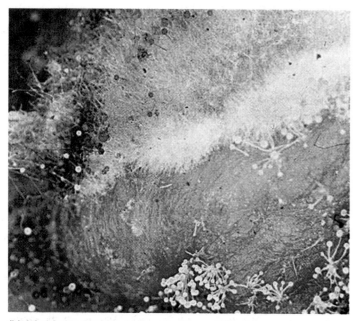

(b) A fruit being decomposed by a common soil fungus.

FIGURE 1.2
Microbial habitats.

underground community of microbes is a significant influence on weathering, mineral extraction, and soil formation.

• Bacteria and fungi live in complex associations with plants that assist the plants in obtaining nutrients and water and may protect them against disease. Microbes form similar interrelationships with animals, notably in the stomach of cattle, where a rich assortment of bacteria digest the complex carbohydrates of the animals' diets.

HUMAN USE OF MICROORGANISMS

It is clear that microorganisms have monumental importance to the earth's operation. It is this very same diversity and versatility that also makes them excellent candidates for solving human problems. By accident or choice, humans have been using microorganisms for thousands of years to improve life and even to shape civilizations. Yeasts, a type of microscopic fungi, cause bread to rise and ferment sugar into alcohol to make wine and beers. These and other "home" uses of microbes have been in use for thousands of years. For example, historical records show that households in ancient Egypt kept moldy loaves of bread to apply directly to wounds and lesions. When humans manipulate microorganisms to make products in an industrial setting, it is called biotechnology. For example, some specialized bacteria have unique capacities to mine precious metals (**figure 1.3***a*) or to produce enzymes that are used in laundry detergents (figure 1.3*b*).

Genetic engineering is a newer area of biotechnology that manipulates the genetics of microbes, plants, and animals for the purpose of creating new products and genetically modified organisms. One powerful technique for designing new organisms is termed **recombinant DNA.** This technology makes it possible to deliberately alter DNA[2] and to switch genetic material from one organism to another. Bacteria and fungi were some of the first organisms to be genetically engineered, because they are so adaptable to changes in their genetic makeup. Recombinant DNA technology has unlimited potential in terms of medical, industrial, and agricultural uses. Microbes can be engineered to synthesize desirable proteins such as drugs, hormones, and enzymes.

Among the genetically unique organisms that have been designed by bioengineers are bacteria that contain a natural pesticide, yeast that produce viral vaccines, pigs that produce human hemoglobin, and plants that do not ripen too rapidly. The techniques also extend to the characterization of human genetic material and diseases.

Another way of tapping into the unlimited potential of microorganisms is the relatively new science of **bioremediation.***
This process involves the introduction of microbes into the environment to restore stability or to clean up toxic pollutants. Bioremediation is required to control the massive levels of pollution from industry and modern living. Microbes have a surprising capacity to break down chemicals that would be harmful to other organisms. Agencies and companies have developed microbes to handle oil spills and detoxify sites contaminated with heavy metals, pesticides, and other chemical wastes (figure 1.3*c*). The solid waste disposal industry is interested in developing methods for degrading the tons of garbage in landfills, especially human-made plastics and paper products. One form of bioremediation that has been in use for some time is the treatment of water and sewage. Since clean freshwater supplies are rapidly dwindling worldwide, it will become even more important to find ways to reclaim polluted water.

---

2. DNA, or deoxyribonucleic acid, the chemical substance that comprises the genetic material of organisms.

*bioremediation (by'-oh-ree-mee-dee-ay"-shun) *bios,* life; *re,* again; *mederi,* to heal. The use of biological agents to remedy environmental problems.

(a) An aerial view of a copper mine looks like a giant quilt pattern. The colored patches are various stages of bacteria extracting metals from the ore.

(c) A bioremediation platform placed in a river for the purpose of detoxifying the water containing industrial pollutants.

FIGURE 1.3
Microbes at work.

(b) Microbes as synthesizers. A large complex fermentor manufactures drugs and enzymes using microbial metabolism.

INFECTIOUS DISEASES AND THE HUMAN CONDITION

One of the most fascinating aspects of the microorganisms with which we share the earth is that, despite all of the benefits they provide, they also contribute significantly to human misery as **pathogens.*** Humanity is plagued by nearly 2,000 different microbes that can infect the human body and cause various types of disease. Infectious diseases still devastate human populations worldwide, despite significant strides in understanding and treating them. The most recent estimates from the World Health Organization (WHO) point to a total of 10 billion new infections across the world every year. There are more infections than people because many people acquire more than one infection. Infectious diseases

---

*pathogens (path′-oh-jenz) Gr. *pathos,* disease, and *gennan,* to produce. Disease-causing agents.

are also among the most common causes of death in much of humanity, and they still kill about one-third of the U.S. population. **Table 1.1** depicts the 10 top causes of death per year (by all causes, infectious and noninfectious) in the United States and worldwide. The worldwide death toll from infections is about 13 million people per year. In **figure 1.4** you can see the top infectious causes of death displayed in a different way. Note that many of these infections are treatable with drugs or preventable with vaccines. Those hardest hit are residents in countries where access to adequate medical care is lacking. One-third of the earth's inhabitants live on less than $1 per day, are malnourished, and are not fully immunized.

Malaria, which kills more than a million people every year worldwide (see figure 1.4), is caused by a microorganism transmitted by mosquitoes. Currently the most effective way for citizens of developing countries to avoid infection with malaria is to sleep under a bed net, since the mosquitoes are most active in the evening. Yet even this inexpensive solution is beyond the reach of many. Mothers in Southeast Asia and elsewhere have to make nightly decisions about which of their children will sleep under the single family bed net, since a second one, priced at about $3 to $5, is too expensive for them.

Adding to the overload of infectious diseases, we are also witnessing an increase in the number of new (emerging) and older (reemerging) diseases. SARS, AIDS, hepatitis C, and viral encephalitis are examples of recently identified diseases that cause severe mortality and morbidity and are currently on the rise. To somewhat balance this trend, there have also been some advances in eradication of diseases such as polio, measles, leprosy, and certain parasitic worms. The WHO is currently on a global push to vaccinate children against the most common childhood diseases.

One of the most eye-opening discoveries in recent years is that many diseases which used to be considered noninfectious probably do involve microbial infection. The most famous of these

TABLE 1.1

Top Causes of Death—All Diseases

United States	No. of Deaths	Worldwide	No. of Deaths
1. Heart disease	725,000	1. Heart disease	11.1 million
2. Cancer	550,000	2. Cancer	7.1 million
3. Stroke	167,000	3. Stroke	5.5 million
4. Chronic lower-respiratory disease	124,000	4. Respiratory infections[1]	3.9 million
5. Unintentional injury (accidents)	97,000	5. Chronic lower-respiratory disease	3.6 million
6. Diabetes	68,000	6. Accidents	3.5 million
7. Influenza and pneumonia	63,000	7. HIV/AIDS	2.9 million
8. Alzheimer disease	45,000	8. Perinatal conditions	2.5 million
9. Kidney problems	35,000	9. Diarrheal diseases	2.0 million
10. Septicemia (bloodstream infection)	30,000	10. Tuberculosis	1.6 million

[1]*Diseases in red are those most clearly caused by microorganisms.*
Data adapted from The World Health Report 2002 (World Health Organization).

is gastric ulcers, now known to be caused by a bacterium called *Helicobacter*. But there are more. A connection has been established between certain cancers and viruses, between diabetes and the Coxsackie virus, and between schizophrenia and a virus called the borna agent. Diseases as disparate as multiple sclerosis, obsessive compulsive disorder, and coronary artery disease have been linked to chronic infections with microorganisms. It seems that the golden age of microbiological discovery, during which all of the "obvious" diseases were characterized, and cures or preventions were devised

for them, should more accurately be referred to as the *first* golden age. We're now discovering the subtler side of microorganisms. Their roles in quiet but slowly destructive diseases are now well known. These include female infertility caused by *Chlamydia* infection, and malignancies such as liver cancer (hepatitis viruses) and cervical cancer (human papillomavirus). Most scientists expect that, in time, many chronic conditions will be found to have some association with microbial agents.

As mentioned earlier, another important development in infectious disease trends is the increasing number of patients with weakened defenses that are kept alive for extended periods. They are subject to infections by common microbes that are not pathogenic to healthy people. There is also an increase in microbes that are resistant to drugs. It appears that even with the most modern technology available to us, microbes still have the "last word," as the great French microbiologist Louis Pasteur observed.

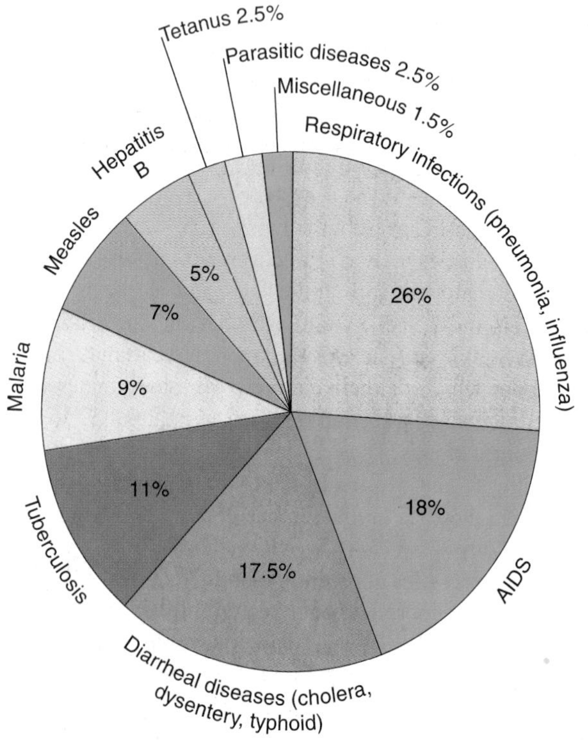

CHAPTER CHECKPOINTS

Microorganisms are defined as "living organisms too small to be seen with the naked eye." Among the members of this huge group of organisms are bacteria, fungi, protozoa, algae, viruses, and parasitic worms.

Microorganisms live nearly everywhere and have impact on many biological and physical activities on earth.

There are many kinds of relationships between microorganisms and humans; most are beneficial, but some are harmful.

The scope of microbiology is incredibly diverse. It includes basic microbial research, research on infectious diseases, study of prevention and treatment of disease, environmental functions of microorganisms, and industrial use of microorganisms for commercial, agricultural, and medical purposes.

In the last 120 years, microbiologists have identified the causative agents for most of the infectious diseases. In addition, they have discovered distinct connections between microorganisms and diseases whose causes were previously unknown.

Microorganisms: We have to learn to live with them because we cannot live without them.

Tetanus 2.5%
Parasitic diseases 2.5%
Miscellaneous 1.5%
Respiratory infections (pneumonia, influenza) 26%
Hepatitis B 5%
Measles 7%
Malaria 9%
Tuberculosis 11%
Diarrheal diseases (cholera, dysentery, typhoid) 17.5%
AIDS 18%

FIGURE 1.4

Worldwide infectious disease statistics. This figure depicts the ten most common infectious causes of death.

The General Characteristics of Microorganisms

CELLULAR ORGANIZATION

As discussed earlier, two basic cell lines appeared during evolutionary history. These lines, termed **procaryotic cells** and **eucaryotic cells,** differ primarily in the complexity of their cell structure (**figure 1.5a**).

In general, procaryotic cells are smaller than eucaryotic cells, and they lack special structures such as a nucleus and **organelles.*** Organelles are small membrane-bound cell structures that perform specific functions in eucaryotic cells. These two cell types and the organisms that possess them (called procaryotes and eucaryotes) are covered in more detail in chapters 2, 4, and 5.

All procaryotes are microorganisms, but only some eucaryotes are microorganisms. The bodies of most microorganisms consist of either a single cell or just a few cells (**figure 1.6**). Because of their role in disease, certain animals such as helminth worms and insects, many of which can be seen with the naked eye, are also considered in the study of microorganisms. Even in its seeming simplicity, the microscopic world is every bit as complex and diverse as the macroscopic one. There is no doubt that microorganisms also outnumber macroscopic organisms by a factor of several million.

A NOTE ON VIRUSES

Viruses are subject to intense study by microbiologists. They are small particles that exist at a level of complexity somewhere between large molecules and cells (see figure 1.5b). Viruses are much simpler than cells; they are composed essentially of a small amount of hereditary material wrapped up in a protein covering. Some biologists refer to viruses as parasitic particles; others consider them to be very primitive organisms. One thing is certain—

they are highly dependent on a host cell's machinery for their activities.

MICROBIAL DIMENSIONS: HOW SMALL IS SMALL?

When we say that microbes are too small to be seen with the unaided eye, what sorts of dimensions are we talking about? This concept is best visualized by comparing microbial groups with the larger organisms of the macroscopic world and also with the molecules and atoms of the molecular world (**figure 1.7**). Whereas the dimensions of macroscopic organisms are usually given in centimeters (cm) and meters (m), those of most microorganisms fall within the range of micrometers (μm) and sometimes, nanometers (nm) and millimeters (mm). The size range of most microbes extends from the smallest viruses, measuring around 20 nm and actually not much bigger than a large molecule, to protozoans measuring 3 to 4 mm and visible with the naked eye.

LIFE-STYLES OF MICROORGANISMS

The majority of microorganisms live a free existence in habitats such as soil and water, where they are relatively harmless and often beneficial. A free-living organism can derive all required foods and other factors directly from the nonliving environment. Some microorganisms require interactions with other organisms. One such group, termed **parasites,** are harbored and nourished by other living organisms, called **hosts.** A parasite's actions cause damage to its host through infection and disease. Although parasites cause important diseases, they make up only a small proportion of microbes.

CHAPTER CHECKPOINTS

Excluding the viruses, there are two types of microorganisms: procaryotes, which are small and lack a nucleus and organelles, and eucaryotes, which are larger and have both a nucleus and organelles.

Viruses are not cellular and are therefore sometimes called particles rather than organisms. They are included in microbiology because of their small size and close relationship with cells.

---

*organelle (or´-gan-el″) L., little organ.

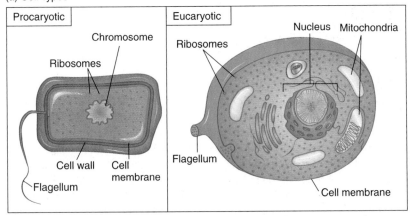

(a) Cell Types

Microbial cells are of the small, relatively simple procaryotic variety (left) or the larger, more complex eucaryotic type (right).

FIGURE 1.5

Cell structure.

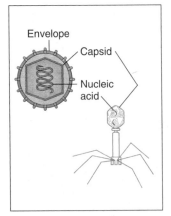

(b) Virus Types

Viruses are tiny particles, not cells, that consist of genetic material surrounded by a protective covering. Shown here are a human virus (top) and bacterial virus (bottom).

(a) Examples of procaryotic organisms

(b) Examples of eucaryotic organisms

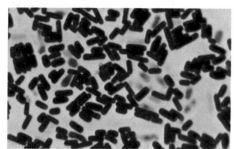

Rod-shaped bacteria, *Clostridium*, found in soil

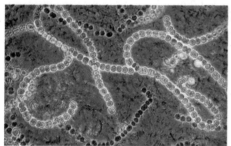

Nostoc, a cyanobacterium that lives in fresh water

The stalked protozoan *Vorticella* is shown in feeding mode. These free-living eucaryotes are common in pond water.

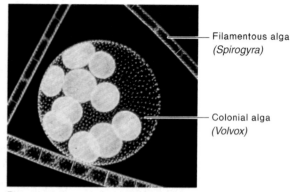

Filamentous alga (*Spirogyra*)

Colonial alga (*Volvox*)

Representatives of algae. *Volvox* is a large, complex colony composed of smaller colonies (spheres) and cells (dots). *Spirogyra* is a filamentous alga composed of elongate cells joined end to end.

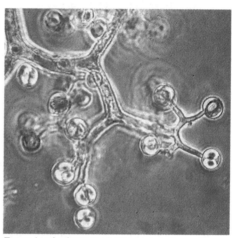

Example of a fungus; shown here is the mold *Thamnidium* displaying its sac-like reproductive vessels.

FIGURE 1.6

The organization of living things and viruses. (Organisms are not shown at the same magnifications.)

Most microorganisms are measured in micrometers, with two exceptions. The helminths are measured in millimeters, and the viruses are measured in nanometers.

Contrary to popular belief, most microorganisms are harmless, free-living species that perform vital functions in both the environment and larger organisms. Comparatively few species are agents of disease.

The Historical Foundations of Microbiology

If not for the extensive interest, curiosity, and devotion of thousands of microbiologists over the last 300 years, we would know little about the microscopic realm that surrounds us. Many of the discoveries in this science have resulted from the prior work of men and women who toiled long hours in dimly lit laboratories with the crudest of tools. Each additional insight, whether large or small, has added to our current knowledge of living things and processes. This section will summarize the prominent discoveries made in the past 300 years: microscopy, the rise of the scientific method, and the development of medical microbiology, including the germ theory and

the origins of modern microbiological techniques. See table B.1 in appendix B, which summarizes some of the pivotal events in microbiology, from its earliest beginnings to the present.

THE DEVELOPMENT OF THE MICROSCOPE: "SEEING IS BELIEVING"

It is likely that from the very earliest history, humans noticed that when certain foods spoiled they became inedible or caused illness, and yet other "spoiled" foods did no harm and even had enhanced flavor. Indeed, several centuries ago, there was already a sense that diseases such as the black plague and smallpox were caused by some sort of transmissible matter. But the causes of such phenomena were vague and obscure because the technology to study them was lacking. Consequently, they remained cloaked in mystery and regarded with superstition—a trend that led even well-educated scientists to believe in spontaneous generation (**Historical Highlights 1.1**). True awareness of the widespread distribution of microorganisms and some of their characteristics was finally made possible by the development of the first microscopes. These devices revealed microbes as discrete entities sharing many of the characteristics of larger, visible plants and animals. Several early scientists fashioned magnifying

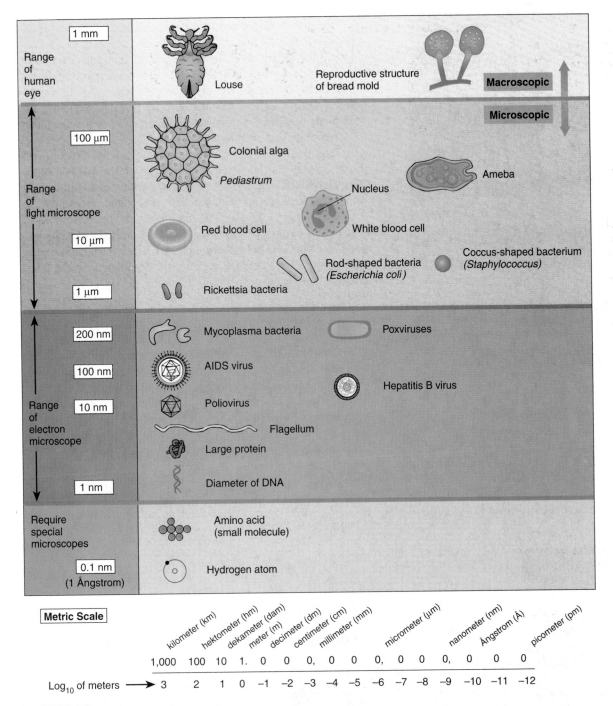

Metric Scale	kilometer (km)	hektometer (hm)	dekameter (dam)	meter (m)	decimeter (dm)	centimeter (cm)	millimeter (mm)	micrometer (µm)	nanometer (nm)	Ångstrom (Å)	picometer (pm)
	1,000	100	10	1.	0	0	0, 0	0 0, 0	0 0, 0	0 0	0

Log$_{10}$ of meters ——→ 3   2   1   0   −1   −2   −3   −4   −5   −6   −7   −8   −9   −10   −11   −12

FIGURE 1.7

The size of things. Common measurements encountered in microbiology and a scale of comparison from the macroscopic to the microscopic, molecular, and atomic. Most microbes encountered in our studies will fall between 100 µm and 10 nm in overall dimensions. The microbes shown are more or less to scale within size zone but not between size zones.

lenses, but their microscopes lacked the optical clarity needed for examining bacteria and other small, single-celled organisms. The first careful and exacting observations awaited the clever single-lens microscope hand-fashioned by Antonie van Leeuwenhoek, a Dutch linen merchant and self-made microbiologist (**figure 1.8**).

Paintings of historical figures like the one of van Leeuwenhoek in figure 1.8 don't always convey a meaningful feeling for the event or person depicted. Imagine a dusty shop in Holland in the late 1600s. Ladies in traditional Dutch garb came in and out, choosing among the bolts of linens for their draperies and upholstery. In between customers, van Leeuwenhoek retired to the workbench in the back of his shop, grinding glass lenses to ever-finer specifications (**figure 1.9**). He could see with increasing clarity the threads in his fabrics. Eventually he became interested in things other than thread counts. He took rainwater from a clay pot and smeared it on his specimen holder, and peered at it through his finest lens. He found "animals appearing to me ten thousand times less than those which may be perceived the water with the naked eye." At this moment van Leeuwenhoek probably became the first person to ever see bacteria.

He didn't stop there. He scraped the plaque from his teeth, and from the teeth of some volunteers who had never cleaned their teeth

FIGURE 1.8

An oil painting of Antonie van Leeuwenhoek (1632–1723) sitting in his laboratory. J.R. Porter and C. Dobell have commented on the unique qualities Leeuwenhoek brought to his craft: "He was one of the most original and curious men who ever lived. It is difficult to compare him with anybody because he belonged to a genus of which he was the type and only species, and when he died his line became extinct."

Leeuwenhoek constructed more than 250 small, powerful microscopes that could magnify up to 300 times (see figure 1.9). Considering that he had no formal training in science and that he was the first person ever to faithfully record this strange new world, his descriptions of bacteria and protozoa (which he called "animalcules") were astute and precise. Because of Leeuwenhoek's extraordinary contributions to microbiology, he is known as the father of bacteriology and protozoology.

From the time of Leeuwenhoek, microscopes evolved into more complex and improved instruments with the addition of refined lenses, a condenser, finer focusing devices, and built-in light sources. The prototype of the modern compound microscope, in use from about the mid-1800s, was capable of magnifications of 1,000 times or more. Even our modern laboratory microscopes are not greatly different in basic structure and function from those early microscopes. The technical characteristics of microscopes and microscopy are the major focus of chapter 3.

THE ESTABLISHMENT OF THE SCIENTIFIC METHOD

A serious impediment to the development of true scientific reasoning and testing was the tendency of early scientists to explain natural phenomena by a mixture of belief, superstition, and argument. The development of an experimental system that answered questions objectively and was not based on prejudice marked the beginning of true scientific thinking. These ideas gradually crept into the consciousness of the scientific community during the 1600s. The general approach taken by scientists to explain a certain natural phenomenon is called the **scientific method.** A primary aim of this method is to formulate a **hypothesis,** a tentative explanation to account for what has been observed or measured (**figure 1.10**).

in their lives and took a good close look at that. He recorded: "In the said matter there were many very little living animalcules, very prettily a-moving. . . . Moreover, the other animalcules were in such enormous numbers, that all the water . . . seemed to be alive." Van Leeuwenhoek started sending his observations to the Royal Society of London, and eventually he was recognized as a scientist of great merit.

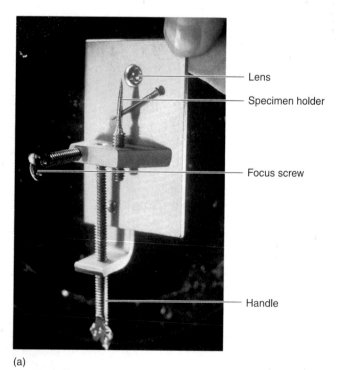

Lens

Specimen holder

Focus screw

Handle

(a)

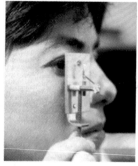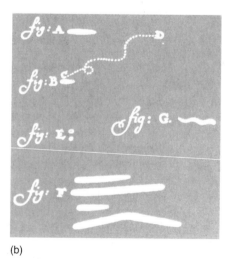

(b)

FIGURE 1.9

Leeuwenhoek's microscope. **(a)** A brass replica of a Leeuwenhoek microscope and how it is held. **(b)** Examples of bacteria drawn by Leeuwenhoek.

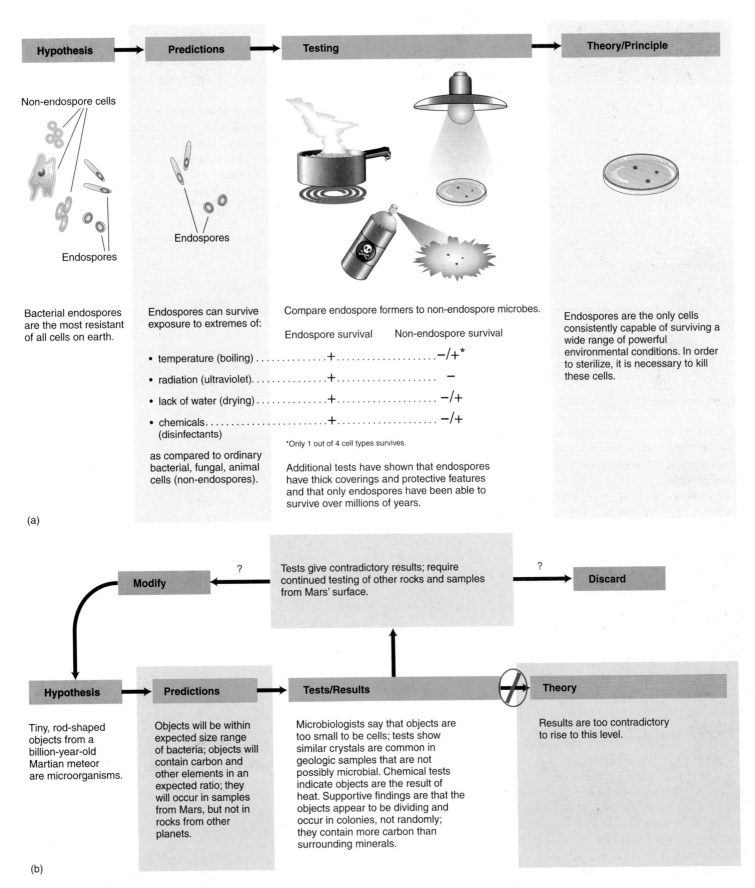

Hypothesis	Predictions	Testing	Theory/Principle

Non-endospore cells

Endospores

Endospores

Bacterial endospores are the most resistant of all cells on earth.

Endospores can survive exposure to extremes of:

as compared to ordinary bacterial, fungal, animal cells (non-endospores).

Compare endospore formers to non-endospore microbes.

	Endospore survival	Non-endospore survival
temperature (boiling)	+	–/+*
radiation (ultraviolet)	+	–
lack of water (drying)	+	–/+
chemicals (disinfectants)	+	–/+

*Only 1 out of 4 cell types survives.

Additional tests have shown that endospores have thick coverings and protective features and that only endospores have been able to survive over millions of years.

Endospores are the only cells consistently capable of surviving a wide range of powerful environmental conditions. In order to sterilize, it is necessary to kill these cells.

(a)

Modify

?   Tests give contradictory results; require continued testing of other rocks and samples from Mars' surface.   ?   Discard

Hypothesis	Predictions	Tests/Results	Theory

Tiny, rod-shaped objects from a billion-year-old Martian meteor are microorganisms.

Objects will be within expected size range of bacteria; objects will contain carbon and other elements in an expected ratio; they will occur in samples from Mars, but not in rocks from other planets.

Microbiologists say that objects are too small to be cells; tests show similar crystals are common in geologic samples that are not possibly microbial. Chemical tests indicate objects are the result of heat. Supportive findings are that the objects appear to be dividing and occur in colonies, not randomly; they contain more carbon than surrounding minerals.

Results are too contradictory to rise to this level.

(b)

FIGURE 1.10

The pattern of deductive reasoning.   The deductive process starts with a general hypothesis that predicts specific expectations. **(a)** This example is based on a well-established principle. **(b)** This example is based on a new hypothesis that has not stood up to critical testing.

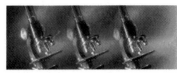

HISTORICAL HIGHLIGHTS 1.1
The Fall of Mysticism and the Rise of Microbiology

For thousands of years, people believed that certain living things arose from vital forces present in nonliving or decomposing matter. This ancient belief, known as **spontaneous generation,** was continually reinforced as people observed that meat left out in the open soon "produced" maggots, that mushrooms appeared on rotting wood, that rats and mice emerged from piles of litter, and other similar phenomena. Though some of these early ideas seem quaint and ridiculous in light of modern knowledge, we must remember that, at the time, mysteries in life were accepted, and the scientific method was not widely practiced.

Even after single-celled organisms were discovered during the mid-1600s, the idea of spontaneous generation continued to exist. Some scientists assumed that microscopic beings were an early stage in the development of more complex ones.

Over the subsequent 200 years, scientists waged an experimental battle over the two hypotheses that could explain the origin of simple life forms. Some tenaciously clung to the idea of **abiogenesis,** * which embraced spontaneous generation. On the other side were advocates of **biogenesis** saying that living things arise only from others of their same kind. There were serious proponents on both sides, and each side put forth what appeared on the surface to be plausible explanations of why their evidence was more correct. Gradually, the abiogenesis hypothesis was abandoned, as convincing evidence for biogenesis continued to mount. The following series of experiments were among the most important in finally tipping the balance. Among the important variables to be considered in challenging the hypotheses were the effects of nutrients, air, and heat and the presence of preexisting life forms in the environment. One of the first people to test the spontaneous generation theory was Francesco Redi of Italy. He conducted a simple experiment in which he placed meat in a jar and covered it with fine gauze. Flies gathering at the jar were blocked from entering and thus laid their eggs on the outside of the gauze. The maggots subsequently developed without access to the

---

*abiogenesis (ah-bee″-oh-jen-uh-sis) Gr. *a,* not, *bios,* living, and *gennan,* to produce. The supposed development of living organisms from nonliving matter.

meat, indicating that maggots were the offspring of flies and did not arise from some "vital force" in the meat. This and related experiments laid to rest the idea that more complex animals such as insects and mice developed through abiogenesis, but it did not convince many scientists of the day that simpler organisms could not arise in that way.

(a)

Frenchman Louis Jablot reasoned that even microscopic organisms must have parents, and his experiments with infusions (dried hay steeped in water) supported that hypothesis. He divided an infusion that had been boiled to destroy any living things into two containers: a heated container that was closed to the air and a heated container that was freely open to the air. Only the open vessel developed microorganisms, which he

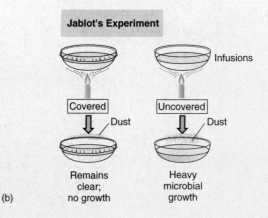

(b)

A good hypothesis must be capable of being either supported or discredited by careful, systematic observation or experimentation. For example, the statement that "microorganisms cause diseases" can be experimentally determined by the tools of science, but the statement that "diseases are caused by evil spirits" cannot.

There are various ways to apply the scientific method, but probably the most common is called the **deductive approach.** In the deductive approach, a scientist constructs a hypothesis, tests its validity by outlining particular events that are predicted by the hypothesis, and then performs experiments to test for those events. The deductive process states: "If the hypothesis is valid, then certain specific events can be expected to occur."

Natural processes have numerous physical, chemical, and biological factors, or *variables,* that can hypothetically affect their outcome. To account for these variables, scientists design

experiments: (1) to thoroughly test for or measure the consequences of each possible variable and (2) to accompany each variable with one or more *control groups.* A control group is designed exactly as the test group but omits only that variable being tested; thus, it may serve as a basis of comparison for the test group. The reasoning is that, if a certain experimental finding occurs only in the test group and not in the control group, the finding must be due to the variable being tested and not to some uncontrolled, untested factor that is not part of the hypothesis **(figure 1.11).**

A lengthy process of experimentation, analysis, and testing eventually leads to conclusions that either support or refute the hypothesis. If experiments do not uphold the hypothesis—that is, if it is found to be flawed—the hypothesis or some part of it is rejected; it is (see figure 1.10*b*) either discarded or modified to fit the presumed had entered in air laden with dust. Regrettably, the validation of biogenesis was temporarily set back by John Needham, an Englishman who did similar experiments using mutton gravy. His results were in conflict with Jablot's because both his heated and unheated test containers teemed with microbes. Unfortunately, his experiments were done before the realization that heat-resistant endospores were not killed by mere boiling. Apparently Jablot had been lucky; his infusions had no endospores.

Additional experiments further defended biogenesis. Franz Shultze and Theodor Schwann of Germany felt sure that air was the source of microbes and sought to prove this by passing air through strong chemicals or hot glass tubes into heat-treated infusions in flasks. When the infusions again remained devoid of living things, the supporters of abiogenesis claimed that the treatment of the air had made it harmful to the spontaneous development of life.

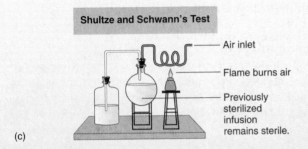

Shultze and Schwann's Test

- Air inlet
- Flame burns air
- Previously sterilized infusion remains sterile.

(c)

Then, in the mid-1800s, the acclaimed microbiologist Louis Pasteur entered the arena. He had recently been studying the roles of microorganisms in the fermentation of beer and wine, and it was clear to him that these processes were brought about by the activities of microbes introduced into the beverage from air, fruits, and grains. The methods he used to discount abiogenesis were simple yet brilliant.

To further clarify that air and dust were the source of microbes, Pasteur filled flasks with broth and fashioned their openings into elongate, swan-neck–shaped tubes. The flasks' openings were freely open to the air but were curved so that gravity would cause any airborne dust particles to deposit in the lower part of the necks. He heated the flasks to sterilize the broth and then incubated them. As long as the flask remained intact, the broth remained sterile, but if the neck was broken off so that dust fell directly down into the container, microbial growth immediately commenced.

Pasteur summed up his findings, "For I have kept from them, and am still keeping from them, that one thing which is above the power of man to make; I have kept from them the germs that float in the air, I have kept from them life."

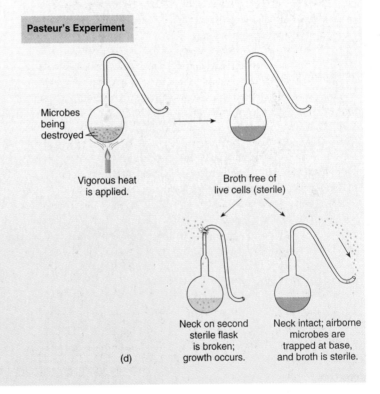

Pasteur's Experiment

Microbes being destroyed

Vigorous heat is applied.

Broth free of live cells (sterile)

Neck on second sterile flask is broken; growth occurs.

Neck intact; airborne microbes are trapped at base, and broth is sterile.

(d)

results of the experiment. If the hypothesis is supported by the results from the experiment, it is not (or should not be) immediately accepted as fact. It then must be tested and retested. Indeed, this is an important guideline in the acceptance of a hypothesis. The results of the experiment must be published and then repeated by other investigators.

In time, as each hypothesis is supported by a growing body of data and survives rigorous scrutiny, it moves to the next level of acceptance—the **theory.** A theory is a collection of statements, propositions, or concepts that explains or accounts for a natural event. A theory is not the result of a single experiment repeated over and over again, but is an entire body of ideas that expresses or explains many aspects of a phenomenon. It is not a fuzzy or weak speculation, as is sometimes the popular notion, but a viable declaration that has stood the test of time and has yet to be disproved by

serious scientific endeavors. Often, theories develop and progress through decades of research and are added to and modified by new findings. At some point, evidence of the accuracy and predictability of a theory is so compelling that the next level of confidence is reached and the theory becomes a law, or principle. For example, although we still refer to the germ *theory* of disease, so little question remains that microbes can cause disease that it has clearly passed into the realm of law.

Science and its hypotheses and theories must progress along with technology. As advances in instrumentation allow new, more detailed views of living phenomena, old theories may be reexamined and altered and new ones proposed. But scientists do not take the stance that theories or even "laws" are ever absolutely proved. The characteristics that make scientists most effective in their work are curiosity, open-mindedness, skepticism, creativity, cooperation, and

Hypothesis: Dental caries (cavities) involve dietary sugar or microbial action or both.

Variables:

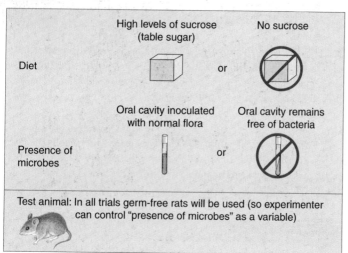

Test animal: In all trials germ-free rats will be used (so experimenter can control "presence of microbes" as a variable)

Experimental Protocol:

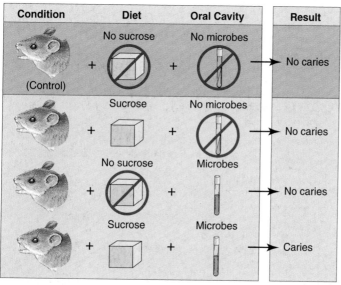

Conclusion: Dental caries will not develop unless both sucrose and microbes are present.

FIGURE 1.11

Variables.   Any factor that can affect the experimental outcome is called a variable, and each combination of variables must be controlled while the hypothesis is being tested.

readiness to revise their views of natural processes as new discoveries are made. The events described on the first page of this chapter provide an important example. See also **Medical Microfile 1.2.**

THE DEVELOPMENT OF MEDICAL MICROBIOLOGY

Early experiments on the sources of microorganisms led to the profound realization that microbes are everywhere: Not only are air and dust full of them, but the entire surface of the earth, its waters, and all objects are inhabited by them. This discovery led to immediate applications in medicine. Thus the seeds of medical microbiology were sown in the mid to latter half of the nineteenth century with the introduction of the germ theory of disease and the resulting use of sterile, aseptic, and pure culture techniques.

The Discovery of Spores and Sterilization
Following Pasteur's inventive work with infusions (see Historical Highlights 1.1), it was not long before English physicist John Tyndall provided the initial evidence that some of the microbes in dust and air have very high heat resistance and that particularly vigorous treatment is required to destroy them. Later, the discovery and detailed description of heat-resistant bacterial endospores by Ferdinand Cohn, a German botanist, clarified the reason that heat would sometimes fail to completely eliminate all microorganisms. The modern sense of the word **sterile,*** meaning completely free of all life forms including spores and viruses, was established from that point on (see chapter 11). The capacity to sterilize objects and materials is an absolutely essential part of microbiology, medicine, dentistry, and some industries.

The Development of Aseptic Techniques
From earliest history, humans experienced a vague sense that "unseen forces" or "poisonous vapors" emanating from decomposing matter could cause disease. As the study of microbiology became more scientific and the invisible was made visible, the fear of such mysterious vapors was replaced by the knowledge and sometimes even the fear of "germs." About 120 years ago, the first studies by Robert Koch clearly linked a microscopic organism with a specific disease. Since that time, microbiologists have conducted a continuous search for disease-causing agents.

At the same time that abiogenesis was being hotly debated, a few budding microbiologists began to suspect that microorganisms could cause not only spoilage and decay but also infectious diseases. It occurred to these rugged individualists that even the human body itself was a source of infection. Dr. Oliver Wendell Holmes, an American physician, observed that mothers who gave birth at home experienced fewer infections than did mothers who gave birth in the hospital, and the Hungarian Dr. Ignaz Semmelweis showed quite clearly that women became infected in the maternity ward after examinations by physicians coming directly from the autopsy room. The English surgeon Joseph Lister took notice of these observations and was the first to introduce **aseptic* techniques** aimed at reducing microbes in a medical setting and preventing wound infections. Lister's concept of asepsis was much more limited than our modern precautions. It mainly involved disinfecting the hands and the air with strong antiseptic chemicals, such as phenol, prior to surgery. It is hard for us to believe, but as recently as the late 1800s surgeons wore street clothes in the operating room and had little idea that hand washing was important. Lister's tech-

*sterile (stair'-il) Gr. *steira,* barren.

*aseptic (ay-sep'-tik) Gr. *a,* no, and *sepsis,* decay or infection. These techniques are aimed at reducing pathogens and do not necessarily sterilize.

FIGURE 1.12

Louis Pasteur (1822–1895), one of the founders of microbiology. Few microbiologists can match the scope and impact of his contributions to the science of microbiology.

FIGURE 1.13

Robert Koch looking through a microscope with colleague Richard Pfeiffer looking on. Robert Koch won the Nobel Prize for Physiology or Medicine in 1905 for his work on tuberculin. Richard Pfeiffer discovered *Haemophilus influenzae* and was a pioneer in typhoid vaccination.

niques and the application of heat for sterilization became the bases for microbial control by physical and chemical methods, which are still in use today.

The Discovery of Pathogens and the Germ Theory of Disease

Two ingenious founders of microbiology, Louis Pasteur of France (**figure 1.12**) and Robert Koch of Germany (**figure 1.13**), introduced techniques that are still used today. Pasteur made enormous contributions to our understanding of the microbial role in wine and beer formation. He invented pasteurization and completed some of the first studies showing that diseases could arise from infection. These studies, supported by the work of other scientists, became known as the **germ theory of disease.** Pasteur's contemporary, Koch, established *Koch's postulates,* a series of proofs that verified the germ theory and could establish whether an organism was pathogenic and which disease it caused (see chapter 13). About 1875, Koch used this experimental system to show that anthrax was caused by a bacterium called *Bacillus anthracis.* So useful were his postulates that the causative agents of 20 other diseases were discovered between 1875 and 1900, and even today, they are the standard for identifying pathogens.

Numerous exciting technologies emerged from Koch's prolific and probing laboratory work. During this golden age of the 1880s, he realized that study of the microbial world would require separating microbes from each other and growing them in culture. It is not an overstatement to say that he and his colleagues invented most of the techniques that are described in chapter 3: inoculation, isolation, media, maintenance of pure cultures, and preparation of specimens for microscopic examination. Other highlights in this era of discovery are presented in later chapters on microbial control (see chapter 11) and vaccination (see chapter 16).

CHAPTER CHECKPOINTS

Our current understanding of microbiology is the cumulative work of thousands of microbiologists, many of whom literally gave their lives to advance knowledge in this field.

The microscope made it possible to see microorganisms and thus to identify their widespread presence, particularly as agents of disease.

Antonie van Leeuwenhoek is considered the father of bacteriology and protozoology because he was the first person to produce precise, correct descriptions of these organisms using microscopes he made himself.

The theory of spontaneous generation of living organisms from "vital forces" in the air was disproved once and for all by Louis Pasteur.

The scientific method is a process by which scientists seek to explain natural phenomena. It is characterized by specific procedures that either support or discredit an initial hypothesis.

Knowledge acquired through the scientific method is rigorously tested by repeated experiments by many scientists to verify its validity. A collection of valid hypotheses is called a theory. A theory supported by much data collected over time is called a law.

Scientific truth changes through time as new research brings new information. Scientists must be able and willing to change theory in response to new data.

Medical microbiologists developed the germ theory of disease and introduced the critically important concept of aseptic technique to control the spread of disease agents.

Koch's postulates are the cornerstone of the germ theory of disease. They are still used today to pinpoint the causative agent of a specific disease.

Louis Pasteur and Robert Koch were the leading microbiologists during the golden age of microbiology (1875–1900). Each had his own research institute.

MEDICAL MICROFILE 1.2
The Serendipity of the Scientific Method: Discovering Drugs

The discoveries in science are not always determined by the strict formulation and testing of a formal hypothesis. Quite often, they involve serendipity* and the luck of being in the right place and time, followed by a curiosity and willingness to change the direction of an experiment. This is especially true in the field of drug discoveries. The first antibiotic, penicillin, was discovered in the late 1920s by Dr. Alexander Fleming, who found a mold colony growing on a culture of bacteria that was wiping out the bacteria. He isolated the active ingredient that eventually launched the era of antibiotics. The search for new drugs to treat infections and cancer has been a continuous focus since that time. Even though the detailed science of testing a drug and working out its chemical structure and action require sophisticated scientific technology, the first and most important part of discovery often lies in a keen eye and an open mind.

In 1987, Dr. Michael Zasloff, a physician and molecular biologist, was doing research in gene expression, using African clawed frogs as a source of eggs. After performing surgery on the frogs and routinely placing them back in a nonsterile aquarium, he was surprised to notice that most of the time the frogs did not get infected or die. If the animal had been a mammal such as a mouse, it would probably not have survived the nonsterile surgery. This led him to conclude that the frog's skin must provide some form of natural protection. He observed that when the skin was stimulated by injury or irritants, it formed a thick white coating in a few moments that reminded him of a self-made "bandage" over the wound. He took a section of skin and extracted the components that were responsible for killing the microbes. His tests showed that they were small proteins called peptides, which he named

magainins, after the Hebrew word for shield. Within 6 months of these findings, Dr. Zasloff made the decision to completely change the subject of his research and started up a new biotechnology company (Magainin Pharmaceuticals) to explore the therapeutic potential for magainins as well as other frog peptides.

The initial tests on this new class of drugs would indicate that they do indeed destroy a variety of bacteria as well as fungi, protozoa, and viruses. Although they are toxic to human cells too, this makes them a possible candidate for cancer treatment. Currently the drugs are being synthesized and tested in the lab for effectiveness and safety. Dr. Zasloff's intriguing observation and subsequent experiments had the impact of opening up a whole new area of biology: isolating antimicrobic peptides from multicellular organisms. Additional studies have shown that these compounds are widespread among amphibians, fish, birds, mammals, and plants. A number of companies are involved in developing applications for animal peptides. This discovery has been well timed, since resistance among microorganisms to traditional drugs is a continuing problem.

An African clawed frog responding to an irritant on its back first forms spots and then a thick opaque blotch of protective chemicals.

*serendipity Making useful discoveries by accident.

Taxonomy: Organizing, Classifying, and Naming Microorganisms

Students just beginning their microbiology studies are often dismayed by the seemingly endless array of new, unusual, and sometimes confusing names for microorganisms. Learning microbial **nomenclature*** is very much like learning a new language, and occasionally its demands may be a bit overwhelming. But paying attention to proper microbial names is just like following a baseball game or a movie plot: You cannot tell the players apart without a program! Your understanding and appreciation of microorganisms will be greatly improved by learning a few general rules about how they are named.

The formal system for organizing, classifying, and naming living things is **taxonomy.*** This science originated more than 250 years ago when Carl von Linné (also known as Linnaeus; 1701–1778), a Swedish botanist, laid down the basic rules for taxonomic categories,

or **taxa.*** Von Linné realized early on that a system for recognizing and defining the properties of living things would prevent chaos in scientific studies by providing each organism with a unique name and an exact "slot" in which to catalogue it. This classification would then serve as a means for future identification of that same organism and permit workers in many biological fields to know if they were indeed discussing the same organism. The von Linné system has served well in categorizing the 2 million or more different types of organisms that have been discovered since that time.

The primary concerns of taxonomy are classification, nomenclature, and identification. These three areas are interrelated and play a vital role in keeping a dynamic inventory of the extensive array of living things. *Classification* is the orderly arrangement of organisms into groups, preferably in a format that shows evolutionary relationships. *Nomenclature* is the process of assigning names to the various taxonomic rankings of each microbial species. *Identification* is the process of discovering and recording the traits of organisms so that they may be placed in an overall taxonomic scheme. A survey of some general methods of identification appears in chapter 3.

*nomenclature (noh′-men-klay″-chur) L. *nomen,* name, and *clare,* to call. A system of naming.

*taxonomy (tacks-on″-uh-mee) Gr. *taxis,* arrangement, and *nomos,* name.

*taxa (tacks′-uh) sing. taxon.

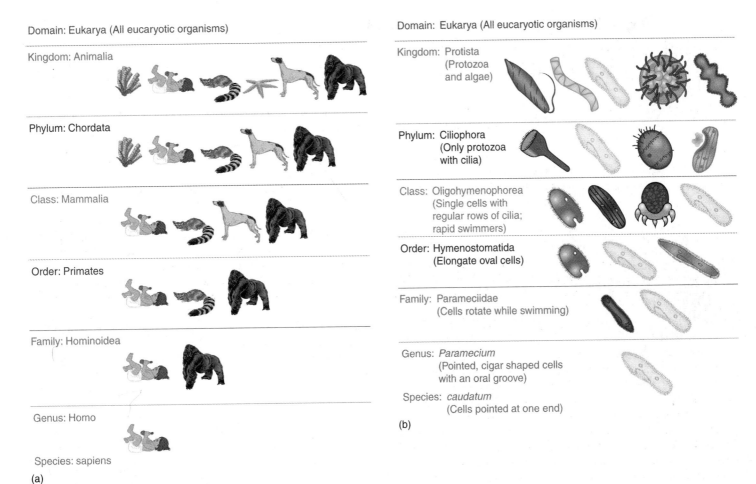

Domain: Eukarya (All eucaryotic organisms)

Kingdom: Animalia

Phylum: Chordata

Class: Mammalia

Order: Primates

Family: Hominoidea

Genus: Homo

Species: sapiens

(a)

Domain: Eukarya (All eucaryotic organisms)

Kingdom: Protista
(Protozoa
and algae)

Phylum: Ciliophora
(Only protozoa
with cilia)

Class: Oligohymenophorea
(Single cells with
regular rows of cilia;
rapid swimmers)

Order: Hymenostomatida
(Elongate oval cells)

Family: Parameciidae
(Cells rotate while swimming)

Genus: *Paramecium*
(Pointed, cigar shaped cells
with an oral groove)

Species: *caudatum*
(Cells pointed at one end)

(b)

FIGURE 1.14

Sample taxonomy.   Two organisms belonging to the Eukarya domain, traced through their taxonomic series. **(a)** Modern humans, *Homo sapiens*. **(b)** A common protozoan, *Paramecium caudatum.*

THE LEVELS OF CLASSIFICATION

The main taxa, or groups, in a classification scheme are organized into several descending ranks, beginning with **domain,** which is a giant, all-inclusive category based on a unique cell type, and ending with a **species,*** the smallest and most specific taxon. All the members of a domain share only one or few general characteristics, whereas members of a species are essentially the same kind of organism—that is, they share the majority of their characteristics. The taxa between the top and bottom levels are, in descending order: **kingdom, phylum*** or **division,**[3] **class, order, family,** and **genus.*** Thus, each domain can be subdivided into a series of kingdoms, each kingdom is made up of several phyla, each phylum contains several classes, and so on. Because taxonomic schemes are to some extent artificial, certain groups of organisms do not exactly fit into the eight taxa. In that case, additional levels can be imposed immediately above (super) or below (sub) a taxon, giving us such categories as superphylum and subclass.

To illustrate the fine points of this system, we compare the taxonomic breakdowns of a human and a protozoan (**figure 1.14**). Humans and protozoa belong to the same domain (Eukarya) but are placed in different kingdoms. To emphasize just how broad the category kingdom is, ponder the fact that we belong to the same kingdom as jellyfish. Of the several phyla within this kingdom, humans belong to the Phylum Chordata, but even a phylum is rather all-inclusive, considering that humans share it with other vertebrates as well as creatures called sea squirts. The next level, Class Mammalia, narrows the field considerably by grouping only those vertebrates that have hair and suckle their young. Humans belong to the Order Primates, a group that also includes apes, monkeys, and lemurs. Next comes the Family Hominoidea, containing only humans and apes. The final levels are our genus, *Homo* (all races of modern and ancient humans), and our species, *sapiens* (meaning wise). Notice that for both the human and the protozoan, the categories become less inclusive and the individual members more closely related. Other examples of classification schemes are provided in sections of chapters 4 and 5 and in several later chapters.

It is important to remember that all taxonomic **hierarchies*** are based on the judgment of scientists with certain expertise in a

---

3. The term *phylum* is used for protozoa and animals; the term *division* is used for bacteria, algae, plants, and fungi.

*species (spee´-sheez) L. *specere,* kind. In biology, this term is always in the plural form.

*phylum (fy´-lum) pl. phyla (fye´-luh) Gr. *phylon,* race.

*genus (jee´-nus) pl. genera (jen´-er-uh) L. birth, kind.

*hierarchy (hy´-ur-ar-kee) L. *hierarchia,* levels of power. Things arranged in the order of rank.

particular group of organisms and that not all other experts may agree with the system being used. Consequently, no taxa are permanent to any degree; they are constantly being revised and refined as new information becomes available or new viewpoints become prevalent. Because this text does not aim to emphasize details of taxonomy, we will usually be concerned with only the most general (kingdom, phylum) and specific (genus, species) levels.

ASSIGNING SPECIFIC NAMES

Many larger organisms are known by a common name suggested by certain dominant features. For example, a bird species might be called a red-headed blackbird or a flowering species a black-eyed Susan. Some species of microorganisms (especially pathogens) are also called by informal names, such as the gonococcus (*Neisseria gonorrhoeae*) or the tubercle bacillus (*Mycobacterium tuberculosis*), but this is not the usual practice. If we were to adopt common names such as the "little yellow coccus"* or the "club-shaped diphtheria bacterium,"* the terminology would become even more cumbersome and challenging than scientific names. Even worse, common names are notorious for varying from region to region, even within the same country. A decided advantage of standardized nomenclature is that it provides a universal language, thereby enabling scientists from all countries on the earth to freely exchange information.

The method of assigning the **scientific,** or **specific name** is called the **binomial (two-name) system of nomenclature.** The scientific name is always a combination of the generic (genus) name followed by the species name. The generic part of the scientific name is capitalized, and the species part begins with a lowercase letter. Both should be italicized (or underlined if italics are not available), as follows:

Staphylococcus aureus

Because other taxonomic levels are not italicized and consist of only one word, one can always recognize a scientific name. An organism's scientific name is sometimes abbreviated to save space, as in *S. aureus,* but only if the genus name has already been stated. The source for nomenclature is usually Latin or Greek. If other languages such as English or French are used, the endings of these words are revised to have Latin endings. In general, the name first applied to a species will be the one that takes precedence over all others. An international group oversees the naming of every new organism discovered, making sure that standard procedures have been followed and that there is not already an earlier name for the organism or another organism with that same name. The inspiration for names is extremely varied and often rather imaginative. Some species have been named in honor of a microbiologist who originally discovered the microbe or who has made outstanding contributions to the field. Other names may designate a characteristic of the microbe (shape, color), a location where it was found, or a disease

it causes. Some examples of specific names, their pronunciations, and their origins are:

- *Staphylococcus aureus* (staf′-i-lo-kok′-us ah′-ree-us) Gr. *staphule,* bunch of grapes, *kokkus,* berry, and Gr. *aureus,* golden. A common bacterial pathogen of humans.
- *Campylobacter jejuni* (cam-pee′-loh-bak-ter jee-joo′-neye) Gr. *kampylos,* curved, *bakterion,* little rod, and *jejunum,* a section of intestine. One of the most important causes of intestinal infection worldwide.
- *Lactobacillus sanfrancisco* (lak″-toh-bass-ill′-us san-fran-siss′-koh) L. *lacto,* milk, and *bacillus,* little rod. A bacterial species used to make sourdough bread.
- *Vampirovibrio chlorellavorus* (vam-py′-roh-vib-ree-oh klor-ell-ah′-vor-us) F. *vampire;* L. *vibrio,* curved cell; *Chlorella,* a genus of green algae; and *vorus,* to devour. A small, curved bacterium that sucks out the cell juices of *Chlorella.*
- *Giardia lamblia* (jee-ar′-dee-uh lam′-blee-uh) for Alfred Giard, a French microbiologist, and Vilem Lambl, a Bohemian physician, both of whom worked on the organism, a protozoan that causes a severe intestinal infection.

Here's a helpful hint: These names may seem difficult to pronounce and the temptation is to simply "slur over them." But when you encounter the name of a microorganism in the chapters ahead it will be extremely useful to you to take the time to sound them out and repeat them until they seem familiar. You are much more likely to remember them that way—and they are less likely to end up in a tangled heap with all of the new language you will be learning.

THE ORIGIN AND EVOLUTION OF MICROORGANISMS

As we indicated earlier, *taxonomy,* the classification of biological species, is a system used to organize all of the forms of life. In biology today there are different methods for deciding on taxonomic categories, but they all rely on the degree of relatedness among organisms. The natural relatedness between groups of living things is called their *phylogeny.* So, biologists use phylogenetic relationships to create a system of taxonomy.

To understand the relatedness among organisms, we must understand some fundamentals of evolution.

Evolution* is an important theme that underlies all of biology, including microbiology. Put simply, evolution states that living things change gradually through hundreds of millions of years and that these evolvements result in various types of structural and functional changes through many generations. The process of evolution is selective: Those changes that most favor the survival of a particular organism or group of organisms tend to be retained, and those that are less beneficial to survival tend to be lost. Space does not permit a detailed analysis of evolutionary theories, but the occurrence of evolution is supported by a tremendous amount of evidence from the fossil record and from the study of **morphology,*** **physiology,*** and **genetics** (inheritance). Evolution accounts

---

Micrococcus luteus (my″-kroh-kok′-us loo′-tee-us) Gr. *micros,* small, and *kokkus,* berry; L. *luteus,* yellow.

Corynebacterium diphtheriae (kor-eye″-nee-bak-ter′-ee-yum dif′-theer-ee-eye) Gr. *coryne,* club, *bacterion,* little rod, and *diphtheriae,* the causative agent of the disease diphtheria.

*evolution (ev-oh-loo′-shun) L. *evolutio,* to roll out.

*morphology (mor-fol′-oh-jee) Gr. *morphos,* form, and *logos,* to study. The study of organismic structure.

*physiology (fiz″-ee-ol′-oh-jee) Gr. physis, nature. The study of the function of organisms.

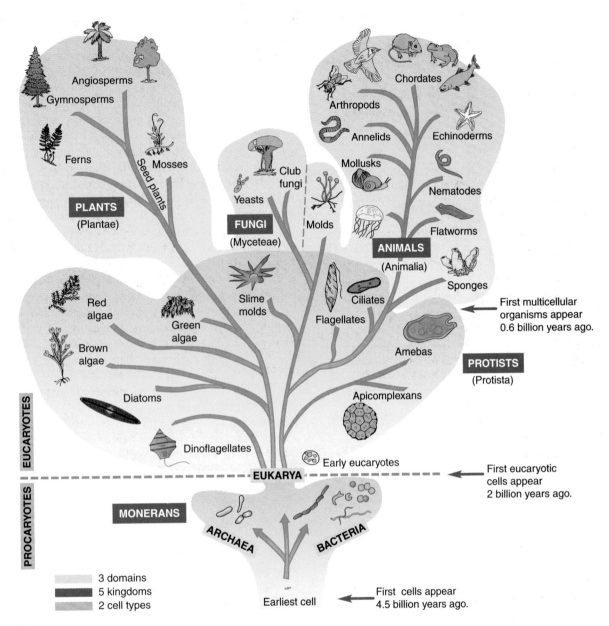

FIGURE 1.15

Traditional Whittaker system of classification.   Kingdoms are based on cell structure and type, the nature of body organization, and nutritional type. Bacteria and Archaea (monerans) have procaryotic cells and are unicellular. Protists have eucaryotic cells and are mostly unicellular. They can be photosynthetic (algae), or they can feed on other organisms (protozoa). Fungi have eucaryotic cells and are unicellular or multicellular; they have cell walls and are not photosynthetic. Plants have eucaryotic cells, are multicellular, have cell walls, and are photosynthetic. Animals have eucaryotic cells, are multicellular, do not have cell walls, and derive nutrients from other organisms.

After Dolphin, Biology Lab Manual, *4th ed., Fig. 14.1, p. 177, McGraw-Hill Companies.*

for the millions of different species on the earth and their adaptation to its many and diverse habitats.

Evolution is founded on two preconceptions: (1) that all new species originate from preexisting species and (2) that closely related organisms have similar features because they evolved from common ancestral forms. Usually, evolution progresses toward greater complexity, and evolutionary stages range from simple, primitive forms that are close to an ancestral organism to more complex, advanced forms. Although we use the terms *primitive* and *advanced* to denote the degree of change from the original set of ancestral traits, it is very important to re-

alize that all species presently residing on the earth are modern, but some have arisen more recently in evolutionary history than others.

The phylogeny, or evolutionary relatedness, of organisms is often represented by a drawing of a tree. The trunk of the tree represents the main ancestral lines and the branches show offshoots into specialized groups of organisms. This sort of arrangement places the more ancient groups at the bottom and the more recent ones at the top. The branches may also indicate origins, how closely related various organisms are, and an approximate timescale for evolutionary history (**figures 1.15** and **1.16**).

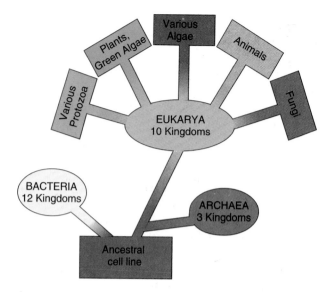

FIGURE 1.16

Woese system. A system for representing the origins of cell lines and major taxonomic groups as proposed by Carl Woese and colleagues. They propose three distinct cell lines placed in superkingdoms called domains. The first primitive cells, called progenotes, were ancestors of both lines of procaryotes (Domains Bacteria and Archaea), and the Archaea emerged from the same cell line as eucaryotes (Domain Eukarya). Some of the traditional kingdoms are still present with this system (see figure 1.15). Protozoa and some algal groups (called various algae here) are lumped into general categories.

SYSTEMS OF PRESENTING A UNIVERSAL TREE OF LIFE

The first phylogenetic trees of life were constructed on the basis of just two kingdoms (plants and animals). In time, it became clear that certain organisms did not truly fit either of those categories, so a third kingdom for simpler organisms that lacked tissue differentiation (protists) was recognized. Eventually, when significant differences became evident even among the protists, Robert Whittaker proposed a fourth kingdom for the bacteria and a fifth one for the fungi.

From that time until very recently, phylogenetic trees of life have looked like the one in figure 1.15. The relationships that were considered in constructing the tree were those based on structural similarities and differences, such as bacterial and eucaryotic cellular organization, and the way the organisms got their nutrition. By 1959 these methods indicated that there were five major subdivisions, or kingdoms: the monera, fungi, protists, plants, and animals. Within these kingdoms were two major cell types, the procaryotic and eucaryotic. The system has proved very useful.

Recently, newer methods for determining phylogeny have led to the development of a differently shaped tree—with important implications for our understanding of evolutionary relatedness. The new techniques are those of *molecular biology,* defined as the study of genes—both their structure and function—at the molecular level. Molecular biological methods have demonstrated that certain types of molecules in cells, called small ribosomal ribonucleic acid (rRNA), provide a "living record" of the evolutionary history of an organism. Analysis of this molecule in procaryotic and eucaryotic cells indicates that certain unusual cells called archaea (originally archaebacteria) are so different from the other two groups that they should be included in a separate superkingdom. Archaea have some structures in common with bacteria and are characterized by their ability to live in extreme environments, such as hot springs or highly salty environments. Under the microscope they resemble bacteria, but molecular biology has revealed that the cells of archaea, though procaryotic in nature, are actually more closely related to eucaryotic cells than to bacterial cells (see table 4.7). To reflect these relationships, Carl Woese and George Fox have proposed a system that assigns all organisms to one of three domains, each described by a different type of cell (see figure 1.16). The procaryotic cell types are placed in the Domains **Archaea** and **Bacteria.** Eucaryotes are all placed in the Domain **Eukarya.** It is believed that these three superkingdoms arose from an ancestor most similar to the archaea. This new system is still undergoing analysis and somewhat complicates the presentation of organisms in that it disposes of some traditional groups, although many of the traditional kingdoms still work within this framework (see figure 1.15). The original Kingdom Protista is now a collection of protozoa and algae that exist in several separate kingdoms (see chapter 5).

This new scheme will not greatly affect our presentation of most microbes, because we will be discussing them at the genus or species level. But you should be aware that biological taxonomy, and more importantly, our view of how organisms evolved on earth, is in a period of transition. And don't forget that our methods of classification reflect our current understanding and are constantly changing as new information is uncovered.

In the interest of balance, we will also present the traditional Whittaker system of classification (see figure 1.15). As we already mentioned this system places all living things in one of five basic kingdoms:

1. the Procaryotae or Monera,
2. the Protista,
3. the Myceteae or Fungi,
4. the Plantae, and
5. the Animalia.

The simple, single-celled organisms at the base of the family tree are in the Kingdom Procaryotae (also called Monera). Because only those organisms with procaryotic cells are placed in it, the nature of cell structure is the main defining characteristic for this kingdom. It includes all of the microorganisms commonly known as **eubacteria,*** cells with typical procaryotic cell structure, and the **archaebacteria,*** cells with atypical cell structure that live in extreme environments (high salt and temperatures).

The other four kingdoms contain organisms composed of eucaryotic cells. The Kingdom **Protista*** contains mostly single-

---

*eubacteria (yoo″-bak-ter′-ee-uh) Gr. *eu,* true, and *bakterion,* little rod. All bacteria besides the archaebacteria.

*archaebacteria (ark″-ee-uh-bak-ter′-ee-uh) Gr. *archaios,* ancient. The same name as archaea.

*Protista (pro-tiss′-tah) Gr. *protos,* the first.

celled microbes that lack more complex levels of organization, such as tissues. Its members include both the microscopic algae, defined as independent photosynthetic cells with rigid walls, and the protozoans, animal-like creatures that feed upon other live or dead organisms and lack cell walls. More information on this group's taxonomy is given in chapter 5. The Kingdom **Myceteae*** contains the fungi, single- or multi-celled eucaryotes that are encased in cell walls and absorb nutrients from other organisms (see chapter 5). With the exception of certain infectious worms and arthropods, the final two kingdoms, Animalia and Plantae, are generally not included in the realm of microbiology because most are large, multicellular organisms with tissues, organs, and organ systems. In general, animals move freely and feed on other organisms, whereas plants grow in an attached state and exhibit a nutritional scheme based on photosynthesis.

Please note that viruses are *not* included in any of the classification or evolutionary schemes, because they are not cells and their position cannot be given with any confidence. Their special taxonomy is discussed in chapter 6.

---

*Myceteae (my-cee′-tee-eye) Gr. *mycos*, the fungi.

CHAPTER CHECKPOINTS

Taxonomy is the formal filing system scientists use to classify living organisms. It puts every organism in its place and makes a place for every living organism.

The taxonomic system has three primary functions: classification, nomenclature, and identification of species.

The eight major taxa, or groups, in the taxonomic system are (in descending order): domain, kingdom, phylum or division, class, order, family, genus, and species.

The binomial system of nomenclature describes each living organism by two names: genus and species.

Taxonomy groups organisms by phylogenetic similarity, which in turn is based on evolutionary similarities in morphology, physiology, and genetics.

Evolutionary patterns show a treelike branching from simple, primitive life forms to complex, advanced life forms.

The Woese classification system places all eucaryotes in the Domain (Superkingdom) Eukarya and subdivides the procaryotes into the two Domains Archaea and Bacteria.

The Whittaker five-kingdom classification system places all bacteria in the Kingdom Procaryotae and subdivides the eucaryotes into Kingdoms Protista, Myceteae, Animalia, and Plantae.

CHAPTER CAPSULE WITH KEY TERMS

I. **Microbiology** is the study of **bacteria, viruses, fungi, protozoa,** and **algae,** which are collectively called **microorganisms,** or microbes. In general, microorganisms are **microscopic** and, unlike **macroscopic** organisms, which are readily visible, they require magnification to be adequately observed or studied.

II. The simplicity, growth rate, and adaptability of microbes are some of the reasons that microbiology is so diverse and has branched out into many subsciences and applications. Important subsciences include **immunology, epidemiology,** public health, food, dairy, aquatic, and industrial microbiology.

III. Microbes live in most of the world's habitats and are indispensable for normal, balanced life on earth. They play many roles in the functioning of the earth's ecosystems.
   A. Microbes are involved in nutrient production and energy flow. Algae and certain bacteria trap the sun's energy to produce food through **photosynthesis.**
   B. Other microbes are responsible for the breakdown and recycling of nutrients through **decomposition.** Microbes are essential to the maintenance of the air, soil, and water.

IV. Microbes have been called upon to solve environmental, agricultural, and medical problems.
   A. **Biotechnology** applies the power of microbes toward the manufacture of industrial products, foods, and drugs.
   B. Microbes form the basis of **genetic engineering** and **recombinant DNA** technology, which alter genetic material to produce new products and modified life forms.
   C. With **bioremediation,** microbes are used to clean up pollutants and wastes in natural environments.

V. Nearly 2,000 microbes are **pathogens** that cause **infectious diseases.** Infectious diseases result in high levels of mortality and

morbidity. Many infections are **emerging,** meaning that they are newly identified pathogens gaining greater prominence. Many older diseases are also increasing.

VI. **Important Historical Events**
   A. Microbiology as a science is about 200 years old. Hundreds of contributors have provided discoveries and knowledge to enrich our understanding.
   B. With his simple microscope, Leeuwenhoek discovered organisms he called animalcules. As a consequence of his findings and the rise of the **scientific method,** the notion of **spontaneous generation,** or **abiogenesis,** was eventually abandoned for **biogenesis.** The scientific method develops rational **hypotheses** and **theories** that can be tested. Theories that withstand repeated scrutiny become law in time.
   C. Early microbiology blossomed with the conceptual developments of **sterilization, aseptic techniques,** and the **germ theory of disease.**

VII. **Characteristics and Classification of Microorganisms**
   A. **Taxonomy** is a hierarchy scheme for the classification, identification, and **nomenclature** of organisms, which are grouped in categories called **taxa,** based on features ranging from general to specific.
      1. Starting with the broadest category, the taxa are **domain, kingdom, phylum** (or **division**), **class, order, family, genus,** and **species.** Organisms are assigned **binomial scientific names** consisting of their genus and species names.
      2. The latest classification scheme for living things is based on the genetic structure of their ribosomes. The Woese-Fox system recognizes three domains: **Archaea,** simple procaryotes that live in extremes; **Bacteria,** typical procaryotes; and **Eukarya,** all types of eucaryotic organisms.

3. An alternative classification scheme uses a five-kingdom organization: Kingdom Procaryotae (Monera), containing the **eubacteria** and the archaea; Kingdom Protista, containing primitive unicellular microbes such as algae and protozoa; Kingdom Myceteae, containing the fungi; Kingdom Animalia, containing animals; and Kingdom Plantae, containing plants.

MULTIPLE-CHOICE QUESTIONS

Select the correct answer from the answers provided. For questions with blanks, choose the combination of answers that most accurately completes the statement.

1. Which of the following is not considered a microorganism?
   a. alga
   b. bacterium
   c. protozoan
   d. mushroom

2. An area of microbiology that is concerned with the occurrence of disease in human populations is
   a. immunology
   b. parasitology
   c. epidemiology
   d. bioremediation

3. Which process involves the deliberate alteration of an organism's genetic material?
   a. bioremediation
   b. biotechnology
   c. decomposition
   d. recombinant DNA

4. A prominent difference between procaryotic and eucaryotic cells is the
   a. larger size of procaryotes
   b. lack of pigmentation in eucaryotes
   c. presence of a nucleus in eucaryotes
   d. presence of a cell wall in procaryotes

5. Which of the following parts was absent from Leeuwenhoek's microscopes?
   a. focusing screw
   b. lens
   c. specimen holder
   d. condenser

6. Abiogenesis refers to the
   a. spontaneous generation of organisms from nonliving matter
   b. development of life forms from preexisting life forms
   c. development of aseptic technique
   d. germ theory of disease

7. A hypothesis can be defined as
   a. a belief based on knowledge
   b. knowledge based on belief
   c. a scientific explanation that is subject to testing
   d. a theory that has been thoroughly tested

8. Which early microbiologist was most responsible for developing sterile laboratory techniques?
   a. Louis Pasteur
   b. Robert Koch
   c. Carl von Linné
   d. John Tyndall

9. Which scientist is most responsible for finally laying the theory of spontaneous generation to rest?
   a. Joseph Lister
   b. Robert Koch
   c. Francesco Redi
   d. Louis Pasteur

10. When a hypothesis has been thoroughly supported by long-term study and data, it is considered
    a. a law
    b. a speculation
    c. a theory
    d. proved

11. Which is the correct order of the taxonomic categories, going from most specific to most general?
    a. domain, kingdom, phylum, class, order, family, genus, species
    b. division, domain, kingdom, class, family, genus, species
    c. species, genus, family, order, class, phylum, kingdom, domain
    d. species, family, class, order, phylum, kingdom

12. By definition, organisms in the same _____ are more closely related than are those in the same _____.
    a. order, family
    b. class, phylum
    c. family, genus
    d. phylum, division

13. Which of the following are procaryotic?
    a. bacteria
    b. archaea
    c. protists
    d. both a and b

14. Order the following items by size, using numbers: 1 = smallest and 8 = largest.
    _7_ AIDS virus
    _4_ ameba
    _5_ rickettsia
    _6_ protein
    _8_ worm
    _2_ coccus-shaped bacterium
    _3_ white blood cell
    _1_ atom

CONCEPT QUESTIONS

These questions are suggested as a *writing-to-learn* experience. For each question, compose a one- or two-paragraph answer that includes the factual information needed to completely address the question. Discuss the concepts in a sequence that allows you to present the subject using clear logic and correct terminology

1. Explain the important contributions microorganisms make in the earth's ecosystems.

2. Describe five different ways in which humans exploit microorganisms for our benefit.

3. Identify the groups of microorganisms included in the scope of microbiology, and explain the criteria for including these groups in the field.

4. Why was the abandonment of the spontaneous generation theory so significant? Using the scientific method, describe the steps you would take to test the theory of spontaneous generation.

5. a. Differentiate between a hypothesis and a theory.
   b. Is the germ theory of disease really a law, and why?

6. a. Differentiate between taxonomy, classification, and nomenclature.
   b. What is the basis for a phylogenetic system of classification?
   c. What is a binomial system of nomenclature, and why is it used?
   d. Give the correct order of taxa, going from most general to most specific. A mnemonic (memory) device for recalling the order is *Did King Philip Come Over For Good Spaghetti?*

7. Compare the new domain system with the five-kingdom system. Does the newer system change the basic idea of procaryotes and eucaryotes? What is the third cell type?

CRITICAL-THINKING QUESTIONS

Critical thinking is the ability to reason and solve problems using facts and concepts. It requires you to apply information to new or different circumstances, to integrate several ideas to arrive at a solution, and to perform practical demonstrations as part of your analysis. These questions can be approached from a number of angles, and in most cases, they do not have a single correct answer.

1. What do you suppose the world would be like if there were cures for all infectious diseases and a means to destroy all microbes? What characteristics of microbes will prevent this from ever happening?

2. a. Where do you suppose the "new" infectious diseases come from?
   b. Name some factors that could cause older diseases to show an increase in the number of cases.
   c. Comment on the sensational ways that some tabloid media portray infectious diseases to the public.

3. Look up each disease shown on figure 1.4 in the index and see which ones could be prevented by vaccines or treated with drugs. How many do you think could have been prevented by modern medicine?

4. Correctly label the types of microorganisms in the drawing at right, using basic characteristics featured in the chapter. (Organisms are not to scale.)

5. What events, discoveries, or inventions were probably the most significant in the development of microbiology and why?

6. List the major variables in abiogenesis outlined in Historical Highlights 1.1 and explain how each was tested and controlled by the scientific method.

7. Can you develop a scientific hypothesis and means of testing the cause of stomach ulcers? (Is it caused by an infection? By too much acid? By a genetic disorder?)

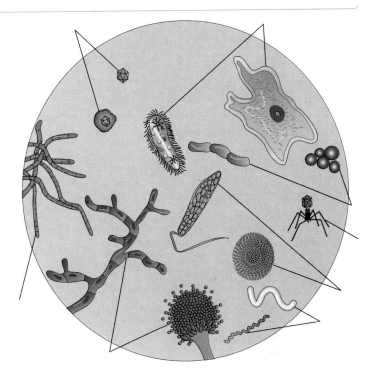

8. Construct the scientific name of a newly discovered species of bacterium, using your name, a pet's name, a place, or a unique characteristic. Be sure to use proper notation and endings.

9. Archaea are found in hot, sulfuric, acidic, salty habitats, much like the early earth's conditions. Speculate on the origins of life, especially as it relates to the archaea.

INTERNET SEARCH TOPICS

1. Using a search engine on the World Wide Web, search for the phrase *emerging diseases*. Adding terms like WHO and CDC will refine your search and take you to several appropriate websites. List the top 10 emerging diseases in the United States and worldwide.

2. Visit the student Online Learning Center at www.mhhe.com/talaro5. Go to chapter 1, Internet Search Topics, and log on to the available websites to:
   a. Explore the "trees of life." Compare the main relationships among the three major domains.
   b. Observe the comparative sizes of microbes arrayed on the head of a pin.

CHAPTER 2

From Atoms to Cells:
A Chemical Connection

I n laboratories all over the world, sophisticated technology is being developed for a wide variety of scientific applications. Refinements in molecular biology techniques now make it possible to routinely identify and cultivate microorganisms, detect genetic disease, diagnose cancer, sequence the genes of organisms, break down toxic wastes, synthesize drugs and industrial products, and genetically engineer microorganisms, plants, and animals. A common thread that runs through new technologies and hundreds of traditional techniques is that, at some point, they involve chemicals and chemical reactions. In fact, if nearly any biological event is traced out to its ultimate explanation, it will invariably involve atoms, molecules, reactions, and bonding.

It is this relationship between the sciences that makes a background in chemistry necessary to biologists and microbiologists. Students with a basic chemistry background will enhance their understanding of and insight into microbial structure and function, metabolism, genetics, drug therapy, immune reactions, and infectious disease. This chapter has been organized to promote a working knowledge of atoms, molecules, bonding, solutions, pH, and biochemistry and to build foundations to later chapters. It concludes with an introduction to cells and a general comparison of procaryotic and eucaryotic cells as a preparation for chapters 4 and 5.

Microbiologists isolate bacteria from caves and other unique sites, in hopes of discovering new microbial species and compounds that may be sources for antibiotics and other drugs.

Chapter Overview

- The understanding of living cells and processes is enhanced by a knowledge of chemistry.
- The structure and function of all matter in the universe is based on atoms.
- Atoms have unique structures and properties that allow chemical reactions to occur.
- Atoms contain protons, neutrons, and electrons in combinations to form elements.
- Living things are composed of approximately 25 different elements.
- Elements interact to form bonds that result in molecules and compounds with different characteristics than the elements that form them.
- Atoms can show variations in charge and polarity.
- Atoms and molecules undergo chemical reactions such as oxidation/reduction, ionization, and dissolution.
- The properties of carbon have been critical in forming macromolecules of life such as proteins, fats, carbohydrates, and nucleic acids.
- The nature of macromolecule structure and shape dictates its functions.

- Cells carry out fundamental activities of life, such as growth, metabolism, reproduction, synthesis, and transport, that are all essentially chemical reactions on a grand scale.

Atoms, Bonds, and Molecules: Fundamental Building Blocks

The universe is composed of an infinite variety of substances existing in the gaseous, liquid, and solid states. All such tangible materials that occupy space and have mass are called **matter.** The organization of matter—whether air, rocks, or bacteria—begins with individual

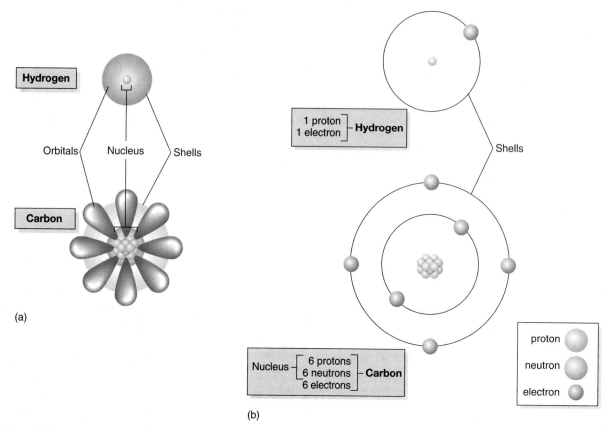

FIGURE 2.1

Models of atomic structure. **(a)** Three-dimensional models of hydrogen and carbon that approximate their actual structure. The nucleus is surrounded by electrons in orbitals that occur in levels called shells. Hydrogen has just one shell and one orbital. Carbon has two shells and four orbitals; the shape of the outermost orbitals is paired lobes rather than circles or spheres. **(b)** Simple models of the same atoms make it easier to show the numbers and arrangements of shells and electrons, and the numbers of protons and neutrons in the nucleus. (Not to accurate scale).

building blocks called atoms. An **atom*** is defined as a tiny particle that cannot be subdivided into smaller substances without losing its properties. Even in a science dealing with very small things, an atom's minute size is striking; for example, an oxygen atom is only 0.0000000013 mm (0.0013 nm) in diameter, and one million of them in a cluster would barely be visible to the naked eye.

Although scientists have not directly observed the detailed structure of an atom, the exact composition of atoms has been well established by extensive physical analysis using sophisticated instruments. In general, an atom derives its properties from a combination of subatomic particles called **protons** (p^+), which are positively charged; **neutrons** (n^0), which have no charge (are neutral); and **electrons** (e^-), which are negatively charged. The relatively larger protons and neutrons make up a central core, or **nucleus,** that is surrounded by one or more electrons **(figure 2.1).** The nucleus makes up the larger mass (weight) of the atom, whereas the electron region accounts for the greater volume. To get a perspective on proportions, consider this: If an atom were the size of a football stadium, the nucleus would be about the size of a marble! The stability of atomic structure is largely maintained by: (1) the mutual attraction of the protons and electrons (opposite charges attract

each other) and (2) the exact balance of proton number and electron number, which causes the opposing charges to cancel each other out. At least in theory then, isolated intact atoms do not carry a charge.

DIFFERENT TYPES OF ATOMS: ELEMENTS AND THEIR PROPERTIES

All atoms share the same fundamental structure. All protons are identical, all neutrons are identical, and all electrons are identical. But when these subatomic particles come together in specific, varied combinations, unique types of atoms called **elements** result. Each element has a characteristic atomic structure and predictable chemical behavior. To date, 92 naturally occurring elements have been described, and 18 have been produced artificially by physicists. By convention, an element is assigned a distinctive name with an abbreviated shorthand symbol. **Table 2.1** lists some of the elements common to biological systems, their atomic characteristics, and some of the natural and applied roles they play.

THE MAJOR ELEMENTS OF LIFE AND THEIR PRIMARY CHARACTERISTICS

The unique properties of each element result from the numbers of protons, neutrons, and electrons it contains, and each element can be identified by certain physical measurements.

*atom (at′-um) Gr. *atomos,* not cut.

TABLE 2.1

The Major Elements of Life and Their Primary Characteristics

Element	Atomic Symbol*	Atomic Number	Atomic Weight	Ionized Form**	Significance in Microbiology
Calcium	Ca	20	40.1	Ca^{++}	Part of outer covering of certain shelled amebas; stored within bacterial spores
Carbon	C	6	12.0	—	Principal structural component of biological molecules
		6	14.0	—	Isotope used in dating fossils
Chlorine	Cl	17	35.5	Cl^-	Component of disinfectants; used in water purification
Cobalt	Co	27	58.9	Co^{++}, Co^{+++}	Trace element needed by some bacteria to synthesize vitamins
		27	60	—	An emitter of gamma rays; used in food sterilization; used to treat cancer
Copper	Cu	29	63.5	Cu^+, Cu^{++}	Necessary to the function of some enzymes; Cu salts are used to treat fungal and worm infections
Hydrogen	H	1	1	H^+	Necessary component of water and many organic molecules; H_2 gas released by bacterial metabolism
		1	3	—	Tritium has 2 neutrons; radioactive; used in clinical laboratory procedures
Iodine	I	53	126.9	I^-	A component of antiseptics and disinfectants; contained in a reagent of the Gram stain
		53	131, 125		Radioactive isotopes for diagnosis and treatment of cancers
Iron	Fe	26	55.8	Fe^{++}, Fe^{+++}	Necessary component of respiratory enzymes; some microbes require it to produce toxin
Magnesium	Mg	12	24.3	Mg^{++}	A trace element needed for some enzymes; component of chlorophyll pigment
Manganese	Mn	25	54.9	Mn^{++}, Mn^{+++}	Trace elements for certain respiratory enzymes
Nitrogen	N	7	14.0	—	Component of all proteins and nucleic acids; the major atmospheric gas
Oxygen	O	8	16.0	—	An essential component of many organic molecules; molecule used in metabolism by many organisms
Phosphorus	P	15	31	—	A component of ATP, nucleic acids, cell membranes; stored in granules in cells
		15	32	—	Radioactive isotope used as a diagnostic and therapeutic agent
Potassium	K	19	39.1	K^+	Required for normal ribosome function and protein synthesis; essential for cell membrane permeability
Sodium	Na	11	23.0	Na^+	Necessary for transport; maintains osmotic pressure; used in food preservation
Sulfur	S	16	32.1	—	Important component of proteins; makes disulfide bonds; storage element in many bacteria
Zinc	Zn	30	65.4	Zn^{++}	An enzyme cofactor; required for protein synthesis and cell division; important in regulating DNA

*Based on the Latin name of the element. The first letter is always capitalized; if there is a second letter, it is always lowercased.

**A dash indicates an element that is usually found in combination with other elements, rather than as an ion.

SPOTLIGHT ON MICROBIOLOGY   2.1
Searching for Ancient Life with Isotopes

Determining the age of the earth and the historical time frame of living things has long been a priority of biologists. Much evidence comes from fossils, geologic sediments, and genetic studies, yet there has always been a need for an exacting scientific reference for tracing samples back in time, possibly even to the beginnings of the earth itself. One very precise solution to this problem comes from patterns that exist in isotopes. The isotopes of an element have the same basic chemical structure, but over billions of years, they have come to vary slightly in the number of neutrons. For example, carbon has 3 isotopes: C12, predominantly found in living things; C13, a less common form associated with nonliving matter; and C14, a radioactive isotope. All isotopes exist in relatively predictable proportions in the earth, solar system, and even universe, so that any variations from the expected ratios would indicate some other factor besides random change.

Isotope chemists use giant machines called microprobes to analyze the atomic structures in fossils and rock samples. These amazing machines can rapidly sort and measure the types and amounts of isotopes, which reflect a sample's age and possibly its origins. The accuracy of this method is such that it can be used like an "atomic clock." It was recently used to verify the dateline for the origins of the first life forms, using 3.85-billion-year-old sediment samples from Greenland. Testing indicated that the content of C12 in the samples was substantially higher than the amount in inorganic rocks, and it was concluded that living cells must have accumulated the C12. This finding shows that the origin of life was 400 million years earlier than the previous estimates.

In a separate study, some ancient Martian meteorites were probed to determine if certain microscopic rods could be some form of microbes (see figure 1.10b). By measuring the ratios of oxygen isotopes in carbonate ions (CO_3^{-2}), chemists were able to detect significant fluctuations in the isotopes from different parts of the same meteorite. Such differences would most likely be caused by huge variations in temperature or other extreme environments that are incompatible with life. From this evidence, they concluded that the tiny rods were not Martian microbes.

Each element is assigned an **atomic number (AN)** based on the number of protons it has. The atomic number is a valuable measurement because an element's proton number does not vary, and knowing it automatically tells you the usual number of electrons (recall that a neutral atom has an equal number of protons and electrons). Another useful measurement is the **mass[1] number (MN),** equal to the number of protons and neutrons. If one knows the mass number and the atomic number, it is possible to determine the numbers of neutrons by subtraction. Hydrogen is a unique element because its common form has only one proton, one electron, and no neutron, making it the only element with the same atomic and mass number.

Isotopes are variant forms of the same element that differ in the number of neutrons and thus have different mass numbers. These multiple forms occur naturally in certain proportions. Carbon, for example, exists primarily as carbon 12 with 6 neutrons (MN=12); but a small amount (about 1%) is carbon 13 with 7 neutrons and carbon 14 with 8 neutrons. Although isotopes have virtually the same chemical properties, some of them have unstable nuclei that spontaneously release energy in the form of radiation. Such *radioactive isotopes* play a role in a number of research and medical applications. Because they emit detectable signs, they can be used to trace the position of key atoms or molecules in chemical reactions, they are tools in diagnosis and treatment, and they are even applied in sterilization procedures (see ionizing radiation in chapter 11). Another application of isotopes is in dating fossils and other ancient materials (**Spotlight on Microbiology 2.1**). An element's **atomic weight** is the average of the mass numbers of all its isotopic forms (table 2.1).

Electron Orbitals and Shells

The structure of an atom can be envisioned as a central nucleus surrounded by a "cloud" of electrons that constantly rotate about the nu-

cleus in pathways (see figure 2.1). The pathways, called **orbitals,** are not actual objects or exact locations, but represent volumes of space in which an electron is likely to be found. Electrons occupy energy shells, proceeding from the lower-level energy electrons nearest the nucleus to the higher-energy electrons in the farthest orbitals.

Electrons fill the orbitals and shells in *pairs,* starting with the shell nearest the nucleus. The first shell contains one orbital and a maximum of 2 electrons; the second shell has four orbitals and up to 8 electrons; the third shell with 9 orbitals can hold up to 18 electrons; and the fourth shell with 16 orbitals contains up to 32 electrons. The number of orbitals and shells and how completely they are filled depends on the numbers of electrons, so that each element will have a unique pattern. For example, helium (AN=2) has only a filled first shell of 2 e^-; oxygen (AN=8) has a filled first shell and a partially filled second shell of 6 e^-; and magnesium (AN=12) has a filled first shell, a filled second one, and a third shell that fills only one orbital, so is nearly empty. As we will see, the chemical properties of an element are controlled mainly by the distribution of electrons in the outermost shell. Figure 2.1 and **figure 2.2** present various simplified models of atomic structure and electron maps.

CHAPTER CHECKPOINTS

Protons (p^+) and neutrons (n^0) make up the nucleus of an atom. Electrons (e^-) orbit the nucleus.

All elements are composed of atoms but differ in the numbers of protons, neutrons, and electrons they possess.

Elements are identified by *atomic weight,* or *mass,* or by *atomic number.*

Isotopes are varieties of one element that contain the same number of protons but different numbers of neutrons.

The number of electrons in an element's outermost orbital (compared with the total number possible) determines its chemical properties and reactivity.

---

1. Mass refers to the amount of matter that a particle contains. The proton and neutron have almost exactly the same mass, which is about 1.7×10^{-24} g, or 1 dalton.

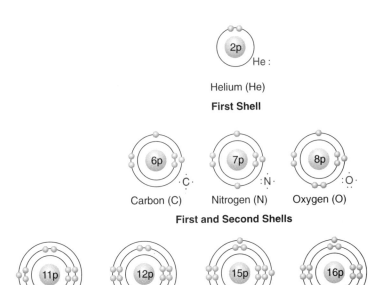

Helium (He)
First Shell

Carbon (C)   Nitrogen (N)   Oxygen (O)
First and Second Shells

Sodium (Na)   Magnesium (Mg)   Phosphorus (P)   Sulfur (S)   Chlorine (Cl)
First, Second, and Third Shells

FIGURE 2.2

Electron orbitals and shells.   Models of several elements show how the shells are filled by electrons as the atomic numbers increase (numbers noted inside nuclei). Electrons tend to appear in pairs, but certain elements have incompletely filled outer shells. Chemists depict elements in shorthand form (red Lewis structures) that indicate only the valence electrons, since these are the electrons involved in chemical bonds.

Bonds and Molecules

Most elements do not exist naturally in pure, uncombined form but are bound together as molecules and compounds. A **molecule*** is a distinct chemical substance that results from the combination of two or more atoms. Some molecules such as oxygen (O_2) and nitrogen gas (N_2) consist of atoms of the same element. Molecules that are combinations of two or more *different* elements are termed **compounds.** Compounds such as water (H_2O) and biological molecules (proteins, sugars, fats) are the predominant substances in living systems. When atoms bind together in molecules, they lose the properties of the atom and take on the properties of the combined substance. In the same way that an atom has an atomic weight, a molecule has a molecular weight (MW), which is calculated from the sum of all of the atomic weights of the atoms it contains.

The **chemical bonds** of molecules and compounds result when two or more atoms share, donate (lose), or accept (gain) electrons **(figure 2.3).** The number of electrons in the outermost shell of an element is known as its **valence.*** The valence determines the degree of reactivity and the types of bonds an element can make. Elements with a filled outer orbital are relatively stable because they have no extra electrons to share with or donate to other atoms. For example, helium has one filled shell, with no tendency either to give up electrons or to take them from other elements, making it a stable, inert (nonreactive) gas. Elements with partially filled outer orbitals are less stable and are more apt to form some sort of bond. Many chemical reactions are based on the tendency of atoms with unfilled outer shells to gain greater stability by achieving, or at least approximating, a filled outer shell. For example, an atom such as oxygen that can accept 2 additional electrons will bond readily with atoms (such as hydrogen) that can share or donate electrons. We ex-

plore some additional examples of the basic types of bonding in the following section.

In addition to reactivity, the number of electrons in the outer shell also dictates the number of chemical bonds an atom can make. For instance, hydrogen can bind with one other atom, oxygen can bind with up to two other atoms, and carbon can bind with four (see figure 2.13).

COVALENT BONDS AND POLARITY: MOLECULES WITH SHARED ELECTRONS

Covalent (cooperative valence) **bonds** form between atoms with valences that suit them to sharing electrons rather than to donating or receiving them. A simple example is hydrogen gas (H_2), which consists of two hydrogen atoms. A hydrogen atom has only a single electron, but when two of them combine, each will bring its electron to orbit about both nuclei, thereby approaching a filled orbital (2 electrons) for both atoms and thus creating a **single covalent bond** (**figure 2.4***a*). Covalent bonding also occurs in oxygen gas (O_2), but with a difference. Because each atom has 2 electrons to share in this molecule, the combination creates two pairs of shared electrons, also known as a **double covalent bond** (figure 2.4*b*). The majority of the molecules associated with living things are composed of single and double covalent bonds between the most common biological elements (carbon, hydrogen, oxygen, nitrogen, sulfur, and phosphorus), which are discussed in more depth in chapter 7. A slightly more complex pattern of covalent bonding is shown for methane gas (CH_4) in figure 2.4*c*.

Other effects of bonding result in differences in polarity. When atoms of different electronegativity form covalent bonds, the electrons are not shared equally and may be pulled more toward one atom than another. This pull causes one end of a molecule to assume a partial negative charge and the other end to assume a partial positive charge. A molecule with such an asymmetrical distribution of charges is termed **polar** and has positive and negative

---

*molecule (mol′-ih-kyool) L. *molecula,* little mass.

*valence (vay′-lents) L. *valentia,* strength. A measure of atomic binding capacity.

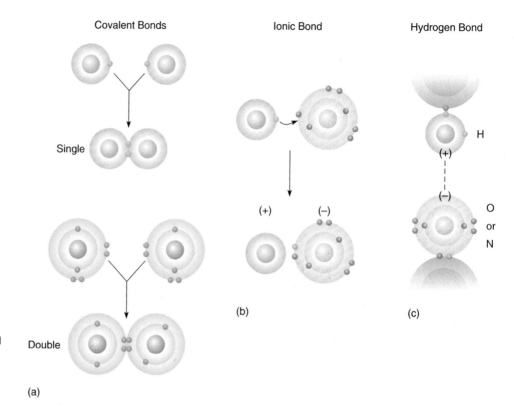

FIGURE 2.3

General representation of three types of bonding. **(a)** Covalent bonds, both single and double. **(b)** ionic bond. **(c)** Hydrogen bond. Note that hydrogen bonds are represented in models and formulas by dotted lines, as shown in **(c)**.

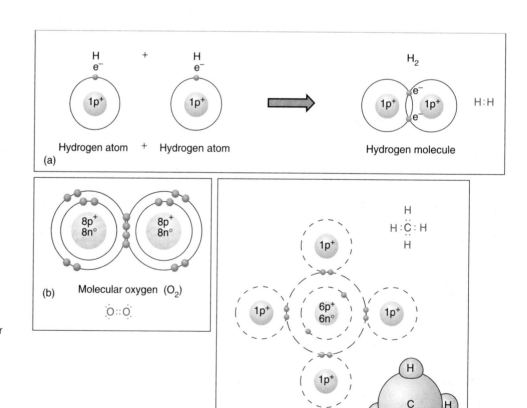

FIGURE 2.4

Examples of molecules with covalent bonding. **(a)** A hydrogen molecule is formed when two hydrogen atoms share their electrons and form a single bond. **(b)** In a double bond, the outer orbitals of two oxygen atoms overlap and permit the sharing of 4 electrons (one pair from each) and the saturation of the outer orbital for both. **(c)** Simple, working, and three-dimensional models of methane. Note that carbon has 4 electrons to share and hydrogens each have one, thereby completing the shells for all atoms in the compound, and creating 4 single bonds.

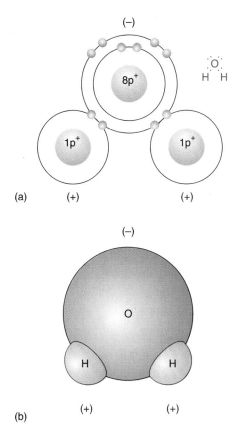

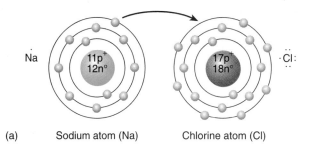

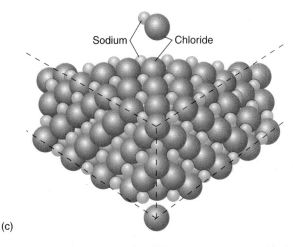

FIGURE 2.5

Polar molecule. **(a)** Simple models and **(b)** a three-dimensional model of a water molecule indicate the polarity, or unequal distribution, of electrical charge, which is caused by the pull of the shared electrons toward the oxygen side of the molecule.

FIGURE 2.6

Ionic bonding between sodium and chlorine. **(a)** When the two elements are placed together, sodium loses its single outer orbital electron to chlorine, thereby filling chlorine's outer shell. **(b)** Simple model of ionic bonding. **(c)** Sodium and chloride ions form large molecules, or crystals, in which the two atoms alternate in a definite, regular, geometric pattern. **(d)** Note the cubic nature of NaCl crystals at the macroscopic level.

poles. Observe the water molecule shown in **figure 2.5** and note that, because the oxygen atom is larger and has more protons than the hydrogen atoms, it will tend to draw the shared electrons with greater force toward its nucleus. This unequal force causes the oxygen part of the molecule to express a negative charge (due to the electrons' being attracted there) and the hydrogens to express a positive charge (due to the protons). The polar nature of water plays an extensive role in a number of biological reactions, which are discussed later. Polarity is a significant property of many large molecules in living systems and greatly influences both their reactivity and their structure.

When covalent bonds are formed between atoms that have the same or similar electronegativity, the electrons are shared equally between the two atoms. Because of this balanced distribution, no part of the molecule has a greater attraction for the electrons. This sort of electrically neutral molecule is termed **nonpolar.**

IONIC BONDS: ELECTRON TRANSFER AMONG ATOMS

In reactions that form **ionic bonds,** electrons are transferred completely from one atom to another and are not shared. These reactions invariably occur between atoms with valences that complement each other, meaning that one atom has an unfilled shell that will readily accept electrons and the other atom has an unfilled shell

that will readily lose electrons. A striking example is the reaction that occurs between sodium (Na) and chlorine (Cl). Elemental sodium is a soft, lustrous metal so reactive that it can burn flesh, and molecular chlorine is a very poisonous yellow gas. But when the two are combined, they form sodium chloride[2] (NaCl)—the familiar nontoxic table salt—a compound with properties quite different from either parent element **(figure 2.6).**

---

2. In general, when a salt is formed, the ending of the name of the negatively charged ion is changed to *-ide.*

The metabolic work of cells, such as synthesis, movement, and digestion, revolves around energy exchanges and transfers. The management of energy in cells is almost exclusively dependent on chemical rather than physical reactions because most cells are far too delicate to operate with heat, radiation, and other more potent forms of energy. The outer-shell electrons are readily portable and easily manipulated sources of energy. It is in fact the movement of electrons from molecule to molecule that accounts for most energy exchanges in cells. Fundamentally, then, a cell must have a supply of atoms that can gain or lose electrons if they are to carry out life processes.

The phenomenon in which electrons are transferred from one atom or molecule to another is termed an **oxidation** and **reduction** (shortened to **redox**) **reaction.** Although the term *oxidation* was originally adopted for reactions involving the addition of oxygen, the term oxidation can include any reaction causing electron release, regardless of the involvement of oxygen. By comparison, reduction is any reaction that causes an atom to receive electrons. All redox reactions occur in pairs. To analyze the phenomenon, let us again review the production of NaCl, but from a different standpoint. Although it is true that these atoms form ionic bonds, the chemical combination of the two is also a type of redox reaction.

When these two atoms react to form sodium chloride, a sodium atom gives up an electron to a chlorine atom. During this reaction, sodium is oxidized because it loses an electron, and chlorine is reduced because it gains an electron. To take this definition further, an atom or molecule, such as sodium, that can donate electrons and thereby reduce another molecule is a reducing agent; one that can receive extra electrons and thereby oxidize another molecule is an oxidizing agent. You may find this concept easier to keep straight if you think of redox agents as partners: The one that gives its electrons away is oxidized; the partner that receives the electrons is reduced. (A mnemonic device to keep track of this is *LEO* says *GER: L*ose *E*lectrons *O*xidized; *G*ain *E*lectrons *R*educed.)

Redox reactions are essential to many of the biochemical processes discussed in chapter 8. In cellular metabolism, electrons alone can be transferred from one molecule to another as described here, but sometimes oxidation and reduction occur with the transfer of hydrogen atoms (which are a proton and an electron) from one compound to another.

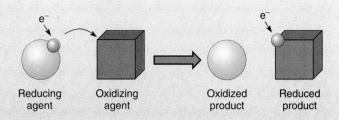

Simplified diagram of the exchange of electrons during an oxidation-reduction reaction.

How does this transformation occur? Sodium has 11 electrons (2 in shell one, 8 in shell two, and only 1 in shell three), so it is 7 short of having a complete outer shell. Chlorine has 17 electrons (2 in shell one, 8 in shell two, and 7 in shell three), making it 1 short of a complete outer shell. These two atoms are very reactive with one another, because a sodium atom will readily donate its single electron and a chlorine atom will avidly receive it. (The reaction is slightly more involved than a single sodium atom's combining with a single chloride atom **(Microbits 2.2),** but this complexity does not detract from the fundamental reaction as described here.) The outcome of this reaction is not many single, isolated molecules of NaCl but rather a solid crystal complex that interlinks millions of sodium and chloride ions (figure 2.6*b*).

Ionization: Formation of Charged Particles

Molecules with intact ionic bonds are electrically neutral, but they can produce charged particles when dissolved in a liquid called a solvent. This phenomenon, called **ionization,** occurs when the ionic bond is broken and the atoms dissociate (separate) into unattached, charged particles called **ions*** **(figure 2.7).** To illustrate what imparts a charge to ions, let us look again at the reaction between sodium and chlorine. When a sodium atom reacts with chlorine and loses one electron, the sodium is left with one more proton than electrons. This imbalance produces a positively charged sodium ion (Na^+). Chlorine, on the other hand, has gained one electron and

now has one more electron than protons, producing a negatively charged ion (Cl^-). Positively charged ions are termed **cations,*** and negatively charged ions are termed **anions.*** (A good mnemonic device is to think of the "t" in cation as a plus (+) sign and the first "n" in anion as a negative (−) sign.) Substances such as salts, acids, and bases that release ions when dissolved in water are termed **electrolytes** because their charges enable them to conduct an electrical current. Owing to the general rule that particles of like charge repel each other and those of opposite charge attract each other, we can expect ions to interact electrostatically with other ions and polar molecules. Such interactions are important in many cellular chemical reactions, in the formation of solutions, and in the reactions microorganisms have with dyes. The transfer of electrons from one molecule to another constitutes a significant mechanism by which biological systems store and release energy (see Microbits 2.2).

Hydrogen Bonding Some types of bonding involve neither sharing, losing, nor gaining electrons but instead are due to attractive forces between nearby molecules or atoms. One such bond is a **hydrogen bond,** a weak type of bond that forms between a hydrogen covalently bonded to one molecule and an oxygen or nitrogen atom on the same molecule or on a different molecule. Because hydrogen in a covalent bond tends to be positively charged, it will attract a

---

*ion (eye′-on) Gr. *ion,* going.

*cation (kat′-eye-on) An ion that migrates toward the negative pole, or cathode, of an electrical field.

*anion (an′-eye-on) An ion that migrates toward the positive pole, or anode.

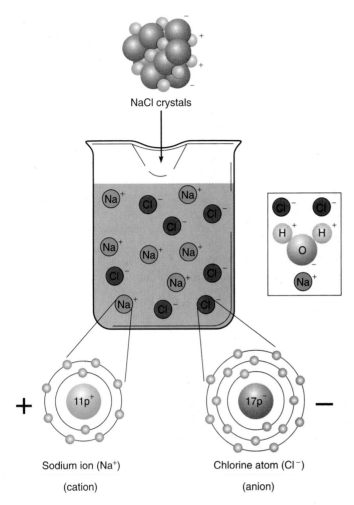

FIGURE 2.7

Ionization. When NaCl in the crystalline form is added to water, the ions are released from the crystal as separate charged particles (cations and anions) into solution. (See also figure 2.11.) In this solution, Cl^- ions are attracted to the hydrogen component of water, and Na^+ ions are attracted to the oxygen (box).

nearby negatively charged atom and form an easily disrupted bridge with it. This type of bonding is usually represented in molecular models with a dotted line. A simple example of hydrogen bonding occurs between water molecules (**figure 2.8**). More extensive hydrogen bonding is partly responsible for the structure and stability of proteins and nucleic acids (see figures 2.22*b* and 2.25).

Chemical Shorthand: Formulas, Models, and Equations

The atomic content of molecules can be represented by a few convenient formulas. We have already been exposed to the molecular formula, which concisely gives the atomic symbols and the number of the elements involved in subscript (CO_2, H_2O). More complex molecules such as glucose ($C_6H_{12}O_6$) can also be symbolized this way, but this formula is not unique, since fructose and galactose also share it. Molecular formulas are useful, but they only summarize the atoms in a compound; they do not show the position of bonds between atoms. For this purpose, chemists use structural formulas illustrating the relationships of the atoms and the number and types of bonds (**figure 2.9**). Other structural models present the

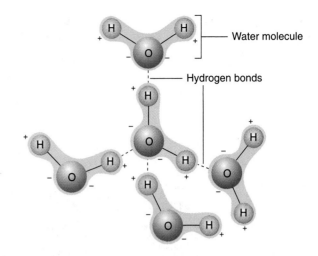

FIGURE 2.8

Hydrogen bonding in water. Because of the polarity of water molecules, the negatively charged oxygen end of one water molecule is weakly attracted to the positively charged hydrogen end of an adjacent water molecule.

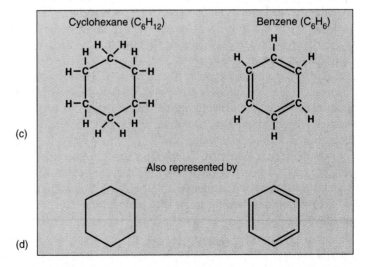

FIGURE 2.9

Comparison of molecular and structural formulas. **(a)** Molecular formulas provide a brief summary of the elements in a compound. **(b)** Structural formulas clarify the exact relationships of the atoms in the molecule, depicting single bonds by a single line and double bonds by two lines. **(c)** In structural formulas of organic compounds, cyclic or ringed compounds may be completely labeled, or **(d)** they may be presented in a shorthand form in which carbons are assumed to be at the angles and attached to hydrogens. See figure 2.14 for structural formulas of three sugars with the same molecular formula, $C_6H_{12}O_6$.

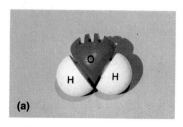

FIGURE 2.10

Three-dimensional, or space-filling, models of (a) water, (b) carbon dioxide, and (c) glucose. By convention, the red atoms are oxygen, the white ones hydrogen, and the black ones carbon.

three-dimensional appearance of a molecule, illustrating the orientation of atoms (differentiated by color codes) and the molecule's overall shape (**figure 2.10**).

The printed page tends to make molecules appear static, but this picture is far from correct, because molecules are capable of changing through chemical reactions. For ease in tracing chemical exchanges between atoms or molecules, and to derive some sense of the dynamic character of reactions, chemists use shorthand equations containing symbols, numbers, and arrows to simplify or summarize the major characteristics of a reaction. Molecules entering or starting a reaction are called **reactants,** and substances left by a reaction are called products. In most instances, summary chemical reactions do not give the details of the exchange, in order to keep the expression simple and to save space.

In a *synthesis* reaction,* the reactants bond together in a manner that produces an entirely new molecule (reactant A plus reactant B yields product AB). An example is the production of sulfur dioxide, a by-product of burning sulfur fuels and an important component of smog:

$$S + O_2 \rightarrow SO_2$$

Some synthesis reactions are not such simple combinations. When water is synthesized, for example, the reaction does not really involve one oxygen atom combining with two hydrogen atoms, because elemental oxygen exists as O_2 and elemental hydrogen exists as H_2. A more accurate equation for this reaction is:

$$2H_2 + O_2 \rightarrow 2H_2O$$

---

*synthesis (sin'-thuh-sis) Gr. *synthesis,* putting together.

The equation for reactions must be balanced—that is, the number of atoms on one side of the arrow must equal the number on the other side to reflect all of the participants in the reaction. To arrive at the total number of atoms in the reaction, multiply the prefix number by the subscript number; if no number is given, it is assumed to be 1.

In *decomposition reactions,* the bonds on a single reactant molecule are permanently broken to release two or more product molecules. One example is the resulting molecules when large nutrient molecules are digested into smaller units; a simpler example can be shown for the common chemical hydrogen peroxide:

$$2H_2O_2 \rightarrow 2H_2O + O_2$$

During *exchange reactions,* the reactants trade portions between each other and release products that are combinations of the two. This type of reaction occurs between acids and bases when they form water and a salt:

$$AB + XY \rightleftharpoons AX + BY$$

The reactions in biological systems can be reversible, meaning that reactants and products can be converted back and forth. These reversible reactions are symbolized with a double arrow, each pointing in opposite directions, as in the exchange reaction above. Whether a reaction is reversible depends on the proportions of these compounds, the difference in energy state of the reactants and products, and the presence of catalysts (substances that increase the rate of a reaction). Additional reactants coming from another reaction can also be indicated by arrows that enter or leave at the main arrow:

$$X + Y \xrightarrow{CD C} XYD$$

SOLUTIONS: HOMOGENEOUS MIXTURES OF MOLECULES

A **solution** is a mixture of one or more substances called **solutes** uniformly dispersed in a dissolving medium called a **solvent.** An important characteristic of a solution is that the solute cannot be separated by filtration or ordinary settling. The solute can be gaseous, liquid, or solid, and the solvent is usually a liquid. Examples of solutions are salt or sugar dissolved in water and iodine dissolved in alcohol. In general, a solvent will dissolve a solute only if it has similar electrical characteristics as indicated by the rule of solubility, expressed simply as "like dissolves like." For example, water is a polar molecule and will readily dissolve an ionic solute such as NaCl, yet a nonpolar solvent such as benzene will not dissolve NaCl.

Water is the most common solvent in natural systems, having several characteristics that suit it to this role. The polarity of the water molecule causes it to form hydrogen bonds with other water molecules, but it can also interact readily with charged or polar molecules. When an ionic solute such as NaCl crystals is added to water, it is dissolved, thereby releasing Na^+ and Cl^- into solution. Dissolution occurs because Na^+ is attracted to the negative pole of the water molecule and Cl^- is attracted to the positive pole; in this way, they are drawn away from the crystal separately into solution.

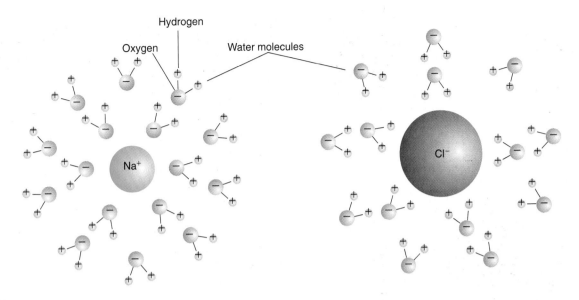

FIGURE 2.11

Hydration spheres formed around ions in solution.   In this example, a sodium cation attracts the negatively charged region of water molecules, and a chloride anion attracts the positively charged region of water molecules. In both cases, the ions become covered with spherical layers of specific numbers and arrangements of water molecules.

As it leaves, each ion becomes **hydrated,** which means that it is surrounded by a sphere of water molecules (**figure 2.11**). Molecules such as salt or sugar that attract water to their surface are termed **hydrophilic.*** Nonpolar molecules, such as benzene, that repel water are considered **hydrophobic.*** A third class of molecules, such as the phospholipids in cell membranes, are considered **amphipathic*** because they have both hydrophilic and hydrophobic properties.

Because most biological activities take place in aqueous (water-based) solutions, the concentration of these solutions can be very important (see chapter 7). The **concentration** of a solution expresses the amount of solute dissolved in a certain amount of solvent. It can be calculated by weight, volume, or percentage. A common way to calculate percentage of concentration is to use the weight of the solute, measured in grams (g), dissolved in a specified volume of solvent, measured in milliliters (ml). For example, dissolving 3 g of NaCl in 100 ml of water produces a 3% solution; dissolving 30 g in 100 ml produces a 30% solution; and dissolving 3 g in 1,000 ml (1 liter) produces a 0.3% solution. A solution with a small amount of solute and a relatively greater amount of solvent (0.3%) is considered dilute or weak. On the other hand, a solution containing significant percentages of solute (30%) is considered concentrated or strong.

A common way to express concentration of biological solutions is by its molar concentration, or *molarity* (M). A standard molar solution is obtained by dissolving one *mole,* defined as the molecular weight of the compound in grams, in 1 L (1,000 ml) of solution. To make a 1 M solution of sodium chloride, we would dissolve 58 g of NaCl to give 1 L of solution; a 0.1 M solution would require 5.8 g of NaCl in 1 L of solution.

ACIDITY, ALKALINITY, AND THE pH SCALE

Another factor with far-reaching impact on living things is the concentration of acidic or basic solutions in their environment. To understand how solutions develop acidity or basicity, we must look again at the behavior of water molecules. Hydrogens and oxygen tend to remain bonded by covalent bonds, but in certain instances, a single hydrogen can break away as the ionic form (H^+), leaving the remainder of the molecule in the form of an OH^- ion. The H^+ ion is positively charged because it is essentially a hydrogen ion that has lost its electron; the OH^- is negatively charged because it remains in possession of that electron. Ionization of water is constantly occurring, but in pure water containing no other ions, H^+ and OH^- are produced in equal amounts, and the solution remains neutral. By one definition, a solution is considered **acidic** when a component dissolved in water (acid) releases excess hydrogen ions[3] (H^+); a solution is **basic** when a component releases excess hydroxyl ions (OH^-), so that there is no longer a balance between the two ions.

To measure the acid and base concentrations of solutions, scientists use the **pH** scale, a graduated numerical scale that ranges from 0 (the most acidic) to 14 (the most basic). This scale is a useful standard for rating relative acidity and basicity; use **figure 2.12** to familiarize yourself with the pH readings of some common substances. It is not an arbitrary scale but actually a mathematical derivation based on the negative logarithm (reviewed in appendix B) of the concentration of H^+ ions in moles per liter (symbolized as $[H^+]$) in a solution, represented as:

$$pH = -\log[H^+]$$

Acidic solutions have a greater concentration of H^+ than OH^-, starting with pH 0, which contains 1.0 moles $H^+/1$. Each of the subsequent whole-number readings in the scale changes in $[H^+]$ by a

---

*hydrophilic (hy-droh-fil′-ik) Gr. *hydros,* water, and *philos,* to love.

*hydrophobic (hy-droh-fob′-ik) Gr. *phobos,* fear.

*amphipathic (am′-fy-path′-ik) Gr. *amphi,* both.

3. Actually, it forms a hydronium ion (H_3O^+), but for simplicity's sake, we will use the notation of H^+.

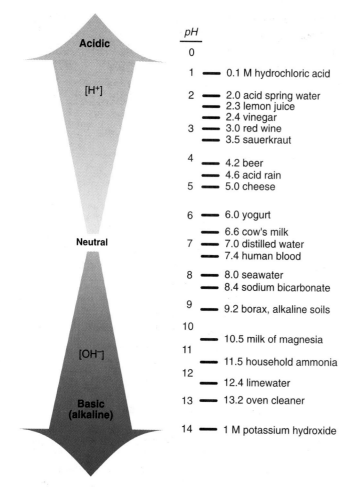

FIGURE 2.12

The pH scale. Shown are the relative degrees of acidity and basicity and the approximate pH readings for various substances.

TABLE 2.2

Hydrogen Ion and Hydroxyl Ion Concentrations at a Given pH

Moles/L of Hydrogen Ions	Logarithm	pH	Moles/L of OH⁻
1.0	10^{-0}	0	10^{-14}
0.1	10^{-1}	1	10^{-13}
0.01	10^{-2}	2	10^{-12}
0.001	10^{-3}	3	10^{-11}
0.0001	10^{-4}	4	10^{-10}
0.00001	10^{-5}	5	10^{-9}
0.000001	10^{-6}	6	10^{-8}
0.0000001	10^{-7}	7	10^{-7}
0.00000001	10^{-8}	8	10^{-6}
0.000000001	10^{-9}	9	10^{-5}
0.0000000001	10^{-10}	10	10^{-4}
0.00000000001	10^{-11}	11	10^{-3}
0.000000000001	10^{-12}	12	10^{-2}
0.0000000000001	10^{-13}	13	10^{-1}
0.00000000000001	10^{-14}	14	10^{-0}

Aqueous solutions containing both acids and bases may be involved in **neutralization** reactions, which give rise to water and other neutral by-products. For example, when equal molar solutions of hydrochloric acid (HCl) and sodium hydroxide (NaOH, a base) are mixed, the reaction proceeds as follows:

$$HCl + NaOH \rightarrow H_2O + NaCl$$

Here the acid and base ionize to H^+ and OH^- ions, which form water, and other ions, Na^+ and Cl^-, which form sodium chloride. Any product other than water that arises when acids and bases react is called a salt. Many of the organic acids (such as lactic and succinic acids) that function in **metabolism*** are available as the acid and the salt form (such as lactate, succinate), depending on the conditions in the cell (see chapter 8).

THE CHEMISTRY OF CARBON AND ORGANIC COMPOUNDS

So far, our main focus has been on the characteristics of atoms, ions, and small, simple substances that play diverse roles in the structure and function of living things. These substances are often lumped together in a category called **inorganic*** chemicals. Examples of inorganic chemicals include NaCl (sodium chloride), $Mg_3(PO_4)_2$ (magnesium phosphate), $CaCO_3$ (calcium carbonate), and CO_2 (carbon dioxide). In reality, however, most of the chemical reactions and structures of living things occur at the level of more complex molecules, termed **organic*** chemicals. These are carbon compounds with a basic framework of the element carbon bonded to other atoms. Organic molecules vary in complexity from the simplest,

tenfold reduction, so that pH 1 contains [0.1 moles H^+/1], pH 2 contains [0.01 moles H^+/1], and so on, continuing in the same manner up to pH 14, which contains [0.00000000000001 moles H^+/1]. These same concentrations can be represented more manageably by exponents: pH 2 has a $[H^+]$ of 10^{-2} moles, and pH 14 has a $[H^+]$ of 10^{-14} moles **(table 2.2)**. It is evident that the pH units are derived from the exponent itself. Even though the basis for the pH scale is $[H^+]$, it is important to note that, as the $[H^+]$ in a solution decreases, the $[OH^-]$ increases in direct proportion. At midpoint—pH 7, or neutrality—the concentrations are exactly equal and neither predominates, this being the pH of pure water previously mentioned.

In summary, the pH scale can be used to rate or determine the degree of acidity or basicity (also called alkalinity) of a solution. On this scale, a pH below 7 is acidic, and the lower the pH, the greater the acidity; a pH above 7 is basic, and the higher the pH, the greater the basicity. Incidentally, although pHs given here in even whole numbers, more often, a pH reading exists in decimal form; for example, pH 4.5 or 6.8 (acidic) and pH 7.4 or 10.2 (basic). Because of the damaging effects of very concentrated acids or bases, most cells operate best under neutral, weakly acidic, or weakly basic conditions (see chapter 7).

***metabolism** (muh-tab′-oh-lizm) A general term referring to the totality of chemical and physical processes occurring in the cell.

***inorganic** (in-or-gan′-ik) Any chemical substances that do not contain both carbon and hydrogen.

***organic** (or-gan′-ik) Gr. *organikos,* instrumental.

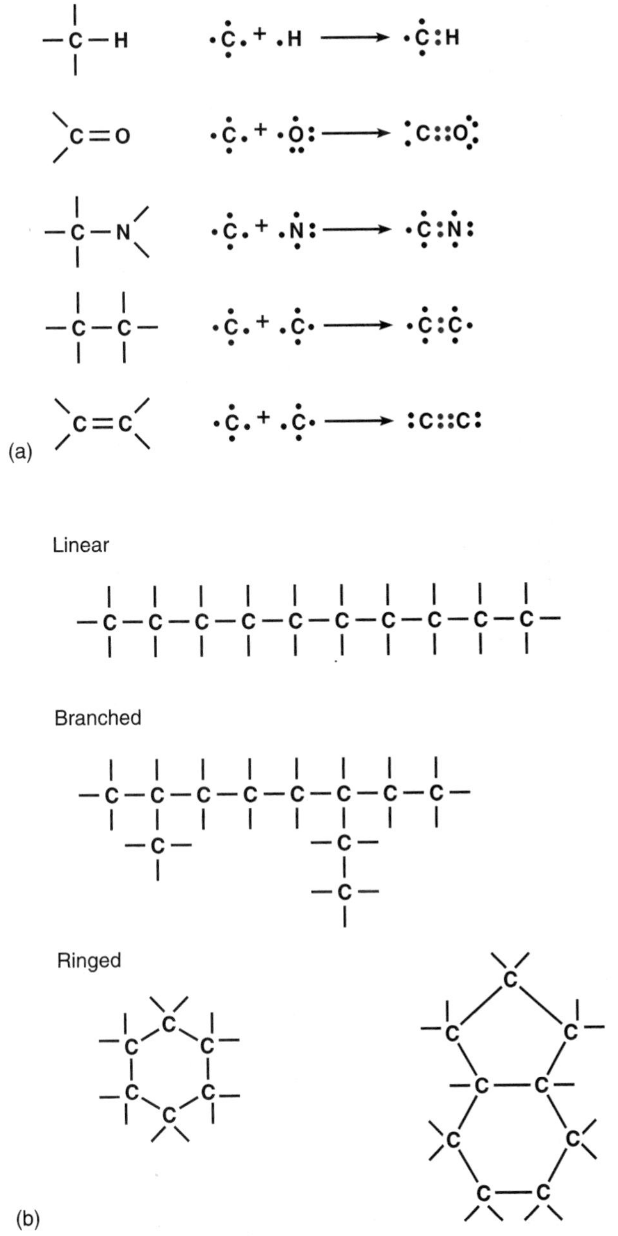

bital to be shared with other atoms (including other carbons) through covalent bonding. As a result, it can form stable chains containing thousands of carbon atoms and still has bonding sites available for forming covalent bonds with numerous other atoms. The bonds that carbon forms are linear, branched, or ringed, and it can form four single bonds, two double bonds, or one triple bond (**figure 2.13**). The atoms with which carbon is most often associated in organic compounds are hydrogen, oxygen, nitrogen, sulfur, and phosphorus.

FUNCTIONAL GROUPS OF ORGANIC COMPOUNDS

One important advantage of carbon's serving as the molecular skeleton for living things is that it is free to bind with an unending array of other molecules. These special molecular groups or accessory molecules that bind to organic compounds are called **functional groups.** Functional groups help define the chemical class of certain groups of organic compounds and confer unique reactive properties on the whole molecule (**table 2.3**). Because each type of functional group behaves in a distinctive manner, reactions of an organic compound can be predicted by knowing the kind of functional group or groups it carries. Many synthesis, decomposition, and transfer reactions rely upon functional groups such as R—OH or R—NH_2. The —**R** designation on a molecule is shorthand for residue, and its placement in a formula indicates that the group attached at that site varies from one compound to another.

CHAPTER CHECKPOINTS

Covalent bonds are chemical bonds in which electrons are shared between atoms. Equally distributed electrons form nonpolar covalent bonds, whereas unequally distributed electrons form polar covalent bonds.

Ionic bonds are chemical bonds in which the outer electron shell either donates or receives electrons from another atom so that the outer shell of each atom is completely filled.

Hydrogen bonds are weak chemical bonds that form between covalently bonded hydrogens and either oxygens or nitrogens on different molecules.

Chemical equations express the chemical exchanges between atoms or molecules. Some arrangements contain more energy than others, and chemical reactions such as synthesis or decomposition may require or release the difference in energy.

Solutions are mixtures of solutes and solvents that cannot be separated by filtration or settling.

The pH, ranging from a highly *acidic* solution to highly *basic* solution, refers to the concentration of hydrogen ions. It is expressed as a number from 0 to 14.

Biologists define organic molecules as those containing both carbon and hydrogen.

Carbon is the backbone of biological compounds because of its ability to form single, double, or triple covalent bonds with itself and many different elements.

Functional (R) groups are specific arrangements of organic molecules that confer distinct properties, including chemical reactivity, to organic compounds.

FIGURE 2.13

The versatility of bonding in carbon. In most compounds, each carbon makes a total of four bonds. **(a)** Both single and double bonds can be made with other carbons, oxygen, and nitrogen; single bonds are made with hydrogen. Simple electron models show how the electrons are shared in these bonds. **(b)** Multiple bonding of carbons can give rise to long chains, branched compounds, and ringed compounds, many of which are extraordinarily large and complex.

methane (CH_4; see figure 2.4c), which has a molecular weight of 16, to certain antibody molecules (produced by an immune reaction) that have a molecular weight of nearly 1,000,000 and are among the most complex molecules on earth.

The role of carbon as the fundamental element of life can best be understood if we look at its chemistry and bonding patterns. The valence of carbon makes it an ideal atomic building block to form the backbone of organic molecules; it has 4 electrons in its outer or-

TABLE 2.3

Representative Functional Groups and Classes of Organic Compounds

Formula of Functional Group	Name	Class of Compounds
R* — O — H	Hydroxyl	Alcohols, carbohydrates
R — C (=O)(OH) (Carboxyl structure)	Carboxyl	Fatty acids, proteins, organic acids
R — C(H)(H) — NH₂	Amino	Proteins, nucleic acids
R — C (=O)(O—R) (Ester structure)	Ester	Lipids
R — C(H)(H) — SH	Sulfhydryl	Cysteine (amino acid), proteins
R — C (=O)(H) (Aldehyde structure)	Carbonyl, terminal end	Aldehydes, polysaccharides
R — C(=O) — C —	Carbonyl, internal	Ketones, polysaccharides
R — O — P(=O)(OH) — OH	Phosphate	DNA, RNA, ATP

The R designation on a molecule is shorthand for residue, and its placement in a formula indicates that what is attached at that site varies from one compound to another.

Macromolecules: Superstructures of Life

The compounds of life fall into the realm of **biochemistry.** Biochemicals are organic compounds produced by (or components of) living things, and they include four main families: carbohydrates, lipids, proteins, and nucleic acids **(table 2.4).** The compounds in these groups are assembled from smaller molecular subunits, or building blocks, and because they are often very large compounds, they are termed **macromolecules.** All macromolecules except lipids are formed by **polymerization,** a process in which repeating subunits termed **monomers*** are bound into chains of various lengths termed **polymers.*** For example, proteins (polymers) are composed of a chain of amino acids (monomers) (see figure 2.22*a*). The large size and complex, three-dimensional shape of macromolecules enables them to function as structural components, molecular messengers, energy sources, enzymes (biochemical catalysts), nutrient stores, and sources of genetic information. In the following section and in later chapters, we will consider numerous concepts relating to the roles of macromolecules in cells.

CARBOHYDRATES: SUGARS AND POLYSACCHARIDES

The term **carbohydrate** originates from the way that most members of this chemical class resemble combinations of carbon and water. Although carbohydrates can be generally represented by the formula $(CH_2O)_n$, in which *n* indicates the number of units of this combination of atoms, some carbohydrates contain additional atoms of sulfur or nitrogen. In molecular configuration, the carbons form chains or rings with two or more hydroxyl groups and either an aldehyde or a ketone group, giving them the technical designation of *polyhydroxy aldehydes* or *ketones* **(figure 2.14).**

Carbohydrates exist in a great variety of configurations. The common term sugar (**saccharide**)* refers to a simple carbohydrate such as a monosaccharide or a disaccharide that has a sweet taste. A **monosaccharide** is a simple polyhydroxy aldehyde or ketone molecule containing from 3 to 7 carbons; a **disaccharide** is a combination of two monosaccharides; and a **polysaccharide** is a polymer of five or more monosaccharides bound in linear or branched chain patterns (see figure 2.14). Monosaccharides and disaccharides are specified by combining a prefix that describes some characteristic of the sugar with the suffix **-ose.** For example, **hexoses** are composed of 6 carbons, and **pentoses** contain 5 carbons. **Glucose** (Gr. sweet) is the most common and universally important hexose; **fructose** is named for fruit (one of its sources); and xylose, a pentose, derives its name from the Greek word for wood. Disaccharides are named similarly: **lactose** (L. milk) is an important component of milk; **maltose** means malt sugar; and **sucrose** (Fr. sugar) is common table sugar or cane sugar.

The Nature of Carbohydrate Bonds

The subunits of disaccharides and polysaccharides are linked by means of **glycosidic bonds,** in which carbons (each is assigned a number) on adjacent sugar units are bonded to the same oxygen atom like links in a chain **(figure 2.15).** For example, maltose is formed when the number 1 carbon on a glucose bonds to the oxygen on the number 4 carbon on a second glucose; sucrose is formed when glucose and fructose bind oxygen between their number 1 and number 2 carbons; and lactose is formed when glucose and galactose connect by their number 1 and number 4 carbons. In order to form this bond, one carbon gives up its OH

---

*monomer (mahn′-oh-mur) Gr. *mono,* one, and *meros,* part.

*polymer (pahl′-ee-mur) Gr. *poly,* many; also the root for polysaccharide and polypeptide.

*saccharide (sak′-uh-ryd) Gr. *sakcharon,* sweet.

TABLE 2.4

Macromolecules and Their Functions

Macromolecule	Description/Basic Structure	Examples/Functions
Carbohydrates		
Monosaccharides	3–7-carbon sugars	Glucose, fructose / Sugars involved in metabolic reactions; building block of disaccharides and polysaccharides
Disaccharides	Two monosaccharides	Maltose (malt sugar) / Composed of two glucoses; an important breakdown product of starch
		Lactose (milk sugar) / Composed of glucose and galactose
		Sucrose (table sugar) / Composed of glucose and fructose
Polysaccharides	Chains of monosaccharides	Starch, cellulose, glycogen / Cell wall, food storage
Lipids		
Triglycerides	Fatty acids + glycerol	Fats, oils / Major component of cell membranes; storage
Phospholipids	Fatty acids + glycerol + phosphate	Membranes
Waxes	Fatty acids, alcohols	Mycolic acid / Cell wall of mycobacteria
Steroids	Ringed structure (not a polymer)	Cholesterol, ergosterol / Membranes of eucaryotes and some bacteria
Proteins		
	Amino acids	Enzymes; part of cell membrane, cell wall, ribosomes, antibodies / Metabolic reactions; structural components
Nucleic acids		
	Pentose sugar + phosphate + nitrogenous base	
	Purines; adenine, guanine	
	Pyrimidines: cytosine, thymine, uracil	
Deoxyribonucleic acid (DNA)	Contains deoxyribose sugar and thymine, not uracil	Chromosomes; genetic material of viruses / Inheritance
Ribonucleic acid (RNA)	Contains ribose sugar and uracil, not thymine	Ribosomes; mRNA, tRNA / Expression of genetic traits

group and the other (the one contributing the oxygen to the bond) loses the H from its OH group. Because a water molecule is produced, this reaction is known as **dehydration synthesis,** a process common to most polymerization reactions (see proteins, page 42). Three polysaccharides (starch, cellulose, and glycogen) are structurally and biochemically distinct, even though all are polymers of the same monosaccharide—glucose. The basis for their differences lies primarily in the exact way the glucoses are bound together, which greatly affects the characteristics of the end product **(figure 2.16).** The synthesis and breakage of each type of bond requires a specialized catalyst called an enzyme (see chapter 8).

The Functions of Polysaccharides

Polysaccharides typically contribute to structural support and protection and serve as nutrient and energy stores. The cell walls in plants and many microscopic algae derive their strength and rigidity from **cellulose,** a long, fibrous polymer (figure 2.16a). Because of this role, cellulose is probably one of the most common organic substances on the earth, yet it is digestible only by certain bacteria, fungi, and protozoa. These microbes, called decomposers, play an essential role in breaking down and recycling plant materials (see figure 7.2). Some bacteria secrete slime layers of a glucose polymer called *dextran.* This substance causes a sticky layer to develop on teeth that leads to plaque (see figure 4.12).

Other structural polysaccharides can be conjugated (chemically bonded) to amino acids, nitrogen bases, lipids, or proteins. **Agar,** an indispensable polysaccharide in preparing solid culture media, is a natural component of certain seaweeds. It is a complex polymer of galactose and sulfur-containing carbohydrates. The exoskeletons of certain fungi contain **chitin,** a polymer of glucosamine (a sugar with an amino functional group). **Peptidoglycan*** is one special class of compounds in which polysaccharides (glycans) are linked to peptide fragments (a short chain of amino acids). This molecule provides the main source of structural support to the bacterial cell wall. The cell wall of gram-negative bacteria also contains **lipopolysaccharide,** a complex of lipid and polysaccharide responsible for symptoms such as fever and shock (see chapters 4 and 13).

The outer surface of many cells has a delicate "sugar coating" composed of polysaccharides bound in various ways to proteins (the combination is called mucoprotein or glycoprotein). This structure, called the **glycocalyx,*** functions in attachment to other cells or as a site for *receptors*—surface molecules that receive and respond to external stimuli. Small sugar molecules account for the differences in human blood types, and carbohydrates

---

*peptidoglycan (pep-tih-doh-gly'-kan).

*glycocalyx (gly"-koh-kay'-lix) Gr. *glycos,* sweet, and *calyx,* covering.

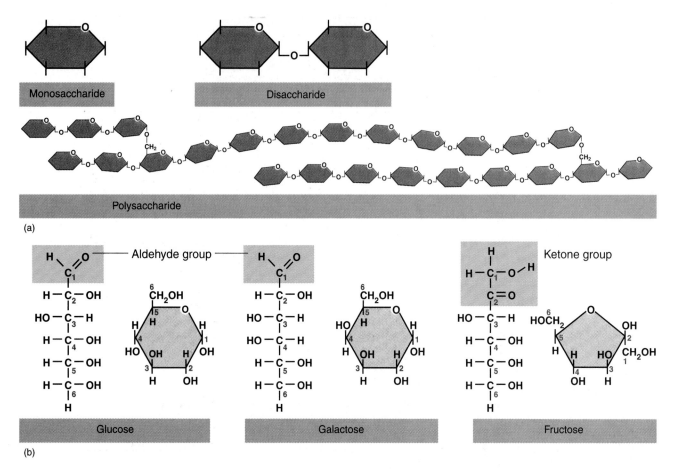

FIGURE 2.14

Common classes of carbohydrates. **(a)** Major saccharide groups, named for the number of sugar units each contains. **(b)** Three hexoses with the same molecular formula and different structural formulas. Both linear and ring models are given. The linear form indicates aldehyde and ketone groups, although in solution the sugars exist in the ring form. Note that the carbons are numbered so as to keep track of reactions within and between monosaccharides.

are a component of large protein molecules called antibodies. Some viruses have glycoproteins on their surface with which they bind to and invade their host cells.

Polysaccharides are usually stored by cells in the form of glucose polymers such as starch (figure 2.16*b*) or **glycogen,** but only organisms with the appropriate digestive enzymes can break them down and use them as a nutrient source. Because a water molecule is required for breaking the bond between two glucose molecules, digestion is also termed **hydrolysis.*** Starch is the primary storage food of green plants, microscopic algae, and some fungi; glycogen (animal starch) is a stored carbohydrate for animals and certain groups of bacteria and protozoa.

LIPIDS: FATS, PHOSPHOLIPIDS, AND WAXES

The term **lipid,** derived from the Greek word *lipos,* meaning fat, is not a chemical designation, but an operational term for a variety of substances that are not soluble in polar solvents such as water (recall that oil and water do not mix) but will dissolve in nonpolar solvents such as benzene and chloroform. This property occurs because the substances we call lipids contain relatively long or complex C—H (hydrocarbon) chains that are nonpolar and thus

hydrophobic. The main groups of compounds classified as lipids are triglycerides, phospholipids, steroids, and waxes.

Important storage lipids are the **triglycerides,** a category that includes fats and oils. Triglycerides are composed of a single molecule of glycerol bound to three fatty acids **(figure 2.17). Glycerol** is a 3-carbon alcohol with three OH groups that serve as binding sites, and fatty acids are long-chain hydrocarbon molecules with a carboxyl group (COOH) at one end that is free to bind to the glycerol. The bond that forms between the —OH group and the —COOH is defined as an **ester bond.** The hydrocarbon portion of a fatty acid can vary in length from 4 to 24 carbons and, depending on the fat, it may be saturated or unsaturated. If all carbons in the chain are single-bonded to 2 other carbons and 2 hydrogens, the fat is saturated; if there is at least one C=C double bond in the chain, it is unsaturated. The structure of fatty acids is what gives fats and oils (liquid fats) their greasy, insoluble nature. In general, solid fats (such as beef tallow) are more saturated, and oils (or liquid fats) are more unsaturated. In most cells, triglycerides are stored in long-term concentrated form as droplets or globules. When the ester linkage is acted on by digestive enzymes called lipases, the fatty acids and glycerol are freed to be used in metabolism. Fatty acids are a superior source of energy, yielding twice as much per gram as other storage molecules (starch). Soaps are K^+ or Na^+ salts of fatty acids whose qualities make them excellent grease removers and cleaners (see chapter 11).

***hydrolysis** (hy-drol′-uh-sis) Gr. *hydros,* water, and *lyein,* to dissolve.

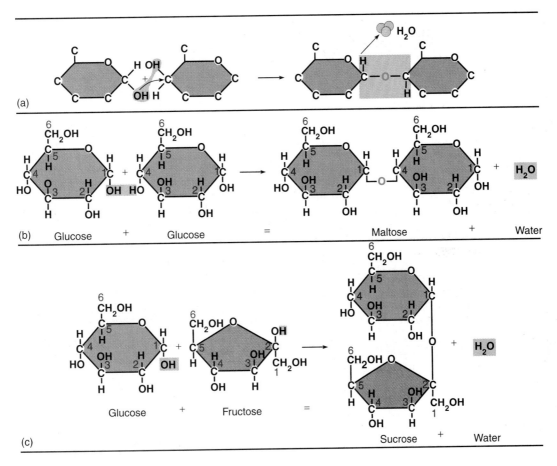

FIGURE 2.15

Glycosidic bond. **(a)** General scheme in the formation of a glycosidic bond by dehydration synthesis. **(b)** Formation of the 1,4 bond between two α glucoses to produce maltose and water. **(c)** Formation of the 1,2 bond between glucose and fructose to produce sucrose and water.

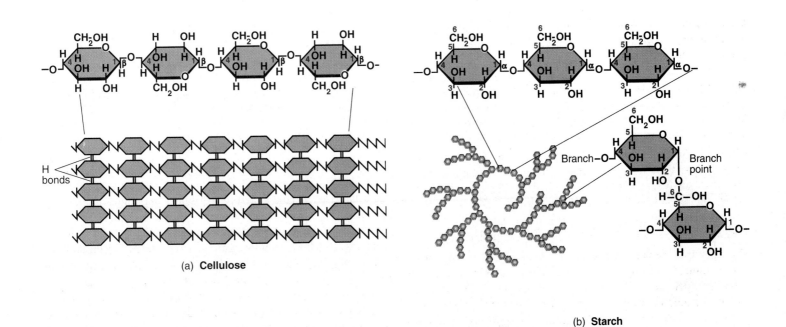

FIGURE 2.16

Polysaccharides. **(a)** Cellulose is composed of β glucose bonded in 1,4 bonds that produce linear, lengthy chains of polysaccharides that are H-bonded along their length. This is the typical structure of wood and cotton fibers. **(b)** Starch is also composed of glucose polymers, in this case α glucose. The main structure is amylose bonded in a 1,4 pattern, with side branches of amylopectin bonded by 1,6 bonds. The entire molecule is compact and granular.

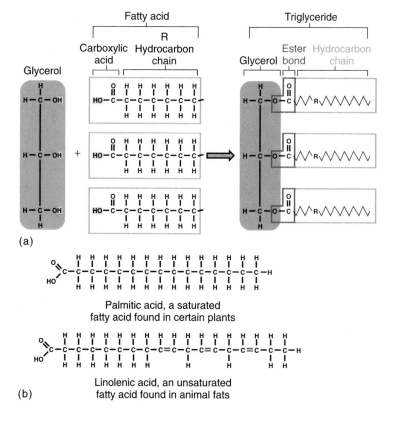

(a)

(b)

Palmitic acid, a saturated
fatty acid found in certain plants

Linolenic acid, an unsaturated
fatty acid found in animal fats

FIGURE 2.17

Synthesis and structure of a triglyceride. **(a)** Because a water molecule is released at each ester bond, this is another form of dehydration synthesis. The jagged lines and R symbol represent the hydrocarbon chains of the fatty acids, which are commonly very long. **(b)** Structural formulas for saturated and unsaturated fatty acids.

Membrane Lipids

A class of lipids that serves as a major structural component of cell membranes is the **phospholipids.** Although phospholipids also contain glycerol and fatty acids, they have some significant differences from triglycerides. Phospholipids contain only two fatty acids attached to the glycerol, and the third glycerol binding site holds a phosphate group. The phosphate is in turn bonded to an alcohol,[4] which varies from one phospholipid to another (**figure 2.18a**). These lipids have a hydrophilic region from the charge on the phosphoric acid–alcohol "head" of the molecule and a hydrophobic region that corresponds to the long, uncharged "tail" (formed by the fatty acids). When exposed to an aqueous solution, the charged heads are attracted to the water phase, and the nonpolar tails are repelled from the water phase (figure 2.18b). This property causes lipids to naturally assume single and double layers (bilayers), which contribute to their biological significance in membranes. When two single layers of polar lipids come together to form a double layer, the outer hydrophilic face of each single layer will orient itself toward the solution, and the hydrophobic portions will become immersed in the core of the bilayer. The structure of lipid bilayers confers characteristics on membranes such as selective permeability and fluid nature (**Microbits 2.3**).

Miscellaneous Lipids

Steroids are complex ringed compounds commonly found in cell membranes and animal hormones. The best known of these is the sterol (meaning a steroid with an OH group) called **cholesterol**

(**figure 2.19**). Cholesterol reinforces the structure of the cell membrane in animal cells and in an unusual group of cell-wall-deficient bacteria called the mycoplasmas (see chapter 4). The cell membranes of fungi also contain a sterol, called ergosterol. *Prostaglandins* are fatty acid derivatives found in trace amounts that function in inflammatory and allergic reactions, blood clotting, and smooth muscle contraction. Chemically, a *wax* is an ester formed between a long-chain alcohol and a saturated fatty acid. The resulting material is typically pliable and soft when warmed but hard and water-resistant when cold (paraffin, for example). Among living things, fur, feathers, fruits, leaves, human skin, and insect exoskeletons are naturally waterproofed with a coating of wax. Bacteria that cause tuberculosis and leprosy produce a wax (wax D) that repels ordinary stains and contributes to their pathogenicity.

PROTEINS: SHAPERS OF LIFE

The predominant organic molecules in cells are **proteins,** a fitting term adopted from the Greek word *proteios,* meaning first or prime. To a large extent, the structure, behavior, and unique qualities of each living thing are a consequence of the proteins they contain. To best explain the origin of the special properties and versatility of proteins, we must examine their general structure. The building blocks of proteins are **amino acids,** which exist in 20 different naturally occurring forms (**table 2.5**). Various combinations of these amino acids account for the nearly infinite variety of proteins. Amino acids have a basic skeleton consisting of a carbon (called the α carbon) linked to an amino group (NH_2), a carboxyl group (COOH), a hydrogen atom (H), and a variable R group. The variations among the amino acids occur at the R group, which is different in each amino acid and imparts the unique characteristics to the

---

4. Alcohols are hydrocarbons containing OH groups.

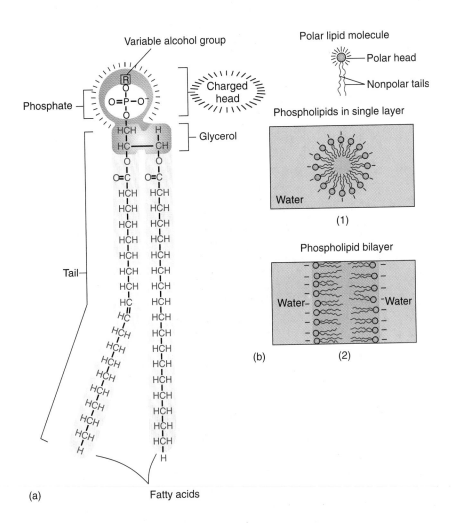

(a)

Variable alcohol group

Phosphate

Charged head

Glycerol

Tail

Fatty acids

Polar lipid molecule

Polar head

Nonpolar tails

Phospholipids in single layer

Water

(1)

Phospholipid bilayer

Water

Water

(b)

(2)

FIGURE 2.18

Phospholipids—membrane molecules.
(a) A complex model of a single molecule of a phospholipid. The phosphate-alcohol head lends a charge to one end of the molecule; its long, trailing hydrocarbon chain is uncharged. **(b)** The behavior of phospholipids in water-based solutions causes them to become arranged **(1)** in single layers called micelles, with the charged head oriented toward the water phase and the hydrophobic nonpolar tail buried away from the water phase, or **(2)** in double-layered phospholipid systems with the hydrophobic tails sandwiched between two hydrophilic layers.

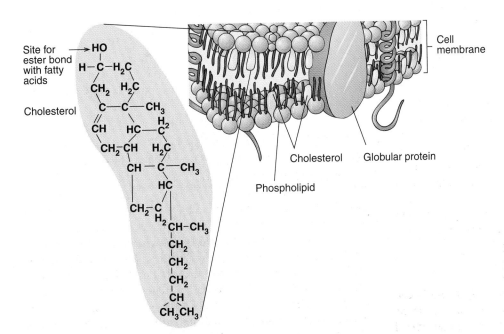

Site for ester bond with fatty acids

Cholesterol

Cell membrane

Cholesterol

Globular protein

Phospholipid

FIGURE 2.19

Formula for cholesterol, an alcoholic steroid that is inserted in some membranes.
Cholesterol can become esterified with fatty acids at its OH group, imparting a polar quality similar to that of phospholipids.

TABLE 2.5

Twenty Amino Acids and Their Abbreviations

Acid	Abbreviation	Characteristic of R Groups*
Alanine	Ala	NP
Arginine	Arg	+
Asparagine	Asn	P
Aspartic acid	Asp	−
Cysteine	Cys	P
Glutamic acid	Glu	−
Glutamine	Gln	P
Glycine	Gly	P
Histidine	His	+
Isoleucine	Ile	NP
Leucine	Leu	NP
Lysine	Lys	+
Methionine	Met	NP
Phenylalanine	Phe	NP
Proline	Pro	NP
Serine	Ser	P
Threonine	Thr	P
Tryptophan	Trp	NP
Tyrosine	Tyr	P
Valine	Val	NP

*NP, nonpolar; P, polar; +, positively charged; −, negatively charged.

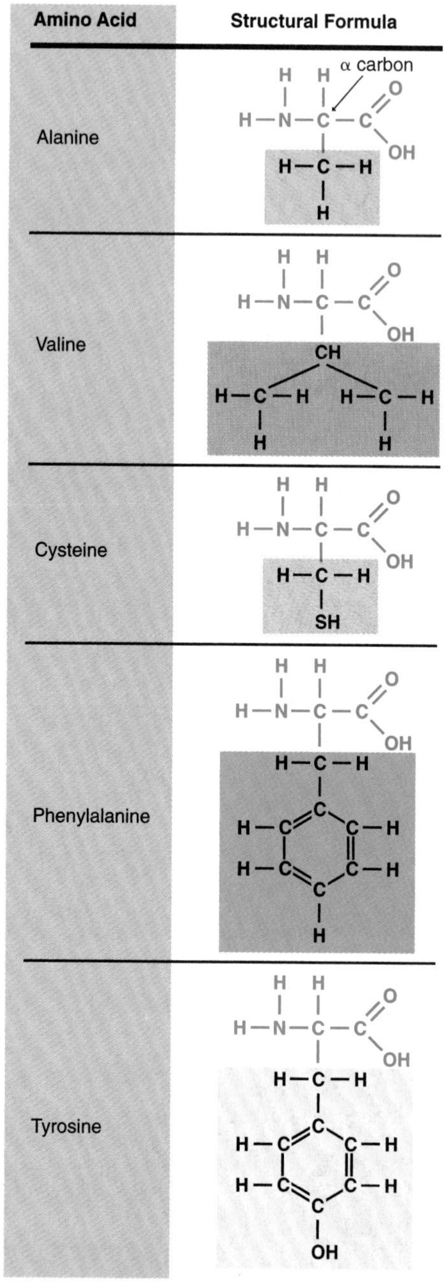

molecule and to the proteins that contain it **(figure 2.20).** A covalent bond called a **peptide bond** forms between the amino group on one amino acid and the carboxyl group on another amino acid. As a result of peptide bond formation, it is possible to produce molecules varying in length from two amino acids to chains containing thousands of them.

Various terms are used to denote the nature of compounds containing peptide bonds. **Peptide*** usually refers to a molecule composed of short chains of amino acids, such as a dipeptide (two amino acids), a tripeptide (three), and a tetrapeptide (four) **(figure 2.21).** A **polypeptide** contains an unspecified number of amino acids, but usually has more than 20, and is often a smaller subunit of a protein. A protein is the largest of this class of compounds and usually contains a minimum of 50 amino acids. It is common for the terms *polypeptide* and *protein* to be used interchangeably, though not all polypeptides are large enough to be considered proteins. In chapter 9 we see that protein synthesis is not just a random connection of amino acids; it is directed by information provided in DNA.

Protein Structure and Diversity

The reason that proteins are so varied and specific is that they do not function in the form of a simple straight chain of amino acids (called the primary structure). A protein has a natural tendency to assume more complex levels of organization, called the secondary, tertiary, and quaternary structures **(figure 2.22).** The **primary (1°) structure** is more correctly described as the type, number,

*peptide (pep′-tyd) Gr. *pepsis,* digestion.

FIGURE 2.20

Structural formulas of selected amino acids. The basic structure common to all amino acids is shown in blue and the variable group, or R group, is placed in a colored box. Note the variations in structure of this reactive component.

and order of amino acids in the chain, which varies extensively from protein to protein. The **secondary (2°) structure** arises when various functional groups exposed on the outer surface of the molecule interact by forming hydrogen bonds. This interaction causes the amino acid chain to twist into a coiled configuration called the α *helix* or to fold into an accordion pattern called a β-*pleated sheet.* Some proteins contain both types of secondary configurations. Proteins at the secondary level undergo a third degree of torsion called the **tertiary (3°) structure** created by

MICROBITS   2.3
Membranes: Cellular Skins

The word **membranes** appears frequently in descriptions of cells in this chapter and in chapters 4 and 5. The word itself describes any lining or covering, including such multicellular structures as the mucous membranes of the body. From the perspective of a single cell, however, a membrane is a thin, double-layered sheet composed of lipids such as phospholipids and sterols (averaging about 40% of membrane content) and protein molecules (averaging about 60%). The primary role of membranes is as a cell membrane that completely encases the cytoplasm. Membranes are also components of eucaryotic organelles such as nuclei, mitochondria, and chloroplasts, and they appear in internal pockets of certain procaryotic cells. Even some viruses, which are not cells at all, can have a membranous protective covering.

Cell membranes are so thin—on the average, just 0.0070 μm (7 nm) thick—that they cannot actually be seen with an optical microscope. Even at magnifications made possible by electron microscopy (500,000×), very little of the precise architecture can be visualized, and a cross-sectional view has the appearance of railroad tracks. Following detailed microscopic and chemical analysis, S. J. Singer and C. K. Nicholson proposed a simple and elegant theory for membrane structure

called the **fluid mosaic* model.** According to this theory, a membrane is a continuous bilayer formed by lipids that are oriented with the polar lipid heads toward the outside and the nonpolar heads toward the center of the membrane. Embedded at numerous sites in this bilayer are various-sized globular proteins. Some proteins are situated only at the surface; others extend fully through the entire membrane. The configuration of the inner and outer sides of the membrane can be quite different because of the variations in protein shape and position.

Membranes are dynamic and constantly changing because the lipid phase is in motion and many proteins can migrate freely about, somewhat as icebergs do in the ocean. This fluidity is essential to such activities as engulfment of food and discharge or secretion by cells. The structure of the lipid phase provides an impenetrable barrier to many substances. This property accounts for the selective permeability and capacity to regulate transport of molecules. It also serves to segregate activities within the cell's cytoplasm. Membrane proteins function in receiving molecular signals (receptors), in binding and transporting nutrients, and in acting as enzymes, topics to be discussed in chapters 7 and 8.

*mosaic (moh-zay′-ik) An intricate design made up of many small fragments.

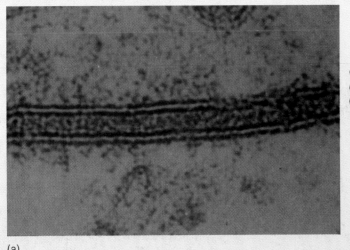

(a)

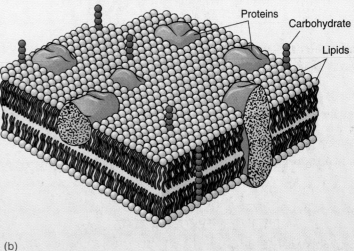

Proteins
Carbohydrate
Lipids

(b)

Extreme magnification of **(a)** a cross section of a cell membrane, which appears as double tracks. **(b)** A generalized version of the fluid mosaic model of a cell membrane indicates a bilayer of lipids with globular proteins embedded to some degree in the lipid matrix. This structure explains many characteristics of membranes, including flexibility, solubility, permeability, and transport.

additional bonds between functional groups (figure 2.22c). In proteins with the sulfur-containing amino acid **cysteine,*** considerable tertiary stability is achieved through covalent disulfide bonds between sulfur atoms on two different parts of the molecule. Some complex proteins assume a **quaternary (4°) structure,** in which more than one polypeptide forms a large, multiunit protein. This is typical of antibodies (see chapter 15) and some enzymes that act in cell synthesis.

The most important outcome of intrachain[5] bonding and folding is that each different type of protein develops a unique shape, and its surface displays a distinctive pattern of pockets and bulges. As a result, a protein can react only with molecules that complement or fit its particular surface features like a lock and key. Such a degree of specificity can provide the functional diversity required for many thousands of different cellular activities. **Enzymes** serve as the catalysts for all chemical reactions in cells, and nearly

*cysteine (sis′-tuh-yeen) Gr. *kystis,* sac. An amino acid first found in urine stones.

5. **Intra**chain means within the chain; **inter**chain would be between two chains.

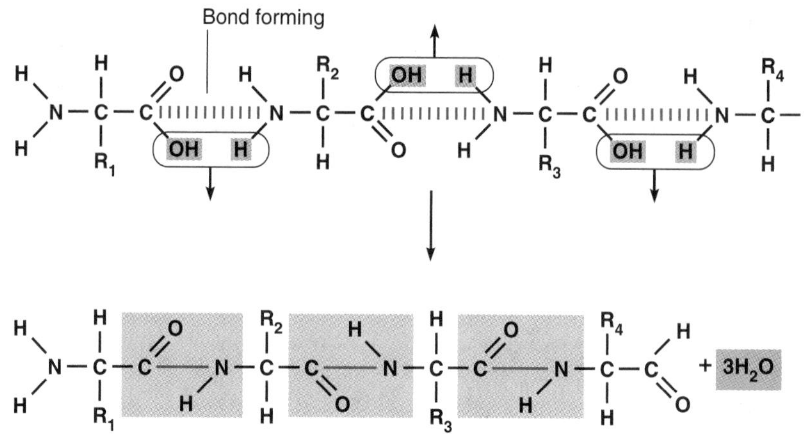

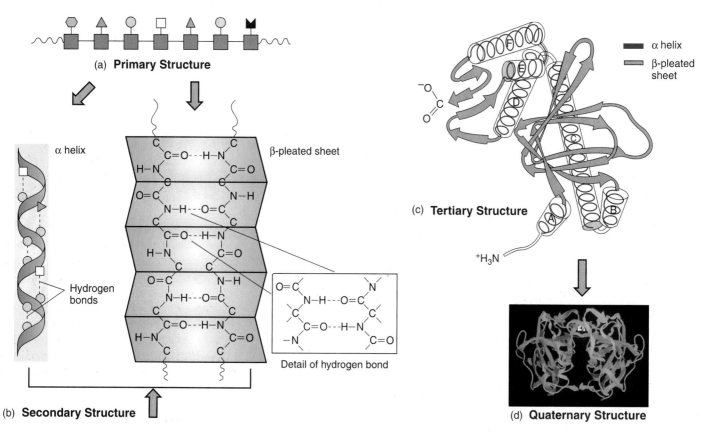

FIGURE 2.21

The formation of peptide bonds in a tetrapeptide.

FIGURE 2.22

Stages in the formation of a functioning protein. **(a)** Its primary structure is a series of amino acids bound in a chain. **(b)** Its secondary structure develops when the chain forms hydrogen bonds that fold it into one of several configurations such as an alpha helix or beta-pleated sheet. Some proteins have several configurations in the same molecule. **(c)** A protein's tertiary structure is due to further folding of the molecule into a three-dimensional mass that is stabilized by hydrogen, ionic, and disulfide bonds between functional groups. The letters and arrows denote the order and direction of the folded chain. **(d)** The quaternary structure exists only in proteins that consist of more than one polypeptide chain. Shown here is a computer model of the nitrogenase iron protein, with the two polypeptide chains arranged symmetrically.

every reaction requires a different enzyme (see chapter 8). **Antibodies** are complex glycoproteins with specific regions of attachment for bacteria, viruses, and other microorganisms; certain bacterial toxins (poisonous products) react with only one specific organ or tissue; and proteins embedded in the cell membrane have reactive sites restricted to a certain nutrient. Some proteins function as

receptors to receive stimuli from the environment. The functional three-dimensional form of a protein is termed the *native state,* and if it is disrupted by some means, the protein is said to be *denatured.* Such agents as heat, acid, alcohol, and some disinfectants disrupt the stabilizing intrachain bonds and cause the molecule to become nonfunctional (see figure 11.4).

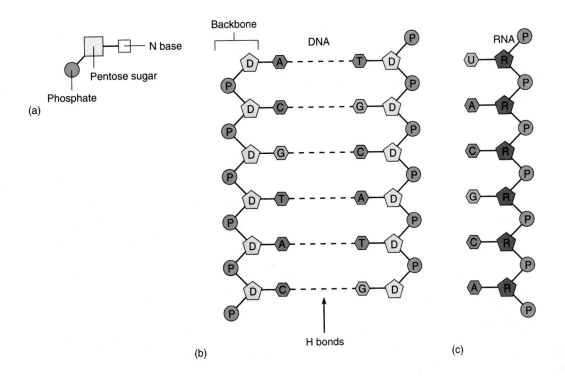

FIGURE 2.23

The general structure of nucleic acids. **(a)** A nucleotide, composed of a phosphate, a pentose sugar, and a nitrogen base, is the monomer of both DNA and RNA. **(b)** In DNA, the polymer is composed of alternating deoxyribose (D) and phosphate (P) with nitrogen bases (A, T, C, G) attached to the deoxyribose. Two of these polynucleotide strands are oriented so that the bases are paired across the central axis of the molecule. **(c)** In RNA, the polymer is composed of alternating ribose (R) and phosphate (P) attached to nitrogen bases (A, U, C, G), but it is only a single strand.

THE NUCLEIC ACIDS: A CELL COMPUTER AND ITS PROGRAMS

The nucleic acids, **deoxyribonucleic* acid (DNA)** and **ribonucleic* acid (RNA),** were originally isolated from the cell nucleus. Shortly thereafter, they were also found in other parts of nucleated cells, in cells with no nuclei (bacteria), and in viruses. The universal occurrence of nucleic acids in all known cells and viruses emphasizes their important roles as informational molecules. DNA, the master computer of cells, contains a special coded genetic program with detailed and specific instructions for each organism's heredity. It transfers the details of its program to RNA, operator molecules responsible for carrying out DNA's instructions and translating the DNA program into proteins that can perform life functions. For now, let us briefly consider the structure and some functions of DNA, RNA, and a close relative, adenosine triphosphate (ATP).

Both nucleic acids are polymers of repeating units called **nucleotides,*** each of which is composed of three smaller units: a **nitrogen base,** a **pentose** (5-carbon) sugar, and a *phosphate* **(figure 2.23a).** The nitrogen base is a cyclic compound that comes in two forms: *purines* (two rings) and *pyrimidines* (one ring). There are two types of purines—**adenine (A)** and **guanine (G)**—and three types of pyrimidines—**thymine (T), cytosine (C),** and

uracil **(U) (figure 2.24).** A characteristic that differentiates DNA from RNA is that DNA contains all of the nitrogen bases except uracil, and RNA contains all of the nitrogen bases except thymine. The nitrogen base is covalently bonded to the sugar *ribose* in RNA and *deoxyribose* (because it has one less oxygen than ribose) in DNA. Phosphate (PO_4^{-3}), a derivative of phosphoric acid (H_3PO_4), provides the final covalent bridge that connects sugars in series. Thus, the backbone of a nucleic acid strand is a chain of alternating phosphate-sugar-phosphate-sugar molecules, and the nitrogen bases branch off the side of this backbone (see figure 2.23b,c).

The Double Helix of DNA

DNA is a huge molecule formed by two very long polynucleotide strands linked along their length by hydrogen bonds between complementary pairs of nitrogen bases. The pairing of the nitrogen bases occurs according to a predictable pattern: Adenine ordinarily pairs with thymine, and cytosine with guanine. The bases are attracted in this way because each pair shares oxygen, nitrogen, and hydrogen atoms exactly positioned to align perfectly for hydrogen bonds **(figure 2.25).**

For ease in understanding the structure of DNA, it is sometimes compared to a ladder, with the sugar-phosphate backbone representing the rails and the paired nitrogen bases representing the steps. Owing to the manner of nucleotide pairing and stacking of the bases, the actual configuration of DNA is a *double helix* that looks somewhat like a spiral staircase (see figures 2.25 and 9.4). As is true of protein, the structure of DNA is intimately related to its

---

*deoxyribonucleic (dee-ox″-ee-ry″-boh-noo-klay′-ik).

*ribonucleic (ry″-boh-noo-klay′-ik) It is easy to see why the abbreviations are used!

*nucleotide (noo′-klee-oh-tyd) From nucleus and acid.

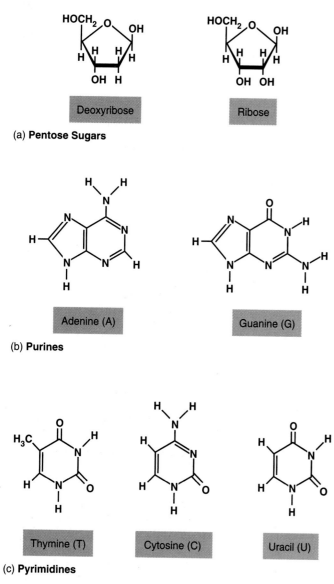

(a) **Pentose Sugars**

(b) **Purines**

(c) **Pyrimidines**

FIGURE 2.24

The sugars and nitrogen bases that make up DNA and RNA. (a) DNA contains deoxyribose, and RNA contains ribose. (b) A and G purines are found in both DNA and RNA. (c) C pyrimidine is found in both DNA and RNA, but T is found only in DNA, and U is found only in RNA.

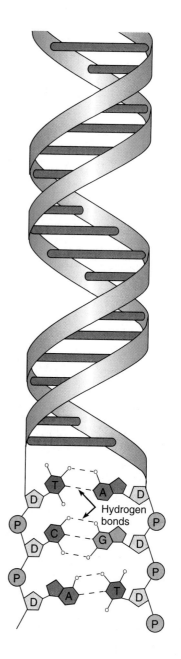

FIGURE 2.25

A structural representation of the double helix of DNA. Shown are the details of hydrogen bonds between the nitrogen bases of the two strands.

function. DNA molecules are usually extremely long, a feature that satisfies a requirement for storing genetic information in the sequence of base pairs the molecule contains. The hydrogen bonds between pairs can be disrupted when DNA is being copied, and the fixed complementary base pairing is essential to maintain the genetic code.

Making New DNA: Passing on the Genetic Message

The biological properties of cells and viruses are ultimately programmed by a master code composed of nucleic acids. This code is in the form of DNA in all cells and many viruses; other viruses are based on RNA alone. Regardless of the exact genetic program, both cells and viruses will continue to exist only if they can duplicate their genetic material and pass it on to subsequent generations.

Figure 2.26 summarizes the main steps in this process and how it differs between cells and viruses.

During its division cycle, the cell has a mechanism for making a copy of its DNA by **replication,*** using the original strand as a pattern (figure 2.26*a*). Note that replication is guided by the double-stranded nature of DNA and the precise pairing of bases that create the master code. Replication requires the separation of the double strand into two single strands by an enzyme that helps to split the hydrogen bonds along the length of the molecule. This event exposes the base code and makes it available for copying.

***replication** (reh″-plih-kay′-shun) A process of making an exact copy of something.

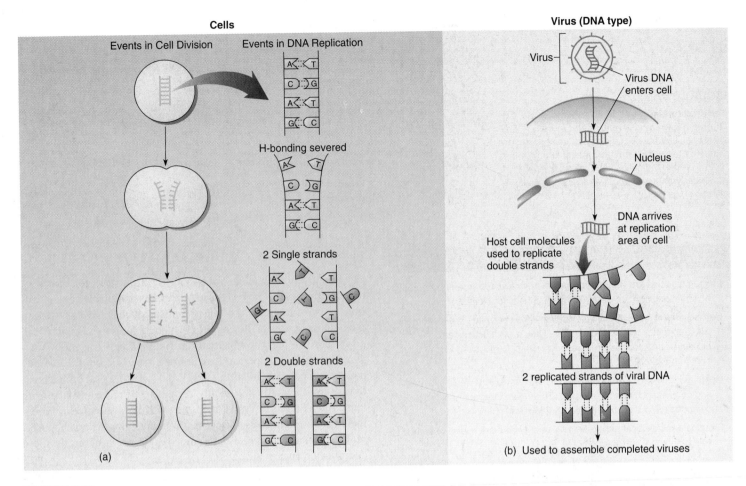

FIGURE 2.26

Simplified view of DNA replication in cells and viruses.   **(a)** The DNA in the cell's chromosome must be duplicated as the cell is dividing. This duplication is accomplished through the separation of the double DNA strand into two single strands. New strands are then synthesized using the original strands as guides to assemble the correct complementary new bases. **(b)** In DNA viruses, the DNA is inserted into its host cell and enters the replication area. Here it is separated into two strands and copied by the machinery of the host cell. Thousands of copies of viral nucleic acid will become part of the completed viruses later released by the infected host cell.

Free nucleotides are used to synthesize matching strands that complement the bases in the code by adhering to the pairing requirements of A–T and C–G. The end result is two separate double strands with the same order of bases as the original molecule. Viruses have a related mode of replication, but being genetic parasites, they must first enter a host cell and take over the cell's regular synthetic machinery and program it to make copies of the virus nucleic acid and, eventually, create whole viruses (figure 2.26*b*). The details of nucleic acid chemistry of cells and viruses will be covered in greater detail in chapter 9.

RNA: Organizers of Protein Synthesis

Like DNA, RNA consists of a long chain of nucleotides. However, RNA is a single strand containing ribose sugar instead of deoxyribose and uracil instead of thymine (see figure 2.23). Several functional types of RNA are formed using the DNA template through a replication-like process. Three major types of RNA are important for protein synthesis. Messenger RNA (mRNA) is a copy of a gene from DNA that provides the order and type of amino acids in a protein; transfer RNA (tRNA) is a carrier that delivers the cor-

rect amino acids for protein assembly; and ribosomal RNA (rRNA) is a major component of ribosomes (see figure 4.19). More information on these important processes is presented in chapter 9.

ATP: The Energy Molecule of Cells

A relative of RNA involved in an entirely different cell activity is **adenosine triphosphate (ATP).** ATP is a nucleotide containing adenine, ribose, and three phosphates rather than just one **(figure 2.27).** It belongs to a category of high-energy compounds (also including guanosine triphosphate, GTP) that give off energy when the bond is broken between the second and third (outermost) phosphate. The presence of these high-energy bonds makes it possible for ATP to release and store energy for cellular chemical reactions. Breakage of the bond of the terminal phosphate releases energy to do cellular work and also generates adenosine diphosphate (ADP). ADP can be converted back to ATP when the third phosphate is restored, thereby serving as an energy depot. Carriers for oxidation-reduction activities (nicotinamide adenine dinucleotide [NAD], for instance) are also derivatives of nucleotides (see chapter 8).

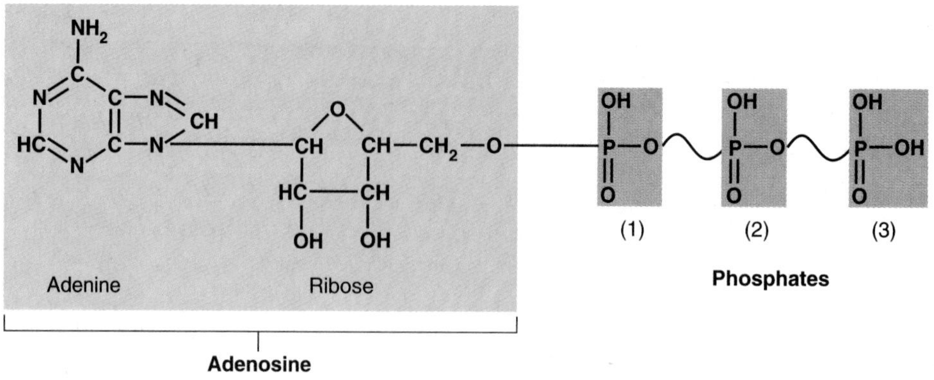

FIGURE 2.27

The structural formula of an ATP molecule, the chemical form of energy transfer in cells. The wavy lines that connect the phosphates represent bonds that release large amounts of energy when broken. Within the cell, several of the hydroxyl groups on the phosphates would be negatively charged oxygens.

CHAPTER CHECKPOINTS ✓✓✓

Macromolecules are very large organic molecules (polymers) built up by polymerization of smaller molecular subunits (monomers).

Carbohydrates are biological molecules whose polymers are monomers linked together by glycosidic bonds. Their main functions are protection and support (in organisms with cell walls) and also nutrient and energy stores.

Lipids are biological molecules such as fats that are insoluble in water and contain special ester linkages. Their main functions are cell components, cell secretions, and nutrient and energy stores.

Proteins are biological molecules whose polymers are chains of amino acid monomers linked together by peptide bonds.

Proteins are called the "shapers of life" because of the many biological roles they play in cell structure and cell metabolism.

Protein shape determines protein function. Shape is dictated by amino acid composition and by the pH and temperature of the protein's immediate environment.

Nucleic acids are biological molecules whose polymers are chains of nucleotide monomers linked together by phosphate–pentose sugar covalent bonds. Double-stranded nucleic acids are linked together by hydrogen bonds. Nucleic acids are information molecules that direct cell metabolism and reproduction. Nucleotides such as ATP also serve as energy transfer molecules in cells.

Cells: Where Chemicals Come to Life

As we proceed in this chemical survey from the level of simple molecules to increasingly complex levels of macromolecules, at some point we cross a line from the realm of lifeless molecules and arrive at the fundamental unit of life called a **cell**.[6] A cell is indeed a huge aggregate of carbon, hydrogen, oxygen, nitrogen, and many other atoms, and it follows the basic laws of chemistry and physics, but it is much more. The combination of these atoms produces characteristics, reactions, and products that can only be described as **living.**

---

6. The word *cell* was originally coined from an Old English term meaning "small room" because of the way plant cells looked to early microscopists.

FUNDAMENTAL CHARACTERISTICS OF CELLS

The bodies of living things such as bacteria and protozoa consist of only a single cell, whereas those of animals and plants contain trillions of cells. Regardless of the organism, all cells have a few common characteristics. They tend to be spherical, polygonal, cubical, or cylindrical, and their protoplasm (internal cell contents) is encased in a cell or cytoplasmic membrane (see Microbits 2.3). They have chromosomes containing DNA and ribosomes for protein synthesis, and they are exceedingly complex in function. Aside from these few similarities, most cell types fall into one of two fundamentally different lines (discussed in chapter 1): the small, seemingly simple procaryotic cells and the larger, structurally more complicated eucaryotic cells.

Eucaryotic cells are found in animals, plants, fungi, and protists. They contain a number of complex internal parts called organelles that perform useful functions for the cell involving growth, nutrition, or metabolism. By convention, organelles are defined as cell components that perform specific functions and are enclosed by membranes. Organelles also partition the eucaryotic cell into smaller compartments. The most visible organelle is the nucleus, a roughly ball-shaped mass surrounded by a double membrane that contains the DNA of the cell. Other organelles include the Golgi apparatus, endoplasmic reticulum, vacuoles, and mitochondria (**table 2.6**).

Procaryotic cells are possessed only by the bacteria and archaea. Sometimes it may seem that procaryotes are the microbial "have-nots" because, for the sake of comparison, they are described by what they lack. They have no nucleus or other organelles. This apparent simplicity is misleading, because the fine structure of procaryotes is complex. Overall, procaryotic cells can engage in nearly every activity that eucaryotic cells can, and many can function in ways that eucaryotes cannot.

PROCESSES THAT DEFINE LIFE

To lay the groundwork for a detailed coverage of cells in chapters 4 and 5, this section provides an overview of cell structure and function and introduces the primary characteristics of life. The biological activities or properties that help define and characterize cells as living entities are:

1. growth;
2. reproduction and heredity;

TABLE 2.6

A General Comparison of Procaryotic and Eucaryotic Cells and Viruses*

Function or Structure	Characteristic	Procaryotic Cells	Eucaryotic Cells	Viruses**
Genetics	Nucleic acids	+	+	+
	Chromosomes	+	+	−
	True nucleus	−	+	−
	Nuclear envelope	−	+	−
Reproduction	Mitosis	−	+	−
	Production of sex cells	+/−	+	−
	Binary fission	+	+	−
Biosynthesis	Independent	+	+	−
	Golgi apparatus	−	+	−
	Endoplasmic reticulum	−	+	−
	Ribosomes	+★★★	+	−
Respiration	Enzymes	+	+	−
	Mitochondria	−	+	−
Photosynthesis	Pigments	+/−	+/−	−
	Chloroplasts	−	+/−	−
Motility/locomotor structures	Flagella	+/−★★★	+/−	−
	Cilia	−	+/−	−
Shape/protection	Cell wall	+★★★	+/−	−
	Capsule	+/−	+/−	−
Complexity of function		+	+	+/−
Size (in general)		0.5−3 μm	2−100 μm	< 0.2 μm

★+ *means most members of the group exhibit this characteristic;* − *means most lack it;* +/− *means only some members have it.*

★★ *Viruses cannot participate in metabolic or genetic activity outside their host cells.*

★★★ *The procaryotic type is functionally similar to the eucaryotic, but structurally unique.*

3. metabolism, including cell synthesis and the release of energy;
4. movement and/or irritability;
5. the capacity to transport substances into and out of the cell; and
6. cell support, protection, and storage mechanisms.

Although eucaryotic cells have specific organelles to perform these functions, procaryotic cells must rely on a few simple, multipurpose cell components. As indicated in chapter 1, viruses are not cells, are not generally considered living things, and show certain signs of life only when they invade a host cell. Table 2.6 indicates their relative simplicity compared with cells.

Reproduction: Bearing Offspring

A cell's **genome,*** its complete set of genetic material, is composed of elongate strands of DNA. The DNA is packed into discrete bodies called **chromosomes,*** In eucaryotic cells, the chromosomes are located entirely within a nuclear membrane. Procaryotic DNA occurs in a special type of circular chromosome that is not enclosed by a membrane of any sort.

Living things devote a portion of their life cycle to producing offspring that will carry on their particular genetic line for many generations. In *sexual reproduction,* offspring are produced through the union of sex cells from two parents. In *asexual[7] reproduction,* offspring originate through the division of a single parent cell into two daughter cells. Sexual reproduction occurs in most eucaryotes, and eucaryotic cells also reproduce asexually by several processes. One is a type of cell division called *binary fission,* a simple process in which the cell splits equally in two. Many eucaryotic cells engage in **mitosis,*** an orderly division of chromosomes that usually accompanies cell division (see figure 5.6). In contrast, procaryotic cells reproduce primarily by binary fission. They have no mitotic apparatus, nor do they reproduce by typical sexual means.

Metabolism: Chemical and Physical Life Processes

Cells synthesize proteins using hundreds of tiny particles called **ribosomes.** In eucaryotes, ribosomes are dispersed throughout the cell or inserted into membranous sacs known as the **endoplasmic reticulum** (see figure 5.8). Procaryotes have smaller ribosomes scattered throughout the protoplasm, since they lack an

---

*genome (jee′-nohm) A combination of the words *gene* and *chromosome.* Refers to an organism's entire set of hereditary factors.

*chromosome (kro′-moh-sohm) Gr. *chroma,* colored, and *soma,* body. The name is derived from the fact that chromosomes stain readily with dyes.

7. *Asexual* refers to the absence of sexual union. (The prefix *a* or *an* means not or without.)

*mitosis (my-toh′-sis) Gr. *mitos,* thread, and *osis,* a condition. Often the term *mitosis* is used synonymously with eucaryotic cell division.

endoplasmic reticulum. Eucaryotes generate energy by chemical reactions in the **mitochondria,** whereas procaryotes use their cell membrane for this purpose. Photosynthetic microorganisms (algae and some bacteria) trap solar energy by means of pigments and convert it to chemical energy in the cell. Algae (eucaryotes) have compact, membranous bundles called **chloroplasts,** which contain the pigment and perform the photosynthetic reactions. Photosynthetic reactions and pigments of procaryotes do not occur in chloroplasts, but in specialized areas of the cell membrane.

Irritability or Motility

All cells have the capacity to respond to chemical, mechanical, or light stimuli. This quality, called **irritability,** helps cells adapt to the environment and obtain nutrients. Although not present in all cells, true **motility,** or self-propulsion, is a notable sign of life. Eucaryotic cells move by one of the following locomotor organelles: cilia, which are short, hairlike appendages; flagella, which are longer, whiplike appendages; or pseudopods, fingerlike extensions of the cell membrane. Motile procaryotes move by means of unusual, propeller-like flagella unique to bacteria or by special fibrils that produce a gliding form of motility. They have no cilia or pseudopods.

Protection and Storage

Many cells are supported and protected by rigid cell walls, which prevent them from rupturing while also providing support and shape. Among eucaryotes, cell walls occur in plants, microscopic algae, and fungi, but not in animals or protozoa. The majority of procaryotes have cell walls, but they differ in composition from the

eucaryotic varieties. As protection against depleted nutrient sources, many microbes store nutrients intracellularly. Eucaryotes store nutrients in membranous sacs called vacuoles, and procaryotes concentrate them in crystals called granules or inclusions.

Transport: Movement of Nutrients and Wastes

Cell survival depends on drawing nutrients from the external environment and expelling waste and other metabolic products from the internal environment. This two-directional transport is accomplished in both eucaryotes and procaryotes by the cell membrane. This membrane, described in Microbits 2.3, has a very similar structure in both eucaryotic and procaryotic cells. Eucaryotes have an additional organelle, the **Golgi apparatus,** that assists in sorting and packaging molecules for transport and removal from the cell.

Table 2.6 summarizes the differences and similarities among procaryotes, eucaryotes, and viruses. You will probably want to review this table after you have completed chapters 4 and 5.

CHAPTER CHECKPOINTS

As the atom is the fundamental unit of matter, so is the cell the fundamental unit of life.

All true cells contain biological molecules that carry out the processes that define life: metabolism and reproduction; these two basic processes are supported by the functions of irritability and motility, protection, storage, and transport.

The cell membrane is of critical importance to all cells because it controls the interchange between the cell and its environment.

CHAPTER CAPSULE WITH KEY TERMS

I. **Atoms, Bonds, and Molecules: The Building Blocks**
  A. Atomic Structure and Elements
    1. All **matter** in the universe is composed of minute particles called **atoms**—the simplest form of matter not divisible into a simpler substance by chemical means. Atoms are composed of smaller particles called protons, neutrons, and electrons. Protons and neutrons comprise the nucleus of an atom, and electrons move about the nucleus in **orbitals,** with the orbitals arranged in shells.
    2. Atoms that differ in numbers of the protons, neutrons, and electrons are elements. Elements can be described by **mass number (MN)** and **atomic number (AN),** and each is known by a distinct name and symbol. Elements may exist in variant forms called **isotopes.**
    3. The major elements of life can be described by a unique symbol, by their atomic number, and by the **atomic weight,** which averages the mass numbers of the isotopes. An important key to understanding the characteristics of an element is the filling of the electron orbitals. Electrons fill the orbitals in pairs, and the outermost shell becomes the focus of reactivity with other atoms.
  B. Bonds and Molecules
    1. Atoms interact to form **chemical bonds** and **molecules.** If the atoms combining to make a molecule are different elements, then the substance is termed a **compound.**

  2. The type of bond is dictated by the electron makeup (**valence**) of the outer orbitals of the atoms. Bond types include:
    a. **Covalent bonds,** with shared electrons. The molecule shares the electrons; the balance of charge will be **polar** if unequal or **nonpolar** if equally shared.
    b. **Ionic bonds,** where electrons are transferred to an atom that can come closer to filling up the outer orbital. Dissociation of these compounds leads to the formation of charged **cations** and **anions.**
    c. **Hydrogen bonds** involve weak covalent bonds between hydrogen and nearby electronegative oxygens and nitrogens.
  3. Molecules can be represented by chemical formulas, which summarize the atoms that make up the molecule, or by structural models that represent the bonding or three-dimensional shape.
  4. The interaction of elements and molecules to form different bonds involving reactants (starting chemicals) and products (end results) can be shown by chemical equations. Some of the reactions that occur include synthesis, decomposition, and exchange reactions. Many reactions are reversible.
  5. Of particular importance are **oxidation-reduction (redox)** reactions between pairs of atoms of molecules whereby electrons are transferred.

C. Solutions, Acids, Bases, and pH
   1. A **solution** is a combination of a solid, liquid, or gaseous chemical (the **solute**) dissolved in a liquid medium (the **solvent**).
   2. Water is the most common solvent in natural systems, and the characteristics of dissolved solutes in water are important to life. The solutes interact with the water molecules and the resulting ions become **hydrated.** Whether a molecule is **hydrophobic, hydrophilic,** or amphipathic will affect their function.
   3. Ionization of water leads to the release of hydrogen ions (H^+) and hydroxyl ions. The **pH** scale expresses the concentration of [H^+] such that a pH of less than 7.0 is considered **acidic,** and a pH of more than that, indicating fewer [H^+], is considered **basic.**

II. **Macromolecules**
   A. Biochemistry studies those molecules that are found in living things. These are based on **organic** compounds, which consist of carbon and hydrogen covalently bonded in various combinations. Examples includes methane and glucose. **Inorganic** compounds do not contain both carbon and hydrogen in combination. Examples include carbon dioxide and sodium chloride.
   B. **Macromolecules** are very large compounds and are generally assembled from single units called **monomers** by **polymerization.**
   C. Macromolecules of life fall into basic categories of **carbohydrates, lipids, proteins,** and nucleic acids.
      1. **Carbohydrates** are composed of carbon, hydrogen, and oxygen and contain aldehyde or ketone groups.
         a. **Monosaccharides** such as glucose are the simplest carbohydrates with 3 to 7 carbons; these are the monomers of carbohydrates.
         b. **Disaccharides** such as lactose consist of two monosaccharides joined by **glycosidic** bonds.
         c. **Polysaccharides** such as starch and **peptidoglycan** are chains of five or more monosaccharides.
      2. **Lipids** contain long hydrocarbon chains and are not soluble in polar solvents such as water due to their nonpolar, hydrophobic character.
         a. **Triglycerides,** including fats and oils, consist of a **glycerol** molecule bonded to three fatty acid molecules. They are important storage lipids.
         b. **Phospholipids** are composed of a glycerol bound to two fatty acids and a phosphoric acid–alcohol group; these molecules have hydrophilic heads and long hydrophobic fatty acid tails. They form single or double lipid layers in the presence of water and are important constituents of all cell membranes.
         c. Other important lipids include sterols and waxes.
      3. **Proteins** are highly complex macromolecules that are crucial in most, if not all, life processes.
         a. **Amino acids** are the basic building blocks of proteins. They all share a basic structure of an amino group, a carboxyl group, an R group, and hydrogen bonded to a carbon atom. There are 20 different R groups, which define the basic set of 20 amino acids, found in all of life.
         b. A **peptide** is a short chain of amino acids bound by **peptide bonds;** a dipeptide has two, and a tripeptide has three; a polypeptide is usually 20 to 50 amino acids; a **protein** contains more than 50 amino acids. Larger proteins predominate in cells.
         c. The structure of a protein is very important to the function it has. This is described by the **primary structure** (the chain of amino acids), the **secondary structure** (formation of helices and sheets due to hydrogen bonding within the chain), **tertiary structure** (cross-links, especially disulfide bonds, between secondary structures), and **quaternary structure** (formation of multisubunit proteins). The incredible variation in shapes is the basis for the diverse roles proteins play as enzymes, antibodies, and structural components.
      4. Nucleic acids
         a. **Nucleotides** are the building blocks of nucleic acids. They are composed of a **nitrogen base, a pentose** sugar, and phosphate, Nitrogen bases are ringed compounds: **adenine (A), guanine (G), cytosine (C), thymine (T),** and **uracil (U).** Pentose sugars may be deoxyribose or ribose.
         b. **Deoxyribonucleic acid (DNA)** is a polymer of nucleotides that occurs as a double-stranded helix with hydrogen bonding in pairs between the helices. It has all of the bases except uracil, and the pentose sugar is deoxyribose. DNA is the master code for a cell's life processes and must be transmitted to the offspring. During cell division, it is **replicated** such that identical copies are passed on to the progeny.
         c. **Ribonucleic acid (RNA)** is a polymer of nucleotides where the sugar is ribose and the uracil is used instead of thymine. It is almost always found single stranded and is used to express the DNA code into proteins.
         d. **Adenosine triphosphate (ATP)** is a nucleotide involved in the transfer and storage of energy in cells. It contains adenine, ribose, and three phosphates in a series.

III. **Cells: Where Chemicals Come to Life**
   A. All living things are composed of **cells,** which are aggregates macromolecules that carry out living processes.
   B. Cells can be divided into two basic types: procaryotes and eucaryotes.
      1. Procaryotic cells are the basic structural unit of bacteria and archaea. They do not contain a nucleus or organelles. They are highly successful and adaptable single-cell life forms.
      2. Eucaryotic cells contain a membrane-surrounded nucleus and a number of organelles that function in specific ways. A wide variety of organisms, from single-celled protozoans to humans, are composed of eucaryotic cells.
      3. A membrane that contains the cytoplasm and controls cell permeability and transport surrounds both basic cell types.
      4. Viruses are not generally considered living, but also rely on host cells to replicate.
   C. Cells show the basic essentials characteristics of life. Parts of cells and macromolecules do not show these characteristics:
      1. Growth: Living entities are able to grow and increase in size, often renewing and rebuilding themselves over time.
      2. **Reproduction** and heredity: Cells must pass on genetic information to their offspring, whether *asexually* (with one parent) or *sexually* (with two parents).
      3. **Metabolism:** This refers to the chemical reactions in the cell, including the synthesis of proteins on ribosomes and the capture and release of energy using ATP.
      4. Movement: Motility originates from special locomotor structures such as flagella and cilia; **irritability** is responsiveness to external stimuli.
      5. Transport: Nutrients must be brought into the cell through the membrane and wastes expelled from the cell.
      6. Cell protection, support, and storage: Cells tend to certain shapes, which must be maintained by modifications to the cell membrane or additional internal and external supports. Varying means of storage protect against nutrient depletion and keep cells alive until more nutrients are available.

MULTIPLE-CHOICE QUESTIONS

1. The smallest unit of matter with unique characteristics is
   a. an electron
   b. a molecule
   c. an atom
   d. a proton

2. The _____ charge of a proton is exactly balanced by the _____ charge of a (an) _____.
   a. negative, positive, electron
   b. positive, neutral, neutron
   c. positive, negative, electron
   d. neutral, negative, electron

3. Electrons move around the nucleus of an atom in pathways called
   a. shells
   b. orbitals
   c. circles
   d. rings

4. Which part of an element does not vary in number?
   a. electron
   b. neutron
   c. proton
   d. all of these vary

5. The number of electrons of a neutral atom is automatically known if one knows the
   a. atomic number
   b. atomic weight
   c. number of orbitals
   d. valence

6. Elements
   a. exist in 92 natural forms
   b. have distinctively different atomic structures
   c. vary in atomic weight
   d. a and b are correct
   e. a, b, and c are correct

7. If a substance contains two or more elements of different types, it is considered
   a. a compound
   b. a monomer
   c. a molecule
   d. organic

8. Bonds in which atoms share electrons are defined as _____ bonds.
   a. hydrogen
   b. ionic
   c. double
   d. covalent

9. What kind of bond would you expect potassium to form with chlorine?
   a. ionic
   b. covalent
   c. polar
   d. nonpolar

10. When a compound carries a positive charge on one end and a negative charge on the other end, it is said to be
    a. ionized
    b. hydrophilic
    c. polar
    d. oxidized

11. Hydrogen bonds can form between _____ adjacent to each other.
    a. two hydrogen atoms
    b. two oxygen atoms
    c. a hydrogen atom and an oxygen atom
    d. negative charges

12. Ions with the same charge will be _____ each other, and ions with opposite charges will be _____ each other.
    a. repelled by, attracted to
    b. attracted to, repelled by
    c. hydrated by, dissolved by
    d. dissolved by, hydrated by

13. An atom that can donate electrons during a reaction is called
    a. an oxidizing agent
    b. a reducing agent
    c. an ionic agent
    d. an electrolyte

14. In a solution of NaCl and water, NaCl is the _____ and water is the _____.
    a. acid, base
    b. base, acid
    c. solute, solvent
    d. solvent, solute

15. A substance that releases H^+ into a solution
    a. is a base
    b. is ionized
    c. is an acid
    d. has a high pH

16. A solution with a pH of 2 _____ than a solution with a pH of 8.
    a. has less H^+
    b. has more H^+
    c. has more OH^-
    d. is less concentrated

17. Fructose is a type of
    a. disaccharide
    b. monosaccharide
    c. polysaccharide
    d. amino acid

18. Bond formation in polysaccharides and polypeptides is accompanied by the removal of a
    a. hydrogen atom
    b. hydroxyl ion
    c. carbon atom
    d. water molecule

19. The monomer unit of polysaccharides such as starch and cellulose is
    a. fructose
    b. glucose
    c. ribose
    d. lactose

20. A phospholipid contains
    a. three fatty acids bound to glycerol
    b. three fatty acids, a glycerol, and a phosphate
    c. two fatty acids and a phosphate bound to glycerol
    d. three cholesterol molecules bound to glycerol

21. Proteins are synthesized by linking amino acids with _____ bonds.
    a. disulfide
    b. glycosidic
    c. peptide
    d. ester

22. The amino acid that accounts for disulfide bonds in the tertiary structure of proteins is
    a. tyrosine
    b. glycine
    c. cysteine
    d. serine

23. DNA is a hereditary molecule that is composed of
    a. deoxyribose, phosphate, and nitrogen bases
    b. deoxyribose, a pentose, and nucleic acids
    c. sugar, proteins, and thymine
    d. adenine, phosphate, and ribose

24. What is meant by DNA replication?
    a. duplication of the sugar-phosphate backbone
    b. matching of base pairs
    c. formation of the double helix
    d. the exact copying of the DNA code into two new molecules

CONCEPT QUESTIONS

1. How are the concepts of an atom and an element related? What causes elements to differ?

2. a. How are mass number and atomic number derived? What is the atomic weight?
   b. Using data in table 2.1, give the electron number of nitrogen, sulfur, calcium, phosphorus, and iron.
   c. What is distinctive about isotopes of elements, and why are they important?

3. a. How is the concept of molecules and compounds related?
   b. Compute the molecular weight of oxygen and methane.

4. a. Why is an isolated atom neutral?
   b. Describe the concept of the atomic nucleus, electron orbitals, and shells.
   c. What causes atoms to form chemical bonds?
   d. Why do some elements not bond readily?
   e. Draw the atomic structure of magnesium and predict what kinds of bonds it will make.

5. Distinguish between the general reactions in covalent, ionic, and hydrogen bonds.

6. a. Which kinds of elements tend to make covalent bonds?
   b. Distinguish between a single and a double bond.
   c. What is polarity?
   d. Why are some covalent molecules polar and others nonpolar?
   e. What is an important consequence of the polarity of water?

7. a. Which kinds of elements tend to make ionic bonds?
   b. Exactly what causes the charges to form on atoms in ionic bonds?
   c. Verify the proton and electron numbers for Na^+ and Cl^-.
   d. Differentiate between an anion and a cation.
   e. What kind of ion would you expect magnesium to make, on the basis of its valence?

8. Differentiate between an oxidizing agent and a reducing agent.

9. Why are hydrogen bonds relatively weak?

10. a. Compare the three basic types of chemical formulas.
    b. Review the types of chemical reactions and the general ways they can be expressed in equations.

11. a. Define solution, solvent, and solute.
    b. What properties of water make it an effective biological solvent, and how does a molecule like NaCl become dissolved in it?
    c. How is the concentration of a solution determined?
    d. What is molarity? Tell how to make a 1M solution of $Mg_3(PO_4)_2$ and a 0.1 M solution of $CaSO_4$.

12. a. What determines whether a substance is an acid or a base?
    b. Briefly outline the pH scale.
    c. How can a neutral salt be formed from acids and bases?

13. a. What atoms must be present in a molecule for it to be considered organic?
    b. What characteristics of carbon make it ideal for the formation of organic compounds?
    c. What are functional groups?
    d. Differentiate between a monomer and a polymer.
    e. How are polymers formed?
    f. Name several inorganic compounds.

14. a. What characterizes the carbohydrates?
    b. Differentiate between mono-, di-, and polysaccharides, and give examples of each.
    c. What is a glycosidic bond, and what is dehydration synthesis?
    d. What are some of the functions of polysaccharides in cells?

15. a. Draw simple structural molecules of triglycerides and phospholipids to compare their differences and similarities.
    b. What is an ester bond?
    c. How are saturated and unsaturated fatty acids different?
    d. What characteristic of phospholipids makes them essential components of cell membranes?
    e. Why is the hydrophilic end of phospholipids attracted to water?

16. a. Describe the basic structure of an amino acid.
    b. What makes the amino acids distinctive, and how many of them are there?
    c. What is a peptide bond?
    d. Differentiate between a peptide, a polypeptide, and a protein.
    e. Explain what causes the various levels of structure of a protein molecule.
    f. What functions do proteins perform in a cell?

17. a. Describe a nucleotide and a polynucleotide, and compare and contrast the general structure of DNA and RNA.
    b. Name the two purines and the three pyrimidines.
    c. Why is DNA called a double helix?
    d. What is the function of RNA?
    e. What is ATP, and what is its function in cells?

18. a. Outline the general structure of a cell, and describe the characteristics of cells that qualify them as living.
    b. Why are viruses not considered living?
    c. Compare the general characteristics of procaryotic and eucaryotic cells.
    d. What are cellular membranes, and what are their functions?
    e. Explain the fluid mosaic model of a membrane.

CRITICAL-THINKING QUESTIONS

1. The "octet rule" in chemistry helps predict the tendency of atoms to acquire or donate electrons from the outer shell. It says that those with fewer than 4 tend to donate electrons and those with more than 4 tend to accept additional electrons; those with exactly 4 can do both. Using this rule, determine what category each of the following elements falls into: N, S, C, P, O, H, Ca, Fe, and Mg. (You will need to work out the valence of the atoms.)

2. Predict the kinds of bonds that occur in ammonium (NH_3), phosphate (PO_4), disulfide (S–S), and magnesium chloride ($MgCl_2$). (Use simple models such as those in figure 2.3.)

3. Work out the following problems:
   a. What is the number of protons in helium?
   b. Will an H bond form between H_3C—CH=O and H_2O? Why or why not?
   c. Draw the following molecules and determine which are polar: Cl_2, NH_3, CH_4.
   d. What is the pH of a solution with a concentration of 0.00001 moles/ml (M) of H^+?
   e. What is the pH of a solution with a concentration of 0.00001 moles/ml (M) of OH^-?

4. a. Describe how hydration spheres are formed around cations and anions.
   b. Which substances will be expected to be hydrophilic and hydrophobic, and what makes them so?
   c. Distinguish between polar and ionic compounds, using your own words.

5. In what way are carbon-based compounds like children's Tinker Toys or Lego blocks?

6. Is galactose an aldehyde or a ketone sugar?

7. a. How many water molecules are released when a triglyceride is formed?
   b. How many peptide bonds are in a tetrapeptide?

8. a. Use pipe cleaners to help understand the formation of the 2° and 3° structures of proteins.
   b. Note the various ways that your pipe cleaner structure can be folded and the diversity of shapes that can be formed.

9. a. Looking at figure 2.25, can you see why adenine forms hydrogen bonds with thymine and why cytosine forms them with guanine?
   b. Show on paper the steps in replication of the following segment of DNA:

$$\text{A T G T T C C C G A T C G G C}$$
$$| \; | \; | \; | \; | \; | \; | \; | \; | \; | \; | \; | \; | \; | \; |$$
$$\text{T A C A A G G G C T A G C C G}$$

10. A useful mnemonic (memory) device for recalling the major characteristics of life is: *Giant Rats Have Many Colored Teeth*. Can you list possible characteristics for which each letter stands?

INTERNET SEARCH TOPICS

1. Use a search engine to explore the topic of isotopes and dating ancient rocks. How can isotopes be used to determine if rocks contain evidence of life?

2. Visit the student Online Learning Center at www.mhhe.com/talaro5. Go to chapter 2, Internet Search Topics, and log on to the available websites to make a search for basic information on elements. Click on the icons for C, H, N, O, P, S and list the source, biological importance, and other useful information about these elements.

Tools of the Laboratory:
The Methods for Studying Microorganisms

The world has recently been on high alert, not only because of disasters, violence, and war, but increasingly because of threats from microorganisms. Acts of terrorism using biological agents such as anthrax have been a major focus of concern, especially since events have now proved beyond a doubt that these agents are readily available and can be released inconspicuously into public places. To complicate matters, microbes from natural sources continue to pose emerging threats to public health. Over the past 30 years, nearly 20 new infectious agents have appeared suddenly and unexpectedly. Examples such as the agents of AIDS and SARS (severe acute respiratory syndrome) are new viruses that apparently originated from animals and have become infectious to humans. At the outset of increased reports of a new or suspected infectious disease, the pressure is immediately on medical science to discover the causative agent and possibly to find a treatment for the disease. The response to such threats invariably falls to the laboratories around the world that have the technical expertise to detect and identify the microbe rapidly and accurately. The reality is that without their basic tools—microscopes, media, biochemical and genetic tests—tracking down the agents of disease or developing treatments for them would be an impossible endeavor.

Any area that deals with microorganisms is faced with this same challenge. We need to be able to observe and analyze not just agents of disease, but those involved in such diverse areas as biotechnology, genetics, microbial ecology, research, and agriculture. In this chapter we address the unique strategies that have grown up around the care, feeding, and study of microbes in the laboratory.

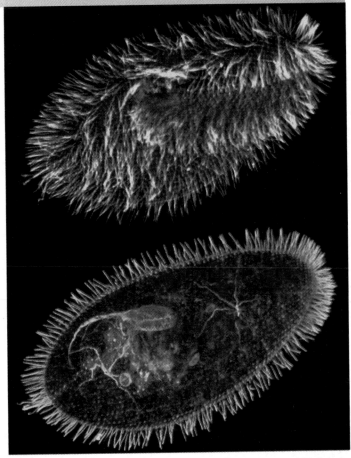

A display of a living protozoan *(Paramecium)* is produced by a combination of two types of microscopy. The confocal microscope provides dramatic three-dimensional images of cilia and internal cell structure. The fluorescent microscope uses dyes to highlight specific cell parts.

Chapter Overview

- A driving force of microbiology has been to find ways to visualize and handle microorganisms.
- Microbes are managed and characterized by implementing the Five I's—inoculation, incubation, isolation, inspection, and identification.
- Cultures are made by removing a sample from a desired source and placing it in containers of media.
- Media can be varied in chemical, physical, and functional purposes, depending on the intention.
- Growth and isolation of microbes leads to pure cultures that permit the study and testing of single species.
- Cultures can be used to provide information on microbial morphology, biochemistry, and genetic characteristics.
- Unknown, invisible samples can become known and visible.
- The microscope is a powerful tool for magnifying and resolving cells and their parts.
- Microscopes exist in several forms, using light, radiation, and electrons to form images.
- Specimens and cultures are prepared for study in fresh (live) or fixed (dead) form.
- Staining procedures highlight cells and allow them to be described and identified.

Methods of Culturing Microorganisms— The Five I's

Biologists studying large organisms such as animals and plants can, for the most part, immediately see and differentiate their experimental subjects from the surrounding environment and from one another. In fact, they can use their senses of sight, smell, hearing, and even touch to detect and evaluate identifying characteristics and to keep track of growth and developmental changes. Because microbiologists cannot rely as much as other scientists on senses other than sight, they are confronted by some unique problems. First, most habitats (such as the soil and the human mouth) harbor microbes in complex associations, so it is often necessary to separate the species from one another. Second, to maintain and keep track of such small research subjects, microbiologists usually have to grow them under artificial conditions. A third difficulty in working with microbes is that they are invisible and widely distributed, and undesirable ones can be introduced into an experiment and cause misleading results. These impediments motivated the development of techniques to control microbes and their growth, primarily sterile, aseptic, and pure culture techniques.[1]

Microbiologists use five basic techniques to manipulate, grow, examine, and characterize microorganisms in the laboratory: inoculation, incubation, isolation, inspection, and identification (the Five I's; **figure 3.1**). Some or all of these procedures are performed by microbiologists, whether the beginning laboratory student, the researcher attempting to isolate drug-producing bacteria from soil, or the clinical microbiologist working with a specimen from a patient's infection. These procedures make it possible to handle and maintain microorganisms as discrete entities whose detailed biology can be studied and recorded.

---

1. *Sterile* relates to the complete absence of viable microbes; *aseptic* relates to prevention of infection; *pure culture* refers to growth of a single species of microbe.

An Overview of Major Techniques Performed by Microbiologists to Locate, Grow, Observe, and Characterize Microorganisms.

Specimen Collection:
Nearly any object or material can serve as a source of microbes. Common ones are body fluids and tissues, foods, water, or soil. Specimens are removed by some form of sampling device. This may be a swab, syringe, or a special transport system that holds, maintains, and preserves the microbes in the sample.

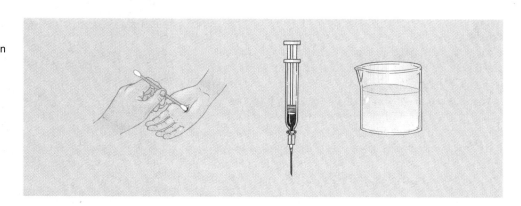

A GUIDE TO THE FIVE I'S: How the Sample Is Processed and Profiled

1. Inoculation:
The sample is placed into a container of sterile **medium** that provides microbes with the appropriate nutrients to sustain growth. Inoculation involves using a sterile tool to spread the sample on the surface of a solid medium or to introduce the sample into a flask or tube. Selection of media with specialized functions can improve later steps of isolation and identification. Some microbes may require a live organism (animal, egg) as the inoculation medium.

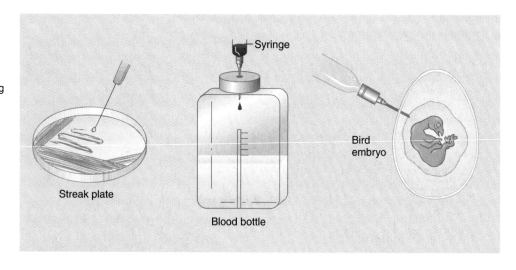

FIGURE 3.1

A summary of the general laboratory techniques carried out by microbiologists. It is not necessary to perform all the steps shown or to perform them exactly in this order, but all microbiologists participate in at least some of these activities. In some cases, one may proceed right from the sample to inspection, and in others, only inoculation and incubation on special media are required.

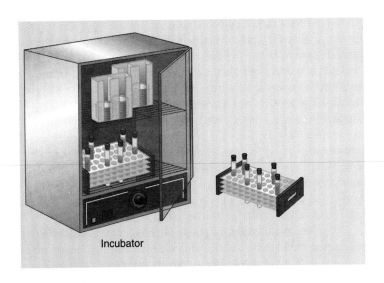

Incubator

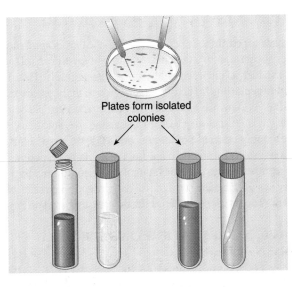

Plates form isolated colonies

2. Incubation:
An incubator can be used to adjust the proper growth conditions of a sample. Setting the optimum temperature and gas content promotes multiplication of the microbes over a period of hours, days, and even weeks. Incubation produces a culture—the visible growth of the microbe in the medium.

3. Isolation:
The end result of inoculation and incubation is **isolation** of the microbe in macroscopic form. The isolated microbes may take the form of separate colonies (discrete mounds of cells) on solid media, or turbidity in broths. Further isolation, also known as subculturing, involves taking a tiny bit of growth from an isolated colony and inoculating a separate medium. This is one way to make a pure culture that contains only a single species of microbe.

4. Inspection:
The cultures are observed macroscopically for obvious growth characteristics (color, texture, size) that could be useful in analyzing the specimen contents. Slides are made to assess microscopic details such as cell shape, size, and motility. Staining techniques may be used to gather specific information on microscopic morphology.

5. Identification:
A major purpose of the 5 I's is to determine the type of microbe, usually to the level of species. Summaries of characteristics are used to develop profiles of the microbe or microbes isolated from the sample. Information can include relevant data already taken during inspection and additional tests that further describe and differentiate the microbes. Specialized tests include biochemical tests to determine metabolic activities specific to the microbe, immunologic tests, and genetic analysis.

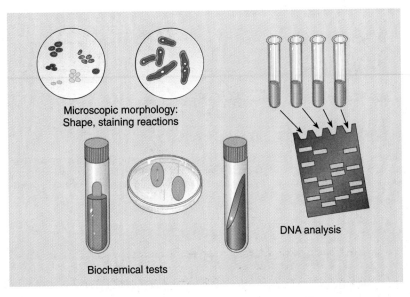

Microscopic morphology: Shape, staining reactions

DNA analysis

Biochemical tests

INOCULATION: PRODUCING A CULTURE

To cultivate, or **culture,*** microorganisms, one introduces a tiny sample (the inoculum) into a container of nutrient **medium*** (pl. media), which provides an environment in which they multiply. This process is called **inoculation.*** The observable growth that appears in or on the medium is known as a culture. The nature of the sample being cultured depends on the objectives of the analysis. Clinical specimens for determining the cause of an infectious disease are obtained from body fluids (blood, cerebrospinal fluid), discharges (sputum, urine, feces), or diseased tissue. Other samples subject to microbiological analysis are soil, water, sewage, foods, air, and inanimate objects. Procedures for proper specimen collection are discussed on page 537.

ISOLATION: SEPARATING ONE SPECIES FROM ANOTHER

Certain **isolation** techniques are based on the concept that if an individual bacterial cell is separated from other cells and provided adequate space on a nutrient surface, it will grow into a discrete

---

***culture** (kul´-chur) Gr. *cultus,* to tend or cultivate. It can be used as a verb or a noun.

***medium** (mee´-dee-um) pl. media; L., middle.

***inoculation** (in-ok″-yoo-lay´-shun) L. *in,* and *oculus,* bud.

mound of cells called a **colony (figure 3.2).** Because it was formed from a single cell, a colony consists of just that one species and no other. Proper isolation requires that a small number of cells be inoculated into a relatively large volume or over an expansive area of medium. It generally requires the following materials: a medium that has a relatively firm surface (see agar in "Physical States of Media," page 63), a Petri plate (a clear, flat dish with a cover), and inoculating tools. In the streak plate method, a small droplet of culture or sample is spread over the surface of the medium according to a pattern that gradually thins out the sample and separates the cells spatially over several sections of the plate **(figure 3.3a,b).** Because of its effectiveness and economy of materials, the streak plate is the method of choice for most applications.

In the loop dilution, or pour plate, technique, the sample is inoculated serially into a series of cooled but still liquid agar tubes so as to dilute the number of cells in each successive tube in the series (figure 3.3c,d). Inoculated tubes are then plated out (poured) into sterile Petri plates and are allowed to solidify (harden). The end result (usually in the second or third plate) is that the number of cells per volume is so decreased that cells have ample space to grow into separate colonies. One difference between this and the streak plate method is that in this technique some of the colonies will develop in the medium itself and not just on the surface.

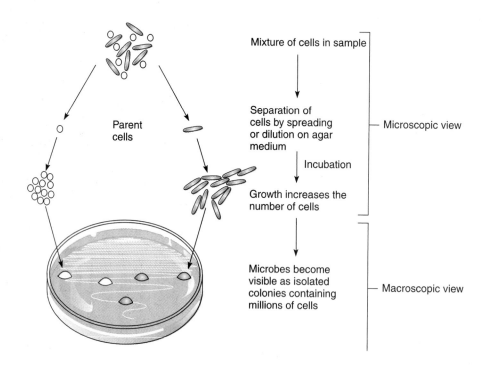

FIGURE 3.2

Isolation technique. Stages in the formation of an isolated colony, showing the microscopic events and the macroscopic result. Separation techniques such as streaking can be used to isolate single cells. After numerous cell divisions, a macroscopic, clinging mound of cells, or a colony, will be formed. This is a relatively simple yet successful way to separate different types of bacteria in a mixed sample.

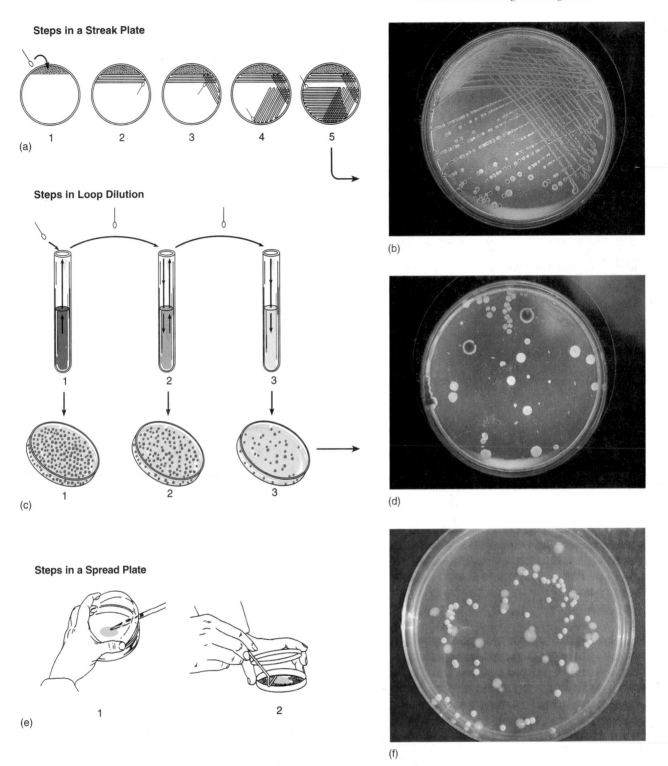

FIGURE 3.3

Methods for isolating bacteria.   **(a)** Steps in a quadrant streak plate and **(b)** resulting isolated colonies of bacteria. **(c)** Steps in the loop dilution method and **(d)** the appearance of plate 3. **(e)** Spread plate and **(f)** its result.

MICROBITS 3.1
Animal Inoculation: "Living Media"

A great deal of attention has been focused on the uses of animals in biology and medicine. Animal rights activists are vocal about practically any experimentation with animals and have expressed their outrage quite forcefully. Certain kinds of animal testing may seem trivial and unnecessary, but many times it is absolutely necessary to use animals bred for experimental purposes, such as guinea pigs, mice, chickens, and even armadillos. Such animals can be an indispensable aid for studying, growing, and identifying microorganisms. One special use of animals involves inoculation of the early life stages (embryos) of birds. Vaccines for influenza are currently produced in duck embryos. Here is a summary of major rationales for live animal inoculation:

1. Animal inoculation is an essential step in testing the effects of drugs and the effectiveness of vaccines before they are administered to humans. It makes progress toward prevention, treatment, and cure possible without risking the lives of humans.
2. Researchers develop animal models for evaluating new diseases or for studying the cause or process of a disease. Koch's postulates is a series of proofs to determine the causative agent of a disease and requires a controlled experiment with an animal that can develop a typical case of the disease. So far, there has not been a completely successful model for HIV infection and human AIDS, but a similar disease in monkeys called simian AIDS has clarified some aspects of human AIDS and is serving as a model for vaccine development.
3. Animals are an important source of antibodies, antisera, antitoxins, and other immune products that can be used in therapy or testing.
4. Animals are sometimes required to determine the pathogenicity or toxicity of certain bacteria. One such test is

the mouse neutralization test for the presence of botulism toxin in food. This test can help identify even very tiny amounts of toxin and thereby can avert outbreaks of this disease. Occasionally, it is necessary to inoculate an animal to distinguish between pathogenic or nonpathogenic strains of *Listeria* or *Candida* (a yeast).
5. Some microbes will not grow on artificial media but will grow in a suitable animal and can be recovered in a more or less pure form. These include animal viruses, the spirochete of syphilis, and the leprosy bacillus (grown in armadillos).

The nude or athymic mouse has genetic defects in hair formation and thymus development. It is widely used to study cancer, immune function, and infectious diseases.

With the spread plate technique, a small volume of liquid, diluted sample is pipetted onto the surface of the medium and spread around evenly by a sterile spreading tool (sometimes called a "hockey stick"). Like the streak plate, cells are pushed into separate areas on the surface so that they can form individual colonies (figure 3.3*e,f*).

Before we continue to cover information on the Five I's, we will take a side trip to look at media in more detail.

MEDIA: PROVIDING NUTRIENTS IN THE LABORATORY

A major stimulus to the rise of microbiology in the late 1800s was the development of techniques for growing microbes out of their natural habitats and in pure form in the laboratory. This milestone enabled the close examination of a microbe and its morphology, physiology, and genetics. It was evident from the very first that for successful cultivation, each microorganism had to be provided with all of its required nutrients in an artificial medium.

Nutritional requirements of microbes vary from a few very simple inorganic compounds to a complex list of specific inorganic and organic compounds. This tremendous diversity is evident in the types of media that can prepared. At least 500 different types of media are used in culturing and identifying microorganisms. Culture media are contained in test tubes, flasks, or Petri plates, and they are inoculated by such tools as loops, needles, pipettes, and swabs. Media are extremely varied in nutrient content and consistency and can be specially formulated for a particular purpose. Culturing parasites that cannot grow on artificial media (all viruses and certain bacteria) requires cell cultures or host animals (**Microbits 3.1**). For an experiment to be properly controlled, sterile technique is necessary. This means that the inoculation must start with a sterile medium and inoculating tools with sterile tips must be used. Measures must be taken to prevent introduction of nonsterile materials such as room air and fingers directly into the media.

Types of Media

Media can be classified on three primary levels:

1. physical form,
2. chemical composition, and
3. functional type **(table 3.1).**

Most media discussed here are designed for bacteria and fungi, though algae and some protozoa can be propagated in media.

Physical States of Media

Liquid media are defined as water-based solutions that do not solidify at temperatures above freezing and that tend to flow freely when the container is tilted **(figure 3.4).** These media, termed broths, milks, or infusions, are made by dissolving various solutes in distilled water. Growth occurs throughout the container and can then present a dispersed, cloudy or particulate appearance. A common laboratory medium, *nutrient broth,* contains beef extract and peptone dissolved in water. Methylene blue milk and litmus milk are opaque liquids containing whole milk and dyes. Fluid thioglycollate is a slightly viscous broth used for determining patterns of growth in oxygen.

At ordinary room temperature, **semisolid media** exhibit a clotlike consistency **(figure 3.5)** because they contain an amount of solidifying agent (agar or gelatin) that thickens them but does not produce a firm substrate. Semisolid media are used to determine the motility of bacteria and to localize a reaction at a specific site. Both motility test medium and SIM contain a small amount (0.3–0.5%) of agar. The medium is stabbed carefully in the center and later observed for the pattern of growth around the stab line. In addition to motility, SIM can test for physiological characteristics used in identification (hydrogen sulfide production and indole reaction).

Solid media provide a firm surface on which cells can form discrete colonies (see figure 3.3) and are advantageous for isolating and culturing bacteria and fungi. They come in two forms: liquefiable and nonliquefiable. Liquefiable solid media, sometimes called

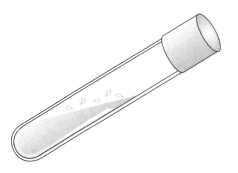

(a)

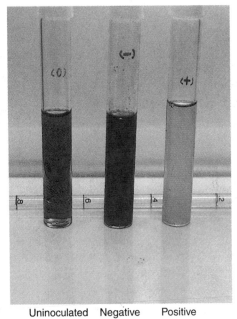

(b)   Uninoculated   Negative   Positive

FIGURE 3.4

Sample liquid media. **(a)** Liquid media tend to flow freely when the container is tilted. **(b)** *Enterococcus faecalis* broth, a selective medium for identifying this species. On the left (0) is a clear, uninoculated broth. The broth in the center is growing without a color change (−), and on the right is a broth with growth and color change (+).

reversible solid media, contain a solidifying agent that changes its physical properties in response to temperature. By far the most widely used and effective of these agents is **agar,*** a complex polysaccharide isolated from the red alga *Gelidium.* The benefits of agar are numerous. It is solid at room temperature, and it melts (liquefies) at the boiling temperature of water (100°C). Once liquefied, agar does not resolidify until it cools to 42°C, so it can be inoculated and poured in liquid form at temperatures (45°–50°C) that will not harm the microbes or the handler. Agar is flexible and moldable, and it provides a basic framework to hold moisture and nutrients, though it is not itself a digestible nutrient for most microorganisms.

Any medium containing 1% to 5% agar usually has the word *agar* in its name. *Nutrient agar* is a common one. Like nutrient broth, it contains beef extract and peptone, as well as 1.5% agar by weight. Many of the examples covered in the section on functional categories of media contain agar. Although gelatin is not nearly as

TABLE 3.1		
Three Categories of Media Classification		
Physical State (Medium's Normal Consistency)	**Chemical Composition (Type of Chemicals Medium Contains)**	**Functional Type (Purpose of Medium)***
1. Liquid 2. Semisolid 3. Solid (can be converted to liquid) 4. Solid (cannot be liquefied)	1. Synthetic (chemically defined) 2. Nonsynthetic (not chemically defined)	1. General purpose 2. Enriched 3. Selective 4. Differential 5. Anaerobic growth 6. Specimen transport 7. Assay 8. Enumeration

*Some media can serve more than one function. For example, a medium such as brain-heart infusion is general purpose and enriched; mannitol salt agar is both selective and differential; and blood agar is both enriched and differential.

*agar (ah′-gur) The Malay word for this seaweed product is *agar-agar.*

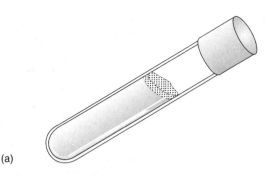

(a)

(b)

FIGURE 3.5

Sample semisolid media. **(a)** Semisolid media have more body than liquid media but less body than solid media. They do not flow freely and have a soft, clotlike consistency. **(b)** Sulfur indole motility (SIM) medium. The growth patterns that develop in this medium can be used to determine various characteristics. Possible results starting from far left are: no growth; no motility; motility; motility and production of hydrogen sulfide gas (black precipitate).

satisfactory as agar, it will create a reasonably solid surface in concentrations of 10% to 15% (but it probably will not remain solid). Agar and gelatin media are illustrated in **figure 3.6.**

Nonliquefiable solid media have less versatile applications than agar media because they do not melt. They include materials such as rice grains (used to grow fungi), cooked meat media (good for anaerobes), and potato slices; all of these media start out solid and remain solid after heat sterilization. Other solid media contain-

ing egg and serum start out liquid and are permanently coagulated or hardened by moist heat.

Chemical Content of Media

Media whose compositions are chemically defined are termed synthetic. Such media contain pure organic and inorganic compounds that vary little from one source to another and have a molecular content specified by means of an exact formula. Synthetic media come in many forms. Some media, such as minimal media for fungi, contain nothing more than a few essential compounds such as salts and amino acids dissolved in water. Others contain a variety of defined organic and inorganic chemicals **(table 3.2).** Such standardized and reproducible media are most useful in research and cell culture when the exact nutritional needs of the test organisms are known. If even one component of a given medium is not chemically definable, the medium belongs in the next category.

Complex, or nonsynthetic, media contain at least one ingredient that is *not* chemically definable—not a simple, pure compound and not representable by an exact chemical formula. Most of these substances are extracts of animals, plants, or yeasts, including such materials as ground-up cells, tissues, and secretions. Examples are blood, serum, and meat extracts or infusions. Infusions are high in vitamins, minerals, proteins, and other organic nutrients. Other nonsynthetic ingredients are milk, yeast extract, soybean digests, and peptone. Peptone is a partially digested protein, rich in amino acids, that is often used as a carbon and nitrogen source. Nutrient broth, blood agar, and MacConkey agar, though different in function and appearance, are all nonsynthetic media. They present a rich mixture of nutrients for microbes that have complex nutritional needs.

A specific example can be used to compare what differentiates a synthetic medium from a nonsynthetic one. Both synthetic *Euglena* medium (table 3.2) and nonsynthetic nutrient broth contain amino acids. But *Euglena* medium has three known amino acids in known amounts, whereas nutrient broth contains amino acids (in peptone) in variable types and amounts. Pure inorganic salts and organic acids are added in precise quantities for *Euglena,* whereas those components are provided by undefined beef extract in nutrient broth.

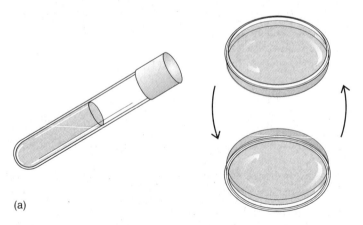

(a)

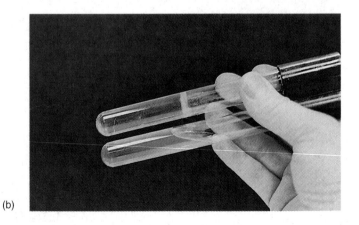

(b)

FIGURE 3.6

Solid media that are reversible to liquids. **(a)** Media containing 1–5% agar-agar are solid enough to remain in place when containers are tilted or inverted. They are liquefiable by heat but generally not by bacterial enzymes. **(b)** Nutrient gelatin contains enough gelatin (12%) to take on a solid consistency. The top tube shows it as a solid. The bottom tube indicates a result when microbial enzymes digest the gelatin and liquefy it.

TABLE 3.2

Medium for the Growth and Maintenance of the Green Alga *Euglena*

Glutamic acid (aa)	6 g
Aspartic acid (aa)	4 g
Glycine (aa)	5 g
Sucrose (c)	30 g
Malic acid (oa)	2 g
Succinic acid (oa)	1.04 g
Boric acid	1.14 mg
Thiamine hydrochloride (v)	12 mg
Monopotassium phosphate	0.6 g
Magnesium sulfate	0.8 g
Calcium carbonate	0.16 g
Ammonium carbonate	0.72 g
Ferric chloride	60 mg
Zinc sulfate	40 mg
Manganese sulfate	6 mg
Copper sulfate	0.62 mg
Cobalt sulfate	5 mg
Ammonium molybdate	1.34 mg

Note: These ingredients are dissolved in 1,000 ml of water.
aa, amino acid; c, carbohydrate; oa, organic acid; v, vitamin; g, gram; mg, milligram.

Media to Suit Every Function

Microbiologists have many types of media at their disposal, with new ones being devised all the time. Depending upon what is added, a microbiologist can fine-tune a medium for nearly any purpose. As a result, only a few species of bacteria or fungi cannot yet be cultivated artificially. Media are used for primary isolation, to maintain cultures in the lab, to determine biochemical and growth characteristics, and for numerous other functions.

General-purpose media are designed to grow as broad a spectrum of microbes as possible. As a rule, they are nonsynthetic and contain a mixture of nutrients that could support the growth of pathogens and nonpathogens alike. Examples include nutrient agar and broth, brain-heart infusion, and trypticase soy agar (TSA). TSA contains partially digested milk protein (casein), soybean digest, NaCl, and agar.

An **enriched medium** contains complex organic substances such as blood, serum, hemoglobin, or special **growth factors** (specific vitamins, amino acids) that certain species must have in order to grow. Bacteria that require growth factors and complex nutrients are termed **fastidious.*** Blood agar, which is made by adding sterile sheep, horse, or rabbit blood to a sterile agar base (**figure 3.7a**) is widely employed to grow fastidious streptococci and other pathogens. Pathogenic *Neisseria* (one species causes gonorrhea) are grown on Thayer-Martin medium or chocolate agar, which is made by heating blood agar (figure 3.7b).

Selective and Differential Media. Some of the cleverest and most inventive media belong to the categories of selective and differential media (**figure 3.8**). These media are designed for special

*****fastidious** (fass-tid′-ee-us) L. *fastidium*, loathing or disgust.

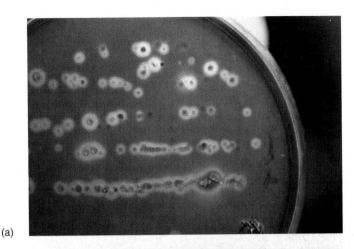

(a)

(b)

FIGURE 3.7

Examples of enriched media. **(a)** Blood agar plate growing bacteria from the human throat. Note that this medium also differentiates among different colonies by their appearance. **(b)** Chocolate agar, a medium that gets its brown color from heated blood, not from chocolate. It is commonly used to culture the fastidious gonococcus *Neisseria gonorrhoeae*.

microbial groups, and they have extensive applications in isolation and identification. They can permit, in a single step, the preliminary identification of a genus or even a species.

A **selective medium (table 3.3)** contains one or more agents that inhibit the growth of a certain microbe or microbes (A, B, C) but not others (D) and thereby encourages, or *selects,* microbe D and allows it to grow. Selective media are very important in primary isolation of a specific type of microorganism from samples containing dozens of different species—for example, feces, saliva, skin, water, and soil. They hasten isolation by suppressing the unwanted background organisms and favoring growth of the desired ones.

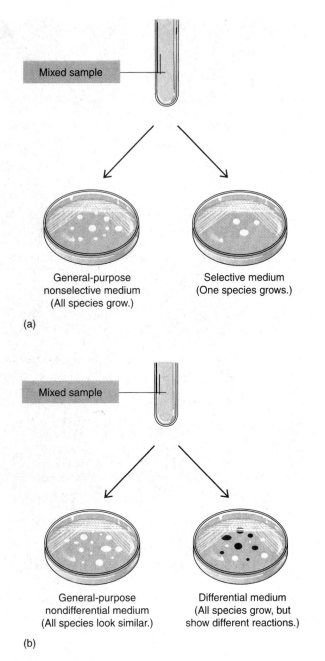

(a)

(b)

FIGURE 3.8

Comparison of selective and differential media with general-purpose media. **(a)** The same mixed sample containing five different species is streaked onto plates of general-purpose nonselective medium and selective medium. Note the results. **(b)** Another mixed sample containing three different species is streaked onto plates of general-purpose nondifferential medium and differential medium. Note the results.

TABLE 3.3

Selective Media, Agents, and Functions

Medium	Selective Agent	Used For
Mueller tellurite	Potassium tellurite	Isolation of *Corynebacterium diphtheriae*
Enterococcus faecalis broth	Sodium azide Tetrazolium	Isolation of fecal enterococci
Phenylethanol agar	Phenylethanol chloride	Isolation of staphylococci and streptococci
Tomato juice agar	Tomato juice, acid	Isolation of lactobacilli from saliva
MacConkey agar	Bile, crystal violet	Isolation of gram-negative enterics
Salmonella/Shigella (SS) agar	Bile, citrate, brilliant green	Isolation of *Salmonella* and *Shigella*
Lowenstein-Jensen	Malachite green dye	Isolation and maintenance of *Mycobacteria*
Sabouraud's agar	pH of 5.6 (acid)	Isolation of fungi—inhibits bacteria

let also inhibit certain gram-positive bacteria. Other agents that have selective properties are antimicrobic drugs and acid. Some selective media contain strongly inhibitory agents to favor the growth of a pathogen that would otherwise be overlooked because of its low numbers in a specimen. Selenite and brilliant green dye are used in media to isolate *Salmonella* from feces, and sodium azide is used to isolate enterococci from water and food (see EF broth, figure 3.4*b*).

Differential media grow several types of microorganisms and are designed to display visible differences among those microorganisms. Differentiation shows up as variations in colony size or color, in media color changes, or in the formation of gas bubbles and precipitates **(table 3.4).** These variations come from the type of chemicals these media contain and the ways that microbes react to them. For example, when microbe X metabolizes a certain substance not used by organism Y, then X will cause a visible change in the medium and Y will not. The simplest differential media show two reaction types such as the use or nonuse of a particular nutrient or a color change in some colonies but not in others. Some media are sufficiently complex to show three or four different reactions **(figure 3.10).**

Dyes can be used as differential agents because many of them are pH indicators that change color in response to the production of an acid or a base. For example, MacConkey agar contains neutral red, a dye that is yellow when neutral and pink or red when acidic. A common intestinal bacterium such as *Escherichia coli* that gives off acid when it metabolizes the lactose in the medium develops red to pink colonies, and one like *Salmonella* that does not give off acid remains its natural color (off-white). Spirit blue agar is used to

Mannitol salt agar (MSA) **(figure 3.9*a*)** contains a concentration of NaCl (7.5%) that is quite inhibitory to most human pathogens. One exception is the genus *Staphylococcus,* which grows well in this medium and consequently can be amplified in very mixed samples. Bile salts, a component of feces, inhibit most gram-positive bacteria while permitting many gram-negative rods to grow. Media for isolating intestinal pathogens (MacConkey agar, Hektoen enteric [HE] agar) contain bile salts as a selective agent (figure 3.9*b*). Dyes such as methylene blue and crystal vio-

(a)

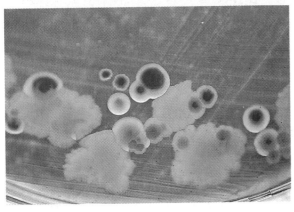

(b)

FIGURE 3.9

Example of selective media. **(a)** Mannitol salt agar is used to isolate members of the genus *Staphylococcus*. It is selective for this genus because its members can grow in the presence of 7.5% sodium chloride, whereas many other species are inhibited by this high concentration. It contains a dye that also pinpoints those species of *Staphylococcus* that produce acid from mannitol and turn the phenol red dye to a bright yellow. **(b)** MacConkey agar differentiates between lactose-fermenting bacteria (indicated by a pink-red reaction in the center of the colony) and lactose-negative bacteria (indicated by an off-white colony with no dye reaction).

TABLE 3.4

Differential Media

Medium	Substances That Facilitate Differentiation	Differentiates Between
Blood agar	Intact red blood cells	Types of hemolysis
Mannitol salt agar	Mannitol, phenol red, and 7.5% NaCl	Species of *Staphylococcus* NaCl also inhibits the salt-sensitive species
Hektoen enteric (HE) agar	Brom thymol blue, acid fuchsin, sucrose, salicin, thiosulfate, ferric ammonium citrate, and bile	*Salmonella, Shigella*, other lactose fermenters from nonfermenters Dyes and bile also inhibit gram-positive bacteria
Spirit blue agar	Spirit blue dye and oil	Bacteria that use fats from those that do not
Urea broth	Urea, phenol red	Bacteria that hydrolyze urea to ammonia
Sulfur indole motility (SIM)	Thiosulfate, iron	H_2S gas producers from nonproducers
Triple-sugar iron agar (TSIA)	Triple sugars, iron, and phenol red dye	Fermentation of sugars, H_2S production
XLD agar	Lysine, xylose, iron, thiosulfate, phenol red	*Enterobacter, Escherichia, Proteus, Providencia, Salmonella*, and *Shigella*
Birdseed agar	Seeds from thistle plant	*Cryptococcus neoformans* and other fungi

detect the hydrolysis (digestion) of fats by lipase enzyme. Positive hydrolysis is indicated by the dark blue color that develops in colonies.

Miscellaneous Media   A reducing medium contains a substance (thioglycollic acid or cystine) that absorbs oxygen or slows the penetration of oxygen in a medium, thus reducing its availability. Reducing media are important for growing anaerobic bacteria or determining oxygen requirements (see figure 7.12). Carbohydrate fermentation media contain sugars that can be fermented (converted to acids) and a pH indicator to show this reaction (figure 3.9*a* and **figure 3.11**). Media for other biochemical reactions that provide the basis for identifying bacteria and fungi are presented in the second half of this book.

Transport media are used to maintain and preserve specimens that have to be held for a period of time before clinical analysis or to sustain delicate species that die rapidly if not held under stable conditions. Stuart's and Amies transport media contain

salts, buffers, and absorbants to prevent cell destruction by enzymes, pH changes, and toxic substances, but will not support growth. Assay media are used by technologists to test the effectiveness of antimicrobial drugs (see chapter 12) and by drug manufacturers to assess the effect of disinfectants, antiseptics, cosmetics, and preservatives on the growth of microorganisms. Enumeration media are used by industrial and environmental microbiologists to count the numbers of organisms in milk, water, food, soil, and other samples.

CHAPTER CHECKPOINTS

Most microorganisms can be cultured on artificial media, but some can be cultured only in living tissue or cells.

Artificial media are classified by their *physical state* as either liquid, semisolid, liquefiable solid, or nonliquefiable solid.

(a)

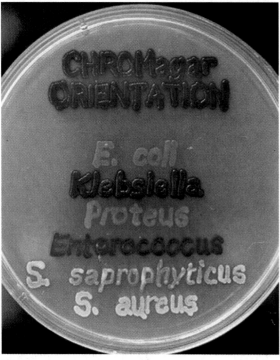

(b)

FIGURE 3.10

Media that differentiate characteristics. **(a)** Triple-sugar iron agar (TSIA) in a slant tube. This medium contains three fermentable carbohydrates, phenol red to indicate pH changes, and a chemical (iron) that indicates H_2S gas production. Reactions (from left to right) are: no growth; growth with no acid production; acid production in the bottom (butt) only; acid production all through the medium; and acid production in the butt with H_2S gas formation (black). **(b)** A state-of-the-art medium developed for culturing and identifying the most common urinary pathogens. CHROMagar Orientation™ uses color-forming reactions to distinguish at least seven species and permits rapid identification and treatment.

Artificial media are classified by their *chemical content* as either *synthetic* or *nonsynthetic,* depending on whether the exact chemical composition is known.

Artificial media are classified by their *function* as either general-purpose media or media with one or more specific purposes. Enriched, selective, differential, transport, assay, and enumerating media are all examples of media designed for specific purposes.

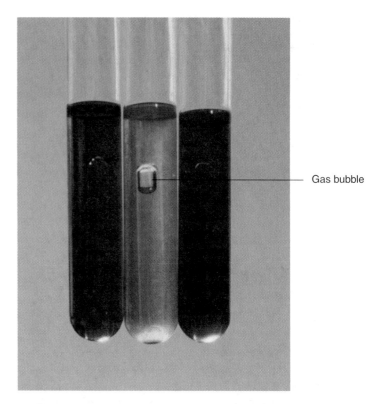

— Gas bubble

FIGURE 3.11

Carbohydrate fermentation in broths. This medium is designed to show fermentation (acid production) and gas formation by means of a small, inverted Durham tube for collecting gas bubbles. The tube on the left is an uninoculated negative control; the center tube is positive for acid (yellow) and gas (open space); the tube on the right shows growth but neither acid nor gas.

INCUBATION, INSPECTION, AND IDENTIFICATION

Once a container of medium has been inoculated, it is **incubated.*** This means it is placed in a temperature-controlled chamber or incubator to encourage multiplication. Although microbes have adapted to growth at temperatures ranging from freezing to boiling, the usual temperatures used in laboratory propagation fall between 20° and 40°C. Incubators can also control the content of atmospheric gases such as oxygen and carbon dioxide that may be required for the growth of certain microbes. During the incubation period (ranging from a day to several weeks), the microbe multiplies and produces growth that is observable macroscopically.[2] Microbial growth in a liquid medium materializes as cloudiness, sediment, scum, or color. A common manifestation of growth on solid media is the appearance of colonies, especially in bacteria and fungi. Colonies are actually large masses of clinging cells with distinctions in size, shape, color, and texture (see figure 4.12*a,b*).

In some ways, culturing microbes is analogous to gardening on a microscopic scale. Cultures are formed by "seeding" tiny plots (media) with microbial cells to grow into separate rows. Extreme

---

2. For all intents and purposes, the macroscopic level is synonymous with the cultural level.

*incubate (in'-kyoo-bayt) Gr. *incubatus,* to lie in or upon.

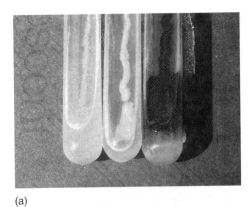
(a)

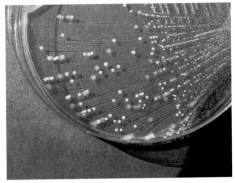
(b)

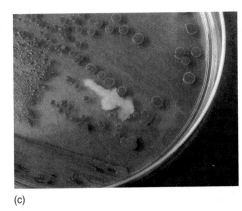
(c)

FIGURE 3.12

Various conditions of cultures.   **(a)** Three tubes containing pure cultures of *Escherichia coli* (white), *Micrococcus luteus* (yellow), *and Serratia marcescens* (red). **(b)** A mixed culture of *M. luteus* and *E. coli* readily differentiated by their colors. **(c)** This plate of *S. marcescens* was overexposed to room air, and it has developed a large, white colony. Because this intruder is not desirable and not identified, the culture is now contaminated.

care is taken to exclude weeds. A **pure culture** is a container of medium that grows only a single known species or type of microorganism (**figure 3.12a**). This type of culture is most frequently used for laboratory study, because it allows the systematic examination and control of one microorganism by itself. Instead of the term *pure culture,* some microbiologists prefer the term **axenic,*** meaning that the culture is free of other living things except for the one being studied. A standard method for preparing a pure culture is to **subculture,** or make a second-level culture from a well-isolated colony. A tiny bit of cells is transferred into a separate container of media and incubated (see figure 3.1, step 3).

A **mixed culture** (figure 3.12b) is a container that holds two or more *identified,* easily differentiated species of microorganisms, not unlike a garden plot containing both carrots and onions. A **contaminated*** **culture** (figure 3.12c) was once pure or mixed (and thus a known entity) but has since had **contaminants** (unwanted microbes of uncertain identity) introduced into it, like weeds into a garden. Because contaminants have the potential for causing disruption, constant vigilance is required to exclude them from microbiology laboratories, as you will no doubt witness from your own experience.

How does one determine what sorts of microorganisms have been isolated in cultures? Certainly the combination of microscopic and macroscopic appearance can be valuable in differentiating the smaller, simpler procaryotic cells from the larger, more complex eucaryotic cells. Appearance can be especially useful in identifying eucaryotic microorganisms to the level of genus or species because of their distinctive morphological features; however, bacteria are generally not identifiable by these methods because very different species may appear quite similar. For them, we must include techniques that characterize their cellular metabolism. These methods, called biochemical tests, can determine fundamental chemical characteristics such as nutrient requirements, products given off during growth, presence of enzymes, and mechanisms for deriving energy.

Several modern analytical and diagnostic tools that focus on genetic and molecular characteristics can detect the exact nature of microbial DNA. In the case of certain pathogens, further information on a microbe is obtained by inoculating a suitable laboratory animal. A profile is prepared by compiling physiological testing results with both macroscopic and microscopic traits. The profile then becomes the raw material used in final identification. In several subsequent chapters, we present more detailed examples of identification methods (see figure 4.28 for an example).

Maintenance and Disposal of Cultures

In most medical laboratories, the cultures and specimens constitute a potential hazard and will require immediate and proper disposal. Both steam sterilizing (see autoclave, chapter 11) and incineration (burning) are used to destroy microorganisms. On the other hand, many teaching and research laboratories maintain a line of stock cultures that represent "living catalogues" for study and experimentation. The largest culture collection can be found at the American Type Culture Collection in Rockville, Maryland, which maintains a voluminous array of frozen and freeze-dried fungal, bacterial, viral, and algal cultures.

CHAPTER CHECKPOINTS

The Five I's—inoculation, incubation, isolation, inspection, and identification—summarize the kinds of laboratory procedures used in microbiology.

Following *inoculation,* cultures are *incubated* at a specified temperature and time to encourage growth.

Isolated colonies that originate from single cells are composed of large numbers of cells clinging together.

A culture may exist in one of the following forms: A *pure culture* contains only one species or type of microorganism. A *mixed culture* contains two or more known species. A *contaminated culture* contains both known and unknown (unwanted) microorganisms.

During *inspection,* the cultures are examined and evaluated macroscopically and microscopically.

---

*axenic (ak-zee′-nik) Gr. *a,* no, and *xenos,* stranger.

*contaminated (kon-tam′-ih-nay-tid) Gr. *con,* together, and L. *tangere,* to touch.

Microorganisms are *identified* in terms of their macroscopic, or colony, morphology; their microscopic morphology; and their biochemical reactions and genetic characteristics.

Microbial cultures are disposed of in two ways: steam sterilization or incineration.

The Microscope: Window on an Invisible Realm

Imagine Leeuwenhoek's excitement and wonder when he first viewed a drop of rainwater and glimpsed an amazing microscopic world teeming with unearthly creatures. Beginning microbiology students still experience this sensation, and even experienced microbiologists never forget their first view. The microbial existence is indeed another world, but it would remain largely uncharted without an essential tool: the microscope. Your efforts in exploring microbes will be more meaningful if you understand some essentials of **microscopy*** and specimen preparation.

MAGNIFICATION AND MICROSCOPE DESIGN

The two key characteristics of a reliable microscope are magnification, or the ability to enlarge objects, and resolving power, or the ability to show detail.

A discovery by early microscopists that spurred the advancement of microbiology was that clear, glass spheres could act as a lens to **magnify*** small objects. Magnification in most microscopes results from a complex interaction between visible light waves and the curvature of the lens. When a beam or ray of light transmitted through air strikes and passes through the convex surface of glass, it experiences some degree of **refraction,*** defined as the bending or change in the angle of the light ray as it passes through a medium such as a lens (see figure 3.15). The greater the difference in the composition of the two substances the light passes between, the more pronounced is the refraction. When an object is placed a certain distance from the spherical lens and **illuminated*** with light, an optical replica, or image, of it is formed by the refracted light. Depending upon the size and curvature of the lens, the image appears enlarged to a particular degree, which is called its power of magnification and is usually identified with a number combined with × (read "times"). This behavior of light is evident if one looks through an everyday object such as a glass ball or a magnifying glass (**figure 3.13**). It is basic to the function of all optical, or light, microscopes, though many of them have additional features that define, refine, and increase the size of the image.

The first microscopes were simple, meaning they contained just a single magnifying lens and a few working parts. Examples of this type of microscope are a magnifying glass, a hand lens, and

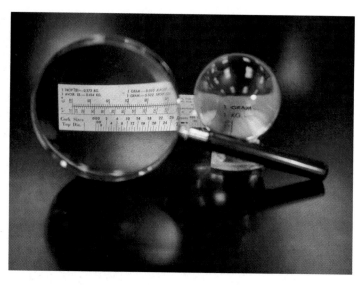

FIGURE 3.13

Effects of magnification. Demonstration of the magnification and image-forming capacity of clear glass "lenses." Given a proper source of illumination, this magnifying glass and crystal ball magnify a ruler two to three times.

Leeuwenhoek's basic little tool shown in figure 1.9a. Among the refinements that led to the development of today's compound microscope were the addition of a second magnifying lens system, a lamp in the base to give off visible light and illuminate the specimen, and a special lens called the condenser that converges or focuses the rays of light to a single point on the object. The fundamental parts of a modern compound light microscope are illustrated in **figure 3.14.**

Principles of Light Microscopy

To be most effective, a microscope should provide adequate magnification, resolution, and clarity of image. Magnification of the object or specimen by a compound microscope occurs in two phases. The first lens in this system (the one closest to the specimen) is the objective lens, and the second (the one closest to the eye) is the ocular lens, or eyepiece (**figure 3.15**). The objective forms the initial image of the specimen, called the **real image.** When this image is projected up through the microscope body to the plane of the eyepiece, the ocular lens forms a second image, the **virtual image.** The virtual image is the one that will be received by the eye and converted to a retinal and visual image. The magnifying power of the objectives alone usually ranges from 4× to 100×, and the power of the ocular alone ranges from 10× to 20×. The total power of magnification of the final image formed by the combined lenses is a product of the separate powers of the two lenses:

Power of objective	×	Usual power of ocular	=	Total magnification
10× low power objective		10×	=	100×
40× high dry objective		10×	=	400×
100× oil immersion objective		10×	=	1,000×

***microscopy** (mye-kraw′-skuh-pee) Gr. The science that studies microscope techniques.

***magnify** (mag′-nih-fye) L. *magnus,* great, and *ficere,* to make.

***refract, refraction** (ree-frakt′, ree-frak′-shun) L. *refringere,* to break apart.

***illuminate** (ill-oo′-mih-nayt) L. *illuminatus,* to light up.

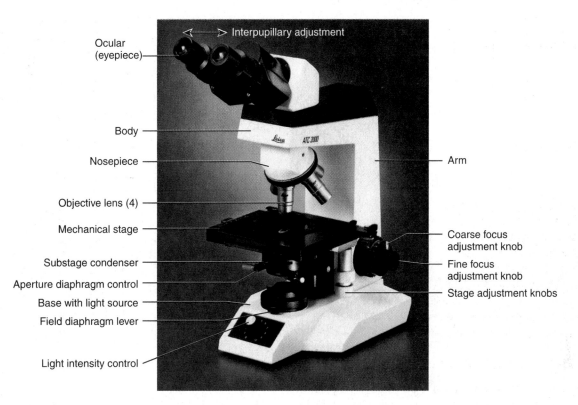

Interpupillary adjustment

Ocular (eyepiece)

Body

Nosepiece

Objective lens (4)

Mechanical stage

Substage condenser

Aperture diaphragm control

Base with light source

Field diaphragm lever

Light intensity control

Arm

Coarse focus adjustment knob

Fine focus adjustment knob

Stage adjustment knobs

FIGURE 3.14

The parts of a student laboratory microscope.   This microscope is a compound light microscope with two oculars (called binocular). It has four objective lenses, a mechanical stage to move the specimen, a condenser, an iris diaphragm, and a built-in lamp.

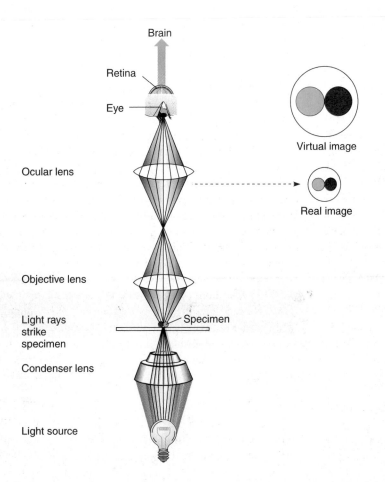

Brain

Retina

Eye

Virtual image

Ocular lens

Real image

Objective lens

Light rays strike specimen

Specimen

Condenser lens

Light source

FIGURE 3.15

The pathway of light and the two stages in magnification of a compound microscope.   As light passes through the condenser, it forms a solid beam that is focused on the specimen. Light leaving the specimen that enters the objective lens is refracted so that an enlarged primary image, the real image, is formed. One does not see this image, but its degree of magnification is represented by the lower circle. The real image is projected through the ocular, and a second image, the virtual image, is formed by a similar process. The virtual image is the final magnified image that is received by the retina and perceived by the brain. Notice that the lens systems cause the image to be reversed.

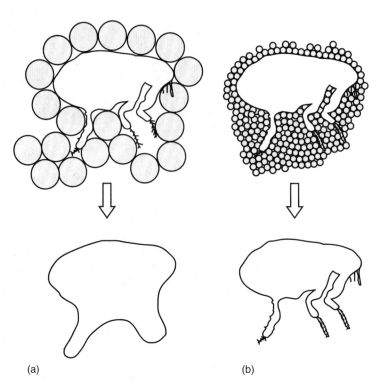

FIGURE 3.16

Effect of wavelength on resolution.   A simple model demonstrates how the wavelength influences the resolving power of a microscope. Here an outline of a flea represents the object being illuminated, and two different-sized circles represent the wavelengths of light. In **(a)**, the longer waves are too large to penetrate between the finer spaces and produce a fuzzy, undetailed image. In **(b)**, shorter waves are small enough to enter small spaces and produce a much more detailed image that is recognizable as a flea.

(a)   (b)

Microscopes are equipped with a nosepiece holding three or more objectives that can be rotated into position as needed. The power of the ocular usually remains constant for a given microscope. Depending on the power of the ocular, the total magnification of standard light microscopes can vary from 40× with the lowest power objective (called the scanning objective) to 2,000× with the highest power objective (the oil immersion objective).

Resolution: Distinguishing Magnified Objects Clearly   Despite the importance of magnification in visualizing tiny objects or cells, an additional optical property is essential for seeing clearly. That property is resolution, or **resolving power.** Resolution is the capacity of an optical system to distinguish or separate two adjacent objects or points from one another. For example, at a certain fixed distance, the lens in the human eye can resolve two small objects as separate points just as long as the two objects are no closer than 0.2 mm apart. The eye examination given by optometrists is in fact a test of the resolving power of the human eye for various-sized letters read at a distance of 20 feet. Because microorganisms are extremely small and usually very close together, they will not be seen with clarity or any degree of detail unless the microscope's lenses can resolve them.

A simple equation in the form of a fraction expresses the main determining factors in resolution:

$$\text{Resolving power (R.P.)} = \frac{\text{Wavelength of light in nm}}{2 \times \text{Numerical aperture of objective lens}}$$

From this equation it is evident that the resolving power is a function of the wavelength of light that forms the image, along with certain characteristics of the objective. The light source for optical microscopes consists of a band of colored wavelengths in the visible spectrum (see figure 11.7). The shortest visible wavelengths are in the violet-blue portion of the spectrum (400 nm), and the longest are in the red portion (750 nm). Because the wavelength must pass between the objects that are being resolved, shorter wavelengths (in the 400–500 nm range) will provide better resolution **(figure 3.16).** Some microscopes have a special blue filter placed over the lamp to limit the longer wavelengths of light from entering the specimen.

The other factor influencing resolution is the **numerical aperture,** a mathematical constant that describes the relative efficiency of a lens in bending light rays. Without going into the mathematical derivation of this constant, it is sufficient to say that each objective has a fixed numerical aperture reading that is determined by the microscope design and ranges from 0.1 in the lowest power lens to approximately 1.25 in the highest power (oil immersion) lens. The most important thing to remember is that a higher numerical aperture number will provide better resolution. In order for the oil immersion lens to arrive at its maximum resolving capacity, a drop of oil must be inserted between the tip of the lens and the specimen on the glass slide. Because oil has the same optical qualities as glass, it prevents refractive loss that normally occurs as peripheral light passes from the slide into the air; this property effectively increases the numerical aperture **(figure 3.17).** There is an absolute limitation to resolution in optical microscopes, which can be demonstrated by calculating the resolution of the oil immersion lens using a blue-green wavelength of light:

$$\text{R.P.} = \frac{500 \text{ nm}}{2 \times 1.25}$$

$$= 200 \text{ nm (or 0.2 } \mu\text{m)}$$

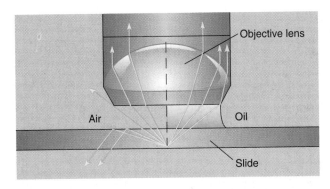

FIGURE 3.17

Workings of an oil immersion lens.   To maximize its resolving power, an oil immersion lens (the one with highest magnification) must have a drop of oil placed at its tip. This forms a continuous medium to transmit a beam of light from the condenser to the objective and effectively increase the numerical aperture. Without oil, some of the peripheral light that passes through the specimen is scattered into the air or onto the glass slide; this scattering decreases resolution.

resolved by the optical microscope and require electron microscopy (discussed later in this chapter). In summary then, the factor that most limits the clarity of a microscope's image is its resolving power. Even if a light microscope were designed to magnify several thousand times, its resolving power could not be increased, and the image it produced would simply be enlarged and fuzzy.

Other constraints to the formation of a clear image are the quality of the lens and light source and the lack of contrast in the specimen. No matter how carefully a lens is constructed, flaws remain. A typical problem is spherical aberration, a distortion in the image caused by irregularities in the lens, which creates a curved, rather than flat, image (see figure 3.13). Another is chromatic aberration, a rainbowlike image that is caused by the lens acting as a prism and separating visible light into its colored bands. Brightness and direction of illumination also affect image formation. Because too much light can reduce contrast and burn out the image, an iris diaphragm on most microscopes controls the amount of light entering the condenser. The lack of contrast in cell components is compensated for by using special lenses (the phase-contrast microscope) and by adding dyes.

VARIATIONS ON THE OPTICAL MICROSCOPE

Optical microscopes that use visible light can be described by the nature of their field, meaning the circular area viewed through the ocular lens. With special adaptations in lenses, condenser, and light sources, four special types of microscopes can be described: bright-field, dark-field, phase-contrast, and interference. A fifth type of optical microscope, the fluorescence microscope, uses ultraviolet radiation as the illuminating source. Each of these microscopes is adapted for viewing specimens in a particular way, as described in the next sections and summarized in parts of **table 3.5.**

In practical terms, this means that the oil immersion lens can resolve any cell or cell part as long as it is at least 0.2 μm in diameter, and that it can resolve two adjacent objects as long as they are at least 0.2 μm apart **(figure 3.18).** In general, organisms that are 0.5 μm or more in diameter are readily seen. This includes fungi and protozoa and some of their internal structures, and most bacteria. However, a few bacteria and most viruses are far too small to be

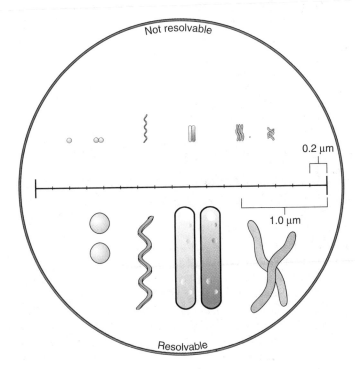

FIGURE 3.18

Effect of magnification.   Comparison of cells that would not be resolvable versus those that would be resolvable under oil immersion at 1,000× magnification. Note that in addition to differentiating two adjacent things, good resolution also means being able to observe an object clearly.

TABLE 3.5		
Comparison of Light Microscopes and Electron Microscopes		
Characteristic	**Light or Optical**	**Electron (Transmission)**
Useful magnification	2,000×	1,000,000 or more
Maximum resolution	200 nm	0.5 nm
Image produced by	Visible light rays	Electron beam
Image focused by	Glass objective lens	Electromagnetic objective lenses
Image viewed through	Glass ocular lens	Fluorescent screen
Specimen placed on	Glass slide	Copper mesh
Specimen may be alive	Yes	No
Specimen requires special stains or treatment	Not always	Yes
Colored images produced	Yes	No

Bright-Field Microscopy

The bright-field microscope is the most widely used type of light microscope. Although we ordinarily view objects like the words on this page with light reflected off the surface, a bright-field microscope forms its image when light is transmitted through the specimen. The specimen, being denser and more opaque than its surroundings, absorbs some of this light, and the rest of the light is transmitted directly up through the ocular into the field. As a result, the specimen will produce an image that is darker than the surrounding brightly illuminated field. The bright-field microscope is a multipurpose instrument that can be used for both live, unstained material and preserved, stained material. The bright-field image is compared with that of other microscopes in **figure 3.19.**

Dark-Field Microscopy

A bright-field microscope can be adapted as a dark-field microscope by adding a special disc called a *stop* to the condenser. The stop blocks all light from entering the objective lens except peripheral light that is reflected off the sides of the specimen itself. The resulting image is a particularly striking one: brightly illuminated specimens surrounded by a dark (black) field (figure 3.19*b*). Some of Leeuwenhoek's more successful microscopes probably operated with dark-field illumination. The most effective use of dark-field microscopy is to visualize living cells that would be distorted by drying or heat or cannot be stained with the usual methods. It can outline the organism's shape and permit rapid recognition of swimming cells that might appear in dental and other infections (**figure 3.20**), but it does not reveal fine internal details.

Phase-Contrast and Interference Microscopy

If similar objects made of clear glass, ice, cellophane, or plastic were immersed in the same container of water, an observer would have difficulty telling them apart because they have similar optical properties. Internal components of a live, unstained cell also lack contrast and can be difficult to distinguish. But cell structures do differ slightly in density, enough that they can alter the light that passes through them in subtle ways. The phase-contrast microscope has been constructed to take advantage of this characteristic. This microscope contains devices that transform the subtle changes in light waves passing through the specimen into differences in light intensity. For example, denser cell parts such as organelles alter the pathway of light more than less dense regions (the cytoplasm). Light patterns coming from these regions will vary in contrast. The amount of internal detail visible by this method is greater than by either bright-field or dark-field methods. The phase-contrast microscope is most useful for observing intracellular structures such as bacterial spores, granules, and organelles, as well as the locomotor structures of eucaryotic cells (figure 3.19*c* and **figure 3.21*a***).

Like the phase-contrast microscope, the differential interference contrast (DIC) microscope provides a detailed view of unstained, live specimens by manipulating the light. But this microscope has additional refinements, including two prisms that add contrasting colors to the image and two beams of light rather than a single one. DIC microscopes produce extremely well-defined images that are vividly colored and appear three-dimensional (figure 3.21*b*).

Most optical microscopes have difficulty forming a clear image of cells at higher magnifications, because cells are often too

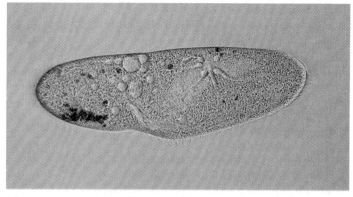

(a)

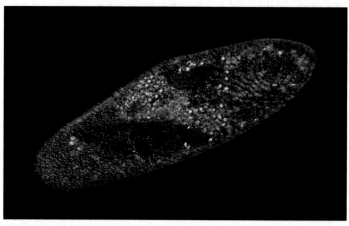

(b)

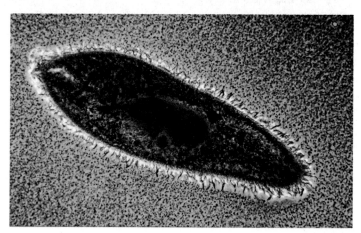

(c)

FIGURE 3.19

Three views of a basic cell.   A live cell of *Paramecium* viewed with **(a)** bright-field (400×), **(b)** dark-field (400×), and **(c)** phase-contrast (400×). Note the difference in the appearance of the field and the degree of detail shown by each method of microscopy. Only in phase-contrast are the cilia (fine hairs) on the cells noticeable. Can you see the nucleus? The oral groove?

thick for conventional lenses to focus all levels of the cell simultaneously. This is especially true of larger cells with complex internal structures. A newer type of microscope that overcomes this impediment is called the *confocal scanning optical microscope*. This microscope uses a beam of light to scan various depths in the

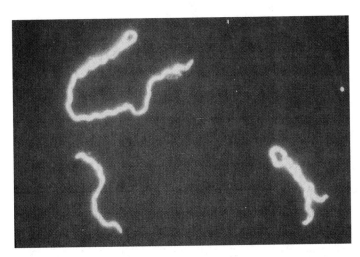

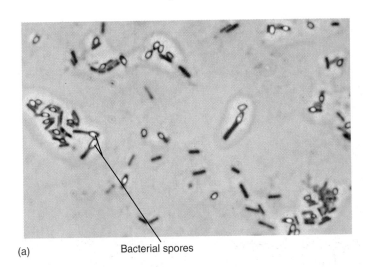

(a)   Bacterial spores

FIGURE 3.20

Dark-field photomicrograph of a spiral-shaped oral bacterium called *Treponema vincenti* **(1,100×).**   This species lives in the space between the gums and the teeth and may be involved in a deteriorating infection called necrotizing gingivitis.

specimen and deliver a sharp image focusing on just a single plane. It is thus able to capture a highly focused view at any level, ranging from the surface to the middle of the cell. It is versatile because it works on live or preserved, stained specimens and it does not require sectioning of the specimen **(see chapter opener).**

Fluorescence Microscopy

The fluorescence microscope is a specially modified compound microscope furnished with an ultraviolet (UV) radiation source and a filter that protects the viewer's eye from injury by these dangerous rays. The name of this type of microscopy originates from the use of certain dyes (acridine, fluorescein) and minerals that show **fluorescence.** This means that the dyes emit visible light when bombarded by shorter ultraviolet rays. For an image to be formed, the specimen must first be coated or placed in contact with a source of fluorescence. Subsequent illumination by ultraviolet radiation causes the specimen to give off light that will form its own image, usually an intense yellow, orange, or red against a black field.

Fluorescence microscopy has its most useful applications in diagnosing infections caused by specific bacteria, protozoans, and viruses. A staining technique with fluorescent dyes is commonly used to detect *Mycobacterium tuberculosis* (the agent of tuberculosis) in patients' specimens (see figure 19.20). In a number of diagnostic procedures, fluorescent dyes are affixed to specific antibodies. These *fluorescent antibodies* can be used to detect the causative agents in such diseases as syphilis, chlamydiosis, trichomoniasis, herpes, and influenza (see figure 6.24b). A newer technology using fluorescent nucleic acid stains can differentiate between live and dead cells in mixtures **(figure 3.22).** A fluorescence microscope can be handy for locating microbes in complex mixtures because only those cells targeted by the technique will fluoresce.

ELECTRON MICROSCOPY

If conventional light microscopes are our windows on the microscopic world, then the electron microscope (EM) is our window on

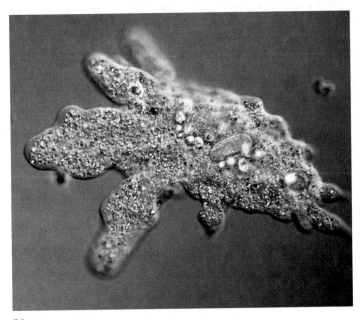

(b)

FIGURE 3.21

Visualizing internal structures.   **(a)** Phase-contrast micrograph of a bacterium containing spores. The relative density of the spores causes them to appear as bright, shiny objects against the darker cell parts (600×). **(b)** Differential interference micrograph of *Amoeba proteus*, a common protozoan. Note the outstanding internal detail, the depth of field, and the bright colors, which are not natural (160×).

the tiniest details of that world. Although this microscope was originally conceived and developed for studying nonbiological materials such as metals and small electronics parts, biologists immediately recognized the importance of the tool and began to use it in the early 1930s. One of the most impressive features of the electron microscope is the resolution it provides.

Unlike the light microscope, the electron microscope forms an image with a beam of electrons that can be made to travel in wavelike patterns when accelerated to high speeds. These waves are 100,000 times shorter than the waves of visible light. Because resolving power is a function of wavelength, electrons have tremendous power to resolve minute structures. Indeed, it is possible to

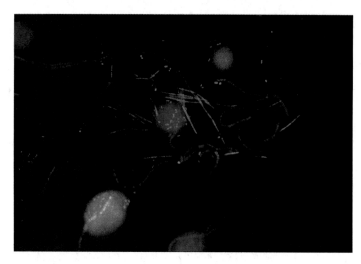

FIGURE 3.22

Fluorescent staining on a fresh sample of cheek scrapings from the oral cavity.   Cheek epithelial cells are the larger unfocused red or green cells. Bacteria appearing here are streptococci (tiny spheres in long chains) and filamentous rods. This technique also indicates whether cells are alive or dead; live cells fluoresce green, and dead cells fluoresce red.

resolve atoms with an electron microscope, though the practical resolution for biological applications is approximately 0.5 nm. Because the resolution is so substantial, it follows that magnification can also be extremely high—usually between 5,000× and 1,000,000× for biological specimens and up to 5,000,000× in some applications. Its capacity for magnification and resolution makes the EM an invaluable tool for seeing the finest structure—or ultrastructure—of cells and viruses. If not for electron microscopes, our understanding of biological structure and function would still be in its early theoretical stages.

In fundamental ways, the electron microscope is a derivative of the compound microscope. It employs components analogous to, but not necessarily the same as, those in light microscopy (**figure 3.23**). For instance, it magnifies in stages by means of two lens systems, and it has a condensing lens, a specimen holder, and focusing apparatus. Otherwise, the two types have numerous differences (**table 3.6**). An electron gun aims its beam through a vacuum to ring-shaped electromagnets that focus this beam on the specimen. Specimens must be pretreated with chemicals or dyes to increase contrast and cannot be observed in a live state. The enlarged image is displayed on a viewing screen or photographed for further study

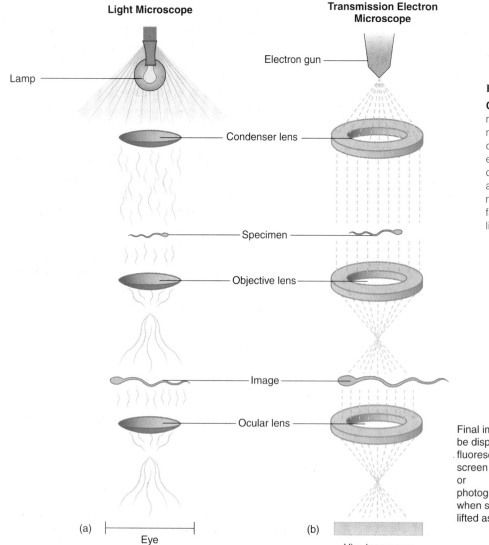

(a) Eye

(b) Viewing screen

FIGURE 3.23

Comparison of two microscopes: **(a)** the light microscope and **(b)** one type of electron microscope (EM; transmission type). These diagrams are highly simplified, especially for the electron microscope, to indicate the common components. Note that the EM's image pathway is actually upside down compared with that of a light microscope. **(c)** The EM is a larger machine with far more complicated working parts than most light microscopes.

Final image can be displayed on fluorescent screen or photographed when screen is lifted aside.

(c)

TABLE 3.6

Comparisons of Types of Microscopy

Microscope	Maximum Practical Magnification	Resolution	Important Features
Visible light as source of illumination Bright-field	2,000×	0.2 μm (200 nm)	Common multipurpose microscope for live and preserved stained specimens; specimen is dark, field is white; provides fair cellular detail
Dark-field	2,000×	0.2 μm	Best for observing live, unstained specimens; specimen is bright, field is black; provides outline of specimen with reduced internal cellular detail
Phase-contrast	2,000×	0.2 μm	Used for live specimens; specimen is contrasted against gray background; excellent for internal cellular detail
Differential interference	2,000×	0.2 μm	Provides brightly colored, highly contrasting, three-dimensional images of live specimens
Ultraviolet rays as source of illumination Fluorescent	2,000×	0.2 μm	Specimens stained with fluorescent dyes or combined with fluorescent antibodies emit visible light; specificity makes this microscope an excellent diagnostic tool
Electron beam forms image of specimen Transmission electron microscope (TEM)	1,000,000×	0.5 nm	Sections of specimen are viewed under very high magnification; finest detailed structure of cells and viruses is shown; used only on preserved material
Scanning electron microscope (SEM)	100,000×	10 nm	Scans and magnifies external surface of specimen; produces striking three-dimensional image

rather than being observed directly through an eyepiece. Because images produced by electrons lack color, electron micrographs (a micrograph is a photograph of a microscopic object) are always shades of black, gray, and white. The color-enhanced micrographs used in this and other textbooks have computer-added color (yes, they are colorized!).

Two general forms of EM are the transmission electron microscope (TEM) and the scanning electron microscope (SEM) (see table 3.6). Transmission electron microscopes are the method of choice for viewing the detailed structure of cells and viruses. This microscope produces its image by transmitting electrons through the specimen. Because electrons cannot readily penetrate thick preparations, the specimen must be sectioned into extremely thin slices (20–100 nm thick) and stained or coated with metals that will increase image contrast. The darkest areas of TEM micrographs represent the thicker (denser) parts, and the lighter areas indicate the more transparent and less dense parts **(figure 3.24).** The TEM can also be used to produce negative images and shadow casts of whole microbes (see figure 6.2).

The scanning electron microscope provides some of the most dramatic and realistic images in existence. This instrument is designed to create an extremely detailed three-dimensional view of all kinds of objects—from plaque on teeth (see figure 21.30) to tapeworm heads (see figure 23.25a). To produce its images, the SEM does not transmit electrons; it bombards the surface of a whole, metal-coated specimen with electrons while scanning back and forth over it. A shower of electrons deflected from the surface is picked up with great fidelity by a sophisticated detector, and the electron pattern is displayed as an image on a television screen. The contours of the specimens resolved with scanning electron micrography are very revealing and often surprising. Areas that look smooth and flat with the light microscope display intriguing surface features with the SEM **(figure 3.25).** Improved technology has continued to refine electron microscopes and to develop variations on the basic plan. One of the most inventive relatives of the EM is the scanning probe microscope **(Microbits 3.2).**

PREPARING SPECIMENS FOR OPTICAL MICROSCOPES

A specimen for optical microscopy is generally prepared by mounting a sample on a suitable glass slide that sits on the stage between the condenser and the objective lens. The manner in which a slide specimen, or mount, is prepared depends upon: (1) the condition of the specimen, either in a living or preserved state; (2) the aims of the examiner, whether to observe overall structure, identify the microorganisms, or see movement; and (3) the type of microscopy available, whether it is bright-field, dark-field, phase-contrast, or fluorescence.

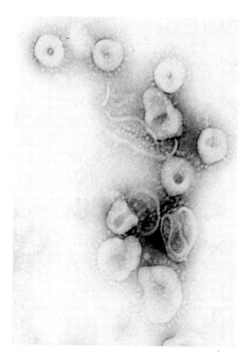

(a)

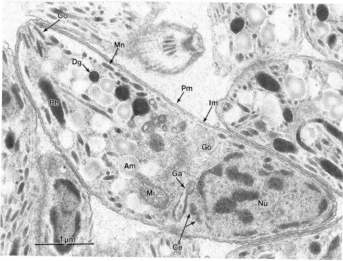

(b)

FIGURE 3.24

Transmission Electron Micrographs.   **(a)** A sample from the respiratory tract reveals coronaviruses (corona for the crownlike envelope) that cause infectious bronchitis in animals (100,000×). A new form of this virus is responsible for severe acute respiratory syndrome in humans (SARS). **(b)** A section through an infectious stage of *Toxoplasma gondii*, the cause of toxoplasmosis. Labels indicate fine structures such as cell membrane (Pm), Golgi complex (Go), nucleus (Nu), mitochondrion (Mi), centrioles (Ce), and granules (Am, Dg).

Fresh, Living Preparations

Live samples of microorganisms are placed in wet mounts or in hanging drop mounts so that they can be observed as near to their natural state as possible. The cells are suspended in a suitable fluid (water, broth, saline) that temporarily maintains viability and provides space and a medium for locomotion. A wet mount consists

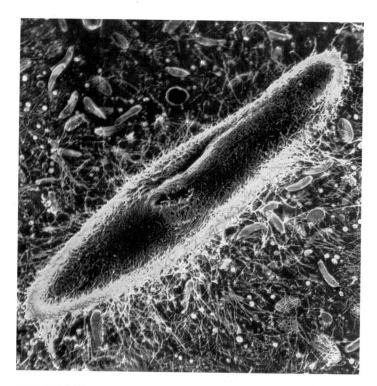

FIGURE 3.25

A false-color scanning electron micrograph (SEM) of *Paramecium*, covered in masses of fine hairs (100×).   These are actually its locomotor and feeding structures—the cilia. Cells in the surrounding medium are bacteria that serve as the protozoan's "movable feast." Compare this with figure 3.20 to appreciate the outstanding three-dimensional detail shown by an SEM.

of a drop or two of the culture placed on a slide and overlaid with a cover glass. Although this type of mount is quick and easy to prepare, it has certain disadvantages. The cover glass can damage larger cells, and the slide is very susceptible to drying and can contaminate the handler's fingers. A more satisfactory alternative is the hanging drop preparation made with a special concave (depression) slide, a Vaseline adhesive or sealant, and a coverslip from which a tiny drop of sample is suspended (see figure 4.4). These types of short-term mounts provide a true assessment of the size, shape, arrangement, color, and motility of cells. Greater cellular detail can be observed with phase-contrast or interference microscopy.

Fixed, Stained Smears

A more permanent mount for long-term study can be obtained by preparing fixed, stained specimens. The smear technique, developed by Robert Koch more than 100 years ago, consists of spreading a thin film made from a liquid suspension of cells on a slide and air-drying it. Next, the air-dried smear is usually heated gently by a process called heat fixation that simultaneously kills the specimen and secures it to the slide. Another important action of fixation is to preserve various cellular components in a natural state with minimal distortion. Fixation of some microbial cells is performed with chemicals such as alcohol and formalin.

MICROBITS   3.2
The Evolution in Resolution: Probing Microscopes

In the past, chemists, physicists, and biologists had to rely on indirect methods to provide information on the structures of the smallest molecules. But technological advances have created a new generation of microscopes that "see" atomic structure by actually feeling it. *Scanning probe microscopes* operate with a minute needle tapered to a tip that can be as narrow as a single atom! This probe scans over the exposed surface of a material and records an image of its outer texture. These revolutionary microscopes have such profound resolution that they have the potential to image single atoms (but not subatomic structure yet) and to magnify 100 million times. The scanning tunneling microscope (STM) was the first of these microscopes. It uses a tungsten probe that hovers near the surface of an object and follows its topography while simultaneously giving off an electrical signal of its pathway, which is then imaged on a screen. The STM is used primarily for detecting defects on the surfaces of electrical conductors and computer chips composed of silicon, but it has also provided the first incredible close-up views of DNA, the genetic material (see Spotlight on Microbiology 9.1).

Another exciting new variant is the atomic force microscope (AFM), which gently forces a diamond and metal probe down onto the surface of a specimen like a needle on a record. As it moves along the surface, any deflection of the metal probe is detected by a sensitive device that relays the information to an imager. The AFM is very useful in viewing the detailed functions of biological molecules such as antibodies and enzymes.

These powerful new microscopes, along with tools that can move and position atoms, have spawned a field called *nanotechnology*—the science of the "small." Scientists in this area use physics, chemistry, biology, and engineering to explore and manipulate small molecules and atoms. Working at these dimensions, they hope to create tiny molecular tools to miniaturize computers and other electronic devices. In the future, it may be possible to use microstructures to deliver drugs, analyze DNA, and treat disease.

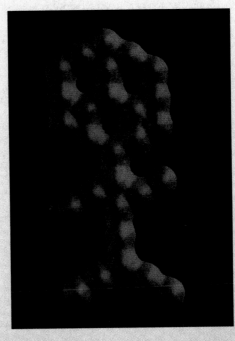

"Carbon monoxide man." An atomic force microscope image of a single carbon monoxide molecule, with its single carbon and oxygen atoms magnified several million times.

Like images on undeveloped photographic film, the unstained cells of a fixed smear are quite indistinct, no matter how great the magnification or how fine the resolving power of the microscope. The process of "developing" a smear to create contrast and make inconspicuous features stand out requires staining techniques. Staining is any procedure that applies colored chemicals called dyes to specimens. Dyes impart a color to cells or cell parts by becoming affixed to them through a chemical reaction. In general, they are classified as basic (cationic) dyes, which have a positive charge, or acidic (anionic) dyes, which have a negative charge. Because chemicals of opposite charge are attracted to each other, cell parts that are negatively charged will attract basic dyes and those that are positively charged will attract acidic dyes **(table 3.7).** Many cells, especially those of bacteria, have numerous negatively charged acidic substances and thus stain more readily with basic dyes. Acidic dyes, on the other hand, tend to be repelled by cells, so they are good for negative staining (discussed in the next section). The chemistry of dyes and their staining reactions is covered in more detail in **Microbits 3.3.**

Negative Versus Positive Staining   Two basic types of staining technique are used, depending upon how a dye reacts with the specimen (summarized in table 3.7). Most procedures involve a **positive stain,** in which the dye actually sticks to the specimen and gives it color. A **negative stain,** on the other hand, is just the reverse (like a photographic negative). The dye does not stick to the specimen but settles around its outer boundary, forming a silhouette. In a sense, negative staining "stains" the glass slide to produce a dark background around the cells. Nigrosin (blue-black) and India ink (a black suspension of carbon particles) are the dyes most commonly used for negative staining. The cells themselves do not stain because these dyes are negatively charged and are repelled by the negatively charged surface of the cells. The value of negative staining is its relative simplicity and the reduced shrinkage or distortion of cells, as the smear is not heat-fixed. A quick assessment can thus be made regarding cellular size, shape, and arrangement. Negative staining is also used to accentuate the capsule that surrounds certain bacteria and yeasts **(figure 3.26,** page 82).

Because many microbial cells lack contrast, it is necessary to use dyes to observe their detailed structure and identify them. Dyes are colored compounds related to or derived from the common organic solvent benzene. When certain double-bonded groups (C=O, C=N, N=N) are attached to complex ringed molecules, the resultant compound gives off a specific color. Most dyes are in the form of a sodium or chloride salt of an acidic or basic compound that ionizes when dissolved in a compatible solvent. The color-bearing ion, termed a **chromophore,** is charged and has an affinity for certain cell parts that are of the opposite charge.

Dyes that have a negatively charged chromophore are termed acidic. An example is eosin, a bright red dye that is used in staining blood cells. Acidic chromophores are attracted to the positively charged molecules of red blood cells and some parts of white blood cells. Because bacterial cells have numerous acidic substances and carry a slightly negative charge on their surface, they do not stain well with acidic dyes.

Basic dyes such as methylene blue have a positively charged chromophore that is attracted to negatively charged cell components (nucleic acids and proteins). Since bacteria have a preponderance of negative ions, they stain readily with other basic dyes, including crystal violet, fuchsin, malachite green, and safranin.

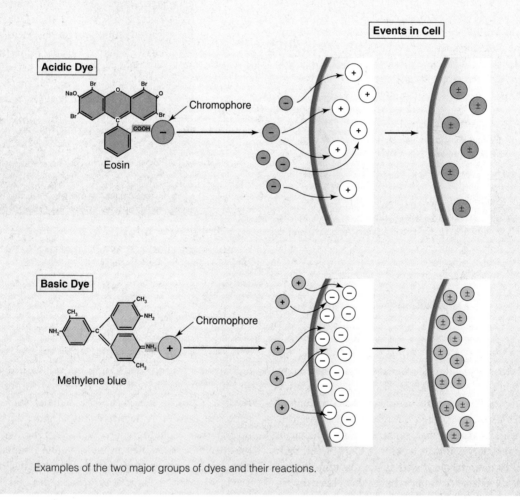

Examples of the two major groups of dyes and their reactions.

Simple Versus Differential Staining   Positive staining methods are classified as simple, differential, or special (figure 3.26). Whereas **simple stains** require only a single dye and an uncomplicated procedure, **differential stains** use two different-colored dyes, called the *primary dye* and the *counterstain,* to distinguish between cell types or parts. These staining techniques tend to be more complex and sometimes require additional chemical reagents to produce the desired reaction.

Most simple staining techniques take advantage of the ready binding of bacterial cells to dyes like malachite green, crystal violet, basic fuchsin, and safranin. Simple stains cause all cells in a smear to appear more or less the same color, regardless of type, but they can still reveal bacterial characteristics such as shape, size, and arrangement. A simple stain with Loeffler's methylene blue is significant because it also reveals the internal granules of *Corynebacterium diphtheriae,* a bacterium that is responsible for diphtheria (see chapter 4).

TABLE 3.7

Comparison of Positive and Negative Stains

	Positive Staining	Negative Staining
Appearance of cell	Colored by dye	Clear and colorless
Background	Not stained (generally white)	Stained (dark gray or black)
Dyes employed	Basic dyes: Crystal violet Methylene blue Safranin Malachite green	Acidic dyes: Nigrosin India ink
Subtypes of stains	Several types: Simple stain Differential stains Gram stain Acid-fast stain Spore stain Special stains Capsule Flagella Spore Granules Nucleic acid	Few types: Capsule Spore (Dorner)

Types of Differential Stains   A satisfactory differential stain uses differently colored dyes to clearly contrast two cell types or cell parts. Common combinations are red and purple, red and green, or pink and blue. Differential stains can also pinpoint other characteristics, such as the size, shape, and arrangement of cells. Typical examples include Gram, acid-fast, and endospore stains. Some staining techniques (spore, capsule) fall into more than one category.

Gram staining, a century-old method named for its developer, Hans Christian Gram, remains the most universal diagnostic staining technique for bacteria. It permits ready differentiation of major categories based upon the color reaction of the cells: **gram positive,** which stain purple, and **gram negative,** which stain pink (red). The Gram stain is the basis of several important bacteriological topics, including bacterial taxonomy, cell wall structure, and identification and diagnosis of infection; in some cases, it even guides the selection of the correct drug for an infection. Gram staining is discussed in greater detail in Historical Highlights 4.2.

The **acid-fast stain,** like the Gram stain, is an important diagnostic stain that differentiates acid-fast bacteria (pink) from non-acid-fast bacteria (blue). This stain originated as a specific method to detect *Mycobacterium tuberculosis* in specimens. It was determined that these bacterial cells have a particularly impervious outer wall that holds fast (tightly or tenaciously) to the dye (carbol fuchsin) even when washed with a solution containing acid or acid alcohol. This stain is used for other medically important mycobacteria such as the leprosy bacillus and for *Nocardia,* an agent of lung or skin infections.

The endospore stain (spore stain) is similar to the acid-fast method in that a dye is forced by heat into resistant bodies called spores or endospores (their formation and significance are discussed in chapter 4). This stain is designed to distinguish between spores and the cells that they come from (so-called **vegetative cells**). Of significance in medical microbiology are the gram-positive, spore-forming members of the genus *Bacillus* (the cause of anthrax) and *Clostridium* (the cause of botulism and tetanus)—dramatic diseases of universal fascination that we consider in later chapters.

Special stains are used to emphasize certain cell parts that are not revealed by conventional staining methods. Capsule staining is a method of observing the microbial capsule, an unstructured protective layer surrounding the cells of some bacteria and fungi. Because the capsule does not react with most stains, it is often negatively stained with India ink, or it may be demonstrated by special positive stains. The fact that not all microbes exhibit capsules is a useful feature for identifying pathogens. One example is *Cryptococcus,* which causes a serious fungal infection in AIDS patients.

Flagellar staining is a method of revealing flagella, the tiny, slender filaments used by bacteria for locomotion (see chapter 4). Because the width of bacterial flagella lies beyond the resolving power of the light microscope, in order to be seen, they must be enlarged by depositing a coating on the outside of the filament and then staining it. This stain works best with fresh, young cultures, because flagella are delicate and can be lost or damaged on older cells. Their presence, number, and arrangement on a cell are taxonomically useful.

CHAPTER CHECKPOINTS

Magnification, resolving power, lens quality, and illumination source all influence the clarity of specimens viewed through the optical microscope.

The maximum resolving power of the optical microscope is 200 nm, or 0.2 μm. This is sufficient to see the internal structures of eucaryotes and the morphology of most bacteria.

There are five types of optical microscopes. Four types use visible light for illumination: bright-field, dark-field, phase-contrast, and interference microscopes. The fifth type, the fluorescence microscope, uses UV light for illumination, but it has the same resolving power as the other optical microscopes.

Electron microscopes (EM) use electrons, not light waves, as an illumination source to provide high magnification ($5,000\times$ to $1,000,000\times$) and high resolution (0.5 nm). Electron microscopes can visualize cell ultrastructure (TEM) and three-dimensional images of cell and virus surface features (SEM).

Specimens viewed through optical microscopes can be either alive or dead, depending on the type of specimen preparation, but all EM specimens are dead because they must be treated with metals for effective viewing.

Stains are important diagnostic tools in microbiology because they can be designed to differentiate cell shape, structure, and biochemical composition of the specimens being viewed.

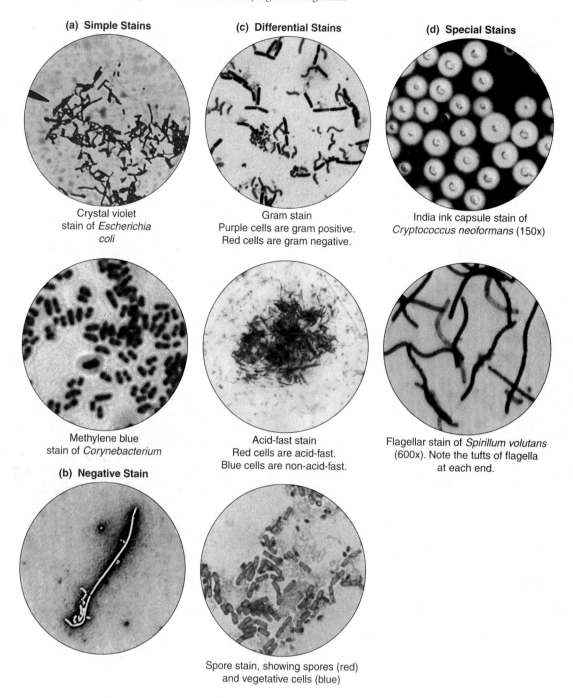

(a) Simple Stains

Crystal violet
stain of *Escherichia
coli*

Methylene blue
stain of *Corynebacterium*

(b) Negative Stain

(c) Differential Stains

Gram stain
Purple cells are gram positive.
Red cells are gram negative.

Acid-fast stain
Red cells are acid-fast.
Blue cells are non-acid-fast.

Spore stain, showing spores (red)
and vegetative cells (blue)

(d) Special Stains

India ink capsule stain of
Cryptococcus neoformans (150x)

Flagellar stain of *Spirillum volutans*
(600x). Note the tufts of flagella
at each end.

FIGURE 3.26

Types of microbiological stains. **(a)** Simple stains. **(b)** Negative stain. **(c)** Differential stains: Gram, acid-fast, and spore. **(d)** Special stains: capsule and flagellar.

CHAPTER CAPSULE WITH KEY TERMS

I. Methods of Culturing Microorganisms
A. Microbiology as a science is very dependent on a number of specialized laboratory techniques. Laboratory steps routinely employed in microbiology are inoculation, incubation, isolation, inspection, and identification.
  1. Initially, a specimen must be collected from a source, whether environmental or a patient. The mode of collection will vary depending on the source and the type of microbe.

  2. **Inoculation** of a **medium** is the first step in obtaining a **culture** of the microorganisms present.
  3. **Isolation** of the microorganisms, such that each microbial cell present is separated from the others and forms discrete **colonies,** is aided by inoculation techniques such as streak plates, pour plates, and spread plates.
  4. **Incubation** of the medium with the microbes under the right conditions allows growth to visible colonies. Generally,

isolated colonies would be **subcultured** for further testing at this point. The goal is a **pure culture** in most cases, or a **mixed culture. Contaminated cultures** can ruin correct analysis and study.

5. Inspection begins with macroscopic characteristics of the colonies, and continues with microscopic analysis, often using microscopy.

6. Identification correlates the various morphological, physiological, genetic, and serological methods as needed to be able to pinpoint the actual species or even strain of microbe.

B. Media: Providing Nutrients in the Laboratory

1. Artificial media allow the growth and isolation of microorganisms in the laboratory, and can be classified by their physical form, chemical composition, and functional types.

2. Physical types of media include those that are liquid, such as broths and milk, those that are **semisolid,** and those that are solid. Solid media may be liquefiable, containing a solidifying agent such as **agar** or gelatin.

3. Chemical composition of a medium may be completely chemically defined, thus synthetic. Nonsynthetic, or complex, media contain ingredients that are not completely definable.

4. Functional types of media serve different purposes, often allowing biochemical tests to be performed at the same time. Types include general-purpose, **enriched, selective, differential,** anaerobic (reducing), assay, and enumeration **media.** Transport media are important for conveying certain clinical specimens to the laboratory.

5. In certain instances, microorganisms have to be grown in animals and bird embryos.

II. **Microscopy: Viewing Microorganisms at the Cellular and Subcellular Level**

A. Optical, or light, microscopy depends on lenses that **refract** light rays, drawing the rays to a focus to produce a magnified image. Common principles underlie microscope design.

1. A simple microscope consists of a single magnifying lens, whereas a compound microscope relies on two lenses: the ocular lens and the objective lens.

2. The total power of magnification is calculated from the product of the ocular and objective magnifying powers.

3. Resolution, or the **resolving power,** is a measure of a microscope's capacity to make clear images of very small objects. Resolution is improved with shorter wavelengths of illumination and with a higher **numerical aperture** of the lens. Light microscopes are limited to magnifications around $2,000\times$ by the resolution. A student microscope with a $10\times$ ocular and $100\times$ oil objective is sufficient to see most bacteria.

4. Modifications in the lighting or the lens system give rise to the bright-field, dark-field, phase-contrast, interference, and fluorescence microscopes.

B. Electron microscopy depends on electromagnets that serve as lenses to focus electron beams. A transmission electron microscope (TEM) projects the electrons through prepared sections of the specimen, providing detailed structural images of cells, cell parts and viruses. A scanning electron microscope (SEM) is more like dark-field microscopy, bouncing the electrons off the surface of the specimen to detectors. This results in lower magnification but rich images of the external surfaces of microbes.

C. Specimen preparation in optical microscopy is governed by the condition of the specimen, the purpose of the inspection, and the type of microscope being used.

1. Wet mounts and hanging drop mounts permit examination of the characteristics of live cells, such as motility, shape, and arrangement.

2. Fixed mounts are made by drying and heating a film of the specimen called a smear. This is then stained using dyes to permit visualization of cells or cell parts.

D. Staining uses either basic (cationic) dyes with positive charges or acidic (anionic) dyes with negative charges on the **chromophore.** The surfaces of microbes are negatively charged and attract basic dyes. This is the basis of positive staining. In **negative staining,** the microbe repels the dye and it stains the background. Dyes may be used alone and in combination.

1. Simple stains use just one dye, and highlight cell morphology.

2. Differential stains require a primary dye and a contrasting counterstain in order to distinguish cell types or parts. Important differential stains include the **Gram stain, acid-fast stain,** and the endospore stain.

3. Special stains are designed to bring out distinctive characteristics. Examples include capsule stains and flagellar stains.

MULTIPLE-CHOICE QUESTIONS

1. Which of the following is not one of the Five I's?
   a. inspection
   b. identification
   c. induction
   d. incubation
   e. inoculation

2. The term *culture* refers to the _____ growth of microorganisms in _____.
   a. rapid, an incubator
   b. macroscopic, media
   c. microscopic, the body
   d. artificial, colonies

3. A mixed culture is
   a. the same as a contaminated culture
   b. one that has been adequately stirred
   c. one that contains two or more known species
   d. a pond sample containing algae and protozoa

4. Agar is superior to gelatin as a solidifying agent because agar
   a. does not melt at room temperature
   b. solidifies at 75°C
   c. is not usually decomposed by microorganisms
   d. both a and c

5. The process that most accounts for magnification is
   a. a condenser
   b. refraction of light rays
   c. illumination
   d. resolution

6. A subculture is a
   a. colony growing beneath the media surface
   b. culture made from a contaminant
   c. culture made in an embryo
   d. culture made from an isolated colony

7. Resolution is _____ with a longer wavelength of light.
   a. improved
   b. worsened
   c. not changed
   d. not possible

8. A real image is produced by the
   a. ocular
   b. objective
   c. condenser
   d. eye

9. A microscope that has a total magnification of 1,500× with the oil immersion lens has an ocular of what power?
   a. 150×
   b. 1.5×
   c. 15×
   d. 30×

10. The specimen for an electron microscope is always
   a. stained with dyes
   b. sliced into thin sections
   c. killed
   d. viewed directly

11. Motility is best observed with a
   a. hanging drop preparation
   b. negative stain
   c. streak plate
   d. flagellar stain

12. Bacteria tend to stain more readily with cationic (positively charged) dyes because bacteria
   a. contain large amounts of alkaline substances
   b. contain large amounts of acidic substances
   c. are neutral
   d. have thick cell walls

13. The primary difference between a TEM and SEM is in
   a. magnification capability
   b. colored versus black and white images
   c. preparation of the specimen
   d. type of lenses

14. **Multiple Matching.** For each type of medium, select all descriptions that fit. For media that fit more than one description, briefly explain why this is the case.
   _____ mannitol salt agar
   _____ chocolate agar
   _____ MacConkey agar
   _____ nutrient broth
   _____ Sabouraud's agar
   _____ triple-sugar iron agar
   _____ *Euglena* agar
   _____ SIM medium
   _____ Stuart's medium
   a. selective medium
   b. differential medium
   c. chemically defined (synthetic) medium
   d. enriched medium
   e. general-purpose medium
   f. complex medium
   g. transport medium

CONCEPT QUESTIONS

1. a. Describe briefly what is involved in the Five I's.
   b. Name three basic differences between inoculation and contamination.

2. a. Name two ways that pure, mixed, and contaminated cultures are similar and two ways that they differ from each other.
   b. What must be done to avoid contamination?

3. a. Explain what is involved in isolating microorganisms and why it is necessary to do this.
   b. Compare and contrast three common laboratory techniques for separating bacteria in a mixed sample.
   c. Describe how an isolated colony forms.
   d. Explain why an isolated colony and a pure culture are not the same thing.

4. a. Explain the two principal functions of dyes in media.
   b. Differentiate among the ingredients and functions of enriched, selective, and differential media.

5. Differentiate between microscopic and macroscopic methods of observing microorganisms, citing a specific example of each method.

6. a. Contrast the concepts of magnification, refraction, and resolution.
   b. Briefly explain how an image is made and magnified.
   c. Trace the pathway of light from its source to the eye, explaining what happens as it passes through the major parts of the microscope.

7. a. On the basis of the formula for resolving power, explain why a smaller R. P. value is preferred to a larger one.
   b. What does it mean in practical terms if the resolving power is 1.0 μm?
   c. How does a value greater than 1.0 μm compare? (Is it better or worse?)
   d. How does a value less than 1.0 μm compare?
   e. What can be done to a microscope to improve resolution?

8. a. Compare bright-field, dark-field, phase-contrast, and fluorescence microscopy as to field appearance, specimen appearance, light source, and uses.
   b. Of what benefit is the dark-field microscope if the field is black?

9. a. Compare and contrast the optical compound microscope with the electron microscope.
   b. Why is the resolution so superior in the electron microscope?
   c. What will you never see in an unretouched electron micrograph?
   d. Compare the way that the image is formed in the TEM and SEM.

10. Evaluate the following preparations in terms of showing microbial size, shape, motility, and differentiation: spore stain, negative stain, simple stain, hanging drop slide, and Gram stain.

11. a. Itemize the various staining methods, and briefly characterize each.
   b. For a stain to be considered a differential stain, what must it do?
   c. Explain what happens in positive staining to cause the reaction in the cell.
   d. Explain what happens in negative staining that causes the final result.

CRITICAL-THINKING QUESTIONS

1. Describe the steps you would take to isolate, cultivate, and identify a microbial pathogen from a urine sample. (Hint: Look at the Five I's.)

2. A certain medium has the following composition:

Glucose	15 g
Yeast extract	5 g
Peptone	5 g
KH_2PO_4	2 g
Distilled water	1,000 ml

   a. To what chemical category does this medium belong?
   b. How could you convert *Euglena* agar (table 3.2) into a nonsynthetic medium?

3. a. Name four categories that blood agar fits into.
   b. Name four differential reactions that TSIA shows.
   c. Can you tell what functional kind of medium *Enterococcus faecalis* medium is?

4. a. What kind of medium might you make to selectively grow a bacterium that lives in the ocean?
   b. One that lives in the human stomach?
   c. What characteristic of dyes makes them useful in differential media?
   d. Why are intestinal bacteria able to grow on media containing bile?

5. a. When buying a microscope, what features are most important to check for?
   b. What is probably true of a $20 microscope that claims to magnify 1,000×?

6. How can one obtain 2,000× magnification with a 100× objective?

7. a. In what ways are dark-field microscopy and negative staining alike?
   b. How is the dark-field microscope like the scanning electron microscope?

8. Biotechnology companies have engineered hundreds of different types of mice, rats, pigs, goats, cattle, and rabbits to have genetic diseases similar to diseases of humans or to synthesize drugs and other biochemical products. They have patented these animals and sell them to researchers for study and experimentation.
   a. What do you think of creating new life forms just for experimentation?
   b. Comment or start a class discussion on the benefits, safety, and humanity of this trend.

9. This is a test of your living optical system's resolving power. Prop your book against a wall about 20 inches away and determine the line that is no longer resolvable by your eye. See if you can determine your actual resolving power, using a millimeter ruler.

So, Naturalists observe,

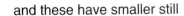

a flea has smaller

fleas that on him prey;

and these have smaller still

to bite 'em; and so proceed,

ad infinitum.

Source: Poem by Jonathan Swift.

INTERNET SEARCH TOPICS

1. Search through several websites using the words "electron micrograph." Find examples of TEM and SEM micrographs and their applications in science and technology.

2. Search using the words "laboratory identification of anthrax" to make an outline of the basic techniques used in analysis of the microbe, under the headings of the Five I's.

3. Visit the student Online Learning Center at www.mhhe.com/talaro5. Go to chapter 3, Internet Search Topics, and log on to the available websites to:
   a. Explore information and images on microscopes and microscopy. Visit the photo gallery to compare different types of microscope images.
   b. View how the numerical aperture changes with magnification.

CHAPTER 4

Procaryotic Profiles:
The Bacteria and Archaea

S mall and deceptively simple, procaryotes are among nature's most abundant and ubiquitous microorganisms. If it were somehow possible to eradicate all bacteria in the world, humans would notice the effects immediately and, for a while, might find it a favorable change. We would not have to be as careful about preparing and refrigerating foods; plaque would no longer develop on our teeth; and there would be fewer cleaning chores around the house. Quite suddenly, the medical community's goal of eradicating certain infectious diseases, such as tuberculosis, cholera, and tetanus, could be a reality. But there are other considerations to take into account. We would also have to do without certain foods, like sauerkraut, yogurt, Swiss cheese, and sourdough bread. At first, this may seem a small price to pay, but are these slight inconveniences the only sacrifices to be expected? Within a few days, industrial processes that produce vitamins, drugs, and solvents would lie abandoned, and most molecular biology research labs and biotechnology companies would be shut down. In a few months and years, humus containing dead animal and plant matter would build up and trap the very elements needed to sustain the living world. Clearly, bacteria play vital roles in all aspects of our existence. We literally can't live without them!

In order to explore the roles of bacteria in nutrition, genetics, drug therapy, infection, and microbial ecology, we must first understand several aspects of the structure and behavior of procaryotic cells. The primary topics to be covered in this chapter are elements of microscopic anatomy, physiology, identification, and classification and a survey of selected bacterial groups.

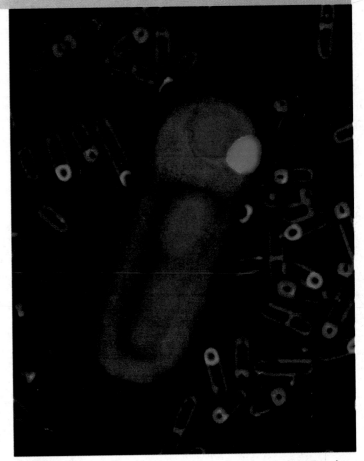

These biological "safety pins" are actually stages in endospore formation of *Bacillus subtilis*, stained with fluorescent proteins. The large red and blue cell is a vegetative cell in the early stages of sporulating. The developing spores are shown in green and orange.

Chapter Overview

- Procaryotic cells are the smallest, simplest, and most abundant cells on earth.
- Representative procaryotes include bacteria and archaea, both of which lack a nucleus and organelles but are functionally complex.
- The structure of bacterial cells is compact and capable of adaptations to a multitude of habitats.
- The cell is encased in an envelope that protects, supports, and regulates transport.
- Bacteria have special structures for motility and adhesion to the environment.
- Bacterial cells contain genetic material in a single chromosome, and ribosomes for synthesizing proteins.
- Bacteria have the capacity for reproduction, nutrient storage, dormancy, and resistance to adverse conditions.
- Shape, size, and arrangement of bacterial cells are extremely varied.
- Bacterial taxonomy and classification is based on their structure, metabolism, and genetics.
- Archaea are procaryotes related to eucaryotic cells that possess unique biochemistry and genetics.
- Archaea are adapted to the most extreme habitats on earth.

Procaryotic Form and Function: External Structure

The evolutionary history of procaryotic cells extends back at least 3.5 billion years (see figure 4.29). It is now generally thought that the very first cells to appear on the earth were a type of archaea possibly related to modern forms that live on sulfur compounds in geothermal ocean vents. The fact that these organisms have endured for so long in such a variety of habitats indicates a cellular structure and function that are amazingly versatile and adaptable. The general cellular organization of a procaryotic cell can be represented with this flowchart:

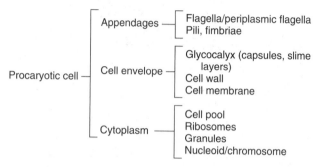

All bacterial cells invariably have a cell membrane, cytoplasm, ribosomes, and a chromosome; the majority have a cell wall and some form of surface coating or glycocalyx. Specific structures that are found in some, but not all, bacteria are flagella, pili, fimbriae, capsules, slime layers, and granules.

The Structure of a Generalized Procaryotic Cell

Bacterial cells appear featureless and two-dimensional when viewed with an ordinary microscope. Not until they are subjected to the scrutiny of the electron microscope and biochemical studies does their intricate and functionally complex nature become evident. The descriptions of bacterial structure, except where otherwise noted, refer to the **bacteria,*** a category of procaryotes with peptidoglycan in their cell walls. **Figure 4.1** presents a three-dimensional anatomical view of a generalized (rod-shaped)

---

*bacteria Formerly known as eubacteria (yoo'-bak-ter-ee-uh) Gr. *eu,* true, and *bakterion,* rod. Includes all procaryotes besides the archaea.

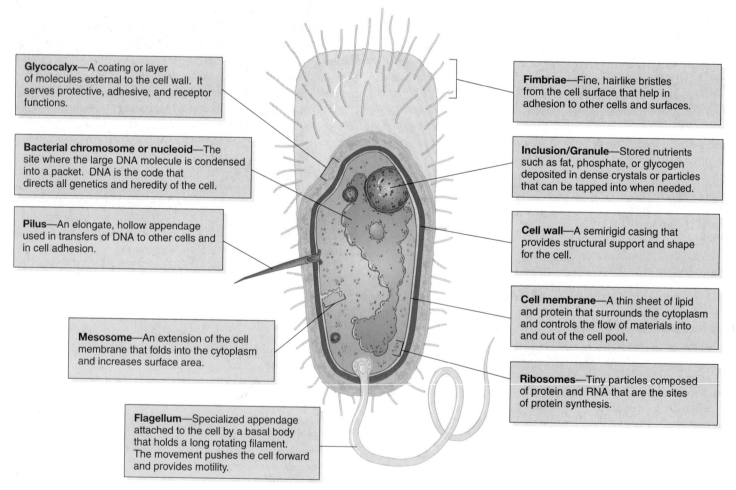

FIGURE 4.1

Structure of a procaryotic cell. Cutaway view of a typical rod-shaped bacterium, showing major structural features. Note that not all components are found in all cells.

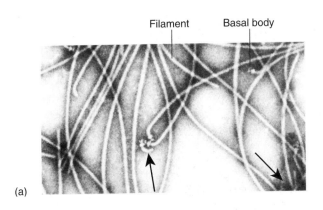

Filament   Basal body

(a)

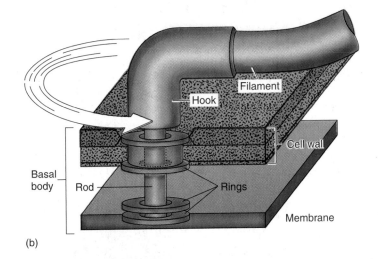

Filament

Hook

Cell wall

Basal
body   Rod   Rings

Membrane

(b)

FIGURE 4.2

Structure of flagella. **(a)** Electron micrograph of basal bodies and filaments (at arrows) of bacterial flagella (66,000×). **(b)** Details of the basal body in a gram-negative cell. The hook, rings, and rod function together as a tiny device that rotates the filament 360°.

bacterial cell. As we survey the principal anatomical features of this cell, we will perform a microscopic dissection of sorts, following a course that begins with the outer cell structures and proceeds to the internal contents.

APPENDAGES: CELL EXTENSIONS

Several discrete types of accessory structures sprout from the surface of bacteria. These elongate **appendages*** are common but are not present on all species. Appendages can be divided into two major groups: those that provide motility (flagella and axial filaments) and those that provide attachments (fimbriae and pili).

Flagella—Bacterial Propellers

The procaryotic **flagellum*** is an appendage of truly amazing construction, certainly unique in the biological world. The primary function of flagella is to confer **motility,** or self-propulsion—that is, the capacity of a cell to swim freely through an aqueous habitat. The extreme thinness of a bacterial flagellum necessitates high magnification to reveal its special architecture, which occurs in three distinct parts: the filament, the hook (sheath), and the basal body (**figure 4.2**). The **filament,** a helical structure composed of proteins, is approximately 20 nm in diameter and varies from 1 to 70 nm in length. It is inserted into a curved, tubular hook. The hook is anchored to the cell by the basal body, a stack of rings firmly anchored through the cell wall, to the cell membrane. This arrangement permits the hook with its filament to rotate 360°, rather than undulating back and forth like a whip as was once thought. As the flagellum rotates in a counterclockwise direction, it causes the cell body to swim in a direct forward path (see figure 4.5).

One can generalize that all spirilla, about half of the bacilli, and a small number of cocci are flagellated (these bacterial shapes are shown in figure 4.22). Flagella vary both in number and arrangement according to two general patterns: (1) In a *polar* arrangement, the fla-

gella are attached at one or both ends of the cell. Three subtypes of this pattern are: **monotrichous,*** with a single flagellum; **lophotrichous,*** with small bunches or tufts of flagella emerging from the same site; and **amphitrichous,*** with flagella at both poles of the cell. (2) In a **peritrichous*** arrangement, flagella are dispersed randomly over the surface of the cell (**figure 4.3**). The type of arrangement has some bearing on the swimming speed of the bacterium. The speediest forms are polar, flagellated cells such as *Thiospirillum,* which can zip along at 5.2 mm/minute, and *Pseudomonas aeruginosa,* which can swim 4.4 mm/minute. Taking into account the small dimensions of these bacteria, such speeds are comparable to some protozoa and animals. Peritrichous rods such as *Escherichia coli* tend to swim at a relatively slower pace (1 mm/minute).

The presence of motility is one piece of information used in the laboratory identification or diagnosis of pathogens. Special stains or electron microscope preparations must be used to see arrangement, since flagella are too minute to be seen in live preparations with a light microscope. Often it is sufficient to know simply whether a bacterial species is motile. One way to detect motility is to stab a tiny mass of cells into a soft (semisolid) medium. Growth spreading rapidly through the entire medium is indicative of motility (see figure 3.5). Alternatively, cells can be observed microscopically with a hanging drop slide (**figure 4.4**). A truly motile cell will flit, dart, or wobble around the field, making some progress, whereas one that is nonmotile jiggles about in one place but makes no progress.

Fine Points of Flagellar Function Flagellated bacteria can perform some rather sophisticated feats. They can detect and move in response to chemical signals—a type of behavior called **chemotaxis.*** Positive chemotaxis is movement of a cell in the direction

---

***monotrichous** (mah″-noh-trik′-us) Gr. *mono*, one, and *tricho*, hair.

***lophotrichous** (lo″-foh-trik′-us) Gr. *lopho*, tuft or ridge.

***amphitrichous** (am″-fee-trik′-us) Gr. *amphi*, on both sides.

***peritrichous** (per″-ee-trik′-us) Gr. *peri*, around.

***chemotaxis** (ke″-moh-tak′-sis) pl. chemotaxes; Gr. *chemo*, chemicals, and *taxis*, an ordering or arrangement.

***appendage** (uh-pen′-dij) Any external projection of a body.

***flagellum** (flah-jel′-em) pl. flagella; L., a whip.

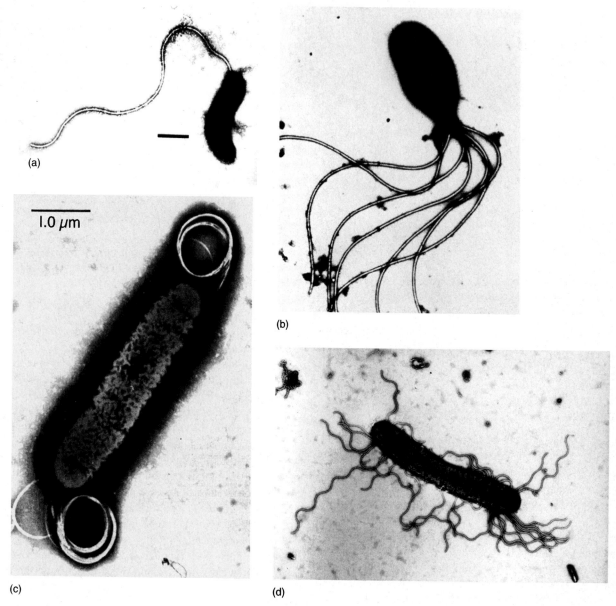

FIGURE 4.3

Electron micrographs depicting types of flagellar arrangements. **(a)** Monotrichous flagellum on the predatory bacterium *Bdellovibrio*.
(b) Lophotrichous flagella on *Vibrio fischeri*, a common marine bacterium (23,000×). **(c)** Unusual flagella on *Aquaspirillum* are amphitrichous
(and lophotrichous) in arrangement and coil up into tight loops. **(d)** An unidentified bacterium discovered inside the cells of *Paramecium* exhibits
peritrichous flagella.

(b) From Reichelt and Baumann, Arch. Microbiol. 94:283–330. © Springer-Verlag, 1973.

of a favorable chemical stimulus (usually a nutrient); negative chemotaxis is movement away from a repellent (potentially harmful) compound.

The flagellum is effective in guiding bacteria through the environment primarily because the system for detecting chemicals is linked to the mechanisms that drive the flagellum. Located in the cell membrane are clusters of receptors[1] that bind specific molecules coming from the immediate environment. The attachment of sufficient numbers of these molecules transmits signals to the flagellum and sets it into rotary motion. If several flagella are present, they become aligned and rotate as a group **(figure 4.5)**. As a flagellum rotates counterclockwise, the cell itself swims in a smooth linear direction toward the stimulus, called a run. Runs are interrupted at various intervals by tumbles, during which the flagellum reverses direction and causes the cell to stop and change its course. It is believed that attractant molecules inhibit tumbles and permit progress toward the stimulus. Repellents cause numerous tumbles, allowing the bacterium to redirect itself away from the stimulus **(figure 4.6)**. Some photosynthetic bacteria exhibit *phototaxis*, a type of movement in response to light rather than chemicals.

---

1. Cell surface molecules that bind specifically with other molecules.

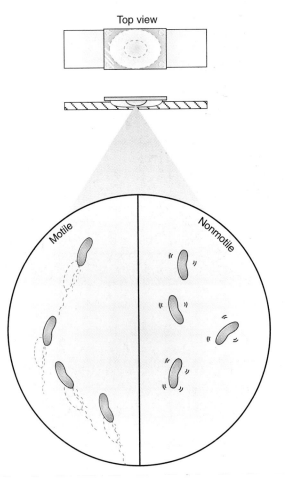

Top view

Motile

Nonmotile

FIGURE 4.4

Motility detection. A hanging drop slide can be used to detect motility in bacteria by observing the microscopic behavior of the cell. In true motility, the cell swims and progresses from one point to another. It is assumed that a motile cell has one or more flagella even though the flagella are not visible. Nonmotile cells oscillate in the same relative space because of bombardment by molecules, a physical process called Brownian movement, but they do not swim from one point to another.

Periplasmic Flagella: Internal Flagella

Corkscrew-shaped bacteria called **spirochetes*** show an unusual, wriggly mode of locomotion caused by two or more long, coiled threads, the periplasmic flagella or *axial filaments.* A periplasmic flagellum is a type of internal flagellum that is enclosed in the space between the cell wall and the cell membrane **(figure 4.7).** The filaments curl closely around the spirochete coils yet are free to contract and impart a twisting or flexing motion to the cell. This form of locomotion must be seen in live cells such as the spirochete of syphilis to be truly appreciated.

Appendages for Attachment and Mating

The structures termed **pilus*** and **fimbria*** both refer to bacterial surface appendages that provide some type of adhesion, but not locomotion. We will use the term *fimbriae* to refer to the shorter, nu-

---

*spirochete (spy′-roh-keet) Gr. *speira,* coil, and *chaite,* hair.

*pilus (py′-lus) pl. pili; L., hair.

*fimbria (fim′-bree-ah) pl. fimbriae; L., a fringe.

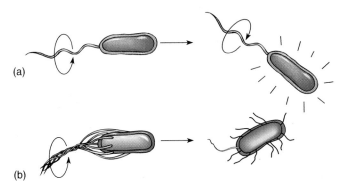

(a)

(b)

FIGURE 4.5

The operation of flagella and the mode of locomotion in bacteria with polar and peritrichous flagella. **(a)** In general, when a polar flagellum rotates in a counterclockwise direction, the cells swims forward. When the flagellum reverses direction and rotates clockwise, the cell stops and tumbles. **(b)** In peritrichous forms, all flagella sweep toward one end of the cell and rotate as a single group. During tumbles, the flagella lose coordination.

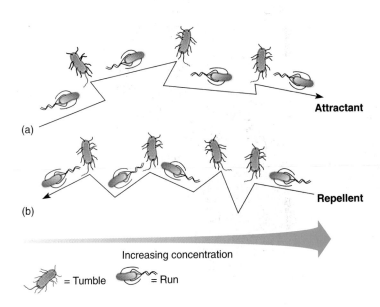

Attractant

(a)

(b)

Repellent

Increasing concentration

= Tumble   = Run

FIGURE 4.6

Chemotaxis in bacteria. The cell shows a primitive mechanism for progressing **(a)** toward positive stimuli and **(b)** away from irritants by swimming in straight runs or by tumbling. Bacterial runs allow straight, undisturbed progress toward the stimulus, whereas tumbles interrupt progress to allow the bacterium to redirect itself away from the stimulus after sampling the environment.

merous strands and the term *pili* to refer to the longer, sparser appendages.

Fimbriae are small, bristlelike fibers sprouting off the surface of many bacterial cells **(figure 4.8).** Their exact composition varies, but most of them contain protein. Fimbriae have an inherent tendency to stick to each other and to surfaces. They may be responsible for the mutual clinging of cells that leads to biofilms and other thick aggregates of cells on the surface of liquids and for the microbial colonization of inanimate solids such as rocks and glass **(Spotlight on Microbiology 4.1).** Some pathogens can colonize and infect host tissues because of a tight adhesion between their fimbriae and epithelial cells (figure 4.8*b*). For example, the gonococcus

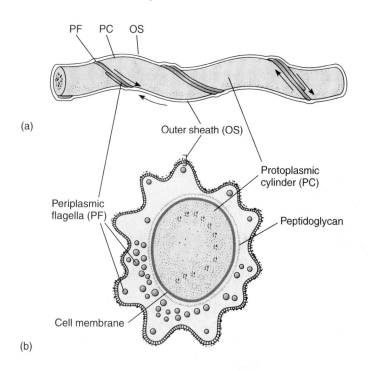

(a)

(b)

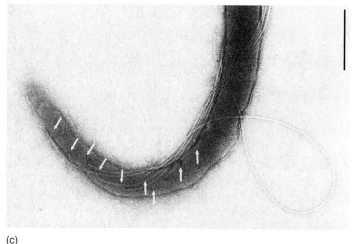

(c)

FIGURE 4.7

The orientation of periplasmic flagella on the spirochete cell.
(a) Longitudinal section. **(b)** Cross section. Contraction of the filaments imparts a spinning and undulating pattern of locomotion. **(c)** Electron micrograph captures the details of periplasmic flagella and their insertion points (arrows) in *Borrelia burgdorferi*. (Bar = 0.2 μm)

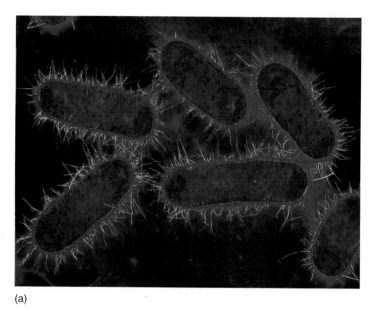

(a)

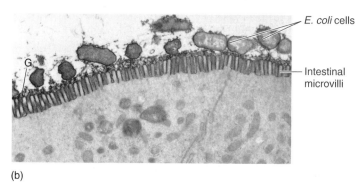

(b)

FIGURE 4.8

Form and function of bacterial fimbriae.   **(a)** Several cells of pathogenic *Escherichia coli* covered with numerous stiff fibers called fimbriae (30,000×). Note also the dark blue granules, which are the chromosomes. **(b)** A row of *E. coli* cells tightly adheres by their fimbriae to the surface of intestinal cells (12,000×). This is how the bacterium clings and gains access to the body during an infection. (G = glycocalyx)

(agent of gonorrhea) invades the genitourinary tract, and *Escherichia coli* invades the intestine by this means. Mutant forms of these pathogens that lack fimbriae are unable to cause infections.

A pilus (also called a *sex pilus*) is an elongate, rigid tubular structure made of a special protein, *pilin*. So far, true pili have been found only on gram-negative bacteria, where they are involved primarily in a mating process between cells called **conjugation,*** which involves partial transfer of DNA from one cell to another **(figure 4.9).** A pilus from the donor cell unites with a recipient cell,

thereby providing a cytoplasmic connection for making the transfer. Production of pili is controlled genetically, and conjugation takes place only between compatible gram-negative cells. Conjugation in gram-positive bacteria does occur, but involves aggregation proteins rather than sex pili. The roles of pili and conjugation are further explored in chapter 9.

THE CELL ENVELOPE: THE OUTER WRAPPING OF BACTERIA

The majority of bacteria have a chemically complex external covering, termed the cell envelope, that lies outside of the cytoplasm. It is composed of three basic layers known as the glycocalyx, the cell wall, and the cell membrane **(figure 4.10).** The layers of the envelope are stacked one upon another and are often tightly bonded together like the outer husk and casings of a coconut. Although each envelope layer performs a distinct function, together they act as a single protective unit. The envelope is extensive and can account for one-tenth to one-half of a cell's volume.

---

*conjugation (kon-joo-gay'-shun) L. *conjugatus,* linked together.

Pili

Fimbriae

Viruses

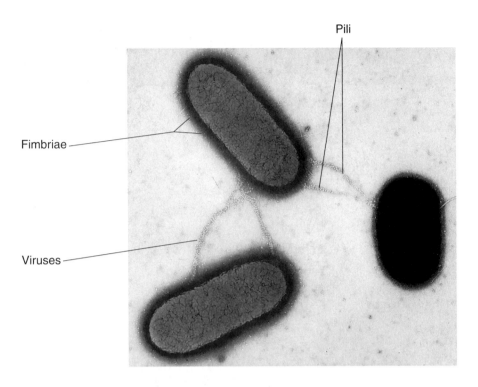

FIGURE 4.9

Three bacteria in the process of conjugating.
Clearly evident are the sex pili forming mutual
conjugation bridges between a donor (upper cell)
and two recipients (two lower cells). Interesting
points to observe are the fimbriae on the donor
cell and the bacterial viruses (phages) forming
tiny spots on the pili.

Glycocalyx
(varies in
structure)

Cell wall
(varies in
structure)

Cell membrane

Cell envelope

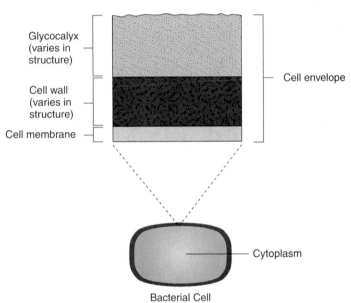

Cytoplasm

Bacterial Cell

FIGURE 4.10

The relationship of the three layers of the cell envelope.

Slime Layer

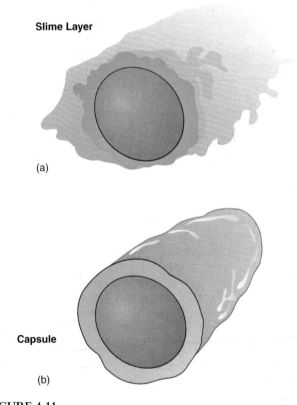

(a)

Capsule

(b)

FIGURE 4.11

Bacterial cells sectioned to show the types of glycocalyces.   (a) The
slime layer is a loose structure that is easily washed off. **(b)** The capsule is
a thick, structured layer that is not readily removed.

The Bacterial Surface Coating, or Glycocalyx

The bacterial cell surface is frequently exposed to severe environmental conditions. The **glycocalyx** develops as a coating of macromolecules to protect the cell and, in some cases, help it adhere to its
environment. Glycocalyces differ among bacteria in thickness, organization, and chemical composition. Some bacteria are covered
with a loose, soluble shield called a slime layer that evidently protects them from loss of water and nutrients (**figure 4.11a**). Other
bacteria produce **capsules** of repeating polysaccharide units, of
protein, or of both (figure 4.11b). A capsule is bound more tightly

to the cell than a slime layer is, and it has a thicker, gummy consistency that gives a prominently sticky (mucoid) character to the
colonies of most encapsulated bacteria (**figure 4.12**).

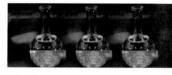

SPOTLIGHT ON MICROBIOLOGY 4.1
Biofilms—The Glue of Life

Being aware of the widespread existence of microorganisms on earth, we should not be surprised that, when left undisturbed, they gather in masses, cling to various surfaces, and capture available moisture and nutrients. The formation of these living layers, called **biofilms,** is actually a universal phenomenon that all of us have observed. Consider the scum that builds up in toilet bowls and shower stalls in a short time if they are not cleaned; or the algae that collect on the walls of swimming pools; and, more intimately—the constant deposition of plaque on teeth. Microbes making biofilms is a primeval tendency that has been occurring for billions of years as a way to create stable habitats with adequate access to food, water, atmosphere, and other essential factors. Biofilms are often cooperative associations among several microbial groups (bacteria, fungi, algae, and protozoa) as well as plants and animals.

Substrates are most likely to accept a biofilm if they are moist and have developed a thin layer of organic material such as polysaccharides or glycoproteins on their exposed surface (see figure at right). This depositing process occurs within a few minutes to hours, making a slightly sticky texture that attracts primary colonists, usually bacteria. These early cells attach (adsorb to) and begin to multiply on the surface. As they grow, various secreted substances in their glycocalyx (receptors, fimbriae, slime layers, capsules) increase the binding of cells to the surface and thicken the biofilm. As the biofilm evolves, it undergoes specific adaptations to the habitat in which it forms. In many cases, the earliest colonists contribute nutrients and create microhabitats that serve as a matrix for other microbes to attach and grow into the film, forming complete communities. The biofilm varies in thickness and complexity, depending upon where it occurs and how long it keeps developing. Complexity ranges from single cell layers to thick microbial mats with dozens of dynamic interactive layers.

Biofilms are a profoundly important force in the development of terrestrial and aquatic environments. They dwell permanently in bedrock and the earth's sediments, where they play an essential role in recycling elements, leaching minerals, and participating in soil formation. Biofilms associated with plant roots promote the mutual exchange of nutrients between the microbes and roots. The human body contains biofilms in the form of normal flora that live in the skin and mucous membranes, and on structures such as teeth (see the description of plaque formation in figure 21.29). Bacteria can also persistently colonize medical devices such as catheters, artificial heart valves, and other inanimate objects placed in the body (see figure 4.13). Invasive biofilms can wreak havoc with human-made structures such as cooling towers, storage tanks, air conditioners, and even stone buildings. Additional information on biofilms is found in chapter 26.

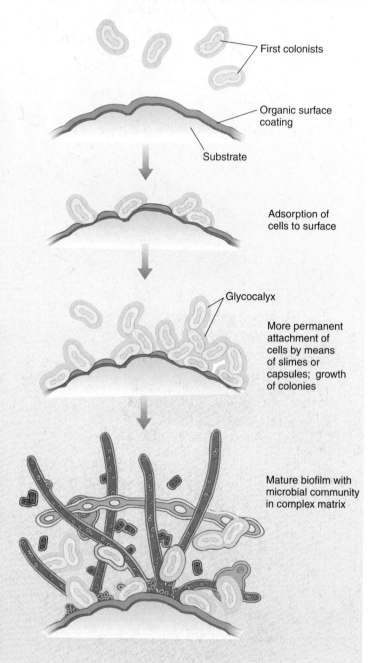

First colonists

Organic surface coating

Substrate

Adsorption of cells to surface

Glycocalyx

More permanent attachment of cells by means of slimes or capsules; growth of colonies

Mature biofilm with microbial community in complex matrix

Specialized Functions of the Glycocalyx Capsules are formed by a few pathogenic bacteria, such as *Streptococcus pneumoniae* (a cause of pneumonia, an infection of the lung), *Haemophilus influenzae* (one cause of meningitis), and *Bacillus anthracis* (the cause of anthrax). Encapsulated bacterial cells generally have greater pathogenicity because capsules protect the bacteria against white blood cells called phagocytes. Phagocytes are a natural body defense that can engulf and destroy foreign cells

through phagocytosis, thus preventing infection. A capsular coating blocks the mechanisms that phagocytes use to attach to and engulf bacteria. By escaping phagocytosis, the bacteria are free to multiply and infect body tissues. Encapsulated bacteria that mutate to nonencapsulated forms usually lose their pathogenicity.

Other types of glycocalyces can be important in formation of biofilms. The thick, white plaque that forms on teeth comes in part from the surface slimes produced by certain streptococci in the oral

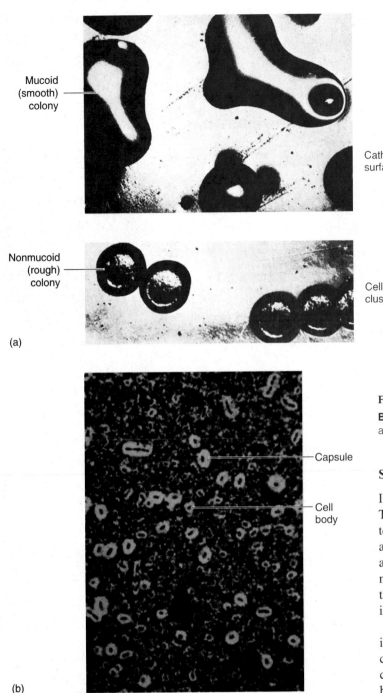

Mucoid (smooth) colony

Nonmucoid (rough) colony

(a)

Capsule

Cell body

(b)

FIGURE 4.12

Structure of encapsulated bacteria. **(a)** The appearance of colonies composed of encapsulated cells (mucoid) compared with those lacking capsules (nonmucoid). Even at the macroscopic level, the slippery, gel-like character of the capsule is evident. **(b)** Staining reveals the microscopic appearance of a large, well-developed capsule.

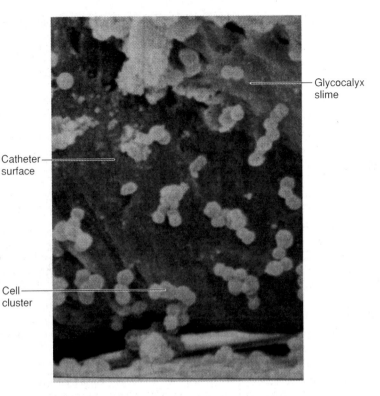

Glycocalyx slime

Catheter surface

Cell cluster

FIGURE 4.13

Biofilm. Scanning electron micrograph of *Staphylococcus aureus* cells attached to a catheter by a slime secretion.

STRUCTURE OF THE CELL WALL

Immediately below the glycocalyx lies a second layer, the cell wall. This structure accounts for a number of important bacterial characteristics. In general, it determines the shape of a bacterium, and it also provides the kind of strong structural support necessary to keep a bacterium from bursting or collapsing because of changes in osmotic pressure. In this way, the cell wall functions like a bicycle tire that maintains the necessary shape and prevents the more delicate inner tube from bursting when it is expanded.

The cell walls of most bacteria gain their relatively rigid quality from a unique macromolecule called **peptidoglycan (PG).** This compound is composed of a repeating framework of long glycan* chains cross-linked by short peptide fragments to provide a strong but flexible support framework **(figure 4.14).** Peptidoglycan is only one of several materials found in cell walls, and its amount and exact composition vary among the major bacterial groups.

Because many bacteria live in aqueous habitats with a low solute concentration, they are constantly absorbing excess water by osmosis. Were it not for the strength and relative rigidity of the peptidoglycan in the cell wall, they would rupture from internal pressure. Understanding this function of the cell wall has been a tremendous boon to the drug industry. Several types of drugs used to treat infection (penicillin, cephalosporins) are effective because they target the peptide cross-links in the peptidoglycan, thereby disrupting its integrity. With their cell walls incomplete or missing, such cells have very little protection from **lysis*** (see figure 12.2).

cavity. This slime initially allows them to adhere to the teeth and provides a niche for other oral bacteria that, in time, can lead to dental disease. The glycocalyx of some bacteria is so highly adherent that it is responsible for persistent colonization of nonliving materials such as plastic catheters, intrauterine devices, and metal pacemakers that are in common medical use **(figure 4.13).**

---

*glycan (gly´-kan) Gr., sugar. These are large polymers of simple sugars.

*lysis (ly´-sis) Gr., to loosen. A process of cell destruction, as occurs in bursting.

(a) The peptidoglycan of a cell wall can be presented as a crisscross network pattern similar to a chain-link fence, forming a single massive molecule that molds the outer structure of the cell into a tight box.

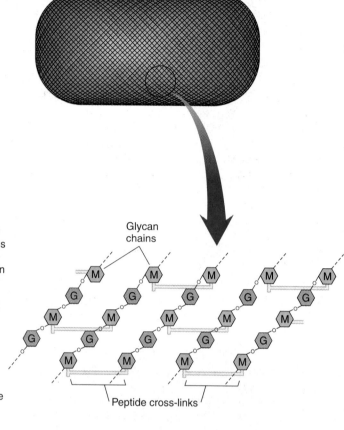

(b) An idealized view of the molecular pattern of peptidoglycan. It contains alternating glycans (G and M) bound together in long strands. The G stands for *N*-acetyl glucosamine, and the M stands for *N*-acetyl muramic acid. A muramic acid molecule binds to an adjoining muramic acid on a parallel chain by means of a cross-linkage of peptides.

Glycan chains

Peptide cross-links

(c) A detailed view of the links between the muramic acids. Tetrapeptide chains branching off the muramic acids connect by interbridges also composed of amino acids. The types of amino acids in the interbridge can vary and it may be lacking entirely (gram-negative cells). It is this linkage that provides rigid yet flexible support to the cell and that may be targeted by drugs like penicillin.

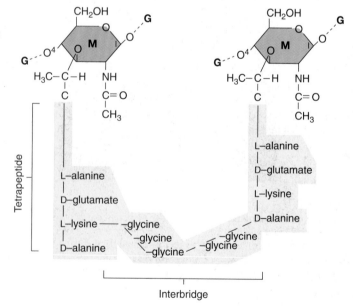

FIGURE 4.14

Structure of peptidoglycan in the cell wall.

Some disinfectants (alcohol, detergents) also kill bacterial cells by damaging the cell wall. Lysozyme, an enzyme contained in tears and saliva, provides a natural defense against certain bacteria by hydrolyzing the bonds in the glycan chains and causing the wall to break down.

Differences in Cell Wall Structure

More than a hundred years ago, long before the detailed anatomy of bacteria was even remotely known, a Danish physician named Hans Christian Gram developed a staining technique, the **Gram stain,** that delineates two generally different groups of bacteria (**Historical

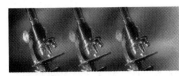

HISTORICAL HIGHLIGHTS   4.2
The Gram Stain: A Grand Stain

In 1884, Hans Christian Gram discovered a staining technique that could be used to make bacteria in infectious specimens more visible. His technique consisted of timed, sequential applications of crystal violet (the primary dye), Gram's iodine (IKI, the mordant), an alcohol rinse (decolorizer), and a contrasting counterstain. The initial counterstain used was yellow or brown and later replaced by the red dye, safranin. Since that substitution, bacteria that stained purple are called gram positive, and those that stained red are called gram negative.

Although these staining reactions involve an attraction of the cell to a charged dye (see chapter 3), it is important to note that the terms *gram positive* and *gram negative* are not used to indicate the electrical charge of cells or dyes but whether or not a cell retains the primary dye-iodine complex after decolorization. There is nothing specific in the reaction of gram-positive cells to the primary dye or in the reaction of gram-negative cells to the counterstain. The different results in the Gram stain are due to differences in the structure of the cell wall and how it reacts to the series of reagents applied to the cells.

In the first step, crystal violet is attracted to the cells in a smear and stains them all the same purple color. The second and key differentiating step is the addition of the mordant—Gram's iodine. The mordant is a stabilizer that causes the dye to form large crystals in the peptidoglycan meshwork of the cell wall. Because the peptidoglycan layer in gram-positive cells is thicker, the entrapment of the dye is far more extensive in them than in gram-negative cells. Application of alcohol in the third step dissolves lipids in the outer membrane and removes the dye from the peptidoglycan layer and the gram-negative cells. By contrast, the crystals of dye tightly embedded in the peptidoglycan of gram-positive bacteria are relatively inaccessible and resistant to removal. Because gram-negative bacteria are colorless after decolorization, their presence is demonstrated by applying the counterstain safranin in the final step.

This century-old staining method remains the universal basis for bacterial classification and identification. It permits differentiation of four major categories based upon color reaction and shape: gram-positive rods, and gram-positive cocci, gram-negative rods, and gram-negative cocci (see table 4.6). The Gram stain can also be a practical aid in diagnosing infection and in guiding drug treatment. For example, gram staining a fresh urine or throat specimen can help pinpoint the possible cause of infection, and in some cases it is possible to begin drug therapy on the basis of this stain. Even in this day of elaborate and expensive medical technology, the Gram stain remains an important and unbeatable first tool in diagnosis.

Step	Microscopic Appearance of Cell		Chemical Reaction in Cell Wall (very magnified view)	
	Gram (+)	Gram (−)	Gram (+)	Gram (−)
1. Crystal violet			Both cell walls affix the dye	
2. Gram's iodine			Dye crystals trapped in wall	No effect of iodine
3. Alcohol			Crystals remain in cell wall	Cell wall partially dissolved, loses dye
4. Safranin (red dye)			Red dye has no effect	Red dye stains the colorless cell

Gram stain technique and theory.

Highlights 4.2). We now know that the contrasting staining reactions that occur with this stain are due entirely to some very fundamental differences in the structure of bacterial cell walls. The two major groups shown by this technique are the gram-positive bacteria and the gram-negative bacteria. Because the Gram stain does not actually reveal the nature of these physical differences, we must turn to the electron microscope and to biochemical analysis.

The extent of the differences between gram-positive and gram-negative bacteria is evident in the physical appearance of their cell envelopes **(figure 4.15)**. In gram-positive cells, a microscopic section resembles an open-faced sandwich with two layers: the thick outer cell wall, composed primarily of peptidoglycan, and the cell membrane. A similar section of a gram-negative cell envelope shows a complete sandwich with three layers: the cell wall, composed of an outer membrane and a thin layer of peptidoglycan, and the cell membrane. See **table 4.1** for a further comparison of cell wall types.

The Gram-Positive Cell Wall   The bulk of the gram-positive cell wall is a thick, homogeneous sheath of peptidoglycan ranging from 20 to 80 nm in thickness. It also contains tightly bound acidic polysaccharides, including teichoic acid and lipoteichoic acid **(figure 4.16)**. Teichoic acid is a polymer of ribitol or glycerol and phosphate embedded in the peptidoglycan sheath. Lipoteichoic acid is similar in structure but is attached to the lipids in the plasma membrane. These molecules appear to function in cell wall maintenance and

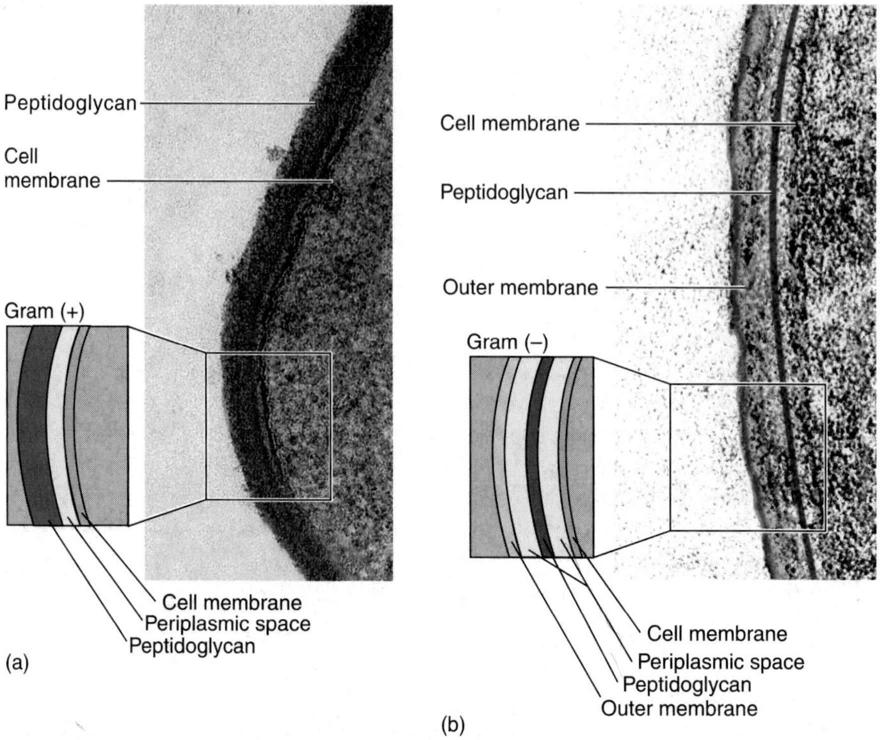

FIGURE 4.15

A comparison of the envelopes of gram-positive and gram-negative cells. **(a)** A photomicrograph of a gram-positive cell wall/membrane and an artist's interpretation of its open-faced sandwich–style layering with two layers. **(b)** A photomicrograph of a gram-negative cell wall/membrane and an artist's interpretation of its complete sandwich–style layering with three distinct layers.

TABLE 4.1		
Comparison of Gram-Positive and Gram-Negative Cell Walls		
Characteristic	**Gram-Positive**	**Gram-Negative**
Number of major layers	1	2
Chemical composition	Peptidoglycan Teichoic acid Lipoteichoic acid	Lipopolysaccharide Lipoprotein Peptidoglycan
Overall thickness	Thicker (20–80 nm)	Thinner (8–11 nm)
Outer membrane	No	Yes
Periplasmic space	Narrow	Extensive
Porin proteins	No	Yes
Permeability to molecules	More penetrable	Less penetrable

enlargement during cell division, and they also contribute to the acidic charge on the cell surface. In some cases, the cell wall of gram-positive bacteria is pressed tightly against the cell membrane with very little space between them, but in other cells, a thin **periplasmic* space** is evident between the cell membrane and cell wall.

The Gram-Negative Cell Wall The gram-negative cell wall is more complex in morphology because it contains an outer membrane (OM), has a thinner shell of peptidoglycan, and has an extensive space surrounding the peptidoglycan (figure 4.16). The outer membrane is somewhat similar in construction to the cell membrane, except that it contains specialized types of polysaccharides and proteins. The uppermost layer of the OM contains *lipopolysaccharide* (LPS). The polysaccharide chains extending off the surface function as antigens and receptors. The innermost layer of the OM is a phospholipid layer anchored by means of lipoproteins to the peptidoglycan layer below. The outer membrane serves as a partial chemical sieve by allowing only relatively small molecules to penetrate. Access is provided by special membrane channels formed by *porin proteins* that completely span the outer membrane. The size of these porins can be altered so as to block the entrance of harmful chemicals, making them one defense of gram-negative bacteria against certain antibiotics.

The bottom layer of the gram-negative wall is a single, thin (1–3 nm) sheet of peptidoglycan. Although it acts as a somewhat rigid protective structure as previously described, its thinness gives gram-negative bacteria a relatively greater flexibility and sensitivity to lysis. There is a well-developed *periplasmic space* surrounding the peptidoglycan. This space is an important reaction site for a large and varied pool of substances that enter and leave the cell.

Practical Considerations of Differences in Cell Wall Structure Variations in cell wall anatomy contribute to several other differences between the two cell types. The outer membrane contributes an extra barrier in gram-negative bacteria that makes them more impervious to some antimicrobic chemicals such as dyes and disinfectants, so they are generally more difficult to inhibit or kill than are gram-positive bacteria. One exception is for alcohol-based compounds, which can dissolve the lipids in the outer membrane and disturb its integrity. Treating infections caused by gram-negative bacteria often requires different drugs from gram-positive infections, especially drugs that can cross the outer membrane.

*periplasmic (per″-ih-plaz′-mik) Gr. *peri,* around, and *plastos,* formed.

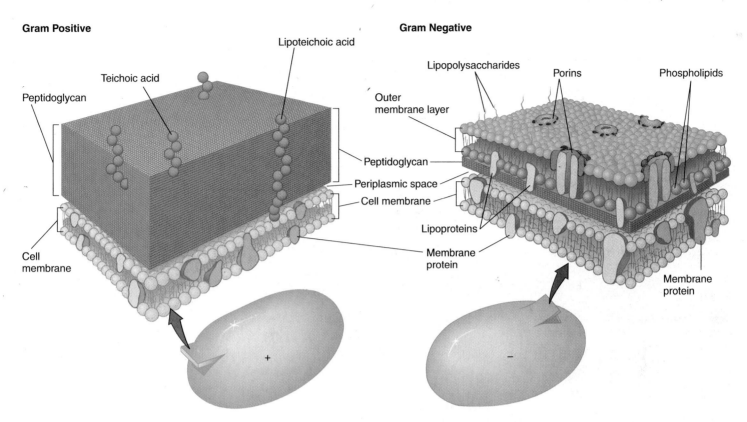

FIGURE 4.16

A comparison of the detailed structure of gram-positive and gram-negative cell walls.

The cell wall or its parts can interact with human tissues and contribute to disease. The lipids have been referred to as *endotoxins* because they stimulate fever and shock reactions in gram-negative infections such as meningitis and typhoid fever. Proteins attached to the outer portion of the cell wall of several gram-positive species, including *Corynebacterium diphtheriae* (the agent of diphtheria) and *Streptococcus pyogenes* (the cause of strep throat), also have toxic properties. The lipids in the cell walls of certain *Mycobacterium* species are harmful to human cells as well. Because most macromolecules in the cell walls are foreign to humans, they stimulate antibody production by the immune system (see chapter 15).

Nontypical Cell Walls

Several bacterial groups lack the cell wall structure of gram-positive or gram-negative bacteria, and some bacteria have no cell wall at all. Although these exceptional forms can stain positive or negative in the Gram stain, examination of their fine structure and chemistry shows that they do not really fit the descriptions for typical gram-negative or -positive cells. For example, the cells of *Mycobacterium* and *Nocardia* contain peptidoglycan and stain gram positive, but the bulk of their cell wall is composed of unique types of lipids. One of these is a very-long-chain fatty acid called *mycolic acid*, or cord factor, that contributes to the pathogenicity of this group (see chapter 19). The thick, waxy nature imparted to the cell wall by these lipids is also responsible for a high degree of resistance to certain chemicals and dyes. Such resistance is the basis for the **acid-fast stain** used to diagnose tuberculosis and leprosy. In this stain, hot carbol fuchsin dye becomes tenaciously attached (is held fast)

to these cells so that an acid-alcohol solution will not remove the dye (see chapter 3).

Because they are from a more ancient and primitive line of procaryotes, the archaea exhibit unusual and chemically distinct cell walls. In some, the walls are composed almost entirely of polysaccharides, and in others, the walls are pure protein; but as a group, they all lack the true peptidoglycan structure described previously. Since a few archaea and all mycoplasmas (discussed in a later section of this chapter) lack a cell wall entirely, their cell membrane must serve the dual functions of support as well as transport.

Some bacteria that ordinarily have a cell wall can lose it during part of their life cycle. These wall-deficient forms are referred to as **L forms** or L-phase variants (for the Lister Institute, where they were discovered). L forms arise naturally from a mutation in the wall-forming genes, or they can be induced artificially by treatment with a chemical such as lysozyme or penicillin that disrupts the cell wall. When a gram-positive cell is exposed to either of these two chemicals, it will lose the cell wall completely and become a **protoplast,*** a fragile cell bounded only by a membrane that is highly subject to lysis **(figure 4.17a).** A gram-negative cell exposed to these same substances loses it peptidoglycan but retains its outer membrane, leaving a less fragile but nevertheless weakened **spheroplast*** (figure 4.17*b*). Evidence points to a role for L forms in certain infections (see chapter 21).

---

*protoplast (proh′-toh-plast) Gr. *proto*, first, and *plastos*, formed.

*spheroplast (sfer′-oh-plast) Gr. *sphaira*, sphere.

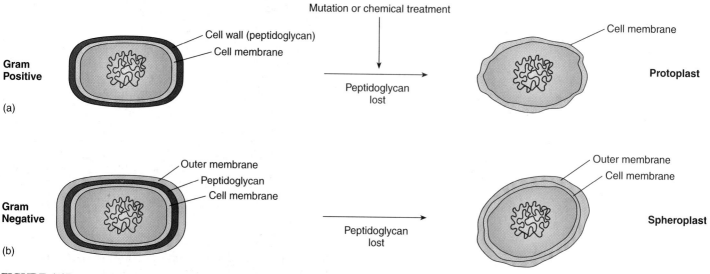

FIGURE 4.17

The conversion of walled bacterial cells to L forms: **(a)** gram-positive bacteria; **(b)** gram-negative bacteria.

CELL MEMBRANE STRUCTURE

Appearing just beneath the cell wall is the cell, or cytoplasmic, membrane, a very thin (5–10 nm), flexible sheet molded completely around the cytoplasm. Its general composition was described in chapter 2 as a lipid bilayer with proteins embedded to varying degrees (see Microbits 2.3). Bacterial cell membranes have this typical structure, containing primarily phospholipids (making up about 30–40% of the membrane mass) and proteins (contributing 60–70%). Major exceptions to this description are the membranes of mycoplasmas, which contain high amounts of sterols—rigid lipids that stabilize and reinforce the membrane—and the membranes of archaea, which contain unique branched hydrocarbons rather than fatty acids.

In some locations, the cell membrane forms internal folds in the cytoplasm called **mesosomes*** (see figure 4.1). These are prominent in gram-positive bacteria but are harder to see in gram-negative bacteria because of their relatively small size. Mesosomes presumably increase the internal surface area available for membrane activities. It has been proposed that mesosomes function in cell wall synthesis and to guide the duplicated bacterial chromosomes into the two daughter cells during cell division (see figure 7.14). Some scientists are not convinced that mesosomes exist in live cells, and believe they are artifacts that appear in bacteria fixed for electron microscopy. Support for the existence of functioning internal membranes comes from specialized procaryotes such as cyanobacteria and even mitochondria, considered as a type of procaryotic cell.

Photosynthetic procaryotes such as cyanobacteria contain dense stacks of internal membranes that carry the photosynthetic pigments (see figure 4.33).

Functions of the Cell Membrane

Since bacteria have none of the eucaryotic organelles, the cell membrane provides a site for functions such as energy reactions,

nutrient processing, and synthesis. A major action of the cell membrane is to regulate *transport,* that is, the passage of nutrients into the cell and the discharge of wastes. Although water and small uncharged molecules can diffuse across the membrane unaided, the membrane is a *selectively permeable* structure with special carrier mechanisms for passage of most molecules (see chapter 7). The glycocalyx and cell wall can bar the passage of large molecules, but they are not the primary transport apparatus. The cell membrane is also involved in *secretion,* or the discharge of a metabolic product into the extracellular environment.

The membranes of procaryotes are an important site for a number of metabolic activities. Most enzymes of respiration and ATP synthesis reside in the cell membrane since procaryotes lack mitochondria (see chapter 8). Enzyme structures located in the cell membrane also help synthesize structural macromolecules to be incorporated into the cell envelope and appendages. Other products (enzymes and toxins) are secreted by the membrane into the extracellular environment.

CHAPTER CHECKPOINTS

Bacteria are the oldest form of cellular life. They are also the most widely dispersed, occupying every conceivable microclimate on the planet.

The appendages of bacteria provide motility (flagella), attachment (pili and fimbriae) and in some bacteria, a means of DNA transfer (sex pili).

Flagella vary in number and arrangement as well as in the type and rate of motion they produce.

The cell envelope is the outermost covering of bacteria. It consists of three basic layers: (1) the glycocalyx, (2) the cell wall, and (3) the cell membrane.

The composition of the procaryotic cell wall is used to classify bacteria into four major divisions: gram-positive bacteria, gram-negative bacteria, bacteria with no cell walls, and bacteria with chemically unique cell walls.

---

*mesosome (mes'-oh-sohm) Gr. *mesos,* middle, and *soma,* body.

Gram-positive bacteria retain the crystal violet and stain purple. Gram-negative bacteria lose the crystal violet and stain red from the safranin counterstain.

Gram-positive bacteria have thick cell walls of peptidoglycan and acidic polysaccharides such as teichoic acid, and they have a thin periplasmic space. The cell walls of gram-negative bacteria are thinner but contain an additional outer membrane and have a wide periplasmic space.

The bacterial cell membrane is typically composed of phospholipids and proteins, and it performs many metabolic functions as well as transport activities.

Bacterial Form and Function: Internal Structure

CONTENTS OF THE CELL CYTOPLASM

Encased by the cell membrane is a dense, gelatinous solution referred to as **cytoplasm,** which is another prominent site for many of the cell's biochemical and synthetic activities. Its major component is water (70–80%), which serves as a solvent for the cell pool, a complex mixture of nutrients including sugars, amino acids, and salts. The components of this pool serve as building blocks for cell synthesis or as sources of energy. The cytoplasm also contains larger, discrete cell masses such as the chromatin body, ribosomes, mesosomes, and granules.

Bacterial Chromosomes and Plasmids: The Sources of Genetic Information

The hereditary material of bacteria exists in the form of a single circular strand of DNA designated as the **bacterial chromosome.*** By definition, bacteria do not have a nucleus; that is, their DNA is not enclosed by a nuclear membrane but instead is aggregated in a dense area of the cell called the **nucleoid.*** The chromosome is actually an extremely long molecule of DNA that is tightly coiled around special basic protein molecules so as to fit inside the cell compartment (see Microbits 9.2). Arranged along its length are genetic units (genes) that carry information required for bacterial maintenance and growth. When exposed to special stains or observed with an electron microscope, chromosomes have a granular or fibrous appearance (**figure 4.18**).

Although the chromosome is the minimal genetic requirement for bacterial survival, many bacteria contain other, nonessential pieces of DNA called **plasmids.*** These tiny strands can be free or integrated into the chromosome; they may be duplicated and passed on to offspring. They are not essential to bacterial growth and metabolism, but they often confer protective traits such as resisting drugs and producing toxins and enzymes (see chapter 9). Because they can be readily manipulated in the laboratory and transferred from one bacterial cell to another, plasmids are an important agent in modern genetic engineering techniques.

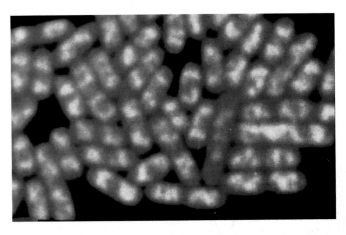

FIGURE 4.18

Chromosome structure. Fluorescent staining highlights the chromosomes of the bacterial pathogen *Salmonella enteriditis*. The cytoplasm is orange, and the chromosome fluoresces bright yellow. Some bacteria appear to have more than one chromosome because they are in the process of dividing.

Ribosomes: Sites of Protein Synthesis

A bacterial cell contains thousands of tiny, discrete units called **ribosomes.*** When viewed even by very high magnification, ribosomes show up as fine, spherical specks dispersed throughout the cytoplasm that often occur in chains (polysomes). They are also attached to the cell membrane. Chemically, a ribosome is a combination of a special type of RNA called ribosomal RNA, or rRNA (about 60%), and protein (40%). One method of characterizing ribosomes is by S, or Svedberg,[2] units, which rate the molecular sizes of various cell parts that have been spun down and separated by molecular weight and shape in a centrifuge. Heavier, more compact structures sediment faster and are assigned a higher S rating. Combining this method of analysis with high-resolution electron micrography has revealed that the procaryotic ribosome, which has an overall rating of 70S, is actually composed of two smaller subunits. The 30S unit looks something like a heart; the 50S unit bears a resemblance to a crown (**figure 4.19**). They fit together to form a miniature platform upon which protein synthesis is performed. We examine the more detailed functions of ribosomes in chapter 9.

Inclusions, or Granules: Storage Bodies

Most bacteria are exposed to severe shifts in the availability of food. During periods of nutrient abundance, they compensate by laying down nutrients intracellularly in **inclusion bodies,** or **inclusions,*** of varying size, number, and content. As the environmental source of these nutrients becomes depleted, the bacterial cell can mobilize its own storehouse as required. Some inclusion bodies enclose condensed, energy-rich organic substances, including glycogen and poly β-hydroxybutyrate (PHB), within special single-layered membranes (**figure 4.20**). A unique type of inclusion found in some aquatic bacteria are gas vesicles that provide buoyancy and

---

***chromosome** (kroh′-mah-som) Gr. *chromato,* color, and *soma,* body.

***nucleoid** (noo′-klee-oid) L. *nucis,* nut, and *oid,* form or like.

***plasmid** (plaz′-mid) Gr. *plasma,* a form, and *id,* belonging to.

---

2. Named in honor of T. Svedberg, the Swedish chemist who developed the ultracentrifuge in 1926.

***ribosome** (reye′-boh-sohm) Gr. *ribose,* a pentose sugar, and *soma,* body.

***inclusion** (in-kloo′-zhun). Any substance held in the cell cytoplasm in an insoluble state.

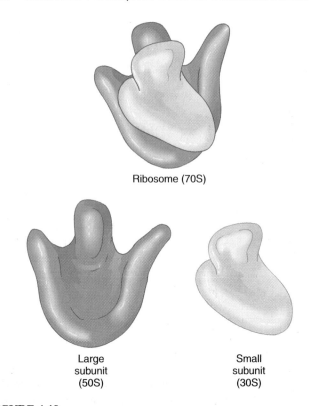

Ribosome (70S)

Large
subunit
(50S)

Small
subunit
(30S)

FIGURE 4.19

A model of a procaryotic ribosome, showing the small (30S) and large (50S) subunits, both separate and joined.

flotation. Other inclusions, also called granules, contain crystals of inorganic compounds and are not enclosed by membranes. Sulfur granules of photosynthetic bacteria (see figure 4.34*b*) and polyphosphate granules of *Corynebacterium* (see figure 4.22) and *Mycobacterium* are of this type. The latter represent an important source of building blocks for nucleic acid and ATP synthesis. They have been termed **metachromatic* granules** because they stain a contrasting color (red, purple) in the presence of methylene blue dye.

Perhaps the most unique cell granule is not involved in cell nutrition but rather in cell orientation. Magnetotactic bacteria contain crystalline particles of iron oxide (magnetosomes) that have magnetic properties. Evidently the bacteria use these granules to be pulled by the polar and gravitational fields into deeper habitats with a lower oxygen content.

BACTERIAL ENDOSPORES: AN EXTREMELY RESISTANT STAGE

Ample evidence indicates that the anatomy of bacteria helps them adjust rather well to adverse habitats. But of all microbial structures, nothing can compare to the bacterial **endospore*** (or simply spore) for withstanding hostile conditions and facilitating survival.

Endospores are dormant bodies produced by the gram-positive genera *Bacillus, Clostridium,* and *Sporosarcina.* These bacteria have a two-phase life cycle—a vegetative cell and an endospore **(figure 4.21*a*).** The vegetative cell is a metabolically active and

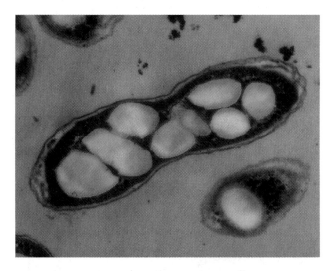

FIGURE 4.20

An example of a storage inclusion in a bacterial cell (32,500×). Substances such as polyhydroxybutyrate can be stored in an insoluble, concentrated form that provides an ample, long-term supply of that nutrient.

growing entity that can be induced by environmental conditions to undergo spore formation, or **sporulation.** Once formed, the spore exists in an inert, resting condition that shows up prominently in a spore or Gram stain (figure 4.21*b*). The chapter-opening photo also shows stages in this cycle, stained with differential fluorescent dyes. Features of spores, including size, shape, and position in the vegetative cell, are somewhat useful in identifying some species.

Endospore Formation and Resistance

The depletion of nutrients, especially an adequate carbon or nitrogen source, is the stimulus for a vegetative cell to begin spore formation. Once this stimulus has been received by the vegetative cell, it undergoes a conversion to a committed sporulating cell called a **sporangium.*** Complete transformation of a vegetative cell into a sporangium and then into a spore requires 6 to 8 hours in most spore-forming species. **Table 4.2** illustrates some major physical and chemical events in this process. Bacterial endospores are the hardiest of all life forms, capable of withstanding extremes in heat, drying, freezing, radiation, and chemicals that would readily kill vegetative cells. Their survival under such harsh conditions is due

A Note on Terminology

The word spore can have more than one usage in microbiology. It is a generic term that refers to any tiny compact cells that are produced by vegetative or reproductive structures of microorganisms. Spores can be quite variable in origin, form, and function. The bacterial type discussed here is called an endospore, because it is produced inside a cell. It functions in *survival,* not in reproduction, because no increase in cell numbers is involved in its formation. In contrast, the fungi produce many different types of spores for both survival and reproduction (see chapter 5).

---

*metachromatic (met″-uh-kroh-mah′-tik) Gr. *meta,* other, and *chromo,* color.

*endospore (en′-doh-spor) Gr. *endo,* inside, and *sporos,* seed.

*sporangium (spor-anj′-yum) L. *sporos,* and Gr. *angeion,* vessel.

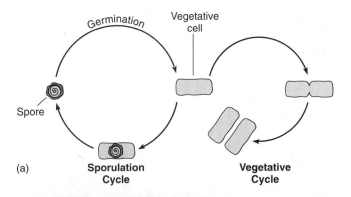

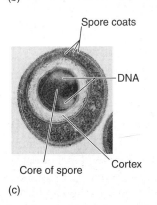

Sporeformers can exist in an active, growing, vegetative state, or they can enter a dormant survival state through sporulation. Their actual fate depends upon the availability of nutrients.

(b)

A fluorescent stain of *Bacillus subtilis* is designed to show stages in its cycle. The enlarged red/blue cell (3,000×) is a vegetative cell in the early stages of sporulation. Smaller (1,000×) cells with orange and green terminal spots highlight later stages in sporulation.

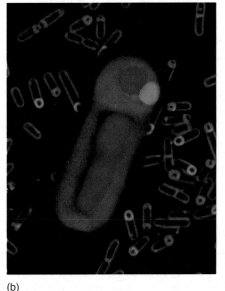

(c)

High magnification (10,000×) cross section of a single spore reveals the dense protective layers that surround the central core.

FIGURE 4.21
The general life cycle of a spore-forming bacterium.

to several factors. The heat resistance of spores has been linked to their high content of calcium and *dipicolinic acid,* although the exact role of these chemicals is not yet clear. We know, for instance, that heat destroys cells by inactivating proteins and DNA and that this process requires a certain amount of water in the protoplasm. Because the deposition of calcium dipicolinate in the spore removes water and leaves the spore very dehydrated, it is less vulnerable to the effects of heat. It is also metabolically inactive and highly resistant to damage from further drying. The thick, impervious cortex and spore coats also protect against radiation and chemicals (figure 4.21c). The longevity of bacterial spores verges on immortality. One record describes the isolation of viable spores from a fossilized bee that was 25 million years old. More recently, microbiologists unearthed a viable spore from a 250-million-year-old salt crystal! Initial analysis of this ancient microbe indicates it is a species of *Bacillus* that is genetically different from known species.

The Germination of Endospores

After lying in a state of inactivity for an indefinite time, spores can be revitalized when favorable conditions arise. The breaking of dormancy, or germination, happens in the presence of water and a specific chemical or environmental stimulus (germination agent). Once initiated, it proceeds to completion quite rapidly ($1\frac{1}{2}$ hours). Although the specific germination agent varies among species, it is generally a small organic molecule such as an amino acid or an inorganic salt. This agent stimulates the formation of hydrolytic (digestive) enzymes by the spore membranes. These enzymes digest the cortex and expose the core to water. As the core rehydrates and takes up nutrients, it begins to grow out of the spore coats. In time, it reverts to a fully active vegetative cell, resuming the vegetative cycle.

Medical Significance of Bacterial Spores

Although the majority of spore-forming bacteria are relatively harmless, several bacterial pathogens are sporeformers. In fact, some aspects of the diseases they cause are related to the persistence and resistance of their spores. *Bacillus anthracis* is the agent of anthrax, a skin and lung infection of domestic animals transmissible to humans. The genus *Clostridium* includes even more pathogens, including *C. tetani,* the cause of tetanus (lockjaw), and *C. perfringens,* the cause of gas gangrene. When the spores of these species are embedded in a wound that contains dead tissue, they can germinate, grow, and release potent toxins. Another toxin-forming species, *C. botulinum,* is the agent of botulism, a deadly form of food poisoning.

Because they inhabit the soil and dust, spores are a constant intruder where sterility and cleanliness are important. They resist ordinary cleaning methods that use boiling water, soaps, and disinfectants, and they frequently contaminate cultures and media. Hospitals and clinics must take precautions to guard against the potential harmful effects of spores in wounds. Spore destruction is a particular concern of the food-canning industry. Several endospore-forming species cause food spoilage or poisoning. Ordinary boiling (100°C) will usually not destroy such spores, so canning is carried out in pressurized steam at 120°C for 20 to 30 minutes. Such rigorous conditions will ensure that the food is sterile and free from viable bacteria.

TABLE 4.2

General Stages in Endospore Formation

Stage	State of Cell	Process/Event
1	Vegetative cell	Cell in early stage of binary fission doubles chromosome.
2	Vegetative cell becomes **sporangium** in preparation for sporulation	One chromosome and a small bit of cytoplasm are walled off as a protoplast at one end of the cell. This core contains the minimum structures and chemicals necessary for guiding life processes. During this time, the sporangium remains active in synthesizing compounds required for spore formation.
3	Sporangium	The protoplast is engulfed by the sporangium to continue the formation of various protective layers around it.
4	Sporangium with prospore	Special peptidoglycan is laid down to form a cortex around the spore protoplast, now called the prospore; calcium and dipicolinic acid are deposited; core becomes dehydrated and metabolically inactive.
5	Sporangium with prospore	Three heavy and impervious protein spore coats are added.
6	Mature endospore	Endospore becomes thicker, and heat resistance is complete; sporangium is no longer functional and begins to deteriorate.
7	Free spore	Complete lysis of sporangium frees spore; it can remain dormant yet viable for thousands of years.
8	Germination	Addition of nutrients and water reverses the dormancy. The spore then swells and liberates a young vegetative cell.
9	Vegetative cell	Restored vegetative cell.

CHAPTER CHECKPOINTS

- The cytoplasm of bacterial cells serves as a solvent for materials (the cell pool) used in all cell functions.

- The genetic material of bacteria is DNA. Genes are arranged in a single circular chromosome. Additional genes are carried on plasmids.

- Bacterial ribosomes are dispersed in the cytoplasm in chains (polysomes) and are also embedded in the cell membrane.

- Bacteria store nutrients in their cytoplasm in structures called inclusions. Inclusions vary in structure and the materials that are stored.

- A few families of bacteria produce dormant bodies called endospores, which are the hardiest of all life forms, surviving for centuries.

- The genera *Bacillus* and *Clostridium* are sporeformers, and both contain deadly pathogens.

Bacterial Shapes, Arrangements, and Sizes

For the most part, bacteria function as independent single-celled, or unicellular, organisms. Although it is true that an individual bacterial cell can live attached to others in colonies or other such groupings, each one is fully capable of carrying out all necessary life activities, such as reproduction, metabolism, and nutrient processing (unlike the more specialized cells of a multicellular organism).

Bacteria exhibit considerable variety in shape, size, and colonial arrangement. It is convenient to describe most bacteria by one of three general shapes as dictated by the configuration of the cell wall **(figure 4.22).** If the cell is spherical or ball-shaped, the bacterium is described as a **coccus.*** Cocci can be perfect spheres, but they also can exist as oval, bean-shaped, or even pointed variants. A cell that is cylindrical (longer than wide) is termed a rod, or **bacillus.*** There is also a genus named *Bacillus*. As might be expected, rods are also quite varied in their actual form. Depending on the bacterial species, they can be blocky, spindle-shaped, round-ended, long and threadlike (filamentous), or even clubbed or drumstick-shaped. When a rod is short and plump, it is called a **coccobacillus;** if it is gently curved, it is a **vibrio.*** A bacterium having the shape of a curviform or spiral-shaped cylinder is called a **spirillum,*** a rigid helix, twisted twice or more along its axis (like a corkscrew). Another spiral cell mentioned earlier in conjunction with periplasmic flagella is the **spirochete,** a more flexible form that resembles a spring. Refer to **table 4.3** for a comparison of other features of the

---

*coccus (kok′-us) pl. cocci (kok′-seye) Gr. *kokkos,* berry.

*bacillus (bah-sil′-lus) pl. bacilli (bah-sil′-eye) L. *bacill,* small staff or rod.

*vibrio (vib′-ree-oh) L. *vibrare,* to shake.

*spirillum (spy-ril′-em) pl. spirilla; L. *spira,* a coil.

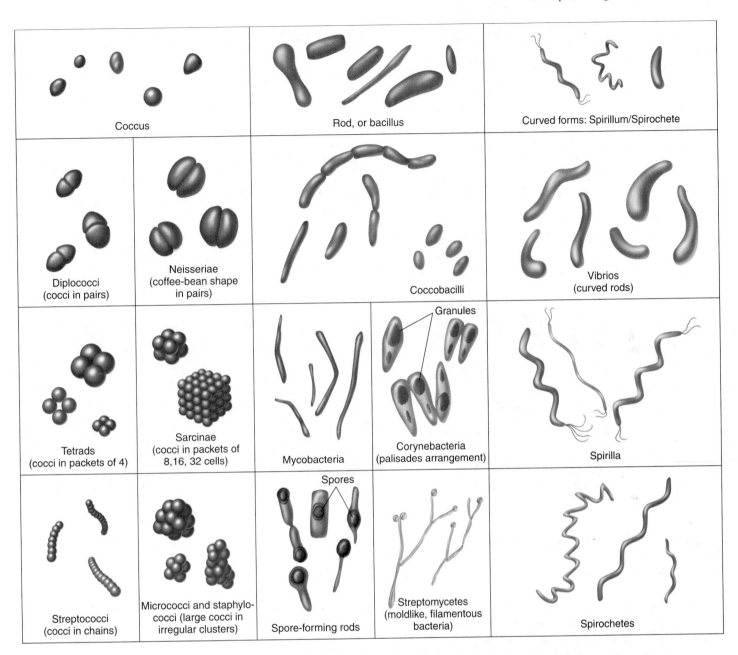

FIGURE 4.22

Bacterial shapes and arrangements.   May not be shown to exact scale.

two helical bacterial forms. Because bacterial cells look rather two-dimensional and flat with traditional staining and microscope techniques, they are seen to best advantage with a scanning electron microscope that emphasizes their striking three-dimensional forms **(figure 4.23).**

It is rather common for cells of the same species to vary to some extent in shape and size. This phenomenon, called **pleomorphism*** **(figure 4.24a),** is due to individual variations in cell wall structure caused by nutritional or slight hereditary differences. For example, although the cells of *Corynebacterium diphtheriae* are generally considered rod-shaped, in culture they display variations such as club-shaped, swollen, curved, filamentous, and coccoid.

Pleomorphism reaches an extreme in the mycoplasmas, which entirely lack cell walls and thus display extreme variations in shape (see figure 4.32).

The cells of bacteria can also be categorized according to arrangement, or style of grouping (see figure 4.22). The main factors influencing the arrangement of a particular cell type are its pattern of division and how the cells remain attached afterward. The greatest variety in arrangement occurs in cocci, which can be single, in pairs **(diplococci),*** in **tetrads** (groups of four), in irregular clusters (both **staphylococci*** and **micrococci),*** or in chains

---

***pleomorphism** (plee″-oh-mor′-fizm) Gr. *pleon,* more, and *morph,* form or shape.

***diplococci**; Gr. *diplo,* double.

***staphylococci** (staf″-ih-loh-kok′-seye) Gr. *staphyle,* a bunch of grapes.

***micrococci** Gr. *mikros,* small.

TABLE 4.3

Comparison of the Two Spiral-Shaped Bacteria

	Spirilla	Spirochetes		Spirilla	Spirochetes
Overall appearance	Rigid helix	Flexible helix	**Number of helical turns**	Varies from 1 to 20	Varies from 3 to 70
Mode of locomotion	Polar flagella; cells swim by rotating around like corkscrews; do not flex	Periplasmic flagella within sheath; cells flex; can swim by rotation or by creeping on surfaces	**Gram reaction** (cell wall type)	Gram negative	Gram negative
	1 to several flagella; can be in tufts	2 to 100 periplasmic flagella	**Examples of important types**	Most are harmless saprobes; one species, *Spirillum minor,* causes rat bite fever	*Treponema pallidum* cause of syphilis; *Borrelia* and *Leptospira,* important pathogens

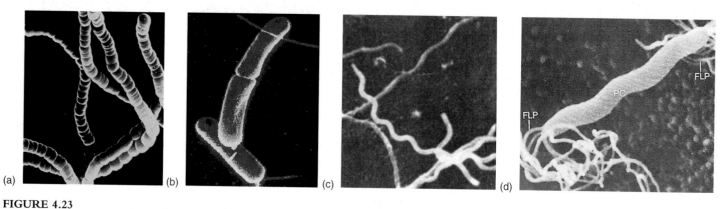

FIGURE 4.23

SEM photographs of basic bacterial shapes reveals their three dimensions and surface features. **(a)** Cocci in chains. **(b)** A rod-shaped bacterium (*Escherichia coli*) in a diplobacillus arrangement. **(c)** A spirochete (*Borrelia burgdorferi,* the cause of Lyme disease) is a long, thin cell with irregular coils and no external flagella. **(d)** A spirillum is thicker with a few even coils (PC) and external flagella (FLP). Can you tell what the flagellar arrangement is?

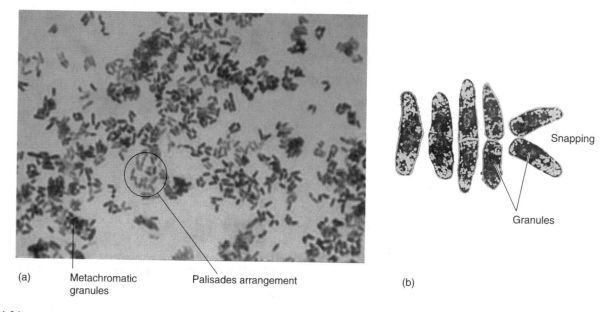

FIGURE 4.24

Pleomorphism in *Corynebacterium.* **(a)** Cells occur in a great variety of shapes and sizes (800×). This genus typically exhibits an unusual formation called a palisades arrangement that is caused by snapping, shown in **(b).** Close examination will also reveal darkly stained granules inside the cells. (3,000×).

of a few to hundreds of cells (**streptococci**).* An even more complex grouping is a cubical packet of eight, sixteen, or more cells called a **sarcina** (sar´-sih-nah). These different coccal groupings are the result of the division of a coccus in a single plane, in two perpendicular planes, or in several intersecting planes; after division, the resultant daughter cells remain attached (**figure 4.25**).

Bacilli are less varied in arrangement because they divide only in the transverse plane (perpendicular to the axis). They occur either as single cells, as a pair of cells with their ends attached (diplobacilli), or as a chain of several cells (streptobacilli). A **palisades*** arrangement, typical of the corynebacteria, is formed when the cells of a chain remain partially attached by a small hinge region at the ends. The cells tend to fold (snap) back upon each other, forming a row of cells oriented side by side (see figure 4.24b). The reaction can be compared to the behavior of boxcars on a jackknifed train, and the result looks su-

---

*streptococci (strep″-toh-kok´-seye) Gr. *streptos,* twisted.

*palisades (pal´-ih-saydz) L. *pale,* a stake. A fence made of a row of stakes.

perficially like an irregular picket fence. Spirilla are occasionally found in short chains, but spirochetes rarely remain attached after division. Comparative sizes of typical cells are presented in **figure 4.26.**

CHAPTER CHECKPOINTS

Most bacteria have one of three general shapes: coccus (round), bacillus (rod), or spiral, based on the configuration of the cell wall. Two types of spiral cells are spirochetes and spirilla.

Shape and arrangement of cells are key means of describing bacteria. Arrangements of cells are based on the number of planes in which a given species divides.

Cocci can divide in many planes to form pairs, chains, packets, or clusters. Bacilli divide only in the transverse plane. If they remain attached, they form chains or palisades.

Bacteria range in size from the smallest rickettsias to the largest spiral forms.

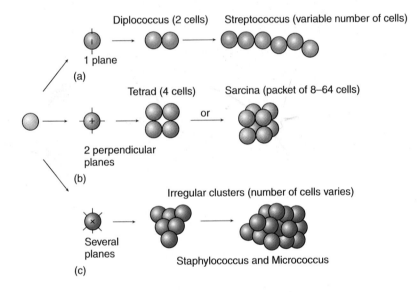

FIGURE 4.25

Arrangement of cocci resulting from different planes of cell division. **(a)** Division in one plane produces diplococci and streptococci. **(b)** Division in two planes at right angles produces tetrads and packets. **(c)** Division in several planes produces irregular clusters.

Viruses	Bacterial Cells		Other Cells

Rickettsias

Coccus

Bacillus

Spirillum

Spirochete

Human red blood cell

Ameba

| 0 | 5 | 10 | 15 | 20 | 25 | 30 | 35 | 40 | 45 | 50 | 55 | 60 | 65 | 70 | 75 | µm |

FIGURE 4.26

The dimensions of bacteria. The sizes of bacteria range from those just barely visible with light microscopy (0.2 μm) to those measuring a thousand times that size. Cocci measure anywhere from 0.5 to 3.0 μm in diameter; bacilli range from 0.2 to 2.0 μm in diameter and from 0.5 to 20 μm in length; vibrios and spirilla vary from 0.2 to 2.0 μm in diameter and from 0.5 to 100 μm in length. Spirochetes range from 0.1 to 3.0 μm in diameter and from 0.5 to 250 μm in length. Note the range of sizes as compared with eucaryotic cells and viruses. Comparisons are given as average sizes.

Bacterial Identification and Classification Systems

Every speck of soil, dab of saliva, and droplet of pond water is teeming with a rich variety of bacteria. Although study of such mixed populations can be useful, in most laboratory studies, one must prepare pure cultures of the individual members of the population (see chapter 3). This is especially critical in medical labs, which are charged with gathering the characteristics of an infectious agent and identifying it so that proper treatment can be given. Isolation and laboratory growth are also necessary for discovery and verification of new species.

The methods that a microbiologist uses to identify bacteria to the level of genus and species fall into the main categories of morphology (microscopic and macroscopic), bacterial physiology or biochemistry, serological analysis, and genetic techniques. Data from a cross section of such tests can produce a unique profile of each bacterium. Final differentiation of any unknown species is accomplished by comparing its profile with the characteristics of known bacteria in tables, charts, and keys (see figure 20.8). Many of the identification systems are automated and incorporate computers to process data and provide a "best fit" identification. However, not all methods are used on all bacteria. A few bacteria can be identified by placing them in an automated machine that analyzes only the kind of fatty acids they contain; in contrast, some are identifiable by a Gram stain and a few physiological tests; others may require a diverse spectrum of morphological, biochemical, and genetic tests. The following list summarizes some of the general areas of bacteriological testing (see also figure 3.1).

METHODS USED IN BACTERIAL IDENTIFICATION

Microscopic Morphology   Traits that can be valuable aids to identification are combinations of cell shape and size, Gram stain reaction, acid-fast reaction, and special structures, including endospores, granules, and capsules. Electron microscope studies can pinpoint additional structural features (such as the cell wall, flagella, pili, and fimbriae).

Macroscopic Morphology   Appearance of colonies, including texture, size, shape, pigment, speed of growth and patterns of growth in broth and gelatin media.

Physiological/Biochemical Characteristics   These have been the traditional mainstay of bacterial identification. Enzymes and other biochemical properties of bacteria are fairly reliable and stable expressions of the chemical identity of each species. Dozens of diagnostic tests exist for determining the presence of specific enzymes and to assess nutritional and metabolic activities. Examples include tests for fermentation of sugars; capacity to digest or metabolize complex polymers such as proteins and polysaccharides; production of gas; presence of enzymes such as catalase, oxidase, and decarboxylases; and sensitivity to antimicrobic drugs. Special rapid identification test systems that record the major biochemical reactions of a culture have streamlined data collection (**figure 4.27**).

Chemical Analysis   Analyzing the types of specific structural substances that the bacterium contains, such as the chemical composition of peptides in the cell wall and lipids in membranes.

Serological Analysis   Bacteria have surface and other molecules called antigens that are recognized by the immune system. One immune response to antigens is the production of molecules called antibodies that are designed to bind tightly to the antigens. This response is so specific that antibodies can be used as a means of identifying bacteria in specimens and cultures. Laboratory kits based on this technique are available for immediate identification of a number of pathogens.

Genetic and Molecular Analysis   Examining the genetic material itself has revolutionized the identification and classification of bacteria.

> **G + C base composition.** The overall percentage of guanine and cytosine (the G + C content as compared with A + T content) in DNA is a general indicator of relatedness because it is a trait that does not change rapidly. Bacteria with a significant difference in G + C percentage are likely to be genetically distinct species or genera. For

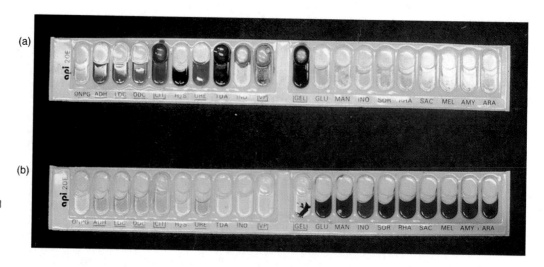

FIGURE 4.27

Rapid tests.   The API 20E manual biochemical system for microbial identification. **(a)** Positive and **(b)** negative results.

example, although superficially similar in Gram reaction, shape, and other morphological characteristics, *Escherichia* has a G + C base composition of 48–52% and *Pseudomonas* has a composition of 58–70%, indicating that they probably are not closely related. This technique is most applicable for clarifying the taxonomic position of a bacterium, but it is too nonspecific to be applicable as a precise identification tool.

DNA analysis using genetic probes. The exact order and arrangement of the DNA code is unique to each organism (see figure 2.26). With a technique called *hybridization*, it is possible to identify a bacterial species by analyzing segments of its DNA. This requires small fragments of single-stranded DNA (or RNA) called **probes** that are known to be complementary to the specific sequences of DNA from a particular microbe (see figure 10.3, page 292). The test is conducted by extracting unknown test DNA from cells in specimens or cultures and binding it to special blotter paper. After several different probes have been added to the blotter, it is observed for visible signs that the probes have become fixed (hybridized) to the test DNA. The binding of probes onto several areas of the test DNA indicates close correspondence and makes positive identification possible.

Nucleic acid sequencing and rRNA analysis. One of the most viable indicators of evolutionary relatedness and affiliation is comparison of the sequence of nitrogen bases in ribosomal RNA, a major component of ribosomes (**figure 4.28**). Ribosomes have the same function (protein synthesis) in all cells, and they tend to remain more or less stable in their nucleic acid content over long periods. Thus, any major differences in the sequence, or "signature," of the rRNA is likely to indicate some distance in ancestry. This technique is powerful at two levels: It is effective for differentiating general group differences (it was used to separate the three superkingdoms of life discussed in chapter 1), and it can be fine-tuned to identify at the species level (for example in *Mycobacterium* and *Legionella*). Elements of these and other identification methods are presented in more detail in chapters 10, 16, 20, and 21.

CLASSIFICATION SYSTEMS IN THE PROCARYOTAE

Classification systems serve both practical and academic purposes. They aid in differentiating and identifying unknown species in medical and applied microbiology. They are also useful in organizing bacteria and as a means of studying their relationships and origins. Since the classification was started around 200 years ago, several thousand species of bacteria and archaea have been identified, named, and catalogued.

For years there has been intense interest in tracing the origins of and evolutionary relationships among bacteria, but doing so has not been an easy task. As a rule, tiny, relatively soft organisms do not form fossils very readily. Several times since the 1960s, however, scientists have discovered microscopic fossils of procaryotes that look very much like modern bacteria. Some of the rocks that contain these fossils have been dated back billions of years (**figure 4.29**). One of the questions that has plagued taxonomists is, What characteristics are the most indicative of closeness in ancestry? Early bacteriologists found it convenient to classify bacteria

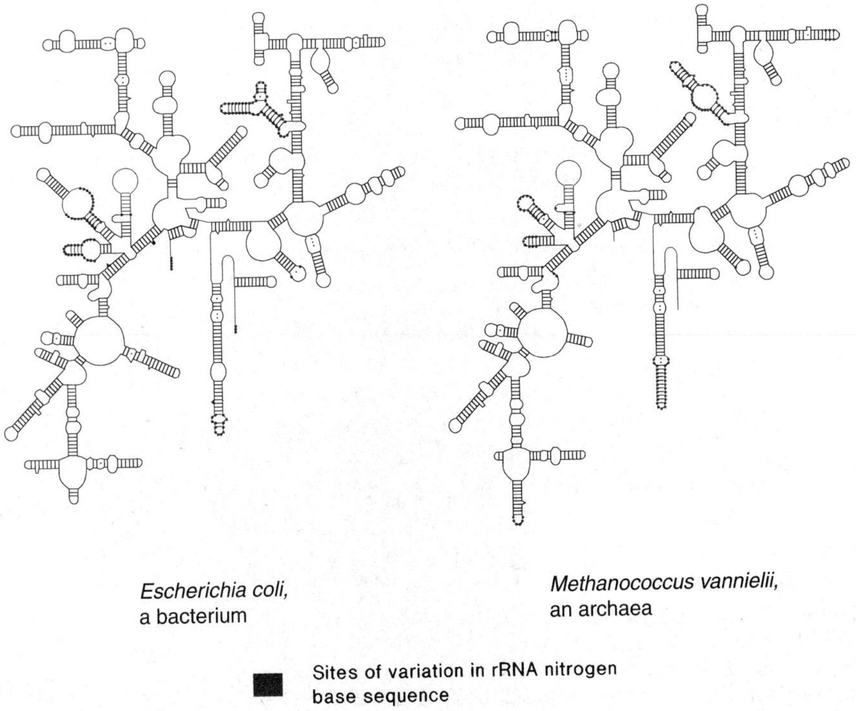

FIGURE 4.28

A modern molecular technique that identifies the cell type or subtype by analyzing the nitrogen base sequence of its ribosomal RNA. The segments of rRNA shown here are from the small subunit of *Escherichia coli*, a common intestinal bacterium, and *Methanococcus vannielii*, an archaea that inhabits the deep sediments of marshes. The shaded regions indicate areas of rRNA that differ significantly between bacteria (*Escherichia*) and archaea (*Methanococcus*).

Escherichia coli, a bacterium

Methanococcus vannielii, an archaea

■ Sites of variation in rRNA nitrogen base sequence

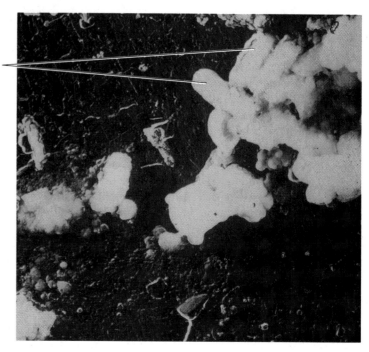

Rod-shaped cells

FIGURE 4.29

Ancient procaryotes.   Electron micrograph of
fossilized rod-shaped bacterial cells preserved in
rock sediment. This specimen is more than 2
billion years old (24,000×).

according to shape, variations in arrangement, growth characteristics, and habitat. However, as more species were discovered and as techniques for studying their biochemistry were developed, it soon became clear that similarities in cell shape, arrangement, and staining reactions do not automatically indicate relatedness. Even though the gram-negative rods look alike, there are hundreds of different species, with highly significant differences in biochemistry and genetics. If we attempted to classify them on the basis of Gram stain and shape alone, we could not assign them to a more specific level than class. Increasingly, classification schemes are turning to genetic and molecular traits that cannot be visualized under a microscope or in culture.

The most functional classification schemes make use of current knowledge to show natural relationships. Not only must they be flexible enough to add newly discovered species, but they must also complement a number of microbiology disciplines. In general, schemes for organizing bacteria can be phylogenetic, based on evolutionary relationships, or **phenetic,*** based on their morphology or biochemistry. One system of classification has not been permanent or universally accepted to date; indeed, most systems are in a state of flux as new information and methods of analysis become available. Of the current proposed systems, we present three:

1. the overall scheme of classification from the ninth edition of *Bergey's Manual of Systematic Bacteriology,* a manual of bacterial descriptions and classification published continuously since 1923 **(table 4.4);**
2. a method that groups bacteria by comparing rRNA sequence; and
3. a practical system that uses a few morphological and physiological traits to categorize the major, medically important bacterial families (see table 4.6).

*phenetic (fuh-neh'-tik) Referring to the phenotype or expression of traits.

Taxonomic Scheme

The ninth edition of *Bergey's Manual* organizes the Kingdom Procaryotae into four major divisions. These somewhat natural divisions are based upon the nature of the cell wall. The **Gracilicutes*** have gram-negative cell walls and thus are thin-skinned;

*Gracilicutes (gras"-ih-lik'-yoo-teez) L. *gracilus,* thin, and *cutis,* skin.

TABLE 4.4

Major Taxonomic Groups of Bacteria per *Bergey's Manual*

Division I. Gracilicutes: Gram-Negative Bacteria
Class I.   Scotobacteria: Gram-negative non-photosynthetic bacteria (examples in table 4.5)
Class II.   Anoxyphotobacteria: Gram-negative photosynthetic bacteria that do not produce oxygen (purple and green bacteria)
Class III. Oxyphotobacteria: Gram-negative photosynthetic bacteria that evolve oxygen (cyanobacteria)

Division II. Firmicutes: Gram-Positive Bacteria
Class I.   Firmibacteria: Gram-positive rods or cocci (examples in table 4.6)
Class II.   Thallobacteria: Gram-positive branching cells (the actinomycetes)

Division III. Tenericutes
Class I.   Mollicutes: Bacteria lacking a cell wall (the mycoplasmas)

Division IV. Mendosicutes
Class I.   Archaebacteria: Bacteria with atypical compounds in the cell wall and membranes

Source: Data from Bergey's Manual of Systematic Bacteriology, *9th ed. Williams & Wilkins Company, Baltimore, 1984.*

TABLE 4.5

Bergey's Manual Taxonomic Rankings
for *Borrelia Burgdorferi*

Taxonomic Rank	Includes
Kingdom: Monera	All bacteria
Division: Gracilicutes	All bacteria with gram-negative cell wall
Class: Scotobacteria	Non-photosynthetic gram-negative bacteria
Order: Spirochaetales	Helical, flexible, motile bacteria (spirochetes)
Family: Spirochaetaceae	Helical, motile bacteria lacking hooked ends
Genus: *Borrelia*	Tiny, loose, irregularly coiled spirochetes
Species: *burgdorferi*	Causative agent of Lyme disease

the **Firmicutes*** have gram-positive cell walls that are thick and strong; the **Tenericutes*** lack a cell wall and thus are soft; and the **Mendosicutes*** are the archaea (also called archaebacteria), primitive procaryotes with unusual cell walls and nutritional habits. The first two divisions contain the greatest number of species. The 200 or so species that cause human and animal diseases can be found in four classes: the Scotobacteria, Firmibacteria, Thallobacteria, and Mollicutes. The system used in *Bergey's Manual* further organizes bacteria into subcategories such as classes, orders, and families, but these are not available for all groups. An example of the entire classification of one bacterial species is shown in **table 4.5.**

Ribosomal RNA Scheme

The classification scheme developed by analyzing the similarities in base sequence of rRNA revealed 12 distinct branches on the bacterial "tree" (**figure 4.30**). It just so happens that some groupings are similar to those used in Bergey's system (gram-positive bacteria), but others are genetically different enough to be placed into their own separate categories (spirochetes).

1. Gram-positive eubacteria: Selected representatives are *Bacillus, Clostridium, Mycobacterium, Staphylococcus, Actinomyces,* and the cell-wall-free mycoplasmas.
2. Gram-negative eubacteria (Proteobacteria) includes purple photosynthetic bacteria (*Chromatium*) and non-photosynthetic relatives represented by *Pseudomonas, Vibrio, Neisseria,* and the rickettsias.
3. Cyanobacteria: photosynthetic bacteria with chlorophyll *a* that evolve (give off) oxygen; includes *Oscillatoria* and *Spirulina.*
4. Spirochetes: flexible helical cells with periplasmic flagella such as *Treponema* and *Borrelia.*
5. Walled, budding bacteria that lack peptidoglycan in their cell walls: includes *Planctomyces.*

6. The *Bacteroides, Flavobacterium, Fusobacterium,* and *Cytophaga:* a mixed group morphologically and physiologically.
7. Chlamydias: unusual obligate parasites of vertebrates; lack ability to complete metabolism independently; lack peptidoglycan; one genus—*Chlamydia.*
8. Green sulfur bacteria: anaerobic bacteria that contain bacteriochlorophyll and use sulfur in metabolism; do not give off oxygen during photosynthesis; includes *Chlorobium.*
9. Green nonsulfur bacteria: filamentous, gliding, thermophilic, photosynthetic bacteria that contain bacteriochlorophyll, do not evolve oxygen; includes *Chloroflexus.*
10. Unique bacteria with extreme resistance to electromagnetic radiation: *Deinococcus,* gram-positive cocci, and *Thermus,* thermophilic rods.
11. Unusual thermophilic bacteria inhabiting hot oceanic vents: *Thermotoga.*
12. Primitive bacteria in the genera *Aquiflex* and *Hydrogenobacter* that inhabit very hot habitats and derive energy from hydrogen gas and sulfur compounds.

Ribosomal RNA analysis of the archaea detected three well-defined groups:

1. Representatives that synthesize methane from simple inorganic compounds (methanogens)
2. Extremely halophilic (salt-loving) archaea
3. Extremely thermophilic (heat-loving) archaea that use sulfur in their metabolism

The last section this chapter covers features of archaeal form, function, and ecology.

Diagnostic Scheme

Many medical microbiologists prefer an informal working system that outlines the major families and genera (**table 4.6**). This system is more applicable for their purposes because it is restricted to bacterial disease agents, depends less on nomenclature, and is based on readily accessible morphological and physiological tests rather than on phylogenetic relationships. It also divides the bacteria into gram positive, gram negative, and those without cell walls and then subgroups them according to cell shape, arrangement, and certain physiological traits such as oxygen usage: *Aerobic* bacteria use oxygen in metabolism; *anaerobic* bacteria do not use oxygen in metabolism; and facultative bacteria may or may not use oxygen. Further tests not listed on the table would be required to separate closely related genera and species. Many of these are included in later chapters on specific bacterial groups.

Species and Subspecies in Bacteria

Among most organisms, the species level is a distinct, readily defined, and natural taxonomic category. In animals, for instance, a species is a distinct type of organism that can produce viable offspring only when it mates with others of its own kind. This definition does not work for bacteria primarily because they do not exhibit a typical mode of sexual reproduction. They can accept genetic information from unrelated forms, and they can also alter

*Firmicutes (fer-mik′-yoo-teez) L. *firmus,* strong.

*Tenericutes (ten″-er-ik′-yoo-teez) L. *tener,* soft

*Mendosicutes (men-doh-sik′-yoo-teez) L. *mendosus,* to be false.

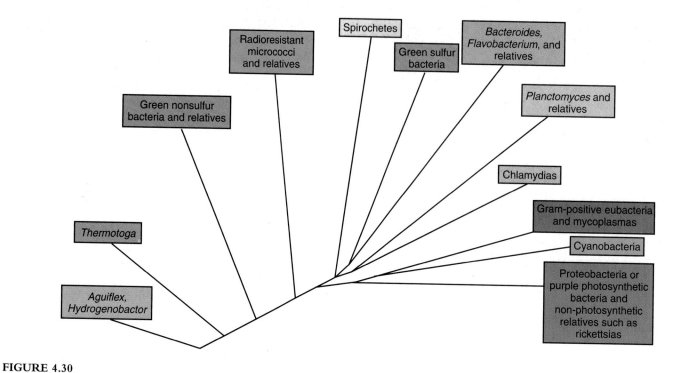

FIGURE 4.30

Separation chart for the eubacteria proposing phylogenetic relationships based on rRNA sequences. This scheme divides them into 11 genetically discrete groups. The branches suggest evolutionary origins, and branches that have greater similarities appear closer to each other.

Source: Data from C. R. Woese, Microbiol. Rev. 51(1987):221–71. American Society for Microbiology, Washington, D.C.

their genetic makeup by a variety of mechanisms. Thus, it is necessary to hedge a bit when we define a bacterial species. Theoretically, it is a collection of bacterial cells, all of which share an overall similar pattern of traits, in contrast to other groups whose pattern differs significantly. Although the boundaries that separate two closely related species in a genus are in some cases very arbitrary, this definition still serves as a method to separate the bacteria into various kinds that can be cultured and studied. As additional information on bacterial genomes is discovered, it may be possible to define species according to specific combinations of genetic codes found only in a particular isolated culture.

Since the individual members of given species can show variations, we must also define levels within species (subspecies) called **strains** and types. A strain or variety of bacteria is a culture derived from a single parent that differs in structure or metabolism from other cultures of that species (also called biovars or morphovars). For example, there are pigmented and nonpigmented strains of *Serratia marcescens* and flagellated and nonflagellated strains of *Pseudomonas fluorescens*. A type is a subspecies that can show differences in antigenic makeup (serotype or serovar), in susceptibility to bacterial viruses (phage type), and in pathogenicity (pathotype).

CHAPTER CHECKPOINTS

A bacterial species can be properly identified only if it is grown in pure culture—that is, in isolation from all other forms of life.

Key traits that are used to identify a bacterial species include (1) morphology, (2) Gram stain or other stain characteristics,

(3) presence of specialized structures, (4) macroscopic appearance of colonies, (5) biochemical reactions, and (6) nucleotide composition of both DNA and rRNA.

Bacteria are formally classified by phylogenetic relationships and phenotypic characteristics.

Medical identification of pathogens uses a more informal system of classification based on Gram stain, morphology, biochemical reactions, and metabolic requirements.

A bacterial species is loosely defined as a collection of bacterial cells that shares an overall similar pattern of traits different from other groups of bacteria.

Variant forms within a species (subspecies) include strains and types.

Survey of Procaryotic Groups with Unusual Characteristics

The bacterial world is so diverse that we cannot do complete justice to it in this introductory chapter. This variety extends into all areas of bacterial biology, including nutrition, mode of life, and behavior. Certain types of bacteria exhibit such unusual qualities that they deserve special mention. In this minisurvey, we will consider some medically important groups and some more remarkable representatives of bacteria living free in the environment that are ecologically important. Many of the bacteria mentioned here do not have the morphology typical of bacteria discussed previously, and in a few cases, they are vividly different (**Spotlight on Microbiology 4.3**).

TABLE 4.6

Medically Important Families and Genera of Bacteria, with Notes on Some Diseases*

I. Bacteria with gram-positive cell wall structure

Cocci in clusters or packets that are aerobic or facultative
Family Micrococcaceae: *Staphylococcus* (members cause boils, skin infections)

Cocci in pairs and chains that are facultative
Family Streptococcaceae: *Streptococcus* (species cause strep throat, dental caries)

Anaerobic cocci in pairs, tetrads, irregular clusters
Family Peptococcaceae: *Peptococcus, Peptostreptococcus* (involved in wound infections)

Spore-forming rods
Family Bacillaceae: *Bacillus* (anthrax), *Clostridium* (tetanus, gas gangrene, botulism)

Non-spore-forming rods
Family Lactobacillaceae: *Lactobacillus, Listeria* (milk-borne disease), *Erysipelothrix* (erysipeloid)
Family Propionibacteriaceae: *Propionibacterium* (involved in acne)

Family Corynebacteriaceae: *Corynebacterium* (diphtheria)

Family Mycobacteriaceae: *Mycobacterium* (tuberculosis, leprosy)

Family Nocardiaceae: *Nocardia* (lung abscesses)

Family Actinomycetaceae: *Actinomyces* (lumpy jaw), *Bifidobacterium*

Family Streptomycetaceae: *Streptomyces* (important source of antibiotics)

II. Bacteria with gram-negative cell wall structure

Family Neisseriaceae
Aerobic cocci
Neisseria (gonorrhea, meningitis), *Branhamella*
Aerobic coccobacilli
Moraxella, Acinetobacter
Anaerobic cocci
Family Veillonellaceae
Veillonella (dental disease)
Miscellaneous rods
Brucella (undulant fever), *Bordetella* (whooping cough), *Francisella* (tularemia)
Aerobic rods
Family Pseudomonadaceae: *Pseudomonas* (pneumonia, burn infections)
Miscellaneous: *Legionella* (Legionnaires' disease)
Facultative or anaerobic rods and vibrios
Family Enterobacteriaceae: *Escherichia, Edwardsiella, Citrobacter, Salmonella* (typhoid fever), *Shigella*
(dysentery), *Klebsiella, Enterobacter, Serratia, Proteus, Yersinia* (one species causes plague)

Family Vibronaceae: *Vibrio* (cholera, food infection), *Campylobacter, Aeromonas*

Miscellaneous genera: *Chromobacterium, Flavobacterium, Haemophilus* (meningitis), *Pasteurella,*
Cardiobacterium, Streptobacillus
Anaerobic rods
Family Bacteroidaceae: *Bacteroides, Fusobacterium* (anaerobic wound and dental infections)
Helical and curviform bacteria
Family Spirochaetaceae: *Treponema* (syphilis), *Borrelia* (Lyme disease), *Leptospira* (kidney infection)
Obligate intracellular bacteria
Family Rickettsiaceae: *Rickettsia* (Rocky Mountain spotted fever), *Coxiella* (Q fever)
Family Bartonellaceae: *Bartonella* (trench fever, cat scratch disease)
Family Chlamydiaceae: *Chlamydia* (sexually transmitted infection)

III. Bacteria with no cell walls
Family Mycoplasmataceae: *Mycoplasma* (pneumonia), *Ureaplasma* (urinary infection)

*Details of pathogens and diseases in chapters 18, 19, 20, and 21.

SPOTLIGHT ON MICROBIOLOGY 4.3
Redefining Bacterial Size

Many microbiologists believe we are still far from having a complete assessment of the bacterial world, mostly because the world is so large and bacteria are so small. This fact becomes evident in the periodic discoveries of exceptional bacteria that are reported in newspaper headlines. Among the most remarkable are giant and dwarf bacteria.

Big Bacteria Break Records

In 1985, biologists discovered a new bacterium living in the intestine of surgeonfish that at the time was a candidate for the *Guinness Book of World Records*. The large cells, named *Epulopiscium fishelsoni* ("guest at a banquet of fish"), measure around 100 μm in length, although some specimens were as large as 300 μm. This record was recently broken when marine microbiologist Heide Schultz discovered an even larger species of bacteria living in ocean sediments near the African country of Namibia. These gigantic cocci are arranged in strands that look like pearls and contain hundreds of golden sulfur granules, inspiring their name, *Thiomargarita namibia* ("sulfur pearl of Namibia") (see photo at right). The size of the individual cells ranges from 100 up to 750 μm (3/4 mm), and many are large enough to see with the naked eye. By way of comparison, if the average bacterium were the size of a mouse, *Thiomargarita* would be as large as a blue whale!

Closer study revealed that they are indeed procaryotic and have bacterial ribosomes and DNA, but that they also have some unusual adaptations to their life cycle. They live an attached existence embedded in sulfide sediments (H_2S) that are free of gaseous oxygen. They obtain energy through oxidizing these sulfides using dissolved nitrates (NO_3). Since the quantities of these substances can vary with the seasons, they must be stored in cellular depots. The sulfides are borne as granules in the cytoplasm, and the nitrates occupy a giant, liquid-filled vesicle that takes up a major proportion of cell volume. Because of their morphology and physiology, the cells can survive for up to three months without an external source of nutrients by tapping into their "storage tanks." These bacteria are found in such large numbers in the sediments that it is thought that they are essential to the ecological cycling of H_2S gas in this region, converting it to less toxic substances.

Miniature Microbes—The Smallest of the Small

At the other extreme, microbiologists are being asked to reevaluate the lower limits of bacterial size. Up until now it has been generally accepted that the smallest cells on the planet are some form of mycoplasma with dimensions of 0.2 to 0.3 μm, which is right at the limit of resolution with light microscopes. A new controversy is brewing over the discovery of tiny cells that look like dwarf bacteria but are ten times smaller than mycoplasmas and a hundred times smaller than the average bacterial cell. These minute cells have been given the name **nanobacteria** or **nanobes** (Gr. *nanos,* one-billionth).

Nanobacteria-like forms were first isolated from blood and serum samples. The tiny cells appear to grow in culture, have cell walls, and contain protein and nucleic acids, but their size range is only from 0.05 μm to 0.2 μm. Similar nanobes have been extracted by minerologists studying sandstone rock deposits in the ocean at temperatures of 100°–170°C and deeply embedded in billion-year-old minerals. The minute filaments were able to grow and are capable of depositing minerals in a test tube. Many geologists are convinced that these nanobes are real, that they are probably similar to the first microbes on earth, and that they play a strategic role in the evolution of the earth's crust. Microbiologists tend to be more skeptical. It has been postulated that the minimum cell size to contain a functioning genome and reproductive and synthetic machinery is approximately 0.14 μm. They believe that the nanobes are really just artifacts or bits of larger cells that have broken free. It is partly for this reason that most bacteriologists rejected the idea that small objects found in a Martian meteor were microbes but were more likely caused by chemical reactions. Additional studies are needed to test this curious question of nanobes, and possibly to answer some questions about the origins of life on earth and even other planets.

Thiomargarita namibia—giant cocci.

UNUSUAL FORMS OF MEDICALLY SIGNIFICANT BACTERIA

Most bacteria are free-living or parasitic forms that can metabolize and reproduce by independent means. Two groups of bacteria—the rickettsias and chlamydias—have adapted to life inside their host cells, where they are considered **obligate intracellular parasites.** These unusual bacteria are covered in greater detail in chapter 21, but here we provide brief descriptions.

Rickettsias Rickettsias[3] are distinctive, very tiny, gram-negative bacteria **(figure 4.31).** Although they have a somewhat typical bacterial morphology, they are atypical in their life cycle and other adaptations. Most are pathogens that alternate between a mammalian host and blood-sucking arthropods,[4] such as fleas,

---

3. Named for Howard Ricketts, a physician who first worked with these organisms and later lost his life to typhus.

4. An arthropod is an invertebrate with jointed legs, such as an insect, tick, or spider.

Rickettsial cells

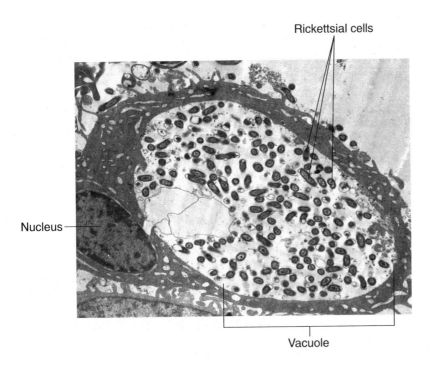

Nucleus

Vacuole

FIGURE 4.31

Transmission electron micrograph of the rickettsia *Coxiella burnetii*, the cause of Q fever. Its mass growth inside a host cell has filled a vacuole and displaced the nucleus to one side.

lice, or ticks. Rickettsias cannot survive or multiply outside a host cell and cannot carry out metabolism completely on their own, so they are closely attached to their hosts. Several important human diseases are caused by rickettsias. Among these are Rocky Mountain spotted fever, caused by *Rickettsia rickettsii* (transmitted by ticks), and epidemic typhus, caused by *Rickettsia prowazekii* (transmitted by lice). An exceptionally resistant rickettsia, *Coxiella burnetti* (the cause of Q fever), is transmitted in air and dust by arthopods.

Chlamydias   Bacteria of the genus *Chlamydia* are similar to the rickettsias in that they require host cells for growth and metabolism, but they are not closely related and are not transmitted by arthropods. Because of their tiny size and obligately parasitic lifestyle, they were at one time considered a type of virus. Later studies indicated that their structure was that of a gram-negative cell and that their mode of cell division (binary fission) was clearly procaryotic. Species that carry the greatest medical impact are *Chlamydia trachomatis,* the cause of both a severe eye infection (trachoma) that can lead to blindness and one of the most common sexually transmitted diseases; *Chlamydia psittaci,* the agent of ornithosis, or parrot fever, a disease of birds that can be transmitted to humans; and *Chlamydia pneumoniae,* an agent in lung infections (see chapter 21).

Mycoplasmas and Other Cell-Wall-Deficient Bacteria
Mycoplasmas are bacteria that naturally lack a cell wall. Although other bacteria require an intact cell wall to prevent the bursting of the cell, the mycoplasmal cell membrane is stabilized by sterols and is resistant to lysis. These extremely tiny, pleomorphic cells are considered the smallest cells, ranging from 0.1 to 0.5 μm in size. They range in shape from filamentous to coccus or doughnut-shaped. They are *not* obligate parasites and can be grown on artificial media, although added sterols are required for the cell mem-

branes of some species. Mycoplasmas are found in many habitats, including plants, soil, and animals. The most important medical species is *Mycoplasma pneumoniae* **(figure 4.32),** which adheres to the epithelial cells in the lung and causes an atypical form of pneumonia in humans.

FREE-LIVING NONPATHOGENIC BACTERIA

Photosynthetic Bacteria
The nutrition of most bacteria is heterotrophic, meaning that they derive their nutrients from other organisms. Photosynthetic bacteria, however, are independent cells that contain special light-trapping pigments and can use the energy of sunlight to synthesize all required nutrients from simple inorganic compounds. The two general types of photosynthetic bacteria are those that produce oxygen during photosynthesis and those that produce some other substance, such as sulfur granules or sulfates.

Cyanobacteria: Blue-Green Bacteria   The cyanobacteria were called blue-green algae for many years and were grouped with the eucaryotic algae. However, further study verified that they are indeed bacteria with a gram-negative cell wall (Gracilicutes) and general procaryotic structure. These bacteria range in size from 1 μm to 10 μm, and they can be unicellular or can occur in colonial or filamentous groupings **(figure 4.33*b,c*).** Some species occur in packets surrounded by a gelatinous sheath (figure 4.33*c*). A specialized adaptation of cyanobacteria are extensive internal membranes called **thylakoids,** * which contain granules of chlorophyll *a* and other photosynthetic pigments (figure 4.33*a*). They also have gas inclusions, which permit them to float on the water surface and increase their light exposure, and cysts that convert gaseous nitrogen (N_2) into a form usable by plants. This group is sometimes

---

*thylakoid (theye′-luh-koid) Gr. *thylakon,* and *eidos,* form.

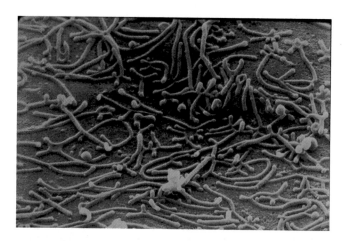

FIGURE 4.32

Scanning electron micrograph of *Mycoplasma pneumoniae* (62,000×). Cells like these that naturally lack a cell wall exhibit extreme pleomorphism.

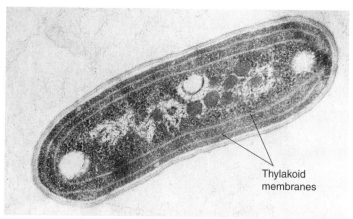

(a)

(b)

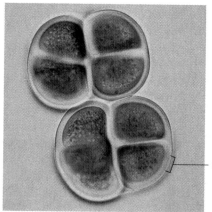

(c)

FIGURE 4.33

Structure and examples of cyanobacteria. **(a)** Electron micrograph of a cyanobacterial cell (80,000×) reveals folded stacks of membranes that contain the photosynthetic pigments and increase surface area for photosynthesis. **(b)** Two species of *Oscillatoria*, a gliding, filamentous form (100×). **(c)** *Chroococcus*, a colonial form surrounded by a gelatinous sheath (600×).

called the blue-green bacteria in reference to their content of **phycocyanin*** pigment that tints some members a shade of blue, although other members are colored yellow and orange. Some representatives glide or sway gently in the water from the action of filaments in the cell envelope that cause wavelike contractions.

Cyanobacteria are very widely distributed in nature. They grow profusely in fresh water and seawater and are thought to be responsible for periodic blooms that kill off fish by depleting the available oxygen. Some members are so pollution-resistant that they serve as biological indicators of polluted water. Cyanobacteria inhabit and flourish in hot springs (see Spotlight on Microbiology 7.1) and have even exploited a niche in dry desert soils and rock surfaces. Some types of lichens are symbiotic associations between fungi and cyanobacteria that assist in the breakdown of rock and soil formation (see chapter 26).

Green and Purple Sulfur Bacteria The green and purple bacteria are also photosynthetic and contain pigments. They differ from the cyanobacteria in having a different type of chlorophyll called *bacteriochlorophyll* and by not giving off oxygen as a product of photosynthesis. They live in sulfur springs, freshwater lakes, and swamps that are deep enough for the anaerobic conditions they require yet where their pigment can still absorb wavelengths of light **(figure 4.34*a*).** These bacteria are named for their predominant colors, but they can also develop brown, pink, purple, blue, and orange coloration. They exist as single cells of many different shapes and frequently are motile. Both groups utilize sulfur compounds (H_2S, S) in their metabolism, and some can deposit intracellular granules of sulfur or sulfates (figure 4.34*b*).

Gliding, Fruiting Bacteria The gliding bacteria are a mixed collection of gram-negative bacteria that live in water and soil. The name is derived from the tendency of members to glide over moist surfaces. The gliding property evidently involves rotation of filaments or fibers

---

*phycocyanin (feye-koh´seye´-an-in) Gr. *phykos*, seaweed, and *cyan*, blue. A bluish pigment unique to this group.

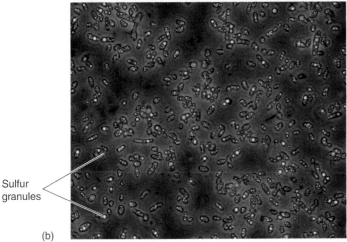

Sulfur granules

(a)

(b)

FIGURE 4.34

Behavior of purple sulfur bacteria. **(a)** Floating purple mats are huge masses of purple sulfur bacteria blooming in the Baltic Sea. Photosynthetic bacteria can have significant effects on the ecology of certain habitats. **(b)** Microscopic view of purple sulfur bacteria (*Chromatium vinosum*) carrying large, yellow sulfur granules internally.

just under the outer membrane of the cell wall. They do not have flagella. There are several morphological forms, including slender rods, long filaments, cocci, and some miniature, tree-shaped fruiting bodies. Probably the most intriguing and exceptional members of this group are the slime bacteria, or myxobacteria **(figure 4.35).** What sets the myxobacteria apart from other bacteria are the complexity and advancement of their life cycle. During this cycle, the vegetative cells swarm together and differentiate into a many-celled, colored structure called the fruiting body. The fruiting body is a survival structure that makes spores by a method very similar to that of certain fungi. These fruiting structures are often large enough to be seen with the unaided eye on tree bark and plant debris.

Appendaged Bacteria

The appendaged bacteria are quite varied in their structure and life cycles, but all of them produce an extended process of the cell wall in the form of a bud, a stalk, or a long thread **(figure 4.36).** The stalked bacteria live attached to the surface of objects in aquatic en-

vironments. One type can even grow in distilled water or tap water. The stalks evidently help them trap minute amounts of organic materials present in the water. Budding bacteria reproduce entirely by budding; that is, they form a tiny bulb (bud) at the end of a thread. The bud then breaks off, enlarges, develops a flagellum, and swarms to another area to start its own cycle. These bacteria can also grow in very low-nutrient habitats.

ARCHAEA: THE OTHER PROCARYOTES

The discovery and characterization of novel procaryotic cells that have unusual anatomy, physiology, and genetics changed our views of microbial taxonomy and classification (see chapter 1). These single-celled, simple organisms, called **archaea*** are now considered a third cell type in a separate superkingdom (the Domain Archaea). We include them in this chapter because they are procaryotic in general structure and they do share many bacterial characteristics. Evidence is accumulating that they are actually more closely related to Domain Eukarya than to bacteria. For example, archaea and eucaryotes share a number of ribosomal RNA sequences that are not found in bacteria, and their protein synthesis and ribosomal subunit structures are similar. **Table 4.7** outlines selected points of comparison of the three domains.

Among the ways that the archaea differ significantly from other cell types are that certain genetic sequences (CACACACCG) are found only in their rRNA (see figure 4.28), and that they have unique membrane lipids and cell wall construction. It is clear that the archaea are the most primitive of all life forms and are most closely related to the first cells that originated on the earth 4 billion years ago. The early earth is thought to have contained a hot, anaerobic "soup" with sulfuric gases and salts in abundance. The modern archaea still live in the remaining habitats on the earth that have these same ancient conditions—the most extreme habitats in nature. It is for this reason that they are often called extremophiles, meaning that they "love" extreme conditions in the environment.

Metabolically, the archaea exhibit nearly incredible adaptations to what would be deadly conditions for other organisms. These hardy microbes have adapted to multiple combinations of heat, salt, acid, pH, pressure, and atmosphere. Included in this group are methane producers, hyperthermophiles, extreme halophiles, and sulfur reducers.

Members of the group called *methanogens* can convert CO_2 and H_2 into methane gas (CH_4) through unusual and complex pathways. These archaea are common inhabitants of anaerobic swamp mud, the bottom sediments of lakes and oceans, and even the digestive systems of animals. The gas they produce collects in swamps and may become a source of fuel. Methane may also contribute to the "greenhouse effect," which maintains the earth's temperature and can contribute to global warming (see chapter 26).

Other types of archaea—the extreme halophiles—require salt to grow and may have such a high salt tolerance that they can multiply in sodium chloride solutions (36% NaCl) that would destroy most cells. They exist in the saltiest places on the earth—inland seas, salt lakes, salt mines, and salted fish. They are not

---

***archaea** (ark′-ee-uh) Gr. *arche,* beginning.

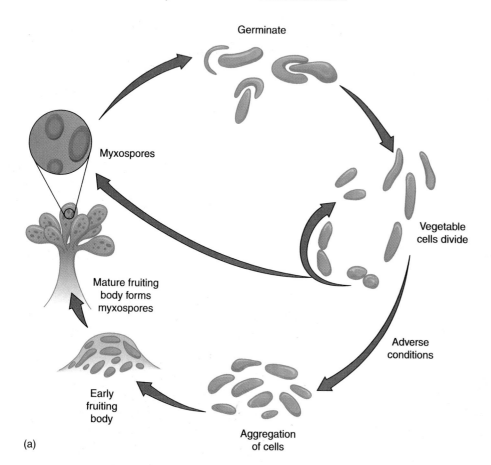

(b)

(a)

FIGURE 4.35

Myxobacterium. (a) The life cycle of a myxobacterium (*Chondromyces*). **(b)** A photograph of an actual mature fruiting body.

FIGURE 4.36

Budding bacteria. The life cycle of *Caulobacter,* showing the development of the flagellated swarm cell and stalked phases.

particularly common in the ocean because the salt content is not high enough. Many of the "halobacteria" use a red pigment to synthesize ATP in the presence of light. These pigments are responsible for "red herrings," the color of the Red Sea, and the red color of salt ponds **(figure 4.37).**

Archaea adapted to growth at very high temperatures are defined as hyperthermophilic (they love high temperatures). These organisms flourish at temperatures between 80° and 105°C and cannot grow at 50°C. They live in volcanic waters and soils and submarine vents and are also often salt- and acid-tolerant as well. One member, *Thermoplasma,* lives in hot, acidic habitats in the waste piles around coal mines that regularly sustain a pH of 1 and a temperature of nearly 60°C. Researchers sampling sulfur vents in the deep ocean discovered thermophilic archaea flourishing at temperatures up to 250°C—150° above the temperature of boiling water! Not only were these archaea growing prolifically at this high temperature, but they were also living at 265 atmospheres of pressure. (On the earth's surface, pressure is about one atmosphere.) For additional discussion of the unusual adaptations of archaea, see chapter 7.

CHAPTER CHECKPOINTS

The rickettsias are a group of bacteria that are intracellular parasites, dependent on their eucaryote host for energy and nutrients. Most are pathogens that alternate between arthropods and mammalian hosts.

The chlamydias are also small, intracellular parasites that infect humans, mammals, and birds. They do not require arthropod vectors.

TABLE 4.7

Comparison of Three Cellular Domains

Characteristic	Bacteria	Archaea	Eukarya
Cell type	Procaryotic	Procaryotic	Eucaryotic
Chromosomes	Single, circular	Single, circular	Several, linear
Types of ribosomes	70S	70S but structure is more similar to 80S	80S
Contains unique ribosomal RNA signature sequences	+	+	+
Number of sequences shared with Eukarya	1	3	
Protein synthesis similar to Eukarya	−	+	
Presence of peptidoglycan in cell wall	+	−	−
Cell membrane lipids	Fatty acids with ester linkages	Long-chain, branched hydrocarbons with ether linkages	Fatty acids with ester linkages
Sterols in membrane	− (some exceptions)	−	+

The mycoplasmas are bacteria that lack cell walls and therefore lack a definite cell shape. Although they are not obligate parasites, many require a eucaryote host and have specialized growth requirements when grown in culture.

Many bacteria are free-living, rather than parasitic. The photosynthetic bacteria, gliding bacteria, and appendaged bacteria encompass many subgroups that colonize specialized habitats, not other living organisms.

Archaea are another type of procaryotic cell that constitute the third domain of life. They exhibit unusual biochemistry and genetics that make them different from bacteria. Members are adapted to extreme habitats with high temperature, salt, pressure, or acid.

(a)

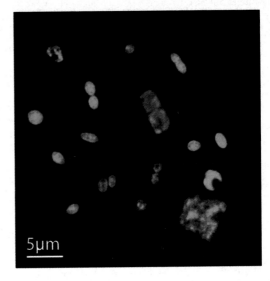

5µm

(b)

FIGURE 4.37

Halophiles around the world. **(a)** A solar evaporation pond in Owens Lake, California, is extremely high in salt and mineral content. The archea that dominate in this hot saline habitat produce brilliant red pigments with which they absorb light to drive cell synthesis. **(b)** A sample taken from a saltern in Australia viewed by fluorescent microscopy (1,000×). Note the range of cell shapes (cocci, rods, and square) found in this community.

CHAPTER CAPSULE WITH KEY TERMS

I. **Procaryotic Structures and Their Functions**
   A. General Features of Procaryotes
      1. Procaryotes consist of two major groups, the bacteria and the archaea. Life on earth would not be possible without them.
      2. Procaryotic cells lack the membrane-surrounded organelles and nuclear compartment of eucaryotic cells but are still complex in their structure and function. All procaryotes have a cell membrane, cytoplasm, ribosomes, and a chromosome.
   B. External Features of Bacteria
      1. Appendages: Some bacteria have projections that extend from the cell. Flagella (and internal **axial filaments** found in

spirochetes) are used for motility. **Fimbriae** and **pili** function in adhering to the environment; sex pili provide a means of genetic exchange.
      2. The cell envelope: Most procaryotes are surrounded by a protective envelope that is bounded by the **glycocalyx** and the cytoplasmic membrane, with a cell wall in between. The glycocalyx may be a slime layer or a **capsule** with important defensive functions.
      3. The cell wall: The cell wall functions to keep the cell shape and provide osmotic protection. The **Gram stain** differentiates two types of cells on the basis of the structure: gram-positive

bacteria have a thick layer of **peptidoglycan** while gram-negative bacteria have a thin layer encased in an outer membrane. Important variations in cell walls or the lack thereof are seen in archaea and various bacterial species (e.g., acid-fast species, *Mycoplasma*).

4. Cell membrane: All cells are surrounded by a cytoplasmic membrane that functions in regulating transport into the cell and secretion out of the cell. In addition, procaryotes utilize it as a site of metabolic activities, and **mesosomes** may provide additional sites of function.

C. Internal Features of Bacteria

1. The contents of the cytoplasm include the cell pool, a single circular **chromosome** condensed in the **nucleoid,** and **ribosomes.** The latter are sites of protein synthesis and are 70S in size. Extra genetic information may be present as **plasmids.** Nutrients may be stored in **inclusions** and **granules.**

2. Bacterial **endospores:** A few bacteria are able to make a highly resistant structure for survival, and resume growth at a later time.

D. Bacterial Shapes, Arrangements, and Sizes

1. Most bacteria are unicellular and are found in a great variety of shapes, arrangements and sizes. General shapes include **cocci, bacilli,** and helical forms such as **spirilla** and **spirochetes.** Some show great variation within the species in shape and size and are **pleomorphic.** Other variations include **coccobacilli, vibrios,** and filamentous forms.

2. Procaryotes divide by binary fission and do not utilize mitosis. Various arrangements result from cell division and are termed **diplococci, streptococci, staphylococci, tetrads,** and **sarcina** for cocci; bacilli may form pairs, chains, or **palisades.**

II. **Classification and Survey of Procaryotes**

A. Identification and Classification of Bacteria

1. Bacteria are identified by means of microscopic and macroscopic morphology as well as by biochemical, serological, molecular, and genetic characteristics.

2. Classification schemes utilize the information just mentioned for grouping together bacterial species that are related by evolutionary relationships or by **phenetic** relationships.

3. An important taxonomic system is standardized by *Bergey's Manual of Systematic Bacteriology,* which divides procaryotes into four major groups:

   a. **Gracilicutes:** Bacteria with gram-negative cell walls.

   b. **Firmicutes:** Bacteria with gram-positive cell walls.

   c. **Tenericutes:** Bacteria without cell walls.

   d. **Mendosicutes:** Archaebacteria.

4. Bacterial species may be divided into **strains** and types.

B. Unusual Procaryotic Groups

1. Medically important bacteria: Several groups of bacteria are so different that they have not always fit well in classification schemes. **Rickettsias** and chlamydias are within the gram-negative group but are small **obligate intracellular parasites** that replicate within cells of the hosts they invade. **Mycoplasmas** are able to survive without cell walls by strengthening their cell membrane with sterols.

2. Nonpathogenic bacterial groups: The majority of bacterial species are free-living and not involved in disease. Unusual groups include photosynthetic bacteria such as cyanobacteria, which provide oxygen to the environment, and the green and purple bacteria.

3. Archaea, the other major procaryote group: Archaea share the major characteristics of procaryotes but do have some differences with bacteria in certain genetic aspects and some cell components. Many are adapted to extreme environments, much as may have been found originally on earth. They are not considered medically important, but are of ecological and potential economic importance.

MULTIPLE-CHOICE QUESTIONS

1. Which of the following is not found in all bacterial cells?
   a. cell membrane   c. ribosomes
   b. a nucleoid   d. capsule

2. The major locomotor structures in bacteria are
   a. flagella   c. fimbriae
   b. pili   d. cilia

3. Pili are tubular shafts in ____ bacteria that serve as a means of ____.
   a. gram-positive, genetic exchange
   b. gram-positive, attachment
   c. gram-negative, genetic exchange
   d. gram-negative, protection

4. An example of a glycocalyx is
   a. a capsule   c. outer membrane
   b. pili   d. a cell wall

5. Which of the following is a primary bacterial cell wall function?
   a. transport   c. support
   b. motility   d. adhesion

6. Which of the following is present in both gram-positive and gram-negative cell walls?
   a. an outer membrane   c. teichoic acid
   b. peptidoglycan   d. lipopolysaccharides

7. Mesosomes are internal extensions of the
   a. cell wall   c. cell membrane
   b. chromosome   d. capsule

8. Metachromatic granules are concentrated crystals of ____ that are found in ____.
   a. fat, *Mycobacterium*
   b. dipicolinic acid, *Bacillus*
   c. sulfur, *Thiobacillus*
   d. PO_4, *Corynebacterium*

9. Bacterial endospores function in
   a. reproduction   c. protein synthesis
   b. survival   d. storage

10. A bacterial arrangement in packets of eight cells is described as a ____.
    a. micrococcus   c. tetrad
    b. diplococcus   d. sarcina

11. The major difference between a spirochete and a spirillum is
    a. presence of flagella   c. the nature of motility
    b. a cell with coils   d. size

12. Genetic analysis of bacteria would include
    a. fermentation testing
    b. ability to digest complex nutrients
    c. presence of oxidase
    d. G + C content

13. Which division of bacteria has a gram-positive cell wall?
    a. Gracilicutes   c. Firmicutes
    b. Archaea   d. Tenericutes
14. To which division of bacteria do cyanobacteria belong?
    a. Tenericutes   c. Firmicutes
    b. Gracilicutes   d. Mendosicutes

15. Archaea differ from bacteria on the basis of
    a. structure of envelope
    b. size
    c. the archaea having a nucleus
    d. type of locomotor structures

CONCEPT QUESTIONS

1. a. Name several general characteristics that could be used to define the procaryotes.
   b. Do any other microbial groups besides bacteria have procaryotic cells?
   c. What does it mean to say that bacteria are ubiquitous? In what habitats are they found? Give some general means by which bacteria derive nutrients.
2. a. Describe the structure of a flagellum and how it operates. What are the four main types of flagellar arrangement?
   b. How does the flagellum dictate the behavior of a motile bacterium? Differentiate between flagella and periplasmic flagella.
   c. List some direct and indirect ways that one can determine bacterial motility.
3. a. List the components of the cell envelope.
   b. Explain the position of the glycocalyx.
   c. What are the functions of slime layers and capsules?
   d. How is the presence of a slime layer evident even at the level of a colony?
4. a. Differentiate between pili and fimbriae.
   b. How do their structures differ?
   c. How do their functions differ?
5. a. Compare the cell envelopes of gram-positive and gram-negative bacteria.
   b. What function does peptidoglycan serve?
   c. To which part of the cell envelope does it belong?
   d. Give a simple description of its structure.
   e. What happens to a cell that has its peptidoglycan disrupted or removed?
   f. What functions does the LPS layer serve?
6. a. What is the Gram stain?
   b. What is there in the structure of bacteria that causes some to stain purple and others to stain red?
   c. How does the precise structure of the cell walls differ in gram-positive and gram-negative bacteria?
   d. What other properties besides staining are different in gram-positive and gram-negative bacteria?
   e. What is the periplasmic space, and how does it function?
   f. What characteristics does the outer membrane confer on gram-negative bacteria?
7. a. List five functions that the cell membrane performs in bacteria.
   b. What are mesosomes and some of their possible functions?
8. a. Compare the composition of the bacterial chromosome (nucleoid) and plasmids.
   b. What are the functions of each?

9. a. What is unique about the structure of bacterial ribosomes?
   b. How do they function?
   c. Where are they located?
10. a. Compare and contrast the structure and function of inclusions and granules.
    b. What are metachromatic granules, and what do they contain?
11. a. Describe the vegetative stage of a bacterial cell.
    b. Describe the structure of an endospore, and explain its function.
    c. Describe the endospore-forming cycle.
    d. Explain why an endospore is not considered a reproductive body.
    e. Why are endospores so difficult to destroy?
12. a. Draw the three bacterial shapes.
    b. How are spirochetes and spirilla different?
    c. What is a vibrio? A coccobacillus?
    d. What is pleomorphism?
    e. What is the difference between the use of the term bacillus and the name *Bacillus*? *Staphylococcus* and staphylococcus?
13. a. Rank the size ranges in bacteria according to shape.
    b. Rank the bacteria in relationship to viruses and eucaryotic cell size.
    c. Use the size bars to measure the cells in figure 4.37 and Spotlight 4.3.
14. a. What characteristics are used to classify bacteria?
    b. What are the most useful characteristics for categorizing bacteria into families?
    c. In what ways is ribosomal RNA an important method for differentiating bacteria and groups of organisms?
15. a. How is the species level in bacteria defined?
    b. Name at least three ways bacteria are grouped below the species level.
    c. In what ways are they important?
16. a. Describe at least two circumstances that give rise to L forms.
    b. How do L forms survive?
    c. In what ways are they important?
17. Name several ways in which bacteria are medically and ecologically important.
18. a. Explain the characteristics of archaea that indicate it is a unique domain of living things that is neither a bacterium nor a eucaryote.
    b. What leads microbiologists to believe the archaea are more closely related to eucaryotes than to bacteria?
    c. What is meant by the term *extremophile?* Describe some archaeal adaptations to extreme habitats.

CRITICAL-THINKING QUESTIONS

1. What would happen if one stained a gram-positive cell with safranin? A gram-negative cell with crystal violet? What would happen to the two types if the mordant were omitted?

2. What is required to kill endospores? How do you suppose archaeologists were able to date some spores as being 3,000 years old?

3. Using clay, demonstrate how cocci can divide in several planes and show the outcome of this division. Show how the arrangements of bacilli occur, including palisades.

4. Using a corkscrew and a spring to compare the flexibility and locomotion of spirilla and spirochetes, explain which cell type is represented by each object.

5. Under the microscope, you see a rod-shaped cell that is swimming rapidly forward.
   a. What do you automatically know about that bacterium's structure?
   b. How would a bacterium use its flagellum for phototaxis?
   c. Can you think of another function of flagella besides locomotion?

6. Can you think of ways in which an evolutionary tree built on rRNA sequences might be misleading?

7. a. Name an unusual type of bacterium that lives in extreme habitats.
   b. What adaptations enable it to survive there?

8. a. Name a bacterium that has no cell walls.
   b. How is it protected from osmotic destruction?

9. a. Name a bacterium that is aerobic, gram positive, and spore-forming.
   b. What habitat would you expect this species to occupy?

10. a. Name an acid-fast bacterium.
    b. What characteristics make this bacterium different from other gram-positive bacteria?

11. a. Name two main groups of obligate intracellular parasitic bacteria.
    b. Why can't these groups live independently?

12. a. Name a bacterium that contains sulfur granules.
    b. What is the advantage in storing these granules?

13. a. Name a bacterium that uses chlorophyll to photosynthesize.
    b. Describe the two major groups of photosynthetic bacteria.
    c. How are they similar?
    d. How are they different?

14. a. What are some possible adaptations that the giant bacterium *Thiomargarita* has had to make because of its large size?
    b. If a regular bacterium were the size of an elephant, estimate the size of a nanobe at that scale.

15. Propose a hypothesis to explain how bacteria and archaea could have, together, given rise to eucaryotes.

16. Explain or illustrate exactly what will happen to the cell wall if the synthesis of the interbridge is blocked by penicillin. What if the glycan is hydrolyzed by lysozyme?

17. Ask your lab instructor to help you make a biofilm and examine it under the microscope. One possible technique is to suspend a glass slide in an aquarium for a few weeks, then carefully air-dry, fix, and gram stain it. Observe the diversity of cell types.

INTERNET SEARCH TOPICS

1. Go to a search engine and type in "Martian Microbes." Look for papers and information that support or reject the idea that fossil structures discovered in an ancient meteor from Mars could be bacteria. What are some of the reasons that microbiologists are skeptical of this possibility?

2. Search the Internet for information on nanobacteria. Give convincing reasons why these are or are not real organisms.

3. Visit the student Online Learning Center at www.mhhe.com/talaro5. Go to chapter 4, Internet Search Topics, and log on to the available website to observe short clips on dividing bacteria and bacterial motility.

CHAPTER 5

Eucaryotic Cells and Microorganisms

The eucaryotic cell is a complex, compartmentalized unit that differs from the procaryotic cell by containing a nucleus and several other specialized structures called organelles. Although exact cell structures differ somewhat among the several groups of eucaryotic organisms, the eucaryotic cell is the typical cell of certain microbial groups (fungi, algae, protozoa, and helminth worms) as well as all animals and plants. In this chapter, we examine the overall structure and function of eucaryotic cells in preparation for later chapters that deal with related microbiological concepts, including metabolism, genetics, nutrition, drug therapy, immunology, and disease. Because of the tremendous variety of eucaryotic microorganisms and their practical importance in medicine, industry, and agriculture, this chapter will also cover the major characteristics of each group of eucaryotic microorganisms.

Chapter Overview

- Eucaryotic cells are large complex cells divided into separate compartments by membrane-bound components called organelles.
- Major organelles—the nucleus, mitochondria, chloroplasts, endoplasmic reticulum, Golgi apparatus, and locomotor appendages—each serve an essential function to the cell, such as heredity, production of energy, synthesis, transport, and movement.
- Eucaryotic cells are found in fungi, protozoa, algae, plants, and animals, and they exhibit single-celled, colonial, and multicellular body plans.
- Fungi and Protista (algae and protozoa) are the major kingdoms that contain eucaryotic microorganisms.
- Fungi are eucaryotes that feed on organic substrates, have cell walls, reproduce asexually and sexually by spores, and exist in macroscopic or microscopic forms.
- Most fungi are free-living decomposers that are beneficial to biological communities; some may cause infections in animals and plants.
- Microscopic fungi include yeasts with spherical budding cells and molds with elongate filamentous hyphae in mycelia.
- Algae are aquatic photosynthetic protists with rigid cell walls and chloroplasts containing chlorophyll and other pigments.
- Algae belong to several groups based on their type of pigments, cell wall, stored food materials, and body plan.
- Protozoa are protists that feed by engulfing other cells, lack a cell wall, usually have some type of locomotor organelle, and may form dormant cysts.
- Subgroups of protozoa differ in their organelles of motility (flagella, cilia, pseudopods, nonmotile).

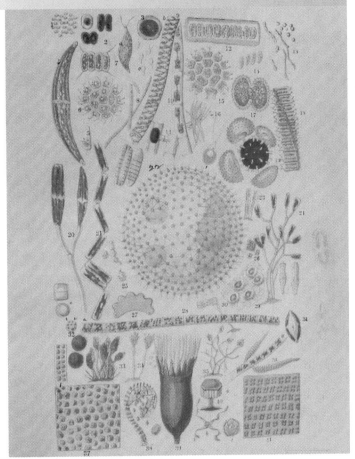

These drawings are from a book published more than a century ago, "Common Objects of the Microscope." The book displays hundreds of well-known algae, molds, and yeasts found in water and soil samples.

- Most protozoa are free-living aquatic cells that feed on bacteria and algae, and a few are animal parasites.
- The infective helminths are flatworms and roundworms that have greatly modified body organs so as to favor their parasitic life style.

The Nature of Eucaryotes

Evidence from paleontology indicates that the first eucaryotic cells appeared on the earth approximately 2 billion years ago. Some fossilized cells that look remarkably like algae or protozoa appear in

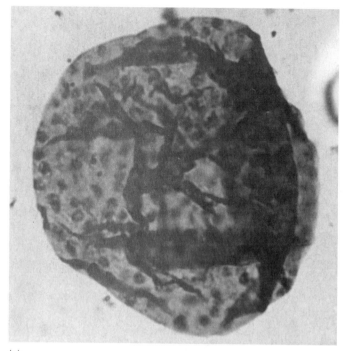

(a)

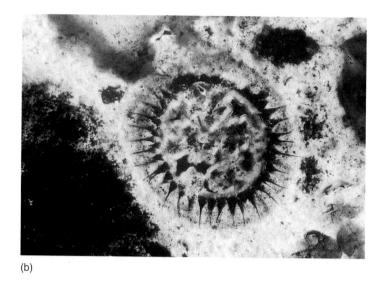

(b)

FIGURE 5.1

Ancient eucaryotic protists caught up in fossilized rocks. Both forms
are called acritarchs. **(a)** An ornamented algalike cell found in Siberian
shale deposits and dated from 850 million to 950 million years ago. **(b)** A
large, disclike cell bearing a crown of spines is from Chinese rock dated
590 million to 610 million years ago.

shale sediments from China, Russia, and Australia that date from
850 million to 950 million years ago (**figure 5.1**). Biologists have
discovered profound evidence to suggest that the eucaryotic cell
evolved from procaryotic organisms by a process of intracellular
symbiosis* (**Spotlight on Microbiology 5.1**). It now seems clear
that some of the **organelles** that distinguish eucaryotic cells origi-

---

*symbiosis (sim-beye-oh′-sis) Gr. *sym*, together, and *bios*, to live.

nated from procaryotic cells that became trapped inside them. The
structure of these first eucaryotic cells was so versatile that eucaryotic microorganisms soon spread out into available habitats and
adopted greatly diverse styles of living.

The first primitive eucaryotes were probably single-celled
and independent, but, over time, some forms began to aggregate,
forming colonies and filaments. With further evolution, some of
the cells within colonies became *specialized,* or adapted to perform a particular function advantageous to the whole colony, such
as locomotion, feeding, or reproduction. Complex multicellular
organisms evolved as individual cells in the organism lost the ability to survive apart from the intact colony. Although a multicellular organism is composed of many cells, it is more than just a disorganized assemblage of cells like a colony. Rather, it is composed
of distinct groups of cells that cannot exist independently of the
rest of the body. The cell groupings of multicellular organisms that
have a specific function are termed *tissues,* and groups of tissues
make up *organs.*

Looking at modern eucaryotic organisms, we find examples
of many levels of cellular organization. All protozoa, as well as

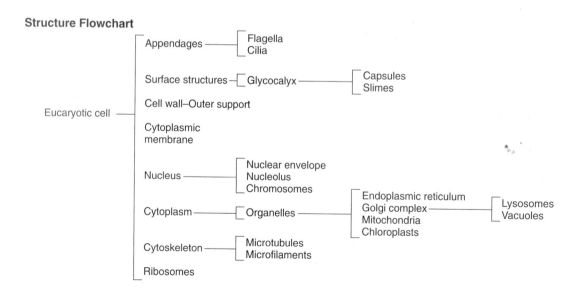

Structure Flowchart

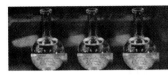

SPOTLIGHT ON MICROBIOLOGY 5.1
The Extraordinary Emergence of Eucaryotic Cells

For years, biologists have grappled with the problem of how a cell as complex as the eucaryotic cell originated. One of the most fascinating explanations is that of **endosymbiosis,*** which proposes that eucaryotic cells arose when a much larger procaryotic cell engulfed smaller bacterial cells that began to live and reproduce inside the procaryotic cell rather than being destroyed. As the smaller cells took up permanent residence, they came to perform specialized functions for the larger cell, such as food synthesis and oxygen utilization, that enhanced the cell's versatility and survival. Over time, when the cells evolved into a single functioning entity, the relationship became obligatory. Although the theory of endosymbiosis has been greeted with some controversy, this phenomenon has been demonstrated in the laboratory with amebas infected with bacteria that gradually became dependent upon the bacteria for survival.

The biologist most responsible for validation of the theory of endosymbiosis is Dr. Lynn Margulis. Using modern molecular techniques, she has accumulated convincing evidence of the relationships between the organelles of modern eucaryotic cells and the structure of bacteria. In many ways, the mitochondrion of eucaryotic cells is something like a tiny cell within a cell. It is capable of independent division, contains a circular chromosome that has bacterial DNA sequences, and has ribosomes that are clearly procaryotic. Mitochondria also have bacterial membranes and can be inhibited by drugs that affect only bacteria. This link is so well established that recent phylogenetic trees of bacteria show mitochondria and chloroplasts as two of the bacterial branches.

Chloroplasts likely arose when endosymbiotic cyanobacteria provided their host cells with a built-in feeding mechanism. Evidence is seen in a modern flagellated protist that harbors specialized chloroplasts with cyanobacterial chlorophyll and thylakoids. Margulis also has convincing evidence that eucaryotic cilia and flagella are the consequence of endosymbiosis between spiral bacteria and the cell membrane of early eucaryotic cells.

It is tempting to envision the development of the eucaryotic cell through a fusion of archaeal and bacterial cells. The archaea would have served as a source of ribosomes and certain aspects of protein synthesis, and bacteria would have given rise to mitochondria and chloroplasts. The first eukarya probably resembled modern protists such as *Giardia* or *Plagiopyla*. Researchers are pursuing further genetic analysis of these microbes and their organelles in order to add further support to the theory.

*endosymbiosis (en′-doh-sym-bee-oh′-sis) Gr. *endo,* within; *syn,* together, and *bios,* life. An intimate association between two cells, in which one is living inside another.

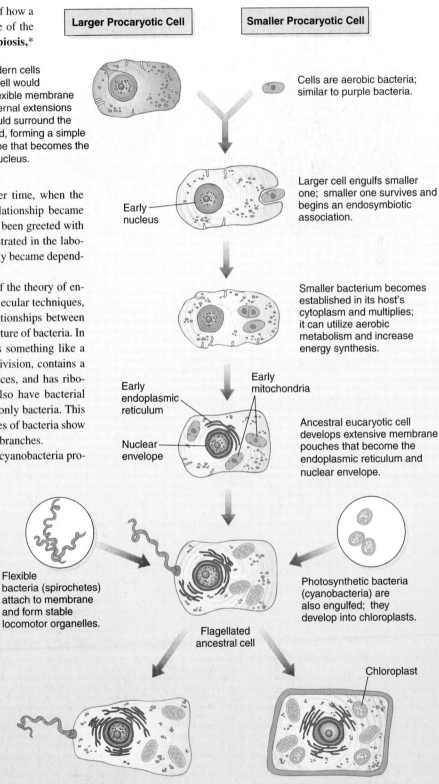

Larger Procaryotic Cell

Smaller Procaryotic Cell

No modern cells exist; cell would have flexible membrane and internal extensions that could surround the nucleoid, forming a simple envelope that becomes the early nucleus.

Cells are aerobic bacteria; similar to purple bacteria.

Early nucleus

Larger cell engulfs smaller one; smaller one survives and begins an endosymbiotic association.

Smaller bacterium becomes established in its host's cytoplasm and multiplies; it can utilize aerobic metabolism and increase energy synthesis.

Early endoplasmic reticulum

Early mitochondria

Nuclear envelope

Ancestral eucaryotic cell develops extensive membrane pouches that become the endoplasmic reticulum and nuclear envelope.

Flexible bacteria (spirochetes) attach to membrane and form stable locomotor organelles.

Flagellated ancestral cell

Photosynthetic bacteria (cyanobacteria) are also engulfed; they develop into chloroplasts.

Chloroplast

Protozoa, fungi, animals

Algae, higher plants

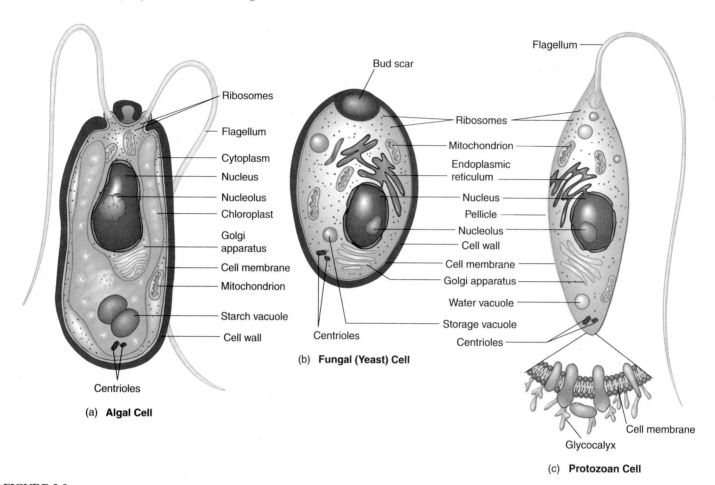

FIGURE 5.2

The structure of three representative eucaryotic cells. **(a)** *Chlamydomonas*, a unicellular motile alga. **(b)** A yeast cell (fungus). **(c)** *Peranema*, a flagellated protozoan.

numerous algae and fungi, live a unicellular or colonial existence. Truly multicellular organisms are found only among plants and animals and some of the fungi (mushrooms) and algae (seaweeds). Only certain eucaryotes are traditionally studied by microbiologists—primarily the protozoa, the microscopic algae and fungi, and animal parasites.

Form and Function of the Eucaryotic Cell: External Structures

The cells of eucaryotic organisms are so varied that no one member can serve as a typical example. **Figure 5.2** presents the generalized structure of typical algal, fungal, and protozoan cells. The outline at the bottom of page 124 shows the organization of a eucaryotic cell. (Only motile algae contain all categories of eucaryotic organelles.)

In general, eucaryotic microbial cells have a cytoplasmic membrane, nucleus, mitochondria, endoplasmic reticulum, Golgi apparatus, vacuoles, cytoskeleton, and glycocalyx. A cell wall, locomotor appendages, and chloroplasts, are found only in some groups. In the following sections, we cover the microscopic structure and functions of the eucaryotic cell. As with the procaryotes, we begin on the outside and proceed inward through the cell.

LOCOMOTOR APPENDAGES: CILIA AND FLAGELLA

Motility allows a microorganism to locate life-sustaining nutrients and to migrate toward positive stimuli such as sunlight; it also permits avoidance of harmful substances and stimuli. Locomotion by means of flagella or cilia is a common property of protozoa, many algae, and a few fungal and animal cells.

Although they share the same name, eucaryotic **flagella** are much different from those of procaryotes. The eucaryotic flagellum is thicker (by a factor of 10), structurally more complex, and covered by an extension of the cell membrane. A single flagellum is a long, sheathed cylinder containing regularly spaced hollow tubules—microtubules—that extend along its entire length **(figure 5.3a)**. A cross section reveals nine pairs of closely attached microtubules surrounding a single central pair. This scheme, called the 9 + 2 arrangement, is a universal pattern of flagella and cilia (figure 5.3b). During locomotion, the adjacent microtubules slide past each other, whipping the flagellum back and forth. Although details of this process are too complex to discuss here, it involves expenditure of energy and a coordinating mechanism in the cell membrane. Flagella can move the cell by pushing it forward like a fishtail or by pulling it by a lashing or twirling motion (figure 5.3c). The placement and number of flagella can be useful in identifying flagellated protozoa and certain algae.

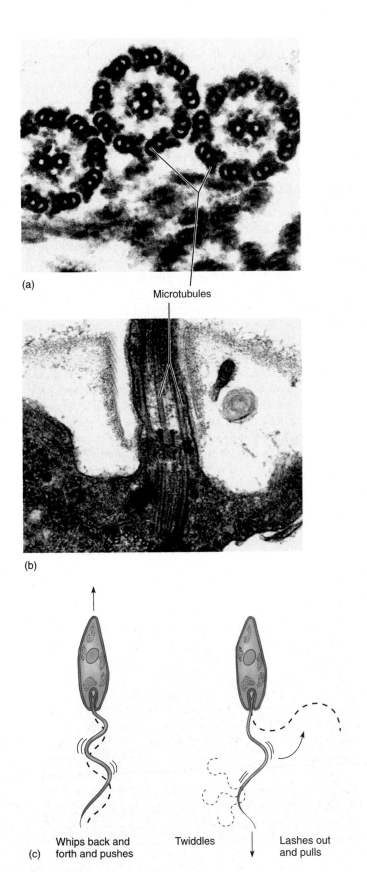

(a)

Microtubules

(b)

(c) Whips back and forth and pushes   Twiddles   Lashes out and pulls

FIGURE 5.3

The structures of microtubules. **(a)** A cross section (circular area) that reveals the typical 9 + 2 arrangement found in both flagella and cilia. **(b)** Longitudinal section through a flagellum, showing microtubules. **(c)** Locomotor patterns seen in flagellates.

Cilia* are very similar in overall architecture to flagella, but they are shorter and more numerous (some cells have several thousand). They are found only on a single group of protozoa and certain animal cells. In the ciliated protozoa, the cilia occur in rows over the cell surface, where they beat back and forth in regular oar-like strokes (see figure 5.30). Such protozoa are among the fastest of all motile cells. The fastest ciliated protozoon can swim up to 2,500 μm/s—a meter and a half per minute! On some cells, cilia also function as feeding and filtering structures.

SURFACE STRUCTURES: THE GLYCOCALYX

Most eucaryotic cells have a **glycocalyx,** an outermost boundary that comes into direct contact with the environment (see figure 5.2). This structure is usually composed of polysaccharides and appears as a network of fibers, a slime layer, or a capsule much like the glycocalyx of procaryotes. Because of its positioning, the glycocalyx contributes to protection, adherence of cells to surfaces, and reception of signals from other cells and from the environment. We will explore the function of receptors that are part of the glycocalyx in chapters 14 and 15 (host defenses and immune interactions). The nature of the layer beneath the glycocalyx varies among the several eucaryotic groups. Fungi and most algae have a thick, rigid cell wall, whereas protozoa, a few algae, and all animal cells lack a cell wall and have only a cell membrane.

THE CELL WALL

The cell walls of the fungi and algae are rigid and provide structural support and shape, but they are different in chemical composition from procaryotic cell walls. Fungal cell walls have a thick, inner layer of polysaccharide fibers composed of chitin or cellulose and a thin outer layer of mixed glycans **(figure 5.4).** The cell walls of algae are quite varied in chemical composition. Substances commonly found among various algal groups are cellulose, pectin,[1] mannans,[2] and minerals such as silicon dioxide and calcium carbonate.

THE CELL MEMBRANE

The cell membrane of eucaryotic cells is a typical bilayer of phospholipids in which protein molecules are embedded. In addition to phospholipids, eucaryotic membranes also contain *sterols* of various kinds. Sterols are different from phospholipids in both structure and behavior (see figure 2.19). Their relative rigidity confers stability on eucaryotic membranes. This strengthening feature is extremely important in cells that lack a cell wall. Cytoplasmic membranes of eucaryotes are functionally similar to those of procaryotes, serving as selectively permeable barriers in transport (see chapter 7). Unlike procaryotes, eucaryotic cells also contain a number of individual membrane-bound organelles that are extensive enough to account for 60% to 80% of their volume.

---

1. A polysaccharide composed of galacturonic acid subunits.

2. A polymer of the sugar known as mannose.

   *cilia (sil'-ee-uh) sing. cilium; L. *siliam,* eyelid.

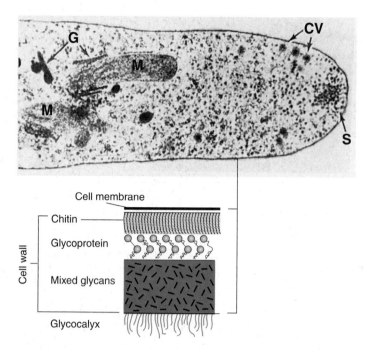

FIGURE 5.4

Glycocalyx structure. Cross section through the tip of a fungal cell to show the general structure of the cell wall in relationship to the cell membrane and glycocalyx. Top: Photomicrograph. (S, growing tip; CV, coated vesicles; G, Golgi apparatus; M, mitochondrion.) Bottom: The cell wall is a thick, rigid structure composed of complex layers of polysaccharides and proteins.

Survey of Major Organelles

Eucaryotic cells accomplish their cellular work with unique intracellular membrane-wrapped organelles. Each type of organelle carries out one or two functions, and it can partition the cell into many small compartments, thereby serving to organize and separate metabolic events. Functions such as synthesis and secretion often proceed in an assembly-line fashion, with each organelle in the series affecting the outcome of the final product or process.

CHAPTER CHECKPOINTS

Eucaryotes are cells with organelles and a nucleus compartmentalized by membranes. They might have originated from procaryote ancestors about 2 billion years ago. Eucaryotic cell structure enabled eucaryotes to diversify from single cells into a huge variety of complex multicellular forms.

The cell structures common to most eucaryotes are the cell membrane, vacuoles, mitochondria, endoplasmic reticulum, Golgi apparatus, and a cytoskeleton. Cell walls, chloroplasts, and locomotor organs are present in some eucaryote groups.

Microscopic eucaryotes use locomotor organs such as flagella or cilia for moving themselves or their food.

The glycocalyx is the outermost boundary of most eucaryotic cells. Its functions are protection, adherence, and reception of chemical signals from the environment or from other organisms. The glycocalyx is supported by either a cell wall or a cell membrane.

The cytoplasmic membrane of eucaryotes is similar in function to that of procaryotes, but it differs in composition, possessing sterols as additional stabilizing agents.

Eucaryotic cells have membrane-bound organelles, which are never present in procaryotes. Such compartmentalization allows a wide variety of cellular functions to occur in specialized areas of the cell.

Form and Function of the Eucaryotic Cell: Internal Structures

THE NUCLEUS: THE CONTROL CENTER

The nucleus is a compact sphere that is the most prominent organelle of eucaryotic cells. It is separated from the cell cytoplasm by an external boundary called a **nuclear envelope.** The envelope has a unique architecture. It is composed of two parallel membranes separated by a narrow space, and it is perforated with small, regularly spaced openings, or pores, formed at sites where the two membranes unite (**figure 5.5**). The nuclear pores are passageways through which macromolecules migrate from the nucleus to the cytoplasm and vice versa. The nucleus contains a matrix called the **nucleoplasm** and a granular mass, the **nucleolus,*** that can stain more intensely than the immediate surroundings because of its RNA content. The nucleolus is the site for ribosomal RNA synthesis and a collection area for ribosomal subunits. The subunits are transported through the nuclear pores into the cytoplasm for final assembly into ribosomes.

A prominent feature of the nucleoplasm in stained preparations is a network of dark fibers known as **chromatin*** because of its attraction for dyes. Analysis has shown that chromatin actually

---

*nucleolus (noo-klee′-oh-lus) pl. nucleoli; L. "a little nucleus."

*chromatin (kroh′-muh-tin) Gr. *chromos*, color.

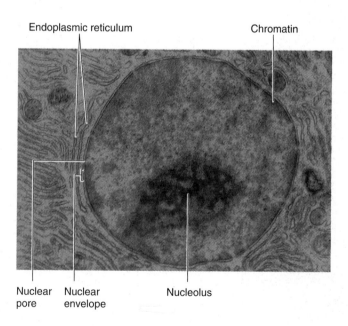

FIGURE 5.5

The nucleus. Electron micrograph section of an interphase nucleus, showing its most prominent features.

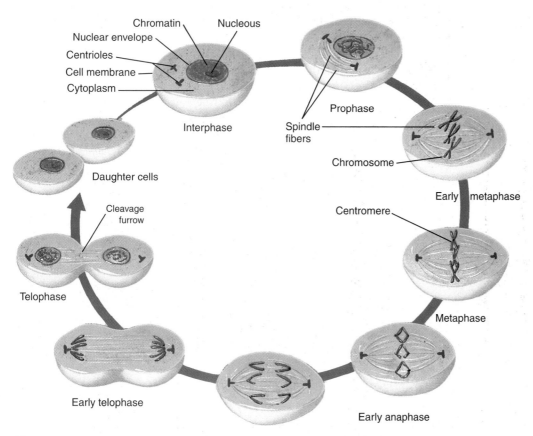

(a)

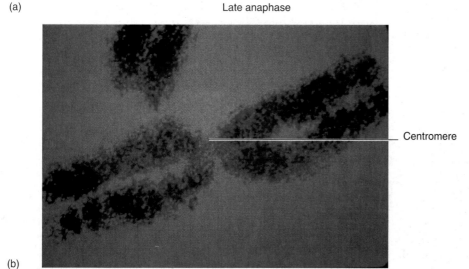

(b)

FIGURE 5.6

Changes in the cell and nucleus that accompany mitosis in a eucaryotic cell such as a yeast. **(a)** Before mitosis (at interphase), chromosomes are visible only as chromatin. As mitosis proceeds (early prophase), chromosomes take on a fine, threadlike appearance as they condense, and the nuclear membrane and nucleolus are temporarily disrupted. **(b)** By metaphase, the chromosomes are fully visible as X-shaped structures. The shape is due to duplicated chromosomes attached at a central point, the centromere. Spindle fibers attach to these and facilitate the separation of individual chromosomes during metaphase. Later phases serve in the completion of chromosomal separation and division of the cell proper into daughter cells.

comprises the eucaryotic **chromosomes,** large units of genetic information in the cell. The chromosomes in the nucleus of most cells are not readily visible because they are long, linear DNA molecules bound in varying degrees to **histone**[3] proteins, and they are far too fine to be resolved as distinct structures without extremely high magnification. During **mitosis,*** however, when the duplicated chromosomes are separated equally into daughter cells, the chromosomes themselves finally become readily visible as discrete bodies **(figure 5.6).** This appearance arises when the DNA becomes highly condensed by forming coils and supercoils around the histones to prevent the chromosomes from tangling as they are separated into new cells (see Microbits 9.2).

The Relationship of Chromosome Number to Reproductive Mode

To ensure its continued existence, a eucaryotic cell maintains a specified number of chromosomes that is typical for its species. The set of chromosomes exist in the **haploid*** (single, unpaired) state or in the

---

3. A simple type of protein containing a high proportion of lysine and arginine.

*mitosis (my-toh′-sis) Gr. *mitos,* thread, and *osis,* a condition of.

*haploid (hap′-loyd) Gr. *haplos,* simple, and *eidos,* form.

diploid* (matched pair) state. Most fungi, many algae, and some protozoa are examples of organisms whose cells occur in the haploid state during most of the life cycle. For example, the chromosome number of one type of ameba is 25, and one species of the fungus *Penicillium* possesses 5 chromosomes. Regardless of the chromosome state of the adult organism, all gametes (sex cells) are haploid.

---

*diploid (dip′-loyd) Gr. *di,* two, and *eidos,* form.

The cells of animals and plants, and of some protozoa, fungi, and algae, spend the greater part of their life cycle in the diploid state. This state comes about when a female haploid gamete (egg) fuses with a male haploid gamete (sperm) and produces a double (diploid) set of chromosomes in the offspring during sexual reproduction. Examples are humans, with 46 chromosomes (23 pairs), a parasitic blood fluke, with 16 chromosomes (8 pairs), and a redwood tree, with 22 chromosomes (11 pairs). Keep in mind that in all eucaryotic cells, whether haploid or diploid, the normal chromosome number is maintained through mitosis (figure 5.6 and **figure 5.7**).

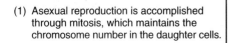

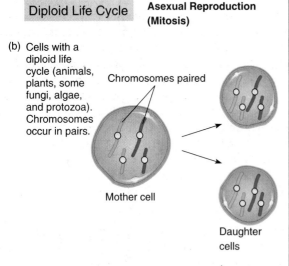

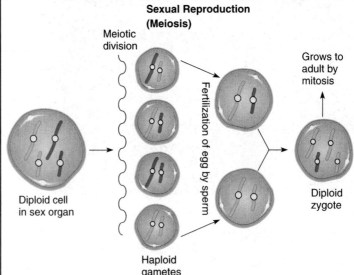

Haploid Life Cycle

Asexual Reproduction (Mitosis)

(a) Cells with haploid life cycle (algae, fungi, some protozoa). Chromosomes are unpaired.

Chromosomes unpaired
Mother cell
Daughter cells

(1) Asexual reproduction is accomplished through mitosis, which maintains the chromosome number in the daughter cells.

Sexual Reproduction (Meiosis)

Fusion
Haploid gametes
Diploid zygote
Meiotic division
Haploid offspring

(2) Sexual reproduction involves the fusion of specialized sex cells (already haploid), forming a diploid zygote. The haploid state in the offspring is restored by meiosis of the zygote. Note that reassortment of chromosomes can produce haploid offspring unlike either original gamete.

Diploid Life Cycle

Asexual Reproduction (Mitosis)

(b) Cells with a diploid life cycle (animals, plants, some fungi, algae, and protozoa). Chromosomes occur in pairs.

Chromosomes paired
Mother cell
Daughter cells

(3) Asexual reproduction through mitosis maintains the chromosome number and content.

Sexual Reproduction (Meiosis)

Meiotic division
Diploid cell in sex organ
Haploid gametes
Fertilization of egg by sperm
Diploid zygote
Grows to adult by mitosis

(4) In order to reproduce sexually, the normally diploid cells must produce haploid gametes through meiotic division. During fertilization, these haploid gametes (egg, sperm) form a zygote that restores the diploid number. Depending upon which gametes fuse, combinations of chromosomes unlike the original parents are possible in offspring.

FIGURE 5.7

Schematic of haploid versus diploid life cycles. Shown are the roles of mitosis and meiosis in reproduction and maintaining an organism's chromosome number.

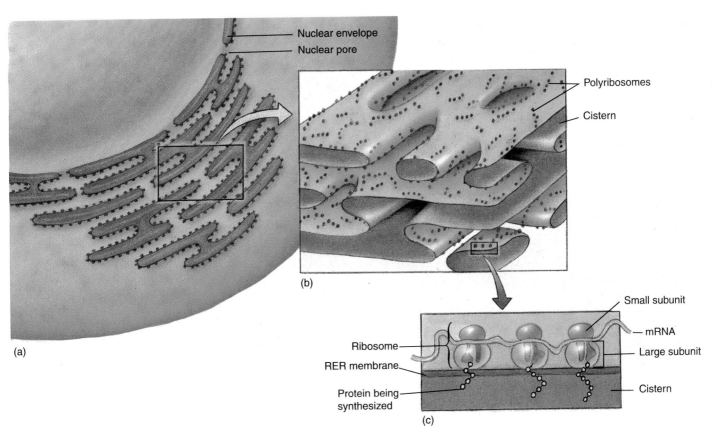

FIGURE 5.8

The origin and detailed structure of the rough endoplasmic reticulum (RER). **(a)** Schematic view of the origin of the RER from the outer membrane of the nuclear envelope. **(b)** Three-dimensional projection of the RER. **(c)** Detail of the orientation of a ribosome on the RER membrane.

When it comes to sexual reproduction in diploid organisms, however, the diploid chromosome number must be reduced to haploid, because gametes must be haploid for proper fertilization. This reduction is accomplished through reduction division, or **meiosis*** (figure 5.7*b*). During this process, diploid cells in the sex organs undergo two sets of divisions that result in haploid sex cells. Formation of a diploid zygote following fertilization then restores the chromosome number characteristic of that organism. Another example of the function of meiosis occurs in the sexual reproductive cycles of haploid fungi. Normal mitosis alone can produce their gametes, but after these gametes unite, they form a diploid zygote. In order to restore the fungus to the haploid state, the diploid nucleus undergoes meiosis and produces haploid offspring (see figure 5.7*a*, step 2, and figure 5.20).

ENDOPLASMIC RETICULUM: A PASSAGEWAY IN THE CELL

The **endoplasmic reticulum*** **(ER)** is a microscopic series of tunnels used in transport and storage. In sections, it appears as parallel, flat pouches bounded by membranes. Two kinds of endoplasmic reticulum are the **rough endoplasmic reticulum (RER) (figure 5.8)** and the **smooth endoplasmic reticulum (SER).** Electron micrographs show that the RER originates from the outer membrane of the nuclear envelope and extends in a continuous network through the cytoplasm, even out to the cell membrane. This architecture permits the spaces in the RER, or cisternae, to transport materials from the nucleus to the cytoplasm and ultimately to the cell's exterior. The RER appears rough because of large numbers of ribosomes partly attached to its membrane to act as secretory factories. Proteins synthesized on the ribosomes are shunted into the cavity of the reticulum and held there for later packaging and transport. The SER is a closed tubular network without ribosomes that functions in nutrient processing and in synthesis and storage of nonprotein macromolecules such as lipids.

GOLGI APPARATUS: A PACKAGING MACHINE

The **Golgi*** **apparatus,** also called the Golgi **complex** or body, is a discrete organelle consisting of a stack of several flattened, disc-shaped sacs called cisternae. These sacs have outer limiting membranes and cavities like those of the endoplasmic reticulum, but they

---

*meiosis (my-oh´-sis) Gr. *meiosis,* diminution. A process that permits paired chromosome sets to be reduced to single sets.

*endoplasmic (en˝-doh-plas´-mik) Gr. *endo,* within, and *plasm,* something formed; reticulum (rih-tik´-yoo-lum) L. *rete,* net or network.

*Golgi (gol´-jee) Named for C. Golgi, an Italian histologist who first described the apparatus in 1898.

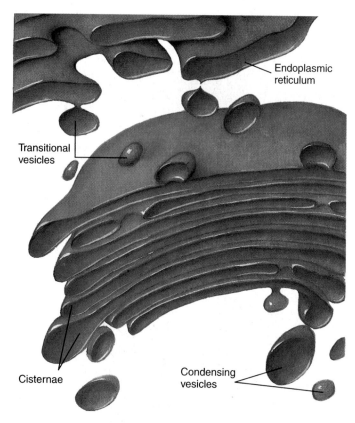

FIGURE 5.9

Detail of the Golgi apparatus. The flattened layers are cisternae. Vesicles enter the upper surface and leave the lower surface.

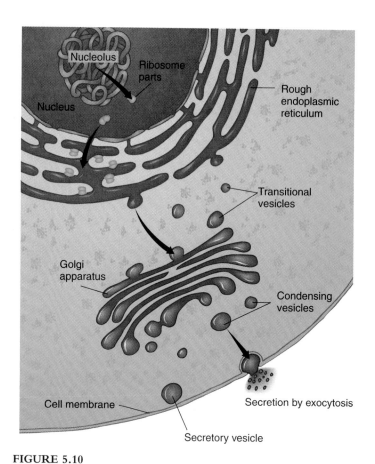

FIGURE 5.10

The transport process. The cooperation of organelles in protein synthesis and transport: nucleus → RER → Golgi apparatus → vesicles → secretion.

do not form a continuous network **(figure 5.9).** This organelle is always closely associated with the endoplasmic reticulum both in its location and function. At a site where it meets the Golgi apparatus, the endoplasmic reticulum buds off tiny membrane-bound packets of protein called *transitional vesicles** that are picked up by the forming face of the Golgi apparatus. Once in the complex itself, the proteins are often modified by the addition of polysaccharides and lipids. The final action of this apparatus is to pinch off finished *condensing vesicles* that will be conveyed to organelles such as lysosomes or transported outside the cell as secretory vesicles **(figure 5.10).** The production of antibodies and receptors in human white blood cells involves this packaging and transport system (see chapter 15).

Nucleus, Endoplasmic Reticulum, and Golgi Apparatus: Nature's Assembly Line

As the keeper of the eucaryotic genetic code, the nucleus ultimately governs and regulates all cell activities. But, because the nucleus remains fixed in a specific cellular site, it must direct these activities through a structural and chemical network (figure 5.10). This network includes ribosomes, which originate in the nucleus, and the rough endoplasmic reticulum, which is continuously connected with the nuclear envelope. Initially, a segment of the genetic code of DNA is copied into RNA and passed out through the nuclear pores directly to the ribosomes on the endoplasmic reticulum. Here,

specific proteins are synthesized from the RNA code and deposited in the lumen (space) of the endoplasmic reticulum. After being transported to the Golgi apparatus, the protein products are chemically modified and packaged into vesicles that can be used by the cell in a variety of ways. Some of the vesicles contain enzymes to digest food inside the cell; other vesicles are secreted to digest materials outside the cell, and yet others are important in the enlargement and repair of the cell wall and membrane.

A **lysosome*** is one type of vesicle originating from the Golgi apparatus that contains a variety of enzymes. Lysosomes are involved in intracellular digestion of food particles and in protection against invading microorganisms. They also participate in digestion and removal of cell debris in damaged tissue. Other types of vesicles include **vacuoles,*** which are membrane-bound sacs containing fluids or solid particles to be digested, excreted, or stored. They are formed in phagocytic cells (certain white blood cells and protozoa) in response to food and other substances that have been engulfed. The contents of a food vacuole are digested through the merger of the vacuole with a lysosome **(figure 5.11).** Other types of vacuoles are used in storing reserve food such as fats and glycogen. Protozoa living in freshwater habitats regulate osmotic pressure by means of contractile vacuoles, which regularly expel excess water that has diffused into the cell (see figure 5.29).

---

*vesicle (ves'-ik-l) L. *vesios*, bladder. A small sac containing fluid.

*lysosome (ly'-soh-sohm) Gr. *lysis*, dissolution, and *soma*, body.

*vacuole (vak'-yoo-ohl) L. *vacuus*, empty. Any membranous space in the cytoplasm.

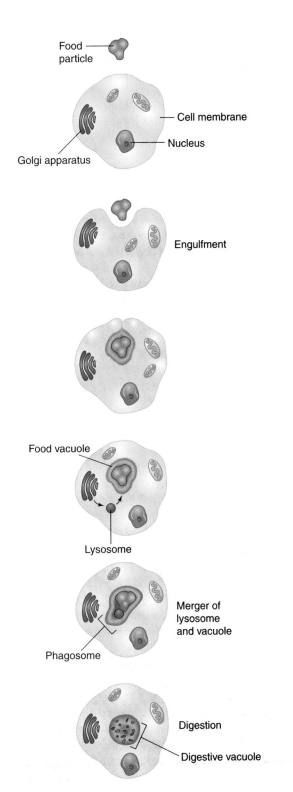

FIGURE 5.11
The origin and action of lysosomes in phagocytosis.

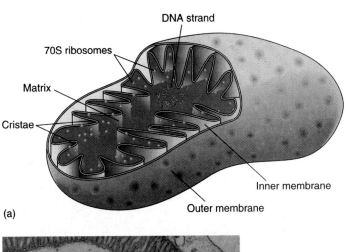

(a)

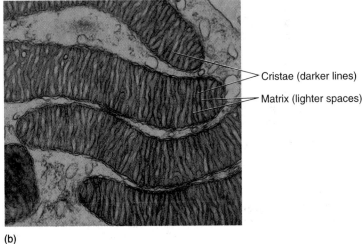

(b)

FIGURE 5.12

General structure of a mitochondrion. **(a)** in three-dimensional projection and **(b)** by electron microscopy. In most cells, mitochondria are elliptical or spherical, although in certain fungi, algae, and protozoa, they are long and filament-like.

MITOCHONDRIA: ENERGY GENERATORS OF THE CELL

Although the nucleus is the cell's control center, none of the cellular activities it commands could proceed without a constant supply of energy, the bulk of which is generated in most eucaryotes by **mitochondria.*** When viewed with light microscopy, mitochondria appear as round to elongate particles scattered throughout the cytoplasm. The internal ultrastructure reveals that a single mitochondrion consists of a smooth, continuous outer membrane that forms the external contour, and an inner, folded membrane nestled neatly within the outer membrane (**figure 5.12a**). The folds on the inner membrane, called **cristae,*** may be tubular, like fingers, or folded into shelflike bands.

The cristae membranes hold the enzymes and electron carriers of aerobic respiration. This is an oxygen-using process that extracts chemical energy contained in nutrient molecules and stores it in the form of high-energy molecules, or ATP. More detailed functions of mitochondria are covered in chapter 8. The spaces around the cristae are filled with a chemically complex fluid called the **matrix,*** which holds ribosomes, DNA, and the pool of enzymes and other compounds involved in the metabolic cycle. Mitochondria (along with chloroplasts) are unique among organelles in that they divide independently of the cell, contain circular strands of DNA, and have procaryotic-sized 70S ribosomes. These findings

---

*mitochondria (my″-toh-kon′-dree-uh) sing. mitochondrion; Gr. *mitos,* thread, and *chondrion,* granule.

*cristae (kris′-te) sing. crista; L. *crista,* a comb.

*matrix (may′-triks) L. *mater,* mother or origin.

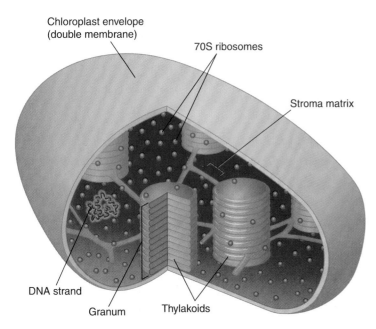

Chloroplast envelope (double membrane)

70S ribosomes

Stroma matrix

DNA strand

Granum

Thylakoids

FIGURE 5.13
Detail of an algal chloroplast.

have prompted some intriguing speculations on their evolutionary origins (see Spotlight on Microbiology 5.1).

CHLOROPLASTS: PHOTOSYNTHESIS MACHINES

Chloroplasts* are remarkable organelles found in algae and plant cells that are capable of converting the energy of sunlight into chemical energy through photosynthesis. The photosynthetic role of chloroplasts makes them the primary producers of organic nutrients upon which all other organisms (except certain bacteria) ultimately depend. Another important photosynthetic product of chloroplasts is oxygen gas. Although chloroplasts resemble mitochondria, chloroplasts are larger, contain special pigments, and are much more varied in shape.

There are differences among various algal chloroplasts, but most are generally composed of two membranes, one enclosing the other. The smooth, outer membrane completely covers an inner membrane folded into small, disclike sacs called **thylakoids*** that are stacked upon one another into **grana.** These structures carry the green pigment chlorophyll and sometimes additional pigments as well. Surrounding the thylakoids is a ground substance called the **stroma*** (**figure 5.13**). The role of the photosynthetic pigments is to absorb and transform solar energy into chemical energy, which is then converted by reactions in the stroma to carbohydrates. We further explore some important aspects of photosynthesis in chapters 7 and 26.

THE CYTOSKELETON: A SUPPORT NETWORK

The cytoplasm of a eucaryotic cell is penetrated by a flexible framework of molecules called the cytoskeleton (**figure 5.14**). This framework appears to have several functions, such as anchoring

*chloroplast (klor′-oh-plast) Gr. *chloros,* green, and *plastos,* to form.

*thylakoid (thy′-lah-koid) Gr. *thylakon,* a small sac, and *eidos,* form.

*stroma (stroh′-mah) Gr. *stroma,* mattress or bed.

organelles, providing support, and permitting shape changes and movement in some cells. The two main types of cytoskeletal elements are **microfilaments** and microtubules. Microfilaments are thin protein strands that attach to the cell membrane and form a network through the cytoplasm. Some microfilaments are responsible for movements of the cytoplasm, often made evident by the streaming of organelles around the cell in a cyclic pattern. Other microfilaments are active in *ameboid motion,* a type of movement typical of cells such as amebas and phagocytes that produces extensions of the cell membrane (pseudopods) into which the cytoplasm flows (see figure 5.29). Microtubules are long, hollow tubes that maintain the shape of eucaryotic cells without walls and transport substances from one part of a cell to another. The spindle fibers that play an essential role in mitosis are actually microtubules that attach to chromosomes and separate them into daughter cells. As indicated earlier, microtubules are also responsible for the movement of cilia and flagella.

RIBOSOMES: PROTEIN SYNTHESIZERS

In an electron micrograph of a eucaryotic cell, ribosomes are numerous, tiny particles that give a stippled appearance to the cytoplasm. Ribosomes are distributed in two ways: Some are scattered freely in the cytoplasm and cytoskeleton; others are intimately associated with the rough endoplasmic reticulum as previously described. Multiple ribosomes are often found arranged in short chains called polyribosomes (polysomes). The basic structure of eucaryotic ribosomes is similar to that of procaryotic ribosomes. Both are composed of large and small subunits of ribonucleoprotein (see figure 5.8). By contrast, however, the eucaryotic ribosome is the larger 80S variety that is a combination of 60S and 40S subunits. As in the procaryotes, eucaryotic ribosomes are the staging areas for protein synthesis.

CHAPTER CHECKPOINTS

The genome of eucaryotes is located in the nucleus, a spherical structure surrounded by a double membrane. The nucleus contains the nucleolus, the site of ribosome synthesis. DNA is organized into chromosomes in the nucleus and during meiosis and mitosis the chromatin is condensed.

All eucaryotes alternate between a haploid state and a diploid state at some point in their life cycle. Most fungi, many algae, and some protozoa are haploid except in their zygote stage.

Plants and animals, as well as some protozoa, fungi, and algae, are diploid for most of their life cycles, but their reproductive cells (gametes) are always haploid.

Microscopic eucaryotes reproduce asexually by mitosis and sexually by meiosis.

The endoplasmic reticulum (ER) is an internal network of membranous passageways extending throughout the cell.

The Golgi apparatus is a packaging center that receives materials from the ER and then forms vesicles around them for storage or for transport to the cell membrane for secretion.

The mitochondria generate energy in the form of ATP to be used in numerous cellular activities.

Chloroplasts, membranous packets found in plants and algae, are used in photosynthesis.

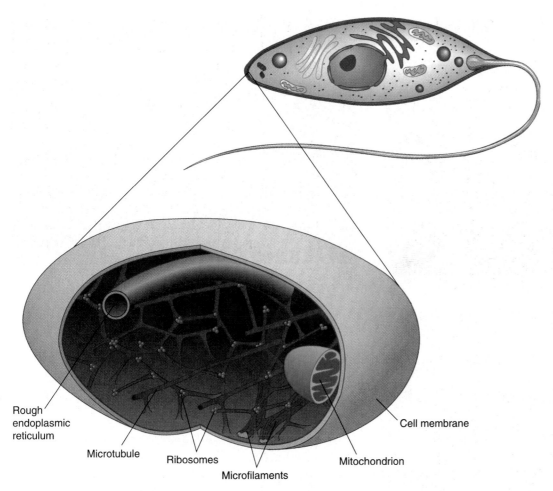

FIGURE 5.14

A model of the cytoskeleton.
Depicted is the relationship between microtubules, microfilaments, and organelles.

Rough endoplasmic reticulum

Microtubule Ribosomes Microfilaments Mitochondrion Cell membrane

The cytoskeleton maintains the shape of cells and produces movement of cytoplasm within the cell, movement of chromosomes at cell division, and, in some groups, movement of the cell as a unit.

Ribosomes are components of protein synthesis present in both eucaryotes and procaryotes.

Survey of Eucaryotic Microorganisms

With the general structure of the eucaryotic cell in mind, let us next examine the amazingly wide range of adaptations that this cell type has undergone. The following sections contain a general survey of the principal eucaryotic microorganisms—fungi, algae, protozoa, and parasitic worms—while simultaneously introducing elements of their structure, life history, classification, identification, and importance.

The Kingdom of the Fungi

The position of the **fungi*** in the biological world has been debated for many years. Although they were originally classified with the green plants (along with algae and bacteria), they were later separated from plants and placed in a group with algae and protozoa (the

Protista). Even at that time, however, many **mycologists*** were struck by several unique qualities of fungi that warranted their being placed into their own separate kingdom, and eventually they were.

The Kingdom Fungi, or Myceteae, is large and filled with forms of great variety and complexity. For practical purposes, the approximately 100,000 species of fungi can be divided into two groups: the *macroscopic fungi* (mushrooms, puffballs, gill fungi) and the *microscopic fungi* (molds, yeasts). Although the majority of fungi are either unicellular or colonial, a few complex forms such as mushrooms and puffballs do have a limited form of cellular specialization. Cells of the microscopic fungi exist in two basic morphological types: yeasts and hyphae. A yeast cell is distinguished by its round to oval shape and by its mode of asexual reproduction. It grows swellings on its surface called buds. **Hyphae*** are long, threadlike cells found in the bodies of filamentous fungi, or molds **(figure 5.15)**. Some species form a **pseudohypha,*** a chain of yeasts formed when buds remain attached in a row **(figure 5.16)**. Because of the manner of formation, it is not a true hypha like that of molds. While some fungal cells exist only in a yeast form

*fungi (fun´-jy) sing. fungus; Gr. *fungos,* mushroom.

*mycologist (my-kol´-uh-jist) Gr. *mykes,* standard root for fungus. Scientists who study fungi.

*hypha (hy´-fuh) pl. hyphae (hy´-fee) Gr. *hyphe,* a web.

*pseudohypha (soo˝-doh-hy´-fuh) pl. pseudohyphae; Gr. *pseudo,* false, and *hyphe.*

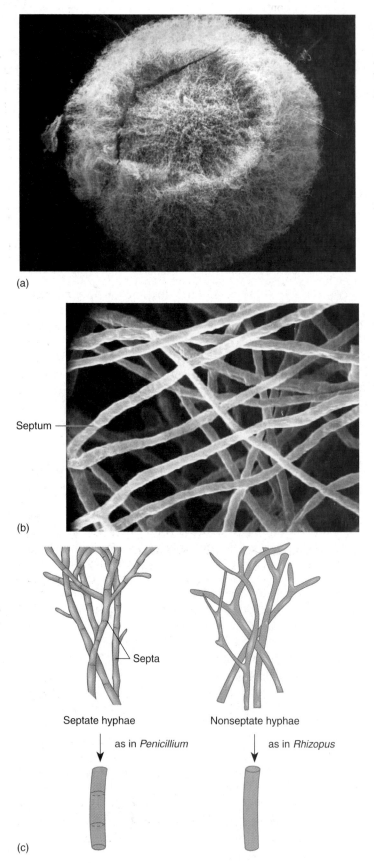

(a)

(b)

Septum

(c)

Septa

Septate hyphae

as in *Penicillium*

Nonseptate hyphae

as in *Rhizopus*

FIGURE 5.15

***Diplodia maydis*, a pathogenic fungus of corn plants.** **(a)** Scanning electron micrograph of a single colony showing its filamentous texture (24×). **(b)** Close-up of hyphal structure (1,200×). **(c)** Basic structural types of hyphae.

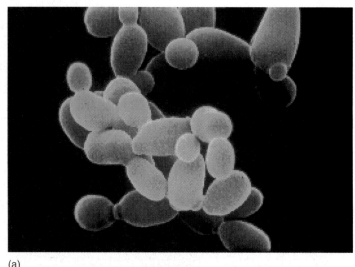

(a)

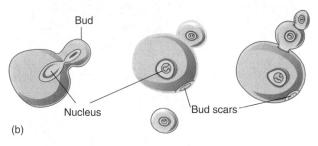

Bud

Nucleus

Bud scars

(b)

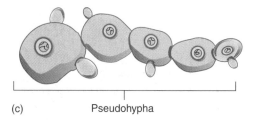

(c)   Pseudohypha

FIGURE 5.16

Microscopic morphology of yeasts. **(a)** Scanning electron micrograph of the brewer's, or baker's, yeast *Saccharomyces cerevisiae* (21,000×). **(b)** Formation and release of yeast buds. **(c)** Formation of pseudohypha (a chain of budding yeast cells).

and others occur primarily as hyphae, a few, called **dimorphic,*** can take either form, depending upon growth conditions, especially changing temperature. This variability in growth form is particularly characteristic of some pathogenic molds (see figure 22.1).

FUNGAL NUTRITION

All fungi are **heterotrophic.*** They acquire nutrients from a wide variety of organic materials called **substrates (figure 5.17).** Most fungi are **saprobes,*** meaning that they obtain these substrates

---

*dimorphic (dy-mor′-fik) Gr. *di,* two, and *morphe,* form.

*heterotrophic (het-ur-oh-tro′-fik) Gr. *hetero,* other, and *troph,* to feed. A type of nutrition that relies on an organic nutrient source.

*saprobe (sap′-rohb) Gr. *sapros,* rotten, and *bios,* to live. Also called saprotroph or saprophyte.

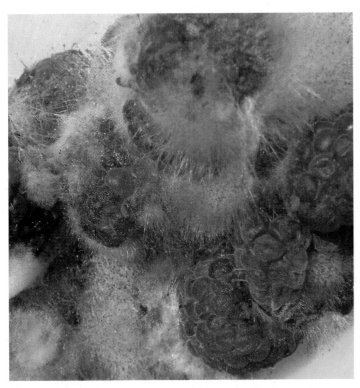

(a)

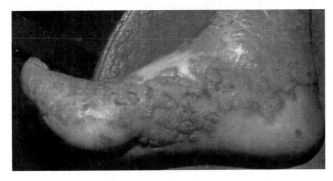

(b)

FIGURE 5.17

A small assortment of nutritional sources (substrates) of fungi. **(a)** A fungal mycelium growing on raspberries. The fine hyphal filaments and black sporangia are typical of *Rhizopus*. **(b)** The skin of the foot infected by a soil fungus, *Fonsecaea pedrosoi*.

from the remnants of dead plants and animals in soil or aquatic habitats. Fungi can also be parasites on the bodies of living animals or plants, although very few fungi absolutely require a living host. In general, the fungus penetrates the substrate and secretes enzymes that reduce it to small molecules that can be absorbed by the cells. Fungi have enzymes for digesting an incredible array of substances, including feathers, hair, cellulose, petroleum products, wood, and rubber. It has been said that every naturally occurring organic material on the earth can be attacked by some type of fungus. Fungi are often found in nutritionally poor or adverse environments. In our laboratory, we occasionally discover fungi in mineral and weak acid solutions ("bottle imps") and on specimens preserved in formaldehyde. Various fungi thrive in substrates with high salt or sugar content, at relatively high temperatures, and even in snow and glaciers. Their medical and agricultural impact is extensive. A number of species cause **mycoses** (fungal infections) in animals, and thousands of species are important plant pathogens. Fungal toxins may cause disease in humans, and airborne fungi are a frequent cause of allergies and other medical conditions (**Spotlight on Microbiology 5.2**).

ORGANIZATION OF MICROSCOPIC FUNGI

The cells of most microscopic fungi grow in loose associations or colonies. The colonies of yeasts are much like those of bacteria in that they have a soft, uniform texture and appearance. The colonies of filamentous fungi are noted for the striking cottony, hairy, or velvety textures that arise from their microscopic organization and morphology. The woven, intertwining mass of hyphae that makes up the body or colony of a mold is called a **mycelium*** (see figure 5.16).

Although hyphae contain the usual eucaryotic organelles, they also have some unique organizational features. In most fungi, the hyphae are divided into segments by cross walls, or **septa,*** a condition called septate. The nature of the septa varies from solid partitions with no communication between the compartments to partial walls with small pores that allow the flow of organelles and nutrients between adjacent compartments. Nonseptate hyphae consist of one long, continuous cell *not* divided into individual compartments by cross walls. With this construction, the cytoplasm and organelles move freely from one region to another, and each hyphal element can have several nuclei.

Hyphae can also be classified according to their particular function. **Vegetative hyphae** (mycelia) are responsible for the visible mass of growth that appears on the surface of a substrate and penetrates it to digest and absorb nutrients. During the development of a fungal colony, the vegetative hyphae give rise to structures called **reproductive,** or **fertile, hyphae** which branch off vegetative mycelium. These hyphae are responsible for the production of fungal reproductive bodies called **spores.** Other specializations of hyphae are illustrated in **figure 5.18.**

REPRODUCTIVE STRATEGIES AND SPORE FORMATION

Fungi have many complex and successful reproductive strategies. Most can propagate by the simple outward growth of existing hyphae or by fragmentation, in which a separated piece of mycelium can generate a whole new colony. But the primary reproductive mode of fungi involves the production of various types of spores. Do not confuse fungal spores with the more resistant, nonreproductive bacterial spores. Fungal spores are responsible not only for multiplication but also for survival, producing genetic variation, and dissemination. Because of their compactness and relatively light weight, spores are dispersed widely through the environment by air, water, and living things. Upon encountering a favorable

---

*mycelium (my-see′-lee-um) pl. mycelia; Gr. *mykes*, fungus, and *helos*, nail.

*septa (sep′-tuh) sing. septum; L. *saepio*, to fence or wall.

FIGURE 5.18

Functional types of hyphae using the mold Rhizopus as an example. **(a)** Vegetative hyphae are those surface and submerged filaments that digest, absorb, and distribute nutrients from the substrate. This species also has special anchoring structures called rhizoids. **(b)** Later, as the mold matures, it sprouts reproductive hyphae that produce asexual spores. **(c)** During the asexual life cycle, the free mold spores settle on a substrate and send out germ tubes that elongate into hyphae. Through continued growth and branching, an extensive mycelium is produced. So prolific are the fungi that a single colony of mold can easily contain 5,000 spore-bearing structures. If each of these released 2,000 single spores and if every spore released from every sporangium on every mold were able to germinate, we would soon find ourselves in a sea of mycelia. Most spores do not germinate, but enough are successful to keep the numbers of fungi and their spores very high in most habitats.

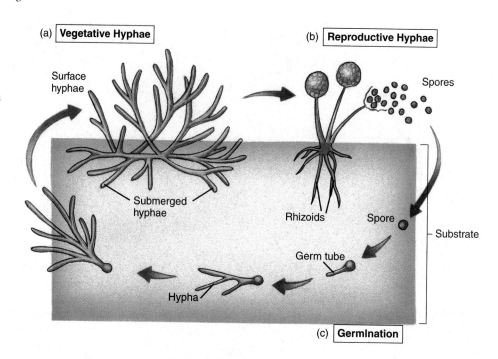

(a) **Vegetative Hyphae**

Surface hyphae

Submerged hyphae

(b) **Reproductive Hyphae**

Spores

Rhizoids   Spore

Germ tube

Hypha

Substrate

(c) **Germination**

substrate, a spore will germinate and produce a new fungus colony in a very short time (figure 5.18).

The fungi exhibit such a marked diversity in spores that they are largely classified and identified by their spores and spore-forming structures. Although there are some elaborate systems for naming and classifying spores, we will present a basic overview of the principal types. The most general subdivision is based on the way the spores arise. **Asexual spores** are the products of mitotic division of a single parent cell, and **sexual spores** are formed through a process involving the fusing of two parental nuclei followed by meiosis.

Asexual Spore Formation

On the basis of the nature of the reproductive hypha and the manner in which the spores originate, there are two subtypes of asexual spore **(figure 5.19):**

1. **Sporangiospores** (figure 5.19*a*) are formed by successive cleavages within a saclike head called a **sporangium,*** which is attached to a stalk, the sporangiophore. These spores are initially enclosed but are released when the sporangium ruptures.

2. **Conidia*** (conidiospores) are free spores not enclosed by a spore-bearing sac (figure 5.19*b*). They develop either by the pinching off of the tip of a special fertile hypha or by the segmentation of a preexisting vegetative hypha. Conidia are the most common asexual spores, and they occur in these forms:

   arthrospore (ar′-thro-spor) Gr. *arthron,* joint. A rectangular spore formed when a septate hypha fragments at the cross walls.

   chlamydospore (klam-ih′-doh-spor) Gr. *chlamys,* cloak. A spherical conidium formed by the thickening of a hyphal cell. It is released when the surrounding hypha fractures, and it serves as a survival or resting cell.

   blastospore. A spore produced by budding from a parent cell that is a yeast or another conidium; also called a bud.

   phialospore (fy′-ah-lo-spor) Gr. *phialos,* a vessel. A conidium that is budded from the mouth of a vase-shaped spore-bearing cell called a phialide or *sterigma,* leaving a small collar.

   microconidium and *macroconidium.* The smaller and larger conidia formed by the same fungus under varying conditions. Microconidia are one-celled, and macroconidia have two or more cells.

   porospore. A conidium that grows out through small pores in the spore-bearing cell; some are composed of several cells.

Sexual Spore Formation

If fungi can propagate themselves successfully with millions of asexual spores, what is the function of their sexual spores? The answer lies in important variations that occur when fungi of different genetic makeup combine their genetic material. Just as in plants and animals, this linking of genes from two parents creates offspring with combinations of genes different from that of either parent. The offspring from such a union can have slight variations in form and function that are potentially advantageous in the adaptation and survival of their species. Because fungal cells spend most of their life cycle in the haploid state, their cells already have a chromosome number compatible for sexual union. In general, during sexual reproduction, the haploid nuclei from two compatible parental cells fuse, forming a diploid nucleus that subsequently participates in sexual spore development (see figure 5.7*a,* step 2).

---

*sporangium (spo-ran′-jee-um) pl. sporangia; Gr. *sporos,* seed, and *angeion,* vessel.

*conidia (koh-nid′-ee-uh) sing. conidium; Gr. *konidion,* a particle of dust.

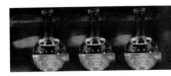

The Mother of All Fungi

Far from being microscopic, one of the largest organisms in the world is a massive basidiomycete growing in a forest in eastern Oregon. The mycelium of a single colony of *Armillaria ostoye* covers an area of 2,200 acres and stretches 3.5 miles across and 3 feet into the ground. Most of the fungus lies hidden beneath the ground, where it periodically fruits into edible mushrooms. Experts with the forest service took random samples over a large area and found that the mycelium is from genetically identical stock. They believe this fungus was started from a single mating pair of sexual hyphae around 2,400 years ago. Another colossal species from Washington state covers 1,500 acres and weighs around 100 tons. Both fungi penetrate the root structure of trees and spread slowly from tree to tree, sapping their nutrients and gradually destroying whole forests.

A Fungus in Your Future

Biologists are developing some rather imaginative uses for fungi as a way of controlling both the life and death of plants. Dutch and Canadian researchers studying ways to control a devastating fungus infection of elm trees (Dutch elm disease) have come up with a brand new use of an old method—they actually vaccinate the trees. Ordinarily, the disease fungus invades the plant vessels and chokes off the flow of water. The natural tendency of the elm to defend itself by surrounding and inactivating the fungus is too slow to save it from death. But treating the elm trees before they get infected helps them develop an immunity to the disease. Plants are vaccinated somewhat like humans and animals: nonpathogenic spores or proteins from fungi are injected into the tree over a period of time. So far it appears that the symptoms of disease and the degree of damage can be significantly reduced. This may be the start of a whole new way to control fungal pests.

At the other extreme, government biologists working for narcotic control agencies have unveiled a recent plan to use fungi to kill unwanted plants. The main targets would be plants grown to produce illegal drugs like cocaine and heroin in the hopes of cutting down on these drugs right at the source. A fungus infection (*Fusarium*) that wiped out 30% of the coca crop in Peru dramatically demonstrated how effective this might be. Since then, at least two other fungi that could destroy opium poppies and marijuana plants have been isolated. Purposefully releasing plant pathogens such as *Fusarium* into the environment has stirred a great deal of controversy. Critics in South America emphasize that even if the fungus appears specific to a particular plant, there is too much potential for it to switch hosts to food and ornamental plants and wreak havoc with the ecosystem. United States biologists who support the plan of using fungal control agents say that it is not as dangerous as massive spraying with pesticides, and that extensive laboratory tests have proved that the species of fungi being used will be very specific to the illegal drug plants and will not affect close relatives. Limited field tests will be started in the near future, paid for by a billion-dollar fund created by the U.S. government as part of its war on drugs.

Fungi, Fungi, Everywhere

The importance of fungi in the ecological structure of the earth is well founded. They are essential contributors to complex environments such as soil, and they play numerous beneficial roles as decomposers of organic debris and as partners to plants. Fungi also have great practical importance due to their metabolic versatility. They are productive sources of drugs (penicillin) to treat human infections and other diseases, and they are used in industry to ferment foods and synthesize organic chemicals.

The fact that they are so widespread also means that they frequently share human living quarters, especially in locations that provide ample moisture and nutrients. Often their presence is harmless and limited to a film of mildew on shower stalls or other moist environments. In some cases, depending on the amount of contamination and the type of mold, these indoor fungi can also give rise to various medical problems. Such common air contaminants as *Penicillium, Aspergillus, Cladosporium,* and *Stachybotrys* (see figure 5.25) all have the capacity to give off airborne spores and toxins that, when inhaled, cause a whole spectrum of symptoms sometimes referred to as "sick building syndrome." The usual source of harmful fungi is the presence of chronically water-damaged walls, ceilings, and other building materials that have come to harbor these fungi. Such materials contain plant products that serve as a rich source of nutrients. People exposed to these houses or buildings report symptoms that range from skin rash, flulike reactions, sore throat, and headaches to fatigue, diarrhea, allergies, and immune suppression. Because they can mimic other diseases, the association with indoor toxic fungi may be missed. Recent reports of sick buildings have been on the rise, affecting thousands of people, and some deaths have been reported in small children. The control of indoor fungi requires correcting the moisture problem, removing the infested materials, and decontaminating the living spaces. Mycologists are currently studying the mechanisms of toxic effects with an aim to develop better diagnosis and treatment.

Armillaria (honey) mushrooms sprouting from the base of a tree in Oregon provide only a tiny hint of the massive fungus from which they arose.

140 CHAPTER 5 Eucaryotic Cells and Microorganisms

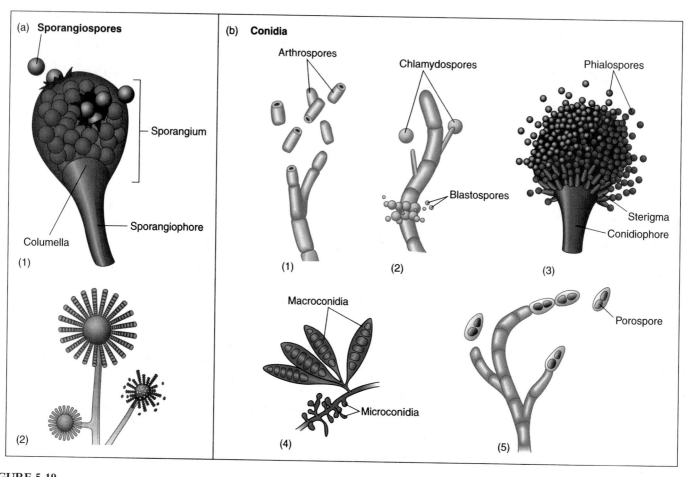

FIGURE 5.19

Types of asexual mold spores. **(a)** Sporangiospores: (*1*) *Absidia*, (*2*) *Syncephalastrum*, **(b)** Conidia: (*1*) arthrospores (e.g., *Coccidioides*), (*2*) chlamydospores and blastospores (e.g., *Candida albicans*), (*3*) phialospores (e.g., *Aspergillus*), (*4*) macroconidia and microconidia (e.g., *Microsporum*), and (*5*) porospores (e.g., *Alternaria*).

The majority of fungi produce sexual spores at some point. The nature of this process varies from the simple fusion of fertile hyphae of two different strains to a complex union of differentiated male and female structures and the development of special fruiting structures. We will consider the three most common sexual spores: zygospores, ascospores, and basidiospores. These spore types provide an important basis for classifying the major fungal divisions.

Zygospores* are sturdy diploid spores formed when hyphae of two opposite strains (called the plus and minus strains) fuse and create a diploid zygote that swells and becomes covered by strong, spiny walls **(figure 5.20).** When its wall is disrupted, and moisture and nutrient conditions are suitable, the zygospore germinates and forms a mycelium that gives rise to a sporangium. Meiosis of diploid cells of the sporangium results in haploid nuclei that develop into sporangiospores. Both the sporangia and the sporangiospores that arise from sexual processes are outwardly

identical to the asexual type, but because the spores arose from the union of two separate fungal parents, they are not genetically identical.

In general, haploid spores called **ascospores*** are created inside a special fungal sac, or **ascus** (pl. asci) **(figure 5.21).** Although details can vary among types of fungi, the ascus and ascospores are formed when two different strains or sexes join together to produce offspring. In many species, the male sexual organ fuses with the female sexual organ. The end result is a number of terminal cells, each containing a diploid nucleus. Through differentiation, each of these cells enlarges to form an ascus, and its diploid nucleus undergoes meiosis (often followed by mitosis) to form four to eight haploid nuclei that will mature into ascospores. A ripe ascus breaks open and releases the ascospores. Some species form an elaborate fruiting body to hold the asci (figure 5.21).

---

*zygospores (zy'-goh-sporz) Gr. *zygon*, yoke, to join.

*ascospores (as'-koh-sporz) Gr. *ascos*, a sac.

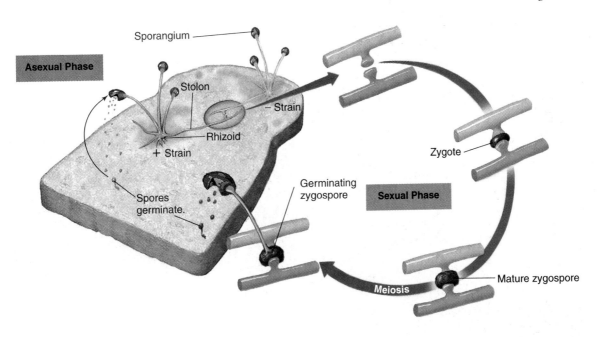

FIGURE 5.20

Formation of zygospores in *Rhizopus stolonifer*. Sexual reproduction occurs when two mating strains of hyphae grow together, fuse, and form a mature zygospore. Germination of the zygospore involves meiotic division and production of a haploid sporangium that looks just like the asexual one.

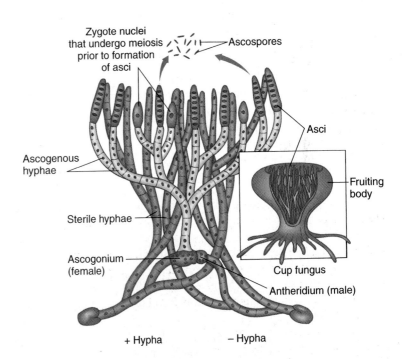

FIGURE 5.21

Production of ascospores in a cup fungus.
Inset shows the cup-shaped fruiting body that houses the asci.

Basidiospores* are haploid sexual spores formed on the outside of a club-shaped cell called a **basidium (figure 5.22).** In general, spore formation follows the same pattern of two mating types coming together, fusing, and forming terminal cells with diploid nuclei. Each of these cells becomes a basidium, and its nucleus produces, through meiosis, four haploid nuclei. These nuclei are ex-truded through the top of the basidium, where they develop into basidiospores. Notice the location of the basidia along the gills in mushrooms, which are often dark from the spores they contain. It may be a surprise to discover that the fleshy part of a mushroom is actually a fruiting body designed to protect and help disseminate its sexual spores.

---

*basidiospores (bah-sid′-ee-oh-sporz) Gr. *basidi,* a pedestal.

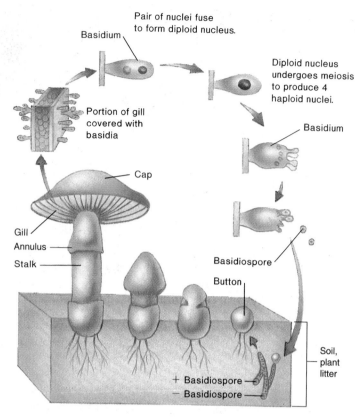

FIGURE 5.22
Formation of basidiospores in a mushroom.

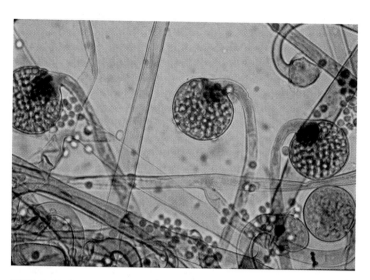

FIGURE 5.23

A representative Zygomycota, *Circinella*. Note the sporangia, sporangiospores, and nonseptate hyphae.

FUNGAL CLASSIFICATION

It is often difficult for microbiologists to assign logical and useful classification schemes to microorganisms while at the same time preserving their relationships. This difficulty is due to the fact that the organisms do not always perfectly fit the neat categories made for them, and even experts cannot always agree on the nature of the categories. The fungi are no exception, and there are several ways to classify them. For our purposes, we will adopt a classification scheme with a medical mycology emphasis, in which the Kingdom Fungi is subdivided into two subkingdoms, the Amastigomycota and the Mastigomycota. The Amastigomycota are common inhabitants of terrestrial habitats. Some of the characteristics used to separate them into subgroups are the type of sexual reproduction and their hyphal structure. The Mastigomycota are primitive filamentous fungi that live primarily in water and may cause disease in potatoes and grapes.

The next section outlines the four divisions of Amastigomycota, including major characteristics and important members.

Amastigomycota: Fungi That Produce Sexual and Asexual Spores (Perfect)

Division I—Zygomycota (also Phycomycetes) Sexual spores: zygospores; asexual spores: mostly sporangiospores, some conidia. Hyphae are usually nonseptate. If septate, the septa are complete. Most species are free-living saprobes; some are animal parasites. Can be obnoxious contaminants in the laboratory and on food and vegeta-

bles. Examples are mostly molds: *Rhizopus,* a black bread mold; *Mucor; Syncephalastrum; Circinella* (**figure 5.23,** asexual phase only).

Division II—Ascomycota (also Ascomycetes) Sexual spores: produce ascospores in asci, asexual spores: many types of conidia, formed at the tips of conidiophores. Hyphae with porous septa. Many important species. Examples: *Histoplasma,* the cause of Ohio Valley fever; *Microsporum,* one cause of ringworm (a common name for certain fungal skin infections that often grow in a ringed pattern); *Penicillium,* one source of antibiotics (**figure 5.24,** asexual phase only); and *Saccharomyces,* a yeast used in making bread and beer. Most of the species in this division are either molds or yeasts; it also includes many human and plant pathogens, such as *Pneumocystis carinii,* a pathogen of AIDS patients.

Division III—Basidiomycota (also Basidiomycetes) Sexual reproduction by means of basidia and basidiospores; asexual spores: conidia. Incompletely septate hyphae. Some plant parasites and one human pathogen. Fleshy fruiting bodies are common. Examples: mushrooms, puffballs, bracket fungi, and plant pathogens called rusts and smuts. The one human pathogen, the yeast *Cryptococcus neoformans,* causes an invasive systemic infection in several organs, including the skin, brain, and lungs (see figure 22.26).

Amastigomycota: Fungi That Produce Only Asexual Spores (Imperfect)

From the beginnings of fungal classification, any fungus that lacked a sexual state was called "imperfect" and was placed in a catchall category, the **Fungi Imperfecti,** or **Deuteromycota.** A species would remain classified in that category until its sexual state was described (if ever). Gradually, many species of Fungi Imperfecti were found to make sexual spores, and they were assigned to the taxonomic grouping that best fit those spores. This system created a quandary, because several very important species, especially pathogens, had to be regrouped and renamed. In many cases, the

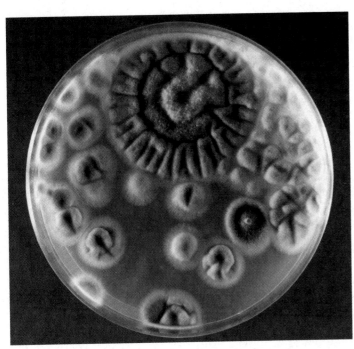

(a)

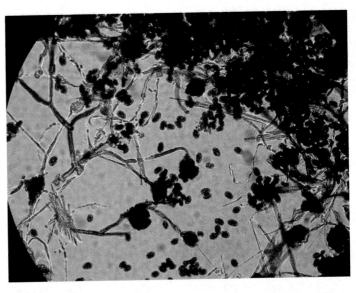

FIGURE 5.25

Mycelium and spores of a representative of Deuteromycota called *Stachybotrys.* This black mold is implicated in building contamination that leads to toxic diseases (see Spotlight on Microbiology 5.2).

(b)

FIGURE 5.24

A common Ascomycota, *Penicillium*. **(a)** Macroscopic view of a typical blue-green colony. **(b)** Microscopic view shows the brush arrangement of phialospores (220×).

especially the imperfect states of *Blastomyces* and *Microsporum.* Other species are *Coccidioides immitis,* the cause of valley fever; *Candida albicans,* the cause of various yeast infections; and *Cladosporium,* a common mildew fungus, and *Strachybotrys,* a toxic mold **(figure 5.25).**

FUNGAL IDENTIFICATION AND CULTIVATION

Fungi are identified in medical specimens by first being isolated on special types of media and then being observed macroscopically and microscopically. Examples of media for cultivating fungi are cornmeal, blood, and Sabouraud's agar. The latter medium is useful in isolating fungi from mixed samples because of its low pH, which inhibits the growth of bacteria but not most fungi. Because the fungi are classified into general groups by the presence and type of sexual spores, it would seem logical to identify them in the same way, but sexual spores are rarely if ever demonstrated in the laboratory setting. As a result, the asexual spore-forming structures and spores are usually used to identify organisms to the level of genus and species. Other characteristics that contribute to identification are hyphal type, colony texture and pigmentation, physiological characteristics, and genetic makeup (see figure 22.3).

THE ROLES OF FUNGI IN NATURE AND INDUSTRY

Nearly all fungi are free-living and do not require a host to complete their life cycles. Even among those fungi that are pathogenic, most human infection occurs through accidental contact with an environmental source such as soil, water, or dust. Humans are generally quite resistant to fungal infection, except for two main types of fungal pathogens: the true pathogens, which can infect even healthy persons, and the opportunistic pathogens, which attack persons who are already weakened in some way.

older names were so well entrenched in the literature that it was easier to retain them and assign a new generic name for their sexual stage. Consequently, *Blastomyces* and *Histoplasma* are also known as *Ajellomyces* (sexual phase).

Division IV—Deuteromycota* Asexual spores: conidia of various types. Hyphae septate. Majority are yeasts or molds, some dimorphic. Saprobes and a few animal and plant parasites. Examples: Several human pathogens were originally placed in this group,

---

***Deuteromycota** (doo´-ter-oh-my-koh˝-tuh) Gr. *deutero,* second, and *mykes,* fungus.

TABLE 5.1

Major Fungal Infections of Humans

Degree of Tissue Involvement and Area Affected	Name of Infection	Name of Causative Fungus
Superficial (not deeply invasive)		
Outer epidermis	Pityriasis versicolor	*Malassezia furfur*
Epidermis, hair, and dermis can be attacked	Dermatophytosis, also called tinea or ringworm of the scalp, body, feet (athlete's foot), toenails	*Microsporum, Trichophyton,* and *Epidermophyton*
Mucous membranes, skin, nails	Candidiasis, or yeast infection	*Candida albicans*
Subcutaneous (invades just beneath the skin)		
Nodules beneath the skin; can invade lymphatics	Sporotrichosis	*Sporothrix schenckii*
Large fungal tumors of limbs and extremities	Mycetoma, madura foot	*Pseudallescheria boydii, Madurella mycetomatis*
Systemic (deep; organism enters lungs; can invade other organs)		
Lung	Coccidioidomycosis (San Joaquin Valley fever)	*Coccidioides immitis*
	North American blastomycosis (Chicago disease)	*Blastomyces dermatitidis*
	Histoplasmosis (Ohio Valley fever)	*Histoplasma capsulatum*
	Cryptococcosis (torulosis)	*Cryptococcus neoformans*
Lung, skin	Paracoccidioidomycosis (South American blastomycosis)	*Paracoccidioides brasiliensis*

Mycoses (fungal infections) vary in the way the agent enters the body and the degree of tissue involvement (**table 5.1**). The list of opportunistic fungal pathogens has been increasing in the past few years because of newer medical techniques that keep immunocompromised patients alive. Even so-called harmless species found in the air and dust around us may be able to cause opportunistic infections in patients who already have AIDS, cancer, or diabetes.

Fungi are involved in other medical conditions besides infections (see Spotlight on Microbiology 5.2). Fungal cell walls give off chemical substances that can cause allergies. The toxins produced by poisonous mushrooms can induce neurological disturbances and even death. The mold *Aspergillus flavus* (see figure 22.27) synthesizes a potentially lethal poison called aflatoxin,[4] which is the cause of a disease in domestic animals that have eaten grain infested with the mold and is also a cause of liver cancer in humans.

Fungi pose an ever-present economic hindrance to the agricultural industry. A number of species are pathogenic to field plants such as corn and grain, and fungi also rot harvested fresh produce during shipping and storage. It has been estimated that as much as 40% of the yearly fruit crop is consumed not by humans but by fungi. On the beneficial side, however, fungi play an essential role in decomposing organic matter and returning essential minerals to the soil. They form stable associations with plant roots (mycorrhizae) that increase the ability of the roots to absorb water and nutrients. Industry has tapped the biochemical potential of fungi to produce large quantities of antibiotics, alcohol, organic acids, and vitamins. Some fungi are eaten or used to impart flavorings to food. The yeast *Saccharomyces* produces the alcohol in beer and wine and the gas that causes bread to rise. Blue cheese, soy sauce, and cured meats derive their unique flavors from the actions of fungi (see chapter 26).

The Protists

The algae and protozoa have been traditionally combined into the Kingdom Protista. The two major taxonomic categories of this kingdom are Subkingdom Algae and Subkingdom Protozoa. Although these general types of microbes are now known to occupy several kingdoms, it is still useful to retain the concept of a protist as any unicellular or colonial organism that lacks true tissues. This concept is especially helpful in our survey of major groups of microorganisms, because many algae are multicellular and macroscopic, and we cannot present an exhaustive coverage of all forms.

THE ALGAE: PHOTOSYNTHETIC PROTISTS

The **algae*** are a group of photosynthetic organisms usually recognized by their larger members, such as seaweeds and kelps. In addition to being beautifully colored and diverse in appearance, they vary in length from a few micrometers to 100 meters. Algae occur in unicellular, colonial, and filamentous forms, and the larger forms can possess tissues and simple organs. **Table 5.2** lists the major characteristics of the various subgroups of algae.

Algal cells as a group exhibit all of the eucaryotic organelles. The most noticeable of these are the chloroplasts, which contain, in addition to the green pigment chlorophyll, a number of other pigments that create the yellow, red, and brown coloration of some groups (table 5.2). The chloroplasts can be of such unique character that they are used in identification (for example, *Spirogyra* in **figure 5.26b**). Most algal cells are enclosed by a complex cell wall, which accounts for the distinctive appearance of unicellular members such as diatoms (figure 5.26c) and dinoflagellates. The outermost structure in a group called the euglenids (for example, *Euglena* in figure 5.26a) is a thick, flexible membrane called a **pellicle.** Motility by flagella or gliding is common among the algae, and many members contain tiny light-sensitive areas (eyespots) that coordinate with the flagella to guide the cell toward the light it requires for photosynthesis.

Algae are widespread inhabitants of fresh and marine waters. They are one of the main components of the large floating

---

4. From **a**spergillus, **fl**avus, **tox**in.

*algae (al'-jee) sing. alga; L. seaweeds.

TABLE 5.2

Summary of Algal Characteristics

Group	Organization	Primary Habitat	Cell Wall	Pigmentation	Ecology/ Importance	Example(s)
Euglenophyta (euglenids)	Mainly unicellular; motile by flagella	Fresh water	None; pellicle instead	Chlorophyll, carotenoids, xanthophyll	Some are close relatives of Mastigophora	*Euglena*
Pyrrophyta (dinoflagellates)	Unicellular, dual flagella	Marine plankton	Cellulose or atypical wall	Chlorophyll, carotenoids	Cause of "red tide"	*Gonyaulax*
Chrysophyta (diatoms or golden-brown algae)	Mainly unicellular, some filamentous forms, unusual form of motility	Fresh water and marine	Silicon dioxide	Chlorophyll, fucoxanthin	Diatomaceous earth, major component of plankton	*Navicula,* other diatoms
Phaeophyta (brown algae— kelps)	Multicellular, vascular system, holdfasts	Marine, subtidal forests	Cellulose, alginic acid	Chlorophyll, carotenoids, fucoxanthin	Source of an emulsifier, alginate	*Fucus, Sargassum*
Rhodophyta (red seaweeds)	Multicellular	Marine, intertidal forests	Cellulose	Chlorophyll, carotenoids, xanthophyll, phycobilin	Source of agar and carrageenan, a food additive	*Gelidium*
Chlorophyta (green algae, grouped with plants)	Varies from unicellular, colonial, filamentous, to multicellular	Fresh water and salt water	Cellulose	Chlorophyll, carotenoids, xanthophyll	Precursor of higher plants	*Chlamydomonas, Spirogyra, Volvox*

community of microscopic organisms called **plankton.** In this capacity, they play an essential role in the aquatic food web and produce most of the earth's oxygen. Other algal habitats include the surface of soil, rocks, and plants, and several species are even hardy enough to live in hot springs or snowbanks (see figure 7.10).

Algae are often described by common names such as green algae, brown algae, golden brown algae, and red algae in reference to their predominant color. A more technical system divides the microscopic algae into divisions or kingdoms based on the types of chlorophyll and other pigments, the type of cell covering, the nature of their stored foods, and genetic factors. The common names for these groups are:

1. Euglenophyta, or euglenids;
2. Pyrrophyta, or dinoflagellates;
3. Chrysophyta, or diatoms;
4. Phaeophyta, or brown algae;
5. Rhodophyta, or red seaweeds; and
6. Chlorophyta, or green algae.

Algae reproduce asexually through fragmentation, binary fission, and mitosis, and some produce motile spores. Their sexual reproductive cycles can be highly complex, with stages similar to those of fungi. The capacity of algae to photosynthesize is extremely important to the earth. They form the basis of aquatic food webs and play an essential role in the earth's oxygen and carbon dioxide balance. Some products of algae have industrial applications. For example, fossilized marine diatoms yield an abrasive

powder called diatomaceous earth that is used in polishes, bricks, and filters, and certain seaweeds are a source of agar and algin used in microbiology, dentistry, and the food and cosmetics industries. We consider the role of algae in microbial ecology in chapter 26.

Animal tissues would be rather inhospitable to algae, so algae are rarely infectious. One exception is *Prototheca,* an unusual nonphotosynthetic alga, which has been associated with skin and subcutaneous infections in humans and animals.

The primary medical threat from algae is due to a type of food poisoning caused by the toxins of certain marine **dinoflagellates.** During particular seasons of the year, the overgrowth of these motile algae imparts a brilliant red color to the water, which is referred to as a "red tide." When intertidal animals feed, their bodies accumulate toxins given off by the dinoflagellates that can persist for several months (see figure 26.19). Paralytic shellfish poisoning is caused by eating exposed clams or other invertebrates. It is marked by severe neurological symptoms and can be fatal. Ciguatera is a serious intoxication caused by dinoflagellate toxins that have accumulated in fish such as bass and mackerel. Cooking does not destroy the toxin, and there is no antidote.

Several episodes of a severe infection caused by *Pfiesteria piscicida,* a toxic dinoflagellate, have been reported over the past several years in the United States. The disease was first reported in fish and was later transmitted to humans. This newly identified species occurs in at least 20 forms, including spores, cysts, and amebas (see figure 5.26*d*), that can release potent toxins. Both fishes and humans develop neurological symptoms and bloody

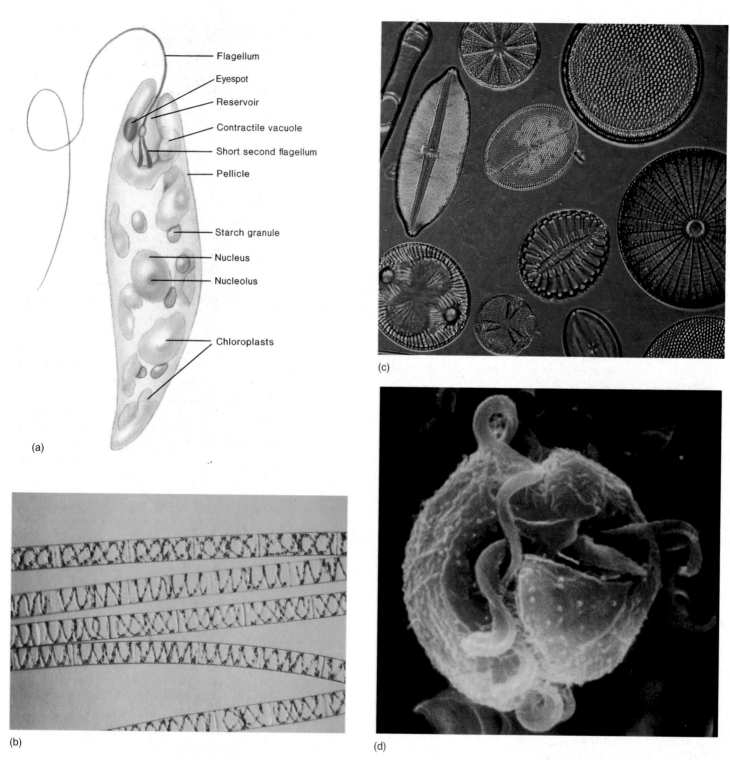

FIGURE 5.26

Representative microscopic algae. **(a)** *Euglena*, a unicellular motile example (150×). **(b)** *Spirogyra*, a colonial filamentous form with spiral chloroplasts. **(c)** A strew of beautiful chrysophyta—diatoms—shows the intricate and varied structure of their silica cell wall. **(d)** Zoospore of the dinoflagellate *Pfiesteria piscicida*. Although it is free-living, it is known to parasitize fish and release potent toxins that kill fish and sicken humans.

skin lesions. The cause of the epidemic has been traced to nutrient-rich agricultural runoff water that promoted the sudden "bloom" of dinoflagellates. Those microbes first attacked and killed millions of fish and later people whose occupations exposed them to fish and contaminated water.

BIOLOGY OF THE PROTOZOA

If a poll were taken to choose the most engrossing and vivid group of microorganisms, many people would choose the protozoa. Although their name comes from the Greek for "first animals," they are far from being simple, primitive organisms. The protozoa constitute a very large group (about 65,000 species) of creatures that although single-celled, have startling properties when it comes to movement, feeding, and behavior. Although most members of this group are harmless, free-living inhabitants of water and soil, a few species are parasites collectively responsible for hundreds of millions of infections of humans each year. Before we consider a few examples of important pathogens, let us examine some general aspects of protozoan biology.

Protozoan Form and Function

Most protozoan cells are single cells containing the major eucaryotic organelles except chloroplasts (see figure 5.2c). Their organelles can be highly specialized for feeding, reproduction, and locomotion. The cytoplasm is usually divided into a clear outer layer called the **ectoplasm** and a granular inner region called the *endoplasm.* Ectoplasm is involved in locomotion, feeding and protection. Endoplasm houses the nucleus, mitochondria, and food and contractile vacuoles. Some ciliates and flagellates[5] even have organelles that work somewhat like a primitive nervous system to coordinate movement. Because protozoa lack a cell wall, they have a certain amount of flexibility. Their outer boundary is a cell membrane that regulates the movement of food, wastes, and secretions. Cell shape can remain constant (as in most ciliates) or can change constantly (as in amebas). Certain amebas (foraminiferans) encase themselves in hard shells made of calcium carbonate. The size of most protozoan cells falls within the range of 3 to 300 μm. Some notable exceptions are giant amebas and ciliates that are large enough (3–4 mm in length) to be seen swimming in pond water.

Nutritional and Habitat Range   Protozoa are heterotrophic and usually require their food in a complex organic form. Free-living species scavenge dead plant or animal debris and even graze on live cells of bacteria and algae. Some species have special feeding structures such as oral grooves, which carry food particles into a passageway or gullet that packages the captured food into vacuoles for digestion. A remarkable feeding adaptation can be seen in the ciliate *Didinium,* which can easily devour another microbe that is nearly its size. Some protozoa absorb food directly through the cell membrane. Parasitic species live on the fluids of their host, such as plasma and digestive juices, or they can actively feed on tissues.

Although protozoa have adapted to a wide range of habitats, their main limiting factor is the availability of moisture. Their predominant habitats are fresh and marine water, soil, plants, and animals. Even extremes in temperature and pH are not a barrier to their existence; hardy species are found in hot springs, ice, and habitats with low or high pH. Many protozoa can convert to a resistant, dormant stage called a cyst.

Styles of Locomotion   Except for one group (the Apicomplexa), protozoa are motile by **pseudopods,** * **flagella,** or **cilia.** A few species have both pseudopods (also called pseudopodia) and flagella. Some unusual protozoa move by a gliding or twisting movement that does not appear to involve any of these locomotor structures.

Pseudopods are blunt, branched, or long and pointed, depending on the particular species. As previously described, the flowing action of the pseudopods results in ameboid motion, and pseudopods also serve as feeding structures in many amebas (see figure 5.29). The structure and behavior of flagella and cilia were discussed in the first section of this chapter. Flagella vary in number from one to several, and in certain species they are attached along the length of the cell by an extension of the cytoplasmic membrane called the *undulating membrane* (see figure 5.28). In most ciliates, the cilia are distributed over the entire surface of the cell in characteristic patterns. Because of the tremendous variety in ciliary arrangements and functions, ciliates are among the most diverse and awesome cells in the biological world. In certain protozoa, cilia line the oral groove and function in feeding; in others, they fuse together to form stiff props (cirri) that serve as primitive rows of walking legs.

Life Cycles and Reproduction   Most protozoa are recognized by a motile feeding stage called the **trophozoite*** that requires ample food and moisture to remain active. A large number of species are also capable of entering into a dormant, resting stage called a **cyst** when conditions in the environment become unfavorable for growth and feeding. During *encystment,* the trophozoite cell rounds up into a sphere, and its ectoplasm secretes a tough, thick cuticle around the cell membrane **(figure 5.27).** Because cysts are more resistant than ordinary cells to heat, drying, and chemicals, they can survive adverse periods. They can be dispersed by air currents and may even be an important factor in the spread of diseases such as amebic dysentery. If provided with moisture and nutrients, a cyst breaks open and releases the active trophozoite.

The life cycles of protozoans vary from simple to complex. Several protozoan groups exist at all times in the trophozoite state. Many alternate between a trophozoite and a cyst stage, depending on the conditions of the habitat. The life cycle of a parasitic protozoon dictates its mode of transmission to other hosts. For example, the flagellate *Trichomonas vaginalis* causes a common sexually transmitted disease. Because it does not form cysts, it is more delicate, and must be transmitted by intimate contact between sexual partners. In contrast, intestinal pathogens such as *Entamoeba histolytica* and *Giardia lamblia* form cysts and are readily transmitted in contaminated water and foods.

All protozoa reproduce by relatively simple, asexual methods, usually mitotic cell division. Several parasitic species, including the agents of malaria and toxoplasmosis, reproduce asexually inside a host cell by multiple fission. Sexual reproduction also occurs during the life cycle of most protozoa. Ciliates participate in **conjugation,** a form of genetic exchange in which members of two different mating types fuse temporarily and exchange micronuclei. This process of sexual recombination yields new and different genetic combinations that can be advantageous in evolution.

Classification of Selected Medically Important Protozoa

Taxonomists have not escaped problems classifying protozoa. They, too, are very diverse and frequently frustrate attempts to generalize or place them in neat groupings. The most recent system

---

5. The terms *ciliate* and *flagellate* are common names of protozoan groups that move by means of cilia and flagella.

*pseudopod (soo'-doh-pod) pl. pseudopods or pseudopodia; Gr. *pseudo,* false, and *pous,* feet.

*trophozoite (trof"-oh-zoh'-yte) Gr. *trophonikos,* to nourish, and *zoon,* animal.

FIGURE 5.27

The general life cycle exhibited by many protozoa. All protozoa have a trophozoite, but not all produce cysts.

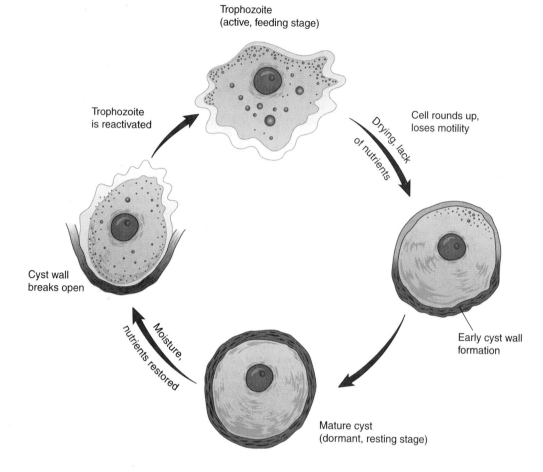

Trophozoite
(active, feeding stage)

Trophozoite
is reactivated

Cell rounds up,
loses motility

Drying, lack
of nutrients

Cyst wall
breaks open

Early cyst wall
formation

Moisture,
nutrients restored

Mature cyst
(dormant, resting stage)

places them in all five of the Kingdoms on the Eukarya tree (see figure 1.16), but this method may be more complex than is necessary for our survey. We will simplify this system by presenting four groups, based on method of motility, mode of reproduction, and stages in the life cycle, summarized as follows:

The Mastigophora (Flagellata)[6] Motility is primarily by flagella alone or by both flagellar and ameboid motion. Single nucleus. Sexual reproduction, when present, by syngamy; division by longitudinal fission. Several parasitic forms lack mitochondria and Golgi apparatus. Most species form cysts and are free-living; the group also includes several parasites. Some species are found in loose aggregates or colonies, but most are solitary. Members include: *Trypanosoma* and *Leishmania,* important blood pathogens spread by insect vectors; *Giardia,* an intestinal parasite spread in water contaminated with feces; *Trichomonas,* a parasite of the reproductive tract of humans spread by sexual contact **(figure 5.28).**

The Sarcodina[6] Cell form is primarily an ameba **(figure 5.29).** Major locomotor organelles are pseudopods, although some species have flagellated reproductive states. Asexual reproduction by fission. Two groups have an external shell; mostly uninucleate; usually encyst. Most amebas are free-living and not infectious; *Entamoeba* is a pathogen or parasite of humans; shelled amebas called foraminifera and radiolarians are responsible for chalk deposits in the ocean.

The Cilophora (Ciliata) Trophozoites are motile by cilia **(figure 5.30);** some have cilia in tufts for feeding and attachment; most develop cysts; have both macronuclei and micronuclei; division by transverse fission; most have a definite mouth and feeding organelle; show relatively advanced behavior **(figure 5.31).** The majority of ciliates are free-living and harmless. One important pathogen, *Balantidium coli,* lives in vertebrate intestines and can infect humans.

The Apicomplexa (Sporozoa) Motility is absent in most cells except male gametes. Life cycles are complex, with well-developed asexual and sexual stages. Sporozoa produce special sporelike cells called **sporozoites* (figure 5.32)** following sexual reproduction, which are important in transmission of infections; most form thick-walled zygotes called oocysts; entire group is parasitic. *Plasmodium,* the most prevalent protozoan parasite, causes 100 million to 300 million cases of malaria each year worldwide. It is an intracellular parasite with a complex cycle alternating between humans and mosquitoes (see figure 23.11). *Toxoplasma gondii* causes an acute infection (toxoplasmosis) in humans, which is acquired from cats and other animals.

Protozoan Identification and Cultivation

The unique appearance of most protozoa makes it possible for a knowledgeable person to identify them to the level of genus and often species by microscopic morphology alone. Characteristics to consider in identification include the shape and size of the cell; the

---

6. Some biologists prefer to combine Mastigophora and Sarcodina into the phylum Sarcomastigophora. Because the algal group Euglenophyta has flagella, it may also be included in the Mastigophora.

---

*sporozoite (spor″-oh-zoh′-yte) Gr. *sporos,* seed, and *zoon,* animal.

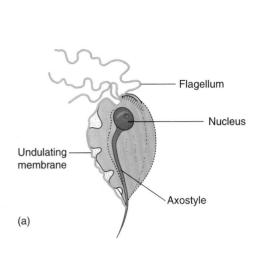

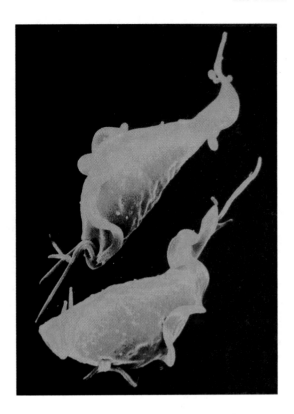

FIGURE 5.28

The structure of a typical mastigophoran, *Trichomonas vaginalis*.
This genital tract pathogen, is shown in **(a)** a drawing and **(b)** a scanning
electron micrograph brings out the details of flagella and undulating
membrane.

(b)

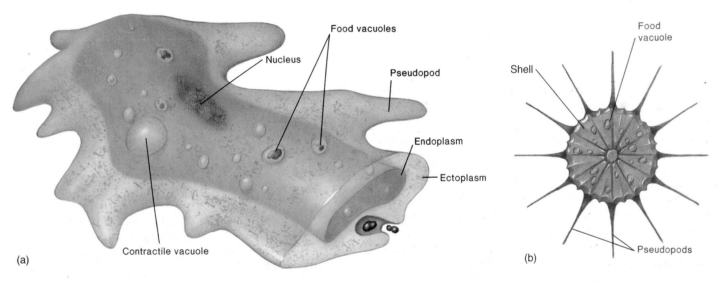

FIGURE 5.29

Examples of sarcodinians. **(a)** The structure of an ameba. **(b)** Radiolarian, a shelled ameba with long, pointed pseudopods. Pseudopods are used in both movement and feeding.

type, number, and distribution of locomotor structures; the presence of special organelles or cysts; and the number of nuclei. Medical specimens taken from blood, sputum, cerebrospinal fluid, feces, or the vagina are smeared directly onto a slide and observed with or without special stains. Occasionally, protozoa are cultivated on artificial media or in laboratory animals for further identification or study.

Important Protozoan Parasites
Although protozoan infections are very common, they are actually caused by only a small number of species often restricted geographically to the tropics and subtropics. In this survey, we first in-troduce some concepts of the host-protozoan relationship and then look at examples that illustrate some of the main features of protozoan diseases. Details of several other medically important protozoa are given in chapter 23.

Parasitic protozoa are traditionally studied along with the helminths in the science of parasitology. Although a parasite is generally defined as an organism that obtains food and other requirements at the expense of a host, the range of host-parasite relationships can be very broad. At one extreme are the so-called good parasites, which occupy their host with little harm. An example is certain amebas that

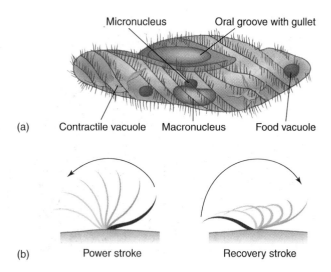

(a)

Micronucleus — Oral groove with gullet

Contractile vacuole — Macronucleus — Food vacuole

(b) Power stroke — Recovery stroke

FIGURE 5.30

Structure and locomotion in ciliates. **(a)** The structure of a typical representative, *Paramecium*. **(b)** Cilia beat in coordinated waves, driving the cell forward and backward. View of a single cilium shows that it has a pattern of movement like a swimmer, with a power forward stroke and a repositioning stroke.

live in the human intestine and feed off organic matter there. At the other extreme are parasites (*Plasmodium, Trypanosoma*) that multiply in host tissues such as the blood or brain, causing severe damage and disease. Between these two extremes are parasites of varying pathogenicity, depending on their particular adaptations.

Most human parasites go through three general stages:

1. The microbe is transmitted to the human host from a source such as soil, water, food, other humans, or animals.
2. The microbe invades and multiplies in the host, producing more parasites that can infect other suitable hosts.
3. The microbe leaves the host in large numbers by a specific means and, to survive, must find and enter a new host.

There are numerous variations on this simple theme. For instance, the microbe can invade more than one host species (alternate hosts) and undergo several changes as it cycles through these hosts, such as sexual reproduction or encystment. Some microbes are spread from human to human by means of **vectors,*** defined as animals such as insects that carry diseases. Others can be spread through bodily fluids and feces.

---

*vector (vek′-tur) L. *vectur*, one who carries.

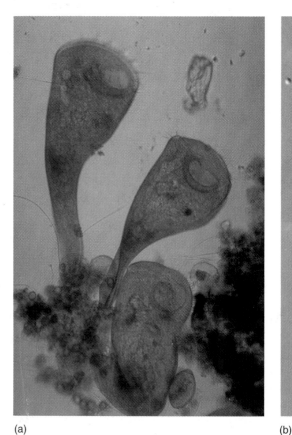

(a)

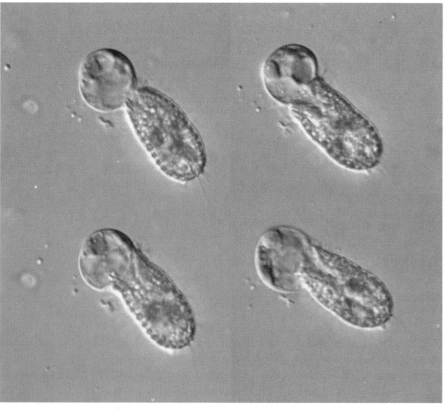

(b)

FIGURE 5.31

Selected ciliate representatives. **(a)** Large, funnel-shaped *Stentor* with a rotating row of cilia around its oral cavity. Currents produced by the cilia sweep food particles into the gullet. **(b)** Stages in the process of *Coleps* feeding on an alga (round cell). The predaceous ciliate gradually pulls its prey into a large oral groove. Yummy!

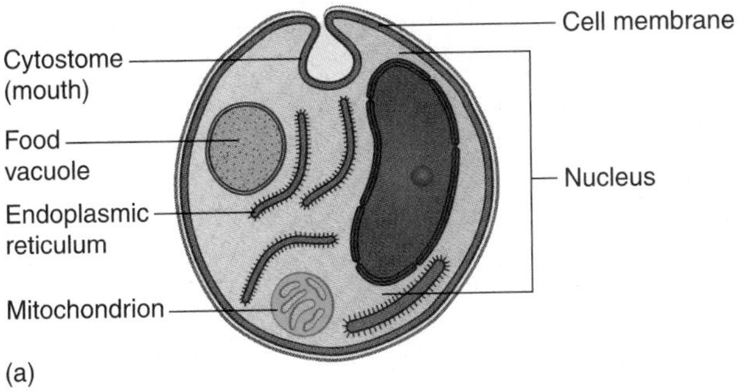

Cytostome (mouth)
Food vacuole
Endoplasmic reticulum
Mitochondrion
Cell membrane
Nucleus
(a)

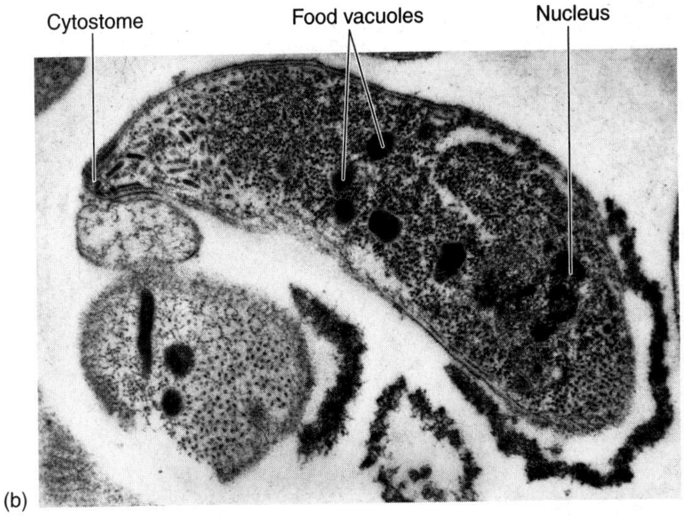

Cytostome
Food vacuoles
Nucleus
(b)

FIGURE 5.32

Sporozoan parasites. (a) General cell structure. Note the lack of specialized locomotor organelles. **(b)** Scanning electron micrograph of the sporozoite of *Cryptosporidium,* an intestinal parasite of humans and other mammals.

Pathogenic Flagellates: Trypanosomes Trypanosomes are parasites belonging to the genus *Trypanosoma.** The two most important representatives are *T. brucei* and *T. cruzi,* species that are closely related but geographically restricted. *Trypanosoma brucei* occurs in Africa, where it causes approximately 10,000 new cases of sleeping sickness each year. *Trypanosoma cruzi,* the cause of Chagas disease,[7] is endemic to South and Central America, where it infects several million people a year. Both species have long, crescent-shaped cells with a single flagellum that is sometimes attached to the cell body by an undulating membrane **(figure 5.33).** Both occur in the blood during infection and are transmitted by blood-sucking vectors. We will use *T. cruzi* to illustrate the phases of a trypanosomal life cycle and to demonstrate the complexity of parasitic relationships.

The trypanosome of Chagas disease relies on the close relationship of a warm-blooded mammal and an insect that feeds on mammalian blood. The mammalian hosts are numerous, including dogs, cats, opossums, armadillos, and foxes. The vector is the *reduviid* bug,* an insect that is sometimes called the "kissing bug" because of its habit of biting its host at the corner of the mouth. Transmission occurs from bug to mammal and from mammal to bug, but usually not from mammal to mammal, except across the placenta during pregnancy. The general phases of this cycle are presented in figure 5.33.

The trypanosome trophozoite multiplies in the intestinal tract of the reduviid bug and is harbored in the feces. The bug seeks a host and bites the mucous membranes, usually of the eye, nose, or lips. As it fills with blood, the bug soils the bite with feces containing the trypanosome. Ironically, the victims themselves inadvertently contribute to the entry of the microbe by scratching the bite wound. The trypanosomes ultimately become established and multiply in muscle and white blood cells. Periodically, these parasitized cells rupture, releasing large numbers of

(a) Infective Trypanosome
Reduviid bug
Cycle in Human Dwellings
(b) Mode of infection
Cycle in the Wild

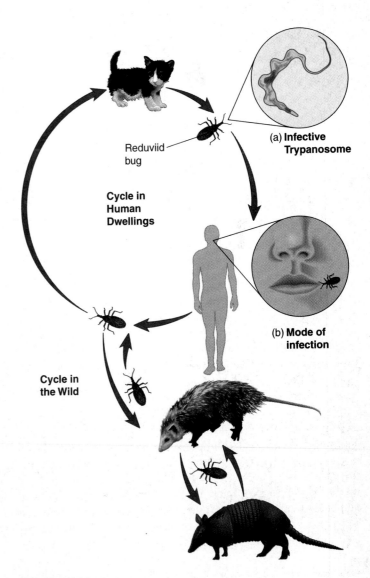

FIGURE 5.33

Cycle of transmission in Chagas disease. Trypanosomes (inset *a*) are transmitted among mammalian hosts and human hosts by means of a bite from the kissing bug (inset *b*).

7. Named for Carlos Chagas, the discoverer of *T. cruzi.*

**Trypanosoma* (try″-pan-oh-soh′-mah) Gr. *trypanon,* borer, and *soma,* body.

**reduviid* (ree-doo′-vee-id) A member of a large family of flying insects with sucking, beaklike mouths.

FIGURE 5.34

Stages in the infection and transmission of amebic dysentery. Arrows show the route of infection; insets show the appearance of *Entamoeba histolytica*. **(a)** Cysts are eaten. **(b)** Trophozoites (amebas) emerge from cysts. **(c)** Trophozoites invade the large intestinal wall. **(d)** Mature cysts are released in the feces, and may be spread through contaminated food and water.

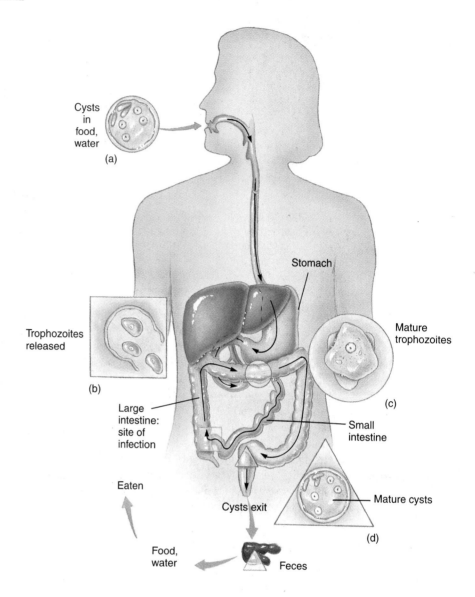

Cysts in food, water
(a)

Stomach

Trophozoites released
(b)

Mature trophozoites

Large intestine: site of infection

Small intestine
(c)

Eaten

Cysts exit

Mature cysts
(d)

Food, water

Feces

new trophozoites into the blood. Eventually, the trypanosome can spread to many systems, including the lymphoid organs, heart, liver, and brain. Manifestations of the resultant disease range from mild to very severe, and include fever, inflammation, and heart and brain damage. In many cases, the disease has an extended course and can cause death.

Infective Amebas: Entamoeba Several species of amebas cause disease in humans, but probably the most common disease is amebiasis, or amebic dysentery,* caused by *Entamoeba histolytica*. This microbe is widely distributed in the world, from northern zones to the tropics, and is nearly always associated with humans. Amebic dysentery is the fourth most common protozoan infection in the world. This microbe has a life cycle quite different from the trypanosomes in that it does not involve multiple hosts and a blood-sucking vector. It lives part of its cycle as a trophozoite and part as a cyst. Because the cyst is the more resistant

form and can survive in water and soil for several weeks, it is the more important stage for transmission. The primary way that people become infected is by ingesting food or water contaminated with human feces.

Figure 5.34 shows the major features of the amebic dysentery cycle, starting with the ingestion of cysts. The viable, heavy-walled cyst passes through the stomach unharmed. Once inside the small intestine, the cyst germinates into a large multinucleate ameba that subsequently divides to form small amebas (the trophozoite stage). These trophozoites migrate to the large intestine and begin to feed and grow. From this site, they can penetrate the lining of the intestine and invade the liver, lungs, and skin. Common symptoms include gastrointestinal disturbances such as nausea, vomiting, and diarrhea, leading to weight loss and dehydration. Untreated cases with extensive damage to the organs experience a high death rate. The cycle is completed in the infected human when certain trophozoites in the feces begin to form cysts, which then pass out of the body with fecal matter. Knowledge of the amebic cycle and role of cysts has been helpful in controlling the disease. Important preventive measures include sewage treatment,

*dysentery (dis'-en-ter"-ee) Any inflammation of the intestine accompanied by bloody stools. It can be caused by a number of factors, both microbial and nonmicrobial.

curtailing the use of human feces as fertilizers, and adequate sanitation of food and water.

The Parasitic Helminths

Tapeworms, flukes, and roundworms are collectively called **helminths,** from the Greek word meaning worm. Adult animals are usually large enough to be seen with the naked eye, and they range from the longest tapeworms, measuring up to about 23 m in length, to roundworms less than 1 mm in length. Nevertheless, they are included among microorganisms because the microscope is necessary to see the details of smaller helminths and to identify their eggs and larvae.

On the basis of morphological form, the two major groups of parasitic helminths are the **flatworms** (Phylum Platyhelminthes), with a very thin, often segmented body plan **(figure 5.35)**, and the **roundworms** (Phylum Aschelminthes, also called **nematodes***), with an elongate, cylindrical, unsegmented body plan **(figure 5.36)**. The flatworm group is subdivided into the **cestodes,*** or tapeworms, named for their long, ribbonlike arrangement, and the **trematodes,*** or flukes, characterized by flat, ovoid bodies. Not all flatworms and roundworms are parasites by nature; many live free in soil and water. Both here and in chapter 23, we concern ourselves with the medically important worms (see table 23.4).

GENERAL WORM MORPHOLOGY

All helminths are multicellular animals equipped to some degree with organs and organ systems. In parasites, the most developed organs are those of the reproductive tract, with some degree of

---

*nematode (neem′-ah-tohd) Gr. *nemato,* thread, and *eidos,* form.

*cestode (sess′-tohd) L. *cestus,* a belt, and *ode,* like.

*trematode (treem′-a-tohd) Gr. *trema,* hole. Named for the appearance of having tiny holes.

---

reduction in the digestive, excretory, nervous, and muscular systems. In particular groups, such as the cestodes, reproduction is so dominant that the worms are reduced to little more than a series of flattened sacs filled with ovaries, testes, and eggs (see figure 5.35*a*). Not all worms have such extreme adaptations as cestodes, but most have a highly developed reproductive potential, thick cuticles for protection, and mouth glands for breaking down the host's tissue.

LIFE CYCLES AND REPRODUCTION

Many worms have complex life cycles that alternate between hosts. Reproduction of individuals is primarily sexual, involving the production of eggs and sperm in the same worm or in separate male and female worms. Fertilized eggs are usually released to the environment and are provided with a protective shell and extra food to aid their development into larvae. Even so, most eggs and larvae are vulnerable to heat, cold, drying, and predators and are destroyed or unable to reach a new host. To counteract this formidable mortality rate, certain worms have adapted a reproductive capacity that borders on the incredible: A single female *Ascaris*[8] can lay 200,000 eggs a day, and a large female can contain over 25,000,000 eggs at varying stages of development! If only a tiny number of these eggs makes it to another host, the parasite will have been successful in completing its life cycle.

A HELMINTH CYCLE: THE PINWORM

To illustrate a helminth cycle in humans, we will use the example of a roundworm, *Enterobius vermicularis,* the pinworm or seatworm. This worm causes a very common infestation of the large intestine (see figure 5.36). Worms range from 2 to 12 mm long and have a

---

8. *Ascaris* is a genus of parasitic intestinal roundworms.

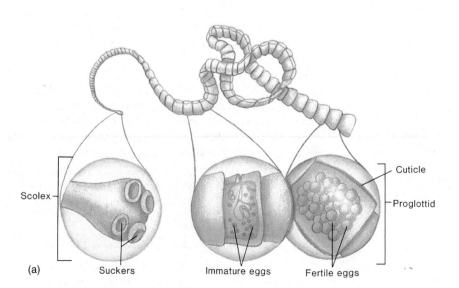

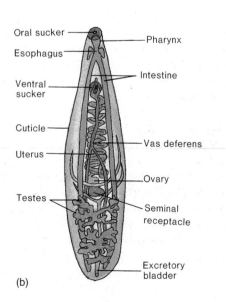

FIGURE 5.35

Parasitic flatworms. **(a)** A cestode (beef tapeworm), showing the scolex; long, tapelike body; and magnified views of immature and mature proglottids. **(b)** The structure of a trematode (liver fluke). Note the suckers that attach to host tissue and the dominance of reproductive and digestive organs.

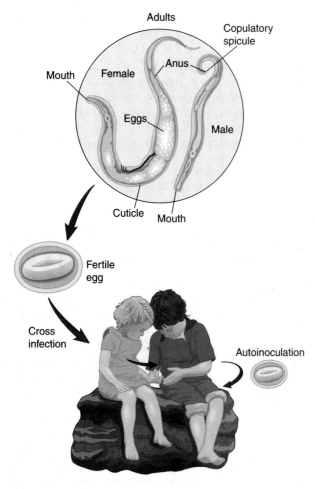

FIGURE 5.36

The life cycle of the pinworm, a roundworm. Eggs are the infective stage and are transmitted by unclean hands. Children frequently reinfect themselves and also pass the parasite on to others.

tapered, curved cylinder shape. The condition they cause, enterobiasis, is usually a simple, uncomplicated infection that does not spread beyond the intestine.

A cycle starts when a person swallows microscopic eggs picked up from another infected person by direct contact or by touching articles that person has touched. The eggs hatch in the intestine and then release larvae that mature into adult worms in the appendix (within about one month). There, male and female worms mate, and the female migrates out to the anus to deposit eggs, which cause intense itchiness that is relieved by scratching. Herein lies a significant means of dispersal: Scratching contaminates the fingers, which, in turn, transfer eggs to bedclothes and other inanimate objects. This person becomes a host and a source of eggs and can spread them to others in addition to reinfesting himself. Enterobiasis occurs most often among families and in other close living situations. Its distribution is worldwide among all socioeconomic

groups, but it seems to attack younger people more frequently than older ones.

HELMINTH CLASSIFICATION AND IDENTIFICATION

The helminths are classified according to their shape; their size; the degree of development of various organs; the presence of hooks, suckers, or other special structures; the mode of reproduction; the kinds of hosts; and the appearance of eggs and larvae. They are identified in the laboratory by microscopic detection of the adult worm or its larvae and eggs, which often have distinctive shapes or external and internal structures (see chapter 23). Occasionally, they are cultured in order to verify all of the life stages.

DISTRIBUTION AND IMPORTANCE OF PARASITIC WORMS

About 50 species of helminths parasitize humans. They are distributed in all areas of the world that support human life. Some worms are restricted to a given geographic region, and many have a higher incidence in tropical areas. This knowledge must be tempered with the realization that jet-age travel, along with human migration, are gradually changing the patterns of worm infections, especially of those species that do not require alternate hosts or special climatic conditions for development. The yearly estimate of worldwide cases numbers in the billions, and these are not confined to developing countries. A conservative estimate places 50,000,000 helminth infections in North America alone. The primary targets are malnourished children.

CHAPTER CHECKPOINTS

The eucaryotic microorganisms include the Fungi (Myceteae), the Protista (algae and protozoa), and the Helminths (Kingdom Animalia).

The Kingdom Fungi (Myceteae) is composed of non-photosynthetic haploid species with cell walls. The fungi are either saprobes or parasites, and may be unicellular, colonial, or multicellular. Forms include yeasts (unicellular budding cells) and molds (filamentous cells called hyphae). Their primary means of reproduction involves asexual and sexual spores.

The protists are mostly unicellular or colonial eucaryotes that lack specialized tissues. There are two major organism types: the Algae and the Protozoa. Algae are photosynthetic organisms that contain chloroplasts with chlorophyll and other pigments. Protozoa are heterotrophs that usually display some form of locomotion. Most are single-celled trophozoites, and many produce a resistant stage, or cyst.

The Kingdom Animalia has only one group that contains microscopic members. These are the helminths or worms. Parasitic members include flatworms and roundworms, that are able to invade and reproduce in human tissues.

CHAPTER CAPSULE WITH KEY TERMS

I. **Eucaryotic Structure**
   A. Major organelles and other structural features include:
   Appendages (cilia, flagella), **glycocalyx,** cell wall, cytoplasmic
   (or cell) membrane, ribosomes, **organelles** (nucleus, **nucleolus,
   endoplasmic reticulum, Golgi complex, mitochondria,**
   chloroplasts, cytoskeleton, **microfilaments,** microtubules).
   A review comparing the major differences between eucaryotic
   and procaryotic cells is provided in table 2.6, page 51.

II. **The Kingdom Fungi (Myceteae)**
   Common names of two particular types: macroscopic fungi
   (mushrooms, bracket fungi, puffballs) and microscopic fungi
   (yeasts, molds).
   A. *Overall Morphology:* At the cellular (microscopic) level, typical
   eucaryotic cell, with thick cell walls. Yeasts are single cells that
   form buds and **pseudohyphae. Hyphae** are long, tubular
   filaments that can be septate or nonseptate and grow in a network
   called a **mycelium;** hyphae are characteristic of the filamentous
   fungi called molds. Some fungi (mushrooms) produce
   multicellular structures such as fleshy fruiting bodies. Fungal cells
   are largely nonmotile, except for some motile gametes.
   B. *Nutritional Mode/Distribution:* All are **heterotrophic.** The
   majority are harmless **saprobes** living off organic **substrates** such
   as dead animal and plant tissues. A few are **parasites,** living on
   the tissues of other organisms, but none are obligate. Preferred
   temperature of growth is 20°–40°C. Distribution is extremely
   widespread in many habitats.
   C. *Reproduction:* Primarily through **spores** formed on special
   reproductive hyphae. In asexual reproduction, spores are formed
   through budding, partitioning of a hypha, or in special
   sporogenous structures; examples are **conidia** and
   sporangiospores. In sexual reproduction, spores are formed
   following fusion of male and female strains and the formation of a
   sexual structure; sexual spores are one basis for classification.
   D. *Major Groups:* The four main divisions among the terrestrial
   fungi, given with sexual spore type, are Zygomycota
   (**zygospores**), Ascomycota (**ascospores**), Basidiomycota
   (**basidiospores**), and Deuteromycota (no sexual spores).
   E. *Importance:* Essential decomposers of plant and animal detritis in
   the environment with return of valuable nutrients to the ecosystem.
   Economically beneficial as sources of antibiotics; used in making
   foods and in genetic studies. Adverse impact: decomposition of
   fruits and vegetables; several fungi cause infections, or **mycoses;**
   some produce substances that are toxic if eaten.

III. **Microscopic Protists**
   A. The Algae
   Include plantlike protists, kelps, and seaweeds. Specific groups
   are the euglenids, green algae, diatoms, **dinoflagellates,** brown
   algae, and red seaweeds.
   1. *Overall Morphology:* Contain **chloroplasts** with **chlorophyll**
   and other pigments; cell wall; may or may not have flagella.
   Microscopic forms are unicellular, colonial, filamentous;
   macroscopic forms are colonial and multicellular.
   2. Nutritional Mode/Distribution:
   Photosynthetic; most are free-living in the aquatic
   environment, both fresh water and marine (common
   component of **plankton**).
   3. *Major Groups:* Classification according to types of pigments
   and cell walls. Microscopic algae: Euglenophyta, Chlorophyta,

Chrysophyta, Pyrrophyta: Macroscopic algae: Phaephyta
(kelps) and Rhodophyta (seaweeds) are multicellular algae.
   4. *Importance:* Algae provide the basis of the food web in most
   aquatic habitats, and they produce a large proportion of
   atmospheric O_2 through photosynthesis. Some are harvested as
   a source of cosmetics, food, and medical products.
   Dinoflagellates cause red tides and give off toxins that are
   harmful to humans and animals.
   B. The Protozoa
   Include animal-like protists; unicellular animals such as amebas,
   flagellates, ciliates, sporozoa.
   1. *Overall Morphology:* Most are unicellular, colonies rare, no
   multicellular forms; most have locomotor structures, such as
   flagella, cilia, pseudopods; special feeding structures can be
   present; lack a cell wall; extreme variations in shape. Can exist
   in **trophozoite,** a motile, feeding stage, or **cyst,** a dormant
   resistant stage.
   2. *Nutritional Mode/Distribution:* All are heterotrophic. Most are
   free-living in a moist habitat (water, soil); feed by engulfing
   other microorganisms and organic matter. A number of animal
   parasites; can be spread from host to host by insect **vectors.**
   3. *Reproduction:* Asexual by binary fission and **mitosis,** budding;
   sexual by fusion of free-swimming gametes, conjugation.
   4. *Major Groups:* Protozoa are subdivided into four groups based
   upon mode of locomotion and type of reproduction:
   Mastigophora, the flagellates, motile by flagella; Sarcodina, the
   amebas, motile by pseudopods; Ciliophora, the ciliates, motile
   by cilia; Apicomplexa, all parasites; motility not well
   developed; produce unique reproductive structures.
   5. *Importance:* Ecologically important in food webs and
   decomposing organic matter. Medical significance: hundreds
   of millions of people are afflicted with one of the many
   protozoan infections (malaria, trypanosomiasis, amebiasis).

IV. **The Helminth Parasites**
   Includes parasitic worms, tapeworms, flukes, nematodes.
   A. *Overall Morphology:* Animal cells; multicellular; individual
   organs specialized for reproduction, digestion, movement,
   protection, though some of these are reduced.
   B. *Nutritional/Reproductive Mode:* Parasitize host tissues; have
   mouthparts for attachment to or digestion of host tissues. Most
   have well-developed sex organs that produce eggs and sperm.
   Fertilized eggs go through larval period in or out of host body.
   C. *Major Groups:*
   1. **Flatworms** have highly flattened body; no definite body
   cavity; digestive tract a blind pouch; simple excretory and
   nervous systems. **Cestodes** (tapeworms) are long chains of
   segments; attach to host's intestine by hooked mouthpart.
   Trematodes, or flukes, are flattened, nonsegmented worms
   with sucking mouthparts.
   2. **Roundworms (nematodes)** have round bodies, a complete
   digestive tract, a protective surface cuticle, spines and hooks on
   mouth; excretory and nervous systems poorly developed.
   D. *How Transmitted and Acquired:* Through ingestion of larvae or
   eggs in food; from soil or water; some are carried by insect
   vectors.
   E. *Importance:* Afflict billions of humans. Because these worms are
   so common, they have medical and economic impact of
   staggering proportions.

MULTIPLE-CHOICE QUESTIONS

1. Both flagella and cilia are found primarily in
   a. algae
   b. protozoa
   c. fungi
   d. both b and c

2. Features of the nuclear envelope include
   a. ribosomes
   b. a double membrane structure
   c. pores that allow communication with the cytoplasm
   d. b and c
   e. all of these

3. In general, if two haploid cells fuse, _____ will result.
   a. a germ cell
   b. a diploid zygote
   c. mitosis
   d. meiosis

4. The cell wall is found in which eucaryotes?
   a. fungi
   b. algae
   c. protozoa
   d. a and b

5. What is embedded in rough endoplasmic reticulum?
   a. ribosomes
   b. Golgi apparatus
   c. chromatin
   d. vesicles

6. Yeasts are _____ fungi, and molds are _____ fungi.
   a. macroscopic, microscopic
   b. unicellular, filamentous
   c. motile, nonmotile
   d. water, terrestrial

7. In general, fungi derive nutrients through
   a. photosynthesis
   b. engulfing bacteria
   c. digesting organic substrates
   d. parasitism

8. A hypha divided into compartments by cross walls is called
   a. nonseptate
   b. imperfect
   c. septate
   d. perfect

9. A conidium is a/an _____ spore, and a zygospore is a/an _____ spore.
   a. sexual, asexual
   b. free, endo
   c. ascomycete, basidiomycete
   d. asexual, sexual

10. Algae generally contain some types of
    a. spore
    b. chlorophyll
    c. locomotor organelle
    d. toxin

11. All protozoa have a
    a. locomotor organelle
    b. cyst stage
    c. pellicle
    d. trophozoite

12. The protozoan trophozoite is the
    a. active feeding stage
    b. inactive dormant stage
    c. infective stage
    d. spore-forming stage

13. All mature sporozoa are
    a. parasitic
    b. nonmotile
    c. carried by vectors
    d. both a and b

14. Helminth parasites reproduce with
    a. spores
    b. eggs and sperm
    c. mitosis
    d. cysts
    e. all of these

15. Mitochondria likely originated from
    a. archaea
    b. invaginations of the cell membrane
    c. purple bacteria
    d. cyanobacteria

16. **Single Matching.** Select the description that best fits the word in the left column.
    _____ diatom   a. the cause of malaria
    _____ *Rhizopus*   b. single-celled alga with silica in its cell wall
    _____ *Histoplasma*   c. fungal cause of Ohio Valley fever
    _____ *Cryptococcus*   d. the cause of amebic dysentery
    _____ euglenid   e. genus of black bread mold
    _____ dinoflagellate   f. helminth worm involved in pinworm infection
    _____ *Trichomonas*   g. motile flagellated alga with eyespots
    _____ *Entamoeba*   h. a yeast that infects the lungs
    _____ *Plasmodium*   i. flagellated protozoan genus that causes an STD
    _____ *Enterobius*   j. alga that causes red tides

CONCEPT QUESTIONS

1. Construct a chart that reviews the major similarities and differences between procaryotic and eucaryotic cells.

2. a. Which kingdoms of the five-kingdom system contain eucaryotic microorganisms? How do unicellular, colonial, and multicellular organisms differ from each other?
   b. Give examples of each type.

3. a. Describe the anatomy and functions of each of the major eucaryotic organelles.
   b. How are flagella and cilia similar? How are they different?
   c. Compare and contrast the smooth ER, the rough ER, and the Golgi apparatus in structure and function.

4. Trace the synthesis of cell products, their processing, and their packaging through the organelle network.

5. a. Describe the detailed structure of the nucleus.
   b. Why can one usually not see the chromosomes?
   c. When are the chromosomes visible?
   d. What causes them to be visible?

6. a. Define mitosis and explain its function.
   b. What happens to the chromosome number during this process?
   c. How does a diploid organism remain diploid and a haploid organism remain haploid?

7. a. Define meiosis and explain its function.
   b. When does it occur in diploid organisms?
   c. In haploid organisms?
   d. How does it differ from mitosis?

8. Describe some of the ways that organisms use lysosomes.

9. For what reasons would a cell need a "skeleton?"

10. a. Differentiate between the yeast and hypha types of fungal cell.
    b. What is a mold?
    c. What does it mean if a fungus is dimorphic?

11. a. How does a fungus feed?
    b. Where would one expect to find fungi?

12. a. Describe the functional types of hyphae.
    b. Describe the two main types of asexual fungal spores and how they are formed.
    c. What are some types of conidia?
    d. What is the reproductive potential of molds in terms of spore production?
    e. How do mold spores differ from procaryotic spores?

13. a. Explain the importance of sexual spores to fungi.
    b. Describe the three main types of sexual spores, and show how each is formed by means of a simple diagram.

14. How are fungi classified? Give an example of a member of each fungus division and describe its structure and importance.

15. What is a mycosis? What kind of mycosis is athlete's foot? What kind is coccidioidomycosis?

16. What is a working definition of a "protist?"

17. a. Describe the principal characteristics of algae that separate them from protozoa.
    b. How are algae important?
    c. What causes the many colors in the algae?
    d. Are there any algae of medical importance?

18. a. Explain the general characteristics of the protozoan life cycle.
    b. Describe the protozoan adaptations for feeding.
    c. Describe protozoan reproductive processes.

19. a. Briefly outline the characteristics of the four protozoan groups.
    b. What is an important pathogen in each group?

20. a. Which protozoan group is the most complex in structure and behavior?
    b. In life cycle?
    c. What characteristics set the apicomplexa apart from the other protozoan groups?

21. a. Construct a chart that compares the four groups of eucaryotic microorganisms (fungi, algae, protozoa, helminths) in cellular structure (see figure 5.2).
    b. Indicate whether the group has a cell wall, chloroplasts, motility, or some other distinguishing feature.
    c. Include also the manner of nutrition and body plan (unicellular, colonial, filamentous, or multicellular).

22. Discuss the adaptations of parasitic worms to their lifestyles, and explain why these adaptations are necessary or advantageous to the worms' survival.

CRITICAL-THINKING QUESTIONS

1. Suggest some ways that one would go about determining if mitochondria and chloroplasts are a modified procaryotic cell.

2. Give the common name of a eucaryotic microbe that is unicellular, walled, non-photosynthetic, nonmotile, and bud-forming.

3. Give the common name of a microbe that is unicellular, nonwalled, motile with flagella, and has chloroplasts.

4. Which group of microbes has long, thin pseudopods and is encased in a hard shell?

5. What general type of multicellular parasite is composed primarily of thin sacs of reproductive organs?

6. a. Name two parasites that are transmitted in the cyst form.
   b. How must a non-cyst-forming pathogenic protozoan be transmitted? Why?

7. You just found an old container of food in the back in your refrigerator. You open it and see a mass of multicolored fuzz. As a budding microbiologist, describe how you would determine what types of organisms are growing on the food.

8. Explain what factors could cause opportunistic mycoses to be a growing medical problem.

9. a. How are bacterial endospores and cysts of protozoa alike?
   b. How do they differ?

10. You have gone camping in the mountains and plan to rely on water present in forest pools and creeks for drinking water. Certain encysted pathogens often live in this type of water, but you do not discover this until you arrive at the campground. How might you treat the water to prevent becoming infected?

11. a. Explain the two levels of parasitism at work in Chagas disease.
    b. Do you suppose the trypanosome has a parasite, too?
    c. If so, could its parasite have a parasite?
    d. What does this tell you about the prevalence of parasitism and its success as a mode of life?
    e. What is a potential weakness in parasitism?

12. Can you think of a way to determine if a child is suffering from pinworms? Hint: Scotch tape is involved.

13. How could the cycle of a parasite worm be interrupted to stop the spread of infection?

INTERNET SEARCH TOPICS

Use the World Wide Web to explore these topics:

1. The endosymbiotic theory of eucaryotic cell evolution. List data from studies that support this idea.

2. The medical problems caused by mycotoxins and "sick building" syndrome.

3. Several excellent websites provide information and animations on eucaryotic cells and protists. Visit the student Online Learning Center at www.mhhe.com/talaro5. Go to chapter 5, Internet Search Topics, and log on to the available websites to observe the varied styles of feeding, reproduction, and locomotion seen among these microbes.

CHAPTER 6

An Introduction to the Viruses

The concept of viruses can inspire a sense of mystery and awe. At times we assign to them imaginary powers that they do not have, and yet, at other times, they seem to have powers beyond imagination. Despite success in eradicating viral diseases like smallpox and polio, humans continue to experience outbreaks of new viral infections every few months. In fact, viruses are the most prominent emerging microbes, accounting for over 50% of new outbreaks of infectious diseases worldwide.

We need no more powerful reminder of these realities than to consider the sudden onset of a new disease SARS (severe acute respiratory syndrome) in late 2002. The disease began in mainland China and soon spread worldwide, causing great fear and panic. In several countries, infected people were quarantined and tested at travel terminals for signs of infection (see **chapter-opening photo**). The virus was rapidly identified as a new coronavirus, but its origins are yet to be determined.

A different virus—the West Nile virus—has also caused a great deal of concern in the United States. Several hundred cases have been reported, with a number of deaths. This virus has somehow migrated from Africa or Asia and become established in American birds and mosquitoes. At about the same time, the Nipah virus jumped hosts from bats to pigs to humans, causing a deadly epidemic among agricultural workers in parts of Asia.

Viruses are often blamed for unexplained illnesses and symptoms, and are usually considered the culprit when other infectious agents have been ruled out. They are tied intimately to their host cells, where they often linger and even become part of their host's genetic material. Because of their roles in disease and genetics, it is very important to have a working knowledge of the basic characteristics of viruses. The primary aim of this chapter is to familiarize you with their many unique properties and to provide a survey of their structure, physiology, multiplication, and diversity.

Travelers from Hong Kong are screened for symptoms of SARS before being allowed to enter Thailand. All arrivals at Bangkok international airport were evaluated for illness in spring of 2003, at the height of the SARS epidemic.

Chapter Overview

Viruses:

- Are a unique group of tiny infectious particles that are obligate parasites of cells.
- Do not exhibit the characteristics of life, but can regulate the functions of host cells.
- Infect all groups of living things and produce a variety of diseases.
- Are not cells but resemble complex molecules composed of protein and nucleic acid.
- Are encased in an outer shell or envelope and contain either DNA or RNA as their genetic material.
- Are genetic parasites that take over the host cell's metabolism and synthetic machinery.
- Can instruct the cell to manufacture new virus parts and assemble them.
- Are released in a mature, infectious form, followed by destruction of the host cell.
- May persist in cells, leading to slow progressive diseases and cancer.
- Are identified by structure, host cell, type of nucleic acid, outer coating, and type of disease.
- Are among the most common infectious agents, causing serious medical and agricultural impact.

The Search for the Elusive Viruses

The discovery of the light microscope made it possible to see first-hand the agents of many bacterial, fungal, and protozoan diseases. But the techniques for observing and cultivating these relatively large microorganisms were virtually useless for viruses. For many years, the cause of viral infections such as smallpox and polio was unknown, even though it was clear that the diseases were transmitted from person to person. The French bacteriologist Louis Pasteur was certainly on the right track when he postulated that rabies was caused by a "living thing" smaller than bacteria, and in 1884 he was able to develop the first vaccine for rabies. Pasteur also proposed the term **virus** (L. poison) to denote this special group of infectious agents.

The first substantial revelations about the unique characteristics of viruses occurred in the 1890s. First, D. Ivanovski and M. Beijerinck showed that a disease in tobacco was caused by a virus (tobacco mosaic virus). Then, Friedrich Loeffler and Paul Frosch discovered an animal virus that causes foot-and-mouth disease in cattle. These early researchers found that when infectious fluids from host organisms were passed through porcelain filters designed to trap bacteria, the filtrate still remained infectious. This result proved that an infection could be caused by a cell-free fluid containing agents smaller than bacteria and thus first introduced the concept of a *filterable virus.*

Over the succeeding decades, a remarkable picture of the physical, chemical, and biological nature of viruses began to take form. Years of experimentation were required to show that viruses were noncellular particles with a definite size, shape, and chemical composition. Like bacteria, they could be cultured in the laboratory. By the 1950s, virology had grown into a multifaceted discipline that promised to provide much information on disease, genetics, and even life itself.

The Position of Viruses in the Biological Spectrum

Viruses are a unique group of biological entities known to infect every type of cell, including bacteria, algae, fungi, protozoa, plants, and animals. Although the emphasis in this chapter is on animal viruses, much credit for our knowledge must be given to experiments with bacterial and plant viruses. The exceptional and curious nature of viruses prompts numerous questions, including:

1. Are they organisms; that is, are they alive?
2. What are their distinctive biological characteristics?
3. How can particles so small, simple, and seemingly insignificant be capable of causing disease and death? and
4. What is the connection between viruses and cancer?

In this chapter, we address these questions and many others.

The unusual structure and behavior of viruses have led to debates about their connection to the rest of the microbial world. One viewpoint holds that viruses are unable to exist independently

TABLE 6.1
Novel Properties of Viruses

- Are obligate intracellular parasites of bacteria, protozoa, fungi, algae, plants, and animals.
- Ultramicroscopic size, ranging from 20 nm up to 450 nm (diameter).
- Are not cells; structure is very compact and economical.
- Do not independently fulfill the characteristics of life (see chapter 2).
- Are inactive macromolecules outside of the host cell and active only inside host cells.
- Are geometric; can form crystal-like masses.
- Basic structure consists of protein shell (capsid) surrounding nucleic acid core.
- Nucleic acid can be either DNA or RNA but not both.
- Nucleic acid can be double-stranded DNA, single-stranded DNA, single-stranded RNA, or double-stranded RNA.
- Molecules on virus surface impart high specificity for attachment to host cell.
- Multiply by taking control of host cell's genetic material and regulating the synthesis and assembly of new viruses.
- Lack enzymes for most metabolic processes.
- Lack machinery for synthesizing proteins.

from the host cell, so they are not living things but are more akin to large, infectious molecules. Another viewpoint proposes that even though viruses do not exhibit most of the life processes of cells (discussed in chapter 2), they can direct them and thus are certainly more than inert and lifeless molecules. Depending upon the circumstances, both views are defensible. In applied virology, this debate has greater philosophical than practical importance because viruses are agents of disease and must be dealt with through control, therapy, and prevention, whether we regard them as living or not. In keeping with their special position in the biological spectrum, it is best to describe viruses as *infectious particles* (rather than organisms) and as either *active* or *inactive* (rather than alive or dead).

Viruses are different from their host cells in size, structure, behavior, and physiology. They are a type of **obligate intracellular parasites** that cannot multiply unless they invade a specific host cell and instruct its genetic and metabolic machinery to make and release quantities of new viruses. Because of this characteristic, viruses are capable of causing serious damage and disease. Other unique properties of viruses are summarized in **table 6.1.**

CHAPTER CHECKPOINTS

Viruses are noncellular entities whose properties have been identified through technological advances in microscopy and tissue culture.

Viruses are infectious particles that invade every known type of cell. They are not alive, yet they are able to redirect the metabolism of living cells to reproduce virus particles.

Viral replication inside a cell usually causes death or disease of that cell.

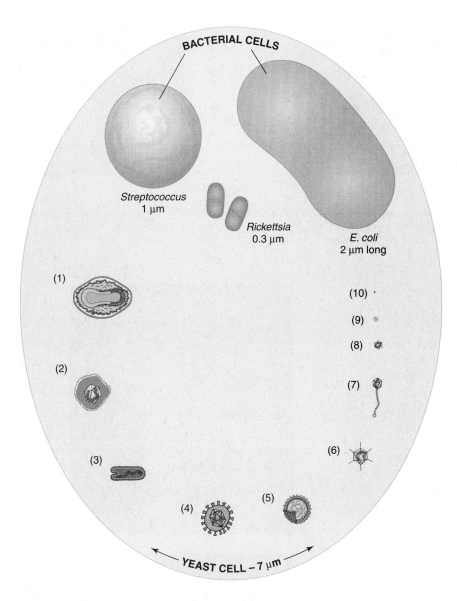

BACTERIAL CELLS

Streptococcus
1 μm

Rickettsia
0.3 μm

E. coli
2 μm long

(1)

(2)

(3)

(4)

(5)

(6)

(7)

(8)

(9)

(10)

YEAST CELL – 7 μm

Viruses	
1. Poxvirus	250 nm
2. Herpes simplex	150 nm
3. Rabies	125 nm
4. HIV	110 nm
5. Influenza	100 nm
6. Adenovirus	75 nm
7. T2 bacteriophage	65 nm
8. Poliomyelitis	30 nm
9. Yellow fever	22 nm
Protein Molecule	
10. Hemoglobin molecule	15 nm

FIGURE 6.1

Size comparison of viruses with a eucaryotic cell (yeast) and bacteria.   Viruses range from largest (1) to smallest (9). A molecule of protein (10) is included to indicate proportion of macromolecules.

The General Structure of Viruses

SIZE RANGE

As a group, viruses represent the smallest infectious agents (with some unusual exceptions to be discussed later in this chapter). Their size relegates them to the realm of the ultramicroscopic. This term means that most of them are so minute (< 0.2 μm) that an electron microscope is necessary to detect them or to examine their fine structure. They are dwarfed by their host cells: More than 2,000 bacterial viruses could fit into an average bacterial cell, and more than 50 million polioviruses could be accommodated by an average human cell. Animal viruses range in size from the small parvoviruses[1] (around 20 nm [0.02 μm] in diameter) to poxviruses[2] that are as large as small bacteria (up to 450 nm [0.4 μm] in length) (**figure 6.1**). Some cylindrical viruses are relatively long (800 nm [0.8 μm] in length) but so narrow in diameter (15 nm [0.015 μm]) that their visibility is still limited without the high magnification and resolution of an electron microscope. Figure 6.1 compares the sizes of several viruses with procaryotic and eucaryotic cells and molecules.

Viral architecture is most readily observed through special stains in combination with electron microscopy (**figure 6.2**). Negative staining uses very thin layers of an opaque salt to outline the shape of the virus against a dark background and to enhance textural features on the viral surface. Internal details are revealed by positive staining of specific parts of the virus such as protein or nucleic acid. The *shadowcasting* technique attaches a virus preparation to a surface and showers it with a dense metallic vapor directed from a certain angle. The thin metal coating over the surface of the virus enhances its contours, and a shadow is cast on the unexposed side.

---

1. DNA viruses that cause respiratory infections in humans.

2. A group of large, complex viruses, including smallpox, that cause raised skin swellings called pox.

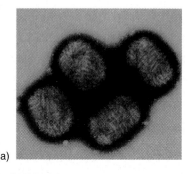

(a)

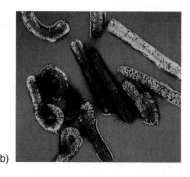

(b)

(c)

FIGURE 6.2

Methods of viewing viruses. **(a)** Negative staining of an orfvirus (a type of poxvirus), revealing details of its outer coat (122,000×). **(b)** Positive stain of the Ebola virus, a type of filovirus, so named because of its tendency to form long strands. Note the textured capsid. **(c)** Shadowcasting image of a vaccinia virus (17,500×).

UNIQUE VIRAL CONSTITUENTS: CAPSIDS, NUCLEIC ACIDS, AND ENVELOPES

It is important to realize that viruses bear no real resemblance to cells and that they lack any of the protein-synthesizing machinery found in even the simplest cells. Their molecular structure is composed of regular, repeating subunits that give rise to their crystalline appearance. Indeed, many purified viruses can form large aggregates or crystals if subjected to special treatments **(figure 6.3).** The general plan of virus organization is the utmost in its simplicity and compactness. Viruses contain only those parts needed to invade and control a host cell: an external coating and a core containing one or more nucleic acid strands of either DNA or RNA. This pattern of organization can be represented with a flowchart:

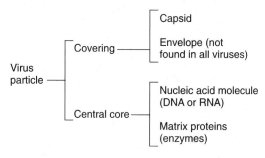

All but one group of viruses have a protein **capsid,*** or shell, that surrounds the nucleic acid in the central core. Together the capsid and the nucleic acid are referred to as the **nucleocapsid (figure 6.4).** Members of 13 of the 20 families of animal viruses possess an additional covering external to the capsid called an **envelope,** which is actually a modified piece of the host's cell membrane (figure 6.4*b*). Viruses that consist of only a nucleocapsid are considered **naked viruses** (figure 6.4*a*). As we shall see later, the enveloped viruses also differ from the naked viruses in the way that they enter and leave a host cell.

The Viral Capsid: The Protective Outer Shell

When a virus particle is magnified several hundred thousand times, the capsid appears as the most prominent geometric feature (see figure 6.3*b*). In general, each capsid is constructed from identical

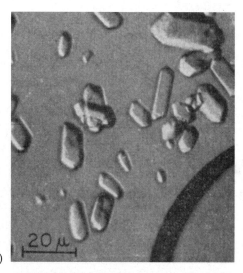

(a)

(b)

FIGURE 6.3

The crystalline nature of viruses. **(a)** Light microscope magnification (1,200×) of purified poliovirus crystals. **(b)** Highly magnified (150,000×) electron micrograph of the capsids of this same virus, demonstrating their highly geometric nature.

*capsid (kap'-sid) L. *capsa,* box.

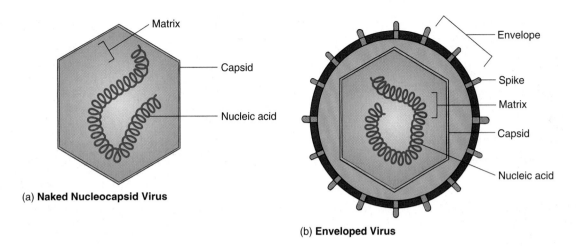

Matrix — Capsid — Nucleic acid

(a) **Naked Nucleocapsid Virus**

Envelope — Spike — Matrix — Capsid — Nucleic acid

(b) **Enveloped Virus**

FIGURE 6.4

Generalized structure of viruses.   **(a)** The simplest virus is a naked virus (nucleocapsid) consisting of a geometric capsid assembled around a nucleic acid strand or strands. **(b)** An enveloped virus is composed of a nucleocapsid surrounded by a flexible membrane called an envelope. The envelope usually has special receptor spikes inserted into it.

subunits called **capsomers*** that are constructed from protein molecules. The capsomers spontaneously self-assemble into the finished capsid. Depending on how the capsomers are shaped and arranged, this binding results in two different types: helical and icosahedral.

The simpler **helical capsids** have rod-shaped capsomers that bond together to form a series of hollow discs resembling a bracelet. During the formation of the nucleocapsid, these discs link with other discs to form a continuous helix into which the nucleic acid strand is coiled (**figure 6.5**). In electron micrographs, the appearance of a helical capsid varies with the type of virus. The nucleocapsids of naked helical viruses are very rigid and tightly wound into a cylinder-shaped package (**figure 6.6a,b**). An example is the *tobacco mosaic virus,* which attacks tobacco leaves. Enveloped helical nucleocapsids are more flexible and tend to be arranged as a looser helix within the envelope (figure 6.6c,d). This type of morphology is found in several enveloped human viruses, including those of influenza, measles, and rabies.

The capsids of a number of major virus families are arranged in an **icosahedron***—a three-dimensional, 20-sided figure with 12 evenly spaced corners. The arrangements of the capsomers vary from one virus to another. Some viruses construct the capsid from a single type of capsomer while others may contain several types of capsomers (**figure 6.7**). Although the capsids of all icosahedral viruses have this sort of symmetry, they can have major variations in the number of capsomers; for example, a poliovirus has 32, and an adenovirus has 240 capsomers. Individual capsomers can look either ring- or dome-shaped,

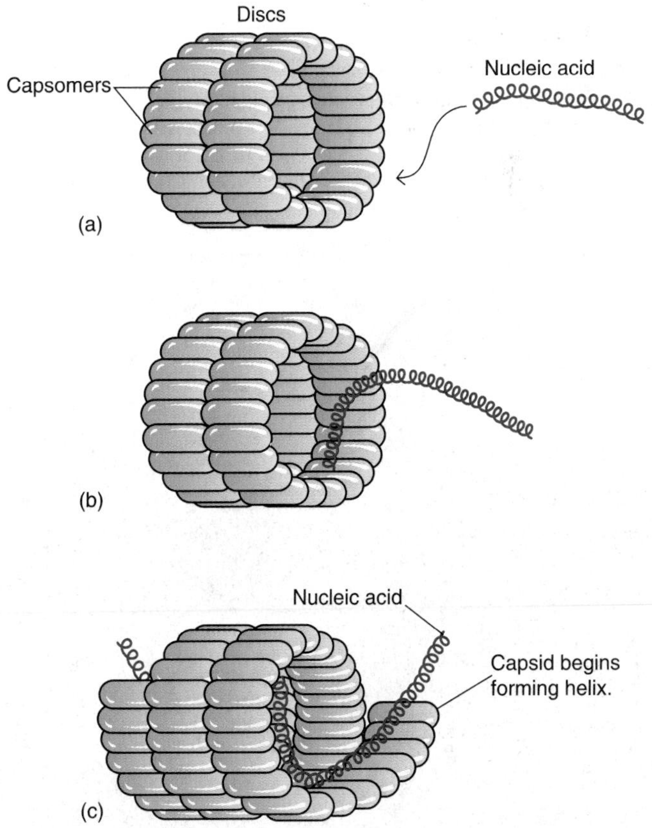

Discs

Capsomers

Nucleic acid

(a)

(b)

Nucleic acid

Capsid begins forming helix.

(c)

FIGURE 6.5

Assembly of helical nucleocapsids.   **(a)** Capsomers assemble into hollow discs. **(b)** The nucleic acid is inserted into the center of the disc. **(c)** Elongation of the nucleocapsid progresses from both ends, as the nucleic acid is wound "within" the lengthening helix.

***capsomer** (kap′-soh-meer) L. *capsa,* box, and *mer,* part.

***icosahedron** (eye″-koh-suh-hee′-drun) Gr. *eikosi,* twenty, and *hedra,* side. A type of polygon.

(a)

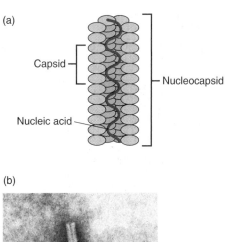

Capsid

Nucleocapsid

Nucleic acid

(b)

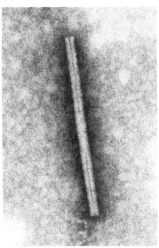

(c)

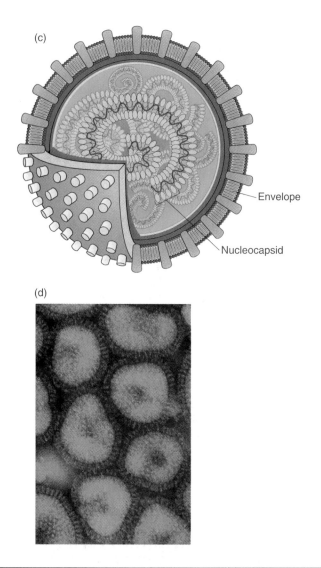

Envelope

Nucleocapsid

(d)

FIGURE 6.6

Typical variations of viruses with helical nucleocapsids. Naked helical virus (tobacco mosaic virus): **(a)** a schematic view and **(b)** a greatly magnified micrograph. Note the overall cylindrical morphology. Enveloped helical virus (influenza virus): **(c)** a schematic view and **(d)** an electron micrograph of the same virus (350,000×).

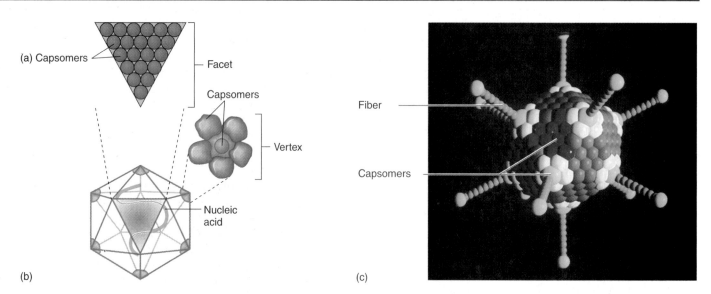

(a) Capsomers

Facet

Capsomers

Vertex

Nucleic acid

(b)

Fiber

Capsomers

(c)

FIGURE 6.7

The structure and formation of an icosahedral virus (adenovirus is the model). **(a)** A facet or "face" of the capsid is composed of 21 identical capsomers arranged in a triangular shape. The vertices or "points" consist of 5 capsomers arranged with a single penton in the center. Other viruses can vary in the number, types, and arrangement of capsomers. **(b)** An assembled virus shows how the facets and vertices come together to form a shell around the nucleic acid. **(c)** A three-dimensional model of this virus shows fibers attached to the pentons.

Capsomers   Capsid

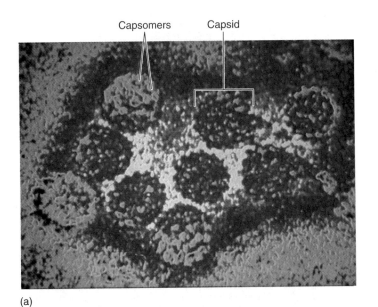

(a)

Envelope

Capsid

DNA core

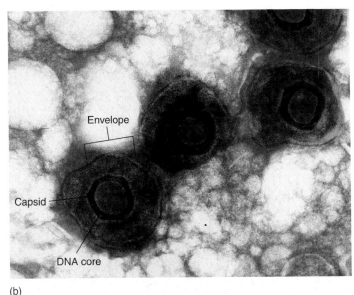

(b)

FIGURE 6.8

Two types of icosahedral viruses, highly magnified. **(a)** False-color micrograph of papillomaviruses with unusual, ring-shaped capsomers. **(b)** Herpesvirus, an enveloped icosahedron (300,000×).

and the capsid itself can appear spherical or cubical **(figure 6.8).** During assembly of the virus, the nucleic acid is packed into the center of this icosahedron, forming a nucleocapsid. Another factor that alters the appearance of icosahedral viruses is whether or not they have an outer envelope; contrast a papillomavirus (warts) and its naked nucleocapsid with herpes simplex (cold sores) and its enveloped nucleocapsid (figure 6.10).

The Viral Envelope

When **enveloped viruses** (mostly animal) are released from the host cell, they take with them a bit of its membrane system in the form of an envelope (see figure 6.20). Some viruses bud off the cell membrane; others leave via the nuclear envelope or the endoplasmic reticulum. Whichever avenue of escape, the viral envelope differs significantly from the host's membranes. In the envelope, some or all of the regular membrane proteins are replaced with special viral proteins. Some proteins form a binding layer between the envelope and capsid of the virus, and glycoproteins (proteins bound to a carbohydrate) remain exposed on the outside of the envelope. These protruding molecules, called **spikes** or **peplomers,*** are essential for the attachment of viruses to the next host cell. Because the envelope is more supple than the capsid, enveloped viruses are pleomorphic and range from spherical to filamentous in shape.

Functions of the Viral Capsid/Envelope

The outermost covering of a virus is indispensable to viral function because it protects the nucleic acid from the effects of various enzymes and chemicals when the virus is outside the host cell. For example, the capsids of enteric (intestinal) viruses such as polio and hepatitis A are resistant to the acid- and protein-digesting enzymes of the gastrointestinal tract. Capsids and envelopes are also responsible for helping to introduce the viral DNA or RNA into a suitable host cell, first by binding to the cell surface and then by assisting in penetration of the viral nucleic acid (to be discussed in more detail later in the chapter). In addition, parts of viral capsids and envelopes stimulate the immune system to produce antibodies that can neutralize viruses and protect the host's cells against future infections (see chapters 15, 24, and 25).

Complex Viruses: Atypical Viruses

Two special groups of viruses, termed complex viruses **(figure 6.9),** are more intricate in structure than the helical, icosahedral, naked, or enveloped viruses just described. The **poxviruses** (including the agent of smallpox) are very large DNA viruses that lack a regular capsid and have in its place several layers of lipoproteins and coarse surface fibrils. Another group of very complex viruses, the **bacteriophages,*** have a polyhedral head, a helical tail, and fibers for attachment to the host cell. Their mode of multiplication is covered in a later section of this chapter. **Figure 6.10** summarizes the morphological types of some common viruses.

Nucleic Acids: At the Core of a Virus

So far, one biological constant is that the genetic information of living cells is carried by nucleic acids (DNA, RNA). Viruses, although neither alive nor cells, are no exception to this rule, but there is a significant difference. Unlike cells, which contain both DNA and RNA, viruses contain either DNA or RNA *but not both.* Because viruses must pack into a tiny space all of the genes necessary to instruct the host cell to make new viruses, the number of viral genes is quite small compared with that of a cell. It varies from four genes in hepatitis B virus to hundreds of genes in some herpesviruses. By comparison, the bacterium *Escherichia coli* has approximately 4,000 genes, and a human cell has approximately 40,000 genes.

---

*peplomer (pep'-loh-meer) Gr. *peplos,* envelope, and *mer,* part.

*bacteriophage (bak-teer'-ee-oh-fayj") From *bacteria,* and Gr. *phagein,* to eat. These viruses parasitize bacteria.

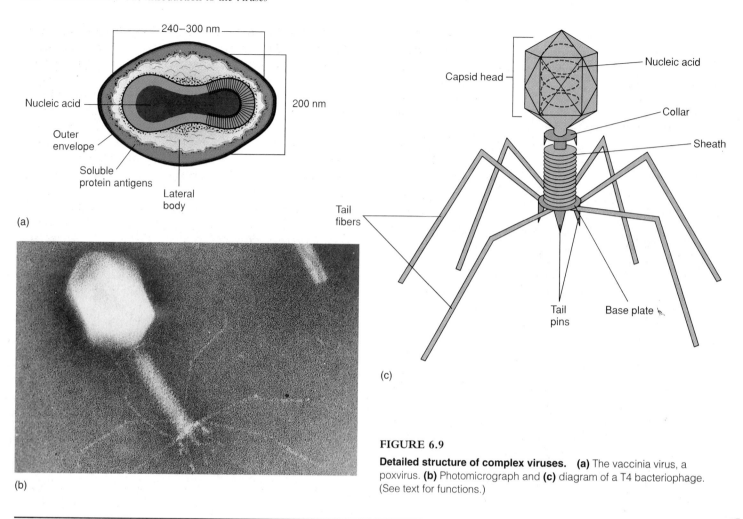

(a)

240–300 nm

200 nm

Nucleic acid

Outer envelope

Soluble protein antigens

Lateral body

(b)

Capsid head

Nucleic acid

Collar

Sheath

Tail fibers

Tail pins

Base plate

(c)

FIGURE 6.9

Detailed structure of complex viruses.   **(a)** The vaccinia virus, a poxvirus. **(b)** Photomicrograph and **(c)** diagram of a T4 bacteriophage. (See text for functions.)

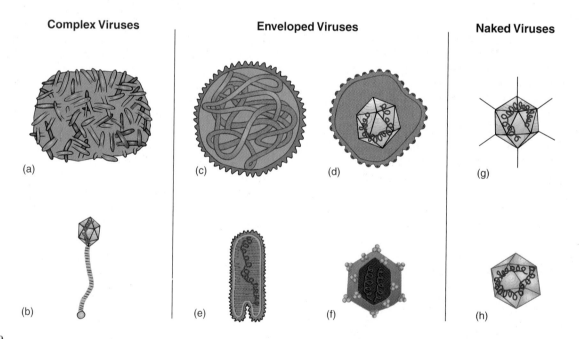

Complex Viruses

Enveloped Viruses

Naked Viruses

(a)

(b)

(c)

(d)

(e)

(f)

(g)

(h)

FIGURE 6.10

Complex viruses: **(a)** poxvirus, a large DNA virus; **(b)** flexible-tailed bacteriophage. Enveloped viruses: **(c)** mumps virus, an enveloped RNA virus with a helical nucleocapsid; **(d)** herpesvirus, an enveloped DNA virus with an icosahedral nucleocapsid; **(e)** rhabdovirus, an RNA virus with a bullet-shaped envelope; **(f)** HIV (AIDS), an RNA retrovirus with an icosahedral capsid. Naked viruses: **(g)** adenovirus, a DNA virus with fibers on the capsid; **(h)** papillomavirus, a DNA virus that causes warts.

In chapter 2 we learned that DNA usually exists as a double-stranded molecule and that RNA is single-stranded. Although most viruses follow this same pattern, a few exhibit distinctive and exceptional forms. Notable examples are the parvoviruses (the cause of erythema infectiosum[3]), which contain single-stranded DNA, and reoviruses (a cause of respiratory and intestinal tract infections), which contain double-stranded RNA. In all cases, these tiny strands of genetic material carry the blueprint for viral structure and functions. In a very real sense, viruses are genetic parasites because they cannot multiply until their nucleic acid has reached the internal habitat of the host cell. At the minimum, they must carry genes for synthesizing the viral capsid and genetic material, for regulating the actions of the host, and for packaging the mature virus.

Other Substances in the Virus Particle

In addition to the protein of the capsid, the proteins and lipids of envelopes, and the nucleic acid of the core, viruses can contain enzymes for specific operations within their host cell. They may come with pre-formed enzymes that are required for viral replication. Examples include *polymerases** that assemble DNA and RNA and replicases that copy RNA. The AIDS virus comes equipped with reverse transcriptase for synthesizing DNA from RNA. Viruses completely lack the genes for synthesis of metabolic enzymes. As we shall see, this deficiency has little consequence, because viruses have adapted to completely take over their hosts' metabolic resources. Some viruses can actually carry away substances from their host cell. For instance, arenaviruses pack along host ribosomes, and retroviruses "borrow" the host's tRNA molecules.

How Viruses Are Classified and Named

Although viruses are not classified as members of the kingdoms discussed in chapter 1, they are diverse enough to require their own classification scheme to aid in their study and identification. In an informal and general way, we have already begun classifying viruses—as animal, plant, or bacterial viruses; enveloped or naked viruses; DNA or RNA viruses; and helical or icosahedral viruses. These introductory categories are certainly useful in organization and description, but the study of specific viruses requires a more standardized method of nomenclature. For many years, the animal viruses were classified mainly on the basis of their hosts and the kind of diseases they caused. Newer systems for naming viruses take into account the actual nature of the virus particles themselves, with only partial emphasis on host and disease. The main criteria presently used to group viruses are structure, chemical composition, and similarities in genetic makeup.

A widely used scheme for classifying animal viruses first assigns them to one of two superfamilies, either those containing DNA or those containing RNA. DNA viruses can be further subdivided into six families, and RNA viruses into 13 families, for a total of 19 families of animal viruses. Virus families are given a name composed of a Latin root followed by *-viridae*. Characteristics used for placement in a particular family include type of capsid, nucleic acid strand number, presence and type of envelope, overall viral size, and area of the host cell in which the virus multiplies. Some virus families are named for their microscopic appearance (shape and size). Examples include **rhabdoviruses,** which have a bullet-shaped envelope, and *togaviruses,** which have a cloaklike envelope. Anatomical or geographic areas have also been used in naming. For instance, **adenoviruses** were first discovered in adenoids (one type of tonsil), and hantaviruses were originally isolated in an area in Korea called Hantaan. Viruses can also be named for their effects on the host. *Lentiviruses** tend to cause slow, chronic infections. Acronyms made from blending several characteristics include *picornaviruses,** which are tiny RNA viruses, and reoviruses (or **r**espiratory **e**nteric **o**rphan viruses), which inhabit the respiratory tract and the intestine and are not yet associated with any known disease state.

Each different type of virus is also assigned genus status according to its host, target tissue, and the type of disease it causes **(table 6.2).** Viral genera are denoted by a special Latinized root followed by the suffix *-virus* (for example, *Enterovirus* and *Hantavirus*). Because the use of standardized species names has not been widely accepted, the genus or common English vernacular names (for example, poliovirus and rabies virus) predominate in discussions of specific viruses in this text.

A Note on Terminology

Although the terms *virus* and *virus particle* are interchangeable, virologists find it convenient to distinguish between the various states in which a virus can exist. A fully formed, extracellular particle that is virulent (able to establish infection in a host) is called a **virion** (vir'-ee-on). Once its genetic material has entered a host cell, a virus may undergo a multiplication phase called the **lytic** (lih'-tik) **cycle** in which the host cell is disrupted to release more virions. Some viruses remain in an inactive, or **latent** (lay'-tunt), stage in which the virus does not lyse the host cell. This stage is called the **lysogenic phase.**

Modes of Viral Multiplication

Viruses tend to remain closely associated with their hosts. In addition to providing the viral habitat, the host cell is absolutely necessary for viral multiplication. The process of viral multiplication is an extraordinary biological phenomenon. Viruses have often been aptly described as minute parasites that seize control of the synthetic and genetic machinery of cells. The nature of this cycle

---

3. A common childhood disease described in chapter 24.

*polymerase (pol-im'-ur-ace) An enzyme that synthesizes a large molecule from smaller subunits.

*rhabdovirus (rab"-doh-vy'-rus) Gr. *rhabdo,* little rod.

*togavirus (toh"-guh-vy'-rus) L. *toga,* covering or robe.

*adenovirus (ad"-uh-noh-vy'-rus) G. *aden,* gland.

*lentivirus (len"-tee-vy'-rus) Gr. *lente,* slow. HIV, the AIDS virus, belongs in this group.

*picornavirus (py-kor"-nah-vy'-rus) Sp. *pico,* small, plus RNA.

TABLE 6.2

Important Human Virus Families, Genera, Common Names, and Types of Diseases*

Family	Genus of Virus	Common Name of Genus Members	Name of Disease
DNA Viruses			
Poxviridae	*Orthopoxvirus*	Variola major and minor	Smallpox, cowpox
Herpesviridae	*Simplexvirus*	Herpes simplex (HSV) 1 virus	Fever blister, cold sores
		Herpes simplex (HSV) 2 virus	Genital herpes
	Varicellovirus	Varicella zoster virus (VZV)	Chickenpox, shingles
	Cytomegalovirus	Human cytomegalovirus (CMV)	CMV infections
Adenoviridae	*Mastadenovirus*	Human adenoviruses	Adenovirus infection
Papovaviridae	*Papillomavirus*	Human papillomavirus (HPV)	Several types of warts
	Polyomavirus	JS virus (JCV)	Progressive multifocal leukoencephalopathy (PML)
Hepadnaviridae	*Hepadnavirus*	Hepatitis B virus (HBV or Dane particle)	Serum hepatitis
Parvoviridae	*Erythrovirus*	Parvovirus B19	Erythema infectiosum
RNA Viruses			
Picornaviridae	*Enterovirus*	Poliovirus	Poliomyelitis
		Coxsackievirus	Hand-foot-mouth disease
	Hepatovirus	Hepatitis A virus (HAV)	Short-term hepatitis
	Rhinovirus	Human rhinovirus	Common cold, bronchitis
Calciviridae	*Calicivirus*	Norwalk virus	Viral diarrhea, Norwalk virus syndrome
Togaviridae	*Alphavirus*	Eastern equine encephalitis virus	Eastern equine encephalitis (EEE)
		Western equine encephalitis virus	Western equine encephalitis (WEE)
		Yellow fever virus	Yellow fever
		St. Louis encephalitis virus	St. Louis encephalitis
	Rubivirus	Rubella virus	Rubella (German measles)
Flaviviridae	*Flavivirus*	Dengue fever virus	Dengue fever
		West Nile fever virus	West Nile fever
Bunyaviridae	*Bunyavirus*	Bunyamwera viruses	California encephalitis
	Hantavirus	Sin Nombre virus	Respiratory distress syndrome
	Phlebovirus	Rift Valley fever virus	Rift Valley fever
	Nairovirus	Crimean–Congo hemorrhagic fever virus (CCHF)	Crimean–Congo hemorrhagic fever
Filoviridae	*Filovirus*	Ebola, Marburg virus	Ebola fever
Reoviridae	*Coltivirus*	Colorado tick fever virus	Colorado tick fever
	Rotavirus	Human rotavirus	Rotavirus gastroenteritis
Orthomyxoviridae	*Influenza virus*	Influenza virus, type A (Asian, Hong Kong, and swine influenza viruses)	Influenza or "flu"
Paramyxoviridae	*Paramyxovirus*	Parainfluenza virus, types 1–5	Parainfluenza
		Mumps virus	Mumps
	Morbillivirus	Measles virus	Measles (red)
	Pneumovirus	Respiratory syncytial virus (RSV)	Common cold syndrome
Rhabdoviridae	*Lyssavirus*	Rabies virus	Rabies (hydrophobia)
Retroviridae	*Oncornavirus*	Human T-cell leukemia virus (HTLV)	T-cell leukemia
	Lentivirus	HIV (human immunodeficiency viruses 1 and 2)	Acquired immunodeficiency syndrome (AIDS)
Arenaviridae	*Arenavirus*	Lassa virus	Lassa fever
Coronaviridae	*Coronavirus*	Infectious bronchitis virus (IBV)	Bronchitis
		Enteric corona virus	Coronavirus enteritis
		SARS virus	Severe acute respiratory syndrome

*See also figures 24.1 and 25.1 for additional information.

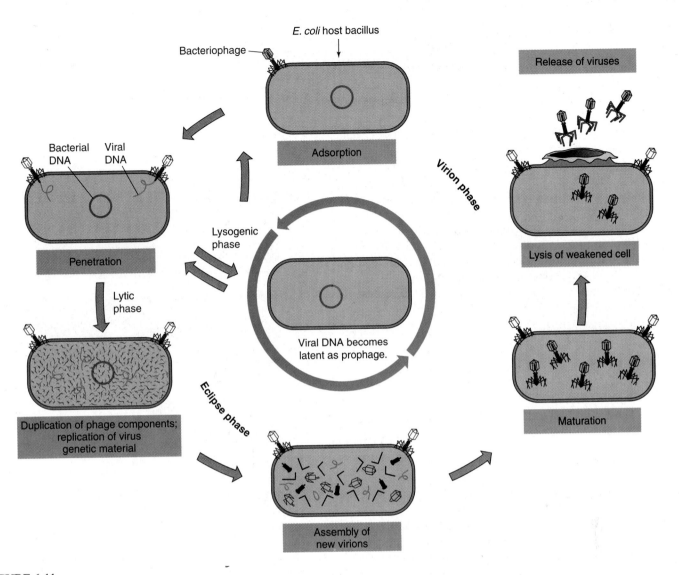

FIGURE 6.11

Events in the multiplication cycle of T-even bacteriophages.   The cycle is divided into the eclipse phase (during which the phage is developing but is not yet infectious) and the virion phase (when the virus matures and is capable of infecting a host). (The text provides further details of this cycle. See figure 6.15 for the steps in the lysogenic phase.)

dictates viral pathogenicity, transmission, the responses of the immune defenses, and human measures to control viral infections. From these perspectives, we cannot overemphasize the importance of a working knowledge of the relationship between viruses and their host cells.

The multiplication cycles of viruses are similar enough that virologists use certain viruses (such as bacteriophages and enveloped animal viruses) as general models. Although a given cycle occurs continuously and not in discrete steps, it is helpful to demarcate its major sequential events. These events are **adsorption,*** a recognition process between a virus and host cell that results in

virus attachment to the external surface of the host cell; **penetration,** entrance of the virion (either a whole virus or just its nucleic acid) into the host cell; **replication,*** copying and expression of the viral genome by the host's synthetic equipment, resulting in the production of the various virus components; **assembly** and **maturation** of these individual viral parts into whole, intact virions; and **release,** escape from the host cell of the active, infectious viral particles **(figure 6.11).** We present an overview of the genetic events of virus cycles here, but this topic is covered in greater detail in chapter 9. The following section compares and contrasts the multiplication stages for bacterial and human viruses.

*adsorption (ad-sorp′-shun) L. *ad,* to, and *sorbere,* to suck. The attachment of one thing onto the surface of another.

*replication (rep-lih-kay′-shun) L. *replicare,* to reply. To make an exact duplicate.

THE MULTIPLICATION CYCLE
IN BACTERIOPHAGES

When Frederick Twort and Felix d'Herelle discovered bacterial viruses in 1915, it first appeared that the bacterial host cells were being eaten by some unseen parasite, hence the name bacteriophage was used. Most bacteriophages (often shortened to *phage*) contain double-stranded DNA, though single-stranded DNA and RNA types exist as well. So far as is known, every bacterial species is parasitized by various specific bacteriophages. Probably the most widely studied bacteriophages are those of the intestinal bacterium *Escherichia coli*—especially the T-even (for even-numbered type) phages. Their complex structure has been previously described. They have an icosahedral capsid head containing DNA, a central tube (surrounded by a sheath), collar, base plate, tail pins, and fibers, which in combination make an efficient package for infecting a bacterial cell (see figure 6.9c). Momentarily setting aside a strictly scientific and objective tone, it is tempting to think of these extraordinary viruses as minute spacecrafts docking on an alien planet, ready to unload their genetic cargo.

Adsorption: Docking onto the Host Cell Surface

For a successful infection, a phage must meet and stick to a susceptible host cell. Adsorption takes place when certain molecules on the tail fibers bind to specific molecules (receptors) on the cell envelope of the host bacterium (see figure 6.11). Among the bacterial structures that serve as phage receptors are surface molecules of the cell wall, pili, and flagella. Attachment is a function of correct fit and chemical attraction; that is, the phage's tail molecules must interact specifically with those of the bacterial receptors. Once the phage is firmly affixed to the cell, it is positioned for penetration.

Bacteriophage Penetration: Entry of the Nucleic Acid

After adsorption, a phage is still on the outside of its host cell and remains inactive unless it can gain entrance to the cell's cytoplasm. The strong, rigid bacterial cell wall is quite an impenetrable barrier, and the entire virus particle is unable to cross it. But the T-even bacteriophages have an exquisite mechanism for injecting nucleic acid across this barrier and into the cell. Like a miniscule syringe, the sheath constricts, pushing the inner tube through the host's cell wall and membrane, forming a passageway for its nucleic acid into the interior of the cell (**figure 6.12**). Once inside, the viral nucleic acid alone can complete the multiplication cycle. The spent virus shell has fulfilled its principal functions—protection, adsorption, and injection of DNA—and it will remain attached outside to the cell wall fragment even after the cell has lysed.

The Bacteriophage Assembly Line

Entry of the nucleic acid into the cell results in sweeping changes in the bacterial cell's activities. Within a few minutes, the bacterium stops synthesizing its own molecules, and its metabolism shifts to the expression of genes on the viral nucleic acid strand. In T-even bacteriophages, the viral DNA redirects the genetic and metabolic activity of the cell, blocking the utilization of host DNA and ensuring that viral DNA is copied and used to synthesize new viral components (see replication of DNA in chapter 2). Expression of that phage's genetic information gives rise to these viral molecules:

1. proteins to "seal" the cell (because the host cell is punctured by the virus during injection, and it is advantageous for the

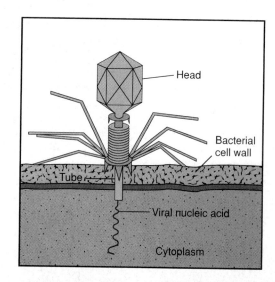

(a)

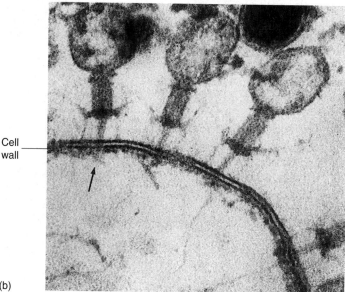

(b)

FIGURE 6.12

Penetration of a bacterial cell by a T-even bacteriophage. **(a)** After adsorption, the phage plate becomes embedded in the cell wall, and the sheath contracts, pushing the tube through the cell wall and releasing the nucleic acid into the interior of the cell. **(b)** Section through *Escherichia coli* with attached phages. Note that these phages have injected their nucleic acid through the cell wall and now have empty heads.

virus to repair the hole before the host cell's integrity is destroyed);
2. enzymes for copying the virus genome;
3. proteins that make up the capsid head and parts of the tail; and
4. enzymes that help weaken the cell wall so that the new phages can easily escape.

This period of viral synthesis exploits the host's cytoplasmic nutrient resources, its ribosomes, and its energy supplies. The early stages of viral replication are known as *eclipse,* a period during which no mature virions can be detected within the host cell. This corresponds with the period of penetration, replication, and early assembly, and if the cell were disrupted at this time, no infective virus would be released.

As the host cell rapidly produces new phage parts, these parts spontaneously assemble into bacteriophages, similar to a mass production assembly line **(figure 6.13).** The first parts to assemble are the capsid and tail. Viral DNA is inserted into the capsid before the capsomers are completely joined and the collar and sheath unite with the tail pins. Mature phage particles are formed when the capsid and tail fit together and the fibers are anchored to the base plate. An average-sized *Escherichia coli* cell can contain up to 200 new phage units at the end of this period. Eventually, the host cell becomes so packed with viruses that it **lyses**—splits open—thereby liberating the mature virions **(figure 6.14).** This process is hastened by viral enzymes produced late in the infection cycle that digest the cell envelope, thereby weakening it. Upon release, the virulent phages can spread to other susceptible bacterial cells and begin a new cycle of infection.

Lysogeny: The Silent Virus Infection

The lethal effects of a virulent phage on the host cell present a dramatic view of virus-host interaction. Not all bacteriophages complete the lytic cycle, however. Special DNA phages, called **temperate* phages,** undergo adsorption and penetration into the bacterial host but are not replicated or released. Instead, the viral DNA enters an inactive **prophage*** state, during which it is inserted into the bacterial chromosome. This viral DNA will be retained by the bacterial cell and copied during its normal cell division so that the cell's progeny will also have the temperate phage DNA **(figure 6.15).** This condition, in which the host chromosome carries bacteriophage DNA, is termed **lysogeny.*** Because the viral genome is not expressed, the bacterial cells carrying temperate phages do not lyse, and they appear entirely normal. On occasion, the prophage in a lysogenic cell will be activated and progress directly into viral replication and the lytic cycle. Lysogeny is a less deadly form of parasitism than the full lytic cycle and is thought to be an advancement that allows the virus to spread without killing the host. In a later section and in chapters 9, 17, and 24, we describe a similar relationship that exists between certain animal viruses and human cells.

*temperate (tem'-pur-ut) A reduction in intensity.

*prophage (pro'-fayj) L. *pro*, before, plus phage.

*lysogeny (ly-soj'-uhn-ee) The potential ability to produce phage.

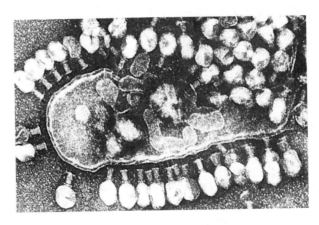

FIGURE 6.14

A weakened bacterial cell, crowded with viruses. The cell has ruptured and released numerous virions that can then attack nearby susceptible host cells. Note the empty heads of "spent" phages lined up around the ruptured wall.

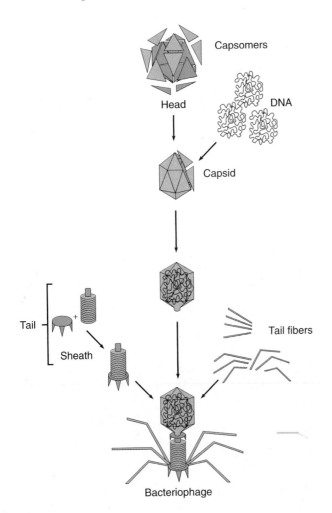

FIGURE 6.13

Bacteriophage assembly line. First the protein subunits (head, sheath, tail fibers) are synthesized by the host cell. A strand of viral nucleic acid also replicated by the host's machinery is inserted during the later stages of capsid formation. During final assembly, the prefabricated components fit together into whole parts and finally into the finished products.

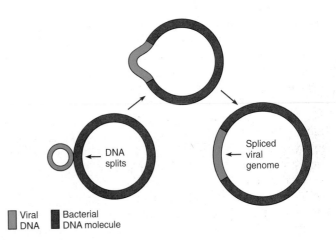

FIGURE 6.15

The lysogenic state in bacteria. A bacterial DNA molecule can accept and insert viral DNA molecules at specific sites on its genome. This additional viral DNA is duplicated along with the regular genome and can provide adaptive characteristics for the host bacterium.

The cycle of bacterial viruses illustrates general features of viral multiplication in a very concrete and memorable way. It is fascinating to realize that viruses are capable of lying "dormant" in their host cells, possibly becoming active at some later time. Because of the intimate association between the genetic material of the virus and host, phages occasionally serve as transporters of

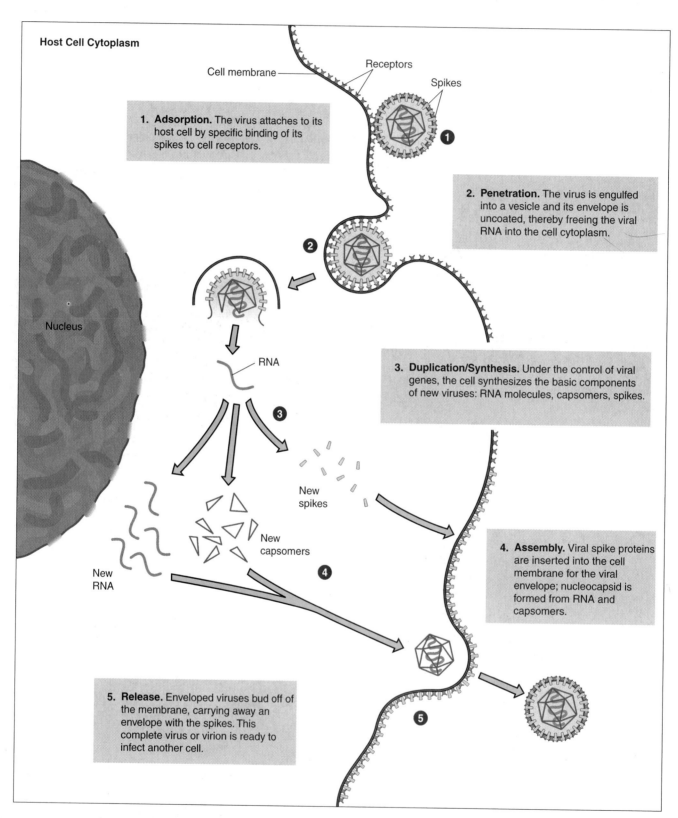

Host Cell Cytoplasm

1. **Adsorption.** The virus attaches to its host cell by specific binding of its spikes to cell receptors.

2. **Penetration.** The virus is engulfed into a vesicle and its envelope is uncoated, thereby freeing the viral RNA into the cell cytoplasm.

3. **Duplication/Synthesis.** Under the control of viral genes, the cell synthesizes the basic components of new viruses: RNA molecules, capsomers, spikes.

4. **Assembly.** Viral spike proteins are inserted into the cell membrane for the viral envelope; nucleocapsid is formed from RNA and capsomers.

5. **Release.** Enveloped viruses bud off of the membrane, carrying away an envelope with the spikes. This complete virus or virion is ready to infect another cell.

Cell membrane — Receptors — Spikes — Nucleus — RNA — New spikes — New capsomers — New RNA

FIGURE 6.16

General features in the multiplication cycle of an enveloped animal virus. Using an RNA virus (rubella virus), the major events are outlined, although other viruses will vary in exact details of the cycle.

bacterial genes from one bacterium to another and consequently can play a profound role in bacterial genetics. This phenomenon, called transduction, is one way that genes for toxin production and drug resistance are transferred between bacteria (see chapters 9 and 12).

MULTIPLICATION CYCLES IN ANIMAL VIRUSES

Now that we have a working concept of viral multiplication in bacteria, let us turn to the animal viruses for yet another model system. Although the basic cycle is very similar to that of bacteriophages, several significant differences exist between the host cells, their viruses, and the multiplication cycles **(table 6.3).** The general phases in the cycle of animal viruses are adsorption, penetration, uncoating, replication, assembly, and exit from the host cell. The length of the entire multiplication cycle varies from 8 hours in polioviruses to 36 hours in herpesviruses. See **figure 6.16** for the major phases of one type of animal virus.

Adsorption and Host Range

Invasion begins when the virus encounters a susceptible host cell and adsorbs specifically to receptor sites on the cell membrane. The membrane receptors that viruses attach to are usually glycoproteins

the cell requires for its normal function. For example, the rabies virus affixes to the acetylcholine receptor of nerve cells, and the human immunodeficiency virus (HIV or AIDS virus) attaches to the CD4 protein on certain white blood cells. The mode of attachment varies between the two general types of viruses. In enveloped forms such as influenza virus and HIV, glycoprotein spikes bind to the cell membrane receptors. Viruses with naked nucleocapsids (poliovirus, for example), possess surface molecules that adhere to cell membrane receptors **(figure 6.17).**

Because a virus can invade its host cell only through making an exact fit with a specific host molecule, the scope of hosts it can infect in a natural setting is limited. This limitation, known as the **host range,** may be as restricted as hepatitis B, which infects only liver cells of humans; intermediate like the poliovirus, which infects intestinal and nerve cells of primates (humans, apes, and

TABLE 6.3

Comparison of Bacteriophage and Animal Virus Multiplication

	Bacteriophage	Animal Virus
Adsorption	Precise attachment of special tail fibers to cell wall	Attachment of capsid or envelope to cell surface receptors
Penetration	Injection of nucleic acid through cell wall; no uncoating of nucleic acid	Whole virus is engulfed and uncoated, or virus surface fuses with cell membrane, nucleic acid is released
Replication and Maturation	Occurs in cytoplasm Cessation of host synthesis	Occurs in cytoplasm and nucleus Cessation of host synthesis
	Viral DNA or RNA is replicated and begins to function Viral components synthesized	Viral DNA or RNA is replicated and begins to function Viral components synthesized
Viral Persistence	Lysogeny	Latency, chronic infection, cancer
Exit from Host Cell	Cell lyses when viral enzymes weaken it	Some cells lyse; enveloped viruses bud off host cell membrane
Cell Destruction	Immediate	Immediate; delayed in some

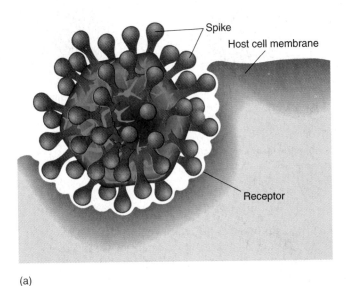

(a)

(b)

FIGURE 6.17

The mode by which animal viruses adsorb to the host cell membrane. **(a)** An enveloped coronavirus with prominent spikes. The configuration of the spike has a complementary fit for cell receptors. The process in which the virus lands on the cell and plugs into receptors is termed docking. **(b)** An adenovirus has a naked capsid that adheres to its host cell by nestling surface molecules on its capsid into the receptors on the host cell's membrane.

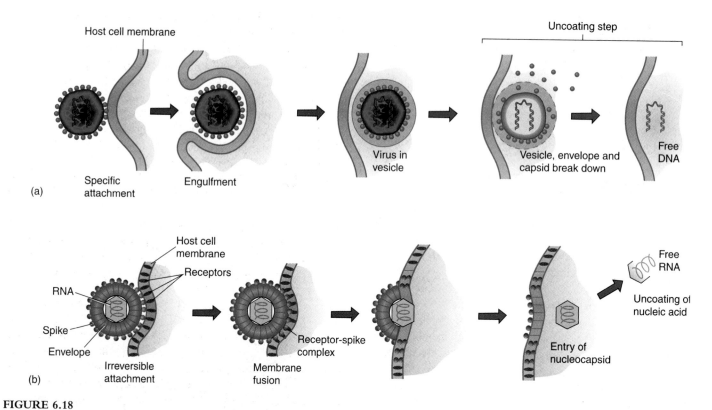

FIGURE 6.18

Two principal means by which animal viruses penetrate. **(a)** Endocytosis (engulfment) and uncoating of a herpesvirus. **(b)** Fusion of the cell membrane with the viral envelope (mumps virus).

monkeys); or as broad as the rabies virus, which can infect various cells of all mammals. Cells that lack compatible virus receptors are resistant to adsorption and invasion. This explains why, for example, human liver cells are not infected by the canine hepatitis virus and dog liver cells cannot host the human hepatitis A virus. It also explains why viruses usually have tissue specificities called *tropisms** for certain cells in the body. The hepatitis B virus targets the liver, and the mumps virus targets salivary glands. However, the fact that most viruses can be induced to infect cells that they would not infect naturally makes it possible to cultivate them in the laboratory.

Penetration/Uncoating of Animal Viruses

Animal viruses exhibit some impressive mechanisms for entering a host cell. Unlike bacteriophages, they have no mechanism to inject their nucleic acids. Instead, the flexible cell membrane of the host is penetrated by the whole virus or its nucleic acid **(figure 6.18).** In penetration by **endocytosis** (figure 6.18*a*), the entire virus is engulfed by the cell and enclosed in a vacuole or vesicle. When enzymes in the vacuole dissolve the envelope and capsid, the virus is said to be **uncoated,** a process that releases the viral nucleic acid into the cytoplasm. The exact manner of uncoating varies, but in most cases, the virus fuses with the wall of the vesicle. Another means of entry involves direct fusion of the viral envelope with the host cell membrane (as in influenza and mumps viruses) (figure 6.18*b*). In this

form of penetration, the envelope merges directly with the cell membrane, thereby liberating the nucleocapsid into the cell's interior.

Replication and Maturation of Animal Viruses: Host Cell As Factory

The synthetic and replicative phases of animal viruses are highly regulated and extremely complex at the molecular level. They are too detailed to cover at this point. A more thorough coverage of viral genetics is included in chapters 9, 24, and 25. As with bacteriophages, the free viral nucleic acid exerts control over the host's synthetic and metabolic machinery. How this control proceeds will vary, depending on whether the virus is a DNA or an RNA virus. In general, the DNA viruses (except poxviruses) enter the host cell's nucleus and are replicated and assembled there (see figure 9.19). With few exceptions (such as retroviruses), RNA viruses are replicated and assembled in the cytoplasm (see figure 9.20).

For an overview of the phases of duplication and assembly, we will use RNA viruses as a model. Almost immediately upon entry, the viral nucleic acid alters the genetic expression of the host and instructs it to synthesize the building blocks for new viruses. First, the RNA of the virus becomes a message for synthesizing viral proteins (translation). The viruses with *positive sense** RNA molecules already contain the correct message for translation into proteins. Viruses with *negative sense* RNA molecules must first be

---

*tropism** (troh'-pizm) Gr. *trope,* a turn. Having a special affinity for an object or substance.

---

*sense The "sense" of the strand relates to its readability into a protein.

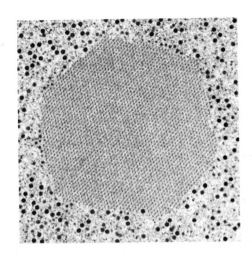

FIGURE 6.19
Nucleus of a cell, containing a crystalline mass of adenovirus (35,000×).

converted into a positive sense message. Some viruses come equipped with the necessary enzymes for synthesis of viral components; others utilize those of the host.

In the next phase, new RNA is synthesized using host nucleotides. Proteins for the capsid, spikes, and viral enzymes are synthesized on the host's ribosomes using its amino acids. Toward the end of the cycle, mature virus particles are constructed from the growing pool of parts. In most instances, the capsid is first laid down as an empty shell that will serve as a receptacle for the nucleic acid strand. Electron micrographs taken during this time show cells with masses of viruses, often in crystalline packets (**figure 6.19**). One important event in enveloped viruses is the insertion of viral spikes into the host's cell membrane so they can be picked up as the virus buds off with its envelope (see figures 6.16 and 6.20).

Release of Mature Viruses

To complete the cycle, assembled viruses leave their host in one of two ways. Nonenveloped and complex viruses that reach maturation in the cell nucleus or cytoplasm are released when the cell lyses or ruptures. Enveloped viruses are liberated by **budding** or **exocytosis**[4] from the membranes of the cytoplasm, nucleus, endoplasmic reticulum, or vesicles. During this process, the nucleocapsid binds to the membrane, which curves completely around it and forms a small pouch. Pinching off the pouch releases the virus with its envelope (**figure 6.20**). Budding of enveloped viruses causes them to be shed gradually, without the sudden destruction of the cell. Regardless of how the virus leaves, most active viral infections are ultimately lethal to the cell because of accumulated damage. Lethal damages include a permanent shutdown of metabolism and genetic expression, destruction of cell membrane and organelles, toxicity of virus components, and release of lysosomes.

The number of viruses released by infected cells is variable, controlled by factors such as the size of the virus and the health of the host cell. About 3,000 to 4,000 virions are released from a single cell infected with poxviruses, whereas a poliovirus-infected cell can release over 100,000 virions. If even a small number of these virions happens to meet another susceptible cell and infect it, the potential for rapid viral proliferation is immense.

Damage to the Host Cell and Persistent Infections

The short- and long-term effects of viral infections on animal cells are well documented. **Cytopathic* effects** (CPEs) are defined as virus-induced damage to the cell that alters its microscopic appearance. Individual cells can become disoriented, undergo gross changes in shape or size, or develop intracellular changes (**figure 6.21a**). It is common to note *inclusion bodies,* or compacted masses of viruses or damaged cell organelles, in the nucleus and cytoplasm (figure 6.21b). Examination of cells and tissues for cytopathic effects is an important part of the diagnosis of viral infections. **Table 6.4** summarizes some prominent cytopathic effects associated with specific viruses.

Although accumulated damage from a virus infection kills most host cells, some cells maintain a carrier relationship, in which the cell harbors the virus and is not immediately lysed. These so-called *persistent infections* can last from a few weeks to the rest of one's life. One of the more serious complications occurs with the measles virus. It may remain hidden in brain cells for many years, causing progressive damage and loss of function. Several viruses remain in a *chronic latent state,*[5] periodically becoming reactivated. Examples of this are herpes simplex viruses (fever blisters and genital herpes) and herpes zoster virus (chickenpox and shingles). Both viruses can go into latency in nerve cells and later emerge under the influence of various stimuli to cause recurrent infections. Specific damage that occurs in viral diseases is covered more completely in chapters 24 and 25.

Some animal viruses enter their host cell and permanently alter its genetic material, leading to cancer. These viruses are termed *oncogenic,* and their effect on the cell is called **transformation.** A startling feature of these viruses is that their nucleic acid is consolidated into the host DNA like a lysogenic phage. Transformed cells have an increased rate of growth; alterations in chromosomes; changes in the cell's surface molecules; and the capacity to divide for an indefinite period, unlike normal animal cells. Mammalian viruses capable of initiating tumors are called *oncoviruses.* Some of these are DNA viruses such as papillomavirus (genital warts are associated with cervical cancer), herpesviruses (Epstein-Barr virus causes Burkitt's lymphoma), and hepatitis B virus. Retroviruses are an unusual family of RNA viruses that can program the synthesis of DNA using their single-stranded RNA as a template. They can then insert their DNA into a host chromosome, where it can remain for the life of the cell (see figure 25.16a). One type of retrovirus is HIV, the cause of AIDS. Other viruses related to HIV—HTLV I and II[6]—are involved in human cancers. These findings have spurred a great deal of speculation on the possible involvement of viruses in cancers whose cause is still unknown. Additional information on the connection between viruses and cancer is found in chapters 9, 17, 24, and 25.

---

4. For enveloped viruses, these terms are interchangeable. They mean the release of a virus from an animal cell by enclosing it in a portion of membrane derived from the cell.

5. Meaning that they exist in an inactive state over long periods.

6. Human T-cell lymphotropic viruses; cause types of leukemia.

*cytopathic (sy″-toh-path′-ik) Gr. *cyto,* cell, and *pathos,* disease.

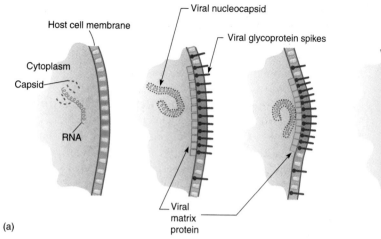

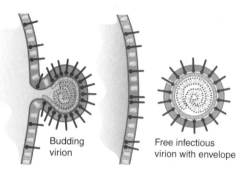

(a)

(b)

FIGURE 6.20

Process of maturation of an enveloped virus (parainfluenza virus).
(a) As the virus is budded off the membrane, it simultaneously picks up an envelope and spikes. **(b)** AIDS viruses (HIV) leave their host white blood cell by budding off its surface. (Note early and mature stages.)

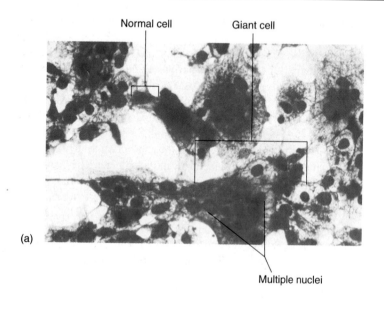

(a)

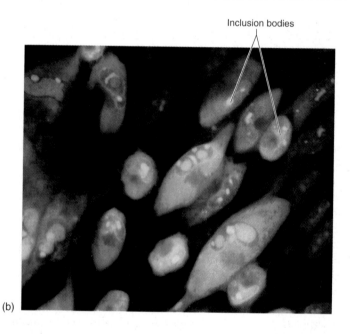

(b)

FIGURE 6.21

Cytopathic changes in cells and cell cultures infected by viruses. **(a)** Human epithelial cells infected by herpes simplex virus demonstrate multinucleate giant cells. **(b)** Fluorescent-stained human cells infected with cytomegalovirus. Note the inclusion bodies. Note also that both viruses disrupt the cohesive junctions between cells.

TABLE 6.4

Cytopathic Changes in Selected Virus-Infected Animal Cells

Virus	Response in Animal Cell
Smallpox virus	Cells round up; inclusions appear in cytoplasm
Herpes simplex	Cells fuse to form multinucleated giant cells; nuclear inclusions
Adenovirus	Clumping of cells; nuclear inclusions
Poliovirus	Cell lysis; no inclusions
Reovirus	Cell enlargement; vacuoles and inclusions in cytoplasm
Influenza virus	Cells round up; no inclusions
Rabies virus	No change in cell shape; cytoplasmic inclusions (Negri bodies)
HIV	Giant cells with numerous nuclei (multinucleate)

CHAPTER CHECKPOINTS

Virus size range is from 20 nm to 450 nm (diameter). Viral particles are also called virions.

Viruses are composed of an outer protein capsid enclosing either DNA or RNA plus a variety of enzymes. Some viruses also exhibit an envelope around the capsid.

Viruses go through a multiplication cycle that generally involves adsorption, penetration of their nucleic acid, viral synthesis and assembly, and viral release by lysis or budding.

These events turn the host cell into a factory solely for making and shedding new viruses. This results in the ultimate destruction of the cell.

Bacterial viruses, or bacteriophages, sometimes inadvertently incorporate bacterial DNA into the virion in place of viral DNA, thus providing a means of genetic transfer in bacteria.

Lysogeny is a condition in which viral DNA is inserted into the bacterial chromosome and remains inactive for an extended period. It is replicated right along with the chromosome every time the bacterium divides.

Animal viruses vary significantly from bacteriophage in their methods of adsorption, penetration, site of replication, and method of exit from host cells.

Animal viruses can persist in host tissues as chronic latent infections that can reactivate periodically throughout the host's life. Some persistent animal viruses are oncogenic.

Techniques in Cultivating and Identifying Animal Viruses

One problem hampering earlier animal virologists was their inability to propagate specific viruses routinely in pure culture and in sufficient quantities for their studies. Virtually all of the pioneering attempts at cultivation had to be performed in an organism that was the usual host for the virus. But this method had its limitations. How could researchers have ever traced the stages of viral multiplication if they had been restricted to the natural host, especially in the case of human viruses? Fortunately, systems of cultivation with broader applications were developed, including *in vivo** inoculation of laboratory-bred animals and embryonic bird tissues and *in vitro** cell (or tissue) culture methods. Such use of substitute host systems permits greater control, uniformity, and wide-scale harvesting of viruses.

The primary purposes of viral cultivation are:

1. to isolate and identify viruses in clinical specimens;
2. to prepare viruses for vaccines; and
3. to do detailed research on viral structure, multiplication cycles, genetics, and effects on host cells.

USING LIVE ANIMAL INOCULATION

Specially bred strains of white mice, rats, hamsters, guinea pigs, and rabbits are the usual choices for animal cultivation of viruses. Invertebrates (insects) or nonhuman primates are occasionally used as well. Because viruses can exhibit some host specificity, certain animals can propagate a given virus more readily than others. Depending on the particular experiment, tests can be performed on adult, juvenile, or newborn animals. The animal is exposed to the virus by injection of a viral preparation or specimen into the brain, blood, muscle, body cavity, skin, or footpads.

USING BIRD EMBRYOS

An **embryo** is an early developmental stage of animals marked by rapid differentiation of cells. Birds undergo their embryonic period within the closed protective case of an egg, which makes an incubating bird egg a nearly perfect system for viral propagation. It is an intact and self-supporting unit, complete with its own sterile environment and nourishment. Furthermore, it furnishes several embryonic tissues that readily support viral multiplication.

Chicken, duck, and turkey eggs are the most common choices for inoculation. The egg must be injected through the shell, usually by drilling a hole or making a small window. Rigorous sterile techniques must be used to prevent contamination by bacteria and fungi from the air and the outer surface of the shell. The exact tissue that is inoculated is guided by the type of virus being cultivated and the goals of the experiment (**figure 6.22**).

Viruses multiplying in embryos may or may not cause effects visible to the naked eye. The signs of viral growth include death of the embryo, defects in embryonic development, and localized areas of damage in the membranes, resulting in discrete, opaque spots called pocks (a variant of pox). If a virus does not produce overt changes in the developing embryonic tissue, virologists have other methods of detection. Embryonic fluids and tissues can be prepared for direct examination with an electron microscope. Certain viruses can also be detected by their ability to agglutinate* red blood cells or by their reaction with an antibody of known specificity that will affix to its corresponding virus, if it is present.

in vivo (in vee′-voh) L. *vivos*, life. Experiments performed in a living body.

in vitro (in vee′-troh) L. *vitros*, glass. Experiments performed in test tubes or other artificial environments.

agglutinate To form large masses or clumps.

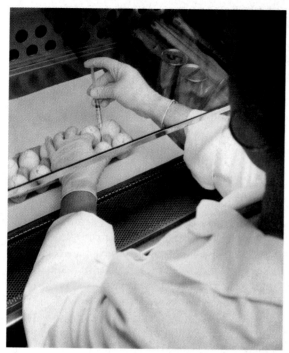

(a) A technician inoculates fertilized chicken eggs with viruses in the first stage of preparing vaccines. This process requires the highest levels of sterile and aseptic precautions.

FIGURE 6.22
Cultivating animal viruses in a developing bird embryo.

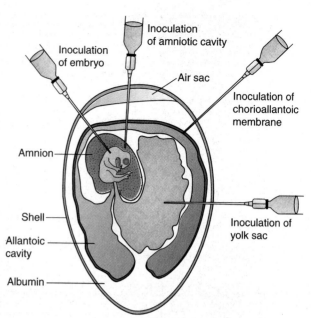

(b) The shell is perforated using sterile techniques, and a virus preparation is injected into a site selected to grow the viruses. Targets include the allantoic cavity, a fluid-filled sac that functions in embryonic waste removal; the amniotic cavity, a sac that cushions and protects the embryo itself; the chorioallantoic membrane, which functions in embryonic gas exchange; the yolk sac, a membrane that mobilizes yolk for the nourishment of the embryo; and the embryo itself.

USING CELL (TISSUE) CULTURE TECHNIQUES

The most important early discovery that led to cultivation of viruses was the development of a simple and effective way to grow populations of isolated animal cells in culture. These types of in vitro cultivation systems are termed **cell culture** or **tissue culture.** (Although these terms are used interchangeably, cell culture is probably a more accurate description.) So prominent is this method that most viruses are propagated in some sort of cell culture, and much of the virologist's work involves developing and maintaining these cultures. Animal cell cultures are grown in sterile chambers with special media that contain the correct nutrients required by animal cells to survive. The cultured cells grow in the form of a *monolayer,* a single, confluent sheet of cells that supports viral multiplication and permits close inspection of the culture for signs of infection **(figure 6.23).**

Cultures of animal cells usually exist in the primary or continuous form. *Primary cell cultures* are prepared by placing freshly isolated animal tissue in a growth medium. The cells undergo a series of mitotic divisions to produce a monolayer. Embryonic, fetal, adult, and even cancerous tissues have served as sources of primary cultures. A primary culture retains several characteristics of the original tissue from which it was derived, but this original line generally has a limited existence. Eventually, it will die out or mutate into a line of cells that can grow continuously. These cell lines tend to have altered chromosome numbers, grow rapidly, and show changes in morphology, and they can be continuously subcultured, provided they are routinely transferred to fresh nutrient medium. One very clear advantage of cell culture is that a specific cell line can be available for viruses with a very narrow host range. Strictly human viruses can be propagated in one of several primary or continuous human cell lines, such as embryonic kidney cells, fibroblasts, bone marrow, and heart cells.

One way to detect the growth of a virus in culture is to observe degeneration and lysis of infected cells in the monolayer of cells. The areas where virus-infected cells have been destroyed show up as clear, well-defined patches in the cell sheet called **plaques*** (figure 6.23c). This same technique is used to detect and count bacteriophages, because they also produce plaques when grown in soft agar cultures of their host cells (bacteria). A plaque develops when the viruses released by an infected host cell radiate out to adjacent host cells. As new cells become infected, they die and release more viruses, and so on. As this process continues, the infection spreads gradually and symmetrically from the original point of infection, causing the macroscopic appearance of round, clear spaces that correspond to areas of dead cells.

---

*plaque (plak) Fr. *placke,* patch or spot.

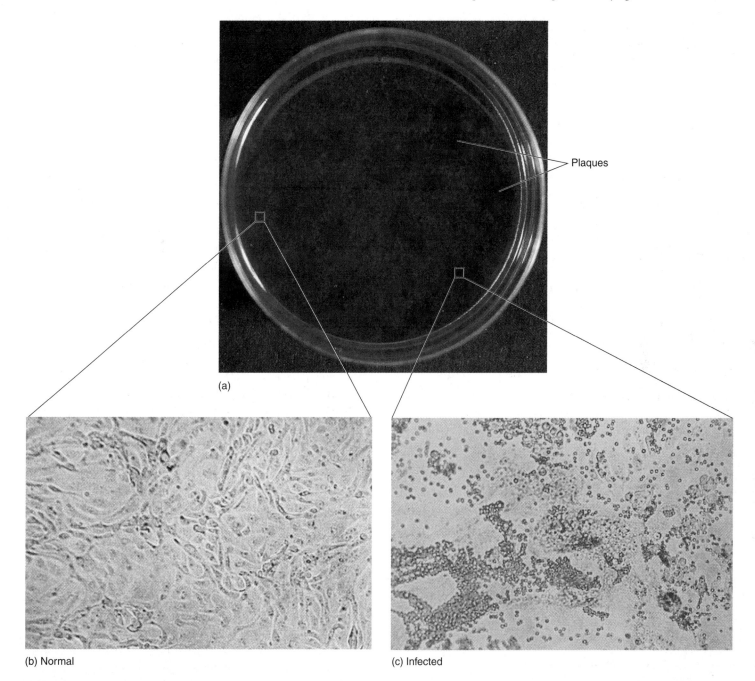

(a)

Plaques

(b) Normal

(c) Infected

FIGURE 6.23

Appearance of normal and infected cell cultures.   **(a)** Macroscopic view of a Petri dish containing a monolayer (single layer of attached cells) of monkey kidney cells. Clear spaces in culture indicate sites of virus growth (plaques). Microscopic views of **(b)** normal, undisturbed cell layer and **(c)** plaques, which consist of cells disrupted by viral infection.

CHAPTER CHECKPOINTS

Animal viruses must be studied in some type of host cell environment such as laboratory animals, bird embryos, or tissue cultures.

Tissue cultures are cultures of host cells grown in special sterile chambers containing correct types and proportions of growth factors using aseptic techniques to exclude unwanted microorganisms.

Virus growth in tissue culture is detected by the appearance of plaques.

Medical Importance of Viruses

[See also chapters 24 and 25]

The number of viral infections that occur on a worldwide basis is nearly impossible to predict accurately. Certainly, viruses are the most common cause of acute infections that do not result in hospitalization, especially when one considers widespread diseases such as colds, hepatitis, chickenpox, influenza, herpes, and warts. If one also takes into account prominent viral infections found only in certain regions of the world, such as Dengue fever, Rift Valley fever, and yellow fever, the total could easily exceed several billion cases each year. Although most viral infections do not result in death, some, such as rabies, AIDS, and Ebola infection have very high mortality rates, and others can lead to long-term debility (polio, neonatal rubella). Current research is focused on the possible connection of viruses to chronic afflictions of unknown cause, such as type I diabetes, multiple sclerosis, and various cancers.

Despite the reputation viruses have for being highly detrimental, in some cases, they may actually show a beneficial side (**Medical Microfile 6.1**).

OTHER NONCELLULAR INFECTIOUS AGENTS

Not all noncellular infectious agents have typical viral morphology. One group of unusual forms, even smaller and simpler than viruses, is implicated in chronic, persistent diseases in humans and animals. These diseases are called spongiform encephalopathies because the brain tissue removed from affected animals resembles a sponge. The infection has a long period of latency (usually several years) before the first clinical signs appear. Signs range from mental derangement to loss of muscle control, and the diseases are progressive and universally fatal.

A common feature of these conditions is the deposition of distinct protein fibrils in the brain tissue. Some researchers have hypothesized that these fibrils are the agents of the disease and have named them **prions.***

Creutzfeldt-Jakob disease afflicts the central nervous system of humans and causes gradual degeneration and death. Cases in which medical workers developed the disease after handling autopsy specimens seem to indicate that it is transmissible, but by an unknown mechanism. Several animals (sheep, mink, elk) are victims of similar transmissible diseases. Bovine spongiform encephalopathy, or "mad cow disease," was recently the subject of fears and a crisis in Europe when researchers found evidence that the disease could be acquired by humans who consumed contaminated beef. This was the first incidence of prion disease transmission from animals to humans. Several hundred Europeans developed symptoms of a variant form of Creutzfeldt-Jakob disease, leading to strict governmental controls on exporting cattle and beef products.

The exact mode of prion infection is currently undergoing analysis. The fact that they are composed primarily of protein (no nucleic acid) has certainly revolutionized our ideas of what can constitute an infectious agent. One of the most compelling questions is just how a prion could be replicated, since all other infectious agents require some nucleic acid (see chapter 25).

Other fascinating viruslike agents in human disease are defective forms called satellite viruses that are actually dependent on other viruses for replication! Two remarkable examples are the adeno-associated virus (AAV), which can replicate only in cells infected with adenovirus, and the delta agent, a naked strand of RNA that is expressed only in the presence of the hepatitis B virus and can worsen the severity of liver damage.

Plants are also parasitized by viruslike agents called **viroids** that differ from ordinary viruses by being very small (about one-tenth the size of an average virus) and being composed of only naked strands of RNA, lacking a capsid or any other type of coating. Viroids are significant pathogens in several economically important plants, including tomatoes, potatoes, cucumbers, citrus trees, and chrysanthemums.

DETECTION AND CONTROL OF VIRAL INFECTIONS

Life-threatening viral diseases such as AIDS, hantavirus pulmonary syndrome, influenza, and those that pose a serious risk to the fetus (rubella or cytomegalovirus) demand rapid detection and correct diagnosis so that appropriate therapy and other control measures can be prescribed. Viruses tend to be more difficult to identify than cellular microbes. Physicians often use the overall clinical picture of the disease (specific signs) to guide diagnosis, but exact virus identification from clinical specimens (blood, respiratory secretions, feces, throat, and vaginal secretions) is more satisfactory (**figure 6.24***a,b*). One direct, rapid method is to examine a specimen for the presence of the virus or signs of cytopathic changes in cells or tissues (see herpesviruses, figure 24.9). This may involve immunofluorescence techniques or direct examination with an electron microscope (figure 6.24*c*). Samples can also be screened for the presence of indicator molecules (antigens) from the virus itself. A new technique that is rapidly becoming standard procedure is the polymerase chain reaction (PCR), which can detect and amplify even minute amounts of viral DNA or RNA in a sample (figure 6.24*e*). In certain infections, definitive diagnosis will require the use of cell culture, embryos, or animals, but this method can be time-consuming and slow to give results (figure 6.24*f*). Frequently, it is helpful to use a screening test to detect specific antibodies that indicate signs of virus infection in a patient's blood. It is the main test for HIV infection (figure 6.24*d*). Details of viral diagnosis are provided in chapters 10, 16, 24, and 25.

TREATMENT OF ANIMAL VIRAL INFECTIONS

The nature of viruses has at times been a major impediment to effective therapy. Because viruses are not bacteria, antibiotics aimed at disrupting procaryotic cells do not work on them. On the other hand, many antiviral drugs block virus replication by targeting the function of host cells, and can cause severe side effects. An example is azidothymidine (AZT), a drug used to treat AIDS, which has severe toxic side effects such as immunosuppression

---

*prion (pree´-on) **prot**einacious **in**fectious particle.

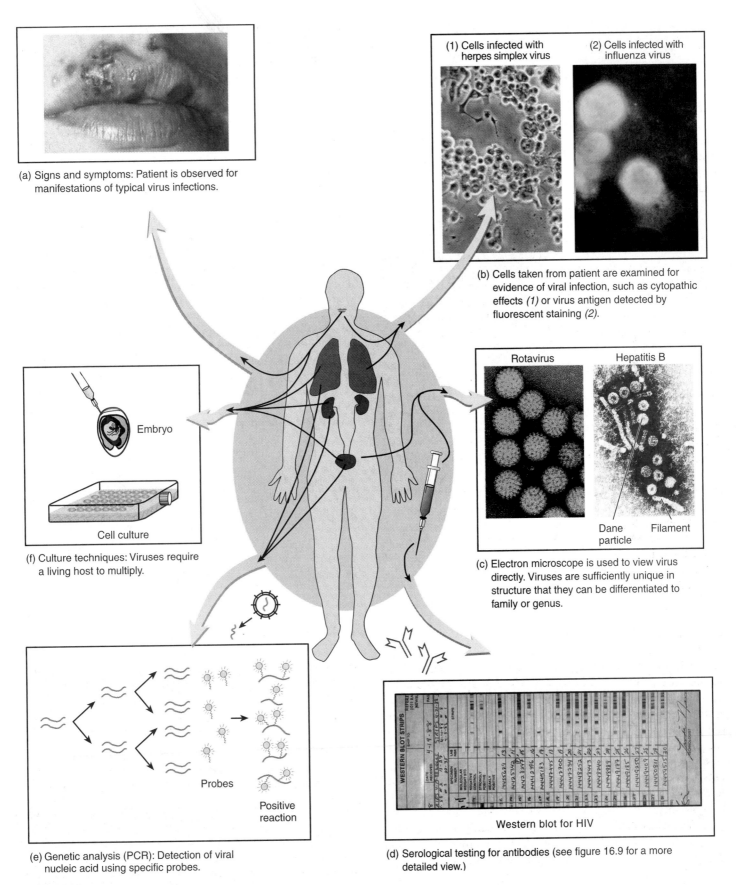

(a) Signs and symptoms: Patient is observed for manifestations of typical virus infections.

(1) Cells infected with herpes simplex virus

(2) Cells infected with influenza virus

(b) Cells taken from patient are examined for evidence of viral infection, such as cytopathic effects *(1)* or virus antigen detected by fluorescent staining *(2)*.

Rotavirus

Hepatitis B

Dane particle

Filament

(c) Electron microscope is used to view virus directly. Viruses are sufficiently unique in structure that they can be differentiated to family or genus.

Embryo

Cell culture

(f) Culture techniques: Viruses require a living host to multiply.

Probes

Positive reaction

(e) Genetic analysis (PCR): Detection of viral nucleic acid using specific probes.

WESTERN BLOT STRIPS

Western blot for HIV

(d) Serological testing for antibodies (see figure 16.9 for a more detailed view.)

FIGURE 6.24

Summary of methods used to diagnose viral infections.

MEDICAL MICROFILE 6.1
A Positive View of Viruses

Looking at this beautiful tulip, one would never guess that it derives its pleasing appearance from a viral infection. It contains tulip mosaic virus, which alters the development of the plant cells and causes complex patterns of colors in the petals. Aside from this, the virus does not cause severe harm to the plants. Despite the reputation of viruses as cell killers, there is another side of viruses—that of being harmless, and in some cases, even beneficial.

Although there is no agreement on the origins of viruses, it is highly likely that they have been in existence for billions of years. Virologists are convinced that viruses have been an important force in the evolution of living things. This is based on the fact that they interact with the genetic material of their host cells and that they carry genes from one host to another (transduction). It is convincing to imagine that viruses arose early in the history of cells as loose pieces of genetic material that became dependent nomads, moving from cell to cell. Viruses are also a significant factor in the functioning of many ecosystems because of the effects they have on their host cells. For example, it is documented that seawater can contain 10 million viruses per milliliter. Since viruses are made of the same elements as living cells, it is estimated that the sum of viruses in the ocean represent 270 million metric tons of organic matter.

Over the past several years, biomedical experts have been looking at viruses as vehicles to treat infections and disease. Viruses are already essential for production of vaccines to treat viral infections such as influenza, polio, and measles. Vaccine experts have also engineered new types of viruses by combining a less harmful virus such as vaccinia or adenovirus with some genetic material from a pathogen such as HIV and herpes simplex (see figure 25.21). This technique creates a vaccine that provides immunity but does not expose the person to the intact pathogen. Several of these types of vaccines are currently in development.

The "harmless virus" approach is also being used to treat genetic diseases such as cystic fibrosis and sickle-cell anemia. With gene therapy, the normal gene is inserted into a retrovirus, such as the mouse leukemia virus, and the patient is infected with this altered virus. It is hoped that the virus will introduce the needed gene into the cells and correct the defect. Dozens of experimental trials are currently underway to develop potential cures for diseases, with some successes (see chapter 10).

Virologists have also created mutant adenoviruses (ONYX) that target cancer cells. These viruses cannot spread among normal cells, but when they enter cancer cells, they immediately cause the cells to self-destruct. One problem has been that infection with these mouse viruses has led to the development of cancer in some patients.

An older therapy getting a second chance involves use of bacteriophages to treat bacterial infections. This technique was tried in the past with mixed success, but was abandoned for more efficient antimicrobic drugs. The basis behind the therapy is that bacterial viruses would seek out only their specific host bacteria and would cause complete destruction of the bacterial cell. Newer experiments with animals have demonstrated that this method can control infections as well as traditional drugs. Some potential applications being considered are adding phage suspension to grafts to control skin infections and to intravenous fluids for blood infections.

and anemia. In addition, most antiviral drugs prevent the virus from replicating or maturing and therefore must be capable of entering human cells to have an effect. Another compound that shows some potential for treating and preventing viral infections is a naturally occurring human cell product called *interferon* (see chapters 12 and 14). Vaccines that stimulate immunity are an extremely valuable tool but are available for only a limited number of viral diseases (see chapter 16).

CHAPTER CHECKPOINTS

Viruses are easily responsible for several billion infections each year. It is conceivable that many chronic diseases of unknown cause will be connected to viral agents.

Other noncellular agents of disease are the prions, which are not viruses at all, but protein fibers; viroids, extremely small lengths of protein-coated nucleic acid; and satellite viruses, which require larger viruses to cause disease.

Diagnosis of viral disease agents is done directly through culture, electron microscopy, immunofluorescence microscopy, and by the PCR technique. Indirect diagnosis is done by testing patient serum for host antibody to viral agents.

Viral infections are difficult to treat because the drugs that attack the viral replication cycle also cause serious side effects in the host.

CHAPTER CAPSULE WITH KEY TERMS

Other terms used to describe viruses are general (virions, viral particles) and specific (bacteriophage, poxvirus, herpesvirus, etc.). They may also be named for host cell, disease, and morphology.

I. **Overall Morphology of Viruses**

   A. **Viruses** are infectious particles and not cells; lack protoplasm, organelles, locomotion of any kind; are large, complex molecules; can be crystalline in form. A virus particle is composed of a nucleic acid core (DNA or RNA, not both) surrounded by a geometric protein shell, or **capsid;** combination called a **nucleocapsid;** capsid is **helical** or **icosahedral** in configuration; many are covered by a membranous **envelope** containing viral protein **spikes;** complex viruses have additional external and internal structures.

   B. *Shapes/Sizes:* Cuboidal, spherical, cylindrical, brick- and bullet-shaped. Smallest infectious forms range from the largest poxvirus (0.45 mm or 450 nm) to the smallest viruses (0.02 mm or 20 nm).

   C. *Nutritional and Other Requirements:* Lack enzymes for processing food or generating energy; are tied entirely to the host cell for all needs (**obligate intracellular parasites**).

II. **Distribution/Host Range**

Viruses are known to parasitize all types of cells, including bacteria, algae, fungi, protozoa, animals, and plants. Each viral type is limited in its **host range** to a single species or group, mostly due to specificity of adsorption of virus to specific host receptors.

III. **Classification**

   A. The two major types of viruses are *DNA* and *RNA viruses.* These are further subdivided into families, depending on shape and size of capsid, presence or absence of an envelope, whether double- or single-stranded nucleic acid, and antigenic similarities.

   B. *Animal Viruses:* Common names of major DNA animal viruses include poxviruses (smallpox), herpesviruses, adenoviruses, papillomavirus (wart), hepatitis B virus, parvoviruses. RNA animal viruses include poliovirus, hepatitis A virus, rhinovirus (common cold), encephalitis viruses, yellow fever virus, rubella virus, influenza virus, mumps virus, measles virus, rabies viruses, and HIV (AIDS virus).

   C. Other noncellular infectious agents cause spongiform encephalopathies such as Creutzfeldt-Jakob disease in humans and bovine spongiform encephalopathy (mad cow disease) in cattle. The agents are probably protein particles called **prions.**

IV. **Multiplication Cycle**

   A. A unique method of multiplication occurs—no division is involved. Viruses make multiple copies of themselves through:

     1. attachment, or **adsorption,** to a host cell; they must have special binding molecules for receptors on the host cell;

     2. **penetration** of a cell by means of whole virus or only nucleic acid;

     3. assuming control of the cell's genetic and metabolic components, thus

     4. programming synthesis of new viral parts;

     5. directing **assembly** of these parts into new virus particles, and

     6. **releasing** complete, mature viruses.

   B. **Bacteriophages** are viruses that attack bacteria; they penetrate by injecting their nucleic acid; they are released as virulent phage upon **lysis** of the cell.

   C. Enveloped viruses penetrate by being engulfed into a vesicle. They are released by **budding** off with an envelope.

   D. Some viruses go into a **latent** or **lysogenic phase** in which they integrate into the DNA of the host cell and later may become active and produce a lytic infection.

V. **Method of Cultivation**

The need for an intracellular habitat makes it necessary to grow viruses in living cells, either in the intact host animal, in bird embryos, or in isolated cultures of host cells (**cell culture**).

VI. **Identification**

Viruses are identified by means of **cytopathic effects** in host cells, direct examination of viruses or their components in samples, analyzing blood for antibodies against viruses, performing genetic analysis of samples to detect virus nucleic acid, growing viruses in culture, and symptoms.

VII. **Importance of Viruses**

   A. *Medical:* Viruses attach to specific target hosts or cells. They cause a variety of infectious diseases, ranging from mild respiratory illness (common cold) to destructive and potentially fatal conditions (rabies, AIDS). Some viruses can cause birth defects and cancer in humans and other animals.

   B. *Agricultural:* Hundreds of cultivated plants and domestic animals are susceptible to viral infections, often with adverse economic and ecologic repercussions.

   C. *Research:* Because of their simplicity, viruses have become an invaluable tool for studying basic genetic principles.

VIII. **Unique Features**

   • Viruses exist at a level between living things and nonliving molecules.

   • They consist of a capsid and nucleic acid, either DNA or RNA, not both.

   • Lack metabolism and respiratory enzymes.

   • Multiply inside host cells using the assembly line of the host's synthetic machinery.

   • Are genetic parasites.

   • Are host-specific.

   • Are ultramicroscopic and crystallizable.

   • Some viruses are enveloped, others are naked.

   • Some have unique nucleic acid structure (single-stranded DNA and double-stranded RNA).

   • Some animal viruses can become latent.

MULTIPLE-CHOICE QUESTIONS

1. A virus is a tiny infectious
   a. cell
   b. living thing
   c. particle
   d. nucleic acid

2. Viruses are known to infect
   a. plants
   b. bacteria
   c. fungi
   d. all organisms

3. The capsid is composed of protein subunits called
   a. spikes
   b. protomers
   c. virions
   d. capsomers

4. The envelope of an animal virus is derived from the _____ of its host cell.
   a. cell wall
   b. cell membrane
   c. glycocalyx
   d. receptors

5. The nucleic acid of a virus is
   a. DNA only
   b. RNA only
   c. both DNA and RNA
   d. either DNA or RNA

6. The general steps in a viral multiplication cycle are
   a. adsorption, penetration, replication, maturation, and release
   b. endocytosis, uncoating, replication, assembly, and budding
   c. adsorption, uncoating, duplication, assembly, and lysis
   d. endocytosis, penetration, replication, maturation, and exocytosis

7. A prophage is an early stage in the development of a/an
   a. bacterial virus
   b. poxvirus
   c. lytic virus
   d. enveloped virus

8. The nucleic acid of animal viruses enters the host cell through
   a. translocation
   b. fusion
   c. endocytosis
   d. all of these

9. In general, RNA viruses multiply in the cell _____, and DNA viruses multiply in the cell _____.
   a. nucleus, cytoplasm
   b. cytoplasm, nucleus
   c. vesicles, ribosomes
   d. endoplasmic reticulum, nucleolus

10. Enveloped viruses carry surface receptors called
    a. buds
    b. spikes
    c. fibers
    d. sheaths

11. Viruses that persist in the cell and cause recurrent disease are considered
    a. oncogenic
    b. cytopathic
    c. latent
    d. resistant

12. Examples of cytopathic effects of viruses are
    a. inclusion bodies
    b. giant cells
    c. multiple nuclei
    d. all of these

13. Viruses cannot be cultivated in
    a. tissue culture
    b. bird embryos
    c. live mammals
    d. blood agar

14. Clear patches in cell cultures that indicate sites of virus infection are called
    a. plaques
    b. pocks
    c. colonies
    d. prions

15. Label the parts of this virus. Identify this virus as to capsid, nucleic acid, and other features. Can you identify it?

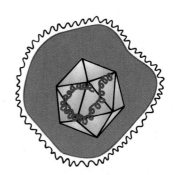

CONCEPT QUESTIONS

1. a. Describe 10 *unique* characteristics of viruses (can include structure, behavior, multiplication).
   b. After consulting table 6.1, what additional statements can you make about viruses, especially as compared with cells?

2. a. Explain what it means to be an obligate intracellular parasite.
   b. What is another way to describe the sort of parasitism exhibited by viruses?

3. a. Characterize the viruses according to size range.
   b. What does it mean to say that they are ultramicroscopic?
   c. That they are filterable?

4. a. Describe the general structure of viruses.
   b. What is the capsid, and what is its function?
   c. How are the two types of capsids constructed?
   d. What is a nucleocapsid?
   e. Give examples of viruses with the two capsid types.
   f. What is an enveloped virus, and how does the envelope arise?
   g. Give an example of a common enveloped human virus.
   h. What are spikes, how are they formed, and what is their function?

5. a. What are bacteriophages, and what is their structure?
   b. What is a tobacco mosaic virus?
   c. How are the poxviruses different from other animal viruses?

6. a. Since viruses lack metabolic enzymes, how can they synthesize necessary components?
   b. What are some enzymes with which the virus is equipped?

7. a. How are viruses classified? What are virus families?
   b. How are generic and common names used?
   c. Look at table 6.3 and count the total number of different viral diseases. How many are caused by DNA viruses? How many are RNA-virus diseases?

8. a. Compare and contrast the main phases in the lytic multiplication cycle in bacteriophages and animal viruses.
   b. When is a virus a virion?
   c. What is necessary for adsorption?
   d. Why is penetration so different in the two groups?
   e. What is eclipse?
   f. In simple terms, what does the virus nucleic acid do once it gets into the cell?
   g. What processes are involved in assembly?

9. a. What is a prophage or temperate phage?
   b. What is lysogeny?

10. a. What dictates the host range of animal viruses?
    b. What are two ways that animal viruses penetrate the host cell?
    c. What is uncoating?
    d. Describe the two ways that animal viruses leave their host cell.

11. a. Describe several cytopathic effects of viruses.
    b. What causes the appearance of the host cell?
    c. How might it be used to diagnose viral infection?

12. a. What does it mean for a virus to be persistent or latent, and how are these events important?
    b. Briefly describe the action of an oncogenic virus.

13. a. Describe the three main techniques for cultivating viruses.
    b. What are the advantages of using cell culture?
    c. The disadvantages of using cell culture?
    d. What is a disadvantage of using live intact animals or embryos?
    e. What are a cell line and a monolayer?
    f. How are plaques formed?

14. a. What is the principal effect of the agent of Creutzfeldt-Jakob disease?
    b. How is the proposed agent different from viruses?
    c. What are viroids?

15. Why are virus diseases more difficult to treat than bacterial diseases?

16. Circle the viral infections from this list: cholera, rabies, plague, cold sores, whooping cough, tetanus, genital warts, gonorrhea, mumps, Rocky Mountain spotted fever, syphilis, rubella, rat bite fever.

CRITICAL-THINKING QUESTIONS

1. a. What characteristics of viruses could be used to characterize them as life forms?
   b. What makes them more similar to lifeless molecules?

2. a. Comment on the possible origin of viruses. Is it not curious that the human cell welcomes a virus in and hospitably removes its coat as if it were an old acquaintance?
   b. How do spikes play a part in the action of the host cell?

3. Go to the following website and open the virus infection animation: http://darwin.bio.uci.edu/~faculty/wagner/hsvbinding.html. Describe the unusual features of the herpesvirus multiplication cycle you see there.

4. a. If viruses that normally form envelopes were prevented from budding, would they still be infectious?
   b. If the RNA of an influenza virus were injected into a cell by itself, could it cause a lytic infection?

5. The end result of most viral infections is death of the host cell.
   a. If this is the case, how can we account for such differences in the damage that viruses do (compare the effects of the cold virus with those of the rabies virus)?
   b. Describe the adaptation of viruses that does not immediately kill the host cell and explain what its function might be.

6. a. Given that DNA viruses can actually be carried in the DNA of the host cell's chromosomes, comment on what this phenomenon means in terms of inheritance in the offspring.
   b. Discuss the connection between viruses and cancers, giving possible mechanisms for viruses that cause cancer.

7. You have been given the problem of constructing a building on top of a mountain. It will be impossible to use huge pieces of building material. Using the virus capsid as a model, can you think of a way to construct the building?

8. The AIDS virus attacks only specific types of human cells, such as certain white blood cells and nerve cells. Can you explain why a virus can enter some types of human cells but not others?

9. a. Consult table 6.3 to determine which viral diseases you have had and which ones you have been vaccinated against.
   b. Which viruses would you investigate as possible oncoviruses?

10. One early problem in cultivating the AIDS virus was the lack of a cell line that would sustain indefinitely *in vitro,* but eventually one was developed. What do you expect were the stages in developing this cell line?

11. a. If you were involved in developing an antiviral drug, what would be some important considerations? (Can a drug "kill" a virus?)
    b. How could multiplication be blocked?

12. a. Is there such a thing as a "good virus"?
    b. Explain why or why not. Consider both bacteriophages and viruses of eucaryotic organisms.

13. Why is an embryonic or fetal viral infection so harmful?

14. How are computer viruses analogous to real viruses?

INTERNET SEARCH TOPICS

1. Visit the student Online Learning Center at www.mhhe.com/talaro5. Go to chapter 6, Internet Search Topics, and log on to the available websites to explore viruses.

2. Look up emerging viral diseases and make note of the newest viruses that have arisen since 1999. What kinds of diseases do they cause, and where did they possibly originate from?

3. Find websites that discuss prions and prion-based diseases. What possible way do the prions replicate?

Elements of Microbial Nutrition, Ecology, and Growth

Carved into a hillside near Butte, Montana, lies the Berkeley Pit, an industrial body of water that stretches about one mile across and contains a volume of close to 30 billion gallons. This small lake was formerly the site of a copper pit mine abandoned in 1982 and left to fill up with water from local springs. Lying at the bottom of the pit was a massive layer of mining waste that had accumulated over many years of mining.

The gradual dissolution of the waste in the water transformed Berkeley Pit into a giant vat of concentrated chemicals so toxic that it was considered uninhabitable to living things. Among the substances found in abundance are heavy metals such as lead, cadmium, aluminum, copper, iron, and zinc, as well as arsenic and sulfides. The pit also has a high level of acidity—10,000 times more than a normal freshwater habitat. The greatest concern is that water from the pit will leak into local streams and contaminate groundwater, river drainages, and eventually the ocean. The Environmental Protection Agency is currently reviewing possible methods of cleaning up the site to avert a possible ecological disaster.

When scientists from the University of Montana began researching the composition of the water in the pit, they were startled by what they found. The samples showed signs of a well-established community of microorganisms that had taken hold over the 20 years the pit lay undisturbed. It included an array of very hardy bacteria, protozoa, fungi, and algae—nearly 50 species in all (see **Chapter-opening photo**). The microbes evidently entered the pit from the water and air, and rather than being killed, they survived, grew, and spread into available **niches.*** In a relatively short time, some of them had actually evolved to depend on the toxic sludge for survival. An important observation made by the Montana researchers is that the microbes appear to be naturally detoxifying the water. They are currently investigating a way to adapt this "self-cleaning" technology to clear the system of toxic chemicals and in time convert the lake to a useful water reservoir.

This example should serve as a dramatic reminder of the overwhelming adaptability of microorganisms, which allows them to prevail in environments that would be toxic to all other living things (see **Spotlight on Microbiology 7.1,** page 190). There are millions of habitats on earth, both of natural and human origin. Microbes are exposed to a tremendous variety in nutrient sources, temperatures, gases, salt, pH, and radiation. With this theme in mind, this chapter will take a closer look at this adaptability in the light of nutrition, responses to environmental factors, transport, and growth.

---

*niche (nitch) Fr. *nichier,* to nest. The role of an organism in its ecological setting.

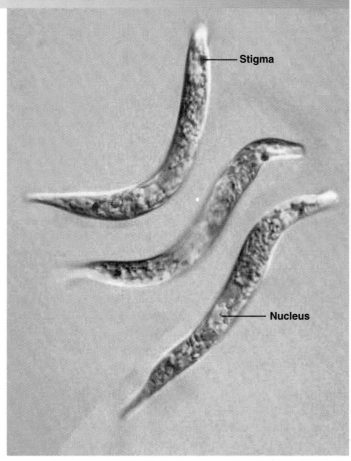

A single-celled clean-up crew. The alga *Euglena mutabilis* isolated from the Berkeley Pit is one of the hardiest microbes on earth. In addition to thriving in very acidic waters high in heavy metals, it can also exist in a wide range of temperatures. Samples of the algae taken from acid mines indicate that it actively removes pollutants from fresh water.

Chapter Overview

- Microbes exist in every known natural habitat on earth.
- Microbes show enormous capacity to adapt to environmental factors.
- Factors that have the greatest impact on microbes are nutrients, temperature, pH, amount of available water, atmospheric gases, light, pressure, and other organisms.
- Nutrition involves absorbing required chemicals from the environment for use in metabolism.
- Autotrophs can exist solely on inorganic nutrients, while heterotrophs require both inorganic and organic nutrients.
- Energy sources for microbes may come from light or chemicals.

- Temperature adaptations may be at cold, moderate, and hot temperatures.
- Oxygen and carbon dioxide are primary gases used in metabolism.
- Transport of materials by cells involves movement by passive or active mechanisms across the cell membrane.
- The water content of the cell versus its environment dictates the osmotic adaptations of cells.
- Microbes interact in a variety of ways with one another and with other organisms that share their habitats.
- The pattern of population growth in simple microbes is to double the number of cells in each generation.
- Growth rate is limited by lack of nutrients and buildup of waste products.

Microbial Nutrition

Nutrition* is a process by which chemical substances called **nutrients** are acquired from the environment and used in cellular activities such as metabolism and growth. With respect to nutrition, microbes are not really so different from humans (**Microbits 7.2,** page 191). Bacteria living in mud on a diet of inorganic sulfur or protozoa digesting wood in a termite's intestine seem to show radical adaptations, but even these organisms require a constant influx of certain substances from their habitat. In general, all living things require a source of elements such as carbon, hydrogen, oxygen, phosphorus, potassium, nitrogen, sulfur, calcium, iron, sodium, chlorine, magnesium, and certain other elements.[1] But the ultimate source of a particular element, its chemical form, and how much of it the microbe needs are all points of variation between different types of organisms. Any substance, whether in elemental or molecular form, that must be provided to an organism is called an **essential nutrient.** Once absorbed, nutrients are processed and transformed into the chemicals of the cell.

Two categories of essential nutrients are **macronutrients** and **micronutrients.** Macronutrients are required in relatively large quantities and play principal roles in cell structure and metabolism. Examples of macronutrients are proteins, carbohydrates, and other molecules that contain carbon, hydrogen, and oxygen. Micronutrients, or **trace elements,** such as manganese, zinc, and nickel are present in much smaller amounts and are involved in enzyme function and maintenance of protein structure. What constitutes a micronutrient can vary from one microbe to another and often must be determined in the laboratory. This determination is made by deliberately omitting the substance in question from a growth medium to see if the microbe can grow in its absence.

Another way to categorize nutrients is according to their carbon content. An **inorganic nutrient** is an atom or simple molecule that contains a combination of atoms other than carbon and hydrogen. The natural reservoirs of inorganic compounds are mineral deposits in the crust of the earth, bodies of water, and the atmosphere. Examples include metals and their salts (magnesium sulfate, ferric nitrate, sodium phosphate), gases (oxygen, carbon dioxide), and water (**table 7.1**).

---

1. A useful mnemonic device for keeping track of major elements required by most organisms is: **C HOPKINS Ca Fe, Na Cl Mg**ood. Read aloud, it becomes, See Hopkin's Cafe, NaCl (salt) M good! Boldfaced letters are short for the atomic symbol of that element.

*nutrition L. *nutrire,* nourishment.

TABLE 7.1

Principal Inorganic Reservoirs of Elements

Element	Inorganic Environmental Reservoir
Carbon	CO_2 in air; CO_3^{-2} in rocks and sediments
Oxygen	O_2 in air, certain oxides, water
Nitrogen	N_2 in air; NO_3^-, NO_2^-, NH_4^+ in soil and water
Hydrogen	Water, H_2 gas, mineral deposits
Phosphorus	Mineral deposits (PO_4^{-3}, H_3PO_4)
Sulfur	Mineral deposits, volcanic sediments (SO_4^{-2}, H_2S, S)
Potassium	Mineral deposits, the ocean (KCl, K_3PO_4)
Sodium	Mineral deposits, the ocean ($NaCl$, $NaSi$)
Calcium	Mineral deposits, the ocean ($CaCO_3$, $CaCl_2$)
Magnesium	Mineral deposits, geologic sediments ($MgSO_4$)
Chloride	The ocean ($NaCl$, NH_4Cl)
Iron	Mineral deposits, geologic sediments ($FeSO_4$)
Manganese, molybdenum, cobalt, nickel, zinc, copper, other micronutrients	Various geologic sediments

In contrast, the molecules of **organic nutrients** contain carbon and hydrogen atoms and are usually the products of living things. They range from the simplest organic molecule, methane (CH_4), to large polymers (carbohydrates, lipids, proteins, and nucleic acids). The source of nutrients is extremely varied: Some microbes obtain their nutrients entirely from inorganic sources, and others require a combination of organic and inorganic sources (see table 7.3). Parasites capable of invading and living on the human body derive all essential nutrients from host tissues, tissue fluids, secretions, and wastes.

CHEMICAL ANALYSIS OF MICROBIAL CYTOPLASM

Examining the chemical composition of a bacterial cell can indicate its nutritional requirements. **Table 7.2** lists the major contents of the intestinal bacterium *Escherichia coli.* Some of these components are absorbed in a ready-to-use form, and others must be synthesized by the cell from simple nutrients. Several important features of cell composition can be summarized as follows:

- Water content is the highest of all the components (70%).
- Proteins are the next most prevalent chemical.
- About 97% of the dry cell weight is composed of organic compounds.
- About 96% of the cell is composed of six elements (represented by CHNOPS).
- Bioelements are needed in the overall scheme of cell growth, but most of them are available to the cell as compounds and not as pure elements (table 7.2).
- A cell as "simple" as *E. coli* contains on the order of 5,000 different compounds, yet it needs to absorb only a few types of nutrients to synthesize this great diversity. These include $(NH_4)_2SO_4$, $FeCl_2$, $NaCl$, trace elements, glucose, KH_2PO_4, $MgSO_4$, $CaHPO_4$, and water.

TABLE 7.2

Analysis of the Chemical Composition of an *Escherichia coli* Cell

	% Total Weight	% Dry Weight
Organic Compounds		
Proteins	15	50
Nucleic acids		
RNA	6	20
DNA	1	3
Carbohydrates	3	10
Lipids	2	Not determined
Miscellaneous	2	Not determined
Inorganic Compounds		
Water	70	
All others	1	3
Elements		
Carbon (C)	—	50
Oxygen (O)	—	20
Nitrogen (N)	—	14
Hydrogen (H)	—	8
Phosphorus (P)	—	3
Sulfur (S)	—	1
Potassium (K)	—	1
Sodium (Na)	—	1
Calcium (Ca)	—	0.5
Magnesium (Mg)	—	0.5
Chloride (Cl)	—	0.5
Iron (Fe)	—	0.2
Manganese (Mn), zinc (Zn), molybdenum (Mo), copper (Cu), cobalt (Co), zinc (Zn)	—	0.3

SOURCES OF ESSENTIAL NUTRIENTS

The elements that make up nutrients ultimately exist in an environmental inorganic reservoir of some type. These reservoirs serve not only as a permanent, longtime source of these elements but also can be replenished by the activities of organisms. In fact, as we shall see in chapter 26, the ability of microbes to keep elements cycling is essential to all life on the earth.

For convenience, this section on nutrients is organized by element. You will no doubt notice that some categories overlap and that many of the compounds furnish more than one element.

Carbon Sources

It seems worthwhile to emphasize a point about the *extracellular source* of carbon as opposed to the *intracellular function* of carbon compounds. Although a distinction is made between the type of carbon compound cells absorb as nutrients (inorganic or organic), the majority of carbon compounds involved in the normal structure and metabolism of all cells are organic.

A **heterotroph*** is an organism that must obtain its carbon in an organic form. Because organic carbon originates from the bodies of other organisms, heterotrophs are dependent on other life forms. Among the common organic molecules that can satisfy this requirement are proteins, carbohydrates, lipids, and nucleic acids. In most cases, these nutrients provide several other elements as well. Some organic nutrients available to heterotrophs already exist in a form that is simple enough for absorption (for example, monosaccharides and amino acids), but many larger molecules must be digested by the cell before absorption. Moreover, heterotrophs vary in their capacities to use various organic carbon sources. Some are restricted to a few substrates, whereas others (certain *Pseudomonas* bacteria, for example) are so versatile that they can metabolize more than 100 different substrates.

---

*heterotroph (het′-uhr-oh-trohf) Gr. *hetero,* other, and *troph,* to feed.

TABLE 7.3

Nutrient Requirements and Sources for Two Ecologically Different Bacterial Species

	Thiobacillus thioxidans	*Mycobacterium tuberculosis*
Habitat	Sulfur springs	Human lung
Nutritional Type	Chemoautotroph (sulfur-oxidizing bacteria)	Chemoheterotroph (parasite; cause of tuberculosis in humans)
Essential Element	**Provided Principally By**	
Carbon	Carbon dioxide (CO_2) in air	Minimum of L-glutamic acid (an amino acid), glucose (a simple sugar), albumin (blood protein), oleic acid (a fatty acid), and citrate
Hydrogen	H_2O, H^+ ions	Nutrients listed above, H_2O
Nitrogen	Ammonium (NH_4)	Ammonium, L-glutamic acid
Oxygen	Phosphate (PO_4), sulfate (SO_4), O_2	Oxygen gas (O_2) in atmosphere
Phosphorus	PO_4	PO_4
Sulfur	Elemental sulfur (S), SO_4	SO_4
Miscellaneous Nutrients		
Vitamins	None required	Pyridoxine, biotin
Inorganic salts	Potassium, calcium, iron, chloride	Sodium, chloride, iron, zinc, calcium, copper, magnesium
Principal energy source	Oxidation of sulfur (inorganic)	Oxidation of glucose (organic)

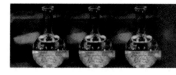

SPOTLIGHT ON MICROBIOLOGY 7.1
Life in the Extremes

Any extreme habitat—whether hot, cold, salty, acidic, alkaline, high pressure, arid, oxygen-free, or toxic—is likely to harbor microorganisms that have made special adaptations to their conditions. Although in most instances the inhabitants are archaea and bacteria, certain fungi, protozoans, and algae are also capable of living in harsh habitats. Microbiologists have termed such remarkable organisms **extremophiles.**

Hot and Cold

Perhaps the most extreme habitats are hot springs, geysers, volcanoes, and ocean vents, all of which support flourishing microbial populations. Temperatures in these regions range from 50°C to well above the boiling point of water, with some ocean vents even approaching 350°C. Many heat-adapted microbes are highly primitive archaea whose genetics and metabolism are extremely modified for this mode of existence. A unique ecosystem based on hydrogen sulfide–oxidizing bacteria exists in the hydrothermal vents lying along deep oceanic ridges (see Spotlight on Microbiology 7.5). Heat-adapted bacteria even plague home water heaters and the heating towers of power and industrial plants.

A large part of the earth exists at cold temperatures. Microbes settle and grow throughout the Arctic and Antarctic, and in the deepest parts of the ocean, in temperatures that hover near the freezing point of water (2°–10°C). Several species of algae and fungi thrive on the surfaces of snow and glaciers (see figure 7.10). More surprising still is a strategy of bacteria and algae adapted to the sea ice of Antarctica. Although the ice appears to be completely solid, it is honeycombed by various-sized pores and tunnels filled with liquid water. These frigid microhabitats harbor a virtual microcosm of planktonic life, including predators (fish and shrimp) that live on these algae and bacteria.

Salt, Acidity, Alkalinity

The growth of most microbial cells is inhibited by high amounts of salt; for this reason, salt is a common food preservative. Yet whole communi-ties of salt-dependent bacteria and algae occupy habitats in oceans, salt lakes, and inland seas, some of which are saturated with salt (30%). Most of these microbes have demonstrable metabolic requirements for high levels of minerals such as sodium, potassium, magnesium, chlorides, or iodides. Because of their salt-loving nature, some species are pesky contaminants in salt-processing plants, pickling brine, and salted fish.

Highly acidic or alkaline habitats are not common, but acidic bogs, lakes, and alkaline soils contain a specialized microbial flora. A few species of algae and bacteria can actually survive at a pH near that of concentrated hydrochloric acid. They not only require such a low pH for growth, but particular bacteria (for example, *Thiobacillus*) actually help maintain the low pH by releasing strong acid.

Other Frontiers to Conquer

It was once thought that the region far beneath the soil and upper crust of the earth's surface was sterile. However, new work with deep core samples (from 330 m down) indicates a vast microbial population in these zones. Myriad bacteria, protozoa, and fungi exist in this moist clay, which is high in minerals and complex organic substrates. Even deep mining deposits two miles into the earth's crust harbor a rich assortment of bacteria. They thrive in mineral deposits that are hot (90°C) and radioactive. Many biologists believe these are very similar to the first ancient microbes to have existed on earth.

Numerous species have carved a niche for themselves in the depths of mud, swamps, and oceans, where oxygen gas and sunlight cannot penetrate. The predominant living things in the deepest part of the oceans (10,000 m or below) are pressure- and cold-loving microorganisms. Even parched zones in sand dunes and deserts harbor a hardy brand of microbes, and thriving bacterial populations can be found in petroleum, coal, and mineral deposits containing copper, zinc, gold, and uranium.

As a rule, a microbe that has adapted to an extreme habitat will die if placed in a moderate one. And, except for rare cases, none of the organisms living in these extremes are pathogens, because the human body is a hostile habitat for them.

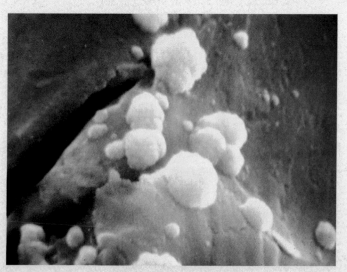

(a)

(b)

(a) Cells of *Sulfolobus,* an archaean that lives in mineral deposits of hot springs and volcanoes. It can survive temperatures of about 90°C and acidity of pH 1.5. **(b)** Clumps of bacteria (dark matter) clinging to life on crystals of ice deep in the Antarctic sediments.

An ameba gorging itself on bacteria could be compared to a person eating a bowl of vegetable soup, because its nutrient needs and uses are fundamentally similar to a human's. Most food is a complex substance that contains many different types of nutrients. Some smaller molecules such as sugars can be absorbed directly by the cell; larger food debris and molecules must be first be ingested and broken down into a size that can be absorbed. As nutrients are taken in, they add to a dynamic pool of inorganic and organic compounds dissolved in the cytoplasm. This pool will provide raw materials to be assimilated into the organism's own specialized proteins, carbohydrates, lipids, and other macromolecules used in growth and metabolism.

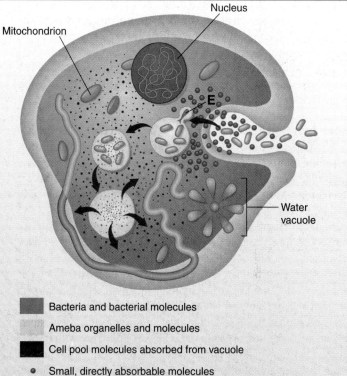

Steps in obtaining and incorporating nutrients. Food particles are phagocytosed into a vacuole that fuses with a lysozyme containing digestive enzymes (E). Smaller subunits of digested macromolecules are transported out of the vacuole into the cell pool and are used in the metabolic and synthetic activities of the cell.

- ▮ Bacteria and bacterial molecules
- ▯ Ameba organelles and molecules
- ▮ Cell pool molecules absorbed from vacuole
- • Small, directly absorbable molecules

An **autotroph*** is an organism that uses CO_2, an inorganic gas, as its carbon source. Because autotrophs have the special capacity to convert CO_2 into organic compounds, they are not nutritionally dependent on other living things. In a later section, we enlarge on the topic of nutritional types as based on carbon and energy sources.

Nitrogen Sources

The main reservoir of nitrogen is nitrogen gas (N_2), which makes up 79% of the earth's atmosphere. This element is indispensable to the structure of proteins, DNA, RNA, and ATP. Such nitrogenous compounds are the primary source for heterotrophs, but to be useful, they must first be degraded into their basic building blocks (proteins into amino acids; nucleic acids into nucleotides). Some bacteria and algae utilize inorganic nitrogenous nutrients (NO_3^-, NO_2^-, or NH_3). A small number of bacteria can transform N_2 into compounds usable by other organisms through the process of nitrogen fixation (see chapter 26). Regardless of the initial form in which the inorganic nitrogen enters the cell, it must first be converted to NH_3, the only form that can be directly combined with carbon to synthesize amino acids and other compounds.

Oxygen Sources

Because oxygen is a major component of organic compounds such as carbohydrates, lipids, and proteins, it plays an important role in the structural and enzymatic functions of the cell. Oxygen is likewise a common component of inorganic salts such as sulfates, phosphates, nitrates, and water. Free gaseous oxygen (O_2) makes up 20% of the atmosphere. It is absolutely essential to the metabolism of many organisms, as we shall see later in this chapter and in chapter 8.

Hydrogen Sources

Hydrogen is a major element in all organic compounds and several inorganic ones, including water (H_2O), salts ($Ca[OH]_2$), and certain naturally occurring gases (H_2S, CH_4, and H_2). These gases are both used and produced by microbes. Hydrogen performs these overlapping roles in the biochemistry of cells:

1. maintaining pH,
2. forming hydrogen bonds between molecules, and
3. serving as the source of free energy in oxidation-reduction reactions of respiration (see chapter 8).

Phosphorus (Phosphate) Sources

The main inorganic source of phosphorus is phosphate (PO_4^{-3}), derived from phosphoric acid (H_3PO_4) and found in rocks and oceanic mineral deposits. Phosphate is a key component of nucleic acids and is thereby essential to the genetics of cells and viruses. Because it is also found in ATP, it also serves in cellular energy transfers. Other phosphate-containing compounds are phospholipids in cell membranes and coenzymes such as NAD (see chapter 8). Certain environments have very little available phosphate for use by organisms and therefore limit the ability for these organisms to grow.

***autotroph** (aw′-toh-trohf) Gr. *auto,* self, and *troph,* to feed.

One method utilized by *Corynebacteriun* is to concentrate and store phosphate in metachromatic granules.

Sulfur Sources

Sulfur is widely distributed throughout the environment in mineral form. Rocks and sediments (such as gypsum) can contain sulfate (SO_4^{-2}), sulfides (FeS), hydrogen sulfide gas (H_2S), and elemental sulfur (S). Sulfur is an essential component of some vitamins (vitamin B_1) and the amino acids methionine and cysteine; the latter determines shape and structural stability of proteins by forming unique linkages called disulfide bonds (see figure 2.22).

Other Nutrients Important in Microbial Metabolism

The remainder of important elements are mineral ions. **Potassium** is essential to protein synthesis and membrane function. **Sodium** is important for certain types of cell transport. **Calcium** is a stabilizer of the cell wall and endospores of bacteria. It is also combined with carbonate (CO_3^{-2}) in the formation of shells by foraminiferans and radiolarians. **Magnesium** is a component of chlorophyll and a stabilizer of membranes and ribosomes. **Iron** is an important component of the cytochrome pigments of cell respiration. Zinc is an essential regulatory element for eucaryotic genetics. It is a major component of "zinc fingers"—binding factors that help enzymes adhere to specific sites on DNA. Copper, cobalt, nickel, molybdenum, manganese, silicon, iodine, and boron are needed in small amounts by some microbes but not others. A discovery with important medical implications is that metal ions can directly influence certain diseases by their effects on microorganisms. For example, the bacteria that cause gonorrhea and meningitis grow more rapidly in the presence of iron ions.

Growth Factors: Essential Organic Nutrients

Few microbes are as versatile as *Escherichia coli* in assembling molecules from scratch. Many fastidious bacteria lack the genetic and metabolic mechanisms to synthesize every organic compound they need for survival. An organic compound such as an amino acid, nitrogen base, or vitamin that cannot be synthesized by an organism and must be provided as a nutrient is a **growth factor.** For example, although all cells require 20 different amino acids for proper assembly of proteins, many cells cannot synthesize them all. Those that must be obtained from food are called **essential amino acids.** A notable example of the need for growth factors occurs in *Haemophilus influenzae,* a bacterium that causes meningitis and respiratory infections in humans. It can grow only when hemin (factor X), NAD (factor V), thiamine and pantothenic acid (vitamins), uracil, and cysteine are provided by another organism or a growth medium.

HOW MICROBES FEED: NUTRITIONAL TYPES

The earth's limitless habitats and microbial adaptations are rivaled by an elaborate menu of nutritional schemes. Fortunately, most organisms show consistent trends and can be described by a few general categories (**table 7.4**) and a few selected terms (see "A Note on Terminology"). The main determinants of a microbe's nutritional type are its sources of carbon and energy. In a previous section, microbes were defined as autotrophs, whose primary carbon source is inorganic carbon (CO_2), and heterotrophs, which are dependent on

TABLE 7.4

Nutritional Categories of Microbes by Carbon and Energy Source

Category	Carbon Source	Energy Source	Example
Autotroph	CO_2	**Nonliving Environment**	
Photoautotroph	CO_2	Sunlight	Photosynthetic organisms, such as algae, plants, cyanobacteria
Chemoautotroph	CO_2	Simple inorganic chemicals	Only certain bacteria, such as methanogens, vent bacteria
Heterotroph	**Organic**	**Other Organisms or Sunlight**	
Photoheterotroph	Organic	Sunlight	Nonsulfur bacteria
Chemoheterotroph	Organic	Metabolic conversion of the nutrients from other organisms	Protozoa, fungi, many bacteria, animals
Saprobe	Organic	Metabolizing the organic matter of dead organisms	Fungi, bacteria (decomposers)
Parasite	Organic	Utilizing the tissues, fluids of a live host	Various parasites and pathogens; can be bacteria, fungi, protozoa, animals

organic carbon compounds. In terms of energy source, microbes that photosynthesize are generally classified as **phototrophs,** and those that oxidize chemical compounds are **chemotrophs.** The terms for carbon and energy source are often merged into a single word for convenience (table 7.4). The categories described here are meant to describe only the major nutritional groups and do not include unusual exceptions.

Autotrophs and Their Energy Sources

Autotrophs derive energy from one of two possible nonliving sources: sunlight (photoautotrophs) and chemical reactions involving simple inorganic chemicals (chemoautotrophs). **Photoautotrophs** are photosynthetic; that is, they capture the energy of light rays and transform it into chemical energy that can be used in cell metabolism (**Microbits 7.3,** page 197). Because photosynthetic organisms (algae, plants, some bacteria) produce organic molecules that can be used by themselves and heterotrophs, they form the basis for most food webs. Their role as primary producers of organic matter is discussed in chapter 26.

A significant type of bacteria called **chemoautotrophs** have an unusual nutritional adaptation that requires neither sunlight nor organic nutrients. Some microbiologists prefer to call them **lithoautotrophs** (rock feeders) in reference to their total reliance on inorganic minerals. These bacteria derive energy in diverse and rather amazing ways. In very simple terms, they remove electrons from inorganic substrates such as hydrogen gas, hydrogen sulfide, sulfur, or iron and combine them with carbon dioxide and hydrogen. This reaction provides simple organic molecules and a modest amount of energy to drive the synthetic

A Note on Terminology

Much of the vocabulary for describing microbial adaptations is based on some common root words. These are combined in various ways that assist in discussing the types of nutritional or ecologic adaptations, as shown in this partial list:

Root	Meaning	Example of Use
troph-	Food, nourishment	Trophozoite—the feeding stage of protozoa
-phile	To love	Extremophile—an organism that has adapted to ("loves") extreme environments
-obe	To live	Microbe—to live "small"
hetero-	Other	Heterotroph—an organism that requires nutrients from other organisms
auto-	Self	Autotroph—an organism that does not need other organisms for food (obtains nutrients from a nonliving source)
photo-	Light	Phototroph—an organism that uses light as an energy source
chemo-	Chemical	Chemotroph—an organism that uses chemicals for energy, rather than light
sapro-	Rotten	Saprobe—an organism that lives on dead organic matter
halo-	Salt	Halophile—an organism that can grow in high-salt environments
thermo-	Heat	Thermophile—an organism that grows best at high temperatures
psychro-	Cold	Psychrophile—an organism that grows best at cold temperatures
aero-	Air (O_2)	Aerobe—an organism that uses oxygen in metabolism

Modifier terms are also used to specify the nature of an organism's adaptations. **Obligate** or **strict** refers to being restricted to a narrow niche or habitat, such as an obligate thermophile that requires high temperatures to grow. By contrast, **facultative*** means not being so restricted, but being able to adapt to a wider range of metabolic conditions and habitats. A facultative halophile can grow with or without high salt concentration.

processes of the cell. Chemoautotrophic bacteria play an important part in recycling inorganic nutrients. For an example of chemoautotrophy and its importance to deep-sea communities, see Spotlight on Microbiology 7.5.

An interesting group of chemoautotrophs are **methanogens,*** which produce methane (CH_4) from hydrogen gas and carbon dioxide (**figure 7.1**).

$$4H_2 + CO_2 \rightarrow CH_4 + 2H_2O$$

Methane, sometimes called "swamp gas," or "natural gas," is formed in anaerobic, hydrogen-containing microenvironments of soil, swamps, mud, and even in the intestines of some animals. Many methanogens are archaea that live in extreme habitats such as ocean vents and hot springs, where temperatures reach up to 125°C. Methane, which is used as a fuel in a large percentage of homes, can also be produced in limited quantities using Biogas generators which are devices primed with a mixed population of microbes (including methanogens) and fueled with various waste materials that can supply enough methane to drive a steam generator. Methane also plays a role as one of the greenhouse gases that is currently an environmental concern (see chapter 26).

Heterotrophs and Their Energy Sources

The majority of heterotrophic microorganisms are **chemoheterotrophs** that derive both carbon and energy from organic compounds. Processing these organic molecules by respiration or fermentation releases energy in the form of ATP. An example of chemoheterotrophy is *aerobic respiration,* the principal energy-yielding reaction in animals, most protozoa and fungi, and aerobic bacteria. It can be simply represented by the equation:

$$\text{Glucose } [(CH_2O)_n] + O_2 \rightarrow CO_2 + H_2O + \text{Energy (ATP)}$$

You might notice that this reaction is complementary to photosynthesis. Here, glucose and oxygen are reactants, and carbon dioxide is given off. Indeed, the earth's balance of both energy and metabolic gases is greatly dependent on this relationship. Chemoheterotrophic microorganisms belong to one of two main categories that differ in how they obtain their organic nutrients: **Saprobes**[2] are free-living microorganisms that feed primarily on organic detritus from dead organisms, and **parasites** ordinarily derive nutrients from the cells or tissues of a host.

Saprobic Microorganisms Saprobes occupy a niche as decomposers of plant litter, animal matter, and dead microbes. If not for the work of decomposers, the earth would gradually fill up with organic material, and the nutrients it contains would not be recycled. Most saprobes, notably bacteria and fungi, have a rigid cell wall and cannot engulf large particles of food. To compensate, they release enzymes to the extracellular environment and digest the food particles into smaller molecules that can pass freely into the cell

---

2. Synonyms are *saprotroph* and *saprophyte.* We prefer to use the terms *saprobe* and *saprotroph* because they are more consistent with other terminology and because the term *saprophyte* is a holdover from the time when bacteria and fungi were grouped with plants.

*methanogen (meth-an-oh-gen) From *methane,* a colorless, odorless gas, and *gennan,* to produce.

---

*facultative (fak′-uhl-tay-tiv) Gr. *facult,* capability or skill, and *tatos,* most.

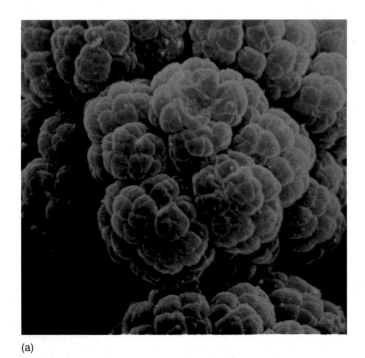

(a)

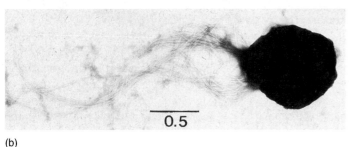

(b)

FIGURE 7.1

Methane-producing archaea. Members of this group are primitive procaryotes with unusual cell walls and membranes. **(a)** SEM of a small colony of *Methanosarcina*. **(b)** *Methanococcus junnuschii,* a motile archaea that inhabits hot vents in the seafloor and uses hydrogen gas as a source of energy.

(figure 7.2). *Obligate saprobes* exist strictly on organic matter in soil and water and are unable to adapt to the body of a live host. This group includes many free-living protozoa, fungi, and bacteria. Apparently, there are fewer of these strict species than was once thought, and many supposedly nonpathogenic saprobes can infect a susceptible host. When a saprobe infects a host, it is considered a *facultative parasite.* Such an infection usually occurs when the host is compromised, and the microbe is considered an *opportunistic pathogen.* For example, although its natural habitat is soil and water, *Pseudomonas aeruginosa* frequently causes infections in hospitalized patients. The yeast *Cryptococcus neoformans* causes a severe lung and brain infection in AIDS patients, yet its natural habitat is the soil.

Parasitic Microorganisms Parasites live in or on the body of a host, which they usually harm to some degree. Parasites inclined to cause damage to tissues (disease) or even death are called **pathogens.** Parasites range from viruses to helminth worms, and they can live on the body (ectoparasites), in the organs and tissues (endoparasites), or even within cells (intracellular parasites, the

Digestion in Bacteria and Fungi

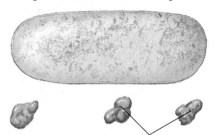

Organic debris

(a) Walled cell is inflexible.

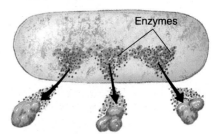

Enzymes

(b) Enzymes are transported across the wall.

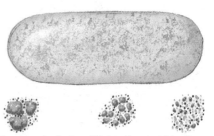

(c) Enzymes hydrolyze the bonds on nutrients.

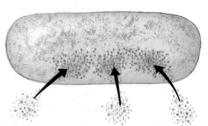

(d) Smaller molecules are transported into the cytoplasm.

FIGURE 7.2

Extracellular digestion in a saprobe with a cell wall (bacterium or fungus). (a) A walled cell is inflexible and cannot engulf large pieces of organic debris. **(b)** In response to a usable substrate, the cell synthesizes enzymes that are transported across the wall into the extracellular environment. **(c)** The enzymes hydrolyze the bonds in the debris molecules. **(d)** Digestion produces molecules small enough to be transported into the cytoplasm.

most extreme type). Although there are several degrees of parasitism, the more successful parasites generally have no fatal effects and eventually evolve to a less harmful relationship with their host. *Obligate parasites* (for example, the leprosy bacillus and the syphilis spirochete) are unable to grow outside of a living host. Parasites that are less strict can be cultured artificially if provided with the correct nutrients and environmental conditions. Bacteria such as *Streptococcus pyogenes* (the cause of strep throat) and *Staphylococcus aureus* can grow on artificial media as saprobes.

TABLE 7.5

Summary of Transport Processes in Cells

General Process	Nature of Transport	Examples	Description	Qualities
Passive	Energy expenditure is not required. Substances exist in a gradient and move from areas of higher concentration towards areas of lower concentration in the gradient.	Diffusion	A fundamental property of atoms and molecules that exist in a state of random motion	Nonspecific Brownian movement
		Osmosis	Diffusion of water molecules across a membrane barrier that is freely permeable to water but selectively permeable to other molecules	Direction depends on osmolarity of cell vs. habitat
		Facilitated diffusion	Molecule binds to a receptor in membrane and is carried across to other side	Molecule specific; transports both ways
Active	Energy expenditure is required. Molecules need not exist in a gradient. Rate of transport is increased. Transport may occur against a concentration gradient.	Carrier-mediated active transport	Atoms or molecules are pumped into or out of the cell by specialized receptors. Driven by ATP	Transports simple sugars, amino acids, inorganic ions (Na^+, K^+)
		Group translocation	Molecule is moved across membrane and simultaneously converted to a metabolically useful substance.	Alternate system for transporting nutrients (sugars, amino acids)
		Bulk transport	Mass transport of large particles, cells, and liquids by engulfment and vesicle formation	Includes endocytosis, exocytosis, pinocytosis

Obligate intracellular parasitism is an extreme but relatively common mode of life. Microorganisms that spend all or part of their life cycle inside a host cell include the viruses, a few bacteria (rickettsias, chlamydias), and certain protozoa (apicomplexa). Contrary to what one might think, the inside of a cell is not completely without hazards, and microbes must overcome some difficult challenges. They must find a way into the cell, keep from being destroyed, not destroy the host cell too soon, multiply, and find a way to infect other cells. Intracellular parasites obtain different substances from the host cell, depending on the group. Viruses are the most extreme, parasitizing the host's genetic and metabolic machinery. Rickettsias are primarily energy parasites, and the malaria protozoan is a hemoglobin parasite.

TRANSPORT MECHANISMS FOR NUTRIENT ABSORPTION

A microorganism's habitat provides necessary nutrients—some abundant, others scarce—that must still be taken into the cell. Survival also requires that cells transport waste materials into the environment. Whatever the direction, transport occurs across the cell membrane, the structure specialized for this role. This is true even in organisms with cell walls (bacteria, algae, and fungi), because the cell wall is usually too nonselective to screen the entrance or exit of molecules. Two general types of transport are **passive transport,** which follows physical laws that are not unique to living systems and do not generally require direct energy input from the cell; **active transport,** which requires carrier proteins in the membranes of living cells and the expenditure of energy (see **table 7.5**).

Passive Transport: Diffusion, Osmosis

The driving force of passive transport is atomic and molecular movement—the natural tendency of atoms and molecules to be in constant random motion. The existence of this motion is evident in Brownian movement (see figure 4.4) of small particles suspended in liquid. It can be demonstrated by a variety of simple observations. A drop of perfume released into one part of a room is soon smelled in another part, or a lump of sugar in a cup of tea spreads through the whole cup without stirring. This phenomenon of molecular movement, in which atoms or molecules move in a gradient from an area of higher density or concentration to an area of lower

How Molecules Diffuse In Aqueous Solutions

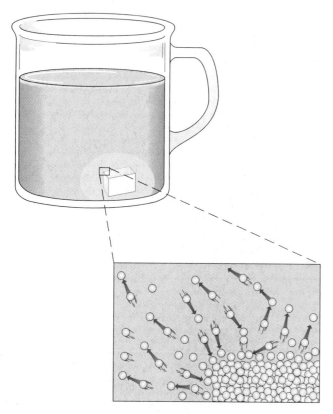

FIGURE 7.3

Diffusion of molecules in aqueous solutions. A high concentration of sugar exists in the cube at the bottom of the liquid. An imaginary molecular view of this area shows that sugar molecules are in a constant state of motion. Those at the edge of the cube diffuse from the concentrated area into more dilute regions. As diffusion continues, the sugar will spread evenly throughout the aqueous phase, and eventually there will be no gradient. At that point, the system is said to be in equilibrium.

density or concentration, is **diffusion*** (**figure 7.3**). Although passive transport requires molecules to be arranged in a gradient, diffusion remains an important way for cells to obtain freely diffusible materials (oxygen, carbon dioxide, and water) and to release wastes to the environment.

Diffusion of water through a selectively permeable membrane, a process called **osmosis,*** is also a physical phenomenon that is easily demonstrated in the laboratory with nonliving materials. It provides a model of how cells deal with various solute concentrations in aqueous solutions (**figure 7.4**). In an osmotic system, the membrane is *selectively,* or *differentially, permeable,* having passageways that allow free diffusion of water but can block certain other dissolved molecules. When this membrane is placed between solutions of differing concentrations and the solute is not diffusible (protein, for example), then under the laws of diffusion, water will diffuse at a faster rate from the side that has more water to the side that has less water. As long as the concentrations of the solutions differ, one side will

*diffusion (dih-few′-zhun) L. *dis,* apart, and *fundere,* to pour.

*osmosis (oz-moh′-sis) Gr. *osmos,* impulsion, and *osis,* a process.

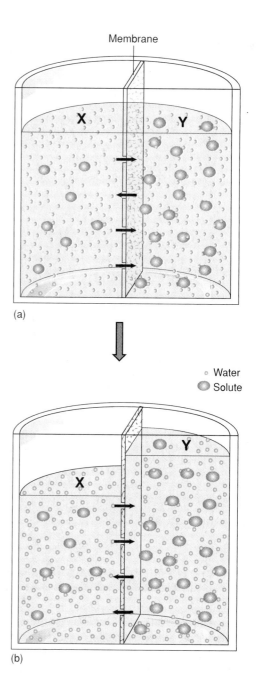

FIGURE 7.4

Osmosis, the diffusion of water through a selectively permeable membrane. **(a)** A membrane has pores that allow the ready passage of water but not large solute molecules from one side to another. Placement of this membrane between solutions of different solute concentrations (X = less concentrated and Y = more concentrated) results in a diffusion gradient for water. Water molecules undergo diffusion and move across the membrane pores in both directions. Because there is more water in solution X, the opportunity for a water molecule to successfully hit and go through a pore is greater for X than for Y. The result will be a net movement of water from X to Y. **(b)** The level of solution on the Y side rises as water continues to diffuse in. This process will continue until equilibration occurs and the rate of diffusion of water is equal on both sides.

experience a net loss of water and the other a net gain of water, until equilibrium is reached and the rate of diffusion is equalized.

Be aware of an important point about the dynamics of osmosis and the solutions involved. Whereas the primary considerations

MICROBITS   7.3
Light-Driven Organic Synthesis

Two equations sum up the reactions of photosynthesis in a simple way. The first equation shows a reaction that results in the production of oxygen:

$$CO_2 + H_2O \xrightarrow[\text{by chlorophyll}]{\text{Sunlight absorbed}} (CH_2O)_n^* + O_2$$

This oxygenic (oxygen-producing) type of photosynthesis occurs in plants, algae, and cyanobacteria (see figure 4.34). The function of chlorophyll is to capture light energy. Carbohydrates produced by the reaction can be used by the cell to synthesize other cell components, and they also become a significant nutrient for heterotrophs that feed on

---

*$(CH_2O)_n$ is shorthand for a carbohydrate.

them. The production of oxygen is vital to maintaining this gas in the atmosphere.

A second equation shows a photosynthetic reaction that does not result in the production of oxygen:

$$CO_2 + H_2S \xrightarrow[\text{by bacteriochlorophyll}]{\text{Sunlight absorbed}} (CH_2O)_n + S + H_2O$$

This anoxygenic (no oxygen produced) type of photosynthesis is found in bacteria such as purple and green sulfur bacteria. Note that the type of chlorophyll (bacteriochlorophyll, a substance unique to this microbes), one of the reactants (hydrogen sulfide gas), and one product (elemental sulfur) are different from those in the first equation. These bacteria live in the anaerobic regions of aquatic habitats.

---

are the concentration and direction of movement of water (solvent), the solute concentration is also involved, because it will dictate how much water is present to diffuse. For example, a more concentrated 50% solution contains 50 parts of solute and 50 parts of water. By comparison, a less concentrated 5% solution contains 5 parts of solute and 95 parts of water. When these two solutions are placed on opposite sides of a membrane, water will diffuse at a faster rate from the 5% solution into the 50% solution.

Osmosis in living systems is similar to the model shown in figure 7.4. Living membranes generally block the entrance and exit of larger molecules and permit free diffusion of water. Because most cells are surrounded by some free water, the amount of water entering or leaving has a far-reaching impact on cellular activities and survival. This osmotic relationship between cells and their environment is determined by the relative concentrations of the solutions on either side of the cell membrane (**figure 7.5**). Such systems can be compared using the terms isotonic, hypotonic, and hypertonic.

Under **isotonic*** conditions, the environment is equal in concentration to the cell's internal environment, and because diffusion of water proceeds at the same rate in both directions, there is no net change in cell volume. Isotonic solutions are the most stable environments for cells, because they are already in an osmotic steady-state with the cell. Parasites living in host tissues are most likely to be living in isotonic habitats.

Under **hypotonic*** conditions, the solute concentration of the external environment is lower than that of the cell's internal environment. Pure water provides the most hypotonic environment for cells because it has no solute. The net direction of osmosis is from the hypotonic solution into the cell, and cells without walls swell and can burst.

Hypertonic* conditions are also out of balance with the tonicity of the cell's cytoplasm, but in this case, the environment has a higher solute concentration than the cytoplasm. Because a hypertonic environment will force water to diffuse out of a cell, it is said to have high *osmotic pressure* or potential. The growth-limiting effect of hypertonic solutions on microbes is the principle behind using concentrated salt and sugar solutions as preservatives for food.

Adaptations to Osmotic Variations in the Environment

Let us now see how specific microbes have adapted osmotically to their environments. In general, isotonic conditions pose little stress on cells, so survival depends on counteracting the adverse effects of hypertonic and hypotonic environments.

A bacterium and an ameba living in fresh pond water are examples of cells that live in constantly hypotonic conditions. The rate of water diffusing across the cell membrane into the cytoplasm is rapid and constant, and the cells would die without a way to adapt. The majority of bacterial cells compensate by having a cell wall that protects them from bursting even as the cytoplasm membrane becomes *turgid** from pressure. The ameba's adaptation is an anatomical and physiological one that requires the constant expenditure of energy. It has a water, or contractile, vacuole that siphons excess water back out into the habitat like a tiny pump (see figure 5.29).

A microbe living in a high-salt environment (hypertonic) has the opposite problem and must either restrict its loss of water to the environment or increase the salinity of its internal environment. Halobacteria living in the Great Salt Lake and the Dead Sea actually absorb salt to make their cells isotonic with the environment, thus they have a physiological need for a high-salt concentration in their habitats (see halophiles on page 119).

---

*isotonic (eye-soh-tahn´-ik) Gr. *iso,* same, and *tonos,* tension.

*hypotonic (hy-poh-tahn´-ik) Gr. *hypo,* under, and *tonos,* tension.

*hypertonic (hy-pur-tahn´-ik) Gr. *hyper,* above, and *tonos,* tension.

*turgid (ter´-jid) A condition of being swollen or congested.

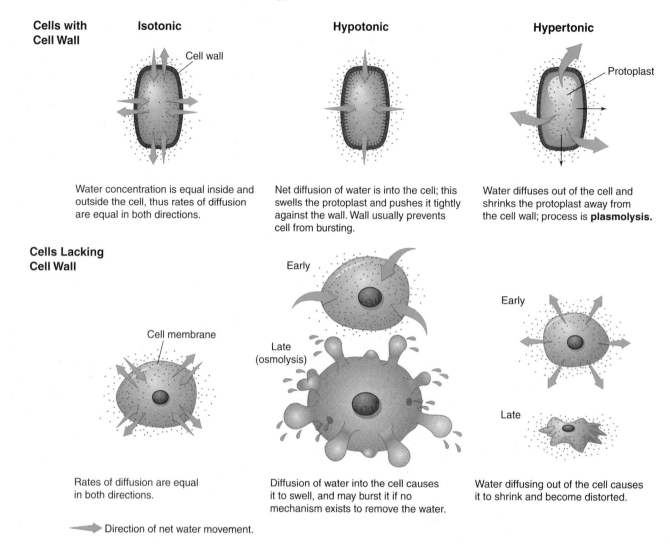

Cells with Cell Wall

Isotonic

Cell wall

Water concentration is equal inside and outside the cell, thus rates of diffusion are equal in both directions.

Hypotonic

Net diffusion of water is into the cell; this swells the protoplast and pushes it tightly against the wall. Wall usually prevents cell from bursting.

Hypertonic

Protoplast

Water diffuses out of the cell and shrinks the protoplast away from the cell wall; process is **plasmolysis.**

Cells Lacking Cell Wall

Cell membrane

Rates of diffusion are equal in both directions.

Early

Late (osmolysis)

Diffusion of water into the cell causes it to swell, and may burst it if no mechanism exists to remove the water.

Early

Late

Water diffusing out of the cell causes it to shrink and become distorted.

Direction of net water movement.

FIGURE 7.5

Cell responses to solutions of differing osmotic content.

Diffusion

All molecules, regardless of being in a solid, liquid, or gas, are in continuous movement, and as the temperature increases, the faster the molecular movement becomes. This is called **"thermal"** movement. In any solution, including cytoplasm, these moving molecules cannot travel very far without having collisions with other molecules and, therefore, will bounce off each other like millions of pool balls every second. As a result of each collision, the directions of the colliding molecules are altered and the direction of any one molecule is unpredictable and is therefore "random." If we start with a solution in which the solute, or dissolved substance, is more concentrated in one area than another then the random thermal movement of molecules in this solution will eventually distribute the molecules from the area of higher concentration to the area of lower concentration, thus evenly distributing the molecules. This net movement of molecules down their concentration gradient by random thermal motion is known as diffusion. Diffusion of molecules across the cell membrane is largely determined by the concentration gradient and permeability of the substance.

So far, the discussion of passive or simple diffusion has not included the added complexity of membranes or cell walls, which hinder simple diffusion by adding a physical barrier. These structures may inhibit the diffusion of a substance several thousand times by decreasing the permeability when compared to that of molecules simply dissolved in water. Therefore, simple diffusion is limited to small nonpolar molecules like oxygen or lipid soluble molecules that may dissolve through the membranes. It is imperative that a cell be able to move polar molecules and ions across the plasma membrane, and given the greatly decreased permeability of these chemicals simple diffusion will not allow this movement. Therefore the concept of **facilitated diffusion** must be introduced **(figure 7.6).** This type of mediated transport mechanism utilizes a carrier protein that will bind a specific substance. This binding will change the conformation of the carrier protein such that the substance is moved across the membrane. Once the substance is transported, the carrier protein resumes its original shape and is ready to transport again. These carrier proteins exhibit **specificity,** which means that they bind and transport only a single type of molecule. For example, a carrier protein that transports sodium will not bind

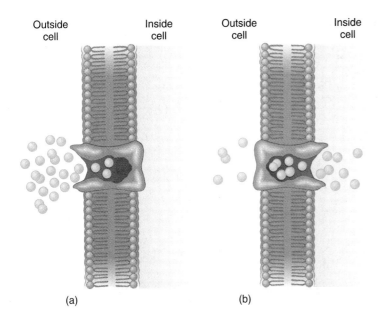

Outside cell — Inside cell — Outside cell — Inside cell

(a)   (b)

FIGURE 7.6

Passive transport. Facilitated diffusion involves the attachment of a molecule to a specific protein carrier. **(a)** Bonding of the molecule causes a conformational change in the protein that facilitates the molecule's passage across the membrane. **(b)** The membrane receptor opens into the cell and releases the molecule.

glucose. A second characteristic exhibited by facilitated diffusion is **saturation.** The rate of transport of a substance is limited by the number of binding sites on the transport proteins. As the substance's concentration increases so does the rate of transport until the concentration of the transported substance is such that all of the transporters' binding sites are occupied. Then the rate of transport reaches a steady state and cannot move faster despite further increases in the substance's concentration. A third characteristic of these carrier proteins is that they exhibit **competition.** This is when two molecules of similar shape can bind the same binding site on a carrier protein. The chemical with the higher binding affinity, or the chemical in the higher concentration, will be transported at a greater rate.

Active Transport: Bringing in Molecules Against a Gradient

Free-living microbes exist under relatively nutrient-starved conditions and cannot rely completely on slow and rather inefficient passive transport mechanisms. To ensure a constant supply of nutrients and other required substances, microbes must capture those that are in extremely short supply and actively transport them into the cell. Features inherent in **active transport** systems are

1. the transport of nutrients against the natural diffusion gradient or in the same direction as the natural gradient but at a rate faster than by diffusion alone,
2. the presence of specific membrane proteins (permeases and pumps; **figure 7.7a**), and
3. the expenditure of energy. Examples of substances transported actively are monosaccharides, amino acids, organic acids, phosphates, and metal ions.

Some freshwater algae have such efficient active transport systems that an essential nutrient can be found in intracellular concentrations 200 times that of the habitat.

An important type of active transport involves specialized pumps, which can rapidly carry ions such as K^+, Na^+, and H^+ across the membrane. This behavior is particularly important in mitochondrial ATP formation (see the discussion of oxidative phosphorylation in chapter 8) and protein synthesis. Another type of active transport, **group translocation,** couples the transport of a nutrient with its conversion to a substance that is immediately useful inside the cell (figure 7.7b). This method is used by certain bacteria to transport sugars (glucose, fructose) while simultaneously adding molecules such as phosphate that prepare them for the next stage in metabolism.

Endocytosis: Eating and Drinking by Cells

Some eucaryotic cells transport large molecules, particles, liquids, or even other cells across the cell membrane. Because the cell usually expends energy to carry out this transport, it is also a form of active transport. The substances transported do not pass physically through the membrane but are carried into the cell by **endocytosis.*** First the cell encloses the substance in its membrane, simultaneously forming a vacuole and engulfing it **(figure 7.8).** Amebas and certain white blood cells ingest whole cells or large solid matter by **phagocytosis.*** Liquids, such as oils or large molecules in solution, enter the cell through **pinocytosis.***

---

*endocytosis (en″-doh-cy-toh′-sis) Gr. *endo,* in; *cyte,* cell; and *osis,* a process.

*phagocytosis (fag″-oh-cy-toh′-sis) Gr. *phagein,* to eat.

*pinocytosis (pin″-oh-cy-toh′-sis) Gr. *pino,* to drink.

CHAPTER CHECKPOINTS

Nutrition is a process by which all living organisms obtain substances from their environment to convert to metabolic uses.

Although the chemical form of nutrients varies widely, all organisms require six bioelements—carbon, hydrogen, oxygen, nitrogen, phosphorus, and sulfur—to survive, grow, and reproduce.

Nutrients are categorized by the amount required (macronutrients or micronutrients), by chemical structure (organic or inorganic), and by their importance to the organism's survival (essential or nonessential).

Microorganisms are classified both by the chemical form of their nutrients and the energy sources they utilize.

Nutrient requirements of microorganisms determine their respective niches in the food webs of major ecosystems.

Nutrients are transported into microorganisms by two kinds of processes: active transport that expends energy and passive transport that occurs independently of energy input.

The molecular size and concentration of a nutrient determine which method of transport is used.

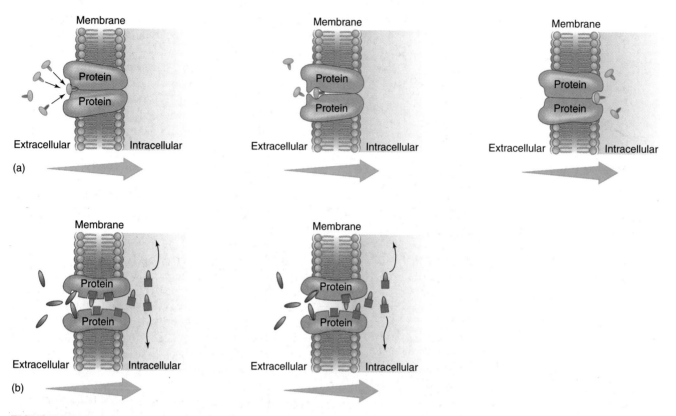

FIGURE 7.7

Active transport. In active transport mechanisms, energy is expended to transport the molecule across the cell membrane. **(a)** Molecular pumps. The membrane proteins (permeases) have attachment sites for essential nutrient molecules. As these molecules bind to the permease, they are pumped into the cell's interior through special membrane protein channels. Microbes have these systems for transporting various ions (sodium, iron) and small organic molecules. **(b)** In group translocation, the molecule is actively captured, but along the route of transport, it is chemically altered. By coupling transport with synthesis, the cell conserves energy.

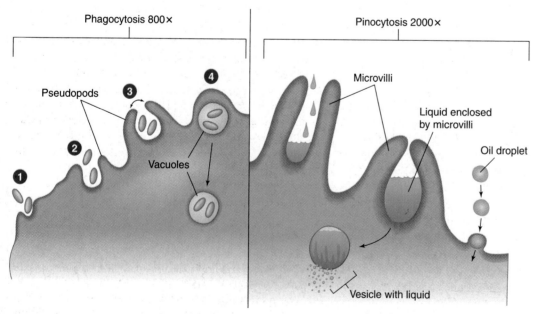

FIGURE 7.8

Endocytosis (phagocytosis and pinocytosis). Solid particles are phagocytosed by large cell extensions called pseudopods, and they are pinocytosed into vesicles by very fine cell protrusions called microvilli. Oil droplets fuse with the membrane and are released directly into the cell.

Environmental Factors That Influence Microbes

Microbes are exposed to a wide variety of environmental factors in addition to nutrients. Microbial ecology focuses on ways that microorganisms deal with or adapt to such factors as heat, cold, gases, acid, radiation, osmotic and hydrostatic pressures, and even other microbes. Adaptation is a complex adjustment in biochemistry or genetics that enables long-term survival and growth. For most microbes, environmental factors fundamentally affect the function of metabolic enzymes. Thus, survival in a changing environment is largely a matter of whether the enzyme systems of microorganisms can adapt to alterations in their habitat. Incidentally, one must be careful to differentiate between growth in a given condition and tolerance, which implies survival without growth.

TEMPERATURE ADAPTATIONS

Microbial cells are unable to control their temperature and therefore assume the ambient temperature of their natural habitats. Their survival is dependent on adapting to whatever temperature variations are encountered in that habitat. The range of temperatures for microbial growth can be expressed as three *cardinal temperatures.* The **minimum temperature** is the lowest temperature that permits a microbe's continued growth and metabolism; below this temperature, its activities are inhibited. The **maximum temperature** is the highest temperature at which growth and metabolism can proceed. If the temperature rises slightly above maximum, growth will stop, but if it continues to rise beyond that point, the enzymes and nucleic acids will eventually become permanently inactivated otherwise known as denaturation, and the cell will die. This is why heat works so well as an agent in microbial control. The **optimum temperature** covers a small range, intermediate between the minimum and maximum, which promotes the fastest rate of growth and metabolism (rarely is the optimum a single point).

Depending on their natural habitats, some microbes have a narrow cardinal range, others a broad one. Some strict parasites will not grow if the temperature varies more than a few degrees below or above the host's body temperature. For instance, the typhus rickettsia multiplies only in the range of 32°–38°C, and rhinoviruses (one cause of the common cold) multiply successfully only in tissues that are slightly below normal body temperature (33°–35°C). Other organisms are not so limited. Strains of *Staphylococcus aureus* grow within the range of 6°–46°C, and the intestinal bacterium *Enterococcus faecalis* grows within the range of 0°–44°C.

Another way to express temperature adaptation is to describe whether an organism grows optimally in a cold, moderate, or hot temperature range. The terms used for these ecological groups are psychrophile, mesophile, and thermophile **(figure 7.9),** respectively.

A **psychrophile** (sy′-kroh-fyl) is a microorganism that has an optimum temperature below 15°C and is capable of growth at 0°C. It is obligate with respect to cold and generally cannot grow above 20°C. Laboratory work with true psychrophiles can be a real challenge. Inoculations have to be done in a cold room because room temperature can be lethal to the organisms. Unlike most laboratory cultures, storage in the refrigerator incubates, rather than inhibits,

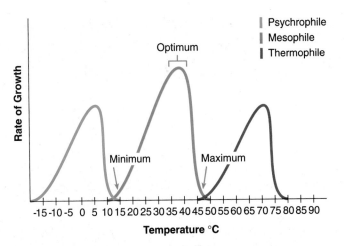

FIGURE 7.9

Ecological groups by temperature of adaptation. Psychrophiles can grow at or near 0°C and have an optimum below 15°C. As a group, mesophiles can grow between 10°C and 50°C, but their optima usually fall between 20°C and 40°C. Generally speaking, thermophiles require temperatures above 45°C and grow optimally between this temperature and 80°C. The cardinal temperatures are labeled for mesophiles. Note that the extremes of the ranges can overlap to an extent.

them. As one might predict, the habitats of psychrophilic bacteria, fungi, and algae are snowfields **(figure 7.10),** polar ice, and the deep ocean. Rarely, if ever, are they pathogenic. True psychrophiles must be distinguished from *psychrotrophs* or *facultative psychrophiles* that grow slowly in cold but have an optimum temperature above 20°C. Bacteria such as *Staphylococcus aureus* and *Listeria monocytogenes* are a concern because they can grow in refrigerated food and cause food-borne illness.

The majority of medically significant microorganisms are **mesophiles** (mez′-oh-fylz), organisms that grow at intermediate temperatures. Although an individual species can grow at the extremes of 10°C or 50°C, the optimum growth temperatures (optima) of most mesophiles fall into the range of 20°–40°C. Organisms in this group inhabit animals and plants as well as soil and water in temperate, subtropical, and tropical regions. Most human pathogens have optima somewhere between 30°C and 40°C (human body temperature is 37°C). *Thermoduric* microbes, which can survive short exposure to high temperatures but are normally mesophiles, are common contaminants of heated or pasteurized foods (see chapter 11). Examples include heat-resistant cysts such as *Giardia* or sporeformers such as *Bacillus* and *Clostridium.*

A **thermophile** (thur′-moh-fyl) is a microbe that grows optimally at temperatures greater than 45°C. Such heat-loving microbes live in soil and water associated with volcanic activity and in habitats directly exposed to the sun. Thermophiles vary in heat requirements, with a general range of growth of 45°–80°C. Most eucaryotic forms cannot survive above 60°C, but a few thermophilic bacteria called hyperthermophiles, grow between 80°C and 110°C (currently thought to be the temperature limit endured by enzymes and cell structures). Strict thermophiles are so heat-tolerant that researchers may use an autoclave to isolate them in culture. Currently, there is intense interest in thermal microorganisms by biotechnology companies **(Spotlight on Microbiology 7.4).**

SPOTLIGHT ON MICROBIOLOGY 7.4
Cashing in on "Hot" Microbes

The smoldering thermal springs in Yellowstone National Park are more than just one of the geologic wonders of the world. They are also a hotbed of some of the most unusual microorganisms in the world. The thermophiles thriving at temperatures near the boiling point are the focus of serious interest from the scientific community. For many years, biologists have been intrigued that any living thing could function at such high temperatures. Such questions as these come to mind: Why don't they melt and disintegrate, why don't their proteins coagulate, and how can their DNA possibly remain intact?

One of the earliest thermophiles to be isolated was *Thermus aquaticus*. It was discovered by Thomas Brock in Yellowstone's Mushroom Pool in 1965 and was registered with the American Type Culture Collection. Interested researchers studied this species and discovered that it has extremely heat-stable proteins and nucleic acids, and its cell membrane does not break down readily at high temperatures. Later, an extremely heat-stable DNA-replicating enzyme was isolated from the species.

What followed is a riveting example of how pure research for the sake of understanding and discovery also offered up a key ingredient in a multimillion-dollar process. Developers of the polymerase chain reaction (PCR), a versatile tool for making multiple copies of DNA fragments, found that the technique would work only if they performed the test at temperatures around 65°–72°C. The mesophilic enzymes they tested were destroyed at such high temperatures, but the Thermus DNA-copying enzyme (Taq polymerase) functioned splendidly. The PCR technique became the basis for a variety of test procedures in forensics and gene detection and analysis.

Spurred by this remarkable success story, biotechnology companies have descended on Yellowstone, which contains over 10,000 hot springs, geysers, and hot habitats. These industries are looking to unusual bacteria and archaea as a source of "extremozymes," enzymes that operate under high

Biotechnology researchers harvesting samples in Yellowstone National Park.

temperatures and acidity. So far, about a dozen other organisms with useful enzymes have been discovered. Some provide applications in the dairy, brewing, and baking industry for high temperature processing and fermentations. Others are being considered for waste treatment and bioremediation.

This quest has also brought attention to questions such as: Who owns these microbes, and can their enzymes be patented? As of the year 2000, the Park Service has gotten a legal ruling that allows them to share in the profits from companies and to add that money to their operating budget. The U.S. Supreme Court has also ruled that a microbe isolated from natural habitats cannot be patented. Only the technology that uses the microbe can be patented.

(a)

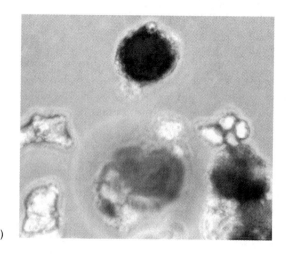

(b)

FIGURE 7.10

Red snow. (a) An early summer snowbank provides a perfect habitat for psychrophilic photosynthetic organisms like *Chlamydomonas nivalis*. **(b)** Microscopic view of this snow alga (actually classified as a "green" alga although a red pigment dominates at this stage of its life cycle).

GAS REQUIREMENTS

The atmospheric gases that most influence microbial growth are O_2 and CO_2. Of these, oxygen gas has the greatest impact on microbial adaptation. Not only is it an important respiratory gas, but it is also a powerful oxidizing agent that exists in many toxic forms. In general, microbes fall into one of three categories: those that use oxygen and can detoxify it; those that can neither use oxygen nor detoxify it; and those that do not use oxygen but can detoxify it.

How Microbes Process Oxygen

As oxygen enters into cellular reactions, it is transformed into several toxic products. Singlet oxygen ($^1O_2$) is an extremely reactive molecule produced by both living and nonliving processes. Notably, it is one of the substances produced by phagocytes to kill invading bacteria (see chapter 14). The buildup of singlet oxygen and the oxidation of membrane lipids and other molecules can damage and destroy a cell. The highly reactive superoxide ion (O_2^-), peroxides (H_2O_2), and hydroxyls (OH^-) are other destructive metabolic by-products of oxygen. To protect themselves against damage, most cells have developed enzymes that go about the business of scavenging and neutralizing these chemicals. The complete conversion of superoxide ion into harmless oxygen requires a two-step process and at least two enzymes:

Step 1. $O_2^- + O_2^- + 2H^+ \xrightarrow{\text{Superoxide dismutase}} H_2O_2 \text{ (hydrogen peroxide)} + O_2$

Step 2. $H_2O_2 + H_2O_2 \xrightarrow{\text{Catalase}} 2H_2O + O_2$

In this series of reactions (essential for aerobic organisms), the superoxide ion is first converted to hydrogen peroxide and normal oxygen by the action of an enzyme called superoxide dismutase. Because hydrogen peroxide is also toxic to cells (it is a disinfectant and antiseptic), it will be degraded by the enzyme catalase into water and oxygen. If a microbe is not capable of dealing with toxic oxygen by these or similar mechanisms, it is forced to live in habitats free of oxygen.

With respect to oxygen requirements, several general categories are recognized. An **aerobe*** (aerobic organism) can use gaseous oxygen in its metabolism and possesses the enzymes needed to process toxic oxygen products. An organism that cannot grow without oxygen is an **obligate aerobe.** Most fungi and protozoa, as well as many bacteria (genera *Micrococcus* and *Bacillus*), have strict requirements for oxygen in their metabolism.

A **facultative anaerobe** is an aerobe that does not require oxygen for its metabolism and is capable of growth in the absence of oxygen. This type of organism metabolizes by aerobic respiration when oxygen is present, but in its absence, it adopts an anaerobic mode of metabolism such as fermentation. Facultative anaerobes usually possess catalase and superoxide dismutase. A large number of bacterial pathogens fall into this group (for example, gram negative *enteric** bacteria and staphylococci). A **microaerophile** (myk″-roh-air′-oh-fyl) does not grow at normal atmospheric tensions of oxygen but requires a small amount of it in metabolism.

Most organisms in this category live in a habitat (soil, water, or the human body) that provides small amounts of oxygen but is not directly exposed to the atmosphere.

An **anaerobe** (anaerobic microorganism) lacks the metabolic enzyme systems for using oxygen in respiration. Because **strict,** or **obligate, anaerobes** also lack the enzymes for processing toxic oxygen, they cannot tolerate any free oxygen in the immediate environment and will die if exposed to it. Strict anaerobes live in highly reduced habitats, such as deep muds, lakes, oceans, and soil. Even though human cells use oxygen and oxygen is found in the blood and tissues, some body sites present anaerobic pockets or microhabitats where colonization or infection can occur. One region that is an important site for anaerobic infections is the oral cavity. Dental caries are partly due to the complex actions of aerobic and anaerobic bacteria, and most gingival infections consist of similar mixtures of oral bacteria that have invaded damaged gum tissues (see chapter 21). Another common site for anaerobic infections is the large intestine, a relatively oxygen-free habitat that harbors a rich assortment of strictly anaerobic bacteria. Anaerobic infections can accompany abdominal surgery and traumatic injuries (gas gangrene and tetanus). Growing anaerobic bacteria usually requires special media, methods of incubation, and handling chambers that exclude oxygen (**figure 7.11***a*).

Aerotolerant anaerobes do not utilize oxygen but can survive and grow to a limited extent in its presence. These anaerobes are not harmed by oxygen, mainly because they possess alternate mechanisms for breaking down peroxides and superoxide. Certain lactobacilli and streptococci use manganese ions or peroxidases to perform this task.

Determining the oxygen requirements of a microbe from a biochemical standpoint can be a very time-consuming process. Often it is illuminating to perform culture tests with reducing media (those that contain an oxygen-absorbing chemical). One such technique demonstrates oxygen requirements by the location of growth in a tube of fluid thioglycollate (**figure 7.12**).

Although all microbes require some carbon dioxide in their metabolism, *capnophiles** grow best at a higher CO_2 tension than is normally present in the atmosphere. This becomes important in the initial isolation of some pathogens from clinical specimens, notably *Neisseria* (gonorrhea, meningitis), *Brucella* (undulant fever), and *Streptococcus pneumoniae*. Incubation is carried out in a CO_2 incubator that provides 3% to 10% CO_2 (see figure 7.11*b*).

EFFECTS OF pH

Microbial growth and survival are also influenced by the pH of the habitat. The pH was defined in chapter 2 as the degree of acidity or alkalinity (basicity) of a solution. It is expressed by the pH scale, a series of numbers ranging from 0 to 14. The pH of pure water (7.0) is neutral, neither acidic nor basic. As the pH value decreases toward 0, the acidity increases, and as the pH increases toward 14, the alkalinity increases. The majority of organisms live or grow in habitats between pH 6 and 8 because strong acids and bases can be highly damaging to enzymes and other cellular substances.

***aerobe** (air′-ohb) Although the prefix means air, it is used in the sense of oxygen.

***enteric** (en-terr′-ik) Gr. *enteron*, intestine. A family of bacteria that live in the large intestines of animals.

***capnophile** (kap′-noh-fyl) Gr. *kapnos*, smoke.

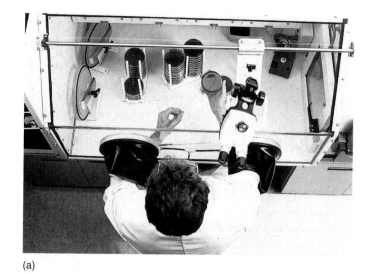

(a)

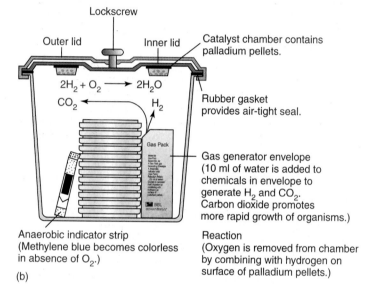

(b)

FIGURE 7.11

Culturing techniques for anaerobes. (a) A special anaerobic environmental chamber makes it possible to handle strict anaerobes without exposing them to air. It also has provisions for incubation and inspection in a completely O_2-free system. **(b)** The anaerobic jar, or CO_2 incubator system. To create an anaerobic environment, a packet is activated to produce hydrogen gas, and the chamber is sealed tightly. The gas reacts with available oxygen to produce water. Carbon dioxide can also be added to the system for growth of capnophiles.

A few microorganisms live at pH extremes. Obligate *acidophiles* include *Euglena mutabilis,* an alga that grows in acid pools between 0 and 1.0 pH, and *Thermoplasma,* an archaea that lacks a cell wall, lives in hot coal piles at a pH of 1 to 2, and will lyse if exposed to pH 7. Because many molds and yeasts tolerate moderate acid, they are the most common spoilage agents of pickled foods. Alkalinophiles live in hot pools and soils that contain high levels of basic minerals (up to pH 10.0). Bacteria that decompose urine create alkaline conditions, since ammonium (NH_4^+, an alkaline ion) can be produced when urea (a component of urine) is digested. Metabolism of urea is one way that Proteus spp. can neutralize the acidity of the urine to colonize and infect the urinary system.

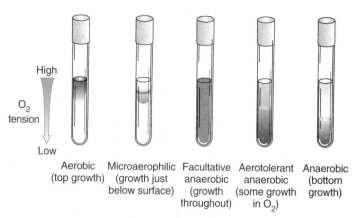

Demonstration of Oxygen Requirements

FIGURE 7.12

Use of thioglycollate broth to demonstrate oxygen requirements. Thioglycollate is a chemical that absorbs O_2 gas from the air. Oxygen at the top of the tube is dissolved in the medium and absorbed; its presence is indicated by the red dye resazurin. When a series of tubes is inoculated with bacteria that differ in O_2 requirements, the relative position of growth provides some indication of their adaptations to oxygen use.

OSMOTIC PRESSURE

Although most microbes exist under hypotonic or isotonic conditions, a few, called **halophiles** (hay′-loh-fylz), live in habitats with a high solute concentration. *Obligate halophiles* such as *Halobacterium* and *Halococcus* inhabit salt lakes, ponds, and other hypersaline habitats. They grow optimally in solutions of 25% NaCl but require at least 9% NaCl (combined with other salts) for growth. These archaea have significant modifications in their cell walls and membranes and will lyse in hypotonic habitats. *Facultative halophiles* are remarkably resistant to salt, even though they do not normally reside in high-salt environments. For example, *Staphylococcus aureus* can grow on NaCl media ranging from 0.1% up to 20%. Although it is common to use high concentrations of salt and sugar to preserve food (jellies, syrups, and brines), many bacteria and fungi actually thrive under these conditions and are common spoilage agents. The term to describe microbes that withstand and grow at high osmotic pressures is **osmophile.**

MISCELLANEOUS ENVIRONMENTAL FACTORS

Various forms of electromagnetic radiation (ultraviolet, infrared, visible light) stream constantly onto the earth from the sun. Some microbes (phototrophs) can use visible light rays as an energy source, but non-photosynthetic microbes tend to be damaged by the toxic oxygen products produced by contact with light. Some microbial species produce yellow carotenoid pigments to protect against the damaging effects of light by absorbing and dismantling toxic oxygen. Other types of radiation that can damage microbes are ultraviolet and ionizing rays (X rays and cosmic rays). In chapter 11, we will see just how these types of energy are applied in microbial control.

Descent into the ocean depths subjects organisms to increasing hydrostatic pressure. Deep-sea microbes called **barophiles** exist under pressures that range from a few times to over 1,000 times the pressure of the atmosphere. These bacteria are so strictly adapted to high pressures that they will rupture when exposed to normal atmospheric pressure.

Because of the high water content of cytoplasm, all cells require water from their environment to sustain growth and metabolism. Water is the solvent for cell chemicals, and it is needed for enzyme function and digestion of macromolecules. A certain amount of water on the external surface of the cell is required for the diffusion of nutrients and wastes. Even in apparently dry habitats, such as sand or dry soil, the particles retain a thin layer of water usable by microorganisms. Dormant, dehydrated cell stages (for example, spores and cysts) tolerate extreme drying because of the inactivity of their enzymes.

ECOLOGICAL ASSOCIATIONS AMONG MICROORGANISMS

Up to now, we have considered the importance of nonliving environmental influences on the growth of microorganisms. Another profound influence comes from other organisms that share (or sometimes are) their habitats. In all but the rarest instances, microbes live in shared habitats, which give rise to complex and fascinating associations. Some associations are between similar or dissimilar types of microbes; others involve multicellular organisms such as animals or plants. Interactions can have beneficial, harmful, or no particular effects on the organisms involved; they can be obligatory or nonobligatory to the members; and they often involve nutritional interactions. This outline provides an overview of the major types of microbial associations:

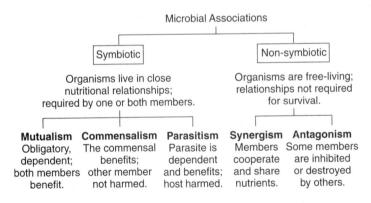

A general term used to denote a situation in which two organisms live together in a close partnership is **symbiosis,** * and the members are termed *symbionts*. Three main types of symbiosis occur. **Mutualism** exists when organisms live in an obligatory but mutually beneficial relationship. This association is rather common in nature because of the survival value it has for the members involved. **Spotlight on Microbiology 7.5** gives several examples to illustrate this concept. In the other symbiotic relationships the relationship tends to be unequal, meaning it benefits one member and not the other, and it can be obligatory.

In a relationship known as **commensalism,** * the member called the **commensal** receives benefits, while its coinhabitant is neither harmed nor benefited. A classic commensal interaction between microorganisms called **satellitism** arises when one member provides

*symbiosis (sim″-bye-oh′-sis) Gr. *syn*, together, and *bios*, to live.

*commensalism (kuh-men′-sul-izm) L. *com*, together, and *mensa*, table.

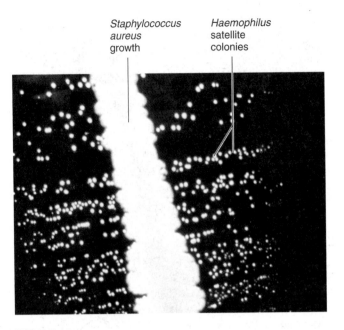

FIGURE 7.13

Satellitism, a type of commensalism between two microbes. In this example *Staphylococcus aureus* provides growth factors to *Haemophilus influenzae*, which grows as tiny satellite colonies near the streak of *Staphylococcus*. By itself, *Haemophilus* could not grow on blood agar. The *Staphylococcus* gives off several nutrients such as vitamins and amino acids that diffuse out to the *Haemophilus*, thereby promoting its growth.

nutritional or protective factors to the other. One example of nutritional satellitism is observed when one microbe provides a growth factor that another one needs **(figure 7.13).** Some microbes can break down a substance that would be toxic or inhibitory to another microbe. Relationships between humans and resident commensals that derive nutrients from the body are discussed in a later section.

In an earlier section, we introduced the concept of **parasitism** as a relationship in which the host organism provides the parasitic microbe with nutrients and a habitat. Multiplication of the parasite usually harms the host to some extent. As this relationship evolves, the host can even develop tolerance for or dependence on a parasite, at which point we call the relationship commensalism or mutualism.

Synergism * is an interrelationship between two or more free-living organisms that benefits them but is not necessary for their survival. Together, the participants cooperate to produce a result that none of them could do alone. This form of shared metabolism can be viewed by the reaction:

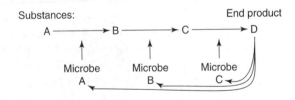

Each microbe performs a specific action on a chemical in the series. The end product is useful to all three microbes.

*synergism (sin′-ur-jizm) Gr. *syn*, together; *erg*, work; and *ism*, process.

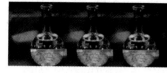

SPOTLIGHT ON MICROBIOLOGY 7.5
Life Together: Mutualism

A tremendous variety of mutualistic partnerships occur in nature. These associations gradually evolve over millions of years as the participating members come to rely on some critical substance or habitat that they share. One of the earliest such associations is thought to have resulted in eucaryotic cells (see Spotlight on Microbiology 5.1).

Protozoan cells often receive growth factors from symbiotic bacteria and algae that, in turn, are nurtured by the protozoan cell. One peculiar ciliate propels itself by affixing symbiotic bacteria to its cell membrane to act as "oars." These relationships become so obligatory that some amebas and ciliates require mutualistic bacteria for survival. This kind of relationship is especially striking in the complex mutualism of termites, which harbor protozoans specialized to live only inside them. The protozoans, in turn, contain endosymbiotic bacteria. Wood eaten by the termite gets processed by the protozoan and bacterial enzymes, and all three organisms thrive.

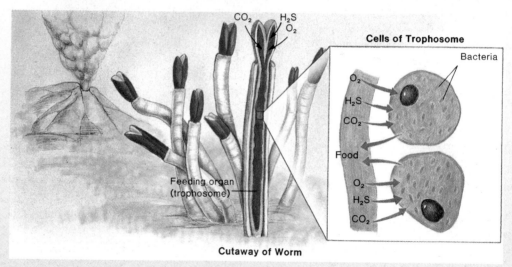

A view of a vent community based on mutualism and chemoautotrophy. The giant tube worm *Riftia* houses bacteria in its specialized feeding organ, the trophosome. Raw materials in the form of dissolved inorganic molecules are provided to the bacteria through the worm's circulation. With these, the bacteria produce usable organic food that is absorbed by the worm.

An example of synergism is observed in the exchange between soil bacteria and plant roots (see chapter 26). The plant provides various growth factors, and the bacteria help fertilize the plant by supplying it with minerals. In synergistic infections, a combination of organisms can produce tissue damage that a single organism would not cause alone. Gum disease, dental caries, and gas gangrene involve mixed infections by bacteria interacting synergistically.

Antagonism* is an association between free-living species that arises when members of a community compete. In this interaction, one microbe secretes chemical substances into the surrounding environment that inhibit or destroy another microbe in the same habitat. The first microbe may gain a competitive advantage by increasing the space and nutrients available to it. Interactions of this type are common in the soil, where mixed communities often compete for space and food. *Antibiosis**—the production of inhibitory compounds called antibiotics—is actually a form of antagonism. Hundreds of naturally occurring antibiotics have been isolated from bacteria and fungi and used as drugs to control diseases (see chapter 12). *Bacteriocins** are another class of antimicrobial proteins that are toxic to bacteria other than the ones that produced them.

INTERRELATIONSHIPS BETWEEN MICROBES AND HUMANS

The human body is a rich habitat for symbiotic bacteria, fungi, and a few protozoa. Microbes that normally live on the skin, in the alimentary tract, and in other sites are called the *normal microbial flora* (see chapter 13). These residents participate in commensal, parasitic, and synergistic relationships with their human hosts. For example, *Escherichia coli* living symbiotically in the intestine produce vitamin K, and species of symbiotic *Lactobacillus* residing in the vagina help maintain an acidic environment that protects against infection by other microorganisms. Hundreds of commensal species "make a living" on the body without either harming or benefiting it. For example, many bacteria and yeasts reside in the outer dead regions of the skin; oral microbes feed on the constant flow of nutrients in the mouth; and billions of bacteria live on the wastes in the large intestine. Because the normal flora and the body are in a constant state of change, these relationships are not absolute, and a commensal can convert to a parasite by invading body tissues and causing disease.

CHAPTER CHECKPOINTS

The environmental factors that control microbial growth are temperature, pH, moisture, radiation, gases, and other microorganisms.

***antagonism** (an-tag'-oh-nizm) Gr. *antagonistes,* an opponent.

***antibiosis** (an"-tee-by-oh'-sis) Gr. *anti,* against, and *bios,* life.

***bacteriocin** (bak-teer'-ee-oh-sin) Gr. *bakterion,* little rod, and *ios,* poison.

Symbiosis Between Microbes and Animals

Microorganisms carry on symbiotic relationships with animals as diverse as sponges, worms, and mammals. Bacteria and protozoa are essential in the operation of the rumen (a complex, four-chambered stomach) of cud-chewing mammals. These mammals produce no enzymes of their own to break down the cellulose that is a major part of their diet, but the microbial population harbored in their rumens does. The complex food materials are digested through several stages, during which time the animal regurgitates and chews the partially digested plant matter (the cud) and occasionally burps methane produced by the microbial symbionts.

Thermal Vent Symbionts

Another fascinating symbiotic relationship has been found in the deep hydrothermal vents in the seafloor, where geologic forces spread the crustal plates and release heat and gas. These vents are a focus of tremendous biological and geologic activity. Discoveries first made in the late 1970s demonstrated that the source of energy in this community is not the sun, because the vents are too deep for light to penetrate (2,600 m). Instead, this ecosystem is based on a massive chemoautotrophic bacterial population that oxidizes the abundant hydrogen sulfide (H_2S) gas given off by the volcanic activity there. As the bottom of the food web, these bacteria serve as the primary producers of nutrients that service a broad spectrum of specialized animals.

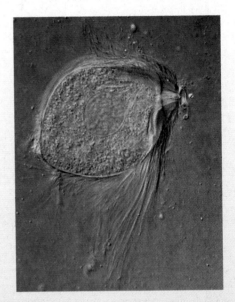

Termites are thought to be responsible for wood damage; however, it is the endosymbiotic relationship of this protozoan that inhabits the gut of the termite that performs the digestion of wood.

Environmental factors control microbial growth by their influence on microbial enzymes.

Three cardinal temperatures for a microorganism describe its temperature range and the temperature at which it grows best. These are the minimum temperature, the maximum temperature, and the optimum temperature.

Microorganisms are classified by their temperature requirements as psychrophiles, mesophiles, or thermophiles.

Most eucaryotic microorganisms are aerobic, but bacteria vary widely in their oxygen requirements from facultative to anaerobic.

Microorganisms live in associations with other species that range from mutually beneficial symbiosis to parasitism and antagonism.

The Study of Microbial Growth

When microbes are provided with nutrients and the required environmental factors, they become metabolically active and grow. Growth takes place on two levels. On one level, a cell synthesizes new cell components and increases its size; on the other level, the number of cells in the population increases. This capacity for multiplication, increasing the size of the population by cell division, has tremendous importance in microbial control, infectious disease, and biotechnology. In the following sections, we will focus primarily on the characteristics of bacterial growth that are generally representative of single-celled microorganisms.

THE BASIS OF POPULATION GROWTH: BINARY FISSION

The division of a bacterial cell occurs mainly through **binary, or transverse, fission;** *binary* means that one cell becomes two, and *transverse* refers to the division plane forming across the width of the cell. During binary fission, the parent cell enlarges, duplicates its chromosome, and forms a central transverse septum that divides the cell into two daughter cells. This process is repeated at intervals by each new daughter cell in turn, and with each successive round of division, the population increases. The stages in this continuous process are shown in greater detail in **figures 7.14** and 7.15.

THE RATE OF POPULATION GROWTH

The time required for a complete fission cycle—from parent cell to two new daughter cells—is called the **generation, or doubling, time.** The term *generation* has a similar meaning as it does in humans. It is the period between an individual's birth and the time of producing offspring. In bacteria, each new fission cycle or generation

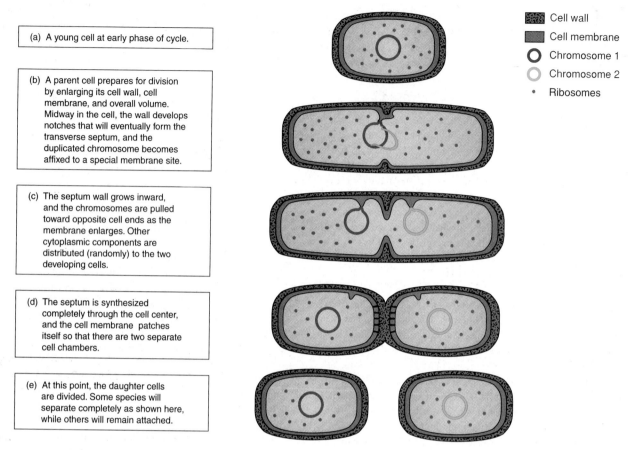

(a) A young cell at early phase of cycle.

(b) A parent cell prepares for division by enlarging its cell wall, cell membrane, and overall volume. Midway in the cell, the wall develops notches that will eventually form the transverse septum, and the duplicated chromosome becomes affixed to a special membrane site.

(c) The septum wall grows inward, and the chromosomes are pulled toward opposite cell ends as the membrane enlarges. Other cytoplasmic components are distributed (randomly) to the two developing cells.

(d) The septum is synthesized completely through the cell center, and the cell membrane patches itself so that there are two separate cell chambers.

(e) At this point, the daughter cells are divided. Some species will separate completely as shown here, while others will remain attached.

Cell wall
Cell membrane
Chromosome 1
Chromosome 2
Ribosomes

FIGURE 7.14

Steps in binary fission of a rod-shaped bacterium.

increases the population by a factor of 2, or doubles it. Thus, the initial parent stage consists of 1 cell, the first generation consists of 2 cells, the second 4, the third 8, then 16, 32, 64, and so on (**figure 7.15**). As long as the environment remains favorable, this doubling effect can continue at a constant rate. With the passing of each generation, the population will double, over and over again.

The length of the generation time is a measure of the growth rate of an organism. Compared with the growth rates of most other living things, bacteria are notoriously rapid. The average generation time is 30 to 60 minutes under optimum conditions. The shortest generation times average 5 to 10 minutes, and the longest generation times require days. For example, *Mycobacterium leprae,* the cause of leprosy, has a generation time of 10 to 30 days—as long as in some animals. Most pathogens have relatively short doubling times. *Salmonella enteritidis* and *Staphylococcus aureus,* bacteria that cause food-borne illness, double in 20 to 30 minutes, which is why leaving food at room temperature even for a short period has caused many a person to be suddenly stricken with an attack of food-borne disease. In a few hours, a population of these bacteria can easily grow from a small number of cells to several million.

Figure 7.15 shows several quantitative characteristics of growth: (A) The cell population size can be represented by the number 2 with an exponent (2^1, 2^2, 2^3, 2^4); (B) the exponent increases by one in each generation; and (C) the number of the exponent is also the number of the generation. This growth pattern is

termed **exponential.** Because these populations often contain very large numbers of cells, it is useful to express them by means of exponents or logarithms (see appendix A). The data from a growing bacterial population are graphed by plotting the number of cells as a function of time. The cell number can be represented logarithmically or arithmetically. Plotting the logarithm number over time provides a straight line indicative of exponential growth. Plotting the data arithmetically gives a constantly curved slope. In general, logarithmic graphs are preferred because an accurate cell number is easier to read, especially during early growth phases.

Predicting the number of cells that will arise during a long growth period (yielding millions of cells) is based on a relatively simple concept. One could use the method of addition $2 + 2 = 4$; $4 + 4 = 8$; $8 + 8 = 16$; $16 + 16 = 32$, and so on, or a method of multiplication (for example, $2^5 = 2 \times 2 \times 2 \times 2 \times 2$), but it is easy to see that for 20 or 30 generations, this calculation could be very tedious. An easier way to calculate the size of a population over time is to use an equation such as:

$$N_f = (N_i)2^n$$

In this equation, N_f is the total number of cells in the population at some point in the growth phase, N_i is the starting number, the exponent n denotes the generation number, and 2^n represents the number of cells in that generation. If we know any two of the values,

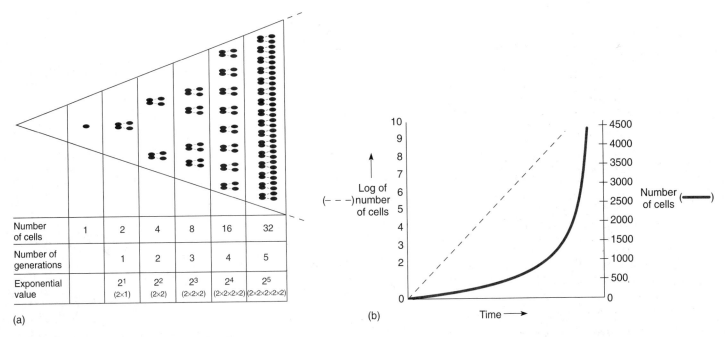

FIGURE 7.15

The mathematics of population growth. (a) Starting with a single cell, if each product of reproduction goes on to divide in a binary fashion, the population doubles with each new division cycle or generation. This process can be represented by logarithms (2 raised to an exponent) or simple numbers. **(b)** Plotting the logarithm of the cells produces a straight line indicative of exponential growth, whereas plotting the cell numbers arithmetically gives a curved slope.

other values can be calculated. Let us use the example of *Staphylococcus aureus* to calculate how many cells (N_f) will be present in an egg salad sandwich after it sits in a warm car for 4 hours. We will assume that N_i is 10 (number of cells deposited in the sandwich while it was being prepared). To derive *n*, we need to divide 4 hours (240 minutes) by the generation time (we will use 20 minutes). This calculation comes out to 12, so 2^n is equal to 2^{12}. Referring to the logarithmic tables for the base 2 in Appendix A, page A-2, we find that 2^{12} is 4,096.

$$\text{Final number } (N_f) = 10 \times 4{,}096$$
$$= 40{,}960 \text{ cells in the sandwich}$$

This same equation, with modifications, is used to determine the generation time, a more complex calculation that requires knowing the number of cells at the beginning and end of a growth period. Such data are obtained through actual testing by a method discussed in the following section.

THE POPULATION GROWTH CURVE

In reality, a population of bacteria does not maintain its potential growth rate and does not double endlessly, because in most systems numerous factors prevent the cells from continuously dividing at their maximum rate. Quantitative laboratory studies indicate that a population typically displays a predictable pattern, or **growth curve,** over time. The method traditionally used to observe the population growth pattern is a viable count technique, in which the total number of live cells is counted over a given time period. In brief, this method entails

1. placing a tiny number of cells into a sterile liquid medium;
2. incubating this culture over a period of several hours;

3. sampling the broth at regular intervals during incubation;
4. plating each sample onto solid media; and
5. counting the number of colonies present after incubation.

Microbits 7.6 gives the details of this process.

STAGES IN THE NORMAL GROWTH CURVE

The system of batch culturing described in Microbits 7.6 is *closed,* meaning that nutrients and space are finite and there is no mechanism for the removal of waste products. Data from an entire growth period of 3 to 4 days typically produce a curve with a series of phases termed the lag phase, the exponential growth (log) phase, the stationary phase, and the death phase (**figure 7.16**).

The **lag phase** is a relatively "flat" period on the graph when the population appears not to be growing or is growing at less than the exponential rate. Growth lags primarily because:

1. The newly inoculated cells require a period of adjustment, enlargement, and synthesis;
2. the cells are not yet multiplying at their maximum rate; and
3. the population of cells is so sparse or dilute that the sampling misses them.

The length of the lag period varies somewhat from one population to another.

The cells reach the maximum rate of cell division during the **exponential growth (log) phase,** a period during which the curve increases geometrically. This phase will continue as long as cells have adequate nutrients and the environment is favorable.

At the **stationary growth phase,** the population enters a survival mode in which cells stop growing or grow slowly. The curve

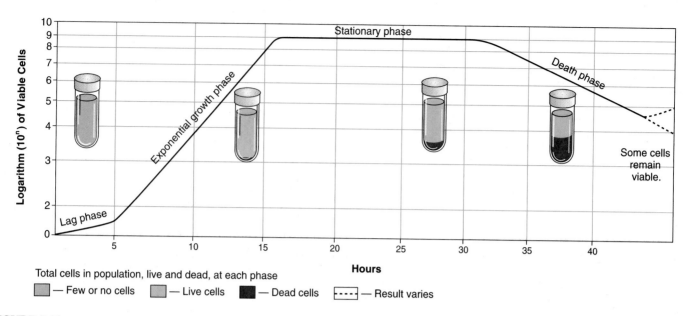

FIGURE 7.16

The growth curve in a bacterial culture. On this graph, the number of viable cells expressed as a logarithm (log) is plotted against time. See text for discussion of the various phases. Note that with a generation time of 30 minutes, the population has risen from 10 (10^1) cells to 1,000,000,000 (10^9) cells in only 16 hours.

levels off because the rate of cell inhibition or death balances out the rate of multiplication. The decline in the growth rate is caused by depleted nutrients and oxygen, excretion of organic acids and other biochemical pollutants into the growth medium, due to the increased density of cells.

As the limiting factors intensify, cells begin to die in exponential numbers (literally perishing in their own wastes), and they are unable to multiply. The curve now dips downward as the **death phase** begins. The speed with which death occurs depends on the relative resistance of the species and how toxic the conditions are, but it is usually slower than the exponential growth phase. Viable cells often remain many weeks and months after this phase has begun. In the laboratory, refrigeration is used to slow the progression of the death phase so that cultures will remain viable as long as possible.

Practical Importance of the Growth Curve

The tendency for populations to exhibit phases of rapid growth, slow growth, and death has important implications in microbial control, infection, food microbiology, and cultural technology. Antimicrobial agents such as heat and disinfectants rapidly accelerate the death phase in all populations, but microbes in the exponential growth phase are more vulnerable to these agents than are those that have entered the stationary phase. In general, actively growing cells are more vulnerable to conditions that disrupt cell metabolism and binary fission.

Growth patterns in microorganisms can account for the stages of infection (see chapter 13). Microbial cells produced during the exponential phase are far more numerous and virulent than those released at a later stage of infection. A person shedding bacteria in the early and middle stages of an infection is more likely to spread it to others than is a person in the late stages. The course of

an infection is also influenced by the relatively faster rate of multiplication of the microbe, which can overwhelm the slower growth rate of the host's own cellular defenses.

Understanding the stages of cell growth is crucial for work with cultures. Sometimes a culture that has reached the stationary phase is incubated under the mistaken impression that enough nutrients are present for the culture to multiply. In most cases, it is unwise to continue incubating a culture beyond the stationary phase, because doing so will reduce the number of viable cells and the culture could die out completely. It is also preferable to do stains (an exception is the spore stain) and motility tests on young cultures, because the cells will show their natural size and correct reaction, and motile cells will have functioning flagella.

For certain research or industrial applications, closed batch culturing with its four phases is inefficient. The alternative is an automatic growth chamber called the **chemostat,** or continuous culture system. This device can admit a steady stream of new nutrients and siphon off used media and old bacterial cells, thereby stabilizing the growth rate and cell number. The chemostat is very similar to the industrial fermenters used to produce vitamins and antibiotics (see chapter 26). It has the advantage of maintaining the culture in a biochemically active state and preventing it from entering the death phase.

OTHER METHODS OF ANALYZING POPULATION GROWTH

Microbiologists have developed several alternative ways of analyzing bacterial growth qualitatively and quantitatively. One of the simplest methods for estimating the size of a population is through turbidometry. This technique relies on the simple observation that a tube of clear nutrient solution loses its clarity and becomes cloudy,

A growing population is established by inoculating a flask containing a known quantity of sterile liquid medium with a few cells of a pure culture. The flask is incubated at that bacteria's optimum temperature and timed. The population size at any point in the growth cycle is quantified by removing a tiny measured sample of the culture from the growth chamber and plating it out on a solid medium to develop isolated colonies. This procedure is repeated at evenly spaced intervals (every hour for 24 hours).

Evaluating the samples involves a common and important principle in microbiology: One colony on the plate represents one cell or colony-forming unit (CFU) from the original sample. Because the CFU of some bacteria is actually composed of several cells (consider the clustered arrangement of *Staphylococcus,* for instance), using a colony count

can underestimate the exact population size to an extent. This is not a serious problem because, in such bacteria, the CFU is the smallest unit of colony formation and dispersal. Multiplication of the number of colonies in a single sample by the container's volume gives a fair estimate of the total population size (number of cells) at any given point. The growth curve is determined by graphing the number for each sample in sequence for the whole incubation period (see figure 7.16).

Because of the scarcity of cells in the early stages of growth, some samples can give a zero reading even if there are viable cells in the culture. The sampling itself can remove enough viable cells to alter the tabulations, but since the purpose is to compare relative trends in growth, these factors do not significantly change the overall pattern.

	60 min	120 min	180 min	240 min	300 min	360 min	420 min	480 min	540 min	600 min
Number of colonies (CFU) per 0.1 ml	0*	1	3	7	13	23	45	80	135	230
Total cell population in flask	0*	5,000	15,000	35,000	65,000	115,000	225,000	400,000	675,000	1,150,000

Flask inoculated

Samples taken at equally spaced intervals (0.1 ml)

500 ml

0.1 ml

Sample is diluted in liquid agar medium and poured or spread over surface of solidified medium

Plates are incubated, colonies are counted

* Zero CFUs only means that too few cells are present to be assayed.

The viable plate count, a technique for determining population size and rate of growth.

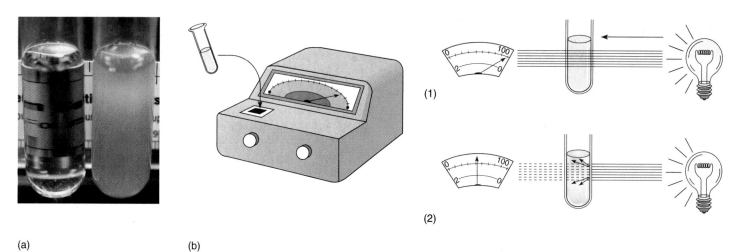

FIGURE 7.17

Turbidity measurements as indicators of growth. (a) Holding a broth to the light is one method of checking for gross differences in cloudiness (turbidity). The broth on the left is transparent, indicating little or no growth; the broth on the right is cloudy and opaque, indicating heavy growth. **(b)** The eye is not sensitive enough to pick up fine degrees in turbidity; more sensitive measurements can be made with a spectrophotometer. (*1*) A tube with no growth will allow light to easily pass. Therefore more light will reach the photodetector and give a higher transmittance value. (*2*) In a tube with growth, the cells scatter the light resulting in less light reaching the photodetector and, therefore, gives a lower transmittance valve.

or **turbid,** as microbes grow in it. In general, the greater the turbidity, the larger the population size, which can be measured by means of sensitive instruments (**figure 7.17b**).

Enumeration of Bacteria

Turbidity readings are useful for evaluating relative amounts of growth, but if a more quantitative evaluation is required, the viable colony count described previously or some other enumeration (counting) procedure is necessary. The **direct,** or **total, cell count** involves counting the number of cells in a sample microscopically (**figure 7.18**). This technique, very similar to that used in blood cell counts, employs a special microscope slide (cytometer) calibrated to accept a tiny sample that is spread over a premeasured grid. The cell count from a cytometer can be used to estimate the total number of cells in a larger sample (for instance, of milk or

water). One inherent inaccuracy in this method is that no distinction can be made between dead and live cells, both of which are included in the count.

Counting can be automated by sensitive devices such as the *Coulter counter,* which electronically scans a culture as it passes through a tiny pipette. As each cell flows by, it is detected and registered on an electronic sensor (**figure 7.19a**). A *flow cytometer* works on a similar principle, but in addition to counting, it can measure cell size and even differentiate between live and dead cells. When used in conjunction with fluorescent dyes and antibodies to tag cells, it has been used to differentiate between gram-positive and gram-negative bacteria. It is being adapted for use as a rapid method to identify pathogens in patient specimens and to differentiate blood cells. Aquatic microbiologists use this tool to separate microbes from mixed plankton samples that would be difficult to monitor by traditional culture methods (figure 7.19b).

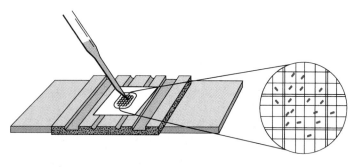

FIGURE 7.18

Direct microscopic count of bacteria. A small sample is placed on the grid under a cover glass. Individual cells, both living and dead, are counted. This number can be used to calculate the total count of a sample.

CHAPTER CHECKPOINTS

Microbial growth refers both to increase in cell size and increase in number of cells in a population.

The generation time is a measure of the growth rate of a microbial species. It varies in length according to environmental conditions.

Microbial cultures in a nutrient-limited environment exhibit four distinct stages of growth: the lag phase, the exponential growth (log) phase, the stationary phase, and the death phase.

Microbial cell populations show distinct phases of growth in response to changing nutrient and waste conditions.

Signs of population growth can be quantified by measuring turbidity, colony counts, and direct cell counts.

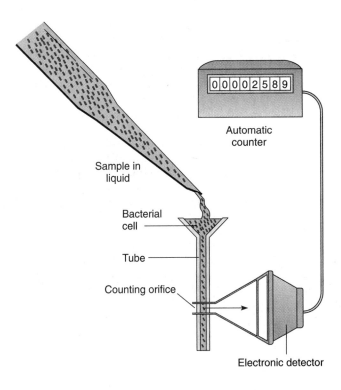

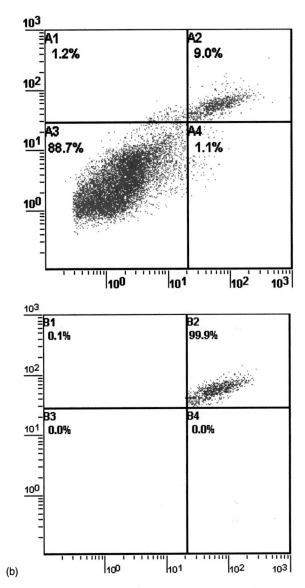

(a) **Coulter counter.** As cells pass through this device, they trigger an electronic sensor that tallies their numbers.

(b) **Flow cytometry.** A variation on the Coulter counter is used to detect, count, and sort the cells in a sample and display a plot of the results. In one version of this procedure, zoospores of the toxic algae *Pfiesteria* (B$_2$) have been separated from a mixed sample (A$_3$). The method is so accurate and precise that the sorted sample is a pure culture of the zoospores.

(b)

FIGURE 7.19

Methods of electronic counting, sorting, and identification.

CHAPTER CAPSULE WITH KEY TERMS

I. Microbial Nutrition, Ecology, and Growth
  A. Source of Nutrients
   Nutrition consists of taking in chemical substances (nutrients) and assimilating and extracting energy from them.
   1. Substances required for survival are **essential nutrients**—usually containing the elements (C, H, N, O, P, S, Na, Cl, K, Ca, Fe, Mg). Essential nutrients are considered **macronutrients** (required in larger amounts) or **micronutrients** (trace elements required in smaller amounts—Zn, Mn, Cu). Nutrients are classed as either **inorganic** or **organic.**
   2. A **growth factor** is an organic nutrient (amino acid and vitamin) that cannot be synthesized and must be provided.

II. Nutritional Categories
  A. An **autotroph** depends on carbon dioxide for its carbon needs. A phototroph that derives energy from light is a **photoautotroph;** one that extracts energy from inorganic substances is a **chemoautotroph. Methanogens** are chemoautotrophs that produce methane.
  B. A **heterotroph** acquires carbon from organic molecules. A **saprobe** is a decomposer that feeds upon dead organic matter, and a **parasite** feeds from a live host and usually causes harm. Disease-causing parasites are pathogens.

III. Environmental Influences on Microbes
   Every organism adapts to a particular habitat and **niche.**
  A. *Temperature:* An organism exhibits **optimum, minimum,** and **maximum temperatures.** Organisms that cannot grow above

20°C but thrive below 15°C and continue to grow even at 0°C are known as **psychrophiles. Mesophiles** grow from 10°C to 50°C, having temperature optima from 20°C to 40°C. The growth range of **thermophiles** is 45°C to 80°C.

B. *Oxygen Requirements:* The ecological need for free oxygen (O_2) is based on whether a cell can handle toxic by-products such as superoxide and peroxide.

1. **Aerobes** grow in normal atmospheric oxygen and have enzymes to handle toxic oxygen by-products. An aerobic organism capable of living without oxygen if necessary is a **facultative anaerobe.** An aerobe that prefers a small amount of oxygen but does not grow under anaerobic conditions is a **microaerophile.**

2. **Strict (obligate) anaerobes** do not use free oxygen and cannot produce enzymes to dismantle reactive oxides. They are actually damaged or killed by oxygen. An **aerotolerant anaerobe** cannot use oxygen for respiration, yet is not injured by it.

C. *Effects of pH:* Acidity and alkalinity affect the activity and integrity of enzymes and the structural components of a cell. Optimum pH for most microbes ranges approximately from 6 to 8. **Acidophiles** prefer lower pH, and **alkalinophiles** prefer higher pH.

D. *Other Environmental Factors:* Radiation and barometric pressure affect microbial growth. A **barophile** is adapted to life under high pressure (bottom dwellers in the ocean, for example).

IV. **Transport Mechanisms**

A. A microbial cell must take on nutrients from its surroundings by transporting them across the cell membrane.

B. **Passive transport** involves the natural movement of substances down a concentration gradient and requires no additional energy **(diffusion).**

C. **Osmosis** is diffusion of water through a selectively permeable membrane. A form of passive transport that can move specific substances is **facilitated diffusion.**

D. Osmotic changes that affect cells are **hypotonic** solutions, which contain a lower solute concentration, and **hypertonic** solutions, which contain a higher solute concentration. **Isotonic** solutions have the same solute concentration as the inside of the cell. A **halophile** thrives in hypertonic surroundings, and an **obligate**

halophile requires a salt concentration of at least 15%, but grows optimally in 25%.

E. In **active transport,** substances are taken into the cell by a process that consumes energy. In **group translocation,** molecules are altered during transport.

F. **Phagocytosis** and **pinocytosis** are forms of active transport in which bulk quantities of solid and fluid material are taken into the cell.

V. **Microbial Interactions**

A. Microbes coexist in varied relationships in nature.

1. Types of **symbiosis** are **mutualism,** a reciprocal, obligatory, and beneficial relationship between two organisms, and **commensalism,** an organism receiving benefits from another without harming the other organism in the relationship. One form of this is *satellitism.* **Parasitism** occurs between a host and an infectious agent.

2. **Synergism** is a mutually beneficial but not obligatory coexistence. **Antagonism** entails competition, inhibition, and injury directed against the opposing organism. A special case of antagonism is antibiotic production.

VI. **Microbial Growth**

A. The splitting of a parent bacterial cell to form a pair of similar-sized daughter cells is known as **binary,** or **transverse, fission.**

B. The duration of each division is called the **generation,** or **doubling, time.** A population theoretically doubles with each generation, so the growth rate is **exponential,** and each cycle increases in **geometric progression.**

C. A **growth curve** is a graphic representation of a closed population over time. Plotting a curve requires an estimate of live cells, called a **viable count.** The initial flat period of the curve is called the **lag phase,** followed by the **exponential growth phase,** in which viable cells increase in logarithmic progression. Adverse environmental conditions combine to inhibit the growth rate, causing a plateau, or **stationary growth phase.** In the **death phase,** nutrient depletion and waste buildup cause increased cell death.

D. Cell numbers can be counted directly by a microscope counting chamber, Coulter counter, or flow cytometer. Cell growth can also be determined by turbidometry and a total cell count.

MULTIPLE-CHOICE QUESTIONS

1. An organic nutrient essential to an organism's metabolism that cannot be synthesized itself is termed a/an
   a. trace element
   b. micronutrient
   c. growth factor
   d. essential nutrient

2. The source of the necessary elements of life is
   a. an inorganic environmental reservoir
   b. the sun
   c. rocks
   d. the air

3. An organism that can synthesize all its required organic components from CO_2 using energy from the sun is a
   a. photoautotroph
   b. photoheterotroph
   c. chemoautotroph
   d. chemoheterotroph

4. An obligate halophile requires high
   a. pH
   b. temperature
   c. salt
   d. pressure

5. Chemoautotrophs can survive on ____ alone.
   a. minerals
   b. CO_2
   c. minerals and CO_2
   d. methane

6. Which of the following statements is true for *all* organisms?
   a. they require organic nutrients
   b. they require inorganic nutrients
   c. they require growth factors
   d. they require oxygen gas

7. A pathogen would most accurately be described as a
   a. parasite
   b. commensal
   c. saprobe
   d. symbiont

8. Which of the following is true of passive transport?
   a. it requires a gradient
   b. it uses the cell wall
   c. it includes endocytosis
   d. it only moves water

9. A cell exposed to a hypertonic environment will ____ by osmosis.
   a. gain water
   b. lose water
   c. neither gain nor lose water
   d. burst

10. Active transport of a substance across a membrane requires
    a. a gradient
    b. the expenditure of ATP
    c. water
    d. diffusion

11. Environmental factors such as temperature and pH exert their effect on the ____ of microbial cells.
    a. membranes
    b. DNA
    c. enzymes
    d. cell wall

12. Psychrophiles would be expected to grow
    a. in hot springs
    b. on the human body
    c. at refrigeration temperatures
    d. at low pH

13. Superoxide ion is toxic to strict anaerobes because they lack
    a. catalase
    b. peroxidase
    c. dismutase
    d. oxidase

14. The time required for a cell to undergo binary fission is called the
    a. exponential growth rate
    b. growth curve
    c. generation time
    d. lag period

15. In a viable plate count, each ____ represents a ____ from the sample population.
    a. cell, colony
    b. colony, cell
    c. hour, generation
    d. cell, generation

16. During the ____ phase, the rate of new cells being added to the population has slowed down.
    a. stationary
    b. death
    c. lag
    d. exponential growth

CONCEPT QUESTIONS

1. Differentiate between micronutrients and macronutrients.
   a. What elements do the letters in the mnemonic device on page 188 stand for?
   b. Briefly describe the general function of CHNOPS in the cell.
   c. Define growth factors, and give examples of them.

2. Name some functions of metallic ions in cells.

3. Compare autotrophs and heterotrophs with respect to the form of carbon-based nutrients they require.

4. Describe the nutritional strategy of two types of chemoautotrophs (lithotrophs) described in the chapter.

5. Briefly fill in the following table:

	Source of Carbon	Usual Source of Energy	Example
Photoautotroph			
Photoheterotroph			
Chemoautotroph			
Chemoheterotroph			
Saprobe			
Parasite			

6. a. Compare and contrast passive and active forms of transport, using examples of what is being transported and the requirements for each.
   b. How are phagocytosis and pinocytosis similar? How are they different?

7. Compare the effects of isotonic, hypotonic, and hypertonic solutions on an ameba and on a bacterial cell. If a cell lives in a hypotonic environment, what will occur if it is placed in a hypertonic one? Answer for the opposite case as well.

8. Look at the following diagrams and predict in which direction osmosis will take place. Use arrows to show the net direction of osmosis. Is one of these microbes a halophile? Which one?

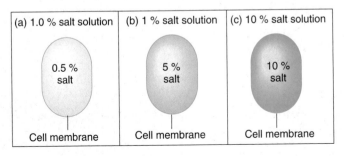

(a) 1.0 % salt solution	(b) 1 % salt solution	(c) 10 % salt solution
0.5 % salt	5 % salt	10 % salt
Cell membrane	Cell membrane	Cell membrane

9. Why are most pathogens mesophilic?

10. What are the ecological roles of psychrophiles and thermophiles?

11. What is the natural habitat of a facultative parasite? Of a strict saprobe? Give examples of both.

12. a. Explain what it means to be an obligate intracellular parasite.
    b. Name three groups of obligate intracellular parasites.

13. a. Classify a human with respect to oxygen requirements.
    b. What might be the habitat of an aerotolerant anaerobe?
    c. Where in the body are anaerobic habitats apt to be found?

14. Where do superoxide ions and hydrogen peroxide originate? What are their toxic effects?

15. a. Define symbiosis and differentiate among mutualism, commensalism, synergism, parasitism, and antagonism, using examples.
    b. How are parasitism and antagonism similar and different?
    c. Are any of these relationships obligatory?

16. Explain the relationship between colony counts and colony-forming units. Why can one use the number of colonies as an index of population size?

17. Why is growth called exponential? What is the size of a population in 20 generations? Explain what is happening to the population at points A, B, C, and D in the following diagram.

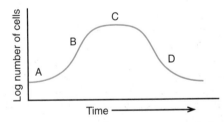

18. Why are we concerned with differentiating live versus dead cells in populations?

19. Contrast the various methods of detecting and counting microbial populations and discuss their advantages and disadvantages.

CRITICAL-THINKING QUESTIONS

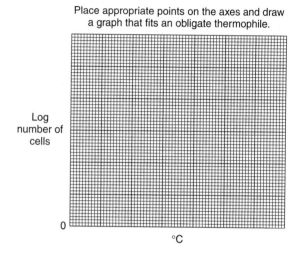

Place appropriate points on the axes and draw a graph that fits an obligate thermophile.

Log number of cells

0

°C

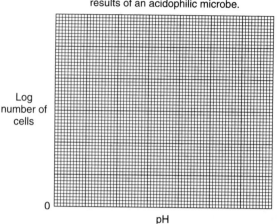

Enter points and show the growth results of an acidophilic microbe.

Log number of cells

0

pH

1. a. Is there a microbe that could grow on a medium that contains only the following compounds dissolved in water: $CaCO_3$, $MgNO_3$, $FeCl_2$, $ZnSO_4$, and glucose? Defend your answer.
   b. Check the last entry in the cell analysis summary on page 189. Are all needed elements present here? Where is the carbon?

2. Describe how one might determine the nutrient requirements of a microbe from Mars. If, after exhausting various nutrient schemes, it still does not grow, what other factors might one take into account?

3. a. The noted microbial ecologist Martin W. Beijerinck has stated: "Everything is everywhere, the environment selects." Use this concept to explain what ultimately determines whether a microorganism can live in a certain habitat.
   b. Give two examples of ways that microbes become modified to survive.

4. a. Explain how osmotic pressure and pH are used in preserving foods.
   b. What effects do they have on microbes?

5. a. How might one effectively treat anaerobic infections using gas?
   b. Explain how it would work.

6. How can you explain the observation that unopened milk will spoil even while refrigerated?

7. What would be the effect of a fever on a thermophilic pathogen?

8. Patients with acidotic diabetes are especially susceptible to fungal infections. Can you explain why?

9. Using the concept of synergism, can you describe a way to grow a fastidious microbe?

10. Describe ways to isolate an antibiotic-producing bacterium or mold.

11. a. If an egg salad sandwich sitting in a warm car for 4 hours develops 40,960 bacterial cells, how many more cells would result with just one more hour of incubation? (Use the same criteria that were stated in the sample problem.)
    b. With 10 hours of incubation?
    c. What would the cell count be after 4 hours if the initial bacterial dose were 100?
    d. What do your answers tell you about using clean techniques in food preparation and storage (other than aesthetic considerations)?

12. Why is an older culture needed for spore staining?

13. Discuss the idea of biotechnology companies being allowed to isolate and own microorganisms taken from the earth's habitats and derive profits from them. Can you come up with a solution that encourages exploration yet also benefits the public?

INTERNET SEARCH TOPICS

1. Look up the term **extremophile** on a search engine. Find examples of microbes that have adapted to various extremes. Report any species that exist in several extremes simultaneously.

2. Find a website that discusses the use of microbes living in high temperatures to bioremediate (clean up) the environment.

3. Research the Rio Tinto River in Spain. What kinds of microbes dominate in this habitat?

4. Visit the student Online Learning Center at www.mhhe.com/talaro5. Go to chapter 7, Internet Search Topics, and log on to the available websites to:
   a. Research the methods used by the laboratory to quantify microbes in water samples.
   b. Discover some very interesting symbiotic relationships.

CHAPTER 8

Microbial Metabolism:
The Chemical Crossroads of Life

Cells maintain a complex internal habitat that is highly structured and controlled. They host a never-ending array of metabolic reactions that are required for functions such as nutrient processing, growth, and release of energy. Among the most remarkable features of cell metabolism are enzymes, specialized proteins that perform key roles in most of these reactions. Recently, biochemists clarified the actions of an enzyme called OMP decarboxylase, used to synthesize a molecule that is found universally in the genetic material of all organisms and viruses. They found that in the presence of the OMP enzyme, this reaction happens very rapidly—about 30 times per second. The chances of the reaction happening without the enzyme would take so long (78 million years) that it is essentially not a realistic possibility. It is clear that without this enzyme, life as we know it could never have evolved. This is but one example of the amazing cellular machinery that is ultimately responsible for the tremendous diversity in the earth's life forms. In addition, it shows us that the metabolic activities of cells often share common patterns. With these themes in mind, this chapter will cover some of the major unifying characteristics of metabolism, enzymes, the flow of energy, and the pathways that govern nutrient processing. Having this background will greatly benefit your understanding of future topics such as genetics, drug therapy, disinfection, biotechnology, diseases, and the identification of microorganisms.

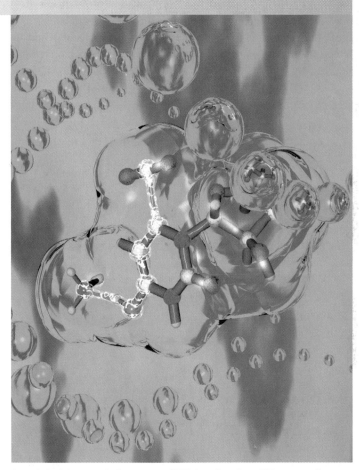

An artist's vision reveals the enzyme OMP decarboxylase. This enzyme is shown assisting in the formation of uracil, a component of RNA, by releasing carbon dioxide gas. It is one of the most efficient enzymes ever discovered.

Chapter Overview

- Cells are constantly involved in an orderly activity called metabolism that encompasses all of their chemical and energy transactions.
- Enzymes are essential metabolic participants that drive cell reactions.
- Enzymes are protein catalysts that speed up chemical processes by lowering the required energy.
- Enzymes have a specific shape tailored to perform their actions on a single type of molecule called a substrate.
- Enzymes derive some of their special characteristics from cofactors such as vitamins, and they show sensitivity to environmental factors.
- Enzymes are involved in activities that synthesize, digest, oxidize, and reduce compounds, and convert one substance to another.
- Enzymes are regulated by several mechanisms that alter the structure or synthesis of the enzyme.
- The energy of living systems resides in the atomic structure of chemicals that can be acted upon and changed.
- Cell energetics involves the release of energy that powers the formation of bonds.

- The energy of electrons is transferred from one molecule to another in coupled redox reactions.
- Electrons are transferred from substrates such as glucose to coenzyme carriers and ultimately captured in high-energy adenosine triphosphate (ATP).
- Cell pathways involved in extracting energy from fuels are glycolysis, the tricarboxylic acid cycle, and electron transport.
- The molecules used in aerobic respiration are glucose and oxygen, and the products are CO_2, H_2O, and ATP.
- Microbes participate in alternate pathways such as fermentation and anaerobic respiration.
- Cells manage their metabolites through linked pathways that have numerous functions and can proceed in more than one direction.

The Metabolism of Microbes

Metabolism, from the Greek term *metaballein,* meaning change, pertains to all chemical reactions and physical workings of the cell. Although metabolism entails thousands of different reactions, most of them fall into one of two general categories. **Anabolism,*** sometimes also called *biosynthesis,* is any process that results in synthesis of cell molecules and structures. It is a building and bond-making process that forms larger macromolecules from smaller ones, and it usually requires the input of energy. **Catabolism*** is the opposite of anabolism. Catabolic reactions are degradative; they break the bonds of larger molecules into smaller molecules, and often produce energy. The linking of anabolism to catabolism ensures the efficient completion of many thousands of cellular processes.

In summary, metabolism performs these functions:

1. degrades macromolecules into smaller molecules and releases energy;
2. energy that is released is converted into ATP or released as heat; and
3. the smaller molecules are assembled to form larger macromolecules specific to the cell and in this process ATP is utilized to form bonds (**figure 8.1**).

It has built-in controls for reducing or stopping a process that is not in demand and other controls for storing excess nutrients. The metabolic

---

***anabolism** (ah-nab′-oh-lizm) Gr. *anabole,* a throwing up.

***catabolism** (kah-tab′-oh-lizm) Gr. *katabole,* a throwing down.

workings of the cell are indeed intricate and complex, but they are also elegant and efficient. It is this very organization that sustains life.

ENZYMES: CATALYZING THE CHEMICAL REACTIONS OF LIFE

A microbial cell could be viewed as a microscopic factory, complete with basic building materials, a source of energy, and a "blueprint" for running its extensive network of metabolic reactions. But the chemical reactions of life, even when highly organized and complex, cannot proceed without a special class of proteins called **enzymes.*** Enzymes are a remarkable example of **catalysts,*** chemicals that increase the rate of a chemical reaction without becoming part of the products or being consumed in the reaction. Do not make the mistake of thinking that an enzyme creates a reaction. Because of the great energy of some molecules, a reaction could occur spontaneously at some point even without an enzyme, but at a very slow rate (**figure 8.2**). A study of the enzyme urease shows that it increases the rate of the breakdown of urea by a factor of 100 trillion as compared to an

---

***enzyme** (en′-zyme) Gr. *en,* in, and *syme,* leaven. Named for catalytic agents first found in yeasts.

***catalyst** (kat′-uh-list) Gr. *katalysis,* dissolution.

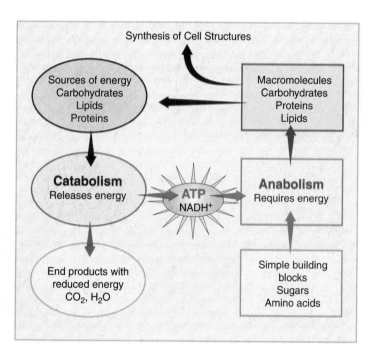

FIGURE 8.1

Simplified Model of Metabolism. Cellular reactions fall into two major categories. Catabolism involves the breakdown of complex organic molecules to extract energy and form simpler end products. Anabolism uses the energy to synthesize necessary macromolecules and cell structures from simple precursors. These processes are interactive and complementary, thereby maintaining chemical and energy balance.

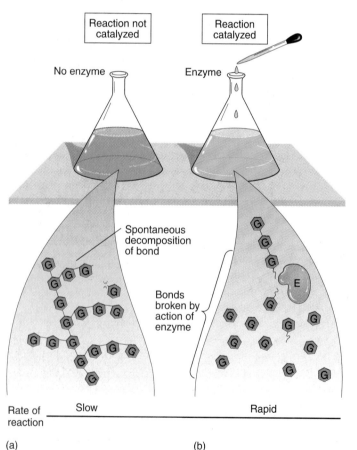

(a)   (b)

FIGURE 8.2

The effects of a catalyst. This example shows the breakdown of starch into glucose. **(a)** A few of the bonds will break in the absence of an enzyme, but it is a very slow process. **(b)** Addition of an enzyme (E) can speed up the reaction so that it happens very rapidly.

MICROBITS 8.1
Enzymes as Biochemical Levers

An analogy will allow us to envision the relationship of enzymes to the energy of activation. A large boulder sitting precariously on a cliff's edge contains a great deal of potential energy, but it will not fall to the ground and release this energy unless something disturbs it. It might eventually be disturbed spontaneously as the cliff erodes below it, but that could take a very long time. Moving it with a crowbar (to overcome the resistance of the boulder's weight) will cause it to tumble down freely by its own momentum. If we relate this analogy to chemicals, the reactants are the boulder on the cliff, the activation energy is the energy needed to move the boulder, the enzyme is the crowbar, and the product is the boulder at the bottom of the cliff. Like the crowbar, an enzyme permits a reaction to occur rapidly by lowering the energy obstacle. (Although this analogy concretely illustrates the concept of the energy of activation, it is imperfect in that the enzyme, unlike the person with the crowbar, does not actually add any energy to the system.)

This phenomenon can be represented by plotting the energy of the reaction against the direction of the reaction. The curve of reaction is shown in the accompanying graph. It is evident that there is an energy "hill" called the energy of activation (E_{act}) that must be overcome for the reaction to proceed. When a catalyst is present, it lowers the E_{act}. It does this by providing the reactants with an energy "shortcut" by which they can progress to the final state. Note that the products are the same final energy state with or without an enzyme.

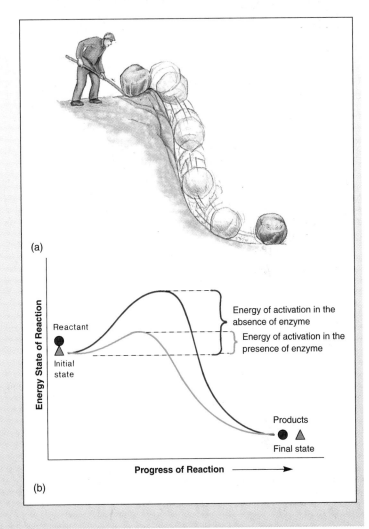

(a)

(b)

Analogy demonstrating the influence of enzymes on chemical reactions. **(a)** A boulder can represent potential energy available for a chemical reaction. **(b)** Graph of chemical reaction, with and without an enzyme. Energy (called energy of activation) is required in both cases to convert a reactant molecule to products. But in an enzyme-catalyzed reaction, the enzyme significantly lowers this energy of activation and allows the reaction to proceed more readily and rapidly.

---

uncatalyzed reaction. Because most uncatalyzed metabolic reactions do not occur fast enough to sustain cell processes, enzymes, which speed up the rate of reactions, are indispensable to life. Other major characteristics of enzymes are summarized in **table 8.1.**

How Do Enzymes Work?

We have said that an enzyme speeds up the rate of a metabolic reaction, but just how does it do this? During a chemical reaction, reactants are converted to products by bond formation or breakage. A certain amount of energy is required to initiate every such reaction, which limits its rate. This resistance to a reaction, which must be overcome for a reaction to proceed, is measurable and is called the **energy of activation** or activation energy, **(Microbits 8.1).** In the laboratory, overcoming this initial resistance can be achieved by

1. increasing thermal energy (heating) to increase molecular velocity,
2. increasing the concentration of reactants, or
3. adding a catalyst.

TABLE 8.1

Checklist of Enzyme Characteristics

- Are composed of protein and may require cofactors
- Act as organic catalysts to speed up the rate of cellular reactions
- Lower the activation energy required for a chemical reaction to proceed (Microbits 8.1)
- Have unique characteristics such as shape, specificity, and function
- Enable metabolic reactions to proceed at a speed compatible with life
- Provide a reactive site for target molecules called substrates
- Are much larger in size than their substrates
- Associate closely with substrates but do not become integrated into the reaction products
- Are not used up or permanently changed by the reaction
- Can be recycled, thus function in extremely low concentrations
- Are limited by particular conditions of temperature and pH
- Can be regulated by feedback and genetic mechanisms

In most living systems, the first two alternatives are not feasible, because elevating the temperature is potentially harmful and higher concentrations of reactants are not practical. This leaves only the action of catalysts, and enzymes fill this need efficiently and potently.

At the molecular level, an enzyme promotes a reaction by serving as a physical site upon which the reactant molecules, called **substrates,** can be positioned for various interactions. The enzyme is much larger in size than its substrate, and it presents a unique niche that fits only that particular substrate. Although an enzyme binds to the substrate and participates directly in changes to the substrate, it does not become a part of the products, is not used up by the reaction, and can function over and over again. Enzyme speed, defined as the number of substrate molecules converted per enzyme per second, is well documented. Speeds range from several million for catalase to a thousand for lactate dehydrogenase. To further visualize the roles of enzymes in metabolism, we must next look at their structure.

Enzyme Structure

The primary structure of all enzymes is protein (with some exceptions—**Microbits 8.2,** page 224), and they can be classified as simple or conjugated. Simple enzymes consist of protein alone, whereas conjugated enzymes (**figure 8.3**) contain protein and nonprotein molecules. A conjugated enzyme, sometimes referred to as a **holoenzyme,*** is a combination of a protein, now called the **apoenzyme,** and one or more **cofactors (table 8.2).** Cofactors are either organic molecules, called **coenzymes,** or inorganic elements (metal ions). In

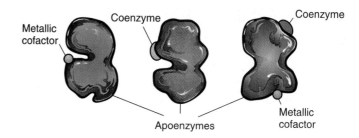

FIGURE 8.3

Conjugated enzyme structure.   All have an apoenzyme (polypeptide or protein) component and one or more cofactors.

some enzymes, the cofactor is loosely associated with the apoenzyme by noncovalent bonds; in others, it is linked by covalent bonds.

Apoenzymes: Specificity and the Active Site

Apoenzymes range in size from small polypeptides with about 100 amino acids and a molecular weight of 12,000 to large polypeptide conglomerates with thousands of amino acids and a molecular weight of over one million. Like all proteins, an apoenzyme exhibits levels of molecular complexity called the primary, secondary, tertiary, and, in larger enzymes, quaternary organization (**figure 8.4**). As we saw in chapter 2, the first three levels of structure arise when a

---

*holoenzyme (hol-oh-en′-zyme) Gr. *holos,* whole.

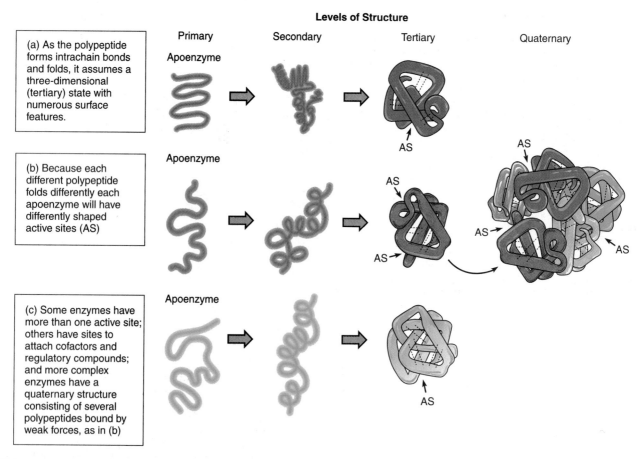

FIGURE 8.4

How the active site and specificity of the apoenzyme arise.

TABLE 8.2

Selected Enzymes, Catalytic Actions, Cofactors, and Functions

Enzyme	Action	Metallic Cofactor
Catalase	Breaks down hydrogen peroxide	Iron (Fe)
Oxidase	Adds electrons to oxygen	Iron, copper (Cu)
Hexokinase	Transfers phosphate to glucose	Magnesium (Mg)
Urease	Splits urea into an ammonium ion	Nickel (Ni)
Nitrate reductase	Reduces nitrate to nitrite	Molybdenum (Mo)
DNA polymerase complex	Synthesis of DNA	Zinc (Zn) and Mg

Enzyme	Coenzyme Involved/Vitamin Required	Function of Coenzyme Complex
Pyruvate dehydrogenase	Coenzyme A/Pantothenic acid (B3)	Recognition site for binding substrate
Pyruvic dehydrogenase	Thiamine pyrophosphate/Thiamine (B1)	Transfer of aldehyde groups
Various dehydrogenases	Nicotinamide adenine dinucleotide (NAD)/Niacin	Carries electrons in redox reactions
Succinate reductase	Flavin adenine dinucleotide (FAD)/Riboflavin (B2)	Electron transfers in mitochondria

single polypeptide chain undergoes an automatic folding process and achieves stability by forming disulfide and other types of bonds. Folding causes the surface of the apoenzyme to acquire three-dimensional features that result in the enzyme's specificity for substrates. The actual site where the substrate binds is a crevice or groove called the **active site,** or **catalytic site,** and there can be from one to several such sites (figure 8.4). Each type of enzyme has a different primary structure (type and sequence of amino acids),[1] variations in folding, and unique active sites.

Enzyme-Substrate Interactions

For a reaction to take place, a temporary enzyme-substrate union must occur at the active site. There are several explanations for the process of attachment **(figure 8.5).** The specificity is often described as a "lock-and-key" fit in which the substrate is inserted into the active site's pocket. This is a useful analogy, but an enzyme is not a rigid lock mechanism. It is likely that the enzyme actually helps pull the substrate into the active site through slight changes in its shape. This is called an *induced fit* (figure 8.5d).

The bonds formed between the substrate and enzyme are weak and, of necessity, easily reversible. Once the enzyme-substrate complex has formed, appropriate reactions occur on the substrate, often with the aid of a cofactor, and a product is formed and released. The enzyme can then attach to another substrate molecule and repeat this action. Although enzymes can potentially catalyze reactions in both directions, most examples in this chapter will depict them working in one direction only.

Cofactors: Supporting the Work of Enzymes

In chapter 7 we learned that microorganisms require specific metal ions called trace elements and certain organic growth factors. In many cases, the need for these substances arises from their roles as cofactors. The metallic cofactors, including iron, copper, magnesium, manganese, zinc, cobalt, selenium, and many others, participate in precise functions between the enzyme and its substrate. In general, metals activate enzymes, help bring the active site and substrate close together, and participate directly in chemical reactions with the enzyme-substrate complex.

Coenzymes are organic compounds that work in conjunction with an apoenzyme to perform a necessary alteration of a substrate. The general function of a coenzyme is to remove a functional group from one substrate molecule and add it to another

---

1. As specified by the genetic blueprint in DNA (see chapter 9).

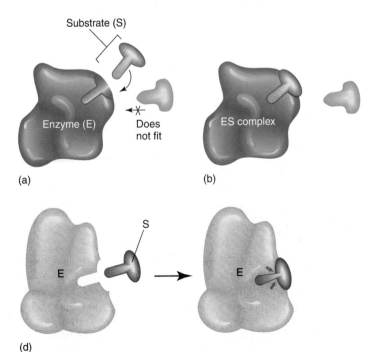

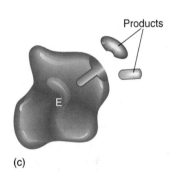

FIGURE 8.5

Enzyme-substrate reactions. **(a)** When the enzyme and substrate come together, the substrate (S) must show the correct fit and position with respect to the enzyme (E). **(b)** When the ES complex is formed, it enters a transition state. During this temporary but tight interlocking union, the enzyme participates directly in breaking or making bonds. **(c)** Once the reaction is complete, the enzyme releases the products. **(d)** The induced fit model proposes that the enzyme recognizes its substrate and adapts slightly to it so that the final binding is even more precise.

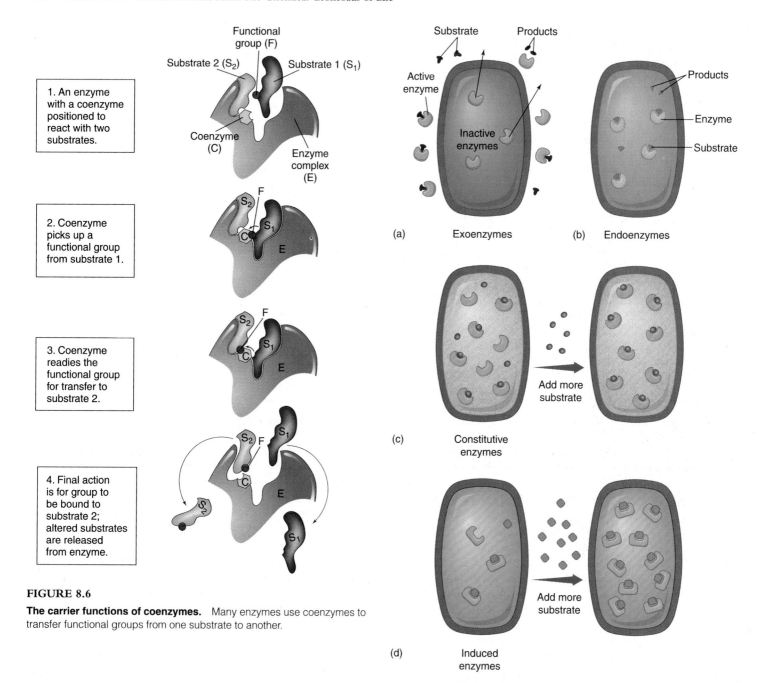

FIGURE 8.6

The carrier functions of coenzymes. Many enzymes use coenzymes to transfer functional groups from one substrate to another.

FIGURE 8.7

Types of enzymes, as described by their location of action and quantity. **(a)** Exoenzymes are released outside the cell to function. **(b)** Endoenzymes remain in the cell and function there. **(c)** Constitutive enzymes are present in relatively constant amounts in a cell. The addition of more substrate does not increase the numbers of these enzymes. **(d)** Induced enzymes are normally present in trace amounts, but their quantity can be increased a thousandfold by the addition of substrate.

substrate, thereby serving as a transient carrier of this group (**figure 8.6**). The specific activities of coenzymes are many and varied. In a later section of this chapter, we shall see that coenzymes carry and transfer hydrogen atoms, electrons, carbon dioxide, and amino groups. One of the most important components of coenzymes are **vitamins** (see table 8.2), which explains why vitamins are important to nutrition and may be required as growth factors for living things. Vitamin deficiencies prevent the complete holoenzyme from forming. Consequently, both the chemical reaction and the structure or function dependent upon that reaction are compromised.

Classification of Enzyme Functions

Enzymes are classified and named according to characteristics such as site of action, type of action, and substrate (**Microbits 8.3**).

Location and Regularity of Enzyme Action Enzymes exhibit several patterns of performance. After initial synthesis in the cell, **exoenzymes** are transported extracellularly, where they break down (hydrolyze) large food molecules or harmful chemicals. Examples of exoenzymes are cellulase, amylase, and penicillinase. By contrast, **endoenzymes** are retained intracellularly and function there. Most enzymes of the metabolic pathways are of this variety (**figure 8.7**).

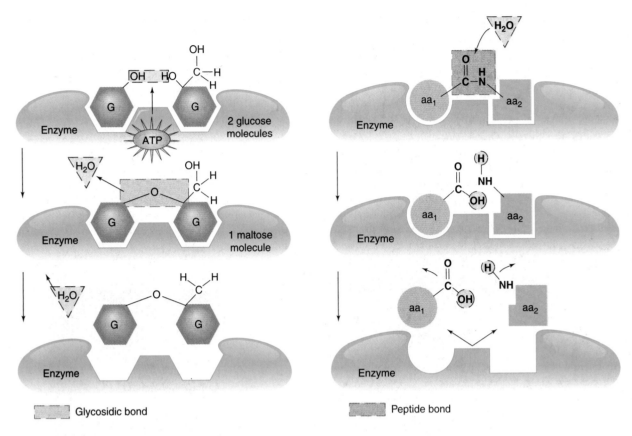

(a) **Condensation Reaction.** Forming a glycosidic bond between two glucose molecules to generate maltose requires the removal of a water molecule and energy from ATP.

(b) **Hydrolysis Reaction.** Breaking a peptide bond between two amino acids requires a water molecule that adds an H and OH to the amino acids.

FIGURE 8.8

Examples of enzyme-catalyzed synthesis and hydrolysis reactions.

In terms of their presence in the cell, enzymes are not all produced in equal amounts or at equal rates. Some, called **constitutive enzymes** (figure 8.7c), are always present and in relatively constant amounts, regardless of the amount of substrate. The enzymes involved in utilizing glucose, for example, are very important in metabolism and thus are constitutive. An **induced** (or inducible) **enzyme** (figure 8.7d) is not constantly present and is produced only when its substrate is present. Induced enzymes are present in amounts ranging from a few molecules per cell to several thousand times that many, depending on the metabolic requirement. This property of selective synthesis of enzymes prevents a cell from wasting resources by making enzymes that will not be used immediately. The induction of enzymes constitutes an important metabolic control discussed later in this section of the chapter and again in chapter 9.

Synthesis and Hydrolysis Reactions   A growing cell is in a frenzy of activity, constantly synthesizing proteins, DNA, and RNA; forming storage polymers such as starch and glycogen; and assembling new cell parts. Such anabolic reactions require enzymes (ligases) to form covalent bonds between smaller substrate molecules. Also known as *condensation reactions,* synthesis reactions typically require ATP and always release one water molecule for each bond made **(figure 8.8a).** Catabolic reactions involving energy transactions, remodeling of cell structure, and digestion of

macromolecules are also very active during cell growth. For example, digestion requires enzymes to break down substrates into smaller molecules so they can be used by the cell. Because the breaking of bonds requires the input of water, digestion is often termed a *hydrolysis* * *reaction* (figure 8.8b).

Transfer Reactions by Enzymes   Other enzyme-driven processes that involve the simple addition or removal of a functional group are important to the overall economy of the cell. Oxidation-reduction and other transfer activities are examples of these types of reactions.

Some atoms and compounds readily give or receive electrons and participate in oxidation (the loss of electrons) or reduction (the gain of electrons). The compound that loses the electrons is **oxidized,** and the compound that receives the electrons is **reduced.** Such oxidation-reduction (redox) reactions are common in the cell and indispensable to the energy transformations discussed later in this chapter (see Microbits 2.2). Important components of cellular redox reactions are oxidoreductases, which remove electrons from one substrate and add them to another, and their coenzyme carriers, nicotinamide adenine dinucleotide (NAD; see figure 8.14) and flavin adenine dinucleotide (FAD).

---

*hydrolysis (hy-drol′-uh-sis) Gr. *hydro,* water, and *lysis,* splitting.

MICROBITS   8.2

Unconventional Enzymes

The molecular reactions of cells are still an active and rich source of discovery, and new findings come along nearly every few months that break older "rules" and change our understanding. It was once an accepted fact that proteins were the only biological molecules that act as catalysts, until biologists found a novel type of RNA termed **ribozymes.** Ribozymes are associated with many types of cells and a few viruses, and they display some of the properties of protein catalysts, such as having a specific active site and interacting with a substrate. But these molecules are remarkable because their substrate is other RNA. Ribozymes are thought to be remnants of the earliest molecules on earth that could have served as both catalysts and genetic material. In natural systems, ribozymes are involved in self-splicing or cutting of RNA molecules during final processing of the genetic code (see chapter 9). Further research has shown that ribozymes can be designed to handle other kinds of activities, such as inhibiting gene expression. Several companies have developed and are testing ribozyme-based therapies for treating cancer and AIDS.

Another intriguing new type of enzyme comes to us from studies of the immune system. Using antibodies as their basic tool, immunologists have developed catalytic molecules called *abzymes.* Antibodies, like enzymes, are proteins that are very specific to their target molecules and attach tightly to them. They can be modified in the lab to create abzymes that are capable of forming a reversible complex with substrate, speeding up reactions, and participating in bond formation and breakage. Abzymes have an additional benefit—they may be fine-tuned to fit almost any substrate shape, and they can be manufactured through monoclonal antibody technology (see chapter 15). The potential of abzymes for use as drugs is also being explored by medicine and industry. Tests are currently under-

way on specific drugs and other treatments for cancer, infections, and medical problems such as blood clots.

Another focus of study that could lead to new applications is the search for *extremozymes,* enzymes from microbes that live under rigorous conditions of temperature, pH, salt, and pressure. Because these enzymes are adapted to working under such extreme conditions, they may well be useful in industrial processes that require work in heat or cold or very low pH. Several enzymes that can work under high temperature and low pH have been isolated and are either currently in use (PCR technique) or in development.

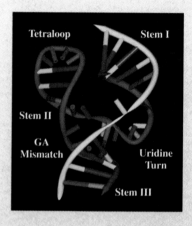

A "hammerhead" ribozyme consisting of a single RNA strand curved around to form an active site (indentation between tetraloop and stem I).

Other enzymes play a role in the molecular conversions necessary for the economical use of nutrients by directing the transfer of functional groups from one molecule to another. For example, *aminotransferases* convert one type of amino acid to another by transferring an amino group (see figure 8.28); *phosphotransferases* participate in the transfer of phosphate groups and are involved in energy transfer; *methyltransferases* move a methyl (CH_3) group from substrate to substrate; and *decarboxylases* (also called carboxylases) catalyze the removal of carbon dioxide from organic acids in several metabolic pathways.

The Role of Microbial Enzymes in Disease   Many pathogens secrete unique exoenzymes that help them avoid host defenses or promote their multiplication in tissues. Because these enzymes contribute to pathogenicity, they are referred to as virulence factors, or toxins in some cases. *Streptococcus pyogenes* (a cause of throat and skin infections) produces a streptokinase that digests blood clots and apparently assists in invasion of wounds. *Pseudomonas aeruginosa,* a respiratory and skin pathogen, produces elastase and collagenase, which digest elastin and collagen. These increase the severity of certain lung

diseases and burn infections. *Clostridium perfringens,* an agent of gas gangrene, synthesizes lecithinase C, a lipase that profoundly damages cell membranes and accounts for the tissue death associated with this disease. Not all enzymes digest tissues; some, such as penicillinase, inactivate penicillin and thereby protect a microbe from its effects.

The Sensitivity of Enzymes to Their Environment

The activity of an enzyme is highly influenced by the cell's environment. In general, enzymes operate only under the natural temperature, pH, and osmotic pressure of an organism's habitat. When enzymes are subjected to changes in these normal conditions, they tend to be chemically unstable, or **labile.** Low temperatures inhibit catalysis, and high temperatures denature the apoenzyme. **Denaturation** is a process by which the weak bonds that collectively maintain the native shape of the apoenzyme are broken. This disruption causes extreme distortion of the enzyme's shape and prevents the substrate from attaching to the active site (see figure 11.4). Such nonfunctional enzymes block metabolic reactions and thereby can lead to cell death. Low or high pH or certain chemicals (heavy metals, alcohol) are also denaturing agents.

Most metabolic reactions require a separate and unique enzyme. Up to the present time, researchers have discovered and named over 5,000 of them. Given the complex chemistry of the cell and the extraordinary diversity of living things, many more enzymes probably remain to be discovered. A standardized system of nomenclature and classification was developed to prevent discrepancies.

In general, an enzyme name is composed of two parts: a prefix or stem word derived from a certain characteristic—usually the substrate acted upon or the type of reaction catalyzed, or both—followed by the ending *-ase.*

The system classifies the enzyme in one of these six classes, on the basis of its general biochemical action:

1. *Oxidoreductases* transfer electrons from one substrate to another, and *dehydrogenases* transfer a hydrogen from one compound to another.
2. *Transferases* transfer functional groups from one substrate to another.
3. *Hydrolases* cleave bonds on molecules with the addition of water.
4. *Lyases* add groups to or remove groups from double-bonded substrates.

5. *Isomerases* change a substrate into its isomeric* form.
6. *Ligases* catalyze the formation of bonds with the input of ATP and the removal of water.

Each enzyme is also assigned a technical name that indicates the specific reaction it catalyses. These names are useful for biochemists but are often too lengthy for routine use. More often, a common name is derived from some notable feature such as the substrate or enzyme action, with *-ase* added to the ending. For example, *hydrogen peroxide oxidoreductase* is more commonly known as *catalase,* and *mucopeptide N-acetylmuramoylhydrolase* is called (thankfully) *lysozyme.* With this system, an enzyme that digests a carbohydrate substrate is a *carbohydrase;* a specific carbohydrase, *amylase,* acts on starch (amylose is a major component of starch). The enzyme *maltase* digests the sugar maltose. An enzyme that hydrolyzes peptide bonds of a protein is a *proteinase, protease,* or *peptidase,* depending on the size of the protein substrate. Some fats and other lipids are digested by *lipases.* DNA is hydrolyzed by *deoxyribonuclease,* generally shortened to *DNase.* A *synthetase* or *polymerase* bonds together many small molecules into large molecules. Other examples of enzymes are presented in table 8.A. (See also table 8.2.)

---

*An isomer is a compound that has the same molecular formula as another compound but differs in arrangement of the atoms.

TABLE 8.A

A Sampling of Enzymes, Their Substrates, and Their Reactions

Common Name	Systematic Name	Enzyme Class	Substrates	Action
Cellulase	1,4 β-glycosidase	Hydrolase	Cellulose	Cleaves 1,4 β-glycosidic linkages
Lactase	β-D-galactosidase	Hydrolase	Lactose	Breaks lactose down into glucose and galactose
Penicillinase	Beta-lactamase	Hydrolase	Penicillin	Hydrolyzes beta-lactam ring
Lipase	Triacylglycerol acylhydrolase	Hydrolase	Triglycerides	Cleaves bonds between glycerol and fatty acids
DNA polymerase	DNA nucleotidyl-transferase	Transferase	DNA nucleosides	Synthesizes a strand of DNA using the complementary strand as a model
Hexokinase	ATP-glucose phosphotransferase	Transferase	Glucose	Catalyzes transfer of phosphate from ATP to glucose
Aldolase	Fructose diphosphate aldolase	Lyase	Fructose diphosphate	Catalyzes the conversion of the substrate to two 3-carbon fragments
Lactate dehydrogenase	Same as common name	Oxidoreductase	Pyruvic acid	Catalyzes the conversion of pyruvic acid to lactic acid
Oxidase	Cytochrome oxidase	Oxidoreductase	Molecular oxygen	Catalyzes the reduction (addition of electrons and hydrogen) to O_2

REGULATION OF ENZYMATIC ACTIVITY AND METABOLIC PATHWAYS

Metabolic reactions proceed in a systematic, highly regulated manner that maximizes the use of available nutrients and energy. The cell responds to environmental conditions by adopting those metabolic reactions that most favor growth and survival. Because enzymes are critical to these reactions, the regulation of metabolism is largely the regulation of enzymes by an elaborate system of checks and balances. Before we examine some of these control systems, let us take a look at some general features of metabolic pathways.

Metabolic Pathways

Metabolic reactions rarely consist of a single action or step. More often, they occur in a multistep series or pathway, with each step catalyzed by an enzyme. An individual reaction is shown in various ways, depending on the purpose at hand **(figure 8.9).** The product of one reaction is often the reactant (substrate) for the next, forming a linear chain of reactions. Many pathways have branches that pro-

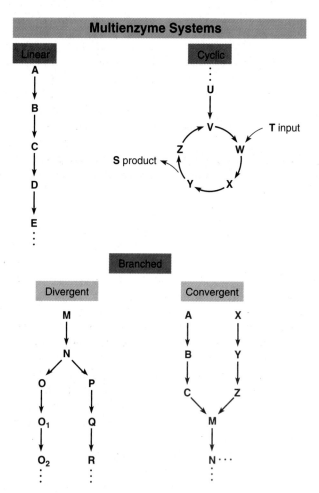

FIGURE 8.9

Patterns of metabolism. In general, metabolic pathways consist of a linked series of individual chemical reactions that produce intermediary metabolites and lead to a final product. These pathways occur in several patterns, including linear, cyclic, and branched. Anabolic pathways involved in biosynthesis result in a more complex molecule, each step adding on a functional group, whereas catabolic pathways involve the dismantling of molecules and can generate energy. Virtually every reaction in a series involves a specific enzyme.

vide alternate methods for nutrient processing. Others take a cyclic form, in which the starting molecule is regenerated to initiate another turn of the cycle (for example, the TCA cycle; see figure 8.21). Pathways generally do not stand alone; they are interconnected and merge at many sites.

Every pathway has one or more enzyme pacemakers (usually the slowest enzyme in the series) that sets the rate of a pathway's progression. These enzymes respond to various control signals and, in so doing, determine whether a pathway proceeds. Regulation of pacemaker enzymes proceeds on two fundamental levels. Either the enzyme itself is directly inhibited or activated, or the amount of the enzyme in the system is altered (decreased or increased). Factors that affect the enzyme directly provide a means for the system to be finely controlled or tuned, whereas regulation at the genetic level (enzyme synthesis) provides a slower, less sensitive control.

Direct Controls on the Actions of Enzymes

Competitive Inhibition   In **competitive inhibition,** other molecules with a structure similar to the normal substrate can occupy the enzyme's active site. Although these molecular mimics have the proper fit for the site, they cannot be further acted upon. The attachment of a mimic effectively prevents an enzyme from attaching to its usual substrate and blocks its activity, output, and possibly the rest of that pathway **(figure 8.10).** This mechanism is common in natural metabolic pathways. It is also an important mode of action by some drugs for treating infections (see chapter 12).

Feedback Control   The term *feedback* is borrowed from the field of engineering, where it describes a process in which the product of a system is *fed back* into the system to keep it in balance. Feedback is positive or negative, depending upon whether the process is stimulated (+) or inhibited (−) by the product. **Negative feedback** is common in electrical appliances such as hot plates or furnaces with automatically controlled temperatures. These devices possess a means of controlling the temperature (a heat source) and a temperature sensor (thermostat) that is set to a desired point and regulates the turning on or off of the heat source. In the case of a hot plate, when the set temperature has been reached, the thermostat switches the heating elements off. A subsequent fall below the set temperature causes the heating element to be turned on again.

Many enzymatic reactions are controlled by negative feedback, whereby the end product being fed back into the system *negates* (cancels) an enzyme's activity:

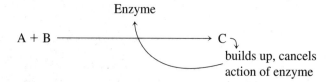

At high levels of substrate and scant levels of end product, the enzyme works freely. In time, the end product builds up to sufficient levels, and it stops the action of that enzyme. It usually targets the first enzyme in a metabolic pathway. The mechanisms of negative feedback—that is, how the product actually stops the action of an enzyme—can best be understood if we look at the allosteric behavior of enzymes.

We have previously shown that globular proteins of enzymes are bulky and possess numerous surface features and that one part

Negative Feedback

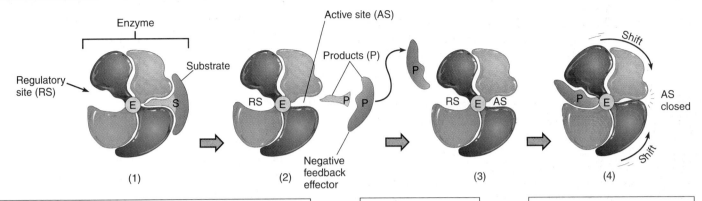

(1)   (2)   (3)   (4)

(1), (2) Allosteric enzymes have a quaternary structure with two different sites of attachment—the active site (AS) and the regulatory site (RS). The enzyme complex normally attaches to the substrate at the AS and releases products (P).

(3) One product can function as a negative-feedback effector by fitting into a regulatory site.

(4) The entrance of P into RS causes a confirmational shift of the enzyme that closes the active site. The enzyme cannot catalyze further reactions with substrate as long as the product is inserted in the regulatory site.

Competitive Inhibition

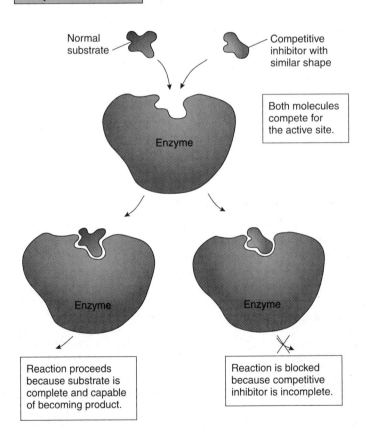

Normal substrate

Competitive inhibitor with similar shape

Both molecules compete for the active site.

Enzyme

Enzyme   Enzyme

Reaction proceeds because substrate is complete and capable of becoming product.

Reaction is blocked because competitive inhibitor is incomplete.

FIGURE 8.10

Examples of two common control mechanisms for enzymes.

*allosteric (al-oh-stair´-ik) Gr. *allos*, other, and *steros*, solid. Literally, "another space."

of an enzyme's terrain (the active site) accommodates the substrate. **Allosteric* enzymes** have an additional **regulatory site** for the attachment of molecules other than substrate. This regulatory site usually exists on a different polypeptide unit of a quaternary structure. When an end-product molecule fits into this regulatory site, the enzyme's active site is so distorted that it can no longer bind to its substrate. Although this distortion does not denature the enzyme and is quite reversible, allosteric inhibitors can temporarily stop the action of that enzyme (figure 8.10). This mechanism is termed **feedback inhibition.** When at some point the end product is used up and more of it is needed, the enzyme will be released from inhibition and can resume catalysis.

Controls on Enzyme Synthesis

Controlling enzymes by controlling their synthesis is another effective mechanism, because enzymes do not last indefinitely. Some wear out, some are deliberately degraded, and others are diluted with each cell division. For catalysis to continue, enzymes eventually must be replaced. This cycle works into the scheme of the cell, where replacement of enzymes can be regulated according to cell demand. The mechanisms of this system are genetic in nature; that is, they require regulation of DNA and the protein synthesis machinery, topics we shall encounter once again in chapter 9.

Enzyme repression is a means to stop further synthesis of an enzyme somewhere along its pathway. As the level of the end product from a given enzymatic reaction has built to excess, the genetic apparatus responsible for replacing these enzymes is automatically suppressed **(figure 8.11).** The response time takes longer than feedback inhibition, but its effects are more enduring.

A response that resembles the inverse of feedback repression is **enzyme induction.** In this process, enzymes appear (are induced) only when suitable substrates are present; that is, the synthesis of enzyme is induced by its substrate (see figures 8.7*d* and 9.21). Both mechanisms are important genetic control systems in bacteria.

A classic model of enzyme induction occurs in the response of *Escherichia coli* to certain sugars. For example, if a particular strain of *E. coli* is inoculated into a medium whose principal carbon source is lactose, it will produce the enzyme lactase to hydrolyze it into glucose and galactose. If the bacterium is subsequently inoculated into

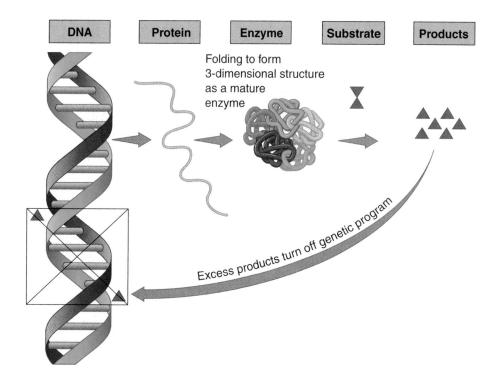

| DNA | Protein | Enzyme | Substrate | Products |

Folding to form
3-dimensional structure
as a mature
enzyme

Excess products turn off genetic program

FIGURE 8.11

One type of genetic control of enzyme synthesis: feedback/enzyme repression.   The enzyme is synthesized continuously until enough product has been made, at which time the excess product reacts with a site on DNA that regulates the enzyme's synthesis, thereby inhibiting further enzyme production.

a medium containing only sucrose as a carbon source, it will cease synthesizing lactase and begin synthesizing sucrase. This response enables the organism to adapt to a variety of nutrients, and it also prevents a microbe from wasting energy, making enzymes for which no substrates are present.

CHAPTER CHECKPOINTS

Metabolism includes all the biochemical reactions that occur in the cell. It is a self-regulating complex of interdependent processes that encompasses many thousands of chemical reactions.

Anabolism is the energy-requiring subset of metabolic reactions, which synthesize large molecules from smaller ones.

Catabolism is the energy-releasing subset of metabolic reactions, which degrade or break down large molecules into smaller ones.

Enzymes are proteins that catalyze all biochemical reactions by forming enzyme-substrate complexes. The binding of the substrate by an enzyme makes possible both bond-forming and bond-breaking reactions, depending on the pathway involved. Enzymes may utilize cofactors as carriers and activators.

Enzymes are classified and named according to the kinds of reactions they catalyze.

To function effectively, enzymes require specific conditions of temperature, pH, and osmotic pressure.

Enzyme activity is regulated by processes of feedback inhibition, induction, and repression, which, in turn, respond to availability of substrate and concentration of end products, as well as to other environmental factors.

The Pursuit and Utilization of Energy

Energy is defined as the capacity to do work or to cause change. In order to carry out the work of an array of metabolic processes, cells require constant input and expenditure of some form of usable energy. Energy commonly exists in various forms:

1. thermal, or heat, energy from molecular motion,
2. radiant (wave) energy from visible light or other rays,
3. electrical energy from a flow of electrons,
4. mechanical energy from a physical change in position,
5. atomic energy from reactions in the nucleus of an atom, and
6. chemical energy present in the bonds of molecules.

Cells are far too fragile to rely in any constant way on thermal or atomic energy for cell transactions. Except for certain chemoautotrophic bacteria, the ultimate source of energy is the sun, but only photosynthetic autotrophs can tap this source directly. This capacity of photosynthesis to convert solar energy into chemical energy provides both a nutritional and an energy basis for all heterotrophic living things. For the most part, only chemical energy can routinely operate cell transactions, and chemical reactions are the universal basis of cellular energetics.

In this section we will further explore the role of energy as it is related to the oxidation of fuels, redox carriers, and the generation of ATP.

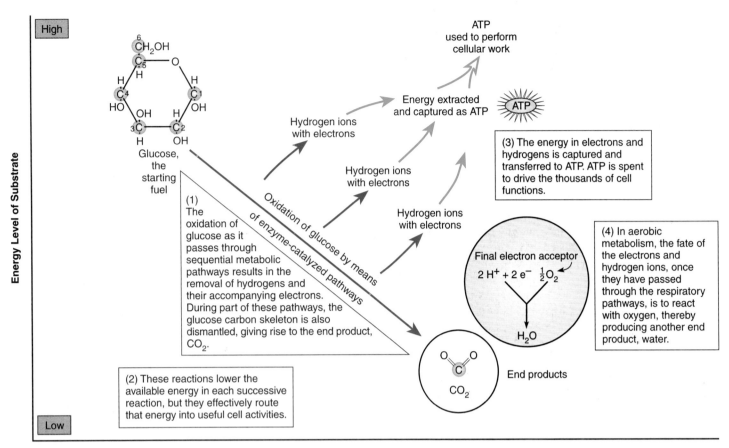

FIGURE 8.12

A simplified model that summarizes the cell's energy machine.   The central events of cell energetics include the release of energy during the systematic dismantling of a fuel such as glucose. This is achieved by the shuttling of hydrogens and electrons to sites in the cell where their energy can be transferred to ATP. The final products are CO_2 and H_2O molecules.

CELL ENERGETICS

Cells manage energy in the form of chemical reactions that change molecules. This often involves activities such as the making or breaking of bonds and the transfer of electrons. Not all cellular reactions are equal with respect to energy. Some release energy, and others require it to proceed. For example, a reaction that proceeds as follows:

$$X + Y \xrightarrow{\text{Enzyme}} Z + \text{Energy}$$

releases energy as it goes forward. This type of reaction is termed **exergonic.*** Energy of this type is considered free—it is available for doing cellular work. Energy transactions such as the following:

$$\overset{\text{Enzyme}}{\text{Energy} + A + B \longrightarrow C}$$

are called **endergonic,*** because they are driven forward with the addition of energy. Exergonic and endergonic reactions are coupled, so that released energy is immediately put to use.

Summaries of metabolism might make it seem that cells "create" energy from nutrients, but they do not. What they actually do is extract chemical energy already present in nutrient fuels and apply that energy toward useful work in the cell, much like a gasoline engine releases energy as it burns fuel. The engine does not actually produce energy, but it converts the potential energy of the fuel to the free energy of work.

The processes by which cells handle energy are well understood, and most of them are explainable in basic biochemical terms. At the simplest level, cells possess specialized enzyme systems that trap the energy present in the bonds of nutrients as they are progressively broken **(figure 8.12).** During exergonic reactions, energy released by electrons is stored in certain high-energy phosphate molecules such as ATP. As we shall see, the ability of ATP to temporarily store and release the energy of chemical bonds fuels endergonic cell reactions. Before discussing ATP, let us examine the process behind electron transfer: redox reactions.

***exergonic** (ex-er-gon′-ik) Gr. *exo,* without, and *ergon,* work. In general, catabolic reactions are exergonic. (energy releasing)

***endergonic** (en-der-gon′-ik) Gr. *endo,* within, and *ergon,* work. Anabolic reactions tend to be endergonic. (energy requiring)

A CLOSER LOOK AT BIOLOGICAL OXIDATION AND REDUCTION

We stated earlier that biological systems often extract energy through **redox reactions.** Such reactions always occur in pairs, with an electron donor and an electron acceptor, which constitute a *conjugate pair,* or *redox pair.* The reaction can be represented as follows (see also figure 8.13):

Electron donor + Electron acceptor $\longrightarrow$ Electron donor + Electron acceptor

•e−

(Reduced)   (Oxidized)   (Oxidized)   (Reduced)

(Review Microbits 2.2 for detailed figure.)

This process salvages electrons along with their inherent energy, and it changes the energy balance, leaving the reduced compound with less energy than the oxidized one. The released energy can be captured to **phosphorylate,** or add an inorganic phosphate, to ADP or to some other compound. This process stores the energy in a high-energy molecule (ATP, for example). In many cases, the cell does not handle electrons as discrete entities but rather as parts of an atom such as hydrogen. For simplicity's sake, we will continue to use the term *electron transfer,* but keep in mind that hydrogens are involved in the transfer process. The removal of hydrogens (a hydrogen atom consists of a single proton and a single electron) from a compound during a redox reaction is called dehydrogenation. The job of handling these protons and electrons falls to one or more carriers, which function as short-term repositories for the electrons until they can be transferred (**figure 8.13**). As we shall see, dehydrogenations are an essential supplier of electrons for the respiratory electron transport system.

Electron Carriers: Molecular Shuttles

Electron carriers resemble shuttles that are alternately loaded and unloaded, repeatedly accepting and releasing electrons and hydrogens to facilitate the transfer of redox energy. Most carriers are coenzymes that transfer both electrons and hydrogens, but some transfer electrons only. The most common carrier is NAD (nicotinamide adenine dinucleotide), which carries hydrogens (and a pair of electrons) from dehydrogenation reactions (**figure 8.14**). Reduced NAD can be represented in various ways. Because 2 hydrogens are removed, the actual carrier state is $NADH + H^+$, but this is somewhat cumbersome, so we will represent it with the shorter NADH. In catabolic pathways, electrons are extracted and carried through a series of redox reactions until the final electron acceptor

at the end of a particular pathway is reached (see figure 8.12). In aerobic metabolism, this acceptor is molecular oxygen; in anaerobic metabolism, it is some other inorganic or organic compound. Other common redox carriers are FAD, NADP (NAD phosphate), coenzyme A, and the compounds of the respiratory chain, which are fixed into membranes.

ADENOSINE TRIPHOSPHATE: METABOLIC MONEY

In what ways do cells extract chemical energy from electrons, store it, and then tap the storage sources? To answer these questions, we must look more closely at the powerhouse molecule, adenosine triphosphate. ATP has also been described as metabolic money because it can be earned, banked, saved, spent, and exchanged. As a temporary energy repository, ATP provides a connection between energy-yielding catabolism and all other cellular activities that require energy. Some clues to its energy-storing properties lie in its unique molecular structure (**figure 8.15**).

The Molecular Structure of ATP

ATP is a three-part molecule consisting of a nitrogen base (adenine) linked to a 5-carbon sugar (ribose), with a chain of three phosphate groups bonded to the ribose (figure 8.15). The type, arrangement, and especially the proximity of atoms in ATP combine to form a compatible but unstable high-energy molecule. The high energy of ATP originates in the orientation of the phosphate groups, which are relatively bulky and carry negative charges. The proximity of these repelling electrostatic charges imposes a strain that is most acute on the bonds between the last two phosphate groups. The strain on the phosphate bonds accounts for the energetic quality of ATP, because removal of the terminal phosphates releases the bond energy.

Breaking the bonds between two successive phosphates yields adenosine diphosphate (ADP) and adenosine monophosphate (AMP). It is worthwhile noting how economically a cell manages its pool of adenosine nucleotides. AMP derivatives help form the backbone of RNA and are also a major component of certain coenzymes (NAD, FAD, and coenzyme A).

The Metabolic Role of ATP

ATP is the primary energy currency of the cell and when it is used in a chemical reaction it must then be replaced. Therefore ATP utilization and replenishment is an ongoing cycle. In many instances,

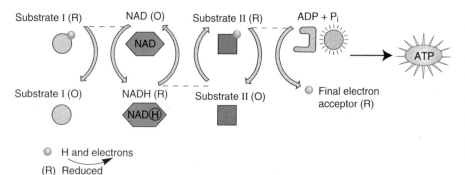

Substrate I (R)   NAD (O)   Substrate II (R)   ADP + P$_i$

Substrate I (O)   NADH (R)   Substrate II (O)   Final electron acceptor (R)   ATP

- H and electrons
(R) Reduced
(O) Oxidized

FIGURE 8.13

The role of electron carriers in redox reactions.
Electrons and hydrogens are transferred by paired or coupled reactions in which the donor compound is oxidized (loses electrons) and the acceptor is reduced (gains electrons). A molecule such as NAD can serve as a transient carrier of the electrons between substrates. During this series of reactions, some energy given off by electron transfer is used to synthesize ATP. To complete the reaction, the electrons are passed to a final electron acceptor.

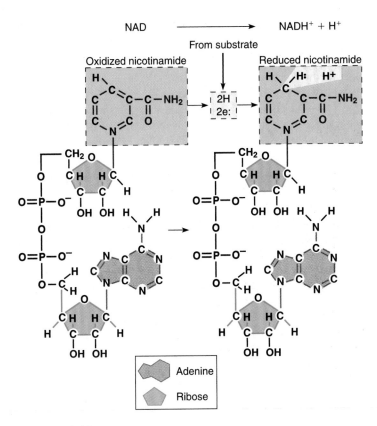

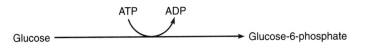

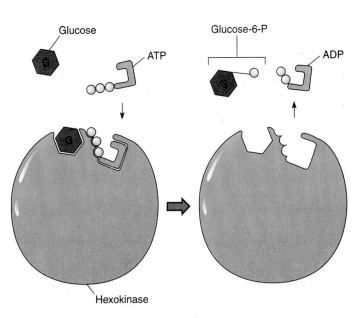

FIGURE 8.14

Details of NAD reduction.   This coenzyme contains the vitamin nicotinamide (niacin) and the purine adenine attached to double ribose phosphate molecules (a dinucleotide). The principal site of action is on the nicotinamide (boxed area). Hydrogens and electrons donated by a substrate interact with a carbon on the top of the ring. One hydrogen bonds there, carrying two electrons (H:), and the other hydrogen is carried in solution as H^+ (a proton).

FIGURE 8.16

An example of phosphorylation of glucose by ATP.   The first step in catabolizing glucose is the addition of a phosphate from ATP by an enzyme called hexokinase. This use of high-energy phosphate as an activator is a recurring feature of many metabolic pathways.

FIGURE 8.15

The structure of adenosine triphosphate (ATP) and its partner compounds, ADP and AMP.

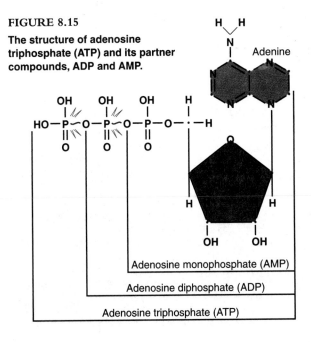

the energy released during ATP hydrolysis powers biosynthesis by activating individual subunits before they are enzymatically linked together. ATP is also used to prepare a molecule for catabolism such as the phosphorylation of a 6-carbon sugar during the early stages of glycolysis (**figure 8.16**).

When ATP is utilized by the removal of the terminal phosphate to release energy plus ADP, ATP then needs to be replenished. The reversal of this process; that is, adding the terminal phosphate to ADP, will replenish ATP.

$$ATP \leftrightarrows ADP + Pi + Energy$$

In heterotrophs, the energy infusion that regenerates a high-energy phosphate comes from certain steps of catabolic pathways, in which nutrients such as carbohydrates are degraded and yield energy. ATP is formed when substrates or electron carriers add a high-energy phosphate bond to ADP. Some ATP molecules are formed through *substrate-level phosphorylation*. In substrate-level phosphorylation, ATP is formed by transfer of a phosphate group from a phosphorylated compound (substrate) directly to ADP to yield ATP (**figure 8.17**).

Substrate

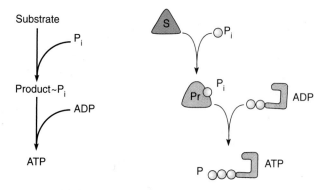

FIGURE 8.17

ATP formation at the substrate level shown in outline and illustrated form. The inorganic phosphate (P_i) and the substrates form a bond with high potential energy. In a reaction catalyzed by ATP synthase, the phosphate is transferred to ADP, thereby producing ATP.

Other ATPs are formed through *oxidative phosphorylation,* a series of redox reactions occurring during the final phase of the respiratory pathway. Phototrophic organisms have a system of *photophosphorylation,* in which the ATP is formed through a series of sunlight-driven reactions (discussed in more detail in chapter 26).

CHAPTER CHECKPOINTS ✓✓✓

All metabolic processes require the constant input and expenditure of some form of usable energy. Chemical energy is the currency that runs the metabolic processes of the cell, but many forms of energy are involved in cell metabolism.

Chemical energy is obtained from the electrons of nutrient molecules through catabolism. It is used to perform the cellular "work" of biosynthesis, movement, membrane transport, and growth.

Energy is extracted from nutrient molecules by redox reactions. A redox pair of substances passes electrons and hydrogens between them. The donor substance loses electrons, becoming oxidized. The acceptor substance gains electrons, becoming reduced.

ATP is the energy molecule of the cell. It donates free energy to anabolic reactions and is continuously regenerated by three phosphorylation processes: substrate phosphorylation, oxidative phosphorylation, and (in certain organisms) photophosphorylation.

Pathways of Bioenergetics

One of the scientific community's greatest achievements was deciphering the biochemical pathways of cells. Initial work with bacteria and yeasts, followed by studies with animal and plant cells, clearly demonstrated metabolic similarities and strongly supported the concept of the universality of metabolism. The study of the production and use of energy by cells is called **bioenergetics,** including catabolic routes that degrade nutrients and anabolic routes that are involved in cell synthesis. Although these pathways are interconnected and interdependent, anabolic pathways are not simply reversals of catabolic ones. At first glance it might seem more economical to use identical pathways, but having different enzymes and divided pathways allows anabolism and catabolism to proceed simultaneously without interference. For simplicity, we shall focus our discussion on the most common catabolic pathways that will illustrate general principles of other pathways as well.

CATABOLISM: AN OVERVIEW OF NUTRIENT BREAKDOWN AND ENERGY RELEASE

The primary catabolism of fuels that results in energy release in many organisms proceeds through a series of three coupled pathways:

1. **glycolysis,**[*] also called the Embden-Meyerhof-Parnas (EMP) pathway;
2. the **tricarboxylic acid cycle (TCA),** also known as the citric acid or Krebs cycle;[2] and
3. the **respiratory chain** (electron transport and oxidative phosphorylation).

Each segment of the pathway is responsible for a specific set of actions on various products of glucose. The interconnections of these pathways in aerobic respiration are represented in figure 8.18, and their reactions are summarized in **table 8.3.** In the following sections, we will observe each of these pathways in greater detail.

---

2. The EMP pathway is named for the biochemists who first outlined its steps. TCA refers to the involvement of several organic acids containing three carboxylic acid groups, citric acid being the first tricarboxylic acid formed; Krebs is in honor of Sir Hans Krebs who, with F. A. Lipmann, delineated this pathway, an achievement for which they won the Nobel Prize in 1953.

[*]glycolysis (gly-kol′-ih-sis) Gr. *glykys,* sweet, and *lysis,* a loosening.

TABLE 8.3

Metabolic Strategies Among Heterotrophic Microorganisms

Scheme	Pathways Involved	Final Electron Acceptor	Products	Chief Microbe Type
Aerobic respiration	Glycolysis, TCA cycle, electron transport	O_2	ATP, CO_2, H_2O	Aerobes; facultative anaerobes
Anaerobic metabolism				
Fermentative	Glycolysis	Organic molecules	ATP, CO_2, ethanol, lactic acid[*]	Facultative, aerotolerant, strict anaerobes
Respiration	Glycolysis, TCA cycle, electron transport	Various inorganic salts (NO_3^-, SO_4^{-2}, CO_3^{-3})	CO_2, ATP, organic acids, H_2S, CH_4, N_2	Anaerobes; some facultatives

[*]*The products of microbial fermentations are extremely varied and include organic acids, alcohols, and gases.*

ENERGY STRATEGIES IN MICROORGANISMS

Nutrient processing is extremely varied, especially in bacteria, yet in most cases it is based on three basic catabolic pathways. In previous discussions, microorganisms were categorized according to their requirement for oxygen gas, and this requirement is related directly to their mechanisms of energy release. As we shall see, **aerobic respiration** is a series of reactions (glycolysis, the TCA cycle, and the respiratory chain) that converts glucose to CO_2 and gives off energy (review figure 8.12). It relies on free oxygen as the final acceptor for electrons and hydrogens and produces a relatively large amount of ATP. Aerobic respiration is characteristic of many bacteria, fungi, protozoa, and animals, and it is the system we will emphasize here. Facultative and aerotolerant anaerobes may use only the glycolysis scheme to incompletely oxidize or **ferment** glucose. In this case, oxygen is not required, organic compounds are the final electron ac-

ceptors, and a relatively small amount of ATP is produced. Some strictly anaerobic microorganisms metabolize by means of **anaerobic respiration.** This system involves the same three pathways as aerobic respiration, but it does not use molecular oxygen as the final electron acceptor but instead, NO_3^-, SO_4^{-2}, CO_3^{-3}, are utilized as the final electron acceptor. Aspects of fermentation and anaerobic respiration are covered in subsequent sections of this chapter.

Aerobic Respiration

Aerobic respiration is a series of enzyme-catalyzed reactions in which electrons are transferred from fuel molecules such as glucose to oxygen as a final electron acceptor. This pathway is the principal energy-yielding scheme for aerobic heterotrophs, and it provides both ATP and metabolic intermediates for many other pathways in the cell, including those of protein, lipid, and carbohydrate synthesis (**figure 8.18**).

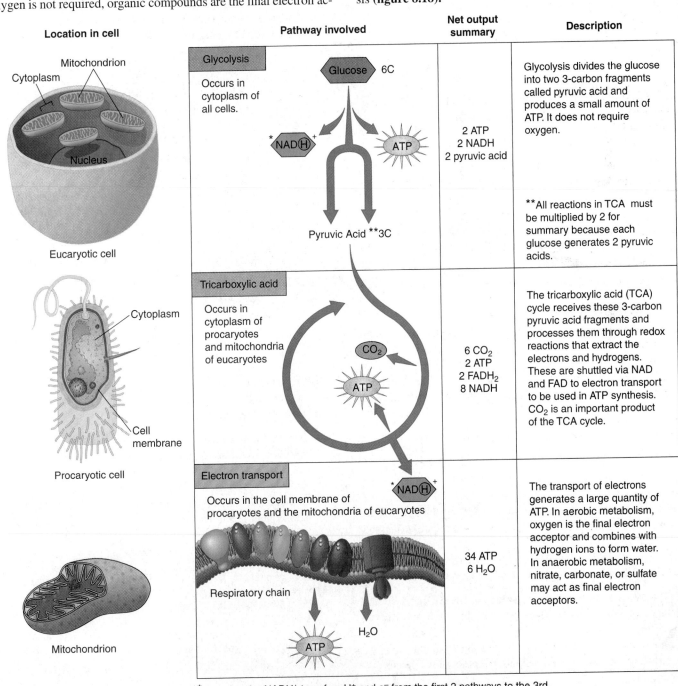

*Note that the NADH⁺ transfers H⁺ and e⁻ from the first 2 pathways to the 3rd.

FIGURE 8.18

Overview of the flow, location, and products of pathways in aerobic respiration. Glucose is degraded through a gradual stepwise process to carbon dioxide and water, while simultaneously extracting energy as ATP and NADH.

Aerobic respiration in microorganisms can be summarized by an equation:

$$\text{Glucose } (C_6H_{12}O_6) + 6\,O_2 + 38\,\text{ADP} + 38\,P_i \rightarrow$$

$$6\,CO_2 + 6\,H_2O + 38\,\text{ATP}$$

The seeming simplicity of the equation for aerobic respiration conceals its complexity. Fortunately, we do not have to present all of the details to address some important concepts concerning its reactants and products, as follows

1. the steps in the oxidation of glucose,
2. the involvement of coenzyme carriers and the final electron acceptor,
3. where and how ATP originates,
4. where carbon dioxide originates,
5. where oxygen is required, and
6. where water originates.

Glucose: The Starting Compound Carbohydrates such as glucose are good fuels because these compounds are readily oxidized; that is, they are superior hydrogen and electron donors. The enzymatic withdrawal of hydrogen from them also removes electrons that can be used in energy transfers. The end products of the conversion of these carbon compounds are energy-rich ATP and energy-poor carbon dioxide and water. Polysaccharides (starch, glycogen) and disaccharides (maltose, lactose) are stored sources of glucose for the respiratory pathways. Although we use glucose as the main starting compound, other hexoses (fructose, galactose) and fatty acid subunits can enter the pathways of aerobic respiration as well, as we will see in a later section of this chapter.

Glycolysis: The Starting Lineup
A process called glycolysis enzymatically converts glucose through several steps into pyruvic acid. Depending on the organism and the conditions, it may be only the first phase of aerobic respiration, or it may serve as the primary metabolic pathway (fermentation). Glycolysis provides a significant means to synthesize a small amount of ATP anaerobically and also to generate pyruvic acid, an essential intermediary metabolite.

Steps in the Glycolytic Pathway Glycolysis, or the EMP pathway, proceeds along nine linear steps, starting with glucose and ending with pyruvic acid. The first portion of glycolysis involves activation of the substrate, and the steps following involve oxidation reactions of the glucose fragments, the synthesis of ATP, and the formation of pyruvic acid. *Although each step of metabolism is catalyzed by a specific enzyme, we will not mention it for most reactions.* The following outline lists the principal steps of glycolysis.

1. **Glucose** is phosphorylated by means of an **ATP** acting with the enzyme hexokinase. The product is **glucose-6-phosphate.** (Numbers in chemical names refer to the position of the phosphate on the carbon skeleton.) This is a way of "priming" the system and keeping the glucose inside the cell.

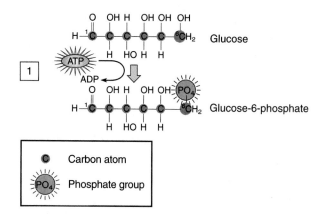

2. Glucose-6-phosphate is converted to its isomer, **fructose-6-phosphate,** by phosphoglucoisomerase.

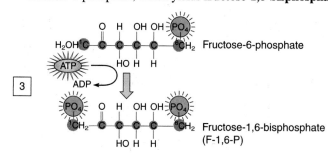

3. Another **ATP** is spent in phosphorylating the first carbon of fructose-6-phosphate, which yields **fructose-1,6-bisphosphate.**

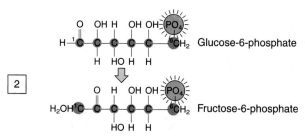

Up to this point, no energy has been released, no oxidation-reduction has occurred, and, in fact, 2 ATPs have been used. In addition, the molecules remain in the 6-carbon state.

4. Now doubly activated, fructose-1,6-bisphosphate is split into two 3-carbon fragments: **glyceraldehyde-3-phosphate (G-3-P)** and **dihydroxyacetone phosphate (DHAP).** These molecules are isomers, and DHAP is enzymatically converted to G-3-P, which is the more reactive form for subsequent reactions.

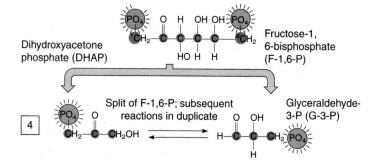

The effect of the splitting of fructose bisphosphate is to double every subsequent reaction, because where there was once a single molecule, there are now two to be fed into the remaining pathways.

5. Each molecule of glyceraldehyde-3-phosphate becomes involved in the single oxidation-reduction reaction of glycolysis, a reaction that sets the scene for ATP synthesis. Two reactions occur simultaneously and are catalyzed by the same enzyme, called glyceraldehyde-3-phosphate dehydrogenase. First, the coenzyme **NAD** picks up hydrogen from G-3-P, forming **NADH.** This step is followed by the addition of an inorganic phosphate (PO_4^{-3}) to form a high-energy bond on the third carbon of the G-3-P substrate. The product of these reactions is **bisphosphoglyceric acid.**

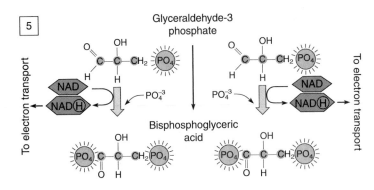

In aerobic organisms, the NADH formed during this step will undergo further reactions in the electron transport system, where the final H acceptor will be oxygen, and an additional 3 ATPs will be generated per NADH. In organisms that ferment glucose anaerobically, the NADH will be oxidized back to NAD, and the hydrogen acceptor will be an organic compound (see figure 8.25).

6. One of the high-energy phosphates of bisphosphoglyceric acid is donated to **ADP** via substrate-level phosphorylation resulting in a molecule of **ATP.** The product of this reaction is **3-phosphoglyceric acid.**

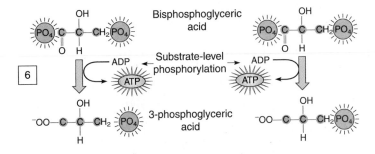

7, 8. During this phase, a substrate for the synthesis of a second ATP is produced in two substeps. First, 3-phosphoglyceric acid is converted to **2-phosphoglyceric acid** through the shift of a phosphate from the third to the second carbon.

Then, the removal of a water molecule from 2-phosphoglyceric acid produces **phosphoenolpyruvic acid** and generates another **high-energy phosphate.**

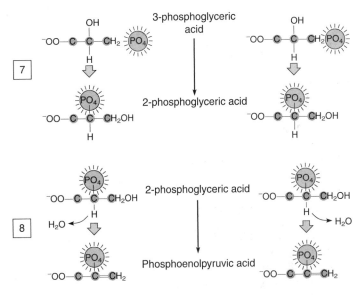

9. In the final reaction of glycolysis, phosphoenolpyruvic acid gives up its high-energy phosphate to form a second **ATP.** Again via substrate-level phosphorylation. This reaction, catalyzed by pyruvate kinase, also produces **pyruvic acid** (pyruvate), a compound with many roles in metabolism.

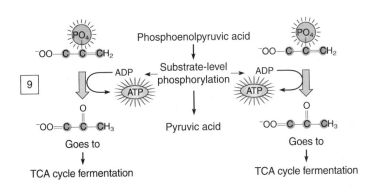

The 2 ATPs formed during steps 6 and 9 are examples of substrate-level phosphorylation, in that the high-energy phosphate is transferred directly from a substrate to ADP. These reactions are catalyzed by kinases, special types of transferase that can phosphorylate a substrate. Because both molecules that arose in step 4 undergo these reactions, an overall total of **4** ATPs is generated in the partial oxidation of a glucose to 2 pyruvic acids. However, 2 ATPs were expended for steps 1 and 3, so the net number of ATPs available to the cell from these reactions is **2.** For a summary refer to **figure 8.19.**

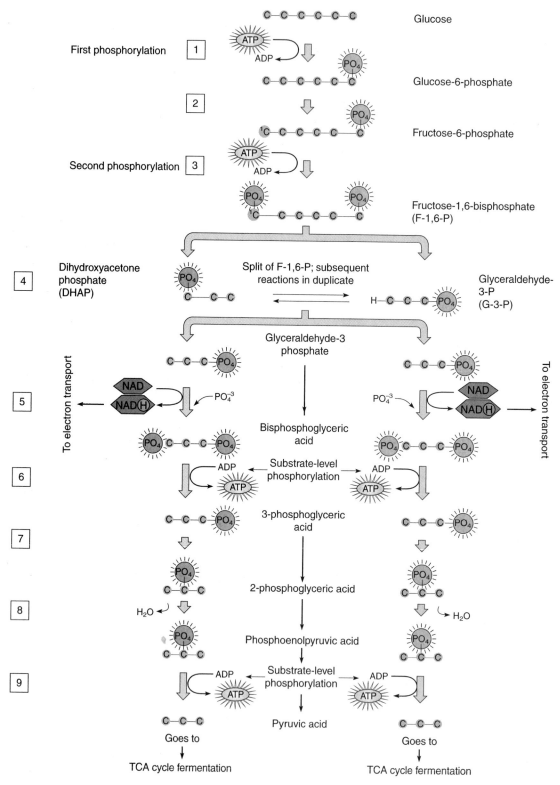

FIGURE 8.19

Summary figure for glycolysis. For more detail of each step see "Glycolysis: The Starting Lineup," page 234.

PYRUVIC ACID—A CENTRAL METABOLITE

Pyruvic acid occupies an important position in several pathways, and different organisms handle it in different ways **(figure 8.20)**. In strictly aerobic organisms and some anaerobes, pyruvic acid enters the TCA cycle for further processing and energy release. Facultative anaerobes can adopt a fermentative metabolism, in which pyruvic acid is further reduced into acids or other products.

THE TRICARBOXYLIC ACID CYCLE—A CARBON AND ENERGY WHEEL

As we have seen, the anaerobic oxidation of glucose yields a comparatively small amount of energy and gives off pyruvic acid. Pyruvic acid is still energy-rich, containing a number of extractable hydrogens and electrons to power ATP synthesis, but this can be achieved only through the work of the second and third phases of respiration, in which pyruvic acid is converted to CO_2 and H_2O (see figure 8.18). In the following section, we will examine the initial phase of this process, that takes place in the mitochondrial matrix in eucaryotes and in the cytoplasm of bacteria.

The Processing of Pyruvic Acid

To connect the glycolysis pathway to the **tricarboxylic acid** cycle, the pyruvic acid is first converted to a starting compound for that cycle **(figure 8.21)**. This step involves the first oxidation-reduction

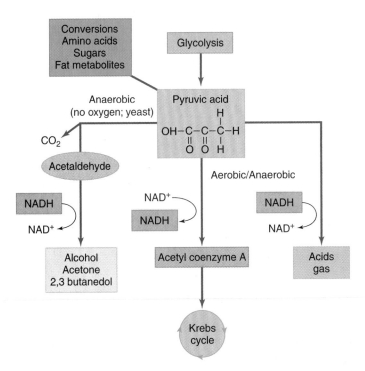

FIGURE 8.20

The fates of pyruvic acid (pyruvate).   This metabolite is an important hub in the processing of nutrients by microbes. It may be fermented anaerobically to several end products or oxidized completely to CO_2 and H_2O through the TCA cycle and the electron transport system. It can also serve as a source of raw material for synthesizing amino acids and carbohydrates.

reaction of this phase of respiration, and it also releases the first carbon dioxide molecule. It involves a cluster of enzymes and **coenzyme A** that participate in the dehydrogenation (oxidation) of pyruvic acid, the reduction of NAD to NADH, and the decarboxylation of pyruvic acid to a 2-carbon **acetyl** group. The acetyl group remains attached to coenzyme A, forming **acetyl coenzyme A** (acetyl CoA) that feeds into the TCA cycle. Pyruvate dehydrogenase, the enzyme complex that makes this reaction possible, is huge. In *E. coli,* it contains four vitamins and has a molecular weight of 6 million—larger than a ribosome!

The NADH formed during this reaction will be shuttled into electron transport and used to generate ATP via oxidative phosphorylation; its formation is one of five dehydrogenations associated with the TCA cycle and accounts for the greater output of ATP in aerobic respiration. **Keep in mind that all reactions described actually happen twice for each glucose because of the two pyruvates that are released during glycolysis.** An important feature of this pathway is that acetyl groups from the breakdown of certain fats can enter the pathway at this same point (see figure 8.27).

The acetyl groups are subsequently fed into the tricarboxylic acid cycle and combined with a 4-carbon oxaloacetate molecule to form citric acid, a 6-carbon molecule. Although this seems a cumbersome way to dismantle such a small molecule, it is necessary in biological systems for extracting larger amounts of energy from the remaining fragment of the acetyl groups.

Steps in the TCA Cycle

As we have seen, a cyclic pathway is one in which the starting compound is regenerated at the end. The tricarboxylic acid cycle has eight steps, beginning with citric acid formation and ending with **oxaloacetic acid**[3] (figure 8.21). As we take a single spin around the TCA cycle, it will be instructive to keep track of

1. the numbers of carbons of each substrate and product (#C),
2. reactions where CO_2 is generated,
3. the involvement of the electron carriers NAD and FAD, and
4. the site of ATP synthesis.

The reactions in the TCA cycle are:

1. **Oxaloacetic acid** (oxaloacetate; 4C) reacts with the **acetyl group** (2C) on acetyl CoA, thereby forming **citric acid** (citrate; 6C) and releasing coenzyme A so it can join with another acetyl group.
2. Citric acid is converted to its isomer, **isocitric acid** (isocitrate; 6C), to prepare this substrate for the decarboxylation and dehydrogenation of the next step.
3. Isocritic acid is acted upon by an enzyme complex including NAD or NADP (depending on the organism), in a reaction that generates NADH or NADPH, splits off a carbon dioxide, and leaves **α-ketoglutaric acid** (α-ketoglutarate; 5C).

---

3. In biochemistry, the terms used for organic acids appear as either the acid form (oxaloacetic acid) or its salt (oxaloacetate).

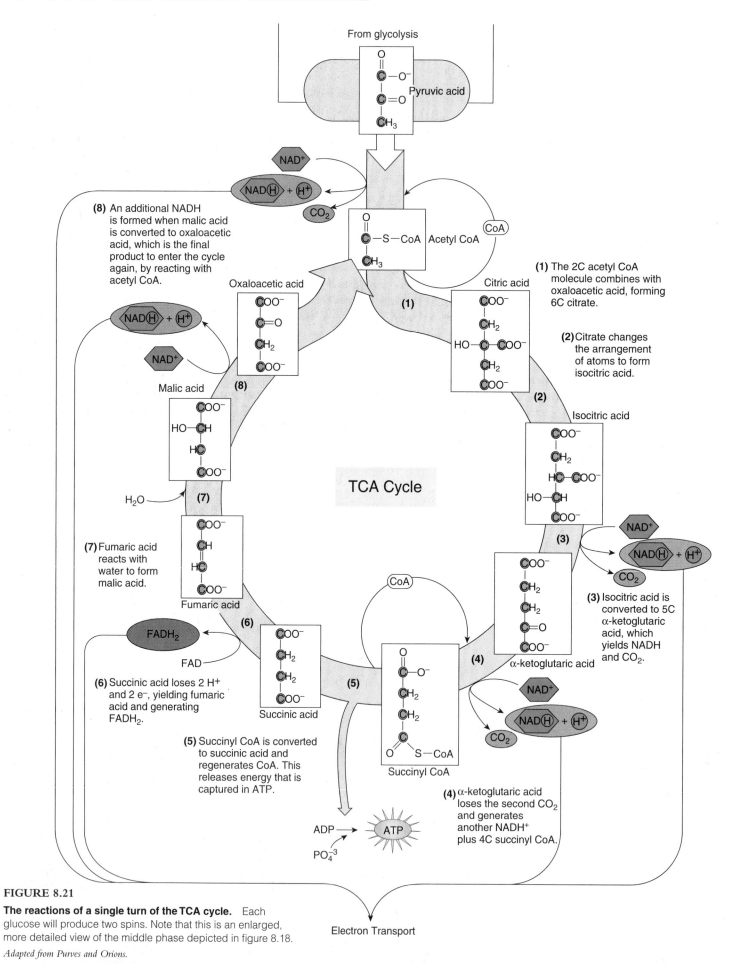

FIGURE 8.21

The reactions of a single turn of the TCA cycle.   Each glucose will produce two spins. Note that this is an enlarged, more detailed view of the middle phase depicted in figure 8.18.

Adapted from Purves and Orions.

4. Alpha-ketoglutaric acid serves as a substrate for the last decarboxylation reaction and yet another redox reaction involving coenzyme A and yielding NADH. The product is the high-energy compound **succinyl CoA** (4C).

At this point, the cycle has completed the formation of 3 CO_2s that balance out the original 3-carbon pyruvic acid that began the TCA. The remaining steps are needed not only to regenerate the oxaloacetic acid to start the cycle again but also to extract more energy from the intermediate compounds leading to oxaloacetic acid.

5. Succinyl CoA is the source of the one substrate level phosphorylation in the TCA cycle. In most microbes, it proceeds with the formation of ATP. The product of this reaction is **succinic acid** (succinate; 4C).
6. Succinic acid next becomes dehydrogenated, but in this case, the electron and H^+ acceptor is **flavin adenine dinucleotide (FAD).** The enzyme that catalyzes this reaction, succinyl dehydrogenase, is found in the bacterial cell membrane and mitochondrial crista of eucaryotic cells. As a consequence, the $FADH_2$ directly enters the electron transport system and bypasses the first step shown in figure 8.23. **Fumaric acid** (fumarate; 4C) is the product of this reaction.
7. The addition of water to fumaric acid (called hydration) results in **malic acid** (malate; 4C). This is one of the few reactions in respiration that directly incorporates water.
8. Malic acid is dehydrogenated (with formation of a final NADH), and **oxaloacetic acid** is formed. This step brings the cycle back to its original starting position, where oxaloacetic acid can react with acetyl coenzyme A.

THE RESPIRATORY CHAIN: ELECTRON TRANSPORT AND OXIDATIVE PHOSPHORYLATION

We now come to the energy chain, which is the final "processing mill" for electrons and hydrogen and the major generator of ATP. Overall, the electron transport system (ETS) consists of a chain of special redox carriers that receive electrons from reduced carriers (NADH, $FADH_2$) generated by glycolysis and the TCA cycle and

shuttle them in a sequential and orderly fashion (see figure 8.18). The flow of electrons down this chain is highly energetic and gives off ATP at various points. The step that finalizes the transport process is the acceptance of electrons and hydrogen by oxygen, producing water. Some variability exists from one organism to another, but the principal compounds that carry out these complex reactions are NADH dehydrogenase, flavoproteins, coenzyme Q **(ubiquinone),*** and **cytochromes.*** The cytochromes contain a tightly bound metal atom at their center that is actively involved in accepting electrons and donating them to the next carrier in the series. The highly compartmentalized structure of the respiratory chain is an important factor in its function. Note in **figure 8.22** that the electron transport carriers and enzymes are embedded in the inner mitochondrial membranes in eucaryotes. The equivalent structure for housing them in bacteria is the cell membrane.

Elements of Electron Transport: The Energy Cascade

The principal questions about the electron transport system are: How are the electrons passed from one carrier to another in the series? How is this progression coupled to ATP synthesis? and, Where and how is oxygen utilized? Although the biochemical details of this process are rather complicated, the basic reactions consist of a number of redox reactions now familiar to us. In general, the seven carrier compounds and their enzymes are arranged in linear sequence and are reduced and oxidized in turn.

The sequence of electron carriers in the respiratory chain of most aerobic organisms is

1. NADH dehydrogenase, which is closely associated in a complex with the adjacent carrier, which is
2. flavin mononucleotide (FMN);
3. coenzyme Q;
4. cytochrome *b;*

---

*ubiquinone (yoo-bik'-wih-nohn) L. *ubique,* everywhere. A type of chemical, similar to vitamin K, that is very common in cells.

*cytochrome (sy'-toh-krohm) Gr. *cyto,* cell, and *kroma,* color. Cytochromes are pigmented, iron-containing molecules similar to hemoglobin.

FIGURE 8.22

The action sites of the mitochondrion.
(a) Location of the electron transport scheme with respect to cristae membranes and matrix.
(b) Exploded view of electron transport chain, showing orientation of electron carriers along the crista membrane.

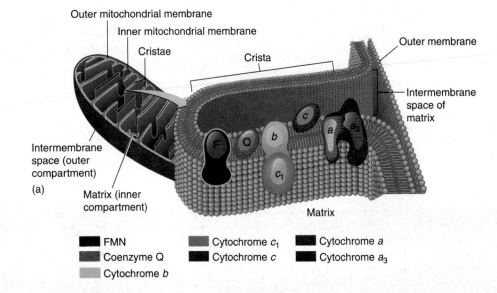

◼ FMN	◼ Cytochrome c_1	◼ Cytochrome a
◼ Coenzyme Q	◼ Cytochrome c	◼ Cytochrome a_3
◼ Cytochrome b		

5. cytochrome c_1;
6. cytochrome c; and
7. cytochromes a and a_3, which are complexed together.

Conveyance of the NADHs from glycolysis and the TCA cycle to the first carrier sets in motion the remaining six steps. With each redox exchange, the energy level of the reactants is lessened. The released energy is captured and used by the **ATP synthase** complex, stationed along the cristae in close association with the ETS carriers. Each NADH that enters electron transport can give rise to 3 ATPs. This coupling of ATP synthesis to electron transport is termed **oxidative phosphorylation.** Since $FADH_2$ from the TCA cycle enters the cycle after the NAD and FMN complex reactions, it has less energy to release, and 2 ATPs result from its processing.

The Formation of ATP and Chemiosmosis

What biochemical processes are involved in coupling electron transport to the production of ATP? We will first look at the system in eucaryotes, which have the components of electron transport embedded in a precise sequence on mitochondrial mem-

branes (figure 8.22). They are stationed between the inner mitochondrial matrix and the outer intermembrane space (**figure 8.23**). According to a widely accepted concept called **chemiosmosis,** as the electron transport carriers shuttle electrons, they actively pump hydrogen ions (protons) into the outer compartment of the mitochondrion. This process sets up a concentration gradient of hydrogen ions called the *proton motive force (PMF)*. The PMF generates a difference in charge between the outer membrane compartment ($+$) and the inner membrane compartment ($-$) (**figure 8.24a**).

Separating the charge has the effect of a battery, which can temporarily store potential energy. This charge will be maintained by the impermeability of the inner cristae membranes to H^+. The only site where H^+ can diffuse into the inner compartment is at the ATP synthase complex, which sets the stage for the final processing of H^+ leading to ATP synthesis.

ATP synthase is a complex enzyme composed of two large units, F_0 and F_1. It is embedded in the membrane but can rotate like a motor and trap chemical energy. As the H^+ ions flow through the F_0 center of the enzyme by diffusion, the F_1 compartments pull in

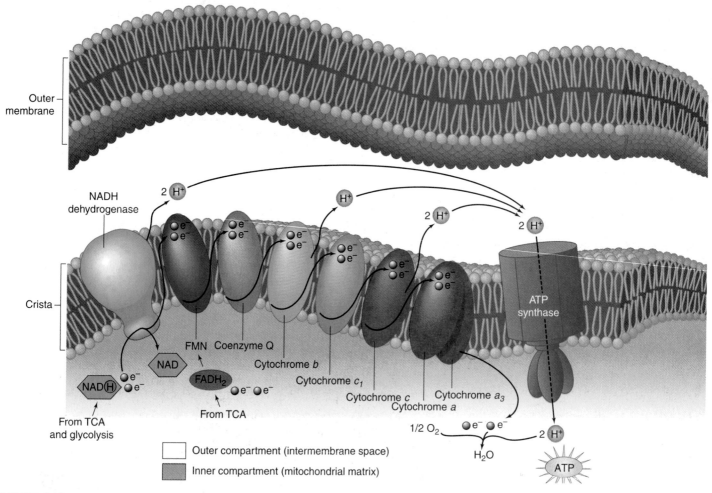

FIGURE 8.23

The electron transport system and oxidative phosphorylation on the mitochondrial crista. Starting at NADH dehydrogenase, electrons brought in from the TCA cycle by NADH are passed along the chain of electron transport carriers. Each adjacent pair of transport molecules undergoes a redox reaction. Coupled to the transport of electrons is the simultaneous active transport of H^+ into the outer compartment by specific carriers. These processes set the scene for ATP synthesis and final H^+ and e^- acceptance by oxygen (see figure 8.24 for details).

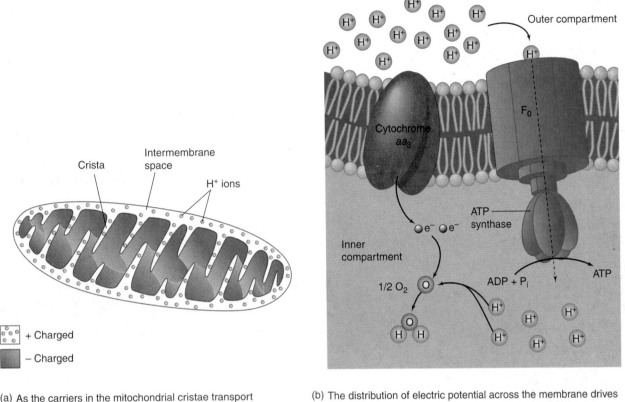

Crista

Intermembrane space

H+ ions

☐ + Charged

■ − Charged

(a) As the carriers in the mitochondrial cristae transport electrons, they also actively pump H+ ions (protons) to the intermembrane space, producing a chemical and charge gradient between the outer and inner mitochondrial compartments.

Outer compartment

F_0

Cytochrome aa_3

ATP synthase

Inner compartment

$1/2\ O_2$

$ADP + P_i$

ATP

(b) The distribution of electric potential across the membrane drives the synthesis of ATP by ATP synthase. The rotation of this enzyme couples diffusion of H+ to the inner compartment with the bonding of ADP and P_i. The final event of electron transport is the reaction of the electrons with the accumulated H+ and O_2 to form metabolic H_2O. This step is catalyzed by cytochrome oxidase (cytochrome aa_3)

Cell wall

Periplasmic space

Cell membrane with ETS

Cytoplasm

ATP synthase

ADP

ATP

(c) Enlarged view of bacterial cell envelope to show the relationship of electron transport and ATP synthesis. Bacteria have the ETS and ATP synthase stationed in the cell membrane. ETS carriers transport H+ and electrons from the cytoplasm to the periplasmic space. Here, it is collected to create a gradient just as it occurs in mitochondria.

FIGURE 8.24

Chemiosmosis—the force behind ATP synthesis.

ADP and P_i. Rotation causes a three-dimensional change in the enzyme that bonds these two molecules, thereby releasing ATP into the inner compartment (figure 8.24b). The enzyme is then rotated back to the start position and will continue the process.

Bacterial ATP synthesis occurs by means of this same overall process. However, bacteria have the ETS stationed in the cell membrane, and the direction of the proton movement is from the cytoplasm to the periplasmic space (figure 8.24c). This difference will affect the amount of ATP produced (discussed in the next section). In both cell types, the chemiosmotic theory has been supported by tests showing that oxidative phosphorylation is blocked if the mitochondrial or bacterial cell membranes are disrupted.

Potential Yield of ATPs from Oxidative Phosphorylation

The total of five NADHs (four from the TCA cycle and one from glycolysis) can be used to synthesize:

15 ATPs for ETS (5 × 2 × 3 per electron pair)

and

15 × 2 = 30 ATPs per glucose

The single FADH produced during the TCA cycle results in:

2 ATPs per electron pair

and

2 × 2 = 4 ATPs per glucose

Table 8.4 summarizes the total of ATP and other products for the entire aerobic pathway. These totals are the potential yields possible but may not be fulfilled by many organisms.

SUMMARY OF AEROBIC RESPIRATION

Originally, we presented a summary equation for respiration. We are now in a position to tabulate the input and output of this equation at various points in the pathways and sum up the final ATP. Close examination of table 8.4 will review several important facets of aerobic respiration:

1. The total possible yield of ATP is 40: 4 from glycolysis, 2 from the TCA cycle, and 34 from electron transport. However, since 2 ATPs were expended in early glycolysis, this leaves a maximum of **38 ATPs.**

The actual totals may be lower in certain eucaryotic cells because energy is expended in transporting the NADH across the mitochondrial membrane. Certain aerobic bacteria come closest to achieving the full total of 38 because they lack mitochondria and thus do not have to use ATP in transport of NADH across the outer mitochondrial membrane.

2. Six carbon dioxide molecules are generated during the TCA cycle.
3. Six oxygen molecules are consumed during electron transport.
4. Six water molecules are produced in electron transport and 2 in glycolysis, but because 2 are used in the TCA cycle, this leaves a net number of 6.

The Terminal Step

The terminal step, during which oxygen accepts the electrons, is catalyzed by cytochrome aa_3, also called cytochrome oxidase. This large enzyme complex is specifically adapted to receive electrons from cytochrome c, pick up hydrogens from solution, and react with oxygen to form a molecule of water (figure 8.24c). This reaction, though in actuality more complex, is summarized as follows:

$$2\,H^+ + 2\,e^- + \tfrac{1}{2}\,O_2 \rightarrow H_2O$$

Most eucaryotic aerobes have a fully functioning cytochrome system, but bacteria exhibit wide-ranging variations in this part of the system. Some species lack one or more of the redox steps; others have several alternative electron transport schemes. Because many bacteria lack cytochrome c oxidase, this variation can be used to differentiate among certain genera of bacteria. The oxidase test (see figure 18.29) can detect this enzyme. It is used to help identify members of the genera *Neisseria* and *Pseudomonas* and some species of *Bacillus*. Another variation in the cytochrome system is evident in certain bacteria (*Klebsiella, Enterobacter*) that can grow even in the presence of cyanide because they lack cytochrome oxidase. Cyanide will cause rapid death in humans and other eucaryotes because it

TABLE 8.4

Summary of Aerobic Respiration for One Glucose Molecule

	Glycolysis*	Net Output	TCA Cycle*	Net Output	Respiratory Chain	Net Output	Total Net Output per Glucose
ATP produced	2 × 2 =	4	1 × 2 =	2	17 × 2 =	34	40 − 2 (used) = 38**
ATP used	2		0		0		
NADH produced	1 × 2 =	2	4 × 2 =	8	0		10
FADH produced	0		1 × 2 =	2	0		2
CO_2 produced	0		3 × 2 =	6	0		6
O_2 used	0		0		3 × 2 =	6	
H_2O produced	2		0		3 × 2 =	6	8 − 2 (used) = 6
H_2O used	0		2		0		

*Products are multiplied by 2 because the first figure represents the amount for only one trip through the pathway, and two molecules make this trip for each glucose.
**This amount can vary among microbes.

blocks cytochrome oxidase, thereby completely terminating aerobic respiration, but it is harmless to these bacteria.

A potential side reaction of the respiratory chain in aerobic organisms is the incomplete reduction of oxygen to superoxide ion (O_2-) and hydrogen peroxide (H_2O_2). Because these toxic oxygen products (previously discussed in chapter 7) can be very damaging to cells, aerobes have various neutralizing enzymes (*superoxide dismutase* and *catalase*). One exception is the genus *Streptococcus,* which can grow well in oxygen yet lacks both cytochromes and catalase. The tolerance of these organisms to oxygen can be explained by the neutralizing effects of a special peroxidase. The lack of cytochromes, catalase, and peroxidases in anaerobes as a rule limits their ability to process free oxygen and contributes to its toxic effects on them.

Alternate Catabolic Pathways

Certain bacteria follow a different pathway in carbohydrate catabolism. The **phosphogluconate pathway** (also called the hexose monophosphate shunt) provides ways to anaerobically oxidize glucose and other hexoses, to release ATP, to produce large amounts of NADPH, and to process pentoses (5-carbon sugars). This pathway, common in heterolactic fermentative bacteria, yields various end products, including lactic acid, ethanol, and carbon dioxide. Furthermore, it is a significant intermediate source of pentoses for nucleic acid synthesis.

ANAEROBIC RESPIRATION

Some bacteria have evolved an anaerobic respiratory system that functions like the aerobic cytochrome system except that it utilizes oxygen-containing salts, rather than free oxygen, as the final electron acceptor. Of these, the nitrate (NO_3^-) and nitrite (NO_2^-) reduction systems are best known. The reaction in species such as *Escherichia coli* is represented as:

$$\text{Nitrate reductase}$$
$$\downarrow$$
$$NO_3^- + NADH \rightarrow NO_2^- + H_2O + NAD^+$$

The enzyme nitrate reductase catalyzes the removal of oxygen from nitrate, leaving nitrite and water as products. A test for this reaction is one of the physiological tests used in identifying bacteria.

Some species of *Pseudomonas* and *Bacillus* possess enzymes that can further reduce nitrite to nitric oxide (NO), nitrous oxide (N_2O), and even nitrogen gas (N_2). This process, called **denitrification,** is a very important step in recycling nitrogen in the biosphere. Other oxygen-containing nutrients reduced anaerobically by various bacteria are carbonates and sulfates. None of the anaerobic pathways produce as much ATP as aerobic respiration.

CHAPTER CHECKPOINTS

Bioenergetics describes metabolism in terms of production, utilization, and transfer of energy by cells.

Catabolic pathways release energy through three pathways: glycolysis, the tricarboxylic acid cycle, and the respiratory electron transport system.

Cellular respiration is described by the nature of the final electron acceptor. Aerobic respiration implies that O_2 is the final hydrogen acceptor. Anaerobic respiration implies that some other molecule is the final hydrogen acceptor. If the final hydrogen acceptor is an organic molecule, the anaerobic process is considered fermentation.

Carbohydrates are preferred cell energy sources because they are superior hydrogen (electron) donors.

Glycolysis is the catabolic processes by which glucose is oxidized and converted into two molecules of pyruvic acid, with a net gain of 2 ATP. This process is also called substrate phosphorylation.

The tricarboxylic acid cycle processes the 3-carbon pyruvic acid and generates three CO_2 molecules. The electrons it releases are transferred to redox carriers for energy harvesting. It also generates 2 ATPs.

The electron transport chain generates free energy through sequential redox reactions collectively called oxidative phosphorylation. This energy is used to generate up to 36 ATP for each glucose molecule catabolized.

THE IMPORTANCE OF FERMENTATION

Of all the results of pyruvate metabolism, probably the most varied is fermentation. Technically speaking, **fermentation*** is the incomplete oxidation of glucose or other carbohydrates in the absence of oxygen. This process uses organic compounds as the terminal electron acceptors and yields a small amount of ATP. Over time, the term *fermentation* has acquired several looser connotations. Originally, Pasteur called the microbial action of yeast during wine production *ferments* **(Historical Highlights 8.4),** and to this day, biochemists use the term in reference to the production of ethyl alcohol by yeasts acting on glucose and other carbohydrates. Fermentation is also what bacteriologists call the formation of acid, gas, and other products by the action of various bacteria on pyruvic acid. The process is a common metabolic strategy among bacteria. Industrial processes that produce chemicals on a massive scale through the actions of microbes are also called fermentations (see chapter 26). Each of these usages is acceptable for one application or another.

It may seem that fermentation would yield only meager amounts of energy (2 ATPs maximum per glucose). What actually happens, however, is that many bacteria can grow as fast as they would in the presence of oxygen. This rapid growth is made possible by an increase in the rate of glycolysis. From another standpoint, fermentation permits independence from molecular oxygen and allows colonization of anaerobic environments. It also enables microorganisms with a versatile metabolism to adapt to variations in the availability of oxygen. For them, fermentation provides a means to grow even when oxygen levels are too low for aerobic respiration.

Bacteria that digest cellulose in the rumens of cattle (as described in chapter 7) are largely fermentative. After initially hydrolyzing cellulose to glucose, they ferment the glucose to organic acids, which are then absorbed as the bovine's principal

---

*fermentation (fur-men-tay′-shun) L. *fervere,* to boil, or *fermentatum,* leaven or yeast.

energy source. Even human muscle cells can undergo a form of fermentation that permits short periods of activity after the oxygen supply in the muscle has been exhausted. Muscle cells convert pyruvic acid into lactic acid, which allows anaerobic production of ATP to proceed for a time. But this cannot go on indefinitely, and after a few minutes, the accumulated lactic acid causes muscle fatigue.

Products of Fermentation in Microorganisms

Alcoholic beverages (wine, beer, whiskey) are perhaps the most prominent among fermentation products; others are solvents (acetone, butanol), organic acids (lactic, acetic), dairy products, and many other foods. Derivatives of proteins, nucleic acids, and other organic compounds are fermented to produce vitamins, antibiotics, and even hormones such as hydrocortisone.

Fermentation products can be grouped into two general categories: alcoholic fermentation products and acidic fermentation products (**figure 8.25**). **Alcoholic fermentation** occurs in yeast species that have metabolic pathways for converting pyruvic acid to ethanol. This process involves a decarboxylation of pyruvic acid to acetaldehyde, followed by a reduction of the acetaldehyde to ethanol. In oxidizing the NADH formed during glycolysis, this step regenerates NAD, thereby allowing the glycolytic pathway to continue. These processes are crucial in the production of beer and wine, though the actual techniques for arriving at the desired amount of ethanol and the prevention of unwanted side reactions are important tricks of the brewer's trade (see Historical Highlights 8.4). Note that the products of alcoholic fermentation are not only ethanol but also CO_2, a gas that accounts for the bubbles in champagne and beer.

Alcohols other than ethanol can be produced during bacterial fermentation pathways. Certain clostridia produce butanol and isopropanol through a complex series of reactions. Although this process was once an important source of alcohols for industrial use, it has been largely replaced by a nonmicrobial petroleum process.

The pathways of **acidic fermentation** are extremely varied (see figure 8.25). Lactic acid bacteria ferment pyruvate in the same way that humans do—by reducing it to lactic acid. If the product of this fermentation is mainly lactic acid, as in certain species of *Streptococcus* and *Lactobacillus*, it is termed *homolactic*. The souring of milk is due largely to the production of this acid by bacteria. When glucose is fermented to a mixture of lactic acid, acetic acid, and carbon dioxide, as is the case with *Leuconostoc* and other species of *Lactobacillus*, the process is termed *heterolactic fermentation*.

Many members of the family Enterobacteriaceae (*Escherichia, Shigella,* and *Salmonella*) possess enzyme systems for converting pyruvic acid to several acids simultaneously (**figure 8.26**). **Mixed acid fermentation** produces a combination of acetic, lactic, succinic, and formic acids, and it lowers the pH of a medium to about 4.0. *Propionibacterium* produces primarily propionic acid, which gives the characteristic flavor to Swiss cheese while fermentation gas (CO_2) produces the holes. Some members also further decompose formic acid completely to carbon dioxide and hydrogen gases. Because enteric bacteria commonly occupy the intestine, this fermentative activity accounts for the accumulation of some types of gas—primarily CO_2 and H_2—in the intestine. Some bacteria reduce the organic acids and produce the neutral end product 2,3-butanediol (**Microbits 8.5**).

We have provided only a brief survey of fermentation products, but it is worth noting that microbes can be harnessed to synthesize a variety of other substances by varying the raw materials provided them. In fact, so broad is the meaning of the word *fermentation* that the large-scale industrial syntheses by microorganisms often utilize entirely different mechanisms from those described here, and they even occur aerobically, particularly in antibiotic, hormone, vitamin, and amino acid production (see chapter 26).

Biosynthesis and the Crossing Pathways of Metabolism

Our discussion now turns from catabolism and energy extraction to anabolic functions and biosynthesis. In this section we will overview aspects of intermediary metabolism, including amphibolic pathways, the synthesis of simple molecules, and the synthesis of macromolecules.

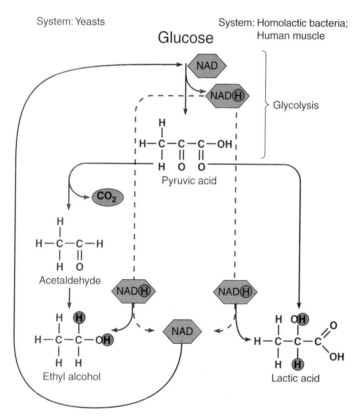

FIGURE 8.25

The chemistry of fermentation systems that produce acid and alcohol. In both cases, the final electron acceptor is an organic compound. In yeasts, pyruvic acid is decarboxylated to acetaldehyde, and the NADH given off in the glycolytic pathway reduces acetaldehyde to ethyl alcohol. In homolactic fermentative bacteria, pyruvic acid is reduced by NADH to lactic acid. Both systems regenerate NAD to feed back into glycolysis or other cycles.

HISTORICAL HIGHLIGHTS 8.4
Pasteur and the Wine-to-Vinegar Connection

The microbiology of alcoholic fermentation was greatly clarified by Louis Pasteur after French wine makers hired him to uncover the causes of periodic spoilage in wines. Especially troublesome was the conversion of wine to vinegar and the resultant sour flavor. Up to that time, wine formation had been considered strictly a chemical process. After extensively studying beer making and wine grapes, Pasteur concluded that wine, both fine and not-so-fine, was the result of microbial action on the juices of the grape and that wine "disease" was caused by contaminating organisms that produced undesirable products such as acid. Although he did not know it at the time, the bacterial contaminants responsible for the acidity of the spoiled wines were likely to be *Acetobacter* or *Gluconobacter* introduced by the grapes, air, or wine-making apparatus. These common gram-negative genera further oxidized ethanol to acetic acid and are presently used in commercial

vinegar production. The following formula shows how this is accomplished:

$$H-\underset{\underset{H}{|}}{\overset{\overset{H}{|}}{C}}-\underset{\underset{H}{|}}{\overset{\overset{H}{|}}{C}}-OH \longrightarrow H-\underset{\underset{H}{|}}{\overset{\overset{H}{|}}{C}}-C\overset{O}{\underset{OH}{\Big\langle}} + 2H^+$$

Ethanol   Acetic acid

Pasteur's far-reaching solution to the problem is still with us today—mild heating, or *pasteurization,* of the grape juice to destroy the contaminants, followed by inoculation of the juice with a pure yeast culture. It also introduced the concept that a single microbe could be responsible for both a desired product and a disease. The topic of wine making is explored further in chapter 26.

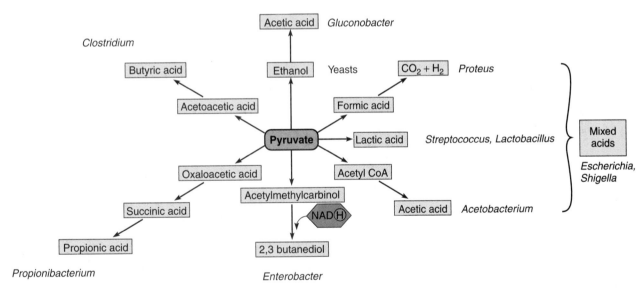

FIGURE 8.26

Miscellaneous products of pyruvate fermentation and the bacteria involved in their production.

THE FRUGALITY OF THE CELL—WASTE NOT, WANT NOT

It must be obvious by now that cells have mechanisms for careful management of carbon compounds. Rather than being dead ends, most catabolic pathways contain strategic molecular intermediates (metabolites) that can be diverted into anabolic pathways. In this way, a given molecule can serve multiple purposes, and the maximum benefit can be derived from all nutrients and metabolites of the cell pool. The property of a system to integrate catabolic and anabolic pathways to improve cell efficiency is termed **amphibolism.***

The amphibolic nature of intermediary metabolism can be better appreciated by looking at an overview of the major metabolic pathways and their points of interaction (**figure 8.27**). The pathways of glucose catabolism are an especially rich "metabolic marketplace." The principal sites of amphibolic interaction occur during glycolysis (glyceraldehyde-3-phosphate and pyruvic acid) and the TCA cycle (acetyl coenzyme A and various organic acids).

***amphibolism** (am-fee-bol'-izm) Gr. *amphi,* two-sided, and *bole,* a throw.

MICROBITS 8.5

Fermentation and Biochemical Testing

The knowledge and understanding of the fermentation products of given species are important not only in industrial production but also in identifying bacteria by biochemical tests. Fermentation patterns in enteric bacteria, for example, are an important identification tool. Specimens are grown in media containing various carbohydrates, and the production of acid or acid and gas is noted (see figure 3.11). For instance, *Escherichia* ferments the milk sugar lactose, whereas *Shigella* and *Proteus* do not. On the basis of its gas production during glucose fermentation,

Escherichia can be further differentiated from *Shigella*, which ferments glucose but does not generate gas. Other enteric bacteria are separated on the basis of whether glucose is fermented to mixed acids or to 2,3-butanediol. *Escherichia coli* produces mixed acids, whereas *Enterobacter* and *Serratia* form primarily 2,3-butanediol. These are part of the IMViC testing described in chapter 20. The M refers to the methyl red test for mixed acids, and the V refers to the Voges-Proskauer test for the butanediol pathway.

Amphibolic Sources of Cellular Building Blocks

Glyceraldehyde-3-phosphate can be diverted away from glycolysis and converted into precursors for amino acid, carbohydrate, and triglyceride (fat) synthesis. (A precursor molecule is a compound that is the source of another compound.) Earlier we noted the numerous directions that pyruvic acid catabolism can take. In terms of synthesis, pyruvate also plays a pivotal role in providing intermediates for amino acids. In the event of an inadequate glucose supply, it serves as the starting point in glucose synthesis from various metabolic intermediates, a process called **gluconeogenesis.***

The acetyl group that starts the TCA cycle is another extremely versatile metabolite that can be fed into a number of synthetic pathways. This 2-carbon fragment can be converted as a single unit into one of several amino acids, or a number of these fragments can be condensed into hydrocarbon chains that are important bases for fatty acid and lipid synthesis. Note that the reverse is also true—fats can be degraded to acetyl and thereby enter the TCA cycle at

---

*gluconeogenesis (gloo″-koh-nee″-oh-gen′-uh-sis) Literally, the formation of "new" glucose.

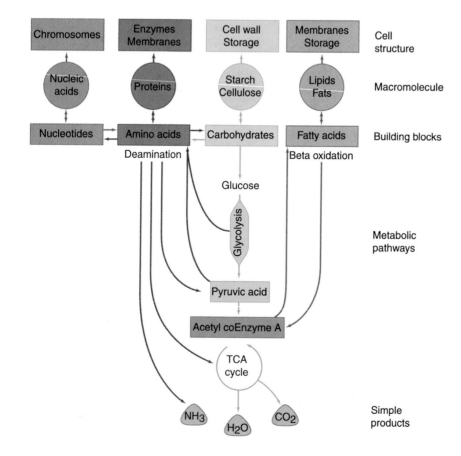

FIGURE 8.27

A summary of metabolic interactions. Intermediate compounds such as pyruvic acid and acetyl coenzyme A serve an amphibolic function. With comparatively small modifications, these compounds can be converted into other compounds and enter a different pathway. Note that catabolism of glucose (center) furnishes numerous intermediates for anabolic pathways that synthesize amino acids, fats, nucleic acids, and carbohydrates. These building blocks can serve in further synthesis of larger molecules to construct various cell components.

acetyl coenzyme A. This aerobic process, called **beta oxidation,** can provide a large amount of energy. Oxidation of a 6-carbon fatty acid yields 50 ATP, compared with 38 for a 6-carbon sugar.

Two metabolites of carbohydrate catabolism that the TCA cycle spins off, oxaloacetic acid and α-ketoglutaric acid, are essential intermediates in the synthesis of certain amino acids. This occurs through **amination,** the addition of an amino group to a carbon skeleton (**figure 8.28a**). A certain core group of amino acids can then be used to synthesize others. Amino acids and carbohydrates can be interchanged through transamination (figure 8.28b).

Pathways that synthesize the nitrogen bases (purines, pyrimidines), which are components of DNA and RNA, originate in amino acids and so can be dependent on intermediates from the TCA cycle as well. Because the coenzymes NAD, NADP, FAD, and others contain purines and pyrimidines similar to the nucleic acids, their synthetic pathways are also dependent on amino acids. During times of carbohydrate deprivation, organisms can likewise convert amino acids to intermediates of the TCA cycle by deamination (removal of an amino group) and thereby derive energy from proteins. Deamination results in the formation of nitrogen waste products such as ammonium ions or urea (figure 8.28c).

Formation of Macromolecules

Monosaccharides, amino acids, fatty acids, nitrogen bases, and vitamins—the building blocks that make up the various macromolecules and organelles of the cell—come from two possible sources. They can enter the cell from the outside as nutrients, or they can be synthesized through various cellular pathways. The degree to which an organism can synthesize its own building blocks is determined by its genetic makeup, a factor that varies tremendously from group to group. In chapter 7 we found that autotrophs require only CO_2 as a carbon source, a few minerals to synthesize all cell substances, and no organic nutrients. Some heterotrophic organisms (*E. coli,* yeasts) are also very efficient in that they can synthesize all cellular substances from minerals and one organic carbon source such as glucose. Compare this with a strict parasite that has few synthetic abilities of its own and drains most precursor molecules from the host.

Whatever their source, once these building blocks are added to the metabolic pool, they are available for synthesis of polymers by the cell. The details of synthesis vary among the types of macromolecules, but all of them involve the formation of bonds by specialized enzymes and the expenditure of ATP (see figure 8.8a).

Amino Acids, Protein Synthesis, and Nucleic Acid Synthesis

Proteins account for a large proportion of a cell's constituents. They are essential components of enzymes, the cell membrane, the cell wall, and cell appendages. As a general rule, 20 amino acids are needed to make these proteins (see chapter 2). Although some organisms (*E. coli,* for example) have pathways that will synthesize all 20 amino acids, others (especially animals) lack some or all of the pathways for amino acid synthesis and must acquire the essential ones from their diets. Protein synthesis itself is a complex process that requires a genetic blueprint and the operation of intricate cellular machinery, as we shall see in chapter 9.

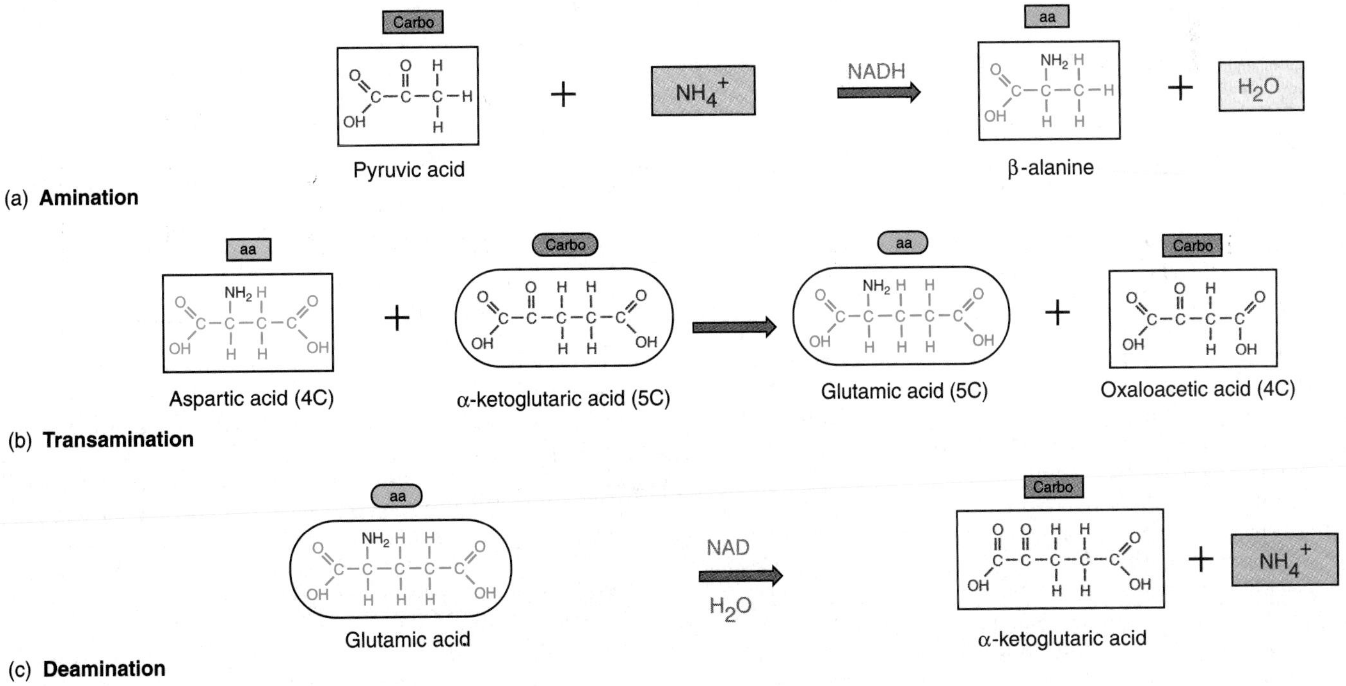

(a) **Amination**

(b) **Transamination**

(c) **Deamination**

FIGURE 8.28

Reactions that produce and convert amino acids. All of them require energy as ATP or NAD and specialized enzymes. **(a)** Through amination (the addition of an ammonium molecule [amino group]), a carbohydrate can be converted to an amino acid. **(b)** Through transamination (transfer of an amino group from an amino acid to a carbohydrate fragment), metabolic intermediates can be converted to amino acids that are in low supply. **(c)** Through deamination (removal of an amino group), an amino acid can be converted to a useful intermediate of carbohydrate catabolism. This is how proteins are used to derive energy. Ammonium is one waste product.

DNA and RNA are important for the hereditary continuity of cells and the overall direction of protein synthesis. Because nucleic acid synthesis is a major topic of genetics and is closely allied to protein synthesis, it will likewise be covered in chapter 9.

Carbohydrate Biosynthesis

The role of glucose in bioenergetics is so crucial that its biosynthesis is ensured by several alternative pathways. Certain structures in the cell depend on an adequate supply of glucose as well. It is the major component of the cellulose cell walls of some eucaryotes and of certain storage granules (starch, glycogen). Monosaccharides other than glucose are important in the synthesis of bacterial cell walls. Peptidoglycan contains a linked polymer of muramic acid and glucosamine. Carbohydrates (deoxyribose, ribose) are also an essential building block in nucleic acids. Polysaccharides are the predominant components of cell surface structures such as capsules and the glycocalyx, and they are commonly found in slime layers (dextran).

CHAPTER CHECKPOINTS

"Intermediary metabolism" refers to the metabolic pathways that use intermediate compounds and connect anabolic and catabolic reactions.

Amphibolic compounds are the "crossroads compounds" of intermediary metabolism. They not only participate in catabolic pathways but also are precursor molecules to biosynthetic pathways.

Biosynthetic pathways utilize building-block molecules from two sources: the environment and the cell's own catabolic pathways. Microorganisms construct macromolecules from these monomers using ATP and specialized enzymes particular to each species.

Proteins are essential macromolecules in all cells because they function as structural constituents, enzymes, and cell appendages.

Carbohydrates are crucial as energy sources, cell wall constituents, and components of nucleotides.

CHAPTER CAPSULE WITH KEY TERMS

I. **Microbial Metabolism**
A. **Metabolism** is the sum of cellular chemical and physical activities; it involves chemical changes to reactants and the release of products using well-established pathways.
B. Metabolism is a complementary process consisting of **anabolism,** synthetic reactions that convert small molecules into large molecules, and **catabolism,** in which large molecules are degraded. Together, they generate thousands of intermediate molecular states, called **metabolites,** which are regulated at many levels.

II. **Enzymes: Metabolic Catalysts**
A. Metabolism is made possible by organic **catalysts,** or **enzymes,** that speed up reactions by lowering the **energy of activation.** Enzymes are not consumed and can be reused. Each enzyme acts specifically upon its matching molecule or **substrate.**
B. *Enzyme Structure:* Depending upon its composition, an enzyme is either **conjugated** or **simple.** A conjugated enzyme consists of a protein component called the **apoenzyme** and one or more activators called **cofactors.** Some cofactors are organic molecules called **coenzymes,** and others are inorganic elements, typically metal ions. To function, a conjugated enzyme must be a complete **holoenzyme** with all its parts. Simple enzymes are composed solely of protein.
C. *Enzyme Specificity:* Substrate attachment occurs in the special pocket called the **active,** or **catalytic, site.** In order to fit, a substrate must conform to the active site of the enzyme. This three-dimensional state is determined by the amino acid content, sequence, and folding of the apoenzyme. Thus, enzymes are usually **substrate-specific.**
D. *Cofactors:* Metallic cofactors impart greater reactivity to the enzyme-substrate complex. Coenzymes such as **NAD** (nicotinamide adenine dinucleotide) are transfer agents that pass functional groups from one substrate to another. Coenzymes usually contain **vitamins.**
E. *Enzyme Classification:* Enzyme names consist of a prefix derived from the type of reaction or the substrate and the ending *-ase.* Thus, an **exoenzyme** is secreted, but an **endoenzyme** is not. Moreover, a **constitutive enzyme** is regularly found in a cell, whereas an **induced enzyme** is synthesized only if its substrate is present.

F. *Types of Enzyme Function:* The release of water that comes with formation of new covalent bonds is a **condensation reaction. Hydrolysis reactions** involve addition of water to break bonds. **Functional groups** may be added, removed, or traded in many reactions. Coupled **redox reactions** transfer electrons and protons (H^+) from one substrate to another. Compounds yielding electrons are another. Compounds yielding electrons are **oxidized,** whereas those gaining electrons are **reduced.**
G. *Enzyme Sensitivity:* Enzymes are **labile** (unstable) and function only within narrow operating ranges of temperature and pH, and they are especially vulnerable to **denaturation.** Enzymes are frequently the targets for physical and chemical agents used in control.

III. **Regulation of Enzymatic Activity**
A. Regulatory controls can act on enzymes directly or on the process that gives rise to the enzymes.
1. A substance that resembles the normal substrate and can occupy the same active site is said to exert **competitive inhibition.**
2. In **feedback** (end product) **control,** the concentration of the product at the end of a pathway blocks the action of a key enzyme by **feedback inhibition.** This is found in **allosteric enzymes,** which have a **regulatory site** different from the active site.
3. Another mechanism, **feedback repression,** inhibits at the genetic level by controlling the synthesis of key enzymes. Enzyme **induction** conserves cell resources by producing enzymes only when the appropriate substrate is present.

IV. **Major Pathways of Bioenergetics**
A. *The Production and Use of Energy:*
1. **Energy** is the capacity of a system to perform work. It is consumed in **endergonic reactions** and is released in **exergonic reactions.** The freed energy is associated with electrons that can be temporarily captured and transferred to high-energy molecules.
2. Extracting energy requires a series of **electron carriers** arrayed in a downhill **redox chain** between electron donors and electron acceptors. In **oxidative phosphorylation,** energy is transferred to high-energy compounds such as ATP.

B. *Principal Pathways in Oxidation of Glucose:* Carbohydrates, such as glucose, are energy-rich because they can yield a large number of electrons per molecule. Glucose is dismantled in stages.
1. **Glycolysis** is a pathway that degrades glucose to **pyruvic acid** in the absence of oxygen.
2. Important intermediates in glycolysis are glucose-6-phosphate, fructose-1,6-bisphosphate, glyceraldehyde-3-phosphate, bisphosphoglyceric acid, phosphoglyceric acids, phosphoenolpyruvate, and pyruvic acid, which enters the second phase of reactions.

C. *Fate of Pyruvic Acid in TCA and Electron Transport:*
1. Pyruvic acid is processed in **aerobic respiration** via the **tricarboxylic acid (TCA) cycle** and its associated **electron transport** chain.
2. **Acetyl coenzyme A** is the product of pyruvic acid processing that undergoes further oxidation and decarboxylation in the TCA cycle, which generates ATP, CO_2, and H_2O.
3. The respiratory chain completes energy extraction. Important redox carriers of the electron transport system are NAD, **FAD** (flavin adenine dinucleotide), **coenzyme Q**, and **cytochromes.**
4. The **chemiosmotic hypothesis** is a conceptual model that explains the origin and maintenance of electropotential gradients across a membrane that leads to ATP synthesis, by **ATP synthase.**
5. The final electron **acceptor** in aerobic respiration is oxygen. In **anaerobic respiration,** sulfate, nitrate, or nitrite serve this function. Bacteria serve as important agents in the nitrogen cycle (**denitrification**).

D. **Fermentation** is anaerobic respiration in which both the electron donor and final electron acceptors are organic compounds.
1. Fermentation enables anaerobic and facultative microbes to survive in environments devoid of oxygen. Production of alcohol, vinegar, and certain industrial solvents relies upon fermentation.
2. The **phosphogluconate pathway** is an alternative anaerobic pathway for hexose oxidation that also provides for the synthesis of NADPH and pentoses.

E. *Versatility of Glycolysis and TCA Cycle:* Many pathways of metabolism are bidirectional, or **amphibolic,** pathways that can be adapted to serving several functions.
1. Metabolites of these pathways double as building blocks and sources of energy. Intermediates such as pyruvic acid are convertible into amino acids through **amination.** Amino acids can be **deaminated** and used as a source of glucose and other carbohydrates (**gluconeogenesis**). Components for purines and pyrimidines are derived from amino acid pathways.
2. Two-carbon acetyl molecules from pyruvate **decarboxylation** can be used in fatty acid synthesis. Combined with glyceraldehyde-3-phosphate, these fatty acids yield triglyceride, a typical storage fat. Alternately, fats can be broken down and fed into the respiratory pathways by **beta oxidation.**

MULTIPLE-CHOICE QUESTIONS

1. _____ is another term for biosynthesis.
   a. catabolism
   b. anabolism
   c. metabolism
   d. catalyst

2. Catabolism is a form of metabolism in which _____ molecules are converted into _____ molecules.
   a. large, small
   b. small, large
   c. amino acid, protein
   d. food, storage

3. An enzyme _____ the activation energy required for a chemical reaction.
   a. increases
   b. converts
   c. lowers
   d. catalyzes

4. An enzyme
   a. becomes part of the final products
   b. is nonspecific for substrate
   c. is consumed by the reaction
   d. is heat and pH labile

5. An apoenzyme is where the _____ is located.
   a. cofactor
   b. coenzyme
   c. redox reaction
   d. active site

6. Many coenzymes are
   a. metals
   b. vitamins
   c. proteins
   d. substrates

7. To digest cellulose in its environment, a fungus produces a/an
   a. endoenzyme
   b. exoenzyme
   c. catalase
   d. polymerase

8. In negative feedback control of enzymes, a buildup in the amount of _____ decreases the activity in the enzyme.
   a. substrate
   b. reactant
   c. product
   d. ATP

9. Energy in biological systems is primarily
   a. electrical
   b. chemical
   c. radiant
   d. mechanical

10. Energy is carried from catabolic to anabolic reactions in the form of _____.
    a. ADP
    b. high-energy ATP bonds
    c. coenzymes
    d. inorganic phosphate

11. Exergonic reactions
    a. release potential energy
    b. consume energy
    c. form bonds
    d. occur only outside the cell

12. A reduced compound is
    a. NAD
    b. FAD
    c. NADH
    d. ADP

13. Most oxidation reactions in microbial bioenergetics involve the
    a. removal of electrons and hydrogens
    b. addition of electrons and hydrogens
    c. addition of oxygen
    d. removal of oxygen

14. A product or products of glycolysis is/are
    a. ATP
    b. H_2O
    c. CO_2
    d. both a and b

15. Fermentation of a glucose molecule has the potential to produce a net number of ____ ATPs.
    a. 4   c. 40
    b. 2   d. 0

16. Complete oxidation of glucose in aerobic respiration can yield a net output of ____ ATP.
    a. 40   c. 38
    b. 6   d. 2

17. The compound that enters the TCA cycle from glycolysis is
    a. citric acid   c. pyruvic acid
    b. oxaloacetic acid   d. acetyl coenzyme A

18. The $FADH_2$ formed during the TCA cycle enters the electron transport system at which site?
    a. NADH dehydrogenase
    b. cytochrome
    c. coenzyme Q
    d. ATP synthase

19. ATP synthase complexes can generate ____ ATPs for each NADH that enters electron transport.
    a. 1   c. 3
    b. 2   d. 4

20. **Matching.** Match the term a, b, or c with the metabolic pathway where it occurs.
    a. glycolysis (Embden-Meyerhof-Parnas)
    b. TCA (Krebs) cycle
    c. electron transport/oxidative phosphorylation
    ____ H^+ and e^- are delivered to O_2 as the final acceptor.
    ____ Pyruvic acid is formed.
    ____ GTP is formed.
    ____ H_2O is produced.
    ____ CO_2 is formed.
    ____ Fructose bisphosphate is split into two 3-carbon fragments.
    ____ $NADH^+$ is oxidized.
    ____ ATP synthase is active.

CONCEPT QUESTIONS

1. a. Describe the chemistry of enzymes and explain how the apoenzyme forms.
   b. Show diagrammatically the interaction of holoenzyme and its substrate and general products that can be formed from a reaction.

2. Give the general name of the enzyme that:
   a. synthesizes ATP; digests RNA
   b. phosphorylates glucose
   c. catalyzes the formation of acetyl from pyruvic acid (just before step 1 in the TCA cycle)
   d. reduces pyruvic acid to lactic acid
   e. reduces nitrate to nitrite

3. Differentiate among the chemical composition and functions of various cofactors. Provide examples of each type.

4. a. Explain what an allosteric enzyme is and how negative feedback works. Two steps in glycolysis are catalyzed by allosteric enzymes. These are: (1) phosphoglucoisomerase, which catabolizes step 2, and (2) pyruvate kinase, which catabolizes step 9.
   b. Suggest what metabolic products might regulate these enzymes.
   c. How might one place these regulators in figure 8.19?

5. Explain how oxidation of a substrate proceeds without oxygen.

6. In the following redox pairs, which compound is reduced and which is oxidized?
   a. NAD and NADH
   b. $FADH_2$ and FAD
   c. lactic acid and pyruvic acid
   d. NO_3^- and NO_2^-
   e. Ethanol and acetaldehyde

7. a. Describe the roles played by ATP and NAD in metabolism.
   b. What particular features of their structure lend them to these functions?

8. Discuss the relationship of:
   a. anabolism to catabolism
   b. ATP to ADP
   c. glycolysis to fermentation
   d. electron transport to oxidative phosphorylation

9. a. What is meant by the concept of the "final electron acceptor"?
   b. What are the final electron acceptors in aerobic, anaerobic, and fermentative metabolism?

10. Name the major ways that substrate-level phosphorylation is different from oxidative phosphorylation.

11. Compare and contrast the location of glycolysis, TCA cycle, and electron transport in procaryotic and eucaryotic cells.

12. a. Outline the basic steps in glycolysis, indicating where ATP is used and given off.
    b. Where does NADH originate, and what is its fate in an aerobe?
    c. What is the fate of NADH in a fermentative organism?

13. a. What is the source of ATP in the TCA cycle?
    b. How many ATPs could be formed from the original glucose molecule?
    c. How does the total of ATPs generated differ between bacteria and many eucaryotes? What causes this difference?

14. Name the sources of oxygen in bacteria that use anaerobic respiration.

15. a. Summarize the chemiosmotic theory of ATP formation.
    b. What is unique about the actions of ATP synthase?

16. How are aerobic and anaerobic respiration different?

17. Compare the general equation for aerobic metabolism with table 8.4 and verify that all figures balance.

18. Water is one of the end products of aerobic respiration. Where in the metabolic cycles is it formed, and where is it used?

19. Briefly outline the use of certain metabolites of glycolysis and TCA in amphibolic pathways.

CRITICAL-THINKING QUESTIONS

1. Using the simplified chart, fill in a summary of the major starting compounds required and the products given off by each phase of metabolism. Use arrows to pinpoint approximately where the reactions take place.

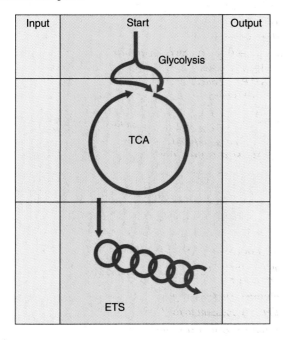

2. Use the graph below to diagram the energetics of a chemical reaction, with and without an enzyme. Be sure to position reactants and products at appropriate points and to indicate the stages in the reaction and the energy levels.

Ames Testing
pg

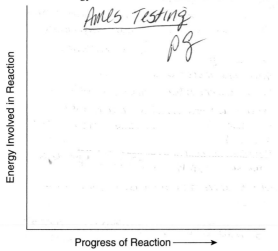

3. Using the concept of fermentation, describe the microbial (biochemical) mechanisms that cause milk to sour.

4. Explain how it is possible for certain microbes to survive and grow in the presence of cyanide, which would kill many other organisms.

5. Explain why allosteric enzyme regulation is considered "fine control" and genetic regulation is "coarse control."

6. Suggest the advantages of having metabolic pathways staged in specific membrane or organelle locations rather than being free in the cytoplasm.

7. What adaptive advantages does a fermentative metabolism confer on a microbe?

8. Explain some of the main functions of vitamins and why they are essential growth factors in human and microbial metabolism.

9. Describe some of the special adaptations of the enzymes found in extremophiles.

10. Beer production requires an early period of rapid aerobic metabolism of glucose by yeast. Given that anaerobic conditions are necessary to produce alcohol, can you explain why this step is necessary?

11. How many ATPs would be formed as a result of aerobic respiration if cytochrome oxidase were missing from the respiratory chain (as is the case with many bacteria)?

12. Draw a model of ATP synthase in three dimensions, showing how it works. Where in the mitochondrion does the ATP supply collect?

13. Microorganisms are being developed to control human-made pollutants and oil spills that are metabolic poisons to animal cells. A promising approach has been to genetically engineer bacteria to degrade these chemicals. What is actually being manipulated in these microbes?

INTERNET SEARCH TOPICS

1. Look up fermentation on the WWW, and outline some of the products made by this process.

2. Find websites that feature three-dimensional views of enzymes. Make simple models of three different enzymes indicating enzyme structure, active site, and substrate.

3. Visit the student Online Learning Center at www.mhhe.com/talaro5. Go to chapter 8, Internet Search Topics, and log on to the available websites to view animations and tutorials of the biochemical pathways and enzymes.

Why are bacteria useful in studying genetics

1) Can be grown rapidly, therefore many generations can be studied in a short period of time. 2) Large populations of essentially identical cells can be cultured from a single parent. 3) Are genetically simple organisms (E. coli = 4,288 genes) 4) Genetic material is readily transferred from one compatible cell to another, Easier to investigate mechanisms of gene function.   Definitions:

Genetics - the study of inheritance or heredity of living things
Gene: - a segment of DNA that encodes for a given trait.
Genotype - sum total of all genes that make up an organism
Phenotype - the visible traits seen when a gene is expressed
Mutation - any permanent change in the nitrogenous base sequence of DNA

Chromosomes
Topoisomers - changes the configuration of DNA
DNA gyrase - produces super coils.
90% of bacterial chro. encode for proteins
remainder tells cells what to do controls + regulates of cellular functions

Extrachro. Circles of DNA
Plasmid - piece of DNA which contains additional genes not found in chro
What organisms have plasmids Some: bacteria, fungi + protozoa
Plasmids in bacteria - Free floating in cytoplasm or connected 2 chrom
Plasmids are not required for cell growth.
genes found in plasmids? Antibiotic resistant genes, F-pili

Tasks performed by DNA
1) storage of genetic information
2) Inheritance to daughter cells
3) Expression of the genetic message (phenotype)

Nucleotides
Consists of nitrogenous base, pentase + phosphate group
Grouped by Structure: 1) Purines - adenine + Guanine
Pyrimidines - Thymine, cytosine and Uracil

DNA	Vs.	RNA
A, T		A, U
double		single
circular		linear
deoxyribose		ribose

DNA Structure
Antiparallel - arrangement is when the phosphase ribose backbone is aligned in opp. direction 5'→3' to 3'→5'
Is a double helix held together by hydrogen bonds
Strands are complimentary
Eg: (Strand 1) 5' ATCCTGCCAT 3'
(Strand 2) 3' TAGGACGGTA 5'

Semiconservative Replication
get 2 identical daughter cells
each has 1 strand of the old chrom. and other strand is new.

DNA Replication
Need to unwind + separate DNA is an enzyme: Helicase
Enzyme which is responsible for making a complementary strand = DNA polymers.

DNA Polymers
Catalyzes the synthesis of a single strand of DNA into double strand in presents of free nucleotides.
highly accurate - one base pairing error 10^6 base pairs
Is able to correct errors
Polymerizes - 1000 nucleotides / second.

Protein Synthesis
Transcription - conversions of DNA master code in mRNA
Enzyme: RNA Polymers
mRNA
contains code in triplets known as codons 1 AA

Translation   includes 2 additional RNA molecules.
Conversion of mRNA code into a polypeptide chain
tRNA carries AA to ribosome - 1st step mRNA binds to 5' ribosomal end
rRNA - functions as the stage for protein synthesis
tRNA - Anticodon is going to bind to the codon   AA attaches to 3'

Mutation
Mutant Chrom. - cells w altered genotype.
wild type Chrom - cells w original non-mutated chrom's

Point Mutation
Replacement of a single nucleotide by a diff nucleotide
A. silent - 3 base codons codes for same AA.
• usually occurs in 3rd base of codons
Eg: GGT + GGA - still code for proline
B. Missense Mutation - changes codon to a different am. Acid
can have mild to server consequences
hydro phobic for H. no change
hydrophobic to hydrophilic - large change (dramatic)   (stop)
C Nonsense Mutation - changes codon to termination codon
always results in termination of protein therefore a shortened
nonfunctional polypeptide will result
will this be a lethal mut. (yes, if protein it was making was essential)
(no if enzyme being produced "non-essential)

Frame shift Mut.
Insertion - addition of 1 or 2 nucleotides into DNA (larger)
Deletion - subtraction of 1 or 2 nucleotides (shorter)
Causes shift in the reading frames, alters transcription
usually results in a non-fuctional protein

Mutagen testing
Mutation - permanent change in nucleotide sequence
Any physical or chemical agent that will ↑'s the rate of mutation

DNA Recombination
An event in which one bacterium donates DNA to another bacterium
A- Conjugation - plasmids, F-pil, mediated
B- Transformation - exchange of free DNA
C- Transduction - bacteriophage mediated transfer of DNA by bacteriophage

Koch's Postulates
1) Infectious agent isolated from diseased host + never from healthy host. 2) Able to isolate infectious agent or artificial media in pure culture.
3) When the infectious agent is innoculated into a susceptible host, it will cause the same disease. 4) Infectious agent than has to be reisolated in pure culture from experimentally infected Host.

Take DNA + Make a protein
Antisense Strand - 5' ATG TTT ATA GCT AGC TAG 3' } Transcription
Sense Strand - 3' TAC AAA TAT CGA TCG ATC 5'
mRNA - 5' AUG UUU AUA GCU AGC UAG 3'
Anticodon - 3' UAC AAA UAU CGA UCG AUC 5'
protein

F-Methionine Phenylalanine F-Methionine Alanine Serine Stop.

MET — PHE
|
peptide bond

Microbial Genetics

1.) Standard concentration of organism using 0.5 McFarland , 100,000,000 orgs/ml. *Kirby Bauer.*
2.) Standard media - Mueller Hinton 4mm
3.) Incubate time 18-24 hrs. @ 36% 1°C UISA, ORSA, MRSA, VRE
4.) Standard disc size, Antibiotic Concentrations.

T he study of modern genetics is really a study of the language of the cell, a special language found in deoxyribonucleic acid—DNA. In this chapter we shall investigate the structure and function of this molecule and explore how it is copied, how its language is interpreted into useful cell products, how it is controlled, how it changes, how it is transferred from one cell to another, and its applications in microbial genetics.

The search for the structure of the genetic material is one of the most compelling stories in biology. See **Historical Highlights 9.1** for a short history of this saga.

Chapter Overview

- Genetics is the study of the transfer of information between biological entities. The molecules most important to this endeavor are DNA, RNA (both of which carry information), and proteins, which carry out most cellular functions and are built using the information in DNA and RNA.
- DNA is a very long molecule composed of small subunits called nucleotides. The sequence of the nucleotides contains the information needed to eventually direct the synthesis of all proteins in the cell.
- Viruses contain various forms of DNA and RNA that are translated by the genetic machinery of their host cells to form functioning viral particles.
- The genetic activities of cells are highly regulated by operons, groups of genes that interact as a unit to control the use or synthesis of metabolic substances.
- The DNA molecule must be replicated for the distribution of genetic material to offspring. When replication is not faithful, permanent changes in the sequence of the DNA, called mutations, can occur. Because mutations may alter the function or expression of genes, they serve as a force in the evolution of organisms.
- Bacteria undergo genetic recombination through the transfer of small pieces of DNA between bacteria as well as the uptake of DNA from the environment.

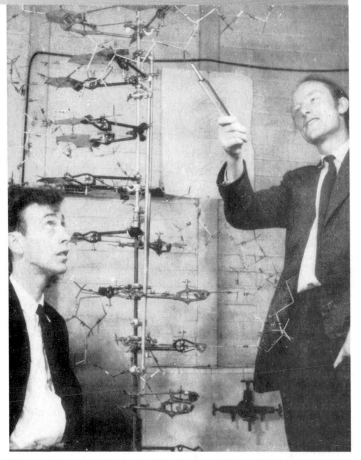

The men who cracked the code of life. Dr. James Watson (left) and Dr. Francis Crick (right) stand next to their model that finally explained the structure of DNA in 1953.

1. the transmission of biological properties (traits) from parent to offspring;
2. the expression and variation of those traits;
3. the structure and function of the genetic material; and
4. how this material changes.

The study of genetics exists on several levels (**figure 9.1**). Organismal genetics observes the heredity of the whole organism or cell; chromosomal genetics examines the characteristics and actions of chromosomes; and molecular genetics deals with the biochemistry of the genes. All of these levels are useful areas of exploration, but in order to understand the expressions of microbial structure, physiology, mutations, and pathogenicity, we need to examine the operation of genes at the cellular and molecular levels. The study of

Introduction to Genetics and Genes: Unlocking the Secrets of Heredity

Genetics* is the study of the inheritance, or **heredity,*** of living things. It is a wide-ranging science that explores

*genetics L. *genesis,* birth, generation.

*heredity L. *hereditas,* heirship. All the characteristics genetically inherited by an organism.

HISTORICAL HIGHLIGHTS   9.1
Deciphering the Structure of DNA

The search for the primary molecules of heredity was a serious focus throughout the first half of the twentieth century. At first many biologists thought that protein was the genetic material. An important milestone occurred in 1944 when Oswald Avery, Colin MacLeod, and Maclyn McCarty purified DNA and demonstrated at last that it was indeed the blueprint for life. This was followed by an avalanche of research, which continues today.

One area of extreme interest concerned the molecular structure of DNA. In 1951, American biologist James Watson and English physicist Francis Crick collaborated on solving the DNA puzzle. Although they did little of the original research, they were intrigued by several findings from other scientists. It had been determined by Erwin Chargaff that any model of DNA structure would have to contain deoxyribose, phosphate, purines, and pyrimidines arranged in a way that would provide variation and a simple way of copying itself. Watson and Crick spent long hours constructing models with cardboard cutouts and kept alert for any and every bit of information that might give them an edge.

Two English biophysicists, Maurice Wilkins and Rosalind Franklin, had been painstakingly collecting data on X-ray crystallographs of DNA for several years. With this technique, molecules of DNA bombarded by X rays produce a photographic image that can predict the three-dimensional structure of the molecule. After being allowed to view certain X-ray data, Watson and Crick noticed an unmistakable pattern: The molecule appeared to be a double helix. Gradually, the pieces of the puzzle fell into place, and a final model was assembled—a model that explained all of the qualities of DNA, including how it is copied (see chapter opening photo). Although Watson and Crick were rightly hailed for the clarity of their solution, it must be emphasized that their success was due to the considerable efforts of a number of English and American scientists. This historic discovery showed that the tools of physics and chemistry have useful applications in biological systems, and it also spawned ingenious research in all areas of molecular genetics.

Since the discovery of the double helix in 1953, an extensive body of biochemical, microscopic, and crystallographic analysis has left little doubt that the model first proposed by Watson and Crick is correct. Newer techniques using scanning tunneling microscopy produce three-dimensional images of DNA magnified 2 million times. These images verify the helical shape and twists of DNA represented by models.

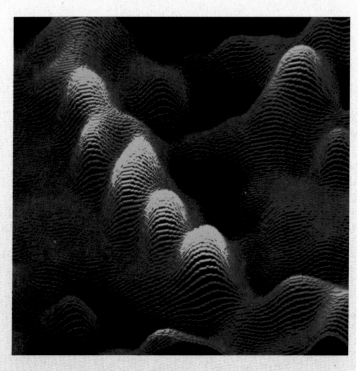

The first direct glimpse at DNA's structure. This false-color scanning tunneling micrograph of calf thymus gland DNA (2,000,000×) brings out the well-defined folds in the helix.

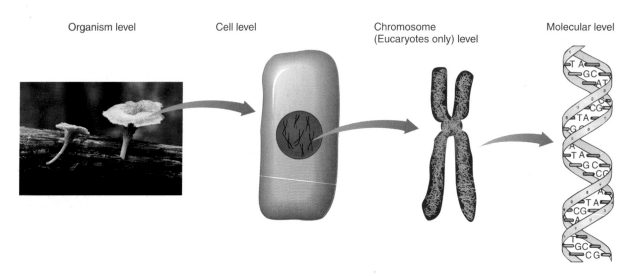

FIGURE 9.1

Levels of genetic study.   The operations of genetics can be observed at the levels of organism, cell, chromosome, and DNA sequence.

microbial genetics provides a greater understanding of human genetics and an increased appreciation for the astounding advances in genetic engineering we are currently witnessing (chapter 10).

THE NATURE OF THE GENETIC MATERIAL

For a species to survive, it must have the capacity of self-replication. In single-celled microorganisms, reproduction involves the division of the cell by means of binary fission or budding, but these forms of reproduction involve a more significant activity than just simple cleavage of the cell mass. Because the genetic material is responsible for inheritance, it must be accurately duplicated and separated into each daughter cell to ensure normal function. This genetic material itself is a long molecule of DNA that can be studied on several levels. Before we look at how DNA is copied, let us explore the organization of this genetic material, proceeding from the general to the specific.

The Levels of Structure and Function of the Genome

The **genome** is the sum total of genetic material of a cell. Although most of the genome exists in the form of chromosomes, genetic material can appear in nonchromosomal sites as well **(figure 9.2)**. For example, bacteria and some fungi contain tiny extra pieces of DNA (plasmids), and certain organelles of eucaryotes (the mitochondria and chloroplasts) are equipped with their own genetic programs. Genomes of cells are composed exclusively of DNA, but viruses contain either DNA or RNA as the principal genetic material. Although the specific genome of an individual organism is unique, the general pattern of nucleic acid structure and function is similar among all organisms.

In general, a **chromosome** is a discrete cellular structure composed of a neatly packaged elongate DNA molecule. The chromosomes of eucaryotes and bacterial cells differ in several respects. The structure of eucaryotic chromosomes consists of a DNA molecule tightly wound around histone proteins (**Microbits 9.2** page 259),

whereas a bacterial chromosome (chromatin body) is condensed and secured into a packet by means of histonelike proteins. Eucaryotic chromosomes are located in the nucleus; they vary in number from a few to hundreds; they can occur in pairs (diploid) or singles (haploid); and they appear elongate. In contrast, most bacteria have a single, circular chromosome, although exceptions exist in a few bacteria that have linear or multiple chromosomes.

The chromosomes of all cells are subdivided into basic informational packets called genes. A **gene** can be defined from more than one perspective. In classical genetics, the term refers to the fundamental unit of heredity responsible for a given trait in an organism. In the molecular and biochemical sense, it is a site on the chromosome that provides information for a certain cell function. More specifically still, it is a certain segment of DNA that contains the necessary code to make a **protein** or RNA molecule. This last definition of a gene will be emphasized in this chapter.

The Size and Packaging of Genomes

Genomes vary greatly in size. The smallest viruses have four or five genes; the bacterium *Escherichia coli* has a single chromosome containing 4,288 genes, and a human cell packs about ten times that many into 46 chromosomes. The exact number is not yet known, but is estimated to be somewhere around 31,000 genes. The total length of DNA relative to cell size is notorious: the chromosome of *E. coli* would measure about 1 mm if unwound and stretched out linearly, and yet this fits within a cell that measures just over 1 μm across, making the DNA 1,000 times longer than the cell **(figure 9.3)**. Still, the bacterial chromosome takes up only about one-third to one-half of the cell's volume. Likewise, if the sum of all DNA contained in the 46 human chromosomes were unraveled and laid end to end, it would measure about 6 feet. This means that the DNA is about 180,000 times longer than a cell 10 μm wide and a million times longer than the width of the nucleus. How can such elongated genomes fit into the miniscule volume of

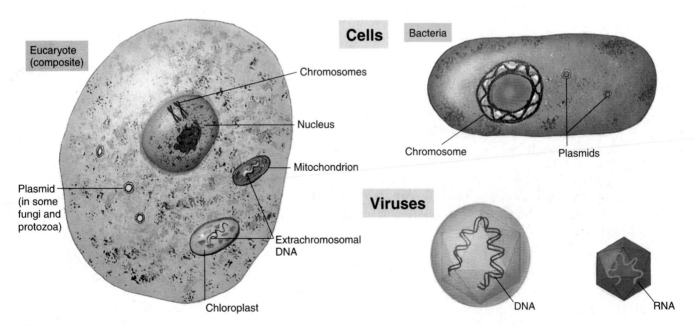

FIGURE 9.2

The general location and forms of the genome in selected cell types and viruses. (not to scale)

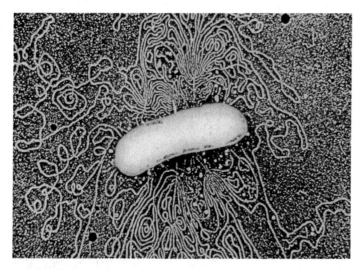

FIGURE 9.3

An *Escherichia coli* cell disrupted to release its DNA molecule. The cell has spewed out its single, uncoiled DNA strand into the surrounding medium.

a cell, and in the case of eucaryotes, into an even smaller compartment, the nucleus? The answer lies in the regular coiling of the DNA chain (Microbits 9.2).

THE DNA CODE: A SIMPLE YET PROFOUND MESSAGE

Examining the function of DNA at the molecular level requires an even closer look at its structure. To do this we will imagine being able to magnify a small piece of a gene about 5 million times. What such fine scrutiny will disclose is one of the great marvels of biology. Our first view of **DNA** in chapter 2 revealed that it is a gigantic molecule, a type of nucleic acid, with two strands combined into a double helix (see figures 2.23 and 2.25). The general structure of DNA is universal, except in some viruses that contain single-stranded DNA. The basic unit of DNA structure is a **nucleotide,** and a gene in a typical bacterium consists of several million nucleotides linked end to end. Each nucleotide is composed of **phosphate, deoxyribose sugar,** and a **nitrogenous base.** The nucleotides covalently bond to form a sugar-phosphate linkage that becomes the backbone of each strand. Each sugar attaches in a repetitive pattern to two phosphates. One of the bonds is to the number 5′ (read "five prime") carbon on deoxyribose, and the other is to the 3′ carbon, which confers a certain order and direction on each strand **(figure 9.4).**

The nitrogenous bases, **purines** and **pyrimidines,** attach by covalent bonds at the 1′ position of the sugar. They span the center of the molecule and pair with appropriate complementary bases from the other side of the helix. The paired bases are so aligned as to be joined by hydrogen bonds. Such weak bonds are easily broken, allowing the molecule to be "unzipped" into its complementary strands. Later we will see that this feature is of great importance in gaining access to the information encoded in the nitrogenous base sequence. Pairing of purines and pyrimidines is not random; it is dictated by the formation of hydrogen bonds between certain bases. Thus, in DNA, the purine **adenine** (A) pairs with the pyrimidine **thymine** (T), and the purine **guanine** (G) pairs with the pyrimidine **cytosine** (C). New research also indicates that the bases are attracted to each other in this pattern because each has a complementary three-dimensional shape that matches its pair. Although the base-pairing partners generally do not vary, the sequence of base pairs along the DNA molecule can assume any order, resulting in an infinite number of possible nucleotide sequences.

Other important considerations of DNA structure concern the nature of the double helix itself. The halves are not parallel or oriented in the same direction. One side of the helix runs in the opposite direction of the other, in an *antiparallel arrangement* (figure 9.4*b*). The order of the bond between the carbon on deoxyribose and the phosphates is used to keep track of the direction of the two sides of the helix. Thus, one helix runs from the 5′ to 3′ direction, and the other runs from the 3′ to 5′ direction. This characteristic is a significant factor in DNA synthesis and translation. As apparently perfect and regular as the DNA molecule may seem, it is not exactly symmetrical. The torsion in the helix and the stepwise stacking of the nitrogen bases produce two different-sized surface features, the major and minor grooves (figure 9.4*c*).

THE SIGNIFICANCE OF DNA STRUCTURE

The arrangement of nitrogenous bases in DNA has two essential effects.

1. **Maintenance of the code during reproduction.** The constancy of base-pairing guarantees that the code will be retained during cell growth and division. When the two strands are separated, each one provides a template (pattern or model) for the replication (exact copying) of a new molecule **(figure 9.5).** Because the sequence of one strand automatically gives the sequence of its partner, the code can be duplicated with fidelity.

2. **Providing variety.** The order of bases along the length of the DNA strand constitutes the genetic program, or the language, of the DNA code. Adding to our earlier definitions, the message present in a gene is a precise arrangement of these bases, and the genome is the collection of all DNA bases that, in an ordered combination, are responsible for the unique qualities of each organism.

It is tempting to ask how such a seemingly simple code can account for the extreme differences among forms as diverse as a virus, *E. coli,* and a human. The English language, based on 26 letters, can create an infinite variety of words, but how can an apparently complex genetic language such as DNA be based on just four nitrogen base "letters"? A mathematical example can explain the possibilities. For a segment of DNA that is 1,000 nucleotides long, there are $4^{1,000}$ different sequences possible. Carried out, this number would approximate 1.5×10^{602}, a number so huge that it literally provides endless degrees of variation.

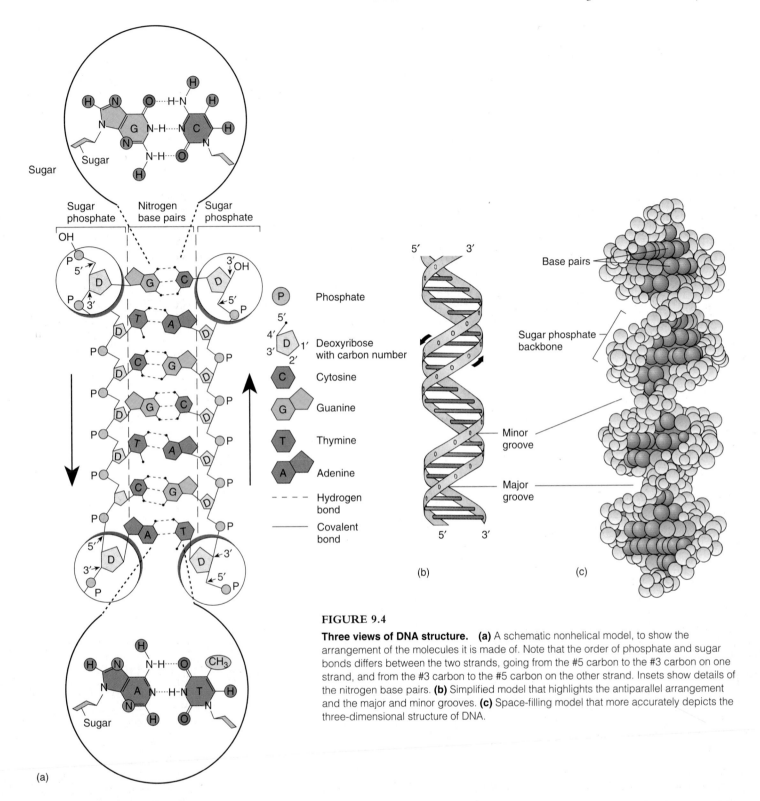

FIGURE 9.4

Three views of DNA structure. **(a)** A schematic nonhelical model, to show the arrangement of the molecules it is made of. Note that the order of phosphate and sugar bonds differs between the two strands, going from the #5 carbon to the #3 carbon on one strand, and from the #3 carbon to the #5 carbon on the other strand. Insets show details of the nitrogen base pairs. **(b)** Simplified model that highlights the antiparallel arrangement and the major and minor grooves. **(c)** Space-filling model that more accurately depicts the three-dimensional structure of DNA.

DNA REPLICATION: PRESERVING THE CODE AND PASSING IT ON

It has been established that the sequence of bases along the length of a gene constitutes the language of DNA. For this language to be preserved for hundreds of generations, it will be necessary for the genetic program to be duplicated and passed on to each offspring. This process of duplication is called **DNA replication.** In the following example, we will show replication in bacteria, but with some exceptions, it also applies to the process as it works in eucaryotes and some viruses. Early in binary fission, the metabolic machinery of a bacterium responds to a message and initiates the duplication of the chromosome. This DNA replication must be completed during a single generation time (around 20 minutes in *E. coli*).

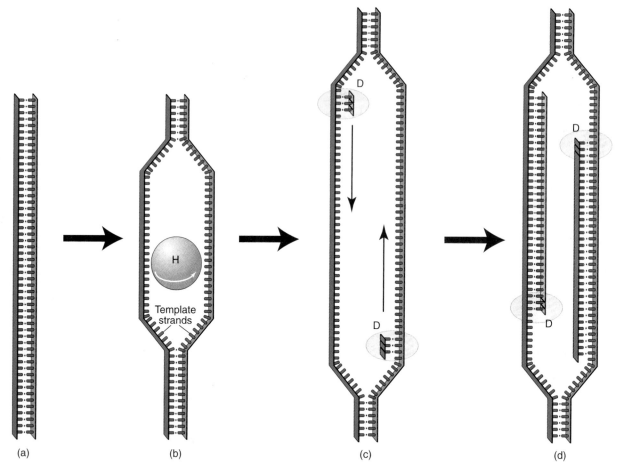

FIGURE 9.5

Simplified steps to show the semiconservative replication of DNA.   **(a, b)** The two strands of the double helix are unwound by a helicase (H), which disrupts the hydrogen bonds and exposes the nitrogen base codes of DNA. Each single strand formed will serve as a template to synthesize a new double strand of DNA. **(c)** A DNA polymerase (D) proceeds along the DNA molecule, attaching the correct nucleotides according to the pattern of the template. An A on the template will pair with a T on the new molecule, and a C will pair with a G. **(d)** The resultant new DNA strands contain one strand of the newly synthesized DNA and the original template strand. The integrity of the code is kept intact because the linear arrangement of the bases is maintained during this process. Note that the actual details of the process are presented in figure 9.6.

The Overall Replication Process

What features allow the DNA molecule to be exactly duplicated, and how is its integrity retained? DNA replication requires a careful orchestration of the actions of 30 different enzymes (partial list in **table 9.1**), which separate the strands of the existing DNA molecule, copy its template, and produce two complete daughter molecules. A simplified version of replication is shown in figure 9.5 and includes the following:

1. uncoiling the parent DNA molecule,
2. unzipping the hydrogen bonds between the base pairs, thus separating the two strands and exposing the nucleotide sequence of each strand (which is normally buried in the center of the helix) to serve as templates, and
3. synthesizing two new strands by attachment of the correct complementary nucleotides to each single-stranded template.

It is worth noting that each daughter molecule will be identical to the parent in composition, but neither one is completely new; the strand that serves as a template is an original parental DNA strand.

TABLE 9.1	
Some Enzymes Involved in DNA Replication and Their Functions	
Enzyme	**Function**
Helicase	Unzipping the DNA helix
Primase	Synthesizing an RNA primer
DNA polymerase III	Adding bases to the new DNA chain; proofreading the chain for mistakes
DNA polymerase I	Removing primer, closing gaps, repairing mismatches
Ligase	Final binding of nicks in DNA during synthesis and repair
Gyrase	Supercoiling

The preservation of the parent molecule in this way, termed **semiconservative replication,** helps explain the reliability and fidelity of replication.

MICROBITS   9.2
The Packaging of DNA: Winding, Twisting, and Coiling

Packing the mass of DNA into the cell involves several levels of DNA structure called supercoils or superhelices. In the simpler system of procaryotes, the circular chromosome is packaged by the action of a special enzyme called a **topoisomerase*** (specifically, DNA **gyrase**).* This enzyme coils the chromosome into a tight bundle by introducing a reversible series of twists into the DNA molecule. The system in eucaryotes is more complex, with three or more levels of coiling. First, the DNA molecule of

a chromosome, which is linear, is wound twice around the histone proteins, creating a chain of **nucleosomes**.* The nucleosomes fold in a spiral formation upon one another. An even greater supercoiling occurs when this spiral arrangement further twists on its radius into a giant spiral with loops radiating from the outside. This extreme degree of compactness is what makes the eucaryotic chromosome visible during mitosis (see figure 5.6). In addition to reducing the volume occupied by DNA, supercoiling solves the problem of keeping the chromosomes from getting tangled during cell division, and it protects the code from massive disruptions due to breakage.

***topoisomerase** (tah"-poh-eye-saw'-mur-ayce) Any enzyme that changes the configuration of DNA.

***gyrase** (jy'-rayce) L. *gyros,* ring or circle. A bacterial enzyme that produces supercoils.

***nucleosomes** "Nucleus bodies" arranged like beads on a chain.

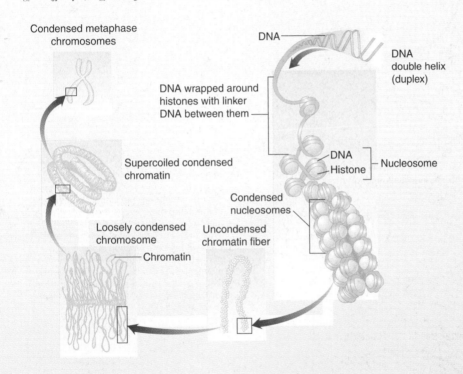

Condensed metaphase chromosomes

DNA wrapped around histones with linker DNA between them

Supercoiled condensed chromatin

Loosely condensed chromosome

Chromatin

Uncondensed chromatin fiber

Condensed nucleosomes

DNA

DNA double helix (duplex)

DNA

Histone

Nucleosome

The packaging of DNA. Eucaryotic DNA undergoes several orders of coiling and supercoiling, which greatly condense it.

Refinements and Details of Replication

The circular bacterial DNA molecule replicates by means of a special configuration called a **replicon,** which is defined as an origin of replication and all of the DNA replicated from that origin. Because a bacterial chromosome contains only a single origin of replication, the entire chromosome consists of a single replicon. The origin of replication is a short sequence rich in adenine and thymine that, you should recall, are held together by only two hydrogen bonds rather than three. Because the origin of replication is AT-rich, less energy is required to separate the two strands than would be required if the origin were rich in guanine and cytosine. Prior to the start of replication enzymes called *helicases* (unzipping enzymes) bind to the DNA at the origin. These enzymes untwist the helix and break the hydrogen bonds holding the two strands together, resulting in two separate strands, each of which will be used as a template for the synthesis of a new strand.

The process of synthesizing a new daughter strand of DNA using the parental strand as a template is carried out by the enzyme DNA polymerase III. The entire process of replication does however depend on several enzymes and can be most easily understood by keeping in mind a few points concerning both the structure of the DNA molecule and the limitations of DNA polymerase III. These points include:

1. The nucleotides that need to be read by DNA polymerase III are buried deep within the double helix. Accessing these nucleotides requires both that the DNA molecule be unwound and that the two strands of the helix be separated from one another.
2. DNA polymerase III is unable to *begin* synthesizing a chain of nucleotides but can only continue to add nucleotides to an already existing chain.

3. DNA polymerase III can only add nucleotides in one direction, so the new strand is always synthesized 5′ to 3′.

With these constraints in mind, the details of replication can be more easily comprehended.

Replication begins when an RNA *primer* is synthesized and enters at the initiation site **(figure 9.6a)**. DNA polymerase III cannot begin synthesis unless it has this short strand of RNA to serve as a starting point for adding nucleotides. Because the bacterial DNA molecule is circular, opening of the circle forms two **replication**

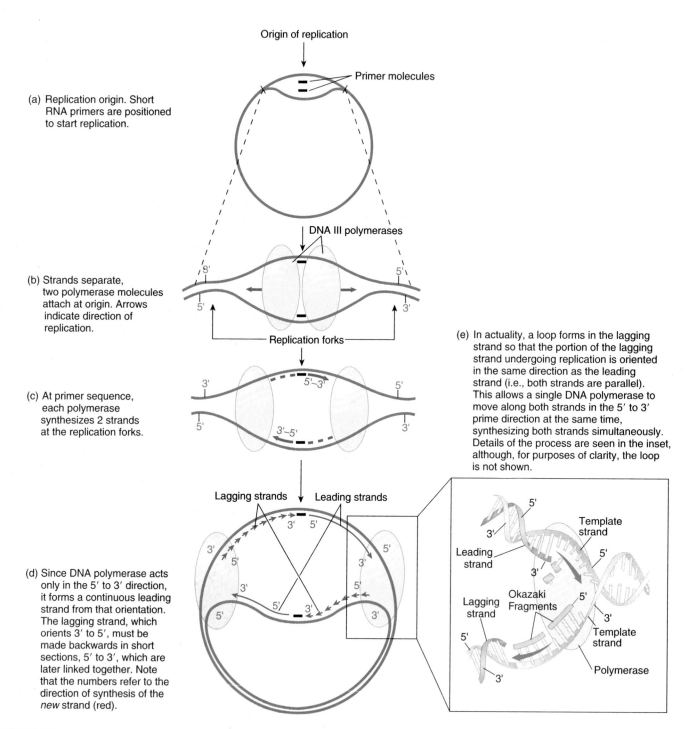

(a) Replication origin. Short RNA primers are positioned to start replication.

(b) Strands separate, two polymerase molecules attach at origin. Arrows indicate direction of replication.

(c) At primer sequence, each polymerase synthesizes 2 strands at the replication forks.

(d) Since DNA polymerase acts only in the 5′ to 3′ direction, it forms a continuous leading strand from that orientation. The lagging strand, which orients 3′ to 5′, must be made backwards in short sections, 5′ to 3′, which are later linked together. Note that the numbers refer to the direction of synthesis of the *new* strand (red).

(e) In actuality, a loop forms in the lagging strand so that the portion of the lagging strand undergoing replication is oriented in the same direction as the leading strand (i.e., both strands are parallel). This allows a single DNA polymerase to move along both strands in the 5′ to 3′ prime direction at the same time, synthesizing both strands simultaneously. Details of the process are seen in the inset, although, for purposes of clarity, the loop is not shown.

FIGURE 9.6

The bacterial replicon: a model for DNA synthesis. **(a)** Circular DNA has a special origin site where replication originates. **(b)** When strands are separated, two replication forks form, and a DNA polymerase III enters at each fork. **(c)** Starting at the primer sequence, both polymerases move along the template strands (blue), synthesizing the new strands (red) at each fork. **(d)** DNA polymerase works only in the 5′ to 3′ direction, necessitating a different pattern of replication at each fork. Because the leading strand orients in the 5′ to 3′ direction, it will be synthesized continuously. The lagging strand, which orients in the opposite direction, can only be synthesized in short sections, 5′ to 3′, which are later linked together. **(e)** Inset presents details of process at one replication fork, and shows the Okazaki fragments and the relationship of the template, leading, and lagging strands.

forks, each containing its own set of replication enzymes. The DNA polymerase III is a huge enzyme complex that encircles the replication fork and adds nucleotides in accordance with the template pattern. As synthesis proceeds, the forks continually open up to expose the template for replication (figure 9.6).

Because DNA polymerase is correctly oriented for synthesis *only* in the 5′ to 3′ direction of the new molecule (red) strand, only one strand, called the **leading strand,** can be synthesized as a continuous, complete strand. The strand with the opposite orientation (3′ to 5′) is termed the **lagging strand** (figure 9.6d,e). Because it cannot be synthesized continuously, the polymerase adds nucleotides a few at a time in the direction away from the fork (5′ to 3′). As the fork opens up a bit, the next segment is synthesized backwards to the point of the previous segment, a process repeated at both forks until synthesis is complete. In this way, the DNA polymerase is able to synthesize the two new strands simultaneously. This manner of synthesis produces one strand containing short fragments of DNA (100 to 1,000 bases long) called **Okazaki fragments.** These fragments are attached to the growing end of the lagging strand by another enzyme called **DNA ligase.**

Elongation and Termination of the Daughter Molecules
The addition of nucleotides proceeds at an astonishing pace, estimated in some bacteria at 750 bases per second at each fork! As replication proceeds, the duplicated strand loops down **(figure 9.7a).**

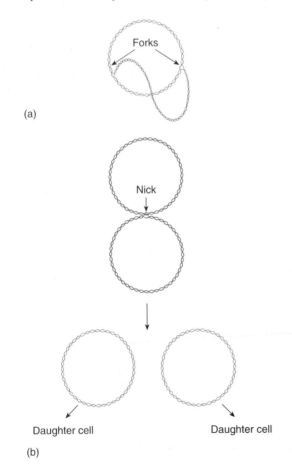

(a)

Nick

(b)

Daughter cell   Daughter cell

FIGURE 9.7

Completion of chromosome replication in bacteria. **(a)** As replication proceeds, one strand loops down. **(b)** Final separation is achieved through repair and the release of two completed molecules. The daughter cells receive these during binary fission.

The DNA polymerase I removes the RNA primers used to initiate DNA synthesis and replaces them with DNA. When the forks come full circle and meet, ligases move along the lagging strand to begin the initial linking of the fragments and to complete synthesis and separation of the two circular daughter molecules (figure 9.7b).

Like any language, DNA is occasionally "misspelled" when an incorrect base is added to the growing chain. Studies have shown that such mistakes are made once in approximately 10^8 to 10^9 bases, but most of these are corrected. If not corrected, they are referred to as mutations and can lead to serious cell dysfunction and even death. Because continued cellular integrity is very dependent on accurate replication, cells have evolved their own proofreading function for DNA. DNA polymerase III, the enzyme that elongates the molecule, can also detect incorrect, unmatching bases, excise them, and replace them with the correct base. DNA polymerase I can also proofread the molecule and repair damaged DNA.

Replication in Other Biological Systems   The replication pattern of eucaryotes is similar to that of procaryotes. It also uses replicons and a variety of DNA polymerases, and replication proceeds both directions from the point of origin. The main difference is that the eucaryotic chromosomes are linear and require hundreds to thousands of replicons acting simultaneously in order to complete replication efficiently.

A novel form of DNA synthesis called **rolling circle** occurs in circular genetic material found in plasmids and some bacterial viruses **(figure 9.8).** In general, it involves the replication of one strand of the parent DNA, forming a single strand that rolls off the circle. This strand is then replicated, thus converting it to a double-stranded duplicate of the original DNA molecule.

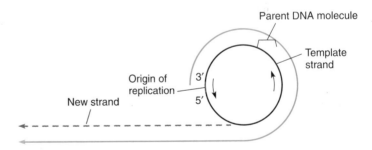

FIGURE 9.8

Simplified model of rolling circle DNA replication.   This pattern occurs in some viruses and plasmids. As the parent DNA rotates, a single strand (green) is synthesized on one template strand, and this new strand is rolled off the circle. A complementary strand (blue) is then formed in sections using this new strand, thus creating a double strand.

CHAPTER CHECKPOINTS

Nucleic acids are molecules that contain the blueprints of life in the form of genes. DNA is the blueprint molecule for all cellular organisms. The blueprints of viruses, however, can be either DNA or RNA.

The total amount of DNA in an organism is termed its genome. The genome of each species contains a unique arrangement of genes that define its appearance (phenotype), metabolic activities, and pattern of reproduction.

The genome of procaryotes is quite small compared with the genomes of eucaryotes. Bacterial DNA consists of a few thousand genes in one circular chromosome. Eucaryotic genomes range from thousands to hundreds of thousands of genes. Their DNA is packaged in tightly wound spirals arranged in discrete chromosomes.

DNA copies itself just before cellular division by the process of semiconservative replication. Semiconservative replication means that each "old" DNA strand is the template upon which each "new" strand is synthesized.

The circular bacterial chromosome is replicated at two forks as directed by DNA polymerase III. At each fork, two new strands are synthesized—one continuously and one in short fragments, and mistakes are proofread and removed.

Applications of the DNA Code: Transcription and Translation

We have explored how the genetic message in the DNA molecule is conserved through replication. Now we must consider the precise role of DNA in the cell. Given that the sequence of bases in DNA is a genetic code, just what is the nature of this code and how is it utilized by the cell? Although the genome is full of critical information, the molecule itself does not perform cell processes directly. Its stored information is conveyed to RNA molecules, which carry out instructions. The concept that genetic information flows from DNA to RNA

to protein is a central theme of molecular biology **(figure 9.9)**. More precisely, it states that the master code of DNA is first used to synthesize an RNA molecule via a process called **transcription,** and the information contained in the RNA is then used to produce proteins in a process known as **translation.** The principal exceptions to this pattern are found in RNA viruses, which convert RNA to other RNA, and in retroviruses, which convert RNA to DNA.

THE GENE-PROTEIN CONNECTION

Genes fall into three basic categories: *structural genes* that code for proteins, genes that code for RNA, and *regulatory genes* that control gene expression. The sum of all of these types of genes constitutes an organism's distinctive genetic makeup, or **genotype.*** The expression of the genotype creates traits (certain structures or functions) referred to as the **phenotype.*** Just as a person inherits a combination of genes (genotype) that gives a certain eye color or height (phenotype), a bacterium inherits genes that direct the formation of a flagellum, and a virus contains genes for its capsid structure.

The Triplet Code and the Relationship to Proteins
Several questions invariably arise concerning the relationship between genes and cell function. For instance, how does gene structure lead to the expression of traits in the individual, and what

*genotype (jee′-noh-typ) Gr. *gennan,* to produce, and *typos,* type.

*phenotype (fee′-noh-typ) Gr. *phainein,* to show. The physical manifestation of gene expression.

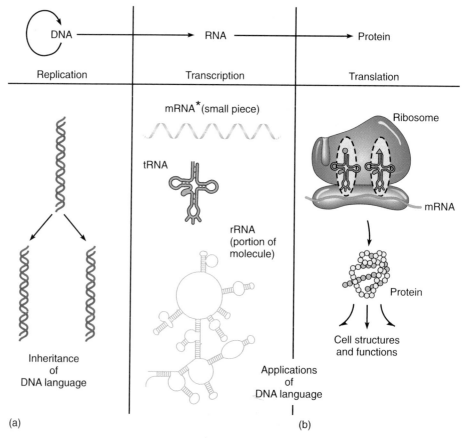

FIGURE 9.9

Summary of the flow of genetic information in cells. DNA is the ultimate storehouse and distributor of genetic information. **(a)** DNA can be replicated and passed to offspring. **(b)** DNA must be deciphered into a usable cell language. It does this by transcribing its code into RNA helper molecules that translate that code into protein.

*The sizes of RNA are not to scale—tRNA and mRNA are enlarged to show details.

features of gene expression cause one organism to be so distinctly different from another? For answers, we must turn to the correlation between gene and protein structure. We know that each structural gene is a linear sequence of nucleotides that codes for a protein. Because each protein is different, each gene must also differ somehow in its composition. In fact, the language of DNA exists in the order of groups of three consecutive bases called triplets on one DNA strand (**figure 9.10**). Thus, one gene differs from another in its composition of triplets. An equally important part of this concept is that each triplet represents a code for a particular amino acid. When the triplet code is transcribed and translated, it dictates the type and order of amino acids in a polypeptide (protein) chain.

The final key points that connect DNA and protein function are:

1. A protein's primary structure—the order and type of amino acids in the chain—determines its characteristic shape and function.
2. Proteins ultimately determine phenotype, the expression of all aspects of cell function and structure. Put more simply, living things are what their proteins make them.
3. DNA is mainly a blueprint that tells the cell which kinds of proteins to make and how to make them.

THE MAJOR PARTICIPANTS IN TRANSCRIPTION AND TRANSLATION

Transcription, the formation of RNA using DNA as a template, and **translation,** the synthesis of proteins using RNA as a template, are highly complex. A number of components participate: most prominently, messenger RNA, transfer RNA, ribosomes, several types of enzymes, and a storehouse of raw materials. After first examining each of these components, we shall see how they come together in the assembly line of the cell.

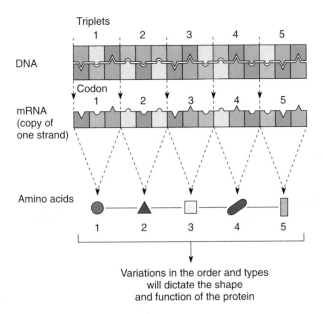

FIGURE 9.10

Simplified view of the DNA-protein relationship. The DNA molecule is a continuous chain of base pairs, but the sequence must be interpreted in groups of three base pairs (a triplet). Each triplet as copied into mRNA codons will translate into one amino acid; consequently, the ratio of base pairs to amino acids is 3:1.

RNAs: Tools in the Cell's Assembly Line

Ribonucleic acid is an encoded molecule like DNA, but its general structure is different in several ways:

1. It is a **single-stranded molecule;** that exists in helical form. This single strand can assume secondary and tertiary levels of complexity due to bonds within the molecule, leading to specialized forms of RNA (tRNA and rRNA—figure 9.9).
2. RNA contains **uracil,** instead of thymine, as the complementary base-pairing mate for adenine. This does not change the inherent DNA code in any way because the uracil still follows the pairing rules.
3. Although RNA, like DNA, contains a backbone that consists of alternating sugar and phosphate molecules, the sugar in RNA is **ribose** rather than deoxyribose.

The many functional types of RNA range from small regulatory pieces to large structural ones (**table 9.2** and **Microbits 9.3**, page 266). All types of RNA are formed through transcription of a DNA gene, but only mRNA is further translated into another type of molecule (protein).

Messenger RNA: Carrying DNA's Message

Messenger RNA (mRNA) is a transcript (copy) of a structural gene or genes complementary to DNA. It is synthesized by a process similar to synthesis of the leading strand during DNA replication,

TABLE 9.2			
Types of Ribonucleic Acid			
RNA Type	Contains Codes For	Function in Cell	Translated
Messenger (mRNA)	Sequence of amino acids in protein	Carries the DNA master code to the ribosome	Yes
Transfer (tRNA)	A cloverleaf tRNA to carry amino acids	Brings amino acids to ribosome during translation	No
Ribosomal (rRNA)	Several large structural rRNA molecules	Forms the major part of a ribosome and participates in protein synthesis	No
Micro (miRNA) and small interfering (siRNA)	Regulatory RNAs	Regulation of gene expression and coiling of chromatin	No
Primer	An RNA that can begin DNA replication	Primes DNA	No
Ribozymes	RNA enzymes, parts of splicer enzymes	Remove introns from other RNAs in eucaryotes	No

and the complementary base-pairing rules ensure that the code will be faithfully copied in the mRNA transcript. The message of this transcribed strand is later read as a series of triplets called **codons** **(figure 9.11a),** and the length of the mRNA molecule varies from about 100 nucleotides to several thousands. The details of transcription and the function of mRNA in translation will be covered shortly.

Transfer RNA: The Key to Translation

Transfer RNA (tRNA) is also a copy of a specific region of DNA; however, it differs from mRNA. It is uniform in length, being 75–95 nucleotides in length, and it contains sequences of bases that form hydrogen bonds with complementary sections of the same tRNA strand. At these points, the molecule bends back upon itself into several *hairpin loops*, giving the molecule a secondary *cloverleaf* structure that folds even further into a complex, three-dimensional helix (figure 9.11*b*). This compact molecule is an adaptor that converts RNA language into protein language. The bottom loop of the cloverleaf exposes a triplet, the **anticodon,** that both designates the specificity of the tRNA and complements mRNA's codons. At the opposite end of the molecule is a binding site for the amino acid that is specific for

that tRNA's anticodon. For each of the 20 **amino acids** (see table 2.5), there is at least one specialized type of tRNA to carry it. Binding of an amino acid to its specific tRNA, a process known as "charging" the tRNA takes place in two enzyme-driven steps: First an ATP activates the amino acid, and then this group binds to the acceptor end of the tRNA. Because tRNA is the molecule that will convert the master code on mRNA into a protein, the accuracy of this step is crucial.

The Ribosome: A Mobile Molecular Factory for Translation

The procaryotic (70S) ribosome is a particle composed of tightly packaged ribosomal RNA (rRNA) and protein. The rRNA component of the ribosome is also a long polynucleotide molecule. It forms complex three-dimensional figures that contribute to the structure and function of ribosomes (see figure 9.9). The interactions of proteins and rRNA create the two subunits of the ribosome that engage in final translation of the genetic code (see figure 9.13). A metabolically active bacterial cell can accommodate up to 20,000 of these miniscule factories—all actively engaged in reading the genetic program, taking in raw materials, and emitting proteins at an impressive rate.

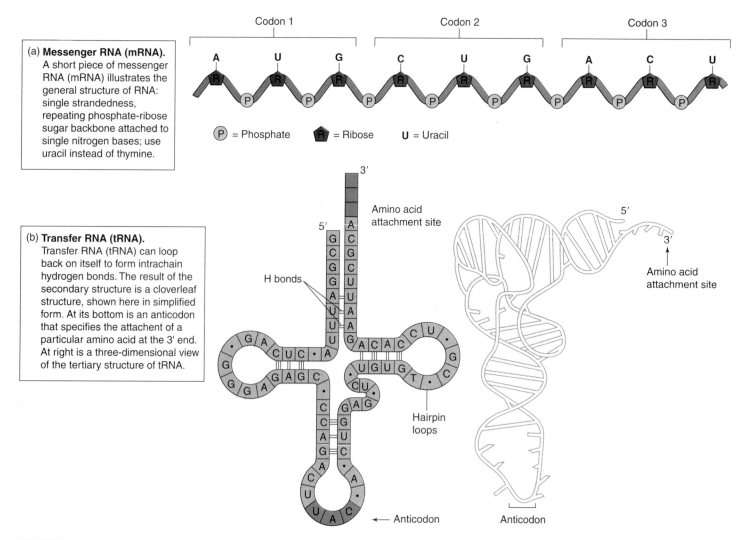

(a) Messenger RNA (mRNA). A short piece of messenger RNA (mRNA) illustrates the general structure of RNA: single strandedness, repeating phosphate-ribose sugar backbone attached to single nitrogen bases; use uracil instead of thymine.

Ⓟ = Phosphate   ⬠ = Ribose   **U** = Uracil

(b) Transfer RNA (tRNA). Transfer RNA (tRNA) can loop back on itself to form intrachain hydrogen bonds. The result of the secondary structure is a cloverleaf structure, shown here in simplified form. At its bottom is an anticodon that specifies the attachent of a particular amino acid at the 3' end. At right is a three-dimensional view of the tertiary structure of tRNA.

FIGURE 9.11

Characteristics of messenger and transfer RNA.

TRANSCRIPTION: THE FIRST STAGE OF GENE EXPRESSION

During transcription, the DNA code is converted to RNA through several stages, directed by a huge and very complex enzyme system, **RNA polymerase (figure 9.12).** Only one strand of the DNA—the *template strand*—contains meaningful instructions for synthesis of a functioning polypeptide. The nontranscribed strand is called the *coding strand*. The strand of DNA that serves as a template varies from one gene to another.

Transcription is initiated when RNA polymerase recognizes a segment of the DNA called the *promoter region*. This region consists of two sequences of DNA just prior to the beginning of the gene to be transcribed. The first sequence, which occurs approximately 35 bases prior to the start of transcription, is tightly bound by RNA polymerase. Transcription is allowed to begin when the DNA helix begins to unwind at the second sequence, which is located about 10 bases prior to the start of transcription. As the DNA helix unwinds, the polymerase advances and begins synthesizing an RNA molecule complementary to the template strand of DNA. The nucleotide sequence of promoters differs only slightly from gene to gene, with all promoters being rich in adenine and thymine.

During elongation, which proceeds in the 5′ to 3′ direction (with regard to the growing RNA molecule), the mRNA is assembled by the addition of nucleotides that are complementary to the DNA template. Be reminded that uracil (U) is placed as adenine's complement. As elongation continues, the part of DNA already transcribed is rewound into its original helical form. At termination, the polymerases recognize another code that signals the separation and release of the mRNA strand, also called the **transcript.** How long is the mRNA? The very smallest mRNA might consist of 100 bases; an average-sized mRNA might consist of 1,200 bases; and a large one, of several thousand.

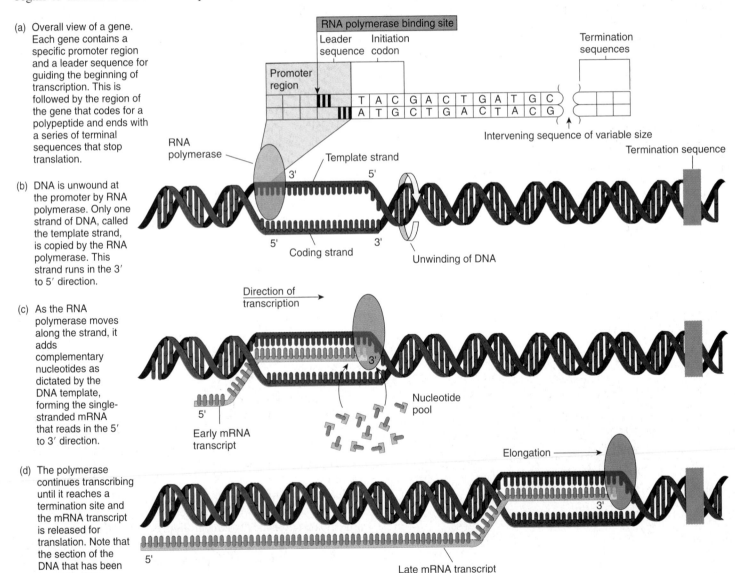

(a) Overall view of a gene. Each gene contains a specific promoter region and a leader sequence for guiding the beginning of transcription. This is followed by the region of the gene that codes for a polypeptide and ends with a series of terminal sequences that stop translation.

(b) DNA is unwound at the promoter by RNA polymerase. Only one strand of DNA, called the template strand, is copied by the RNA polymerase. This strand runs in the 3′ to 5′ direction.

(c) As the RNA polymerase moves along the strand, it adds complementary nucleotides as dictated by the DNA template, forming the single-stranded mRNA that reads in the 5′ to 3′ direction.

(d) The polymerase continues transcribing until it reaches a termination site and the mRNA transcript is released for translation. Note that the section of the DNA that has been transcribed is rewound into its original configuration.

FIGURE 9.12

The major events in mRNA synthesis or transcription.

Since the earliest days of molecular biology, RNA has been the unsung stepchild of the cell, quietly ferrying the information in DNA to ribosomes to direct the formation of proteins. Current research however is showing a new, dynamic role for RNA in the cell that may forever change the reputation of this humble molecule.

Short lengths of RNA, ranging in size from 21 to 28 nucleotides, seem to have the ability to control the expression of certain genes. Experiments in late 2001 identified an enzyme that produced two types of RNAs, dubbed micro RNAs (miRNAs) and small interfering RNAs (siRNAs), by cleaving a larger, double stranded RNA precursor. Once produced, these small RNA molecules, direct a cellular enzyme complex called RISC to degrade specific mRNA molecules, silencing the expression of these genes. The process, known as RNA interference or RNAi is thought to protect cells from foreign DNA that could otherwise damage the integrity of the genome.

A second type of regulation seems to occur when small RNAs alter the structure of chromosomes. As DNA and proteins coil together to form chromatin, small RNAs direct how tightly or loosely the chromatin is constructed (see Microbits 9.2). Just as a closed book cannot be read, DNA sequences contained within tightly coiled chromatin are generally inaccessible to the cell, silencing the expression of those genes. RNAi seems to be the next step toward unraveling the mechanism behind this well-known but poorly understood phenomenon.

One of the most promising roles for RNAi lies in its use as a laboratory tool; enabling researchers to easily down-regulate the expression of genes to study their effects. Currently, such research is carried out by creating "gene knockouts," a time- and labor-intensive process.

Our knowledge of the full role of small RNAs in the cell is just beginning. In the meantime, scientists will keep studying small RNAs while being mindful of the old adage "Good things come in small packages."

TRANSLATION: THE SECOND STAGE OF GENE EXPRESSION

In translation, all of the elements needed to synthesize a protein, from the mRNA to the amino acids, are brought together on the ribosomes **(figure 9.13).** The entire process proceeds through the stages of initiation, elongation, termination, and protein folding and processing.

Initiation of Translation

The mRNA molecule leaves the DNA transcription site and is transported to ribosomal staging sites in the cytoplasm. Ribosomal subunits are specifically adapted to assembling and forming sites to hold the mRNA and tRNAs. The ribosome thus recognizes these

molecules and stabilizes reactions between them. The small subunit binds to the 5′ end of the mRNA, and the large subunit supplies enzymes for making peptide bonds on the protein. The ribosome begins to scan the mRNA by moving in the 5′ to 3′ direction along the mRNA. The first codon it encounters is the START codon, which is almost always AUG (and rarely, GUG).

With the mRNA message in place on the assembled ribosome, the next step in translation involves entrance of tRNAs with their amino acids (figure 9.13). The pool of cytoplasm around the region contains a complete array of tRNAs, previously charged by having the correct amino acid attached. The step in which the complementary tRNA meets with the mRNA code is guided by the two sites on the large subunit of the ribosome called the P site (left) and the A site (right).[1] Think of these sites as shallow depressions in the larger subunit of the ribosome, each of which accommodates a tRNA. The ribosome also has an exit or E site where used tRNAs are released.

The Master Genetic Code: The Message in Messenger RNA

By convention, the master genetic code is represented by the mRNA codons and the amino acids they specify **(figure 9.14).** Except in a very few cases, this code is universal, whether for procaryotes, eucaryotes, or viruses. It is worth noting that once the triplet code on mRNA is known, the original DNA sequence, the complementary tRNA code, and the types of amino acids in the protein are automatically known **(figure 9.15).** One cannot predict from protein structure what the exact mRNA codons are because of a factor called degeneracy, meaning that a particular amino acid can be coded for by more than a single codon.

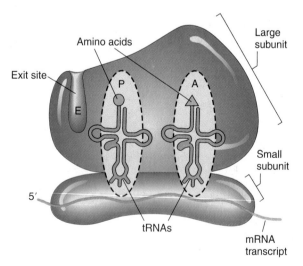

FIGURE 9.13

The "players" in translation. A ribosome serves as the stage for protein synthesis. Assembly of the small and large subunits results in specific sites for holding the mRNA and two tRNAs with their amino acids.

---

1. P stands for peptide site; A stands for aminoacyl site (a charged amino acid); E stands for exit site.

Second Position

First Position	U		C		A		G		Third Position
U	UUU	Phenylalanine	UCU	Serine	UAU	Tyrosine	UGU	Cysteine	U
	UUC		UCC		UAC		UGC		C
	UUA	Leucine	UCA		UAA	STOP**	UGA	STOP**	A
	UUG		UCG		UAG		UGG	Tryptophan	G
C	CUU	Leucine	CCU	Proline	CAU	Histidine	CGU	Arginine	U
	CUC		CCC		CAC		CGC		C
	CUA		CCA		CAA	Glutamine	CGA		A
	CUG		CCG		CAG		CGG		G
A	AUU	Isoleucine	ACU	Threonine	AAU	Asparagine	AGU	Serine	U
	AUC		ACC		AAC		AGC		C
	AUA		ACA		AAA	Lysine	AGA	Arginine	A
	AUG	START Methionine*	ACG		AAG		AGG		G
G	GUU	Valine	GCU	Alanine	GAU	Aspartic acid	GGU	Glycine	U
	GUC		GCC		GAC		GGC		C
	GUA		GCA		GAA	Glutamic acid	GGA		A
	GUG		GCG		GAG		GGG		G

* This codon initiates translation.
**For these codons, which give the orders to stop translation, there are no corresponding tRNAs and no amino acids.

FIGURE 9.14

The Genetic Code: Codons of mRNA That Specify a Given Amino Acid. The master code for translation is found in the mRNA codons.

anti sense - 5" A
sense DNA - 3' T
use mRNA 5 (NO-T)
T RNA 3" A U

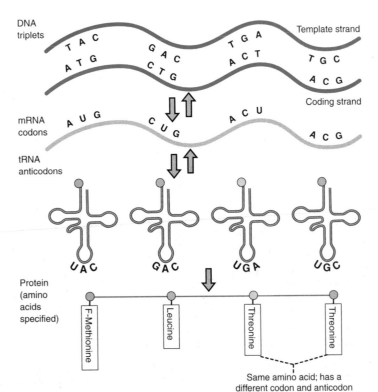

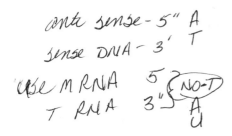

FIGURE 9.15

Interpreting the DNA code. If the DNA code is known, the mRNA codon can be surmised. If a codon is known, the anticodon can be determined, but it may not be possible to determine exact codon sequence from the anticodons because of the degeneracy of the code. It is also not possible to determine the exact codon or anticodon from protein structure.

In figure 9.14, the mRNA codons and their corresponding amino acid specificities are given. Because there are 64 different triplet codes[2] and only 20 different amino acids, it is not surprising that some amino acids are represented by several codons. For example, leucine and serine can each be represented by any of six different triplets, and only tryptophan and methionine are represented by a single codon. In such codons as leucine, only the first two nucleotides are required to encode the correct amino acid, and the third nucleotide does not change its sense. This property, called *wobble,* is thought to permit some variation or mutation without harming the message.

The Beginning of Protein Synthesis

With mRNA serving as the guide, the stage is finally set for actual protein assembly. The correct tRNA (labeled 1 on **figure 9.16**) enters the P site and binds to the **start codon (AUG)** presented by the mRNA. Rules of pairing dictate that the anticodon of this tRNA must be complementary to the mRNA codon AUG, thus the tRNA with anticodon UAC will first occupy site P. It happens that the amino acid carried by the initiator tRNA in bacteria is formyl *methionine* (fMet; see figure 9.14), though in many cases, it may not remain a permanent part of the finished protein.

Continuation and Completion of Protein Synthesis: Elongation and Termination

While reviewing the dynamic process of protein assembly, you will want to remain aware that the ribosome shifts its "reading frame" to the right along the mRNA from one codon to the next. This brings the next codon into place on the ribosome and makes a space for the next tRNA to enter the A position. A peptide bond is formed between the amino acids on the adjacent tRNAs, and the polypeptide grows in length (figure 9.16).

Elongation begins with the filling of the A site by a second tRNA (2 on figure 9.16). The identity of this tRNA and its amino acid is dictated by the second mRNA codon.

The entry of tRNA 2 into the A site brings the two adjacent tRNAs in favorable proximity for a peptide bond to form between the amino acids (aa) they carry. The fMet is transferred from the first tRNA to aa 2, resulting in two coupled amino acids called a dipeptide (figure 9.16b).

For the next step to proceed, some room must be made on the ribosome, and the next codon in sequence must be brought into position for reading. This process is accomplished by *translocation,* the enzyme-directed shifting of the ribosome to the right along the mRNA strand, which causes the blank tRNA (1) to be discharged from the ribosome (figure 9.16c) at the E site. This also shifts the tRNA holding the dipeptide into P position. Site A is temporarily left empty. The tRNA that has been released is now free to drift off into the cytoplasm and become recharged with an amino acid for later additions to this or another protein.

The stage is now set for the insertion of tRNA 3 at site A as directed by the third mRNA codon (figure 9.16d). This insertion is followed once again by peptide bond formation between the dipeptide and aa 3 (making a tripeptide), splitting of the peptide from tRNA 2, and translocation. This releases tRNA 2, shifts mRNA to the next position, moves tRNA 3 to position P, and opens position A for the next

tRNA (4). From this point on, peptide elongation proceeds repetitively by this same series of actions out to the end of the mRNA.

The termination of protein synthesis is not simply a matter of reaching the last codon on mRNA. It is brought about by the presence of at least one special codon occurring just after the codon for the last amino acid. Termination codons—UAA, UAG, and UGA—are codons for which there is no corresponding tRNA. Although they are often called **nonsense codons,** they carry a necessary and useful message: *Stop* here. When this codon is reached, a special enzyme breaks the bond between the final tRNA and the finished polypeptide chain, releasing it from the ribosome.

Before newly made proteins can carry out their structural or enzymatic roles, they often require finishing touches. Even before the peptide chain is released from the ribosome, it begins folding upon itself to achieve its biologically active tertiary conformation. Other alterations, called *posttranslational* modifications, may be necessary. Some proteins must have the starting amino acid (formyl methionine) clipped off; proteins destined to become complex enzymes have cofactors added; and some join with other completed proteins to form quaternary levels of structure.

The operation of transcription and translation is machinelike in its precision. Protein synthesis in bacteria is both efficient and rapid. At 37°C, 12–17 amino acids per second are added to a growing peptide chain. An average protein consisting of about 400 amino acids requires less than half a minute for complete synthesis. Further efficiency is gained when the translation of mRNA starts while transcription is still occurring (**figure 9.17**). A single mRNA is long enough to be fed through more than one ribosome simultaneously. This permits the synthesis of hundreds of protein molecules from the same mRNA transcript arrayed along a chain of ribosomes. This **polyribosomal complex** is indeed an assembly line for mass production of proteins. Protein synthesis consumes an enormous amount of energy. Nearly 1,200 ATPs are required just for synthesis of an average-sized protein.

EUCARYOTIC TRANSCRIPTION AND TRANSLATION: SIMILAR YET DIFFERENT

Eucaryotes and procaryotes share many similarities in protein synthesis. The start codon in eucaryotes is also AUG, but it codes for a different form of methionine. Another difference is that eucaryotic mRNAs code for just one protein, unlike bacterial mRNAs, which often contain several genes in series.

There are a few differences between procaryotic and eucaryotic gene expression. The presence of the DNA in a separate compartment (the nucleus) means that eucaryotic transcription and translation cannot be simultaneous. The mRNA transcript must pass through pores in the nuclear membrane and be carried to the ribosomes in the cytoplasm for translation.

We have given the simplified definition of a gene that works well for procaryotes, but most eucaryotic genes are not colinear[3]—meaning that they do *not* exist as an uninterrupted series of triplets coding for a protein. A eucaryotic gene contains the code for a protein, but located along the gene are one to several intervening

---

2. $64 = 4^3$ (the 4 different codons in all possible combinations of 3).

3. Colinearity means that the base sequence can be read directly into a series of amino acids.

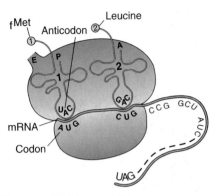

(a) Entrance of tRNAs 1 and 2

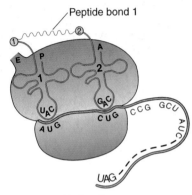

(b) Formation of peptide bond

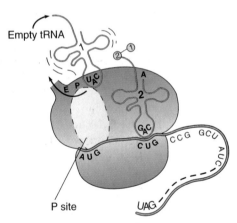

(c) Discharge of tRNA 1 at E site

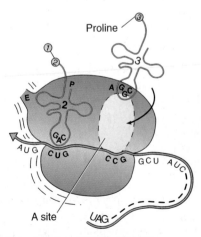

(d) First translocation; tRNA 2 shifts into P site; enter tRNA 3 by ribosome

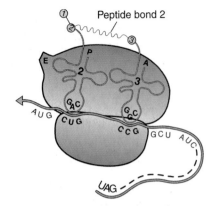

(e) Formation of peptide bond

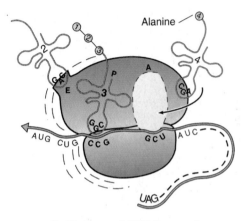

(f) Discharge of tRNA 2; second translocation; enter tRNA 4

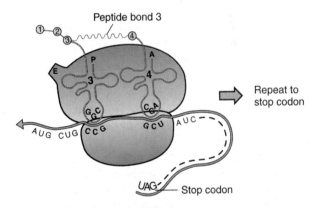

(g) Formation of peptide bond

FIGURE 9.16

The events in protein synthesis.

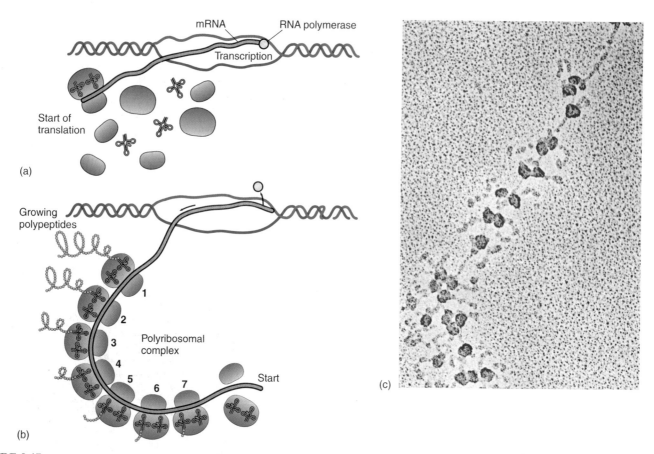

FIGURE 9.17

Speeding up the protein assembly line bacteria. **(a)** The mRNA transcript encounters ribosomal parts immediately as it leaves the DNA. **(b)** The ribosomal factories assemble along the mRNA in a chain, each ribosome reading the message and translating it into protein. Many products will thus be well along the synthetic pathway before transcription has even terminated. **(c)** Photomicrograph of a polyribosomal complex in action. Note that the protein "tails" vary in length depending on the stage of translation.

sequences of bases, called **introns,** that do not code for protein. Introns are interspersed between coding regions, called **exons,** that will be translated into protein **(figure 9.18).** We can use words as examples. A short section of colinear procaryotic gene might read TOM SAW OUR DOG DIG OUT; a eucaryotic gene that codes for the same portion would read TOM SAW XZKP FPL OUR DOG QZWVP DIG OUT. The recognizable words are the exons, and the nonsense letters represent the introns.

This unusual genetic architecture, sometimes called a split gene, requires further processing before translation (figure 9.18). Transcription of the entire gene with both exons and introns occurs first, producing a pre-mRNA. Next a type of RNA and protein called a *spliceosome,* recognizes the exon-intron junctions and enzymatically cuts through them. The action of this splicer enzyme loops the introns into lariat-shaped pieces, excises them, and joins the exons end to end. By this means, a strand of mRNA with no intron material is produced. This completed mRNA strand can then proceed to the cytoplasm to be translated.

At first glance, this system seems to be a cumbersome way to make a transcript, and the value of this extra genetic baggage is still the subject of much debate. Several different types of introns have been discovered, some of which do code for cell substances. One particular intron discovered in yeast gives the code for a reverse transcriptase, and other introns can be translated into endonucleases. Some experts hypothesize that introns represent a "sink" for extra bits of genetic material that could be available for splicing into existing genes, thus promoting genetic change and evolution. There is also evidence that introns serve as genetic regulators and in ribosomal assembly.

THE GENETICS OF ANIMAL VIRUSES

The genetic nature of viruses was described in chapter 6. Viruses essentially consist of one or more pieces of DNA or RNA enclosed in a protective coating. Above all, they are genetic parasites that require access to their host cell's genetic and metabolic machinery to be replicated, transcribed, and translated, and they also have the potential for genetically changing the cells. Because they contain only those genes needed for the production of new viruses, the genomes of viruses tend to be very compact and economical. In fact, this very simplicity makes them excellent subjects for the study of gene function.

The genetics of viruses is quite diverse (see chapters 24 and 25). In many viruses, the nucleic acid is linear in form; in others, it is circular. The genome of most viruses exists in a single molecule, though in a few, it is segmented into several smaller molecules. Most

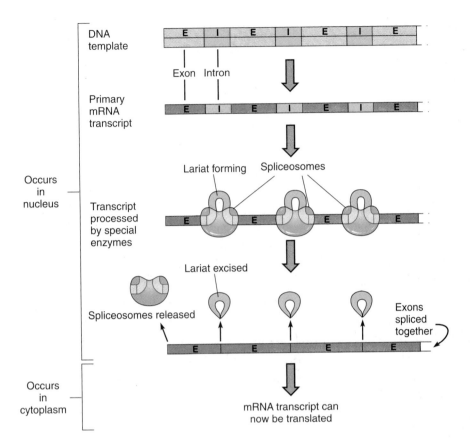

FIGURE 9.18

The split gene of eucaryotes. Eucaryotic genes have an additional complicating factor in their translation. Their coding sequences, or exons (E), are interrupted at intervals by segments called introns (I) that are not part of that protein's code. Introns are transcribed but not translated, which necessitates their removal by RNA splicing enzymes before translation.

viruses contain normal double-stranded (ds) DNA or single-stranded (ss) RNA, but other patterns exist. There are ssDNA viruses, dsRNA viruses, and retroviruses, which work backward by making dsDNA from ssRNA. In some instances, viral genes overlap one another, and in a few DNA viruses, both strands contain a translatable message.

A few generalities can be stated about viral genetics. In all cases, the viral nucleic acid penetrates the cell and is introduced into the host's gene-processing machinery at some point. In successful infection, an invading virus instructs the host's machinery to synthesize large numbers of new virus particles by a mechanism specific to a particular group. With few exceptions, replication of the DNA molecule of DNA animal viruses occurs in the nucleus, where the cell's DNA replication machinery lies and the genome of RNA viruses is replicated in the cytoplasm. In all viruses, viral mRNA is translated into viral proteins on host cell ribosomes using host tRNA. In the next section, we will briefly observe some major patterns of genetic replication (**table 9.3**).

Replication, Transcription, and Translation of dsDNA Viruses

Replication of dsDNA viruses is divided into phases (**figure 9.19**). During the early phase, viral DNA enters the nucleus, where several genes are transcribed into a messenger RNA. The newly synthesized RNA transcript then moves into the cytoplasm to be translated into viral proteins (enzymes) needed to replicate the viral DNA; this replication occurs in the nucleus. The host cell's own DNA polymerase is often involved, though some viruses (herpes, for example) have their own. During the late phase, other parts of the viral genome are transcribed and translated into proteins required to

TABLE 9.3
Patterns of Genetic Duplication in Viruses
DNA Viruses
dsDNA → dsDNA (semiconservative)
ssDNA → dsDNA → ssDNA (only one virus group)
RNA Viruses
(+) ssRNA → (−) ssRNA → (+) ssRNA
(−) ssRNA → (+) ssRNA → (−) ssRNA
ssRNA ssDNA → dsDNA → ssRNA (retroviruses)
dsRNA → ssRNA → dsRNA (conservative)

form the capsid and other structures. The new viral genomes and capsids are assembled, and the mature viruses are released by budding or cell disintegration.

Double-stranded DNA viruses interact directly with the DNA of their host cell. In some viruses, the viral DNA becomes silently *integrated* into the host's genome by insertion at a particular site on the host genome (figure 9.19). This integration may later lead to the transformation[4] of the host cell into a cancer cell and the production of a tumor. Several DNA viruses, including hepatitis B (HBV), the herpesviruses, and papillomaviruses (warts), are known to be initiators of cancers and are thus termed **oncogenic.*** The mechanisms

---

4. The process of genetic change in a cell, leading to malignancy.

*oncogenic (ahn″-koh-jen′-ik) Gr. *onkos*, mass, and *gennan*, to produce. Refers to any cancer-causing process. Viruses that do this are termed oncoviruses

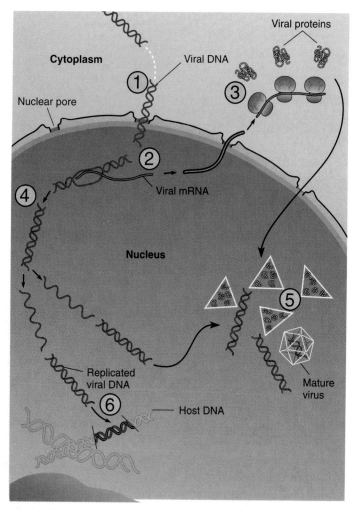

FIGURE 9.19

Genetic stages in the multiplication of double-stranded DNA viruses.
The virus penetrates the host cell and releases DNA, which **(1)** enters the
nucleus and **(2)** is transcribed. Other events are: **(3)** Viral mRNA is
translated into structural proteins; proteins enter the nucleus. **(4)** Viral DNA
is replicated repeatedly in the nucleus. **(5)** Viral DNA and proteins are
assembled into a mature virus in the nucleus. **(6)** Because it is double-
stranded, the viral DNA can insert itself into host DNA (latency).

of **transformation** and oncogenesis involve special genes called
oncogenes that can regulate cellular genomes (see chapter 17).

Replication, Transcription, and Translation of RNA Viruses

RNA viruses exhibit several differences from DNA viruses. Their
genomes are smaller and less stable; they enter the host cell already
in an RNA form; and the virus cycle occurs entirely in the cytoplasm for most viruses. RNA viruses can have one of the following
genetic messages:

1. a positive-sense genome (+) that comes ready to be
   translated into proteins,
2. a negative-sense genome (−) that must be converted to
   positive sense before translation, and
3. a positive-sense genome (+) that can be converted to DNA,
   or a dsRNA genome.

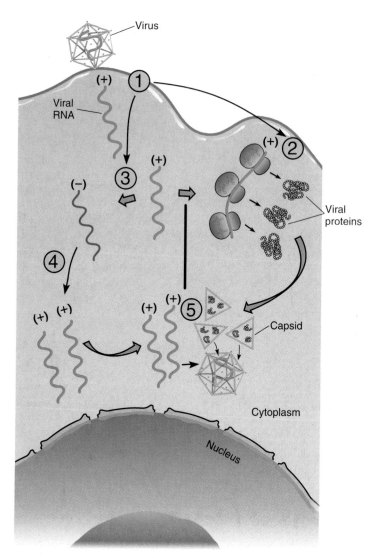

FIGURE 9.20

Replication of positive-sense, single-stranded RNA viruses. In
general, these viruses do not enter the nucleus. **(1)** Penetration and
uncoating of viral RNA. **(2)** Because it is positive in sense and single-
stranded, the RNA can be directly translated on host cell ribosomes into
various necessary viral proteins. **(3)** A negative genome is synthesized
against the positive template to produce large numbers of positive
genomes for final assembly. **(4)** The negative template is then used to
synthesize a series of positive replicates. **(5)** RNA strands and proteins
assemble into mature viruses.

Positive-Sense Single-Stranded RNA Viruses Positive-sense
RNA viruses such as polio and hepatitis A virus must first replicate
a negative strand as a master template to produce more positive
strands. Shortly after the virus uncoats in the cell, its positive strand
is translated into a large protein that is soon cleaved into individual
functional units, one of which is an RNA polymerase that initiates
the replication of the viral strand (**figure 9.20**). Replication of a
single-stranded positive-sense strand is done in two steps. First, a
negative strand is synthesized using the parental positive strand as a
template by the usual base-pairing mechanism. The resultant negative strand becomes a master template against which numerous positive daughter strands are made. Further translation of the viral
genome produces large numbers of structural proteins for final assembly and maturation of the virus.

RNA Viruses with Reverse Transcriptase: Retroviruses A most unusual class of viruses has a unique capability to reverse the order of the flow of genetic information. Thus far in our discussion, all genetic entities have shown the patterns DNA → DNA, DNA → RNA, or RNA → RNA. Retroviruses, including HIV, the cause of AIDS, and HTLV I, a cause of one type of human leukemia, synthesize DNA using their RNA genome as a template (see figure 25.14). They accomplish this by means of an enzyme, **reverse transcriptase,** that comes packaged with each virus particle. This enzyme synthesizes a single-stranded DNA against the viral RNA template and then directs the formation of a complementary strand of this ssDNA, resulting in a double strand of viral DNA. The dsDNA strand enters the nucleus, where it can be integrated into the host genome and transcribed by the usual mechanisms into new viral ssRNA. Translation of the viral RNA yields viral proteins for final virus assembly. The capacity of a retrovirus to become inserted into the host's DNA as a provirus has several possible consequences. In some cases, these viruses are oncogenic and are known to transform cells and produce tumors. It allows the AIDS virus to remain latent in an infected cell until a stimulus activates it to continue a productive cycle.

CHAPTER CHECKPOINTS

Information in DNA is converted to proteins by the process of transcription and translation. These proteins may be structural or functional in nature. Structural proteins contribute to the architecture of the cell while functional proteins (enzymes) control an organism's metabolic activities.

The DNA code occurs in groups of three bases; this code is copied onto RNA as codons; the message provides the types of amino acids in a protein; this code is universal in all cells and viruses.

The processes of transcription and translation are similar but not identical for procaryotes and eucaryotes. Eucaryotes transcribe DNA in the nucleus, remove its introns, and translate it in the cytoplasm. Bacteria transcribe and translate simultaneously because the DNA is not sequestered in a nucleus and the bacterial DNA is free of introns.

Viruses replicate by utilizing the transcription and translation processes of cellular organisms. They invade host cells and "force" them to transcribe and translate viral genes into new virus particles. In some cases, viruses connect themselves to the host genome and are passed on to its progeny, some of which will become virus factories many generations later.

The genetic material of viruses can be either DNA or RNA occurring in single, double, positive-, or negative-sense strands. Viral DNA contains just enough information to force a cell to become a virus factory.

Genetic Regulation of Protein Synthesis and Metabolism

In chapter 8 we surveyed the metabolic reactions in cells and the enzymes involved in those reactions. At that time, we mentioned types of metabolic regulation that are genetic in origin. These control mechanisms ensure that genes are active only when their products are required. In this way, enzymes will be produced to appropriately reflect nutritional status and prevent the waste of energy and materials in dead-end synthesis. Genetic function in procaryotes is regulated by a specific collection of genes called an **operon.*** Operons consist of a coordinated set of genes, all of which are regulated as a single unit. Operons are described as either inducible or repressible. The category each operon falls into is determined by how transcription is affected by the environment surrounding the cell. Many catabolic operons are inducible, meaning that the operon is turned on (induced) by the substrate of the enzyme for which the structural genes code. In this way, the enzymes needed to metabolize a nutrient (lactose, for example) are only produced when that nutrient is present in the environment.

A similar system often controls the genes coding for anabolic enzymes, such as those used to synthesize amino acids. In the case of these repressible operons, several genes in series are turned off (repressed) by the product synthesized by the enzyme.

THE LACTOSE OPERON: A MODEL FOR INDUCIBLE GENE REGULATION IN BACTERIA

The best understood cell system for explaining control through genetic induction is the **lactose *(lac)* operon.** This concept, first postulated in 1961 by François Jacob and Jacques Monod, accounts for the regulation of lactose metabolism in *Escherichia coli.* Many other operons with similar modes of action have since been identified, and together they furnish convincing evidence that the environment of a cell can have great impact on gene expression.

The lactose operon is made up of three segments, or **loci,*** of DNA:

1. the **regulator,** composed of the gene that codes for a protein capable of repressing the operon (a **repressor**);
2. the *control locus,* composed of two genes, the **promoter** (recognized by RNA polymerase) and the **operator,** a sequence where transcription of the structural genes is initiated; and
3. the *structural locus,* made up of three genes, each coding for a different enzyme needed to catabolize lactose (**figure 9.21**).

One of the enzymes, β-galactosidase, hydrolyzes the lactose into its monosaccharides; another, permease, brings lactose across the cell membrane.

In bacteria, structural genes required for the metabolism of a nutrient tend to be arranged in series. This is an efficient strategy that permits genes for a particular metabolic pathway to be induced or repressed in unison by a single regulatory element. The enzymes of the operon are of the inducible sort mentioned in chapter 8. The promoter, operator, and structural components lie adjacent to one another, but the regulator can be at a distant site.

In inductive systems like the *lac* operon, the operon is normally in an *off* mode and does not initiate enzyme synthesis when the appropriate substrate is absent (figure 9.21*a*). How is the operon maintained in this mode? The key is in the repressor protein that is coded by the regulatory gene. This relatively large molecule

---

*operon (op'-ur-on) L. *opera,* exertion. A regulatory site for gene expression.

*loci (loh'-sy) sing. locus (loh'-kus) L. *locus,* a place. The site on a chromosome occupied by a gene.

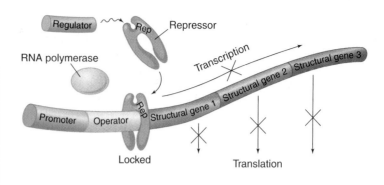

(a) Operon Off. In the absence of lactose, a repressor protein (the product of a regulatory gene located elsewhere on the bacterial chromosome) attaches to the operator gene of the operon. This effectively locks the operator and prevents any transcription of structural genes downstream (to its right). Suppression of transcription (and consequently, of translation) prevents the unnecessary synthesis of enzymes for processing lactose.

(b) Operon On. Upon entering the cell, the substrate (lactose) becomes a genetic inducer by attaching to the repressor, which loses its grip and falls away. The RNA polymerase is now free to initiate transcription, and the enzymes produced by translation of the mRNA perform the necessary reactions on their lactose substrate.

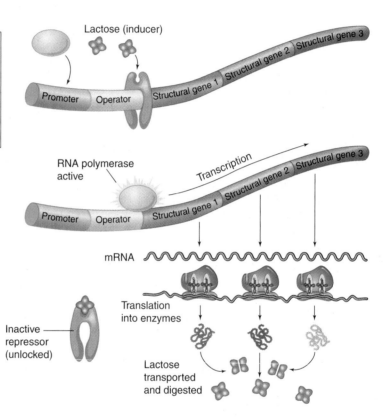

FIGURE 9.21

The lactose operon in bacteria: how inducible genes are controlled by substrate.

has sites for allosteric binding, one for the operator and another for lactose. In the absence of lactose, this repressor binds with the operator locus, thereby blocking the transcription of the structural genes lying downstream. Think of the repressor as a lock on the operator, and if the operator is locked, the structural genes cannot be transcribed.

If lactose is added to the cell's environment, it triggers several events that turn the operon *on*. Because lactose is ultimately responsible for stimulating protein synthesis, it is called an *inducer.* The binding of lactose to the repressor protein causes a conformational change in the repressor that dislodges it from the operator segment (figure 9.21b). The control segment that was previously inactive is opened up, and RNA polymerase can now bind to the promoter. The structural genes are transcribed in a single unbroken

transcript coding for all three enzymes. During translation, however, each gene synthesizes a separate protein.

As lactose is depleted, further enzyme synthesis is not necessary, so the order of events reverses. At this point there is no longer sufficient lactose to inhibit the repressor, hence the repressor is again free to attach to the operator. The operator is locked, and transcription of the structural genes and protein synthesis related to lactose both stop.

A fine but important point about the *lac* operon is that it functions only in the absence of glucose. Glucose is the preferred carbon source because it can be used immediately in growth, and does not require induction of an operon. When glucose is present, a second regulatory system ensures that the *lac* operon is inactive, regardless of lactose levels in the environment.

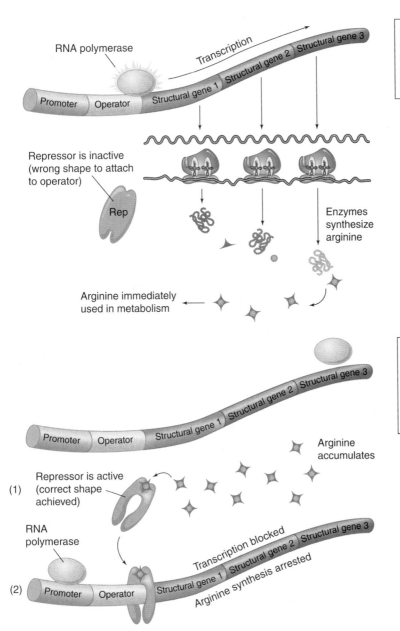

(a) Operon On. A repressible operon remains on when its nutrient products (here, arginine) are in great demand by the cell because the repressor is unable to bind to the operator at low nutrient levels.

(b) Operon Off. The operon is repressed when (1) arginine builds up and, serving as a corepressor, activates the repressor. (2) The repressor complex affixes to the operator and blocks the RNA polymerase and further transcription of genes for arginine synthesis.

FIGURE 9.22

Repressible operon: control of a gene through excess nutrient.

A REPRESSIBLE OPERON

Bacterial systems for synthesis of amino acids, purines and pyrimidines and many other processes work on a slightly different principle—that of repression. Similar factors such as repressor proteins, operators, and a series of structural genes exist for this operon, but with some important differences. Unlike the *lac* operon, this operon is normally in the *on* mode and will be turned *off* only when this nutrient is no longer required. The nutrient plays an additional role as a **corepressor** needed to block the action of the operon.

A growing cell that needs the amino acid arginine (arg) effectively illustrates the operation of a repressible operon. Under these conditions, the *arg* operon is set to *on,* and arginine is being actively synthesized through the action of its enzymatic products **(figure**

9.22a). In an active cell, the arginine will be used immediately, and the repressor will remain inactive (unable to bind the operator) because there is too little free arginine to activate it. As the cell's metabolism begins to slow down, however, the synthesized arginine will no longer be used up and will accumulate. The free arginine is then available to act as a corepressor by attaching to the repressor. This reaction produces a functioning repressor that locks the operator and stops further transcription and arginine synthesis (figure 9.22b).

Analogous gene control mechanisms in eucaryotic cells are not as well understood, but it is known that gene function can be altered by intrinsic regulatory segments similar to operons. Some molecules, called transcription factors, insert on the grooves of the DNA molecule and enhance transcription of specific genes. Examples include zinc "fingers" and leucine "zippers." These transcription factors can regulate gene expression in response to environmental stimuli such as nutrients, toxin levels, or even temperature. Eucaryotic genes are also regulated developmentally, leading to the hundreds of different tissue types found in higher multicellular organism.

ANTIBIOTICS THAT AFFECT TRANSCRIPTION AND TRANSLATION

Naturally occurring cell nutrients are not the only agents capable of modifying gene expression. Some infection therapy is based on the concept that certain drugs react with DNA, RNA, or ribosomes and thereby alter genetic expression (see chapter 12). Treatment with such drugs is based on an important premise: that growth of the infectious agent will be inhibited by blocking its protein-synthesizing machinery selectively, without disrupting the cell synthesis of the patient receiving the therapy.

Drugs that inhibit protein synthesis exert their influence on transcription or translation. For example, the rifamycins used in therapy for tuberculosis bind to RNA polymerase, blocking the initiation step of transcription, and are selectively more active against bacterial RNA polymerase than the corresponding eucaryotic enzyme. Actinomycin D binds to DNA and halts mRNA chain elongation, but its mode of action is not selective for bacteria. For this reason, it is very toxic and never used to treat bacterial infections, though it can be applied in tumor treatment.

The ribosome is a frequent target of antibiotics that inhibit ribosomal function and ultimately protein synthesis. The value and safety of these antibiotics again depend upon the differential susceptibility of procaryotic and eucaryotic ribosomes. One problem with drugs that selectively disrupt procaryotic ribosomes is that the mitochondria of humans contain a procaryotic type of ribosome, and these drugs may inhibit the function of the host's mitochondria. One group of antibiotics (including erythromycin and spectinomycin) prevents translation by interfering with the attachment of mRNA to ribosomes. Chloramphenicol, lincomycin, and tetracycline bind to the ribosome in a way that blocks the elongation of the polypeptide, and aminoglycosides (such as streptomycin) inhibit peptide initiation and elongation. It is interesting to note that these drugs have served as important tools to explore genetic events because they can arrest specific stages in these processes.

CHAPTER CHECKPOINTS

Gene expression must be orchestrated to coordinate the organism's needs with nutritional resources. Genes can be turned "on" and "off" by specific molecules, which expose or hide their nucleotide codes for transcribing proteins. Most, but not all, of these proteins are enzymes.

Operons are collections of genes in bacteria that code for products with a coordinated function. They include genes for regulatory, operational, and structural components of the cell. Nutrients can combine with regulator gene products to turn a set of structural genes on (inducible genes) or off (repressible genes). The *lac* (lactose) operon is an example of an inducible operon. The *arg* (arginine) operon is an example of a repressible operon.

The rifamycins, tetracyclines, and aminoglycosides are classes of antibiotics that are effective because they interfere with transcription and translation processes in microorganisms.

Mutations: Changes in the Genetic Code

As precise and predictable as the rules of genetic expression seem, permanent changes do occur in the genetic code. Indeed, genetic change is the driving force of evolution. In microorganisms, such changes may become evident in altered gene expression such as the appearance or disappearance of anatomical or physiological traits. For example, a normally colored bacterium can lose its ability to form pigment, or a strain of the malarial parasite can develop resistance to a drug. Phenotypic changes of this type are ultimately due to changes in the genotype. Any permanent, inheritable change in the genetic information of the cell is a **mutation.** On a strictly molecular level, a mutation is an alteration in the nitrogen base sequence of DNA. It can involve the loss of base pairs, the addition of base pairs, or a rearrangement in the order of base pairs. Do not confuse this with genetic recombination, in which microbes transfer whole segments of genetic information between themselves.

A microorganism that exhibits a natural, nonmutated characteristic is known as a **wild type,** or wild strain. If a microorganism bears a mutation, it is called a **mutant strain.** Mutant strains can show variance in morphology, nutritional characteristics, genetic control mechanisms, resistance to chemicals, temperature preference, and nearly any type of enzymatic function. Mutant strains are very useful for tracking genetic events, unraveling genetic organization, and pinpointing genetic markers. A classic method of detecting mutant strains involves addition of various nutrients to a culture to screen for its use of that nutrient. For example, in a culture of a wild-type bacterium that is lactose-positive (meaning it has the necessary enzymes for fermenting this sugar), a small number of mutant cells have become lactose-negative, having lost the capacity to ferment this sugar. If the culture is plated on a medium containing indicators for fermentation, each colony can be observed for its fermentation reaction, and the negative strain isolated. Another standard method of detecting and isolating microbial mutants is by replica plating **(figure 9.23).**

CAUSES OF MUTATIONS

A mutation is described as spontaneous or induced, depending upon its origin. A **spontaneous mutation** is a random change in the DNA arising from errors in replication due to unknown causes. The frequency of spontaneous mutations has been measured for a number of organisms. Mutation rates vary tremendously, from one mutation in 10^5 replications (a high rate) to one mutation in 10^{10} replications (a low rate). The rapid rate of bacterial reproduction allows these mutations to be observed more readily in bacteria than in most eucaryotes.

Induced mutations result from exposure to known **mutagens,** which are primarily physical or chemical agents that interact with DNA in a disruptive manner **(table 9.4).** The carefully controlled use of mutagens has proved a useful way to induce mutant strains of microorganisms for study.

Chemical mutagenic agents such as acridine dyes insert completely across the DNA helices between adjacent bases to produce a frameshift mutation and distort the helix. Analogs[5] of the nitrogen

5. An analog is a chemical structured very similarly to another chemical except for minor differences in functional groups.

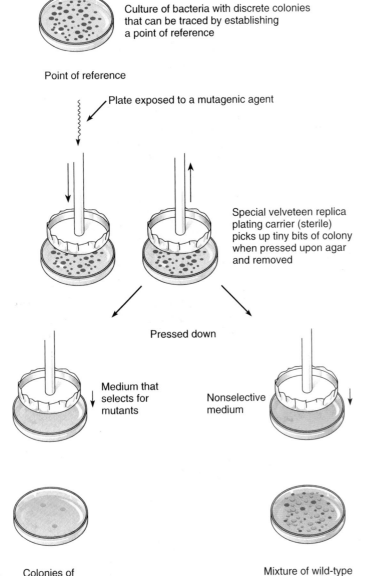

Culture of bacteria with discrete colonies that can be traced by establishing a point of reference

Point of reference

Plate exposed to a mutagenic agent

Special velveteen replica plating carrier (sterile) picks up tiny bits of colony when pressed upon agar and removed

Pressed down

Medium that selects for mutants

Nonselective medium

Colonies of mutant strains

Mixture of wild-type and mutant strains

FIGURE 9.23

The general basis of replica plating.   This method was developed by Joshua Lederberg for detecting and isolating mutant strains of microorganisms.

TABLE 9.4	
Selected Mutagenic Agents and Their Effects	
Agent	**Effect**
Chemical	
Nitrous acid, bisulfite	Removes an amino group from some bases
Ethidium bromide	Inserts between the paired bases
Acridine dyes	Cause frameshifts due to insertion between base pairs
Nitrogen base analogs	Compete with natural bases for sites on replicating DNA
Radiation	
Ionizing (gamma rays, X rays)	Form free radicals that cause single or double breaks in DNA
Ultraviolet	Causes cross-links between adjacent pyrimidines

CATEGORIES OF MUTATIONS

Mutations range from large mutations, in which large genetic sequences are gained or lost (see later section on transposons), to small ones that affect only a single base on a gene. These latter mutations, which involve addition, deletion, or substitution of a few bases, are called **point mutations.**

To understand how a change in DNA influences the cell, remember that the DNA code appears in a particular order of triplets (three bases) that is transcribed into mRNA codons, each of which specifies an amino acid. A permanent alteration in the DNA that is copied faithfully into mRNA and translated can change the structure of the protein. A change in a protein can likewise change the morphology and physiology of a cell. Most mutations have a harmful effect on the cell, leading to cell dysfunction or death; these are called lethal mutations. Neutral mutations produce neither adverse nor helpful changes. A small number of mutations are beneficial in that they provide the cell with a useful change in structure or physiology.

Substitution mutations occur when bases are mismatched but no bases are inserted or deleted. This type of mutation affects only a single amino acid but can affect the protein in one of several ways.

A change in the code that leads to placement of a different amino acid is called a **missense mutation.** A missense mutation can do one of the following:

1. create a faulty, nonfunctional (or less functional) protein,
2. produce a protein which functions in a different manner, or
3. cause no significant alteration in protein function.

A **nonsense mutation,** on the other hand, changes a normal codon into a stop codon that does not code for an amino acid and stops the production of the protein wherever it occurs. A nonsense mutation almost always results in a nonfunctional protein. A **silent mutation** alters a base but does not change the amino acid and thus has no effect. For example, because of the degeneracy of the code, ACU, ACC, ACG, and ACA all code for threonine, so a mutation that changes only the last base will not alter the sense of the message in any way. A **back-mutation** occurs when a gene that has undergone mutation reverses (mutates back) to its original base composition.

Mutations also occur when one or more bases are inserted into or deleted from a newly synthesized DNA strand. This type of

bases (5-bromodeoxyuridine and 2-aminopurine, for example) are chemical mimics of natural bases that are incorporated into DNA during replication. Addition of these abnormal bases leads to mistakes in base-pairing. Many chemical mutagens are also carcinogens, or cancer-causing agents (see the discussion of the Ames test in a later section of this chapter).

Physical agents that alter DNA are primarily types of radiation. High-energy gamma rays and X rays introduce major physical changes into DNA, and it accumulates breaks that may not be repairable. Ultraviolet (UV) radiation induces abnormal bonds between adjacent pyrimidines that prevent normal replication (see figure 11.9). Exposure to large doses of radiation can be fatal, which is why radiation is so effective in microbial control; it can also be carcinogenic in animals.

TABLE 9.5

Classification of Major Types of Mutations

(a) Wild type (original, non-mutated sequence)	THE BIG BAD DOG ATE THE FAT RED CAT
Substitution mutations	
(b) Missense	THE BIG BAD DOG ATE THE FIT RED CAT
(c) Nonsense	THE BIG BAD (stop)
Frameshift mutations	
(d) Insertion	THE BIG BAB DDO GAT ETH EFA TRE DCA T
(e) Deletion	THE BIG BDD OGA TET HEF ATR EDC AT

Categories of Mutations Based on Type of DNA Alteration
(a) The wild-type sequence of a gene is the DNA sequence found in most organisms and is generally considered the "normal," sequence. (b) A missense mutation causes a different amino acid to be incorporated into a protein. Effects range from unnoticeable to severe, based on how different the two amino acids are. (c) A nonsense mutation converts a codon to a stop codon, resulting in premature termination of protein synthesis. Effects of this type of mutation are almost always severe. (d, e) Insertion and deletion mutations cause a change in the reading frame of the mRNA, resulting in a protein in which every amino acid after the mutation is affected. Because of this, frameshift mutations almost always result in a nonfunctional protein.

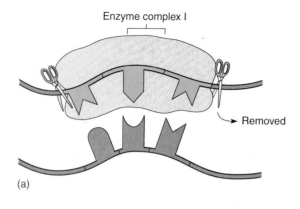
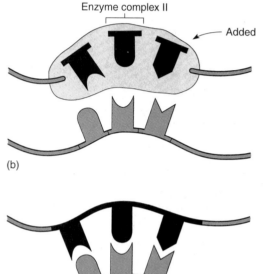

FIGURE 9.24

Excision repair of mutation by enzymes.   **(a)** The first enzyme complex recognizes one or several incorrect bases and removes them. **(b)** The second complex (DNA polymerase I and ligase) places correct bases and seals the gaps. **(c)** Repaired DNA.

mutation, known as a **frameshift,** is so named because the reading frame of the ribosome has been changed. Frameshift mutations nearly always result in a nonfunctional protein because every amino acid after the mutation is different from what was coded for in the original DNA. Also note that insertion or deletion of bases in multiples of three (3,6,9, etc.) results in the addition or deletion of amino acids but does not disturb the reading frame. The effects of all of these types of mutations can be seen in **table 9.5.**

REPAIR OF MUTATIONS

Earlier we indicated that DNA has a proofreading mechanism to repair mistakes in replication that might otherwise become permanent (see page 263). Because mutations are potentially life-threatening, the cell has additional systems for finding and repairing DNA that has been damaged by various mutagenic agents and processes. Most ordinary DNA damage is resolved by enzymatic systems specialized for finding and fixing such defects.

DNA that has been damaged by ultraviolet radiation can be restored by photoactivation or light repair. This repair mechanism requires visible light and a light-sensitive enzyme, DNA photolyase, which can detect and attach to the damaged areas (sites of abnormal pyrimidine binding). Ultraviolet repair mechanisms are successful only for a relatively small number of UV mutations. Cells cannot repair severe, widespread damage and will die. In humans, the genetic disease *xeroderma pigmentosa* is due to non-functioning genes for the enzyme photolyase. Persons suffering from this rare disorder develop severe skin cancers; this relation provides strong evidence for a link between cancer and mutations.

Mutations can be excised by a series of enzymes that remove the incorrect bases and add the correct ones. This process is known as *excision repair.* First, enzymes break the bonds between the bases and the sugar-phosphate strand at the site of the error. A different enzyme subsequently removes the defective bases one at a time, leaving a gap that will be filled in by DNA polymerase I and ligase (**figure 9.24**). A repair system can also locate mismatched bases that were missed during proofreading: for example, C mistakenly paired with A, or G with T. The base must be replaced soon after the mismatch is made, or it will not be recognized by the repair enzymes.

THE AMES TEST

New agricultural, industrial, and medicinal chemicals are constantly being added to the environment, and exposure to them is widespread. The discovery that many such compounds are mutagenic and that up to 83% of these mutagens are linked to cancer is significant. Although animal testing has been a standard method of detecting chemicals with carcinogenic potential, a more rapid

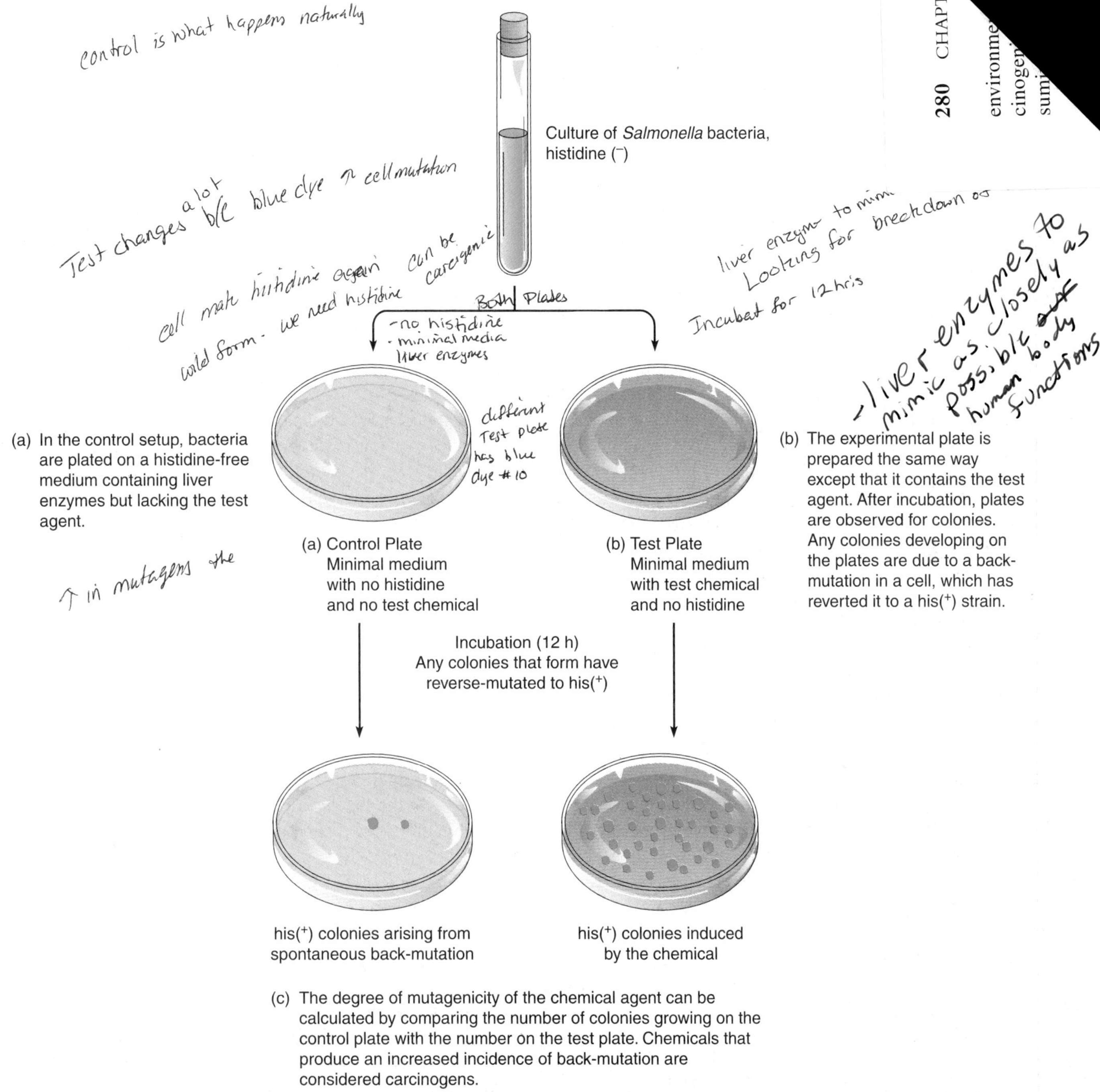

[Handwritten annotations:]

control is what happens naturally

Test changes a lot b/c blue dye → cell mutation

cell make histidine again can be carcinogenic
wild form - we need histidine

↑ in mutagens the

liver enzyme to mim[...]
Looking for breakdown o[...]
Incubat for 12 hrs

- liver enzymes to mimic as closely as possible out human body functions

Culture of *Salmonella* bacteria,
histidine (⁻)

Both Plates
-no histidine
-minimal media
liver enzymes

different Test plate has blue dye #10

(a) In the control setup, bacteria are plated on a histidine-free medium containing liver enzymes but lacking the test agent.

(a) Control Plate
Minimal medium
with no histidine
and no test chemical

(b) Test Plate
Minimal medium
with test chemical
and no histidine

(b) The experimental plate is prepared the same way except that it contains the test agent. After incubation, plates are observed for colonies. Any colonies developing on the plates are due to a back-mutation in a cell, which has reverted it to a his(⁺) strain.

Incubation (12 h)
Any colonies that form have
reverse-mutated to his(⁺)

his(⁺) colonies arising from
spontaneous back-mutation

his(⁺) colonies induced
by the chemical

(c) The degree of mutagenicity of the chemical agent can be calculated by comparing the number of colonies growing on the control plate with the number on the test plate. Chemicals that produce an increased incidence of back-mutation are considered carcinogens.

FIGURE 9.25

The Ames test. This test is based on a strain of *Salmonella typhimurium* that cannot synthesize histidine (his⁻). It lacks the enzymes to repair DNA so that mutations show up readily, and has leaky cell walls that permit the ready entrance of chemicals. Many potential carcinogens (benzanthracene and aflatoxin, for example) are mutagenic agents only after being acted on by mammalian liver enzymes, so an extract of these enzymes is added to the test medium.

screening system called the **Ames test**[6] is also commonly used. In this ingenious test, the experimental subjects are bacteria whose gene expression and mutation rate can be readily observed and monitored. The premise is that any chemical capable of mutating bacterial DNA can similarly mutate mammalian (and thus human) DNA and is therefore potentially hazardous.

One indicator organism in the Ames test is a mutant strain of *Salmonella typhimurium*[7] that has lost the ability to synthesize the amino acid histidine, a defect highly susceptible to back-mutation because the strain also lacks DNA repair mechanisms. Mutations that cause reversion to the wild strain, which is capable of synthesizing histidine, occur spontaneously at a low rate. A test agent is considered a mutagen if it enhances the rate of back-mutation beyond levels that would occur spontaneously. One variation on this testing procedure is outlined in **figure 9.25**. The Ames test has proved invaluable for screening an assortment of

---

6. Named for its creator, Bruce Ames.

7. *S. typhimurium* inhabits the intestine of poultry and causes food poisoning in humans. It is used extensively in genetic studies of bacteria.

...ntal and dietary chemicals for mutagenesis and car-...icity without resorting to more expensive and time-con-...ng animal studies.

POSITIVE AND NEGATIVE EFFECTS OF MUTATIONS

Many mutations are not repaired. How the cell copes with them depends on the nature of the mutation and the strategies available to that organism. Mutations are permanent and heritable and will be passed on to the offspring of organisms and new viruses and become a long-term part of the gene pool. Some mutations are harmful to organisms; others provide adaptive advantages.

If a mutation leading to a nonfunctional protein occurs in a gene for which there is only a single copy, as in haploid or simple organisms, the cell will probably die. This happens when certain mutant strains of *E. coli* acquire mutations in the genes needed to repair damage by UV radiation. Mutations of the human genome affecting the action of a single protein (mostly enzymes) are responsible for more than 3,500 diseases (see chapter 10).

Although most spontaneous mutations are not beneficial, a small number contribute to the success of the individual and the population by creating variant strains with alternate ways of expressing a trait. Microbes are not "aware" of this advantage and do not direct these changes; they simply respond to the environment they encounter. Those organisms with beneficial mutations can more readily adapt, survive, and reproduce. In the long-range view, mutations and the variations they produce are the raw materials for change in the population and, thus, for evolution.

Mutations that create variants occur frequently enough that any population contains mutant strains for a number of characteristics, but as long as the environment is stable, these mutants will never comprise more than a tiny percentage of the population. When the environment changes, however, it can become hostile for the survival of certain individuals, and only those microbes bearing protective mutations will be equipped to survive in the new environment. In this way, the environment naturally selects certain mutant strains that will reproduce, give rise to subsequent generations, and in time, be the dominant strain in the population. Through these means, any change that confers an advantage during selection pressure will be retained by the population. One of the clearest models for this sort of selection and adaptation is acquired drug resistance in bacteria (see chapter 12). One of the fascinating mechanisms bacteria have developed for increasing their adaptive capacity is genetic exchange (called genetic recombination).

DNA Recombination Events

Genetic recombination through sexual reproduction is an important means of genetic variation in eucaryotes. Although bacteria have no exact equivalent to sexual reproduction, they exhibit a primitive means for sharing or recombining parts of their genome. An event in which one bacterium donates DNA to another bacterium is a type of genetic transfer termed **recombination,** the end result of which

is a new strain different from both the donor and the original recipient strain. Recombination in bacteria depends in part on the fact that bacteria contain extrachromosomal DNA and are adept at interchanging genes. Genetic exchanges have tremendous effects on the genetic diversity of bacteria, and unlike spontaneous mutations, are generally beneficial to them. They provide additional genes for resistance to drugs and metabolic poisons, new nutritional and metabolic capabilities, and increased virulence and adaptation to the environment in general.

TRANSMISSION OF GENETIC MATERIAL IN BACTERIA

DNA transfer between bacterial cells typically involves small pieces of DNA in the form of plasmids or chromosomal fragments. Plasmids are small, circular pieces of DNA that contain their own origin of replication and therefore can replicate independently of the bacterial chromosome. Plasmids are found in many bacteria (as well as some fungi) and typically contain, at most, only a few dozen genes. Although plasmids are not necessary for bacterial survival they often confer useful traits, such as antibiotic resistance, upon the bacteria that carry them. Chromosomal fragments, which have escaped from a lysed bacterial cell, are also common vehicles for the transfer of genetic information between cells. An important difference between plasmids and fragments is that while a plasmid has its own origin of replication and is stably replicated and inherited, chromosomal fragments must integrate themselves into the bacterial chromosome in order to be replicated and eventually passed to progeny cells. Both plasmids and integrated DNA fragments are transcribed and translated along with the bacterial chromosome. The process of genetic recombination is rare in nature, but its frequency can be increased in the laboratory, where the ability to shuffle genes between organisms is highly prized.

Depending upon the mode of transmission, the means of genetic recombination is called conjugation, transformation, or transduction. **Conjugation*** requires the attachment of two related species and the formation of a bridge that can transport DNA. **Transformation*** entails the transfer of naked DNA and requires no special vehicle. **Transduction*** is DNA transfer mediated through the action of a bacterial virus (**table 9.6**).

Conjugation: Bacterial Sex

Conjugation is a mode of sexual mating in which a plasmid or other genetic material is transferred by a donor to a recipient cell via a direct connection (**figure 9.26**). It occurs primarily in gram-negative bacteria, but many gram-positive cells can conjugate. In gram-negative cells, the donor has a plasmid (**fertility, or F' factor**) that allows the synthesis of a **pilus,** or **conjugative pilus.** The

---

*conjugation (kahn″-jew-gay′-shun) L. *conjugatus,* yoked together.

*transformation (trans-for-may′-shun) L. *trans,* across, and *formatio,* to form. This term is also used to mean the cancerous (malignant) conversion of cells.

*transduction (trans-duk′-shun) L. *transducere,* to lead across.

TABLE 9.6

Types of Intermicrobial Exchange

Mode	Factors Involved	Direct or Indirect*	Genes Transferred
Conjugation	Donor cell with pilus Fertility plasmid in donor Both donor and recipient alive Bridge forms between cells to transfer DNA	Direct	Drug resistance; resistance to metals; toxin production; enzymes; adherence molecules; degradation of toxic substances; uptake of iron
Transformation	Free donor DNA (fragment) Live, competent recipient cell	Indirect	Polysaccharide capsule; unlimited with cloning techniques
Transduction	Donor is lysed bacterial cell Defective bacteriophage is carrier of donor DNA Live, competent recipient cell of same species as donor	Indirect	Toxins; enzymes for sugar fermentation; drug resistance

Direct means the donor and recipient are in contact during exchange; indirect means they are not.

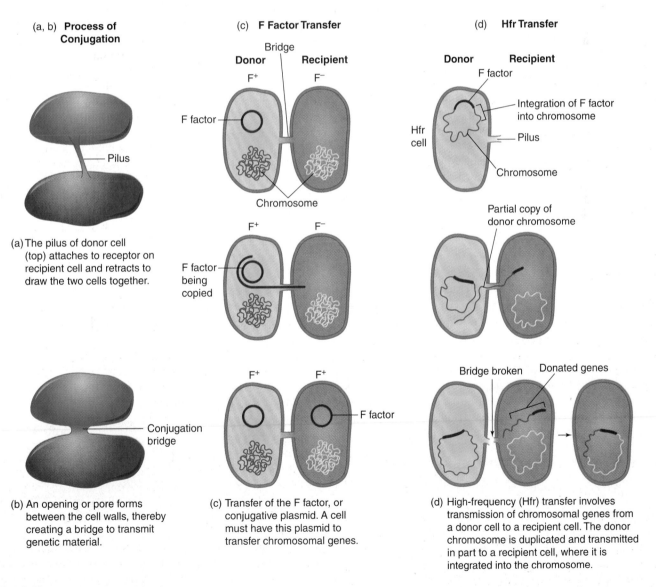

(a, b) **Process of Conjugation**

(a) The pilus of donor cell (top) attaches to receptor on recipient cell and retracts to draw the two cells together.

(b) An opening or pore forms between the cell walls, thereby creating a bridge to transmit genetic material.

(c) **F Factor Transfer**

(c) Transfer of the F factor, or conjugative plasmid. A cell must have this plasmid to transfer chromosomal genes.

(d) **Hfr Transfer**

(d) High-frequency (Hfr) transfer involves transmission of chromosomal genes from a donor cell to a recipient cell. The donor chromosome is duplicated and transmitted in part to a recipient cell, where it is integrated into the chromosome.

FIGURE 9.26

Conjugation: genetic transmission through direct contact between two cells.

recipient cell is a related species or genus that has a recognition site on its surface. A cell's role in conjugation is denoted by F^+ for the cell that has the F plasmid and by F^- for the cell that lacks it. Contact is made when a pilus grows out from the F^+ cell, attaches to the surface of the F^- cell, contracts, and draws the two cells together (see figure 4.9). In both gram-positive and gram-negative cells, an opening is created between the connected cells, and the replicated DNA passes across from one cell to the other. Conjugation is a very conservative process, in that the donor bacterium generally retains a copy of the genetic material being transferred.

There are hundreds of conjugative plasmids with some variations in their properties. One of the best understood plasmids is the F factor in *E. coli,* which exhibits these patterns of transfer:

1. The donor (F^+) cell makes a copy of its F factor and transmits this to a recipient (F^-) cell. The F^- cell is thereby changed into an F^+ cell capable of producing a pilus and conjugating with other cells (figure 9.26c). No additional donor genes are transferred at this time.

2. In high-frequency recombination (Hfr) donors, the fertility factor has been integrated into the F^+ donor chromosome.

The term *high-frequency recombination* was adopted to denote that a cell with an integrated F factor transmits its chromosomal genes at a higher frequency than other cells.

The F factor can direct a more comprehensive transfer of part of the donor chromosome to a recipient cell. This transfer occurs through duplication of the DNA by means of the rolling circle mechanism. One strand of DNA is retained by the donor, and the other strand is transported across to the recipient cell (figure 9.26d). The F factor may not be transferred during this process. The transfer of an entire chromosome takes about 100 minutes, but the pilus bridge between cells is ordinarily broken before this time, and rarely is the entire genome of the donor cell transferred.

Conjugation has great biomedical importance. Special **resistance (R) plasmids,** or **factors,** that bear genes for resisting antibiotics and other drugs are commonly shared among bacteria through conjugation. Transfer of R factors can confer multiple resistance to antibiotics such as tetracycline, chloramphenicol, streptomycin, sulfonamides, and penicillin. This phenomenon is discussed further in chapter 12. Other types of R factors carry genetic codes for resistance to heavy metals (nickel and mercury) or for synthesizing virulence factors (toxins, enzymes, and adhesion molecules) that increase the pathogenicity of the bacterial strain. Conjugation studies have also provided an excellent way to map the bacterial chromosome.

Transformation: Capturing DNA from Solution

One of the cornerstone discoveries in microbial genetics was made in the late 1920s by the English biochemist Frederick Griffith working with *Streptococcus pneumoniae* and laboratory mice. The pneumococcus exists in two major strains based on the presence of the capsule, colonial morphology, and pathogenicity. Encapsulated strains bear a smooth (S) colonial appearance and are virulent; strains lacking a capsule have a rough (R) appearance and are nonvirulent (see figure 4.12). (Recall that the capsule protects a bacterium from the phagocytic host defenses.) To set the groundwork, Griffith showed that when mice were injected with a live, virulent (S) strain, they soon died **(figure 9.27a).** Mice injected with a live, nonvirulent (R) strain remained alive and healthy (figure 9.27b). Next he tried a variation on this theme. First, he heat-killed an S strain and injected it into mice, which remained healthy (figure 9.27c). Then came the ultimate test: Griffith injected both dead S cells and live R cells into mice, with the result that the mice died from pneumococcal blood infection (figure 9.27d). If killed bacterial cells do not come back to life and the nonvirulent live strain was harmless, why did the mice die? Although he did not know it at the time, Griffith had demonstrated that dead S cells, while passing through the body of the mouse, broke open and released some of their DNA (by chance, that part containing the genes for making a capsule). A few of the live R cells subsequently picked up this loose DNA and were transformed by it into virulent, capsule-forming strains.

Later studies supported the concept that a chromosome released by a lysed cell breaks into fragments small enough to be accepted by a recipient cell and that DNA, even from a dead cell, retains its genetic code. This nonspecific acceptance by a bacterial cell of small fragments of soluble DNA from the surrounding environment is termed **transformation.** Transformation is apparently facilitated by special DNA-binding proteins on the cell wall that capture DNA from the surrounding medium. Cells that are capable of accepting genetic material through this means are termed *competent.* The new DNA is processed by the cell membrane and transported into the cytoplasm, where it is inserted into the bacterial chromosome. Transformation is a natural event found in several groups of gram-positive and gram-negative bacterial species. In addition to genes coding for the capsule, bacteria also exchange genes for antibiotic resistance and bacteriocin synthesis in this way.

Because transformation requires no special appendages, and the donor and recipient cells do not have to be in direct contact, the process is useful for certain types of recombinant DNA technology. With this technique, foreign genes from a completely unrelated organism are inserted into a plasmid, which is then introduced into a competent bacterial cell through transformation. These recombinations can be carried out easily in a test tube, and human genes can be experimented upon and even expressed outside the human body by placing them in a microbial cell. This same phenomenon in eucaryotic cells, termed *transfection,* is an essential aspect of genetically engineered yeasts, plants, and mice, and it has been proposed as a future technique for curing genetic diseases in humans. These topics are covered in more detail in chapter 10.

Transduction: The Case of the Piggyback DNA

Bacteriophages (bacterial viruses) have been previously described as destructive bacterial parasites. Infection by a virus does not always kill the host cell, however, and viruses can in fact serve as genetic vectors (an entity that can bring foreign DNA into a cell). The process by which a bacteriophage serves as the carrier of DNA from a donor cell to a recipient cell is **transduction.** Although it occurs naturally in a broad spectrum of bacteria, the participating bacteria in a single transduction event must be the same species because of the specificity of viruses for host cells.

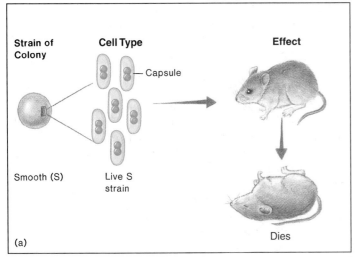

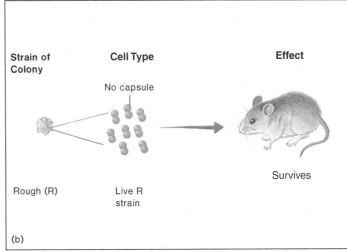

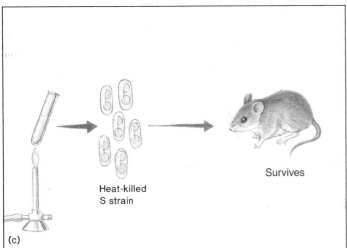

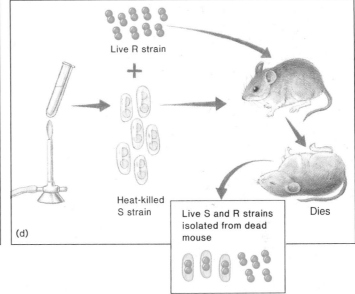

FIGURE 9.27

Griffith's classic experiment in transformation.   In essence, this experiment proved that DNA released from a killed cell can be acquired by a live cell. The cell receiving this new DNA is genetically transformed—in this case, from a nonvirulent strain to a virulent one.

There are two versions of transduction. In *generalized transduction* (**figure 9.28**), random fragments of disintegrating host DNA are taken up by the phage during assembly. Virtually any gene from the bacterium can be transmitted through this means. In *specialized transduction* (**figure 9.29**), a highly specific part of the host genome is regularly incorporated into the virus. This specificity is explained by the prior existence of a temperate prophage inserted in a fixed site on the bacterial chromosome. When activated, the prophage DNA separates from the bacterial chromosome, carrying a small segment of host genes with it. During a lytic cycle, these specific viral-host gene combinations are incorporated by the viral particles and carried to another bacterial cell.

Several cases of specialized transduction have biomedical importance. The virulent strains of bacteria such as *Corynebacterium diphtheriae, Clostridium* spp., and *Streptococcus pyogenes*

all produce toxins with profound physiological effects, whereas nonvirulent strains do not produce toxins. It turns out that toxicity arises from the expression of bacteriophage genes that have been introduced by transduction. Only those bacteria infected with a temperate phage are toxin formers. Other instances of transduction are seen in staphylococcal transfer of drug resistance and in the transmission of gene regulators in gram-negative rods (*Escherichia, Salmonella*).

Transposons: "This Gene Is Jumpin' "   One type of genetic transferral of great interest involves transposable elements, or **transposons.** Transposons have the distinction of shifting from one part of the genome to another and so are termed "jumping genes." When the idea of their existence in corn plants was first postulated by the geneticist Barbara McClintock, it was greeted with some

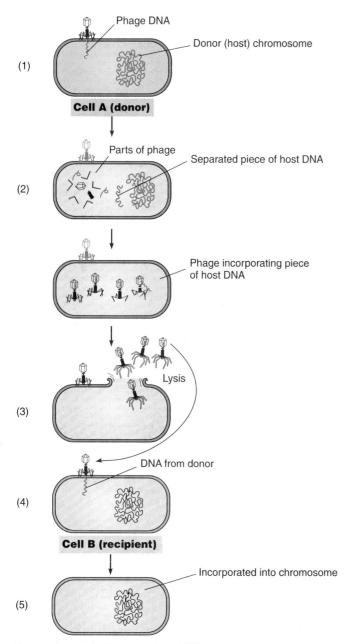

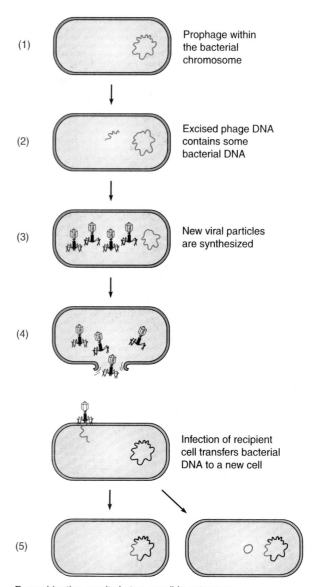

Recombination results in two possible outcomes.

FIGURE 9.28

Generalized transduction: genetic transfer by means of a virus carrier.
(1) A phage infects cell A (the donor cell) by normal means. **(2)** During
replication and assembly, a phage particle incorporates a segment of
bacterial DNA by mistake. **(3)** Cell A then lyses and releases the mature
phages, including the genetically altered one. **(4)** The altered phage
adsorbs to and penetrates another host cell (cell B), injecting the DNA
from cell A rather than viral nucleic acid. **(5)** Cell B receives this donated
DNA, which recombines with its own DNA. Because the virus is defective
(biologically inactive as a virus), it is unable to complete a lytic cycle. The
transduced cell survives and can use this new genetic material.

FIGURE 9.29

**Specialized transduction: transfer of specific genetic material by
means of a virus carrier.** **(1)** Specialized transduction begins with a cell
that contains a prophage (a viral genome integrated into the host cell
chromosome). **(2)** Rarely, the virus enters a lytic cycle and, as it excises
itself from its host cell, inadvertently includes some bacterial DNA.
(3) Replication and assembly results in production of a chimeric virus,
containing some bacterial DNA. **(4)** Release of the recombinant virus and
subsequent infection of a new host results in transfer of bacterial DNA
between cells. **(5)** Recombination can occur between the bacterial
chromosome and the virus DNA, resulting in either bacterial DNA or a
combination of viral and bacterial DNA being incorporated into the
bacterial chromosome.

skepticism, because it had long been believed that the locus of a given gene was set and that genes did not or could not move around. Now it is evident that jumping genes are widespread among procaryotic and eucaryotic cells and viruses.

All transposons share the general characteristic of traveling from one location to another on the genome—from one chromosomal site to another, from a chromosome to a plasmid, or from a plasmid to a chromosome (**figure 9.30**). Because transposons occur in plasmids, they can also be transmitted from one cell to another in bacteria and a few eucaryotes. Some transposons replicate themselves before jumping to the next location, and others simply move without replicating first.

Transposons contain DNA that codes for the enzymes needed to remove and reintegrate the transposon at another site in the genome. Flanking the coding region of the DNA are sequences

called inverted repeats, which mark the point at which the transposon is removed or reinserted into the genome. The smallest transposons consist of only these two genetic sequences and are often referred to as *insertion elements*. Other transposons contain additional genes that provide traits such as antibiotic resistance or toxin production.

The overall effect of transposons—to scramble the genetic language—can be beneficial or adverse, depending upon such variables as where insertion occurs in a chromosome, what kinds of genes are relocated, and the type of cell involved. On the beneficial side, transposons are known to be involved in

1. the creation of different genetic combinations, necessary for the high levels of variation in antibodies and receptor molecules in cells (see chapter 15);
2. changes in traits such as colony morphology, pigmentation, and antigenic characteristics;
3. replacement of damaged DNA; and
4. the intermicrobial transfer of drug resistance (in bacteria).

Some more disruptive and harmful outcomes of transposons can lead to deletions, insertions, translocations, and chromosome breakage. For example, retroviruses such as the AIDS virus behave as transposons by randomly inserting into the genome, thereby leading to severe cell dysfunction (see chapters 17 and 25).

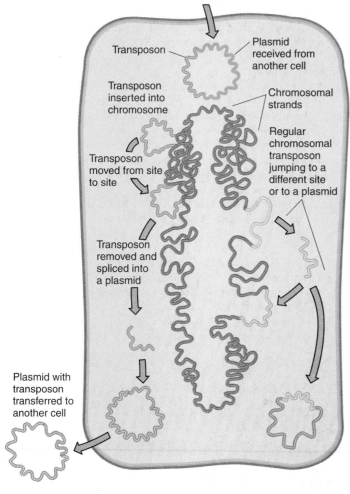

FIGURE 9.30

Transposons: shifting segments of the genome. Potential mechanisms in the movement of transposons in bacterial cells. Some transposons are plasmids that are shifted from site to site; others are regular parts of chromosomes that are moved out of one site and spliced into another.

CHAPTER CHECKPOINTS

Changes in the genetic code can occur by two means: mutation and recombination. Mutation means a change in the nucleotide sequence of the organism's genome.

Recombination means the addition of genes from an outside source, such as a virus or another cell.

Mutations can be either spontaneous or induced by exposure to some external mutagenic agent.

All cells have enzymes that repair damaged DNA. When the degree of damage exceeds the ability of the enzymes to make repairs, mutations occur.

Mutation-induced changes in DNA nucleotide sequencing range from a single nucleotide to addition or deletion of large sections of genetic material.

The Ames test is used to identify potential carcinogens on the basis of their ability to cause back-mutations in bacteria.

Genetic recombination occurs in eucaryotes through sexual reproduction. In bacteria, recombination occurs through the processes of transformation, conjugation, and transduction and the incorporation of lysogenic viruses into the genome.

Transposons are genes that can relocate from one part of the genome to another, causing rearrangement of genetic material. Such rearrangements have both beneficial and harmful consequences for the organism involved.

CHAPTER CAPSULE WITH KEY TERMS

I. **Genes and the Genetic Material**
   A. **Genetics** is the study of **heredity** and can be studied at the level of the organism, **genome, chromosome, gene** and **DNA.** Genes provide the information needed to construct **proteins**, which have structural or catalytic functions in the cell.

II. **Gene Structure and Replication**
   A. **DNA** is a long molecule in the form of a double helix. Each strand of the helix consists of a string of **nucleotides** which form hydrogen bonds with their counterparts on the other strand. **Adenine** base pairs with **thymine** while **guanine** base pairs with **cytosine.** The order of the nucleotides specifies which amino acids will be used to construct proteins during the process of translation.
   B. DNA **replication** is **semiconservative** and requires the participation of several enzymes.

III. **Gene Function**
   A. DNA is used to produce RNA (**transcription**) and RNA is then used to produce protein (**translation**).
   B. RNA: Unlike DNA, RNA is single stranded, contains **uracil** instead of thymine and **ribose** instead of **deoxyribose.**
      1. Major forms of RNA found in the cell include **mRNA, tRNA,** and **rRNA.**
      2. The genetic information contained in DNA is copied to produce an RNA molecule. **Codons** in the mRNA pair with **anticodons** in the tRNA to specify what **amino acids** to assemble on the ribosome during translation.
   C. Transcription and Translation
      Transcription occurs when **RNA polymerase** copies the template strand of a segment of DNA. RNA is always made in the 5′ to 3′ direction.
   D. Translation occurs when the mRNA is used to direct the synthesis of proteins on the ribosome. Codons in the mRNA pair with anticodons in the tRNA to assemble a string of amino acids. This occurs until a stop codon is reached.
   E. Eucaryotic Gene Expression: Eucaryotic genes are composed of **exons** (expressed sequences) and **introns** (intervening sequences). The introns must be removed and the exons spliced together to create the final mRNA.

IV. **The Genetics of Animal Viruses**
   A. Genomes of viruses are found in many physical forms not seen in cells, including dsDNA, ssDNA, dsRNA, and ssRNA.
   B. DNA viruses tend to replicate in the nucleus while RNA viruses replicate in the cytoplasm. Retroviruses synthesize dsDNA from ssRNA.
   C. Some viruses integrate their DNA into the host cell genome. Integration by **oncogenic** viruses can lead to **transformation** of the host cell into an immortal cancerous cell.

V. **Regulation of Genetic Function**
   A. Protein synthesis is regulated through gene induction or repression, as controlled by an **operon.** Operons consist of several structural genes controlled by a common regulatory element.
      1. **Inducible operons** such as the lactose operon are normally off but can be turned on by a lactose inducer.
      2. **Repressible operons** are usually on but can be turned off when their end product is no longer needed.
      3. Many antibiotics prevent bacterial growth by interfering with transcription or translation.

VI. **Gene Mutation**
   A. Permanent changes in the genome of a microorganism are known as **mutations.** Mutations may be **spontaneous** or **induced.**
   B. **Point mutations** entail a change in one or a few bases and are categorized as **missense-, nonsense-, silent-,** or **back-mutations,** based on the effect of the change in nucleotide(s).
   C. Many mutations, particularly those involving mismatched bases or damage from ultraviolet light, can be corrected using enzymes found in the cell.
   D. The **Ames test** measures the mutagenicity of chemicals by determining the ability of a chemical to induce mutations in bacteria.

VII. **DNA Recombination**
   A. Intermicrobial transfer and genetic recombination permit gene sharing between bacteria. Major types of recombination include **conjugation, transformation,** and **transduction.**
   B. **Transposons** are DNA sequences that regularly move to different places within the genome of a cell, as a consequence generating mutations and variations in chromosome structure.

MULTIPLE-CHOICE QUESTIONS

1. What is the smallest unit of heredity?
   a. chromosome   c. codon
   b. gene   d. nucleotide

2. A nucleotide contains which of the following?
   a. 5 C sugar
   b. nitrogen base
   c. phosphate
   d. b and c only
   e. all of these

3. The nitrogen bases in DNA are bonded to the
   a. phosphate   c. ribose
   b. deoxyribose   d. hydrogen

4. DNA replication is semiconservative because the _____ strand will become half of the _____ molecule.
   a. RNA, DNA   c. sense, mRNA
   b. template, finished d. codon, anticodon

5. In DNA, adenine is the complementary base for _____, and cytosine is the complement for _____.
   a. guanine, thymine   c. thymine, guanine
   b. uracil, guanine   d. thymine, uracil

6. The base pairs are held together primarily by
   a. covalent bonds   c. ionic bonds
   b. hydrogen bonds   d. gyrases

7. Why must the lagging strand of DNA be replicated in short pieces?
   a. because of limited space
   b. otherwise, the helix will become distorted
   c. the DNA polymerase can synthesize in only one direction
   d. to make proofreading of code easier

8. Messenger RNA is formed by _____ of a gene on the DNA template strand.
   a. transcription   c. translation
   b. replication   d. transformation

9. Transfer RNA is the molecule that
   a. contributes to the structure of ribosomes
   b. adapts the genetic code to protein structure
   c. transfers the DNA code to mRNA
   d. provides the master code for amino acids

10. As a general rule, the template strand on DNA will always begin with
    a. TAC
    b. AUG
    c. ATG
    d. UAC

11. The *lac* operon is usually in the _____ position and is activated by a/an _____ molecule.
    a. on, repressor
    b. off, inducer
    c. on, inducer
    d. off, repressor

12. The repressible operon is important in regulating _____.
    a. amino acid synthesis
    b. DNA replication
    c. sugar metabolism
    d. ATP synthesis

13. For mutations to have an effect on populations of microbes, they must be
    a. inheritable
    b. permanent
    c. beneficial
    d. a and b
    e. all of the above

14. Which of the following characteristics is *not* true of a plasmid?
    a. It is a circular piece of DNA.
    b. It is required for normal cell function.
    c. It is found in bacteria.
    d. It can be transferred from cell to cell.

15. Which genes can be transferred by all three methods of intermicrobial transfer?
    a. capsule production
    b. toxin production
    c. F factor
    d. drug resistance

16. Which of the following would occur through specialized transduction?
    a. acquisition of Hfr plasmid
    b. transfer of genes for toxin production
    c. transfer of genes for capsule formation
    d. transfer of a plasmid with genes for degrading pesticides

17. **Multiple Matching.** Fill in the blanks with all the letters of the words below that apply.
    E,h ___ genetic transfer that occurs after the donor is dead
    F ___ carries the codon
    B ___ carries the anticodon
    G ___ a process synonymous with mRNA synthesis
    E ___ bacteriophages participate in this transfer
    C ___ a process requiring an F⁺ pilus
    A ___ duplication of the DNA molecule
    I ___ process in which transcribed DNA code is deciphered into a polypeptide
    C,E,H ___ involves plasmids

    a. replication
    b. tRNA
    c. conjugation
    d. ribosome
    e. transduction
    f. mRNA
    g. transcription
    h. transformation
    i. translation
    j. none of these

CONCEPT QUESTIONS

1. Compare the genetic material of eucaryotes, bacteria, and viruses in terms of general structure, size, and mode of replication.

2. Briefly describe how DNA is packaged to fit inside a cell.

3. Describe what is meant by the antiparallel arrangement of DNA.

4. On paper, replicate the following segment of DNA:

   5′ A T C G G C T A C G T T C A C 3′
   3′ T A G C C G A T G C A A G T G 5′

   a. Show the direction of replication of the new strands and explain what the lagging and leading strands are.
   b. Explain how this is semiconservative replication. Are the new strands identical to the original segment of DNA?

5. Name several characteristics of DNA structure that enable it to be replicated with such great fidelity generation after generation.

6. Explain the following relationship: DNA formats RNA, which makes protein.

7. What message does a gene provide? How is the language of the gene expressed?

8. If a protein is 3,300 amino acids long, how many nucleotide pairs long is the gene sequence that codes for it?

9. a. What is a palindrome in DNA and what is one function?
   b. Draw a short sequence of a palindrome.

10. Compare the structure and functions of DNA and RNA.

11. a. Where does transcription begin?
    b. What are the template and coding strands of DNA?
    c. Why is only one strand transcribed, and is the same strand of DNA always transcribed?

12. Compare and contrast the actions of DNA and RNA polymerase.

13. What are the functions of start and stop codons? Give examples of them.

14. The following sequence represents triplets on DNA:

    TAC CAG ATA CAC TCC CCT GCG ACT

    a. Give the mRNA codons and tRNA anticodons that correspond with this sequence, and then give the sequence of amino acids in the polypeptide.
    b. Provide another mRNA strand that can be used to synthesize this same protein.
    c. Using figure 9.14, give the type and order of the amino acids in the peptide.

15. a. Summarize how bacterial and eucaryotic cells differ in gene structure, transcription, and translation.
    b. Discuss the roles of exons and introns.

16. a. Compare DNA viruses with RNA viruses in their general methods of nucleic acid synthesis and viral replication.
    b. Read the section describing retroviruses (p. 273) and simply sketch a step-by-step drawing of its genetic pattern.

17. a. What is an operon? Describe the functions of regulators, promoters, and operators.
    b. Compare and contrast the *lac* operon with a repressible operon system.
    c. How is the *lac* system related to the feedback control of enzymes mentioned in chapter 8?

18. Explain the ideas behind the Ames test and what it is used for.

19. Describe the principal types of mutations. Give an example of a mutation that is beneficial and one that is lethal or harmful.

20. a. Compare conjugation, transformation, and transduction on the basis of general method, nature of donor, and nature of recipient.
    b. Explain the differences between general and specialized transduction, using drawings.

c. List some examples of genes that can be transferred by intermicrobial transfer.

21. By means of a flowchart, show the possible jumps that a transposon can make. Show the involvement of viruses in its movement.

CRITICAL-THINKING QUESTIONS

1. A simple test you can do to demonstrate the coiling of DNA in bacteria is to open a large elastic band, stretch it taut, and twist it. First it will form a loose helix, then a tighter helix, and finally, to relieve stress, it will twist back upon itself. Further twisting will result in a series of knotlike bodies; this is how bacterial DNA is condensed.

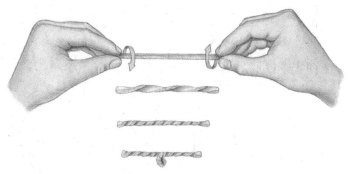

2. Knowing that retroviruses operate on the principle of reversing the direction of transcription from RNA to DNA, propose a drug that might possibly interfere with their replication.

3. Using the piece of DNA in concept question 14, show a deletion, an insertion, a substitution, and nonsense mutations. Which ones are frameshift mutations? Are any of your mutations nonsense? Missense? (Use the universal code to determine this.)

4. Using figure 9.14 and table 9.5, go through the steps in mutation of a codon followed by its transcription and translation that will give the end result in silent, missense, and nonsense mutations. Why is a change in the RNA code alone not really a mutation?

5. Explain the principle of "wobble" and find four amino acids that are encoded by wobble bases (figure 9.14). Suggest some benefits of this phenomenon to microorganisms.

6. Suggest a reason for having only one strand of DNA serve as a source of useful genetic information. What could be some possible functions of the coding strand?

7. The enzymes required to carry out transcription and translation are themselves produced through these same processes. Speculate which may have come first in evolution—proteins or nucleic acids—and explain your choice.

8. Why can one not reliably predict the sequence of nucleotides on mRNA or DNA by observing the amino acid sequence of proteins?

9. Speculate on the manner in which transposons can be involved in cancer.

10. Explain what is meant by the expression: phenotype = genotype + environment.

INTERNET SEARCH TOPICS

1. Find information on introns. Explain at least five current theories as to their possible functions.

2. Do an Internet search under the heading "DNA music." Explore several websites, discovering how this music is made and listening to some examples.

3. Visit the student Online Learning Center at www.mhhe.com/talaro5. Go to chapter 9, Internet Search Topics, and log on to the available websites to locate animations, 3D graphics, and interactive tutorials that help you to visualize replication, transcription, and translation.

CHAPTER 10

Genetic Engineering:
A Revolution in Molecular Biology

In chapter 9 we looked at the ways in which microorganisms duplicate, exchange, and use their genetic information. In scientific parlance this is called *basic science,* because no product or application is directly derived from it. Human beings being what they are, however, it is never long before basic knowledge is used to derive *applied science,* or useful products and applications that owe their invention to the basic research that preceded them. As an example, basic science into the workings of the electron has led to the development of television, computers, and cell phones. None of these staples of modern life were envisioned when early physicists were deciphering the nature of subatomic particles, but without the knowledge of how electrons worked, our ability to harness them for our own uses would never have materialized.

The same scenario can be seen with regard to genetics. The knowledge of how DNA was manipulated within the cell to carry out the goals of a microbe allowed scientists to utilize these processes to accomplish goals more to the liking of human beings. Contrary to being a new idea, the methods of genetic manipulation we will review are simply more efficient ways of accomplishing goals that humans have had for thousands of years.

Examples of human goals that have been more efficiently attained through the use of modern genetic technologies can be seen in each of these scenarios:

1. A farmer mates his two largest pigs in the hopes of producing larger offspring. Unfortunately he quite often ends up with small or unhealthy animals due to other genes that are transferred during mating. Genetic manipulation allows for the transfer of specific genes, so that only advantageous traits are selected.

2. Courts have, for thousands of years, relied on a description of a person's phenotype (eye color, hair color, etc.) as a means of identification. By remembering that a phenotype is the product of a particular sequence of DNA, you can quickly see how looking at someone's DNA (perhaps from a drop of blood) gives a clue as to his or her identification.

3. Lastly, many diseases are the result of a missing or dysfunctional protein, and we have generally treated the disease by replacing the protein as best we can, generally resulting in only temporary relief and limited success. Examples include insulin-dependent diabetes, adenosine deaminase deficiency, and blood clotting disorders. Genetic engineering offers the promise that someday soon fixing the underlying mutation responsible for the lack of a particular protein can treat these diseases far more successfully than we've been able to in the past.

The golden rice grains seen in the photo have been genetically engineered to produce beta-carotene, a precursor to vitamin A. Lack of vitamin A leads to over 1 million deaths and 300,000 cases of blindness a year.

Chapter Overview

- **Genetic engineering*** deals with the ability of scientists to manipulate DNA through the use of an expanding repertoire of recombinant DNA techniques. These techniques allow DNA to be cut, separated by size, and even sequenced to determine the actual composition and order of nucleotides.

- Biotechnology, the use of an organism's biochemical processes to create a product, has been greatly advanced by the introduction of recombinant DNA technologies because organisms can now be

*genetic engineering Sometimes called bioengineering—defined as the direct, deliberate modification of an organism's genome. Biotechnology is a more encompassing term that includes the use of DNA, genes, or genetically altered organisms in commercial production.

genetically modified to accomplish goals that were previously impossible, such as bacteria that have been engineered to produce human insulin.

- Differences in DNA between organisms are being exploited so that they can be used to identify people, animals, and microbes, revolutionizing fields such as police work and epidemiology.
- Genes can be probed to predict the likelihood of a particular genetic disease long before the disease strikes (and becomes less treatable). In a growing number of cases, missing or mutated genes are being replaced to correct inherited defects.
- Recombinant DNA technology is used to create genetically modified food that may have increased nutritional properties, be easier to grow, or may even vaccinate the person eating it against a host of diseases.
- Recombinant DNA techniques have given humans greater power over our own and other species than we've ever had before. Such power is not without risks, and **bioethics*** is an important part of any discussion of bioengineering.

Basic Elements and Applications of Genetic Engineering

Information on genetic engineering and its biotechnological applications is growing at such an expanding rate that some new discovery or product is disclosed almost on a daily basis. To keep this subject somewhat manageable, we will present essential concepts and applications that are both conventional and as up-to-date as possible. As an organizational aid, the topics are divided into the following six major sections, each designated with a roman numeral:

    I. Tools and Techniques of Genetic Engineering;
   II. Methods in Recombinant DNA Technology;
  III. Biochemical Products of Recombinant DNA Technology;
  IV. Genetically Modified Organisms;
   V. Genetic Treatments; and
  VI. Genome Analysis.

I. Tools and Techniques of Genetic Engineering

DNA: AN AMAZING MOLECULE

All of the intrinsic properties of DNA hold true whether the DNA is in a bacterium or test tube. For example, the enzyme helicase is able to unwind the two strands of the double helix just as easily in the lab as it does in a bacterial cell. But in the laboratory we can take advantage of our knowledge of DNA chemistry to make helicase unnecessary. It turns out that when DNA is heated to just below boiling (90°–95°C), the two strands separate, revealing the information contained in their bases. With the nucleotides exposed, DNA can be easily identified, replicated, or transcribed. If heat-denatured DNA is then slowly cooled, complementary nucleotides will hydrogen bond with one another and the strands will renature,

or regain their familiar double-stranded form. As we shall see, this process is a necessary feature of the polymerase chain reaction and nucleic acid probes described later.

Enzymes for Dicing, Splicing, and Reversing Nucleic Acids

The polynucleotide strands of DNA can also be clipped crosswise at selected positions by means of enzymes called **restriction endonucleases.*** These enzymes recognize foreign DNA and are capable of breaking the phosphodiester bonds between adjacent nucleotides on both strands of DNA, leading to a break in the DNA strand. In the bacterial cell, this ability protects against the incompatible DNA of bacteriophages or plasmids. In the biotechnologist's lab, the enzymes can be used to cleave DNA at desired sites and are a must for the techniques of recombinant DNA technology.

So far, hundreds of restriction endonucleases have been discovered in bacteria. Each type has a known sequence of 4 to 10 base pairs as its target, so sites of cutting can be finely controlled. These enzymes have the unique property of recognizing and clipping at base sequences called **palindromes** (figure 10.1b). Palindromes are sequences of DNA that are identical when read from the 5′ to 3′ direction on one strand and the 3′ to 5′ direction on the other strand.

Endonucleases are named by combining the first letter of the bacterial genus, the first two letters of the species, and the endonuclease number. Thus, EcoRI* is the first endonuclease found in *Escherichia coli,* and HindIII* is the third endonuclease discovered in *Haemophilus influenzae* Type d (**figure 10.1b).**

Endonucleases are used in the laboratory to cut DNA into smaller pieces for further study as well as to remove and insert it during recombinant DNA techniques, described in a subsequent section. Endonucleases such as HaeIII make straight, blunt cuts on DNA. But more often (EcoRI and HindIII, for example), the enzymes make staggered symmetrical cuts that leave short tails called "sticky ends." Such adhesive tails will base-pair with complementary tails on other DNA fragments or plasmids. This effect makes it possible to splice genes into specific sites.

The pieces of DNA produced by restriction endonucleases are termed *restriction fragments.* Because DNA sequences vary, even among members of the same species, differences in the cutting pattern of specific restriction endonucleases give rise to restriction fragments of differing lengths, known as *restriction fragment length polymorphisms* (RFLPs). RFLPs allow the direct comparison of the DNA of two different organisms at a specific site, which, as we will see, has many uses (see figures 10.16 and 10.17).

Another enzyme, called a **ligase,** is necessary to seal the sticky ends together by rejoining the phosphate-sugar bonds cut by endonucleases. Its main application is in final splicing of genes into plasmids and chromosomes.

An enzyme called **reverse transcriptase** is best known for its role in the replication of the AIDS virus and other retroviruses. It also provides geneticists with a valuable tool for converting RNA into DNA. Copies called **complementary DNA, or cDNA,** can be

---

***bioethics** A field that relates biological issues to human conduct and moral judgment.

***restriction endonuclease** The meaning of restriction is that the enzymes do not act upon the bacterium's DNA; an endonuclease nicks DNA internally, not at the ends.

***EcoRI** is pronounced "ee′-koh-arr-one."

***HindIII** is pronounced "hindy-three."

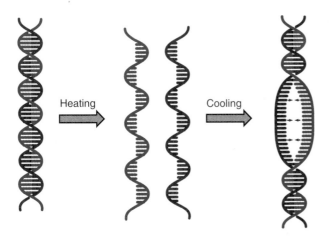

(a) DNA heating and cooling. DNA responds to heat by denaturing—losing its hydrogen bonding, and thereby separating into its two strands. When cooled, the two strands rejoin at complementary sites. The two strands need not be from the same organisms as long as they have matching sites.

Endonuclease	EcoRI	HindIII	HaeIII
Cutting pattern	G A A T T C C T T A A G	A A G C T T T T C G A A	G G C C C C G G

(b) Examples of palindromes and cutting patterns.

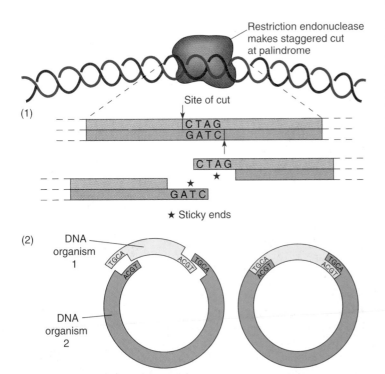

Restriction endonuclease makes staggered cut at palindrome

(1) Site of cut

CTAG
GATC

CTAG
★
★
GATC

★ Sticky ends

(2) DNA organism 1

TGCA ACGT TGCA
ACGT   ACGT

DNA organism 2

TGCA
ACGT

TGCA
ACGT

(c) Action of restriction endonucleases. (1) A restriction endonuclease recognizes and cleaves DNA at the site of a specific palindromic sequence. Cleavage can produce staggered tails called sticky ends that accept complementary tails for gene splicing. (2) The sticky ends can be used to join DNA from different organisms by cutting it with the same restriction enzyme, ensuring that all fragments have complementary ends.

FIGURE 10.1

Some useful properties of DNA.

made from messenger, transfer, ribosomal, and other forms of RNA. The technique provides a valuable means of synthesizing eucaryotic genes from mRNA transcripts. The advantage is that the synthesized gene will be free of the intervening sequences (introns) that can complicate the management of eucaryotic genes in genetic engineering. Complementary DNA can also be used to analyze the nucleotide sequence of RNAs, such as those found in ribosomes and transfer RNAs (see figure 4.29).

Analysis of DNA

One way to produce a readable pattern of DNA fragments is through **gel electrophoresis.** In this technique, samples are placed in compartments (wells) in a soft agar gel and subjected to an electrical current. The phosphate groups in DNA give the entire molecule an overall negative charge, which causes the DNA to move toward the positive pole in the gel. The rate of movement is based primarily on the size of the fragments. The larger fragments move more slowly and remain nearer the top of the gel, whereas the smaller fragments migrate faster and are positioned farther from the wells. The positions of DNA fragments are determined by staining the DNA fragments in the gel (**figure 10.2**). Electrophoresis patterns can be quite distinctive and are very useful in characterizing DNA fragments and comparing the degree of genetic similarities among samples as in a genetic fingerprint (see figure 10.16).

Nucleic Acid Hybridization and Probes

Two different nucleic acids can **hybridize** by uniting at their complementary sites. All different combinations are possible: Single-stranded DNA can unite with other single-stranded DNA or RNA, and RNA can hybridize with other RNA. This property has been the inspiration for specially formulated oligonucleotide tracers called **gene probes.** These probes consist of a short stretch of DNA of a known sequence that will base-pair with a stretch of DNA with a complementary sequence, if one exists in the test sample. Hybridization probes have practical value because they can detect specific nucleotide sequences in unknown samples. So that areas of hybridization can be visualized, the probes carry reporter molecules such as radioactive labels, which are isotopes that emit radiation, or luminescent labels, which give off visible light. Reactions can be revealed by placing photographic film in contact with the test reaction. Fluorescent probes contain dyes that can be visualized with ultraviolet radiation, and enzyme-linked probes react with substrate to release colored dyes (**figure 10.3**).

When probes hybridize with an unknown sample of DNA or RNA, they tag the precise area and degree of hybridization and help determine the nature of nucleic acid present in a sample. In a method called the **Southern blot,**[1] DNA fragments are first separated by electrophoresis and then denatured and transfered to a special filter. A DNA probe is then incubated with the sample, and wherever this probe encounters the segment for which it is complementary, it will attach and form a hybrid. Development of the hybridization pattern will show up as one or more bands (figure 10.4). This method is a sensitive and specific way to isolate fragments

---

1. Named for its developer, E. M. Southern. The "northern" blot is a similar method used to analyze RNA, while the western blot detects proteins. There is a reason scientists became scientists and not comedians.

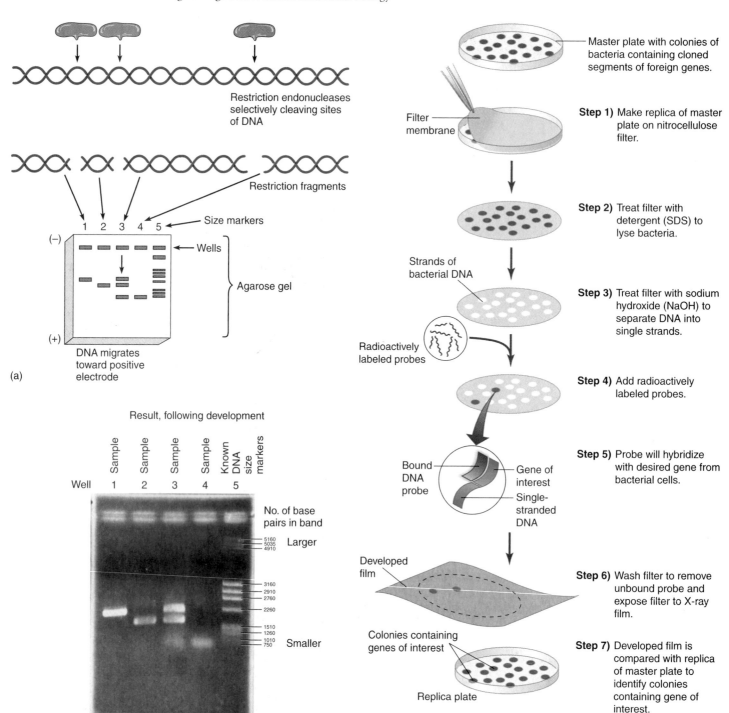

FIGURE 10.2

Revealing the patterns of DNA with electrophoresis. **(a)** After cleavage into fragments, DNA is loaded into wells on one end of an agarose gel. When an electrical current is passed through the gel (from the negative pole to the positive pole), the DNA, being negatively charged, migrates toward the positive pole. The larger fragments, measured in numbers of base pairs, migrate more slowly and remain nearer the wells than the smaller (shorter) fragments. **(b)** An actual developed gel (here stained with ethidium bromide) reveals a separation pattern of the fragments of DNA. The size of a given DNA band can be determined by comparing it to a known set of markers (lane 5) called a ladder. This method can be used for general screening of DNA or for genetic fingerprints.

FIGURE 10.3

A hybridization test relies on the action of microbe-specific probes to identify an unknown bacteria or virus.

from a complex mixture and to find specific gene sequences on DNA. Southern blotting is also one of the important first steps for preparing isolated genes.

Probes are commonly used for diagnosing the cause of an infection from a patient's specimen and identifying a culture of an unknown bacterium or virus. The method for doing this was first outlined in chapter 4. A simple and rapid method called a hybridization test does not require electrophoresis. DNA from a test sample is

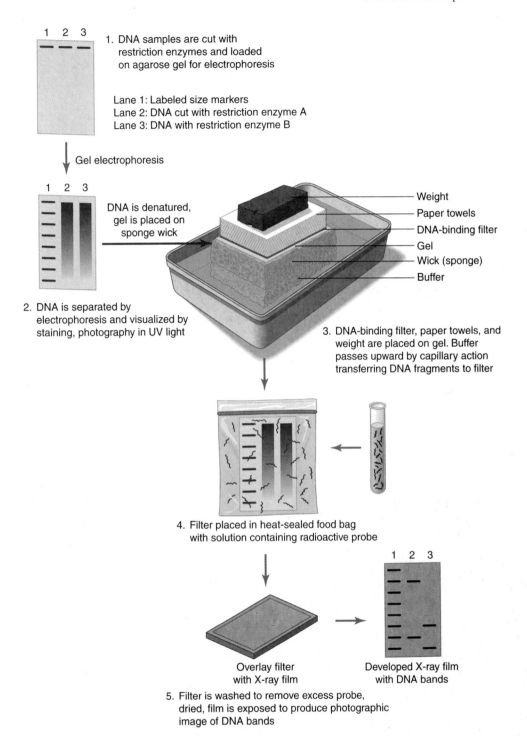

1. DNA samples are cut with restriction enzymes and loaded on agarose gel for electrophoresis

Lane 1: Labeled size markers
Lane 2: DNA cut with restriction enzyme A
Lane 3: DNA with restriction enzyme B

Gel electrophoresis

DNA is denatured, gel is placed on sponge wick

2. DNA is separated by electrophoresis and visualized by staining, photography in UV light

Weight
Paper towels
DNA-binding filter
Gel
Wick (sponge)
Buffer

3. DNA-binding filter, paper towels, and weight are placed on gel. Buffer passes upward by capillary action transferring DNA fragments to filter

4. Filter placed in heat-sealed food bag with solution containing radioactive probe

Overlay filter with X-ray film

Developed X-ray film with DNA bands

5. Filter is washed to remove excess probe, dried, film is exposed to produce photographic image of DNA bands

FIGURE 10.4

Conducting a Southern blot hybridization test.   The 5-step process reveals the hybridized fragments on the X-ray film.

isolated, denatured, placed on an absorbent filter, and combined with a microbe-specific probe. The blot is then developed and observed for areas of hybridization. Commercially available diagnostic kits are now on the market for identifying intestinal pathogens such as *E. coli, Salmonella, Campylobacter, Shigella, Clostridium difficile,* rotaviruses, and adenoviruses. Other bacterial probes exist for *Mycobacterium, Legionella, Mycoplasma, and Chlamydia;* viral probes are available for herpes simplex and

zoster, papilloma (genital warts), hepatitis A and B, and AIDS. DNA probes have also been developed for human genetic markers and some types of cancer.

With another method, called *fluorescent in situ hybridization* (FISH), probes are applied to intact cells and observed microscopically for the presence and location of specific genetic marker sequences on genes. It is a very effective way to locate genes on chromosomes. In situ techniques can also be used to identify

In early 2001 a press conference was held to announce that the human genome had been sequenced. The 3.1 billion base pairs that make up the DNA found in (nearly) every human cell had been identified and put in the proper order. Champagne corks popped, balloons fell, bands played, reporters reported on the significance of the occasion and . . . nothing else seemed to change. What, you may ask, has the Human Genome Project done for me lately? Here for your perusal are a few FAQs.

Q: Who sequenced the genome?

A: Francis Collins was the head of the publicly funded Human Genome Project (HGP) while Craig Venter was the head of Celera Genomics, a private company that developed a new, more powerful method of DNA sequencing and competed with the HGP. A compromise was finally reached whereby both groups took credit for sequencing the genome.

Q: How big was this project?

A: The Human Genome Project was first discussed in the mid 1980s and got underway in 1990. Although the project was to have taken at least 15 years, advances in technology led to its being completed in just over 10. The 3.1 billion base pairs of DNA code for only about 30,000 genes, not the 100,000 or so that the genome was thought to contain only a few years ago.

Q: Will I ever see any benefits from this project?

A: Absolutely. Sequencing the genome was only the fist step. Knowing what proteins are produced in the body, what they do, and how they interact with one another are essential to understanding the workings of the human body, in both health and disease. By knowing the genetic signatures of different diseases we will be able to design extraordinarily sensitive diagnostic tests that detect not only a disease but particular subtypes of each malady. With a precise genetic identification of, for example, a tumor, treatment can be tailored to be as effective as possible, while dramatically reducing side effects.

Q: What's next?

A: While everyone is still digesting the human genome (being the equivalent of a million page book, it's a long read), several groups have gone to work on the mouse genome. The thinking is that areas within genes that are similar in mouse and human (known as regions of homology) are most important for the function of the gene and are the most likely targets for potential drugs or genetic treatments.

Q: Where could I go to check out a really cool website on the human genome?

A: http://www.ornl.gov/hgmis/

J. Craig Venter, left, and Francis Collins mapped the human genome.

unknown bacteria living in natural habitats without having to culture them, and they can be used to detect RNA in cells and tissues.

Methods Used to Size, Synthesize, and Sequence DNA

The relative sizes of nucleic acids are usually denoted by the number of base pairs (bp) or nucleotides they contain. For example, the palindromic sequences recognized by endonucleases are usually 4 to 10 bp in length, an average gene in *E. coli* is approximately 1,300 bp, or 1.3 kilobases (kb), and its entire genome is approximately 4,700,000 bp, 4,700 kb or 4.7 megabases (Mb). The DNA of the human mitochondrion contains 16 kb, and the Epstein-Barr virus (a cause of infectious mononucleosis) has

172 kb. Humans have approximately 3.1 billion base pairs (Bb) arrayed along 46 chromosomes. The recently completed Human Genome Project had as its goal the elucidation of the entire human genome (see **Microbits 10.1**).

DNA Sequencing: Determining the Exact Genetic Code
Analysis of DNA by its size, restriction patterns, and hybridization characteristics is instructive, but the most detailed information comes from determining the actual order and types of bases in DNA. This process, called **DNA sequencing,** provides the identity and order of nucleotides for all types of DNA, including genomic, cDNA, artificial chromosomes, plasmids, and cloned genes. The most common sequencing technique was developed by Frederick

FIGURE 10.5

Steps in a Sanger DNA sequence technique.

(1) Isolated unknown DNA fragment (one strand will be shown for clarity).

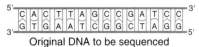

Original DNA to be sequenced

(2) DNA is denatured to produce single template strand.

(3) Strand is labeled with specific primer molecule.

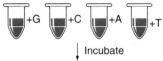

(4) DNA polymerase and regular nucleotide mixture (ATP, CTP, GTP, and TTP) are added; ddG, ddA, ddC, and ddT are placed in separate reaction tubes with the regular nucleotides. The dd nucleotides are labelled with some type of tracer, which allows them to be visualized.

+G   +C   +A   +T

↓ Incubate

(5) Newly replicated strands are terminated at the point of addition of a dd nucleotide.

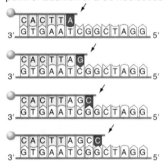

(6) A schematic view of how all possible positions on the fragment are occupied by a labeled nucleotide.

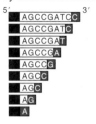

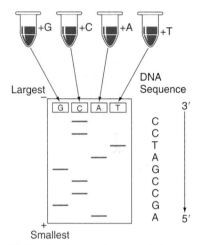

+G   +C   +A   +T

Largest

DNA Sequence

Reading panel:
3′
C
C
T
A
G
C
C
G
A
5′

Smallest

(7) Running the reaction tubes in four separate gel lanes separates them by size and nucleotide type. Reading from bottom to top, one base at a time, provides the correct DNA sequence.

Sanger and is based on the synthesis and analysis of a complementary strand of DNA in a test tube (**figure 10.5**).

Since most DNA being sequenced is very long, it is made more manageable by cutting into a large number of shorter fragments and separated. The test strands, typically several hundred nucleotides long, are then denatured to expose single strands that will serve as templates to synthesize complementary strands. The fragments are divided into four separate tubes that contain primers to set the start point for the synthesis to begin on one of the strands. The primer is labeled with a fluorescent or radioactive tag, which allows it to be detected. The nucleotides will attach at the 3′ end of the primer, using the template strands as a guide, essentially the same as regular DNA synthesis as performed in a cell.

The tubes are incubated with the necessary DNA polymerase and all four of the regular nucleotides needed to carry out the process of elongating the complementary strand. Each tube also contains a single type (A, T, G or C) of dideoxynucleotide (dd) which is critical to the sequencing process. Dideoxynucleotides have no oxygen bound to the 3′ carbon of ribose. This oxygen atom is needed for the chemical attachment of the next nucleotide in the growing DNA strand (see figure 2.24*a* and 9.4*a*). As the reaction proceeds, strands elongate by adding normal nucleotides, but a small percentage of fragments will randomly incorporate the complementary dd nucleotide and be terminated. Eventually, all possible positions in the sequence will incorporate a terminal dideoxynucleotide, thus producing a series of strands that reflect the correct sequence.

The reaction products are placed into four wells (G, C, A, T) of a polyacrylamide gel, which is sensitive enough to separate strands that differ by only a single nucleotide in length. Electrophoresis separates the fragments in order according to both size and lane, and only the fragments carrying the labelled nucleotides will be readable on the gel. The gel indicates the comparative orientation of the bases, which graduate in size from the smallest fragments that terminate early and migrate farthest (bottom of gel) to successively longer fragments (moving stepwise to the top). Reading the order of first appearance of a given gel fragment in the G, C, A, or T lanes provides the correct sequence of bases of the complementary strand, and it also allows one to infer the sequence of the template strand as well. This method of sequencing is remarkably accurate, with only about one mistake in every 1,000 bases. For managing the giant genomes of humans, mice, and other organisms, it has been necessary to automate this method using high-density arrays that can be rapidly sequenced and displayed by a computer.

(a) In cycle 1, the DNA to be amplified is denatured, primed, and replicated by a polymerase that can function at high temperature. The two resulting strands then serve as templates for a second cycle of denaturation, priming, and synthesis.*

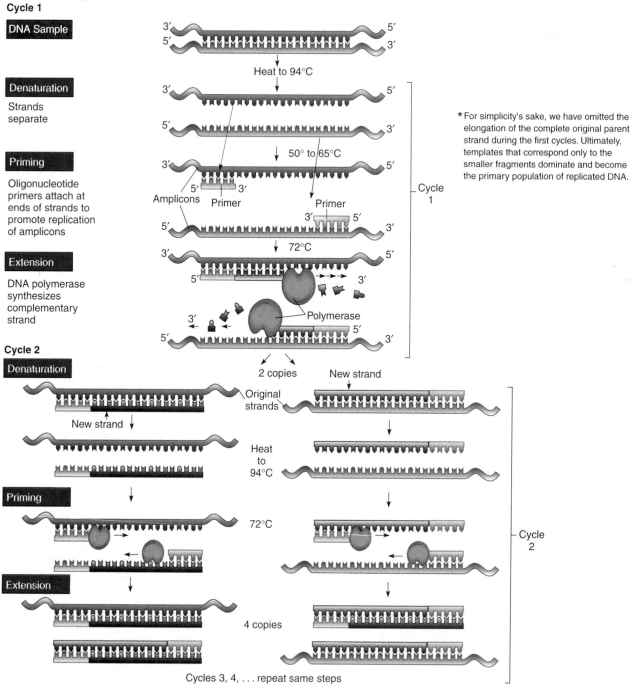

*For simplicity's sake, we have omitted the elongation of the complete original parent strand during the first cycles. Ultimately, templates that correspond only to the smaller fragments dominate and become the primary population of replicated DNA.

Cycle 1

DNA Sample

Denaturation

Strands separate

Priming

Oligonucleotide primers attach at ends of strands to promote replication of amplicons

Extension

DNA polymerase synthesizes complementary strand

Cycle 2

Denaturation

Priming

Extension

Cycles 3, 4, . . . repeat same steps

(b) A view of the process after 6 cycles, with 64 copies of amplified DNA. Continuing this process for 20 to 40 cycles can produce millions of identical DNA molecules.

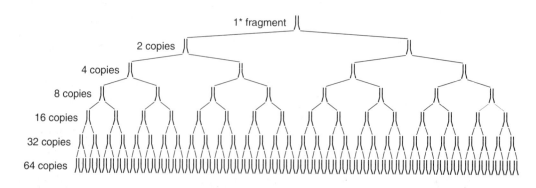

FIGURE 10.6

Schematic of the polymerase chain reaction.

Polymerase Chain Reaction: A Molecular Xerox Machine for DNA

Some of the techniques used to analyze DNA and RNA are limited by the small amounts of test nucleic acid available. This problem was largely solved by the invention of a simple, versatile way to amplify DNA called the **polymerase chain reaction (PCR).** This technique rapidly increases the amount of DNA in a sample without the need for making cultures or carrying out complex purification techniques. It is so sensitive that it holds the potential to detect cancer from a single cell or to diagnose an infection from a single gene copy. It is comparable to being able to pluck a single DNA "needle" out of a "haystack" of other molecules and make unlimited copies of the DNA. The rapid rate of PCR makes it possible to replicate a target DNA from a few copies to billions of copies in a few hours. It can amplify DNA fragments that consist of a few base pairs to whole genomes containing several million base pairs.

To understand the idea behind PCR, it will be instructive to review figure 9.6, which describes synthesis of DNA as it occurs naturally in cells. The PCR method uses essentially the same events, with the opening up of the double strand, using the exposed strands as templates, the addition of primers, and the actions of a DNA polymerase.

Initiating the reaction requires a few specialized ingredients **(figure 10.6).** The **primers** are synthetic oligonucleotides (short DNA strands) of a known sequence of 15 to 30 bases that serve as landmarks to indicate where DNA amplification will begin. These take the place of RNA primers that would normally be synthesized by primase (in the cell). Depending upon the purposes and what is known about the DNA being replicated, the primers can be random, attaching to any sequence they may fit, or they may be highly specific and chosen to amplify a known gene. To keep the DNA strands separated, processing must be carried out at a relatively high temperature. This necessitates the use of special **DNA polymerases** isolated from thermophilic bacteria. Examples of these unique enzymes are Taq polymerase obtained from *Thermus aquaticus* and Vent polymerase from *Thermococcus litoralis.* Enzymes isolated from these thermophilic organisms will remain active at the elevated temperatures used in PCR. Another useful component of PCR is a machine called a thermal cycler that automatically initiates the cyclic temperature changes.

The PCR technique operates by repetitive cycling of three basic steps: denaturation, priming, and extension. The first step involves heating target DNA to 94°C to separate it into two strands. Next, the system is cooled to between 50° and 65°C, depending on the exact nucleotide sequence of the primer. Primers are added in a concentration that favors binding to the complementary strand of test DNA. This reaction prepares the two DNA strands, now called **amplicons,** for synthesis. In the third phase, which proceeds at 72°C, DNA polymerase and raw materials in the form of nucleotides are added. Beginning at the free end of the primers on both strands, the polymerases extend the molecule by adding appropriate nucleotides and produce two complete strands of DNA.

It is through repetition of these same steps that DNA becomes amplified. When the DNAs formed in the first cycle are denatured, they become amplicons to be primed and extended in the second cycle. Each subsequent cycle converts the new DNAs to amplicons and doubles the number of copies. The number of cycles required to produce a million molecules is 20, but the process is usually carried out to 30 or 40 cycles. One significant advantage of this technique has been its natural adaptability to automation. A PCR machine can perform 20 cycles on nearly 100 samples in 2 or 3 hours.

Once the PCR is complete the amplified DNA can be analyzed by any of the techniques discussed earlier. PCR can be adapted to analyze RNA by initially converting an RNA sample to DNA with reverse transcriptase. This cDNA can then be amplified by PCR in the usual manner. It is by such means that ribosomal RNA and messenger RNA are readied for sequencing. The polymerase chain reaction has found prominence as a powerful workhorse of molecular biology, medicine, and biotechnology. It often plays an essential role in gene mapping, the study of genetic defects and cancer, forensics, taxonomy, and evolutionary studies.

For all of its advantages, PCR has some problems. A serious concern is the introduction and amplification of nontarget DNA from the surrounding environment such as a skin cell from the technician carrying out the PCR reaction rather than from the person whose DNA was supposed to be amplified. Such contamination can be minimized by using equipment and rooms dedicated for DNA analysis maintained with the utmost degree of cleanliness. Problems with contaminants can also be reduced by using gene-specific primers and treating samples with special enzymes that can degrade the contaminating DNA before it is amplified.

CHAPTER CHECKPOINTS

The genetic revolution has produced a wide variety of industrial technologies that translate and radically alter the blueprints of life. The potential of biotechnology promises not only improved quality of life and enhanced economic opportunity but also serious ethical dilemmas. Increased public understanding is essential for developing appropriate guidelines for responsible use of these revolutionary techniques.

Genetic engineering utilizes a wide range of methods that physically manipulate DNA for purposes of visualization, sequencing, hybridizing, and identifying specific sequences. The tools of genetic engineering include specialized enzymes, gel electrophoresis, DNA sequencing machines, and gene probes. The polymerase chain reaction (PCR) technique amplifies small amounts of DNA into larger quantities for further analysis.

II. Methods in Recombinant DNA Technology: How to Imitate Nature

The primary intent of **recombinant DNA technology** is to deliberately remove genetic material from one organism and combine it with that of a different organism. Its origins can be traced to 1970, when microbiologists first began to duplicate the clever tricks bacteria do naturally with bits of extra DNA such as plasmids, transposons, and proviruses. As we mentioned earlier, humans have been trying to artificially influence genetic transmission of traits for centuries. The discovery that bacteria can readily accept, replicate, and express foreign DNA made them powerful agents for studying the genes of other organisms in isolation. The practical applications of this work were soon realized by biotechnologists. Bacteria could be genetically engineered to mass-produce substances such as hormones, enzymes, and vaccines that were difficult to synthesize by the usual industrial methods.

Figure 10.7 provides an overview of the recombinant DNA procedure. An important objective of this technique is to form genetic **clones.*** Cloning involves the removal of a selected gene from an animal, plant, or microorganism (the genetic donor) followed by its propagation in a different host organism. Cloning requires that the desired donor gene first be selected, excised by restriction endonucleases, and isolated. The gene is next inserted into a **vector** (usually a plasmid or a virus) that will insert the DNA into a **cloning host.** The cloning host is usually a bacterium or a yeast that can replicate the gene and translate it into the protein product for which it codes. In the next section, we examine the elements of gene isolation, vectors, and cloning hosts and show how they participate in a complete recombinant DNA procedure.

TECHNICAL ASPECTS OF RECOMBINANT DNA AND GENE CLONING

The first hurdles in cloning a target gene are to locate its exact site on the genetic donor's chromosome and to isolate it. Some of the first isolated genes came from viruses, which have extremely small and manageable genomes, and later genes were isolated from bacteria and yeasts. At first, the complexity of the human genome made locating specific genes very difficult. However, this process has been greatly improved by tools for excising, mapping, and identifying genes. Among the most common strategies for obtaining genes in an isolated state are:

1. The DNA is removed from cells and separated into fragments by endonucleases. Each fragment is then inserted into a vector and cloned. The cloned fragments are Southern blotted and probed to identify desired sequences. This is a long and tedious process, because each fragment of DNA must be examined for the cloned gene. For example, if the human genome were separated into 20 kb fragments, there would be at least 150,000 clones to test.
2. A gene can be synthesized from isolated mRNA transcripts using reverse transcriptase (cDNA).
3. A gene can be amplified using PCR in many cases.

Although gene cloning and isolation can be very laborious, a fortunate outcome is that, once isolated, genes can be maintained in a cloning host and vector just like a microbial pure culture. *Genomic libraries,* which are collections of isolated genes, now exist for millions of genes from hundreds of organisms and viruses.

Characteristics of Cloning Vectors

A good recombinant vector has two indispensable qualities: It must be capable of carrying a significant piece of the donor DNA, and it must be readily accepted by the cloning host. Plasmids are excellent vectors because they are small, well characterized, and easy to manipulate, and they can be transferred into appropriate host cells

---

*__*clone__ (klohn) Gr. *klon,* young shoot or twig. Do not confuse the molecular biology use of clone with other uses. For example, a clone could be an organism that is genetically identical to its parent.

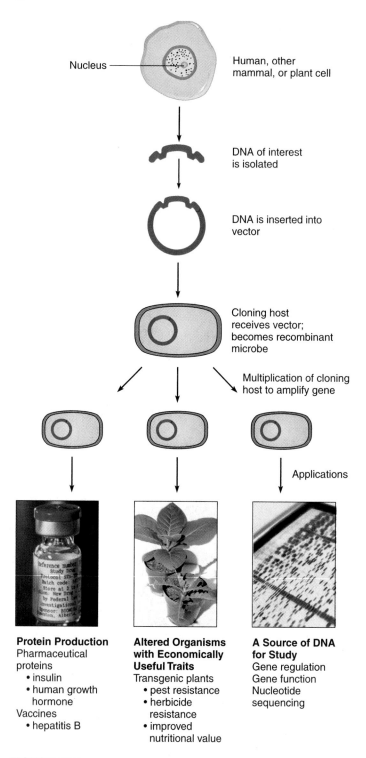

Protein Production
Pharmaceutical proteins
• insulin
• human growth hormone
Vaccines
• hepatitis B

Altered Organisms with Economically Useful Traits
Transgenic plants
• pest resistance
• herbicide resistance
• improved nutritional value

A Source of DNA for Study
Gene regulation
Gene function
Nucleotide sequencing

FIGURE 10.7

Methods and applications of genetic technology. Practical applications of genetic engineering include the development of pharmaceuticals, genetically modified organisms, and forensic techniques.

through transformation. Bacteriophages also serve well because they have the natural ability to inject DNA into bacterial hosts through transduction. A common vector in early work was an *E. coli* plasmid that carries genetic markers for resistance to antibiotics, although it is restricted by the relatively small amount of foreign

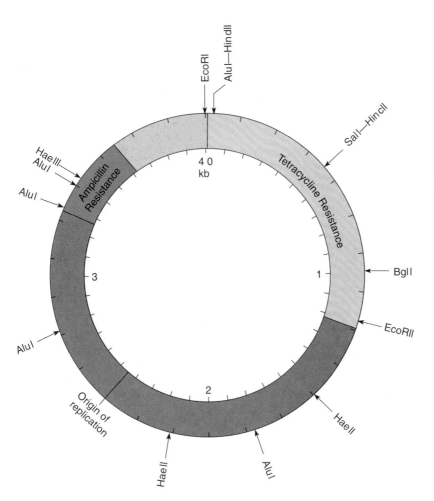

FIGURE 10.8

Partial map of the pBR322 plasmid of *Escherichia coli.* Arrows delineate sites cleaved by various restriction endonucleases. Numbers indicate the size of regions in kilobases (kb). Other important components are genes for ampicillin and tetracycline resistance and the origin of replication.

DNA it can accept. A modified phage vector, the *Charon*[2] phage, is missing large sections of its genome, so it can carry a fairly large segment of foreign DNA. The simple plasmids and bacteriophages that were a staple of early recombinant DNA methodologies evolved into newer, more advanced vectors. Today, a single call to a toll-free number will bring any of hundreds of unique cloning vectors to your door. Although every vector has characteristics that make it ideal for a specific project, all vectors can be thought of as having three primary areas of interest (**figure 10.8**):

1. An origin of replication (ORI) is needed somewhere on the vector so that it will be replicated by the DNA polymerase of the cloning host.
2. The size of the donated DNA that the vector will accept— early plasmids were limited to an insert size of less than 10 kb of DNA, far too small for most eucaryotic genes, with their sizable introns. Newer vectors such as cosmids can hold 100 kb while complex *bacterial artificial chromosomes* (BACs) and *yeast artificial chromosomes* (YACs) can hold as much as 300 kb and 1,000 kb, respectively.
3. Vectors typically contain a gene that confers drug resistance to their cloning host. In this way, cells can be grown on drug-containing media and only those cells that harbor a plasmid will be selected for growth.

2. Named for the mythical boatman in Hades who carried souls across the River Styx.

Characteristics of Cloning Hosts

The best cloning hosts possess several key characteristics (**table 10.1**). The traditional cloning host and the one still used in the majority of experiments is *Escherichia coli.* Because this bacterium was the original recombinant host, the protocols using it are well established, relatively easy, and reliable. Hundreds of specialized cloning vectors have been developed for it. The main disadvantage with this species is that the splicing of mRNA as well as the modification of proteins that normally occurs in the endoplasmic reticulum and Golgi apparatus are unavailable in this procaryotic cloning host. One alternative host for certain industrial processes and research is the yeast *Saccharomyces cerevisiae,* which, being eucaryotic, already possesses mechanisms for processing and modifying eucaryotic

TABLE 10.1

Desirable Features in a Microbial Cloning Host

Rapid overturn, fast growth rate
Can be grown in large quantities using ordinary culture methods
Nonpathogenic
Genome that is well delineated (mapped)
Capable of accepting plasmid or bacteriophage vectors
Maintains foreign genes through multiple generations
Will secrete a high yield of proteins from expressed foreign genes

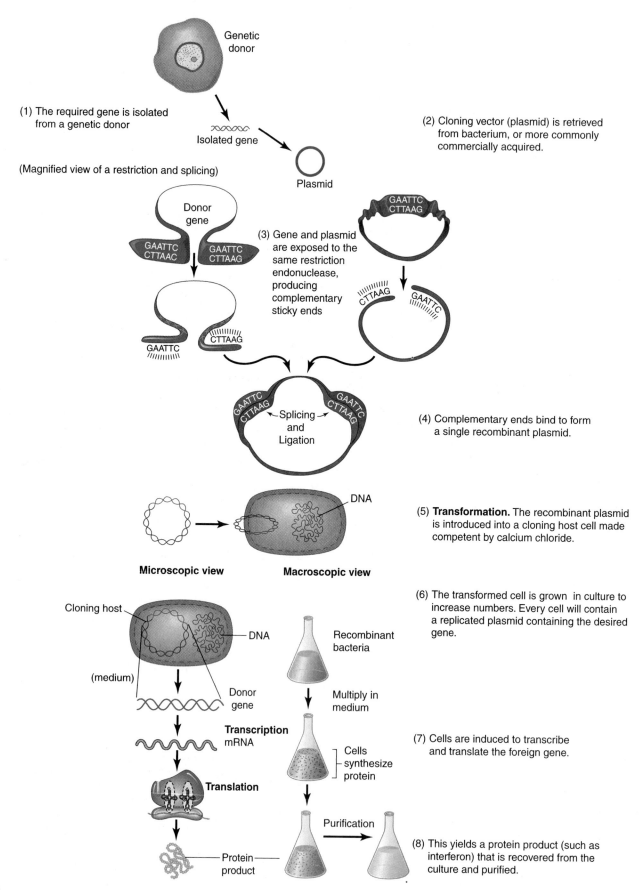

(1) The required gene is isolated from a genetic donor

Genetic donor

Isolated gene

(2) Cloning vector (plasmid) is retrieved from bacterium, or more commonly commercially acquired.

(Magnified view of a restriction and splicing)

Plasmid

Donor gene

GAATTC
CTTAAC

GAATTC
CTTAAG

GAATTC
CTTAAG

(3) Gene and plasmid are exposed to the same restriction endonuclease, producing complementary sticky ends

CTTAAG

GAATTC

CTTAAG

GAATTC

GAATTC
CTTAAG

←Splicing→
and
Ligation

GAATTC
CTTAAG

(4) Complementary ends bind to form a single recombinant plasmid.

DNA

Microscopic view   **Macroscopic view**

(5) **Transformation.** The recombinant plasmid is introduced into a cloning host cell made competent by calcium chloride.

Cloning host

DNA

(medium)

Donor gene

Recombinant bacteria

Multiply in medium

Transcription
mRNA

Cells synthesize protein

Translation

Purification

Protein product

(6) The transformed cell is grown in culture to increase numbers. Every cell will contain a replicated plasmid containing the desired gene.

(7) Cells are induced to transcribe and translate the foreign gene.

(8) This yields a protein product (such as interferon) that is recovered from the culture and purified.

FIGURE 10.9

Steps in recombinant DNA, gene cloning, and product retrieval.

gene products. Certain techniques may also employ different bacteria *(Bacillus subtilis)*, animal cell culture, and even live animals and plants to serve as cloning hosts. In our coverage, we present the recombinant process as it is performed in bacteria and yeasts.

CONSTRUCTION OF A RECOMBINANT, INSERTION INTO A CLONING HOST, AND GENETIC EXPRESSION

This section illustrates the main steps in recombinant DNA technology to produce a drug called alpha-2 interferon (Roferon-A). This form of interferon is used to treat cancers such as hairy-cell leukemia and Kaposi's sarcoma in AIDS patients. The human alpha interferon gene is a linear molecule of approximately 500 bp that codes for a polypeptide of 166 amino acids. It was originally isolated and identified from human blood cells and prepared from processed mRNA transcripts that are free of introns. This is a necessary step because the bacterial cloning host has none of the machinery needed to excise this nontranslated part of a gene.

The first step in cloning is to prepare the isolated interferon gene for splicing into an *E. coli* plasmid (**figure 10.9**). One way this is accomplished is to digest both the gene and the plasmid with the same restriction enzyme, resulting in complementary sticky ends on both the vector and insert DNA. In the presence of the endonuclease, the plasmid's circular molecule is nicked open, and its sticky ends are exposed (figure 10.9). When the gene and plasmid are placed together, their free ends base-pair, and a ligase makes the final covalent seals. The resultant gene and plasmid combination is called a **recombinant.**

Following this procedure, the recombinant is introduced by transformation into the cloning host, a special laboratory strain of *E. coli* that lacks any extra plasmids that could complicate the expression of the gene. Because the recombinant plasmid enters only some of the cloning host cells, it is necessary to locate these recombinant clones. Cultures are plated out on medium containing ampi-

cillin, and only those clones that carry the plasmid with ampicillin resistance can form colonies (**figure 10.10**). These recombinant colonies are selected from the plates and cultured. As the cells multiply, the plasmid is replicated along with the cell's chromosome. In a few hours of growth, there can be billions of cells, each containing the interferon gene. Once the gene has been successfully cloned and tested, this step does not have to be repeated—the recombinant strain can be maintained in culture for production purposes.

The bacteria's ability to express the eucaryotic gene is ensured, because the plasmid has been modified with the necessary transcription and translation recognition sequences. As the *E. coli* culture grows it transcribes and translates the interferon gene, synthesizes the peptide, and secretes it into the growth medium. At the end of the process, the cloning cells and other chemical and microbial impurities are removed from the medium. Final processing to excise a terminal amino acid from the peptide yields the interferon product in a relatively pure form (see figure 10.9). The scale of this procedure can range from test tube size to gigantic industrial vats that can manufacture thousands of gallons of product (see figure 26.40).

Although the process we have presented here produces interferon, some variation of it can be used to mass produce a variety of hormones, enzymes, and agricultural products such as pesticides.

CHAPTER CHECKPOINTS

Recombinant DNA techniques combine DNA from different sources to produce microorganism "factories" that produce hormones, enzymes, and vaccines on an industrial scale. Cloning is the process by which genes are removed from the original host and duplicated for transfer into a cloning host by means of cloning vectors.

Plasmids, bacteriophages, and cosmids are types of cloning vectors used to transfer recombinant DNA into a cloning host. Cloning hosts are simply organisms that readily accept recombinant DNA, grow easily, and synthesize large quantities of specific gene product.

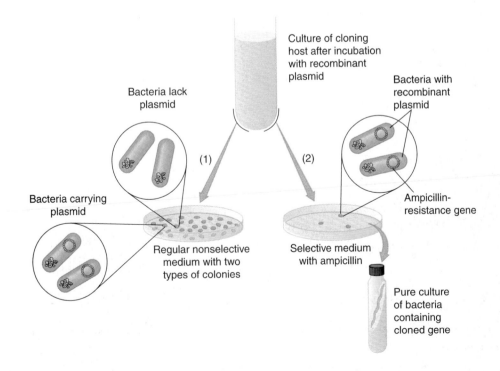

Culture of cloning host after incubation with recombinant plasmid

Bacteria lack plasmid

Bacteria with recombinant plasmid

Bacteria carrying plasmid

(1)   (2)

Ampicillin-resistance gene

Regular nonselective medium with two types of colonies

Selective medium with ampicillin

Pure culture of bacteria containing cloned gene

FIGURE 10.10

One method for screening clones of bacteria that have been transformed with the donor gene. **(1)** Plating the culture on nonselective medium will not separate the transformed cells from normal cells, which lack the plasmid. **(2)** Plating on selective medium containing ampicillin will permit only cells containing the plasmid to multiply. Colonies growing on this medium that carry the cloned gene can be used to make a culture for gene libraries, industrial production, and other processes. (Plasmids are shown disproportionate to size of cell.)

III. Biochemical Products of Recombinant DNA Technology

Recombinant DNA technology is used by pharmaceutical companies to manufacture medications that cannot be manufactured by any other means. Diseases such as diabetes and dwarfism, caused by the lack of an essential hormone, are treated by replacing the missing hormone. Insulin of animal origin was once the only form available to treat diabetes, even though such animal products can cause allergic reactions in sensitive individuals. In contrast, dwarfism cannot be treated with animal growth hormones, so originally the only source of human growth hormone (HGH) was the pituitaries of cadavers. At one time, not enough HGH was available to treat the thousands of children in need. Another serious problem with using natural human products is the potential for infection. For example, infectious agents such as the prion responsible for Creutzfeld-Jakob disease can be transmitted in this manner. Similarly, clotting factor VIII, a protein needed for blood to clot properly, is missing in persons suffering from hemophilia A. Persons lacking factor VIII have historically received periodic injections of the missing protein which had been collected from blood plasma. While the donated protein alleviated the symptoms of factor VIII deficiency, a tragic side effect was seen in the early 1980s, when a large percentage of the patients receiving plasma-derived factor VIII contracted HIV infections as a result of being exposed to the virus through the donated plasma.

Recombinant technology changed the outcome of these and many other conditions by enabling large-scale manufacture of life-saving hormones and enzymes of human origin. Recombinant human insulin can now be prescribed for diabetics, and recombinant HGH can now be administered to children with dwarfism and Turner syndrome. HGH is also used to prevent the wasting syndrome that occurs in AIDS and cancer patients. In all of these applications, recombinant DNA technology has led to both a safer product and one that can be manufactured in quantities previously unfathomable. Other protein-based hormones, enzymes, and vaccines produced through recombinant DNA are summarized in **table 10.2.**

Nucleic acid products also have a number of medical applications. A new development in vaccine formulation involves using microbial DNA as a stimulus for the immune system. So far, animal tests using DNA vaccines for AIDS and influenza indicate that this may be a breakthrough in vaccine design. Recombinant DNA could also be used to produce DNA-based drugs for the types of gene and antisense therapy discussed in the genetic treatments section.

TABLE 10.2

Current Protein Products from Recombinant DNA Technology

Immune Treatments

Interferons—peptides used to treat some types of cancer, multiple sclerosis, and viral infections such as hepatitis and genital warts

Interleukins—types of cytokines that regulate the immune function of white blood cells; used in cancer treatment.

Orthoclone—an immune suppressant in transplant patients

Macrophage colony stimulating factor (GM-CSF)—used to stimulate bone marrow activity after bone marrow grafts

Tumor necrosis factor (TNF)—used to treat cancer

Granulocyte-colony-stimulating factor (Neupogen)—developed for treating cancer patients suffering from low neutrophil counts

Hormones

Erythropoietin (EPO)—a peptide that stimulates bone marrow used to treat some forms of anemia

Tissue plasminogen activating factor (tPA)—can dissolve potentially dangerous blood clots

Hemoglobin A—form of artificial blood to be used in place of real blood for transfusions

Factor VIII—needed as replacement blood-clotting factor in type A hemophilia

Relaxin—an aid to childbirth

Ovidrel and Gonal—female hormones that stimulate egg maturation

Enzymes

rH DNase (pulmozyme)—a treatment that can break down the thick lung secretions of cystic fibrosis

Antitrypsin—replacement therapy to benefit emphysema patients

PEG-SOD—a form of superoxide dismutase that minimizes damage to brain tissue after severe trauma

Vaccines

Vaccines for hepatitis B and *Haemophilus influenzae* Type b meningitis

Experimental malaria and AIDS vaccines based on recombinant surface antigens

Miscellaneous

Bovine growth hormone or bovine somatotropin (BST)—given to cows to increase milk production

Apolipoprotein—to deter the development of fatty deposits in the arteries and to prevent strokes and heart attacks

Spider silk—a light, tough fabric for parachutes and bulletproof vests

IV. Genetically Modified Organisms

The process of artificially introducing foreign genes into organisms is termed *transfection,* and the recombinant organisms produced in this way are called *transgenic* or genetically modified organisms (GMOs). Foreign genes have been inserted into a variety of microbes, plants, and animals through recombinant DNA techniques developed especially for them. Transgenic "designer" organisms are available for a variety of biotechnological applications. Because they are unique life forms that would never have otherwise occurred, they can be patented.

RECOMBINANT MICROBES: MODIFIED BACTERIA AND VIRUSES

One of the first practical applications of recombinant DNA in agriculture was to create a genetically altered strain of the bacterium *Pseudomonas syringae.* The wild strain ordinarily contains an ice nucleation gene that promotes ice or frost formation on moist plant surfaces. Genetic alteration of the frost gene using recombinant plasmids created a different strain that could prevent ice crystals from forming. A commercial product called Frostban has been successfully applied to stop frost damage in strawberry and potato crops. A strain of *Pseudomonas fluorescens* has been engineered with the gene from a bacterium *(Bacillus thuringiensis)* that codes for an insecticide. These recombinant bacteria are released to colonize plant roots and help destroy invading insects. All releases of recombinant microbes must be approved by the Environmental Protection Agency (EPA) and are closely monitored. So far, extensive studies have shown that such microbes do not proliferate in the environment and probably pose little harm to humans.

Although a number of recombinant proteins are produced by transformed bacterial hosts, many of the enzymes, hormones, and antibodies being used in drug therapy are currently being manufactured using mammalian cell cultures as the cloning and expression hosts. One of the primary advantages to this alternate procedure is that these cell cultures can modify the proteins (adding carbohydrates, for example) so that they are biologically more active. Some forms of reproductive hormones, human growth hormone, and interferon are products of cell culture.

Viruses can be genetically engineered for transfection purposes. Several viral vectors have been developed for gene therapy and for an experimental AIDS vaccine (see figure 25.20). In Australia, a mousepox virus was developed to control populations of mice that had become a serious pest. Unfortunately, the virus was so lethal that its use was halted for fear it may jump hosts.

Another very significant bioengineering interest has been to create microbes to bioremediate disturbed environments. Biotechnologists have already developed and tested several types of bacteria that clean up oil spills and degrade pesticides and toxic substances (see chapter 26). The growing power of biotechnology has caused some to wonder about possible malevolent uses of this new ability (see **Spotlight on Microbiology 10.2** and **Microbits 10.3**).

TRANSGENIC PLANTS: IMPROVING CROPS AND FOODS

Two unusual species of bacteria in the genus *Agrobacterium* are the original genetic engineers of the plant world. These pathogens live in soil and can invade injured plant tissues. Inside the wounded tissue, the bacteria transfer a discrete DNA fragment called T-DNA into the host cell. The DNA is integrated into the plant cell's chromosome, where it directs the synthesis of nutrients to feed the bacteria and hormones to stimulate plant growth. In the case of *Agrobacterium tumefaciens,* increased growth causes development of a tumor called *crown gall disease,* a mass of undifferentiated tissue on the stem. The other species, *A. rhizogenes,* attacks the roots and transforms them into abnormally overgrown "hairy roots."

The capacity of these bacteria, especially *A. tumefaciens,* to transfect host cells can be attributed to a large plasmid termed **Ti** (tumor-inducing). This plasmid inserts into the genomes of the infected plant cells and transforms them **(figure 10.11).** Even after the bacteria in a tumor are dead, the plasmid genes remain in the cell nucleus and the tumor continues to grow. This plasmid is a perfect vector for inserting foreign genes into plant genomes. The procedure involves removing the Ti plasmid, inserting a previously isolated gene into it, and returning it to *Agrobacterium.* Infection of the plant by the recombinant bacteria automatically transfers the plasmid with the foreign gene into the plant cells. The insertion of genes with a Ti plasmid works primarily to engineer important dicot plants such as potatoes, tomatoes, cotton, and grapes.

For plants that cannot be transformed with *Agrobacterium,* seed embryos have been successfully implanted with additional genes using gene guns—small "shotguns" that shoot the genes into plant embryos with tiny gold bullets. Agricultural officials had approved more than 130 plants for field testing through the year 2002. This movement toward release of engineered plants into the environment has led to some controversy. Many plant geneticists and ecologists are seriously concerned that transgenic plants will share their genes for herbicide, pesticide, and virus resistance with natural plants, leading to "superweeds" that could flourish and become indestructible. The Department of Agriculture is carefully regulating all releases of transgenic plants. See **table 10.3** for examples of plants that have been engineered.

TRANSGENIC ANIMALS: ENGINEERING EMBRYOS

Animals too can be genetically engineered. In fact, animals are so amenable to transfection that several hundred strains of transgenic animals have been introduced by research and industry. One reason for this movement toward animals is that, unlike bacteria and yeasts, they can express human genes in organs and organ systems that are very similar to those of humans. This advantage has led to the design of animal models to study human genetic diseases and then to use these natural systems to test new genetic therapies before they are used in humans. Animals such as sheep or goats can also be engineered to become "factories" capable of manufacturing proteins useful to humans and excreting them in their milk or semen, a process often referred to (when done to produce medically useful proteins) as "pharming."

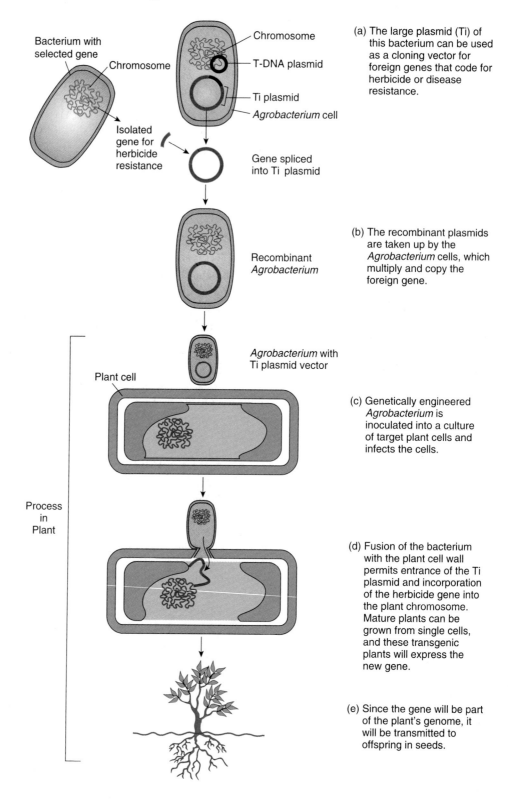

Bacterium with selected gene

Chromosome

Isolated gene for herbicide resistance

Chromosome

T-DNA plasmid

Ti plasmid

Agrobacterium cell

(a) The large plasmid (Ti) of this bacterium can be used as a cloning vector for foreign genes that code for herbicide or disease resistance.

Gene spliced into Ti plasmid

Recombinant *Agrobacterium*

(b) The recombinant plasmids are taken up by the *Agrobacterium* cells, which multiply and copy the foreign gene.

Agrobacterium with Ti plasmid vector

Plant cell

Process in Plant

(c) Genetically engineered *Agrobacterium* is inoculated into a culture of target plant cells and infects the cells.

(d) Fusion of the bacterium with the plant cell wall permits entrance of the Ti plasmid and incorporation of the herbicide gene into the plant chromosome. Mature plants can be grown from single cells, and these transgenic plants will express the new gene.

(e) Since the gene will be part of the plant's genome, it will be transmitted to offspring in seeds.

FIGURE 10.11

Bioengineering of plants. Most techniques employ a natural tumor-producing bacterium called *Agrobacterium tumefaciens*.

The most effective way to insert genes into animals is by using a virus to transfect the fertilized egg or early embryo (germline engineering) **(figure 10.12)**. Foreign genes can be delivered into an egg by pulsing the egg with high voltage or by injection. The success rate does not have to be high, because a transgenic animal will usually pass the genes on to its offspring. The dominant animals in the genetic engineer's laboratory are transgenic forms of mice. Mice are so easy to engineer that in some ways they have become as prominent in genetic engineering as *E. coli.* In some early experiments, fertil-

ized mouse embryos were injected with human genes for growth hormone, producing supermice twice the size of normal mice. Laboratories have developed thousands of genetically modified mice.

The so-called knockout mouse has become a standard way of producing animals with tailor-made genetic defects. The technique involves transfecting a mouse embryo with a defective gene and cross-breeding the progeny through several generations. Eventually, mice will be born with two defective genes and will express the disease. Mouse models exist for cystic fibrosis, hardening of the

TABLE 10.3

Examples of Engineered (Transgenic) Plants

Plant	Trait	Results
Nicotiana tabacum (tobacco)	Herbicide resistance	 Tobacco plants in the upper row have been transformed with a gene that provides protection against Buctril, a systemic herbicide. Plants in the lower row are normal and not transformed. Both groups were sprayed with Buctril and allowed to sit for 6 days. (The control plants at the beginning of each row were sprayed with a blank mixture lacking Buctril.)
Pisum sativum (garden pea)	Pest protection	Pea plants were engineered with a gene that prevents digestion of the seed starch (see seeds on the left). This gene keeps tiny insects called weevils from feeding on the seeds. Seeds on the right are from plants that were not engineered and are suffering from weevil damage (note holes).
Oryza sativa (rice)	Added nutritional value	Golden rice has been engineered to provide beta-carotene. Lack of beta-carotene results in over a million deaths a year (see **chapter opening photo**).

Several embryos recovered from sacrificed female

Embryos transferred to a depression slide containing culture medium

Culture medium
Oil

As embryo is held in place, DNA is injected into pronucleus.

Holding pipette

Pronucleus
DNA to be injected
Injection pipette

Several injected embryos are placed into oviduct of receptive female.

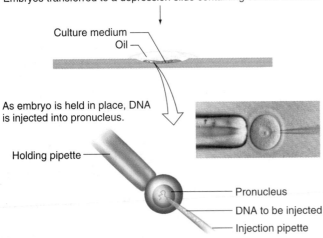

FIGURE 10.12
How transgenic mice are created.

arteries, Gaucher's disease (a lysosomal storage disease), Alzheimer disease, and sickle-cell anemia. See **table 10.4** for a survey of applications of transgenic animals used in animal husbandry and the drug industry.

CHAPTER CHECKPOINTS

Bioengineered hormones, enzymes, and vaccines are safer and more effective than similar substances derived from animals. Recombinant DNA drugs and vaccines are useful alternatives to traditional treatments for disease.

Transfection is the process by which foreign genes are introduced into organisms. The transfected organism is termed *transgenic*. Transgenic microorganisms are genetically designed for medical diagnosis, crop improvement, pest reduction, and bioremediation. Transgenic animals are genetically designed to model genetic therapies, improve meat yield, or synthesize specific biological products.

V. Genetic Treatments: Introducing DNA into the Body

GENE THERAPY

We have known for decades that for certain diseases, the disease phenotype is due to the lack of a single specific protein. For example, type I diabetes is caused by a lack of insulin, leaving those with the disease unable to properly regulate their blood sugar. The initial treatment for this disease was simple; provide diabetics with insulin isolated from a different source, in most cases pigs or cows. While this treatment was adequate for most diabetics (especially in the short term), our increasingly sophisticated genetic engineering abilities

TABLE 10.4

Pharmaceutical Production by Some Transgenic Animals

Treatment For:	Selected Animal	Protein Expressed in Milk	Application
Hemophilia	Pig	Factor VIII and IX	Blood-clotting factor
Emphysema	Sheep	Alpha antitrypsin	Supplemental protein—lack of this protein leads to emphysema
Cancer	Rabbit	Human interleukin-2	Stimulates the production of T lymphocytes to fight selected cancers
Septicemia	Cow	Lactoferrin	Iron-binding protein inhibits the growth of bacteria and viral infection
Surgery, trauma, burns	Cow	Human albumin	Return blood volume to its normal level
GHD (growth hormone deficiency)	Goat	HGH (human growth hormone)	Supplemental protein improves bone metabolism, density, and strength
Heart attack, stroke	Goat	Tissue plasmagen activator (tPA)	Dissolves blood clots to reduce heart damage (if administered within 3 hours)

These two mice are genetically the same strain, except that the one on the right has been transfected with the human growth hormone gene (shown in its circular plasma vector).

Little piglets have been genetically engineered to synthesize human Factor VIII, a clotting agent used by type A hemophiliacs. The females will produce it in their milk when mature. It is estimated that only a few hundred pigs would be required to meet the world's demand for this drug.

made loitering around the slaughterhouse seem a decidedly low-tech solution to a high-tech problem. We've already discussed the first way in which genetic engineering has been used in the treatment of disease, namely producing recombinant proteins in bacteria or yeast rather than isolating the protein from animals or humans. In fact recombinant human insulin was the first genetically engineered drug to be approved for use in humans. The next logical step is to see if we can correct or repair a faulty gene in humans suffering from a fatal or debilitating disease, a process known as **gene therapy.**

The inherent benefit of this therapy is to permanently cure the physiological dysfunction by repairing the genetic defect. There are two strategies for this therapy. In *ex vivo* therapy, the normal gene is cloned in vectors such as retroviruses (mouse leukemia virus) or adenoviruses that are infectious but relatively harmless. Tissues removed from the patient are incubated with these genetically modified viruses to transfect them with the normal gene. The transfected cells are then reintroduced into the patient's body by transfusion (**figure 10.13**). In contrast, the *in vivo* type of therapy skips the intermediate step of incubating excised patient tissue. Instead, the naked DNA or a virus vector is directly introduced into the patient's tissues.

Experimentation with various types of gene therapy, or clinical testing, is performed on human volunteers who manifest the particular genetic condition. Over 600 of these trials have been or are being carried out in the United States and other countries. Most trials target cancer, single gene defects, and infections, and most gene deliveries are carried out by virus vectors. So far the therapeutic trials have been hampered by several difficulties relating to effectiveness and safety.

Many of the systems have a low rate of productive insertion of the gene into the patient cells. In other cases, the genes are not properly translated or are not retained by the cells. Serious concerns have arisen from the potential for virus vectors such as the mouse leukemia virus (a retrovirus) to insert into the DNA of the patient and cause other diseases such as cancer. Late in 2002, a number of therapy trials were suspended when two young boys acquired a type of leukemia following successful therapy for an immunodeficiency disease. Alternate methods of gene delivery are being developed with gene guns similar to those used on plants, and liposomes, tiny lipid spheres that can readily carry molecules across cell membranes.

The first gene therapy experiment in humans was initiated in 1990 by researchers at the National Institutes of Health. The subject was a 4-year-old girl suffering from a severe immunodeficiency disease caused by the lack of the enzyme adenosine deaminase (ADA). She was transfused with her own blood cells that had been engineered to contain a functional ADA gene. Later, other children were given the same type of therapy. So far, the children have shown remarkable improvement and continue to be healthy, but the treatment is not permanent and must be repeated.

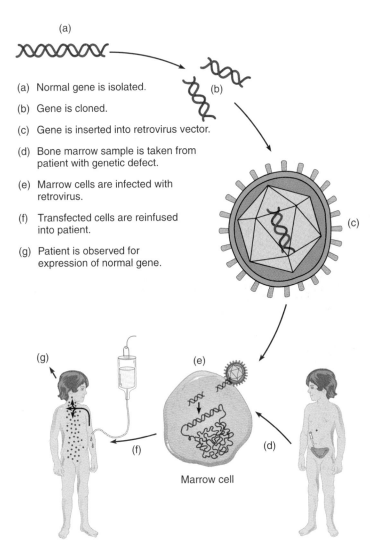

(a) Normal gene is isolated.

(b) Gene is cloned.

(c) Gene is inserted into retrovirus vector.

(d) Bone marrow sample is taken from patient with genetic defect.

(e) Marrow cells are infected with retrovirus.

(f) Transfected cells are reinfused into patient.

(g) Patient is observed for expression of normal gene.

Marrow cell

FIGURE 10.13
Protocol for the *ex vivo* type of gene therapy in humans.

Several patients have been treated successfully for another type of severe combined immunodeficiency syndrome called X-1–linked SCID that is due to a missing enzyme required for mature immune cells. This is the first instance of an apparent cure. Full function has been restored to several hemophilic children, using a similar technique. Medical researchers have been able to correct the damage in certain vascular diseases by infusing growth factor genes into their circulatory systems. In trials for cystic fibrosis (CF) therapy, adenoviruses carrying a normal CF gene are delivered directly into the nose and lungs. So far, the results of this therapy have been less successful, probably because of a low acceptance rate for take-up of the gene by respiratory cells.

Other clinical trials are being carried out for sickle cell anemia, AIDS, and ovarian and melanoma cancers **(table 10.5).**

The ultimate sort of gene therapy is germline therapy, in which genes are inserted into an egg, sperm, or early embryo. In this type of therapy the new gene will be present in all cells of the individual. The therapeutic gene is also heritable (i.e., can be passed to subsequent generations). Because of this last fact, germline gene therapy is not yet being pursued in humans.

ANTISENSE AND TRIPLEX DNA TECHNOLOGY: GENETIC MEDICINES

Up to now we have considered the use of genetic technology to replace a missing or faulty protein that is needed for normal cell function. A different problem arises when a disease results from the inappropriate expression of a protein. For example Alzheimer disease, most viral diseases, and many cancers occur when an unwanted gene is expressed, rather than when a desirable gene is missing. The solution in cases such as this is to prevent transcription or translation of a gene, and scientists, as they often do, look for clues as to how bacteria accomplish this feat. Some genes in bacteria are repressed when an antisense RNA binds to them, preventing translation at the ribosome. This antisense RNA has bases that are complementary to the sense strand of mRNA in the area surrounding the initiation site. When the antisense RNA binds to its particular mRNA, the now double-stranded RNA is inaccessible to the ribosome, resulting in a loss of translation of that mRNA.

As this process was tested in the laboratory, two significant differences between procaryotic and eucaryotic systems emerged. The first was that single-stranded DNA was usually used as the antisense agent, as it was easier to manufacture than RNA. The second was that, for some genes, once the antisense strand bound to the mRNA, not only was the hybrid RNA not translated, it was unable even to leave the nucleus **(figure 10.14*a*).**

Early clinical trials of antisense therapeutics have been promising, with most antisense therapies dramatically lowering the rate of synthesis of the targeted protein. Because the antisense agent can be targeted to a specific sequence of DNA, side effects are rare. Most concern has been centered on the difficulty of getting high enough levels of the drug into the nucleus of cells and the fact that these drugs must be taken for life. In 1998 Vitravene, an antisense DNA used to treat cytomegalovirus retinitis (blindness brought on by infection with cytomegalovirus in immunocompromised persons), became the first antisense therapy approved for use in humans.

Triplex DNA: Adding a Third Strand

Although the usual structure of DNA is a helix composed of two strands, researchers have known for years that DNA can exist in other naturally and artificially induced formats. One of these unusual variations, *triplex DNA,* is a triple helix formed when a third strand of DNA inserts into the major groove of the molecule. The inserted strand forms hydrogen bonds with the purine bases on one of the adjacent strands (see figure 10.14*b*). It seems logical that the template of DNA that has an extra strand wedged into its structure can be relatively inaccessible to normal transcription. Indeed, it is theorized that this may be another natural means for cells to control their genetic expression.

The potential therapeutic applications of triplex DNA are still in their infancy. So far, oligonucleotides have been synthesized to form triplex DNA that interacts with regulatory sequences in genes.

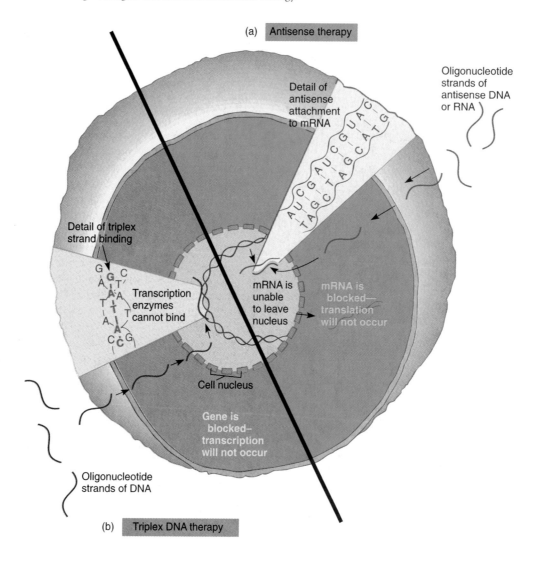

FIGURE 10.14

Mechanisms of antisense DNA and triplex DNA.   These genetic medicines are designed to prevent the expression of an undesirable gene or virus. Both use oligonucleotides of known sequence that can bind with genetic targets that are the cause of disease. The site of action is in the nucleus. **(a)** When antisense strands enter the nucleus, they bind to complementary areas on an existing mRNA molecule. Such bound RNA will be unavailable for translation on the cell's ribosomes, and no protein will be made. **(b)** Triplex DNA strands attach by base-pairing to regulatory sites and other sites on the chromosome itself. This action masks the points of attachment for enzymes of transcription, and mRNA cannot be made. Without mRNA, there will be no translation or protein formed.

It can interrupt the action of transcription factors and prevent RNA polymerase from forming mRNA transcripts. Studies performed on cell cultures revealed that the genes coding for oncogenes, viruses, and the receptor for interleukin-2 (an immune stimulant) could be blocked with triplex DNA. In the clinical trials to come, this therapy will no doubt be plagued with some of the same problems mentioned for the antisense approach.

VI. Genome Analysis: Maps, Fingerprints, and Family Trees

An old geneticist's joke begins: "How do you describe a cat?" The answer "GATCCT. . . " is becoming more and more a reality. As was mentioned in the beginning of the chapter, DNA technology

has allowed us to accomplish many age-old goals by new and improved means. By remembering that phenotypes, whether human (hair color, eye color, blood type), bacterial (spore formation, shape, production of catalase), or even viral (ability to attach to the membrane of human cells), are the result of specific sequences of DNA, we can easily see how DNA can be used to differentiate among organisms in the same way that observing any of these phenotypes can. Additionally, DNA can be used to "see" the phenotype of an organism that is no longer present, as when a rapist is identified by DNA extracted from semen taken from the rape victim. Finally, possession of a particular sequence of DNA may indicate an increased risk of a genetic disease. Detection of this piece of DNA (known as a marker) can identify a person as being at increased risk for cancer or Alzheimer disease long before symptoms arise. The ability to detect diseases before symptoms arise is especially im-

TABLE 10.5

Sample of Human Genetic Defects Targeted for Gene Therapy

Disease	Nature of Defect	Gene Replacement Strategy
Adenosine deaminase (ADA) deficiency	Severe immunodeficiency due to destruction of lymphocytes by toxic metabolic product.	ADA gene inserted into bone marrow white blood cells using virus vector or liposomes.
X-1–linked SCID	Enzyme needed for maturing B and T cells is missing.	Bone marrow is modified by means of engineered virus.
Cystic fibrosis	Lack of necessary protein for conducting chloride ions across membranes causes buildup of mucus and secretions in lungs, pancreas, other organs.	CFTR gene placed in adenovirus or liposomes and inoculated into nasal cavity or lungs.
Sickle-cell anemia, beta-thalassemia	Lack of correct gene to make the normal B globin protein of hemoglobin.	Hemoglobin A (HbA) gene placed in bone marrow cells by vector.
Hemophilia B	Lack of gene to produce factor VIII, needed for blood clotting.	Normal gene for factor VIII incorporated into marrow cells.
Gaucher's disease	Deficiency in an enzyme that is needed to metabolize glucocerebroside leads to its buildup in lysosomes.	Insertion of normal glucocerebrosidase gene into marrow cells.
Muscular dystrophy	Lack of gene for producing dystrophin necessary for normal muscle development.	Dystrophin gene delivered into muscle tissue by one of several methods.
AIDS	Infection by HIV destroys important immune cells (T cells) and causes severe loss of immune function.	Fibroblasts are infected with gene that stimulates an immune response against HIV; other test uses ribozyme enzyme to block virus activity.
Malignant melanoma	Severest form of skin cancer that does not respond well to traditional therapy.	Insertion of B7 gene into tumor cells to boost the natural white blood cell response against the cancer.
Acute myelogenous leukemia	Blood disease caused by overexpression of a gene.	Genetic modification of cells by replacement of defective gene; also addition of antisense DNA to block the actions of cancer genes.

portant for diseases such as cancer, where early treatment is sometimes the very difference between life and death. With examples like this in mind, let's look at several ways in which new DNA technology is allowing us to accomplish goals in ways that were only dreamed of a few years ago.

Cartoon by John Chase. Reprinted by permission.

"We finished the genome map, now we can't figure out how to fold it!"

GENOME MAPPING AND SCREENING: AN ATLAS OF THE GENOME

We have seen a variety of methods for accessing the genomes of organisms, but the most useful information comes from knowing the sequential makeup of the genetic material. Genetic engineers find it very informative to know the **locus,** or exact position of a particular gene on a chromosome, and they also seek information on the types and numbers of **alleles,** which are sites that vary from one individual to another. The process of determining location of loci and other qualities of genomic DNA is called **mapping.** Maps vary in resolution and applications. *Linkage maps* show the relative proximity and order of genes on a chromosome and are relatively low resolution because only a few exact locations are mapped. *Physical maps* are more detailed arrays that depict not only the relative positions of distinct sections of DNA, but give the numerical size in base pairs **(figure 10.15).** This technology uses restriction fragments and fluorescent hybridization probes to visualize the position of a particular selected site along a segment of DNA.

By far the most detailed maps are *sequence maps,* which are produced by the sequencers we discussed earlier. They give an exact order of bases in a plasmid, chromosome, or entire genome. Because this level of resolution is most promising for understanding the nature of the genes, what they code for, and their functions,

SPOTLIGHT ON MICROBIOLOGY 10.2
A Moment to Think

We are embarking on a very potentially troublesome journey, where we begin to reduce all other animals on this planet to genetically engineered products. . . . We will increasingly think of ourselves as just gene codes and blueprints and programs that can be tinkered with.

> Jeremy Rifkin, Foundation on Economic Trends

Never postpone experiments that have clearly defined future benefits for fear of dangers that can't be quantified [because] we can react rationally only to real (as opposed to hypothetical) risks,

> James Watson, Nobel Laureate and first head of the Human Genome Project

I've never been less well equipped intellectually to vote on an issue than I am on this.

> Unnamed U.S. Senator, prior to a vote on stem cell research and human cloning

Of these three statements, the third is without a doubt the most frightening. There are always several sides to every issue and the best we can hope for is that the people regulating genetic technology are as well informed as possible. Those who will determine the limits of genetic technology include not only scientists and politicians but voters and consumers as well. History is rife with both knee-jerk rejections of new technologies that have later proven to be invaluable (vaccinations) as well as complacency as dangerous products were foisted onto an ignorant and unsuspecting public (Fen-Phen, thalidomide). Ethical choices can only be properly made from a standpoint of intellectual awareness, and in this era of advertising, polling and focus groups, people with a stake in genetic technology know that the most effective way to drum up support (both for or against) is not by carefully educating the public as to the uses and limits of our newfound powers but rather by publicizing exaggerated claims of frightening scenarios. Enlightenment is often a casualty of these advertising campaigns, and you as a student, voter, and potential consumer of bioengineered products need to realize that the truth lies somewhere between the photograph seen here and the blissful utopia often portrayed by some proponents of biotechnology.

Throughout this chapter, a common theme has been that biotechnology, genetically modified organisms, and gene-based therapies are

A demonstrator at a biotechnology conference protests the development and sale of genetically modified foods.

simply new ways of achieving long-held goals. This is of course a vast understatement. Society has never before had the ability to do even a fraction of the things we can do today, and at some point we must decide where we should stop, or if we need to stop at all. Where do we as individuals and as a society draw the line? The answer will differ for each person, based on his or her unique experiences and personal philosophies, but should certainly include a realistic knowledge of the probable benefits and possible risks. We live in serious times and these are serious choices requiring serious thought. Nothing worthwhile is ever easy.

Human Chromosome 16

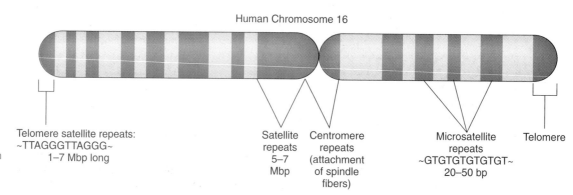

FIGURE 10.15

A physical map of chromosome 16. This map shows areas on the chromosome that are devoted to support DNA, most of which consist of highly repetitive nucleotides.

Telomere satellite repeats: ~TTAGGGTTAGGG~ 1–7 Mbp long

Satellite repeats 5–7 Mbp

Centromere repeats (attachment of spindle fibers)

Microsatellite repeats ~GTGTGTGTGTGT~ 20–50 bp

Telomere

this form of map dominates the research programs. When the Human Genome Project was launched in 1986 to map the human genome, it was expected to require 15 to 20 years and cost 3 billion dollars. Due to advances in technology, it was completed early in 2001 (Microbits 10.1).

Other genome projects have also been highly successful, so much so that the genomes of over 1,000 organisms, viruses, and organelles have been fully sequenced. As of 2001, this includes around 600 viruses, 125 procaryotes (bacteria and archaea), and over 300 eucaryotes (including human, mouse, *C. elegans* (a nematode worm), yeast, fruitfly, *Arabidopsis* (a small flowering plant), and rice. One of the remarkable discoveries in this huge enterprise has been how similar the genomes of relatively unrelated organisms are. Humans share around 80% of their DNA codes with mice, about 60% with rice, and even 30% with the worm *C. elegans.*

Another interesting surprise has been that a very large amount of the human genome contains DNA sequences that do not code for cell protein. Ninety-seven percent of it is made up of support DNA that functions in chromosome stabilizing and division, gene regulation, and ribosome assembly.

Although sequencing provides the ultimate genetic map, it does not automatically identify the exact genes and alleles. Analyzing and storing this massive amount of new data requires specialized computers. Because it is information of both a biological and mathematical content, two whole new disciplines have grown up around managing these data: *genomics* and *bioinformatics.* The job of genomics and bioinformatics will be to analyze and classify genes, determine protein sequences, and ultimately determine the function of the genes. This fact is likely to bring another golden era in biology and microbiology, that of unraveling the complexity of relationships of genes and gene regulation. In time, it will provide a complete understanding of such phenomena as normal cell function, disease, development, aging, and many other issues. In addition, it will allow us to characterize the exact genetic mechanisms behind pathogens, and allow new treatments to be developed against them.

DNA FINGERPRINTING: A UNIQUE PICTURE OF A GENOME

Although DNA is based on a structure of nucleotides, the exact way these nucleotides are combined is unique for each organism. It is now possible to apply DNA technology in a manner that emphasizes these differences and arrays the entire genome in a pattern for comparison. *DNA fingerprinting* (also called DNA typing or profiling) is best known as a tool of forensic science first devised in the mid-1980s by Alex Jeffreys of Great Britain (**figure 10.16**).

Several of the methods discussed previously in this chapter are involved in the creation of a DNA fingerprint. Techniques such as the use of restriction endonucleases for cutting DNA precisely, PCR amplification for increasing the number of copies of a certain genome, electrophoresis to separate fragments, hybridization probes to locate specific loci and alleles, and Southern blotting for producing a visible record are all employed by biotechnologists. Several different methods of DNA fingerprinting are available, but all depend on distinguishing one sequence of DNA from another by comparing the sequence of the strands at specific loci. One type of analysis depends on the ability of a restriction enzyme to cut DNA at a specific

recognition site. If a given strand of DNA possesses the recognition site for a particular restriction enzyme, the DNA strand will be cut, resulting in two smaller pieces of DNA. If the same strand from a different person does not contain the recognition site (perhaps due to a mutation many generations ago), it will not be cut by the restriction enzyme and will remain as a single large piece of DNA. When each of these DNA samples are digested with restriction enzymes and separated on an electrophoretic gel, the first will display two small bands while the second will display one larger band. This is an example of a restriction fragment length polymorphism, which was discussed earlier. All methods of DNA fingerprinting depend on some variation of this strategy to ferret out differences in DNA sequence at the same location in the genome. The type of polymorphism seen previously is also sometimes referred to as an *SNP* or *single nucleotide polymorphism,* because only a single nucleotide is altered. Tens of thousands of these differences at a single locus are known to exist throughout the genome. Other genetic **markers,** or observable variations in DNA structure, depend on small repeated sequences of nucleotides which vary in the number of repeated segments from person to person (and are hence polymorphic). Two of the most common markers of this type are *VNTRs,* or variable number of tandem repeats (which tend to be hundreds of nucleotides in length) and *microsatellite polymorphisms* (which tend toward dozens of nucleotides in length). In either type of marker, the number of nucleotides, rather than the identity of a single nucleotide, varies. This variation can be easily determined by cutting DNA with a restriction enzyme and then looking for changes in the size of the fragments produced on a gel.

The most powerful uses of DNA fingerprinting are in forensics, detecting genetic diseases, determining parentage, analyzing the family trees of humans and animals, identifying microorganisms, and tracing the lineage of ancient organisms.

One of the first uses of the technique of DNA fingerprinting was in forensic medicine. Besides real fingerprints, criminals often leave other evidence at the site of a crime—a hair, a piece of skin or fingernail, semen, blood, or saliva. Because it can be combined with amplifying techniques such as PCR, DNA fingerprinting provides a way to test even specimens that have only minute amounts of DNA, or even old specimens. Regardless of the method used, the primary function of the fingerprint is to provide a snapshot of the genome, displaying it as a unique picture of an individual organism. Markers such as those just described are often used in combination to increase the odds of a correct identification. For example, imagine a sample of blood found at a crime scene is thought to have come from the criminal and has tested positive for markers A, B, C, and D. If marker A is present in 1 person out of 20, marker B in 1 person out of 10, marker C in 1 person out of 25, and marker D in 1 person out of 50, the probability that a single person would have all four markers is $(1/20 \times 1/10 \times 1/25 \times 1/50)$ or 1 out of 250,000. When combined with traditional evidence such as being seen leaving the crime scene, DNA fingerprinting is convincing indeed!

DNA fingerprinting is so widely accepted that it is now routine to use it as trial evidence. It has been used to convict and exonerate thousands of accused defendants in criminal cases. In fact, the technology is so reliable that many older cases for which samples are available are being reopened and solved. The demand has been so high that many forensic units are overwhelmed by requests for testing. The FBI is currently setting up a national database of DNA

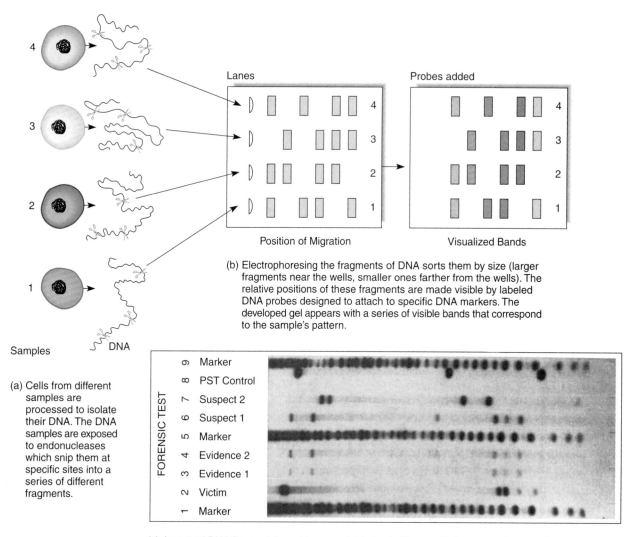

(a) Cells from different samples are processed to isolate their DNA. The DNA samples are exposed to endonucleases which snip them at specific sites into a series of different fragments.

(b) Electrophoresing the fragments of DNA sorts them by size (larger fragments near the wells, smaller ones farther from the wells). The relative positions of these fragments are made visible by labeled DNA probes designed to attach to specific DNA markers. The developed gel appears with a series of visible bands that correspond to the sample's pattern.

(c) An actual DNA fingerprint used in a rape trial. Control lanes with known markers are in lanes 1, 5, 8, and 9. The second lane contains a sample of DNA from the victim's blood. Evidence samples 1 and 2 (lanes 3 and 4) contain semen samples taken from the victim. Suspects 1 and 2 (lanes 6 and 7) were tested. Can you tell by comparing evidence and suspect lanes which individual committed the rape?

FIGURE 10.16

DNA fingerprints: the bar codes of life.

fingerprints, and several states allow DNA fingerprinting of felons in prison (so-called DNA dogtags). In a most amazing application of this knowledge, forensic scientists were able to positively identify the Unabomber, Ted Kaczynski, by DNA recovered from the back of postage stamps he had licked!

Fingerprinting can also be used to detect genes that increase the risk for hereditary diseases and the patterns of inheritance in conditions such as Huntington disease, cystic fibrosis, and Alzheimer disease **(figure 10.17).** Test DNA from various family members is allowed to react with gene-specific probes that will show exactly which restriction sites (genetic markers) the person possesses. The genetic markers are not the actual gene that causes the disease, but they are located nearby and are inherited in exactly the same pattern. The genetic family tree for many diseases can be worked out by

comparing fingerprint patterns in a family that is known to carry and express the gene. Futurists predict that a DNA fingerprint will be made for every child at birth and kept on file, much as footprints and regular fingerprints have been used until now.

This same method of analysis can be used for clarifying human paternity and maternity, the pedigree (genetic ancestry) of domestic animals, and the genetic diversity of animals bred in zoos.

Some insight into the power of DNA typing comes from its use to identify victims of the World Trade Center attack in 2001. This same technology has also been effective in identifying remains of unknown soldiers during the Gulf War and the shuttle *Columbia* astronauts.

Infectious disease laboratories are developing numerous test procedures to profile bacteria and viruses. Profiling is used to

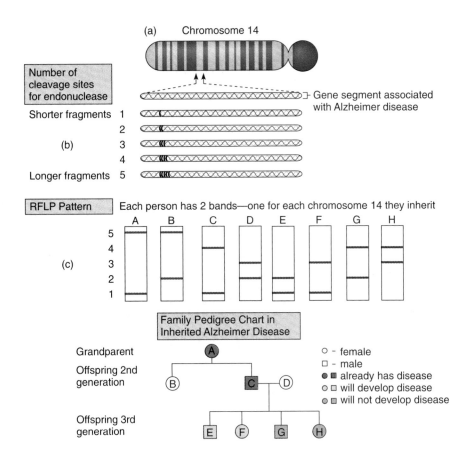

FIGURE 10.17

Pedigree analysis based on genetic screening for familial type Alzheimer disease. **(a)** The site for this disease is on chromosome 14. **(b)** Gene mapping has determined that this chromosome segment has five variations, depending on the number and size of cleavage sites. **(c)** A Southern blot using the restriction fragments (RFLPs) is performed on a family with three generations. Each person will have two bands that correspond with each of the chromosomes they have inherited. The grandmother (A) with pattern 1,5 and her son (C) with pattern 1,4 both already have the disease. Patterns for his sister (B), his wife (D), and their four children (E, F, G, and H) are also shown. Comparing the band patterns of A and B with those of other family members makes it possible to predict which children will inherit the gene for Alzheimer disease. Because the band shared by A and C is 1, we know that this is the variant associated with Alzheimer disease. Children E (1,2) and F (1,3) will develop the disease, and children G (2,4) and H (3,4) will not. It is also evident that this form of Alzheimer disease is inherited as a dominant trait.

identify the gonococcus, *Chlamydia, syphilis spirochete,* and *M. tuberculosis* (see figure F in Introduction to Identification Techniques in Medical Microbiology, page 543). It is also an essential tool for determining genetic relationships between microbes.

Because even fossilized DNA can remain partially intact, ancient DNA is being studied for comparative and evolutionary purposes. Anthropologists have traced the migrations of ancient people by analyzing the mitochondrial DNA found in bone fragments. Mitochondrial DNA (mtDNA) is less subject to degradation than is chromosomal DNA and can be used as an evolutionary time clock. A group of evolutionary biologists have been able to recover Neanderthal mtDNA and use it to show that this group of ancient primates could not have been an ancestor of humans. Biologists studying canine origins have also been able to show that the wolf is the ancestral form for modern canines.

Knowing the complete nucleotide sequence of the human genome is really only half the battle. With very few exceptions all cells in an organism contain the same DNA, so knowing the sequence of that DNA, while certainly helpful, is of very little use when comparing two cells or tissues from the same organism. Recall that genes are expressed in response to both internal needs and external stimuli, and while the DNA content of a cell is static, the mRNA (and hence protein) content at any given time provides scientists with a profile of genes currently being expressed in the cell. What truly distinguishes a liver cell from a kidney cell or a healthy cell from a diseased cell are the genes expressed in each.

Twin advances in biology and electronics have allowed biologists to view the expression of genes in any given cell using a technique called DNA microarray analysis. Prior to the advent of this technology, scientists were able to track the expression of, at most, a few genes at a time. Given the complex interrelationship that exists between genes in a typical cell, tracking two or three genes was of very little practical use. Microarrays are able to track the expression of thousands of genes at once and are able to do so in a single efficient experiment. Microarrays consist of a glass slide (occasionally a silicon chip or nylon membrane is used) onto which have been bound sequences from tens of thousands of different genes. To the chip a solution containing fluorescently labeled cDNA, representing all of the mRNA molecules in a cell at a given time, is added and the labeled cDNA is allowed to bind to any complementary DNA bound to the chip. Bound cDNA is then detected by exciting the fluorescent tag on the cDNA with a laser and recording the fluorescence with a detector linked to a computer. The computer can then interpret this data to determine what mRNAs are present in the cell under a variety of conditions **(figure 10.18).**

Possible uses of microarrays include developing extraordinarily sensitive diagnostic tests that search for a specific pattern of gene expression. As an example, being able to identify a patient's cancer as one of many subtypes (rather than just, for instance, breast cancer) will allow pharmacologists and doctors to treat each cancer with the drug that will be most effective. Again we see genetic technology as a more effective way to reach long-held goals.

Every technological innovation in human history has been used for both benevolence and malevolence. Particle physics brought us both the computer and atomic weapons while airplanes have been used to deliver organs for transplants as well as to drop bombs. Our increasing use of genetic technology to manipulate organisms, combined with an extensive knowledge of the genomes of many pathogens brings to mind a question. Could terrorists use the techniques and information we've developed to create more effective weapons of bioterrorism? The answer is troubling but is fortunately tempered by some good news as well.

Startlingly effective pathogens have been created in the laboratory inadvertently. In one incident, what started as a contraceptive vaccine to help control Australia's mouse population resulted in a recombinant mousepox virus that, by inactivating part of the immune system, quickly killed infected mice, even those who had been previously vaccinated against the virus. As mousepox is related to smallpox, it wasn't a great leap to imagine creating a smallpox virus that was even more virulent than wild-type smallpox. Another easily imagined scenario is the incorporation of antibiotic resistance genes into highly virulent bacterial pathogens. The anthrax scare of late 2001 was tempered somewhat by the fact that anthrax is easily treated with the antibiotic ciprofloxacin. If the anthrax were a strain that had been engineered to carry a gene for ciprofloxacin resistance prior to being used as a biological weapon, its power, both to kill and to induce fear, would have been that much greater. This is not idle speculation; when several scientists from Biopreparat, a network of laboratories in Russia and Kazakhstan defected to the West, they claimed they had created a form of *Yersinia pestis,* the bacterium responsible for plague, that was resistant to 16 different antibiotics.

Several companies have developed techniques to increase the rate of evolutionary change in microbial populations in the hope of producing bacteria with novel, beneficial traits (such as production of new antibiotics). A California company developed and then used this type of technology to create a strain of *E. coli* that was 32,000 times less sensitive to

the antibiotic cefotaxime. After this persuasive demonstration of its new technology, the company destroyed the resultant strain at the request of the American Society of Microbiology, which was concerned about the potential of the resistant strain falling into the wrong hands.

There is, however, some good news. Pathogens engineered in the laboratory would probably struggle to survive in the outside world. It is most likely impossible to engineer a pathogen that both retains its newly enhanced (or acquired) virulence and remains transmissible. Any new trait would immediately be selected against unless it conferred some selective advantage on its bearer. The questions then become how quickly will the pathogen lose its virulence and how many people will be infected before that happens.

Despite the small likelihood of an engineered germ warfare scenario succeeding, in this day and age such threats must be taken absolutely seriously. Scientists at the Center for National Security and Arms Control have developed a system for detecting early signs of bioweapons exposure, regardless of the pathogen responsible. The system, called the Rapid Syndrome Validation Project, collects data from physicians who encounter patients who display any of six syndromes, such as respiratory distress or acute hepatitis. Computer software then identifies suspicious outbreaks of disease without waiting for diagnostic laboratory tests. This epidemiological method of tracking allows the inclusion of data from physicians who may not be familiar with the presenting symptoms of a particular pathogen, an all too likely occurrence when you consider that an engineered bacteria or virus may in fact be something that no one has ever seen before.

Another group, the U.S. Advanced Research Projects Agency, is developing new biosensors to detect many disparate pathogens as well as working to create an assortment of new antibiotics and vaccines aimed at a broad range of as yet unknown infectious agents. So while legitimate methods of genetic engineering are co-opted to produce agents of bioterrorism, scientists are using these same methods to produce defenses against them. What goes around comes around.

FIGURE 10.18

Gene expression analysis using microarrays. The microarray seen here consists of oligonucleotides bound to a glass slide. The gene expression of two cells (for example, one healthy and one diseased) can be compared by labeling the cDNA from each cell with either a red or green fluorescent label and allowing the cDNAs to bind to the microarray. A laser is used to excite the bound cDNAs while a detector records those spots that fluoresce. The color of each spot reveals whether the DNA on the microarray bound to cDNAs are present in the healthy sample, the diseased sample, both samples or neither sample.

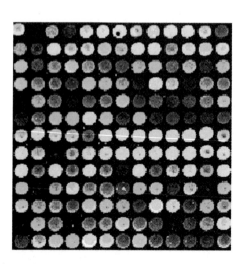

CHAPTER CHECKPOINTS

Gene therapy is the replacement of faulty host genes with functional genes by use of cloning vectors such as viruses and liposomes. This type of transfection can be used to treat genetic disorders and acquired disease. Antisense DNA and triplex DNA are used to block expression of undesirable host genes in plants and animals as well as those of intracellular parasites.

DNA technology has advanced understanding of basic genetic principles that have significant applications in a wide range of disciplines, particularly medicine, evolution, forensics, and anthropology.

The Human Genome Project has not only identified sites of specific human genes but has also clarified the functions of a special type of DNA (support DNA) that does not code for proteins.

DNA fingerprinting is a technique by which individuals are identified for purposes of medical diagnosis, genetic ancestry, and forensics.

Mitochondrial DNA analysis is being used to trace evolutionary origins in animals and plants.

Microarray analysis can determine what genes are transcribed in a given tissue. It is used to identify and devise treatments for diseases based on the genetic profile of the disease.

CHAPTER CAPSULE WITH KEY TERMS

Genetic engineering refers to the manipulation of an organism's genome and is often used in conjunction with biotechnology, the use of an organism's biochemical and metabolic pathways for industrial production of proteins.

I. **Techniques in Genetic Engineering**
  A. **Amazing DNA**
    1. Restriction **endonucleases** are used to cut DNA at specific recognition sites, resulting in small segments of DNA which can be more easily managed in the laboratory.
    2. The enzyme **ligase** is used to covalently join DNA molecules while **reverse transcriptase** is used to obtain a DNA copy of a particular mRNA.
    3. Short strands of DNA called oligonucleotides are used either as probes for specific DNA sequences or as primers for DNA polymerase.
  B. **Nucleic Acid Hybridization and Probes**
    1. The two strands of DNA can be separated (denatured) by heating. These single stranded nucleic acids will bind, or **hybridize,** to their complementary sequence, allowing them to be used as **gene,** or **hybridization,** probes.
      a. In the Southern blot method, DNA isolated from an organism is probed to identify specific DNA sequences.
      b. A technique originally developed by Frederick Sanger allows for the determination of the exact sequence of a stretch of DNA.
    2. The **polymerase chain reaction** is a technique used to amplify a specific segment of DNA several million times in just a few hours.
II. **Methods in Recombinant DNA Technology**
  A. **Technical Aspects of Recombinant DNA Technology**
    1. Recombinant DNA methods have as their goal the transfer of DNA from one organism to another. This commonly involves inserting the DNA into a **vector** (usually a plasmid or virus) and transferring the recombinant vector to a **cloning host** (usually a bacterium or yeast) for expression.
      a. Cloning vectors are differentiated primarily by the host in which they normally reside and the amount of DNA they are capable of holding. Examples include plasmids, bacteriophages, and cosmids, as well as the larger bacterial and yeast artificial chromosomes (BAC's and YAC's).
      b. Cloning hosts are fast growing, nonpathogenic organism that are genetically well understood. Examples include *E. coli* and *Saccharomyces cerevisiae.*

    2. Recombinant organisms are created by cutting both the donor DNA and the plasmid DNA with restriction endonucleases and then joining them together with ligase. The vector is then inserted into a cloning host by transformation, and those cells containing a vector are selected by their resistance to antibiotics.
    3. When the appropriate recombinant cell has been identified, it is grown on an industrial scale. The cloning host expresses the gene it contains, and the resulting protein is excreted from the cell where it can be recovered from the media and purified.
III. **Biochemical Products of Recombinant DNA Technology**
  A. In many instances the final product of recombinant DNA technology is a protein.
    1. Medical conditions such as diabetes, dwarfism, and blood clotting disorders can be treated with recombinant insulin, human growth factor, and clotting factors, respectively.
IV. **Genetically Modified Organisms (GMOs)**
  A. Recombinant organisms (transgenics) include microbes, plants, and animals that have been engineered to express genes they did not formally possess. Examples include genetically engineered *Pseudomonas syringae* bacteria that prevent ice crystals from forming on plants, animals that can produce pharmaceutical proteins in their milk, and crops with resistance to common herbicides.
V. **Genetic Treatments**
  A. **Gene therapy** describes a collection of (for now) experimental techniques in which defective genes are replaced, or, more commonly, have their expression increased or decreased. Therapy is said to be *in vivo* if cells are altered within the body and *ex vivo* if the alteration is done on cells taken from the body.
    1. Antisense and triplex DNA therapy are two methods that specifically attenuate the expression of certain genes and would be useful for treating diseases such as cancer, where the condition can be traced to the inappropriate expression of a gene.
VI. **Genome Analysis**
  A. Gene **mapping** is a method of identifying landmarks throughout the genome. Several types of maps are available, each with a different level of resolution. The ultimate map is a DNA sequence map, which provides the exact order of all of the nucleotides that make up the genome.
    1. The Human Genome Project has completed a sequence map for humans, identifying the sequence of the approximately 30,000 genes found in our DNA.

B. **DNA Fingerprinting**
    1. Because the DNA sequence of an individual is unique, so too is the collection of DNA fragments produced by cutting that DNA with a restriction endonuclease. The collection of specific fragments produced by digestion of the DNA with a restriction endonuclease is commonly referred to as the DNA fingerprint of an organism (more properly as a set of *restriction fragment length polymorphisms*), and serves as a unique identifier of any organism.

C. **Microarray Analysis** is a method of determining which genes are actively transcribed in a cell under a variety of conditions, such as health versus disease or growth versus differentiation.
    1. Rapid advances in both biotechnology and computer analysis allow the detection of all the actively transcribed genes in a cell under a given set of conditions. This allows much more accurate diagnosis of disease and far more specific treatment.

MULTIPLE-CHOICE QUESTIONS

1. Which gene is incorporated into plasmids to detect recombinant cells?
  a. restriction endonuclease
  b. virus receptors
  c. a gene for antibiotic resistance
  d. reverse transcriptase

2. Which of the following is *not* essential to carry out the polymerase chain reaction?
  a. primers   c. gel electrophoresis
  b. DNA polymerase   d. high temperature

3. Which of the following is *not* a part of the Sanger method to sequence DNA?
  a. dideoxy nucleotides   c. electrophoresis
  b. DNA polymerase   d. reverse transcriptase

4. What do we call the synthetic unit of the polymerase chain reaction?
  a. annealer   c. amplicon
  b. ligase   d. primer

5. The function of ligase is to
  a. rejoin segments of DNA
  b. make longitudinal cuts in DNA
  c. synthesize cDNA
  d. break down ligaments

6. The pathogen of plant roots that is used as a cloning host is
  a. *Pseudomonas*
  b. *Agrobacterium*
  c. *Escherichia coli*
  d. *Saccharomyces cerevisiae*

7. Which of the following, that when combined with its complement, could be clipped by an endonuclease?
  a. ATCGATCGTAGCTAGC
  b. AAGCTTTTCGAA
  c. GAATTC
  d. ACCATTGGTA

8. The antisense DNA strand that complements mRNA AUGCGCGAC is
  a. UACGCUCUG
  b. GTCTCGCAT
  c. TACGCGCTG
  d. DNA cannot complement mRNA

9. Which DNA fragment will be closest to the top (negative pole) of an electrophoretic gel?
  a. 450 bp   c. 5 kb
  b. 3,560 bp   d. 1,500 bp

10. Which of the following is a primary participant in cloning an isolated gene?
  a. restriction endonuclease
  b. vector
  c. host organism
  d. all of these

11. For which of the following would a nucleic acid probe *not* be used?
  a. locating a gene on a chromosome
  b. developing a Southern blot
  c. identifying a microorganism
  d. constructing a recombinant plasmid

12. **Single Matching.** Match the term with its description:
  ____ nucleic acid probe
  ____ antisense strand
  ____ template strand
  ____ reverse transcriptase
  ____ Taq polymerase
  ____ triplex DNA
  ____ primer
  ____ restriction endonuclease
  a. enzyme that transcribes RNA into DNA
  b. DNA molecule with an extra strand inserted
  c. the nontranslated strand of DNA or RNA
  d. enzyme that snips DNA at palindromes
  e. oligonucleotide that initiates the PCR
  f. strand of nucleic acid that is transcribed or translated
  g. thermostable enzyme for synthesizing DNA
  h. oligonucleotide used in hybridization

CONCEPT QUESTIONS

1. Define genetic engineering and biotechnology, and summarize the important purposes of these fields. Review the use of the terms *genome, chromosome, gene, DNA,* and *RNA* from chapter 9.

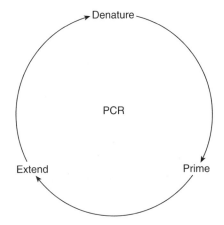

2. a. Describe the processes involved in denaturing and renaturing of DNA.
   b. What is useful about this procedure?
   c. Why is it necessary to denature the DNA in the Southern blot test?
   d. How would the Southern blot be used with PCR?

3. a. Define restriction endonuclease.
   b. Define palindrome and draw three different palindromic sequences.
   c. Using nucleotide letters, show how a piece of DNA can be cut to circularize it.
   d. What are restriction length polymorphisms, and how are they used?

4. Explain how electrophoresis works and the general way that DNA is sized. Estimate the size of the DNA fragment in base pairs in the first lane of the gel in figure 10.3*b*. Define oligonucleotides, explain how they are formed, and give three uses for them.

5. a. Briefly describe the functions of DNA synthesizers and sequencers.
   b. How would you make a copy of DNA from an mRNA transcript?
   c. Show how this process would look, using base notation.
   d. What is this DNA called?
   e. Why would it be an advantage to synthesize eucaryotic genes this way?

6. Go through the basic steps of the Sanger method to sequence DNA, and make a simple drawing to help visualize it.

7. a. Explain the meaning of the shorthand in the next column to represent the polymerase chain reaction.
   b. Describe the effects of temperature change in PCR.
   c. Exactly what are the functions of the primer and Taq polymerase?

   d. Explain why the PCR is unlikely to amplify contaminating bacterial DNA in a sample of human DNA.
   e. Explain how PCR could be used to pick a gene out of a complex genome and amplify it.

8. a. What characteristics of plasmids and bacteriophages make them good cloning vectors?
   b. Name several types of vectors, and explain what benefits they have.
   c. List the types of genes that they can contain.

9. a. Describe the principles behind recombinant DNA technology.
   b. Outline the main steps in cloning a gene.
   c. Once cloned, how can this gene be used?
   d. Characterize several ways that recombinant DNA technology can be used.

10. a. What characteristics of bacteria make them good cloning hosts?
    b. What is one way to determine whether a bacterial culture has received a recombinant plasmid?

11. a. What is transfection, and what are transgenic organisms?
    b. Explain how *Agrobacterium* is used to transfect plants.
    c. Describe a method for transfecting animals.
    d. Summarize some of the uses for transgenic plants and animals.

12. a. Briefly outline the purposes and significant steps in a gene therapy procedure.
    b. What is the main difference between *ex vivo* and *in vivo* gene therapy?
    c. Does the virus vector used in gene therapy replicate itself in the host cell? Would this be desirable or not?
    d. What are some of the main problems with gene therapy?

13. a. Describe the molecular mechanisms by which a DNA antisense molecule could work as a genetic medicine.
    b. Do the same for triplex DNA.
    c. Are the therapies permanent?
    d. Why or why not?

14. a. What is a gene map?
    b. Show by a diagram how chromosomal, physical, and sequence maps are different.
    c. Which organisms are being mapped, and what uses will these maps perform?
    d. Why did the human genome map require 15 years to complete?
    e. What are some possible effects of knowing the genetic map of humans?

15. a. Describe what a DNA fingerprint is and why, and how restriction fragments can be used to form a unique DNA pattern.
    b. Discuss briefly how DNA fingerprinting is being used routinely by biomedicine, the law, the military, and human biology.

CRITICAL-THINKING QUESTIONS

1. a. Give an example of a benefit of genetic engineering to society and a possible adverse outcome.
   b. Give an example of an ecological benefit and a possible adverse side effect.

2. a. In reference to Spotlight on Microbiology 10.2, what is your opinion of the dangers associated with genetic engineering?
   b. Most of us would agree to growth hormone therapy for a child

   with dwarfism, but how do we deal with parents who want to give growth hormones to their 8-year-old son so that he will be "better at sports?"

3. a. If gene probes, fingerprinting, and mapping could make it possible for you to know of future genetic diseases in you or one of your children, would you wish to use this technology to find out?
   b. What if it were used as a screen for employment or insurance?

4. a. Can you think of a reason that bacteria make restriction endonucleases?
   b. What is it about the endonucleases that prevents bacteria from destroying their own DNA?
   c. Look at figure 10.5, and determine the correct DNA sequence for the fragment's template strand. (Be careful of orientation.)
   d. Which suspect was more likely the rapist, according to the fingerprint in figure 10.16c?

5. a. Describe how a virus might be genetically engineered to make it highly virulent.
   b. Can you trace the genetic steps in the development of a tomato plant that has become frost-free from the addition of a flounder's antifreeze genes? (Hint: You need to use *Agrobacterium*.)
   c. What is a different approach to preventing frost on tomatoes?

6. a. Give three different methods to treat or cure hemophilia using the techniques of genetic engineering described in the chapter.
   b. Outline a way that gene therapy could be used to treat a patient with genetic defects due to the presence of extra genes (trisomies as described in chapter 17).
   c. Give a possible explanation for why gene therapy using bone marrow is more successful than, say, that which involves therapy in the lungs.

7. You have obtained a blood sample in which only red blood cells are left to analyze.
   a. Can you conduct a DNA analysis of this blood?
   b. If no, explain why.
   c. If yes, explain what you would use to analyze it and how to do it.

8. The way that PCR amplifies DNA is similar to the doubling in a population of growing bacteria; a single DNA strand is used to synthesize 2 DNA strands, which become 4, then 8, then 16, etc. If a complete cycle takes 3 minutes,
   a. how many strands of DNA would theoretically be present after 10 minutes?
   b. after 30 minutes?
   c. after 1 hour?

9. a. Design an antisense DNA drug for this gene: TACGGCTATATTCCGGGC.
   b. Design an antisense RNA drug for the same gene.
   c. How would a triplex drug for this gene work?
   d. Describe how antisense DNA works to block the replication of viruses.

10. a. How would you regard the release of a bioengineered bacterium in your own backyard?
    b. Describe any moral, ethical, or biological problems associated with eating tomatoes from an engineered plant or pork from a transgenic pig.
    c. What are the moral considerations of using transgenic animals to manufacture various human products?

11. You are on a jury to decide whether a person committed a homicide and you have to weigh DNA fingerprinting evidence. Two different sets of fingerprints were done: one that tested 5 markers and one that tested 10. Both sets match the defendant's profile.
    a. Which one is more reliable and why?
    b. How informative is the fingerprint if you are told that the fingerprint pattern occurs in one person out of 10,000 in the general population?
    c. Would knowing that the defendant lived in the same apartment building as the victim have any effect on your decision?

12. a. How do you suppose the fish and game department, using DNA evidence, could determine whether certain individuals had poached a deer or if a particular mountain lion had eaten part of a body.
    b. Can you think of some reasons that it would *not* be possible to recreate dinosaurs using the technology we have described in this chapter?

13. a. Look at the pedigree chart for Alzheimer disease. How many genes for this disease must be inherited for it to be expressed?
    b. What kinds of offspring could be produced if a person with the genotype of 1,3 married a person with genotype 1,5?
    c. If cystic fibrosis has the same general profile of inheritance as Alzheimer disease, yet it is recessive (requires two genes to be expressed), what would be the future of the offspring in the pedigree chart?

14. Who actually owns the human genome?
    a. Make cogent arguments on various sides of the question.
    b. Explain the steps required in producing a structural map (base sequence) of DNA.

15. It is possible that by 2010, we can have a detailed understanding of normal and pathologic phenotypes. Discuss the idea that the human genome map (genotype) is now a phenotype.

INTERNET SEARCH TOPICS

1. Visit the student Online Learning Center at www.mhhe.com/talaro5. Go to chapter 10, Internet Search Topics, and log on to the available websites to
   a. Discover more about the human genome projects. (Or type "Genbank" into a search engine.)
   b. Research information on all aspects of biotechnology at the website for the National Center for Biotechnology Information. The Science Primer is a great place to start for an easy-to-follow lesson on things biotech.
   c. Review the latest happenings in clinical trials for gene therapy.

2. Type the words "victim identifications world trade center disaster" into a search engine such as Google. Visit the search result sites to discover the remarkable process of identifying people who were killed in the 9/11/2001 attack.

CHAPTER 11

Physical and Chemical Control of Microbes

The natural condition of humanity is to share surroundings with a large, diverse population of microorganisms. The complete exclusion of microbes from the environment is not only impossible but of questionable value. In many instances, however, our health and comfort can depend on the ability to destroy, inhibit, and remove microbes in the habitats we share. These techniques, also known as antimicrobial control, are very broad in scope. They include routine activities such as cleaning, refrigeration, and cooking. They are also of central importance in the medical, dental, and commercial settings to prevent infection and spoilage, and to ensure the safety of food, water, and other products. Both this chapter and chapter 12 will survey important aspects of microbial control.

Chapter Overview

- The control of microbes in the environment is a constant concern of health care and industry since microbes are the cause of infection and food spoilage, among other undesirable events.
- Antimicrobial control is accomplished using both physical techniques and chemical agents to destroy, remove, or reduce microbes in a given area.
- Many factors must be contemplated when choosing an antimicrobial technique, including the material being treated, the type of microbes involved, the microbial load, and the time available for treatment.
- Antimicrobial agents damage microbes by disrupting the structure of the cell wall or cell membrane, preventing synthesis of nucleic acids (DNA and RNA), or altering the function of cellular proteins.
- Microbicidal agents kill microbes by inflicting nonreversible damage to the cell. Microbistatic agents temporarily inhibit the reproduction of microbes but do not inflict irreversible damage. Mechanical antimicrobial agents physically remove microbes from materials but do not necessarily kill or inhibit them.
- Heat is the most important physical agent in microbial control and can be delivered in both moist (steam sterilization, pasteurization) and dry (incinerators, Bunsen burners) forms.
- Radiation exposes materials to high energy waves that can enter and damage microbes. Examples are ionizing and ultraviolet radiation.
- Chemical antimicrobials are available for every level of microbial treatment, from low-level disinfectants to high-level sterilants. Antimicrobial chemicals include halogens, alcohols, phenolics, peroxides, heavy metals, detergents, and aldehydes.

A look down the cleaning products aisle of most any store will confirm our preoccupation (some would say obsession) with microbial control.

Controlling Microorganisms

Much of the time in our daily existence, we take for granted tap water that is drinkable, food that is not spoiled, shelves full of products to eradicate "germs," and drugs to treat infections. Controlling our degree of exposure to potentially harmful microbes is a monumental concern in our lives, and it has a long and eventful history (**Historical Highlights 11.1**).

HISTORICAL HIGHLIGHTS   11.1
Microbial Control in Ancient Times

No one knows for sure when humans first applied methods that could control microorganisms, but perhaps the discovery and use of fire in prehistoric times was the starting point. We do know that records describing simple measures to control decay and disease appear from civilizations that existed several thousand years ago. We know, too, that these ancient people had no concept that germs caused disease, but they did have a mixture of religious beliefs, skills in observing natural phenomena, and possibly, a bit of luck. This combination led them to carry out simple and sometimes rather hazardous measures that contributed to the control of microorganisms.

Salting, smoking, pickling, and drying foods and exposing food, clothing, and bedding to sunlight were prevalent practices among early civilizations. The Egyptians showed surprising sophistication and understanding of decomposition by embalming the bodies of their dead with strong salts and pungent oils. They introduced filtration of wine and water as well. The Greeks and Romans burned clothing and corpses during epidemics, and they stored water in copper and silver containers. The armies of Alexander the Great reportedly boiled their drinking water and buried their wastes. Burning sulfur to fumigate houses and applying sulfur as a skin ointment also date from this approximate era.

During the great plague pandemic of the Middle Ages, it was commonplace to bury corpses in mass graves, burn the clothing of plague victims, and ignite aromatic woods in the houses of the sick in the belief that fumes would combat the disease. In a desperate search for some sort of protection, survivors wore peculiar garments and anointed their bodies with herbs, strong perfume, and vinegar. These attempts may sound foolish and antiquated, but it now appears that they may have had some benefits. Burning wood releases formaldehyde, which could have acted as a disinfectant; herbs, perfume, and vinegar contain mild antimicrobial substances. Each of these early methods, although somewhat crude, laid the foundations for microbial control methods that are still in use today.

Illustration of protective clothing used by doctors in the 1700s to avoid exposure to plague victims. The beaklike portion of the hood contained volatile perfumes to protect against foul odors and possibly inhaling "bad air."

GENERAL CONSIDERATIONS IN MICROBIAL CONTROL

The methods of microbial control belong to the general category of *decontamination* procedures, in that they destroy or remove contaminants. In microbiology, contaminants are microbes present at a given place and time that are undesirable or unwanted. Most decontamination methods employ either physical agents, such as heat or *radiation,* or chemical agents such as disinfectants and antiseptics. This separation is convenient, even though the categories overlap in some cases; for instance, radiation can cause damaging chemicals to form, or chemicals can generate heat. A flowchart (**figure 11.1**) summarizes the major applications and aims in microbial control.

RELATIVE RESISTANCE OF MICROBIAL FORMS

The primary targets of microbial control are microorganisms capable of causing infection or spoilage that are constantly present in the external environment and on the human body. This targeted population is rarely simple or uniform; in fact, it often contains mixtures of microbes with extreme differences in resistance and

harmfulness. Contaminants that can have far-reaching effects if not adequately controlled include bacterial vegetative cells and endospores, fungal hyphae and spores, yeasts, protozoan trophozoites and cysts, worms, insects and their eggs, and viruses. This scheme compares the general resistance these forms have to physical and chemical methods of control:

Highest resistance
   Bacterial endospores

Moderate resistance
   Protozoan cysts; some fungal sexual spores (zygospores); some viruses. In general, naked viruses are more resistant than enveloped forms. Among the most resistant viruses are the hepatitis B virus and the poliovirus. Bacteria with more resistant vegetative cells are *Mycobacterium tuberculosis, Staphylococcus aureus,* and *Pseudomonas* species.

Least resistance
   Most bacterial vegetative cells; (other than zygospores) fungal spores and hyphae; enveloped viruses; yeasts; and protozoan trophozoites

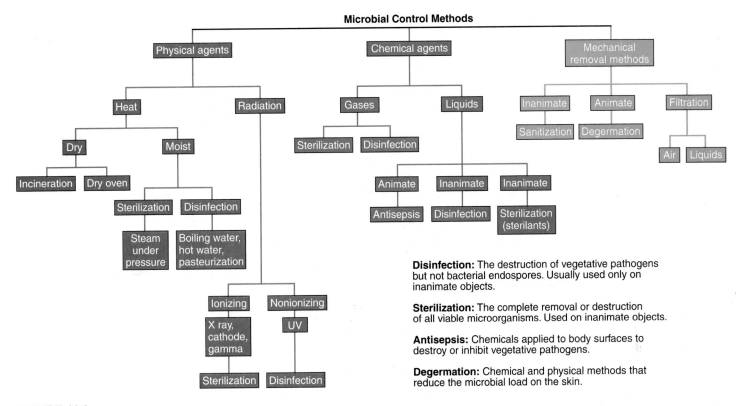

FIGURE 11.1

Microbial Control Methods

Actual comparative figures on the requirements for destroying various groups of microorganisms are shown in **table 11.1.** Bacterial endospores are the most resistant microbial entities, being as much as 18 times harder to destroy than their counterpart vegetative cells. Because of their resistance to microbial control, their destruction is the goal of sterilization (see definition in following section) as any process that kills endospores will invariably kill all less resistant microbial forms. Other methods of control (disinfection, antisepsis) act primarily upon microbes that are less hardy than endospores.

TERMINOLOGY AND METHODS OF MICROBIAL CONTROL

Through the years, a growing terminology has emerged for describing and defining measures that control microbes. To complicate matters, the everyday use of some of these terms can at times be vague and inexact. For example, occasionally one may be directed to "sterilize" or "disinfect" a patient's skin, even though this usage does not fit the technical definition of either term. To lay the groundwork for the concepts in microbial control to follow, we present here a series of concepts, definitions, and usages in antimicrobial control.

Sterilization

Sterilization is a process that destroys or removes all viable microorganisms, including viruses. Any material that has been subjected to this process is said to be **sterile.*** These terms should be

---

TABLE 11.1

Relative Resistance of Bacterial Endospores and Vegetative Cells to Control Agents

Method	Spores*	Vegetative Forms*	Relative Resistance**
Heat (moist)	120°C	80°C	1.5×
Radiation (X-ray) dosage	0.4 Mrad	0.1 Mrad	4×
Ultraviolet rays (exposure time)	1.5 h	10 min	9×
Sterilizing gas (ethylene oxide)	1,200 mg/l	700 mg/l	1.7×
Sporicidal liquid (2% glutaraldehyde)	3 h	10 min	18×

*Values are based on methods (concentration, exposure time, intensity) that are required to destroy the most resistant pathogens in each group.
**The greater resistance of spores versus vegetative cells given as an average figure.

---

used only in the strictest sense for methods that have been proved to sterilize. An object cannot be slightly sterile or almost sterile—it is either sterile or not sterile. Control methods that sterilize are generally reserved for inanimate objects, because sterilizing parts of the human body would call for such harsh treatment that it would be

---

*sterile (ster′ -ill) Gr. *steria*, barren. (This has another, older meaning that connotes the inability to produce offspring.)

highly dangerous and impractical. As we shall see in chapter 13, many internal parts of the body—the brain, muscles, and liver, for example—are naturally free of microbes.

Sterilized products—surgical instruments, syringes, and commercially packaged foods, just to name a few—are essential to human well-being. Although most sterilization is performed with a physical agent such as heat, a few chemicals called *sterilants* can be classified as sterilizing agents because of their ability to destroy spores.

At times, sterilization is neither practicable nor necessary, and only certain groups of microbes need to be controlled. Some antimicrobial agents eliminate only the susceptible vegetative states of microorganisms but do not destroy the more resistant endospore and cyst stages. Keep in mind that the destruction of spores is not always a necessity, because most of the infectious diseases of humans and animals are caused by non-spore-forming microbes.

Microbicidal Agents

The root *-cide,* meaning to kill, can be combined with other terms to define an antimicrobial agent aimed at destroying a certain group of microorganisms. For example, a **bactericide** is a chemical that destroys bacteria except for those in the endospore stage. It may or may not be effective on other microbial groups. A fungicide is a chemical that can kill fungal spores, hyphae, and yeasts. A virucide is any chemical known to inactivate viruses, especially on living tissue. A sporicide is an agent capable of destroying bacterial endospores. A sporicidal agent can also be a sterilant because it can destroy the most resistant of all microbes.

Agents That Cause Microbistasis

The Greek words *stasis* and *static* mean to stand still. They can be used in combination with various prefixes to denote a condition in which microbes are temporarily prevented from multiplying but are not killed outright. Although killing or permanently inactivating microorganisms is the usual goal of microbial control, microbistasis does have meaningful applications. **Bacteriostatic** agents prevent the growth of bacteria on tissues or on objects in the environment, and *fungistatic* chemicals inhibit fungal growth. Materials used to control microorganisms in the body (antiseptics and drugs) have microbistatic effects because many microbicidal compounds can be too toxic to human cells.

Germicides, Disinfection, Antisepsis

A **germicide,*** also called a *microbicide,* is any chemical agent that kills pathogenic microorganisms. A germicide can be used on inanimate (nonliving) materials or on living tissue, but it ordinarily cannot kill resistant microbial cells. Any physical or chemical agent that kills "germs" is said to have **germicidal** properties.

The related term, **disinfection,*** refers to the use of a physical process or a chemical agent (a **disinfectant**) to destroy vegetative pathogens but not bacterial endospores. It is important to note that disinfectants are normally used only on inanimate objects because, in the concentrations required to be effective, they can be toxic to human and other animal tissue. Disinfection processes also remove the harmful products of microorganisms (toxins) from materials. Examples of disinfection include applying a solution of

5% bleach to an examining table, boiling food utensils used by a sick person, and immersing thermometers in an iodine solution between uses.

In modern usage, **sepsis** is defined as the growth of microorganisms in the body or the presence of microbial toxins in blood and other tissues. The term **asepsis*** refers to any practice that prevents the entry of infectious agents into sterile tissues and thus prevents infection. Aseptic techniques commonly practiced in health care range from sterile methods that exclude all microbes to **antisepsis.*** In antisepsis, chemical agents called **antiseptics** are applied directly to exposed body surfaces (skin and mucous membranes), wounds, and surgical incisions to destroy or inhibit vegetative pathogens. Examples of antisepsis include preparing the skin before surgical incisions with iodine compounds, swabbing an open root canal with hydrogen peroxide, and ordinary handwashing with a germicidal soap.

Methods That Reduce the Numbers of Microorganisms

Several applications in commerce and medicine do not require actual sterilization, disinfection, or antisepsis but are based on reducing the levels of microorganisms (the microbial load) so that the possibility of infection or spoilage is greatly decreased. Restaurants, dairies, breweries, and other food industries consistently handle large numbers of soiled utensils that could readily become sources of infection and spoilage. These industries must keep microbial levels to a minimum during preparation and processing. **Sanitization*** is any cleansing technique that mechanically removes microorganisms as well as other debris to reduce contaminations to safe levels. A **sanitizer** is a compound such as soap or detergent used to perform this task.

Cooking utensils, dishes, bottles, cans, and used clothing that have been washed and dried may not be completely free of microbes, but they are considered safe for normal use (sanitary). Air sanitization with ultraviolet lamps reduces airborne microbes in hospital rooms, veterinary clinics, and laboratory installations. It is important to note that some sanitizing processes (such as dishwashing machines) are rigorous enough to sterilize objects, but this is not true of all sanitization methods. Also note that sanitization is often preferable to sterilization. In a restaurant, for example, you could be given a sterile fork with someone else's old food on it and a sterile glass with lipstick on the rim. On top of this, realize that the costs associated with sterilization would lead to the advent of the $50 Happy Meal. In a situation such as this, the advantage of being sanitary as opposed to sterile can be clearly seen.

It is often necessary to reduce the numbers of microbes on the human skin through **degermation.** This process usually involves scrubbing the skin or immersing it in chemicals, or both. It also emulsifies oils that lie on the outer cutaneous layer and mechanically removes potential pathogens on the outer layers of the skin. Examples of degerming procedures are the surgical handscrub, the application of alcohol wipes to the skin, and the cleansing of a wound with germicidal soap and water. The concepts of antisepsis and degermation clearly overlap, since a degerming procedure can simultaneously be antiseptic, and vice versa.

---

*germicide (jer´-mih-syd) L. *germen,* germ, and *caedere,* to kill. Germ is a common term for a pathogenic microbe.

*disinfection (dis˝-in-fek´-shun) L. *dis,* apart, and *inficere,* to corrupt.

*asepsis (ay-sep´-sis) Gr. *a,* no or none, and *sepsis,* decay.

*antisepsis *anti,* against.

*sanitization (san˝-ih-tih-zay´-shun) L. *sanitas,* health.

WHAT IS MICROBIAL DEATH?

Death is a phenomenon that involves the permanent termination of an organism's vital processes. Signs of life in complex organisms such as animals are self-evident, and death is made clear by loss of nervous function, respiration, or heartbeat. In contrast, death in microscopic organisms that are composed of just one or a few cells is often hard to detect, because they reveal no conspicuous vital signs to begin with. Lethal agents (such as radiation and chemicals) do not necessarily alter the overt appearance of microbial cells. Even the loss of movement in a motile microbe cannot be used to indicate death. This fact has made it necessary to develop special qualifications that define and delineate microbial death.

The destructive effects of chemical or physical agents occur at the level of a single cell. As the cell is continuously exposed to an agent such as intense heat or toxic chemicals, various cell structures become dysfunctional, and the entire cell can sustain irreversible damage. At present, the most practical way to detect this damage is to determine if a microbial cell can still reproduce when exposed to a suitable environment. If the microbe has sustained metabolic or structural damage to such an extent that it can no longer reproduce, even under ideal environmental conditions, then it is no longer viable. *The permanent loss of reproductive capability, even under optimum growth conditions, has become the accepted microbiological definition of death.*

Factors That Affect Death Rate

The ability to define microbial death has tremendous theoretical and practical importance. It allows medicine and industry to test the conditions required to destroy microorganisms, to pinpoint the ways that antimicrobial agents kill cells, and to establish standards of sterilization and disinfection in these fields. Hundreds of testing procedures have been developed for evaluating physical and chemical agents. Appendix B presents a general breakdown of certain types of antimicrobial tests, along with the kinds of variables that must be controlled.

The cells of a culture show marked variation in susceptibility to a given microbicidal agent. Death of the whole population is not instantaneous but begins when a certain threshold of microbicidal agent (some combination of time and concentration) is met and continues in a logarithmic manner as the time or concentration of the agent is increased **(figure 11.2)**. Because many microbicidal agents target the cells' metabolic processes, active cells (younger, rapidly dividing) tend to die more quickly than those that are less metabolically active (older, inactive). Eventually, a point is reached at which survival of any cells is highly unlikely; this point is equivalent to sterilization.

The effectiveness of a particular agent is governed by several factors besides time. These additional factors influence the action of antimicrobial agents:

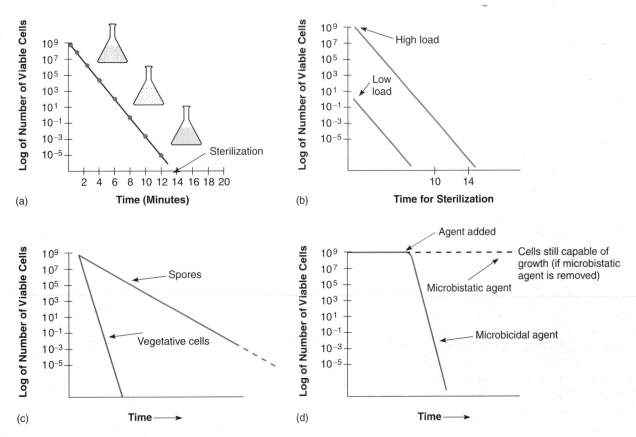

FIGURE 11.2

Factors that influence the rate at which microbes are killed by antimicrobial agents. (a) Length of exposure to the agent. During exposure to a chemical or physical agent, all cells of a microbial population, even a pure culture, do not die simultaneously. Over time, the number of viable organisms remaining in the population decreases logarithmically, giving a straight-line relationship on a graph. The point at which the number of survivors is infinitesimally small is considered sterilization. (b) Effect of the microbial load. (c) Relative resistance of spores versus vegetative forms. (d) Action of the agent, whether microbicidal or microbistatic.

1. The number of microorganisms (figure 11.2*b*). A higher load of contaminants requires more time to destroy.
2. The nature of the microorganisms in the population (figure 11.2*c*). In most actual circumstances of disinfection and sterilization, the target population is not a single species of microbe but a mixture of bacteria, fungi, spores, and viruses, presenting an even greater spectrum of microbial resistance.
3. The temperature and pH of the environment.
4. The concentration (dosage, intensity) of the agent. For example, UV radiation is most microbicidal at 260 nm; most disinfectants are more active at higher concentrations.
5. The mode of action of the agent (figure 11.2*d*). How does it kill or inhibit the microorganism?
6. The presence of solvents, interfering organic matter, and inhibitors. Large amounts of saliva, blood, and feces can inhibit the actions of disinfectants and even of heat.

The influence of these factors will be discussed in greater detail in subsequent sections.

HOW ANTIMICROBIAL AGENTS WORK: THEIR MODES OF ACTION

An antimicrobial agent's adverse effect on cells is known as its *mode* (or *mechanism*) *of action*. Agents affect one or more cellular targets, inflicting damage progressively until the cell is no longer able to survive. Antimicrobials have a range of cellular targets, with the agents that are least selective in their targeting tending to be effective against the widest range of microbes (examples include heat and radiation). More selective agents (drugs, for example) tend to target only a single cellular component and are much more restricted as to the microbes they are effective against.

The cellular targets of physical and chemical agents fall into four general categories:

1. the cell wall,
2. the cell membrane,
3. cellular synthetic processes (DNA, RNA), and
4. proteins.

The Effects of Agents on the Cell Wall

The cell wall maintains the structural integrity of bacterial and fungal cells. Several types of chemical agents damage the cell wall by blocking its synthesis, digesting it, or breaking down its surface. A cell deprived of a functioning cell wall becomes fragile and is lysed very easily. Examples of this mode of action include some antimicrobial drugs (penicillins) that interfere with the synthesis of the cell wall in bacteria (see figure 12.2). Detergents and alcohol can also disrupt cell walls, especially in gram-negative bacteria.

How Agents Affect the Cell Membrane

All microorganisms have a cell membrane composed of lipids and proteins, and even some viruses have an outer membranous envelope. As we learned in previous chapters, a cell's membrane provides a two-way system of transport. If this membrane is disrupted, a cell loses its selective permeability and can neither prevent the loss of vital molecules nor bar the entry of damaging chemicals.

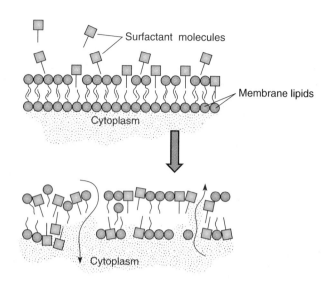

FIGURE 11.3

Mode of action of surfactants on the cell membrane. Surfactants inserting in the lipoidal layers disrupt it and create abnormal channels that alter permeability and cause leakage both into and out of the cell.

Loss of those abilities leads to cell death. Detergents called **surfactants*** work as microbicidal agents because they lower the surface tension of cell membranes. Surfactants are polar molecules with hydrophilic and hydrophobic regions that can physically bind to the lipid layer and penetrate the internal hydrophobic region of membranes. In effect, this process "opens up" the once tight interface, leaving leaky spots that allow injurious chemicals to seep into the cell and important ions to seep out (**figure 11.3**).

Agents That Affect Protein and Nucleic Acid Synthesis

Microbial life depends upon an orderly and continuous supply of proteins to function as enzymes and structural molecules. As we saw in chapter 9, these proteins are synthesized on the ribosomes through a complex process called translation. For example, the antibiotic chloramphenicol binds to the ribosomes of bacteria in a way that stops peptide bonds from forming. In its presence, many bacterial cells are inhibited from forming proteins required in growth and metabolism and are thus inhibited from multiplying. Most of the agents that block protein synthesis are drugs used in antimicrobial therapy. These drugs will be discussed in greater detail in chapter 12.

The nucleic acids are likewise necessary for the continued functioning of microbes. DNA must be regularly replicated and transcribed in growing cells, and any agent that either impedes these processes or changes the genetic code is potentially antimicrobial. Some agents bind irreversibly to DNA, preventing both transcription (formation of RNA) and translation; others are mutagenic agents. Gamma, ultraviolet, or X radiation causes mutations that result in permanent inactivation of DNA. Chemicals such as formaldehyde and ethylene oxide also interfere with DNA and RNA function.

---

*surfactant (sir-fak'-tunt) A word derived from **surface-act**ing age**nt**.

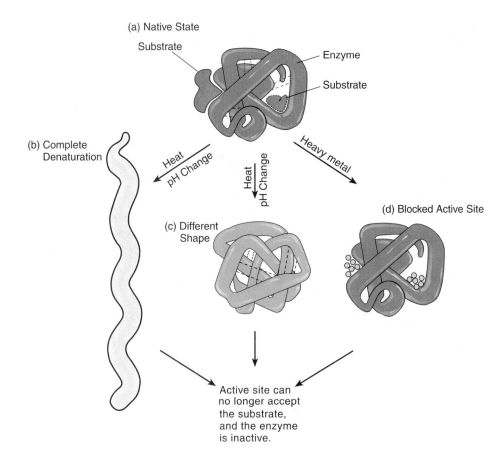

FIGURE 11.4

Modes of action affecting protein function.
(a) The native (functional) state is maintained by bonds that create active sites to fit the substrate. Some agents denature the protein by breaking all or some secondary and tertiary bonds. Results are **(b)** complete unfolding or **(c)** random bonding and incorrect folding. **(d)** Some agents react with functional groups on the active site and interfere with bonding.

Agents That Alter Protein Function

A microbial cell contains large quantities of proteins that function properly only if they remain in a normal three-dimensional configuration called the *native state*. The antimicrobial properties of some agents arise from their capacity to disrupt, or **denature,** proteins. In general, denaturation occurs when the bonds that maintain the secondary and tertiary structure of the protein are broken. Breaking these bonds will cause the protein to unfold or create random, irregular loops and coils **(figure 11.4)**. One way that proteins can be denatured is through coagulation by moist heat (the same reaction seen in the irreversible solidification of the white of an egg when boiled). Chemicals such as strong organic solvents (alcohols, acids) and phenolics also coagulate proteins. Other antimicrobial agents, such as metallic ions, attach to the active site of the protein and prevent it from interacting with its correct substrate. Regardless of the exact mechanism, such losses in normal protein function can promptly arrest metabolism. Most antimicrobials of this type are nonselective as to the microbes they affect.

Practical Concerns in Microbial Control

Numerous considerations govern the selection of a workable method of microbial control. These are among the most pressing concerns:

1. Does the application require sterilization, or is disinfection adequate? In other words, must spores be destroyed, or is it necessary to destroy only vegetative pathogens?
2. Is the item to be reused or permanently discarded? If it will be discarded, then the quickest and least expensive method should be chosen.
3. If it will be reused, can the item withstand heat, pressure, radiation, or chemicals?
4. Is the control method suitable for a given application? (For example, ultraviolet radiation is a good sporicidal agent, but it will not penetrate solid materials.) Or, in the case of a chemical, will it leave an undesirable residue?
5. Will the agent penetrate to the necessary extent?
6. Is the method cost- and labor-efficient, and is it safe?

A remarkable variety of substances can require sterilization. They run the gamut from durable solids such as rubber to sensitive liquids such as serum, and from air to tissue grafts. Hundreds of situations requiring sterilization confront the network of persons involved in health care, be it technician, nurse, doctor, or manufacturer, and no universal method works well in every case.

Considerations such as cost, effectiveness, and method of disposal are all important. For example, the disposable plastic items such as catheters and syringes that are used in invasive medical procedures have the potential for infecting the tissues. These must be sterilized during manufacture by a nonheating method (gas or radiation), because heat can damage delicate plastics. After these items have been used, it is often necessary to destroy or decontaminate them before they are discarded because of the potential risk to the handler (from needlesticks). Steam sterilization, which is quick and sure, is a sensible choice at this point, because it does not matter if the plastic is destroyed. Health care workers are held to very high standards of infection prevention (see universal precautions, page 542).

Methods of Physical Control

Microorganisms have adapted to the tremendous diversity of habitats the earth provides, even severe conditions of temperature, moisture, pressure, and light. For microbes that normally withstand such extreme physical conditions, our attempts at control would probably have little effect. Fortunately for us, we are most interested in controlling microbes that flourish in the same environment in which humans live. The vast majority of these microbes are readily controlled by abrupt changes in their environment. Most prominent among antimicrobial physical agents is heat. Other less widely used agents include radiation, filtration, ultrasonic waves, and even cold. The following sections will examine some of these methods and explore their practical applications in medicine, commerce, and the home.

HEAT AS AN AGENT OF MICROBIAL CONTROL

A sudden departure from a microbe's temperature of adaptation is likely to have a detrimental effect on it. As a rule, elevated temperatures (exceeding the maximum growth temperature) are microbicidal, whereas lower temperatures (below the minimum growth temperature) are microbistatic. The two physical states of heat used in microbial control are moist and dry. *Moist heat* occurs in the form of hot water, boiling water, or steam (vaporized water). In practice, the temperature of moist heat usually ranges from 60° to 135° Celsius. As we shall see, the temperature of steam can be regulated by adjusting its pressure in a closed container. The expression *dry heat* denotes air with a low moisture content that has been heated by a flame or electric heating coil. In practice, the temperature of dry heat ranges from 160°C to several thousand degrees Celsius.

Mode of Action and Relative Effectiveness of Heat

Moist heat and dry heat differ in their modes of action as well as in their efficiency. Moist heat operates at lower temperatures and shorter exposure times to achieve the same effectiveness as dry heat **(table 11.2).** Although many cellular structures are damaged by moist heat, its most microbicidal effect is the coagulation and denaturation of proteins, which quickly and permanently halts cellular metabolism.

Dry heat dehydrates the cell, removing the water necessary for metabolic reactions, and it also denatures proteins. However, the lack of water actually increases the stability of some protein conformations, necessitating the use of higher temperatures when dry heat is employed as a method of microbial control. At very high temperatures, dry heat of course oxidizes cells, reducing them to ashes. This is the method used in the laboratory when a loop is flamed or in industry when medical waste is incinerated.

Heat Resistance and Thermal Death of Spores and Vegetative Cells

Bacterial endospores exhibit the greatest resistance, and vegetative states of bacteria and fungi are the least resistant to both moist and dry heat. Destruction of spores usually requires temperatures above boiling **(table 11.3),** although resistance varies widely. In boiling water (100°C), the spores of *Bacillus anthracis* (the agent of anthrax) can be destroyed in a few minutes, whereas the spores of some thermophilic and anaerobic species can require several hours.

Vegetative cells also vary in their sensitivity to heat, though not to the same extent as spores **(table 11.4).** Among bacteria, the death times with moist heat range from 50°C for 3 minutes *(Neisseria*

TABLE 11.2

Comparison of Times and Temperatures to Achieve Sterilization with Moist and Dry Heat

	Temperature	Time to Sterilize
Moist Heat	121°C	15 min
	125°C	10 min
	134°C	3 min
Dry Heat	121°C	600 min
	140°C	180 min
	160°C	120 min
	170°C	60 min

TABLE 11.3

Thermal Death Times of Various Endospores

Organism	Temperature	Time of Exposure to Kill Spores
Moist Heat		
Bacillus subtilis	121°C	1 min
B. stearothermophilis	121°C	12 min
Clostridium botulinum	120°C	10 min
Cl. tetani	105°C	10 min
Dry Heat		
Bacillus subtilis	121°C	120 min
B. stearothermophilis	140°C	5 min
Clostridium botulinum	120°C	120 min
Cl. tetani	100°C	60 min

TABLE 11.4

Average Thermal Death Times of Vegetative Stages of Microorganisms

Microbial Type	Temperature	Time (Min)
Non-spore-forming pathogenic bacteria	58°C	28
Non-spore-forming nonpathogenic bacteria	61°C	18
Vegetative stage of spore-forming bacteria	58°C	19
Fungal spores	76°C	22
Yeasts	59°C	19
Heat inactivation of viruses		
Nonenveloped	57°C	29
Enveloped	54°C	22
Protozoan trophozoites	46°C	16
Protozoan cysts	60°C	6
Worm eggs	54°C	3
Worm larvae	60°C	10

gonorrhoeae) to 60°C for 60 minutes (*Staphylococcus aureus*). It is worth noting that vegetative cells of sporeformers are just as susceptible as vegetative cells of non-sporeformers and that pathogens are neither more nor less susceptible than nonpathogens. Other microbes, including fungi (yeasts, molds, and some of their spores), protozoa, and worms, are rather similar in their sensitivity to heat. Viruses are surprisingly resistant to heat, with a tolerance range extending from 55°C for 2 to 5 minutes (adenoviruses) to 60°C for 600 minutes (hepatitis A virus). For practical purposes, all non-heat-resistant forms of bacteria, yeasts, molds, protozoa, worms, and viruses are destroyed by exposure to 80°C for 20 minutes.

Practical Concerns in the Use of Heat: Thermal Death Measurements

Adequate sterilization requires that both temperature and length of exposure be considered. As a general rule, higher temperatures allow shorter exposure times, and lower temperatures require longer exposure times. A combination of these two variables constitutes the **thermal death time,** or TDT, defined as the shortest length of time required to kill all test microbes at a specified temperature. The TDT has been experimentally determined for the microbial species that are common or important contaminants in various heat-treated materials. Another way to compare the susceptibility of microbes to heat is the **thermal death point** (TDP), defined as the lowest temperature required to kill all microbes in a sample in 10 minutes.

Many perishable substances are processed with moist heat. Some of these products are intended to remain on the shelf at room temperature for several months or even years. The chosen heat treatment must render the product free of agents of spoilage or disease. At the same time, the quality of the product and the speed and cost of processing must be considered. For example, in the commercial preparation of canned green beans, one of the cannery's greatest concerns is to prevent growth of the agent of botulism. From several possible TDTs (that is, combinations of time and temperature) for *Clostridium botulinum* spores, the cannery must choose one that kills all spores but does not turn the beans to mush. Out of these many considerations emerges an optimal TDT for a given processing method. Commercial canneries heat low-acid foods at 121°C for 30 minutes, a treatment that sterilizes these foods. Because of such strict controls in canneries, cases of botulism due to commercially canned foods are rare.

Common Methods of Moist Heat Control

The four ways that moist heat is employed to control microbes are

1. steam under pressure;
2. live, nonpressurized steam;
3. boiling water; and
4. pasteurization.

Sterilization with Steam Under Pressure At sea level, normal atmospheric pressure is 15 pounds per square inch (psi), or 1 atmosphere. At this pressure water will boil (change from a liquid to a gas) at 100°C, and the resultant steam will remain at exactly that temperature, which is unfortunately too low to reliably kill all microbes. In order to raise the temperature of steam, the pressure at which it is generated must be increased. As the pressure is increased, the temperature at which water boils and the temperature of the steam produced both rise. For example, at a pressure of 20 psi (5 psi above normal), the temperature of steam is 109°C. As the temperature is increased to 10 psi above normal, the steam's temperature rises to 115°C, and at 15 psi above normal (a total of 2 atmospheres), it will be 121°C. It is not the pressure by itself that is killing microbes but the increased temperature it produces.

Such pressure-temperature combinations can be achieved only with a special device that can subject pure steam to pressures greater than 1 atmosphere. Health and commercial industries use an *autoclave* for this purpose, and a comparable home appliance is the pressure cooker. Autoclaves have a fundamentally similar plan: a cylindrical metal chamber with an airtight door on one end and racks to hold materials (**figure 11.5**). Its construction includes a complex network of valves, pressure and temperature gauges, and ducts for regulating and measuring pressure and conducting the steam into the chamber. Sterilization is achieved when the steam condenses against the objects in the chamber and gradually raises their temperature.

Experience has shown that the most efficient pressure-temperature combination for achieving sterilization is 15 psi, which yields 121°C. It is possible to use higher pressure to reach higher temperatures (for instance, increasing the pressure to 30 psi raises the temperature 11°C), but doing so will not significantly reduce the exposure time and can harm the items being sterilized. It is important to avoid overpacking or haphazardly loading the chamber, which prevents steam from circulating freely around the contents and impedes the full contact that is necessary. The duration of the process is adjusted according to the bulkiness of the items in the load (thick bundles of material or large flasks of liquid) and how full the chamber is. The range of holding times varies from 10 minutes for light loads to 40 minutes for heavy or bulky ones; the average time is 20 minutes.

The autoclave is a superior choice to sterilize heat-resistant materials such as glassware, cloth (surgical dressings), rubber (gloves), metallic instruments, liquids, paper, some media, and some heat-resistant plastics. If the items are heat-sensitive (plastic Petri plates) but will be discarded, the autoclave is still a good choice. However, the autoclave is ineffective for sterilizing substances that repel moisture (oils, waxes, powders).

(a)

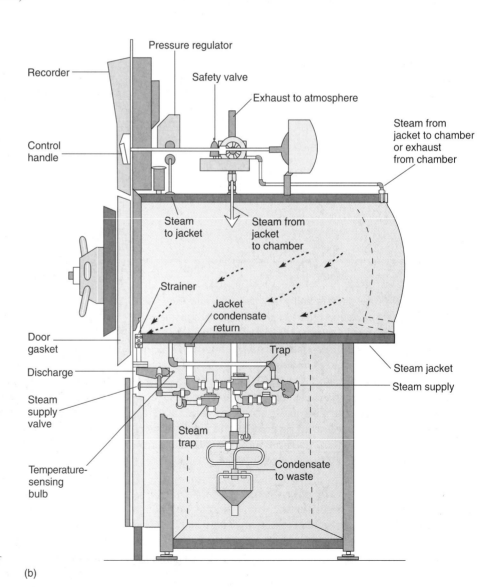

FIGURE 11.5

Steam sterilization with the autoclave.

(a) A large automatic autoclave used in sterilization by drug companies. **(b)** Cutaway section, showing autoclave components.

(b) From John J. Perkins, Principles and Methods of Sterilization in Health Science, *2nd ed., 1969. Courtesy of Charles C. Thomas, Publisher, Springfield, Illinois.*

(b)

Intermittent Sterilization Selected substances that cannot withstand the high temperature of the autoclave can be subjected to *intermittent sterilization,* also called **tyndallization.**[1] This technique requires a chamber to hold the materials and a reservoir for boiling water. Items in the chamber are exposed to free-flowing steam for 30 to 60 minutes. This temperature is not sufficient to reliably kill spores, so a single exposure will not suffice. On the assumption that surviving spores will germinate into less resistant vegetative cells, the items are incubated at appropriate temperatures for 23 to 24 hours, and then again subjected to steam treatment. This cycle is repeated for 3 days in a row. Because the temperature never gets above 100°C, highly resistant spores that do not germinate may survive even after 3 days of this treatment.

Intermittent sterilization is used most often to process heat-sensitive culture media, such as those containing sera, egg, or carbohydrates (which can break down at higher temperatures) and some canned foods. It is probably not effective in sterilizing items such as instruments and dressings that provide no environment for spore germination, but it certainly can disinfect them.

Pasteurization: Disinfection of Beverages Fresh beverages such as milk, fruit juices, beer, and wine are easily contaminated during collection and processing. Because microbes have the potential for spoiling these foods or causing illness, heat is frequently used to reduce the microbial load and destroy pathogens. **Pasteurization** is a technique in which heat is applied to liquids to kill potential agents of infection and spoilage, while at the same time retaining the liquid's flavor and food value.

Ordinary pasteurization techniques require special heat exchangers that expose the liquid to 71.6°C for 15 seconds (flash method) or to 63°–66°C for 30 minutes (batch method). The first method is preferable because it is less likely to change flavor and nutrient content, and it is more effective against certain resistant pathogens such as *Coxiella* and *Mycobacterium*. Although these treatments inactivate most viruses and destroy the vegetative stages of 97–99% of bacteria and fungi, they do not kill endospores or **thermoduric** microbes (mostly nonpathogenic lactobacilli, micrococci, and yeasts). Milk is not sterile after regular pasteurization. In fact, it can contain 20,000 microbes per milliliter or more, which explains why even an unopened carton of milk will eventually spoil. Newer techniques can also produce *sterile milk* that has a storage life of 3 months. This milk is processed with ultrahigh temperature (UHT)—134°C for 1 to 2 seconds (see chapter 26).

One important aim in pasteurization is to prevent the transmission of milk-borne diseases from infected cows or milk handlers. The primary targets of pasteurization are non-spore-forming pathogens: *Salmonella* species (a common cause of food infection), *Campylobacter jejuni* (acute intestinal infection), *Listeria monocytogenes* (listeriosis), *Brucella* species (undulant fever), *Coxiella burnetii* (Q fever), *Mycobacterium bovis, M. tuberculosis,* and several enteric viruses.

Pasteurization also has the advantage of extending milk storage time, and it can also be used by some wineries and breweries to stop fermentation and destroy contaminants.

Boiling Water: Disinfection A simple boiling water bath or chamber can quickly decontaminate items in the clinic and home. Because a single processing at 100°C will not kill all resistant cells, this method can be relied on only for disinfection and not for sterilization. Exposing materials to boiling water for 30 minutes will kill most non-spore-forming pathogens, including resistant species such as the tubercle bacillus and staphylococci. Probably the greatest disadvantage with this method is that the items can be easily recontaminated when removed from the water. Boiling is also a recommended method of disinfecting unsafe drinking water. In the home, boiling water is a fairly reliable way to sanitize and disinfect materials for babies, food preparation, and utensils, bedding, and clothing from the sickroom.

Dry Heat: Hot Air and Incineration

Dry heat is not as versatile or as widely used as moist heat, but it has several important sterilization applications. The temperatures and times employed in dry heat vary according to the particular method, but in general, they are greater than with moist heat. *Incineration* in a flame or electric heating coil is perhaps the most rigorous of all heat treatments. The flame of a Bunsen burner reaches 1,870°C at its hottest point, and furnaces/incinerators operate at temperatures of 800°–6,500°C. Direct exposure to such intense heat ignites and reduces microbes and other substances to ashes and gas.

Incineration of microbial samples on inoculating loops and needles using a Bunsen burner is a very common practice in the microbiology laboratory. This method is fast and effective, but it is also limited to metals and heat-resistant glass materials. Incinerators **(figure 11.6)** are regularly employed in hospitals and research labs for complete destruction and disposal of infectious materials such as syringes, needles, cultural materials, dressings, bandages, bedding, animal carcasses, and pathology samples.

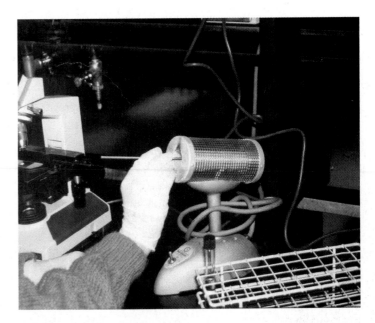

FIGURE 11.6

Dry heat incineration. Infrared incinerator with shield to prevent spattering of microbial samples during flaming.

---

1. Named for the British physicist John Tyndall who did early experiments with sterilizing procedures.

The hot-air oven provides another means of dry-heat sterilization. The so-called *dry oven* is usually electric (occasionally gas) and has coils that radiate heat within an enclosed compartment. Heated, circulated air transfers its heat to the materials in the oven. Sterilization requires exposure to 150°–180°C for 2 to 4 hours, which ensures thorough heating of the objects and destruction of spores.

The dry oven is used in laboratories and clinics for heat-resistant items that do not sterilize well with moist heat. Substances appropriate for dry ovens are glassware, metallic instruments, powders, and oils that steam does not penetrate well. This method is not suitable for plastics, cotton, and paper, which may burn at the high temperatures, or for solutions, which will dry out. Another limitation is the time required for it to work.

THE EFFECTS OF COLD AND DESICCATION

The principal benefit of cold treatment is to slow growth of cultures and microbes in food during processing and storage. *It must be emphasized that cold merely retards the activities of most microbes.* Although it is true that some microbes are killed by cold temperatures, most are not adversely affected by gradual cooling, long-term refrigeration, or deep-freezing. In fact, freezing temperatures, ranging from −70°C to −135°C, provide an environment that can preserve cultures of bacteria, viruses, and fungi for long periods. Some psychrophiles grow very slowly even at freezing temperatures and can continue to secrete toxic products. Unawareness of these facts is probably responsible for numerous cases of food poisoning from frozen foods that have been defrosted at room temperature and then inadequately cooked. Pathogens able to survive several months in the refrigerator are *Staphylococcus aureus, Clostridium* species (sporeformers), *Streptococcus* species, and several types of yeasts, molds, and viruses. Outbreaks of *Salmonella* food infection traced backed to refrigerated foods such as ice cream, eggs, and Tiramisu, are testimony to the inability of freezing temperatures to reliably kill pathogens.

Vegetative cells directly exposed to normal room air gradually become dehydrated, or **desiccated**.* More delicate pathogens such as *Streptococcus pneumoniae,* the spirochete of syphilis, and *Neisseria gonorrhoeae* can die after a few hours of air-drying, but many others are not killed and some are even preserved. Endospores of *Bacillus* and *Clostridium* are viable for millions of years under extremely arid conditions. Staphylococci and streptococci in dried secretions, and the tubercle bacillus surrounded by sputum, can remain viable in air and dust for lengthy periods. Many viruses (especially nonenveloped) and fungal spores can also withstand long periods of desiccation. Desiccation can be a valuable way to preserve foods because it greatly reduces the amount of water available to support microbial growth.

It is interesting to note that a combination of freezing and drying—**lyophilization***—is a common method of preserving mi-croorganisms and other cells in a viable state for many years. Pure cultures are frozen instantaneously and exposed to a vacuum that rapidly removes the water (it goes right from the frozen state into the vapor state). This method avoids the formation of ice crystals that would damage the cells. Although not all cells survive this process, enough of them do to permit future reconstitution of that culture.

As a general rule, chilling, freezing, and desiccation should not be construed as methods of disinfection or sterilization because their antimicrobial effects are erratic and uncertain, and one cannot be sure that pathogens subjected to them have been killed.

RADIATION AS A MICROBIAL CONTROL AGENT

Another way in which energy can serve as an antimicrobial agent is through the use of radiation. **Radiation** is defined as energy emitted from atomic activities and dispersed at high velocity through matter or space. Although radiation exists in many states and can be described and characterized in various ways, we will consider only those types suitable for microbial control: gamma rays, X rays, and ultraviolet radiation.

Modes of Action of Ionizing Versus Nonionizing Radiation

The actual physical effects of radiation on microbes can be understood by visualizing the process of **irradiation**, or bombardment with radiation, at the cellular level (**figure 11.7**). When a cell is bombarded by certain waves or particles, its molecules absorb some of the available energy, leading to one of two consequences: (1) If the radiation ejects orbital electrons from an atom, it causes ions to form. This is the effect of **ionizing radiation.** One of the most sensitive targets for ionizing radiation is the DNA molecule, which can sustain mutations on a broad scale. Secondary lethal effects appear to be chemical changes in organelles and the production of toxic substances. Gamma rays, X rays, and high-speed electrons are all ionizing in their effects. (2) **Nonionizing radiation,** best exemplified by UV, excites atoms by raising them to a higher energy state, but it does not ionize them. This atomic excitation, in turn, leads to the formation of abnormal bonds within molecules such as DNA and is thus a source of mutations (see chapter 9).

Ionizing Radiation: Gamma Rays, X Rays, and Cathode Rays

Over the past several years, ionizing radiation has become safer and more economical to use, and its applications have mushroomed. It is a highly effective alternative for sterilizing materials that are sensitive to heat or chemicals. Because it sterilizes in the absence of heat, irradiation is a type of **cold (or low-temperature) sterilization.** Devices that emit ionizing rays include gamma-ray machines containing radioactive cobalt, X-ray machines similar to those used in medical diagnosis, and cathode-ray machines that operate like the vacuum tube in a television set. Items are placed in these machines and irradiated for a short time with a carefully chosen dosage. The dosage of radiation is measured in Grays (which has replaced the older term, rads). Depending on the application, exposure ranges from 5 to 50 kiloGrays (kGray; a kiloGray is

---

*desiccate (des′-ih-kayt) To dry at normal environmental temperatures.

*lyophilization (ly-off″-il-ih-za′-shun) Gr. *lyein,* to dissolve, and *philein,* to love.

Ionizing Radiation

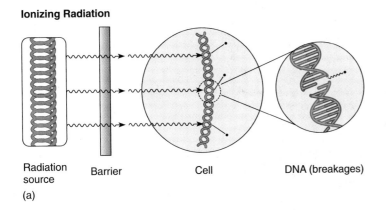

Radiation source   Barrier   Cell   DNA (breakages)

(a)

Nonionizing Radiation

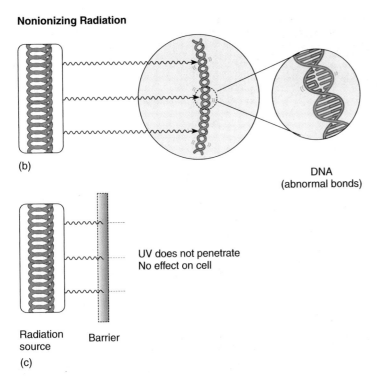

(b)

DNA (abnormal bonds)

UV does not penetrate
No effect on cell

Radiation source   Barrier

(c)

FIGURE 11.7

Cellular effects of irradiation. **(a)** Ionizing radiation can penetrate a solid barrier, bombard a cell, enter it, and dislodge electrons from molecules. Breakage of DNA creates massive mutations. **(b)** Nonionizing radiation enters a cell, strikes molecules, and excites them. The effect on DNA is mutation by formation of abnormal bonds. **(c)** A solid barrier cannot be penetrated by nonionizing radiation.

equal to 1000 Grays). Although all ionizing radiations can penetrate liquids and most solid materials, gamma rays are most penetrating, X rays are intermediate, and cathode rays least penetrating.

Applications of Ionizing Radiation

Foods have, in limited circumstances, been subject to irradiation for more than 50 years. From flour to pork, to ground beef to fruits and vegetables, radiation is used to kill not only bacterial pathogens but also insects and worms and even to inhibit the sprouting of white potatoes. As soon as radiation is mentioned, of course, comes consumer concern that food may be made less nutritious, unpalatable,

or even unsafe by subjecting it to ionizing radiation. Irradiated food has been extensively studied and each of these concerns has been addressed. Irradiation may lead to a small decrease in the amount of thiamine (vitamin B1) in food, but this change is small enough to be inconsequential. The irradiation process does produce short-lived free radical oxidants, which disappear almost immediately (this same type of chemical intermediate is produced through cooking as well). Certain foods do not irradiate well, and are not good candidates for this type of antimicrobial control. The white of eggs becomes milky and liquid, grapefruit gets mushy, and alfalfa seeds do not germinate properly. Lastly, it is important to remember that food is not made radioactive by the irradiation process and many studies, in both animals and humans, have concluded that there are no ill effects from eating irradiated food. In fact, NASA relies on irradiated meat for its astronauts.

While the potential costs of irradiation will always be debated, the potential benefits are enormous. It has been estimated that irradiation of 50% of the meat and poultry in the United States would result in 900,000 fewer cases of infection, 8,500 fewer hospitalizations, and 350 fewer deaths each year. Radiation is currently approved in the United States for the reduction of bacterial pathogens such as *E. coli* and *Salmonella* in beef and chicken, reduction of *Trichinella* worms in pork and the reduction of insects on fruits and vegetables. Officials of the United Nations and World Health Organization are also proponents of food irradiation. An additional benefit of irradiation is that microbes responsible for food spoilage are killed along with pathogens, leading to an increased shelf life. In any event, no irradiated food can be sold to consumers without clear labeling that this method has been used (**figure 11.8**). See chapter 26 for further discussion on irradiation of food.

Sterilizing medical products with ionizing radiation is a rapidly expanding field. Drugs, vaccines, medical instruments (especially plastics), syringes, surgical gloves, and tissues such as bone and skin, and heart valves for grafting all lend themselves to this mode of sterilization. Its main advantages include speed, high penetrating power (it can sterilize materials through outer packages and wrappings), and the absence of heat. Its main disadvantages are potential dangers to machine operators from exposure to radiation and possible damage to some materials.

Nonionizing Radiation: Ultraviolet Rays

Ultraviolet radiation ranges in wavelength from approximately 100 nm to 400 nm. It is most lethal from 240 nm to 280 nm (with a peak at 260 nm). In everyday practice, the source of UV radiation is the germicidal lamp, which generates radiation at 254 nm. Owing to its lower energy state, UV radiation is not as penetrating as ionizing radiation. Because UV radiation passes readily through air, slightly through liquids, and only poorly through solids, the object to be disinfected must be directly exposed to it for full effect.

As UV radiation passes through a cell, it is initially absorbed by DNA. Specific molecular damage occurs on the pyrimidine bases (thymine and cytosine), which form abnormal linkages with each other called **pyrimidine dimers (figure 11.9)**. These bonds occur between adjacent bases on the same DNA strand and interfere with normal DNA replication and transcription. The results are

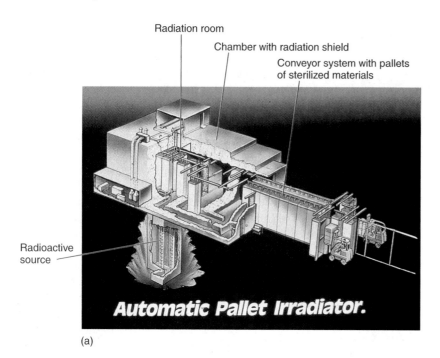

Radiation room

Chamber with radiation shield

Conveyor system with pallets
of sterilized materials

Radioactive
source

Automatic Pallet Irradiator.

(a)

(b)

FIGURE 11.8

Sterilization with ionizing radiation. **(a)** This
irradiation machine uses radioactive cobalt 60 as
a gamma radiation source to sterilize fruits,
vegetables, meats, fish, and spices. Although this
method has stirred some controversy regarding its
safety, it is gaining in acceptance because of its
ability to increase shelf-life and reduce food-borne
infections. **(b)** Regulations dictate that this
universal symbol for radioactivity must be affixed
to all irradiated materials.

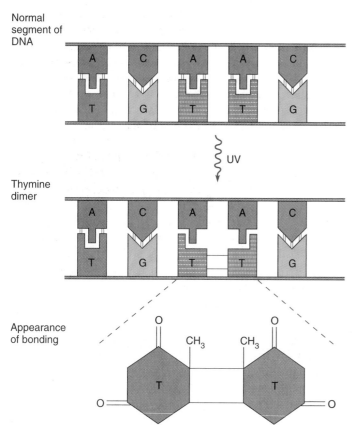

Normal
segment of
DNA

Thymine
dimer

Appearance
of bonding

FIGURE 11.9

**Formation of pyrimidine dimers by the action of ultraviolet (UV)
radiation.** This shows what occurs when two adjacent thymine bases on
one strand of DNA are induced by UV rays to bond laterally with each
other. The result is a thymine dimer shown in greater detail. Dimers can
also occur between adjacent cytosines and thymine and cytosine bases. If
they are not repaired, dimers can prevent that segment of DNA from being
correctly replicated or transcribed. Massive dimerization is lethal to cells.

inhibition of growth and cellular death. In addition to altering DNA
directly, UV radiation also disrupts cells by generating toxic photo-
chemical products called free radicals. These highly reactive molecules interfere with essential cell processes by binding to DNA,
RNA, and proteins. Ultraviolet rays are a powerful tool for destroying fungal cells and spores, bacterial vegetative cells, protozoa, and
viruses. Bacterial spores are about 10 times more resistant to radiation than are vegetative cells, but they can be killed by increasing
the time of exposure.

Applications of UV Radiation UV radiation is usually directed
at disinfection rather than sterilization. Germicidal lamps can cut
down on the concentration of airborne microbes as much as 99%.
They are used in hospital rooms, operating rooms, schools, food
preparation areas, and dental offices. Ultraviolet disinfection of air has
proved effective in reducing postoperative infections, preventing the
transmission of infections by respiratory droplets, and curtailing the
growth of microbes in food-processing plants and slaughterhouses.

Ultraviolet irradiation of liquids requires special equipment
to spread the liquid into a thin, flowing film that is exposed directly
to a lamp. This method can be used to treat drinking water (**figure
11.10**) and to purify other liquids (milk and fruit juices) as an alternative to heat. Ultraviolet treatment has proved effective in freeing
vaccines and plasma from contaminants. The surfaces of solid, nonporous materials such as walls and floors, as well as meat, nuts, tissues for grafting, and drugs, have been successfully disinfected
with UV.

One major disadvantage of UV is its poor powers of penetration through solid materials such as glass, metal, cloth, plastic, and
even paper. Another drawback to UV is the damaging effect of
overexposure on human tissues, including sunburn, retinal damage,
cancer, and skin wrinkles.

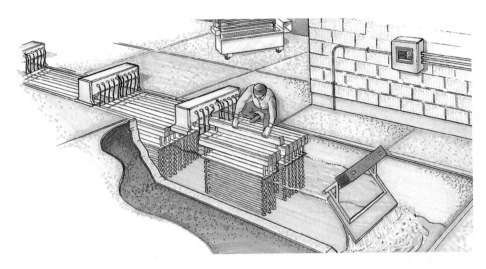

FIGURE 11.10

An ultraviolet (UV) treatment system for disinfection of wastewater. Water flows through racks of UV lamps and is exposed to 254 nm UV radiation. This system has a capacity of several million gallons per day and can be used as an alternative to chlorination.

SOUND WAVES IN MICROBIAL CONTROL

High-frequency sound (sonic) waves beyond the sensitivity of the human ear are known to disrupt cells. These frequencies range from 15,000 to more than 200,000 cycles per second (supersonic to ultrasonic). Sonication transmits vibrations through a water-filled chamber (sonicator) to induce pressure changes and create intense points of turbulence that can stress and burst cells in the vicinity. Gram-negative rods are most sensitive to ultrasonic vibrations, and gram-positive cocci, fungal spores, and bacterial spores are most resistant to them. Sonication also forcefully dislodges foreign matter from objects. Heat generated by sonic waves (up to 80°C) also appears to contribute to an antimicrobial action. Ultrasonic devices are used in dental and some medical offices to clear debris and saliva from instruments before sterilization and to clean dental restorations. However, most sonic machines are not predictable enough to be used in disinfection or sterilization. Other types of ultrasonic devices are available for medical diagnosis and for removing plaque and calculus from teeth.

STERILIZATION BY FILTRATION: TECHNIQUES FOR REMOVING MICROBES

Filtration is an effective method to remove microbes from air and liquids. In practice, a fluid is strained through a filter with openings large enough for the fluid to pass through but too small for microorganisms to pass through (**figure 11.11**).

Most modern microbiological filters are thin membranes of cellulose acetate, polycarbonate, and a variety of plastic materials (Teflon, nylon) whose pore size can be carefully controlled and standardized. Ordinary substances such as charcoal, diatomaceous earth, or unglazed porcelain are also used in some applications. Viewed microscopically, most filters are perforated by very precise, uniform pores (figure 11.11b). The pore diameters vary from coarse (8 μm) to ultrafine (0.02 μm), permitting selection of the minimum particle size to be trapped. Those with the smallest pore diameters permit true sterilization by removing viruses, and some will even remove large proteins. A sterile liquid filtrate is typically produced by suctioning the liquid through a sterile filter into a presterilized container. These filters are also used to separate mixtures of microorganisms and to enumerate bacteria in water analysis (see chapter 26).

Applications of Filtration Sterilization Filtration sterilization is used to prepare liquids that cannot withstand heat, including serum and other blood products, vaccines, drugs, IV fluids, enzymes, and media. Filtration has been employed as an alternative to sterilize milk and beer without altering their flavor. It is also an important step in water purification. Its usage extends to filtering out particulate impurities (crystals, fibers, and so on) that can cause severe reactions in the body. It has the disadvantage of not removing soluble molecules (toxins) that can cause disease. Filtration is also an efficient means of removing airborne contaminants that are a common source of infection and spoilage. High-efficiency particulate air (HEPA) filters are widely used to provide a flow of sterile air to hospital rooms and sterile rooms. A vacuum with a HEPA filter was even used to remove anthrax spores from the Senate offices most heavily contaminated after the terrorist attack in late 2001.

CHAPTER CHECKPOINTS

Physical methods of microbial control include heat, cold, radiation, and drying.

Heat is the most widely used method of microbial control. It is used in combination with water (moist heat) or as dry heat (oven, flames).

The thermal death time (TDT) is the shortest length of time required to kill all microbes at a specific temperature. The TDT is longest for spore-forming bacteria and certain viruses.

The thermal death point (TDP) is the lowest temperature at which all microbes are killed in a specified length of time (10 minutes).

Autoclaving, or steam sterilization, is the process by which steam is heated under pressure to sterilize a wide range of materials in a comparatively short time (minutes to hours). It is effective for most materials except water-resistant substances such as oils, waxes, and powders.

Boiling water and pasteurization of beverages disinfect but do not sterilize materials.

Dry heat is microbicidal under specified times and temperatures. Flame heat, or incineration, is microbicidal. It is used when total destruction of microbes and materials is desired.

Chilling, freezing, and desiccation are microbistatic but not microbicidal. They are not considered true methods of disinfection because they are not consistent in their effectiveness.

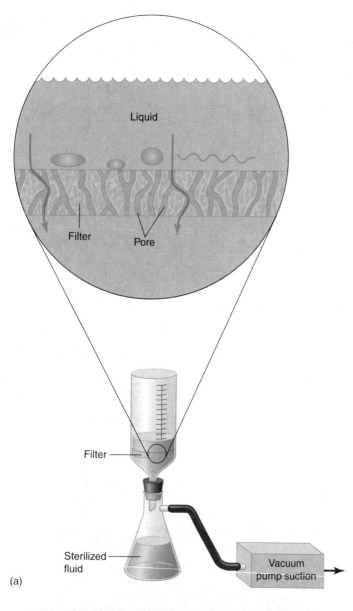

(a)

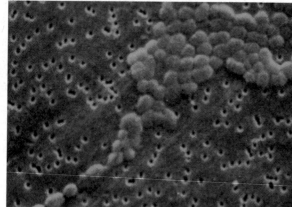

(b)

FIGURE 11.11

Membrane filtration. **(a)** Vacuum assembly for achieving filtration of liquids through suction. Inset shows filter as seen in cross section, with tiny passageways (pores) too small for the microbial cells to enter but large enough for liquid to pass through. **(b)** Scanning electron micrograph of filter, showing relative size of pores and bacteria trapped on its surface (5,900×).

Ionizing radiation (cold sterilization) by gamma rays and X rays is used to sterilize medical products, meats, and spices. It damages DNA and cell organelles by producing disruptive ions.

Ultraviolet light, or nonionizing radiation, has limited penetrating ability. It is therefore restricted to disinfecting air and certain liquids.

Ultrasound is microbistatic to most microbes, but it is microbicidal to gram-negative bacteria. It is used primarily to reduce microbial load from inanimate objects.

Sterilization by filtration removes microbes from heat-sensitive liquids and circulating air. The pore size of the filter determines what kinds of microbes are removed.

Chemical Agents in Microbial Control

Chemical control of microbes probably emerged as a serious science in the early 1800s, when physicians used chloride of lime and iodine solutions to treat wounds and to wash their hands before surgery. At the present time, approximately 10,000 different antimicrobial chemical agents are manufactured; probably 1,000 of them are used routinely in the allied health sciences and the home. There is a genuine need to avoid infection and spoilage, but the abundance of products available to "kill germs, disinfect, antisepticize, clean and sanitize, deodorize, fight plaque, and purify the air" indicates a preoccupation with eliminating microbes from the environment that, at times, seems excessive (**Spotlight on Microbiology 11.2**).

Antimicrobial chemicals occur in the liquid, gaseous, or even solid state and vary from disinfectants and antiseptics to sterilants and preservatives (chemicals that inhibit the deterioration of substances). For the sake of convenience (and sometimes safety) many solid or gaseous antimicrobial chemicals are dissolved in water, alcohol, or a mixture of the two to produce a liquid solution. Solutions containing pure water as the solvent are termed *aqueous,* whereas those dissolved in pure alcohol or water-alcohol mixtures are termed **tinctures.**

CHOOSING A MICROBICIDAL CHEMICAL

The choice and appropriate use of antimicrobial chemical agents is of constant concern in medicine and dentistry. Although actual clinical practices of chemical decontamination vary widely, some desirable qualities in a germicide have been identified, including:

1. rapid action even in low concentrations,
2. solubility in water or alcohol and long-term stability,
3. broad-spectrum microbicidal action without being toxic to human and animal tissues,
4. penetration of inanimate surfaces to sustain a cumulative or persistent action,
5. resistance to becoming inactivated by organic matter,
6. noncorrosive or nonstaining properties,
7. sanitizing and deodorizing properties, and
8. inexpensiveness and ready availability.

As yet, no chemical can completely fulfill all of those requirements, but glutaraldehyde and hydrogen peroxide approach this ideal. At

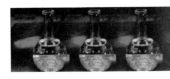

SPOTLIGHT ON MICROBIOLOGY 11.2
Pathogen Paranoia: "The Only Good Microbe Is a Dead Microbe"

The sensational publicity over outbreaks of infections such as influenza, anthrax, and microbial food poisoning has monumentally influenced the public view of microorganisms. Thousands of articles have sprinkled the news services over the past five years. On the positive side, this glut of information has improved people's awareness of the importance of microorganisms. And, certainly, such knowledge can be seen as beneficial when it leads to well-reasoned and sensible choices, such as using greater care in handwashing, food handling, and personal hygiene. But sometimes a little knowledge can be dangerous. The trend also seems to have escalated into an obsessive fear of "germs" lurking around every corner and a fixation on eliminating microbes from the environment and the human body.

As might be expected, commercial industries have found a way to capitalize on those fears. Every year, the number of products that incorporate antibacterial or germicidal "protection" increases dramatically. A widespread array of cleansers and commonplace materials have already had antimicrobic chemicals added. First it was hand soaps and dishwashing detergents, and eventually the list grew to include shampoos, laundry aids, hand lotions, foot pads for shoes, deodorants, sponges and scrub pads, kitty litter, acne medication, cutting boards, garbage bags, toys, and toothpaste.

By far, the prevalent chemical agent routinely added to these products is a phenolic called *triclosan* (Irgasan). This substance is fairly mild and nontoxic and does indeed kill most pathogenic bacteria. However, it does not reliably destroy viruses or fungi and has been linked to cases of skin rashes due to hypersensitivity.

One unfortunate result of the negative news on microbes is how it fosters the feeling that all microbes are harmful, We must not forget that most human beings manage to remain healthy despite the fact that they live in continual intimate contact with microorganisms. We really do not have to be preoccupied with microbes every minute or feel overly concerned that the things we touch, drink, or eat are sterile, as long as they are somewhat clean and free of pathogens. For most of us, resistance to infection is well maintained by our numerous host defenses.

Medical experts are concerned that the widespread overuse of these antibacterial chemicals could favor the survival and growth of resistant strains of bacteria. A study reported in 2000 that many pathogens such as *Mycobacterium tuberculosis* and *Pseudomonas* are naturally resistant to triclosan, and that *E. coli* and *Staphylococcus aureus* have already demonstrated decreased sensitivity to it. The widespread use of this chemical may actually select for "super microbes" that survive ordinary disinfection. Another outcome of overuse of environmental germicides is to reduce the natural contact with microbes that is required to maintain the normal resident flora and stimulate immunities. Constant use of these agents could shift the balance in the normal flora of the body by killing off harmless or beneficial microbes.

Infectious disease specialists urge a happy medium approach. Instead of filling the home with questionable germicidal products, they encourage cleaning with traditional soaps and detergents, reserving more potent products to reduce the spread of infection among household members.

The molecular structure of triclosan, also known as Irgasan and Ster-Zac, a phenol-based chemical that destroys bacteria by disrupting cell walls and membranes.

the same time, we should question the rather overinflated claims made about certain commercial agents such as mouthwashes and disinfectant air sprays.

Germicides are evaluated in terms of their effectiveness in destroying microbes in medical and dental settings. The three levels of chemical decontamination procedures are *high, intermediate,* and *low* (**table 11.5**). High-level germicides kill endospores, and, if properly used, are sterilants. Materials that necessitate high-level control are medical devices—for example, catheters, heart-lung equipment, and implants—that are not heat-sterilizable and are intended to enter body tissues during medical procedures. Intermediate-level germicides kill fungal (but not bacterial) spores, resistant pathogens such as the tubercle bacillus, and viruses. They are used to disinfect items (respiratory equipment, thermometers) that come into intimate contact with the mucous membranes but are noninvasive. Low levels of disinfection eliminate only vegetative bacteria, vegetative fungal cells, and some viruses. They are used to clean materials such as electrodes, straps, and furniture that touch the skin surfaces but not the mucous membranes.

FACTORS THAT AFFECT THE GERMICIDAL ACTIVITY OF CHEMICALS

Factors that control the effect of a germicide include the nature of the microorganisms being treated, the nature of the material being treated, the degree of contamination, the time of exposure, and the strength and chemical action of the germicide (**table 11.6**). Standardized procedures for testing the effectiveness of germicides are summarized in appendix C. The modes of action of most germicides involve the cellular targets discussed in an earlier section of this chapter: proteins, nucleic acids, the cell wall, and the cell membrane.

A chemical's strength or concentration is expressed in various ways, depending upon convention and the method of preparation. The content of many chemical agents can be expressed by more than one notation. In dilutions, a small volume of the liquid chemical (solute) is diluted in a larger volume of solvent to achieve a certain ratio. For example, a common laboratory phenolic disinfectant such as Lysol is usually diluted 1:200; that is, one part of chemical has been added to 200 parts of water by volume. Solutions

---

TABLE 11.5

Qualities of Chemical Agents Used in Health Care

Agent	Target Microbes	Level of Activity	Toxicity	Comments
Chlorine	Sporicidal (slowly)	Intermediate	Gas is highly toxic; solution irritates skin	Inactivated by organics; unstable in sunlight
Iodine	Sporicidal (slowly)	Intermediate	Can irritate tissue; toxic if ingested	Iodophors★ are milder forms
Phenolics	Some bacteria, viruses, fungi	Intermediate to low	Can be absorbed by skin; can cause CNS damage	Poor solubility; expensive
Alcohols	Most bacteria, viruses, fungi	Intermediate	Toxic if ingested; a mild irritant; dries skin	Inflammable, fast-acting
Hydrogen peroxide,★ stabilized	Sporicidal	High	Toxic to eyes; toxic if ingested	Improved stability; works well in organic matter
Quaternary ammonium compounds	Some bactericidal, virucidal, fungicidal activity	Low	Irritating to mucous membranes; poisonous if taken internally	Weak solutions can support microbial growth; easily inactivated
Soaps	Certain very sensitive species	Very low	Nontoxic; few if any toxic effects	Used for removing soil, oils, debris
Mercurials	Weakly microbistatic	Low	Highly toxic if ingested, inhaled, absorbed	Easily inactivated
Silver nitrate	Bactericidal	Low	Toxic, irritating	Discolors skin
Glutaraldehyde★	Sporicidal	High	Can irritate skin; toxic if absorbed	Not inactivated by organic matter; unstable
Formaldehyde	Sporicidal	High to intermediate	Very irritating; fumes damaging, carcinogenic	Slow rate of action; limited applications
Ethylene oxide gas★	Sporicidal	High	Very dangerous to eyes, lungs; carcinogenic	Explosive in pure state; good penetration; materials must be aerated
Dyes	Weakly bactericidal, fungicidal	Low	Low toxicity	Stain materials, skin
Chlorhexidine★	Most bacteria, some viruses, fungi	Low to intermediate	Low toxicity	Fast-acting, mild, has residual effects

★*These forms approach the ideal by having many of the following characteristics: broad spectrum, low toxicity, fast action, penetrating abilities, residual effects, stability, potency in organic matter, and solubility.*

such as chlorine that are effective in very high dilutions are expressed in parts per million (ppm). In percent solutions, the solute is added to water by weight or volume to achieve a certain percentage in the solution. Alcohol, for instance, is used in percentages ranging from 50% to 95%. In general, solutions of low dilution or high percentage have more of the active chemical (are more concentrated) and tend to be more germicidal, but expense and potential toxicity can necessitate using the minimum strength that is effective.

Another factor that contributes to germicidal effectiveness is the length of exposure. Most compounds require adequate contact time to allow the chemical to penetrate and to act on the microbes present. The composition of the material being treated must also be considered. Smooth, solid objects are more reliably disinfected than are those with pores or pockets that can trap soil. An item contaminated with common biological matter such as serum, blood, saliva, pus, fecal material, or urine presents a problem in disinfection. Large amounts of organic material can hinder the penetration of a disinfectant and, in some cases, form bonds that reduce its activity. Adequate cleaning of instruments and other reusable materials ensures that the germicide or sterilant will better accomplish the job for which it was chosen.

GERMICIDAL CATEGORIES ACCORDING TO CHEMICAL GROUP

Several general groups of chemical compounds are widely used for antimicrobial purposes in medicine and commerce (see table 11.5). Prominent agents include halogens, heavy metals, alcohols, phenolic compounds, oxidizers, aldehydes, detergents, and gases. These groups will be surveyed in the following section from the standpoint of each agent's specific forms, modes of action, indications for use, and limitations.

The Halogen Antimicrobial Chemicals

The **halogens**★ are fluorine, bromine, chlorine, and iodine, a group of nonmetallic elements, all of which are found in group VII of the periodic table. Although they can exist in either the ionic (halide) or nonionic state, most halogens exert their antimicrobial effect primarily in the nonionic state, not the halide state (chloride, iodide, for example). Because fluorine and bromine are difficult and dangerous

---

★halogens (hay'-loh-jenz) Gr. *halos,* salt, and *gennan,* to produce.

TABLE 11.6

Required Concentrations and Times for Chemical Destruction of Selected Microbes

Organism	Concentration	Time
Agent: Aqueous Iodine		
Staphylococcus aureus★	2%	2 min
Escherichia coli★★	2%	1.5 min
Enteric viruses	2%	10 min
Agent: Chlorine		
Mycobacterium tuberculosis★	50 ppm	50 sec
Entamoeba cysts (protozoa)	0.1 ppm	150 min
Hepatitis A virus	3 ppm	30 min
Agent: Phenol		
Staphylococcus aureus	1:85 dil	10 min
Escherichia coli	1:75 dil	10 min
Agent: Ethyl Alcohol		
Staphylococcus aureus	70%	10 min
Escherichia coli	70%	2 min
Poliovirus	70%	10 min
Agent: Hydrogen Peroxide		
Staphylococcus aureus	3%	12.5 sec
Neisseria gonorrhoeae	3%	0.3 sec
Herpes simplex virus	3%	12.8 sec
Agent: Quaternary Ammonium Compound		
Staphylococcus aureus	450 ppm	10 min
Salmonella typhi★★	300 ppm	10 min
Agent: Silver Ions		
Staphylococcus aureus	8 µg/ml	48 h
Escherichia coli	2 mg/ml	48 h
Candida albicans (yeast)	14 mg/ml	48 h
Agent: Glutaraldehyde		
Staphylococcus aureus	2%	<1 min
Mycobacterium tuberculosis	2%	<10 min
Herpes simplex virus	2%	<10 min
Agent: Ethylene Oxide Gas		
Streptococcus faecalis	500 mg/l	2–4 min
Influenza virus	10,000 mg/l	25 h
Agent: Chlorhexidine		
Staphylococcus aureus	1:10 dil	15 sec
Escherichia coli	1:10 dil	30 sec

★*Gram-positive vegetative bacterium.*
★★*Gram-negative vegetative bacterium.*

to handle, and are no more effective than chlorine and iodine, only the latter two are used routinely in germicidal preparations. These elements are highly effective components of disinfectants and antiseptics because they are microbicidal and not just microbistatic, and they are sporicidal with longer exposure. For these reasons, halogens are the active ingredients in nearly one-third of all antimicrobial chemicals currently marketed.

Chlorine and Its Compounds   Chlorine has been used for disinfection and antisepsis for approximately 200 years. The major forms used in microbial control are liquid and gaseous chlorine (Cl_2), hypochlorites (OCl), and chloramines (NH_2Cl). In solution, these compounds combine with water and release hypochlorous acid (HOCl), which oxidizes the sulfhydryl (S—H) group on the amino acid cysteine and interferes with disulfide (S—S) bridges on numerous enzymes. The resulting denaturation of the enzymes is permanent and suspends metabolic reactions. Chlorine kills not only bacterial cells and endospores but also fungi and viruses. Chlorine compounds are less effective and relatively unstable, if exposed to light, alkaline pH, and excess organic matter.

Chlorine Compounds in Disinfection and Antisepsis   Gaseous and liquid chlorine are used almost exclusively for large-scale disinfection of drinking water, sewage, and wastewater from such sources as agriculture and industry. Chlorination to a concentration of 0.6 to 1.0 parts of chlorine per million parts of water will ensure that water is safe to drink. This rids the water of most pathogenic vegetative microorganisms without unduly affecting its taste (some persons may debate this).

Hypochlorites are perhaps the most extensively used of all chlorine compounds. The scope of applications is broad, including sanitization and disinfection of food equipment in dairies, restaurants, and canneries and treatment of swimming pools, spas, drinking water, and even fresh foods. Hypochlorites are used in the allied health areas to treat wounds and to disinfect equipment, bedding, and instruments. Common household bleach is a weak solution (5%) of sodium hypochlorite that serves as an all-around disinfectant, deodorizer, and stain remover.

Chloramines (dichloramine, halazone) are being employed more frequently as an alternative to pure chlorine in treating water supplies. Because standard chlorination of water is now believed to produce unsafe levels of cancer-causing substances such as trihalomethanes, some water districts have been directed by federal agencies to adopt chloramine treatment of water supplies. Chloramines also serve as sanitizers and disinfectants and for treating wounds and skin surfaces.

Iodine and Its Compounds   Iodine is a pungent black chemical that forms brown-colored solutions when dissolved in water or alcohol. The two primary iodine preparations are *free iodine* in solution (I_2) and *iodophors*. Iodine rapidly penetrates the cells of microorganisms, where it apparently disturbs a variety of metabolic functions by interfering with the hydrogen and disulfide bonding of proteins (a mode of action similar to chlorine). All classes of microorganisms are killed by iodine if proper concentrations and exposure times are used. Iodine activity is not as adversely affected by organic matter and pH as chlorine is.

Applications of Iodine Solutions   Aqueous iodine contains 2% iodine and 2.4% sodium iodide; it is used as a topical antiseptic before surgery and occasionally as a treatment for burned and infected skin. A stronger iodine solution (5% iodine and 10% potassium iodide) is used primarily as a disinfectant for plastic items, rubber instruments, cutting blades, thermometers, and other inanimate items. Iodine tincture is a 2% solution of iodine and sodium iodide in 70% alcohol that can be used in skin antisepsis. Because iodine can be extremely irritating to the skin and toxic when absorbed, strong aqueous solutions and tinctures (5–7%) are no longer considered safe for routine antisepsis. Iodine tablets are available for disinfecting water during emergencies or destroying pathogens in impure water supplies.

Iodophors are complexes of iodine and a neutral polymer such as a polyvinylalcohol. This formulation allows the slow release of free iodine and increases its degree of penetration. These compounds have largely replaced free iodine solutions in medical antisepsis because they are less prone to staining or irritating tissues. Common iodophor products marketed as Betadine, Povidone (PVP), and Isodine contain 2–10% of available iodine. They are used to prepare skin and mucous membranes for surgery and injections, in surgical handscrubs, to treat burns, and to disinfect equipment and surfaces. A recent study showed that Betadine solution is an effective means of preventing eye infections in newborn infants, and it may replace antibiotics and silver nitrate as the method of choice.

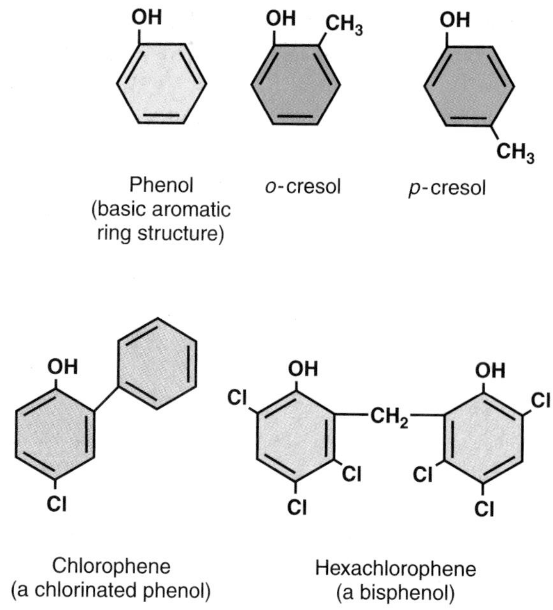

Phenol and Its Derivatives

Phenol (carbolic acid) is an acrid, poisonous compound derived from the distillation of coal tar. First adopted by Joseph Lister in 1867 as a surgical germicide, phenol was the major antimicrobial chemical until other phenolics with fewer toxic and irritating effects were developed. Solutions of phenol are now used only in certain limited cases, but it remains one standard against which other phenolic disinfectants are rated through the use of a *phenol coefficient*, which quantitatively compares a chemical's antimicrobic properties to those of phenol (see appendix C). Substances chemically related to phenol are often referred to as phenolics. Hundreds of these chemicals are now available.

Phenolics consist of one or more aromatic carbon rings with added functional groups **(figure 11.12).** Among the most important are alkylated phenols (cresols), chlorinated phenols, and bisphenols. In high concentrations, they are cellular poisons, rapidly disrupting cell walls and membranes and precipitating proteins; in lower concentrations, they inactivate certain critical enzyme systems. The phenolics are strongly microbicidal and will destroy vegetative bacteria (including the tubercle bacillus), fungi, and most viruses (not hepatitis B), but they are not reliably sporicidal. Their continued activity in the presence of organic matter and their detergent actions contribute to their usefulness. Unfortunately, the toxicity of many of the phenolics makes them too dangerous to use as antiseptics.

Applications of Phenolics   Phenol itself is still used for general disinfection of drains, cesspools, and animal quarters, but it is seldom applied as a medical germicide. The cresols are simple phenolic derivatives that are combined with soap for intermediate or low levels of disinfection in the hospital. Lysol and creolin, in a 1–3% emulsion, are common household versions of this type.

The bisphenols are also widely employed in commerce, clinics, and the home. One type, orthophenyl phenol, is the major ingredient in disinfectant aerosol sprays. This same phenolic is also found in some proprietary compounds (Lysol) often used in hospital and laboratory disinfection. One particular bisphenol,

FIGURE 11.12

Some phenolics.   All contain a basic aromatic ring, but they differ in the types of additional compounds such as Cl and CH_3.

hexachlorophene, was once a common additive of cleansing soaps (pHisoHex) used in the hospital and home. When hexachlorophene was found to be absorbed through the skin and a cause of neurological damage, it was no longer available without a prescription. It is occasionally used to control outbreaks of skin infections.

Perhaps the most widely used phenolic is *triclosan,* chemically known as dichlorophenoxyphenol (see Spotlight on Microbes 11.2). It is the antibacterial compound added to dozens of products, from soaps to kitty litter. It acts as both a disinfectant and antiseptic and is broad-spectrum in its effects.

Chlorhexidine

The compound chlorhexidine (Hibiclens, Hibitane) is a complex organic base containing chlorine and two phenolic rings. Its mode of action targets both cell membranes (lowering surface tension until selective permeability is lost) and protein structure (causing denaturation). At moderate to high concentrations, it is bactericidal for both gram-positive and gram-negative bacteria but inactive against spores. Its effects on viruses and fungi vary. It possesses distinct advantages over many other antiseptics because of its mildness, low toxicity, and rapid action, and it is not absorbed into deeper tissues to any extent. Alcoholic or aqueous solutions of chlorhexidine are now commonly used for handscrubbing, preparing skin sites for surgical incisions and injections, and whole body washing. Chlorhexidine solution also serves as an obstetric antiseptic, a neonatal wash, a wound degermer, a mucous membrane irrigant, and a preservative for eye solutions.

Alcohols As Antimicrobial Agents

Alcohols are colorless hydrocarbons with one or more —OH functional groups. Of several alcohols available, only ethyl and isopropyl are suitable for microbial control. Methyl alcohol is not very microbicidal, and more complex alcohols are either poorly soluble in water or too expensive for routine use. Alcohols are employed alone in aqueous solutions or as solvents for tinctures (iodine, for example). Alcohol's mechanism of action depends in part upon its concentration. Concentrations of 50% and higher dissolve membrane lipids, disrupt cell surface tension, and compromise membrane integrity. Alcohol that has entered the protoplasm denatures proteins through coagulation, but only in alcohol-water solutions of 50–95%. Alcohol is the exception to the rule that higher concentrations of an antimicrobial chemical have greater microbicidal activity. Because water is needed for proteins to coagulate, alcohol shows a greater microbicidal activity at 70% concentration (that is, 30% water) than at 100% (0% water). Absolute alcohol (100%) dehydrates cells and inhibits their growth but is generally not a protein coagulant.

Although useful in intermediate- to low-level germicidal applications, alcohol does not destroy bacterial spores at room temperature. Alcohol can, however, destroy resistant vegetative forms, including tubercle bacilli and fungal spores, provided the time of exposure is adequate. Alcohol is generally more effective in inactivating enveloped viruses than the more resistant nonenveloped viruses such as poliovirus and hepatitis A virus.

Applications of Alcohols Ethyl alcohol, also called ethanol or grain alcohol, is known for being germicidal, nonirritating, and inexpensive. Solutions of 70% to 95% are routinely used as skin degerming agents because the surfactant action removes skin oil, soil, and some microbes sheltered in deeper skin layers. One limitation to its effectiveness is the rate at which it evaporates. Ethyl alcohol is occasionally used to disinfect electrodes, face masks, and thermometers, which are first cleaned and then soaked in alcohol for 15 to 20 minutes. Isopropyl alcohol, sold as rubbing alcohol, is even more microbicidal and less expensive than ethanol, but these benefits must be weighed against its toxicity. It must be used with caution in disinfection or skin cleansing, because inhalation of its vapors can adversely affect the nervous system.

Hydrogen Peroxide and Related Germicides

Hydrogen peroxide (H_2O_2) is a colorless, caustic liquid that decomposes in the presence of light, metals, or catalase into water and oxygen gas. Early formulations were unstable and inhibited by organic matter, but manufacturing methods now permit synthesis of H_2O_2 so stable that even dilute solutions retain activity through several months of storage.

The germicidal effects of hydrogen peroxide are due to the direct and indirect actions of oxygen. Oxygen forms hydroxyl free radicals (—OH), which, like the superoxide radical (see chapter 7), are highly toxic and reactive to cells. Although most microbial cells produce catalase to inactivate the metabolic hydrogen peroxide, it cannot neutralize the amount of hydrogen peroxide entering the cell during disinfection and antisepsis. Hydrogen peroxide is bactericidal, virucidal, and fungicidal and, in higher concentrations, sporicidal.

Applications of Hydrogen Peroxide As an antiseptic, 3% hydrogen peroxide serves a variety of needs, including skin and wound cleansing, bedsore care, and mouthwashing. It is especially useful in treating infections by anaerobic bacteria because of the

lethal effects of oxygen on these forms. Hydrogen peroxide is also a versatile disinfectant for soft contact lenses, surgical implants, plastic equipment, utensils, bedding, and room interiors.

A number of clinical procedures involve delicate reusable instruments such as endoscopes and dental handpieces. Because these devices can become heavily contaminated by tissues and fluids, they really need to be sterilized between patients to prevent transmission of infections such as hepatitis, tuberculosis, and genital warts. These very effective and costly diagnostic tools (a colonoscope may cost up to $30,000) have created another dilemma. They may trap infectious agents where they cannot be easily removed and they are delicate, complex, and difficult to clean. Traditional methods are either too harsh (heat) to protect the instruments from damage or too slow (ethylene oxide) to sterilize them in a timely fashion between patients. This need for effective rapid sterilization has led to the development of low-temperature sterilizing cabinets that contain liquid chemical sterilants (**figure 11.13**). The major types of chemical sterilants used in these machines are powerful oxidizing agents such as hydrogen peroxide (35%) and peracetic acid (35%) that penetrate into delicate machinery, kill the most resistant microbes, and do not corrode or damage the working parts.

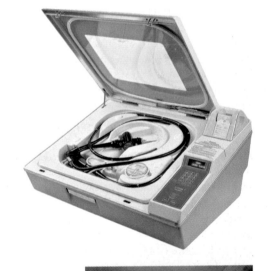

(a) A cabinet for rapid (within 30 minutes) sterile processing of endoscopes and other microsurgical instruments

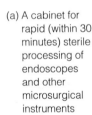

(b) Close-up view of a colonoscope tip, showing ports used in its operation (pencil for size comparison)

FIGURE 11.13

Sterile processing of invasive equipment protects patients.

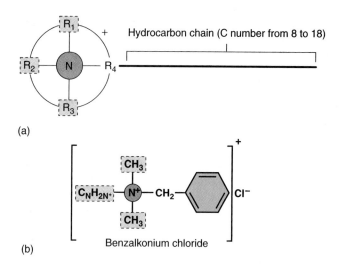

(a)

(b)

FIGURE 11.14

The structure of detergents. **(a)** In general, detergents are polar molecules with a positively charged head and at least one long, uncharged hydrocarbon chain. The head contains a central nitrogen nucleus with various alkyl (R) groups attached. **(b)** A common quaternary ammonium detergent, benzalkonium chloride.

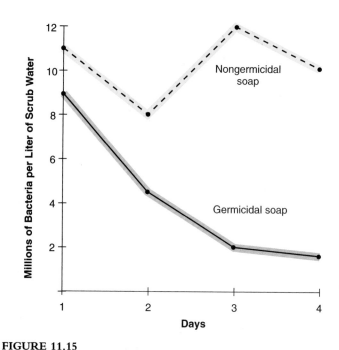

FIGURE 11.15

Graph showing effects of handscrubbing. Comparison of scrubbing over several days with a nongermicidal soap versus a germicidal soap. Germicidal soap has persistent effects on skin over time, keeping the microbial count low. Without germicide, soap does not show this sustained effect.

Vaporized hydrogen peroxide is currently being used as a sterilant in enclosed areas. Hydrogen peroxide plasma sterilizers exist for those applications involving small industrial or medical items. For larger enclosed spaces, such as isolators and passthrough rooms, peroxide generators can be used to fill a room with hydrogen peroxide vapors at concentrations high enough to be sporicidal.

Another compound with effects similar to those of hydrogen peroxide is ozone (O_3), used to disinfect air, water, and industrial air conditioners and cooling towers.

Chemicals with Surface Action: Detergents

Detergents are polar molecules that act as surfactants. Most anionic detergents have limited microbicidal power. This includes most soaps. Much more effective are positively charged (cationic) detergents, particularly the quaternary ammonium compounds (usually shortened to *quats*).

The activity of cationic detergents arises from the amphipathic (two-headed) nature of the molecule. The positively charged end binds well with the predominantly negatively charged bacterial surface proteins while the long, uncharged hydrocarbon chain allows the detergent to disrupt the cell membrane **(figure 11.14)**. Eventually the cell membrane loses selective permeability, leading to the death of the cell. Several other effects are seen but the loss of integrity of the cell membrane is most important.

The effects of detergents are varied. When used at high enough concentrations, quaternary ammonium compounds are effective against some gram-positive bacteria, viruses, fungi, and algae. In low concentrations they exhibit only microbistatic effects. Drawbacks to the quats include their ineffectiveness against the tubercle bacillus, hepatitis virus, *Pseudomonas,* and spores at any concentration. Furthermore, their activity is greatly reduced in the presence of organic matter, and they function best in alkaline solutions. As a result of these limitations, quats are rated only for low-level disinfection in the clinical setting.

Applications of Detergents and Soaps Quaternary ammonium compounds include benzalkonium chloride, Zephiran, and cetylpyridinium chloride (Ceepryn). In dilutions ranging from 1:100 to 1:1,000, quats are mixed with cleaning agents to simultaneously disinfect and sanitize floors, furniture, equipment surfaces, and restrooms. They are used to clean restaurant eating utensils, food-processing equipment, dairy equipment, and clothing. They are common preservatives for ophthalmic solutions and cosmetics. Their level of disinfection is far too low for disinfecting medical instruments.

Soaps are alkaline compounds made by combining the fatty acids in oils with sodium or potassium salts. In usual practice, soaps are only weak microbicides, and they destroy only highly sensitive forms such as the agents of gonorrhea, meningitis, and syphilis. The common hospital pathogen *Pseudomonas* is so resistant to soap that various species grow abundantly in soap dishes. Soaps function primarily as cleansing agents and sanitizers in industry and the home. The superior sudsing and wetting properties of soaps help to mechanically remove large amounts of surface soil, greases, and other debris that contains microorganisms. Soaps gain greater germicidal value when mixed with agents such as chlorhexidine or iodine. They can be used for cleaning instruments before heat sterilization, degerming patients' skin, routine handwashing by medical and dental personnel, and preoperative handscrubbing. Vigorously brushing the hands with germicidal soap over a 15-minute period is an effective way to remove dirt, oil, and surface contaminants as well as some resident microbes, but it will never sterilize the skin **(Historical Highlights 11.3** and **figure 11.15).**

3. large quantities of biological fluids and wastes neutralize their actions; and

4. microbes can develop resistance to metals.

Health and environmental considerations have dramatically reduced the use of metallic antimicrobic compounds in medicine, dentistry, commerce, and agriculture.

Applications of Heavy Metals Weak (0.001% to 0.2%) organic mercury tinctures such as thimerosal (Merthiolate) and nitromersol (Metaphen) are fairly effective antiseptics and infection preventives, but they should never be used on broken skin because they are harmful and can delay healing. The organic mercurials also serve as preservatives in cosmetics and ophthalmic solutions. Mercurochrome, that old staple of the medicine cabinet, is now considered among the poorest of antiseptics.

A silver compound with several applications is silver nitrate ($AgNO_3$) solution. Crede introduced it in the late nineteenth century for preventing gonococcal infections in the eyes of newborn infants who had been exposed to an infected birth canal. This preparation is not used as often now because many pathogens are resistant to it. It has been replaced by antibiotics in most instances. Solutions of silver nitrate (1–2%) can also be used as topical germicides on mouth ulcers and occasionally root canals. Silver sulfadiazine ointment, when added to dressings, effectively prevents infection in second- and third-degree burn patients, and pure silver is now incorporated into catheters to prevent urinary tract infections in the hospital. Colloidal silver preparations are mild germicidal ointments or rinses for the mouth, nose, eyes, and vagina.

For a look at the antimicrobial chemicals found in some common household products, see **table 11.7.**

Aldehydes As Germicides

Organic substances bearing a—CHO functional group (a strong reducing group) on the terminal carbon are called aldehydes. Several common substances such as sugars and some fats are technically aldehydes. The two aldehydes used most often in microbial control are *glutaraldehyde* and *formaldehyde.*

Glutaraldehyde is a yellow acidic liquid with a mild odor. The molecule's two aldehyde groups favor the formation of polymers. The mechanism of activity involves cross-linking protein molecules on the cell surface. In this process, amino acids are alkylated, meaning that a hydrogen atom on an amino acid is replaced by the glutaraldehyde molecule itself (**figure 11.17**). It can also irreversibly disrupt the activity of enzymes within the cell. Glutaraldehyde is rapid and broad-spectrum, and is one of the few chemicals officially accepted as a sterilant and high-level disinfectant. It kills spores in 3 hours and fungi and vegetative bacteria (even *Mycobacterium* and *Pseudomonas*) in a few minutes. Viruses, including the most resistant forms, appear to be inactivated after relatively short exposure times. Glutaraldehyde retains its potency even in the presence of organic matter, is noncorrosive, does not damage plastics, and is less toxic or irritating than formaldehyde. Its principal disadvantage is that it is somewhat unstable, especially with increased pH and temperature.

Formaldehyde is a sharp, irritating gas that readily dissolves in water to form an aqueous solution called **formalin.** Pure formalin

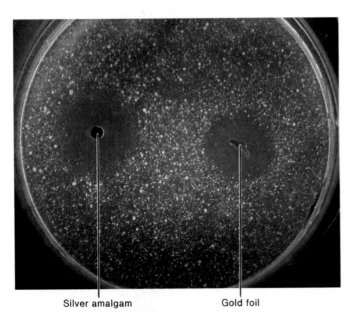

Silver amalgam   Gold foil

FIGURE 11.16

Demonstration of the oligodynamic action of heavy metals. A pour plate inoculated with saliva has small fragments of heavy metals pressed lightly into it. During incubation, clear zones indicating growth inhibition developed around both fragments. The slightly larger zone surrounding the amalgam (used in tooth fillings) probably reflects the synergistic effect of the silver and mercury it contains.

Heavy Metal Compounds

Various forms of the metallic elements mercury, silver, gold, copper, arsenic, and zinc have been applied in microbial control over several centuries. These are often referred to as heavy metals because of their relatively high atomic weight. However, from this list, only preparations containing mercury and silver still have any significance as germicides. Although some metals (zinc, iron) are actually needed in small concentrations as cofactors on enzymes, the higher molecular weight metals (mercury, silver, gold) can be very toxic, even in minute quantities (parts per million). This property of having antimicrobial effects in exceedingly small amounts is called an **oligodynamic* action (figure 11.16).** Heavy metal germicides contain either an inorganic or an organic metallic salt, and they come in the form of aqueous solutions, tinctures, ointments, or soaps.

Mercury, silver, and most other metals exert microbicidal effects by binding onto functional groups of proteins and inactivating them, rapidly bringing metabolism to a standstill (see figure 11.4c). This mode of action can destroy many types of microbes, including vegetative bacteria, fungal cells and spores, algae, protozoa, and viruses (but not endospores).

Unfortunately, there are several drawbacks to using metals in microbial control:

1. metals are very toxic to humans if ingested, inhaled, or absorbed through the skin, even in small quantities, for the same reasons that they are toxic to microbial cells;

2. they commonly cause allergic reactions;

*oligodynamic (ol″-ih-goh-dy-nam′-ik) Gr. oligos, little, and dynamis, power.

TABLE 11.7

Active Ingredients of Various Commercial Antimicrobial Products

Product	Specific Chemical Agent	Antimicrobial Category
Lysol Sanitizing Wipes	Dimethyl benzyl ammonium chloride	Detergent (quat)
Clorox Disinfecting Wipes	Dimethyl benzyl ammonium chloride	Detergent (quat)
Tilex Mildew Remover	Sodium hypochlorites	Halogen
Lysol Mildew Remover	Sodium hypochlorites	Halogen
Ajax Antibacterial Hand Soap	Triclosan	Phenolic
Dawn Antibacterial Hand Soap	Triclosan	Phenolic
Dial Antibacterial Hand Soap	Triclosan	Phenolic
Lysol Disinfecting Spray	Alkyl dimethyl benzyl ammonium Saccharinate/ethanol	Detergent (quats)/alcohol
ReNu Contact Lens Solution	Polyaminopropyl biguanide	Chlorhexidine
Wet Ones Antibacterial Moist Towelettes	Benzethonium chloride	Detergents (quat)
Noxzema Antibacterial Cleanser	Triclosan	Phenolic
Scope Mouthwash	Ethanol	Alcohol
Purell Instant Hand Sanitizer	Ethanol	Alcohol
Pine-Sol	Terpene alcohols and surfactant	Mixed
Allergan Eye Drops	Sodium chlorite	Halogen

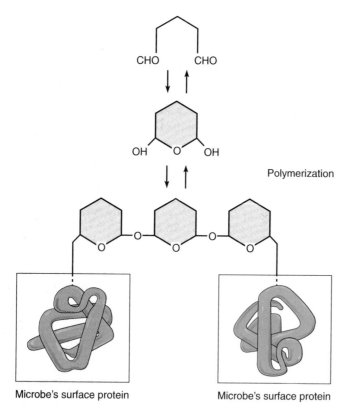

FIGURE 11.17

Actions of glutaraldehyde. The molecule forms a closed ring that can polymerize. When these alkylating polymers react with amino acids, they cross-link and inactivate proteins.

is a 37% solution of formaldehyde gas dissolved in water. The chemical is microbicidal through its attachment to nucleic acids and functional groups of amino acids. Formalin is an intermediate-to high-level disinfectant, although it acts more slowly than glutaraldehyde. Formaldehyde's extreme toxicity (it is classified as a carcinogen) and irritating effects on the skin and mucous membranes greatly limit its clinical usefulness.

A third aldehyde, ortho-phthalaldehyde (OPA) has recently been registered by the EPA as a high-level disinfectant. OPA is pale blue liquid with a barely detectable odor and can be most directly compared to glutaraldehyde. It has a mechanism of activation similar to glutaraldehyde, is stable, nonirritating to the eyes and nasal passages and, for most uses, is much faster acting than glutaraldehyde. It is effective against vegetative bacteria, including *Mycobacterium* and *Pseudomonas,* fungi, and viruses. Chief among its disadvantages is an inability to reliably destroy spores and, on a more practical note, its tendency to stain proteins, including those in human skin.

Applications of the Aldehydes Glutaraldehyde is a milder chemical for sterilizing materials that are damaged by heat. Commercial products (Cidex, Sporicidin) diluted to 2% are used to sterilize respiratory therapy equipment, hemostats, fiberoptic endoscopes (laparoscopes, arthroscopes), and kidney dialysis equipment. Glutaraldehyde is employed on dental instruments (usually in combination with autoclaving) to inactivate hepatitis B and other blood-borne viruses. It also serves to preserve vaccines, sanitize poultry carcasses, and degerm cow's teats.

Formalin tincture (8%) has limited use as a disinfectant for surgical instruments, and formalin solutions have applications in aquaculture to kill fish parasites and control growth of algae and fungi. Any object that comes into intimate contact with the body must be thoroughly rinsed to neutralize the formalin residue. It is, after all, one of the active ingredients in embalming fluid.

HISTORICAL HIGHLIGHTS   11.3
The Quest for Sterile Skin

More than a hundred years ago, before sterile gloves were a routine part of medical procedures, the hands remained bare during surgery. Realizing the danger from microbes, medical practitioners attempted to sterilize the hands of surgeons and their assistants to prevent surgical infections. Several stringent (and probably very painful) techniques involving strong chemical germicides and vigorous scrubbing were practiced. Here are a few examples.

In Schatz's method, the hands and forearms were first cleansed by brisk scrubbing with liquid soap for 3 to 5 minutes, then soaked in a saturated solution of permanganate at a temperature of 110°F until they turned a deep mahogany brown. Next, the limbs were immersed in saturated oxalic acid until the skin became decolorized. Then, as if this were not enough, the hands and arms were rinsed with sterile limewater and washed in warm bichloride of mercury for 1 minute.

Or, there was Park's method (more like a torture). First, the surfaces of the hands and arms were rubbed completely with a mixture of cornmeal and green soap to remove loose dirt and superficial skin. Next, a paste of water and mustard flour was applied to the skin until it began to sting. This potion was rinsed off in sterile water, and the hands and arms were then soaked in hot bichloride of mercury for a few minutes, during which the solution was rubbed into the skin.

Another method once earnestly suggested for getting rid of microorganisms was to expose the hands to a hot-air cabinet to "sweat the germs" out of skin glands. Pasteur himself advocated a quick flaming of the hands to maintain asepsis.

The old dream of sterilizing the skin was finally reduced to some basic realities: The microbes entrenched in the epidermis and skin glands cannot be completely eradicated even with the most intense efforts, and the skin cannot be sterilized without also seriously damaging it. Because this is true for both medical personnel and their patients, the chance always exists that infectious agents can be introduced during invasive medical procedures. Safe surgery had to wait until 1890, when rubber gloves were first made available for placing a sterile barrier around the hands. Of course, this did not mean that skin cleansing and antiseptic procedures were abandoned or downplayed. A thorough scrubbing of the skin, followed by application of an antiseptic, is still needed to remove the most dangerous source of infections—the superficial contaminants constantly picked up from whatever we touch.

(a)

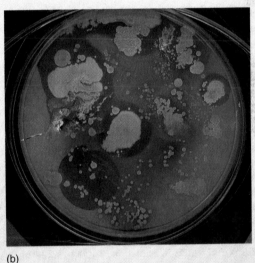

(b)

Microbes on normal unwashed hands. **(a)** A scanning electron micrograph of a piece of skin from a fingertip shows clusters of bacteria perched atop a fingerprint ridge (47,000×). **(b)** Heavy growth of microbial colonies on a plate of blood agar. This culture was prepared by passing an open sterile plate around a classroom of 30 students and having each one touch its surface. After incubation, a mixed population of bacteria and fungi appeared. Clear zones in agar can be indicative of pathogens.

Gaseous Sterilants and Disinfectants

Processing inanimate substances with chemical vapors, gases, and aerosols provides a versatile alternative to heat or liquid chemicals. Currently, those vapors and aerosols having the broadest applications are ethylene oxide (ETO), propylene oxide, and chlorine dioxide.

Ethylene oxide is a colorless substance that exists as a gas at room temperature. It is very explosive in air, a feature that can be eliminated by combining it with a high percentage of carbon dioxide or fluorocarbon. Like the aldehydes, ETO is a very strong alkylating agent, and it reacts vigorously with functional groups of DNA and proteins. Through these actions, it blocks both DNA replication and enzymatic actions. Ethylene oxide is one of a very few gases generally accepted for chemical sterilization because, when employed according to strict procedures, it is a sporicide. A specially designed ETO sterilizer called a *chemiclave,* a variation on the autoclave, is equipped with a chamber, gas ports, and

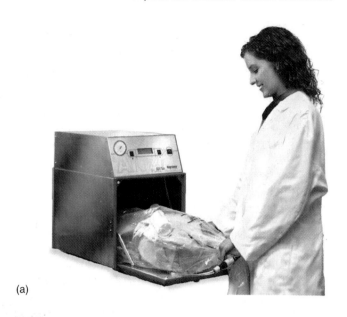

(a)

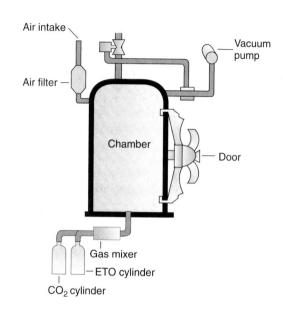

(b)

FIGURE 11.18

Sterilization using gas. **(a)** A chemiclave, an automatic ethylene oxide sterilizer. **(b)** The machine is equipped with gas canisters containing ethylene oxide (ETO) and carbon dioxide, a chamber to hold items, and mechanisms that evacuate gas and introduce air.

temperature, pressure, and humidity controls **(figure 11.18)**. Ethylene oxide is rather penetrating but relatively slow-acting, requiring from 90 minutes to 3 hours. Some items absorb ETO residues and must be aerated with sterile air for several hours after exposure to ensure dissipation of as much residual gas as possible. For all of its effectiveness, ETO has some unfortunate features. Its explosiveness makes it dangerous to handle; it can damage the lungs, eyes, and mucous membranes if contacted directly; and it is rated as a carcinogen by the government.

Chlorine dioxide is another gas that has of late been used as a sterilant. Despite the name, chlorine dioxide works in a completely different way from the chlorine compounds discussed earlier in the chapter. It is a strong alkylating agent, which disrupts proteins and is effective against vegetative bacteria, fungi, viruses, and endospores. Although chlorine dioxide is used for the treatment of drinking water, wastewater, food processing equipment, and medical waste, its most well known use was in the decontamination of the Senate offices after the anthrax attack of 2001 (see **Spotlight on Microbiology 11.4**).

Applications of Gases and Aerosols Ethylene oxide (carboxide, cryoxide) is an effective way to sterilize and disinfect plastic materials and delicate instruments in hospitals and industries. It can safely sterilize prepackaged medical devices, surgical supplies, syringes, and disposable Petri plates. Ethylene oxide has been used extensively to disinfect sugar, spices, dried foods, and drugs.

Propylene oxide is a close relative of ETO, with similar physical properties and mode of action, although it is less toxic. Because it breaks down into a relatively harmless substance, it is safer than ETO for sterilization of foods (nuts, powders, starches, spices).

Betapropiolactone (BPL) is a rapidly microbicidal but somewhat toxic solution that is used for disinfecting whole rooms and instruments, sterilizing bone and arterial grafts, and inactivating viruses in vaccines.

Dyes As Antimicrobial Agents

Dyes are important in staining techniques and as selective and differential agents in media; they are also a primary source of certain drugs used in chemotherapy. Because aniline dyes such as crystal violet and malachite green are very active against gram-positive species of bacteria and various fungi, they are incorporated into solutions and ointments to treat skin infections (ringworm, for example). The yellow acridine dyes, acriflavine and proflavine, are sometimes utilized for antisepsis and wound treatment in medical and veterinary clinics. For the most part, dyes will continue to have limited applications because they stain and have a narrow spectrum of activity.

Acids and Alkalies

Conditions of very low or high pH can destroy or inhibit microbial cells; but they are limited in applications due to their corrosive, caustic, and hazardous nature. Aqueous solutions of ammonium hydroxide remain a common component of detergents, cleansers, and deodorizers. Organic acids are widely used in food preservation because they prevent spore germination and bacterial and fungal growth and because they are generally regarded as safe to eat. Acetic acid (in the form of vinegar) is a pickling agent that inhibits bacterial growth; propionic acid is commonly incorporated into breads and cakes to retard molds; lactic acid is added to sauerkraut and olives to prevent growth of anaerobic bacteria (especially the clostridia), and benzoic and sorbic acids are added to beverages, syrups, and margarine to inhibit yeasts.

SPOTLIGHT ON MICROBIOLOGY 11.4
Decontaminating Congress

Choosing a microbial control technique is usually a straightforward process: cultures are autoclaved, milk is pasteurized, and medical supplies may be irradiated. When letters containing spores of *Bacillus anthracis* were opened in the Hart Office Building of the U.S. Senate, the process got just a bit trickier. Among the many concerns of the Environmental Protection Agency, which was charged with the building's remediation, were these;

- Anthrax is lethal disease.
- *Bacillus anthracis* is a spore-forming bacterium, making eradication difficult.
- The Hart Office Building is populated by thousands of people, who could quickly and easily spread endospores from office to office.
- The area to be decontaminated included heating and air-conditioning vents, carpeting, furniture, office equipment, sensitive papers, artwork, and the various belongings of quickly evacuated workers.

So the goal of the project could be simply stated: Detect and remove all traces of a lethal, highly infectious, spore-forming bacterium from an enormous space filled with all manner of easily damaged material. Easier said than done.

With this goal in mind, the EPA first set to work to determine the extent of contamination. Samples were taken from 25 buildings on Capitol Hill. Nonporous surfaces were swabbed; furniture and carpets were vacuumed into a HEPA filter; air was pumped through a filter to remove any airborne spores. Samples were placed in sterile vials and double bagged before being transferred to the laboratory. Analysis of the sam-

ples revealed the presence of spores in many areas of the Hart Office Building, including the mail-processing areas of 11 senators. Because spores were not found in the entrances to these offices, it was theorized that the spores spread primarily through the mail. The heaviest contamination was found in the office of Senator Tom Daschle, to whom the original anthrax-containing letter was addressed. Additional contamination was found in a conference room, stairwell, elevator, and rest room.

With the scope of the problem identified, the EPA could set out to devise a strategy for remediation, keeping in mind the difficulty of killing spore-forming organisms and the myriad contents of the building, much of it delicate, expensive, and in many cases irreplaceable, It was decided that most areas would be cleaned by a combination of HEPA vacuuming followed by either treatment with liquid chlorine dioxide or Sandia decontamination foam, an antibacterial foam that combines surfactants with oxidizing agents. For the most heavily contaminated areas (Senator Daschle's office and parts of the heating and air-conditioning system), gaseous chlorine dioxide would be used. Chlorine dioxide has been accepted as a sterilant since 1988 and is used for, among other things, treatment of medical waste. Its use in this project was based primarily on the facts that the foam would both work and be unlikely to harm the contents of the building (It was even tested to ensure that it would not damage the ink making up the signature on a document.)

Before fumigation could begin, however, a method of evaluating the success of the remediation effort needed to be devised. Borrowing from a common laboratory technique, 3,000 small slips of paper covered with spores from the organism *Bacillus stearothermophilus* (which is generally considered harder to kill than *B. anthracis*) were dispersed throughout the building. If after the treatment was complete these spores were unable to germinate, then the fumigation could be considered a success.

On December 1, 2001, technicians prepared the 3,000-square-foot office for fumigation. This included constructing barriers to seal off the portion of building being fumigated and raising the humidity in the offices to approximately 75%, to enable the gas to adhere to any lingering spores. Office machines (computers, copiers, for example) were turned on so that the fans inside the equipment would aid in spreading the gas. Finally, at 3:15 A.M. fumigation began. It ended 20 hours later and, after the gas was neutralized and ventilated from the building, technicians entered to collect the test strips. Analyzing the results, it was clear that trace amounts of spores were still present but only in those areas that originally had the worst contamination (these areas were again cleaned with liquid chlorine dioxide). Senator Daschle's office got a makeover with new carpeting, paint, and furniture and shortly thereafter the building was reoccupied. The senator also had his office equipment replaced as it was deemed too contaminated to be used. Just as antibiotics are useless against viruses, computer antivirus software apparently doesn't work against bacteria.

Workers in protective garments prepare the Hart Office Building for decontamination.

CHAPTER CHECKPOINTS

Chemical agents of microbial control are classified by their physical state and chemical nature.

Chemical agents can be either microbicidal or microbistatic. They are also classified as high-, medium-, or low-level germicides.

Factors that determine the effectiveness of a chemical agent include the type and numbers of microbes involved, the material involved, the strength of the agent, and the exposure time.

Halogens are effective chemical agents at both microbicidal and microbistatic levels. Chlorine compounds disinfect water, food, and industrial equipment. Iodine is used as either free iodine or iodophor to disinfect water and equipment. Iodophors are also used as antiseptic agents.

Phenols are strongly microbicidal agents used in general disinfection. Milder phenol compounds, the bisphenols, are also used as antiseptics.

Alcohols dissolve membrane lipids and destroy cell proteins. Their action depends upon their concentration, but they are generally only microbistatic.

Hydrogen peroxide is a versatile microbicide that can be used as an antiseptic for wounds and a disinfectant for utensils. A high concentration is an effective sporicide.

Surfactants are of two types: detergents and soaps. They reduce cell membrane surface tension, causing membrane rupture. Cationic detergents, or quats, are low-level germicides limited by the amount of organic matter present and the microbial load.

Aldehydes are potent sterilizing agents and high-level disinfectants that irreversibly disrupt microbial enzymes.

Ethylene oxide and chlorine dioxide are gaseous sterilants that work by alkylating protein and DNA.

CHAPTER CAPSULE WITH KEY TERMS

I. **Physical methods for controlling microorganisms**
   A. Moist heat denatures proteins and DNA while destroying membranes.
      1. **Sterilization: Autoclaves** utilize steam under pressure to sterilize heat resistant materials while intermittent sterilization can be used to sterilize more delicate items.
      2. **Disinfection:** Pasteurization subjects liquids to temperatures below 100°C and is used to lower the microbial load in liquids. Boiling water can be used to destroy vegetative pathogens in the home.
   B. Dry heat, using higher temperatures than moist heat, can also be used to sterilize.
      1. **Incineration** can be carried out using a Bunsen burner or incinerator. Temperatures range between 600°C and 1800°C.
      2. **Dry ovens** coagulate proteins at temperatures of 15° to 180°C.
   C. Cold temperatures are microbistatic, with refrigeration (0° to 15°C) and freezing (below 0°C) commonly used to preserve food, media and cultures.
   D. Drying and desiccation lead to (often temporary) metabolic inhibition by reducing water in the cell.
   E. Radiation: Energy in the form of radiation is a method of **cold sterilization,** which works by introducing mutations into the DNA of target cells.
      1. **Ionizing radiation,** such as Gamma rays and X-rays, has deep penetrating power and works by causing breaks in the DNA of target organisms.
      2. **Nonionizing radiation** uses **ultraviolet waves** with very little penetrating power and works by creating dimers between adjacent pyrimidines, which interferes with replication.
   F. Filtration involves the physical removal of microbes by passing a gas or liquid though a fine filter and can be used to sterilize air as well as heat sensitive liquids.
II. **Chemical Control of Microorganisms**
   A. Chemicals are divided into **disinfectants, antiseptics, sterilants, sanitizers** and **degermers** based on their level of effectiveness and the surfaces to which they are applied.

   B. Antimicrobic chemicals are found as solids, gases and liquids. Liquids can be either aqueous (water based) or tinctures (alcohol based).
   C. Halogens are chemicals based on elements from group VII of the periodic table.
      1. **Chlorine** is used as chlorine gas, hypochlorites and chloramines. All work by disrupting disulfide bonds and, given adequate time, are sporicidal.
      2. **Iodine** is found both as free iodine (I_2) and iodophors (iodine bound to organic polymers such as soaps). Iodine has a mode of action similar to chlorine and is also sporicidal given enough time.
   D. Phenolics are chemicals based on phenol that work by disrupting cell membranes and precipitating proteins. They are bactericidal, fungicidal and viricidal, but not sporicidal.
      1. Although phenol is now considered too toxic to be used in most circumstances, phenolic compounds (Lysol, Triclosan) are commonly used as, or added to, home and hospital disinfectants.
   E. Chlorhexidine (Hibiclens, Hibitane) is a surfactant and protein denaturant with broad microbicidal properties, although it is not sporicidal. Solutions of chlorhexidine are used as skin degerming agents for preoperative scrubs, skin cleaning and burns.
   F. Ethyl and isopropyl alcohol, in concentrations of 50% to 90%, are useful for microbial control. Alcohols act as **surfactants,** dissolving membrane lipids and coagulating proteins of vegetative bacterial cells and fungi. They are not sporicidal.
   G. Hydrogen peroxide produces highly reactive hydroxyl-free radicals that damage protein and DNA while also decomposing to O_2 gas, which is toxic to anaerobes. Strong solutions of H_2O_2 are sporicidal.
   H. Detergents and soaps
      1. Cationic detergents known as **quaternary ammonium compounds** (quats) act as surfactants that alter the membrane permeability of some bacteria and fungi. They are not sporicidal.
      2. Soaps have little microbicidal activity but rather function by removing grease and soil that contain microbes.

I. Heavy metals: Solutions of silver and mercury kill vegetative cells (but not spores) in exceedingly low concentrations (**oligodynamic action**) by inactivating proteins. Although heavy metal solutions are utilized in specific instances, drawbacks such as their toxicity, ability to cause allergies and neutralization by organic matter generally limits their use.

J. Aldehydes such as glutaraldehyde and formaldehyde kill microbes by alkylating protein and DNA molecules. The use of aldehydes is limited by their toxicity and propensity to irritate living tissue.

K. Gases and aerosols such as **ethylene oxide (ETO),** propylene oxide, betapropiolactone and chlorine dioxide are strong alkylating agents, all of which are sporicidal.

MULTIPLE-CHOICE QUESTIONS

1. A microbicidal agent has what effect?
   a. sterilizes
   b. inhibits microorganisms
   c. is toxic to human cells
   d. destroys microorganisms

2. Microbial control methods that kill _____ are able to sterilize.
   a. viruses
   b. the tubercle bacillus
   c. endospores
   d. cysts

3. Any process that destroys the non-spore-forming contaminants on inanimate objects is
   a. antisepsis
   b. disinfection
   c. sterilization
   d. degermation

4. Sanitization is a process by which
   a. the microbial load on objects is reduced
   b. objects are made sterile with chemicals
   c. utensils are scrubbed
   d. skin is debrided

5. An example of an agent that lowers the surface tension of cells is
   a. phenol
   b. chlorine
   c. alcohol
   d. formalin

6. High temperatures _____ and low temperatures _____ .
   a. sterilize, disinfect
   b. kill cells, inhibit cell growth
   c. denature proteins, burst cells
   d. speed up metabolism, slow down metabolism

7. The temperature-pressure combination for an autoclave is
   a. 100°C and 4 psi
   b. 121°C and 15 psi
   c. 131°C and 9 psi
   d. 115°C and 3 psi

8. Microbe(s) that is/are the target(s) of pasteurization include:
   a. *Clostridium botulinum*
   b. *Mycobacterium* species
   c. *Salmonella* species
   d. both b and c

9. Ionizing radiation removes _____ from atoms.
   a. protons
   b. waves
   c. electrons
   d. ions

10. The primary mode of action of nonionizing radiation is to
    a. produce superoxide ions
    b. make pyrimidine dimers
    c. denature proteins
    d. break disulfide bonds

11. The most versatile method of sterilizing heat-sensitive liquids is
    a. UV radiation
    b. exposure to ozone
    c. beta propiolactone
    d. filtration

12. _____ is the iodine antiseptic of choice for wound treatment.
    a. Eight percent tincture
    b. Five percent aqueous
    c. Iodophor
    d. Potassium iodide solution

13. A chemical with sporicidal properties is
    a. phenol
    b. alcohol
    c. quaternary ammonium compound
    d. glutaraldehyde

14. Silver nitrate is used
    a. in antisepsis of burns
    b. as a mouthwash
    c. to treat genital gonorrhea
    d. to disinfect water

15. Detergents are
    a. high-level germicides
    b. low-level germicides
    c. excellent antiseptics
    d. used in disinfecting surgical instruments

16. Which of the following is an approved sterilant?
    a. chlorhexidine
    b. betadyne
    c. ethylene oxide
    d. ethyl alcohol

CONCEPT QUESTIONS

1. Compare sterilization with disinfection and sanitization. Describe the relationship of the concepts of sepsis, asepsis, and antisepsis.

2. a. Explain the effect of a tuberculocide.
   b. What would be the effect of a pseudomonicide?
   c. What does a virustatic agent do?

3. a. Briefly explain how the type of microorganisms present will influence the effectiveness of exposure to antimicrobial agents.
   b. Explain how the numbers of contaminants can influence the measures used to control them.

4. a. Precisely what is microbial death?
   b. Why does a population of microbes not die instantaneously when exposed to an antimicrobial agent?

5. Why are antimicrobial processes inhibited in the presence of extraneous organic matter?

6. Describe four modes of action of antimicrobial agents, and give a specific example of how each works.

7. a. Summarize the nature, mode of action, and effectiveness of moist and dry heat.
   b. Compare the effects of moist and dry heat on vegetative cells and spores.
   c. Explain the concepts of TDT and TDP, using examples. What are the minimum TDTs for vegetative cells and endospores?

8. How can the temperature of steam be raised above 100°C? Explain the relationship involved.

9. a. What do you see as a basic flaw in tyndallization?
   b. In boiling water devices?
   c. In incineration?
   d. In ultrasonic devices?

10. a. What are several microbial targets of pasteurization?
    b. What are the primary purposes of pasteurization?
    c. What is ultrapasteurization?
11. Explain why desiccation and cold are not reliable methods of disinfection.
12. a. What are some advantages of ionizing radiation as a method of control?
    b. Some disadvantages?
13. a. What is the precise mode of action of ultraviolet radiation?
    b. What are some disadvantages to its use?
14. What are the superior characteristics of iodophors over free iodine solutions?

15. a. What does it mean to lower the surface tension?
    b. What cell parts will be most affected by surface active agents?
16. a. Name one chemical for which the general rule that a higher concentration is more effective is *not* true.
    b. What is a sterilant?
    c. Name the principal sporicidal chemical agents.
17. Why is hydrogen peroxide solution so effective against anaerobes?
18. Give the uses and disadvantages of the heavy metal chemical agents, glutaraldehyde, and the sterilizing gases.
19. What does it mean to say that a chemical has an oligodynamic action?
20. a. Define cold sterilization.
    b. Name three totally different methods that qualify for this definition.

CRITICAL-THINKING QUESTIONS

1. What is wrong with this statement: "Prior to vaccination, the patient's skin was sterilized with alcohol"? What would be the more correct wording?
2. For each item on the following list, give a reasonable method of sterilization. You cannot use the same method more than three times; the method must sterilize, not just disinfect; and the method must not destroy the item or render it useless unless there is no other choice. After considering a workable method, think of a method that would not work. Note: Where an object containing something is given, you must sterilize everything (for example, both the jar and the Vaseline in it). Some examples of methods are autoclave, ethylene oxide gas, dry oven, and ionizing radiation.

room air
blood in a syringe
serum
a pot of soil
plastic Petri plates
heat-sensitive drugs
cloth dressings
leather shoes from a thrift shop
a cheese sandwich
human hair (for wigs)
a flask of nutrient agar
an entire room (walls, floor, etc.)
rubber gloves
disposable syringes

carcasses of cows with "mad cow" disease
inside of a refrigerator
wine
a jar of Vaseline
fruit in plastic bags
talcum powder
milk
orchid seeds
metal instruments
mail contaminated with anthrax spores

3. a. Graph the data on tables 11.3 and 11.4, plotting the time on the Y axis and the temperature on the X axis for three different organisms.
   b. Using pasteurization techniques as a model, compare the TDTs and explain the relationships between temperature and length of exposure.

c. Is there any difference between the graph for a sporeformer and the graph for a non-sporeformer? Explain.
4. Can you think of situations in which the same microbe would be considered a serious contaminant in one case and completely harmless in another?
5. A supermarket/drugstore assignment: With a partner, look at the labels of 25 different products used to control microbes, and make a list of their active ingredients, their suggested uses, and information on toxicity and precautions.
6. Devise an experiment that will differentiate between bacteriocidal and bacteriostatic effects.
7. There is quite a bit of concern that chlorine used as a water purification chemical presents serious dangers. What alternative methods covered by this chapter could be used to purify water supplies and yet keep them safe from contamination?
8. The shelf-life and keeping qualities of fruit and other perishable foods are greatly enhanced through irradiation, saving industry and consumers billions of dollars. How would you personally feel about eating a piece of irradiated fruit? How about spices that had been sterilized with ETO?
9. The microbial levels in saliva are astronomical ($<10^6$ cells per ml, on average). What effect do you think a mouthwash can have against this high number? Comment on the claims made for such products.
10. Can you think of some innovations the health care community can use to deal with medical waste and its disposal that prevent infection but are ecologically sound?

INTERNET SEARCH TOPICS

1. Look for information on the Internet concerning triclosan-resistant bacteria. Based on what you find, do you think the widespread use of this antimicrobial is a good idea? Why or why not?
2. Find websites that discuss problems with sterilizing reusable medical instruments such as endoscopes and the types of diseases that can be transmitted with them.

3 Handwashing is one of our main protections to ensure good health and hygiene. Visit the student Online Learning Center at www.mhhe.com/talaro5. Go to chapter 11, Internet Search Topics, and log on to the available websites to research information, statistics, and other aspects of handwashing.

Drugs, Microbes, Host—The Elements of Chemotherapy

A hundred years ago in the United States, one out of three children was expected to die of an infectious disease before the age of five. Early death or severe lifelong debilitation from scarlet fever, diphtheria, tuberculosis, meningitis, and many other bacterial diseases was a fearsome yet undeniable fact of life to most of the world's population. The introduction of modern drugs to control infections in the 1930s began as a medical revolution that has added significantly to the lifespan and health of humans. It is no wonder that for many years, antibiotics in particular were regarded as the miracle cure-all for infectious diseases. In later discussions, we will evaluate this misconception in light of the shortcomings of chemotherapy. Although antimicrobic drugs have greatly reduced the incidence of certain infections, they have definitely not eradicated infectious disease and probably never will. In fact, in some parts of the world, mortality rates from infectious diseases are as high as before the arrival of antimicrobic drugs. Nevertheless, humans have been taking medicines to try to control diseases for thousands of years **(Historical Highlights 12.1)**.

Chapter Overview

- Antimicrobial chemotherapy involves the use of chemicals to prevent and treat infectious diseases. These chemicals include antibiotics, which are derived from the natural metabolic processes of bacteria and fungi, as well as synthetic drugs.
- To be effective, antimicrobial therapy must disrupt a necessary component of a microbe's structure or metabolism to the extent that microbes are killed or their growth is inhibited. At the same time, the drug must be safe for humans, not being overly toxic, causing allergies or disrupting normal flora.
- Drugs are available to treat infections by all manner of organisms, as well as viruses. Drug choice depends on the microbe's sensitivity, the drug's toxicity, and the health of the patient.
- Adverse side effects of drugs include damage to organs, allergies, and disruption of the host's normal flora, which can lead to other infections.
- A major problem in modern chemotherapy is the development of drug resistance. Many microbes have undergone genetic changes that allow them to circumvent the effects of a drug. The misapplication of drugs and antibiotics has worsened this problem.

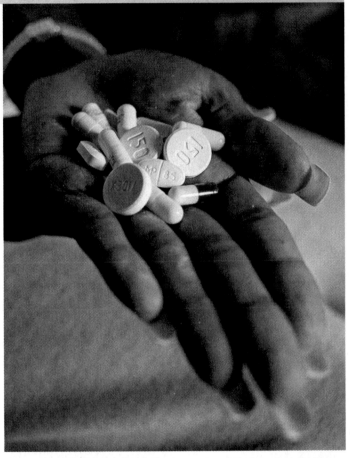

A cocktail for HIV. This handful of pills represents a daily dose of medications taken by AIDS patients to prevent the progression and symptoms of the disease. Expensive and fraught with numerous side effects, this combination of drugs may mean the difference between life and death.

Principles of Antimicrobial Therapy

The goal of antimicrobial chemotherapy is deceivingly simple; administer a drug to an infected person that destroys the infective agent without harming the host's cells. In actuality, this goal is rather difficult to achieve, as many often contradictory factors must be taken into account. The ideal drug should be easily administered, yet be able to reach the infectious agent anywhere in the body, be absolutely toxic to the infectious agent while simultaneously being nontoxic to the host, remain active in the body as long as needed yet

HISTORICAL HIGHLIGHTS 12.1
From Witchcraft to Wonder Drugs

Early human cultures relied on various types of primitive medications such as potions, poultices, and mudplasters. Many were concocted from plant, animal, and mineral products that had been found, usually through trial and error or accident, to have some curative effect upon ailments and complaints. In one ancient Chinese folk remedy, a fermented soybean curd was applied to skin infections. The Greeks used wine and plant resins (myrrh and frankincense), rotting wood, and various mineral salts to treat diseases. Some folk medicines were occasionally effective, but most of them were probably witches' brews that either had no effect or were even harmful. It is interesting that the Greek word *pharmakeutikos* originally meant the practice of witchcraft. These ancient remedies were handed down from generation to generation, but it was not until the Middle Ages that a specific disease was first treated with a specific chemical. Dosing syphilitic patients with inorganic arsenic and mercury compounds may have proved the ancient axiom: Graviora quædum sunt remedia periculus. ("Some remedies are worse than the disease.")

An enormous breakthrough in the science of drug therapy came with the germ theory of infection by Robert Koch (see chapter 1). This allowed disease treatment to focus on a particular microbe, which in turn opened the way for Paul Ehrlich to formulate the first theoretical concepts in chemotherapy in the late 1800s. Ehrlich had observed that certain dyes affixed themselves to specific microorganisms and not to animal tissues. This observation led to the profound idea that if a drug was properly selective in its actions, it would zero in on and destroy a microbial target and leave human cells unaffected. His first discovery was an arsenic-based drug that was very toxic to the spirochete of syphilis but, unfortunately, to humans as well. Ehrlich systematically altered this parent molecule, creating numerous derivatives. Finally, on the 606th try, he arrived at a compound he called *salvarsan*. This drug had some therapeutic merit and was used for a few years, but it eventually had to be discontinued because it was still not selective enough in its toxicity. Ehrlich's work had laid important foundations for many of the developments to come.

Another pathfinder in early drug research was Gerhard Domagk, whose discoveries in the 1930s launched a breakthrough in therapy that marked the true beginning of broad-scale usage of antimicrobic drugs. Domagk showed that the red dye prontosil was chemically changed by the body into a compound with specific activity against bacteria. This substance was sulfonamide—the first *sulfa* drug. In a short time, the structure of this drug was determined, and it became possible to synthesize it on a wide scale and to develop scores of other sulfonamide drugs. Although these drugs had immediate applications in therapy (and still do), still another fortunate discovery was needed before the golden age of antibiotics could really blossom.

The discovery of antibiotics dramatically demonstrates how developments in science and medicine often occur through a combination of accident, persistence, collaboration, and vision. In 1928 in the London

laboratory of Sir Alexander Fleming, a plate of *Staphylococcus aureus* became contaminated with the mold *Penicillium notatum*. Observing these plates, Fleming noted that the colonies of *Staphylococcus* were evidently being destroyed by some activity of the nearby *Penicillium* colonies. Struck by this curious phenomenon, he extracted from the fungus a compound he called penicillin and showed that it was responsible for the inhibitory effects.

Although Fleming understood the potential for penicillin, he was unable to develop it. A decade after his discovery, English chemists Howard Florey and Ernst Chain worked out methods for industrial production of penicillin to help in the war effort. Clinical trials conducted in 1941 ultimately proved its effectiveness, and cultures of the mold were brought to the United States for an even larger-scale effort. When penicillin was made available to the world's population, at first it seemed a godsend. But in time, because of extreme overuse and misunderstanding of its capabilities, it also became the model for one of the most serious drug problems—namely, drug resistance. By the 1950s, the pharmaceutical industry had entered an era of drug research and development that soon made penicillin only one of a large assortment of antimicrobic drugs.

The father of modern antibiotics. A Scottish physician, Sir Alexander Fleming, accidently discovered penicillin when his keen eye noticed that colonies of bacteria were being lysed by a fungal contaminant. He isolated the active ingredient and set the scene for the development of the drug ten years later.

TABLE 12.1

Characteristics of the Ideal Antimicrobial Drug

- Selectively toxic to the microbe but nontoxic to host cells
- Microbicidal rather than microbistatic
- Relatively soluble and functions even when highly diluted in body fluids
- Remains potent long enough to act and is not broken down or excreted prematurely
- Not subject to the development of antimicrobial resistance
- Complements or assists the activities of the host's defenses
- Remains active in tissues and body fluids
- Readily delivered to the site of infection
- Not excessive in cost
- Does not disrupt the host's health by causing allergies or predisposing the host to other infections

be safely and easily broken down and excreted. In short, the perfect drug does not exist, but by balancing drug characteristics against one another, a satisfactory compromise can be achieved (**table 12.1**).

Chemotherapeutic agents are described with regard to their origin, range of effectiveness, and whether they are naturally produced or chemically synthesized. A few of the more important terms you will encounter are found in **table 12.2**.

THE ORIGINS OF ANTIMICROBIAL DRUGS

Nature is undoubtedly the most prolific producer of antimicrobial drugs. Antibiotics, after all, are common metabolic products of aerobic spore-forming bacteria and fungi. By inhibiting the growth of other microorganisms in the same habitat (antagonism), antibiotic producers presumably enjoy less competition for nutrients and space. The greatest numbers of antibiotics are derived from bacteria in the genera *Streptomyces* and *Bacillus* and molds in the genera *Penicillium* and *Cephalosporium*. Not only have chemists created new drugs by altering the structure of naturally occurring antibiotics, they are actively searching for metabolic compounds with antimicrobial effects in species other than bacteria and fungi (**Microbits 12.2**).

CHAPTER CHECKPOINTS

Antimicrobial chemotherapy involves the use of drugs to control infection on or in the body.

Antimicrobial drugs are produced either synthetically or from natural sources. They inhibit or destroy microbial growth in the infected host. Antibiotics are the subset of antimicrobics produced by the natural metabolic processes of microorganisms.

Antimicrobial drugs are classified by their range of effectiveness. Broad-spectrum antimicrobics are effective against many types of microbes. Narrow-spectrum antimicrobics are effective against a limited group of microbes.

Antimicrobial therapy involves three interacting factors: the microbes sensitivity, the drug's toxicity, and the health of the patient.

Spore-forming bacteria and fungi are the primary sources of most antibiotics. The molecular structures of these compounds can be chemically altered to form additional semisynthetic antimicrobics.

	...otherapy
...g	Any chemical used in the treatment, relief, or prophylaxis of a disease
Pr...	Use of a drug to prevent imminent infection of a person at risk
Antim...ial chem...erapy*	The use of chemotherapeutic drugs to control infection
Antimic...bics	All-inclusive term for any antimicrobial drug, regardless of its origin
Antibiotics*	Substances produced by the natural metabolic processes of some microorganisms that can inhibit or destroy other microorganisms
Semisynthetic drugs	Drugs which are chemically modified in the laboratory after being isolated from natural sources
Synthetic drugs	The use of chemical reactions to synthesize antimicrobial compounds in the laboratory
Narrow spectrum (limited spectrum)	Antimicrobics effective against a limited array of microbial types— for example, a drug effective mainly on gram-positive bacteria
Broad spectrum (extended spectrum)	Antimicrobics effective against a wide variety of microbial types—for example, a drug effective against both gram-positive and gram-negative bacteria

*prophylaxis (proh″-fih-lak′-sis) Gr. *prophylassein*, to keep guard before. A process that prevents infection or disease in a person at risk.

*chemotherapy (kee″-moh-ther′-uh-pee) Gr. *chemieia*, chemistry, and *therapeia*, service to the sick. Use of drugs to treat disease.

*antibiotic (an-tee′-by-aw″-tik) Gr. *anti*, against, and *bios*, life.

Interactions Between Drug and Microbe

The goal of antimicrobial drugs is either to disrupt the cell processes or structures of bacteria, fungi, and protozoa or to inhibit virus replication. Most of the drugs used in chemotherapy interfere with the function of enzymes required to synthesize or assemble macromolecules, or they destroy structures already formed in the cell. Above all, drugs should be **selectively toxic,** which means they should kill or inhibit microbial cells without simultaneously damaging host tissues. This concept of selective toxicity is central to chemotherapy, and the best drugs are those that block the actions or synthesis of molecules in microorganisms but not in vertebrate cells. Examples of drugs with selective toxicity are those that block the synthesis of the cell wall in bacteria (penicillins). They have low toxicity and few direct effects on human cells because human cells lack a wall and are thus neutral to this action of the antibiotic. Among the most toxic to human cells are drugs that act upon a structure common to both the infective agent and the host cell, such as the cell membrane

MICROBITS 12.2

A Modern Quest for Designer Drugs

Once the first significant drug was developed, the world immediately witnessed a scientific scramble to find more antibiotics. This search was advanced on several fronts. Hundreds of investigators began the laborious task of screening samples from soil, dust, muddy lake sediments, rivers, estuaries, oceans, plant surfaces, compost heaps, sewage, skin, and even the hair and skin of animals for antibiotic-producing bacteria and fungi. This intense effort has paid off over the past 50 years, because more than 10,000 antibiotics have been discovered (though surprisingly, only a relatively small number have been applicable to chemotherapy). Finding a new antimicrobial substance is only a first step. The complete pathway of drug development from discovery to therapy takes between 12 and 24 years at a cost of billions of dollars.

Antibiotics are products of fermentation pathways that occur in many spore-forming bacteria and fungi. The role of antibiotics in the lives of these microbes must be important since the genes for antibiotic production are preserved in evolution. Some experts theorize that antibiotic-releasing microorganisms can inhibit or destroy nearby competitors or predators; others propose that antibiotics play a part in spore formation. Whatever benefit the microbes derive, these compounds have been extremely profitable for humans. Every year, the pharmaceutical industry farms vast quantities of microorganisms and harvests their products to treat diseases caused by other microorganisms. Researchers have facilitated the work of nature by selecting mutant species that yield more abundant or useful products, by varying the growth medium, or by altering the procedures for large-scale industrial production (see chapter 26).

Another approach in the drug quest is to chemically manipulate molecules by adding or removing functional groups. Drugs produced in this way are designed to have advantages over other, related drugs. Using the **semisynthetic** method, a natural product of the microorganism is joined with various preselected functional groups. The antibiotic is reduced to its basic molecular framework (called the nucleus), and to this nucleus specially selected side chains (R groups) are added. A case in point is the metamorphosis of the semisynthetic penicillins. The nucleus is an inactive penicillin derivative called aminopenicillanic acid, which has an opening on the number 6 carbon for addition of R groups. A particular carboxylic acid (R group) added to this nucleus can "fine-tune" the penicillin, giving it special characteristics. For instance, some R groups will make the product resistant to penicillinase (methicillin), some confer a broader activity spectrum (ampicillin), and others make the product acid-resistant (penicillin V). Cephalosporins and tetracyclines also exist in several semisynthetic versions. The potential for using bioengineering techniques to design drugs seems almost limitless, and, indeed, several drugs have already been produced by manipulating the genes of antibiotic producers. Some investigators are looking for new types of drugs in plants and animals (see Medical Microfile 1.3).

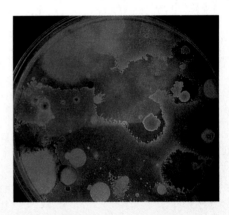

A plate with several discrete colonies of soil bacteria was sprayed with a culture of *Escherichia coli* and incubated. Zones of inhibition (clear areas with no growth) surrounding several colonies indicate species that produce antibiotics.

Synthesizing penicillins. **(a)** The original penicillin G molecule is a fermentation product of *Penicillium chrysogenum* that appears somewhat like a house with a removable patio on the left. This house without the patio is the basic nucleus called aminopenicillanic acid. **(b–d)** Various new fixtures (R groups) can be added to change the properties of the drug. These R groups will produce different penicillins: **(b)** methicillin; **(c)** ampicillin; and **(d)** penicillin V.

(for example, amphotericin B used to treat fungal infections). As the characteristics of the infectious agent become more and more similar to those of the host cell, complete selective toxicity becomes more difficult to achieve, and more and greater side effects are seen. The previous example briefly illustrates this concept. We will examine the subject in more detail in a later section.

MECHANISMS OF DRUG ACTION

If the goal of chemotherapy is to disrupt the structure or function of an organism to the point where it can no longer survive, then the first step toward this goal is to identify the structural and metabolic needs of a living cell. Once the requirements of a living cell have been determined, methods of removing, disrupting, or interfering with these requirements can be employed as potential chemotherapeutic strategies. The metabolism of an actively dividing cell is marked by the production of new cell wall components (in most cells), DNA, RNA, proteins, and cell membrane. Consequently, antimicrobial drugs are divided into categories based on which of these metabolic targets they affect. These categories are outlined in **figure 12.1** and include:

1. inhibition of cell wall synthesis,
2. inhibition of nucleic acid (RNA and DNA) structure and function,
3. inhibition of protein synthesis, and
4. interference with cell membrane structure or function.

As you will see, these categories are not completely discrete, and some effects can overlap.

Antimicrobial Drugs That Affect the Bacterial Cell Wall

The cell walls of most bacteria contain a rigid girdle of peptidoglycan, which protects the cell against rupture from hypotonic environments. Active cells must constantly synthesize new peptidoglycan and transport it to its proper place in the cell envelope. Drugs such as penicillins and cephalosporins react with one or more of the enzymes required to complete this process, causing the cell to develop weak points at growth sites and to become osmotically fragile (**figure 12.2**). Antibiotics that produce this effect are considered bactericidal, because the weakened cell is subject to lysis. It is essential to note that most of these antibiotics are active only in young, growing cells, because old, inactive, or dormant cells do not synthesize peptidoglycan. (One exception is a new class of antibiotics called the penems.)

Cycloserine inhibits the formation of the basic peptidoglycan subunits, and vancomycin hinders the elongation of the peptidoglycan. The beta-lactams (penicillins and cephalosporins) bind and block peptidases that cross-link the glycan molecules, thereby interrupting the completion of the cell wall (**figure 12.3**). Penicillins that do not penetrate the outer membrane are less effective against gram-negative bacteria, but broad-spectrum penicillins and cephalosporins, such as carbenicillin or ceftriaxone, can cross the cell walls of gram-negative species.

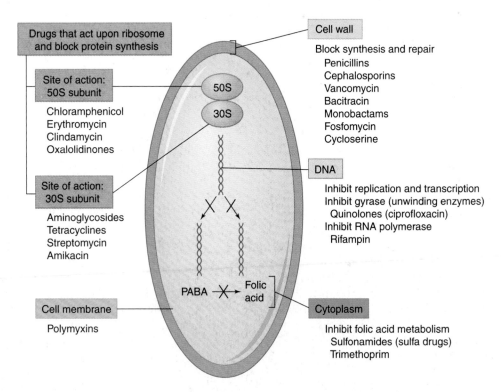

FIGURE 12.1

Primary sites of action of antimicrobic drugs on bacterial cells.   See text for more discussion of mechanisms.

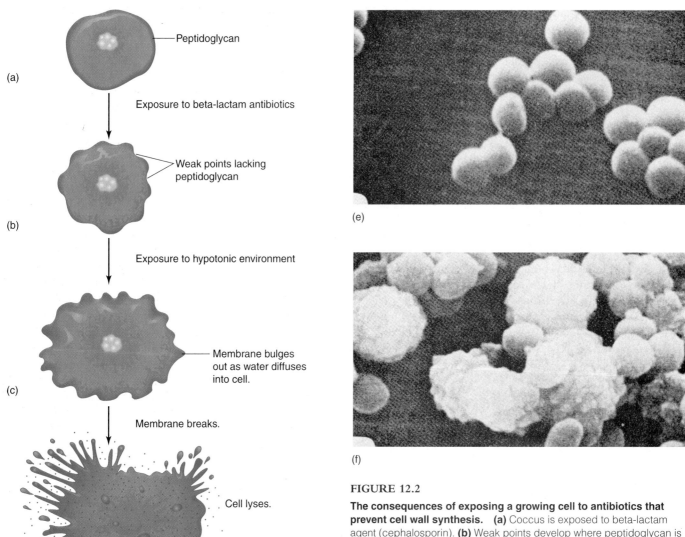

(a) — Peptidoglycan

Exposure to beta-lactam antibiotics

(b) — Weak points lacking peptidoglycan

Exposure to hypotonic environment

(c) — Membrane bulges out as water diffuses into cell.

Membrane breaks.

(d) — Cell lyses.

(e)

(f)

FIGURE 12.2

The consequences of exposing a growing cell to antibiotics that prevent cell wall synthesis. **(a)** Coccus is exposed to beta-lactam agent (cephalosporin). **(b)** Weak points develop where peptidoglycan is incomplete. **(c)** The weakened cell is exposed to a hypotonic environment. **(d)** The cell lyses. **(e)** Scanning electron micrograph of bacterial cells in their normal state (10,000×). **(f)** Scanning electron micrograph of the same cells in the drug-affected state, showing surface bulges (10,000×).

FIGURE 12.3

The mode of action of beta-lactam antibiotics on the bacterial cell wall.
(a) Intact peptidoglycan has chains of NAM (N-acetyl muramic acid) and NAG (N-acetyl glucosamine) glycans cross-linked by peptide bridges.
(b) The beta-lactam antibiotics block the peptidases that link the cross-bridges between NAMs, thereby greatly weakening the cell wall meshwork.

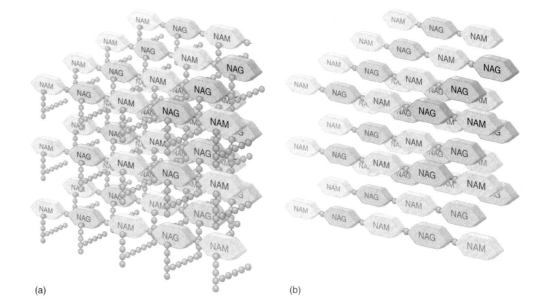

(a)   (b)

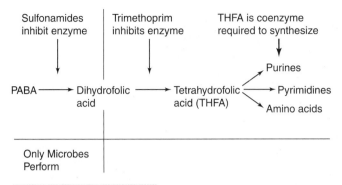

(a) Normal metabolic pathway

(b) Normal folic acid synthesis

Initial folic acid molecule

(c) Inhibition of folic acid synthesis by sulfa drug

Antimicrobial Drugs That Affect Nucleic Acid Synthesis

As you will recall from chapter 9, the metabolic pathway that generates DNA and RNA is a long, enzyme-catalyzed series of reactions. Like any complicated process, it is subject to breakdown at many different points along the way, and inhibition at any given point in the sequence can block subsequent events. Antimicrobial drugs interfere with nucleic acid synthesis by blocking synthesis of nucleotides, inhibiting replication, or stopping transcription. Because functioning DNA and RNA are required for proper translation as well, the effects on protein metabolism can be far-reaching.

Sulfonamides and trimethoprim are drugs that act by mimicking the normal substrate of an enzyme in a process called **competitive inhibition.** They are supplied to the cell in high concentrations to ensure that a needed enzyme is constantly occupied with the **metabolic analog*** rather than the true substrate of the enzyme. As the enzyme is no longer able to produce a needed product, cellular metabolism slows or stops. Sulfonamides and trimethoprim interfere with folate metabolism by blocking enzymes required for the synthesis of tetrahydrofolate, which is needed by bacterial cells for the synthesis of folic acid and the eventual production of DNA and RNA and amino acids. Trimethoprim and sulfonamides are often given simultaneously to achieve a *synergistic effect* which, in pharmacological terms, refers to an additive effect achieved by multiple drugs working together, thus requiring a lower dose of each. **Figure 12.4** illustrates sulfonamides competition with PABA for the active site of the enzyme pteridine synthetase.

---

*analog (an′-uh-log) A compound whose configuration closely resembles another compound required for cellular reactions.

FIGURE 12.4

The mode of action of sulfa drugs. **(a)** The metabolic pathway needed to synthesize tetrahydrofolic acid (THFA) contains two enzymes which are chemotherapeutic targets. **(b)** Under normal circumstances PABA acts as a substrate for the enzyme pteridine synthetase, the product of which is needed for the eventual production of folic acid. **(c)** Sulfonamides act as a competitive inhibitor, constantly filling the active site of the enzyme and preventing PABA from entering the site and being converted to the folic acid molecule.

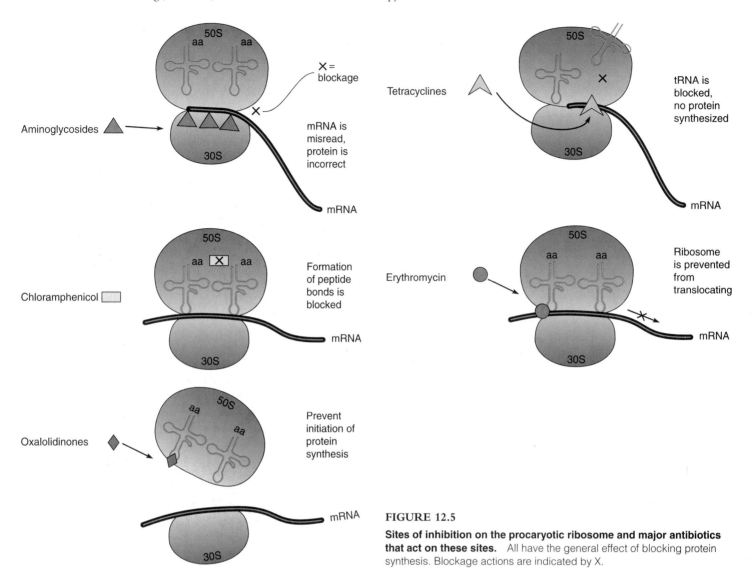

FIGURE 12.5

Sites of inhibition on the procaryotic ribosome and major antibiotics that act on these sites. All have the general effect of blocking protein synthesis. Blockage actions are indicated by X.

The selective toxicity of these compounds is explained by the fact that mammals derive folic acid from their diet and so do not possess this enzyme system. Therefore, inhibition of bacterial and protozoan parasites, which must synthesize folic acid, is easily accomplished while leaving the human host unaffected.

Other antimicrobics inhibit DNA synthesis. Chloroquine (an antimalarial drug) binds and cross-links the double helix. The newer broad-spectrum quinolones inhibit DNA unwinding enzymes or helicases, thereby stopping DNA transcription. Antiviral drugs that are analogs of purines and pyrimidines (including azidothymidine [AZT] and acyclovir) insert in the viral nucleic acid and block further replication.

Drugs That Block Translation

Most inhibitors of translation, or protein synthesis, react with the ribosome-mRNA complex. Although human cells also have ribosomes, the ribosomes of eucaryotes are different in size and structure from those of procaryotes, so these antimicrobics usually have a selective action against bacteria. One potential therapeutic consequence of drugs that bind to the procaryotic ribosome is the damage they can do to eucaryotic mitochondria, which contain a procaryotic type of ribosome. Two possible targets of ribosomal inhibition are the 30S subunit and the 50S subunit (**figure 12.5**). Aminoglycosides (streptomycin, gentamicin, for example) insert on sites on the 30S subunit and cause the misreading of the mRNA, leading to abnormal proteins. Tetracyclines block the attachment of tRNA on the A acceptor site and effectively stop further synthesis. Other antibiotics attach to sites on the 50S subunit in a way that prevents the formation of peptide bonds (chloramphenicol) or inhibits translocation of the subunit during translation (erythromycin).

Drugs That Disrupt Cell Membrane Function

A cell with a damaged membrane invariably dies from disruption in metabolism or lysis and does not even have to be actively dividing to be destroyed. The antibiotic classes that damage cell membranes have specificity for a particular microbial group, based on differences in the types of lipids in their cell membranes.

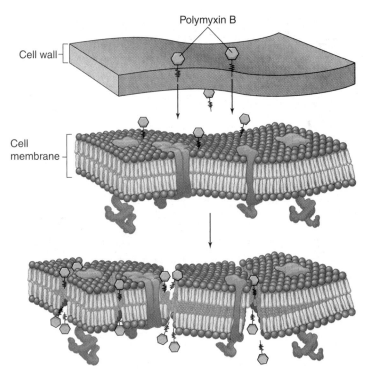

FIGURE 12.6

The detergent action of polymyxin. After passing through the cell wall of gram-negative bacteria, polymyxin binds to the cell membrane and forms abnormal openings that cause the membrane to become leaky.

Polymyxins interact with membrane phospholipids, distort the cell surface, and cause leakage of proteins and nitrogen bases, particularly in gram-negative bacteria (**figure 12.6**). The polyene antifungal antibiotics (amphotericin B and nystatin) form complexes with the sterols on fungal membranes; these complexes cause abnormal openings and seepage of small ions. Unfortunately, this selectivity is not exact, and the universal presence of membranes in microbial and animal cells alike means that most of these antibiotics can be quite toxic to humans.

Survey of Major Antimicrobial Drug Groups

Scores of antimicrobial drugs are marketed in the United States. Although the medical and pharmaceutical literature contains a wide array of names for antimicrobics, most of them are variants of a small number of drug families. About 260 different antimicrobial drugs are currently classified in 20 drug families. Drug reference books may give the impression that there are 10 times that many because various drug companies assign different trade names to the very same generic drug. Ampicillin, for instance, is available under 50 different names. Most antibiotics are useful in controlling bacterial infections, though we shall also consider a number of antifun-

gal, antiviral, and antiprotozoan drugs. **Table 12.3** summarizes some major infectious agents, the diseases they cause, and the drugs indicated to treat them (drugs of choice).

ANTIBACTERIAL DRUGS

Penicillin and Its Relatives

The **penicillin** group of antibiotics, named for the parent compound, is a large, diverse group of compounds, most of which end in the suffix *-cillin*. Although penicillins could be completely synthesized in the laboratory from simple raw materials, it is more practical and economical to obtain natural penicillin through microbial fermentation. The natural product can then be used either in unmodified form or to make semisynthetic derivatives. *Penicillium chrysogenum* is the major source of the drug. All penicillins consist of three parts: a thiazolidine ring, a beta-lactam ring, and a variable side chain that dictates its microbicidal activity (**figure 12.7**).

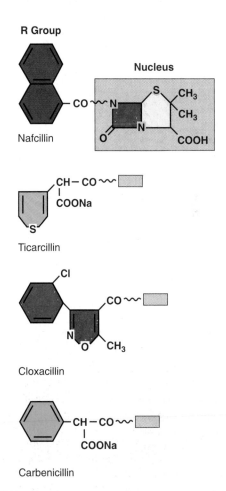

FIGURE 12.7

Chemical structure of penicillins. All penicillins contain a thiazolidine ring (yellow) and a beta-lactam ring (red), but each differs in the nature of the side chain (R group), which is also responsible for differences in biological activity.

TABLE 12.3

Selected Survey of Chemotherapeutic Agents in Infectious Diseases

Infectious Agent	Typical Infection	Drugs of Choice*
Bacteria		
Gram-Positive Cocci		
Staphylococcus aureus	Abscess, skin, toxic shock	Pencillins, vancomycin, cephalosporin
Streptococcus pyogenes	Strep throat, erysipelas, rheumatic fever	Penicillin, cephalosporin, erythromycin
S. pneumoniae	Pneumonia	Penicillin, if sensitive, cephalosporin, erythromycin
Gram-Positive Rods		
Bacillus	Anthrax	Penicillin, doxycycline
Clostridium	Gas gangrene, tetanus	Penicillin, cephalosporin, clindamycin
Corynebacterium	Diphtheria	Erythromycin, penicillin
Acid-Fast Rods		
Mycobacterium	Tuberculosis	(Isoniazid, rifampin, pyrazinamide),★ ethambutol, streptomycin
M. leprae	Leprosy	(Dapsone, rifampin, clofazimine)★★
Nocardia	Nocardiosis	Sulfamethoxazole-Trimethoprim (SxT), cephalosporin
Gram-Negative Cocci		
Neisseria gonorrhoeae	Gonorrhea	Ceftriaxone, quinolones, spectinomycin
N. meningitidis	Meningitis	Penicillin G, ceftriaxone, ampicillin
Gram-Negative Rods		
Bordetella	Pertussis	Erythromycin
Brucella	Brucellosis	Tetracycline, rifampin, gentamicin
Escherichia coli	Sepsis, diarrhea, urinary tract infection	Cephalosporin, quinolone, SxT★★
Haemophilus influenzae	Meningitis	Cefotaxime, cephtriaxone
Legionella	Legionnaires disease	Erythromycin, rifampin
Pseudomonas	Opportunistic lung and burn infections	Ticarcillin, aminoglycoside
Salmonella	Typhoid fever, salmonellosis	Cefalosporin, quinolone
Shigella	Dysentery	Quinolone, SxT, ampicillin
Vibrio cholerae	Cholera	Tetracyclines, SxT
Yersinia pestis	Plague	Streptomycin, gentamicin
Spirochetes		
Borrelia	Lyme disease	Ceftriaxone, doxycycline
Treponema pallidum	Syphilis	Penicillin, tetracyclines
Mycoplasma	Pneumonia, urinary infections	Erythromycin, azithromycin, clarithromycin
Rickettsia	Rocky Mountain spotted fever	Tetracyclines, chloramphenicol
Chlamydia	Urethritis, vaginitis	Azithromycin, doxycycline, erythromycin, quinolones

Subgroups and Uses of Penicillins The characteristics of certain penicillin drugs are shown in **table 12.4** page 360. Penicillins G and V are the most important natural forms. Penicillin is considered the drug of choice for infections by known sensitive, gram-positive cocci (most streptococci) and some gram-negative bacteria (meningococci and the spirochete of syphilis).

Certain semisynthetic penicillins such as ampicillin, carbenicillin, and amoxicillin have broader spectra and thus can be used to treat infections by gram-negative enteric rods. Penicillinase-resistant penicillins such as methicillin, nafcillin, and cloxacillin are useful in treating infections caused by some penicillinase-producing bacteria. Mezlocillin and azlocillin have such an extended spectrum that they can be substituted for combinations of antibiotics. All of the "cillin" drugs are relatively mild and well tolerated because of their specific mode of action on cell walls (which humans lack). The primary problems in therapy include allergy and resistant strains of pathogens. Clavulanic acid is a chemical that inhibits beta-lactamase enzymes, thereby increasing the longevity of beta-lactam antibiotics

in the presence of penicillinase-producing bacteria. For this reason, clavulanic acid is often added to semisynthetic penicillins to augment their effectiveness. For example, clavamox is a combination of amoxicillin and clavulanate and is marketed under the trade name Augmentin (see figure 12.19). Zosyn is a similar combination of tazobactum, a beta-lactamase inhibitor, and piperacillin that is used for a wide variety of systemic infections.

The Cephalosporin Group of Drugs

The **cephalosporins** are a comparatively new group of antibiotics that currently account for a majority of all antibiotics administered. The first compounds in this group were isolated in the late 1940s from the mold *Cephalosporium acremonium*. Cephalosporins are similar to penicillins in their beta-lactam structure that can be synthetically altered **(figure 12.8)** and in their mode of action. The generic names of these compounds are often recognized by the presence of the root *cef, ceph,* or *kef* in their names.

TABLE 12.3

continued

Infectious Agent	Typical Infection	Drugs of Choice★
Fungi		
Systemic Mycoses		
Aspergillus	Aspergillosis	Amphotericin B, azoles, flucytosine
Blastomyces	Blastomycosis	Ketoconazole, amphotericin B
Candida albicans	Candidiasis	Amphotericin B, fluconazole
Coccidioides immitis	Valley fever	Amphotericin B, azoles
Cryptococcus neoformans	Cryptococcosis	Amphotericin B, fluconazole, flucytosine
Pneumocystis (carinii) jiroveci	Pneumonia (PCP)	SxT, pentamidine
Sporothrix schenckii	Sporotrichosis	Iodides, itraconazole
Protozoa		
Entamoeba histolytica	Amebiasis	Metronidazole, tetracycline, paromomycin
Giardia lamblia	Giardiasis	Quinacrine, metronidazole
Plasmodium	Malaria	Chloroquine, primaquine, quinine
Toxoplasma gondii	Toxoplasmosis	Pyrimethamine, sulfadiazine
Trichomonas vaginalis	Trichomoniasis	Metronidazole
Trypanosoma cruzi	Chagas disease	Nitrifurimox★★★
T. brucei	Sleeping sickness	Suramin,★★★ pentamidine
Helminths		
Ascaris	Ascariosis	Mebendazole/pyrantel, piperazine
Cestodes	Tapeworm	Niclosamide, praziquantel
Schistosoma	Schistosomiasis	Praziquantel, metrifonate
Various fluke infections		Praziquantel, tetrachlorethylene, bithionol
Various roundworm infections		Mebendazole, thiabendazol, piperazine
Viruses		
Herpesvirus	Genital herpes, oral herpes, shingles	Acyclovir, ganciclovir
HIV	AIDS	(AZT, protease inhibitors), ddI, ddC, d4T
Orthomyxovirus	Type A influenza	Amantadine, rimantidine

★*More or less in order of preference.*
★★*( ) Usually given in combination.*
★★★*Available in the United States only from the Drug Service of the Centers for Disease Control and Prevention.*

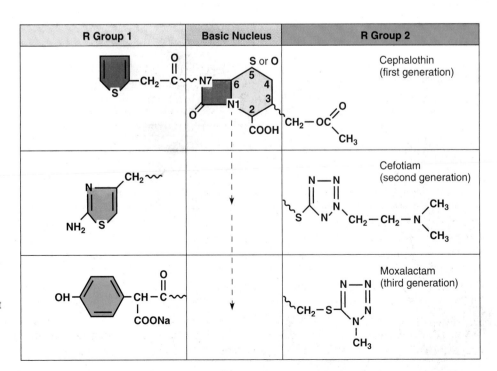

FIGURE 12.8

The structure of cephalosporins. Like penicillin, they have a beta-lactam ring (red), but they have a different main ring (yellow). However, unlike penicillins, they have two sites for placement of R groups (at positions 3 and 7). This makes possible several generations of molecules with greater versatility in function and complexity in structure.

Subgroups and Uses of Cephalosporins The cephalosporins are versatile. They are relatively broad-spectrum, resistant to most penicillinases, and cause fewer allergic reactions than penicillins. Although some cephalosporins are given orally, many are poorly absorbed from the intestine and must be administered **parenterally,*** by injection into a muscle or a vein.

Three generations of cephalosporins exist, based upon their antibacterial activity. First-generation cephalosporins such as cephalothin and cefazolin are most effective against gram-positive cocci and a few gram-negative bacteria. Second-generation forms include cefaclor and cefonacid, which are more effective than the first-generation forms in treating infections by gram-negative bacteria such as *Enterobacter, Proteus,* and *Haemophilus.* Third-generation cephalosporins, such as cephalexin (Keflex) and cefotaxime, are broad-spectrum with especially well-developed activity against enteric bacteria that produce beta-lactamases. Ceftriaxone (rocephin) is a new semisynthetic broad-spectrum drug for treating a wide variety of respiratory, skin, urinary, and nervous system infections.

Other Beta-Lactam Antibiotics

Related antibiotics include imipenem, a broad-spectrum drug for infections with aerobic and anaerobic pathogens. It is active in very small concentrations and can be taken by mouth with few side effects. Aztreonam, isolated from the bacterium *Chromobacterium violaceum,* is a newer narrow-spectrum drug for treating pneumonia, septicemia, and urinary tract infections by gram-negative aerobic bacilli. Aztreonam is especially useful when treating persons who are allergic to penicillin. Because of similarities in their chemical structure, allergies to penicillin often are accompanied by allergies to cephalosporins and carboxypenems (of which imipenem is a member). The structure of aztreonam is chemically distinct so that persons with allergies to penicillin are not usually adversely affected by treatment with aztreonam.

*parenterally (par-ehn′-tur-ah-lee) Gr. para, beyond, and enteron, intestine. A route of drug administration other than the gastrointestinal tract.

The Aminoglycoside Drugs

Antibiotics composed of two or more amino sugars and an aminocyclitol (6-carbon) ring are referred to as **aminoglycosides (figure 12.9).** These complex compounds are exclusively the products of various species of soil **actinomycetes*** in the genera *Streptomyces* (**figure 12.10**) and *Micromonospora.*

*actinomycetes (ak″-tin-oh-my-see′-teez) Gr. actinos, ray, and myces, fungus. A group of filamentous, funguslike bacteria.

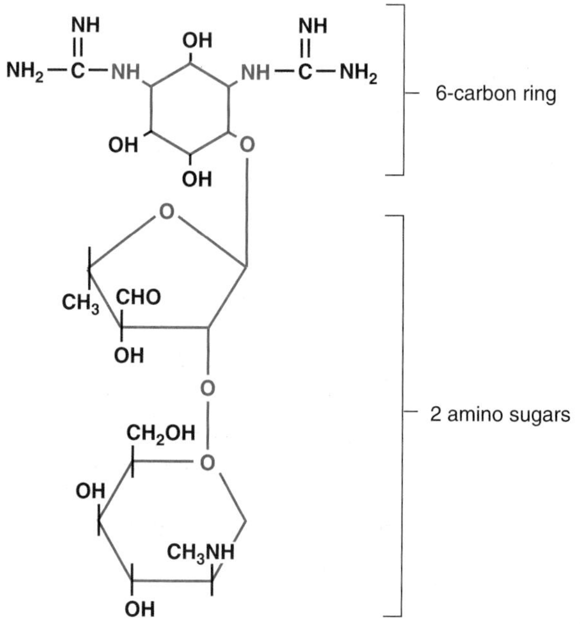

FIGURE 12.9

The structure of streptomycin. Colored portions of the molecule show the general arrangement of an aminoglycoside.

TABLE 12.4

Characteristics of Selected Penicillin Drugs

Name	Spectrum of Action	Uses, Advantages	Disadvantages
Penicillin G	Narrow	Best drug of choice when bacteria are sensitive; low cost; low toxicity	Can be hydrolyzed by penicillinase; allergies occur; requires injection
Penicillin V	Narrow	Good absorption from intestine; otherwise, similar to penicillin G	Hydrolysis by penicillinase; allergies
Oxacillin, dicloxacillin	Narrow	Not susceptible to penicillinase; good absorption	Allergies; expensive
Methicillin, nafcillin	Narrow	Not usually susceptible to penicillinase	Poor absorption; allergies; growing resistance
Ampicillin	Broad	Works on gram-negative bacilli	Can be hydrolyzed by penicillinase; allergies; only fair absorption
Amoxicillin	Broad	Gram-negative infections; good absorption	Hydrolysis by penicillinase; allergies
Carbenicillin	Broad	Same as ampicillin	Poor absorption; used only parenterally
Azlocillin, mezlocillin ticarcillin	Very broad	Effective against *Pseudomonas* species; low toxicity compared with aminoglycosides	Allergies, susceptible to many beta-lactamases

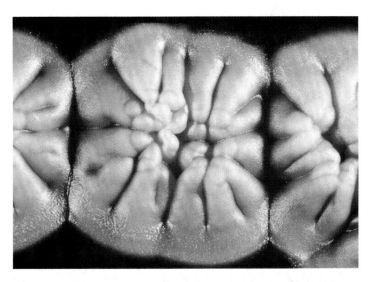

FIGURE 12.10

A colony of Streptomyces, one of nature's most prolific antibiotic producers.

Chloramphenicol

Originally isolated in the late 1940s from *Streptomyces venezuelae*, chloramphenicol is a potent broad-spectrum antibiotic with a unique nitrobenzene structure (figure 12.11*b*). Its primary effect on cells is to block peptide bond formation and protein synthesis. It is one type of antibiotic that is no longer derived from the natural source but is entirely synthesized through chemical processes. Although this drug is fully as broad-spectrum as the tetracyclines, it is so toxic to human cells that its uses are restricted. A small number of people undergoing long-term therapy with this drug incur irreversible damage to the bone marrow that usually results in a fatal form of aplastic anemia.[1] Its administration is now limited to typhoid fever, brain abscesses, and rickettsial and chlamydial infections for which an alternative therapy is not available. Chloramphenicol should never be given in large doses repeatedly over a long time period, and the patient's blood must be monitored during therapy.

---

1. A failure of the blood-producing tissue that results in very low levels of blood cells.

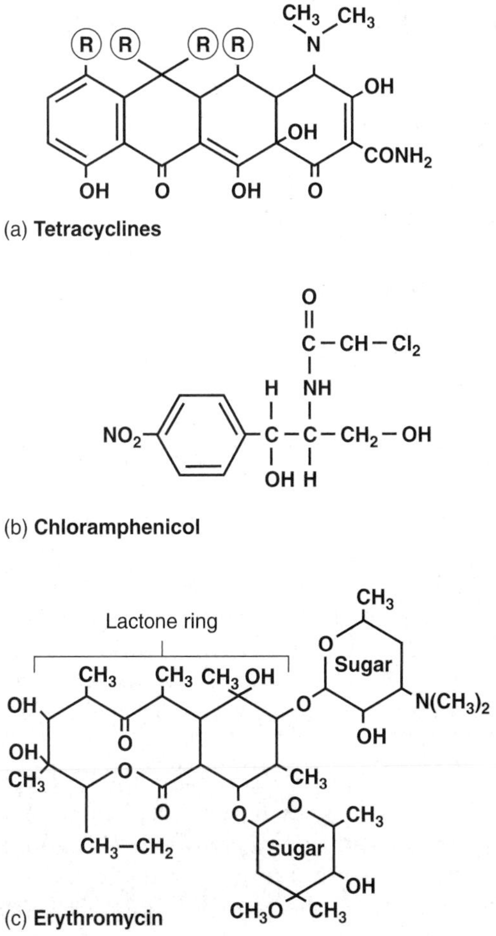

Subgroups and Uses of Aminoglycosides The aminoglycosides have a relatively broad antimicrobial spectrum because they inhibit protein synthesis. They are especially useful in treating infections caused by aerobic gram-negative rods and certain gram-positive bacteria. Streptomycin is among the oldest of the drugs and has gradually been replaced by newer forms with less mammalian toxicity. It is still the antibiotic of choice for treating bubonic plague and tularemia and is considered a good antituberculosis agent. Gentamicin is less toxic and is widely administered for infections caused by gram-negative rods (*Escherichia, Pseudomonas, Salmonella,* and *Shigella*). Two relatively new aminoglycosides, tobramycin and amikacin, are also used for gram-negative bacillary infections and have largely replaced kanamycin.

Tetracycline Antibiotics

In 1948, a colony of *Streptomyces* isolated from a soil sample gave off a substance, aureomycin, with strong antimicrobic properties. This antibiotic was used to synthesize its relatives terramycin and tetracycline. These natural parent compounds and semisynthetic derivatives are known as the **tetracyclines (figure 12.11*a*).** Their action of binding to ribosomes and blocking protein synthesis accounts for the broad-spectrum effects in the group.

Subgroups and Uses of Tetracyclines The scope of microorganisms inhibited by tetracyclines includes gram-positive and gram-negative rods and cocci, aerobic and anaerobic bacteria, mycoplasmas, rickettsias, and spirochetes. Tetracycline compounds such as doxycycline and minocycline are administered orally to treat several sexually transmitted diseases, Rocky Mountain spotted fever, Lyme disease, typhus, *Mycoplasma* pneumonia, cholera, leptospirosis, acne, and even some protozoan infections. Although generic tetracycline is low in cost and easy to administer, its side effects—namely, gastrointestinal disruption due to changes in the normal flora of the gastrointestinal tract and deposition in hard tissues—can limit its use (see table 12.6).

FIGURE 12.11

Structures of miscellaneous broad-spectrum antibiotics.

(a) Tetracyclines. These are named for their regular group of four rings. The several types vary in structure and activity by substitution at the four R groups. **(b)** Chloramphenicol. **(c)** Erythromycin, an example of a macrolide drug. Its central feature is a large lactone ring to which two hexose sugars are attached.

Other *Streptomyces* Antibiotics: Erythromycin, Clindamycin, Vancomycin, Rifamycin

Erythromycin is a macrolide antibiotic first isolated in 1952 from a strain of *Streptomyces*. Its structure consists of a large lactone ring with sugars attached (figure 12.11*c*). This drug is relatively broad-spectrum and of fairly low toxicity. Its mode of action is to block protein synthesis by attaching to the ribosome. It is administered orally as the drug of choice for *Mycoplasma* pneumonia, legionellosis, *Chlamydia* infections, pertussis, and diphtheria and as a prophylactic drug prior to intestinal surgery. It also offers a useful substitute for dealing with penicillin-resistant streptococci and gonococci and for treating syphilis and acne. Newer semisynthetic macrolides include *clarithromycin* and *azithromycin*. Both drugs are useful for middle ear, respiratory, and skin infections and have also been approved for *Mycobacterium* (MAC) infections in AIDS patients. Clarithromycin has additional applications in controlling infectious stomach ulcers.

Clindamycin is a broad-spectrum antibiotic derived from the less effective lincomycin. The tendency of clindamycin to cause adverse reactions in the gastrointestinal tract limits its applications to

1. serious infections in the large intestine and abdomen due to anaerobic bacteria (*Bacteroides* and *Clostridium*) that are unresponsive to other antibiotics,
2. infections with penicillin-resistant staphylococci, and
3. acne medications applied to the skin.

Vancomycin is a narrow-spectrum antibiotic most effective in treating staphylococcal infections in cases of penicillin and methicillin resistance or in patients with an allergy to penicillins. It has also been chosen to treat *Clostridium* infections in children and endocarditis (infection of the lining of the heart) caused by *Enterococcus faecalis*. Because it is very toxic and hard to administer, vancomycin is usually restricted to the most serious, life-threatening conditions.

Another product of the genus *Streptomyces* is rifamycin, which is altered chemically into rifampin. It is somewhat limited in spectrum because the molecule cannot pass through the cell envelope of many gram-negative bacilli. It is mainly used to treat infections by several gram-positive rods and cocci and a few gram-negative bacteria. Rifampin figures most prominently in treating mycobacterial infections, especially tuberculosis and leprosy, but it is usually given in combination with other drugs to prevent development of resistance. Rifampin is also recommended for prophylaxis in *Neisseria meningitidis* carriers and their contacts, and it is occasionally used to treat *Legionella*, *Brucella*, and *Staphylococcus* infections.

The *Bacillus* Antibiotics: Bacitracin and Polymyxin

Bacitracin is a narrow-spectrum peptide antibiotic produced by a strain of the bacterium *Bacillus subtilis*. Since it was first isolated, its greatest claim to fame has been as a major ingredient in a common drugstore antibiotic ointment (Neosporin) for combating superficial skin infections by streptococci and staphylococci. For this purpose, it is usually combined with neomycin (an aminoglycoside) and polymyxin.

Bacillus polymyxa is the source of the **polymyxins**, narrow-spectrum peptide antibiotics with a unique fatty acid component that contributes to their detergent activity (see figure 12.6). Only two polymyxins—B and E (also known as colistin)—have any rou-

tine applications, and even these are limited by their toxicity to the kidney. Either drug can be indicated to treat drug-resistant *Pseudomonas aeruginosa* and severe urinary tract infections caused by other gram-negative rods.

New Classes of Antibiotics

Most new antibiotics are formulated from the traditional drug classes. Two recent additions that are completely new drug types are fosfomycin and synercid. Fosfomycin trimethamine is a phosphoric acid agent effective as alternate treatment for urinary tract infections caused by enteric bacteria. It works by inhibiting an enzyme necessary for cell wall synthesis. Synercid is a combined antibiotic from the streptogramin group of drugs. It is effective against *Staphylococcus* and *Enterococcus* species that cause endocarditis and surgical infections, and against resistant strains of *Streptococcus*. It is one of the main choices when other drugs are ineffective due to resistance. Synercid works by binding to sites on the 50S ribosome, inhibiting translation. Many experts are urging physicians to use these medications only when no other drugs are available to slow the development of drug resistance.

SYNTHETIC ANTIBACTERIAL DRUGS

The synthetic antimicrobics as a group do not originate from bacterial or fungal fermentations. Some were developed from aniline dyes, and others were originally isolated from plants. Although they have been largely supplanted by antibiotics, several types are still useful.

The Sulfonamides, Trimethoprim, and Sulfones

The very first modern antimicrobic drugs were the **sulfonamides,** or sulfa drugs, named for sulfanilamide, an early form of the drug **(figure 12.12).** Although thousands of sulfonamides have been formulated, only a few have gained any importance in chemotherapy. Because of its solubility, sulfisoxazole is the best agent for treating

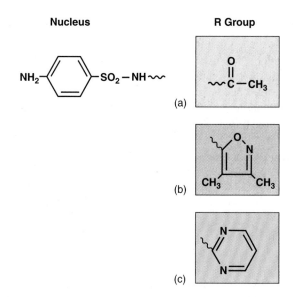

FIGURE 12.12

The structures of some sulfonamides. **(a)** Sulfacetamide, **(b)** sulfadiazine, and **(c)** sulfisoxazole.

shigellosis, acute urinary tract infections, and certain protozoan infections. Silver sulfadiazine ointment and solution are prescribed for treatment of burns and eye infections. Another drug, trimethoprim (Septra, Bactrim), inhibits the enzymatic step immediately following the step inhibited by sulfonamides in the synthesis of folic acid. Because of this, trimethoprim is often given in combination with sulfamethoxazole to take advantage of the synergistic effect of the two drugs (see figure 12.5*a*). This combination is one of the primary treatments for *Pneumocystis (carinii) jiroveci* pneumonia (PCP) in AIDS patients.

Sulfones are compounds chemically related to the sulfonamides but lacking their broad-spectrum effects. This lack does not diminish their importance as key drugs in treating leprosy. The most active form is dapsone, usually given in combination with rifampin and clofazamine (an antibacterial dye) over long periods.

Miscellaneous Antibacterial Agents   Isoniazid (INH) is bactericidal to *Mycobacterium tuberculosis,* but only against growing cells. Although still in use, it has been largely supplanted by rifampicin. Oral doses are indicated for both active tuberculosis and prophylaxis in cases of a positive TB test. Ethambutol, a closely related compound, is effective in treating the early stages of tuberculosis.

A new class of synthetic antibacterial drugs, oxazolidinones was recently developed and the first member of that class, linezolid was approved for use by the FDA. These drugs work by a completely novel mechanism, inhibiting the initiation of protein synthesis. Because this class of drug is not found in nature, it is hoped that resistance among bacteria will be slow to develop. Linezolid (under the name Zyvox) is used to treat infections caused by two of the most difficult clinical pathogens, methicillin-resistant *Staphylococcus aureus* (MRSA) and vancomycin-resistant *Enterococcus* (VRE).

Much excitement has been generated by the new class of synthetic drugs chemically related to quinine called **fluoroquinolones.** These drugs exhibit several ideal traits, including potency and broad spectrum. Even in minimal concentrations, quinolones inhibit a wide variety of gram-positive and gram-negative bacterial species. In addition, they are readily absorbed from the intestine. The principal quinolones, norfloxacin and ciprofloxacin, have been successful in therapy for urinary tract infections, sexually transmitted diseases, gastrointestinal infections, osteomyelitis, respiratory infections, and soft tissue infections. Newer drugs in this category are sparfloxacin and levofloxacin. These agents are especially recommended for pneumonia, bronchitis, and sinusitis. Ciprofloxacin received a great deal of publicity recently as the drug of choice for treating anthrax. Shortly after the first attacks however, the Centers for Disease Control and Prevention changed their recommendation from ciprofloxacin to doxycycline. The change in preferred drug was made not because ciprofloxacin was ineffective but rather to prevent the emergence of ciprofloxacin-resistance bacteria (see Spotlight on Microbiology 12.4). Side effects that limit the use of quinolones include seizures and other brain disturbances.

AGENTS TO TREAT FUNGAL INFECTIONS

Because the cells of fungi are eucaryotic, they present special problems in chemotherapy. For one, the great majority of chemotherapeutic drugs are designed to act on bacteria and are generally ineffective in combating fungal infections. For another, the similarities between fungal and human cells often mean that drugs toxic to fungal cells are also capable of harming human tissues. A few agents with special antifungal properties have been developed for treating systemic and superficial fungal infections. Four main drug groups currently in use are the macrolide polyene antibiotics, griseofulvin, synthetic azoles, and flucytosine (**figure 12.13**).

Polyenes bind to fungal membranes and cause loss of selective permeability. They are specific for fungal membranes because fungal membranes contain a particular sterol component called ergosterol, while human membranes do not (recall that procaryotic organisms do not contain sterols in their membranes). The toxicity of polyenes is not completely selective, however, because mammalian cell membranes contain compounds similar to ergosterol that bind polyenes to a small extent.

Macrolide polyenes, represented by amphotericin B (named for its acidic and basic—amphoteric—properties) and nystatin (for New York State, where it was discovered), have a structure that mimics the lipids in some cell membranes. Amphotericin B (fungizone) is by far the most versatile and effective of all antifungals. Not only does it work on most fungal infections, including skin and mucous membrane lesions caused by *Candida albicans,* but it is one of the few drugs that can be injected to treat systemic fungal infections such as histoplasmosis and cryptococcus meningitis. Nystatin is used only topically or orally to treat candidiasis of the skin and mucous membranes, but it is not useful for subcutaneous or systemic fungal infections or for ringworm.

Griseofulvin is an antifungal product especially active in certain dermatophyte infections such as athlete's foot. The drug is deposited in the epidermis, nails, and hair, where it inhibits fungal growth. Because complete eradication requires several months and griseofulvin is relatively nephrotoxic, this therapy is given only for the most stubborn cases.

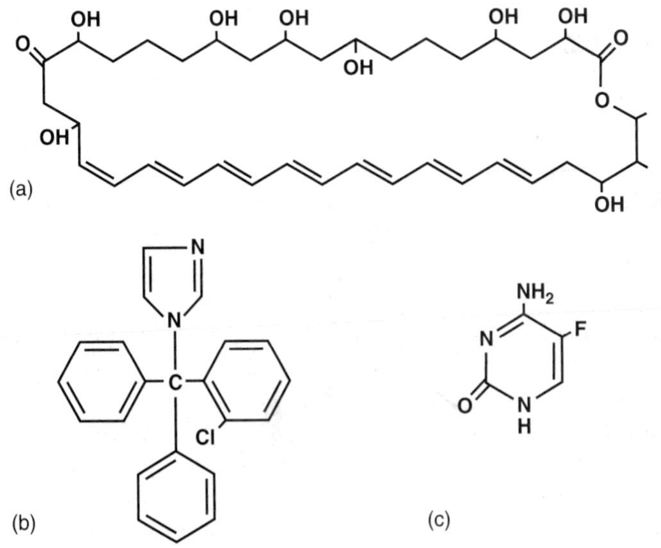

(a)

(b)

(c)

FIGURE 12.13

Some antifungal drug structures.   **(a)** Polyenes. The example shown is amphotericin B (proposed structure), a complex steroidal antibiotic that inserts into fungal cell membranes. **(b)** Clotrimazole, one of the azoles. **(c)** Flucytosine, a structural analog of cystosine that contains fluoride.

The **azoles** are broad-spectrum antifungal agents with a complex ringed structure. The most effective drugs are ketoconazole, fluconazole, clotrimazole, and miconazole. Ketoconazole is used orally and topically for cutaneous mycoses, vaginal and oral candidiasis, and some systemic mycoses. Fluconazole can be used in selected patients for AIDS-related mycoses such as aspergillosis and cryptococcus meningitis. Clotrimazole and miconazole are used mainly as topical ointments for infections in the skin, mouth, and vagina.

Flucytosine is an analog of cytosine that has antifungal properties. It is rapidly absorbed after oral therapy, and it is readily dissolved in the blood and cerebrospinal fluid. Alone, it can be used to treat certain cutaneous mycoses. Now that many fungi are resistant to flucytosine, it must be combined with amphotericin B to effectively treat systemic mycoses.

ANTIPARASITIC CHEMOTHERAPY

The enormous diversity among protozoan and helminth parasites and their corresponding therapies reaches far beyond the scope of this textbook; however, a few of the more common drugs will be surveyed here and again in chapter 23 (see table 23.5). Presently, a small number of approved and experimental drugs are used to treat malaria, leishmaniasis, trypanosomiasis, amebic dysentery, and helminth infections, but the need for new and better drugs has spurred considerable research in this area.

Antimalarial Drugs: Quinine and Its Relatives

Quinine, extracted from the bark of the cinchona tree, was the principal treatment for malaria for hundreds of years, but it has been replaced by the synthesized quinolines, mainly chloroquine and primaquine, which have less toxicity to humans. Because there are several species of *Plasmodium* (the malaria parasite) and many stages in its life cycle, no single drug is universally effective for every species and stage, and each drug is restricted in application. For instance, primaquine eliminates the liver phase of infection, and chloroquine suppresses acute attacks associated with infection of red blood cells. Chloroquine is taken alone for prophylaxis and suppression of acute forms of malaria. Primiquine is administered to patients with relapsing cases of malaria. Mefloquine is a semisynthetic analog of quinine used to treat infections arising from chloroquine-resistant strains of *Plasmodium*.

Chemotherapy for Other Protozoan Infections

A widely used amebicide, metronidazole (Flagyl), is effective in treating mild and severe intestinal infections and hepatic disease caused by *Entamoeba histolytica*. Given orally, it also has applications for infections by *Giardia lamblia* and *Trichomonas vaginalis*. Other drugs with antiprotozoan activities are quinicrine (a quinine-based drug), sulfonamides, and tetracyclines.

Antihelminthic Drug Therapy

Treating helminthic infections has been one of the most difficult and challenging of all chemotherapeutic tasks. Flukes, tapeworms, and roundworms are much larger parasites than other microorganisms and, being animals, have greater similarities to human physiology. Also, the usual strategy of using drugs to block their reproduction is usually not successful in eradicating the adult worms.

The most effective drugs immobilize, disintegrate, or inhibit the metabolism of all stages of the life cycle.

Mebendazole and thiabendazole are broad-spectrum antiparasitic drugs used in several roundworm and tapeworm intestinal infestations. These drugs work locally in the intestine to inhibit the function of the microtubules of worms, eggs, and larvae, which interferes with their glucose utilization and disables them. The compounds pyrantel and piperazine paralyze the muscles of intestinal roundworms. Niclosamide destroys the scolex and the adjoining proglottids of tapeworms, thereby loosening the worm's holdfast. In these forms of therapy, the worms are unable to maintain their grip on the intestinal wall and are expelled along with the feces by the normal peristaltic action of the bowel. Two newer antihelminthic drugs are praziquantel, a treatment for various tapeworm and fluke infections, and ivermectin, a veterinary drug now used for strongyloidiasis and oncocercosis in humans.

ANTIVIRAL CHEMOTHERAPEUTIC AGENTS

The chemotherapeutic treatment of viral infections presents unique problems. Throughout our discussion of infections by bacteria, fungi, protists, and animals, we've relied on differences in structure and metabolism to guide our choice of drug. With viruses, we are dealing with an infectious agent that relies upon the host cell for the vast majority of its metabolic functions. Disrupting viral metabolism requires that we disrupt the metabolism of the host cell to a much greater extent than is desirable. Put another way, selective toxicity with regard to viral infection is almost impossible to achieve because a single metabolic system is responsible for the well-being of both virus and host. Although viral diseases such as measles, mumps, and hepatitis are routinely prevented by the use of effective vaccinations, epidemics of AIDS, influenza, and even the common cold attest to the need for more effective medications for the treatment of viral pathogens.

Over the last few years, several antiviral drugs have been developed that target specific points in the infectious cycle of viruses. Although antiviral drugs are certainly a welcome discovery, the chemotherapeutic treatment of viruses is still in its infancy and most antiviral compounds are quite limited in their usefulness.

Most compounds have their effects on the completion of the virus cycle. Three major modes of action are:

1. barring complete penetration of the virus into the host cell,
2. blocking the transcription and translation of viral molecules, and
3. preventing the maturation of viral particles (**table 12.5**).

Although antiviral drugs protect uninfected cells by keeping viruses from being synthesized and released, most are unable to destroy extracellular viruses or those in a latent state.

Several antiviral agents mimic the structure of nucleotides and compete for sites on replicating DNA. The incorporation of these synthetic nucleotides inhibits further DNA synthesis. **Acyclovir (Zovirax)** and its relatives are synthetic purine compounds that block DNA synthesis in a small group of viruses, particularly the herpesviruses. In the topical form, it is most effective in controlling the primary attack of facial or genital herpes. Intravenous or oral acyclovir therapy can reduce the severity of primary and recurrent genital herpes episodes. Some newer relatives (valacyclovir) are more

TABLE 12.5

Actions of Selected Antiviral Drugs

Category	Example of Drug and Its Structure	Effects of Drug	
Inhibition of Virus Entry: receptor/fusion/uncoating inhibitors	**Fuzeon** Polypeptide of 36 amino acids (trade secret) **Amantidine** and relatives Example: Amantidine **Tamiflu, Relenza** Example: Tamiflu	(1) Blocks HIV infection by preventing the binding of viral GP-41 receptors to cell receptor, thereby preventing fusion of virus with cell (2) Block entry of influenza virus by interfering with fusion of virus with cell membrane and uncoating process (also release) Stop the actions of influenza neuraminidase, required for entry of virus into cell (also assembly)	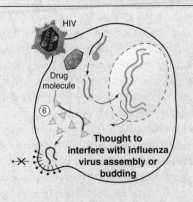 **No infection**

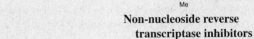

Inhibition of Nucleic Acid Synthesis	**Acyclovir,** other "cyclovirs" Example: Acyclovir **Nucleotide analog reverse transcriptase** (RT) **inhibitors** Example: Zidovudine (AZT) **Non-nucleoside reverse transcriptase inhibitors** Example: Nevirapine	(3) Inactivate viral DNA polymerase and terminates DNA replication in herpesviruses (4) Stop the action of reverse transcriptase in HIV, blocking viral DNA production (5) Attach to HIV RT binding site, stopping its action	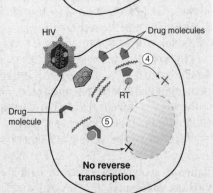 **No viral DNA synthesis** **No reverse transcription**

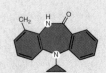

Inhibition of Viral Assembly/Release	**Protease inhibitors** Example: Saquinavir Amantidine and relatives; Tamiflu (see above)	(6) Insert into HIV protease, stopping its action and resulting in inactive noninfectious viruses	**Thought to interfere with influenza virus assembly or budding**

effective and require fewer doses. Famciclovir is used to treat shingles and chickenpox caused by the herpes zoster virus, and ganciclovir is approved to treat cytomegalovirus infections of the eye. An analog of adenine, *vidarabine,* is also effective against the herpesviruses. Ribavirin is a guanine analog used in aerosol form to treat life-threatening infections by the respiratory syncytial virus (RSV) in infants and some types of viral hemorrhagic fever. An interesting aspect of some of these antiviral agents (specifically valacyclovir and famciclovir) is that they are activated by an enzyme encoded by the virus itself, activating the drug only in virally infected cells. The enzyme, thymidine kinase, is used by the virus to process nucleosides before incorporating them into viral RNA or DNA. When the inactive drug enters a virally infected cell, it is activated by the virus' thymidine kinase to produce a working antiviral agent. In cells without viruses, the drug is never activated and DNA replication is allowed to continue unabated.

HIV is classified as a retrovirus, meaning it carries its genetic information in the form of RNA rather than DNA. Upon infection, the RNA genome is used as a template by the enzyme **reverse transcriptase** to produce a DNA copy of the virus' genetic information. Because this particular reaction is not seen outside of the retroviruses, it offers two ideal targets for chemotherapy. The first is interfering with the synthesis of the new DNA strand, which is accomplished using *nucleoside reverse transcriptase inhibitors* (nucleotide analogs) while the second involves interfering with the action of the enzyme responsible for the synthesis, which is accomplished using *non-nucleoside reverse transcriptase inhibitors.*

Azidothymidine (AZT or zidovudine) is a thymine analog that exerts its effect by incorporating itself into the growing DNA chain of HIV and terminating synthesis, in a manner analogous to that seen with acyclovir (see figure 25.19). AZT is used at all stages of HIV infection, including prophylactically with people accidentally exposed to blood or other body fluids. Other approved anti-HIV drugs that act as nucleotide analogs are abacavir, lamivudine (3T3), didanosine (ddI), zalcitabine (ddC), and stavudine (d4T).

Non-nucleoside reverse transcriptase inhibitors accomplish the same goal (preventing reverse transcription of the HIV genome) by binding to the reverse transcriptase enzyme itself, inhibiting its ability to synthesize DNA. Drugs with this mode of action include nevirapine, efavirenz, and delaviridine.

Assembly and release of mature viral particles is also targeted in HIV through the use of protease inhibitors. These drugs (indinavir, saquinavir, nelfinavir, amprenavir), usually used in combination with nucleotide analogs and reverse transcriptase inhibitors, have been shown to reduce the HIV load to undetectable levels by preventing the maturation of virus particles in the cell. Refer to table 12.5 for a summary of HIV drug mechanisms and see chapter 25 for further coverage of this topic.

Amantadine and its relative, rimantidine, are restricted almost exclusively to treating infections by influenza A virus. Relenza and Tamiflu medications are effective treatments for influenza A and B and useful prophylactics as well. Because one action of these drugs is to inhibit the fusion and uncoating of the virus, they must be given rather early in an infection (table 12.5). In addition to these antiviral drugs, dozens of other agents are under investigation. For a novel approach for controlling viruses see **Microbits 12.3.**

A sensible alternative to artificial drugs has been a human-based substance, **interferon (IFN).** Interferon is a carbohydrate-containing protein produced primarily by fibroblasts and leukocytes in response to various immune stimuli. It has numerous biological activities, including antiviral and anticancer properties. Studies have shown that it is a versatile part of animal host defenses, having a great import in natural immunities. (Its mechanism is discussed in chapter 14.)

The first investigations of interferon's antiviral activity were limited by the extremely minute quantities that could be extracted from human blood. Several types of interferon are currently produced by the recombinant DNA technology techniques outlined in chapter 10. Extensive clinical trials have tested its effectiveness in viral infections and cancer. Some of the known therapeutic benefits of interferon include:

1. reducing the time of healing and some of the complications in certain infections (mainly of herpesviruses);
2. preventing or reducing some symptoms of cold and papillomaviruses (warts);
3. slowing the progress of certain cancers, including bone cancer and cervical cancer, and certain leukemias and lymphomas; and
4. treating a rare cancer called hairy-cell leukemia, hepatitis C (a viral liver infection), genital warts, and Kaposi's sarcoma in AIDS patients.

CHAPTER CHECKPOINTS

Antimicrobics are classified into 20 major drug families, based on their chemical composition, source of origin, and their site of action.

The majority of antimicrobics are effective against bacteria, but a limited number are effective against protozoa, helminths, fungi, and viruses.

Penicillins, cephalosporins, bacitracin, vancomycin, and cycloserines block cell wall synthesis, primarily in gram-positive bacteria.

Aminoglycosides and tetracyclines block protein synthesis in procaryotes.

Sulfonamides, trimethoprim, isoniazid, nitrofurantoin, and the fluoroquinolones are synthetic antimicrobics effective against a broad range of microorganisms. They block steps in the synthesis of nucleic acids.

Fungal antimicrobials, such as macrolide polyenes, griseofulvin, azoles, and flucytosine, must be monitored carefully because of the potential toxicity to the infected host. They promote lysis of cell membranes.

There are fewer antiparasitic drugs than antibacterial drugs because parasites are eucaryotes like their human hosts and they have several life stages, some of which can be resistant to the drug.

Antihelminthic drugs immobilize or disintegrate infesting helminths or inhibit their metabolism in some manner.

Antiviral drugs interfere with viral replication by blocking viral entry into cells, blocking the replication process, or preventing the assembly of viral subunits into complete virions.

Many antiviral agents are analogs of nucleotides. They inactivate the replication process when incorporated into viral nucleic acids. HIV antivirals interfere with reverse transcriptase or proteases to prevent the maturation of viral particles.

Although interferon is effective *in vivo* against certain viral infections, commercial interferon is not currently effective as a broad-spectrum antiviral agent.

MICROBITS   12.3

Household Remedies—From Apples to Zinc

Who would have thought that drinking a glass of apple juice, swallowing a clove of garlic, or eating yogurt could be a reliable therapy for infections? A series of research findings from the past few years seems to point to a possible role for these as medicinal aids. Apple juice, fruit juices, and even tea contain natural antiviral substances, thought to be organic acids, that kill the poliovirus and coxsackievirus. Drinking beverages that contain these substances can help prevent the passage of those viruses into the intestine (their usual site of entry). Could this be a reason that "an apple a day keeps the doctor away"?

Another common home remedy has been drinking cranberry juice to reduce the symptoms of urinary tract infections. New research indicates that tannins in this product are excreted through the kidney and can block the attachment of pili by pathogens such as *Escherichia coli*. Controlled studies now support the benefits of zinc ions to control the common cold. It has been suggested that the zinc attaches to cell receptors and blocks the attachment of cold viruses. Various tablets and lozenges are now sold over the counter as cold deterrents.

The therapeutic benefits of certain foods are often surprising. Yogurt made with live cultures has been shown in controlled studies to inhibit *Staphylococcus* infections. Further examination demonstrated that some compound given off by *Lactobacillus* in yogurt prevents the adhesion of pathogens to tissues. This may explain the benefits of eating yogurt to control infections of the gastrointestinal tract and vagina. Research indicates that compounds in garlic extract inactivate viruses and destroy bacteria. If all else fails, one should not overlook the recuperative powers of chicken soup, sometimes known as "Jewish penicillin." This, too, has been found in controlled scientific tests to shorten the length and relieve the symptoms of colds, though the active ingredients have not been isolated. It appears that a timely trip to the kitchen could be as beneficial as one to the medicine cabinet. Perhaps the saying "feeding a cold and starving a fever" has some basis in fact.

THE ACQUISITION OF DRUG RESISTANCE

One unfortunate outcome of the wide-scale use of antimicrobics is the development of microbial drug resistance, an adaptive response in which microorganisms begin to tolerate an amount of drug that would ordinarily be inhibitory. The development of mechanisms for circumventing or inactivating antimicrobic drugs is due largely to the genetic versatility and adaptability of microbial populations. The property of drug resistance can be intrinsic as well as acquired. Intrinsic drug resistance can best be exemplified by the fact that bacteria must, of course, be resistant to any antibiotic that they produce. This type of resistance is, however, limited to a small group of organisms and is generally not a problem with regard to antimicrobial chemotherapy. Of much greater importance is the acquisition of resistance to a drug by a microbe that was previously sensitive to the drug. In our context, the term drug resistance will refer to this last type of acquired resistance.

How Does Drug Resistance Develop?

The genetic events most often responsible for drug resistance are chromosomal mutations or extrachromosomal DNA that are transferred from a resistant species to a sensitive one. Chromosomal drug resistance usually results from spontaneous random mutations in bacterial populations. The chance that such a mutation will be advantageous is minimal, and the chance that it will confer resistance to a specific drug is lower still. Nevertheless, given the huge numbers of microorganisms in any population and the constant rate of mutation, such mutations do occur. The end result varies from slight changes in microbial sensitivity, which can be overcome by larger doses of the drug, to complete loss of sensitivity.

Resistance associated with intermicrobial transfer originates from plasmids called **resistance (R) factors** that are transferred through conjugation, transformation, or transduction. Studies have shown that plasmids encoded with drug resistance are naturally present in microorganisms before they have been exposed to the drug. Such traits are "lying in wait" for an opportunity to be expressed and to confer adaptability on the species. Many bacteria also maintain transposable drug resistance sequences (transposons) that are duplicated and inserted from one plasmid to another or from a plasmid to the chromosome. Chromosomal genes and plasmids containing codes for drug resistance are faithfully replicated and inherited by all subsequent progeny. This sharing of resistance genes accounts for the rapid proliferation of drug-resistant species **(figure 12.14)**. A growing body of evidence points to the ease and

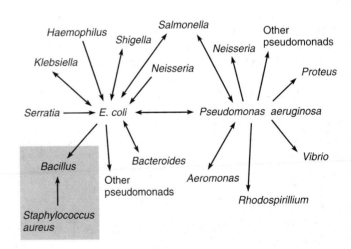

FIGURE 12.14

Spread of resistance factors.   This figure traces documented evidence of known cases in which R factors have been transferred among pathogens. Such promiscuous exchange of drug resistance occurs primarily by conjugation and transduction. Most of the bacteria are gram-negative, but *Bacillus* and *Staphylococcus* are unrelated gram-positive genera. This phenomenon is responsible for the rapid spread of drug-resistant microbes.

Source: Data from Young and Mayer, Review of Infectious Diseases, *1:55, 1979.*

frequency of gene transfers in nature, from totally unrelated bacteria living in the body's normal flora and the environment.

Specific Mechanisms of Drug Resistance

In general, a microorganism becomes resistant to a drug by expressing genes that stop the action of the drug. Gene expression takes the form of:

1. synthesis of enzymes that inactivate the drug,
2. decrease in cell permeability and uptake of the drug,
3. change in the number or affinity of the drug receptor sites, or
4. modification of an essential metabolic pathway.

Some bacteria can become resistant indirectly by lapsing into dormancy, or, in the case of penicillin, by converting to a cell-wall-deficient form (L form) that penicillin cannot affect.

Drug Inactivation Mechanisms Microbes inactivate drugs by producing enzymes that permanently alter drug structure. One example, bacterial exoenzymes called **beta-lactamases,** hydrolyze the *beta-lactam** ring structure of some penicillins and cephalosporins rendering the drugs inactive. Two beta-lactamases—**penicillinase** and cephalosporinase—disrupt the structure of certain penicillin or cephalosporin molecules so their activity is lost **(figure 12.15a).** So many strains of *Staphylococcus aureus* produce penicillinase that regular penicillin is rarely a possible therapeutic choice. Now that some strains of *Neisseria gonorrhoeae,* called PPNG,[2] have also acquired penicillinase, alternative drugs are required to treat gonorrhea. A large number of other gram-negative species are inherently resistant to some of the penicillins and cephalosporins because of naturally occurring beta-lactamases.

---

2. Penicillinase-producing *Neisseria gonorrhoeae.*

*beta-lactam (bay'-tuh-lak'-tam) Molecular structure shown in figure 12.7.

(a) **Drug inactivation**

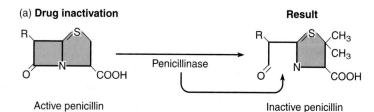

Inactivation of a drug like penicillin by penicillinase, an enzyme that cleaves a portion of the molecule and renders it inactive.

(b) **Decreased permeability**

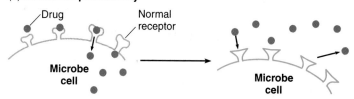

The receptor that transports the drug is altered, so that the drug cannot enter the cell.

(c) **Activation of drug pumps**

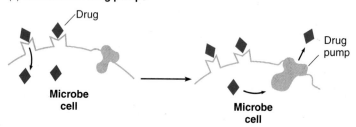

Specialized membrane proteins are activated and continually pump the drug out of the cell.

(d) **Use of alternate metabolic pathway**

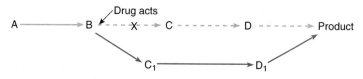

The drug has blocked the usual metabolic pathway, so the microbe circumvents it by using an alternate, unblocked pathway that achieves the required outcome.

FIGURE 12.15

Examples of mechanisms of acquired drug resistance.

Decreased Drug Permeability or Increased Drug Elimination

The resistance of some bacteria can be due to a mechanism that prevents the drug from entering the cell and acting on its target. For example, the outer membrane of the cell wall of certain gram-negative bacteria is a natural blockade for some of the penicillin drugs. Resistance to the tetracyclines can arise from plasmid-encoded proteins that pump the drug out of the cell. Resistance to the aminoglycoside antibiotics is a special case in which microbial cells have lost the capacity to transport the drug intracellularly (figure 12.15*b*).

Many bacteria possess multidrug resistant (MDR) pumps that actively transport drugs and other chemicals out of cells. These pumps are proteins encoded by plasmids or chromosomes. They are stationed in the cell membrane and expel molecules by a proton-motive force similar to ATP synthesis (figure 12.15*c*). They confer drug resistance on many gram-positive pathogens *(Staphylococcus, Streptococcus)* and gram-negative pathogens *(Pseudomonas, E. coli)*. Because they lack selectivity, one type of pump can expel a broad array of antimicrobial drugs, detergents, and other toxic substances.

Change of Drug Receptors

Because most drugs act on a specific target such as protein, RNA, DNA, or membrane structure, microbes can circumvent drugs by altering the nature of this target. In bacteria resistant to rifampin and streptomycin, the structure of key proteins has been altered so that these antibiotics can no longer bind to their cellular targets (figure 12.15*b*). Erythromycin and clindamycin resistance is associated with an alteration on the 50S ribosomal binding site. Penicillin resistance in *Streptococcus pneumoniae* and methicillin resistance in *Staphylococcus aureus* are related to an alteration in the binding proteins in the cell wall. Several species of enterococci have acquired resistance to vancomycin through a similar alteration of cell wall proteins. Fungi can become resistant by decreasing their synthesis of ergosterol, the principal receptor for certain antifungal drugs.

Changes in Metabolic Patterns

The action of antimetabolites can be circumvented if a microbe develops an alternative metabolic pathway or enzyme (figure 12.15*d*). Sulfonamide and trimethoprim resistance develops when microbes deviate from the usual patterns of folic acid synthesis. Fungi can acquire resistance to flucytosine by completely shutting off certain metabolic activities.

Natural Selection and Drug Resistance

So far, we have been considering drug resistance at the cellular and molecular levels, but its full impact is felt only if this resistance occurs throughout the cell population. Let us examine how this might happen and its long-term therapeutic consequences. Recall that any large population of microbes is likely to contain a few individual cells that are already drug-resistant because of prior mutations or transfer of plasmids **(figure 12.16*a*)**. As long as the drug is not present in the habitat, the numbers of these resistant forms will remain low because they have no particular growth advantage. But if the population is subsequently exposed to this drug (figure 12.16*b*), sensitive individuals are inhibited or destroyed, and resistant forms survive and proliferate. During subsequent population growth, offspring of these resistant microbes will inherit this drug resistance. In time, the replacement population will have a preponderance of the drug-resistant forms and can eventually become completely resistant (figure 12.16*c*). In ecological terms, the environmental factor (in this case, the drug) has put selection pressure on the population, allowing the more "fit" microbe (the drug-resistant one) to survive, and the population has evolved to a condition of drug resistance. Natural selection for drug-resistant forms is apparently a common phenomenon. It takes place most frequently in various natural habitats, laboratories, and medical environments, and it even occurs within the bodies of humans and animals during drug therapy **(Spotlight on Microbiology 12.4)**.

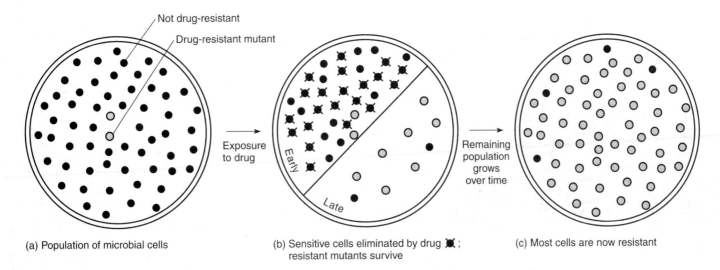

(a) Population of microbial cells

(b) Sensitive cells eliminated by drug ✖; resistant mutants survive

(c) Most cells are now resistant

FIGURE 12.16

The events in natural selection for drug resistance. **(a)** Populations of microbes can harbor some members with a prior mutation that confers drug resistance. **(b)** Environmental pressure (here, the presence of the drug) selects for survival of these mutants. **(c)** They eventually become the dominant members of the population.

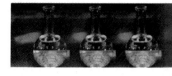

SPOTLIGHT ON MICROBIOLOGY 12.4
The Rise of Drug Resistance

There has been an unrealistic tendency to assume that science would come to the rescue and solve the problem of drug resistance. If drug companies just keep making more and better antimicrobics, soon infectious diseases will be vanquished. This unfortunate attitude has vastly underestimated the extreme versatility and adaptability of microorganisms and the complexity of the task. It is a fact of nature that if you expose a large number of microbes to a variety of drugs, there will always be some genetically favored individuals that survive and thrive. The AIDS virus (HIV) is so prone to drug resistance that it can become resistant during the first few weeks of therapy in one individual. Because it mutates so rapidly, this means that in most cases, it will eventually become resistant to all drugs that have been developed so far.

Ironically, thousands of patients die every year in the United States from infections that lack effective drugs, and 60% of hospital infections are caused by drug-resistant microbes. For many years, concerned observers reported the gradual development of drug resistance in staphylococci, *Salmonella*, and gonococci. But during the past decade, the scope of the problem has escalated. It is now a common event to

TABLE 12.A

Organism	Drug	Year and Prevalence of Resistance		
Pneumococci		**1989**	**2000**	**2002**
	Penicillin	Rare	30%	40%
	Erythromycin	0	15%	40%
	Cephalosporin	0	14%	27%
		1990	**2000**	**2002**
Gonococci	Penicillin	10	70%	74%
	Fluoroquinolone	1.4%	10%	12%
Enterococci	Vancomycin	Rare	50%	52%
		1995	**2000**	**2002**
Campylobacter	Fluoroquinolones	Rare	10%	21%

discover microbes that have become resistant to relatively new drugs in a very short time. In fact, many strains of pathogens have multiple drug resistance, and a few are resistant to all drugs. **Table 12.A** lists some notable examples.

The Hospital Factor

The clinical setting is a prolific source of drug-resistant strains of bacteria. This environment continually exposes pathogens to a variety of drugs, since drugs are inevitably released during therapy. The hospital also maintains patients with weakened defenses, making them highly susceptible to pathogens. A classic example occurred with *Staphylococcus aureus* and penicillin. In the 1950s, hospital strains began to show resistance to this drug, and because of indiscriminate use, they became nearly 100% resistant in 30 years. In a short time, MRSA (methicillin-resistant *S. aureus*) strains appeared, which can tolerate nearly all antibiotics. Up until recently, MRSA has been sensitive to the drug of last resort, vancomycin. In 2002, the first cases of complete resistance to this drug were reported (VRSA). To complicate this problem, strains of MRSA are now being spread into the community.

Drugs in Animal Feeds

Another practice that has contributed significantly to growing drug resistance is the addition of antimicrobics to livestock feed, with the idea of decreasing infections and thereby improving animal health. This practice has had serious impact in both the United States and Europe.

Enteric bacteria such as *Salmonella, Escherichia coli,* and enterococci that live as normal intestinal flora of these animals readily share resistance plasmids and are constantly selected and amplified by exposure to drugs. These pathogens subsequently "jump" to humans and cause drug-resistant infections, oftentimes at epidemic proportions. In a deadly outbreak of *Salmonella* infection in Denmark, the pathogen was found to be resistant to seven different antimicrobics. In the United States, a strain of fluoroquinolone-resistant *Campylobacter* from chickens caused over 5,000 cases of food infection in the late 1990s. The opportunistic pathogen called VRE (vancomycin-resistant enterococcus) has been traced to the use of a vancomycin-like drug in cattle feed. It is now one of the most tenacious of hospital-acquired infections for which there are few drug choices. To attempt to curb this source of resistance, Europe and the United States have banned the use of human drugs in animal feeds.

Worldwide Drug Resistance

The drug dilemma has become a widespread problem, affecting all countries and socioeconomic groups. In general, the majority of infectious diseases, whether bacterial, fungal, protozoan, or viral, are showing increases in drug resistance. In parts of India, the main drugs used to treat cholera (furazolidone, ampicillin) have gone from being highly effective to essentially useless in 10 years. In Southeast Asia 98% of gonococcus infections are multidrug resistant. Malaria, tuberculosis, and typhoid fever pathogens are gaining in resistance, with few alternate drugs to control them. To add to the problem, global travel and globalization of food products means that drug resistance can be rapidly exported.

In countries with adequate money to pay for antimicrobics, most infections will be treated, but at some expense. In the United States alone, the extra cost for treating the drug-resistant variety is around $10 billion per year. In many developing countries, drugs are mishandled by overuse and underuse, either of which can contribute to drug resistance. Many countries that do not regulate the sale of prescription drugs make them readily available to purchase over the counter. For example, the antituberculosis drug INH (isoniazid) is sometimes used as a "lung vitamin" to improve health, and antibiotics are taken in the wrong dose and wrong time for undiagnosed conditions. This means that these countries serve as breeding grounds for drug resistance that can eventually be carried to other countries.

It is clear that we are in a race with microbes and we are falling behind. If the trend is not contained, the world may return to a time when there are few effective drugs left. We simply cannot develop them as rapidly as microbes can develop resistance. In this light, it is essential to fight the battle on more than one front. **Table 12.B** summarizes the several critical strategies to give us an edge in controlling drug resistance.

TABLE 12.B

Strategies to Limit Drug Resistance by Microorganisms

Drug Usage
- Physicians have the responsibility for making an accurate diagnosis and prescribing the correct drug therapy.
- Patients must comply with and carefully follow the physician's guidelines. It is important for the patient to take the correct dosage, by the best route, for the appropriate period. This diminishes the selection for mutants that can resist low drug levels, and ensures elimination of the pathogen.
- Combined therapy: Administering two or more drugs together increases the chances that at least one of the drugs will be effective and that a resistant strain of either drug will not be able to persist. The basis for this method lies in the unlikelihood of simultaneous resistance to several drugs.

Drug Research
- Developing shorter-term, higher-dose antimicrobics that are more effective, less expensive, and have fewer side effects.
- Pharmaceutical companies continue to seek new antimicrobic drugs with structures that are not readily inactivated by microbial enzymes or drugs with modes of action that are not readily circumvented.

Long-Term Strategies
Antimicrobics (especially those for bacterial infections) are overproduced, overprescribed, and used inappropriately on a very wide scale.
- Proposals to reduce the abuses range from educational programs for health workers to requiring written justification from the physician on all antibiotics prescribed.
- Especially valuable antimicrobics may be restricted in their use to only one or two types of infections.
- The addition of antimicrobics to animal feeds must be curtailed worldwide.
- Increase government programs which make effective therapy available to low-income populations.
- Use vaccines whenever possible to provide alternative protection.

Characteristics of Host-Drug Reactions

Although selective antimicrobial toxicity is the ideal constantly being sought, chemotherapy by its very nature involves contact with foreign chemicals that can harm human tissues. In fact, estimates indicate that at least 5% of all persons taking an antimicrobial drug experience some type of serious adverse reaction to it. The major side effects of drugs fall into one of three categories: direct damage to tissues through toxicity, allergic reactions, and disruption in the balance of normal microbial flora. The damage incurred by antimicrobial drugs can be short-term and reversible or permanent, and it ranges in severity from cosmetic to lethal. **Table 12.6** summarizes drug groups and their major side effects.

TOXICITY TO ORGANS

Certain drugs adversely affect the following organs: the liver (hepatotoxic), kidneys (nephrotoxic), gastrointestinal tract, cardiovascular system and blood-forming tissue (hemotoxic), nervous system (neurotoxic), respiratory tract, skin, bones, and teeth.

Because the liver is responsible for metabolizing and detoxifying foreign chemicals in the blood, it can be damaged by a drug or its metabolic products. Injury to liver cells can result in enzymatic abnormalities, fatty liver deposits, hepatitis, and liver failure. The kidney is involved in excreting drugs and their metabolites. Some drugs irritate the nephron tubules, creating changes that interfere with their filtration abilities. Drugs such as sulfonamides crystallize in the kidney pelvis and form stones that can obstruct the flow of urine.

The most common complaint associated with oral antimicrobial therapy is diarrhea, which can progress to severe intestinal irritation or colitis. Although some drugs directly irritate the intestinal lining, the usual gastrointestinal complaints are caused by disruption of the intestinal microflora (discussed in a subsequent section).

Many drugs given for parasitic infections are toxic to the heart, causing irregular heartbeats and even cardiac arrest in extreme cases. Chloramphenicol can severely depress blood-forming cells in the bone marrow, resulting in either a reversible or a permanent (fatal) anemia. Some drugs hemolyze the red blood cells, others reduce white blood cell counts, and still others damage platelets or interfere with their formation, thereby inhibiting blood clotting.

Certain antimicrobics act directly on the brain and cause seizures. Others, such as aminoglycosides, damage nerves (very commonly, the 8th cranial nerve), leading to dizziness, deafness, or motor and sensory disturbances. When drugs block the transmission of impulses to the diaphragm, respiratory failure can result.

The skin is a frequent target of drug-induced side effects. The skin response can be a symptom of drug allergy or a direct toxic effect. Some drugs interact with sunlight to cause photodermatitis, a skin inflammation. Tetracyclines are contraindicated (not advisable) for children from birth to 8 years of age because they bind to the enamel of the teeth, creating a permanent gray to brown discoloration (**figure 12.17**). Pregnant women should avoid tetracyclines because they cross the placenta and can be deposited in the developing fetal bones and teeth.

ALLERGIC RESPONSES TO DRUGS

One of the most frequent drug reactions is heightened sensitivity, or **allergy.** This reaction occurs because the drug acts as an antigen (a foreign material capable of stimulating the immune system) and stimulates an allergic response. This response can be provoked by the intact drug molecule or by substances that develop from the body's metabolic alteration of the drug. In the case of penicillin, for instance, it is not the penicillin molecule itself that causes the allergic response but a product, *benzlpenicilloyl*. Allergic reactions have been reported for every major type of antimicrobic drug, but the penicillins account for the greatest number of antimicrobic allergies, followed by the sulfonamides.

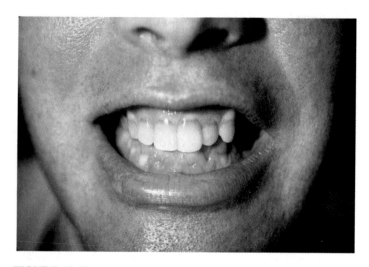

FIGURE 12.17

Drug induced side effect. An adverse effect of tetracycline given to young children is the permanent discoloration of tooth enamel.

TABLE 12.6

Major Adverse Toxic Reactions to Common Drug Groups

Antimicrobic Drug	Primary Tissue Affected	Primary Damage or Abnormality Produced
Antibacterials		
Penicillin G	Skin	Rash
Carbenicillin	Platelets	Abnormal bleeding
Ampicillin	GI tract	Diarrhea and enterocolitis
Cephalosporins	Platelet function	Inhibition of prothrombin synthesis
	White blood cells	Decreased circulation
	Kidney	Nephritis
Tetracyclines	GI tract	Diarrhea and enterocolitis
	Teeth, bones	Discoloration of tooth enamel
	Skin	Reactions to sunlight (photosensitization)
Chloramphenicol	Bone marrow	Injury to red and white blood cell precursors
Aminoglycosides (streptomycin, gentamicin, amikacin)	GI tract, hair cells in cochlea, vestibular cells, neuromuscular, kidney tubules	Diarrhea and enterocolitis; malabsorption; loss of hearing, dizziness, kidney damage
Isoniazid	Liver	Hepatitis
	Brain	Seizures
	Skin	Dermatitis
Sulfonamides	Kidney	Formation of crystals; blockage of urine flow
	Red blood cells	Hemolysis
	Platelets	Reduction in number
Polymyxin	Kidney	Damage to membranes of tubule cells
	Neuromuscular	Weakened muscular responses
Quinolones (ciprofloxacin, norfloxacin)	Nervous system, bones, GI tract	Headache, dizziness, tremors, GI distress
Rifampin	Liver	Damage to hepatic cells
	Skin	Dermatitis
Antifungals		
Amphotericin B	Kidney	Disruption of tubular filtration
Flucytosine	White blood cells	Decreased number
Antiprotozoan Drugs		
Metronidazole	GI tract	Nausea, vomiting
Chloroquine	GI tract	Vomiting
	Brain	Headache
	Skin	Itching
Antihelminthics		
Niclosamide	GI tract	Nausea, abdominal pain
Pyrantel	GI tract	Irritation
	Brain	Headache, dizziness
Antivirals		
Acyclovir	Brain	Seizures, confusion
	Skin	Rash
Amantadine	Brain	Nervousness, light-headedness
	GI tract	Nausea
AZT	Bone marrow	Immunosuppression, anemia

People who are allergic to a drug become sensitized to it during the first contact, usually without symptoms. Once the immune system is sensitized, a second exposure to the drug can lead to a reaction such as a skin rash (hives), respiratory inflammation, and, rarely, anaphylaxis, an acute, overwhelming allergic response that develops rapidly and can be fatal. (This topic is discussed in greater detail in chapter 17.)

SUPPRESSION AND ALTERATION OF THE MICROFLORA BY ANTIMICROBICS

Most normal, healthy body surfaces, such as the skin, large intestine, outer openings of the urogenital tract, and oral cavity, provide numerous habitats for a virtual "garden" of microorganisms. These normal colonists or residents, called the **flora*** or microflora, consist mostly of harmless or beneficial bacteria, but a small number can potentially be pathogens. Although we shall defer a more detailed discussion of this topic to chapter 13, here we focus on the general effects of drugs on this population.

If a broad-spectrum antimicrobic is introduced into the host to treat infection, it will destroy microbes regardless of their roles in the balance, affecting not only the targeted infectious agent but also many others in sites far removed from the original infection (**figure 12.18**). When this therapy destroys beneficial resident species, the pathogens that were once in small numbers begin to overgrow and cause disease. This complication is called a **superinfection.**

Some common examples demonstrate how a disturbance in microbial flora leads to replacement flora and superinfection. A

broad-spectrum cephalosporin used to treat a urinary tract infection by *Escherichia coli* will cure the infection, but it will also destroy the lactobacilli in the vagina that normally maintain a protective acidic environment there. The drug has no effect, however, on *Candida albicans,* a yeast that also resides in normal vaginas. Released from the inhibitory environment provided by lactobacilli, the yeasts proliferate and cause an infection. *Candida* can cause similar superinfections of the oropharynx (thrush) and the large intestine.

Oral therapy with tetracyclines, clindamycin, and broad-spectrum penicillins and cephalosporins is associated with a serious and potentially fatal condition known as *antibiotic-associated colitis* (pseudomembranous colitis). This condition is due to the overgrowth in the bowel of *Clostridium difficile,* a spore-forming bacterium that is resistant to the antibiotic. It invades the intestinal lining and releases toxins that induce diarrhea, fever, and abdominal pain.

*flora (flor′-ah) Gr. *flora,* the goddess of flowers. The microscopic life present in a particular location.

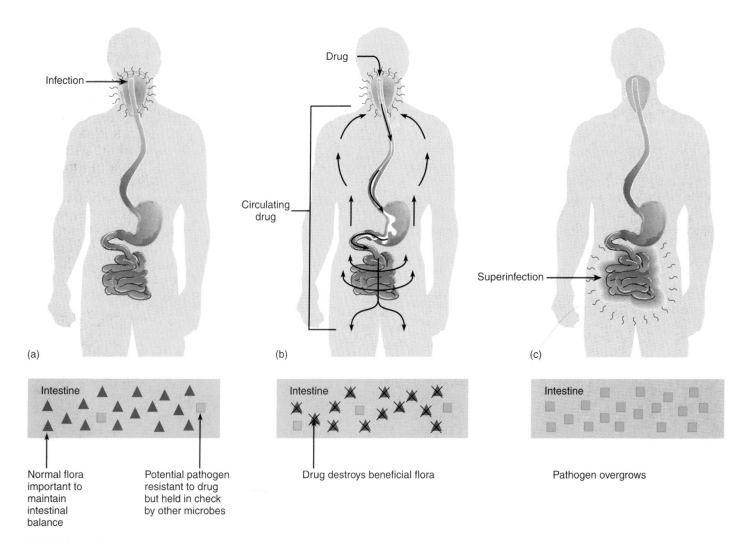

(a)   (b)   (c)

FIGURE 12.18

The role of antimicrobics in disrupting microbial flora and causing superinfections.   **(a)** A primary infection in the throat is treated with an oral antibiotic. **(b)** The drug is carried to the intestine and is absorbed into the circulation. **(c)** The primary infection is cured, but drug-resistant pathogens have survived and create an intestinal superinfection.

CHAPTER CHECKPOINTS

The three major side effects of antimicrobics are toxicity to organs, allergic reactions, and problems resulting from suppression or alteration of normal flora.

Antimicrobics that destroy most but not all normal flora allow the unaffected normal flora to overgrow, causing a superinfection.

Considerations in Selecting an Antimicrobic Drug

Before actual antimicrobic therapy can begin, it is important that at least three factors be known:

1. the nature of the microorganism causing the infection,
2. the degree of the microorganism's susceptibility (also called sensitivity) to various drugs, and
3. the overall medical condition of the patient.

IDENTIFYING THE AGENT

Identification of infectious agents from body specimens should be attempted as soon as possible. It is especially important that such specimens be taken before the antimicrobic drug is given, just in case the drug eliminates the infectious agent. Direct examination of body fluids, sputum, or stool is a rapid initial method for detecting and perhaps even identifying bacteria or fungi. A doctor often begins the therapy on the basis of such immediate findings. The choice of drug will be based on experience with drugs that are known to be effective against the microbe; this is called the "informed best guess." For instance, if a sore throat appears to be caused by *Streptococcus pyogenes,* the physician might prescribe penicillin, because this species seems to be universally sensitive to it so far. If the infectious agent is not or cannot be isolated, epidemiologic statistics may be required to predict the most likely agent in a given infection. For example, *Streptococcus pneumoniae* accounts for the majority of cases of meningitis in children, followed by *Neisseria meningitidis* and *Haemophilus influenzae.*

TESTING FOR THE DRUG SUSCEPTIBILITY OF MICROORGANISMS

Testing is essential in those groups of bacteria commonly showing resistance, primarily *Staphylococcus* species, *Neisseria gonorrhoeae, Streptococcus pneumoniae,* and *Enterococcus faecalis,* and the aerobic gram-negative enteric bacilli. However, not all infectious agents require antimicrobial sensitivity testing. Drug testing in fungal or protozoan infections is difficult and is often unnecessary. When certain groups, such as group A streptococci and all anaerobes (except *Bacteroides*), are known to be uniformly susceptible to penicillin G, testing may not be necessary unless the patient is allergic to penicillin.

Selection of a proper antimicrobial agent begins by demonstrating the *in vitro* activity of several drugs against the infectious agent by means of standardized methods. In general, these tests involve exposing a pure culture of the bacterium to several different drugs and observing the effects of the drugs on growth.

The *Kirby-Bauer* technique is an agar diffusion test that provides useful data on antimicrobic susceptibility. In this test, the surface of a plate of special medium is seeded with the test bacterium, and small discs containing a premeasured amount of antimicrobic are dispensed onto the bacterial lawn. After incubation, the zone of inhibition surrounding the discs is measured and compared with a standard for each drug (**table 12.7** and **figure 12.19**). The profile of antimicrobic sensitivity, or *antibiogram,* provides data for drug selection. The Kirby-Bauer procedure is less effective for bacteria

---

TABLE 12.7

Results of Kirby-Bauer Test

Drug	Zone Sites (mm) Required For:		Actual Result (mm) for: *Staphylococcus aureus*	Evaluation
	Susceptibility (S)	Resistance (R)		
Bacitracin	>13	<8	15	S
Chloramphenicol	>18	<12	20	S
Erythromycin	>18	<13	25	S
Gentamicin	>13	<12	16	S
Kanamycin	>18	<13	20	S
Neomycin	>17	<12	12	R
Penicillin G	>29	<20	10	R
Polymyxin B	>12	<8	10	R
Streptomycin	>15	<11	11	R
Vancomycin	>12	<9	15	S
Tetracycline	>19	<14	25	S

The Disk Diffusion Test

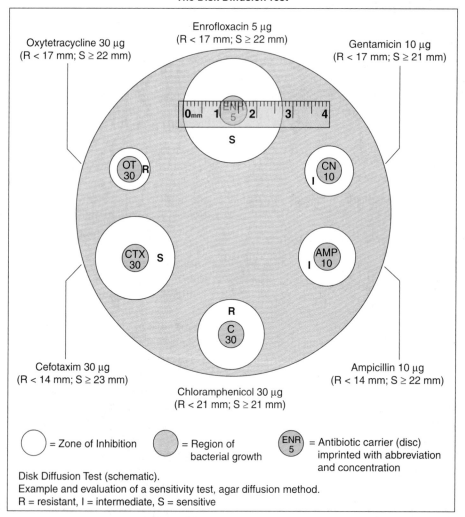

Enrofloxacin 5 µg
(R < 17 mm; S ≥ 22 mm)

Oxytetracycline 30 µg
(R < 17 mm; S ≥ 22 mm)

Gentamicin 10 µg
(R < 17 mm; S ≥ 21 mm)

OT 30 R

CN 10 I

CTX 30 S

AMP 10 I

R
C 30

Cefotaxim 30 µg
(R < 14 mm; S ≥ 23 mm)

Ampicillin 10 µg
(R < 14 mm; S ≥ 22 mm)

Chloramphenicol 30 µg
(R < 21 mm; S ≥ 21 mm)

◯ = Zone of Inhibition   ⬤ = Region of bacterial growth   (ENR 5) = Antibiotic carrier (disc) imprinted with abbreviation and concentration

Disk Diffusion Test (schematic).
Example and evaluation of a sensitivity test, agar diffusion method.
R = resistant, I = intermediate, S = sensitive

(a)

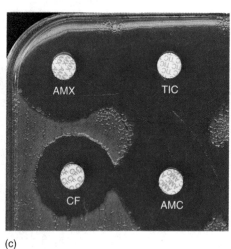

(b)

(c)

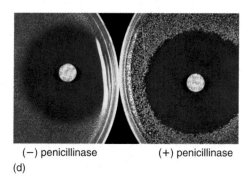

(−) penicillinase   (+) penicillinase

(d)

FIGURE 12.19

Technique for preparation and interpretation of disc diffusion tests. **(a)** Standardized methods are used to seed a lawn of bacteria over the medium. A dispenser delivers several drugs onto a plate, followed by incubation and interpretation of results. During incubation, antimicrobics become increasingly diluted as they diffuse out of the disc into the medium. If the test bacterium is sensitive to a drug, a zone of inhibition develops around its disc. The larger the size of this zone, the greater is the bacterium's sensitivity to the drug. The diameter of each zone is measured in millimeters and evaluated for susceptibility or resistance by means of a comparative standard (see table 12.7). **(b)** Antibiogram of *Escherichia coli* showing resistance to amoxicillin (AMX) and ticarcillin (TIC); when clavulanic acid is combined with amoxicillin (AMC), the result is sensitivity (lower right) due to the combined action of the two drugs. **(c)** Results of test with *Escherichia hermannii* indicate a synergistic effect between ticarcillin (TIC) and AMC (note the expanded zone between these two drugs). **(d)** Plates of *Staphylococcus aureus,* demonstrating one strain with sensitivity to penicillin [(−) penicillinase], and a penicillin-resistant strain [(+) penicillinase]. The resistant culture has an irregular edge at the zone of inhibition.

that are anaerobic, highly fastidious, or slow-growing *(Mycobacterium)*. An alternate diffusion system that provides additional information on drug effectiveness is the E-test **(figure 12.20).**

More sensitive and quantitative results can be obtained with tube dilution tests. First the antimicrobic is diluted serially in tubes of broth, and then each tube is inoculated with a small uniform sample of pure culture, incubated, and examined for growth (turbidity). The smallest concentration (highest dilution) of drug that visibly inhibits growth is called the **minimum inhibitory concentration, or MIC.** The MIC is useful in determining the smallest effective dosage of a drug and in providing a comparative index against other antimicrobics **(figure 12.21** and **table 12.8).** In many clinical laboratories, these antimicrobic testing procedures are performed in automated machines that can test dozens of drugs simultaneously.

THE MIC AND THERAPEUTIC INDEX

The results of antimicrobic sensitivity tests guide the physician's choice of a suitable drug. If therapy has already commenced, it is imperative to determine if the tests bear out the use of that particular drug. Once therapy has begun, it is important to observe the patient's clinical response, because the *in vitro* activity of the drug is not always correlated with its *in vivo* effect. When antimicrobic treatment fails, the failure is due to

1. the inability of the drug to diffuse into that body compartment (the brain, joints, skin);
2. a few resistant cells in the culture that did not appear in the sensitivity test; or
3. an infection caused by more than one pathogen (mixed), some of which are resistant to the drug.

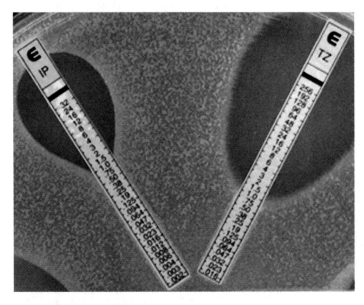

FIGURE 12.20

Alternate to the Kirby-Bauer procedure.   Another diffusion test is the E-test, which uses a strip to produce the zone of inhibition. The advantage of the E-test is that the strip contains a gradient of drug calibrated in μg. This way, the MIC can be measured by observing the mark on the strip that corresponds to the edge of the zone of inhibition. (IP = imepenem and TZ = tazobactam)

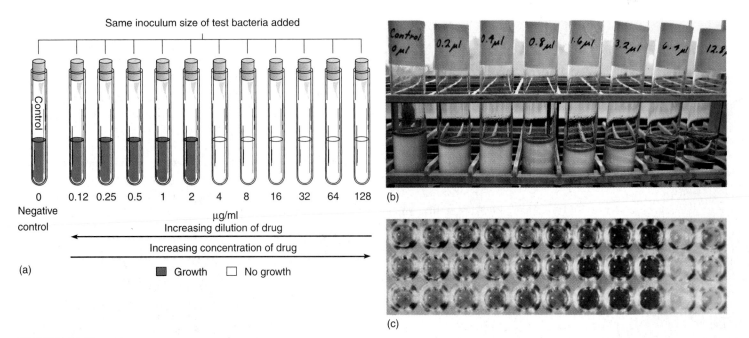

FIGURE 12.21

Tube dilution test for determining the minimum inhibitory concentration (MIC).   **(a)** The antibiotic is diluted serially through tubes of liquid nutrient from right to left. All tubes are inoculated with an identical sample of a test bacterium and then incubated. The first tube on the left is a control that lacks the drug and shows maximum growth. The dilution of the first tube in the series that shows no growth (no turbidity) is the MIC. **(b)** Photograph of MIC tube test using *Escherichia coli* and tetracycline. **(c)** Micro, multiwell plate with an array of three tests. This system can be automated to read the MICs of several drugs simultaneously.

TABLE 12.8					
Comparitive MICs (µg/ml) for Common Drugs and Pathogens					
Bacterium	**Penicillin G**	**Ampicillin**	**Sulfamethoxazole**	**Tetracycline**	**Cefaclor**
Staphylococcus aureus	4.0	0.05	3.0	0.3	4.0
Enterococcus faecalis	3.6	1.6	100.0	0.3	60.0
Neisseria gonorrhoeae	0.5	0.5	5.0	0.8	2.0
Escherichia coli	100.0	12.0	3.0	6–50.0	3.0
Pseudomonas aeruginosa	>500.0	>200.0	NA	>100.0	NA
Salmonella species	12.0	6.0	10.0	1.0	0.8
Clostridium	0.16	NA	NA	3.0	12.0

NA = not available

If therapy does fail, a different drug, combined therapy, or a different method of administration must be considered.

Many factors influence the choice of an antimicrobic drug besides microbial sensitivity to it. The nature and spectrum of the drug, its potential adverse effects, and the condition of the patient can be critically important. When several antimicrobic drugs are available for treating an infection, final drug selection advances to a new series of considerations. In general, it is better to choose the narrowest-spectrum drug of those that are effective if the causative agent is known. This decreases the potential for superinfections and other adverse reactions.

Because drug toxicity is of concern, it is best to choose the one with high selective toxicity for the infectious agent and low human toxicity. The **therapeutic index (TI)** is defined as the ratio of the dose of the drug that is toxic to humans as compared to its minimum effective (therapeutic) dose. The closer these two figures are (the smaller the ratio), the greater is the potential for toxic drug reactions. For example, a drug that has a therapeutic index of:

$$\frac{10 \ \mu g/ml: \text{toxic dose}}{9 \ \mu g/\mu l \ (MIC)} \quad \boxed{TI = 1.1}$$

is a riskier choice than one with a therapeutic index of:

$$\frac{10 \ \mu g/ml}{1 \ \mu g/ml} \quad \boxed{TI = 10}$$

Drug companies recommend dosages that will inhibit the microbes but not adversely affect patient cells. When a series of drugs being considered for therapy have similar MICs, the drug with the highest therapeutic index usually has the widest margin of safety.

The physician must also take a careful history of the patient to discover any preexisting medical conditions that will influence the activity of the drug or the response of the patient. A history of allergy to a certain class of drugs should preclude the administration of that drug. Underlying liver or kidney disease will ordinarily necessitate the modification of drug therapy because these organs play such an important part in metabolizing or excreting the drug. Infants, the elderly, and pregnant women require special precautions. For example, age can diminish gastrointestinal absorption and organ function, and most antimicrobial drugs cross the placenta and could affect fetal development.

The intake of other drugs must be carefully scrutinized, because incompatibilities can result in increased toxicity or failure of one or more of the drugs. For example, the combination of aminoglycosides and cephalosporins increases nephrotoxic effects; antacids reduce the absorption of isoniazid; and the interaction of tetracycline or rifampin with oral contraceptives can abolish the contraceptive's effect. Some drugs (penicillin with certain aminoglycosides, or amphotericin B with flucytosine) act synergistically, so that reduced doses of each can be used in combined therapy. Other concerns in choosing drugs include any genetic or metabolic abnormalities in the patient, the site of infection, the route of administration, and the cost of the drug.

The Art and Science of Choosing an Antimicrobial Drug
Even when all the information is in, the final choice of a drug is not always easy or straightforward. Consider the case of an elderly alcoholic patient with pneumonia caused by *Klebsiella* and complicated by diminished liver and kidney function. All drugs must be given parenterally because of prior damage to the gastrointestinal lining and poor absorption. Drug tests show that the infectious agent is sensitive to third-generation cephalosporins, gentamicin, imipenem and azlocillin. The patient's history shows previous allergy to the penicillins, so these would be ruled out. Drug interactions occur between alcohol and the cephalosporins, which are also associated with serious bleeding in elderly patients, so this may not be a good choice. Aminoglycosides such as gentamicin are nephrotoxic and poorly cleared by damaged kidneys. Imipenem causes intestinal discomfort, but it has less toxicity and would be a viable choice.

In the case of a cancer patient with severe systemic *Candida* infection, there will be fewer criteria to weigh. Intravenous amphotericin B or fluconazole are the only possible choices, despite drug toxicity and other possible adverse side effects. In a life-threatening situation, in which a dangerous chemotherapy is perhaps the only chance for survival, the choices are reduced and the priorities are different.

AN ANTIMICROBIC DRUG DILEMMA

We began this chapter with a view of the exciting strides made in chemotherapy during the past few years, but we must end it on a note of qualification and caution. There is now a worldwide problem in the management of antimicrobic drugs, which rank second only to some nervous system drugs in overall usage. The remarkable progress in treating many infectious diseases has spawned a view of antimicrobics as a "cure-all" for infections as diverse as the common cold and acne. And, although it is true that nothing is as dramatic as curing an infectious disease with the correct antimicrobic drug, in many instances, drugs have no effect or can be harmful. The depth of this problem can perhaps be appreciated better with a few statistics:

1. Roughly 200 million prescriptions for antimicrobics are written in the United States every year. A recent study disclosed that 75% of antimicrobial prescriptions are for pharyngeal, sinus, lung, and upper respiratory infections. A fairly high percentage of these are viral in origin and will have little or no benefit from antibacterial drugs.

    Many drugs are also misprescribed as to type, dosage, or length of therapy. Such overuse of antimicrobics is known to increase the development of antimicrobial resistance, harm the patient, and waste billions of dollars.

2. Drugs are often prescribed without benefit of culture or susceptibility testing, even when such testing is clearly warranted.

3. There is a tendency to use a "shotgun" antimicrobial therapy for minor infections, which involves administering a broad-spectrum drug instead of a more specific narrow-spectrum one. This practice can lead to superinfections and other adverse reactions. Tetracyclines and chloramphenicol are still prescribed routinely for infections that would be treated more effectively with narrower spectrum, less toxic drugs.

4. More expensive newer drugs are chosen when a less costly older one would be just as effective. Among the most expensive drugs are cephalosporins and the longer-acting tetracyclines, yet these are among the most commonly prescribed antibiotics.

5. Tons of excess antimicrobic drugs produced in this country are exported to other countries, where controls are not as strict. Nearly 200 different antibiotics are sold over the counter in Latin America and Asian countries. It is common

to self-medicate without understanding the correct medical indication. Drugs used in this way are largely ineffectual but, worse yet, they are known to be responsible for emergence of drug-resistant bacteria that subsequently cause epidemics.

The medical community recognizes that most physicians are motivated by important and prudent concerns, such as the need for immediate therapy to protect a sick patient and for defensive medicine to provide the very best care possible, but many experts feel that more education is needed for both physicians and patients concerning the proper occasions for prescribing antibiotics. In the final analysis, every allied health professional should be critically aware not only of the admirable and utilitarian nature of antimicrobics but also of their limitations.

CHAPTER CHECKPOINTS

The three major considerations necessary to choose an effective antimicrobic are the nature of the infecting microbe, the microbe's sensitivity to available drugs, and the overall medical status of the infected host.

The Kirby-Bauer test identifies antimicrobics that are effective against a specific infectious bacterial isolate.

The MIC (minimum inhibitory concentration) identifies the smallest effective dose of an antimicrobic toxic to the infecting microbe.

The therapeutic index is a ratio of the amount of drug toxic to the infected host and the MIC. The smaller the ratio, the greater the potential for toxic host-drug reactions.

The effectiveness of antimicrobic drugs is being compromised by several alarming trends: inappropriate prescription, use of broad-spectrum instead of narrow-spectrum drugs, use of higher-cost drugs, sale of over-the-counter antimicrobics in other countries, and lack of sufficient testing before prescription.

CHAPTER CAPSULE WITH KEY TERMS

I. **Antimicrobial Chemotherapy**
   A. Chemotherapeutic drugs are used to control microorganisms in the body. Depending on their source, these drugs are described as **antibiotics, semisynthetic** or synthetic. Based on their mode and spectrum of action, they are described as **broad spectrum** or **narrow spectrum** and microbistatic or microbicidal.
   B. The ideal antimicrobic is **selectively toxic,** highly potent, stable, and soluble in the body's tissues and fluids, does not disrupt the immune system or microflora of the host and is exempt from drug resistance.
   C. Adverse (side) effects of chemotherapy may include toxicity, allergic reactions, disruption of normal microbial **flora, superinfection,** and the acquisition of drug resistance by formerly sensitive microbes.
   D. Strategic approaches to the use of chemotherapeutics include
      1. **Prophylaxis,** where drugs are administered to *prevent* infection in susceptible people.
      2. Combined therapy, where two or more drugs are given simultaneously, either to prevent the emergence of resistant species or achieve synergism.

   E. Proper drug selection involves identification of the microbe involved and its susceptibility to various antimicrobics **(minimum inhibitory concentration,** or **MIC).** Final selection must take into account the **therapeutic index (TI),** spectrum of action, and medical condition of the patient.
   F. The inappropriate use of drugs on a worldwide basis has led to numerous medical and economic problems.

II. **Antibacterial Antibiotics**
   A. **Penicillins** are beta-lactam-based drugs originally isolated from the mold *Penicillium chrysogenum.* The natural form is penicillin G, although various semisynthetic forms, such as ampicillin and methicillin, exist which vary in their spectrum and applications. Penicillin is bactericidal, blocking completion of the cell wall, which causes eventual rupture of the cell.
      1. Major problems encountered in penicillin therapy include allergic reactions and bacterial resistance to the drug through **beta-lactamase.**
      2. Clavulanic acid inhibits beta-lactamases and is sometimes added to semisynthetic penicillins to increase their effectiveness.

B. **Cephalosporins** include both natural and semisynthetic forms initially isolated from the mold *Cephalosporium.* Cephalosporins inhibit peptidoglycan synthesis (like penicillin) but have a much broader spectrum.
  1. Potential problems in the use of cephalosporins include superinfection, allergic reactions, and bacterial resistance due to cephalosporinases
C. **Aminoglycosides** include several narrow-spectrum drugs isolated from unique bacteria found in the genera *Streptomyces.* Examples include streptomycin, gentamicin, tobramycin, and amikacin.
  1. Adverse reactions include damage to the 8th cranial nerve, kidney damage, intestinal disturbance, and drug resistance.
D. **Tetracyclines** and chloramphenicol are very broad-spectrum drugs isolated from *Streptomyces.* Both interfere with translation but their use is limited by adverse effects.
  1. Tetracycline can be hepatotoxic, lead to superinfection, disturbances of the gastrointestinal tract, and discoloration of tooth enamel in children.
  2. Chloramphenicol, in some instances, can cause severe damage to bone marrow.
E. Erythromycin, clindamycin, vancomycin, and rifampin are all isolated from *Streptomyces.*
  1. Erythromycin and clindamycin both affect protein synthesis. Erythromycin is a broad-spectrum alternative for use with penicillin-resistant bacteria, while clindamycin is used primarily for intestinal infection by anaerobes. Gastrointestinal disturbances are a side effect seen with both drugs.
  2. Vancomycin interferes with the early stages of cell wall synthesis and is used for life-threatening, methicillin-resistant staphylococcal infection. Neurotoxicity is a potential problem.
  3. Rifampin interferes with RNA polymerase (thereby affecting transcription) and is primarily used for tuberculosis and leprosy infections.
  4. Resistant bacteria can be found for each of the above drugs.
F. Bacitracin and **polymyxin** are narrow-spectrum antibiotics isolated from the bacterial genera *Bacillus.*
  1. Bacitracin prevents cell wall synthesis in gram-positive organisms. It is used in antibacterial skin ointments, often in combination with neomycin.
  2. Polymyxin is also found in skin ointments and can be used to treat *Pseudomonas* infections. Polymyxin can cause nephrotoxic and neuromuscular reactions in some instances.
G. Fosfomycin and Synercid are newly developed antibiotics that are useful when bacteria have developed resistance to more traditional drugs.
  1. Fosfomycin inhibits cell wall synthesis and is effective against gram-negative organisms.
  2. Synercid inhibits translation and is most effective against gram-positive cocci.

III. **Synthetic Antibacterial Drugs**
A. **Sulfonamides** are broad-spectrum drugs that act as **metabolic analogs, competitively inhibiting** enzymes needed for nucleic acid synthesis. Adverse reactions include allergies and formation of crystals in the kidney.
B. Trimethoprim, dapsone, isoniazid, fluoroquinolones, and oxazolidinones are all synthetic antibacterial drugs.
  1. Trimethoprim inhibits nucleic acid synthesis. Often used in combination with sulfa drugs.

  2. Dapsone is a narrow-spectrum drug related to the sulfonamides. It is used (often in combination with rifampin) to treat leprosy.
  3. Isoniazid (INH) blocks the synthesis of cell wall components in *Mycobacteria.* It is used in the treatment of tuberculosis. Adverse effects include potential liver damage.
  4. Fluoroquinolones (ciprofloxacin) represent a new class of broad-spectrum synthetic drug that has proven useful when the time constraints do not allow an infectious bacterium to be adequately characterized.
  5. Oxazolidinones are another new class of synthetic drug that work by inhibiting the start of protein synthesis. They are useful for treatment of staphylococci that are resistant to other drugs (MRSA and VRE).

IV. **Drugs for Fungal Infection**
A. Polyenes such as amphotericin B and nystatin disrupt fungal membranes by detergent action. Both can cause kidney damage.
B. **Azoles** are synthetic drugs that interfere with membrane synthesis. They include ketoconazole, fluconazole, and miconazole. Adverse reactions include potential liver damage.
C. Flucytosine inhibits DNA synthesis. Because many fungi are now resistant, it must usually be used in conjunction with amphotericin.

V. **Drugs for Protozoan Infections**
A. Quinine, or the related compounds chloraquine, primaquine, or mefloquine are used to treat infections by the malarial parasite *Plasmodium.*
B. Other drugs including metronidazole, suramin, melarsopral, and nitrifurimox are used for other protozoan infections.

VI. **Drugs for Helminth Infections** include mebendazole, thiabendazole, praziquantel, pyrantel, piperizine, and niclosamide, which are all used for antihelminthic chemotherapy. In most cases the cure is not complete until the incapacitated worm and its eggs are expelled from the body in the feces.

VII. **Drugs for Viral Infection** usually act by inhibiting viral penetration, multiplication, or assembly. Because viral and host metabolism are so closely related, toxicity is a potential adverse reaction with all antiviral drugs.
A. **Acyclovir,** valacyclovir, famciclovir, and ribavirin act as nucleoside analogs, inhibiting viral DNA replication, especially in herpesviruses.
B. Amantadine acts early in the cycle of influenza A to prevent viral uncoating.
C. AZT and protease inhibitors such as indinavir are used as anti-HIV drugs.
D. Interferon is a naturally occurring protein with both antiviral and anticancer properties.

VIII. **The Acquisition of Drug Resistance**
Microbes can lose their sensitivity to a drug through the acquisition of **resistance factors.** Drug resistance takes the form of:
A. Drug inactivation.
B. Decreased permeability to drug or increased elimination of drug from the cell.
C. Change of metabolic patterns.
D. Change in drug receptors.

IX. **Host-Drug Reaction**
Side effects of chemotherapy include organ toxicity, allergic responses, alteration of microflora.

MULTIPLE-CHOICE QUESTIONS

1. A compound synthesized by bacteria or fungi that destroys or inhibits the growth of other microbes is a/an
   a. synthetic drug
   b. antibiotic
   c. antimicrobic drug
   d. competitive inhibitor

2. Which statement is *not* an aim in the use of drugs in antimicrobial chemotherapy? The drug should:
   a. have selective toxicity
   b. be active even in high dilutions
   c. be broken down and excreted rapidly
   d. be microbicidal

3. Drugs that prevent the formation of the bacterial cell wall are
   a. quinolones
   b. beta-lactams
   c. tetracyclines
   d. aminoglycosides

4. Sulfonamide drugs initially disrupt which process?
   a. folic acid synthesis
   b. transcription
   c. PABA synthesis
   d. protein synthesis

5. Microbial resistance to drugs is acquired through
   a. conjugation
   b. transformation
   c. transduction
   d. all of these

6. R factors are _____ that contain a code for _____.
   a. genes, replication
   b. plasmids, drug resistance
   c. transposons, interferon
   d. plasmids, conjugation

7. When a patient's immune system becomes reactive to a drug, this is an example of
   a. superinfection
   b. drug resistance
   c. allergy
   d. toxicity

8. An antibiotic that disrupts the normal flora can cause
   a. the teeth to turn brown
   b. aplastic anemia
   c. a superinfection
   d. hepatotoxicity

9. Most antihelminthic drugs function by
   a. weakening the worms so they can be flushed out by the intestine
   b. inhibiting worm metabolism
   c. blocking the absorption of nutrients
   d. inhibiting egg production

10. An example of an antiviral drug that can prevent a viral nucleic acid from being replicated is
    a. azidothymidine
    b. acyclovir
    c. amantadine
    d. both a and b

11. Which of the following effects do antiviral drugs *not* have?
    a. killing extracellular viruses
    b. stopping virus synthesis
    c. inhibiting virus maturation
    d. blocking virus receptors

12. Which of the following modes of action would be most selectively toxic?
    a. interrupting ribosomal function
    b. dissolving the cell membrane
    c. preventing cell wall synthesis
    d. inhibiting DNA replication

13. The MIC is the _____ of a drug that is required to inhibit growth of a microbe.
    a. largest concentration
    b. standard dose
    c. smallest concentration
    d. lowest dilution

14. An antimicrobic drug with a _____ therapeutic index is a better choice than one with a _____ therapeutic index.
    a. low, high
    b. high, low

CONCEPT QUESTIONS

1. Differentiate between antibiotics and synthetic drugs.

2. a. Differentiate between narrow-spectrum and broad-spectrum antibiotics.
   b. Can you determine why some drugs have narrower spectra than others? (Hint: Look at their mode of action.)
   c. How might one determine whether a particular antimicrobic is broad- or narrow-spectrum?

3. a. What is the major source of antibiotics?
   b. What appears to be the natural function of antibiotics?

4. a. Using the diagram at right as a guide, briefly explain how the three factors in drug therapy interact.
   b. What drug characteristics will make treatment most effective?
   c. Why is it better for a drug to be microbicidal than microbistatic?
   d. Which of your answers to question *c* do you think is the most important?

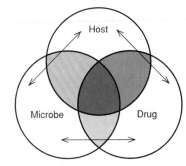

5. a. Explain the major modes of action of antimicrobial drugs, and give an example of each.
   b. What is competitive inhibition?
   c. What is the basic reason that a metabolic analog molecule can inhibit metabolism?
   d. Why do the penicillin group of drugs have milder toxicity than other antibiotics?
   e. What are the long-term effects of drugs that block transcription?
   f. Why would a drug that blocks translation on the ribosomes of bacteria also affect human cells?
   g. Why do drugs that act on bacterial and fungal membranes generally have high toxicity?

6. Construct a chart that summarizes the modes of action and applications of the major groups of antibacterial drugs (antibiotics and synthetics), antifungal drugs, antiparasitic drugs, and antiviral drugs.

7. a. Explain why there are fewer antifungal, antiparasitic, and antiviral drugs than antibacterial drugs.
   b. What effect do nitrogen-base analogs have upon viruses?
   c. Summarize the origins and biological actions of interferon.

8. a. Explain the phenomenon of drug resistance from the standpoint of microbial genetics (include a description of R factors).

b. Multiple drug resistance is becoming increasingly common in microorganisms. Explain how one bacterium can acquire resistance to several drugs.

9. a. Explain four general ways that microbes evade the effects of drugs.
   b. What is the effect of beta-lactamase?

10. What causes mutated or plasmid-altered strains of drug-resistant microbes to persist in a population?

11. a. Generally overview the adverse effects of antimicrobic drugs on the host.

b. On what basis can one explain allergy to drugs?
   c. Describe the stages in a superinfection.

12. a. Outline the steps in antimicrobic susceptibility testing.
    b. Compare the interpretation of the Kirby-Bauer technique and the E test with the MIC technique.
    c. What is the therapeutic index, and how is it used?

13. Summarize the primary considerations in choosing an antimicrobic drug.

CRITICAL-THINKING QUESTIONS

1. Occasionally, one will hear the expression that a microbe has become "immune" to a drug.
   a. What is a better way to explain what is happening?
   b. Explain a simple test one could do to determine if drug resistance was developing in a culture.

2. a. Can you think of additional ways that drug resistance can be prevented?
   b. What can health care workers do?
   c. What can one do on a personal level?

3. Drugs are often given to surgical patients, to dental patients with heart disease, or to healthy family members exposed to contagious infections.
   a. What word would you use to describe this use of drugs?
   b. What is the purpose of this form of treatment?
   c. Explain some potential undesired effects of this form of therapy.

4. a. Your pregnant neighbor has been prescribed a daily dose of oral tetracycline for acne. Do you think this therapy is advisable for her? Why or why not?
   b. A woman has been prescribed a broad-spectrum oral cephalosporin for a strep throat. What are some possible consequences in addition to cure of the infected throat?
   c. A man has a severe case of gastroenteritis that is negative for bacterial pathogens. A physician prescribes an oral antibacterial drug in treatment. What are your opinions of this therapy?

5. You have been directed to take a sample from a growth-free portion of the zone of inhibition in the Kirby-Bauer test and inoculate it onto a plate of nonselective medium.
   a. What does it mean if growth occurs on the new plate?
   b. What if there is no growth?

6. In cases in which it is not possible to culture or drug test an infectious agent (such as middle ear infection), how would the appropriate drug be chosen?

7. Using the results in tables 12.7 and 12.8 and reviewing drug characteristics, choose an antimicrobic for each of the following situations (explain your choice):
   a. for an adult patient suffering from *Mycoplasma* pneumonia
   b. for a child with meningitis (drug must enter into cerebrospinal fluid)
   c. for a patient with allergy to erythromycin
   d. for a urinary tract infection by *Enterococcus*
   e. for gonorrhea

8. What factors can play a part in drug synergism?

9. How would you personally feel about being told by a physician that your infection cannot be cured by an antibiotic, and that the best thing to do is go home, drink a lot of fluids, and take aspirin or other symptom-relieving drugs?

10. a. Refer to figure 12.19a and interpret the results.
    b. Give the MICs for the tests in figure 12.20.
    c. Refer to the tube dilution tests shown in figure 12.21 and give the MIC of the drug being tested.

11. a. Explain the basis for combined therapy.
    b. Give reasons why it could be helpful to use combined therapy in treating HIV infection.

INTERNET SEARCH TOPICS

1. Visit the student Online Learning Center at www.mhhe.com/talaro5. Go to chapter 12, Internet Search Topics, and log on to the available websites to
   a. Look up examples of several types of anti-HIV drugs, comparing information on costs, side effects, and problems in therapy.
   b. Study the Department of Health website. Use the information provided there to discuss the problems of antibiotic resistance with family and friends.

   c. Review an excellent source of case studies in the uses of antimicrobic drugs.

2. Locate information on new types of antibacterial drugs called linezolid, evevnimycin, daptomycin, and Zyvox. Determine their mode of action and indications for use.

Microbe-Human Interactions:
Infection and Disease

T he human body exists in a state of dynamic equilibrium with microorganisms. In the healthy individual, this balance is maintained as a peaceful coexistence and lack of disease. But on occasion, the balance tips in favor of the microorganism, and an infection or disease results. In this chapter, we explore each component of the host-parasite relationship, beginning with the nature and function of normal flora, moving to the stages of infection and disease, and closing with a study of epidemiology and the patterns of disease in populations. These topics will set the scene for chapters 14 and 15, which deal with the ways the host defends itself against assault by microorganisms.

Chapter Overview

- The normal flora of humans includes bacteria, fungi, and protozoa that live on the body without causing disease. Theses microbes can be found in areas exposed to the outside environment, such as the gastrointestinal tract, skin, and respiratory tract, and are generally beneficial to humans.
- Pathogens are those microbes that infect the body and cause disease. Disease results when an adequate number of pathogenic cells enter the body through a specific route, grow, and disrupt tissues.
- Pathogens produce virulence factors such as toxins and enzymes that help them invade and damage the cells of their host. The effects of infection and disease are seen in the host as signs and symptoms, which may include both short- and long-term damage.
- Pathogens may be spread by direct or indirect means involving overtly infected people, carriers, vectors, and vehicles. A significant source of infection is exposure to the hospital environment.
- The field of epidemiology is concerned with the patterns of disease occurrence in a population. Epidemiologists are concerned with monitoring the number of cases; their geographic distribution; the sex, age, and ethnicity of affected people; and the mortality rate.
- Pathogens may be found residing in humans, animals, food, soil, and water. A specific laboratory methodology is used to determine the causative agent of a disease.

The Human Host

In chapter 7, several of the basic interrelationships between humans and microorganisms were considered. Most of the microbes inhabiting the human body benefit from the nutrients and protective habi-

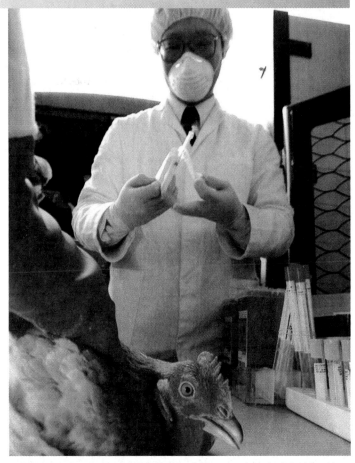

A virologist prepares to take blood from a chicken to screen for infection with a strain of influenza. This deadly virus can cause widespread illness in chickens and is readily transmissible to humans.

tat it provides. From the human point of view, these relationships run the gamut from mutualism to commensalism to parasitism and can have beneficial, neutral, or harmful effects (see page 206). A common characteristic of all microbe-human relationships, regardless of where they lead, is that they begin with contact.

CONTACT, INFECTION, DISEASE—A CONTINUUM

The body surfaces are constantly exposed to microbes. Some microbes become implanted there as colonists (normal flora), some are rapidly lost (transients), and others invade the tissues. Such intimate contact with microbes inevitably leads to **infection,** a condition in which pathogenic microorganisms penetrate the host defenses, enter the tissues, and multiply. When the cumulative effects of the

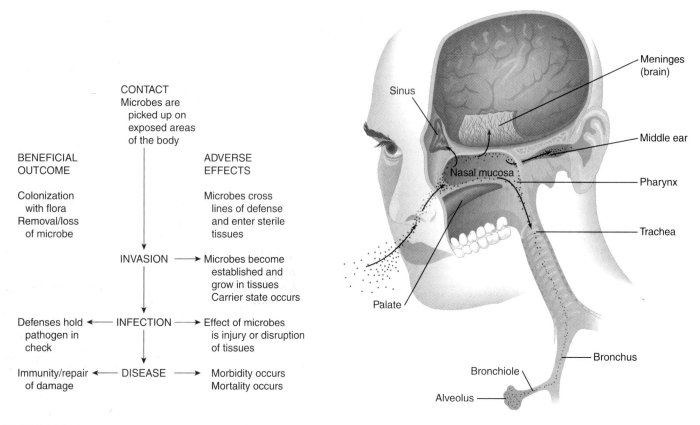

FIGURE 13.1

Associations between microbes and humans.   Effects of exposure can progress in a variety of directions, ranging from no effect to colonization, and from infection to immunity to disease. The example shown here follows the possible events in the case of contact with a pathogen such as *Streptococcus pneumoniae* (the pneumococcus). This bacterium can be harbored harmlessly in the upper respiratory tract, but it may also invade and infect various sites in the head and respiratory tract.

infection damage or disrupt tissues and organs, the **pathologic*** state that results is known as a disease. A disease is defined as any deviation from health. There are hundreds of different diseases caused by such factors as infections, diet, genetics, and aging. In this chapter, however, we will discuss only **infectious disease**—the disruption of a tissue or organ caused by microbes or their products.

The pattern of the host-parasite relationship can be viewed as a series of stages that begins with contact, progresses to infection, and ends in disease. Because of numerous factors relating to host resistance and degree of pathogenicity, not all contacts lead to infection and not all infections lead to disease. In fact, contamination without infection and infection without disease are the rule. Before we consider further details of infection and disease, let us examine what happens when microbes colonize the body, thereby establishing a long-term, usually beneficial, relationship (**figure 13.1**).

RESIDENT FLORA: THE HUMAN AS A HABITAT

With its constant source of nourishment and moisture, relatively stable pH and temperature, and extensive surfaces upon which to settle, the human body provides a favorable habitat for an abun-

dance of microorganisms. In fact, it is so favorable that, cell-for-cell, microbes outnumber human cells ten to one! The large and mixed collection of microbes adapted to the body has been variously called the normal **resident flora,** or **indigenous*** **flora,** though some microbiologists prefer to use the terms *microflora* and *commensals.* The normal residents include an array of bacteria, fungi, protozoa, and, to a certain extent, viruses and arthropods.

Acquiring Resident Flora

The human body offers a seemingly endless variety of environmental niches, with wide variations in temperature, pH, nutrients, and oxygen tension occurring from one area to another. Because the body provides such a range of habitats, it should not be surprising that the body supports a wide range of microbes. As shown in **tables 13.1** and **13.2,** most areas of the body in contact with the outside environment harbor resident microorganisms, while internal organs and tissue, along with the fluids they contain, are generally microbe-free.

The vast majority of microbes that come in contact with the body are removed or destroyed by the host's defenses long before they are able to colonize a particular area. Of those microbes able to establish an ongoing presence, an even smaller number are able to remain without attracting the unwanted attention of the body's

---

***pathologic** (path″-uh-loj′-ik) Any process causing structural and functional damage to the body.

***indigenous** (in-dih′-juh-nus) Belonging or native to.

TABLE 13.1

Sites That Harbor a Normal Flora

Skin and its contiguous mucous membranes
Upper respiratory tract
Gastrointestinal tract (various parts)
Outer opening of urethra
External genitalia
Vagina
External ear canal
External eye (lids, conjunctiva)

TABLE 13.2

Sterile (Microbe-Free) Anatomical Sites and Fluids

All Internal Tissues and Organs
  Heart and circulatory system
  Liver
  Kidneys and bladder
  Lungs
  Brain and spinal cord
  Muscles
  Bones
  Ovaries/testes
  Glands (pancreas, salivary, thyroid)
  Sinuses
  Middle and inner ear
  Internal eye

Fluids Within an Organ or Tissue
  Blood
  Urine in kidneys, ureters, bladder
  Cerebrospinal fluid
  Saliva prior to entering the oral cavity
  Semen prior to entering the urethra
  Amniotic fluid surrounding the embryo and fetus

defenses. This last group of organisms have evolved, along with their human hosts, to produce a complex relationship where the effects of normal flora are generally not deleterious to the host.

Although generally stable, the flora can fluctuate to a limited extent with general health, age, variations in diet, hygiene, hormones, and drug therapy. In many cases, bacterial flora actually benefits the human host by preventing the overgrowth of harmful microorganisms. A common example is the fermentation of glycogen by lactobacilli, which keep the pH in the vagina quite acidic and prevent the overgrowth of the yeast *Candida albicans*. A second example is seen in the large intestine where a protein produced by *E. coli* prevents the growth of pathogenic bacteria such as *Salmonella* and *Shigella*.

Of course, characterizing the normal flora as beneficial or, at worst, commensal to the host presupposes that the host is in good health, with a fully functioning immune system and that the flora is present only in its natural microhabitat within the body. Hosts with a compromised immune system could very easily be infected by

their own flora (see table 13.4). This is seen when AIDS patients contract pneumococcal pneumonia, the causative agent of which (*Streptococcus pneumoniae*) is often carried as normal flora in the nasopharynx. **Endogenous*** infections can also occur when normal flora is introduced to a site that was previously sterile, as when *E. coli* enters the bladder, resulting in a urinary tract infection.

Initial Colonization of the Newborn

The uterus and its contents are normally sterile during embryonic and fetal development and remain essentially germ-free until just before birth. The event that first exposes the infant to microbes is the breaking of the fetal membranes, at which time microbes from the mother's vagina can enter the womb. Comprehensive exposure occurs during the birth process itself, when the baby unavoidably comes into intimate contact with the birth canal (**figure 13.2**). Within 8 to 12 hours after delivery, the newborn typically has been colonized by bacteria such as streptococci, staphylococci, and lactobacilli, acquired primarily from its mother. The nature of the flora initially colonizing the large intestine depends upon whether the baby is bottle- or breast-fed. Bottle-fed infants (receiving milk or a milk-based formula) tend to acquire a mixed population of coliforms, lactobacilli, enteric streptococci, and staphylococci. In contrast, the intestinal flora of breast-fed infants consists primarily of

*endogenous (en-doj-en-us) Growing from within.

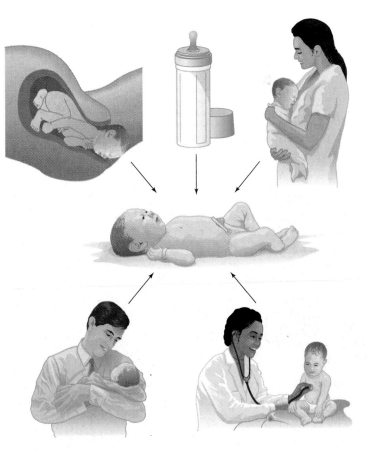

FIGURE 13.2
The origins of flora in newborns.

Bifidobacterium species whose growth is favored by a growth factor from the milk. This bacterium metabolizes sugars into acids that protect the infant from infection by certain intestinal pathogens. The skin, gastrointestinal tract, and portions of the respiratory and genitourinary tract all continue to be colonized as contact continues with family members, health care personnel, the environment, and food.

Milestones that contribute to development of the adult pattern of flora are eruption of teeth, weaning, and introduction of the first solid food. Although exposure to microbes is unavoidable and even necessary for the maturation of the infant's flora, contact with pathogens is dangerous, because the neonate is not yet protected by a full complement of flora, and owing to its immature immune defenses, is extremely susceptible to infection.

INDIGENOUS FLORA OF SPECIFIC REGIONS

Although we tend to speak of the flora as a single unit, it is a complex mixture of hundreds of species, differing somewhat in quality and quantity from one individual to another. Studies of the flora have shown that most people harbor certain specially adapted bacteria, fungi, and protozoa **(table 13.3)**.

Flora of the Human Skin

The skin is the largest and most accessible of all organs. Its major layers are the epidermis, an outer layer of dead cells continually being sloughed off and replaced, and the dermis, which lies atop the subcutaneous layer of tissue **(figure 13.3*a*)**. Depending on its location, skin also contains hair follicles and several types of glands, and the outermost surface is covered with a protective, waxy cuticle that can help microbes adhere. The normal flora resides only in or on the dead cell layers, and except for in follicles and glands, it does not extend into the dermis or subcutaneous levels. The nature of the population varies according to site. Oily, moist skin supports a more prolific flora than dry skin. Humidity, occupational exposure, and clothing also influence its character. Transition zones where the skin joins with the mucous membranes of the nose, mouth, and external genitalia harbor a particularly rich flora.

TABLE 13.3		
Life on Humans: Sites Containing Well-Established Flora and Representative Examples		
Anatomic Sites	**Common Genera**	**Remarks**
Skin	**Bacteria:** *Staphylococcus, Micrococcus, Corynebacterium, Propionibacterium, Mycobacterium*	Microbes live only in upper dead layers of epidermis, glands, and follicles; dermis and layers below are sterile.
	Fungi: *Malassezia* yeast	Dependent on skin lipids for growth.
	Arthropods: *Demodix* mite	Present in sebaceous glands and hair follicles.
Gastrointestinal Tract		
Oral cavity	**Bacteria:** *Streptococcus, Neisseria, Veillonella, Staphylococcus, Fusobacterium, Lactobacillus, Bacteroides, Corynebacterium, Actinomyces, Eikenella, Treponema, Haemophilus*	Colonize the epidermal layer of cheeks, gingiva, pharynx; surface of teeth; found in saliva in huge numbers.
	Fungi: *Candida* species	Can cause thrush.
	Protozoa: *Trichomonas tenax, Entamoeba gingivalis*	Frequent the gingiva of persons with poor oral hygiene.
Large intestine and rectum	**Bacteria:** *Bacteroides, Fusobacterium, Eubacterium, Bifidobacterium, Clostridium,* fecal streptococci, *Peptococcus, Lactobacillus,* coliforms (*Escherichia, Enterobacter*)	Sites of lower gastrointestinal tract other than large intestine and rectum have sparse or nonexistent flora. Flora consists predominantly of strict anaerobes; other microbes are aerotolerant or facultative.
	Fungi: *Candida*	
	Protozoa: *Entamoeba coli, Trichomonas hominis*	Feed on waste materials in the large intestine.
Upper Respiratory Tract	Microbial population exists in the nasal passages, throat, and pharynx; owing to proximity, flora is similar to that of oral cavity	Trachea and bronchi have a sparse population; smaller breathing tubes and alveoli have no normal flora and are essentially sterile.
Genital Tract	**Bacteria:** *Lactobacillus, Streptococcus, Corynebacterium, Escherichia, Mycobacterium*	In females, flora occupies the external genitalia and vaginal and cervical surfaces; internal reproductive structures normally remain sterile. Flora responds to hormonal changes during life.
	Fungi: *Candida*	Cause of yeast infections.
Urinary Tract	**Bacteria:** *Staphylococcus, Streptococcus,* coliforms	In females, flora exists only in the first portion of the urethral mucosa; the remainder of the tract is sterile. In males, the entire reproductive and urinary tract is sterile except for a short portion of the anterior urethra.

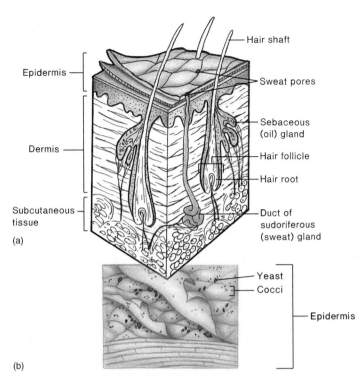

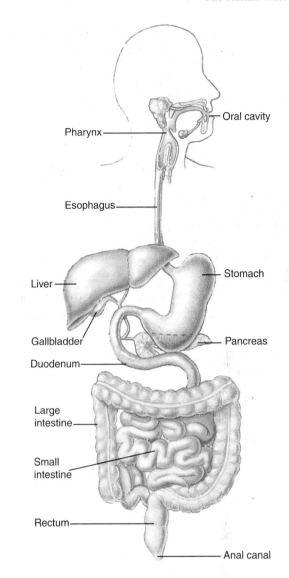

FIGURE 13.3

The landscape of the skin. **(a)** The epidermis, along with associated glands and follicles (colored), provides rich and diverse habitats. Noncolored regions (dermis and subcutaneous layers) are free of microbial flora. **(b)** A highly magnified view of the skin surface reveals beds of rod- and coccus-shaped bacteria and yeasts beneath peeled-back skin flakes.

FIGURE 13.4

Distribution of flora. Areas of the gastrointestinal tract that shelter resident flora are highlighted in color. Noncolored areas do not regularly harbor residents.

Ordinarily, there are two cutaneous populations. The **transient** population, or exposed flora, clings to the skin surface but does not ordinarily grow there. It is acquired by routine contact, and it varies markedly from person to person and over time. The transient flora includes any microbe a person has picked up, often species that do not ordinarily live on the body, and it is greatly influenced by the hygiene of the individual.

The **resident** population lives and multiplies in deeper layers of the epidermis and in glands and follicles (figure 13.3*b*). The composition of the resident flora is more stable, predictable, and less influenced by hygiene than is the transient flora. The normal skin residents consist primarily of bacteria (notably *Staphylococcus, Corynebacterium,* and *Propionibacterium*) and yeasts. Moist skin folds, especially between the toes, tend to harbor fungi, whereas lipophilic mycobacteria and staphylococci are prominent in sebaceous[1] secretions of the axilla, external genitalia, and external ear canal. One species, *Mycobacterium smegmatis,* lives in the cheesy secretion, or *smegma,* on the external genitalia of men and women.

Flora of the Gastrointestinal Tract

The gastrointestinal (GI) tract receives, moves, digests, and absorbs food; it also removes waste. It encompasses the oral cavity, esophagus, stomach, small intestine, large intestine, rectum, and anus. Stating that the GI tract harbors flora may seem to contradict our

earlier statement that internal organs are sterile, but it is not an exception to the rule. How can this be true? In reality, the GI tract is a long, hollow tube (with numerous pockets and curves), bounded by the mucous membranes of the oral cavity on one extreme and those of the anus on the other. Because the innermost surface of this tube is exposed to the environment, it is topographically outside the body, so to speak.

The shifting conditions of pH, oxygen tension, and differences in microscopic anatomy of the GI tract are reflected by the variations in or distribution of the flora (**figure 13.4**). Some microbes remain attached to the mucous epithelium or its associated structures, and others dwell in the **lumen.*** Although the abundance of nutrients invites microbial growth, the only areas that harbor appreciable permanent flora are the oral cavity, large intestine, and rectum. The esophagus

---

1. From sebum, a lipid material secreted by glands in the hair follicles.

*lumen (loo′-men) The space within a tubular structure.

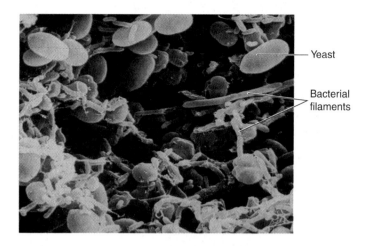

Yeast

Bacterial filaments

FIGURE 13.5

Electron micrograph of the mucous membrane of the large intestine.   Microcolonies of yeasts and long, filamentous bacteria located in pockets are visible.

undergoes wavelike contractions (peristalsis), a process that constantly flushes microorganisms; the stomach acid inhibits most microbes; and peristalsis and digestive enzymes help exclude flora from all but the terminal segment of the small intestine.

Flora of the Mouth   The oral cavity has a unique flora that is among the most diverse and abundant of the body. Microhabitats, including the cheek epithelium, gingiva, tongue, floor of the mouth, and tooth enamel, provide numerous adaptive niches for hundreds of different species to colonize. The most common residents are aerobic *Streptococcus* species—*S. sanguis, S. salivarius, S. mitis*—that colonize the smooth superficial epithelial surfaces. Two species, *S. mutans* and *S. sanguis,* make a major contribution to dental caries by forming sticky dextran slime layers in the presence of simple sugars. The adherence of dextrans to the tooth surface establishes a medium that attracts other bacteria (see figure 21.30).

Once the teeth have erupted, an anaerobic habitat is established in the gingival crevice that favors the colonization of anaerobic bacteria that can be involved in dental caries and periodontal[2] infections. The instant that saliva is secreted from ducts into the oral cavity, it becomes laden with resident and transient flora. Saliva normally has a high bacterial count (up to 5×10^9 cells per milliliter), a fact that tends to make mouthwashes rather ineffective and a human bite very dangerous.

Flora of the Large Intestine   The flora of the intestinal tract has complex and profound interactions with the host. The large intestine (cecum and colon) and the rectum harbor a huge population of microbes ($10^8–10^{11}$ per gram of feces) **(figure 13.5).** So abundant and prolific are these microbes that they constitute 10–30% of the fecal volume. Even an individual on a long-term fast passes feces consisting primarily of bacteria.

Because of the state of the intestinal environment, the predominant fecal flora are strictly anaerobic bacteria *(Bacteroides,*

Bifidobacterium, Fusobacterium, and *Clostridium).* **Coliforms*** such as *Escherichia coli, Enterobacter,* and *Citrobacter* are present in smaller numbers. Many species ferment waste materials in the feces, generating vitamins (B_{12}, vitamin K, pyridoxine, riboflavin, and thiamine) and acids (acetic, butyric, and propionic acids) of potential value to the host. Occasionally significant are bacterial digestive enzymes that convert disaccharides to monosaccharides or promote steroid metabolism.

Intestinal bacteria contribute to intestinal odor by producing **skatole,*** amines, and gases (CO_2, H_2, CH_4, and H_2S). Intestinal gas is known in polite circles as *flatus,* and the expulsion of it as *flatulence.* Some of the gas arises through the action of bacteria on dietary carbohydrate residues from vegetables such as cabbage, corn, and beans. The bacteria produce an average of 8.5 liters of gas daily, but only a small amount is ejected in flatus. Combustible gases occasionally form an explosive mixture in the presence of oxygen that has reportedly ignited during intestinal surgery and ruptured the colon!

Late in childhood, members of certain ethnic groups lose the ability to secrete the enzyme lactase. When they ingest milk or other lactose-containing dairy products, lactose is acted upon instead by intestinal bacteria, and severe intestinal distress can result. The recommended treatment for this deficiency is avoiding these foods or eating various lactose-free substitutes.

Flora of the Respiratory Tract

The first microorganisms to colonize the upper respiratory tract (nasal passages and pharynx) are predominantly oral streptococci. Inhaled air regularly contains microbes that are filtered out, destroyed, or expelled, although some can adapt to specific regions of this habitat. *Staphylococcus aureus* preferentially resides in the nasal entrance, nasal vestibule, and anterior nasopharynx, and *Neisseria* species take up residence in the mucous membranes of the nasopharynx behind the soft palate **(figure 13.6).** Lower still are assorted streptococci and species of *Haemophilus* that colonize the tonsils and lower pharynx. Conditions lower in the respiratory tree (bronchi and lungs) are unfavorable habitats for permanent residents.

Flora of the Genitourinary Tract

The regions of the genitourinary tract that harbor microflora are the vagina and outer opening of the urethra in females and the anterior urethra in males **(figure 13.7).** The internal reproductive organs are kept sterile through physical barriers such as the cervical plug and other host defenses. The kidney, ureter, bladder, and upper urethra are presumably kept sterile by urine flow and regular bladder emptying. Because the urethra in women is so short (about 3.5 cm long), it can form a passage for bacteria to the bladder and lead to urinary tract infections. The principal residents of the urethra are nonhemolytic streptococci, staphylococci, corynebacteria, and occasionally, coliforms.

The vagina presents a notable example of how changes in physiology can greatly influence the composition of the normal flora. An important factor influencing these changes in women is the

---

2. Situated or occurring around the tooth.

---

*coliform (koh′-lih-form) L. *colum,* a sieve. Gram-negative, facultatively anaerobic, and lactose-fermenting microbes.

*skatole (skat′-ohl) Gr. *skatos,* dung. One chemical that gives feces its characteristic stench.

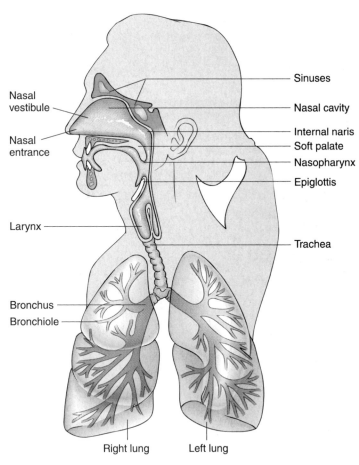

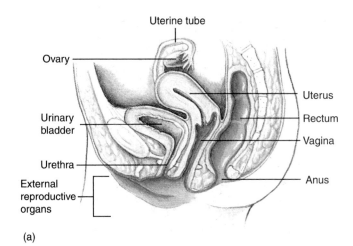

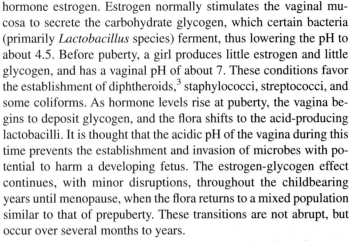

FIGURE 13.6

Colonized regions of the respiratory tract. The moist mucous blanket of the nasopharynx has a well-entrenched flora (indicated by color). Some colonization occurs in the upper trachea, but lower regions of bronchi, bronchioles, and lungs lack resident microbes.

FIGURE 13.7

Location of (a) female and (b) male genitourinary flora (indicated by color).

hormone estrogen. Estrogen normally stimulates the vaginal mucosa to secrete the carbohydrate glycogen, which certain bacteria (primarily *Lactobacillus* species) ferment, thus lowering the pH to about 4.5. Before puberty, a girl produces little estrogen and little glycogen, and has a vaginal pH of about 7. These conditions favor the establishment of diphtheroids,[3] staphylococci, streptococci, and some coliforms. As hormone levels rise at puberty, the vagina begins to deposit glycogen, and the flora shifts to the acid-producing lactobacilli. It is thought that the acidic pH of the vagina during this time prevents the establishment and invasion of microbes with potential to harm a developing fetus. The estrogen-glycogen effect continues, with minor disruptions, throughout the childbearing years until menopause, when the flora returns to a mixed population similar to that of prepuberty. These transitions are not abrupt, but occur over several months to years.

For a look into the laboratory study of resident flora see **Microbits 13.1.**

3. Any nonpathogenic species of *Corynebacterium.*

CHAPTER CHECKPOINTS

Humans are contaminated with microorganisms from the moment of birth onward. An infection is a condition in which contaminating microorganisms overcome host defenses, multiply, and cause disease, damaging tissues and organs.

The resident or normal flora includes bacteria, fungi, and protozoa.

There are two types of cutaneous populations of microbes: the transients, which cling to but do not grow on the superficial layers of the skin, and the more permanent residents, which reside in the deeper layers of the epidermis and its glands.

The flora of the alimentary canal is confined primarily to the mouth, large intestine, and rectum.

The flora of the respiratory tract extends from the nasal cavity to the lower pharynx.

The flora of the genitourinary tract is restricted to the urethral opening in males and to the urethra and vagina in females.

MICROBITS 13.1

Life Without Flora

For years, questions lingered about how essential the microbial flora is to normal life and what functions various members of the flora might serve. The need for animal models to further investigate these questions led eventually to development of laboratory strains of *germ-free,* or **axenic,** mammals and birds. The techniques and facilities required for producing and maintaining a germ-free colony are exceptionally rigorous. After the young mammals are taken from the mother aseptically by cesarian section, they are immediately transferred to a sterile isolator or incubator. The newborns must be fed by hand through gloved ports in the isolator until they can eat on their own, and all materials entering their chamber must be sterile. Rats, mice, rabbits, guinea pigs, monkeys, dogs, hamsters, and cats are some of the mammals raised in the germ-free state.

Experiments with germ-free animals are of two basic varieties: (1) general studies on how the lack of normal microbial flora influences the nutrition, metabolism, and anatomy of the animal, and (2) **gnotobiotic*** studies, in which the germ-free subject is inoculated either with a single type of microbe to determine its individual effect or with several known microbes to determine interrelationships. Results are validated by comparing the germ-free group with a conventional, normal control group. **Table 13.A** summarizes some major conclusions arising from studies with germ-free animals.

A dramatic characteristic of germ-free animals is that they live longer and have fewer diseases than normal controls, as long as they remain in a sterile environment. From this standpoint, it is clear that the flora is not needed for survival and may even be the source of infectious agents. At the same time, it is also clear that axenic life is highly impractical. Additional studies have revealed important facts about the effect of the flora on various organs and systems. For example, the flora contributes significantly to the development of the immune system. When germ-free animals are placed in contact with normal control animals,

---

***gnotobiotic** (noh"-toh-by-ah'-tik) Gr. *gnotos,* known, and *biota,* the organisms of a region.

TABLE 13.A

Effects and Significance of Experiments with Germ-Free Subjects

Effect in Germ-Free Animal	Significance
Enlargement of the cecum; other degenerative diseases of the intestinal tract of rats, rabbits, chickens	Microbes are needed for normal intestinal development
Vitamin deficiency in rats	Microbes are a significant nutritional source of vitamins
Underdevelopment of immune system in most animals	Microbes are needed to stimulate development of certain host defenses
Absence of dental caries and periodontal disease in dogs, rats, hamsters	Normal flora are essential in caries formation and gum disease
Heightened sensitivity to enteric pathogens *(Shigella, Salmonella, Vibrio cholerae)* and to fungal infections	Normal flora are antagonistic against pathogens
Lessened susceptibility to amebic dysentery	Normal flora facilitate the completion of the life cycle of the ameba in the gut

they gradually develop a flora similar to that of the controls. However, germ-free subjects are less tolerant of microorganisms and can die from infections by relatively harmless species. This susceptibility is due to the immature character of the immune system of germ-free animals. These animals have a reduced number of certain types of white blood cells and slower antibody response.

Gnotobiotic experiments have clarified the dynamics of several infectious diseases. Perhaps the most striking discoveries were made in the case of oral diseases. For years, the precise involvement of microbes in dental caries had been ambiguous. Studies with germ-free rats, hamsters, and beagles confirmed that caries development is influenced by heredity, a diet high in sugars, and poor oral hygiene. Even when all these predisposing factors are present, however, germ-free animals still remain free of caries unless they have been inoculated with specific bacteria. Further discussion on dental diseases is found in chapter 21.

The ability of known pathogens to cause infection can also be influenced by normal flora, sometimes in opposing ways. Studies have indicated that germ-free animals are highly susceptible to experimental infection by the enteric pathogens *Shigella* and *Vibrio,* whereas normal animals are less susceptible, presumably because of their protective flora. In marked contrast, *Entamoeba histolytica* (the agent of amebic dysentery) is more pathogenic in the normal animal than in the germ-free animal. One explanation for this phenomenon is that *E. histolytica* must feed on intestinal bacteria to complete its life cycle.

Sterile enclosure for rearing and handling germ-free laboratory animals.

The Progress of an Infection

A microbe whose relationship with its host is parasitic and results in infection and disease is termed a **pathogen.** The type and severity of infection depends both on the pathogenicity of the organism and the condition of the host **(figure 13.8).** Pathogenicity, you will recall, is a broad concept that describes an organism's potential to cause infection or disease, and is used to divide pathogenic microbes into one of two groups. **True pathogens** (primary pathogens) are capable of causing disease in healthy persons with normal immune defenses. They are generally associated with a specific, recognizable disease, which may vary in severity from mild (colds) to severe (malarial) to fatal (rabies). Examples of true pathogens include influenza virus, plague bacillus, and malarial protozoan.

Opportunistic pathogens cause disease when the host's defenses are compromised or when they become established in a part of the body that is not natural to them. Opportunists are not considered pathogenic to a normal healthy person and, unlike primary pathogens, do not generally possess well-developed virulence properties. Examples of opportunistic pathogens include *Pseudomonas* species and *Candida albicans.* Factors that greatly predispose a person to infections, both primary and opportunistic, are shown in **table 13.4.**

Recognizing that classifying pathogens into only two categories may be unduly restrictive, the Centers for Disease Control and Prevention has adopted a system of biosafety categories for pathogens based on their degree of pathogenicity and the relative danger in handling them. This system assigns microbes to one of four levels or classes. Microbes not known to cause disease in humans are assigned to level 1, highly virulent, contagious viruses that pose an extreme risk to humans are classified as level 4. Microbes of intermediate virulence are assigned to levels 2 or 3. This system is explained in more detail in table M.1 on page 544.

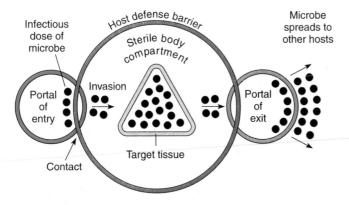

FIGURE 13.8

An overview of the events in infection. An adequate dose of an infectious agent overcomes a defense barrier and enters the sterile tissues of a host through one of several portals of entry. From here, it moves or is carried to a specific organ or tissue, called the target tissue, where it multiplies and usually causes some degree of damage. The exit of the pathogen through the same or another portal can facilitate its transmission to another host.

TABLE 13.4

Factors That Weaken Host Defenses and Increase Susceptibility to Infection*

- Old age and extreme youth (infancy, prematurity)
- Genetic defects in immunity and acquired defects in immunity (AIDS)
- Surgery and organ transplants
- Organic disease: cancer, liver malfunction, diabetes
- Chemotherapy/immunosuppressive drugs
- Physical and mental stress
- Other infections

**These conditions compromise defense barriers or immune responses.*

Differences in pathogenicity can be accounted for on the basis of **virulence.*** Virulence takes into account the ability of microbes to invade a host (invasiveness) and produce toxins (toxigenicity). These properties are called **virulence factors.** Virulence can be due to single or multiple factors. In some microbes, the causes of virulence are clearly established, but in others they are not. Although virulence is sometimes used interchangeably with pathogenicity, it is actually not a synonym. Pathogenicity is a more general term for comparing degrees of microbial infectiousness. In the following section, we examine the effects of virulence factors while simultaneously outlining the stages in the progress of an infection.

THE PORTAL OF ENTRY: GATEWAY TO INFECTION

To initiate an infection, a microbe enters the tissues of the body by a characteristic route, the **portal of entry,** usually a cutaneous or membranous boundary. The source of the infectious agent can be **exogenous,** originating from a source outside the body (the environment or another person or animal), or **endogenous,** already existing on or in the body (normal flora or latent infection).

For the most part, the portals of entry are the same anatomical regions that also support normal flora: the skin, gastrointestinal tract, respiratory tract, and urogenital tract. The majority of pathogens have adapted to a specific portal of entry, one that provides a habitat for further growth and spread. This adaptation can be so restrictive that if certain pathogens enter the "wrong" portal, they will not be infectious. For instance, inoculation of the nasal mucosa with the influenza virus invariably gives rise to the flu, but if this virus contacts only the skin, no infection will result. Likewise, contact with athlete's foot fungi in small cracks in the toe webs can induce an infection, but inhaling the fungus spores will not infect a healthy individual. Occasionally, an infective agent can enter by more than one portal. For instance, *Mycobacterium tuberculosis* enters through both the respiratory and gastrointestinal tracts, and pathogens in the genera *Streptococcus* and *Staphylococcus* have adapted to invasion through several portals of entry such as the skin, urogenital tract, and respiratory tract.

*virulence (veer-yoo-lents) L. *virulentia,* virus, poison.

Infectious Agents That Enter the Skin

The skin is a very common portal of entry. The actual sites of entry are usually nicks, abrasions, and punctures (many of which are tiny and inapparent) rather than smooth, unbroken skin. *Staphylococcus aureus* (the cause of boils), *Streptococcus pyogenes* (an agent of impetigo), the fungal dermatophytes, and agents of gangrene and tetanus gain access through damaged skin. The viral agent of cold sores (herpes simplex, type 1) enters through the mucous membranes near the lips.

Some infectious agents create their own passageways into the skin using digestive enzymes. For example, certain helminth worms burrow through the skin directly to gain access to the tissues. Other infectious agents enter through bites. The bites of insects, ticks, and other animals offer an avenue to a variety of viruses, rickettsias, and protozoa. An artificial means for breaching the skin barrier is contaminated hypodermic needles by intravenous drug abusers. Users who inject drugs are predisposed to a disturbing list of well-known diseases: hepatitis, AIDS, tetanus, tuberculosis, osteomyelitis, and malaria. A resurgence of some of these infections is directly traceable to drug use. Contaminated needles often contain bacteria from the skin or environment that induce heart disease (endocarditis), lung abscesses, and chronic infections at the injection site.

Although the conjunctiva, the outer protective covering of the eye, is ordinarily a relatively good barrier to infection, bacteria such as *Haemophilus aegyptius* (pinkeye), *Chlamydia trachomatis* (trachoma), and *Neisseria gonorrhoeae* have special affinity for this membrane.

The Gastrointestinal Tract as Portal

The gastrointestinal tract is the portal of entry for pathogens contained in food, drink, and other ingested substances. They are adapted to survive digestive enzymes and abrupt pH changes. Most enteric pathogens possess specialized mechanisms for entering and localizing in the mucosa of the small or large intestine. The best-known enteric agents of disease are gram-negative rods in the genera *Salmonella, Shigella, Vibrio,* and certain strains of *Escherichia coli* (see Spotlight on Microbiology 20.3). Viruses that enter through the gut are poliovirus, hepatitis A virus, echovirus, and rotavirus. Important enteric protozoans are *Entamoeba histolytica* (amebiasis) and *Giardia lamblia* (giardiasis). Although the anus is not a typical portal of entry, it becomes one in people who practice anal sex.

The Respiratory Portal of Entry

The oral and nasal cavities are also the gateways to the respiratory tract, the portal of entry for the greatest number of pathogens. The extent to which an agent is carried into the respiratory tree is based primarily on its size. In general, small cells and particles are inhaled more deeply than larger ones. Infectious agents with this portal of entry include the bacteria of streptococcal sore throat, meningitis, diphtheria, and whooping cough and the viruses of influenza, measles, mumps, rubella, chickenpox, and the common cold. Pathogens that are inhaled into the lower regions of the respiratory tract (bronchioles and lungs) can cause **pneumonia,** an inflammatory condition of the lung. Bacteria (*Streptococcus pneumoniae, Klebsiella, Mycoplasma*) and fungi (*Cryptococcus* and *Pneumocystis*) are a few of the agents involved in pneumonias. All types of pneumonia are on the increase owing to the greater susceptibility of AIDS patients to them. Other agents causing unique recognizable lung diseases are *Mycobacterium tuberculosis* and fungal pathogens such as *Histoplasma.*

TABLE 13.5	
Incidence of Common Sexually Transmitted Diseases	
STD	**Estimated Number of New Cases**
Human papillomavirus	5,500,000
Trichomoniasis	5,000,000
Chlamydiosis	783,000
Herpes simplex	1,000,000
Gonorrhea	361,000
Hepatitis B	77,000
Syphilis	32,200
AIDS	41,002

Urogenital Portals of Entry

The urogenital tract is the portal of entry for pathogens that are contracted by sexual means (intercourse or intimate direct contact). In the past, a judgmental attitude toward women and a somewhat puritan attitude toward sex led to these diseases being referred to as venereal, from Venus, the Latin name of the goddess of love. The more contemporary description, **sexually transmitted disease (STD),** reflects the mode of transmission of these diseases more accurately. STDs account for an estimated 4% of infections worldwide, with approximately 13 million new cases occurring in the United States each year. The most recent available statistics (2002) for the estimated incidence of common STDs is provided in **table 13.5.**

The microbes of STDs enter the skin or mucosa of the penis, external genitalia, vagina, cervix, and urethra. Some can penetrate an unbroken surface; others require a cut or abrasion. The once predominant sexual diseases syphilis and gonorrhea have been supplanted by a large and growing list of STDs headed by genital warts, chlamydia, and herpes. Evolving sexual practices have increased the incidence of STDs that were once uncommon, and diseases that were not originally considered STDs are now so classified.[4] Other common sexually transmitted agents are HIV (AIDS virus), *Trichomonas* (a protozoan), *Candida albicans* (a yeast), and hepatitis B virus.

Pathogens That Infect During Pregnancy and Birth

The placenta is an exchange organ formed by maternal and fetal tissues that separates the blood of the developing fetus from that of the mother, yet permits diffusion of dissolved nutrients and gases to the fetus. The placenta is ordinarily an effective barrier against microorganisms in the maternal circulation. However, a few microbes such as the syphilis spirochete can cross the placenta, enter the umbilical vein, and spread by the fetal circulation into the fetal tissues (**figure 13.9**).

Other infections, such as herpes simplex, occur perinatally when the child is contaminated by the birth canal. The common infections of fetus and neonate are grouped together in a unified cluster, known by the acronym **STORCH,** that medical personnel must monitor. STORCH stands for **s**yphilis, **t**oxoplasmosis, **o**ther diseases (hepatitis B, AIDS, and chlamydia), **r**ubella, **c**ytomegalovirus, and **h**erpes simplex virus. The most serious complications of STORCH infections are spontaneous abortion, congenital abnormalities, brain damage, prematurity, and stillbirths.

---

4. Amebic dysentery, scabies, salmonellosis, and *Strongyloides* worms are examples.

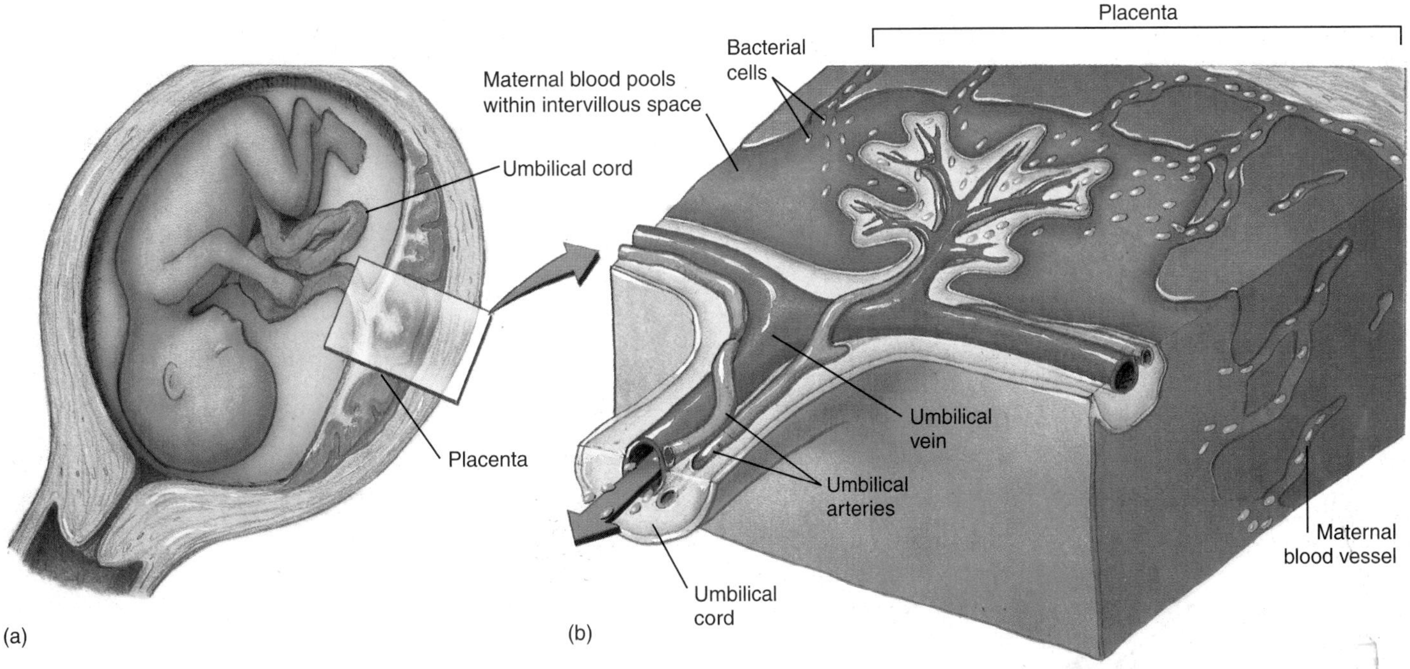

FIGURE 13.9

Transplacental infection of the fetus. **(a)** Fetus in the womb. **(b)** In a closer view, microbes are shown penetrating the maternal blood vessels and entering the blood pool of the placenta. They then invade the fetal circulation by way of the umbilical vein.

THE SIZE OF THE INOCULUM

Another factor crucial to the course of an infection is the quantity of microbes in the inoculating dose. For most agents, infection will proceed only if a minimum number, called the *infectious dose* (ID), is present. This number has been determined experimentally for many microbes. In general, microorganisms with smaller infectious doses have greater virulence. The ID varies from the astonishing number of 1 rickettsial cell in Q fever to about 10 infectious cells in tuberculosis, giardiasis, and coccidioidomycosis. The ID is 1,000 cells for gonorrhea and 10,000 cells for typhoid fever in contrast to 1,000,000,000 cells in cholera. Lack of an infectious dose will generally not result in an infection. But if the quantity is far in excess of the ID, the onset of disease can be extremely rapid. Even weakly pathogenic species can be rendered more virulent with a large inoculum.

MECHANISMS OF INVASION AND ESTABLISHMENT OF THE PATHOGEN

Following entry of the pathogen, the next stage in infection requires that the pathogen

1. bind to the host,
2. penetrate its barriers, and
3. become established in the tissues.

How the pathogen achieves these ends greatly depends upon its specific biochemical and structural characteristics.

How Pathogens Attach

Adhesion is a process by which microbes gain a more stable foothold at the portal of entry. Because adhesion is dependent on binding between specific molecules on both the host and pathogen, a particular pathogen is limited to only those cells (and organisms) to which

it can bind. Once attached, the pathogen is poised advantageously to invade the sterile body compartments. Bacterial pathogens attach most often by mechanisms such as fimbriae (pili), flagella, and adhesive slimes or capsules; viruses attach by means of specialized receptors (**figure 13.10**). Protozoa can infiltrate by means of their organelle of locomotion; for example, *Balantidium coli* is said to penetrate by the boring action of its cilia. In addition, parasitic worms are mechanically fastened to the portal of entry by suckers, hooks, and barbs (see chapter 23). Adhesion methods of various microbes and the diseases they lead to are shown in **table 13.6.**

CHAPTER CHECKPOINTS

Microbial infections result when a microorganism penetrates host defenses, multiplies, and damages host tissue. The *pathogenicity* of a microbe refers to its ability to cause infection or disease. The *virulence* of a pathogen refers to the degree of damage it inflicts on the host tissues.

True pathogens cause infectious disease in healthy hosts, whereas *opportunistic pathogens* become infectious only when the host immune system is compromised in some way.

The site at which a microorganism first contacts host tissue is called the *portal of entry.* Most pathogens have one preferred portal of entry, although some have more than one.

The respiratory system is the portal of entry for the greatest number of pathogens.

The *infectious dose,* or ID, refers to the minimum number of microbial cells required to initiate infection in the host. The ID varies widely among microbial species.

Fimbriae, flagella, hooks, and adhesive capsules are types of adherence factors by which pathogens physically attach to host tissues.

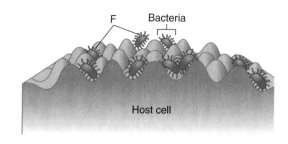

(a) **Fimbriae**

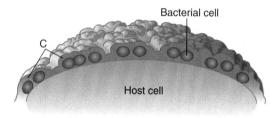

(b) **Capsules**

(c) **Spikes**

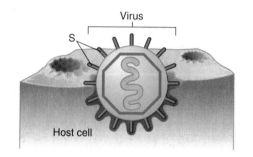

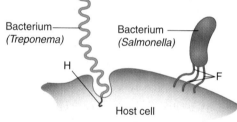

(d) **Hooks or flagella**

FIGURE 13.10

Mechanisms of adhesion by pathogens. **(a)** Fimbriae (F), minute bristlelike appendages. **(b)** Adherent extracellular capsules (C) made of slime or other sticky substances. **(c)** Viral envelope spikes (S). **(d)** Specialized cell hooks (H) or flagella (F). (See table 13.6 for specific examples.)

How Virulence Factors Contribute to Tissue Damage

Virulence factors from a microbe's perspective are simply adaptations it uses to invade and establish itself in the host. These same factors determine the degree of tissue damage that occurs. The effects of pathogen's virulence factors on tissues vary greatly. Cold viruses, for example, invade and multiply but cause relatively little damage to their host. At the other end of the spectrum, pathogens such as *Clostridium tetani* or HIV severely damage or kill their host. For convenience, we will divide the virulence factors into exoenzymes, toxins, and antiphagocytic factors (**figure 13.11**). Although this distinction is useful, there is often a very fine line between enzymes and toxins because many substances called toxins actually function as enzymes.

TABLE 13.6

Adhesion Properties of Microbes

Microbe	Disease	Adhesion Mechanism
Neisseria gonorrhoeae	Gonorrhea	Fimbriae attach to genital epithelium
Escherichia coli	Diarrhea	Well-developed K antigen capsule
Shigella	Dysentery	Fimbriae can attach to intestinal epithelium
Vibrio	Cholera	Glycocalyx anchors microbe to intestinal epithelium
Treponema	Syphilis	Tapered hook embeds in host cell
Mycoplasma	Pneumonia	Specialized tip at ends of bacteria fuse tightly to lung epithelium
Pseudomonas aeruginosa	Burn, lung infections	Fimbriae and slime layer
Streptococcus pyogenes	Pharyngitis, impetigo	Lipotechoic acid and M-protein anchor cocci to epithelium
Streptococcus mutants, S. sobrinus	Dental caries	Dextran slime layer glues cocci to tooth surface
Influenza virus	Influenza	Viral spikes react with receptor on cell surface
Poliovirus	Polio	Capsid proteins attach to receptors on susceptible cells
HIV	AIDS	Viral spikes adhere to white blood cell receptor
Giardia lamblia (protozoan)	Giardiasis	Small suction disc on underside attaches to intestinal surface

Extracellular Enzymes Many pathogenic bacteria, fungi, protozoa, and worms secrete **exoenzymes** that break down and inflict damage on tissues. Other enzymes dissolve the host's defense barriers and promote the spread of microbes to deeper tissues.

Examples of enzymes are:

1. mucinase, which digests the protective coating on mucous membranes and is a factor in amebic dysentery;
2. keratinase, which digests the principal component of skin and hair, and is secreted by fungi that cause ringworm;
3. collagenase, which digests the principal fiber of connective tissue and is an invasive factor of *Clostridium* species and certain worms; and
4. hyaluronidase, which digests hyaluronic acid, the ground substance that cements animal cells together. This enzyme is an important virulence factor in staphylococci, clostridia, streptococci, and pneumococci.

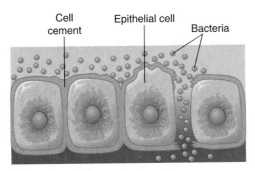

(a) **Exoenzymes**

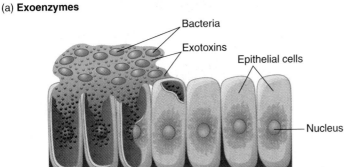

(b) **Toxins**

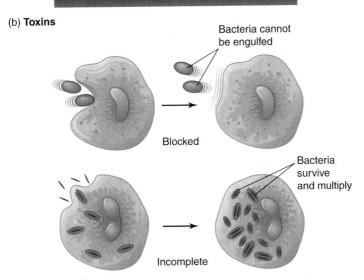

(c) **Phagocytosis**

FIGURE 13.11

The function of exoenzymes, toxins, and phagocyte blockers in invasiveness. **(a)** Exoenzymes. Bacteria produce extracellular enzymes that dissolve extracellular barriers and penetrate through or between cells to underlying tissues. **(b)** Toxins (primarily exotoxins) secreted by bacteria diffuse to target cells, which are poisoned and disrupted. **(c)** Blocked (top) or incomplete (bottom) phagocytosis. Phagocytosis is not effective for several reasons. For example, some pathogens have a protective coating that makes them hard to engulf, or the phagocyte ingests them but the microbes can still multiply.

Some enzymes react with components of the blood. Coagulase, an enzyme produced by pathogenic staphylococci, causes clotting of blood or plasma. By contrast, the bacterial kinases (streptokinase, staphylokinase) do just the opposite, dissolving fibrin clots and expediting the invasion of damaged tissues. In fact, one form of streptokinase (streptase) is marketed as a therapy to dissolve blood clots in patients with problems with thrombi and emboli.[5]

Bacterial Toxins: A Potent Source of Cellular Damage A **toxin** is a specific chemical product of microbes, plants, and some animals that is poisonous to other organisms. **Toxigenicity,** the power to produce toxins, is a genetically controlled characteristic of many species and is responsible for the adverse effects of a variety of diseases generally called **toxinoses.** A type of toxinosis in which the toxin is spread by the blood from the site of infection is a **toxemia** (tetanus and diphtheria, for example), whereas one caused by ingestion of toxins is an **intoxication** (botulism). A toxin is named according to its specific target of action: Neurotoxins act on the nervous system; enterotoxins act on the intestine; hemotoxins lyse red blood cells; and nephrotoxins damage the kidneys.

A more traditional scheme classifies toxins according to their origins **(figure 13.12).** A toxin molecule secreted by a living bacterial cell into the infected tissues is an **exotoxin.** A toxin that is not secreted but is released only after the cell is damaged or lysed is an **endotoxin.** Other important differences between the two groups, summarized in **table 13.7,** are generally chemical and medical in nature.

Exotoxins are proteins with a strong specificity for a target cell and extremely powerful, sometimes deadly, effects. They generally affect cells by damaging the cell membrane and initiating lysis or by disrupting intracellular function. **Hemolysins*** are a class of bacterial exotoxin that disrupts the cell membrane of red blood cells (and some other cells too). This damage causes the red blood cells to **hemolyze**—to burst and release hemoglobin pigment. Hemolysins that increase pathogenicity include the streptolysins of *Streptococcus pyogenes* and the alpha (α) and beta (β) toxins of *Staphylococcus aureus.* When colonies of bacteria growing on blood agar produce hemolysin, distinct zones appear around the colony. The pattern of hemolysis is often used to identify bacteria and determine their degree of pathogenicity (see chapter 18).

The toxins of diphtheria, tetanus, and botulism, among others, attach to a particular target cell, become internalized, and interrupt an essential cell pathway. The consequences of cell disruption depend upon the target. One toxin of *Clostridium tetani* blocks the action of certain spinal neurons; the toxin of *Clostridium botulinum* prevents the transmission of nerve-muscle stimuli; pertussis toxin inactivates the respiratory cilia; and cholera toxin provokes profuse salt and water loss from intestinal cells. More details of the pathology of exotoxins are found in later chapters on specific diseases.

Endotoxins belong to a class of chemicals called lipopolysaccharides (LPS), which are part of the outer membrane of gramnegative cell walls (see Medical Microfile 20.1). When gramnegative bacteria cause infections, some of them eventually lyse and release these LPS molecules into the infection site or into the circulation. Endotoxins differ from exotoxins in having a variety of systemic effects on tissues and organs. Depending upon the amounts present, endotoxins can cause fever, inflammation, hemorrhage, and diarrhea. Blood infection by gram-negative bacteria such as *Salmonella, Shigella, Neisseria meningitidis,* and *Escherichia coli* are particularly dangerous, in that it can lead to fatal endotoxic shock.

---

5. These conditions are intravascular blood clots that can cause circulatory obstructions.

*hemolysin (hee-mahl'-uh-sin) Gr. *haima,* blood, and *lysis,* dissolution. A substance that causes hemolysis.

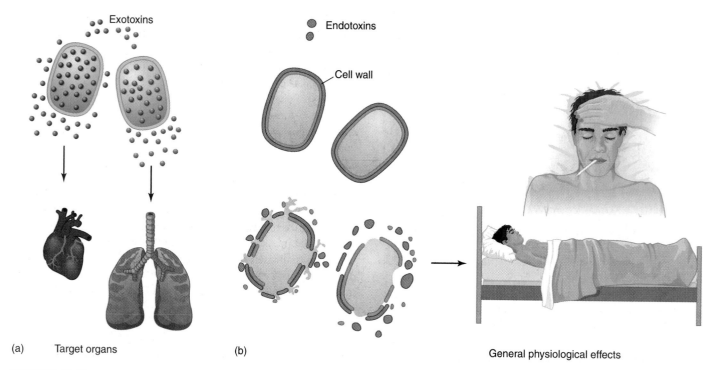

Exotoxins

(a)   Target organs

(b)

General physiological effects

FIGURE 13.12

The origins and effects of circulating exotoxins and endotoxins.   **(a)** Exotoxins, given off by live cells, have highly specific targets and physiological effects. **(b)** Endotoxins, given off when the cell wall of gram-negative bacteria disintegrates, have more generalized physiological effects.

TABLE 13.7

Differential Characteristics of Bacterial Exotoxins and Endotoxins

Characteristic	Exotoxins	Endotoxins
Toxicity	Toxic in minute amounts	Toxic in high doses
Effects on the Body	Specific to a cell type (blood, liver, nerve)	Systemic: fever, inflammation
Chemical Composition	Polypeptides	Lipopolysaccharide of cell wall
Heat Denaturation at 60°C	Unstable	Stable
Toxoid Formation	Convert to toxoid*	Do not convert to toxoid
Immune Response	Stimulate antitoxins**	Do not stimulate antitoxins
Fever Stimulation	Usually not	Yes
Manner of Release	Secreted from live cell	Released by cell during lysis
Typical Sources	A few gram-positive and gram-negative	All gram-negative bacteria

*A toxoid is an inactivated toxin used in vaccines.
**An antitoxin is an antibody that reacts specifically with a toxin.

How Microbes Escape Phagocytosis   *Antiphagocytic factors* are another type of virulence factor used by some pathogens to avoid certain white blood cells called phagocytes. These cells would ordinarily engulf and destroy pathogens by means of enzymes and other antibacterial chemicals (see chapter 14). The antiphagocytic factors of resistant microorganisms help them to circumvent some part of the phagocytic process (see figure 13.11c). The most aggressive strategy involves bacteria that kill phagocytes outright. Species of both *Streptococcus* and *Staphylococcus* produce **leukocidins**, substances that are toxic to white blood cells. Some microorganisms secrete an extracellular surface layer (slime or capsule) that makes it physically difficult for the phagocyte to engulf them. *Streptococcus pneumoniae, Salmonella typhi, Neisseria meningitidis,* and *Cryptococcus neoformans* are notable examples. Some bacteria are well adapted to survival inside phagocytes after ingestion. For instance, pathogenic species of *Legionella, Mycobacterium,* and many rickettsias are readily engulfed but are capable of avoiding further destruction. The ability to survive intracellularly in phagocytes has special significance because it provides a place for the microbes to hide, grow, and be spread throughout the body.

Establishment, Spread, and Pathologic Effects

Aided by virulence factors, microbes eventually settle in a particular target organ and continue to cause damage at the site. The type and scope of injuries inflicted during this process account for the typical stages of an infection (**Medical Microfile 13.2**), the patterns of the infectious disease, and its manifestations in the body.

In addition to the adverse effects of enzymes, toxins, and other factors, multiplication by a pathogen frequently weakens host tissues. Pathogens can obstruct tubular structures such as blood vessels, lymphatic channels, fallopian tubes, and bile ducts. Accumulated damage can lead to cell and tissue death, a condition called **necrosis.** Although viruses do not produce toxins or destructive enzymes, they destroy cells by multiplying in and lysing them. Many of the cytopathic effects of viral infection arise from the impaired metabolism and death of cells (see chapter 6).

Patterns of Infection   Patterns of infection are many and varied (**figure 13.13**). In the simplest situation, a **localized infection,** the microbe enters the body and remains confined to a specific tissue (figure 13.13*a*). Examples of localized infections are boils, fungal skin infections, and warts.

Many infectious agents do not remain localized but spread from the initial site of entry to other tissues. In fact, spreading is necessary for pathogens, such as rabies and hepatitis A virus, whose target tissue is some distance from the site of entry. The rabies virus travels from a bite wound along nerve tracts to its target in the brain, and the hepatitis A virus moves from the intestine to the liver via the circulatory system. When an infection spreads to several sites and tissue fluids, usually in the bloodstream, it is called a **systemic infection** (figure 13.13*b*). Examples of systemic infections are viral diseases (measles, rubella, chickenpox, and AIDS); bacterial diseases (brucellosis, anthrax, typhoid fever, and syphilis); and fungal diseases (histoplasmosis and cryptococcosis). Infectious agents can also travel to their targets by means of nerves (as in rabies) or cerebrospinal fluid (as in meningitis).

A **focal infection** is said to exist when the infectious agent breaks loose from a local infection and is carried into other tissues (figure 13.13*c*). This pattern is exhibited by tuberculosis or by streptococcal pharyngitis, which gives rise to scarlet fever. In the condition called toxemia,[6] the infection itself remains localized at the portal of entry, but the toxins produced by the pathogens are carried by the blood to the actual target tissue. In this way, the target of the bacterial cells can be different from the target of their toxin (see discussions of tetanus and diphtheria in chapter 19).

An infection is not always caused by a single microbe. In a **mixed infection,** several agents establish themselves simultaneously at the infection site (figure 13.13*d*). In some mixed or synergistic infections, the microbes cooperate in breaking down a tissue. In other mixed infections, one microbe creates an environment that enables another microbe to invade. Gas gangrene, wound infections, dental caries, and human bite infections tend to be mixed.

Some diseases are described according to a sequence of infection. When an initial, or **primary,** infection is complicated by another infection caused by a different microbe, the second infection is termed a **secondary infection** (figure 13.13*e*). This pattern often occurs in a child with chickenpox (primary infection) who may scratch his pox and infect them with *Staphylococcus aureus* (secondary

---

6. Not to be confused with toxemia of pregnancy, which is a metabolic disturbance and not an infection.

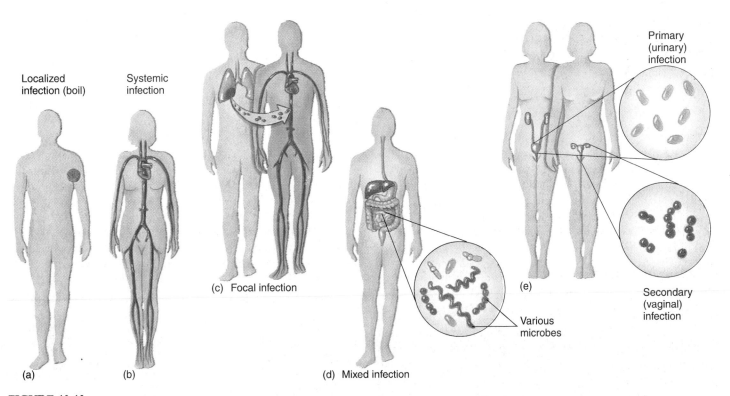

(c) Focal infection

Localized infection (boil)

Systemic infection

(a)   (b)

Primary (urinary) infection

Various microbes

(d) Mixed infection

(e)

Secondary (vaginal) infection

FIGURE 13.13

The occurrence of infections with regard to location, type of microbe, and length of time. **(a)** A localized infection, in which the pathogen is restricted to one specific site. **(b)** Systemic infection, in which the pathogen spreads through circulation to many sites. **(c)** A focal infection occurs initially as a local infection, but circumstances cause the microbe to be carried to other sites systemically. **(d)** A mixed infection, in which the same site is infected with several microbes at the same time. **(e)** In a primary-secondary infection, an initial infection is complicated by a second one in a different location (usually) and caused by a different microbe.

MEDICAL MICROFILE   13.2
The Classic Stages of Clinical Infections

As the body of the host responds to the invasive and toxigenic activities of a parasite, it passes through four distinct phases of infection and disease: the incubation period, the prodromium, the period of invasion, and the convalescent period.

The *incubation period* is the time from initial contact with the infectious agent (at the portal of entry) to the appearance of the first symptoms. During the incubation period, the agent is multiplying at the portal of entry but has not yet caused enough damage to elicit symptoms. Although this period is relatively well defined and predictable for each microorganism, it does vary according to host resistance, degree of virulence, and distance between the target organ and the portal of entry (the farther apart, the longer the incubation period). Overall, an incubation period can range from several hours in pneumonic plague to several years in leprosy. The majority of infections, however, have incubation periods ranging between 2 and 30 days.

The earliest notable symptoms of infection appear as a vague feeling of discomfort, such as head and muscle aches, fatigue, upset stomach, and general malaise. This short period (1–2 days) is known as the *prodromal stage.* The infectious agent next enters a *period of invasion,* during which it multiplies at high levels, exhibits its greatest toxicity, and becomes well established in its target tissue. This period is often marked by fever and other prominent and more specific signs and symptoms, which can include cough, rashes, diarrhea, loss of muscle control, swelling, jaundice, discharge of exudates, or severe pain, depending on the particular infection. The length of this period is extremely variable.

As the patient begins to respond to the infection, the symptoms decline—sometimes dramatically, other times slowly. During the recovery that follows, called the *convalescent period,* the patient's strength and health gradually return owing to the healing nature of the immune response. An infection that results in death is called terminal. This term is particularly applicable if the infection is the immediate cause of death in a patient already suffering from a degenerative disease or cancer. Thus, an alcoholic might succumb to terminal pneumonia, or an AIDS patient will actually die from a secondary infection.

The transmissibility of the microbe during these four stages must be considered on an individual basis. A few agents are released mostly during incubation (measles, for example); many are released during the invasive period (*Shigella*); and others can be transmitted during all of these periods (hepatitis B).

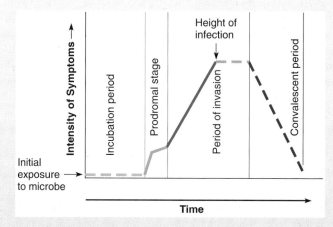

Stages in the course of infection and disease: Dashed lines represent periods with a variable length.

infection). The secondary infection need not be in the same site as the primary infection, and it usually indicates altered host defenses.

Infections that come on rapidly, with severe but short-lived effects, are called **acute infections.** Infections that progress and persist over a long period of time are **chronic infections. Microbits 13.3** illustrates other common terminology used to describe infectious diseases.

SIGNS AND SYMPTOMS: WARNING SIGNALS OF DISEASE

When an infection causes **pathologic** changes leading to disease, it is often accompanied by a variety of signs and symptoms. A **sign** is any objective evidence of disease as noted by an observer; a **symptom** is the subjective evidence of disease as sensed by the patient. In general, signs are more precise than symptoms, though both can have the same underlying cause. For example, an infection of the brain might present with the sign of bacteria in the spinal fluid and symptom of headache. Or a streptococcal infection might produce a sore throat (symptom) and inflamed pharynx (sign). Disease indicators that can be sensed and observed can qualify as either a sign or a symptom. When a disease can be identified or defined by a certain complex of signs and symptoms, it is termed a **syndrome.** Signs and symptoms with considerable importance in diagnosing infectious diseases are shown in **table 13.8.**

TABLE 13.8	
Common Signs and Symptoms of Infectious Diseases	
Signs	**Symptoms**
Fever	Chills
Septicemia	Pain, ache, soreness, irritation
Microbes in tissue fluids	Nausea
Chest sounds	Malaise, fatigue
Skin eruptions	Chest tightness
Leukocytosis	Itching
Leukopenia	Headache
Swollen lymph nodes	Nausea
Abscesses	Abdominal cramps
Tachycardia (increased heart rate)	Anorexia (lack of appetite)
Antibodies in serum	Sore throat

Words in medicine have great power and economy. A single technical term can often replace a whole phrase or sentence, thereby saving time and space in patient charting. The beginning student may feel overwhelmed by what seems like a mountain of new words. However, having a grasp of a few root words and a fair amount of anatomy can help you learn many of these words and even deduce the meaning of unfamiliar ones. Some examples of medical shorthand follow.

The suffix **-itis** means an inflammation and, when affixed to the end of an anatomical term, indicates an inflammatory condition in that location. Thus, meningitis is an inflammation of the meninges surrounding the brain; encephalitis is an inflammation of the brain itself; hepatitis involves the liver; vaginitis, the vagina; gastroenteritis, the intestine; and otitis media, the middle ear. Although not all inflammatory conditions are caused by infections, many infectious diseases inflame their target organs.

The suffix **-emia** is derived from the Greek word *haeima,* meaning blood. When added to a word, it means "associated with the blood." Thus, septicemia means sepsis (infection) of the blood; bacteremia, bacteria in the blood; viremia, viruses in the blood; and fungemia, fungi in the blood. It is also applicable to specific conditions such as toxemia, gonococcemia, and spirochetemia.

The suffix **-osis** means "a disease or morbid process." It is frequently added to the names of pathogens to indicate the disease they cause: for example, listeriosis, histoplasmosis, toxoplasmosis, shigellosis, salmonellosis, and borreliosis. A variation of this suffix is *-iasis,* as in trichomoniasis and candidiasis.

The suffix **-oma** comes from the Greek word *onkomas* (swelling) and means tumor. Although the root is often used to describe cancers (sarcoma, melanoma), it is also applied in some infectious diseases that cause masses or swellings (tuberculoma, leproma).

Signs and Symptoms of Inflammation

The earliest symptoms of disease result from the activation of the body defense process called **inflammation.*** The inflammatory response includes cells and chemicals that respond nonspecifically to disruptions in the tissue. This subject is discussed in greater detail in chapter 14, but it is worth noting here that many signs and symptoms of infection are caused by the mobilization of this system. Some common symptoms of inflammation include fever, pain, soreness, and swelling. Signs of inflammation include **edema,*** the accumulation of fluid in an afflicted tissue; **granulomas** and **abscesses,** walled-off collections of inflammatory cells and microbes in the tissues; and **lymphadenitis,** swollen lymph nodes.

Rashes and other skin eruptions are common symptoms and signs in many diseases, and because they tend to mimic each other, it can be difficult to differentiate among diseases on this basis alone. The general term for the site of infection or disease is **lesion.*** Skin lesions can be restricted to the epidermis and its glands and follicles, or they can extend into the dermis and subcutaneous regions. The lesions of some infections undergo characteristic changes in appearance during the course of disease and thus fit more than one category (see Medical Microfile, page 538, and figure 24.1).

Signs of Infection in the Blood

Changes in the number of circulating white blood cells, as determined by special counts, are considered to be signs of possible infection. **Leukocytosis*** is an increase in the level of white blood cells, whereas **leukopenia*** is a decrease. Other signs of infection revolve around the occurrence of a microbe or its products in the blood. The clinical term for blood infection, **septicemia,** refers to a general state in which microorganisms are multiplying in the blood and are present in large numbers. When small numbers of bacteria or viruses are found in the blood, the correct terminology is **bacteremia** or **viremia,** which means that these microbes are present in the blood but are not necessarily multiplying.

During infection, a normal host will invariably show signs of an immune response in the form of antibodies in the serum or some type of sensitivity to the microbe. This fact is the basis for several serological tests used in diagnosing infectious diseases such as AIDS or syphilis. Such specific immune reactions indicate the body's attempt to develop specific immunities against pathogens. We will concentrate on this role of the host defenses in chapters 14 and 15.

Infections That Go Unnoticed

It is rather common for an infection to produce no noticeable symptoms, even though the microbe is active in the host tissue. In other words, although infected, the host does not manifest the disease. Infections of this nature are known as **asymptomatic, subclinical,** or *inapparent* because the patient experiences no symptoms or disease and does not seek medical attention. However, it is important to note that most infections are attended by some sort of sign. In the section on epidemiology, we further address the significance of subclinical infections in the transmission of infectious agents.

*inflammation (in-flam'-uh-tor"-ee) L. *inflammatio,* to set on fire.

*edema (uh-dee'-muh) Gr. *oidema,* swelling.

*lesion (lee'-zhun) L. *laesio,* to hurt.

*leukocytosis (loo"-koh'-sy-toh'-sis) From *leukocyte,* a white blood cell, and the suffix-*osis.*

*leukopenia (loo"-koh-pee'-nee-uh) From *leukocyte* and *penia,* a loss or lack of.

THE PORTAL OF EXIT: VACATING THE HOST

Earlier, we introduced the idea that a parasite is considered *unsuccessful* if it kills its host. A parasite is equally unsuccessful if it does not have a provision for leaving its host and moving to other susceptible hosts. With few exceptions, pathogens depart by a specific avenue called the **portal of exit (figure 13.14)**. In most cases, the pathogen is shed or released from the body through secretion, excretion, discharge, or sloughed tissue. The usually very high number of infectious agents in these materials increases both virulence and the likelihood that the pathogen will reach other hosts. In many cases, the portal of exit is the same as the portal of entry, but a few pathogens use a different route. As we see in the next section, the portal of exit concerns epidemiologists because it greatly influences the dissemination of infection in a population.

Respiratory and Salivary Portals

Mucus, sputum, nasal drainage, and other moist secretions are the media of escape for the pathogens that infect the lower or upper respiratory tract. The most effective means of releasing these secretions are coughing and sneezing (see figure 13.20), although they can also be released during talking and laughing. Tiny particles of liquid released into the air form aerosols or droplets that can spread the infectious agent to other people. The agents of tuberculosis, influenza, measles, and chickenpox most often leave the host through airborne droplets. Droplets of saliva are the exit route for several viruses, including those of mumps, rabies, and infectious mononucleosis.

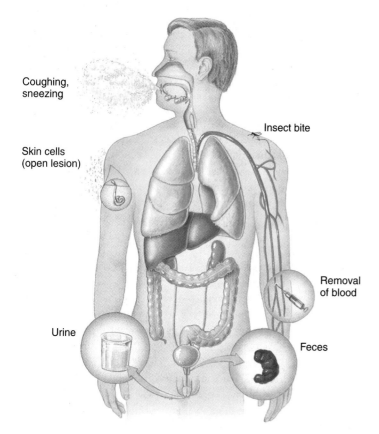

Coughing, sneezing

Skin cells (open lesion)

Insect bite

Urine

Removal of blood

Feces

FIGURE 13.14

Major portals of exit of infectious diseases.

Skin Scales

The outer layer of the skin and scalp are constantly being shed into the environment. A large proportion of household dust is actually composed of skin scales. A single person can shed several billion skin cells a day, and some persons, called shedders, disseminate massive numbers of bacteria into their immediate surroundings. Skin lesions and their exudates can serve as portals of exit in warts, fungal infections, boils, herpes simplex, smallpox, and syphilis.

Fecal Exit

Feces is a very common portal of exit. Some intestinal pathogens grow in the intestinal mucosa and create an inflammation that increases the motility of the bowel. This increased motility speeds up peristalsis, resulting in diarrhea, and the more fluid stool provides a rapid exit for the pathogen. A number of helminth worms release cysts and eggs through the feces. Feces containing pathogens are a public health problem when allowed to contaminate drinking water or when used to fertilize crops (see chapter 23).

Urogenital Tract

A number of agents involved in sexually transmitted infections leave the host in vaginal discharge or semen. This is also the source of neonatal infections such as herpes simplex, *Chlamydia,* and *Candida albicans,* which infect the infant as it passes through the birth canal. Less commonly, certain pathogens that infect the kidney are discharged in the urine: for instance, the agents of leptospirosis, typhoid fever, tuberculosis, and schistosomiasis.

Removal of Blood or Bleeding

Although the blood does not have a direct route to the outside, it can serve as a portal of exit when it is removed or released through a vascular puncture made by natural or artificial means. Blood-feeding animals such as ticks and fleas are common transmitters of pathogens (see Microbits 21.5). The AIDS and hepatitis viruses are transmitted by shared needles or through small gashes in a mucous membrane caused by sexual intercourse. Blood donation is also a means for certain microbes to leave the host, though this means of exit is now unusual because of close monitoring of the donor population and blood used for transfusions.

THE PERSISTENCE OF MICROBES AND PATHOLOGIC CONDITIONS

The apparent recovery of the host does not always mean that the microbe has been completely removed or destroyed by the host defenses. After the initial symptoms in certain chronic infectious diseases, the infectious agent retreats into a dormant state called **latency.** Throughout this latent state, the microbe can periodically become active and produce a recurrent disease. The viral agents of herpes simplex, herpes zoster, hepatitis B, AIDS, and Epstein-Barr can persist in the host for long periods. The agents of syphilis, typhoid fever, tuberculosis, and malaria also enter into latent stages. The person harboring a persistent infectious agent may or may not shed it during the latent stage. If it is shed, such persons are chronic carriers who serve as sources of infection for the rest of the population.

Some diseases leave **sequelae*** in the form of long-term or permanent damage to tissues or organs. For example, meningitis can result in deafness, a strep throat can lead to rheumatic heart disease, Lyme disease can cause arthritis, and polio can produce paralysis.

CHAPTER CHECKPOINTS

Exoenzymes, toxins, and antiphagocytic factors are the three main types of *virulence factors* pathogens utilize to combat host defenses and damage host tissue.

Exotoxins and endotoxins differ in their chemical composition and tissue specificity.

Antiphagocytic factors produced by microorganisms include leukocidins, capsules, and factors that resist digestion by white blood cells.

Patterns of infection vary with the pathogen or pathogens involved. They range from local and focal to systemic.

A mixed infection is caused by two or more microorganisms simultaneously.

Infections can be characterized by their sequence as primary or secondary and by their duration as either acute or chronic.

An infectious disease is characterized by both objective signs and subjective symptoms.

Infectious diseases that are asymptomatic or subclinical nevertheless often produce clinical signs.

The portal of exit by which a pathogen leaves its host is usually but not always the same as the portal of entry.

The portals of exit and entry determine how pathogens spread in a population.

Some pathogens persist in the body in a latent state; others cause long-term diseases called *sequelae*.

Epidemiology: The Study of Disease in Populations

So far, our discussion has revolved primarily around the impact of an infectious disease in a single individual. Let us now turn our attention to the effects of diseases on the community—the realm of **epidemiology.*** By definition, this term involves the study of the frequency and distribution of disease and other health-related factors in defined human populations. It involves many disciplines—not only microbiology, but anatomy, physiology, immunology, medicine, psychology, sociology, ecology, and statistics—and it considers many diseases other than infectious ones, including heart disease, cancer, drug addiction, and mental illness. The epidemiologist is a medical sleuth who collects clues on the causative agent, pathology, sources and modes of transmission, and tracks the numbers and distribution of cases of disease in the community. In fulfilling these demands, the epidemiologist asks who, when, where, how, why, and what questions about diseases. The outcome of these studies helps public health departments develop prevention and treatment programs and establish a basis for predictions.

*sequelae (suh-kwee´-lee) L. *sequi,* to follow.

*epidemiology (ep´´-ih-dee-mee-ahl´-uh-gee) Gr. *epidemios,* prevalent.

WHO, WHEN, AND WHERE? TRACKING DISEASE IN THE POPULATION

Epidemiologists are concerned with all of the factors covered earlier in this chapter: virulence, portals of entry and exit, and the course of disease. But they are also interested in surveillance—that is, collecting, analyzing, and reporting data on the rates of occurrence, mortality, morbidity, and transmission of infections. Surveillance involves keeping data for a large number of diseases seen by the medical community and reported to public health authorities. By law, certain **reportable,** or notifiable, diseases must be reported to authorities; others are reported on a voluntary basis.

A well-developed network of individuals and agencies at the local, district, state, national, and international levels keeps track of infectious diseases. Physicians and hospitals report all notifiable diseases that are brought to their attention. Case reporting can focus on a single individual or collectively on group data.

Local public health agencies first receive the case data and determine how they will be handled. In most cases, health officers investigate the history and movements of patients to trace their prior contacts and to control the further spread of the infection as soon as possible through drug therapy, immunization, and education. In sexually transmitted diseases, patients are asked to name their partners so that these persons can be notified, examined, and treated. It is very important to maintain the confidentiality of the persons in these reports. The principal government agency responsible for keeping track of infectious diseases nationwide is the Centers for Disease Control and Prevention (CDC) in Atlanta, Georgia, which is a part of the United States Public Health Service. The CDC publishes a weekly notice of diseases (the *Morbidity and Mortality Report*) that provides weekly and cumulative summaries of the case rates and deaths for about 50 notifiable diseases, highlights important and unusual diseases, and presents data concerning disease occurrence in the major regions of the United States. Ultimately, the CDC shares its statistics on disease with the World Health Organization (WHO) for worldwide tabulation and control.

Epidemiologic Statistics: Frequency of Cases

The **prevalence** of a disease is the total number of existing cases with respect to the entire population. It is a cumulative statistic, usually represented as the percentage of the population having a particular disease at any given time. Disease **incidence** measures the number of new cases over a certain time period, as compared with the general healthy population. This statistic, also called the case, or morbidity, rate, indicates both the rate and the risk of infection. The equations used to figure these rates are:

$$\text{Prevalence} = \frac{\text{Total number of cases in population}}{\text{Total number of persons in population}} \times 100 = \%$$

$$\text{Incidence} = \frac{\text{Number of new cases}}{\text{Number of healthy persons}} = \text{Ratio}$$

As an example, let us use a classroom of 50 students exposed to a new strain of influenza. Before exposure, the prevalence and incidence in this population are both zero (0/50). If in one week, 5 out of the 50 people contract the disease, the prevalence is 5/50 = 10%, and the incidence is 1 in 9 (5 cases compared with 45 healthy persons). If after 2 weeks, 5 more students contract the flu, the prevalence becomes 5 + 5 = 10/50 = 20%, and the incidence becomes 1 in 8 (5/40). When dealing with large populations, the incidence is usually given in numbers of cases per 1,000 or 100,000 population.

The changes in incidence and prevalence are usually followed over a seasonal, yearly, and long-term basis and are helpful in predicting trends (**figure 13.15**). Statistics of concern to the epidemiologist are the rates of disease with regard to sex, race, or geographic region (figure 13.15). Also of importance is the **mortality rate,** which measures the total number of deaths in a population due to a certain disease. Over the past century, the overall death rate from infectious diseases has dropped, although the number of persons afflicted with infectious diseases (the **morbidity* rate**) has remained relatively high.

Monitoring statistics also makes it possible to define the frequency of a disease in the population (**figure 13.16**). An infectious disease that exhibits a relatively steady frequency over a long time period in a particular geographic locale is **endemic** (figure 13.16*a*). For example, Lyme disease is endemic to certain areas of the United States where the tick vector is found. A certain number of new cases are expected in these areas every year. When a disease is **sporadic,** occasional cases are reported at irregular intervals in random locales (figure 13.16*b*). Tetanus and diphtheria are reported sporadically in the United States (fewer than 50 cases a year). When statistics indicate that the prevalence of an endemic or sporadic disease is increasing beyond what is expected for that population, the pattern is described as an **epidemic** (figure 13.16*c*). (See figure 13.15 for an idea of how epidemics look on graphs.) An epidemic exists when an increasing trend is observed in a particular population. The time period is not defined—it can range from hours in food poisoning to years in syphilis—nor is an exact percentage of increase needed before an outbreak can qualify as an epidemic. Several epidemics occur every year in the United States, most recently among STDs such as chlamydia and gonorrhea. The spread of an epidemic across continents is a **pandemic,** as exemplified by AIDS and influenza (figure 13.16*d*).

One important epidemiologic truism might be called the "iceberg effect," which refers to the fact that only a small portion of an iceberg is visible above the surface of the ocean, with a much more massive part lingering unseen below the surface. Regardless of case reporting and public health screening, a large number of cases of infection in the community go undiagnosed and unreported. (For a list of reportable diseases in the United States, see **table 13.9.**) In the instance of salmonellosis, approximately 40,000 cases are reported each year. Epidemiologists estimate that the actual number is more likely somewhere between 400,000 and 4,000,000 (see figure 20.15). The iceberg effect can be even more lopsided for sexually transmitted diseases or for infections that are not brought to the attention of reporting agencies.

Investigative Strategies of the Epidemiologist

Initial evidence of a new disease or an epidemic in the community is fragmentary. A few sporadic cases are seen by physicians and are eventually reported to authorities. However, it can take several reports before any alarm is registered. Epidemiologists and public health departments must piece together odds and ends of data from a series of

---

*morbidity (mor-bih′-dih-tee) L. *morbidis,* sick. A condition of being diseased.

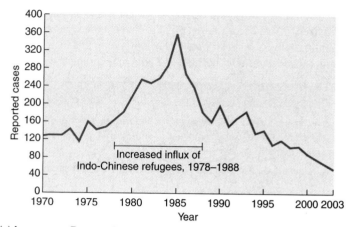

(a) **Leprosy — Reported cases by year, United States, 1970–2003**

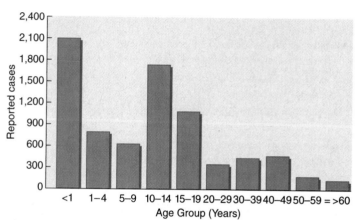

(b) **Pertussis — Reported cases by age group, United States, 2000**

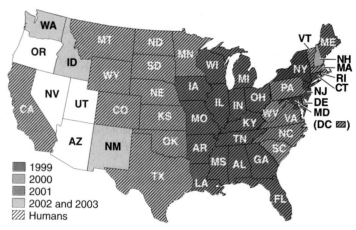

(c) West Nile virus cases were first reported in 1999 among a variety of animals and humans. Colors track its rapid progress across the United States in just four years.

FIGURE 13.15

Graphical representation of epidemiological data. The Centers for Disease Control and Prevention collects epidemiological data which is analyzed with regard to **(a)** time frame, **(b)** age, and **(c)** geographic region.

Source: Centers for Disease Control and Prevention. *Morbidity and Mortality Weekly Report.* United States, 2003.

• Outbreaks

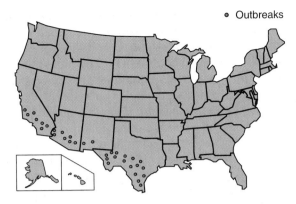

(a) **Endemic Occurrence**

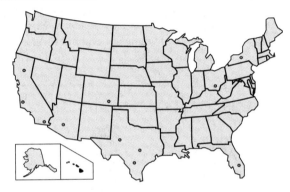

(b) **Sporadic Occurrence**

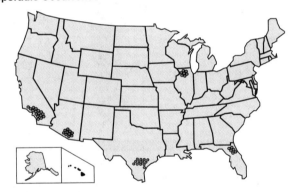

(c) **Epidemic Occurrence**

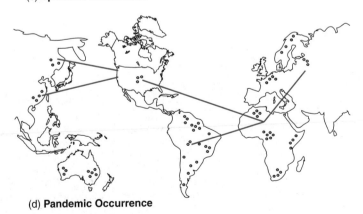

(d) **Pandemic Occurrence**

FIGURE 13.16

Patterns of infectious disease occurrence. **(a)** In endemic occurrence, cases are concentrated in one area at a relatively stable rate. **(b)** In sporadic occurrence, a few cases occur randomly over a wide area. **(c)** An epidemic is an increased number of cases that often appear in geographic clusters. **(d)** Pandemic occurrence means that an epidemic ranges over more that one continent.

TABLE 13.9

Notifiable Diseases in the United States★

AIDS	Lyme disease
Amebiasis	Lymphogranuloma venereum
Anthrax	Malaria
Arboviral infections	Measles (rubeola)
Aseptic meningitis	Meningococcal infections
Botulism	Mumps
Brucellosis	Pertussis
Chancroid	Plague
Chickenpox	Poliomyelitis
Cholera	Psittacosis
Chlamydiosis	Rabies
Cryptosporidiosis	Rheumatic fever
Diphtheria	Rocky Mountain spotted fever
Encephalitis	Rubella
Enterovirus	Salmonellosis
Escherichia coli O157:H7	Shigellosis
Gonorrhea	Syphilis
Hansen's disease (leprosy)	Tetanus
Haemophilus influenzae meningitis	Toxic shock syndrome
Hepatitis A	Trichinosis
Hepatitis B	Tuberculosis
Hepatitis C	Tularemia
Hepatitis, other	Typhoid fever
Influenza	Typhus
Legionellosis	Yellow fever
Leptospirosis	

★*Depending on the state, some of these diseases are reported only if they occur at epidemic levels; others must be reported on a case-by-case basis. You can request a list of reportable infectious diseases in your state by calling the State Department of Health.*

apparently unrelated cases and work backward to reconstruct the epidemic pattern. A completely new disease requires even greater preliminary investigation, because the infectious agent must be isolated and linked directly to the disease (see the discussion of Koch's postulates in a later section of this chapter). All factors possibly impinging on the disease are scrutinized. Investigators search for *clusters* of cases indicating spread between persons or a public (common) source of infection; they also look at possible contact with animals, contaminated food, water, and public facilities, and at human interrelationships or changes in community structure. Out of this maze of case information, the investigators hope to recognize a pattern that indicates the source of infection so that they can quickly move to control it.

RESERVOIRS: WHERE PATHOGENS PERSIST

In order for an infectious agent to continue to exist and be spread, it must have a permanent place to reside. The **reservoir** is the primary habitat in the natural world from which a pathogen originates. Often it is a human or animal carrier, although soil, water, and plants are also reservoirs. The reservoir can be distinguished from the infection **source,** which is the individual or object from which an infection is actually acquired. In diseases such as syphilis, the reservoir and the source are the same (the human body). In the case of hepatitis A, the reservoir (a human carrier) is usually different from the source of infection (contaminated food).

Living Reservoirs

Many pathogens continue to exist and spread because they are harbored by members of a host population. Persons or animals with frank symptomatic infection are obvious sources of infection, but a **carrier** is, by definition, an individual who *inconspicuously* shelters a pathogen and spreads it to others without any notice. Although human carriers are occasionally detected through routine screening (blood tests, cultures) and other epidemiologic devices, they are unfortunately very difficult to discover and control. As long as a pathogenic reservoir is maintained by the carrier state, the disease will continue to exist in that population, and the potential for epidemics will be a constant threat. The duration of the carrier state can be short- or long-term, and actual infection of the carrier may or may not be involved.

Several situations can produce the carrier state. **Asymptomatic** (apparently healthy) **carriers** are indeed infected, but as previously indicated, they show no symptoms (**figure 13.17a**). A few asymptomatic infections (gonorrhea and genital warts, for instance) can carry out their entire course without overt manifestations. Other asymptomatic carriers, called *incubation carriers,* spread the infectious agent during the incubation period. For example, AIDS patients can harbor and spread the virus for months and years before their first symptoms appear. Recuperating patients without symptoms are considered *convalescent carriers* when they continue to shed viable microbes and convey the infection to others. Diphtheria patients, for example, spread the microbe for up to 30 days after the disease has subsided.

An individual who shelters the infectious agent for a long period after recovery because of the latency of the infectious agent is a *chronic carrier.* Patients who have recovered from tuberculosis, hepatitis, and herpes infections frequently carry the agent chronically. About one in 20 victims of typhoid fever continues to

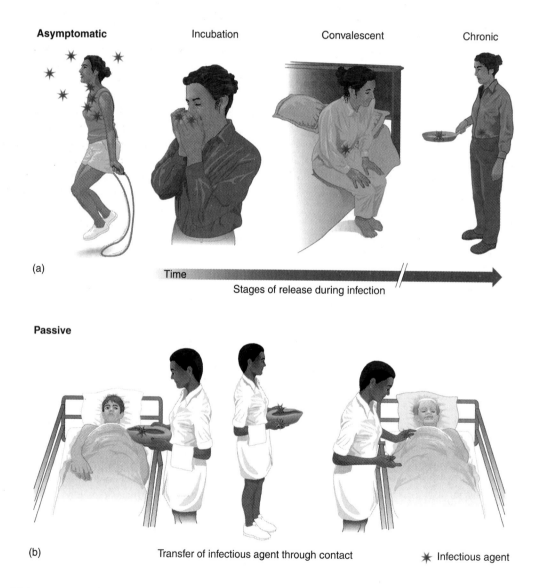

(a)

Asymptomatic   Incubation   Convalescent   Chronic

Time   Stages of release during infection

(b)

Passive

Transfer of infectious agent through contact   ✳ Infectious agent

FIGURE 13.17

Types of carriers. **(a)** An asymptomatic carrier is infected without symptoms. Incubation carriers are in the early stages of infection; convalescent carriers are in the late stages of recovery; chronic carriers sequester the microbe for long periods after the infection is over. **(b)** A passive carrier is contaminated but not infected.

harbor *Salmonella typhi* in the gallbladder for several years, and sometimes for life. The most infamous of these was "Typhoid Mary," a cook who spread the infection to hundreds of victims in the early 1900s.

The **passive carrier** state is of great concern during patient care (see a later section on nosocomial infections). Medical and dental personnel who must constantly handle materials that are heavily contaminated with patient secretions and blood are at risk for picking up pathogens mechanically and accidently transferring them to other patients (figure 13.17*b*). Proper handwashing, handling of contaminated materials, and aseptic techniques greatly reduce this likelihood.

Animals As Reservoirs and Sources   Up to now, we have lumped animals with humans in discussing living reservoirs or carriers, but animals deserve special consideration as vectors of infections. The word **vector** is used by epidemiologists to indicate a live animal that transmits an infectious agent from one host to another. (The term is sometimes misused to include any object that spreads disease.) The majority of vectors are arthropods such as fleas, mosquitoes, flies, and ticks, although larger animals can also spread infection—for example, mammals (rabies), birds (psittacosis), or lower vertebrates (salmonellosis).

By tradition, vectors are placed into one of two categories, depending upon the animal's relationship with the microbe (**figure 13.18**). A **biological vector** actively participates in a pathogen's life cycle, serving as a site in which it can multiply or complete its life cycle. A biological vector communicates the infectious agent to the human host by biting, aerosol formation, or touch. In the case of biting vectors, the animal can

1. inject infected saliva into the blood (the mosquito),
2. defecate around the bite wound (the flea) (figure 13.18*a*), or
3. regurgitate blood into the wound (the tsetse fly).

More detailed discussions of the roles of biological vectors are found in chapters 20, 21, 23, and 25.

Mechanical vectors are not necessary to the life cycle of an infectious agent and merely transport it without being infected. The external body parts of these animals become contaminated when they come into physical contact with a source of pathogens. The agent is subsequently transferred to humans indirectly by an intermediate such as food or, occasionally, by direct contact (as in certain eye infections). Houseflies (figure 13.18*b*) are noxious mechanical vectors. They feed on decaying garbage and feces, and while they are feeding, their feet and mouthparts easily become contaminated. They also regurgitate juices onto food to soften and digest it. Flies spread more than 20 bacterial, viral, protozoan, and worm infections. Other nonbiting flies transmit tropical ulcers, yaws, and trachoma (see chapter 21). Cockroaches, which have similar unsavory habits, play a role in the mechanical transmission of fecal pathogens as well as contributing to allergy attacks in asthmatic children.

Many vectors and animal reservoirs spread their own infections to humans. An infection indigenous to animals but naturally transmissible to humans is a **zoonosis**.* In these types of infections, the human is essentially a dead-end host and does not contribute to the natural persistence of the microbe. Some zoonotic infections (rabies, for instance) can have multihost involvement, and others can have very complex cycles in the wild (see plague in chapter 20). Zoonotic spread of disease is promoted by close associations of humans with animals, and people in animal-oriented or outdoor professions are at greatest risk. At least 150 zoonoses exist worldwide; the most common ones are listed in **table 13.10**. Zoonoses make up a full 70% of all new emerging diseases worldwide. It is worth noting that zoonotic infections are impossible to completely eradicate without also eradicating the animal reservoirs. Attempts have been made to eradicate mosquitoes and certain rodents, but it is inconceivable that such extreme measures would ever be tried on wild birds or mammals.

---

*zoonosis (zoh″-uh-noh′-sis) Gr. *zoion*, animal, and *nosos*, disease.

(a) Biological vectors are infected.

(b) Mechanical vectors are not infected.

FIGURE 13.18

Two types of vectors (magnified views).   **(a)** Biological vectors serve as hosts during pathogen development. One example is the flea, a carrier of bubonic plague and murine typhus that infect by its bite. **(b)** Mechanical vectors such as the housefly ingest filth and transport pathogens on their feet and mouthparts.

TABLE 13.10

Common Zoonotic Infections

Disease	Primary Animal Reservoirs
Viruses	
Rabies	All mammals
Yellow fever	Wild birds, mammals, mosquitoes
Viral fevers	Wild mammals
Hantavirus	Rodents
Influenza	Chickens, swine
West Nile virus	Wild birds, mosquitoes
Bacteria	
Rocky Mountain spotted fever	Dogs, ticks
Psittacosis	Birds
Leptospirosis	Domestic animals
Anthrax	Domestic animals
Brucellosis	Cattle, sheep, pigs
Plague	Rodents, fleas
Salmonellosis	Variety of mammals, birds, and rodents
Tularemia	Rodents, birds, arthropods
Miscellaneous	
Ringworm	Domestic mammals
Toxoplasmosis	Cats, rodents, birds
Trypanosomiasis	Domestic and wild mammals
Trichinosis	Swine, bears
Tapeworm	Cattle, swine, fish
Scabies (mange)	Domestic animals

One technique that can provide an early warning signal for the occurrence of certain mosquito-borne zoonoses has been the use of *sentinel animals.* These are usually domestic animals (most often chickens or horses) that can serve as hosts for diseases such as West Nile fever, various viral encephalitides, and malaria. Sentinel animals are placed at various sites throughout the community, and their blood is monitored periodically for antibodies to the infectious agents that would indicate a recent infection by means of a mosquito bite (See **chapter-opening photo**). The presence of infected animals provides useful data on the potential for human exposure, and it also helps establish the epidemiologic pattern of the zoonosis including where it may have spread.

Nonliving Reservoirs

It is clear that microorganisms have adapted to nearly every habitat in the biosphere. They thrive in soil and water and often find their way into the air. Although most of these microbes are saprobic and cause little harm and considerable benefit to humans, some are opportunists and a few are regular pathogens. Because human hosts are in regular contact with these environmental sources, acquisition of pathogens from natural habitats is of diagnostic and epidemiologic importance.

Soil harbors the vegetative forms of bacteria, protozoa, helminths, and fungi, as well as their resistant or developmental stages such as spores, cysts, ova, and larvae. Regular bacterial

pathogens include the anthrax bacillus and species of *Clostridium* that are responsible for gas gangrene, botulism, and tetanus. Pathogenic fungi in the genera *Coccidioides* and *Blastomyces* are spread by spores in the soil and dust. The invasive stages of the hookworm *Necator* occur in the soil. Natural bodies of water carry fewer nutrients than soil does but still support a number of pathogenic species such as *Legionella, Cryptosporidium,* and *Giardia.*

HOW AND WHY? THE ACQUISITION AND TRANSMISSION OF INFECTIOUS AGENTS

Infectious diseases can be categorized on the basis of how they are acquired. A disease is **communicable** when an infected host can transmit the infectious agent to another host and establish infection in that host. (Although this is standard terminology, one must realize that it is not the disease that is communicated, but the microbe. Also be aware that the word *infectious* is sometimes used interchangeably with the word *communicable,* but this is not precise usage.) The transmission of the agent can be direct or indirect, and the ease with which the disease is transmitted varies considerably from one agent to another. If the agent is highly transmissible, especially through direct contact, the disease is **contagious.** Influenza and measles move readily from host to host and thus are contagious, whereas leprosy is only weakly communicable. Because they can be spread through the population, communicable diseases will be our main focus in the following sections.

In contrast, a **non-communicable** infectious disease does *not* arise through transmission of the infectious agent from host to host. The infection and disease are acquired through some other, special circumstance. Non-communicable infections occur primarily when a compromised person is invaded by his own microflora (as with certain pneumonias, for example) or when an individual has accidental contact with a facultative parasite that exists in a nonliving reservoir such as soil. Some examples are certain mycoses, acquired through inhalation of fungal spores, and tetanus, in which *Clostridium tetani* spores from a soiled object enter a cut or wound. Persons thus infected do not become a source of disease to others.

Patterns of Transmission in Communicable Diseases

The routes or patterns of disease transmission are many and varied. The spread of diseases is by direct or indirect contact with animate or inanimate objects and can be horizontal or vertical. The term *horizontal* means the disease is spread through a population from one infected individual to another; *vertical* signifies transmission from parent to offspring via the ovum, sperm, placenta, or milk. The extreme complexity of transmission patterns among microorganisms makes it very difficult to generalize. However, for easier organization, we will divide microorganisms into two major groups, as shown in **figure 13.19**: transmission by direct contact or transmission by indirect routes (with the latter category divided into vehicle and airborne transmission).

Modes of Direct Transmission In order for microbes to be directly transferred, some type of contact must occur between the skin or mucous membranes of the infected person and that of the

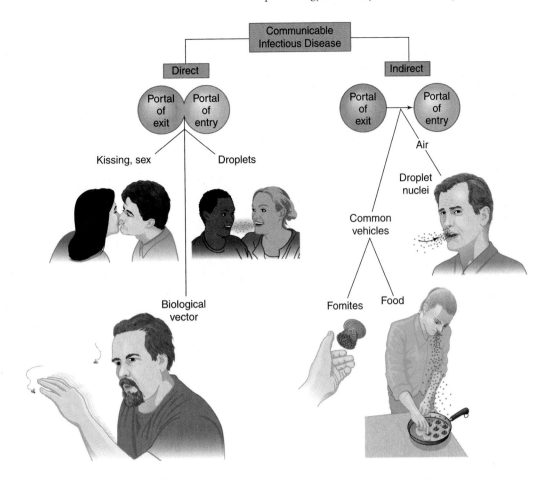

FIGURE 13.19

Summary of how communicable infectious diseases are acquired.
Direct mechanisms of transmission involve physical contact between two hosts. Indirect modes of transmission require some object or material to transfer the infectious agent between hosts.

infectee. It may help to think of this route as the portal of exit meeting the portal of entry without the involvement of an intermediate object or substance (figure 13.19). Included in this category are fine droplets sprayed directly upon a person during sneezing or coughing (as distinguished from droplet nuclei that are transmitted some distance by air). Most sexually transmitted diseases are spread directly. In addition, infections that result from kissing, nursing, placental transfer, or bites by biological vectors are direct. Most obligate parasites are far too sensitive to survive for long outside the host and can be transmitted only through direct contact.

Routes of Indirect Transmission   For microbes to be indirectly transmitted, the infectious agent must pass from an infected host to an intermediate conveyor and from there to another host. This form of communication is especially pronounced when the infected individuals contaminate inanimate objects, food, or air through their activities (figure 13.19). The transmitter of the infectious agent can be either openly infected or a carrier.

Indirect Spread by Vehicles: Contaminated Materials   The term **vehicle** specifies any inanimate material commonly used by humans that can transmit infectious agents. A *common vehicle* is a single material that serves as the source of infection for many individuals. Some specific types of vehicles are food, water, various biological products (such as blood, serum, and tissue), and fomites.

A **fomite*** is an inanimate object that harbors and transmits pathogens. The list of possible fomites is as long as your imagination allows. Probably highest on the list would be objects commonly in contact with the public such as doorknobs, telephones, push buttons, and faucet handles that are readily contaminated by touching. Shared bed linens, handkerchiefs, toilet seats, toys, eating utensils, clothing, personal articles, and syringes are other examples. Although paper money is impregnated with a disinfectant to inhibit microbes, pathogens are still isolated from bills as well as coins.

Outbreaks of food poisoning often result from the role of food as a common vehicle. The source of the agent can be soil, the handler, or a mechanical vector. In the type of transmission termed the *oral-fecal route,* a fecal carrier with inadequate personal hygiene contaminates food during handling, and an unsuspecting person ingests it. Hepatitis A, amebic dysentery, shigellosis, and typhoid fever are often transmitted this way. Because milk provides a rich growth medium for microbes, it is a significant means of transmitting pathogens from diseased animals, infected milk handlers, and environmental sources of contamination. The agents of brucellosis, tuberculosis, Q fever, salmonellosis, and listeriosis are transmitted by contaminated milk. Water that has been contaminated by feces or urine can carry *Salmonella, Vibrio* (cholera) viruses (hepatitis A, polio), and pathogenic protozoans (*Giardia, Cryptosporidium*).

*****fomite** (foh′-myt) L. *fomes,* tinder.

FIGURE 13.20

The explosiveness of a sneeze. Special photography dramatically captures droplet formation in an unstifled sneeze. Even the merest attempt to cover a sneeze with one's hand will reduce this effect considerably. When such droplets dry and remain suspended in air, they are droplet nuclei.

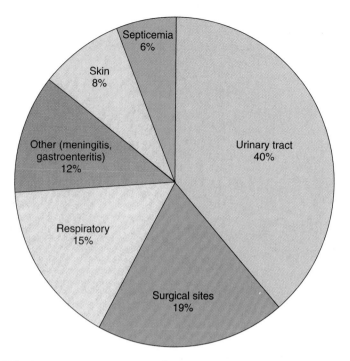

FIGURE 13.21

Most common nosocomial infections. Relative frequency by target area.

Indirect Spread by Airborne Route: Droplet Nuclei and Aerosols Unlike soil and water, outdoor air cannot provide nutritional support for microbial growth and seldom transmits airborne pathogens. On the other hand, indoor air (especially in a closed space) can serve as an important medium for the suspension and dispersal of certain respiratory pathogens via droplet nuclei and aerosols. **Droplet nuclei** are dried microscopic residues created when microscopic pellets of mucus and saliva are ejected from the mouth and nose. They are generated forcefully in an unstifled sneeze or cough (**figure 13.20**) or mildly during other vocalizations. Although the larger beads of moisture settle rapidly, smaller particles evaporate and remain suspended for longer periods. Droplet nuclei are implicated in the spread of hardier pathogens such as the tubercle bacillus and the influenza virus. **Aerosols** are suspensions of fine dust or moisture particles in the air that contain live pathogens. Q fever is spread by dust from animal quarters, and psittacosis by aerosols from infected birds. An unusual outbreak of coccidioidomycosis (a lung infection) occurred during the 1994 southern California earthquake. Epidemiologists speculate that disturbed hillsides and soil gave off clouds of dust containing the spores of *Coccidioides*.

NOSOCOMIAL INFECTIONS: THE HOSPITAL AS A SOURCE OF DISEASE

Infectious diseases that are acquired or develop during a hospital stay are known as **nosocomial*** infections. This concept seems incongruous at first thought, because a hospital is regarded as a place to get treatment for a disease, not a place to acquire a disease. Yet it

---

*nosocomial (nohz″-oh-koh′-mee-al) Gr. *nosos*, disease, and *komeion*, to take care of. Originating from a hospital or infirmary.

is not uncommon for a surgical patient's incision to become infected or a burn patient to develop a case of pneumonia in the clinical setting. The rate of nosocomial infections can be as low as 0.1% or as high as 20% of all admitted patients depending on the clinical setting, with an average of about 5%. In light of the number of admissions, this amounts to from 2 to 4 million cases a year, which result in nearly 90,000 deaths. Nosocomial infections cost time and money. By one estimate, they amount to 8 million additional days of hospitalization a year and an increased cost of $5 to $10 billion.

So many factors unique to the hospital environment are tied to nosocomial infections that a certain number of infections are virtually unavoidable. After all, the hospital both attracts and creates compromised patients, and it serves as a collection point for pathogens. Some patients become infected when surgical procedures or lowered defenses permit resident flora to invade their bodies. Other patients acquire infections directly or indirectly from fomites, medical equipment, other patients, medical personnel, visitors, air, and water.

The health care process itself increases the likelihood that infectious agents will be transferred from one patient to another. Treatments using reusable instruments such as respirators and thermometers constitute a possible source of infectious agents. Indwelling devices such as catheters, prosthetic heart valves, grafts, drainage tubes, and tracheostomy tubes form a ready portal of entry and habitat for infectious agents. Because such a high proportion of the hospital population receives antimicrobial drugs during their stay, drug-resistant microbes are selected for at a much greater rate than is the case outside the hospital.

The most common nosocomial infections involve the urinary tract, the respiratory tract, and surgical incisions (**figure 13.21**). Gram-negative intestinal flora (*Escherichia coli, Klebsiella, Pseudomonas*) are cultured in more than half of patients with nosocomial infections (see figure 20.7). Gram-positive bacteria (staphylococci and streptococci) and yeasts make up most of the remainder. True pathogens such as *Mycobacterium tuberculosis, Salmonella,* hepatitis B, and influenza virus can be transmitted in the clinical setting as well.

The potential seriousness and impact of nosocomial infections have required hospitals to develop committees that monitor infectious outbreaks and develop guidelines for infection control and aseptic procedures.

Medical asepsis includes practices that lower the microbial load in patients, caregivers, and the hospital environment. These practices include proper handwashing, disinfection and sanitization, as well as patient isolation (**table 13.11** summarizes guidelines for the major types of isolation). The goal of these procedures is to limit the spread of infectious agents from person to person. An even higher level of stringency is seen with *surgical asepsis,* which involves all of the strategies listed previously plus ensuring that all surgical procedures are conducted under sterile conditions. This includes sterilization of surgical instruments, dressings, sponges and the like, as well as clothing personnel in sterile garments and scrupulously disinfecting the room surfaces and air.

Hospitals generally employ an *infection control officer* who not only implements proper practices and procedures throughout the hospital but is also tasked with tracking potential outbreaks, identifying breaches in asepsis, and training other health care workers in aseptic technique. Among those most in need of this training are nurses and other caregivers whose work, by its very nature, exposes them to needlesticks, infectious secretions, blood, and physical contact with the patient. The same practices that interrupt the

routes of infection in the patient can also protect the health care worker. It is for this reason that most hospitals have adopted universal precautions that recognize that all secretions from all persons in the clinical setting are potentially infectious and that transmission can occur in either direction. See the section "Introduction to Identification Techniques," that follows chapter 17, for a full description of these precautions.

WHICH AGENT IS THE CAUSE? USING KOCH'S POSTULATES TO DETERMINE ETIOLOGY

An essential aim in the study of infection and disease is determining the precise **etiologic,** or causative, agent. In our modern technological age, we take for granted that a certain infection is caused by a certain microbe, but such has not always been the case. More than a century ago, Robert Koch realized that in order to prove the germ theory of disease he would have to develop a standard for determining causation that would stand the test of scientific scrutiny. Out of his experimental observations on the transmission of anthrax in cows came a series of proofs, called **Koch's postulates,** that established the principal criteria for etiologic studies (**figure 13.22**). These postulates direct an investigator to

1. find evidence of a particular microbe in every case of a disease,
2. isolate that microbe from an infected subject and cultivate it artificially in the laboratory,
3. inoculate a susceptible healthy subject with the laboratory isolate and observe the resultant disease, and
4. reisolate the agent from this subject.

TABLE 13.11
Levels of Isolation Used in Clinical Settings

Type of Isolation*	Protective Measures**	To Prevent Spread of
Enteric Precautions	Gowns and gloves must be worn by all persons having direct contact with patient; masks not required; special precautions taken for disposing of feces and urine	Diarrheal diseases; *Shigella, Salmonella,* and *Escherichia coli* gastroenteritis; cholera; hepatitis A; rotavirus; and giardiasis
Respiratory Precautions	Private room with closed door is necessary; gowns and gloves not required; masks usually indicated; items contaminated with secretions must be disinfected	Tuberculosis, measles, mumps, meningitis, pertussis, rubella, chickenpox
Drainage and Secretion Precautions	Gowns and gloves required for all persons; masks not needed; contaminated instruments and dressings require special precautions	Staphylococcal and streptococcal infections; gas gangrene; herpes zoster; burn infections
Strict Isolation	Private room with closed door required; gowns, masks, and gloves must be worn by all persons; contaminated items must be wrapped and sent to central supply for decontamination	Mostly highly virulent or contagious microbes; includes diphtheria, some types of pneumonia, extensive skin and burn infections, disseminated herpes simplex and zoster
Reverse Isolation (Also Called Protective Isolation)	Same guidelines as for strict isolation; room may be ventilated by unidirectional or laminar airflow filtered through a high-efficiency particulate air (HEPA) filter that removes most airborne pathogens; infected persons must be barred	Used to protect patients extremely immunocompromised by cancer therapy, surgery, genetic defects, burns, prematurity, or AIDS and therefore vulnerable to opportunistic pathogens

*Precautions are based upon the primary portal of entry and communicability of the pathogen.
**In all cases, visitors to the patient's room must report to the nurses' station before entering the room; all visitors and personnel must wash their hands upon entering and leaving the room.

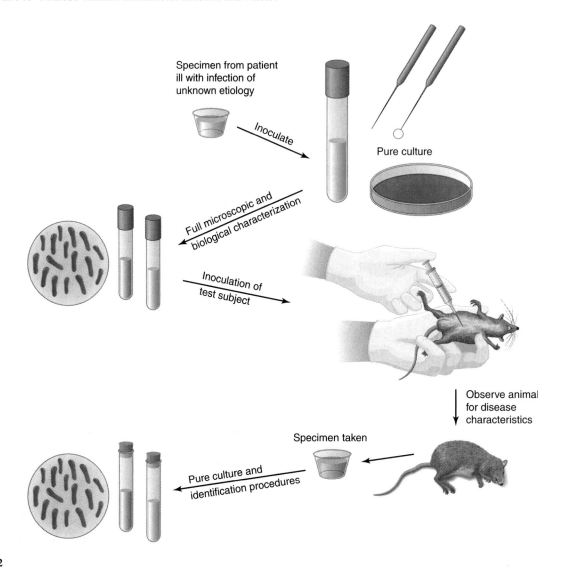

FIGURE 13.22

Koch's postulates: Is this the etiologic agent? The microbe in the initial and second isolations and the disease in the patient and experimental animal must be identical for the postulates to be satisfied.

Valid application of Koch's postulates requires attention to several critical details. Each isolated culture must be pure, observed microscopically, and identified by means of characteristic tests; the first and second isolate must be identical; and the pathologic effects, signs, and symptoms of the disease in the first and second subject must be the same. Once established, these postulates were rapidly put to the test, and within a short time, they had helped determine the causative agents of tuberculosis, diphtheria, and plague. Today, most infectious diseases have been directly linked to a known infectious agent.

Koch's postulates continue to play an essential role in modern epidemiology. Every decade, new diseases challenge the scientific community and require application of the postulates. Prominent examples are toxic shock syndrome, AIDS, Lyme disease, and Legionnaires disease (named for the American Legion members who first contracted a mysterious lung infection in Philadelphia).

Koch's postulates are reliable for most infectious diseases, but they cannot be completely fulfilled in certain situations. For example, some infectious agents are not readily isolated or grown in the laboratory. If one cannot elicit a similar infection by inoculating it into an animal, it is very difficult to prove the etiology. In the past, scientists have attempted to circumvent this problem by using human subjects (**Historical Highlights 13.4).**

A small but vocal group of critics has claimed that the postulates have not been adequately carried out for AIDS and thus, that HIV cannot be claimed as the causative agent, despite overwhelming evidence from observing infected humans and primates that it is. Cases of accidental infection through exposure to blood have yielded much proof. In all cases, subjects developed an early virus syndrome, carried high virus levels, and later developed AIDS. One study with baboons revealed that they do indeed develop symptoms of AIDS when infected with a variant of HIV.

HISTORICAL HIGHLIGHTS   13.4
The History of Human Guinea Pigs

In this day and age, human beings are not used as subjects for determining the cause of infectious disease, but in earlier times they were. A long tradition of human experimentation dates well back into the eighteenth century, with the subject frequently the experimenter himself. In some studies, mycologists inoculated their own skin and even that of family members with scrapings from fungal lesions to demonstrate that the disease was transmissible. In the early days of parasitology, it was not uncommon for a brave researcher to swallow worm eggs in order to study the course of his infection and the life cycle of the worm.

In a sort of reverse test, a German colleague of Koch's named Max von Petenkofer believed so strongly that cholera was *not* caused by a bacterium that he and his assistant swallowed cultures of the vibrio. Fortunately for them, they did not acquire serious infection. Many self-experimenters have not been so fortunate.

One of the most famous cases is that of Jesse Lazear, a Cuban physician who worked with Walter Reed on the etiology of yellow fever in 1900. Dr. Lazear was convinced that mosquitoes were directly involved in the spread of yellow fever, and by way of proof, he allowed himself and two volunteers to be bitten by mosquitoes infected with the blood of yellow fever patients. Although all three became ill as a result of this exposure, Dr. Lazear's sacrifice was the ultimate one—he died of yellow fever. Years later, paid volunteers were used to completely fulfill the postulates.

Dr. Lazear was not the first martyr in this type of cause. Fifteen years previously, a young Peruvian medical student, Daniel Carrion, attempted to prove that a severe blood infection, Oroya fever, had the same etiology as *verucca peruana,* an ancient disfiguring skin disease. After inoculating himself with fluid from a skin lesion, he developed the severe form and died, becoming a national hero. In his honor, the disease now identified as bartonellosis is also sometimes called Carrion's disease. Eventually, the microbe (*Bartonella bacilliformis*) was isolated, a monkey model was developed, and the sand fly was shown to be the vector for this disease.

For many years, syphilis and gonorrhea were thought to be different stages of the same disease because of an unfortunate experiment. In an ironic twist of fate, Fritz Schaudinn eventually proved the causation of syphilis, only to die later from amebic dysentery—which he had acquired from self-experimentation.

The incentive for self-experimentation has continued, but present-day researchers are more likely to test experimental vaccines than dangerous microbes. One exception to this was Dr. J. Robert Warren and Barry Marshall who tested the effects of *Helicobacter pylori* by swallowing a culture. Although they didn't get ulcers, they helped to establish the pathogenicity of the microbe. Even more recently, a parasitologist fed several of his patients live worm eggs to treat their inflammatory bowel disease, and it actually improved their condition significantly.

CHAPTER CHECKPOINTS

Epidemiology is the study of the determinants and distribution of all diseases in populations. The study of infectious disease in populations is just one aspect of this field.

Data on specific, reportable diseases is collected by local, national, and worldwide agencies.

The *prevalence* of a disease is the percentage of existing cases in a given population. The disease *incidence,* or *morbidity rate,* is the ratio of newly infected to uninfected members of a population.

The disease frequency is described as sporadic, epidemic, pandemic, or endemic.

The primary habitat of a pathogen is called its reservoir. A human reservoir is also called a carrier.

Animals can be either reservoirs or vectors of pathogens. An infected animal is a biological vector. Uninfected animals, especially insects, that transmit pathogens mechanically are called mechanical vectors.

Soil and water are nonliving reservoirs for pathogenic bacteria, protozoa, fungi, and worms.

A communicable disease can be transmitted from an infected host to others, but not all infectious diseases are communicable.

The spread of infectious disease from person to person is called horizontal transmission. The spread of infectious disease from parent to offspring is called vertical transmission.

Infectious diseases are spread by either direct or indirect routes of transmission. Vehicles of indirect transmission include soil, water, food, droplet nuclei, and fomites (inanimate objects).

Nosocomial infections are acquired in a hospital from surgical procedures, equipment, personnel, and exposure to drug-resistant microorganisms.

Causative agents of infectious disease must be isolated and identified according to Koch's postulates.

CHAPTER CAPSULE WITH KEY TERMS

I. The Host-Parasite Relationship
The human body is in constant contact with microbes, some of which invade the body and multiply. Microbes are classified as either part of the normal flora of the body or as **pathogens,** depending on whether or not their colonization of the body results in disease.

A. **The Body As a Habitat**
   1. **Resident flora** or microflora consists of a huge and varied population of microbes that reside permanently on all surfaces of the body exposed to the environment. Internal organs and the fluid they contain are free of flora.

2. Colonization begins just prior to birth and continues throughout life. Variations in flora occur in response to changes in an individual's age, diet, hygiene, and health
3. Flora often provide some benefit to the host but are also capable of causing disease, especially when the immune system of the host is compromised or when flora invades a normally sterile area of the body.

B. **Factors Affecting the Course of Infection and Disease**
1. **Pathogenicity** is the property of microorganisms to cause infection and disease; **virulence** refers to the factors used by the microbe to invade and damage host tissues.
2. The pathogenicity of a microbe varies, with greater pathogenicity being ascribed to microbes with more or greater virulence factors.
3. **True pathogens** are able to cause disease in a normal healthy host with intact immune defenses.
4. **Opportunistic pathogens** can cause disease only in persons whose host defenses are compromised by *predisposing conditions*.

C. **Mechanisms of Infection and Disease**
1. Microbes typically enter the body through a specific **portal of entry**. Portals of entry are generally the same as those areas of the body that harbor microflora.
2. The minimum number of microbes needed to produce an infection is termed the infectious dose, and these pathogens may be classified as **exogenous** or **endogenous,** based on their source.
3. Microbes attach to the host cell, a process known as **adhesion,** by means of fimbriae, flagella, capsules, or receptors.

D. **Virulence Factors**
1. **Exoenzymes** digest epithelial tissues and permit invasion of pathogens.
2. **Toxigenicity** refers to a microbe's capacity to produce **toxins** at the site of multiplication. Toxins are divided into **endotoxins** and **exotoxins,** and diseases caused by toxins are classified as **intoxications, toxemias,** or **toxinoses.**
3. Antiphagocytic factors include **leukocidins** and capsules.

E. **Types and Patterns of Infection**
1. **Infectious diseases** follow a typical pattern consisting of an *incubation period, prodromal stage, period of invasion,* and *convalescent period.* Each stage is marked by symptoms of a specific intensity.
2. Infections are classified by:
   a. whether they remain localized at the site of inoculation (**localized infection, systemic infection, focal infection**).
   b. number of microbes involved in an infection, and the order in which they infect the body (**mixed infection, primary infection, secondary infection**).

c. persistence of the infection (**acute infection, chronic infection,** subacute infection).
3. **Signs** of infection refer to *objective* evidence of infection, and **symptoms** refer to *subjective* evidence. A **syndrome** is a disease that manifests as a predictable complex of symptoms.
4. After the initial infection pathogens may remain in the body in a **latent** state and may later cause recurrent infections. Long-term damage to host cells is referred to as **sequelae.**
5. Microbes are released from the body through a specific **portal of exit** that allows them access to another host.

II. **Epidemiology**
A. Epidemiologists are involved in the surveillance of reportable diseases in populations. Diseases are tracked with regard to their **prevalence** and **incidence,** while populations are tracked with regard to **morbidity** and **mortality.**
B. Diseases are described as being **endemic, sporadic, epidemic,** or **pandemic** based on the frequency of their occurrence in a population.
C. The **reservoir** of a microbe refers to its natural habitat; a pathogen's **source** refers to the immediate origin of an infectious agent.
D. **Carriers** are individuals who inconspicuously shelter a pathogen and spread it to others. Carriers are divided into **asymptomatic carriers,** incubation carriers, convalescent carriers, chronic carriers, and **passive carriers.**
E. **Vectors** refer to animals that transmit pathogens
1. **Biological vectors** are animals that are involved in the life cycle of the pathogen they transmit; **mechanical vectors** transmit pathogens without being involved in the life cycle.
2. **Zoonoses** are diseases that can be transmitted to humans but are normally found in animal populations.
F. Agents of disease are either **communicable** or **non-communicable,** with highly communicable diseases referred to as **contagious.**
1. Direct transmission of a disease occurs when a portal of exit meets a portal of entrance.
2. Indirect transmission occurs when an intermediary such as a **vehicle, fomite,** or **droplet nuclei** connect portals of entrance and exit.
G. **Nosocomial** infections are infectious diseases acquired in a hospital setting. Special vigilance is required to reduce their occurrence.
H. **Koch's postulates** are a series of proofs used to determine the **etiologic** agent of a specific disease.

MULTIPLE-CHOICE QUESTIONS

1. The best descriptive term for the resident flora is
   a. commensals   c. pathogens
   b. parasites   d. mutualists

2. Resident flora is commonly found in the
   a. stomach   c. salivary glands
   b. kidney   d. urethra

3. Resident flora is absent from the
   a. pharynx   c. intestine
   b. lungs   d. hair follicles

4. Virulence factors include
   a. toxins   c. capsules
   b. enzymes   d. all of these

5. The specific action of hemolysins is to
   a. damage white blood cells
   b. cause fever
   c. damage red blood cells
   d. cause leukocytosis

6. The _____ is the time that lapses between encounter with a pathogen and the first symptoms.
   a. prodromium   c. period of convalescence
   b. period of invasion   d. period of incubation

7. A short period early in a disease that manifests with general malaise and achiness is the
   a. period of incubation   c. sequela
   b. prodromium   d. period of invasion

8. The presence of a few bacteria in the blood is termed
   a. septicemia   c. bacteremia
   b. toxemia   d. a secondary infection

9. A ____ infection is acquired in a hospital.
   a. subclinical   c. nosocomial
   b. focal   d. zoonosis

10. A/an ____ is a passive animal transporter of pathogens.
    a. zoonosis   c. mechanical vector
    b. biological vector   d. asymptomatic carrier

11. An example of a non-communicable infection is:
    a. measles   c. tuberculosis
    b. leprosy   d. tetanus

12. A general term that refers to an increased white blood cell count is
    a. leukopenia   c. leukocytosis
    b. inflammation   d. leukemia

13. A positive antibody test for HIV would be a ____ of infection.
    a. sign   c. syndrome
    b. symptom   d. sequela

14. Which of the following would *not* be a portal of entry?
    a. the meninges   c. skin
    b. the placenta   d. small intestine

15. Which of the following is *not* a condition of Koch's postulates?
    a. isolate the causative agent of a disease
    b. cultivate the microbe in a lab
    c. inoculate a test animal to observe the disease
    d. test the effects of a pathogen on humans

CONCEPT QUESTIONS

1. Differentiate between contamination, infection, and disease. What are the possible outcomes in each?

2. How are infectious diseases different from other diseases?

3. Name the general body areas that are sterile. Why is the inside of the intestine not sterile like many other organs?

4. What causes variations in the flora of the newborn intestine?

5. What factors influence how the flora of the vagina develops?

6. Why must axenic young be delivered by cesarian section?

7. Explain several ways that true pathogens differ from opportunistic pathogens.

8. a. Distinguish between pathogenicity and virulence.
   b. Define virulence factors, and give examples of them in gram-positive and gram-negative bacteria, viruses, and parasites.

9. Describe the course of infection from contact with the pathogen to its exit from the host.

10. a. Explain why most microbes are limited to a single portal of entry.
    b. For each portal of entry, give a vehicle that carries the pathogen and the course it must travel to invade the tissues.
    c. Explain how the portal of entry could differ from the site of infection.

11. Differentiate between exogenous and endogenous infections.

12. a. What factors possibly affect the size of the infectious dose?
    b. Name five factors involved in microbial adhesion.

13. Which body cells or tissues are affected by hemolysins, leukocidins, hyaluronidase, kinases, tetanus toxin, pertussis toxin, and enterotoxin?

14. Compare and contrast: systemic versus local infections; primary versus secondary infections; infection versus intoxication.

15. What are the differences between signs and symptoms? (First put yourself in the place of a patient with an infection and then in the place of a physician examining you. Describe what you would feel and what the physician would detect upon examining the affected area.)

16. a. What are some important considerations about the portal of exit?
    b. Name some examples of infections and their portals of exit.

17. Complete the table:

	Exotoxins	Endotoxins
Chemical makeup		
General source		
Degree of toxicity		
Effects on cells		
Symptoms in disease		
Examples		

18. a. Explain what it means to be a carrier of infectious disease.
    b. Describe four ways that humans can be carriers.
    c. What is epidemiologically and medically important about carriers in the population?

19. a. Outline the science of epidemiology and the work of an epidemiologist.
    b. Using the following statistics, based on number of reported cases, determine which show endemic, sporadic, or epidemic patterns. Explain how you can determine each type.

United States Region	Disease Statistics	
Chlamydiosis	**2000**	**2001**
Northeast	93,116	115,467
Midwest	168,332	184,111
South	286,136	307,405
West	161,868	176,259
Total cases	709,452	783,242
Lyme disease		
Northeast	14,932	14,435
Midwest	1,343	1,260
South	1,319	1,198
West	136	136
Total cases	17,730	17,029
Rubella		
Northeast	27	9
Midwest	3	5
South	123	5
West	15	4
Total cases	168	23

   c. Explain what would have to occur for these diseases to have a pandemic distribution.

20. Distinguish between mechanical and biological vectors, giving one example of each.

21. a. Explain the precise difference between communicable and non-communicable infectious diseases.
    b. Between direct and indirect modes of transmission.
    c. Between vectors and vehicles as modes of transmission.

22. a. Nosocomial infections can arise from what two general sources?
    b. From this chapter and figure 20.7, outline the major agents involved in nosocomial infections.

c. Do they tend to be true pathogens or opportunists?
d. Outline the two types of hospital asepsis and define isolation.
e. What is the work of an infection control officer?

23. a. List the main features of Koch's postulates.
    b. Why is it so difficult to prove them for some diseases?

CRITICAL-THINKING QUESTIONS

1. a. Discuss the relationship between the vaginal residents and the colonization of the newborn.
   b. Can you think of some serious medical consequences of this relationship?
   c. Why would normal flora cause some infections to be more severe and other infections to be less severe?

2. If the following patient specimens produced positive cultures when inoculated and grown on appropriate media, indicate whether this result indicates a disease state and why or why not:

Urine	Throat
Lung biopsy	Feces
Saliva	Blood
Cerebrospinal fluid	Urine from bladder
Liver biopsy	Semen

   What are the important clinical implications of positive blood or cerebrospinal fluid?

3. Pretend that you have been given the job of developing a colony of germ-free cockroaches.
   a. What will be the main steps in this process?
   b. What possible experiments can you do with these animals?

4. If healthy persons are resistant to infection with opportunists or weak pathogens, what is the expected result if a compromised person is exposed to a true pathogen?

5. Explain how the endotoxin gets into the bloodstream of a patient with endotoxic shock.

6. You are a physician following the course of an infection in a small child who has been exposed to scarlet fever. Describe the main events that occur, what is happening during each stage, and the causes of each sign and symptom.

7. Describe each of the following infections using correct technical terminology. (Descriptions may fit more than one category.) Use terms such as primary, secondary, nosocomial, STD, mixed, latent, toxemia, chronic, zoonotic, asymptomatic, local, systemic, -itis, -emia.
   Caused by needlestick in dental office
   Pneumocystis pneumonia in AIDS patient
   Bubonic plague from rat flea bite
   Diphtheria
   Undiagnosed chlamydiosis
   Acute necrotizing gingivitis
   Syphilis of long duration
   Large numbers of gram-negative rods in the blood
   A boil on the back of the neck
   An inflammation of the meninges
   Scarlet fever

8. Using statistics in the endpapers of this book, find several reportable diseases that show an increase in numbers of cases that could indicate an outbreak or epidemic.

9. Name 10 fomites that you came into contact with today.

10. Examine the zoonotic infectious diseases in table 13.10. Look up each of these infections in the index and find which animals are the primary reservoirs of the pathogens involved.

11. Describe what parts of Koch's postulates were unfulfilled by Dr. Lazear's experiment described in Historical Highlights 13.4.

12. a. Suggest several reasons why urinary tract, respiratory tract, and surgical infections are the most common nosocomial infections.
    b. Name several measures that health care providers must exercise at all times to prevent or reduce nosocomial infections.

INTERNET SEARCH TOPICS

1. Use an Internet search engine to look up the following topics. Write a short summary of what you discover on your searches.
   Sentinel chickens
   Tuskegee syphilis experiment
   Koch's postulates verified for AIDS and HIV
   Vessel sanitation program
   Newcastle virus in chicken

2. To gain additional insight into the science of epidemiology, visit the student Online Learning Center at www.mhhe.com/talaro5. Go to chapter 13, Internet Search Topics, and log on to the available websites. The American Museum of Natural History offers an outstanding overview of infection, disease, and their monitoring, while another site provides an opportunity to act as a detective solving an epidemiologic mystery.

The Nature of
Host Defenses

The survival of the host depends upon an elaborate network of defenses that keeps harmful microbes and other foreign materials from penetrating the body. Should they penetrate, additional host defenses are summoned to prevent them from becoming established in tissues. Defenses involve barriers, cells, and chemicals, and they range from nonspecific to specific and from inborn to acquired. This chapter introduces the main lines of defense intrinsic to all humans. Topics included in this survey are the anatomical and physiological systems that detect, recognize, and destroy foreign substances and the general adaptive responses that account for an individual's long-term immunity or resistance to infection and disease.

Chapter Overview

- The human body possesses a complex series of overlapping defenses that protect it against invasion. The first two of the body's three lines of defense provide nonspecific protection against anything (living or not) regarded as foreign to the body, with the first line consisting of barriers to foreign matter, and the second is charged with protecting the body once foreign matter has entered.

- White blood cells formed in the bone marrow are responsible for most of the reactions of the immune system, including antibody production, phagocytosis, and many aspects of inflammation. Lymphoid organs such as the spleen, lymph nodes, and thymus are also intimately involved in these defense mechanisms.

- Cells of the immune system are able to travel freely between different areas of the body because of the interrelationship between the blood, the lymphatic system, and the reticuloendothelial system.

- Communication between cells of the immune system is facilitated by the use of chemical messengers such as cytokines, leukokines, and interferons. These chemicals are released by cells of the immune system to increase blood flow, stimulate the migration of white blood cells, initiate fever, or induce the death of virally infected cells.

- The inflammatory response is a complex reaction to infection that works to fight foreign agents and limit further damage to the body. As part of the inflammatory response, white blood cells known as phagocytes help to clear foreign organisms from the body. The complement system acts to lyse cells that have been identified as foreign.

- The induction of B and T lymphocytes (the third line of defense) occurs if the body's innate immunities are incapable of resolving an infection. B and T cells provide acquired, long-term protection against specific microbes.

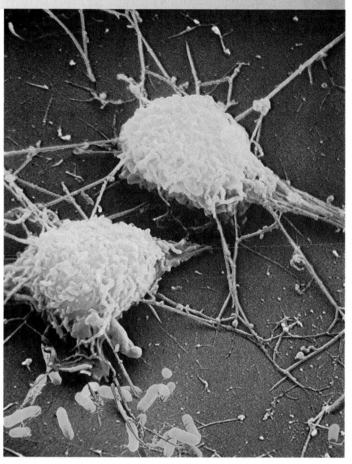

Like an octopus, a macrophage sends out long, tentacled pseudopods to capture its *Escherichia coli* prey (blue cells). False-color scanning electron micrograph (22,000×).

Defense Mechanisms of the Host in Perspective

In chapter 13 we explored the host-parasite relationship, with emphasis on the role of microorganisms in disease. In this chapter we examine the other side of the relationship—that of the host defending itself against microorganisms. As previously stated, in light of the unrelenting contamination and colonization of humans, it is something of a miracle that we are not constantly infected and diseased. Our ability to overcome this continual assault on our bodies is the result of a fascinating, highly complex system of defense. In the war against all sorts of invaders, microbial and otherwise, the body erects a series of barriers, sends in an army of cells, and emits a flood of chemicals to protect tissues from harm.

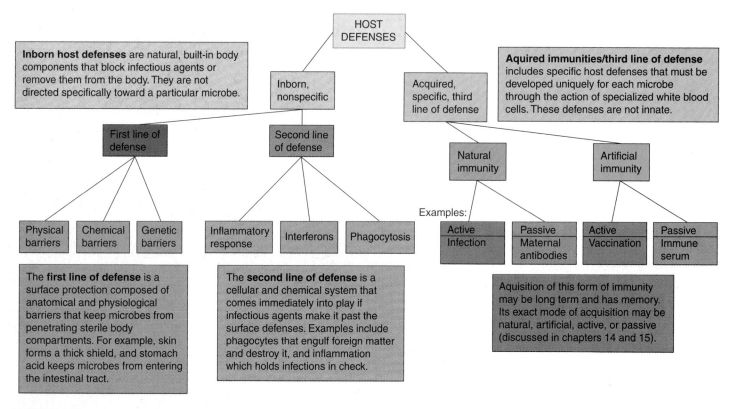

FIGURE 14.1

Flowchart summarizing the major components of the host defenses.   Defenses are classified into one of two general categories: (1) inborn and nonspecific or (2) acquired and specific. These can be further subdivided into the first, second, and third lines of defense, each being characterized by a different level and type of protection. The third line of defense is the most complex and is responsible for specific immunity.

The host defenses embrace a multilevel network of innate, nonspecific protections and specific **immunities*** referred to as the first, second, and third lines of defense (**figure 14.1**). The interaction and cooperation of these three levels of defense normally provide complete protection against infection. The *first line of defense* includes any barrier that blocks invasion at the portal of entry. This mostly nonspecific line of defense limits access to the internal tissues of the body. However, it is not considered a true immune response because it does not involve recognition of a specific foreign substance but is very general in action. The *second line of defense* is a slightly more internalized system of protective cells and fluids that includes inflammation and phagocytosis. It acts rapidly at both the local and systemic levels once the first line of defense has been circumvented. The highly specific *third line of defense* is acquired on an individual basis as each foreign substance is encountered by white blood cells called lymphocytes. The reaction with each different microbe produces unique protective substances and cells that can come into play if that microbe is encountered again. The third line of defense provides long-term immunity.

The human systems are armed with various levels of defense that do not operate in a completely separate fashion; most defenses

overlap and are even redundant in some of their effects. This "immunological overkill" literally bombards microbial invaders with an entire assault force, making their survival unlikely. Because of the interwoven nature of host defenses, we will introduce basic concepts of structure and function that will prepare you for later information on specific reactions of the immune system (see chapter 15).

BARRIERS AT THE PORTAL OF ENTRY: A FIRST LINE OF DEFENSE

A number of defenses are a normal part of the body's anatomy and physiology. These inborn, nonspecific defenses can be divided into physical, chemical, and genetic barriers that impede the entry of not only microbes but any foreign agent, whether living or not. (**figure 14.2**).

Physical or Anatomical Barriers at the Body's Surface

The skin and mucous membranes of the respiratory and digestive tracts have several built-in defenses. The outermost layer (stratum corneum) of the skin is composed of epithelial cells that have become compacted, cemented together, and impregnated with an insoluble protein, keratin. The result is a thick, tough layer that is highly impervious and waterproof. Few pathogens can penetrate this unbroken barrier, especially in regions such as the soles of the feet or the palms of the hands, where the stratum corneum is much thicker than on other parts of the body. Other cutaneous barriers

---

*immunity (im-yoo'-nih-tee) Gr. *immunis*, free, exempt. A state of resistance to infection.

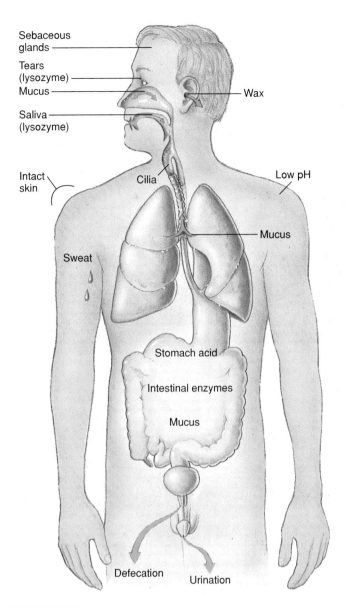

FIGURE 14.2
The primary physical and chemical defense barriers.

Labels on figure:
Sebaceous glands
Tears (lysozyme)
Mucus
Saliva (lysozyme)
Wax
Intact skin
Cilia
Low pH
Mucus
Sweat
Stomach acid
Intestinal enzymes
Mucus
Defecation
Urination

include hair follicles and skin glands. The hair shaft is periodically extruded, and the follicle cells are **desquamated.*** The flushing effect of sweat glands also helps remove microbes.

The mucocutaneous membranes of the digestive, urinary, and respiratory tracts and of the eye are moist and permeable. Despite the normal wear and tear upon these epithelia, damaged cells are rapidly replaced. The mucous coat on the free surface of some membranes impedes the entry and attachment of bacteria. Blinking and tear production (lacrimation) flush the eye's surface with tears and rid it of irritants. The constant flow of saliva helps carry microbes into the harsh conditions of the stomach. Vomiting and defecation also evacuate noxious substances or microorganisms from the body.

---

*desquamate (des′-kwuh-mayt) L. *desquamo,* to scale off. The casting off of epidermal scales.

The respiratory tract is constantly guarded from infection by elaborate and highly effective adaptations. Nasal hair traps larger particles. Rhinitis, the copious flow of mucus and fluids that occurs in allergy and colds, exerts a flushing action. In the respiratory tree (primarily the trachea and bronchi), a ciliated epithelium (called the ciliary escalator) conveys foreign particles entrapped in mucus toward the pharynx to be removed **(figure 14.3).** Irritation of the nasal passage reflexly initiates a sneeze, which expels a large volume of air at high velocity. Similarly, the acute sensitivity of the bronchi, trachea, and larynx to foreign matter triggers coughing, which ejects irritants.

The genitourinary tract derives partial protection from the continuous trickle of urine through the ureters and from periodic bladder emptying that flushes the urethra.

The composition and protective effect exerted by flora were discussed in chapter 13. Even though the resident flora does not constitute an anatomical barrier, its presence can block the access by pathogens to epithelial surfaces and can create an unfavorable environment for pathogens by competing for limited nutrients, or altering the local pH.

Nonspecific Chemical Defenses

The skin and mucous membranes offer a variety of chemical defenses. Sebaceous secretions exert an antimicrobial effect, and specialized glands such as the meibomian glands of the eyelids lubricate the conjunctiva with an antimicrobial secretion. An additional defense in tears and saliva is **lysozyme,** an enzyme that hydrolyzes the peptidoglycan in the cell wall of bacteria. The high lactic acid and electrolyte concentrations of sweat and the skin's acidic pH and fatty acid content are also inhibitory to many microbes. Likewise, the hydrochloric acid in the stomach renders protection against many pathogens that are swallowed, and the intestine's digestive juices and bile are potentially destructive to microbes. Even semen contains an antimicrobial chemical that inhibits bacteria, and the vagina has a protective acidic pH maintained by normal flora.

Genetic Defenses

Some hosts are genetically immune to the diseases of other hosts. One explanation for this phenomenon is that some pathogens have such great specificity for one host species that they are incapable of infecting other species. One way of putting it is: "Humans can't acquire distemper from cats, and cats can't get mumps from humans." This specificity is particularly true of viruses, which can invade only by attaching to a specific host receptor. But it does not hold true for zoonotic infectious agents that attack a broad spectrum of animals. Genetic differences in susceptibility can also exist within members of one species. Humans carrying a gene or genes for sickle-cell anemia are resistant to malaria. Genetic differences also exist in susceptibility to tuberculosis, leprosy, and certain systemic fungal infections.

The vital contribution of barriers is clearly demonstrated in people who have lost them or never had them. Patients with severe skin damage due to burns are extremely susceptible to infections; those with blockages in the salivary glands, tear ducts, intestine, and urinary tract are also at greater risk for infection. But as important as it is, the first line of defense alone is not sufficient to protect against infection. Because many pathogens find a way to circumvent the

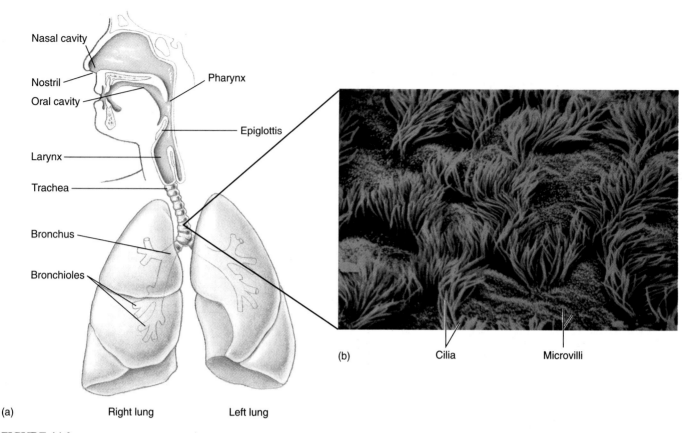

Nasal cavity
Nostril
Oral cavity
Pharynx
Epiglottis
Larynx
Trachea
Bronchus
Bronchioles

(a)   Right lung   Left lung

(b)   Cilia   Microvilli

FIGURE 14.3

The ciliary defense of the respiratory tree. **(a)** The epithelial lining of the airways contains a brush border of cilia to entrap and propel particles upward toward the pharynx. **(b)** Tracheal mucosa (5,000×).

barriers by using their virulence factors (discussed in chapter 13), a whole new set of defenses—inflammation, phagocytosis, specific immune responses—are brought into play.

CHAPTER CHECKPOINTS

The multilevel, interconnecting network of host protection against microbial invasion is organized into three lines of defense.

• The first line consists of physical and chemical barricades provided by the skin and mucous membranes.

• The second line encompasses all the nonspecific cells and chemicals found in the tissues and blood.

• The third line, the specific immune response, is customized to react to specific antigens of a microbial invader. This response immobilizes and destroys the invader every time it appears in the host.

Introducing the Immune System

Immunology encompasses the study of all features of the body's immune system. While the focus of this chapter is, not surprisingly, concerned with infectious microbial agents, be aware that immunology is central to the study of fields as diverse as cancer (at least partly the result of an underactive immune system) and allergy (an overactive system). In chapter 16 we will see that many of the most powerful aspects of immunology have been developed for use in health care, laboratory, or commercial settings.

In the body, the mandate of the immune system can be easily stated. A healthy functioning immune system is responsible for

1. surveillance of the body,
2. recognition of foreign material, and
3. destruction of entities deemed to be foreign **(figure 14.4).**

Because infectious agents could potentially enter through any number of portals, the cells of the immune system constantly move about the body, searching for potential pathogens. This process is carried out primarily by white blood cells, which have been trained to recognize body cells (so-called **self**) and differentiate them from any foreign material in the body, such as invading bacterial cells **(nonself).** The ability to evaluate cells as either self or nonself is central to the functioning of the immune system. While foreign cells must be recognized as a potential threat and dealt with appropriately, self cells must not come under attack by the immune defenses.

The immune system evaluates cells by the **markers**[1] on their surface. These markers, which generally consist of proteins and/or sugars, can be thought of as the cellular equivalent of facial characteristics in humans and allow the cells of the immune system to

---

1. The term *marker* is also employed in genetics in a different sense—that is, to denote a detectable characteristic of a particular genetic mutant. A genetic marker may or may not be a surface marker.

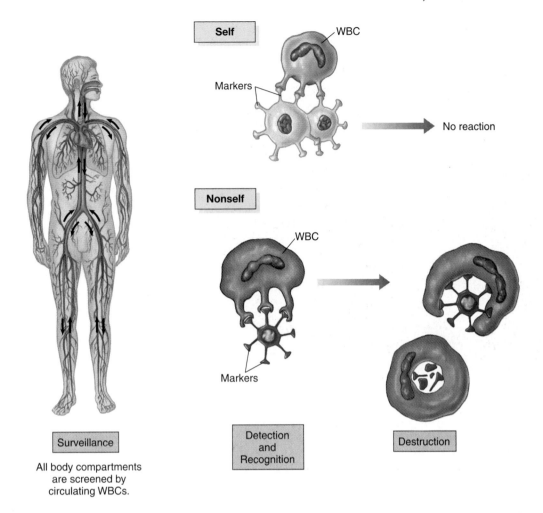

FIGURE 14.4

Search, recognize, and destroy is the mandate of the immune system. White blood cells are equipped with a very sensitive sense of "touch." As they sort through the tissues, they feel surface markers that help them determine what is self and what is not. When self markers are recognized, no response occurs. However, when nonself is detected, a reaction to destroy it is mounted.

identify whether or not a newly discovered cell poses a threat. While self cells are left alone, cells and other objects designated as foreign are marked for destruction by a number of methods, the most common of which is phagocytosis.

CHAPTER CHECKPOINTS

The immune system operates first as a surveillance system that discriminates between the host's self identity markers and the nonself identity markers of foreign cells. When it recognizes that a marker or antigen is foreign, or nonself, the immune system tailors its response specifically to each different antigen. As far as the immune system is concerned, if an antigen is not self, it is foreign, does not belong, and must be destroyed.

Systems Involved in Immune Defenses

Unlike many systems, the immune system does not exist in a single, well-defined site; rather, it encompasses a large, complex, and diffuse network of cells and fluids that permeate every organ and tissue.

It is this very arrangement that promotes the surveillance and recognition processes that help screen the body for harmful substances.

The body is partitioned into several fluid-filled spaces called the intracellular, extracellular, lymphatic, cerebrospinal, and circulatory compartments. Although these compartments are physically separated, they have numerous connections. Their structure and position permit extensive interchange and communication (**figure 14.5**). Among the body compartments that participate in immune function are

1. the **reticuloendothelial* system (RES),**
2. the spaces surrounding tissue cells that contain *extracellular fluid (ECF),*
3. the *bloodstream,* and
4. the **lymphatic system.**

In the following section, we consider the anatomy of these main compartments and how they interact in the second and third lines of defense.

---

*reticuloendothelial (reh-tik″-yoo-loh-en″-doh-thee′-lee-al) L. *reticulum,* a small net, and *endothelium,* lining of the blood vessel.

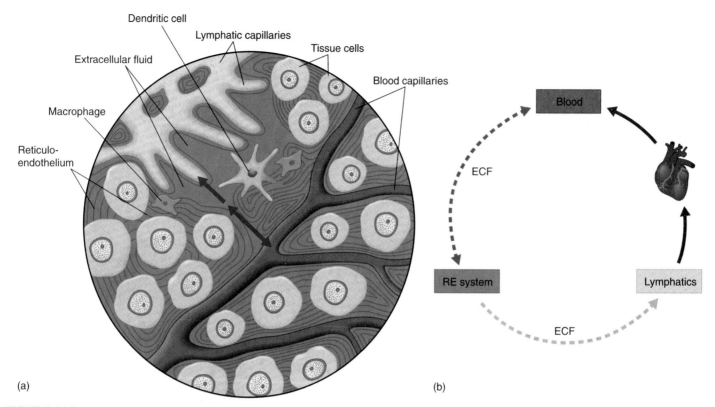

FIGURE 14.5

Connections between the body compartments. **(a)** The meeting of the major fluid compartments at the microscopic level. **(b)** Schematic view of the main fluid compartments and how they form a continuous cyclic system of exchange. Positioning of the compartments favors rapid circulation of microbes, cells, and fluids from one site to another, either through direct (→) or indirect (--→) connections.

THE COMMUNICATING BODY COMPARTMENTS

For effective immune responsiveness, the activities in one fluid compartment must be conveyed to other compartments. Let us see how this occurs by viewing tissue at the microscopic level (figure 14.5*a*). At this level, clusters of tissue cells are in direct contact with the reticuloendothelial system (RES) and the extracellular fluid (ECF). Other compartments (vessels) that penetrate at this level are blood and lymphatic capillaries. This close association allows cells and chemicals that originate in the RES and ECF to diffuse or migrate into the blood and lymphatics; any products of a lymphatic reaction can be transmitted directly into the blood through the connection between these two systems; and certain cells and chemicals originating in the blood can move through the vessel walls into the extracellular spaces and migrate into the lymphatic system.

The flow of events among these systems depends on where an infectious agent or foreign substance first intrudes. A typical progression might begin in the extracellular spaces and RES, move to the lymphatic circulation, and ultimately end up in the bloodstream. Regardless of which compartment is first exposed, an immune reaction in any one of them will eventually be communicated to the others at the microscopic level. An obvious benefit of such an integrated system is that no cell of the body is far removed from competent protection, no matter how isolated. Let us take a closer look at each of these compartments.

Immune Functions of the Reticuloendothelial System

The tissues of the body are permeated by a support network of connective tissue fibers, or a *reticulum,* that originates in the cellular basal lamina, interconnects nearby cells, and meshes with the massive connective tissue network surrounding all organs. This network, called the **reticuloendothelial system (figure 14.6)** is intrinsic to the immune function because it provides a passageway within and between tissues and organs. It also coexists with and helps form a niche for a collection of phagocytic cells termed the **mononuclear phagocyte system.** The RES is heavily endowed with white blood cells called macrophages waiting to attack passing foreign intruders as they arrive in the skin, lungs, liver, lymph nodes, spleen, and bone marrow.

Origin, Composition, and Functions of the Blood

The circulatory system consists of the circulatory system proper, which includes the heart, arteries, veins, and capillaries that circulate the blood, and the lymphatic system, which includes lymphatic vessels and lymphatic organs (lymph nodes) that circulate lymph (see figure 14.12). As we shall see, these two circulations parallel, interconnect with, and complement one another.

The substance that courses through the arteries, veins, and capillaries is **whole blood,** a liquid connective tissue consisting of **blood cells** (formed elements) suspended in **plasma** (ground substance). One can visualize these two components with the naked eye when a tube of *unclotted* blood is allowed to sit or is spun in a centrifuge. The cells' density causes them to settle into an opaque

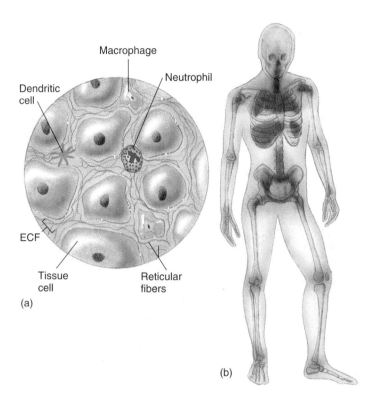

FIGURE 14.6

The reticuloendothelial system occurs as a pervasive, continuous connective tissue framework throughout the body. **(a)** This system begins at the microscopic level with a fibrous support network (reticular fibers) enmeshing each cell. This web connects one cell to another within a tissue or organ and provides a niche for phagocytic white blood cells which can crawl within and between tissues. **(b)** The degrees of shading in the body indicate variations in phagocyte concentration (darker = greater).

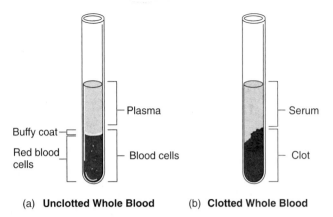

(a) **Unclotted Whole Blood**   (b) **Clotted Whole Blood**

FIGURE 14.7

The macroscopic composition of whole blood. **(a)** When blood containing anticoagulants is allowed to sit for a period, it stratifies into a clear layer of plasma, a thin layer of off-white material called the buffy coat (which contains the white blood cells), and a layer of red blood cells in the bottom, thicker layer. **(b)** Serum is the clear fluid that separates from clotted blood.

and lymphatic organs, and it is finally assumed entirely and permanently by the red bone marrow **(figure 14.8).** Although much of a newborn's red marrow is devoted to hemopoietic function, the active marrow sites gradually recede, and by the age of 4 years, only the ribs, sternum, pelvic girdle, flat bones of the skull and spinal column, and proximal portions of the humerus and femur are devoted to blood cell production.

The relatively short life of blood cells demands a rapid turnover that is continuous throughout a human lifespan. The primary precursor of new blood cells is a pool of undifferentiated cells called pluripotential **stem cells**[2] maintained in the marrow. During development, these stem cells proliferate and *differentiate*—meaning that immature or unspecialized cells develop the specialized form and function of mature cells. The primary lines of cells that arise from this process produce red blood cells (RBCs, or erythrocytes), white blood cells (WBCs, or leukocytes), and platelets (thrombocytes). The white blood cell lines are programmed to develop into several secondary lines of cells during the final process of differentiation **(figure 14.9).** These committed lines of WBCs are largely responsible for immune function.

The *white blood cells,* or **leukocytes,*** are traditionally evaluated by their reactions with hematologic stains that contain a mixture of dyes and can differentiate cells by color and morphology. When this stain used on blood smears is evaluated by the light microscope, the leukocytes appear either with or without noticeable colored granules in the cytoplasm and, on that basis, are divided into two groups: **granulocytes** and **agranulocytes.** Greater magnification reveals that even the agranulocytes have tiny granules in their cytoplasm, so some hematologists also use the appearance of the nucleus to distinguish them. Granulocytes have a lobed nucleus, and agranulocytes have an unlobed, rounded nucleus. Note both of these characteristics in circulating leukocytes (see figure 14.9).

layer at the bottom of the tube, leaving the plasma, a clear, yellowish fluid, on top **(figure 14.7).** In chapters 15 and 16, we will introduce the concept of **serum.** This substance is essentially the same as plasma, except it is the clear fluid from clotted blood. Serum is often used in immune testing and therapy.

Fundamental Characteristics of Plasma   Plasma contains hundreds of different chemicals produced by the liver, white blood cells, endocrine glands, and nervous system and absorbed from the digestive tract. The main component of this fluid is water (92%), and the remainder consists of proteins such as albumin and globulins (including antibodies), other immunochemicals, fibrinogen and other clotting factors, hormones, nutrients (glucose, amino acids, fatty acids), ions (sodium, potassium, calcium, magnesium, chloride, phosphate, bicarbonate), dissolved gases (O_2 and CO_2), and waste products (urea). These substances support the normal physiological functions of nutrition, development, protection, homeostasis, and immunity. We return to the subject of plasma and its function in immune interactions later in this chapter and in chapter 15.

A Survey of Blood Cells   The production of blood cells, or **hemopoiesis,*** begins early in embryonic development in the yolk sac (an embryonic membrane). Later it is taken over by the liver

---

*hemopoiesis (hee″-moh-poy-ee′-sis) Gr. *haima,* blood, and *poiesis,* a making. Also called hematopoiesis.

---

2. Pluripotential stem cells can develop into several different types of blood cells; unipotential cells have already committed to a specific line of development.

*leukocyte (loo′-koh-syte) Gr. *leukos,* white, and *kytos,* cell. The whiteness of unstained WBCs is best seen in the white layer, or buffy coat, of sedimented blood.

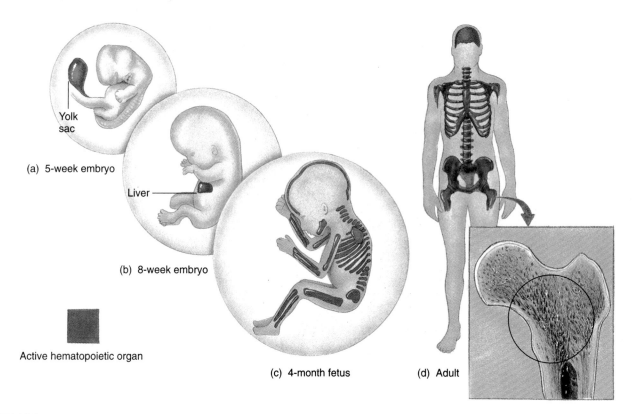

(a) 5-week embryo

Yolk sac

Liver

(b) 8-week embryo

Active hematopoietic organ

(c) 4-month fetus

(d) Adult

FIGURE 14.8

Stages in hemopoiesis. The sites of blood cell production change as development progresses from **(a, b)** yolk sac and liver in the embryo, to **(c)** extensive bone marrow sites in the fetus and **(d)** selected bone marrow sites in the child and adult. **(Inset)** Red marrow occupies the spongy bone (circle) in these areas.

Granulocytes The types of granular leukocytes present in the bloodstream are neutrophils, eosinophils, and basophils. All three are known for prominent cytoplasmic granules that stain with some combination of acidic dye (eosin) or basic dye (methylene blue). Although these granules are useful diagnostically, they also function in numerous physiological events.

Neutrophils* are distinguished from other leukocytes by their conspicuous, lobed nuclei and by their fine, pale lavender granules. In cells newly released from the bone marrow, the nuclei are horseshoe-shaped, but as they age, they form multiple lobes (up to five). These cells, also called **polymorphonuclear*** **neutrophils (PMNs),** make up 55% to 90% of the circulating leukocytes—about 25 billion cells in the circulation at any given moment. The main work of the neutrophils is in phagocytosis.[3] Their high numbers in both the blood and tissues suggest a constant challenge from resident microflora and environmental sources. Most of the cytoplasmic granules carry digestive enzymes and other chemicals that degrade the phagocytosed materials (see the discussion of phagocytosis later in this chapter). The average neutrophil lives only about 8 days, spending much of this time in the tissues and only about 6 to 12 hours in circulation.

Eosinophils* are readily distinguished in a stain preparation by their larger, orange to red (eosinophilic) granules and bilobed nucleus. They are much more numerous in the bone marrow and the spleen than in the circulation, contributing only 1% to 3% of the total WBC count. The role of the eosinophil in the immune system is not fully defined, though several functions have been suggested. Their granules contain peroxidase, lysozyme, and other digestive enzymes, as well as toxic proteins and inflammatory chemicals. The protective action of eosinophils is to attack and destroy large eucaryotic pathogens, but they are also involved in inflammation and allergic reactions. Among their most important targets are helminth worms and fungi. They physically bind to these parasites, release their toxic granules, and eliminate them through lysis. Eosinophils are among the earliest cells to accumulate near sites of inflammation and allergic reactions, where they attract other leukocytes and release chemical mediators (see chapter 17).

Basophils* are characterized by pale-stained, constricted nuclei and very prominent dark blue to black granules. They are the scarcest type of leukocyte, making up less than 0.5% of the total circulating WBCs in a normal individual. Basophils share some morphological and functional similarities with widely distributed tissue cells called **mast*** **cells.** Although these two cell types were once

---

3. The neutrophil is sometimes called a microphage, or "small eater."

*neutrophil (noo'-troh-fil) L. *neuter,* neither, and Gr. *philos,* to love. The granules are neutral and do not react markedly with either acidic or basic dyes. In clinical reports, they are often called "polys" or PMNs for short.

*polymorphonuclear Gr. *poly,* many; *morph,* shape; and *nuclear,* nucleus.

*eosinophil (ee"-oh-sin'-oh-fil) Gr. *eos,* dawn, rosy, and *philos,* to love. Eosin is a red, acidic dye attracted to the granules.

*basophil (bay'-soh-fil) Gr. *basis,* foundation. The granules attract to basic dyes.

*mast From Ger. *mast,* food. Early cytologists thought these cells were filled with food vacuoles.

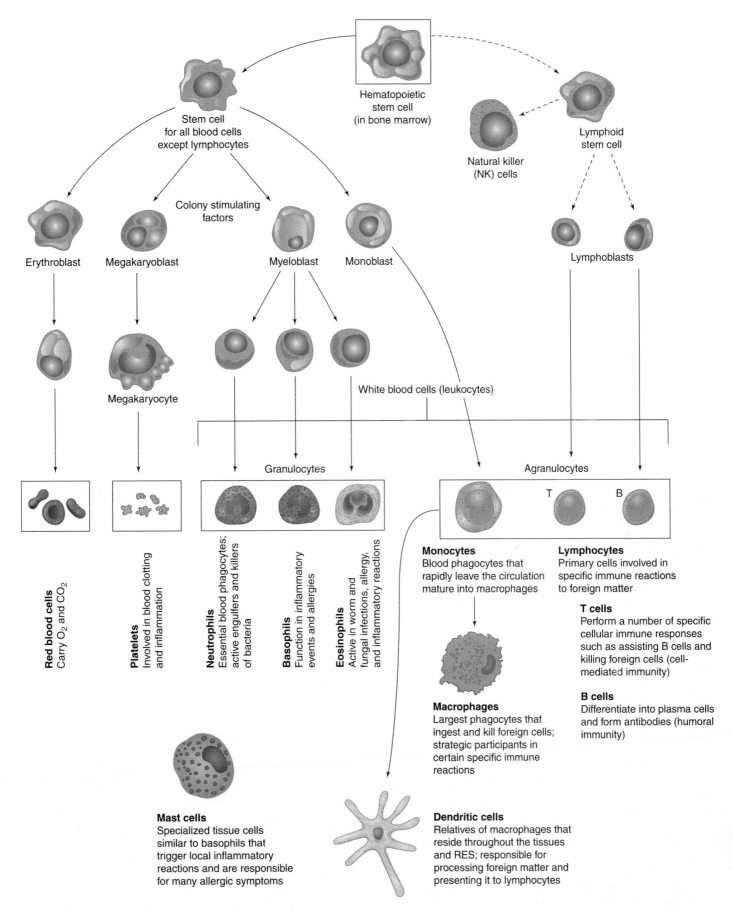

FIGURE 14.9

The development of blood cells and platelets.   Undifferentiated stem cells in the red marrow differentiate to give rise to several different cell lines that become increasingly specialized until mature cells (boxed) are released into circulation. Some cells migrate into the tissues to achieve fully functional status.

regarded as identical, mast cells are nonmotile elements bound to connective tissue around blood vessels, nerves, and epithelia, and basophils are motile elements derived from bone marrow.

Basophils parallel eosinophils in many of their actions, since they also contain granules with potent chemical mediators. Mast cells are first line defenders against the local invasion of pathogens, they recruit other inflammatory cells, and are directly responsible for the release of histamine and other allergic stimulants during immediate allergies (see chapter 17).

Agranulocytes Agranular leukocytes have globular, nonlobed nuclei and lack prominent cytoplasmic granules when viewed with the light microscope. The two general types are monocytes and lymphocytes.

Lymphocytes are the second most common WBC in the blood, comprising 20% to 35% of the total circulating leukocytes. The fact that their overall number throughout the body is among the highest of all cells indicates how important they are to immunity. One estimate suggests that about one-tenth of all adult body cells are lymphocytes, exceeded only by erythrocytes and fibroblasts. In a stained blood smear, most lymphocytes are small, spherical cells with a uniformly dark blue, rounded nucleus surrounded by a thin fringe of clear, pale blue cytoplasm, although in tissues, they can become much larger and can even mimic monocytes in appearance (see figure 14.9). Lymphocytes exist as two functional types—the bursal-equivalent, or **B lymphocytes** (**B cells,** for short), and the thymus-derived, or **T lymphocytes** (**T cells,** for short). B cells were first demonstrated in and named for a special lymphatic gland of chickens called the *bursa of Fabricius,* the site for their maturation in birds. In humans, B cells mature in special bone marrow sites, but humans do not have a bursa. T cells mature in the thymus gland in all birds and mammals. Both populations of cells are transported by the bloodstream and lymph and move about freely between lymphoid organs and connective tissue.

Lymphocytes are the key cells of the third line of defense and the specific immune response **(figure 14.10).** When stimulated by foreign substances (antigens), lymphocytes are transformed into activated cells that neutralize and destroy that foreign substance. The contribution of B cells is mainly in **humoral immunity,**[4] defined as protective molecules carried in the fluids of the body. When activated B cells divide, they form specialized **plasma cells,** which produce **antibodies,*** large protein molecules that interlock with an antigen and participate in its destruction. Activated T cells engage in a spectrum of immune functions characterized as **cell-mediated immunity** in which T cells modulate immune functions and kill foreign cells. The action of both classes of lymphocytes accounts for the recognition and memory typical of immunity. So important are lymphocytes to the defense of the body that a large portion of chapter 15 is devoted to their reactions.

Monocytes* are generally the largest of all white blood cells and the third most common in the circulation (3–7%). As they mature, the nucleus becomes oval- or kidney-shaped—indented on one side, off-center, and often contorted with fine wrinkles. The

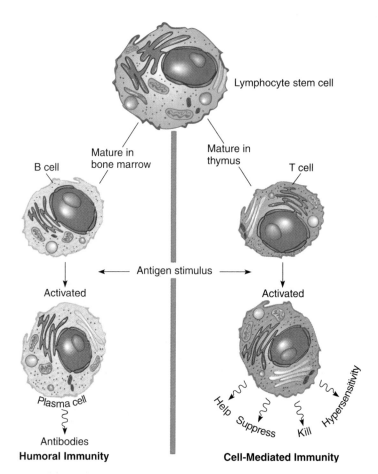

FIGURE 14.10

Summary of the general development and functions of lymphocytes, which are the cornerstone of specific immune reactions. B cells and T cells arise from the same stem cell but later diverge into two cell lines. Their appearances are similar, and one cannot differentiate them on the basis of staining. Note the relatively large nucleus—cytoplasm ratio and the lack of granules.

pale blue cytoplasm holds many fine vacuoles containing digestive enzymes. Monocytes are discharged by the bone marrow into the bloodstream, where they live as phagocytes for a few days. Later they leave the circulation to undergo final differentiation into **macrophages*** (see figure 14.18). Unlike many other WBCs, the monocyte-macrophage series is relatively long-lived and retains an ability to multiply. Macrophages are among the most versatile and important of cells. In general, they are responsible for

1. many types of specific and nonspecific phagocytic and killing functions (they assume the job of cellular housekeepers, "mopping up the messes" created by infection and inflammation);
2. processing foreign molecules and presenting them to lymphocytes; and
3. secreting biologically active compounds that assist, mediate, attract, and inhibit immune cells and reactions.

We touch upon these functions in several ensuing sections.

---

4. In reference to the humors, the liquids of the body. Humoral immunity includes antibodies, complement, and interferon.

*antibody (an′-tih-bahd″-ee) Gr. *anti,* against, and O.E., *bodig,* body.

*monocyte (mon′-oh-syte) From *mono,* one, and *cytos,* cell.

*macrophage (mak′-roh-fayj) Gr. *macro,* large, and *phagein,* to eat. They are the "large eaters" of the tissues.

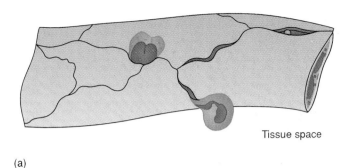

(a)

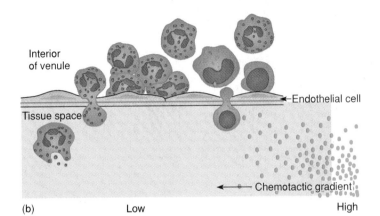

(b)

Interior
of venule

←Endothelial cell

Tissue space

←— Chemotactic gradient

Low   High

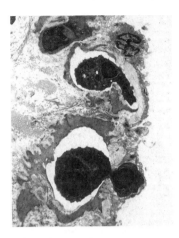

(c)

FIGURE 14.11

Diapedesis and chemotaxis of leukocytes.
(a) View of a venule depicts white blood cells
squeezing themselves between spaces in the
blood vessel wall through diapedesis. **(b)** This
process, shown in cross section, indicates how
the pool of leukocytes adheres to the endothelial
wall. From this site, they are poised to migrate out
of the vessel into the tissue space. **(c)** This
photograph captures neutrophils in the process
of diapedesis.

Another product of the monocyte cell line are **dendritic cells,**
named for their long, thin cell processes (figure 14.9). Immature
dendritic cells move from the blood to the RES and lymphatic tissues, where they trap pathogens. Ingestion of bacteria and viruses
stimulates dendritic cells to migrate to lymph nodes and the spleen.
Here, they mature into highly effective processors and presenters of
foreign proteins (see chapter 15).

Erythrocyte and Platelet Lines   Erythrocytes* develop from
stem cells in the bone marrow and lose their nucleus just prior to
entering the circulation. The resultant red blood cells are simple,
biconcave sacs of hemoglobin that transport oxygen and carbon
dioxide to and from the tissues (see figure 14.9). These are the
most numerous of circulating blood cells, appearing in stains as
small pink circles. Red blood cells do not ordinarily have immune
functions, though they can be the target of immune reactions (see
chapter 17).

Platelets,* or **thrombocytes,*** are formed elements in circulating blood that are *not* whole cells. They originate in the bone
marrow when a giant multinucleate cell called a *megakaryocyte* disintegrates into numerous tiny, irregular-shaped pieces, each containing bits of the cytoplasm and nucleus (see figure 14.9). In stains,
platelets are blue-gray with fine red granules and are readily distinguished from cells by their small size. Platelets function primarily

in hemostasis (plugging broken blood vessels to stop bleeding) and
in releasing chemicals that act in blood clotting and inflammation.

Unique Dynamic Characteristics of White Blood Cells
Many lymphocytes and phagocytes make regular journeys from the
blood and lymphatics to the tissues and back again to the circulation as part of the constant surveillance of the compartments. In order for these WBCs to complete this circuit, they adhere to the inner walls of the smaller blood vessels. From this position, they are
poised to migrate out of the blood into the tissue spaces by a
process called **diapedesis.***

Diapedesis, also known as transmigration, is aided by several
related characteristics of WBCs. For example, they are actively motile
and readily change shape. This phenomenon is also assisted by the
nature of the endothelial cells lining the venules. They contain complex adhesive receptors that capture the WBCs and participate in their
transport from the venules into the extracellular spaces **(figure 14.11).**

Another factor in the migratory habits of these WBCs is
chemotaxis.* This is defined as the tendency of cells to migrate in
response to a specific chemical stimulus given off at a site of injury
or infection (see inflammation and phagocytosis later in this chapter). Through this means, cells swarm from many compartments to
the site of infection and remain there to perform general and specific immune functions. These basic properties are absolutely essential for the sort of intercommunication and deployment of cells
required for most immune reactions (figure 14.11).

***erythrocyte** (eh-rith′-roh-syte) Gr. *erythros,* red. The red color comes from
hemoglobin.

***platelet** (playt′-let) Gr. *platos,* flat, and *let,* small.

***thrombocyte** (throm′-boh-syte) Gr. *thrombos,* clot.

***diapedesis** (dye″-ah-puh-dee′-sis) Gr. *dia,* through, and *pedan,* to leap.

***chemotaxis** (kee-moh-tak′-sis) NL *chemo,* chemical, and *taxis,* arrangement.

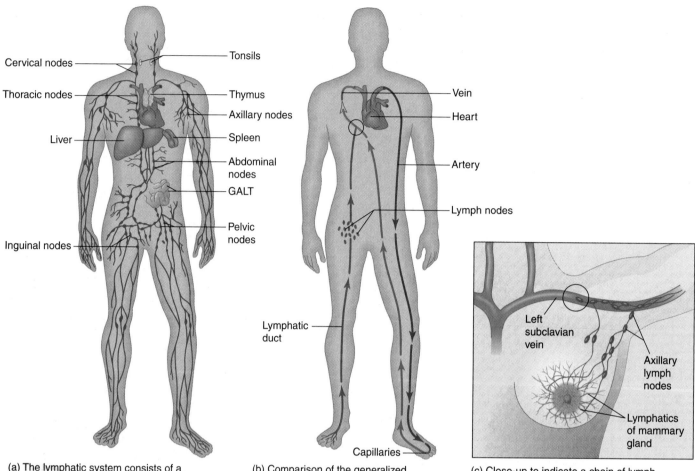

Cervical nodes

Tonsils

Thoracic nodes

Thymus

Axillary nodes

Liver

Spleen

Abdominal nodes

GALT

Inguinal nodes

Pelvic nodes

Vein

Heart

Artery

Lymph nodes

Lymphatic duct

Capillaries

Left subclavian vein

Axillary lymph nodes

Lymphatics of mammary gland

(a) The lymphatic system consists of a branching network of vessels which extend into most body areas. Note the higher density of lymphatic vessels in the "dead-end" areas of the hands, feet, and breast, which are frequent contact points for infections. Other lymphatic organs include the lymph nodes, spleen, gut-associated lymphoid tissue (GALT), the thymus gland, and the tonsils.

(b) Comparison of the generalized circulation of the lymphatic system and the blood. Although the lymphatic vessels parallel the regular circulation, they transport in only one direction unlike the cyclic pattern of blood. Direct connection between the two circulations occurs at points near the heart where large lymph ducts empty their fluid into veins (circled area).

(c) Close-up to indicate a chain of lymph nodes near the axilla and breast and another point of contact between the two circulations (circled area).

FIGURE 14.12

General components of the lymphatic system.

Components and Functions of the Lymphatic System

The lymphatic system is a compartmentalized network of vessels, cells, and specialized accessory organs (**figure 14.12**). It begins in the farthest reaches of the tissues as tiny capillaries that transport a special fluid (lymph) through an increasingly larger tributary system of vessels and filters (lymph nodes), and it leads to major vessels that drain back into the regular circulatory system. Some major functions of the lymphatic system are

1. to provide an auxiliary route for the return of extracellular fluid to the circulatory system proper,[5]
2. to act as a "drain-off" system for the inflammatory response, and
3. to render surveillance, recognition, and protection against foreign materials through a system of lymphocytes, phagocytes, and antibodies.

Lymphatic Fluid Lymph is a plasmalike liquid carried by the lymphatic circulation. It is formed when certain blood components move out of the blood vessels into the extracellular spaces and diffuse or migrate into the lymphatic capillaries. Thus, the composition of lymph parallels that of plasma in many ways. It is made up of water, dissolved salts, and 2% to 5% protein (especially antibodies and albumin). Like blood, it also transports numerous white blood cells (especially lymphocytes) and miscellaneous materials such as fats, cellular debris, and infectious agents that have gained access to the tissue spaces. Unlike blood, red blood cells are not normally found in lymph.

Lymphatic Vessels The system of vessels that transports lymph is constructed along the lines of blood vessels. As the lymph is never subjected to high pressure, the lymphatic vessels appear most similar to thin-walled veins rather than thicker-walled arteries. The tiniest vessels, lymphatic capillaries, accompany the blood capillaries and permeate all parts of the body except the central nervous

---

5. The importance of this function is most evident in cases of impaired lymphatic drainage as seen in patients with filariasis, a roundworm infection (see figure 23.20).

system and certain organs such as bone, placenta, and thymus. Their thin walls are easily permeated by extracellular fluid that has escaped from the circulatory system. Lymphatic vessels are found in particularly high numbers in the hands, feet, and around the areola of the breast.

Two overriding differences between the bloodstream and the lymphatic system should be mentioned. First, because one of the main functions of the lymphatic system is returning lymph to the circulation, the flow of lymph is in one direction only with lymph moving from the extremities toward the heart. Eventually lymph will be returned to the bloodstream through two large veins at the base of the neck. The second difference concerns how lymph travels through the vessels of the lymphatic system. While blood is transported through the body by means of a dedicated pump (the heart), lymph is moved only through the contraction of the skeletal muscles through which the lymphatic ducts wind their way. This dependence on muscle movement helps to explain the swelling of the hands and feet that sometimes occurs during the night (when muscles are inactive) yet dissipates soon after waking.

Lymphoid Organs and Tissues   Other organs and tissues that perform lymphoid functions are the lymph nodes (glands), thymus, spleen, and clusters of tissues in the gastrointestinal tract (gut-associated lymphoid tissue; GALT) and the pharynx (the tonsils, for example). A trait common to these organs is a loose connective tissue framework that houses aggregations of lymphocytes, the important class of white blood cells mentioned previously.

Lymph Nodes   Lymph nodes are small, encapsulated, bean-shaped organs stationed, usually in clusters, along lymphatic channels and large blood vessels of the thoracic and abdominal cavities (see figure 14.12). Major aggregations of nodes occur in the loose connective tissue of the armpit (axillary nodes), groin (inguinal nodes), and neck (cervical nodes). Both the location and architecture of these nodes clearly specialize them for filtering out materials that have entered the lymph and providing appropriate cells and niches for immune reactions.

A view of a single sectioned lymph node reveals its filtering and cellular response systems (**figure 14.13**). Incoming lymphatic vessels (afferent lymphatic ducts) transport the lymph into sinuses

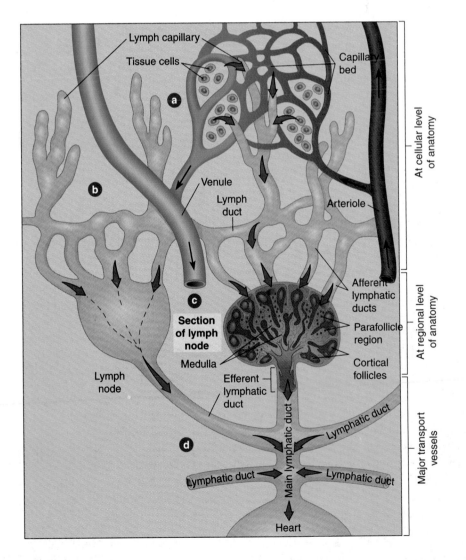

(a) The finest level of lymphatic circulation begins with blind capillaries that pick up fluid, white blood cells, and microbes or other foreign matter from the surrounding tissues and transport this liquid mixture (lymph) away from the extremities via a system of small ducts.

(b) The ducts carry lymph into a circuit of larger ducts that ultimately flow into clusters of specialized filtering organs, the lymph nodes.

(c) The center diagram shows a section through a lymph node to reveal the afferent ducts draining into sinuses that house several types of white blood cells, primarily T lymphocytes, B lymphocytes, macrophages, and dendritic cells. Here, foreign material is filtered out, processed, and becomes the focus of various immune responses.

(d) Lymph continues to trickle from the lymph nodes via efferent ducts into a system of larger drainage vessels, which ultimately connect with large veins near the heart. In this way, cells and products of immunity continually enter the regular circulation.

FIGURE 14.13

Scheme of circulation in the lymphatic vessels and lymph nodes.

in the node that contain segregated populations of lymphocytes. The central zone, or **medulla,*** is an accumulation point for lymph and cells passing through the node. The surrounding germinal centers in the **cortex*** are packed with T and B cells. This system of sinuses and discrete lymphocyte zones filters out particulate materials (microbes, for instance) and contributes WBCs to the lymph as it passes through. Many of the initial encounters between lymphocytes and microbes that result in specific immune responses occur in the lymph nodes.

Spleen The spleen is a lymphoid organ in the upper left portion of the abdominal cavity. It is somewhat similar to a lymph node except that it serves as a filter for blood instead of lymph. While the spleen's primary function is to remove worn-out red blood cells from circulation, its most important immunological function centers on the filtering of pathogens from the blood and their subsequent phagocytosis by resident macrophages. Although adults whose spleens have been surgically removed can live a relatively normal life, asplenic children are severely immunocompromised.

The Thymus: Site of T-Cell Maturation The **thymus*** originates in the embryo as two lobes in the pharyngeal region that fuse into a triangular structure. The size of the thymus is greatest proportionately at birth **(figure 14.14),** and it continues to exhibit high rates of activity and growth until puberty, after which it begins to shrink gradually through adulthood. During the last few decades of life, its function is greatly diminished, because its primary work has been completed. The thymus is sectioned into a medulla, composed of special epithelial cells, and a cortex, containing undifferentiated lymphocytes called thymocytes. Under the influence of thymic hormones, thymocytes develop specificity and are released into the circulation as mature T cells. The T cells subsequently migrate to and settle in other lymphoid organs (for example, the lymph nodes and spleen), where they occupy the specific sites described previously.

The thymus gland was once thought to have no important function. Medical science was so mistaken about its significance that children's necks were sometimes irradiated to "cure" a condition called "enlarged thymus." Experiments in the early 1960s finally clarified its link to lymphocyte development. Children born without a thymus (DiGeorge syndrome, see p. 525), or who have had their thymus surgically removed, are severely immunodeficient and fail to thrive. Adults have developed enough mature T cells that removal of the thymus or reduction in its function has milder effects. Do not confuse the thymus with the thyroid gland, which is located nearby but has an entirely different function.

Miscellaneous Lymphoid Tissue At many sites on or just beneath the mucosa of the gastrointestinal and respiratory tracts lie discrete bundles of lymphocytes. The positioning of this diffuse system provides an effective first-strike potential against the constant influx of microbes and other foreign materials in food and air. In the pharynx, a ring of tissues called the tonsils provides an active

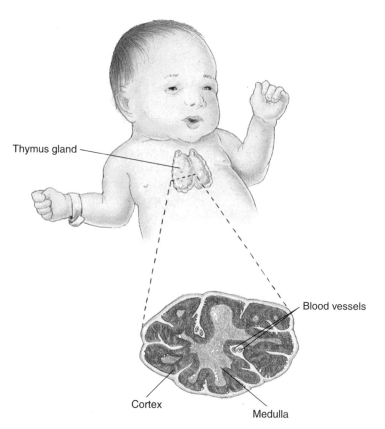

Thymus gland

Blood vessels

Cortex

Medulla

FIGURE 14.14

The thymus gland. Immediately after birth, the thymus is a large organ that nearly fills the region over the midline of the upper thoracic region. In the adult, however, it is proportionately smaller (to compare, see figure 14.12*a*). Section shows the main anatomical regions of the thymus. Immature T cells enter through the cortex and migrate into the medulla as they mature.

source of lymphocytes. The breasts of pregnant and lactating women also become temporary sites of antibody-producing lymphoid tissues (see colostrum, Medical Microfile 15.3). The intestinal tract houses the best-developed collection of lymphoid tissue, called **gut-associated lymphoid tissue,** or **GALT.** Examples of GALT include the appendix, the lacteals (special lymphatic vessels stationed in each intestinal villus), and **Peyer's patches,** compact aggregations of lymphocytes in the ileum of the small intestine. GALT provides immune functions against intestinal pathogens and is a significant source of some types of antibodies. Other, less well-organized collections of secondary lymphoid tissue include the *mucosal-associated lymphoid tissue (MALT), skin-associated lymphoid tissue (SALT), and bronchial-associated lymphoid tissue (BALT).*

CHAPTER CHECKPOINTS

The immune system is a complex collection of fluids and cells that penetrate every organ, tissue space, fluid compartment, and vascular network of the body. The four major subdivisions of this system are the RES, the ECF, the blood vascular system, and the lymphatic system.

---

*medulla (meh-dul′-ah) L. *medius,* middle or marrow.

*cortex (kor′-teks) L. *cortex,* bark, rind, shell.

*thymus (thigh′-mus) Gr. *thymos,* soul, mind.

The RES, or reticuloendothelial system, is a network of connective tissue fibers inhabited by macrophages ready to attack and ingest microbes that have managed to bypass the first line of defense.

The ECF, or extracellular fluid, compartment surrounds all tissue cells and is penetrated by both blood and lymph vessels, which bring all components of the second and third line of defense to attack infectious microbes.

The blood contains both specific and nonspecific defenses. Nonspecific cellular defenses include the granulocytes, macrophages, and dendritic cells. The two components of the specific immune response are the T lymphocytes, which provide specific cell-mediated immunity, and the B lymphocytes, which produce specific antibody or humoral immunity.

The lymphatic system has three functions: (1) It returns tissue fluid to general circulation; (2) it carries away excess fluid in inflamed tissues; (3) it concentrates and processes foreign invaders and initiates the specific immune response. Important sites of lymphoid tissues are lymph nodes, spleen, thymus, tonsils, and GALT.

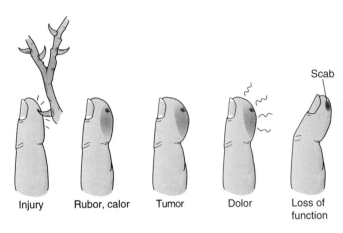

Injury   Rubor, calor   Tumor   Dolor   Loss of function

FIGURE 14.15

The response to injury.   This classic checklist encapsulates the reactions of the tissues to an assault. Each of the events is an indicator of one of the mechanisms of inflammation described in this chapter.

Nonspecific Immune Reactions of the Body's Compartments

Now that we have introduced the principal anatomical and physiological framework of the immune system, let us address some mechanisms that play important roles in host defenses: inflammation, phagocytosis, interferon, and complement. Because of the generalized nature of these defenses, they are primarily nonspecific in their effects, but they also support and interact with the specific immune responses described in chapter 15.

THE INFLAMMATORY RESPONSE: A COMPLEX CONCERT OF REACTIONS TO INJURY

At its most general level, the inflammatory response is a reaction to any traumatic event in the tissues. It is so very commonplace that most of us manifest inflammation in some way every day. It appears in the nasty flare of a cat scratch, the blistering of a burn, the painful lesion of an infection, and the symptoms of allergy. It is readily identifiable by a classic series of signs and symptoms characterized succinctly by four Latin terms: *rubor, calor, tumor,* and *dolor.* Rubor (redness) is caused by increased circulation and vasodilation in the injured tissues; calor (warmth) is the heat given off by the increased flow of blood; tumor (swelling) is caused by increased fluid escaping into the tissues; and dolor (pain) is caused by the stimulation of nerve endings (**figure 14.15**). Although these manifestations can be unpleasant, they serve an important warning that injury has taken place and set in motion responses that save the body from further injury.

Factors that can elicit inflammation include trauma from infection (the primary emphasis here), tissue injury or necrosis due to physical or chemical agents, and specific immune reactions. Although the details of inflammation are very complex, its chief functions can be summarized as follows:

1. to mobilize and attract immune components to the site of the injury,
2. to set in motion mechanisms to repair tissue damage and localize and clear away harmful substances, and
3. to destroy microbes and block their further invasion (**figure 14.16**).

The inflammatory response is a powerful defensive reaction, a means for the body to maintain stability and restore itself after an injury. But when it is chronic, it has the potential to actually *cause* tissue injury, destruction, and disease (**Medical Microfile 14.1**).

THE STAGES OF INFLAMMATION

The process leading to inflammation is a dynamic, predictable sequence of events that can be acute, lasting from a few minutes or hours, to chronic, lasting for days, weeks, or years. Once the initial injury has occurred, a chain reaction takes place at the site of damaged tissue, summoning beneficial cells and fluids into the injured area. As an example, we will look at an injury at the microscopic level and observe the flow of major events (figure 14.16).

Vascular Changes: Early Inflammatory Events
Following an injury, some of the earliest changes occur in the vasculature (arterioles, capillaries, venules) in the vicinity of the damaged tissue. These changes are controlled by nervous stimulation, **chemical mediators,** and **cytokines*** released by blood cells, tissue cells, and platelets in the injured area. Some mediators are *vasoactive*—that is, they affect the endothelial cells and smooth muscle cells of blood vessels, and others are **chemotactic factors,** also called **chemokines,** that affect white blood cells. Inflammatory mediators cause fever, stimulate lymphocytes, prevent virus spread, and cause allergic symptoms (**Medical Microfile 14.2** and **figure 14.17**). Although the constriction of arterioles is stimulated first, it lasts for only a few seconds or minutes and is followed in quick succession by the

---

*cytokine (sy′-toh-kyne) Gr. *cytos,* cell, and *kinein,* to move. A protein or polypeptide produced by WBCs that regulates host defenses.

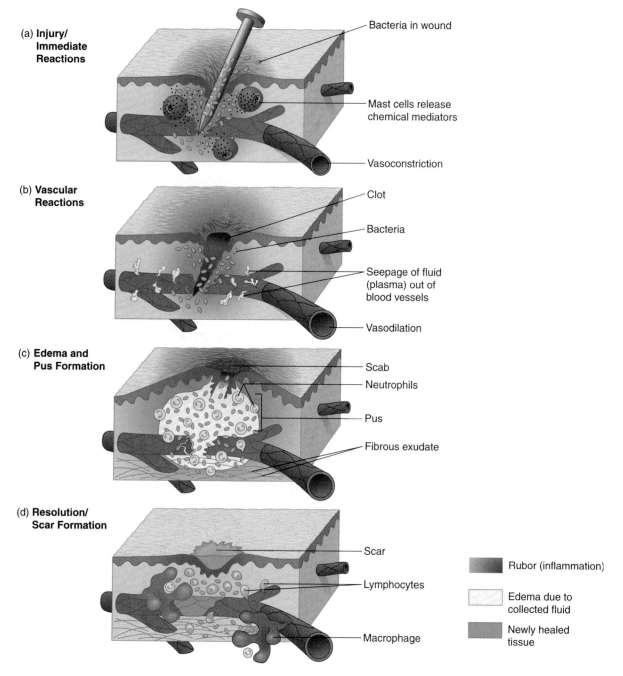

(a) **Injury/Immediate Reactions**
- Bacteria in wound
- Mast cells release chemical mediators
- Vasoconstriction

(b) **Vascular Reactions**
- Clot
- Bacteria
- Seepage of fluid (plasma) out of blood vessels
- Vasodilation

(c) **Edema and Pus Formation**
- Scab
- Neutrophils
- Pus
- Fibrous exudate

(d) **Resolution/Scar Formation**
- Scar
- Lymphocytes
- Macrophage

- Rubor (inflammation)
- Edema due to collected fluid
- Newly healed tissue

FIGURE 14.16

The major events in inflammation. **(a)** Injury → Reflex narrowing of the blood vessels (vasoconstriction) lasting for a short time → Release of chemical mediators into area. **(b)** Increased diameter of blood vessels (vasodilation) → Increased blood flow → Increased vascular permeability → Leakage of fluid (plasma) from blood vessels into tissues (exudate formation). **(c)** Edema → Infiltration of site by neutrophils and accumulation of pus. **(d)** Macrophages and lymphocytes → Repair, either by complete resolution and return of tissue to normal state or by formation of scar tissue.

opposite reaction, vasodilation. The overall effect of vasodilation is to increase the flow of blood into the area, which facilitates the influx of immune components and also causes redness and warmth.

Edema: Leakage of Vascular Fluid into Tissues

Some vasoactive substances cause the endothelial cells surrounding postcapillary venules to contract and form gaps through which blood-borne components exude into the extracellular spaces. The fluid part that escapes is called the *exudate*. Accumulation of this fluid in the tissues gives rise to local swelling and hardness called **edema.** The edematous exudate contains varying amounts of plasma proteins, such as globulins, albumin, the clotting protein fibrinogen, blood cells, and cellular debris. Depending upon its content, the exudate varies from serous (clear) to serosanguinous (containing red blood cells) to purulent (containing pus). In some types of edema, the fibrinogen is

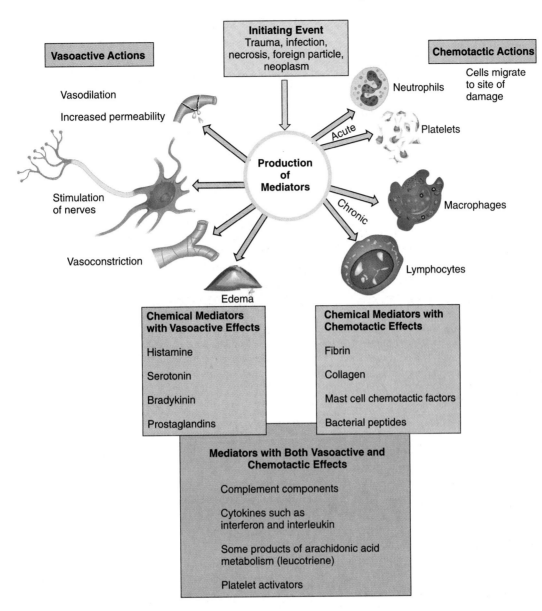

FIGURE 14.17
Chemical mediators of the inflammatory response and their effects.

converted to fibrin threads that enmesh the injury site. Within an hour, multitudes of neutrophils responding chemotactically to special signaling molecules converge on the injured site (see figure 14.16c).

The Benefits of Edema and Chemotaxis   Both the formation of edematous exudate and the infiltration of neutrophils are physiologically beneficial activities. The influx of fluid dilutes toxic substances, and the fibrin clot can effectively trap microbes and prevent their further spread. The neutrophils that aggregate in the inflamed site are immediately involved in phagocytosing and destroying bacteria, dead tissues, and particulate matter (by mechanisms discussed in a later section on phagocytosis). In some types of inflammation, accumulated phagocytes contribute to **pus,** a whitish mass of cells, liquefied cellular debris, and bacteria. Certain bacteria

(streptococci, staphylococci, gonococci, and meningococci) are especially powerful attractants for neutrophils and are thus termed **pyogenic,** or pus-forming, bacteria.

Late Reactions of Inflammation   Sometimes a mild inflammation can be resolved by edema and phagocytosis. Inflammatory reactions that are more long-lived attract a collection of monocytes, lymphocytes, and macrophages to the reaction site. Clearance of pus, cellular debris, dead neutrophils, and damaged tissue is performed by macrophages, the only cells that can engulf and dispose of such large masses. At the same time, B lymphocytes react with foreign molecules and cells by producing specific antimicrobial proteins (antibodies), and T lymphocytes kill intruders directly. Late in the process the tissue is completely repaired, if possible, or replaced by connective tissue in the form of a scar (see

Not every aspect of inflammation is protective or results in the proficient resolution of tissue damage. As one looks over a list of diseases, it is rather striking how many of them are due in part or even completely to an overreactive or dysfunctional inflammatory response.

Some "itis" reactions mentioned in chapter 13 are a case in point (see Microbits 13.4). Inflammatory exudates that build up in the brain in African trypanosomiasis, cryptococcosis, and other brain infections can be so injurious to the nervous system that impairment is permanent. Frequently, an inflammatory reaction that walls off the pathogen leads to an abscess, a swollen mass of neutrophils and dead, liquefied tissue that can harbor live pathogens in the center. Abscesses are a prominent feature of staphylococcal, amebic, and enteric infections.

Other pathologic manifestations of chronic diseases—for example, the tubercles of tuberculosis, the lesions of late syphilis, the disfiguring nodules of leprosy, and the cutaneous ulcers of leishmaniasis—are due to an aberrant tissue response called *granuloma formation* (see figure 19.16). Granulomas develop not only in response to microbes but also in response to inanimate foreign bodies (sutures and

mineral grains that are difficult to break down). This condition is initiated when neutrophils ineffectively and incompletely phagocytose the pathogens or materials involved in an inflammatory reaction. The macrophages then enter to clean up and attempt to phagocytose the dead neutrophils and foreign substances, but they fail to completely manage them. They respond by storing these ingested materials in vacuoles and becoming inactive. Over a given time period, large numbers of adjacent macrophages fuse into giant, inactive multinucleate cells called foreign body giant cells. These sites are further infiltrated with lymphocytes. The resultant collections make the tissue appear granular—hence, the name. A granuloma can exist in the tissue for months, years, or even a lifetime.

Medical science is rapidly searching for new applications for the massive amount of new information on inflammatory mediators. One highly promising area appears to be the use of chemokine inhibitors that could reduce chemotaxis and the massive, destructive influx of leukocytes. Such therapy could ultimately be used for certain cancers, hardening of arteries, and Alzheimer disease.

figure 14.16*d*). If the inflammation cannot be relieved or resolved in this way, it can become chronic and create a long-term pathologic condition.

Fever: An Adjunct to Inflammation

An important systemic component of inflammation is fever, defined as an abnormally elevated body temperature. Although fever is a nearly universal symptom of infection, it is also associated with certain allergies, cancers, and other organic illnesses. Fevers whose causes are unknown are called fevers of unknown origin, or FUO.

The body temperature is normally maintained by a control center in the hypothalamus. This thermostat regulates the body's heat production and heat loss and sets the core temperature at around 37°C (98.6°F), with slight fluctuations (1°F) during a daily cycle. Fever is initiated when a circulating substance called **pyrogen*** resets the hypothalamic thermostat to a higher setting. This change signals the musculature to increase heat production and peripheral arterioles to decrease heat loss through vasoconstriction (**Medical Microfile 14.3**). Fevers range in severity from low-grade (37.7°–38.3°C or 100°–101°F) to moderate (38.8°–39.4°C, or 102°–103°F) to high (40.0°–41.1°C, or 104°–106°F). Pyrogens are described as *exogenous* (coming from outside the body) or *endogenous* (originating internally). Exogenous pyrogens are products of infectious agents such as viruses, bacteria, protozoans, and fungi. One well-characterized exogenous pyrogen is endotoxin, the lipopolysaccharide found in the cell walls of gram-negative

bacteria. Blood, blood products, vaccines, or injectable solutions can also contain exogenous pyrogens. Endogenous pyrogens are liberated by monocytes, neutrophils, and macrophages during the process of phagocytosis and appear to be a natural part of the immune response. Two potent pyrogens released by macrophages are interleukin-1 and tumor necrosis factor.

Benefits of Fever The association of fever with infection strongly suggests that it serves a beneficial role, a view still being debated but gaining acceptance. Aside from its practical and medical importance as a sign of a physiological disruption, increased body temperature has additional benefits:

- Fever inhibits multiplication of temperature-sensitive microorganisms such as the poliovirus, cold viruses, herpes zoster virus, systemic and subcutaneous fungal pathogens, *Mycobacterium* species, and the syphilis spirochete.
- Fever impedes the nutrition of bacteria by reducing the availability of iron. It has been demonstrated that during fever, the macrophages stop releasing their iron stores, which could retard several enzymatic reactions needed for bacterial growth.
- Fever increases metabolism and stimulates immune reactions and naturally protective physiological processes. It speeds up hematopoiesis, phagocytosis, and specific immune reactions.

Treatment of Fever With this revised perspective on fever, whether to suppress it or not can be a difficult decision. Some advocates feel that a slight to moderate fever in an otherwise healthy person should be allowed to run its course, in light of its potential

---

*pyrogen (py′-roh-jen) Gr. *pyr,* fire, and *gennan,* produce. As in funeral pyre and pyromaniac.

MEDICAL MICROFILE 14.2
The Dynamics of Inflammatory Mediators

Just as the nervous system is coordinated by a complex communications network, so too is the immune system. Hundreds of small, active molecules are constantly being secreted to regulate, stimulate, suppress, and otherwise control the many aspects of cell development, inflammation, and immunity. These substances are the products of several types of cells, including monocytes, macrophages, lymphocytes, fibroblasts, mast cells, platelets, and endothelial cells of blood vessels. Their effects may be local or systemic, short-term or long-lasting, nonspecific or specific, protective or pathologic.

In recent times, the field of cytokines has become so increasingly complex that we can include here only an overview of the major groups of important cytokines and other mediators. The major functional types can be categorized into

1. cytokines that mediate nonspecific immune reactions such as inflammation and phagocytosis,
2. cytokines that regulate the growth and activation of lymphocytes,
3. cytokines that activate immune reactions during inflammation,
4. hemopoiesis factors for white blood cells,
5. vasoactive mediators, and
6. miscellaneous inflammatory mediators.

Nonspecific Mediators of Inflammation and Immunity

- *Tumor necrosis factor (TNF),* a substance from macrophages, lymphocytes, and NK cells that increases chemotaxis and phagocytosis and stimulates other cells to secrete inflammatory cytokines. It also serves as an endogenous pyrogen that induces fever, increases blood coagulation, suppresses bone marrow, and causes wasting of the body called cachexia.
- *Interferon (IFN), alpha and beta,* produced by leukocytes and fibroblasts, inhibits virus replication and cell division and increases the action of certain lymphocytes that kill other cells.
- **Interleukin* (IL) 1,** a product of macrophages and epithelial cells that has many of the same biological activities as TNF, such as inducing fever and activation of certain white blood cells.
- *Interleukin-6,* secreted by macrophages, lymphocytes, and fibroblasts. Its primary effects are to stimulate the growth of B cells and to increase the synthesis of liver proteins.
- *Various chemokines.* By definition, chemokines are cytokines that stimulate the movement and migration of white blood cells (chemotactic factors). Included among these are complement C5A, interleukin 8, and platelet factor.

---

*Interleukin is a term that refers to a group of small peptides originally isolated from leukocytes. There are currently 20 known interleukins. We now know that other cells besides leukocytes can synthesize them and that they have a variety of biological activities. Functions of some selected examples will be presented in chapter 15.

Cytokines That Regulate Lymphocyte Growth in Activation

- *Interleukin-2,* the primary growth factor from T cells. Interestingly, it acts on the same cells that secrete it. It stimulates mitosis and secretion of other cytokines. In B cells, it is a growth factor and stimulus for antibody synthesis.
- *Interleukin-4,* a stimulus for the production of allergy antibodies; inhibits macrophage actions; favors development of T cells.
- *Granulocyte colony-stimulating factor (G-CSF),* produced by T cells, macrophages, and neutrophils. It stimulates the activation and differentiation of neutrophils.
- *Macrophage colony-stimulating factor (M-CSF),* produced by a variety of cells. M-CSF promotes the growth and development of macrophages from undifferentiated precursor cells.

Cytokines That Activate Specific Immune Reactions

- *Gamma interferon,* a T-cell derived mediator whose primary function is to activate macrophages. It also promotes the differentiation of T and B cells, activates neutrophils, and stimulates diapedesis.
- *Interleukin-5* activates eosinophils and B cells; *interleukin-10* inhibits macrophages and B cells; and *interleukin-12* activates T cells and killer cells.

Vasoactive Mediators

- **Histamine,** a vasoactive mediator produced by mast cells and basophils that causes vasodilation, increased vascular permeability, and mucus production. It functions primarily in inflammation and allergy.
- **Serotonin,** a mediator produced by platelets and intestinal cells that causes smooth muscle contraction, inhibits gastric secretion, and acts as a neurotransmitter.
- **Bradykinin,** a vasoactive amine from the blood or tissues that stimulates smooth muscle contraction and increases vascular permeability, mucus production, and pain. It is particularly active in allergic reactions.

Miscellaneous Inflammatory Mediators

- **Prostaglandins,** produced by most body cells; complex chemical mediators that can have opposing effects (for example, dilation or constriction of blood vessels) and are powerful stimulants of inflammation and pain.
- **Leukotrienes** stimulate the contraction of smooth muscle and enhance vascular permeability. They are implicated in the more severe manifestations of immediate allergies (constriction of airways).
- **Platelet-activating factor,** a substance released from basophils, causes the aggregation of platelets and the release of other chemical mediators during immediate allergic reactions.

Fever is such a prevalent reaction that it is a prominent symptom of hundreds of diseases. For thousands of years, people believed fever was part of an innate protective response. Hippocrates offered the idea that it was the body's attempt to burn off a noxious agent. Sir Thomas Sydenham wrote in the seventeenth century: "Why, fever itself is Nature's instrument!" So widely held was the view that fever could be therapeutic that pyretotherapy (treating disease by inducing an intermittent fever) was once used to treat syphilis, gonorrhea, leishmaniasis (a protozoan infection), and cancer. This attitude fell out of favor when drugs for relieving fever (aspirin) first came into use in the early 1900s, and an adverse view of fever began to dominate.

Changing Views of Fever

In recent times, the medical community has returned to the original concept of fever as more healthful than harmful. Experiments with vertebrates indicate that fever is a universal reaction, even in cold-blooded animals such as lizards and fish. A study with **febrile*** mice and frogs

---

*febrile (fee´-bril) L. *febris*, fever. Feverish.

indicated that fever increases the rate of antibody synthesis. Work with tissue cultures showed that increased temperatures stimulate the activities of T cells and increase the effectiveness of interferon. Artificially infected rabbits and pigs allowed to remain febrile survive at a higher rate than those given suppressant drugs. Fever appears to enhance phagocytosis of staphylococci by neutrophils in guinea pigs and humans.

Hot and Cold: Why Do Chills Accompany Fever?

Fever almost never occurs as a single response; it is usually accompanied by chills. What causes this oddity—that a person flushed with fever periodically feels cold and trembles uncontrollably? The explanation lies in the natural physiological interaction between the thermostat in the hypothalamus and the temperature of the blood. For example, if the thermostat has been set (by pyrogen) at 102°F but the blood temperature is 99°F, the muscles are stimulated to contract involuntarily (shivering) as a means of producing heat. In addition, the vessels in the skin constrict, creating a sensation of cold, and the piloerector muscles in the skin cause "goose bumps" to form.

---

benefits and minimal side effects. All medical experts do agree that high and prolonged fevers, or fevers in patients with cardiovascular disease, seizures, and respiratory ailments, are risky and must be treated immediately with suppressant drugs. The classic therapy for fever is an **antipyretic*** drug such as aspirin or acetaminophen (Tylenol) that lowers the setting of the hypothalamic center and restores normal temperature. Any physical technique that increases heat loss (tepid baths, for example) can also help reduce the core temperature.

PHAGOCYTES: THE EVER-PRESENT BUSYBODIES OF INFLAMMATION AND SPECIFIC IMMUNITY

By any standard, a phagocyte represents an impressive piece of living machinery, meandering through the tissues to seek, capture, and destroy a target. The general activities of phagocytes are

1. to survey the tissue compartments and discover microbes, particulate matter (dust, carbon particles, antigen-antibody complexes), and injured or dead cells;
2. to ingest and eliminate these materials; and
3. to extract immunogenic information (antigens) from foreign matter.

It is generally accepted that all cells have some capacity to engulf materials, but *professional phagocytes* do it for a living. The three main types of phagocytes are neutrophils, monocytes, and macrophages.

---

*antipyretic (an″-tih-py-reh´-tik) L. *anti*, against, and *pyretos*, fever. An agent that relieves fever.

Neutrophils: Granulocytic Phagocytes

As previously stated, neutrophils are general-purpose phagocytes that react early in the inflammatory response to bacteria and other foreign materials and to damaged tissue (see figure 14.16). A common sign of bacterial infection is a high neutrophil count in the blood (neutrophilia), and neutrophils are also a primary component of pus. Eosinophils are attracted to sites of parasitic infections and antigen-antibody reactions, though they play only a minor phagocytic role.

Macrophage: King of the Phagocytes

After emigrating out of the bloodstream into the tissues, monocytes are transformed by various inflammatory mediators into macrophages. This process is marked by an increase in size and by enhanced development of lysosomes and other organelles (**figure 14.18**). At one time, macrophages were classified as either fixed (adherent to tissue) or wandering, but this terminology can be misleading. All macrophages retain the capacity to move about. Whether they reside in a specific organ or wander depends upon their stage of development and the immune stimuli they receive. Specialized macrophages called **histiocytes** migrate to a certain tissue and remain there during their lifespan. Examples are alveolar (lung) macrophages, the Kupffer cells in the liver, Langerhans cells in the skin (**figure 14.19**), and macrophages in the spleen, lymph nodes, bone marrow, kidney, bone, and brain. Other macrophages do not reside permanently in a particular tissue and drift nomadically throughout the RES. Not only are macrophages dynamic scavengers, but they also process foreign substances and prepare them for reactions with B and T lymphocytes (see chapter 15).

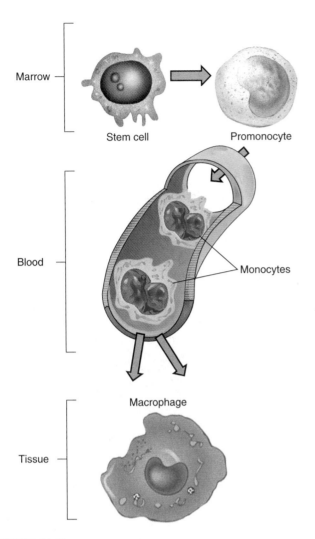

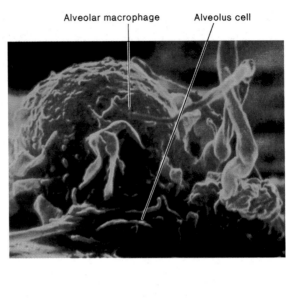

(a)

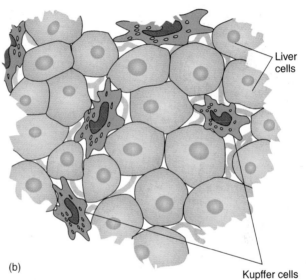

(b)

FIGURE 14.18

The developmental stages of monocytes and macrophages. The cells progress through maturational stages in the bone marrow and peripheral blood. Once in the tissues, a macrophage can remain nomadic or take up residence in a specific organ.

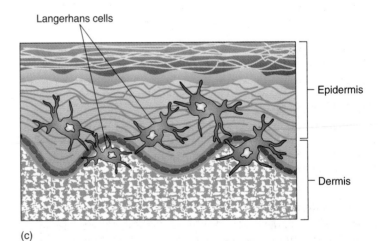

(c)

FIGURE 14.19

Sites containing macrophages. **(a)** Scanning electron micrograph view of a lung with an alveolar macrophage. **(b)** Liver tissue with Kupffer cells. **(c)** Langerhans cells deep in the epidermis.

Mechanisms of Phagocytic Discovery, Engulfment, and Killing

Although the term **phagocytosis** literally means the engulfment of particles by cells, phagocytes actually endocytose both particulate and liquid substances. But phagocytosis is more than just the physical process of engulfment, because phagocytes also actively attack and dismantle foreign cells with a wide array of antimicrobial substances. The events in phagocytosis include chemotaxis, ingestion, phagolysosome formation, destruction, and excretion (**figure 14.20**).

Chemotaxis and Ingestion Phagocytes migrate into a region of inflammation with a deliberate sense of direction, attracted by a gradient of stimulant products from the parasite and host tissue at the site of injury. On the scene of an inflammatory reaction, they often trap cells or debris against the fibrous network of connective tissue or the wall of blood and lymphatic vessels. Phagocytosis is often accompanied by **opsonization** (discussed again in

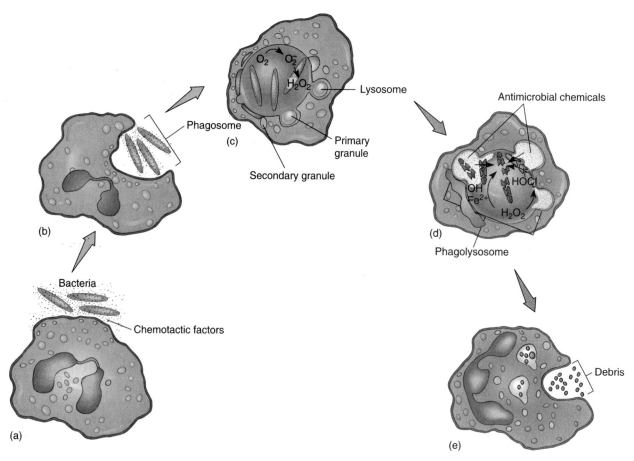

FIGURE 14.20

The phases in phagocytosis. **(a)** Chemotaxis. **(b)** Contact and ingestion (forming a phagosome). **(c)** Formation of phagolysosome (granules fuse with phagosome). **(d)** Killing, digestion of the microbe. **(e)** Release of debris.

chapter 15). This is a process that coats the surface of microorganisms with antibodies or complement, thereby facilitating recognition and engulfment. Once the phagocyte has made contact with its prey, it extends pseudopods that enclose the cells or particles in a pocket and internalize them in a vacuole called a *phagosome.*

Phagolysosome Formation and Killing In a short time, **lysosomes** migrate to the scene of the phagosome and fuse with it to form a **phagolysosome.** Other granules containing antimicrobial chemicals are released into the phagolysosome, forming a potent brew designed to poison and then dismantle the ingested material. The destructiveness of phagocytosis is evident by the death of bacteria within 30 minutes after contacting this battery of antimicrobial substances.

Destruction and Elimination Systems Two separate systems of destructive chemicals await the microbes in the phagolysosome. The oxygen-dependent system elaborates several substances that were described in chapters 7 and 11. Myeloperoxidase, an enzyme found in granulocytes, forms halogen ions

(OCl^-) that are strong oxidizing agents. Other products of oxygen metabolism such as hydrogen peroxide, the superoxide anion (O_2^-), activated or so-called singlet oxygen $(^1O_2)$, and the hydroxyl free radical $(.OH)$ separately and together have formidable killing power. Other mechanisms that come into play are the liberation of lactic acid, lysozyme, and *nitric oxide* (NO), a powerful mediator that kills bacteria and inhibits viral replication. Cationic proteins that injure bacterial cell membranes and a number of proteolytic and other hydrolytic enzymes complete the job. The small bits of undigestible debris are released from the macrophage by exocytosis.

CONTRIBUTORS TO THE BODY'S CHEMICAL IMMUNITY

Interferon: Antiviral Cytokines and Immune Stimulants

Interferon (IFN) was described in chapter 12 as a small protein produced naturally by certain white blood and tissue cells that is used in therapy against certain viral infections and cancer. Although the interferon system was originally thought to be directed exclusively against viruses, it is now known to be involved also in defenses against other microbes and in immune regulation and intercommunication. Three major types are *alpha interferon,* a product of lymphocytes and macrophages; *beta interferon,* a product of fibroblasts and epithelial cells; and *gamma interferon,* a product of T cells.

All three classes of interferon are produced in response to viruses, RNA, immune products, and various antigens. Their biological activities are extensive. In all cases, they bind to cell surfaces and induce changes in genetic expression, but the exact results vary. In addition to antiviral effects discussed in the next section, all three IFNs can inhibit the expression of cancer genes and have tumor suppressor effects. Alpha and beta IFN stimulate phagocytes, and gamma IFN is an immune regulator of macrophages and T and B cells.

Characteristics of Antiviral Interferon

When a virus binds to the receptors on a host cell, a signal is sent to the nucleus that directs the cell to synthesize interferon. After transcribing and translating the interferon gene, newly synthesized interferon molecules are rapidly secreted by the cell into the extracellular space, where it binds to interferon receptors on other cells. The binding of interferon to a second cell induces the production of proteins in that cell that inhibit viral multiplication either by degrading the viral RNA or by preventing the translation of viral proteins **(figure 14.21).** Interferon is not virus-specific, so its synthesis in response to one type of virus will also protect against other types. Because this protein is the direct inhibitor of viruses, it has been a valuable treatment for a number of virus infections.

Other Roles of Interferon

Interferons are also important immune regulatory cytokines that activate or instruct the development of white blood cells. For example, alpha interferon produced by T lymphocytes activates a subset of cells called natural killer (NK) cells. In addition, one type of beta interferon plays a role in the maturation of B and T lymphocytes and in inflammation. Gamma interferon inhibits cancer cells, stimulates B lymphocytes, activates macrophages, and enhances the effectiveness of phagocytosis.

Complement: A Versatile Backup System

Among its many overlapping functions, the immune system has another complex and multiple-duty system called **complement (C factor)** that, like inflammation and phagocytosis, is brought into play at several levels. The complement system, named for its property of "complementing" immune reactions, consists of at least 26 blood proteins that work in concert to destroy bacteria and certain viruses. The sources of complement factors are liver hepatocytes, lymphocytes, and monocytes. Some knowledge of this important system will help in your understanding of topics in chapters 15 and 16.

The concept of a cascade reaction is helpful in understanding how complement functions. A cascade reaction is a sequential physiological response like that of blood clotting, in which the first substance in a chemical series activates the next substance, which activates the next, and so on, until a desired end product is reached. Complement exhibits two schemes, the *classical pathway* and the *alternative pathway,* which differ in how they are activated and in speed and efficiency. The two pathways merge at the final stages into a common pathway with a similar end result **(figure 14.22).**

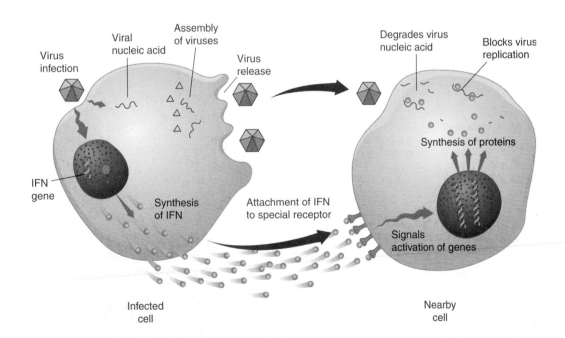

FIGURE 14.21

The antiviral activity of interferon. When a cell is infected by a virus, its nucleus is triggered to transcribe and translate the interferon (IFN) gene. Interferon diffuses out of the cell and binds to IFN receptors on nearby uninfected cells where it induces production of proteins which eliminate viral genes and block viral replication. Note that the original cell is not protected by IFN and that IFN does not prevent viruses from invading the protected cells.

(a) **Initiation.** The classical pathway begins when C1 components bind to receptors on a foreign cell membrane.

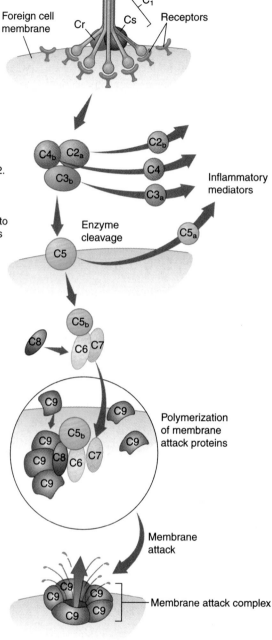

(b) **Amplification and cascade.** The C1 complex is an enzyme that activates a second series of components, C4 and C2. When these have been enzymatically cleaved into separate molecules, they become a second enzyme complex that activates C3. At this same site, C3 binds to C5 and cleaves it to form a product that is tightly bound to the membrane.

(c) **Polymerization.** $C5_b$ is a reactive site for the final assembly of an attack complex. In series, C6, C7, and C8 aggregate with C5 and become integrated into the membrane. They form a substrate upon which the final component, C9 can bind. Up to 15 of these C9 units ring the central core of the complex.

(d) **Membrane attack.** The final product of these reactions is a large, donut-shaped enzyme that punctures small pores through the membrane, leading to cell lysis.

(e) An electron micrograph (187,000$\times$) of a cell reveals multiple puncture sites over its surface. The lighter, ringlike structures are the actual enzyme complex.

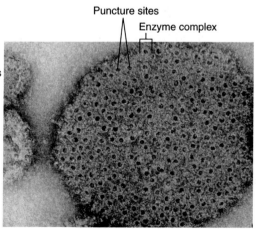

FIGURE 14.22

Steps in the classical complement pathway at a single site. See text for more description.

Since the complement numbers (C1–C9) are based on the order of their discovery, be aware that factors C1–C4 do not appear in numerical order during activation **(table 14.1).**

Overall Stages in the Complement Cascade   In general, the complement cascade includes the three stages of *initiation, amplification and cascade,* and *membrane attack.* At the outset, an initiator (such as microbes, cytokines, and antibodies; **table 14.2**) reacts with the first complement chemical, which propels the reaction on its course. This process develops a recognition site on the surface of the target cell where the initial C components will bind. Through a stepwise series, each component reacts with another on or near the recognition site. In the C2–C5 series, enzymatic cleavage produces several inflammatory cytokines. Other

details of the pathways differ, but whether classical or alternative, the functioning end product is a large ring-shaped protein termed the *membrane attack complex.* This complex can digest holes in the cell membranes of bacteria, cells, and enveloped viruses, thereby destroying them (figure 14.22*a–e*). The two pathways are described in more detail in **Microbits 14.4.**

CHAPTER CHECKPOINTS

Nonspecific immune reactions are generalized responses to invasion, regardless of the type. These include inflammation, phagocytosis, interferon, and complement.

The four symptoms of inflammation are rubor (redness), calor (heat), tumor (edema), and dolor (pain).

Fever is another component of nonspecific immunity. It is caused by both endogenous and exogenous pyrogens. Fever increases the rapidity of the host immune responses and reduces the viability of many microbial invaders.

Macrophages are activated monocytes. Along with neutrophils (PMNs), they are the key phagocytic agents of nonspecific response to disease.

The plasma contains complement, a nonspecific group of chemicals that works with the third line of defense to attack foreign cells.

TABLE 14.1

Complement Proteins

Pathway	Component
Classical	C1q
(Initial Portion)	C1r
Rapid and efficient	C1s
	C4
	C2
	C3
Lectin	**Membrane Attack Components**
(Proteins That	**(Common to Both Pathways)**
Bind Mannans)	C5
	C6
	C7
	C8
	C9
Alternative	Properdin
or Properdin	Factor B
(Initial Portion)	Factor D
Slower and	Factor C3b
less efficient	and Mg

TABLE 14.2

Substances That Activate the Complement Pathways

Activators in the Classical Pathways	Activators in the Alternative Pathway
Complement-fixing antibodies: IgG, IgM	Cell wall components; e.g., yeast and bacteria
Bacterial lipopolysaccharide	Viruses; e.g., influenza virus
Pneumococcal C-reactive protein	Parasites; e.g., *Schistosoma*
Retroviruses	Fungi; e.g., *Cryptococcus*
Polynucleotides	Some tumor cells
Mitochondrial membranes	X-ray opaque media, dialysis membranes

Specific Immunities: The Third and Final Line of Defense

When host barriers and nonspecific defenses fail to control an infectious agent, a person with a normal functioning immune system has an extremely substantial mechanism to resist the pathogen—the third, specific line of immunity. This aspect of immunity is the resistance developed after contracting childhood ailments such as chickenpox or measles that provides long-term protection against future attacks. This sort of immunity is not innate, but adaptive; it is acquired only after an immunizing event such as an infection. The absolute need for acquired or adaptive immunity is impressively documented in children who have genetic defects in this system or in AIDS patients who have lost it. Even with heroic measures to isolate the patient, combat infection, or restore lymphoid tissue, the victim is constantly vulnerable to life-threatening infections.

Acquired specific immunity is the product of a dual system that we have previously mentioned—the B and T lymphocytes. During fetal development, these lymphocytes undergo a selective process that specializes them for reacting only to one specific antigen. During this time, **immunocompetence,** the ability of the body to react with myriad foreign substances, develops. An infant is born with the theoretical potential to acquire millions of different immunities.

Two features that most characterize this third line of defense are specificity and **memory.** Unlike mechanisms such as anatomical barriers or phagocytosis, acquired immunity is highly selective.

For example, the antibodies produced during an infection against the chickenpox virus will function against that virus and not against the measles virus (**figure 14.23*a***). The property of memory pertains to the rapid mobilization of lymphocytes that have been programmed to "recall" their first engagement with the invader and rush to the attack once again (figure 14.23*b*). This complex and fascinating response will be covered more extensively in chapter 15.

CHAPTER CHECKPOINTS

B and T lymphocytes are the agents of the specific immune response. Unlike the nonspecific responses, it develops after exposure to a specific antigen and is tailor-made to it. In addition, the B and T cells "remember" the antigen and respond much more rapidly on subsequent exposures.

(a) Specificity: Viruses and other infectious agents contain antigen molecules that are specific to a single type of lymphocyte. One result of binding will be the production of virus-specific antibodies.

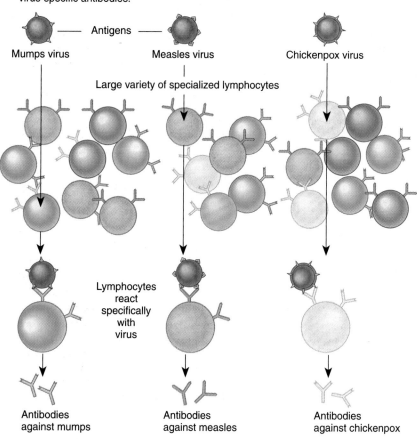

(b) Memory: First contact with antigen creates a unique programmed memory cell that provides quick recall upon second and other future contacts with that antigen.

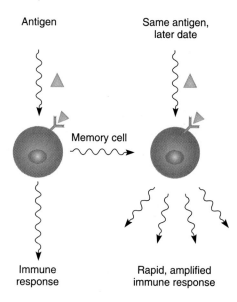

FIGURE 14.23

The characteristics of acquired immunity: **(a)** specificity and **(b)** memory.

MICROBITS 14.4

How Complement Works

The **classical pathway** is a part of the specific immune response covered in chapter 15. It is initiated either by the foreign cell membrane of a parasite or a surface antibody. The first chemical, C1, is a large complex of three molecules, C1q, C1r, and C1s (see table 14.1). When the C1q subunit has recognized and bound to surface receptors on the membrane, the C1r subunit cleaves the C1s proenzyme, and an activated enzyme emerges. During amplification, the C1s enzyme has as its primary targets proenzymes C4 and C2. Through the enzyme's action, C4 is converted into C4a and C4b, and C2 is converted into C2a and C2b. C4b and C2a fragments remain attached as an enzyme, C3 convertase, whose substrate is factor C3. The cleaving of C3 yields subunits C3a and C3b. C3b has the property of binding strongly with the cell membrane in close association with the component C5, and it also forms an enzyme complex with C4b–C2b that converts C5 into two fragments, C5a and C5b. C5b will form the nucleus for the membrane attack complex. This is the point at which the two pathways merge. From this point on, C5b reacts with C6 and C7 to form a stable complex inserted in the membrane. Addition of C8 to the complex causes the polymerization of several C9 molecules into a giant ring-shaped membrane attack complex that bores ring-shaped holes in the membrane. If the target is a cell, this reaction causes it to disintegrate (figure 14.22d,e). If the target is an enveloped virus, the envelope is perforated and the virus inactivated.

The **alternative pathway,** sometimes called the **properdin* pathway,** is not specific to a particular microbe. It can be initiated by a wide variety of microbes, tumors, and cell walls. It requires a different group of serum proteins—factors B, D, and P (properdin), C3b, and magnesium in the initiation and amplification phases rather than C1, C2, or C4 components (see table 14.1). The remainder of the steps occur as in the classical pathway. The principal function of the alternative system is to provide a slower but less specific means of lysing foreign cells (especially gram-negative bacteria) and viruses.

You will notice that at many of the steps of the classical pathway, two molecules are given off. One of these continues in the formation of the membrane attack complex, and the other (C2a, C4a, C3a, or C5a) goes on to become a cytokine or stimulant of inflammation and other immune reactions. Complement components also behave as one type of opsonin that promotes phagocytosis. Complement can participate in inflammation and allergy by causing liberation of vasoactive substances from mast cells and basophils. C3a and C5a are so potent in this response that injecting only one quadrillionth of a gram elicits an immediate flare-up at the site.

*properdin (proh'-pur-din) L. *pro,* before, and *perdere,* to destroy.

CHAPTER CAPSULE WITH KEY TERMS

I. **The Three Levels of Host Defenses**
   A. The *first line of defense* is an inborn, nonspecific system composed of anatomical, chemical, and genetic barriers that block microbes at the portal of entry. The *second line of defense* is also inborn and nonspecific and includes protective cells and fluids in tissues, and the *third line of defense* is acquired and specific and is dependent on the function of T and B cells.

II. **Immunity/Immunology**
   A. **Immunology** is the study of **immunity,** which in turn refers to the development of resistance to infectious agents by the body. Immunity is marked by white blood cells (WBCs), which conduct surveillance of the body, recognizing and differentiating **self** from **nonself** (foreign) cells by virtue of the **markers** they carry on their surface. Under normal conditions self cells are left alone, but nonself cells are destroyed.
   B. For purposes of immunologic study, the body is divided into three compartments: the blood lymphatics, and **reticuloendothelial system.** A fourth component, the extracellular fluid, surrounds the first three and allows constant communication between all areas of the body.

III. **Circulatory System: Blood and Lymphatics**
   A. **Whole blood** consists of **blood cells** (formed by **hemopoiesis** in the bone marrow) dispersed in **plasma. Stem cells** in the bone marrow differentiate to produce white blood cells **(leukocytes),** red blood cells **(erythrocytes),** and megakaryocytes, which give rise to **platelets.**
   B. Leukocytes are the primary mediators of immune function and can be divided into **granulocytes (neutrophils, eosinophils,** and **basophils)** and **agranulocytes (monocytes** and **lymphocytes)** based on their appearance. Monocytes later differentiate to become **macrophages.** Lymphocytes are divided into **B cells,** which produce **antibodies** as part of **humoral immunity,** and **T cells,** which participate in **cell-mediated immunity.** Leukocytes display both **chemotaxis** and **diapedesis** in response to chemical mediators of the immune system.

IV. **Lymphatic System**
   A. The **lymphatic system** parallels the circulatory system and transports lymph while also playing host to cells of the immune system. Lymphoid organs and tissues include lymph nodes, the spleen, **thymus,** as well as areas of less well organized immune tissues such as **GALT,** SALT, MALT, and BALT.

V. **Generalized Immune Reactions**
   A. The inflammatory response is a complex reaction to tissue injury marked by *redness, heat, swelling,* and *pain.*
      1. Blood vessels narrow and then dilate in response to **chemical mediators** and **cytokines.**
      2. **Edema** swells tissues, helping prevent the spread of infection.
      3. WBCs, microbes, debris, and fluid collect to form **pus.**
      4. **Pyrogens** may induce fever.
      5. Macrophages and neutrophils engage in **phagocytosis,** engulfing microbes in a phagosome. Uniting the phagosome with a **lysosome** results in destruction of the phagosome contents.

VI. **Chemical Defenses**
   A. **Interferon (IFN)** is a family of proteins produced by leukocytes and fibroblasts that inhibit the reproduction of viruses by degrading viral RNA or blocking the synthesis of viral proteins.
   B. **Complement** is a complex defense system that results, by way of a cascade mechanism, in the formation of a membrane attack complex that kills cells by creating holes in their membranes.

VII. **Characteristics of Acquired Immunities**
   A. **Immunocompetent** individuals possess a third line of defense that is acquired only after direct exposure to an infectious agent. This type of immunity is mediated by **T** and **B lymphocytes** and is marked by being extremely **specific** and creating **immunologic memory.**

MULTIPLE-CHOICE QUESTIONS

1. An example of a nonspecific chemical barrier to infection is
   a. unbroken skin
   b. lysozyme in saliva
   c. cilia in respiratory tract
   d. all of these

2. Which nonspecific host defense is associated with the trachea?
   a. lacrimation
   b. ciliary lining
   c. desquamation
   d. lactic acid

3. Which of the following blood cells function primarily as phagocytes?
   a. eosinophils
   b. basophils
   c. lymphocytes
   d. neutrophils

4. Which of the following is not a lymphoid tissue?
   a. spleen
   b. thyroid gland
   c. lymph nodes
   d. GALT

5. What is included in GALT?
   a. thymus
   b. Peyer's patches
   c. tonsils
   d. breast lymph nodes

6. Monocytes are ____ leukocytes that develop into ____.
   a. granular, phagocytes
   b. agranular, mast cells
   c. agranular, macrophages
   d. granular, T cells

7. Which of the following inflammatory signs specifies pain?
   a. tumor
   b. dolor
   c. calor
   d. rubor

8. An example of an inflammatory mediator that stimulates vasodilation is
   a. histamine
   b. collagen
   c. complement C5a
   d. interferon

9. ____ is an example of an inflammatory mediator that stimulates chemotaxis.
   a. Endotoxin
   b. Serotonin
   c. Fibrin clot
   d. Interleukin-2

10. An example of an exogenous pyrogen is
   a. interleukin-1
   b. complement
   c. interferon
   d. endotoxin

11. ____ interferon is secreted by ____ and is involved in destroying viruses.
   a. Gamma, fibroblasts
   b. Beta, lymphocytes
   c. Alpha, natural killer cells
   d. Beta, fibroblasts

12. Which of the following substances is *not* produced by phagocytes to destroy engulfed microorganisms?
   a. hydroxyl radicals
   b. superoxide anion
   c. hydrogen peroxide
   d. bradykinin

13. Which of the following is the end product of the complement system?
   a. properdin
   b. cascade reaction
   c. membrane attack complex
   d. complement factor C9

14. Immunologic memory refers to the ability of the immune system to
   a. recognize millions of different antigens
   b. react with millions of different antigens
   c. migrate from the blood vessels into the tissues
   d. recall a previous immune response

15. Which subset of WBCs accounts for acquired, specific immunity?
   a. monocytes
   b. B cells
   c. T cells
   d. both b and c

CONCEPT QUESTIONS

1. a. Explain the functions of the three lines of defense.
   b. Which is the most essential to survival?
   c. What is the difference between nonspecific host defenses and immune responses?

2. a. Use the pointers to describe the major components of the first line of defense.
   b. What effects do these defenses have on microbes?

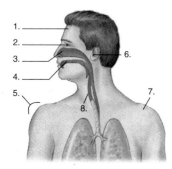

3. a. Describe the main elements of the process through which the immune system distinguishes self from nonself.
   b. How is surveillance of the tissues carried out?
   c. What is responsible for it?
   d. What does the term *foreign* mean in reference to the immune system?

4. a. Trace the complete cycle of a bacterium through the immune compartments, starting with the blood, the RES, the lymphatics, and the ECF. (You will start and end at the same point.)
   b. Describe the direct connection between the bloodstream and the lymphatic circulation?

5. a. What are the main components of the reticuloendothelial system?
   b. Why is it also called the mononuclear phagocyte system?
   c. How does it communicate with tissues and the vascular system?

6. a. Prepare a simplified outline of the cell lines of hemopoiesis.
   b. What is a stem cell?
   c. Review the anatomical locations of hemopoiesis at various stages of development.

7. a. Differentiate between granulocytes and agranulocytes.
   b. Describe the main cell types in each group, their functions, and their incidence in the circulation.

8. a. What is the principal function of lymphocytes?
   b. Differentiate between the two lymphocyte types and between humoral and cell-mediated immunity.

9. a. Why are platelets considered formed elements and not cells?
   b. What are their functions?
   c. How are they associated with inflammation?

10. a. Explain the processes of diapedesis and chemotaxis, and show how they interrelate.
    b. Referring to question 4, explain how these processes can account for the movements of leukocytes through the fluid compartments.

11. a. What is lymph, and how is it formed?
    b. Why are white cells but not red cells normally found in it?
    c. What are the functions of the lymphatic system?
    d. Explain the filtering action of a lymph node.
    e. What is GALT, and what are its functions?

12. Differentiate between the terms *pyogenic* and *pyrogenic*. Use examples to explain the impact they have on the immune response.

13. Briefly account for the origins and actions of the major types of inflammatory mediators and cytokines.

14. a. Describe the events that give rise to macrophages.
    b. What types of macrophages are there, and what are their principal functions?

15. a. Outline the major phases of phagocytosis.
    b. In what ways is a phagocyte a tiny container of disinfectants?

16. a. Briefly describe the three major types of interferon, their sources, and their biological effects.
    b. Describe the mechanism by which interferon acts as an antiviral compound.

17. a. Describe the general complement reaction in terms of a cascade.
    b. Briefly outline what leads to the result shown here and the action of this structure. What is it called?
    c. What are some other functions of complement components?

18. What are immunocompetence, immunologic specificity, and immunologic memory?

CRITICAL-THINKING QUESTIONS

1. Suggest some reasons that there is so much redundancy of action and there are so many interacting aspects of immune responses.

2. a. What are some possible elements missing in children born without a functioning lymphocyte system?
   b. What is the most important component extracted in bone marrow transplants?

3. a. What is the likelihood that plants have some sort of immune protection?
   b. Explain your reasoning.

4. A patient's chart shows an increase in eosinophil levels.
   a. What does this cause you to suspect?
   b. What does it mean if the basophil levels are very high?
   c. What if the neutrophil levels are very high?

5. How can adults continue to function relatively normally after surgery to remove the thymus, tonsils, spleen, or lymph nodes?

6. a. What actions of the inflammatory and immune defenses account for swollen lymph nodes and leukocytosis?
   b. What is pus, and what does it indicate?
   c. In what ways can edema be beneficial?
   d. In what ways is it harmful?

7. An obsolete treatment for syphilis involved inducing fever by deliberately infecting patients with the agent of relapsing fever. A recent but failed AIDS treatment involved applying heat to the body to induce hyperthermia. Can you provide some possible explanations behind these peculiar forms of treatment?

8. Patients with a history of tuberculosis often show scars in the lungs and experience recurrent infection. Account for these effects on the basis of the inflammatory response.

9. Suggest some reasons that our complement system generally does not react against our own cell membranes (especially the alternative pathway).

10. *Shigella, Mycobacterium,* and numerous other pathogens have developed mechanisms that prevent them from being killed by phagocytes.
    a. Suggest two or three factors that help them avoid destruction by the powerful antiseptics in macrophages.
    b. In addition, suggest the potential implications that these infected macrophages can have in the development of disease.

11. Macrophages perform the final job of removing tissue debris and other products of infection. Indicate some of the possible effects when these scavengers cannot successfully complete the work of phagocytosis.

12. Account for the several inflammatory symptoms that occur in the injection site when one has been vaccinated against influenza and tetanus.

13. a. Knowing that fever is potentially both harmful and beneficial, what are some possible guidelines for deciding whether to suppress it or not?
    b. What is the specific target organ of a fever-suppressing drug?

INTERNET SEARCH TOPICS

1. Find a website that deals with the uses of interferon in therapy. Name viral infections and cancers it is used to treat. Are there any adverse side effects?

2. Visit the student Online Learning Center at www.mhhe.com/talaro5. Go to chapter 14, Internet Search Topics, and log on to the available websites to:

   a. Observe movies showing chemotaxis and phagocytosis; or as an alternate, type the terms "chemotaxis" and "phagocytosis" plus "animation" or "movie" into a search engine.
   b. Review a good overview of basic immune function.

3. Use a search engine to locate information on dendritic cells. Outline their origins, distribution, and functions. How are they involved in HIV infection?

Host Defenses - Non-specific

Is not targeting a specific organism

A- Inflammation (wound or injury site)

B- Fever (↑ temp)

C- Skin

D- stomach

E- ciliated cells in resp. sys. flow of urine

      Host Defenses - Specific

Is referred to as aquired

Important cell for CMI?

      memory cells!

Targets for CMI (cell mediated immun.)

Protozoa

Fungi

virus infected cells

Intracellular bacteria

   Tumor cells

Clonal selection theory - B-cells

B-cell is stimulated by specific antigen
by binding to surface receptor (IgD)

clones
Production
of memory B-cells
to specific to
Ag (Z)

clones of
plasma cells
which make
IgM anti(Z)

which ever he gives

O - cell wall somatic

H - Flagellar

K - capsular

The Acquisition of Specific Immunity and Its Applications

The primary focus of this chapter is the remarkable system of lymphocytes that are responsible for specific, acquired immunities. In our more detailed examination, we will discover the complex adaptations for defense against microbes, cancer, and toxic substances. This background will prepare you for coverage of vaccination, immune testing, allergy, and immune deficiency in chapters 16 and 17.

Chapter Overview

- The most specific host defenses are derived from a dual system of lymphocytes that are genetically programmed to react with foreign substances (antigens) found in microbes and other organisms.
- These cells carry glycoprotein receptors that dictate their specificity and reactivity.
- B lymphocytes have antibody receptors, T lymphocytes have T-cell receptors, and macrophages have histocompatibility receptors such as MHC and HLA.
- B cells and T cells arise in the bone marrow, where they proliferate and develop extreme variations in the expression of receptor genes.
- Differentiation of lymphocytes creates billions of genetically different clones that each have a unique specificity for antigen.
- The B cells reach final maturity in special bone marrow sites, and the T cells reach final maturity in the thymus gland.
- Both types of lymphocytes home (migrate) to separate sites in lymphoid tissue where they serve as a constant source of immune cells primed to respond to their correct antigen.
- Antigens are foreign cells, viruses, and molecules that meet a required size and complexity. They are capable of triggering immune reactions by lymphocytes.
- The B and T cells react with antigens through a complex series of cooperative events that involve the presentation of antigens by macrophages and the assistance of helper T cells and cytokine stimulants (interleukins).
- B cells activated by antigen enter the cell cycle and mitosis, giving rise to plasma cells that secrete antibodies (humoral immunity) and long-lived memory cells.
- Antibodies have binding sites that affix tightly to an antigen and hold it in place for agglutination, opsonization, complement fixation, and neutralization.
- The amount of antibodies increases during the initial contact with antigen and rises rapidly during subsequent exposures due to memory cells ready for immediate reactions.

An immune system huddle. The large central dendritic cell (blue) communicates with a team of smaller T lymphocytes, giving the signal for the next play. From this deceptively simple interaction springs the massive power of the immune system.

- T cells have various receptors that signal their ability to respond as helper cells, suppressor cells, and cytotoxic cells that kill complex pathogens and cancer cells.
- Acquired immunities fall into the categories of natural, artificial, active, and passive.

Further Explorations into the Immune System

In chapter 14 we described the capacity of the immune system to survey, recognize, and react to foreign cells and molecules, and we overviewed the characteristics of nonspecific host defenses, blood cells, phagocytosis, inflammation, and complement. In addition, we introduced the concepts of acquired immunity and specificity. In this chapter we take a closer look at those topics.

The elegance and complexity of immune function are largely due to lymphocytes working closely together with macrophages. To simplify and clarify the network of immunologic development and interaction, we present it here as a series of five stages, with each stage covered in a separate section (**figure 15.1**). The principal stages include these:

  I. lymphocyte development and differentiation;
 II. the presentation of antigens;
III. the challenge of B and T lymphocytes by antigens;
 IV. B lymphocytes and the production and activities of antibodies; and
  V. T-lymphocyte responses.

Notice that as each stage is covered, we will simultaneously be following the sequence of an immune response.

The Dual Nature of Specific Immune Responses

In the following overview, color coded Roman numerals correspond with the subsequent text sections that cover these topics in more detail.

I. DEVELOPMENT OF THE DUAL LYMPHOCYTE SYSTEM

Lymphocytes are central to immune responsiveness. They undergo a sequential development that begins in the embryonic yolk sac and shifts to the liver and bone marrow. Although all lymphocytes arise from the same basic stem cell type, at some point in development, they diverge into two distinct types. Final maturation of B cells occurs in specialized bone marrow sites and that of T cells occurs in the thymus. This process commits each individual B cell or T cell to one specificity. Both cell types subsequently migrate to precise, separate areas in the lymphoid organs (for instance, nodes and spleen, as described in chapter 14) to serve as a lifelong continuous source of immune responsiveness.

II. ENTRANCE AND PRESENTATION OF ANTIGENS AND CLONAL SELECTION

When foreign cells, or antigens, enter a fluid compartment of the body, they are immediately swept up in an interconnected network of the lymphatics, blood, and the reticuloendothelial system (RES). In these sites the antigenic materials are met by a battery of cells that work together to screen, entrap, and eliminate them. Certain specialized phagocytes are usually the first cells to recognize and react. They ingest and process antigens and present them to the lymphocytes that are specific for that antigen. This selection process is the trigger that activates the lymphocytes. In most cases, the response of B cells also requires the additional assistance of special classes of T cells called helper T cells.

III. B. AND III. T. ACTIVATION OF LYMPHOCYTES AND CLONAL EXPANSION

When challenged by antigen, both B cells and T cells further differentiate and proliferate. The multiplication of a particular lymphocyte creates a clone, or group of genetically identical cells, some of which are memory cells that will ensure future reactiveness against that antigen. Because the B-cell and T-cell responses depart notably from this point in the sequence, they will be summarized separately.

IV. PRODUCTS OF B LYMPHOCYTES: ANTIBODY STRUCTURE AND FUNCTIONS

The active progeny of a dividing B-cell clone are called plasma cells. These cells are programmed to synthesize and secrete antibodies into the tissue fluid. When these antibodies attach to the antigen for which they are specific, the antigen is marked for destruction or neutralization. Because secreted antibody molecules circulate freely in the tissue fluids, lymph, and blood, the immunity they provide is humoral.

V. HOW T CELLS RESPOND TO ANTIGEN: CELL-MEDIATED IMMUNITY (CMI)

T-cell types and responses are extremely varied. When activated (sensitized) by antigen, a T cell gives rise to one of several types of progeny, each involved in a cell-mediated immune function. The four main functional types of T cells are:

1. helper cells that assist in immune reactions;
2. suppressor cells that suppress and regulate immune reactions;
3. cytotoxic, or killer, cells that destroy specific target cells; and
4. delayed hypersensitivity cells that function in certain allergic reactions.

Although T cells secrete cytokines that help destroy antigen or regulate immune responses, they do not produce antibodies.

CHAPTER CHECKPOINTS

Acquired specific immunity is an elegant but complex matrix of interrelationships between lymphocytes and macrophages consisting of several stages.

Stage I. Lymphocytes originate in hemopoietic tissue but go on to diverge into two distinct types: B cells, which produce antibody, and T cells, which produce cytokines that initiate and coordinate the entire immune response.

Stage II. Antigen-presenting cells detect invading foreign antigens and present them to lymphocytes, which recognize the antigen and initiate the specific immune response.

Stage III. Lymphocytes proliferate, producing clones of progeny that include groups of responder cells and memory cells.

Stage IV. Activated B lymphocytes become plasma cells that produce and secrete large quantities of antibodies.

Stage V. Activated T lymphocytes differentiate into one of four subtypes, which regulate and participate directly in the specific immune responses.

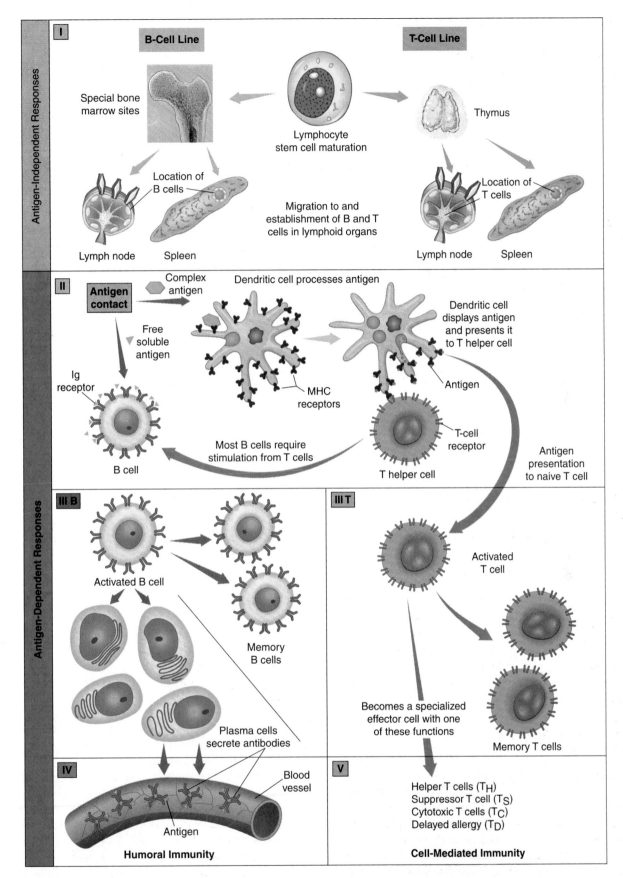

FIGURE 15.1

Overview of the stages of lymphocyte development and function. **I.** Development of B- and T-lymphocyte specificity and migration to lymphoid organs. **II.** Antigen processing by dendritic cell and presentation to lymphocytes; assistance to B cells by T cells. **III B** and **III T.** Lymphocyte activation, clonal expansion, and formation of memory B and T cells. **IV.** Humoral immunity, B-cell line produces antibodies to react with the original antigen. **V.** Cell-mediated immunity. Activated T cells perform various functions, depending on the signal and type of antigen. Details of these processes are covered in each corresponding section heading.

Essential Preliminary Concepts for Understanding Immune Reactions of Sections I–V

Before we examine lymphocyte development and function in greater detail, we must initially review concepts such as the unique structure of molecules (especially proteins), the characteristics of cell surfaces (membranes and envelopes), the ways that genes are expressed, and immune recognition and identification of self and nonself. Ultimately, the shape and function of protein receptors and markers protruding from the surfaces of cells are the result of genetic expression, and these molecules are responsible for specific immune recognition and, thus, immune reactions.

MARKERS ON CELL SURFACES INVOLVED IN RECOGNITION OF SELF AND NONSELF

Chapter 14 touched on the fundamental idea that cell markers or receptors confer specificity and identity. A given cell can express several different receptors, each type playing a distinct and significant role in detection, recognition, and cell communication. Major functions of receptors are:

1. to perceive and attach to nonself or foreign molecules (antigens),
2. to promote the recognition of self molecules,
3. to receive and transmit chemical messages among other cells of the system, and
4. to aid in cellular development.

Because of their importance in the immune response, we will concentrate here on the major receptors of lymphocytes and macrophages.

How Are Receptors Formed?

The nature of cell receptors is dictated by which specific elements of the genome are active during development. As a cell matures — be it liver or brain cell, lymphocyte or macrophage—certain genes that code for the cell receptors will be transcribed and translated into protein products with a distinctive shape, specificity, and function. This receptor is modified and packaged by the endoplasmic reticulum and Golgi complex. It is ultimately inserted into the cell membrane so as to be accessible to antigens, other cells, and chemical mediators **(figure 15.2).** Note that the receptors of many cells, including lymphocytes and macrophages, are glycoproteins with additional carbohydrate fragments added.

Major Histocompatibility Complex

One set of genes that codes for human cell receptors is the **major histocompatibility complex (MHC).** This gene complex gives rise to a series of glycoproteins (called MHC antigens) found on all cells except red blood cells. Because these markers were first identified in humans on the surface of white blood cells, the MHC is also known as the **human leukocyte antigen (HLA)** system. This receptor complex plays a vital role in recognition of self by the immune system and in rejection of foreign tissue. Genes that regulate and code for the MHC of humans are located on the sixth chromosome, clustered in a multigene complex of three subgroups called class I, class II, and class III **(figure 15.3).**

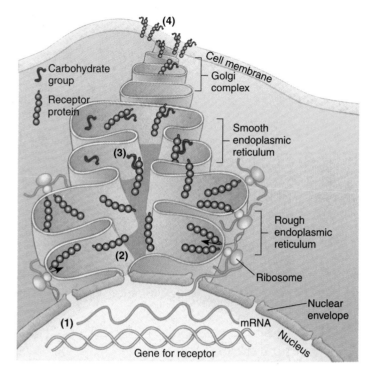

FIGURE 15.2

Receptor formation in a developing cell. **(1)** Gene coding for the receptor is transcribed. **(2)** In the endoplasmic reticulum, mRNA is translated into the protein portion of the receptor. **(3)** A carbohydrate side chain is added, forming glycoprotein. **(4)** The finished receptor is transported to and inserted into the cell membrane.

The functions of the three MHC groups have been identified. Class I genes code for markers that display unique characteristics of self and allow for the recognition of self molecules and the regulation of immune reactions. This class is required for T lymphocytes. The system is rather complicated in its details, but in general, each human being inherits a particular combination of class I MHC (HLA) genes in a relatively predictable fashion. Although millions of different combinations and variations of these genes are possible among humans, the closer the relationship, the greater the probability for similarity in MHC profile (see figure 17.20). Individual differences in the exact inheritance of MHC genes, however, make it highly unlikely that even closely related persons will express an identical MHC profile. This fact introduces an important recurring theme: Although humans are genetically the same species, the cells of each individual express molecules that are foreign (antigenic) to other humans, which is how the term *histo-* (tissue) compatibility (acceptance) originated. This fact necessitates testing for HLA and other antigens when blood is transfused and organs are transplanted (see graft rejection, chapter 17).

Class II MHC genes also code for immune regulatory receptors, but in this case, for receptors that recognize and react with foreign antigens. This system is located primarily on macrophages and B cells, and it is involved in presenting antigens to T cells during cooperative immune reactions.

Unlike the other two classes, which code for molecules that are inserted into cell surfaces, class III MHC genes code for certain secreted complement components such as C2 and C4 (described in chapter 14).

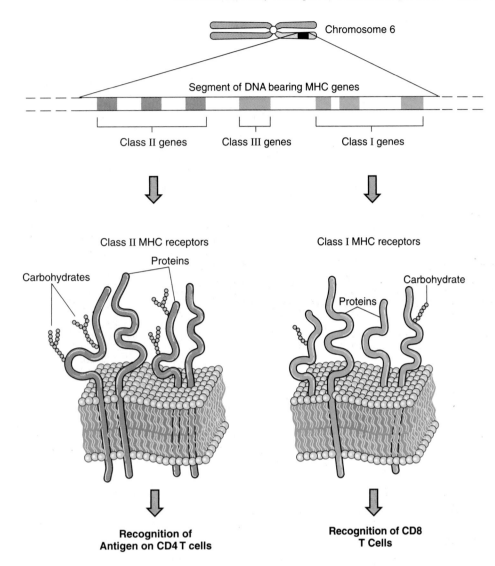

FIGURE 15.3

Glycoprotein receptors of the human major histocompatibility (human leukocyte antigen) gene complex (MHC).

Class I receptors occur on all nucleated cells. Class II receptors occur on white blood cells.

Lymphocyte Receptors and Specificity to Antigen

The part lymphocytes play in immune surveillance and recognition emphasizes the essential role of their receptors. Although they possess MHC antigens for recognizing self, they also carry receptors on their membranes that recognize foreign antigens. Antigen molecules exist in great diversity; there are potentially millions and even billions of unique types. The many sources of antigens include microorganisms as well as an awesome array of chemical compounds in the environment. One of the most fascinating questions in immunology is: How can the lymphocyte receptors be varied to react with such a large number of different antigens? After all, it is generally accepted that there will have to be a different lymphocyte receptor for each unique antigen. Some questions that naturally follow are: How can a cell accommodate enough genetic information to respond to millions or even billions of antigens? When, where, and how does the capacity to distinguish native from foreign tissue arise? To answer these questions, we must first introduce a central theory of immunity.

THE ORIGIN OF DIVERSITY AND SPECIFICITY IN THE IMMUNE RESPONSE

The Clonal Selection Theory and Lymphocyte Development

Research findings have shown that lymphocytes use slightly more than 500 genes to produce the tremendous repertoire of specific receptors they must display for antigens. The most widely-accepted explanation for how this diversity is generated is called the **clonal selection theory.** According to this theory, early undifferentiated lymphocytes in the embryo and fetus undergo a continuous series of divisions and genetic changes that generate hundreds of millions of different cell types, each carrying a particular receptor specificity **(figure 15.4).**

The mechanism, generally true for both B and T cells, can be summarized as follows: As undifferentiated stem cells in the embryo or fetus undergo mitosis, gene segments that code for the receptor

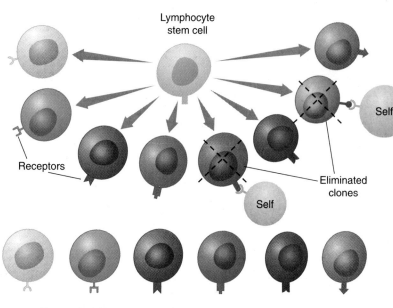

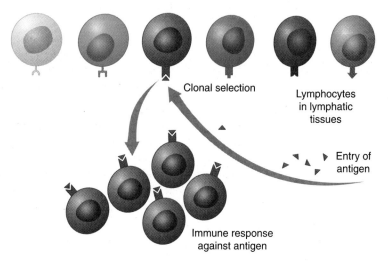

Repertoire of lymphocyte clones, each with unique receptor display

(a) Antigen-Independent Period

1. During development of early lymphocytes from stem cells, a given stem cell undergoes rapid cell division to form numerous progeny.

During this period of cell differentiation, random rearrangements of the genes that code for cell surface protein receptors occur. The result is a large array of genetically distinct cells, called clones, each clone bearing a different receptor that is specific to react with only a single type of foreign molecule or antigen.

2. At this same time, any lymphocyte clones that develop a specificity for self molecules and could be harmful are eliminated or deleted from the pool of diversity. This is called immune tolerance.

3. The specificity for a single antigen molecule is programmed into the lymphocyte and is set for the life of a given clone. The end result is an enormous pool of immature or naive lymphocytes that are ready to further differentiate under the influence of certain organs and immune stimuli.

(b) Antigen-Dependent Period

4. Lymphocytes come to populate the lymphatic organs, where they will finally encounter antigens. These antigens will become the stimulus for the lymphocytes' final activation and immune function. Entry of a specific antigen selects only the lymphocyte clone or clones that carries matching surface receptors. This will trigger an immune response, which varies according to the type of lymphocyte involved.

FIGURE 15.4

Overview of the clonal selection theory of lymphocyte development and diversity. (Simplified for clarity.)

are randomly rearranged so as to create daughter cells that have a different genetic make-up from each other and the original cell (see figure 15.6). In time, every possible recombination occurs, leading to a huge assortment of lymphocytes.[1] Each genetically unique line of lymphocytes arising from these recombinations is termed a **clone.** Be reminded that the genetic code will be expressed as a protein receptor of unique configuration on the surface of the lymphocyte, something like a "sign post" announcing its specificity and reactivity for an antigen. This *proliferative* stage of lymphocyte development does not require the actual presence of foreign antigens, and it is essentially completed within the first year of life.

The second stage of development—*clonal selection and expansion*—does require stimulation by an antigen such as a microbe. When this antigen enters the immune surveillance system, it encoun-

ters specific lymphocytes ready to recognize it. Such contact stimulates that clone to undergo mitotic divisions and expands it into a larger population of lymphocytes all bearing the same specificity. This large increase in the numbers of specific lymphocytes increases the capacity of the immune response to that antigen. Two important generalities one can derive from the clonal selection theory are (1) lymphocyte specificity is preprogrammed, existing in the genetic makeup before an antigen has ever entered the tissues, and (2) each genetically distinct lymphocyte expresses only a single specificity and can react to only one type of antigen. Other important features of the lymphocyte response system will be expanded in later sections.

One potentially problematic outcome of random genetic assortment is the development of clones of lymphocytes able to react to *self.* This outcome could lead to severe damage when the immune system actually perceives self molecules as foreign and mounts a harmful response against the host's tissues. According to a corollary of the clonal selection theory, any such clones are destroyed during

---

1. Estimates of the theoretical number of possible variations that may be created vary from 10^{14} to 10^{18} different specificities.

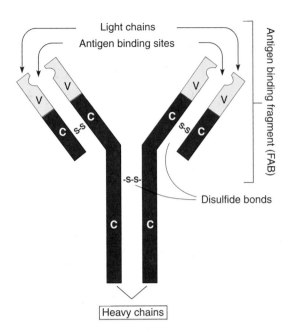

FIGURE 15.5

Simplified structure of an immunoglobulin molecule. The main components are four polypeptide chains—two identical light chains and two identical heavy chains bound by disulfide bonds as shown. Each chain consists of a variable region (V) and a constant region (C). The variable regions of light and heavy chains form a binding site for antigen.

development through clonal *deletion.* The removal of such potentially harmful clones, essential to life, is the basis of **immune tolerance** or **tolerance to self.** Some diseases (autoimmunity) are thought to be caused by the loss of immune tolerance and the survival of certain "forbidden clones" (see chapter 17).

The Specific B-Cell Receptor: An Immunoglobulin Molecule

In the case of B lymphocytes, the receptor genes that undergo the recombination described are those governing **immunoglobulin*** **(Ig)** synthesis. Immunoglobulins are large glycoprotein molecules that serve as the specific receptors of B cells and as antibodies. The basic immunoglobulin molecule is a composite of four polypeptide chains: a pair of identical heavy (H) chains and a pair of identical light (L) chains **(figure 15.5).** One light chain is bonded to one heavy chain, and the two heavy chains are bonded to one another with disulfide bonds, creating a symmetrical, Y-shaped arrangement. The ends of the forks formed by the light and heavy chains contain pockets, called the **antigen binding sites.** It is these sites that can be highly variable in shape to fit a wide range of antigens. This extreme versatility is due to **variable regions (V)** where amino acid composition is highly varied from one clone of B lymphocytes to another. The remainder of the light chains and heavy chains consist of constant regions (C) whose amino acid content does not vary greatly from one antibody to another. Although we will subsequently discuss immunoglobulins and their function as antibodies, for now, we will concentrate on the genetics that explain the origins of lymphocyte specificity.

Development of the Receptors During Lymphocyte Maturation

The genes that code for immunoglobulins lie on three different chromosomes. An undifferentiated lymphocyte has about 150 different genes that code for the variable region of light chains and a total of about 250 genes for the variable and diversity regions (D) of the heavy chains. It has only a few genes for coding for the constant regions and a few for the **joining regions (J)** that join segments of the molecule together. Owing to genetic recombination during development, only the selected (V and D) receptor genes are active in the mature cell, and all the other V and D genes are inactive **(figure 15.6).** This is how the singular specificity of lymphocytes arises.

It is helpful to think of the immunoglobulin genes as blocks of information that code for a polypeptide lying in sequence along a chromosome. During development of each lymphocyte the genetic blocks are independently segregated, randomly selected, and assembled as follows:

- For a heavy chain, a variable region gene and diversity region gene are selected from among the hundreds available and spliced to one joining region gene and one constant region gene.
- For a light chain, one variable, one joining, and one constant gene are spliced together.
- After transcription and translation of each gene complex into a polypeptide, a heavy chain combines with a light chain to form half an immunoglobulin; two of these combine to form a completed monomer (figure 15.6).

Once synthesized, the immunoglobulin product is transported to the cell membrane and inserted there to act as a receptor that expresses the specificity of that cell and to react with an antigen as shown in figure 15.2. The first receptor on most B cells is a small form of IgM, and mature B cells carry IgD receptors (see table 15.2, page 461). It is notable that for each lymphocyte, the genes that were selected for the variable region, and thus for its specificity, will be locked in for the rest of the life of that lymphocyte and its progeny. We will discuss the different classes of Ig molecules further in section IV.

T-Cell Receptors

The T-cell receptor for antigen belongs to the same protein family as the B-cell receptor. It is similar to B cells in being formed by genetic modification, having variable and constant regions, being inserted into the membrane, and having an antigen binding site formed from two parallel polypeptide chains **(figure 15.7).** Unlike the immunoglobulins, the T-cell receptor is relatively small and does not appear to have a humoral function. The several other receptors of T cells are described in a later section.

CHAPTER CHECKPOINTS

The surfaces of all cell membranes contain identity markers known as protein receptors. These cell markers function in identification, communication, and cell development. They are also self identity markers.

Human cell markers are genetically determined by MHCs. MHC I codes for identity markers on all host cells. MHC II codes for immune receptors on macrophages and B cells. MHC III codes for secretion of complement C2 and C4.

***immunoglobulin** (im″-yoo-noh-glahb′-yoo-lin) The technical name for an antibody.

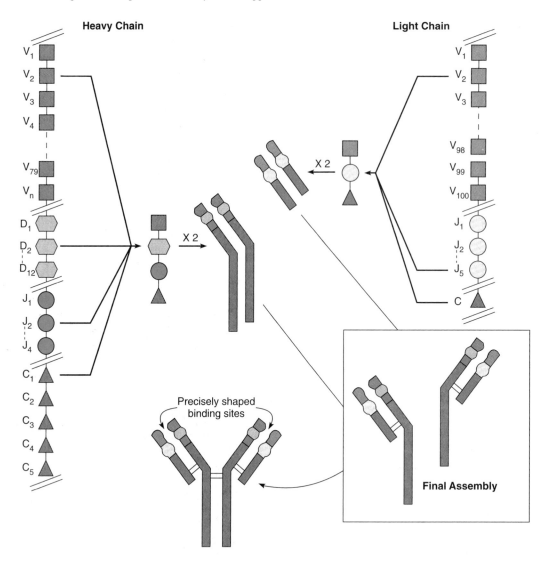

FIGURE 15.6

A simplified look at immunoglobulin genetics. The phenomenon of B-cell differentiation could be compared to a "cutting and pasting" process, in which the final gene that codes for a heavy or light chain is assembled by splicing blocks of genetic material from several regions. *(Left)* The heavy-chain gene is composed of genes from four separate segments (V, D, J, and C) that are transcribed and translated to form a single polypeptide chain. *(Right)* The light-chain genes are put together like heavy ones, except that the final gene is spliced from three gene groups (V, J, and C). During final assembly, first the heavy and light chains are bound, and then the heavy-light combinations are connected to form the immunoglobulin molecule.

The clonal selection theory explains that during prenatal development, both B and T cells develop millions of genetically different clones through independent segregation, random reassortment, and mutation. Together these clones possess enough genetic variability to respond to many millions of different antigens. Each clone, however, can respond to only one specific antigen.

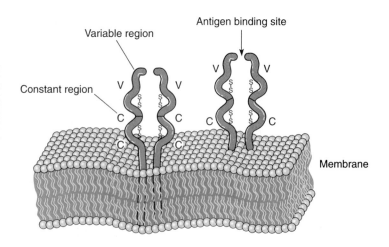

FIGURE 15.7

Proposed structure of the T-cell receptor for antigen. The structure of this polypeptide is very similar to the Fab portion of an immunoglobulin.

The Lymphocyte Response System in Depth

Now that you have a working knowledge of some factors in the development of immune specificity, let us look at each stage of an immune response as originally outlined in figure 15.1.

I. DEVELOPMENT OF THE DUAL LYMPHOCYTE SYSTEM: STAGES IN ORIGIN, DIFFERENTIATION, AND MATURATION

Development generally follows a similar general pattern in both types of lymphocytes (**figure 15.8**). Starting in embryonic and fetal stages, stem cells in the yolk sac, liver, and bone marrow give rise to immature lymphocytes that are released into the circulation. Because these undifferentiated cells cannot yet react with antigens, they must first undergo developmental changes at some specific anatomical location (**table 15.1**). This maturation occurs along two separate lines that will characterize all future responses. The fully mature B and T lymphocytes are released and ultimately take up residence in various lymphoid organs. Lymphocyte differentiation and immunocompetence are basically complete by the late fetal or early neonatal period.

Specific Happenings in B-Cell Maturation

The site of B-cell maturation was first discovered in birds, which have an organ in the intestine called the bursa. For some time, the human bursal equivalent was not established. Now it is known to be certain bone marrow sites that harbor *stromal cells*. These huge cells nurture the lymphocyte stem cells and provide hormonal signals that initiate B-cell development. As a result of gene modification and selection, hundreds of millions of distinct B cells develop. These naive lymphocytes "home" to specific sites in the lymph nodes, spleen, and gut-associated lymphoid tissue (GALT), where they adhere to specific binding molecules. Here they will come into contact with antigens throughout life. In addition to having immunoglobulins as surface receptors, a fully differentiated B cell has a distinctively rough appearance under high magnification, with numerous microvillus projections.

Specific Happenings in T-Cell Maturation

The maturation of T cells and the development of their specific receptors are directed by the thymus gland and its hormones. Due to the complexity of T-cell function, there are at least seven classes of T-cell receptors or markers, termed the CD cluster (abbreviated from cluster of differentiation—CD1, CD2, etc.). Some receptors (see figure 15.7) recognize antigen molecules bound to cells on MHC receptors; others recognize receptors on B cells, other T cells, and macrophages. Like B cells, T cells also migrate to lymphoid organs and occupy specific sites. Additional characteristics typical of

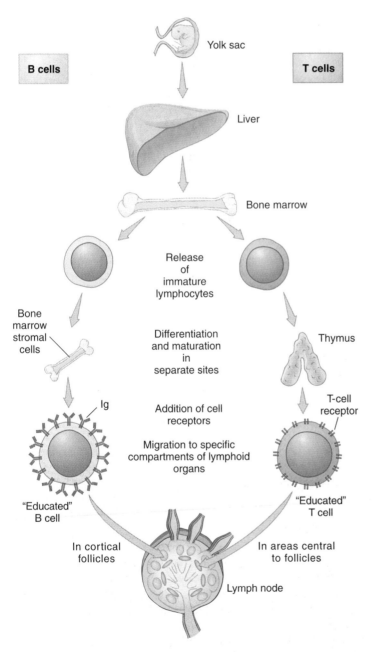

FIGURE 15.8

Major stages in the development of B and T cells.

TABLE 15.1		
Contrasting Properties of B-Cell and T-Cell Lines		
	B Cells	**T Cells**
Site of Maturation	Bone marrow	Thymus
Specific Surface Markers	Immunoglobulin	Several CD receptors T-cell receptor
Texture of Surface	Rough	Smoother
Circulation in Blood	Low numbers	High numbers
Receptors for Antigen	Immunoglobulin	T-cell receptor
Distribution in Lymphatic Organs	Cortex (in follicles)	Paracortical sites (interior to the follicles)
Require MHC receptor binding	No	Yes
Product of Antigenic Stimulation	Plasma cells and memory cells	Several types of sensitized T cells and memory cells
General Functions	Production of antibodies to inactivate, neutralize, target antigens	Cells function in helping other immune cells, suppressing, killing abnormal cells; hypersensitivity; synthesize cytokines

mature T cells are smaller and fewer microvilli under high magnification and the property of rosette formation when mixed with normal sheep red blood cells (see figure 16.13). It has been estimated that 25×10^9 T cells pass between the lymphatic and general circulation per day.

II. ENTRANCE AND PROCESSING OF ANTIGENS AND CLONAL SELECTION

Having reviewed the characteristics of lymphocytes, let us now examine the properties of antigens, the substances that cause them to react. As we reported in chapter 14, an **antigen (Ag)**[2] is a substance that provokes an immune response in specific lymphocytes. The property of behaving as an antigen is called **antigenicity.** The term **immunogen** is another term of reference for a substance that can elicit an immune response. To be perceived as an antigen or immunogen, a substance must meet certain requirements in foreignness, shape, size, and accessibility.

Characteristics of Antigens

One important characteristic of an antigen is that it be perceived as foreign, meaning that it is not a normal constituent of the body. Whole microbes or their parts, cells, or substances that arise from other humans, animals, plants, and various molecules all possess this quality of foreignness and thus are potentially antigenic to the immune system of an individual **(figure 15.9).** Molecules of complex composition such as proteins and protein-containing compounds prove to be more immunogenic than repetitious polymers composed of a single type of unit. Most materials that serve as antigens fall into these chemical categories:

- Proteins and polypeptides (enzymes, albumin, antibodies, hormones, exotoxins)
- Lipoproteins (cell membranes)
- Glycoproteins (blood cell markers)
- Nucleoproteins (DNA complexed to proteins, but not pure DNA)
- Polysaccharides (certain bacterial capsules) and lipopolysaccharides

Effects of Molecular Shape and Size To initiate an immune response, a substance must also be large enough to "catch the attention" of the surveillance cells. Molecules with a molecular weight (MW) of less than 1,000 are seldom complete antigens, and those between 1,000 MW and 10,000 MW are weakly so. Complex macromolecules approaching 100,000 MW are the most immunogenic, a category also dominated by large proteins. Note that large size alone is not sufficient for antigenicity; glycogen, a polymer of glucose with a highly repetitious structure, has an MW over 100,000 and is not normally antigenic, whereas insulin, a protein with an MW of 6,000, can be antigenic.

A lymphocyte's capacity to discriminate differences in molecular shape is so fine that it recognizes and responds to only a portion of the antigen molecule. This molecular fragment, called the **antigenic determinant,** is the primary signal that the molecule is foreign **(figure 15.10).** The particular tertiary structure and shape of this de-

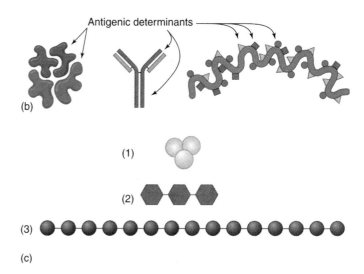

FIGURE 15.9

Characteristics of antigens. **(a)** Whole cells and viruses make good immunogens. **(b)** Complex molecules with several antigenic determinants make good immunogens. **(c)** Poor immunogens include small molecules not attached to a carrier molecule *(1)*, simple molecules *(2)*, and large but repetitive molecules *(3)*.

terminant must conform like a key to the receptor "lock" of the lymphocyte, which then responds to it. Certain amino acids accessible at the surface of proteins or protruding carbohydrate side chains are typical examples. Many foreign cells and molecules are very complex antigenically, with numerous determinants, each of which will elicit a separate and different lymphocyte response. Examples of these multiple, or *mosaic,* antigens include bacterial cells containing cell wall, membrane, flagellar, capsular, and toxin antigens; and viruses, which express various surface and core antigens (figure 15.10).

Small foreign molecules that consist only of a determinant group and are too small by themselves to elicit an immune response are termed **haptens.** However, if such an incomplete antigen is linked to a larger carrier molecule, the combination develops immunogenicity **(figure 15.11).** The carrier group contributes to the size of the complex and enhances the proper spatial orientation of the determinative group, while the hapten serves as the antigenic determinant. Haptens include such molecules as drugs, metals, and ordinarily innocuous household, industrial, and environmental chemicals. Many haptens develop antigenicity in the body by combining with large carrier molecules such as serum proteins (see allergy in chapter 17).

---

2. Originally, **antibody generator.**

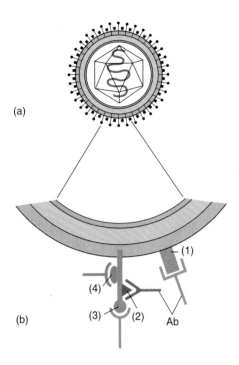

FIGURE 15.10

Mosaic antigens. **(a)** Microbes such as viruses present various sites that serve as separate antigenic determinants. **(b)** Inset indicates that each determinant *(1, 2, 3, 4)* will stimulate a different lymphocyte and antibody response.

Special Types of Antigens   So far, we have emphasized the role of microbial antigens, but tissues and proteins (including enzymes and antibodies) from other humans and animals are also antigenic. On occasion, even a part of the body can take on the character of an antigen. During lymphocyte differentiation, immune tolerance to self tissue occurs, but a few anatomical sites can contain sequestered (hidden) molecules that escape this assessment. Such molecules, called **autoantigens,** can occur in tissues (eye, thyroid gland, for example) that are walled off early in embryonic development before the surveillance system is in complete working order. Because tolerance to these substances has not yet been established, they can subsequently be mistaken as foreign; this mechanism appears to account for some types of autoimmune diseases such as rheumatoid arthritis (chapter 17).

Because each human being is genetically and biochemically unique (except for identical twins), the proteins and other molecules of one person can be antigenic to another. **Alloantigens*** are cell surface markers and molecules that occur in some members of the same species but not in others. Alloantigens are the basis for an individual's blood group (see chapter 17) and major histocompatibility profile, and they are responsible for incompatibilities that can occur in blood transfusion or organ grafting.

On the other hand, different organisms can possess some molecules, called **heterophilic,*** or heterogenetic, antigens, with the same or a similar determinant group. These antigens stimulate a response from the same lymphocyte clone even when they are from totally different sources. Antibodies raised against a heterophilic antigen from one organism will cross-react with a similar or identical antigen from another source. Examples of antigens of different origins with heterophilic determinants are carbohydrate residues on the surfaces of bacteria and red blood cells, the antigens of Group A streptococci and human heart tissue, and cardiolipin, a phospholipid present in a wide assortment of living things. Heterophilic antigens may play a part in diseases such as rheumatic fever and in false positive diagnostic tests (as occur in syphilis).

Some bacterial toxins, called **superantigens,** are potent stimuli for T cells. Their presence in an infection activates T cells at a rate 100 times greater than ordinary antigens. The result can be an overwhelming release of cytokines and cell death. Such diseases as toxic shock syndrome (see chapter 18) and certain autoimmune diseases (see chapter 17) are associated with this class of antigens.

Antigens that evoke allergic reactions, called **allergens,** will be characterized in detail in chapter 17.

Host Response to Antigens: A Cooperative Affair

The basis for most immune responses is the encounter between antigens and white blood cells. Microbes and other foreign substances enter most often through the respiratory or gastrointestinal mucosa and less frequently through other mucous membranes, the skin, or across the placenta. Antigens introduced intravenously

---

*alloantigen (al′-oh) Gr. *allos*, other.

*heterophile (het′-ur-oh-fyl) Gr. *hetero*, other, and *phile*, to love.

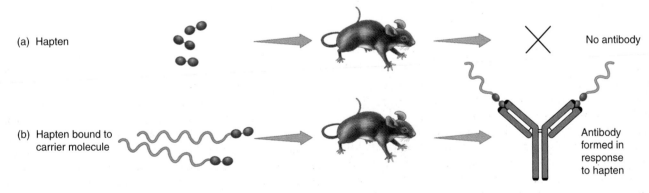

FIGURE 15.11

The hapten-carrier phenomenon.   **(a)** Haptens are too small to be discovered by an animal's immune system; no response. **(b)** A hapten bound to a large molecule will serve as an antigenic determinant and stimulate a response and an antibody that is specific for it.

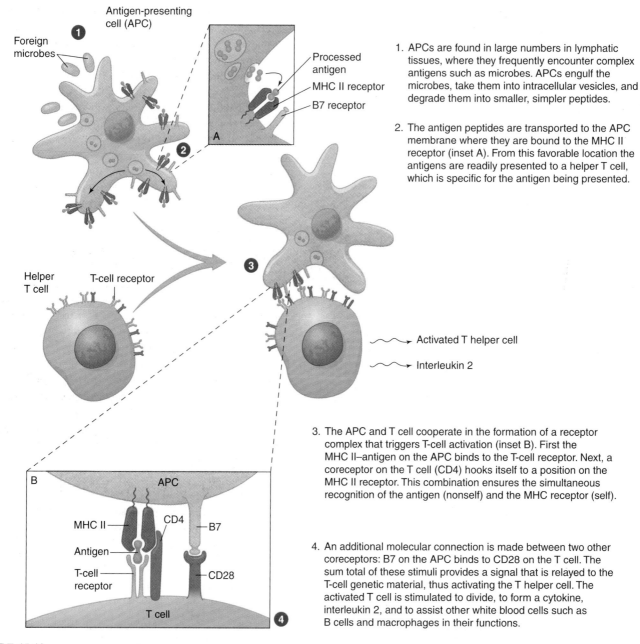

1. APCs are found in large numbers in lymphatic tissues, where they frequently encounter complex antigens such as microbes. APCs engulf the microbes, take them into intracellular vesicles, and degrade them into smaller, simpler peptides.

2. The antigen peptides are transported to the APC membrane where they are bound to the MHC II receptor (inset A). From this favorable location the antigens are readily presented to a helper T cell, which is specific for the antigen being presented.

3. The APC and T cell cooperate in the formation of a receptor complex that triggers T-cell activation (inset B). First the MHC II–antigen on the APC binds to the T-cell receptor. Next, a coreceptor on the T cell (CD4) hooks itself to a position on the MHC II receptor. This combination ensures the simultaneous recognition of the antigen (nonself) and the MHC receptor (self).

4. An additional molecular connection is made between two other coreceptors: B7 on the APC binds to CD28 on the T cell. The sum total of these stimuli provides a signal that is relayed to the T-cell genetic material, thus activating the T helper cell. The activated T cell is stimulated to divide, to form a cytokine, interleukin 2, and to assist other white blood cells such as B cells and macrophages in their functions.

FIGURE 15.12

Interactions between antigen-presenting cells (APCs) and T helper (CD4) cells required for T-cell activation. For T cells to recognize foreign antigens, they must have the antigen processed and presented by a professional APC such as a dendritic cell.

become localized in the liver, spleen, bone marrow, kidney, and lung. If introduced by some other route, antigens are carried in lymphatic fluid and concentrated by the lymph nodes. The lymph nodes and spleen are important in concentrating the antigens and circulating them thoroughly through all areas populated by lymphocytes so that they come into contact with the proper clone.

The Role of Antigen Processing and Presentation

In most immune reactions, the antigen must be further acted upon and formally presented to lymphocytes by dendritic cells or special

macrophages called **antigen presenting cells (APCs).** One of the most prominent APCs is a large **dendritic*** cell that engulfs the antigen and modifies it so that it will be more immunogenic and recognizable to lymphocytes. After processing is complete, the antigen is moved to the surface of the APC and bound to the MHC receptor so that it will be readily accessible to the lymphocytes during presentation **(figure 15.12).**

_____

*dendritic (den'-drih-tik) Gr. *dendron,* tree. In reference to the long, branchlike extensions of the cell membrane. These cells are similar to macrophages in origin and function.

Presentation of Antigen to the Lymphocytes and Its Early Consequences

For lymphocytes to respond to the APC-bound antigen, certain conditions must be met. T-cell-dependent antigens, usually protein-based, require recognition steps between the APC, antigen, and lymphocytes. The first cells on the scene to assist in activating B cells and other T cells are a special class of **helper T cells (T_H).** This class of T cell bears a receptor that binds simultaneously with the class II MHC receptor on the APC and with one site on the antigen (figure 15.12). Once identification has occurred, a cytokine, **interleukin* -1 (IL-1),** produced by the APC, activates this T helper cell. The T_H cell, in turn, produces a different cytokine, **interleukin-2 (IL-2),** that stimulates a general increase in activity of committed B and T cells (see Spotlight on Microbiology 15.2, page 468). The manner in which B and T cells subsequently become activated by the APC–T helper cell complex and their individual responses to antigen will be addressed separately in sections III B and IV, and III T and V.

A few antigens can trigger a response from B lymphocytes without the cooperation of APCs or T helper cells. These T-cell-independent antigens are usually simple molecules such as carbohydrates with many repeating and invariable determinant groups. Examples include lipopolysaccharide from the cell wall of *Escherichia coli,* polysaccharide from the capsule of *Streptococcus pneumoniae,* and molecules from rabies and Epstein-Barr virus. Because so few antigens are of this type, most B-cell reactions require helper T cells.

III. B. ACTIVATION OF B LYMPHOCYTES: CLONAL EXPANSION AND ANTIBODY PRODUCTION

The immunologic activation of most B cells requires a series of events (**figure 15.13**):

1. **Clonal selection and binding of antigen.** In this case, a precommitted B cell of a particular clonal specificity picks up the antigen on its Ig receptors and processes it into small peptide determinants. The antigen is then bound to the MHC II receptors on the B cell. The MHC/Ag receptor on the B cell is bound to the T_H cell, thereby ensuring self recognition.
2. **Instruction by chemical mediators.** The B cell receives developmental signals from macrophages and T cells (interleukins 2, 4 and 6) and various other growth factors, such as IL-4 and -5.
3. The combination of these stimuli on the membrane receptors causes a signal to be transmitted internally to the B-cell nucleus.
4. These events trigger B-cell activation. An activated B cell undergoes an increase in DNA synthesis, organelle bulk, and size in preparation for entering the cell cycle and mitosis.
5.–6. **Clonal expansion.** A stimulated B cell multiplies through successive mitotic divisions and produces a large population of genetically identical daughter cells. Some cells that stop short of becoming fully differentiated are **memory cells,** which remain for long periods to react with that same antigen at a later time. This reaction also expands the clone size, so that subsequent exposure to that antigen provides more cells with that specificity. This expansion of the clone size accounts for the increased memory response. By far the most numerous progeny are large, specialized, terminally differentiated B cells called **plasma cells.**

7. **Antibody production and secretion.** The primary action of plasma cells is to secrete into the surrounding tissues copious amounts of antibodies with the same specificity as the original receptor (figure 15.13). Although an individual plasma cell can produce around 2,000 antibodies per second, production does not continue indefinitely because of regulation from the T suppressor (T_S) class of cells. The plasma cells do not survive for long and deteriorate after they have synthesized antibodies.

CHAPTER CHECKPOINTS

Immature lymphocytes released from hemopoietic tissue migrate (home) to one of two sites for further development. B cells mature in the stromal cells of the bone marrow. T cells mature in the thymus.

Antigens or immunogens are proteins or other complex molecules of high molecular weight that trigger the immune response in the host.

Lymphocytes respond to a specific portion of an antigen called the antigenic determinant. A given microorganism has many such determinants, all of which stimulate individual specific immune responses.

Haptens are molecules that are too small to trigger an immune response alone but can be immunogenic when they attach to a larger substance, such as host serum protein.

Autoantigens and allergens are types of antigens that cause damage to host tissue as a consequence of the immune response.

Antigen-presenting cells (APCs) such as dendritic cells and macrophages bind foreign antigen to their cell surfaces for presentation to lymphocytes. Physical contact between the APC, T cells and B cells activates these lymphocytes to proceed with their respective immune responses.

IV. PRODUCTS OF B LYMPHOCYTES: ANTIBODY STRUCTURE AND FUNCTIONS

The Structure of Immunoglobulins

Earlier we saw that a basic immunoglobulin (Ig) molecule contains four polypeptide chains connected by disulfide bonds. Let us view this structure once again, using an IgG molecule as a model. Two functionally distinct segments called *fragments* can be differentiated. The two "arms" that bind antigen are termed antigen binding fragments (Fabs), and the rest of the molecule is the crystallizable fragment (Fc), so called because it has been crystallized in pure form. The distal end of each Fab fragment (consisting of the variable regions of the heavy and light chains) folds into a groove that will accommodate one antigenic determinant. The presence of a special *hinge* region at the site of attachment between the Fab and Fc fragments allows swiveling of the Fab fragments. In this way, they can change their angle to accommodate nearby antigen sites that vary slightly in distance and position. The Fc fragment is involved in binding to various cells and molecules of the immune system itself. **Figure 15.14** shows three views of antibody structure.

*interleukin (in″-tur-loo′-kin) A peptide that carries signals between white blood cells.

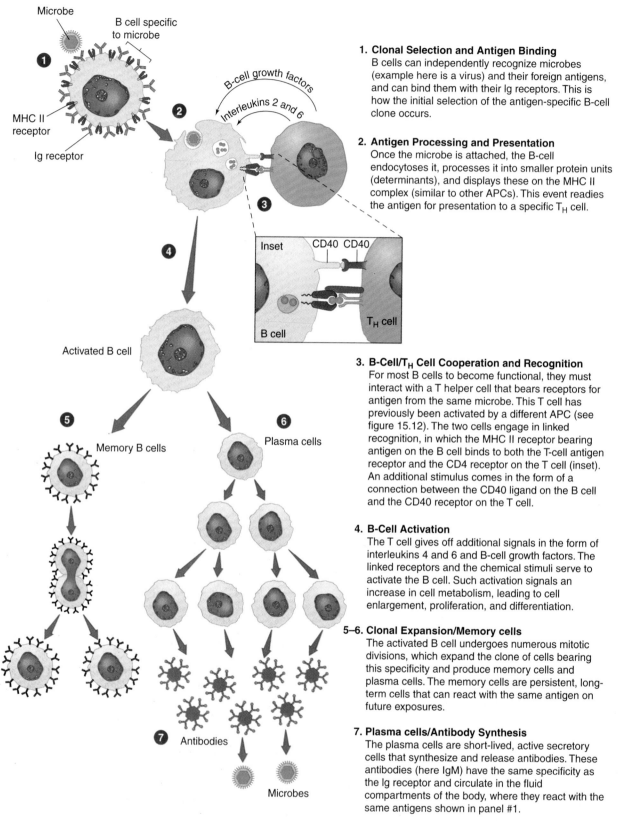

1. **Clonal Selection and Antigen Binding**
   B cells can independently recognize microbes (example here is a virus) and their foreign antigens, and can bind them with their Ig receptors. This is how the initial selection of the antigen-specific B-cell clone occurs.

2. **Antigen Processing and Presentation**
   Once the microbe is attached, the B-cell endocytoses it, processes it into smaller protein units (determinants), and displays these on the MHC II complex (similar to other APCs). This event readies the antigen for presentation to a specific T_H cell.

3. **B-Cell/T_H Cell Cooperation and Recognition**
   For most B cells to become functional, they must interact with a T helper cell that bears receptors for antigen from the same microbe. This T cell has previously been activated by a different APC (see figure 15.12). The two cells engage in linked recognition, in which the MHC II receptor bearing antigen on the B cell binds to both the T-cell antigen receptor and the CD4 receptor on the T cell (inset). An additional stimulus comes in the form of a connection between the CD40 ligand on the B cell and the CD40 receptor on the T cell.

4. **B-Cell Activation**
   The T cell gives off additional signals in the form of interleukins 4 and 6 and B-cell growth factors. The linked receptors and the chemical stimuli serve to activate the B cell. Such activation signals an increase in cell metabolism, leading to cell enlargement, proliferation, and differentiation.

5–6. **Clonal Expansion/Memory cells**
   The activated B cell undergoes numerous mitotic divisions, which expand the clone of cells bearing this specificity and produce memory cells and plasma cells. The memory cells are persistent, long-term cells that can react with the same antigen on future exposures.

7. **Plasma cells/Antibody Synthesis**
   The plasma cells are short-lived, active secretory cells that synthesize and release antibodies. These antibodies (here IgM) have the same specificity as the Ig receptor and circulate in the fluid compartments of the body, where they react with the same antigens shown in panel #1.

FIGURE 15.13

Events in B-cell activation and antibody synthesis.

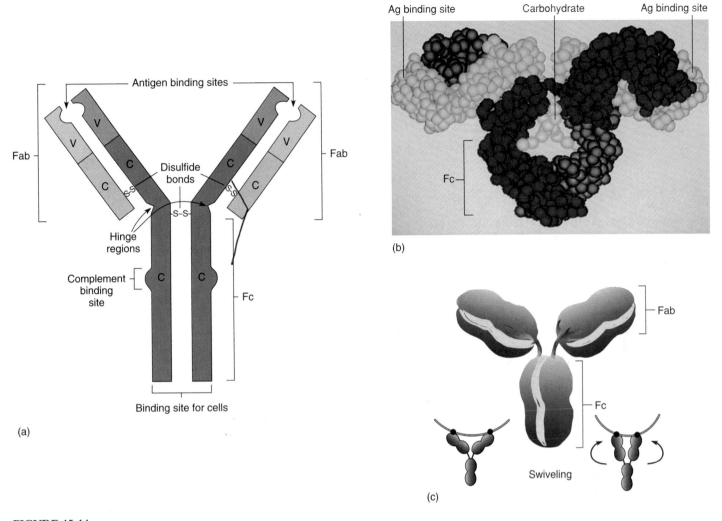

FIGURE 15.14

Working models of antibody structure.   **(a)** Diagrammatic view of IgG depicts the principal functional areas (Fabs and Fc) of the molecule. **(b)** Realistic model of immunoglobulin shows the tertiary and quaternary structure achieved by additional intrachain and interchain bonds and the position of the carbohydrate component. **(c)** The "peanut" model of IgG helps illustrate swiveling of Fabs relative to one another and to Fc.

Antibody-Antigen Interactions and the Function of the Fab

The site on the antibody where the antigenic determinant inserts is composed of a *hypervariable region* whose amino acid content can be extremely varied. Antibodies differ somewhat in the exactness of this groove for antigen, but a certain complementary fit is necessary for the antigen to be held effectively **(figure 15.15).** The specificity of antigen binding sites for antigens is very similar to enzymes and substrates (in fact, some antibodies are used as enzymes; see Microbits 8.2). So specific are some immunoglobulins for antigen that they can distinguish between a single functional group of a few atoms. Because the specificity of the Fab sites is identical, an Ig molecule can bind antigenic determinants on the same cell or on two separate cells and thereby link them.

The principal activity of an antibody is to unite with, immobilize, call attention to, or neutralize the antigen for which it was formed **(figure 15.16).** Antibodies called opsonins stimulate

opsonization,* a process in which microorganisms or other particles are coated with specific antibodies so that they will be more readily recognized by phagocytes, which dispose of them. Opsonization has been likened to putting handles on a slippery object to provide phagocytes a better grip. The capacity for antibodies to aggregate, or **agglutinate,** antigens is the consequence of their cross-linking cells or particles into large clumps. This is a principle behind certain immune tests discussed in chapter 16. The interaction of an antibody with complement can result in the specific rupturing of cells and some viruses. In **neutralization** reactions, antibodies fill the surface receptors on a virus or the active site on a molecule to prevent it from functioning normally. **Antitoxins** are a special type of antibody that neutralize bacterial exotoxins. It should be noted that not all antibodies are protective; some neither benefit nor harm, and a few actually cause diseases (see autoimmunity in chapter 17).

---

*opsonization (ahp″-son-uh-zay′-shun) Gr. *opsonein,* to prepare food.

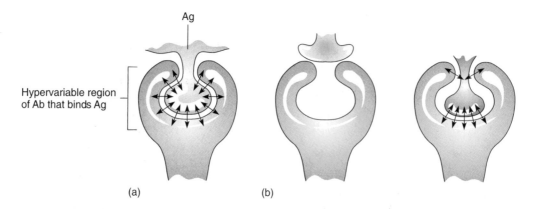

FIGURE 15.15

Antigen-antibody binding. The union of antibody (Ab) and antigen (Ag) is characterized by a certain degree of fit and is supported by weak linkages such as hydrogen bonds and electrostatic attraction. **(a)** In a snug fit such as that shown here, there is great opportunity for attraction and strong attachment. The strength of this union confers high affinity. **(b)** Examples of the relationship of other antigens with this same antibody. The first Ag clearly cannot be accommodated. The second (purple) antigen is not a perfect fit, but it can bind to the antibody.

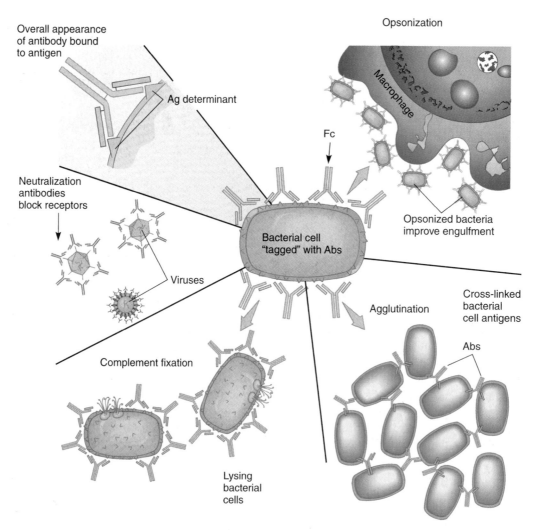

FIGURE 15.16
Summary of antibody functions.

Functions of the Crystallizable Fragment: Interactions with Self

Although the Fab fragments bind antigen, the Fc fragment has a different binding function. In most classes of immunoglobulin, the proximal end of Fc contains an effector molecule that can bind to certain receptors on the membrane of cells, such as macrophages, neutrophils, eosinophils, mast cells, basophils, and lymphocytes. The effect on an antibody's Fc fragment binding to a cell receptor depends upon that cell's role. In the case of opsonization, the attachment of antibody to foreign cells and viruses exposes the Fc fragments to phagocytes. Certain antibodies have receptors on the Fc portion for fixing complement, and in some immune reactions, the binding of Fc causes the release of cytokines. For example, the antibody of allergy (IgE) binds to basophils and mast cells, which causes the release of allergic mediators such as histamine (see chapter 17). The size and amino acid composition of Fc also determine an antibody's permeability, its distribution in the body, and its class.

Accessory Molecules on Immunoglobulins

All antibodies contain molecules in addition to the basic polypeptides. Varying amounts of carbohydrates are affixed to the constant regions in most instances (**table 15.2**). Two additional accessory molecules are the *J chain* that joins the monomers of IgA and IgM, and the *secretory component*, which helps move Ig across mucous membranes. These proteins occur only in certain immunoglobulin classes.

The Classes of Immunoglobulins

Immunoglobulins exist as structural and functional classes called *isotypes* (compared and contrasted in table 15.2). The differences in these classes are due primarily to variations in the Fc fragment and its accessory molecules. The classes are differentiated with shorthand names (Ig, followed by a letter: IgG, IgA, IgM, IgD, IgE).

The structure of **IgG** has already been presented. It is a monomer produced by memory cells responding the second time to a given antigenic stimulus. It is by far the most prevalent antibody circulating throughout the tissue fluids and blood. It has numerous functions: It neutralizes toxins, opsonizes, and fixes complement, and it is the only antibody capable of crossing the placenta.

The two forms of **IgA** are: (1) a monomer that circulates in small amounts in the blood and (2) a dimer that is a significant component of the mucous and serous secretions of the salivary glands, intestine, nasal membrane, breast, lung, and genitourinary tract. The dimer, called secretory IgA, is formed in a plasma cell by two monomers attached by a J piece. To facilitate the transport of IgA

TABLE 15.2

Characteristics of the Immunoglobulin (Ig) Classes

	IgG	IgA (dimer only)	IgM	IgD	IgE
	Monomer	Dimer, Monomer	Pentamer	Monomer	Monomer
Number of Antigen Binding Sites	2	4 2	10	2	2
Molecular Weight	150,000	170,000–385,000	900,000	180,000	200,000
Percent of Total Antibody in Serum	80%	13%	6%	1%	0.002%
Average Life in Serum (Days)	23	6	5	3	2.5
Crosses Placenta?	Yes	No	No	No	No
Fixes Complement?	Yes	No	Yes	No	No
Fc Binds To	Phagocytes	Phagocytes	B lymphocytes	B lymphocytes	Mast cells and basophils
Biological Function	Long-term immunity; memory antibodies	Secretory antibody; on mucous membranes	Produced at first response to antigen; can serve as B-cell receptor	Receptor on B cells	Antibody of allergy; worm infections

C = carbohydrate.
J = J chain.

across membranes, a secretory piece is later added by the gland cells themselves. IgA coats the surface of these membranes and appears free in saliva, tears, colostrum, and mucus. It confers the most important specific local immunity to enteric, respiratory, and genitourinary pathogens. Its contribution in protecting newborns who derive it passively from nursing is mentioned in Medical Microfile 15.3, page 469.

IgM (M for *macro*) is a huge molecule composed of five monomers (making it a pentamer) attached by the Fc receptors to a central J chain. With its 10 binding sites, this molecule has tremendous avidity for antigen (*avidity* means the capacity to bind antigens). It is the first class synthesized by a plasma cell following its first encounter with antigen. Its complement-fixing and opsonizing qualities make it an important antibody in many immune reactions. It circulates mainly in the blood and is far too large to cross the placental barrier.

IgD is a monomer found in minuscule amounts in the serum, and it does not fix complement, opsonize, or cross the placenta. Its main function is to serve as a receptor for antigen on B cells, usually along with IgM. It seems to be the triggering molecule for B-cell activation, and it can also play a role in immune suppression.

IgE is also an uncommon blood component unless one is allergic or has a parasitic worm infection. Its Fc region interacts with receptors on mast cells and basophils. Its biological significance is to stimulate an inflammatory response through the release of potent physiological substances by the basophils and mast cells. Because inflammation would enlist blood cells such as eosinophils and lymphocytes to the site of infection, it would certainly be one defense against parasites. Unfortunately, IgE has another, more insidious effect—that of mediating anaphylaxis, asthma, and certain other allergies (see chapter 17).

Evidence of Antibodies in Serum

Regardless of the site where antibodies are first secreted, a large quantity eventually ends up in the blood by way of the body's communicating networks. If one submits a sample of **antiserum** (serum containing specific antibodies) to electrophoresis, the major groups of proteins migrate in a pattern consistent with their mobility and size **(figure 15.17).** The albumins show up in one band, and the globulins in four bands called alpha-1 (α_1), alpha-2 (α_2), beta (β), and gamma (γ) globulins. Most of the globulins represent antibodies, which explains how the term *immunoglobulin* was derived. **Gamma globulin** is composed primarily of IgG, whereas β and α_2 globulins are a mixture of IgG, IgA, and IgM. As we will see in chapter 16, the gamma globulin fraction of serum is important in immune therapies.

Monitoring Antibody Production over Time: Primary and Secondary Responses to Antigens

We can learn a great deal about how the immune system reacts to an antigen by studying the levels of antibodies in serum over time **(figure 15.18).** This level is expressed quantitatively as the **titer,*** or concentration of antibodies. Upon the first exposure to an antigen, the system undergoes a **primary response.** The earliest part of this response, the *latent period,* is marked by a lack of antibodies for that antigen, but much activity is occurring. During this time, the antigen is being concentrated in lymphoid tissue, and is being processed by the correct clones of B lymphocytes. As plasma cells synthesize antibodies, the serum titer increases to a certain plateau and then tapers off to a low level over a few weeks or months. When the class of antibodies produced during this response is tested, an important characteristic of the response is uncovered. It turns out that, early in the primary response, most of the antibodies are the IgM type, which is the first class to be secreted by naive plasma cells. Later, the class of the antibodies (but not their specificity) is switched to IgG or some other class (IgA or IgE).

When the immune system is exposed again to the same immunogen within weeks, months, or even years, a **secondary response** occurs. The rate of antibody synthesis, the peak titer, and the length of antibody persistence are greatly increased over the primary response. The rapidity and amplification seen in this response are attributable to the memory B cells that were formed during the primary response. Because of its association with recall, the secondary response is also called the **anamnestic*** **response.** The advantage of this response is evident: It provides a quick and potent strike against subsequent exposures to infectious agents. This memory effect forms the basis for giving **boosters**—additional doses of vaccine—to increase the serum titer.

Monoclonal Antibodies: Useful Products from Cancer Cells

The value of antibodies as tools for locating or identifying antigens is well established. For many years, antiserum extracted from human or animal blood was the main source of antibodies for tests and therapy, but most antiserum has a basic problem. It contains **polyclonal antibodies,** meaning that it is a mixture of different antibodies because it reflects dozens of immune reactions from a wide variety of B-cell clones. This characteristic is to be expected, because several immune reactions may be occurring simultaneously, and even a single species of microbe can stimulate several different types of antibodies. Certain applications in immunology require a pure preparation of **monoclonal antibodies (MABs)** that originate from a single clone and have a single specificity for antigen.

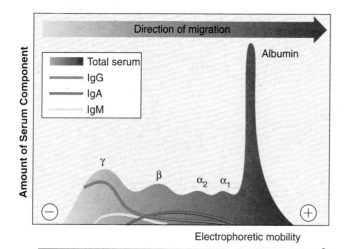

FIGURE 15.17

Pattern of human serum after electrophoresis. When antiserum is subjected to electrical current, the various proteinaceous components are separated into bands. This is a means of separating the different antibodies and serum proteins as well as quantifying them.

*titer (ty′-tur) Fr. *titre,* standard. One method for determining titer is shown in figure 16.5.

*anamnestic (an-am-ness′-tik) Gr. *anamnesis,* a recalling.

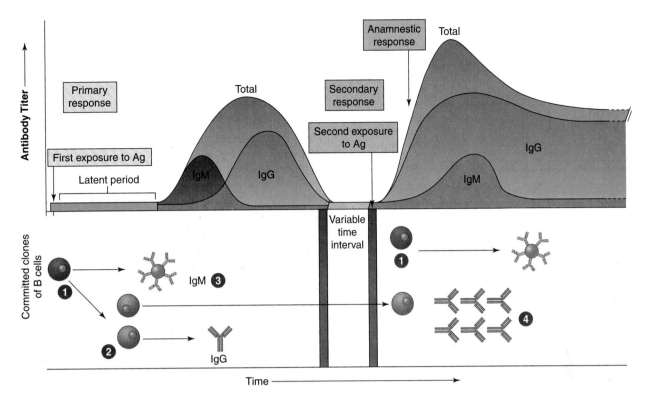

FIGURE 15.18

Primary and secondary responses to antigens. *(Top)* The pattern of antibody titer and subclasses as monitored during initial and subsequent exposure to the same antigen. *(Bottom)* A view of the B-cell responses that account for the pattern. Depicted are clonal selection (1) production of memory cells (2) and the predominant antibody class occurring at first (3) and second contact with antigen (4). Note that residual memory cells remaining from the primary response are ready to act immediately, which produces the rapid rise of antibody levels early in the secondary period.

The technology for producing monoclonal antibodies is possible by hybridizing cancer cells and plasma cells *in vitro* (**figure 15.19**). This technique began with the discovery that tumors isolated from multiple **myelomas*** in mice consist of identical plasma cells. These monoclonal plasma cells secrete a strikingly pure form of antibodies with a single specificity and continue to divide indefinitely. Immunologists recognized the potential in these plasma cells and devised a **hybridoma** approach to creating MABs. The basic idea behind this approach is to hybridize or fuse a myeloma cell with a normal plasma cell from a mouse spleen to create an immortal cell that secretes a supply of functional antibodies with a single specificity.

The introduction of this technology has the potential for numerous biomedical applications. Monoclonal antibodies have provided immunologists with excellent standardized tools for studying the immune system and for expanding disease diagnosis and treatment. Most of the successful applications thus far use MABs in *in vitro* diagnostic testing and research. Although injecting monoclonal antibodies to treat human disease is an exciting prospect, so far this therapy has been stymied because most MABs are of mouse origin, and many humans will develop hypersensitivity to them. The development of human MABs and other novel approaches using genetic engineering is currently under way (**Medical Microfile 15.1**).

*myeloma (my-uh-loh′-muh) Gr. *myelos,* marrow, and *oma,* tumor. A malignancy of the bone marrow.

CHAPTER CHECKPOINTS

B cells produce five classes of antibody: IgM, IgG, IgA, IgD, and IgE. IgM and IgG predominate in plasma. IgA predominates in body secretions. IgD binds to B cells as an antigen receptor. IgE binds to tissue cells, promoting inflammation.

Antibodies bind physically to the specific antigen that stimulates their production, thereby immobilizing the antigen and enabling it to be destroyed by other components of the immune system.

The anamnestic response means that the second exposure to antigen calls forth a much faster and more vigorous response than the first.

Monoclonal, or pure, antibodies can be produced commercially by fusing a plasma cell with a myeloma cell to produce an immortal hybridoma.

III. T ACTIVATION OF T LYMPHOCYTES AND V. HOW T CELLS RESPOND TO ANTIGEN: CELL-MEDIATED IMMUNITY (CMI)

During the time that B cells have been actively responding to antigens, the T-cell limb of the system has been similarly engaged. The responses of T cells, however, are **cell-mediated** immunities, which require the direct involvement of T lymphocytes throughout the course of the reaction. These reactions are among the most complex and diverse in the immune system and involve several subsets

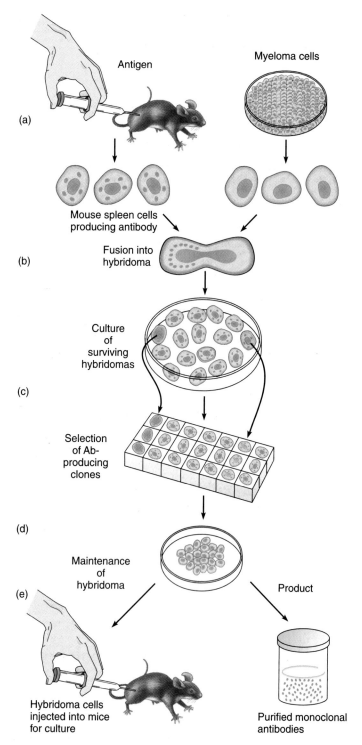

FIGURE 15.19

Summary of the technique for producing monoclonal antibodies by hybridizing myeloma tumor cells with normal plasma cells. **(a)** A normal mouse is inoculated with an antigen having the desired specificity, and plasma cells are isolated from its spleen. A special strain of mouse provides the myeloma cells. **(b)** The two cell populations are mixed with polyethylene glycol, which causes some cells in the mixture to fuse and form hybridomas. **(c)** Surviving cells are cultured and separated into individual wells. **(d)** Tests are performed on each hybridoma to determine the specificity of the antibody (Ab) it secretes. **(e)** A hybridoma with the desired specificity is grown in tissue culture; antibody product is then isolated and purified. The hybridoma is maintained in a susceptible mouse for future use.

of T cells whose particular actions are dictated by CD receptors. All mature T cells have CD2 but **CD4** and **CD8** are found only on certain classes (table 15.3). T cells are restricted; that is, they require some type of MHC (self) recognition before they can be activated, and all produce cytokines with a spectrum of biological effects **(Spotlight on Microbiology 15.2).**

T cells have notable differences in function from B cells. Rather than making antibodies to control foreign antigens, the whole T cell acts directly in contact with the antigen. They also stimulate other T cells, B cells, and phagocytes.

The Activation of T Cells and Their Differentiation Into Subsets

The mature T cells in lymphoid organs are primed to react with antigens that have been processed and presented to them by dendritic cells and macrophages. They recognize an antigen only when it is presented in association with an MHC carrier (see figure 15.12). T cells with CD4 receptors recognize ingested peptides presented on MHC II and T cells with CD8 receptors receive peptides presented on MHC I.

A T cell is initially sensitized when an antigen/MHC complex is bound to its receptors. By mechanisms not yet fully characterized, sensitization leads to the final differentiation of the cell into one of four functionally specialized subsets: helper, suppressor, cytotoxic, or delayed hypersensitivity T cells **(table 15.3** and **figure 15.20).** As with B cells, activated T cells transform in preparation for mitotic divisions, and they divide into one of the subsets of effector cells and memory cells that can interact with the antigen upon subsequent contact. Memory T cells are some of the longest-

TABLE 15.3

Characteristics of Subsets of T Cells

Functional Types	Primary Receptor	Functions/ Important Features
T helper cells (CD4, T_H)	CD4	Assist B cells in recognition of antigen; assist other subsets of T cells in recognition and reaction to antigen; activate macrophages to kill microbes; require MHC II for function
Cytotoxic (killer) cells (T_C)	CD8	Destroy a target foreign cell by lysis; important in destruction of complex microbes, cancer cells, virus-infected cells; graft rejection; allergy; require MHC I for function
Delayed hypersensitivity cells (T_D)	CD4 and CD8	Responsible for allergies occurring several hours or days after contact; skin reactions as in tuberculin test
T suppressor cells (T_S)	Both CD4 and CD8	Regulate immune reactions; cells limit the extent of antibody production; block some T-cell activity; carry CD5 and CD8 receptors

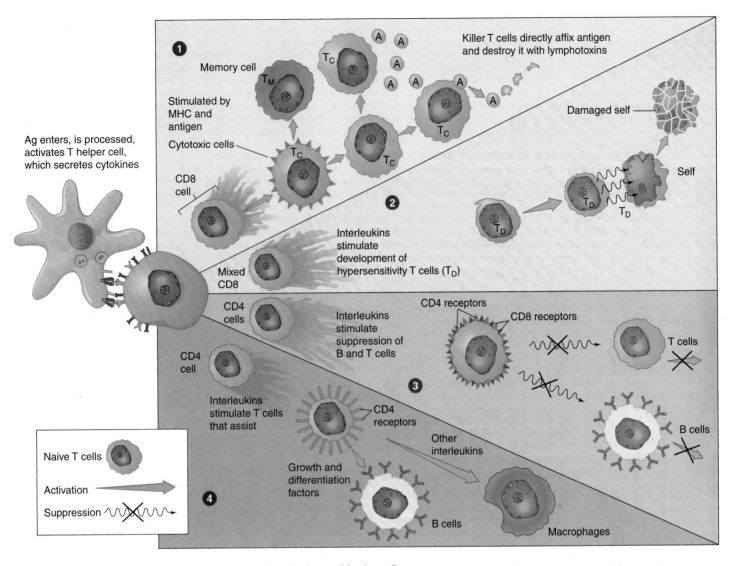

1. T_C cells destroy certain microbes and foreign cells.
2. T_D cells react with allergens and cause a type of hypersensitivity that damages self.
3. T_S cells limit the actions of B and T cells.
4. T_H cells assist in the actions of B and T cells.

FIGURE 15.20

Overall scheme of T-cell activation and differentiation into different types of T cells. T cells secrete specific interleukins that initiate differentiation into one of the classes of T cells listed.

lived blood cells known (70 years in one well-documented case). The functional categories of T cells vary in their CD4 receptors and sensitivity to cytokines.

T Helper (T_H) Cells   Helper cells play a central role in assisting with immune reactions to antigens, including those of B cells and other T cells. They are also involved in activating macrophages and improving opsonization. They do this directly by receptor contact and indirectly by releasing cytokines such as interleukin-2, which stimulates the primary growth and activation of B and T cells, and interleukins-4, -5, and -6, which stimulate various activities of B cells. T helper cells are the most prevalent type of

T cell in the blood and lymphoid organs, making up about 65% of this population. The severe depression of this class of T cells (with CD4 receptors) by HIV is what largely accounts for the immunopathology of AIDS.

Cytotoxic T (T_C) Cells: Cells That Kill Other Cells
Cytotoxicity is the capacity of certain T cells to kill a specific target cell. It is a fascinating and powerful property that accounts for much of our immunity to foreign cells and cancer, and yet, under some circumstances, it can lead to disease. For a **killer T cell** to become activated, it must recognize foreign receptors presented to it, and mount a direct attack upon the target cell. After activation, the T_C cell

Imagine releasing millions of tiny homing pigeons into a molecular forest and having them navigate directly to their proper roost, and you have some sense of what monoclonal antibodies can do. Laboratories use them to identify antigens, receptors, and antibodies; to differentiate cell types (T cells versus B cells) and cell subtypes (different sets of T cells); to diagnose diseases such as cancer and AIDS; and to identify bacteria and viruses.

A number of promising techniques have been directed toward using monoclonals as drugs. When genetic engineering is combined with hybridoma technology, the potential for producing antibodies of almost any desired specificity and makeup is possible. For instance, *chimeric* MABs (monoclonal antibodies) that are part human and part mouse have been produced through splicing antibody genes. It is even possible to design an antibody molecule that has antigen binding sites with different specificities. Monoclonals can be hybridized with plant or bacterial toxins to form immunotoxin complexes that attach to a target cell and poison it. The most exciting prospect of this therapy is that it can destroy a specified cancer cell and not harm normal cells. Monoclonals are currently employed in treatment of cancers such as breast and colon cancers. Drugs such as Rituxan and herceptin bind to cancer cells and trigger their death.

Such antibodies could also be used to suppress allergies, autoimmunities, and graft rejection. Drugs such as OKT3 and Orthoclone are currently used to prevent the rejection of organ transplants by incapacitating cytotoxic T cells. Now that researchers have genetically engineered plants and mice to produce human MABs, therapeutic uses will continue to expand. MABs are undergoing clinical trials to treat heart, lung, and inflammatory diseases.

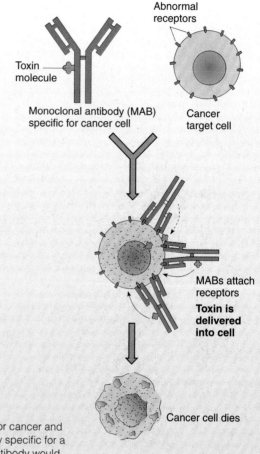

The mechanism of an immunotoxin. A potential therapy for cancer and immune dysfunctions is the use of a monoclonal antibody specific for a certain tumor that carries a potent toxin molecule. The antibody would circulate to cancer cells and deliver the toxin to them. Normal body cells would be unharmed.

---

delivers a dose of several cytokines that severely injures the target cell (**figure 15.21**). This process involves the formation of pore-forming proteins, or **perforins**,* that attack the target cell membrane and create pores through which ions can leak. The loss of selective permeability is followed by target cell death through a process called *apoptosis* (ah-poh-toh′-sis). The apoptosis is genetically programmed and results in destruction of the nucleus and complete cell lysis.

Target Cells That TC Cells Can Destroy Include the Following:

- Fungi, protozoans, and complex bacteria (mycobacteria).
- Virally infected cells (figure 15.21). Cytotoxic cells recognize these because of telltale virus receptors expressed on their surface. Cytotoxic defenses are an essential protection against viruses.

- Cancer cells. T cells constantly survey the tissues and immediately attack any abnormal cells they encounter (**figure 15.22**). The importance of this function is clearly demonstrated in the susceptibility of T-cell-deficient people to cancer (chapter 17).
- Cells from other animals and humans. Cytotoxic CMI is the most important factor in graft rejection. In this instance, the T_C cells attack the foreign tissues that have been implanted into a recipient's body.

Other Types of Killer Cells   Natural killer (NK) cells are a type of lymphocyte related to T cells that lack specificity for antigens. They circulate through the spleen, blood, and lungs and are probably the first killer cells to attack cancer cells and virus-infected cells. They destroy such cells by similar mechanisms as T cells. Their activities are acutely sensitive to cytokines such as interleukin-12 and interferon.

*perforin From the term *perforate* or to penetrate with holes.

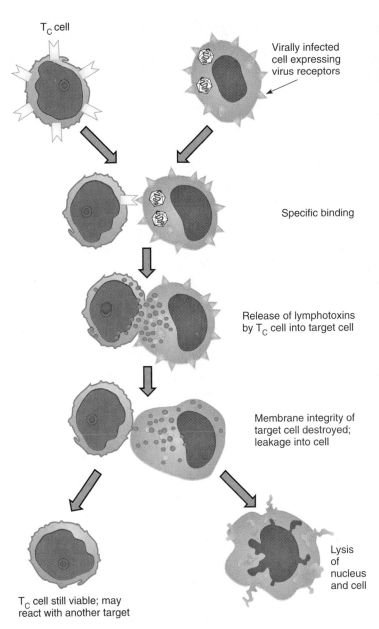

T$_C$ cell

Virally infected
cell expressing
virus receptors

Specific binding

Release of lymphotoxins
by T$_C$ cell into target cell

Membrane integrity of
target cell destroyed;
leakage into cell

Lysis
of
nucleus
and cell

T$_C$ cell still viable; may
react with another target

FIGURE 15.21

**Stages of cell-mediated cytotoxicity and the action of lymphotoxins
on virus-infected target cells.**   Destruction of the infected cell rapidly
terminates the viral cycle.

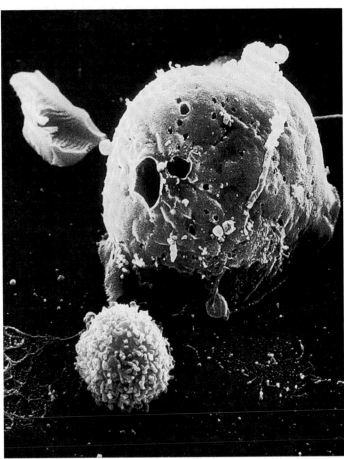

FIGURE 15.22

**A cytotoxic T cell (lower blue cell) has mounted a successful attack on
a tumor cell (larger yellow cell).**   These small killer cells perforate their
cellular targets with holes that lead to lysis and death.

T Suppressor (T$_S$) Cells   Not all T cells are involved in activating immune responses. The T suppressor (T$_S$) cells actually do just
the opposite—they inhibit them. These T cells appear to be a special class of lymphocytes that carry the CD8 receptor and are considered to be primarily regulatory in function. Although the precise
mechanisms are not yet understood. T$_S$ cells are known to inhibit
the functions of B cells and other T cells. They do this by producing various specific protein inhibitors that keep antigen processing
cells and lymphocytes from reacting to antigens. The benefit of this
suppressor function is to restrict random immune responses that
could become destructive or uncontrolled. T$_S$ cells also appear to
play a role in development of tolerance to self and foreign antigens.
It seems likely that abnormalities in the suppressor cells are an underlying basis for certain autoimmune diseases and cancers.

Delayed Hypersensitivity T (T$_D$) Cells   Although immediate
allergies such as hay fever and anaphylaxis are mediated by antibodies, certain delayed responses to allergens (the tuberculin reactions, for example) are initiated by special T cells. These reactions
are discussed in chapter 17.

A PRACTICAL SCHEME FOR CLASSIFYING
SPECIFIC IMMUNITIES

The means by which humans acquire immunities can be conve-
niently encapsulated within four interrelated categories: active, passive, natural, and artificial.

> **Active immunity** occurs when an individual receives an
> immune stimulus (antigen) that activates the B and T
> cells, causing the body to produce immune substances
> such as antibodies. Active immunity is marked by several
> characteristics: (1) It is an essential attribute of an
> immunocompetent individual; (2) it creates a memory
> that renders the person ready for quick action upon
> reexposure to that same antigen; (3) it requires several
> days to develop; and (4) it lasts for a relatively long time,
> sometimes for life. Active immunity can be stimulated by
> natural or artificial means.

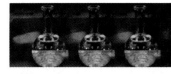

SPOTLIGHT ON MICROBIOLOGY 15.2
Lymphokines: Chemical Products of T Cells

Although the immunities of T cells are usually thought of as cell-mediated, one must not overlook the fact that T cells are also prolific chemical factories. The products of sensitized T cells that communicate with and act upon other cells are a type of cytokine called **lymphokines.** Lymphokines have diverse effects. Some (interferon and interleukin) regulate immune reactions; some mediate inflammation (chemotactic factors); and others (lymphotoxins) kill whole cells. In the previous chapter, we mentioned the role of **gamma interferon** in regulating B cells and T cells. It also activates natural killer cells (NK) and stimulates macrophages. Interleukin-2 was discussed in conjunction with T helper cells and its role as a growth promoter and general stimulus for lymphocyte activity. Interleukin-3 is a powerful stimulus for stem cell development in the bone marrow. Several other interleukins (IL-4, -5, -6, -7) promote proliferation, differentiation, and secretion of B and T cells.

The mechanisms by which cytotoxic lymphocytes destroy their whole cell targets is just being understood. It appears that they seek out the foreign cell's membrane and secrete various metabolic products into it. The compounds that damage the target cell are termed **lymphotoxins.** These molecular poisons can disrupt the cell membrane by means of specialized proteins called perforins. They also precipitate the lysis of the cell nucleus and initiate cell death (see figure 15.21).

Passive immunity occurs when an individual receives immune substances (antibodies) that were produced actively in the body of another human or animal donor. The recipient is protected for a time even though he or she has not had prior exposure to the antigen. It is characterized by: (1) lack of memory for the original antigen, (2) lack of production of new antibodies against that disease, (3) immediate onset of protection, and (4) short-term effectiveness, because antibodies have a limited period of function, and ultimately, the recipient's body disposes of them. Passive immunity can also be natural or artificial in origin.

Natural immunity encompasses any immunity that is acquired during the normal biological experiences of an individual rather than through medical intervention.

Artificial immunity is protection from infection obtained through medical procedures. This type of immunity is induced by immunization with vaccines and immune serum.

Figure 15.23 illustrates the various possible combinations of acquired immunities.

Natural Active Immunities: Getting the Infection

After recovering from infectious disease, a person may be actively resistant to reinfection for a period that varies according to the disease. In the case of childhood viral infections such as measles, mumps, and rubella, this natural active stimulus provides lifelong immunity. Other diseases result in a less extended immunity of a few months to years (such as pneumococcal pneumonia and shigellosis), and reinfection is possible. Even a subclinical infection can stimulate natural active immunity. This probably accounts for the fact that some people are immune to an infectious agent without ever having been noticeably infected with or vaccinated for it.

Natural Passive Immunity: Mother to Child

Natural, passively acquired immunity occurs only as a result of the prenatal and postnatal, mother-child relationship. During fetal life, IgG antibodies circulating in the maternal bloodstream are small enough to pass or be actively transported across the placenta. Antibodies against tetanus, diphtheria, pertussis, and several viruses regularly cross the placenta. This natural mechanism provides an infant with a mixture of many maternal antibodies that can protect it for the first few critical months outside the womb, while its own immune system is gradually developing active immunities. Depending upon the microbe, passive protection lasts anywhere from a few months to a year. But eventually, the infant's body clears the antibody. Most childhood vaccinations are timed so that there is no lapse in protection against common childhood infections.

Another source of natural passive immunity comes to the baby by way of mother's milk (**Medical Microfile 15.3**). Although the human infant acquires 99% of natural passive immunity in utero and only about 1% through nursing, the milk-borne antibodies provide a special type of intestinal protection that is not forthcoming from transplacental antibodies.

Artificial Immunity: Immunization

Immunization is any clinical process that produces immunity in a subject. Because it is often used to give advance protection against infection, it is also called *immunoprophylaxis.* The use of these terms is sometimes imprecise, thus it should be stressed that active immunization, in which a person is administered antigen, is synonymous with vaccination, and that passive immunization, in which a person is given antibodies, is a type of immune therapy.

Vaccination: Artificial Active Immunization

The term *vaccination* originated from the Latin word *vacca* (cow), because the cowpox virus was used in the first preparation for active immunization against smallpox (see Historical Highlights 16.1). Vaccination exposes a person to a specially prepared microbial (antigenic) stimulus, which then triggers the immune system to produce antibodies and lymphocytes to protect the person upon future exposure to that microbe. As with natural active immunity, the degree and length of protection vary. Commercial vaccines are currently available for about 26 diseases. Methods of vaccine antigen selection and modes of vaccination are discussed more fully in chapter 16.

Acquired Immunity

Natural Immunity
is acquired through the normal life experiences of a human and is not induced through medical means.

Artificial Immunity
is that produced purposefully through medical procedures (also called immunization).

Active Immunity
is the consequence of a person developing his own immune response to a microbe.

Passive Immunity
is the consequence of one person receiving a performed immunity made by another person.

Active Immunity
is the consequence of a person developing his own immune response to a microbe.

Passive Immunity
is the consequence of one person receiving a performed immunity made by another person.

  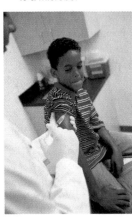 

FIGURE 15.23

Categories of acquired immunities. Natural immunities, which occur during the normal course of life, are either active (acquired from an infection and then recovering) or passive (antibodies donated by the mother to her child). Artificial immunities are acquired through medical practices and can be active (vaccinations with antigen, to stimulate an immune response) or passive (immune therapy with a serum containing antibodies).

MEDICAL MICIROFILE 15.3
Breast Feeding: The Gift of Antibodies

An advertising slogan from the past claims that cow's milk is "nature's most nearly perfect food." One could go a step further and assert that human milk is nature's *perfect* food for young humans. Clearly, it is loaded with essential nutrients, not to mention being available on demand from a readily portable, hygienic container that does not require refrigeration or warming. But there is another and perhaps even greater benefit. During lactation, the breast becomes a site for the proliferation of lymphocytes that produce IgA, a special class of antibody that protects the mucosal surfaces from local invasion by microbes. The very earliest secretion of the breast, a thin, yellow milk called **colostrum,** is very high in IgA. These antibodies form a protective coating in the gastrointestinal tract of a nursing infant that guards against infection by a number of enteric pathogens (*Escherichia coli, Salmonella,* poliovirus, rotavirus). Protection at this level is especially critical because an infant's own IgA and natural intestinal bar-

riers are not yet developed. As with immunity in utero, the necessary antibodies will be donated only if the mother herself has active immunity to the microbe through a prior infection or vaccination.

In recent times, the ready availability of artificial formulas and the changing life-styles of women have reduced the incidence of breast feeding. Where adequate hygiene and medical care prevail, bottle-fed infants get through the critical period with few problems, because the foods given them are relatively sterile and they have received protection against some childhood infections in utero. Mothers in developing countries with untreated water supplies or poor medical services are strongly discouraged from using prepared formulas, because they can actually inoculate the baby's intestine with pathogens from the formula. Millions of neonates suffer from severe and life-threatening diarrhea that could have been prevented by the hygienic practice of nursing.

Immunotherapy: Artificial Passive Immunization

In immunotherapy, a patient at risk for acquiring a particular infection is administered a preparation that contains specific antibodies against that infectious agent. In the past, these therapeutic substances were obtained by vaccinating animals (horses in particular), then taking blood and extracting the serum. However, horse serum is now used only in limited situations because of the potential for hypersensitivity to it. Pooled human serum from donor blood (gamma globulin) and immune serum globulins containing high quantities of antibodies are more frequently used. Immune serum globulins are used to protect people who have been exposed to hepatitis, measles, and rubella. More specific immune serum, obtained from patients recovering from a recent infection, is useful in preventing and treating hepatitis B, rabies, pertussis, and tetanus.

An outline summarizing the system of host defenses covered in chapters 14 and 15 was presented in figure 14.1. You may want to use this resource to review major aspects of immunity and to guide you in answering certain questions (see concept question 17).

CHAPTER CHECKPOINTS

T cells do not produce antibodies. Instead, they produce different cytokines that play diverse roles in the immune response. Each subset of T cell produces a particular cytokine that stimulates lymphocytes or destroys foreign cells.

Active immunity means that your body produces antibodies to a disease agent. If you contract the disease, you can develop natural active immunity. If you are vaccinated, your body will produce artificial active immunity.

In passive immunity, you receive antibodies from another person. Natural passive immunity comes from the mother. Artificial passive immunity is administered medically.

CHAPTER CAPSULE WITH KEY TERMS

Specific Immunity*

I. Development of Lymphocyte Specificity/Receptors

A. Acquired immunity involves the reactions of B and T lymphocytes to foreign molecules, or **antigens.** Before they can react, each lymphocyte must undergo differentiation into its final functional type by developing protein receptors for antigen, the specificity of which is genetically controlled and unique for each type of lymphocyte.

B. The **clonal selection theory** explains this process. Genetic recombination and mutation during embryonic and fetal development produce billions of different lymphocyte clones, each bearing a different receptor. This provides a huge lymphocyte repertoire required to react with antigens.

C. *Tolerance to self,* the elimination of any lymphocyte clones that can attack self, occurs during this time.

D. The receptors on B cells are **immunoglobulin (Ig)** molecules, and receptors on T cells are smaller glycoprotein molecules.

E. Other receptors needed in recognition are governed by the **major histocompatibility (MHC)** gene complex, which is also referred to as the **human leukocyte antigen (HLA)** complex.

F. Expression of these genes gives rise to receptors on most cells that govern cell communication and recognition of self and antigens.

G. *B-Cell Maturation:* Immature B stem cells originate in the yolk sac, liver, and bone marrow and differentiate into mature cells under the influence of special stromal cells in the bone marrow. Mature cells acquire Ig receptors and home to predetermined sites in lymphoid organs, ready to react with antigen.

H. *T-Cell Maturation:* Immature T stem cells originate in the same areas of the embryo and fetus but mature under the influence of the thymus gland. Specificity is acquired through addition of CD receptors, and mature cells home to different sites in lymphoid organs.

II. Introduction of Antigens/Immunogens

A. An antigen (Ag) is any substance that stimulates an immune response.

1. Requirements for **antigenicity** include foreignness (recognition as nonself), large size, and complexity of cell or molecule.

2. Foreign cells and large complex molecules (over 10,000 MW) are most antigenic.

3. Foreign molecules less than 1,000 MW (**haptens**) are not antigenic unless attached to a larger carrier molecule.

4. The **antigenic determinant** is the small molecular group of the foreign substance that is recognized by lymphocytes. Cells, viruses, and large molecules can have numerous antigenic determinants.

B. Special categories of antigens include:

1. **Autoantigens,** molecules on self tissues for which tolerance is inadequate.

2. **Alloantigens,** cell surface markers of one individual that are antigens to another of that same species.

3. **Heterophilic antigens,** molecules from unrelated species that bear similar antigenic determinants.

4. **Superantigens,** complex bacterial toxins.

5. **Allergen,** the antigen that provokes allergy.

III. Cooperation in Immune Reactions to Antigen

A. T-cell-dependent antigens must be processed by large phagocytes such as dendritic cells or macrophages called **antigen-presenting cells (APCs).** An APC alters the antigen and attaches it to its MHC receptor for presentation to lymphocytes.

B. The presentation of a single antigen involves a direct collaboration between an APC, a T helper (T_H), and an antigen specific B or T cell.

C. Cytokines involved are **interleukin-1** from the APC which activates the T_H cells, and **interleukin-2** produced by the T_H cell, which activates B and other T cells.

IV. B-Cell Activation and Antibody Production

A. Once B cells process the antigen, interact with T_H cells, and are stimulated by B-cell growth and differentiation factors, they enter the cell cycle in preparation for mitosis and clonal expansion.

B. Divisions give rise to **plasma cells** that secrete antibodies and **memory cells** that can react to that same antigen later.

1. *Nature of Antibodies* (Immunoglobulins): A single immunoglobulin molecule (monomer) is a large Y-shaped protein molecule consisting of four polypeptide chains. It contains two identical fragments (**Fab**) with ends that form the

---

*Also review figure 15.1.

active site that binds with a unique specificity to an antigen. The single fragment of the antibody (**Fc**) binds to self.

2. *Antigen-Antibody (Ag-Ab) Reactions* include **opsonization, neutralization** by **antitoxin, agglutination,** and complement fixation.
   a. The Fc portion can bind to various body cells and mediate inflammation and allergy.
3. The five **antibody classes,** which differ in size and function, are **IgG, IgA** (secretory Ab), **IgM, IgD,** and **IgE.**
4. *Antibodies in Serum* (**Antiserum**):
   a. Serum antibodies can be identified through electrophoresis and quantified by testing the **titer** (levels of antibodies) over time.
   b. The first introduction of an Ag to the immune system produces a **primary response,** with a gradual increase in Ab titer.
   c. The second contact with the same Ag produces a **secondary,** or **anamnestic, response,** due to memory cells produced during initial response.
5. **Monoclonal antibodies (MAB)** are single specificity antibodies formed by fusing a mouse B cell with a cancer cell. MABs are used in diagnosis of disease, identification of microbes, and therapy.

V. **T Cells and Cell-Mediated Immunity (CMI)**
   A. T cells act directly against antigens and foreign cells. T cells secrete cytokines (**interleukin, interferon, lymphotoxins**) that act on other cells. *Sensitized T cells* proliferate into long-lasting memory

T cells and one of the following depending upon its **CD** receptor:
1. **T helper** (T_H) or CD4 cells assist other T cells and B cells.
2. **Cytotoxic** (T_C) or CD8 cells destroy complex foreign or abnormal cells by forming **perforins** that lyse the cell.
3. Delayed hypersensitivity cells (T_D) cause a form of hypersensitivity.
4. T suppressor cells (T_S) limit the actions of other T cells and B cells.

VI. **Classification of Acquired Immunity**
   A. Immunities acquired through B and T lymphocytes can be classified by a simple system.
   1. **Natural immunity** is acquired as part of normal life experiences.
   2. **Artificial immunity** is acquired through medical procedures such as immunization.
   3. **Active immunity** results when a person is challenged with antigen that stimulates production of antibodies. It creates memory, takes time, and is lasting.
   4. In **passive immunity,** preformed antibodies are donated to an individual. It does not create memory, acts immediately, and is short term.
   B. Combinations of acquired immunity are:
   1. Natural active, acquired upon infection and recovery, and natural passive, acquired by a child through placenta and breast milk.
   2. Artificial active (vaccination), acquired through inoculation with a selected antigen, and artificial passive, administration of **immune serum** or globulin.

MULTIPLE-CHOICE QUESTIONS

1. The primary B-cell receptor is
   a. IgD
   b. IgA
   c. IgE
   d. IgG

2. Which of the MHC classes is required by T cells to recognize antigen?
   a. MHC I
   b. MHC II
   c. MHC III
   d. a and b

3. In humans, B cells mature in the _____, and T cells mature in the _____.
   a. GALT, liver
   b. bursa, thymus
   c. bone marrow, thymus
   d. lymph nodes, spleen

4. Small, simple molecules are _____ antigens.
   a. poor
   b. never
   c. good
   d. heterophilic

5. An example of a mosaic antigen is
   a. albumin
   b. a lipopolysaccharide molecule
   c. a virus
   d. a hapten

6. Which type of cell actually secretes antibodies?
   a. T cells
   b. macrophages
   c. plasma cells
   d. monocytes

7. The cross-linkage of antigens by antibodies is known as
   a. opsonization
   b. a cross-reaction
   c. agglutination
   d. complement fixation

8. The greatest concentration of antibodies is found in the _____ fraction of the serum.
   a. gamma globulin
   b. albumin
   c. beta globulin
   d. alpha globulin

9. _____ is a measurement of the relative amount of immunoglobulin in the serum.
   a. Gamma globulin
   b. Secretory antibody
   c. Antibody titer
   d. Complement fixation

10. A cytokine that stimulates the activity of B and T cells is
   a. interleukin-2
   b. opsonin
   c. lymphotoxin
   d. interleukin-1

11. _____ T cells assist in the functions of certain B cells and other T cells.
   a. Sensitized
   b. Cytotoxic
   c. Helper
   d. Natural killer

12. T$_C$ cells are important in controlling
  a. virus infections
  b. allergy
  c. autoimmunity
  d. all of these

13. Vaccination is synonymous with _____ immunity.
  a. natural active
  b. artificial passive
  c. artificial active
  d. natural passive

14. Fusion between a plasma cell and a tumor cell creates a
  a. lymphoblast
  b. hybridoma
  c. natural killer cell
  d. myeloma

15. T cells are the source of which cytokines?
  a. interleukin
  b. interferon
  c. lymphotoxin
  d. a and b
  e. all of the choices

16. Which of the following can serve as antigen-presenting cells (APCs)?
  a. T cells
  b. B cells
  c. macrophages
  d. dendritic cells
  e. b, c, and d

17. **Multiple matching.** Place all possible matches in the space at the left.

_____ IgG	a. found in mucous secretions
_____ IgA	b. a monomer
_____ IgD	c. a dimer
_____ IgE	d. has greatest number of Fabs
_____ IgM	e. is primarily a surface receptor for B cells
	f. major Ig of primary response to Ag
	g. major Ig of secondary response to Ag
	h. crosses the placenta
	i. fixes complement
	j. involved in allergic reactions

CONCEPT QUESTIONS

1. a. What function do receptors play in specific immune responses?
   b. How can receptors be made to vary so widely?

2. Describe the major histocompatibility complex, and explain how it participates in immune reactions.

3. a. Evaluate the following statement: Each different lymphocyte type must have a unique receptor to react with antigen.
   b. How many different Ags might one be expected to meet up with during life?

4. a. What constitutes a clone of lymphocytes?
   b. Explain the clonal selection theory of antibody specificity and diversity.
   c. During development, when is antigen not needed, and why is it not needed?
   d. When is antigen needed?
   e. Why must the body develop tolerance to self?

5. a. Trace the development of the B-cell receptor from gene to cell surface.
   b. What is the structure of the receptor?
   c. What is the function of the variable regions?

6. a. Trace the origin and development of B lymphocytes; of T lymphocytes.
   b. What is happening during lymphocyte maturation?

7. Describe three ways that B cells and T cells are similar and at least five major ways in which they are different.

8. a. What is an antigen or immunogen?
   b. What is the antigenic determinant?
   c. How do foreignness, size, and complexity contribute to antigenicity?
   d. What is a mosaic antigen?
   e. Why are haptens by themselves not antigenic, even though foreign?
   f. How can they be made to behave as antigens?

9. a. Differentiate among autoantigens, alloantigens, and heterophile antigens.
   b. Explain briefly what importance each has in immune reactions.

10. a. Describe the actions of an antigen-presenting cell.
    b. What is the difference between a T-cell-dependent and T-cell-independent response?

11. a. Trace the immune response system, beginning with the entry of a T-cell-dependent antigen, antigen processing, presentation, the cooperative response among the macrophage and lymphocytes, and the reactions of activated B and T cells.
    b. What are the actions of interleukins-1 and -2?

12. a. On what basis is a particular B-cell clone selected?
    b. How are B cells activated, and what events are involved in this process?
    c. What happens after B cells are activated?
    d. What are the functions of plasma cells, clonal expansion, and memory cells?

13. a. Describe the structure of immunoglobulin.
    b. What are the functions of the Fab and Fc portions?
    c. Describe four or five ways that antibodies function in immunity.
    d. Describe the attachment of Abs to Ags. (What eventually happens to the Ags?)

14. a. Contrast the primary and secondary response to Ag.
    b. Explain the type, order of appearance, and amount of immunoglobulin in each response and the reasons for them.
    c. What causes the latent period? The anamnestic response?
    d. Explain how monoclonal and polyclonal antibodies are different.
    e. Outline the basic steps in production of monoclonal antibodies.
    f. Describe several possible applications of monoclonals in medicine.

15. a. Why are the immunities involving T cells called cell-mediated?
    b. How do T cells become sensitized?
    c. Summarize the function of each category of T cell and the types of receptors with which they are associated. Define cytokines and interleukins, and provide some examples of them.
    d. How do cytotoxic cells kill their target?
    e. Why would the immune system naturally require suppression?
    f. What is a natural killer cell, and what are its functions?

16. a. Contrast active and passive immunity in terms of how each is acquired, how long it lasts, whether memory is triggered, how soon it becomes effective, and what immune cells and substances are involved.

    b. Name at least two major ways that natural and artificial immunities are different.

17. **Multiple matching.** (Summarizes information from chapters 14 and 15 [see figure 14.1].) In the blanks on the left, place the letters of all of the host defenses and immune responses in the right column that can fit the description.

    ____ vaccination for tetanus
    ____ lysozyme in tears
    ____ immunization with horse serum
    ____ in utero transfer of antibodies

    a. active
    b. passive
    c. natural
    d. artificial

    ____ booster injection for diphtheria
    ____ recovery from a case of mumps
    ____ colostrum
    ____ interferon
    ____ action of neutrophils
    ____ injection of gamma globulin
    ____ recovery from a case of mumps
    ____ edema
    ____ humans having protection from canine distemper virus
    ____ stomach acid
    ____ cilia in trachea
    ____ asymptomatic chickenpox
    ____ complement

    e. acquired
    f. innate, inborn
    g. chemical barrier
    h. mechanical barrier
    i. genetic barrier
    j. specific
    k. nonspecific
    l. inflammatory response
    m. second line of defense
    n. none of these

CRITICAL-THINKING QUESTIONS

1. What is the advantage of having lymphatic organs screen the body fluids, directly and indirectly?

2. Cells contain built-in suicide genes to self-destruct by apoptosis under certain conditions. Can you explain why development of the immune system might depend in part on this sort of adaptation?

3. a. Give some possible explanations for the need to have immune tolerance.

   b. Why would it be necessary for the T cells to bind both antigen and self (MHC) receptors?

4. Double-stranded DNA is a large, complex molecule, but it is not generally immunogenic unless it is associated with proteins or carbohydrates. Can you think why this might be so? (Hint: How universal is DNA?)

5. How would you go about producing monoclonal antibodies that would participate in the destruction of cancer cells but would not kill normal human cells?

6. Explain how it is possible for people to give a false positive reaction in blood tests for syphilis, AIDS, and infectious mononucleosis.

7. Describe the cellular/microscopic pathology in the immune system of AIDS patients that results in opportunistic infections and cancers.

8. Explain why most immune reactions result in a polyclonal collection of antibodies.

9. a. Why does rising titer of Abs indicate infection?

   b. Why do some vaccinations require three or four boosters?

10. a. What event must occur for passive immunity to exist at all?

    b. Armed with this knowledge, suggest an effective way to immunize a fetus.

    c. Is this method uniformly safe?

    d. What would be an even safer way to ensure that fetuses get necessary antibodies?

11. a. Combine information on the functions of different classes of Ig to explain the exact mechanisms of natural passive immunity (both transplacental and colostrum-induced).

    b. Why are these sorts of immunity short-term?

12. Use a football game or warfare analogy to produce a scenario depicting the major activities of the immune system from this chapter and from chapter 14.

13. Using words and arrows, complete a flow outline of an immune response, beginning with entrance of antigen; include processing, cell interaction, involvement of cytokines, and the end results for B and T cells.

INTERNET SEARCH TOPICS

1. Find information on new monoclonal antibodies being developed and used for medical purposes. Write a short description summarizing your research into this topic.

2. Access Doc Kaiser's microbiology home page (use any search engine). Click on the immune system section and observe animations of T- and B-cell activation.

3. Visit the student Online Learning Center at www.mhhe.com/talaro5. Go to chapter 15, Internet Search Topics, and log on to the available websites to discover useful diagrams, animations, and information on the immune system. Or, use a search engine and type in the words "immune response and interactive images."

CHAPTER 16

Immunization and Immune Assays

A n expanded knowledge of immune function has yielded significant breakthroughs in medical technology for manipulating and monitoring the immune system. A highly practical benefit of immunology is administering vaccines and other immune treatments against common infectious diseases such as hepatitis and tetanus that once caused untold sickness and death. Another valuable application of this technology is testing the blood for signs of infection or disease. It is routine medical practice to diagnose such infections as HIV, syphilis, hepatitis B, and rubella by means of immunologic analysis. Both separately and in combination, these methods have made sweeping contributions to individual and community health by improving diagnosis and treatment, and controlling the spread of disease. Many hopes for the survival of humankind lie in our ability to harness the amazing workings of the immune system.

Chapter Overview

- Discoveries in basic immune function have created powerful medical tools to artificially induce protective immunities and to determine the status of the immune system.
- Immunization may be administered by means of passive and active methods.
- Passive immunity is acquired by infusing antiserum taken from other patients' blood that contains high levels of protective antibodies. This form of immunotherapy is short-lived.
- Active methods involve administering a vaccine against an infectious agent that may be encountered in the future. This form of protection provides a longer-lived immunity.
- Vaccines contain some form of microbial antigen that has been altered so that it stimulates a protective immune response without causing the disease.
- Vaccines can be made from whole dead or live cells and viruses, parts of cells or viruses, or by recombinant DNA techniques.
- Reactions between antibodies and antigens provide specific and sensitive tests that can be used in diagnosis of disease and identification of pathogens.
- Serology involves the testing of a patient's blood serum for antibodies that can indicate a current or past infection and the degree of immunity.
- Tests that produce visible interactions between antibodies and antigens include agglutination, precipitation, and complement fixation.

Detail from "The Cowpock," an 1808 etching that caricatured the worst fears of the English public concerning Edward Jenner's smallpox vaccine.

- Assays are tests that separate antigens and antibodies and visualize them with radioactivity or fluorescence. These include immunoelectrophoresis, Western blot, and direct and indirect immunoassays.

Practical Applications of Immunologic Function

A knowledge of the immune system and its responses to antigens has provided extremely valuable biomedical applications in two major areas: (1) use of antiserum and vaccination to provide artificial protection against disease and (2) diagnosis of disease through immunologic testing.

IMMUNIZATION: METHODS OF MANIPULATING IMMUNITY FOR THERAPEUTIC PURPOSES

The concept of artificially induced immunity was introduced in chapters 14 and 15. Methods that actively or passively immunize people are widely used in disease prevention and treatment. In the case of passive immunization, a patient is given preformed antibodies, which is actually a form of **immunotherapy.** In the case of active immunization, a patient is vaccinated with a microbe or its antigens, providing a form of advance protection.

Immunotherapy: Artificial Passive Immunity

The first attempts at passive immunization involved the transfusion of horse serum containing antitoxins to prevent tetanus and to treat patients exposed to diphtheria. Since then, antisera from animals have been replaced with products of human origin that function with various degrees of specificity. Immune serum globulin (ISG), sometimes called *gamma globulin,* contains immunoglobulin extracted from the pooled blood of at least 1,000 human donors. The method of processing ISG concentrates the antibodies to increase potency and eliminates potential pathogens (such as the hepatitis B and HIV viruses). It is a treatment of choice in preventing measles and hepatitis A and in replacing antibodies in immunodeficient pa-

tients. Most forms of ISG are injected intramuscularly to minimize adverse reactions, and the protection it provides lasts 2 to 3 months.

A preparation called specific immune globulin (SIG) is derived from a more defined group of donors. Companies that prepare SIG obtain serum from patients who are convalescing and in a hyperimmune state after such infections as pertussis, tetanus, chickenpox, and hepatitis B. These globulins are preferable to ISG because they contain higher titers of specific antibodies obtained from a smaller pool of patients. Although useful for prophylaxis in persons who have been exposed or may be exposed to infectious agents, these sera are often limited in availability.

When a human immune globulin is not available, antisera and antitoxins of animal origin can be used. Sera produced in horses are available for diphtheria, botulism, and spider and snake bites. Unfortunately, the presence of horse antigens can stimulate allergies such as serum sickness or anaphylaxis (see chapter 17). Although donated immunities only last a relatively short time, they act immediately and can protect patients for whom no other useful medication or vaccine exists.

Artificial Active Immunity: Vaccination

Active immunity can be conferred artificially by **vaccination**—exposing a person to material that is antigenic but not pathogenic. The discovery of vaccination was one of the farthest reaching and most important developments in medical science **(Historical Highlights 16.1).** The basic principle behind vaccination is to stimulate

HISTORICAL HIGHLIGHTS 16.1
The Lively History of Active Immunization

The basic notion of immunization has existed for thousands of years. It probably stemmed from the observation that persons who had recovered from certain communicable diseases rarely if ever got a second case. Undoubtedly, the earliest crude attempts involved bringing a susceptible person into contact with a diseased person or animal. The first recorded attempt at immunization occurred in sixth century China. It consisted of drying and grinding up smallpox scabs and blowing them with a straw into the nostrils of vulnerable family members. By the tenth century, this practice had changed to the deliberate inoculation of dried pus from the smallpox pustules of one patient into the arm of a healthy person, a technique later called **variolation** (variola is the smallpox virus). This method was used in parts of the Far East for centuries before Lady Mary Montagu brought it to England in 1721. Although the principles of the technique had some merit, unfortunately, many recipients and their contacts died of smallpox. This outcome vividly demonstrates a cardinal rule for a workable vaccine: It must contain an antigen that will provide protection but not cause the disease. Variolation was so controversial that any English practitioner caught doing it was charged with a felony.

Eventually, this human experimentation paved the way for the first really effective vaccine, developed by the English physician Edward Jenner in 1796 (see chapter 24). Jenner conducted the first scientifically controlled study, one that had a tremendous impact on the advance of medicine. His work gave rise to the words **vaccine** and **vaccination** (from L., *vacca,* cow), which now apply to any immunity obtained by inoculation with selected antigens. Jenner was inspired by the case of a dairymaid

who had been infected by a pustular infection called cowpox. This is a related virus that afflicts cattle but causes a milder condition in humans. She explained that she and other milkmaids had remained free of smallpox. Other residents of the region expressed a similar confidence in the cross-protection of cowpox. To test the effectiveness of this new vaccine, Jenner prepared material from human cowpox lesions and inoculated a young boy. When challenged 2 months later with an injection of crusts from a smallpox patient, the boy proved immune.

Jenner's discovery—that a less pathogenic agent could confer protection against a more pathogenic one—is especially remarkable in view of the fact that microscopy was still in its infancy and the nature of viruses was unknown. At first, the use of the vaccine was regarded with some fear and skepticism (see chapter opening illustration). When Jenner's method proved successful and word of its significance spread, it was eventually adopted in many other countries. Eventually, the original virus mutated into a unique strain (*vaccinia* virus) that became the basis of the current vaccine. In 1973 the World Health Organization declared that smallpox had been eradicated. As a result, smallpox vaccination had been discontinued until recently, due to the threat of bioterrorism.

Other historical developments in vaccination included using heat-killed bacteria in vaccines for typhoid fever, cholera, and plague and techniques for using neutralized toxins for diphtheria and tetanus. Throughout the history of vaccination, there have been vocal opponents and minimizers, but numbers do not lie. Whenever a vaccine has been introduced, the prevalence of that disease has declined (see figure 19.12 for a dramatic example).

TABLE 16.1

Checklist of Requirements for an Effective Vaccine

- It should have a low level of adverse side effects or toxicity and not cause serious harm.
- It should protect against exposure to natural, wild forms of pathogen.
- It should stimulate both antibody (B-cell) response and cytotoxic (T-cell) response.
- It should have long-term, lasting effects (produce memory).
- It should not require numerous doses or boosters.
- It should be inexpensive, have a relatively long shelf life, and be easy to administer.

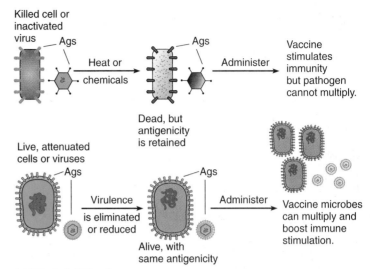

(a) **Whole Cell Vaccines**

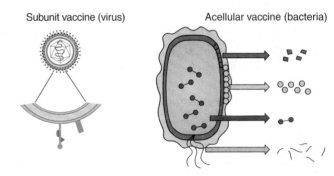

(b) **Vaccines from Microbe Parts**

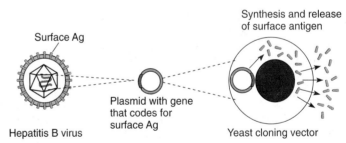

(c) **Recombinant Vaccine**

FIGURE 16.1

Strategies in vaccine design. **(a)** Whole cells or viruses, killed or attenuated. **(b)** Acellular or subunit vaccines are made by disrupting the microbe to release various molecules or cell parts that can be isolated and purified. **(c)** Recombinant vaccines are made by isolating a gene for antigenicity from the pathogen (here a hepatitis virus) and splicing it into a plasmid. Insertion of the recombinant plasmid into a cloning host (yeast) results in the production of large amounts of viral surface antigen to use in vaccine preparation.

a primary and secondary anamnestic response (see figure 15.18) that primes the immune system for future exposure to a virulent pathogen. If this pathogen enters the body, the immune response will be immediate, powerful, and sustained.

Vaccines have profoundly reduced the prevalence and impact of many infectious diseases that were once common and often deadly. In this section, we survey the principles of vaccine preparation and important considerations surrounding vaccine indication and safety. (Vaccines are also given specific consideration in later chapters on bacterial and viral diseases.)

Principles of Vaccine Preparation   A vaccine must be considered from the standpoints of antigen selection, effectiveness, ease in administration, safety, and cost. In natural immunity, an infectious agent stimulates a relatively long-term protective response. In artificial active immunity, the objective is to obtain this same response with a modified version of the microbe or its components. Qualities of an effective vaccine are listed in **table 16.1.** Most vaccine preparations are based on one of the following antigen preparations (**figure 16.1**):

1. killed whole cells or inactivated viruses;
2. live, attenuated cells or viruses;
3. antigenic molecules derived from bacterial cells or viruses, or
4. genetically engineered microbes or microbial antigens.

A survey of the major licensed vaccines and their indications is presented in **table 16.2.**

Large, complex antigens such as whole cells or viruses are very effective immunogens. Depending on the vaccine, these are either killed or attenuated. **Killed** or **inactivated vaccines** are prepared by cultivating the desired strain or strains of a bacterium or virus and treating them with formalin, radiation, heat, or some other agent that does not destroy antigenicity. One type of vaccine for the bacterial disease cholera is of this type (see chapter 20). Salk polio vaccine and one form of influenza vaccine contain inactivated viruses. Because the microbe does not multiply, killed vaccines often require a larger dose and more boosters to be effective.

A number of vaccines are prepared from live, **attenuated*** microbes. **Attenuation** is any process that substantially lessens or negates the virulence of viruses or bacteria. It is usually achieved by modifying the growth conditions or manipulating microbial genes in a way that eliminates virulence factors. Attenuation methods include long-term cultivation, selection of mutant strains that grow at colder temperatures (cold mutants), passage of the microbe through unnatural hosts or tissue culture, and removal of virulence genes. The vaccine for tuberculosis (BCG) was obtained after 13

---

*attenuated (ah-ten′-yoo-ayt-ed) L. *attenuare,* to thin. Able to multiply, but nonvirulent.

TABLE 16.2

Currently Approved Vaccines

Disease/Vaccine Preparation	Route of Administration	Recommended Usage/Comments
Contain Killed Whole Bacteria		
Cholera	Subcutaneous (SQ) injection	For travelers; effect not long-term
Plague	SQ	For exposed individuals and animal workers; variable protection
Contain Live, Attenuated Bacteria		
Tuberculosis (BCG)	Intradermal (ID) injection	For high-risk occupations only; protection variable
Typhoid	Oral	For travelers only; low rate of effectiveness
Acellular Vaccines (Capsular Polysaccharides)		
Meningitis (meningococcal)	SQ	For protection in high-risk infants, military recruits; short duration
Meningitis (*Haemophilus influenzae*)	IM	For infants and children; may be administered with DTaP
Pneumococcal pneumonia	IM or SQ	Important for people at high risk: the young, elderly, and immunocompromised; moderate protection
Pertussis (aP)	IM	For newborns and children; contains recombinant protein antigens
Toxoids (Formaldehyde-Inactivated Bacterial Exotoxins)		
Diphtheria	IM	A routine childhood vaccination; highly effective in systemic protection
Tetanus	IM	A routine childhood vaccination; highly effective
Pertussis	IM	A routine childhood vaccination; highly effective
Botulism	IM	Only for exposed individuals such as laboratory personnel
Contain Inactivated Whole Viruses		
Poliomyelitis (Salk)	IM	Routine childhood vaccine; now used as first choice
Rabies	IM	For victims of animal bites or otherwise exposed; effective
Influenza	IM	For high-risk populations; requires constant updating for new strains; immunity not durable
Hepatitis A	IM	Protection for travelers, institutionalized people
Contain Live, Attenuated Viruses		
Adenovirus infection	Oral	For immunizing military recruits
Measles (rubeola)	SQ	Routine childhood vaccine; very effective
Mumps (parotitis)	SQ	Routine childhood vaccine; very effective
Poliomyelitis (Sabin)	Oral	Routine childhood vaccine; very effective, but can cause polio
Rubella	SQ	Routine childhood vaccine; very effective
Chickenpox (varicella)	SQ	Routine childhood vaccine; immunity can diminish over time
Yellow fever	SQ	Travelers, military personnel in endemic areas
Influenza	Inhaled	See inactivated above
Subunit Viral Vaccines		
Hepatitis B	IM	Recommended for all children, starting at birth; also for health workers and others at risk
Influenza	IM	See influenza above
Recombinant Vaccines		
Hepatitis B	IM	Used more often than subunit, but for same groups
Pertussis	IM	See acellular above

years of subculturing the agent of bovine tuberculosis (see chapter 19). Vaccines for measles, mumps, polio (Sabin), and rubella contain live, nonvirulent viruses. The advantages that favor live preparations are:

1. viable microorganisms can multiply and produce infection (but not disease) like the natural organism;
2. they confer long-lasting protection; and
3. they usually require fewer doses and boosters than other types of vaccines.

Disadvantages of using live microbes in vaccines are that they require special storage facilities, can be transmitted to other people, and can mutate back to a virulent strain (see polio, chapter 25).

If the exact antigenic determinants that stimulate immunity are known, it is possible to produce a vaccine based on a selected component of a microorganism. These vaccines for bacteria are called **acellular** or **subcellular vaccines.** For viruses, they are called **subunit vaccines.** The antigen used in these vaccines may be taken from cultures of the microbes, produced by rDNA technology, or synthesized chemically.

Examples of component antigens currently in use are the capsules of the pneumococcus and meningococcus, the protein surface antigen of anthrax, and the surface proteins of hepatitis B virus. A special type of vaccine is the **toxoid,*** which consists of a purified bacterial exotoxin that has been chemically denatured. By eliciting the production of antitoxins that can neutralize the natural toxin, toxoid vaccines provide protection against toxinoses such as diphtheria and tetanus.

Development of New Vaccines

Despite considerable successes, dozens of bacterial, viral, protozoan, and fungal diseases still remain without a functional vaccine. At the present time, no reliable vaccines are available for malaria, HIV/AIDS, various diarrheal diseases (*E. coli, Shigella*), respiratory diseases, and worm infections that affect over 200 million people per year worldwide. Of all of the challenges facing vaccine specialists, probably the most difficult has been choosing a vaccine antigen that is safe and that properly stimulates immunity. Currently, much attention is being focused on newer strategies for vaccine preparation that employ antigen synthesis, recombinant DNA, and gene cloning technology.

When the exact composition of an antigenic determinant is known, it is possible to synthesize it. This ability permits preservation of antigenicity while greatly increasing antigen purity and concentration. The malaria vaccine currently being used in areas of South America and Africa is composed of three synthetic peptides from the parasite. Several biotechnology companies are using plants to mass produce vaccine antigens. Tests are underway to grow tomatoes, potatoes, and bananas that synthesize proteins from cholera, hepatitis, papillomavirus, and *E. coli* pathogens.

Genetically Engineered Vaccines   Some of the genetic engineering concepts introduced in chapter 10 offer novel approaches to vaccine development. These methods are particularly effective in designing vaccines for obligate parasites that are difficult or expensive to culture, such as the syphilis spirochete or the malaria parasite. This technology provides a means of isolating the genes that encode various microbial antigens, inserting them into plasmid vectors, and cloning them in appropriate hosts. The outcome of recombination can be varied as desired. For instance, the cloning host can be stimulated to synthesize and secrete a protein product (antigen), which is then harvested and purified (figure 16.1c). This is how certain vaccines for hepatitis are currently being prepared. Antigens from the agents of syphilis, *Schistosoma,* and influenza have been similarly isolated and cloned and are currently being considered as potential vaccine material.

Another ingenious technique using genetic recombination has been nicknamed the *Trojan horse* vaccine. The term derives from an ancient legend in which the Greeks sneaked soldiers into the fortress of their Trojan enemies by hiding them inside a large, mobile wooden horse. In the microbial equivalent, genetic material from a selected infectious agent is inserted into a live carrier microbe that is nonpathogenic. In theory, the recombinant microbe will multiply and express the foreign genes, and the vaccine recipient will be immunized against the microbial antigens. Vaccinia, the virus originally used to vaccinate for smallpox, and adenoviruses

have proved practical agents for this technique. Vaccinia is used as the carrier in one of the experimental vaccines for AIDS (see figure 25.19), herpes simplex 2, leprosy, and tuberculosis.

DNA vaccines are being hailed as the most promising of all of the newer approaches to immunization. The technique in these formulations is very similar to gene therapy as described in figure 10.13, except in this case, microbial (not human) DNA is inserted into a plasmid vector and inoculated into a recipient (**figure 16.2a**). The expectation is that the human cells will take up some of the plasmids and express the microbial DNA in the form of proteins. Because these proteins are foreign, they will be recognized during immune surveillance and cause B and T cells to be sensitized and form memory cells.

Experiments with animals have shown that these vaccines are very safe and that only a small amount of the foreign antigen need be expressed to produce effective immunity. Another advantage to this method is that any number of potential microbial proteins can be expressed, making the antigenic stimulus more complex and improving the likelihood that it will stimulate both antibody and cell mediated immunity. At the present time, over 30 DNA-based vaccines are being tested in animals. Vaccines for Lyme disease, hepatitis C, herpes simplex, influenza, tuberculosis, papillomavirus, and malaria are undergoing animal trials, most with encouraging results.

Vaccine effectiveness relies, in part, on the production of antibodies that closely fit the natural antigen. Realizing that such reactions are very much like a molecular jigsaw puzzle, researchers have developed a unique approach to vaccine preparation. The **anti-idiotype** vaccine is based on the principle that the antigen binding (variable) region, or **idiotype,*** of a given antibody (A) can be antigenic to a genetically different recipient and can cause that recipient's immune system to produce antibodies (B; also called anti-idiotypic antibodies) specific for the variable region on antibody A (figure 16.2b). The purpose for making anti-idiotypic antibodies is that they will display an identical configuration as the desired antigen and can be used in vaccines. This method avoids administering a microbial antigen, thus reducing the potential for dangerous side effects. Using monoclonal antibodies, this approach has been used to mimic the surface antigen of hepatitis B virus and *Trypanosoma* with some success.

Concerns about potential terrorist release of Class A bioterrorism agents have created pressure to re-introduce or improve vaccines for smallpox, anthrax, botulism, plague, and tularemia. Smallpox vaccination (see chapter 24) is now being given again to first responders (healthcare providers, the military). Earlier vaccines already exist for anthrax, tularemia, botulism and plague, but they have been difficult to give or not effective. New vaccines are being developed and tested to immunize military and other personnel in the event of a disaster.

Route of Administration and Side Effects of Vaccines

Most vaccines are injected by subcutaneous, intramuscular, or intradermal routes. Oral vaccines are available for only two diseases (table 16.2), but they have some distinct advantages. An oral dose of a vaccine can stimulate protection (IgA) on the mucous membrane of the portal of entry. Oral vaccines are also easier to give, more readily accepted, and well tolerated. Some vaccines require the addition of a special binding substance, or **adjuvant.*** An adjuvant is any compound that enhances immunogenicity and prolongs

*toxoid (tawks'-oyd) Toxinlike.

*idiotype (id'-ee-oh-type) Gr. *idios,* own, peculiar. Another term for the antigen binding site.
*adjuvant (ad'-joo-vunt) L. *adjuvare,* to help.

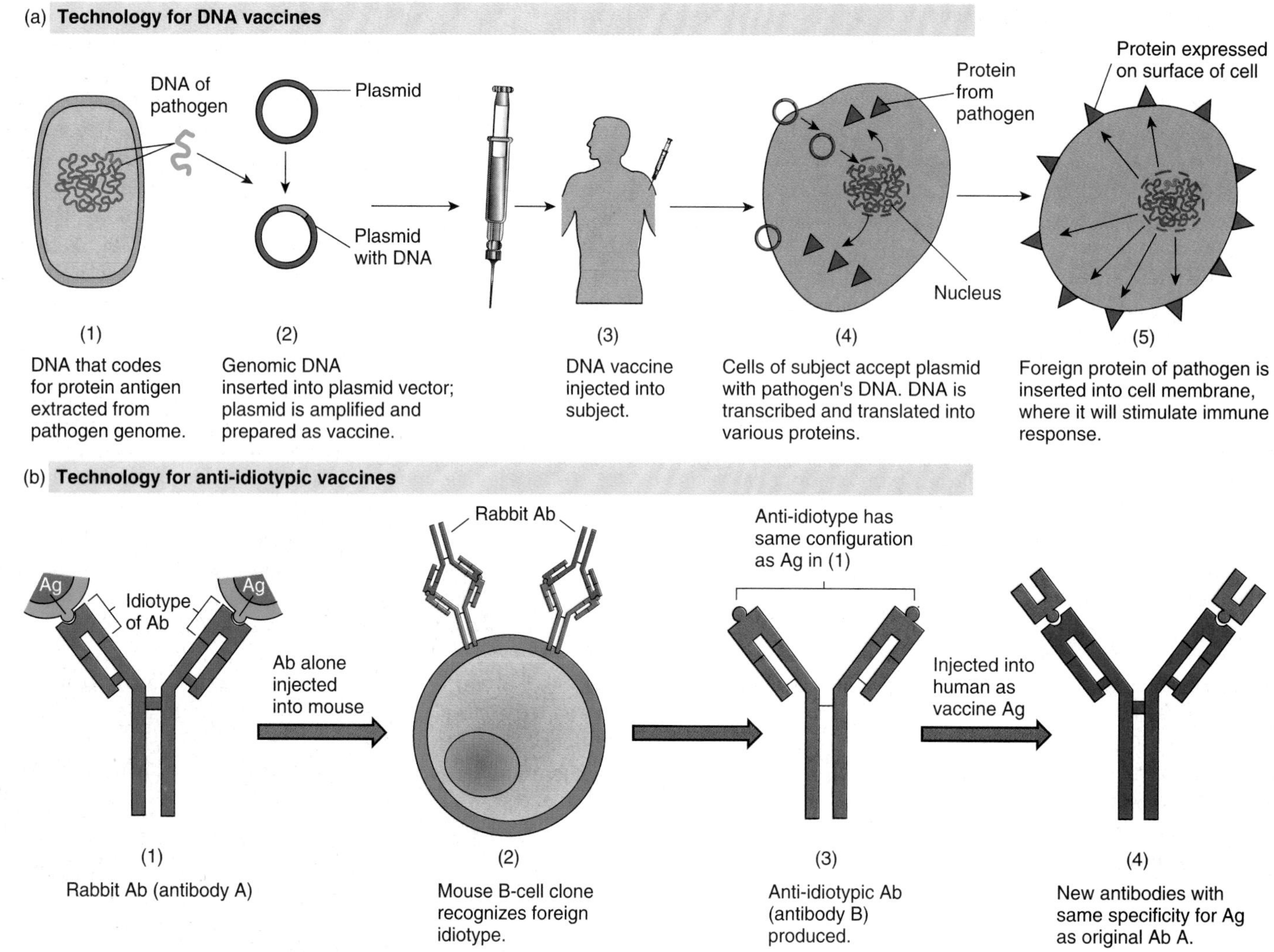

(a) Technology for DNA vaccines

(1)	(2)	(3)	(4)	(5)
DNA that codes for protein antigen extracted from pathogen genome.	Genomic DNA inserted into plasmid vector; plasmid is amplified and prepared as vaccine.	DNA vaccine injected into subject.	Cells of subject accept plasmid with pathogen's DNA. DNA is transcribed and translated into various proteins.	Foreign protein of pathogen is inserted into cell membrane, where it will stimulate immune response.

(b) Technology for anti-idiotypic vaccines

(1)	(2)	(3)	(4)
Rabbit Ab (antibody A)	Mouse B-cell clone recognizes foreign idiotype.	Anti-idiotypic Ab (antibody B) produced.	New antibodies with same specificity for Ag as original Ab A.

FIGURE 16.2

Other technologies useful in preparing vaccines. **(a)** DNA vaccines contain all or part of the pathogen's DNA, which is used to "infect" recipient's cells. Processing of the DNA leads to production of an antigen protein that can stimulate a specific response against that pathogen. **(b)** Anti-idiotypic vaccines use antibodies as foreign proteins to mimic the structure of the natural antigen. See text for details.

antigen retention at the injection site. The adjuvant precipitates the antigen and holds it in the tissues so that it will be released gradually. Its gradual release, presumably, facilitates contact with antigen presenting cells and lymphocytes. Common adjuvants are alum (aluminum hydroxide salts), Freund's adjuvant (emulsion of mineral oil, water, and extracts of mycobacteria), and beeswax.

Vaccines must go through many years of trials in experimental animals and humans before they are licensed for general use. Even after they have been approved, like all therapeutic products, they are not without complications. The most common of these are local reactions at the injection site, fever, allergies, and other adverse reactions. Relatively rare reactions (about 1 case out of 220,000 vaccinations) are panencephalitis (from measles vaccine), back-mutation to a virulent strain (from polio vaccine), disease due to contamination with dangerous viruses or chemicals, and neurological effects of unknown cause (from pertussis and swine flu vaccines). Some patients experience allergic reactions to the medium (eggs or tissue culture) rather than to vaccine antigens. Some recent studies have attempted to link childhood vaccinations to later development of diabetes, asthma, and autism. After thorough examination of records, epidemiologists have found no convincing evidence for a vaccine connection to these diseases.

When known or suspected adverse effects have been detected, vaccines are altered or withdrawn. Most recently, the whole-cell pertussis vaccine was replaced by the acellular capsule (aP) form when it was associated with adverse neurological effects. The live oral rotavirus vaccine had to be withdrawn when children experienced intestinal blockage. Polio vaccine was switched from live, oral to inactivated when too many cases of paralytic disease occurred from back-mutated vaccine stocks. Vaccine companies have also phased out certain preservatives such as thimerosal that are thought to cause allergies and other potential side effects.

Professionals involved in giving vaccinations must understand their inherent risks but also realize that the risks from the infectious disease almost always outweigh the chance of an adverse vaccine reaction. The greatest caution must be exercised in giving live vaccines to immunocompromised or pregnant patients, the latter because of possible risk to the fetus.

To Vaccinate: Why, Whom, and When?

Vaccination confers long-lasting, sometimes lifetime, protection in the individual, but an equally important effect is to protect the public health. Vaccination is an effective method of establishing **herd immunity** in the population. According to this concept, individuals immune to a communicable infectious disease will not harbor it, thus reducing the occurrence of that pathogen. With a larger number of immune individuals in a population (herd), it will be less likely that an unimmunized member of the population will encounter the agent. In effect, collective immunity through mass immunization confers indirect protection on the nonimmune members (such as children). Herd immunity maintained through immunization is an important force in preventing epidemics.

Vaccination is recommended for all typical childhood diseases for which a vaccine is available and for people in certain special circumstances (health workers, travelers, military personnel). **Table 16.3** outlines a general schedule for childhood immunization and the indication for special vaccines. Some vaccines are mixtures

---

TABLE 16.3

Recommended Regimen and Indications for Routine Vaccinations

Vaccine ▼ / Age ►	Birth	1 mo	2 mos	4 mos	6 mos	12 mos	15 mos	18 mos	24 mos	4–6 Yrs	11–12 Yrs	13–18 Yrs	Comments
					range of recommended ages		catch-up vaccination			preadolescent assessment			
Hepatitis B	HB #1	only if mother HBsAg (−)								HepB series			Option depends upon condition of infant; 3 doses given
		HepB #2			HepB #3								
Diphtheria, Tetanus, Pertussis[1]			DTaP	DTaP	DTaP		DTaP		DTaP	Td			Td is tetanus/diphtheria; T is tetanus alone; either one should be given as booster every 10 years.
Haemophilus influenzae Type b			Hib	Hib	Hib	Hib							Schedule depends upon source of vaccine; given with DTaP as TriHIBit or with IPV and HB as Pediatrix
Inactivated Polio[2]			IPV	IPV	IPV				IPV				Similar schedule to DTaP; injected vaccine
Measles, Mumps, Rubella[3]						MMR #1			MMR #2	MMR #2			First dose varies with disease incidence; booster given at either 4–6 or 11–12 years
Varicella						Varicella			Varicella				Used to prevent or limit disease severity of chickenpox in susceptible persons
Pneumococcal[4]			PCV	PCV	PCV	PCV			PCV	PPV			Used to protect children against otitis media
— Vaccines below this line are for selected populations —													
Hepatitis A									Hepatitis A series				
Influenza					Influenza (yearly)								

[1]DTaP. The diphtheria–tetanus–acellular pertussis vaccine has replaced the DTP.
[4]PCV is pneumococcal conjugate vaccine. PPV is pneumococcal polysaccharide vaccine. Infants require the more immunogenic conjugate vaccine.

[2]polio vaccine—is the recommended vaccine for all schedules

[3]Measles vaccine (Attenuvax) can be given alone to children during epidemics or to adults immunized before 1970.

Used in Cases of Specific Risk Due to Occupational or Other Exposure

Vaccine	Group Targeted
Hepatitis B (Recombivax)	Health care personnel; people exposed through life-style
Hepatitis A (Havrix)	Children 2–14 years who live in areas of high prevalence
Pneumococcus (Pneumovax)	Elderly patients, children with sickle-cell anemia
Influenza, polio, tuberculosis (BCG)	Hospital, laboratory, health care workers
Rabies, plague, tularemia	People whose jobs involve contact with animals (veterinarians, forest rangers); known or suspected exposure to rabid animal; living in areas of high incidence
Cholera, hepatitis B, hepatitis A, measles, yellow fever, meningococcal meningitis, polio, rabies, typhoid, plague, botulism	Travelers to endemic regions, including military recruits (varies with geographic destination), laboratory workers

This schedule indicates the recommended ages for routine administration of currently licensed childhood vaccines, as of December 1, 2002, for children through age 18 years. Any dose not given at the recommended age should be given at any subsequent visit when indicated and feasible. ▨ Indicates age groups that warrant special effort to administer those vaccines not previously given. Additional vaccines may be licensed and recommended during the year. Licensed combination vaccines may be used whenever any components of the combination are indicated and the vaccine's other components are not contraindicated. Providers should consult the manufacturers' package inserts for detailed recommendations.

of antigens from several pathogens, notably Pediatrix (DTaP, IPV, and HB). It is also common for several vaccines to be given simultaneously, as occurs in military recruits who receive as many as 15 injections within a few minutes and children who receive boosters for DTaP and polio at the same time they receive the MMR vaccine. Experts doubt that immune interference (inhibition of one immune response by another) is a significant problem in these instances, and the mixed vaccines are carefully balanced to prevent this eventuality. The main problem with simultaneous administration is that side effects can be amplified.

CHAPTER CHECKPOINTS

Knowledge of the specific immune response has two practical applications: (1) commercial production of antisera and vaccines and (2) development of rapid, sensitive methods of disease diagnosis.

Artificial passive immunity usually involves administration of antiserum, and occasionally B and T cells. Antibodies collected from donors (human or otherwise) are injected into people who need protection immediately. Examples include ISG (immune serum globulin) and SIG (specific immune globulin).

Artificial active agents are vaccines that provoke a protective immune response in the recipient but do not cause the actual disease. Vaccination is the process of challenging the immune system with a specially selected antigen. Examples are (1) killed or inactivated microbes, (2) live, attenuated microbes, (3) subunits of microbes, and (4) genetically engineered microbes or microbial parts.

Vaccination programs seek to protect the individual directly through raising the antibody titer and indirectly through the development of herd immunity.

SEROLOGICAL AND IMMUNE TESTS: MEASURING THE IMMUNE RESPONSE *IN VITRO*

The antibodies formed during an immune reaction are important in combating infection, but they hold additional practical value. Characteristics of antibodies (such as their quantity or specificity) can reveal the history of a patient's contact with microorganisms or other antigens. This is the underlying basis of serological testing. **Serology** is the branch of immunology that traditionally deals with *in vitro* diagnostic testing of serum. Serological testing is based on the familiar concept that antibodies have extreme specificity for antigens, so when a particular antigen is exposed to its specific antibody, it will fit like a hand in a glove. The ability to visualize this interaction by some means provides a powerful tool for detecting, identifying, and quantifying antibodies—or for that matter, antigens. The scheme works both ways, depending on the situation. One can detect or identify an unknown antibody using a known antigen, or one can use an antibody of known specificity to help detect or identify an unknown antigen (**figure 16.3**). Modern serological testing has grown into a field that tests more than just serum. Urine, cerebrospinal fluid, whole tissues, and saliva can also be used to determine the immunologic status of patients. These and other immune tests are helpful in confirming a suspected diagnosis or in screening a certain population for disease.

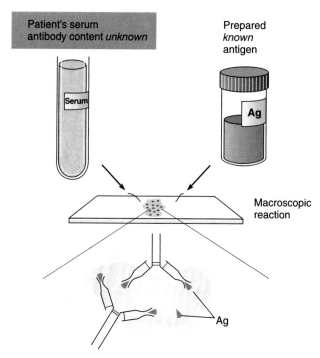

(a) In serological diagnosis of disease, a blood sample is scanned for the presence of antibody using an antigen of known specificity. A positive reaction is usually evident as some visible sign, such as color change or clumping, that indicates a specific interaction between antibody and antigen. (The reaction at the molecular level is rarely observed.)

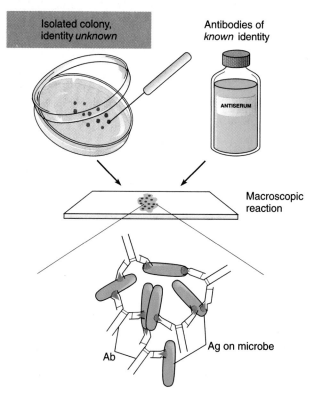

(b) An unknown microbe is mixed with serum containing antibodies of known specificity, a procedure known as serotyping. Microscopically or macroscopically observable reactions indicate a correct match between antibody and antigen and permit identification of the microbe.

FIGURE 16.3

Basic principles of serological testing using antibodies and antigens.

General Features of Immune Testing

The strategies of immunologic tests are diverse, and they underline some of the brilliant and imaginative ways that antibodies and antigens can be used as tools. We will summarize them under the headings of agglutination, precipitation, immunodiffusion, complement fixation, fluorescent antibody tests, and immunoassay tests. First we will overview the general characteristics of immune testing, and we will then look at each type separately.

The most effective serological tests have a high degree of specificity and sensitivity (**figure 16.4**). Specificity is the property of a test to focus upon only a certain antibody or antigen and not to react with unrelated or distantly related ones. Sensitivity means that the test can detect even very small amounts of antibodies or antigens that are the targets of the test. New systems using monoclonal antibodies have greatly improved specificity, and those using radioactivity, enzymes, and electronics have improved sensitivity.

Visualizing Antigen–Antibody Interactions   The primary basis of most tests is the binding of an antibody (Ab) to a specific molecular site on an antigen (Ag). Because this reaction cannot be readily seen without an electron microscope, tests involve some type of endpoint reaction visible to the naked eye or with regular magnification that tells whether the result is positive or negative. In the case of large antigens such as cells, Ab binds to Ag and creates large clumps or aggregates that are visible macroscopically or microscopically (**figure 16.5a**). Smaller Ag-Ab complexes that do not result in readily observable changes will require special indicators in order to be visualized. Endpoints are often revealed by dyes or fluorescent reagents that can tag molecules of interest. Similarly, radioactive isotopes incorporated into antigens or antibodies constitute sensitive tracers that are detectable with photographic film.

An antigen-antibody reaction can be used to read **titer,** or the quantity of antibodies in the serum. Titer is determined by serially diluting a sample in tubes or in a multiple-welled microtiter plate and mixing it with antigen (figure 16.5b). It is expressed as the highest dilution of serum that produces a visible reaction with an antigen. The more a sample can be diluted and yet still react with antigen, the greater is the concentration of antibodies in that sample and the higher is its titer. Interpretation of testing results is discussed in **Medical Microfile 16.2.**

Agglutination and Precipitation Reactions

The essential differences between agglutination and precipitation are in size, solubility, and location of the antigen. In agglutination, the antigens are whole cells such as red blood cells or bacteria with determinant groups on the surface. In precipitation, the antigen is a soluble molecule. In both instances, when Ag and Ab are optimally combined so that neither is in excess, one antigen is interlinked by several antibodies to form an insoluble, three-dimensional aggregate so large that it cannot remain suspended and it settles out (figure 16.5a).

Agglutination Testing   Agglutination is discernible because the antibodies cross-link the antigens to form visible clumps. **Agglutination** tests are performed routinely by blood banks to determine ABO and Rh (Rhesus) blood types in preparation for transfusions.

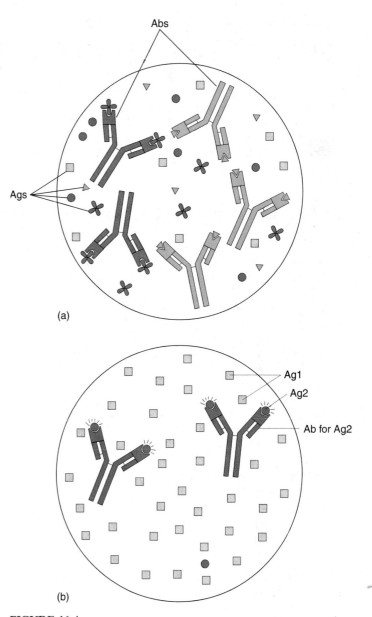

(a)

(b)

FIGURE 16.4

Specificity and sensitivity in immune testing.   **(a)** This test shows specificity in which an antibody (Ab) attaches with great exactness with only one type of antigen (Ag). **(b)** Sensitivity is demonstrated by the fact that Ab can pick up antigens even when the antigen is greatly diluted.

In this type of test, antisera containing antibodies against the blood group antigens on red blood cells are mixed with a small sample of blood and read for the presence or absence of clumping (see figure 17.10). The **Widal test** is an example of a tube agglutination test for diagnosing salmonelloses and undulant fever. In addition to detecting specific antibody, it also gives the serum titer (figure 16.5b).

Numerous variations of agglutination testing exist. The Rapid Plasma Reagin (RPR) test is one of several tests commonly used to test for antibodies to syphilis. The cold agglutinin test, named for antibodies that react only at lower temperatures (4°–20°C), was developed to diagnose *Mycoplasma* pneumonia. The *Weil-Felix reaction* is an agglutination test sometimes used in diagnosing rickettsial infections.

MEDICAL MICROFILE 16.2
When Positive Is Negative: How to Interpret Serological Test Results

What if a patient's serum gives a positive reaction—is **seropositive**—in a serological test? In most situations, it means that antibodies specific for a particular microbe have been detected in the sample. But one must be cautious in proceeding to the next level of interpretation. The mere presence of antibodies does not necessarily indicate that the patient has a disease but only that he or she has possibly had contact with a microbe or its antigens through infection or vaccination. In screening tests for determining a patient's history (rubella, for instance), knowing that a certain titer of antibodies is present can be significant, because it shows that the person has some protection. However, when the test is being used to diagnose current disease, a series of tests to show a rising titer of antibodies is necessary. The accompanying figure indicates how such a test can be used to diagnose patients who have nonspecific symptoms that could fit several diseases. Lyme disease, for instance, can be mistaken for arthritis or viral infections. In the first group, note that the antibody titer against *Borrelia burgdorferi* increased steadily over a 6-week period. A control group that shared similar symptoms did not exhibit a rise in titer for antibodies to this microbe.

Another important consideration in testing is the occasional appearance of biological false positives. These are results in which a patient's serum shows a positive reaction, even though, in reality, he is not or has not been infected by the microbe. False positives such as those in syphilis and AIDS testing arise when antibodies or other substances present in the serum cross-react with the test reagents, producing a positive result. Such false results may require retesting by a method that greatly minimizes cross-reactions.

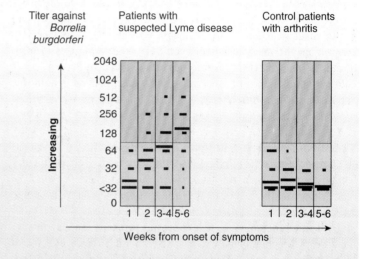

Agglutination

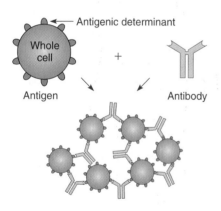

Microscopic appearance of clumps

Precipitation

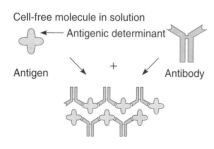

Microscopic appearance of precipitate

(a)

The Tube Agglutination Test

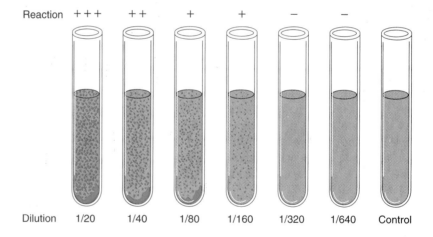

A sample of patient's serum is serially diluted with saline. The dilution is made in a way that halves the number of antibodies in each subsequent tube. An equal amount of the antigen (here, blue bacterial cells) is added to each tube. The control tube has antigen, but no serum. After incubation and centrifugation, each tube is examined for agglutination clumps as compared with the control, which will be cloudy and clump-free. The titer is defined as the dilution of the last tube in the series that shows agglutination.

(b)

FIGURE 16.5

Cellular/molecular view of agglutination and precipitation reactions that produce visible antigen-antibody complexes. Although IgG is shown as the Ab, IgM is also involved in these reactions.

In some tests, special agglutinogens have been prepared by affixing antigen to the surface of an inert particle. In *latex agglutination* tests, the inert particles are tiny latex beads. Kits using latex beads are available for assaying pregnancy hormone in the urine, identifying *Candida* yeasts and bacteria (staphylococci, streptococci, and gonococci), and diagnosing rheumatoid arthritis.

In *viral hemagglutination* testing, agglutinogen is a red blood cell that reacts naturally with certain viral antigens. The RBCs are mixed with a known virus and a patient's serum of unknown content (**figure 16.6**). The test is interpreted differently than other agglutination tests because it is based on a competition between the RBCs and the antibodies for the virus antigens. If the patient's serum does *not* contain antibodies specific to the virus, the virus reacts with the RBCs instead and agglutinates them. Thus, agglutination here indicates no antibodies and a seronegative result. On the other hand, if the specific antibodies for the virus are present, they attach to the virus particles, the RBCs remain free, and there is no agglutination. As performed in the wells of microtiter plates, the difference is apparent to the naked eye (figure 16.6). Several viral diseases (measles, rubella, mumps, mononucleosis, and influenza) can be diagnosed with this test.

Precipitation Tests   In precipitation reactions, the soluble antigen is precipitated (made insoluble) by an antibody. This reaction is observable in a test tube in which antiserum has been carefully laid over an antigen solution (**figure 16.7a**). At the point of contact, a cloudy or opaque zone forms.

One example of this technique is the VDRL (Veneral Disease Research Lab) test that also detects antibodies to syphilis. Although it is a good screening test, it contains a heterophilic antigen (cardiolipin) that may give rise to false positive results. Although precipitation is a useful detection tool, the precipitates are so easily disrupted in liquid media that most precipitation reactions are carried out in agar gels. These substrates are sufficiently soft to allow the reactants (Ab and Ag) to freely diffuse, yet firm enough to hold the Ag-Ab precipitate in place. One technique with applications in microbial identification and diagnosis of disease is the double diffusion (Ouchterlony) method. It is called double diffusion because it involves diffusion of both antigens and antibodies. The test is performed by punching a pattern of small wells into an agar medium and filling them with test antigens and antibodies. A band forming between two wells indicates that antibodies from one well have met and reacted with antigens from the other well. Variations on this technique provide a means of identifying unknown antibodies or antigens (figure 16.7).

Immunoelectrophoresis constitutes yet another refinement of diffusion and precipitation in agar. With this method, a serum sample is first electrophoresed to separate the serum proteins as previously shown (see figure 15.17). Antibodies that react with specific serum proteins are placed in a trough parallel to the direction of

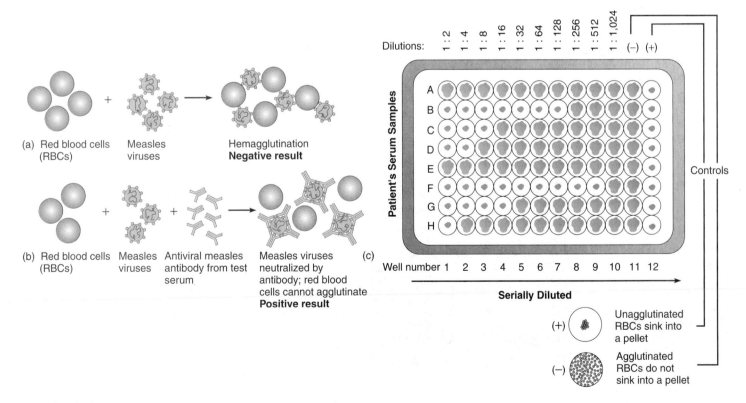

FIGURE 16.6

Theory and interpretation of viral hemagglutination.   **(a)** In an antibody-free system, the natural tendency of measles virus to combine and interlink red blood cells causes an agglutination reaction. **(b)** In a test system using antibodies specific to the measles virus, these Abs will react with the viruses and prevent them from binding to red blood cells, thereby blocking agglutination. Thus, no agglutination means a positive reaction for antibody. **(c)** In an actual test, a multiwell microtiter plate is set up to run dilutions of several patients' sera. In a negative test, the agglutinated RBCs fill up most of the well; in a positive test, the unagglutinated RBCs fall into a small pile (pellet) in the well's bottom. This test allows the reading of titer, expressed as the highest dilution of serum that shows a positive result. For example, Row A is 0; row B is 1:128, etc.

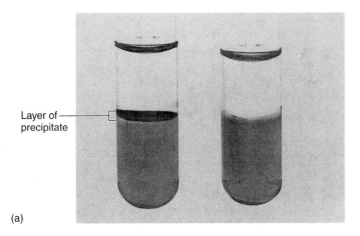

Layer of precipitate

(a)

I.

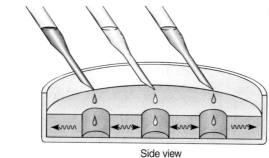

Side view

I. In one method of setting up a double-diffusion test, wells are punctured in soft agar, and antibodies (Ab) and antigens (Ag) are added in a pattern. As the contents of the wells diffuse toward each other, a number of reactions can result, depending on whether antibodies meet and precipitate antigens.

II.

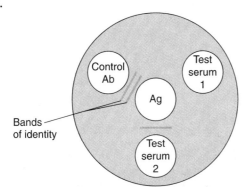

Control Ab

Ag

Test serum 1

Bands of identity

Test serum 2

II. Example of test pattern and results. Antigen (Ag) is placed in the center well and antibody (Ab) samples are placed in outer wells. The control contains known Abs to the test Ag. Note bands that form where Ab/Ag meet. The other wells (1, 2) contain unknown test sera. One is positive and the other is negative. Double bands indicate more than one antigen and antibody that can react.

III.

III. Actual test results for detecting infection with the fungal pathogen *Histoplasma*. Numbers 1 and 4 are controls and 2, 3, 5, 6 are patient test sera. Determine which patients have the infection and which do not.

(b)

FIGURE 16.7

Precipitation reactions. **(a)** A tube precipitation test for streptococcal group antigens. Specific antiserum has been placed in the bottom of the tubes and antigen solution carefully overlaid to form a zone of contact. The left-hand tube has developed a heavy band of precipitate indicative of a positive reaction between antibody and antigen. The right-hand tube is negative. **(b)** Double-diffusion (Ouchterlony) tests in a semisolid matrix.

(a) From Gillies and Dodds, Bacteriology Illustrated, *5th ed., fig. 14. Reprinted by permission of Longman Group Ltd.*

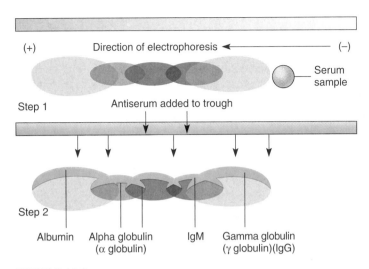

FIGURE 16.8

Immunoelectrophoresis of normal human serum. *Step 1.* Proteins are separated by electrophoresis on a gel. *Step 2.* To identify the bands and increase visibility, antiserum containing antibodies specific for serum proteins is placed in a trough and allowed to diffuse toward the bands. This diffusion produces a pattern of numerous arcs representing major serum components.

migration, forming reaction arcs specific for each protein (**figure 16.8**). This test is widely used to detect disorders in the production of antibodies. *Counterimmunoelectrophoresis,* which uses an electrical current to speed up the migration of antibody and antigen, is a newer technique for identifying bacterial and viral antigens in blood.

The Western Blot for Detecting Proteins

The **Western blot** test is somewhat similar to the previous tests because it involves the electrophoretic separation of proteins, followed by an immunoassay to detect these proteins. This test is a counterpart of the Southern blot test for identifying DNA (see figure 10.4). It is a highly specific and sensitive way to identify or verify a particular protein (antibody or antigen) in a sample (**figure 16.9**). First, the test material is electrophoresed in a gel to separate out particular bands. The gel is then transferred to a special blotter that binds the reactants in place. The blot is developed by incubating it with a solution of antigen or antibody that has been labeled with radioactive, fluorescent, or luminescent labels. Sites of specific binding will appear as a pattern of bands that can be compared with known positive and negative samples. This is currently the verification test for people who are antibody-positive for HIV in the ELISA test (described in a later section), because it tests more types of antibodies and is less subject to misinterpretation than are other antibody tests. The technique has significant applications for detecting microbes and their antigens in specimens.

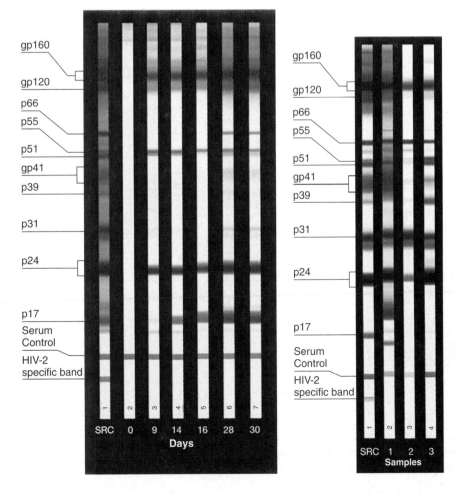

FIGURE 16.9

The Western blot procedure.   The example shown here tests for antibodies to specific HIV antigens. The test strips are prepared by electrophoresing several of the major HIV surface and core antigens and then blotting them onto special filters. The test strips are incubated with a patient's serum and developed with a radioactive or colorimetric label. Sites where HIV antigens have bound antibodies show up as bands. Patients' sera are then compared with a positive control strip (SRC) containing antibodies for all HIV antigens. Certain criteria must be met to consider the result positive.

Interpretation of Bands

Labels correspond with glycoproteins (GP) or proteins (P) that are part of HIV-1 antigen structure.

- The test is considered positive if bands occur at two locations: gp160 or gp120 and g31 or g24.

- The test is considered negative if no bands are present for any HIV antigen.

- The test is considered indeterminate if bands are present, but not at the criteria locations. This result may require retesting at a later date.

Complement Fixation

An antibody that requires (fixes) complement to complete the lysis of its antigenic target cell is termed a **lysin** or cytolysin. When lysins act in conjunction with the intrinsic complement system on red blood cells, the cells hemolyze (lyse and release their hemoglobin). This lysin-mediated hemolysis is the basis of a group of tests called complement fixation, or CF (**figure 16.10**).

Complement fixation testing uses four components—antibody, antigen, complement, and sensitized sheep red blood cells—and it is conducted in two stages. In the first stage, the test antigen is allowed to react with the test antibody (at least one must be of known identity) in the absence of complement. If the Ab-Ag are specific for each other, they form complexes. To this mixture, purified complement proteins from guinea pig blood are added. If antibody and antigen have complexed during the previous step, they attach, or fix, the complement to them, thus preventing it from participating in further reactions. The extent of this complement fixation is determined in the second stage by means of sheep RBCs with surface lysin molecules. The sheep RBCs serve as an indicator complex that can also fix complement. Contents of the stage 1 tube are mixed with the stage 2 tube and observed for hemolysis, which can be observed with the naked eye as a clearing of the solution. If hemolysis *does not* occur, it means that the complement was used

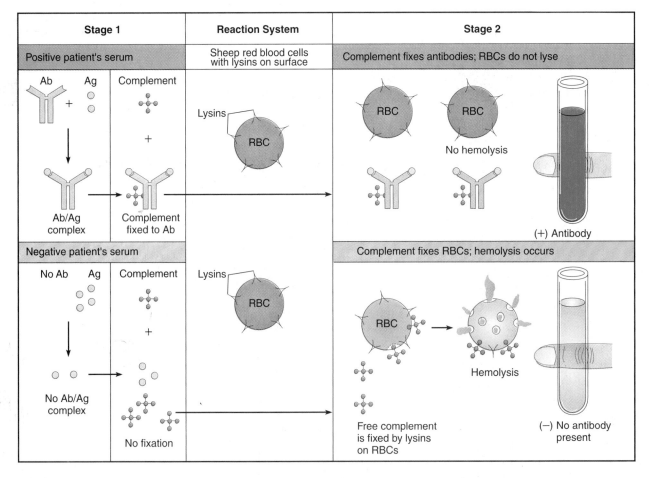

FIGURE 16.10

Complement fixation test. In this example, two serum samples are being tested for antibodies to a certain infectious agent. In reading this test one observes the cloudiness of the tube. If it is cloudy, the RBCs are not hemolyzed and the test is positive. If it is clear and pink, the RBCs are hemolyzed and the test is negative.

up by the first stage Ab-Ag complex and that the unknown antigen or antibody was indeed present. This result is considered positive. If hemolysis *does* occur, it means that unfixed complement from tube 1 reacted with the RBC complex instead, thereby causing lysis of the sheep RBCs. This result is negative for the antigen or antibody that was the target of the test. Complement fixation tests are invaluable in diagnosing influenza, polio, and various fungi.

The antistreptolysin O (ASO) titer test measures the levels of antibody against the streptolysin toxin, an important hemolysin of group A streptococci. It employs a technique related to complement fixation. A serum sample is exposed to known suspensions of streptolysin and then allowed to incubate with RBCs. Lack of hemolysis indicates antistreptolysin antibodies in the patient's serum that have neutralized the streptolysin and prevented hemolysis. This is an important verification procedure for scarlet fever, rheumatic fever, and other related streptococcal syndromes (see chapter 18).

Miscellaneous Serological Tests

A test that relies on changes in cellular activity as seen microscopically is the *Treponema pallidum immobilization* (TPI) test for syphilis. The impairment or loss of motility of the *Treponema* spirochete in the presence of test serum and complement indicates that the serum contains anti-*Treponema pallidum* antibodies (see chapter 21). In *toxin neutralization* tests, a test serum is incubated with the microbe that produces the toxin. If the serum inhibits the growth of the microbe, one can conclude that antitoxins are present.

Serotyping is an antigen-antibody technique for identifying, classifying, and subgrouping certain bacteria into categories called serotypes, using antisera for cell antigens such as the capsule, flagellum, and cell wall. It is widely used in typing *Salmonella* species and strains and is the basis for identifying the numerous serotypes of streptococci. The Quellung test, which identifies serotypes of the pneumococcus involves a precipitation reaction in which antibodies react with the capsular polysaccharide. Although the reaction makes the capsule seem to swell, it is actually creating a zone of Ab-Ag complex on the cell's surface (see figure 18.20).

Fluorescent Antibodies and Immunofluorescence Testing

The property of dyes such as fluorescein and rhodamine to emit visible light in response to ultraviolet radiation was discussed in chapter 3. This property of fluorescence has found numerous applications in diagnostic immunology. The fundamental tool in immunofluorescence

testing is a fluorescent antibody—a monoclonal antibody labeled by a fluorescent dye (fluorochrome).

The two ways that fluorescent antibodies (FABs) can be used are shown in **figure 16.11.** In *direct testing,* an unknown test specimen or antigen is fixed to a slide and exposed to a fluorescent antibody solution of known composition. If the antibodies are complementary to antigens in the material, they will bind to it. After the slide is rinsed to remove unattached antibodies, it is observed with

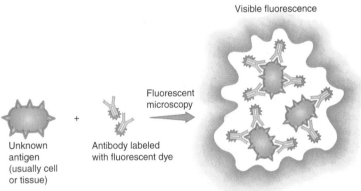
(a) **Direct Testing**

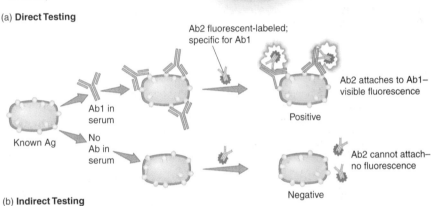
(b) **Indirect Testing**

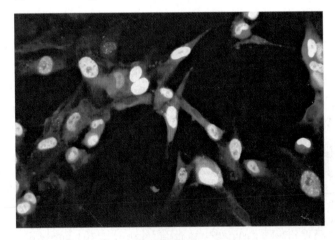
(c) **Indirect Immunofluorescence Testing**

FIGURE 16.11

Immunofluorescence testing. The success of these techniques is contingent upon the accumulation of sufficient labeled antibody (Ab) in one site that it will show up as fluorescing cells or masses. **(a)** Direct: Unidentified antigen (Ag) is directly tagged with fluorescent Ab. **(b)** Indirect: Ag of known identity is used to assay unknown Ab; a positive reaction occurs when the second Ab (with fluorescent dye) affixes to the first Ab. **(c)** An indirect immunofluorescent stain of cells infected with two different viruses. Cells fluorescing green contain cytomegalovirus; cells fluorescing yellow contain adenovirus.

the fluorescent microscope. Fluorescing cells or specks indicate the presence of Ab-Ag complexes and a positive result. These tests are valuable for identifying and locating antigens on the surfaces of cells or in tissues and in identifying the disease agents of syphilis (see figure D, page 541), gonorrhea, chlamydiosis, whooping cough, Legionnaires' disease, plague, trichomoniasis, meningitis, and listeriosis.

In *indirect testing* methods, the fluorescent antibodies are *anti-isotypic* antibodies made to react with the Fc region of another antibody (remember that antibodies can be antigenic). In this scheme, an antigen of known character (a bacterial cell, for example) is combined with a test serum of unknown antibody content. The fluorescent antibody solution that can react with the unknown antibody is applied and rinsed off to visualize whether the serum contains antibodies that have affixed to the antigen. A positive test shows fluorescing aggregates or cells, indicating that the fluorescent antibodies have combined with the unlabeled antibodies. In a negative test, no fluorescent complexes will appear. This technique is frequently used to diagnose syphilis (FTA-ABS) and various viral infections discussed in chapters 24 and 25.

Immunoassays: Tests of Great Sensitivity

The elegant tools of the microbiologist and immunologist are being used increasingly in athletics, criminology, government, and business to test for trace amounts of substances such as hormones, metabolites, and drugs. But traditional techniques in serology are not refined enough to detect a few molecules of these chemicals. Extremely sensitive alternative methods that permit rapid and accurate measurement of trace antigen or antibody are called **immunoassays.** Examples of the technology for detecting an antigen or antibody in minute quantities include radioactive isotope labels, enzyme labels, and sensitive electronic sensors. Many of these tests are based on specifically formulated monoclonal antibodies.

Radioimmunoassay (RIA) Antibodies or antigens labeled with a radioactive isotope can be used to pinpoint minute amounts of a corresponding antigen or antibody. Although very complex in practice, these assays compare the amount of radioactivity present in a sample before and after incubation with a known, labeled antigen or antibody. The labeled substance competes with its natural, non-labeled partner for a reaction site. Large amounts of a bound radioactive component indicate that the unknown test substance was not present. The amount of radioactivity is measured with an isotope counter or a photographic emulsion (autoradiograph). Radioimmunoassay has been employed to measure the levels of insulin and other hormones and to diagnose allergies, chiefly by the radioimmunosorbent test (RIST) for measurement of IgE in allergic patients and the radioallergosorbent test (RAST) to standardize allergenic extracts (chapter 17).

Enzyme-Linked Immunosorbent Assay (ELISA) The **ELISA test,** also known as enzyme immunoassay (EIA), contains an enzyme-antibody complex that can be used as a color

tracer for antigen-antibody reactions. The enzymes used most often are horseradish peroxidase and alkaline phosphatase, both of which release a dye (chromogen) when exposed to their substrate. This technique also relies on a solid support such as a plastic microtiter plate that can *adsorb* (attract on its surface) the reactants (**figure 16.12**).

The *indirect ELISA* test can detect antibodies in a serum sample. As with other indirect tests, the final positive reaction is achieved by means of an antibody-antibody reaction. The indicator antibody is complexed to an enzyme that produces a color change with positive serum samples (figure 16.12*a, b*). The starting reactant is a known antigen that is adsorbed to the surface of a well. To this, an unknown serum is added. After rinsing, an enzyme-Ab reagent that can react with the unknown test antibody is placed in the well. The substrate to the enzyme is then added, and the wells are scanned for color changes. Color development indicates that all the components reacted and that the antibody was present in the patient's serum. This is the common screening test for the antibodies to HIV (AIDS virus), various rickettsial species, hepatitis A and C, the cholera vibrio, and *Helicobacter,* a cause of gastric ulcers. Because false positives can occur, a verification test may be necessary (such as Western blot for HIV).

In *capture ELISA,* or sandwich, tests, a known antibody is adsorbed to the bottom of a well and incubated with a solution containing unknown antigen (figure 16.12*c*). After excess unbound components have been rinsed off, an enzyme-antibody indicator that can react with the antigen is added. If antigen is present, it will attract the indicator-antibody and hold it in place. Next, the substrate to the enzyme is placed in the wells and incubated. Enzymes affixed to the antigen will hydrolyze the substrate and release a colored dye. Thus, any color developing in the wells is a positive result. Lack of color means that the antigen was not present and that the subsequent rinsing removed the enzyme-antibody complex. The capture technique is used to detect antibodies to hantavirus, rubella virus, and *Toxoplasma.*

A newer technology uses electronic monitors that directly read out antibody-antigen reactions. Without belaboring the technical aspects, these systems contain computer chips that sense the minute changes in electrical current given off when an antibody binds to antigen. The potential for sensitivity is extreme; it is thought that amounts as small as 12 molecules of a substance can be detected in a sample. In another procedure, antibody substrate molecules are incubated with sample and then exposed to the enzyme alkaline phosphatase. If the antibody is bound, the enzyme reacts with the substrate and causes visible light to be emitted. The light can be detected by machines or photographic films.

Tests That Differentiate T Cells and B Cells

So far we have concentrated on tests that identify antigens and antibodies in samples, but techniques also exist that differentiate between B cells and T cells and can quantify subsets of each. Information on the types and numbers of lymphocytes in blood and other samples is a common way to evaluate immune dysfunctions such as those in AIDS, immunodeficiencies, and cancer. A simple method for identifying T cells is to mix them with untreated sheep red

(a) **Indirect ELISA,** comparing a positive vs negative reaction. This is the basis for HIV screening tests.

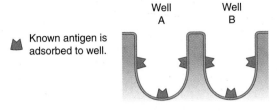

Well A   Well B

Known antigen is adsorbed to well.

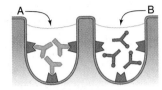

Serum samples with unknown antibodies.

A

B

A   B

Well is washed to remove unbound (nonreactive) antibodies.

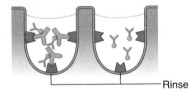

Indicator antibody linked to enzyme attaches to any bound antibody.

Rinse

Wells are rinsed to remove unbound indicator antibody; A colorless substrate for enzyme is added.

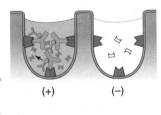

Enzymes linked to indicator Ab hydrolyze the substrate, which releases a dye. Wells that develop color are positive for the antibody: colorless wells are negative.

(+)   (−)

(b) **Microtiter ELISA Plate with 96 Tests for HIV Antibodies.** Colored wells indicate a positive reaction.

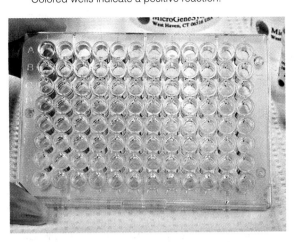

(c) **Capture or Antibody Sandwich ELISA method.** Note that an antigen is trapped between two antibodies. This test is used to detect hantavirus and measles virus.

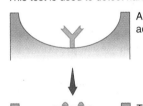
Antibody is adsorbed to well.

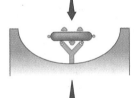
Test antigen is added; if complimentary, antigen binds to antibody.

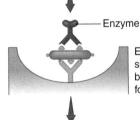
Enzyme

Enzyme-linked antibody specific for test antigen then binds to another antigen, forming a sandwich.

Enzyme's substrate (□) is added, and reaction produces a visible color change (●).

FIGURE 16.12

Methods of ELISA testing.

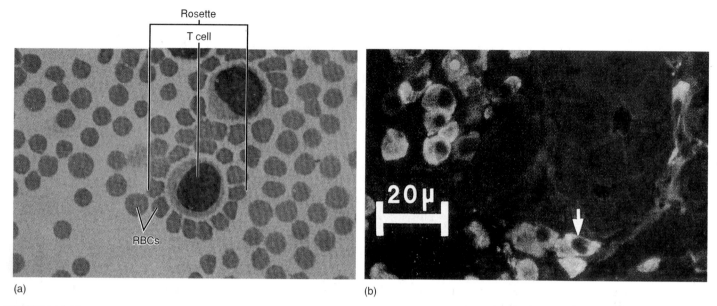

(a)

(b)

FIGURE 16.13

Tests for characterizing T cells and B cells. **(a)** Photomicrograph of rosette formation that identifies T cells. **(b)** Plasma cells (arrow) highlighted by fluorescent antibodies.

blood cells. Receptors on the T cells bind the RBCs into a flower-like cluster called a **rosette formation (figure 16.13a).** Rosetting can also occur in B cells if one uses Ig-coated bovine RBCs or mouse erythrocytes.

For routine identification of lymphocytes, rosette formation has been replaced by fluorescent techniques. These techniques can differentiate between T cells and B cells as well as to subgroup them (figure 16.13b). These subgroup tests utilize monoclonal antibodies produced in response to specific cell markers. B-cell tests categorize different stages in B-cell development and are very useful in characterizing B-cell cancers. Tests that can help differentiate the CD4, CD8, and other T-cell subsets are important in monitoring AIDS and other immunodeficiency diseases.

In Vivo Testing

Probably the first immunologic tests were performed not in a test tube but on the body itself. A classic example of one such technique is the **tuberculin test,** which uses a small amount of purified protein derivative (PPD) from *Mycobacterium tuberculosis* injected into the skin. The appearance of a red, raised, thickened lesion in 48–72 hours can indicate previous exposure to tuberculosis (see figure 17.18a). In practice, *in vivo* tests employ principles similar to serological tests, except in this case an antigen or an antibody is introduced into a patient to elicit some sort of visible reaction. Like the tuberculin test, some of these diagnostic skin tests are useful for evaluating infections due to fungi (coccidioidin and histoplasmin tests, for example) or allergens (see skin testing in chapter 17).

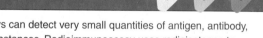

CHAPTER CHECKPOINTS

Complement fixation involves a two-part procedure in which complement fixes to a specific antibody if present, or to red blood cell antigens, if antibody is absent. Lack of RBC hemolysis is indicative of a positive test.

Serological tests can measure the degree to which host antibody binds directly to disease agents or toxins. This is the principle behind tests for syphilis and rheumatic fever.

Direct fluorescent antibody tests indicate presence of an antigen and are useful in identifying infectious agents. Indirect fluorescent tests indicate the presence of a particular antibody and can diagnose infection.

Immunoassays can detect very small quantities of antigen, antibody, or other substances. Radioimmunoassay uses radioisotopes to detect trace amounts of biological substances.

The ELISA test uses enzymes and dyes to detect antigen-antibody complexes. It is widely used to detect viruses, bacteria, and antibodies in HIV infection.

Technicians use precise assays to differentiate between B and T cells and to identify subgroups of these cells for disease diagnosis.

In vivo serological testing, such as the tuberculin test, involves subcutaneous injection of antigen to elicit a visible immune response in the host.

CHAPTER CAPSULE WITH KEY TERMS

Biomedical Applications of Specific Immunity

I. *Immunization:* **Producing immunity by medical intervention**
  A. Passive immunotherapy includes administering immune serum globulin and specific immune globulins pooled from donated serum to prevent infection and disease in those at risk; antisera and antitoxins from animals are occasionally used.
  B. Active immunization is synonymous with **vaccination;** provides an antigenic stimulus that does not cause disease but can produce long-lasting, protective immunity. **Vaccines** are made with:
    1. **Killed** whole cells or **inactivated viruses** that do not reproduce but are antigenic.
    2. Live, **attenuated** cells or viruses that are able to reproduce but have lost virulence.
    3. **Acellular** or **subunit** components of microbes such as surface antigen or neutralized toxins (**toxoids**).
    4. Genetic engineering techniques, including cloning of antigens, recombinant attenuated microbes, and **DNA**-based **vaccines.**
  C. Boosters (additional doses) are often required.
  D. Vaccination increases herd immunity, protection provided by mass immunity in a population.

II. *Serological/Immune Testing:* **Serology** is a science that attempts to detect signs of infection in a patient's serum such as antibodies specific for a microbe.
  A. The basis of serological tests is that Abs specifically bind to Ag *in vitro.* An Ag of known identity will react with antibodies in an unknown serum sample. The reverse is also true; known antibodies can be used to detect and type antigens.
  B. These Ag-Ab reactions are visible in the form of obvious clumps and precipitates, color changes, or the release of radioactivity. Test results are read as positive or negative.
  C. Desirable properties of tests are high specificity and sensitivity.
  D. Types of Tests
    1. In **agglutination** tests, antibody cross-links whole-cell antigens, forming complexes that settle out and form visible clumps in the test chamber; examples are tests for blood type, some bacterial diseases, and viral diseases.

  2. Double diffusion precipitation tests involve the diffusion of Ags and Abs in a soft agar gel, forming zones of precipitation where they meet.
  3. In immunoelectrophoresis, migration of serum proteins in gel is combined with precipitation by antibodies.
  4. The **Western blot** test separates antigen into bands. After the gel is affixed to a blotter, it is reacted with a test specimen and developed by radioactivity or with dyes.
  5. Complement fixation tests detect **lysins**—antibodies that fix complement and can lyse target cells. It involves first mixing test Ag and Ab with complement and then with sensitized sheep RBCs. If the complement is fixed by the Ag-Ab, the RBCs remain intact, and the test is positive. If RBCs are hemolyzed, specific antibodies are lacking.
  6. In direct assays, known marked Ab is used to detect unknown Ag (microbe).
    a. In indirect testing, known Ag reacts with unknown Ab, and the reaction is made visible by a second Ab that can affix to and identify the unknown Ab.
    b. Immunofluorescence testing uses fluorescent antibodies (FABs tagged with fluorescent dye) either directly or indirectly to visualize cells or cell aggregates that have reacted with the FABs.
  7. **Immunoassays** are highly sensitive tests for Ag and Ab.
    a. In **radioimmunoassay,** Ags or Abs are labeled with radioactive isotopes and traced.
    b. The **enzyme-linked immunosorbent assay (ELISA)** can detect unknown Ag or Ab by direct or indirect means. A positive result is visualized when a colored product is released by an enzyme-substrate reaction.
    c. Tests are also available to differentiate B cells from T cells and their subtypes.
  8. With *in vivo* testing, Ags are introduced into the body directly to determine the patient's immunologic history.

MULTIPLE-CHOICE QUESTIONS

1. A living microbe with reduced virulence that is used for vaccination is considered
   a. a toxoid
   b. attenuated
   c. denatured
   d. an adjuvant

2. A vaccine that contains parts of viruses is called
   a. acellular
   b. recombinant
   c. subunit
   d. attenuated

3. Widespread immunity that protects the population from the spread of disease is called
   a. seropositivity
   b. cross-reactivity
   c. epidemic prophylaxis
   d. herd immunity

4. DNA vaccines contain ____ DNA that stimulates cells to make ____ antigens.
   a. human, RNA
   b. microbial, protein
   c. human, protein
   d. microbial, polysaccharide

5. Administration of immune serum globulin is a form of ____ immunization that ____.
   a. active, prevents infection
   b. passive, provides long-term immunity
   c. therapeutic, prevents disease
   d. prophylactic, stimulates the immune system

6. What is the purpose of an adjuvant?
   a. to kill the microbe
   b. to stop allergic reactions
   c. to improve the contact between the antigen and lymphocytes
   d. to make the antigen more soluble in the tissues

7. An example of a recombinant DNA vaccine is
   a. tetanus
   b. MMR
   c. polio
   d. hepatitis B

8. In agglutination reactions, the antigen is a _____; in precipitation reactions, it is a _____.
   a. soluble molecule, whole cell
   b. whole cell, soluble molecule
   c. bacterium, virus
   d. protein, carbohydrate

9. Which reaction requires complement?
   a. hemagglutination
   b. precipitation
   c. hemolysis
   d. toxin neutralization

10. A patient with a _____ titer of antibodies to an infectious agent generally has greater protection than a patient with a _____ titer.
   a. high, low   c. negative, positive
   b. low, high   d. old, new

11. Direct immunofluorescence tests use a labeled antibody to identify _____.
   a. an unknown microbe
   b. an unknown antibody
   c. fixed complement
   d. agglutinated antigens

12. The Western blot test can be used to identify
   a. unknown antibodies
   b. unknown antigens
   c. specific DNA
   d. both a and b

13. An example of an *in vivo* serological test is
   a. indirect immunofluorescence
   b. radioimmunoassay
   c. tuberculin test
   d. complement fixation

14. Which of the following is NOT an important criterion in vaccine development?
   a. low toxicity
   b. requires several boosters
   c. stimulates antibodies and cytotoxic T cells
   d. provides long-lasting immunity

15. **Multiple Matching.** For the following list of vaccines, first look up the disease for which the vaccine is intended (if not apparent). Then list the letters of all the descriptions that fit. (There may be more than one possible answer.)
   _____ BCG
   _____ measles
   _____ pertussis
   _____ tetanus
   _____ mumps
   _____ diphtheria
   _____ hepatitis B
   _____ Hib
   _____ polio
   _____ rubella
   _____ pneumococcus
   _____ hepatitis A
   a. uses live, attenuated microbes
   b. based on a toxoid
   c. subunit/acellular vaccine
   d. used in mixed vaccine
   e. does not require boosters
   f. uses killed, whole cells
   g. uses killed, whole viruses
   h. routinely given in childhood
   i. mainly for people at high risk
   j. made by genetic engineering

CONCEPT QUESTIONS

1. a. Name three products used in passive artificial immunization.
   b. What are the primary reasons for using these substances?
   c. What are some disadvantages of this form of immunization?
   d. What is the difference between immunization used for prophylaxis and that used for treatment?

2. a. Outline the strategies for developing vaccines, and give specific examples for each method.
   b. By what means are microorganisms attenuated?
   c. What is the purpose of an adjuvant?
   d. Describe the way that a Trojan horse vaccine works.

3. a. What are the advantages and disadvantages of a killed vaccine; a live, attenuated vaccine; a subunit vaccine; a recombinant vaccine; and a DNA vaccine?
   b. Use an outline to explain how an inoculation with tetanus toxoid will protect a person the next time he or she steps on a dirty piece of glass.

4. a. Describe the concept of herd immunity.
   b. How does vaccination contribute to its development in a community?
   c. Give some possible explanations for recent epidemics of diphtheria and whooping cough.

5. a. What is the basis of serology and serological testing?
   b. Differentiate between specificity and sensitivity.
   c. Describe several general ways that Ag-Ab reactions are detected.

6. a. What does seropositivity mean?
   b. What is a false positive test result and what are some possible causes?
   c. What is meant by a false negative result and what might account for it?
   d. What does the titer of serum tell us about the immune status?

7. a. Explain how agglutination and precipitation reactions are alike.
   b. In what ways are they different?
   c. Make a drawing of the manner in which antibodies cross-link the antigens in agglutination and precipitation reactions.
   d. Give examples of several tests that employ the two reactions.

8. a. What is meant by complement fixation? What are cytolysins?
   b. What is the purpose of using sheep red blood cells in this test?

9. a. Explain the differences between direct and indirect procedures in serological or immunoassay tests.
   b. How is fluorescence detected?
   c. How is the reaction in a radioimmunoassay detected?
   d. How does a positive reaction in an ELISA test appear? How many wells are positive in figure 16.12(b)?

10. a. Briefly describe the principles and give an example of the use of a specific test using immunoelectrophoresis, Western blot, complement fixation, fluorescence testing (direct and indirect), and immunoassays (direct and indirect ELISA).
    b. Explain a rapid microscopic method for differentiating T cells from B cells.

CRITICAL-THINKING QUESTIONS

1. Describe the relationship between an antitoxin, a toxin, and a toxoid.

2. It is often said that a vaccine does not prevent infection; rather, it primes the immune system to undergo an immediate response to prevent an infection from spreading. Explain what is meant by this statement, and outline what is happening at the cellular/molecular level from the time of vaccination until subsequent contact with the infectious agent actually occurs.

3. At least three boosters are given for DTaP vaccines.
   a. Explain what each subsequent booster does, and why more than one is needed.
   b. Which features of the immune system allow it to react efficiently with 10 to 15 different vaccine antigens simultaneously?

4. Explain how to design a vaccine that could:
   a. induce protective IgA in the intestine
   b. give immunity to dental caries
   c. protect against the liver phase of the malaria parasite
   d. be derived from a single microbe and immunize against two different infectious diseases

5. a. Suggest several reasons that it could be risky to administer a vaccine containing a live, attenuated DNA virus.
   b. Explain what is involved in making a DNA vaccine.
   c. Explain why measles vaccine does not reliably protect infants if it is given before 12 months of age.

6. a. Determine the vaccines you have been given and those for which you will require periodic boosters.
   b. Suggest vaccines you may need in the future.

7. When traders and missionaries first went to the Hawaiian Islands, the natives there experienced severe disease and high mortality rates from smallpox, measles, and certain STDs.
   a. Explain what factors are involved in the sudden outbreaks of disease in previously unexposed populations.
   b. Explain the ways in which vaccination has been responsible for the worldwide eradication of diseases such as smallpox and polio.

8. Why do some tests for antibody in serum (such as for HIV and syphilis) require backup verification with additional tests at a later date?

9. a. Look at figure 16.5b. What is the titer as shown?
   b. If the titer had been 1:40, what interpretation would be made as to the immune status of the patient?
   c. What would it mean if a test 2 weeks later revealed a titer of 1:1280?
   d. What would it mean if no agglutination had occurred in any tube?
   e. From figure 16.6c, can you tell which patients have measles antibody and which do not?
   f. What are the titers of the Ab-positive patients?

10. Why do we interpret positive hemolysis in the complement fixation test to mean negative for the test substance?

11. Observe figure 16.12 and make note of the several steps in the indirect ELISA test. What four essential events are necessary to develop a positive reaction (besides having antibody A)? Hint: what would happen without rinses?

12. Using the criteria for band interpretation in figure 16.9, tell which test strips are consistent with a positive test, negative test, or indeterminate test result.

13. Explain how an immunoassay method could use monoclonal antibodies to differentiate between B and T cells and between different subsets of T cells.

INTERNET SEARCH TOPICS

1. Find information on experimental vaccines for herpes simplex, human papillomavirus, group B *Streptococcus,* cancer, and dental diseases.

2. Visit the student Online Learning Center at www.mhhe.com/talaro5. Go to chapter 16, Internet Search Topics, and log on to the available websites to:
   a. Explore the latest information on vaccines and their development.
   b. Study excellent graphics and explanations of immune tests, such as immunoassays, diagnostic serological tests, and the Western blot test.

[Handwritten notes]

Clonial selection theory

undetermined # of different B-cells in peripheral blood w/ IgD surface receptors
surface receptors recognize specific antigen
antigen binds to IgD surface receptor of B cell
makes clone of plasma cells (identical) make
IgM antibodies for anti ?
clone memory B cells that are specific to the antigen

Bacteria in genetics

1. Can be rapidly grown, therefore many generations can be studied in a short period of time.

2. Large populations of essentially identical cells can be cultivated from a single parent

3. are genetically simple organisms

4. genetic material is readily transferred from one cell to another. Easier to investigate mechanisms of gene function

agar

1. Standard type of agar Mueller Hinton w/ standard depth of 4mm

2. " concentration of antibiotic w/ standard disk size of 6mm

3. " concentration of organism using turbidity of 0.5 McFarland

4. Standard incub. time of 18-24hrs & standard temp of 35-37°C

Type I	atopy chronic local allergic system expose mucus axillary	excess IgE needs sensitizing dose	mast + basil cells responsible for inflam. response
Type II	cytotoxic - Target + destroy cell		
Type III	immune complex reactions	delayed symptoms (non specific cytotoxic)	
Type IIII	delayed hypersensitivity (TB) Type cell mediated		

Type I Hypersensitivity diseases - Hay fever   drug allergen
                 asthma   anaphylaxis
                 dermatitis   food allerg.

Type II - performed IgM/or IgG react w/ foreign cell
            ex. Rh -- mothers - babies

Type III - autoimmune disorder - Lupus - Butterfly rash
     anti body (Ab) against organs + tissues
     Rheumatoid arthritis
     Graves disease
     Hashimois Thyroids
     diabetes / myasthenia gravis, MS

Type IV - no antibodies T-cell
        ex. TB reaction
        contact dermatitis - poison ivy

     Graft vs Host   killer cells + cytotoxic T lymph.
           MHC genes + receptors

ABO	RBC anti	naturally occurring anti
A	A anti	anti B
B	B anti	anti A
O	none	anti A + B
AB	A + B anti	—

H antigen (- hand waving) flagella
O antigen - somatic cell wall
K " - capsular antigen

3- classes of MHC - major Histocompatibility complex
1. Responsible for rejection or acceptance of graft tissue - (T-lymph)
2. Regulates immune response macrophage + B-cell
3. Regulate compl. componenes

B-cells - made in Bone marrow
     antibody on surface
     make plasma cells as by product

T-cells - made in thymus. High in CD-4 recept.   circulation

Classes of Immunoglobin

1. IgG - smallest antibody
     crosses placenta
     activates complement
     long term immunity

2. IgM - Biggest one
     Does not cross placenta
     primary response to antigen
     Fixes complement IgG + IgM

3. IgA - secretory
     immunoglobulin dimer only
     found in saliva, mucus, tears
     Two forms

4. IgE - responsible for allergies to much cause allergies
     - very low concent. in blood (0.002%)
     - responsible for allergic reaction
     - protects against parasites (roundworms)

5. IgD -
     - very low concent. in blood (1%)
     - Bound to surface of B-cells

CHAPTER 17

Disorders in Immunity

[handwritten notes:]
1. ergosterols found in fungi
2. 70S Ribosomes because prokaydic cells are 70S + we are 80s
3. Peptidoglycan kills bacteria

Humans possess a powerful and intricate system of defense, which by its very nature also carries the potential to cause injury and disease. In most instances, a defect in immune function is expressed in commonplace, but miserable, symptoms such as those of hay fever and dermatitis. But abnormal or undesirable immune functions are also actively involved in debilitating or life-threatening diseases such as asthma, anaphylaxis, rheumatoid arthritis, graft rejection, and cancer.

Chapter Overview

- The immune system is subject to several types of dysfunctions termed immunopathologies.
- Some dysfunctions are due to abnormally heightened responses (allergies, hypersensitivities, and autoimmunities).
- Some dysfunctions are due to the reduction or loss in protective immune reactions (immunodeficiencies and cancer).
- Some immune damage is caused by normal actions that are directed at foreign tissues placed in the body for therapy (transfusions and transplants).
- Hypersensitivities are divided into immediate, antibody-mediated, immune complex, and delayed allergies.
- Allergens are the foreign molecules that cause a hypersensitive or allergic response.
- Most hypersensitivities require an initial sensitizing event, followed by a later contact that causes symptoms.
- The immediate type of allergy is mediated by special types of B cells, IgE, and mast cells that release allergic chemicals such as histamine that stimulate symptoms.
- Examples of immediate allergies are atopy, asthma, food allergies, and anaphylaxis.
- Hypersensitivity can arise from the action of other antibodies (IgG and IgM) that fix complement and lyse foreign cells.
- Immune complex reactions are caused by large amounts of circulating antibodies accumulating in tissues and organs.
- Autoimmune diseases, such as rheumatoid arthritis and multiple sclerosis, are due to B and T cells that are abnormally sensitized to react with the body's natural molecules and thus can damage cells and tissues.
- T cell responses to certain allergens and foreign molecules cause the diseases known as delayed-type hypersensitivities and graft rejection.

This delicate poison ivy plant, along with its close relatives poison oak and sumac, is one of the most common causes of allergy in the United States. Its leaves contain an oil that many people are sensitive to. The nature of its effects are covered in this chapter.

- Immunodeficiencies occur when B and T cells and other immune cells are missing or destroyed. They may be inborn and genetic or acquired.
- The primary outcome of immunodeficiencies is manifest in recurrent infections and lack of immune competence.
- Cancer is an abnormal overgrowth of cells due to a genetic defect and the lack of effective immune surveillance, recognition, and destruction of these cells.

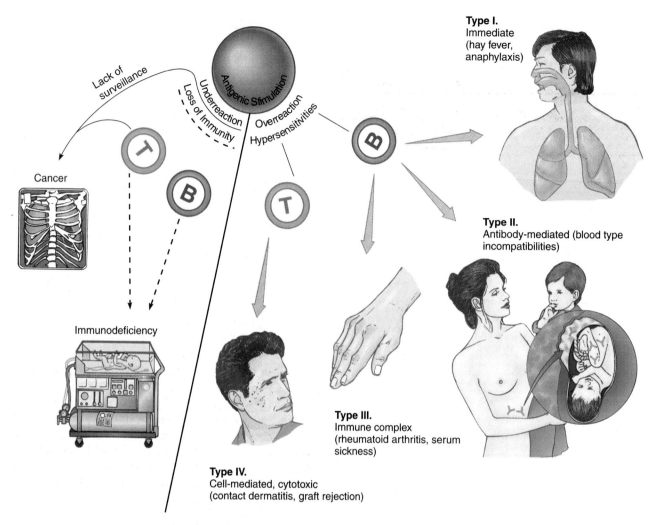

FIGURE 17.1

Overview of diseases of the immune system.   Just as the system of T cells and B cells provides necessary protection against infection and disease, the same system can cause serious and debilitating conditions by overreacting or underreacting to immune stimuli.

The Immune Response: A Two-Sided Coin

With few exceptions, our previous discussions of the immune response have centered around its numerous beneficial effects. The precisely coordinated system that seeks out, recognizes, and destroys an unending array of foreign materials is clearly protective, but it also presents another side—a side that promotes rather than prevents disease. In this chapter, we will survey **immunopathology,** the study of disease states associated with overreactivity or underreactivity of the immune response (**figure 17.1**). In the cases of allergies and **autoimmunity,** the tissues are innocent bystanders attacked by immunologic functions that can't distinguish one's own tissues from those expressing foreign material. In **grafts** and **transfusions,** a recipient reacts to the foreign tissues and cells of another individual. In **immunodeficiency** diseases, immune function is incompletely developed, suppressed, or destroyed. Cancer falls into a special category, because it is both a cause and an effect of immune dysfunction. As we shall see, one fascinating by-product of studies of immune disorders has been our increased understanding of the basic workings of the immune system.

Overreactions to Antigens: Allergy/Hypersensitivity

The term **allergy*** means a condition of altered reactivity or exaggerated immune response that is manifested by inflammation. Although it is sometimes used interchangeably with hypersensitivity, some experts refer to immediate reactions such as hay fever as allergies and to delayed reactions as hypersensitivities. Allergic individuals are acutely sensitive to repeated contact with antigens, called **allergens,** that do not noticeably affect nonallergic individuals. Although the general effects of hyperactivity are detrimental, we must be aware that it involves the very same types of immune reactions as those at work in protective immunities. These include humoral and cell-mediated actions, the inflammatory response, phagocytosis, and complement. Such an association means that all humans have the potential to develop hypersensitivity under particular circumstances.

---

*allergy (al′-er-jee) Gr. *allos,* other, and *ergon,* work.

TABLE 17.1

Hypersensitivity States

Type		Systems and Mechanisms Involved	Examples
I.	Immediate hypersensitivity	IgE-mediated; involves mast cells, basophils, and allergic mediators	Anaphylaxis, atopic allergies such as hay fever, asthma
II.	Antibody-mediated	IgG, IgM antibodies act upon cells with complement and cause cell lysis; includes some autoimmune diseases	Blood group incompatibility, pernicious anemia; myasthenia gravis
III.	Immune complex–mediated	Antibody-mediated inflammation; circulating IgG complexes deposited in basement membranes of target organs; includes some autoimmune diseases	Systemic lupus erythematosus; rheumatoid arthritis; serum sickness; rheumatic fever
IV.	T cell–mediated	Delayed hypersensitivity and cytotoxic reactions in tissues	Infection reactions; contact dermatitis; graft rejection; some types of autoimmunity

Originally, allergies were defined as either immediate or delayed, depending upon the time lapse between contact with the allergen and onset of symptoms. Subsequently, they were differentiated as humoral versus cell-mediated. But as information on the nature of the allergic immune response accumulated, it became evident that, although useful, these schemes oversimplified what is really a very complex spectrum of reactions. The most widely accepted classification, first introduced by immunologists P. Gell and R. Coombs, includes four major categories: type I (atopy and anaphylaxis), type II (IgG- and IgM-mediated cell damage), type III (immune complex), and type IV (delayed hypersensitivity) **(table 17.1).** In general, types I, II, and III involve a B-cell–immunoglobulin response, and type IV involves a T-cell response (figure 17.1). The antigens that elicit these reactions can be exogenous, originating from outside the body (microbes, pollen grains, and foreign cells and proteins), or endogenous, arising from self tissue (autoimmunities).

One of the reasons allergies are easily mistaken for infections is that both involve damage to the tissues and thus trigger the inflammatory response (see figure 14.16). Many symptoms and signs of inflammation (redness, heat, skin eruptions, edema, and granuloma) are prominent features of allergies.

CHAPTER CHECKPOINTS

Immunopathology is the study of diseases associated with excesses and deficiencies of the immune response. Such diseases include allergies, autoimmunity, grafts, transfusions, immunodeficiency disease, and cancer.

An allergy or hypersensitivity is an exaggerated immune response that injures or inflames tissues.

There are four categories of hypersensitivity reactions: type I (atopy and anaphylaxis), type II (transfusion reactions), type III (immune complex reactions), and type IV (delayed hypersensitivity reactions).

Antigens that trigger hypersensitivity reactions are allergens. They can be either exogenous (originate outside the host) or endogenous (involve the host's own tissue).

Type I Allergic Reactions: Atopy and Anaphylaxis

All type I allergies share a similar physiological mechanism, are immediate in onset, and are associated with exposure to specific antigens. However, it is convenient to recognize two subtypes: **Atopy*** is any chronic local allergy such as hay fever or asthma; **anaphylaxis*** is a systemic, often explosive reaction that involves airway obstruction and circulatory collapse. In the following sections, we will consider the epidemiology of type I allergies, allergens and routes of inoculation, mechanisms of disease, and specific syndromes.

EPIDEMIOLOGY AND MODES OF CONTACT WITH ALLERGENS

Allergies exert profound medical and economic impact. Allergists (physicians who specialize in treating allergies) estimate that about 10% to 30% of the population is prone to atopic allergy. It is generally acknowledged that self-treatment with over-the-counter medicines accounts for significant underreporting of cases. The 35 million people afflicted by hay fever (15–20% of the population) spend about half a billion dollars annually for medical treatment. The monetary loss due to employee debilitation and absenteeism is immeasurable. The majority of type I allergies are relatively mild, but certain forms such as asthma and anaphylaxis may require hospitalization and cause death. About 2 million people in the United States suffer from asthma.

The predisposition for type I allergies is inherited. Be aware that what is hereditary is a generalized *susceptibility,* not the allergy to a specific substance. For example, a parent who is allergic to ragweed pollen can have a child who is allergic to cat hair. The prospect of a child's developing atopic allergy is at least 25% if one parent is atopic, increasing up to 50% if grandparents or siblings are also afflicted. The actual basis for atopy appears to be a genetic

---

*atopy (at′-oh-pee) Gr. *atop,* out of place.

*anaphylaxis (an″-uh-fih-lax′-us) Gr. *ana,* excessive, and *phylaxis,* protection.

program that favors allergic antibody (IgE) production, increased reactivity of mast cells, and increased susceptibility of target tissue to allergic mediators. Allergic persons often exhibit a combination of syndromes, such as hay fever, eczema, and asthma.

Other factors that affect the presence of allergy are age, infection, and geographic locale. New allergies tend to crop up throughout an allergic person's life, especially as new exposures occur after moving or changing life-style. In some persons, atopic allergies last for a lifetime; others "outgrow" them, and still others suddenly develop them later in life. Some features of allergy are not yet completely explained.

THE NATURE OF ALLERGENS AND THEIR PORTALS OF ENTRY

As with other antigens, allergens have certain immunogenic characteristics. Not unexpectedly, proteins are more allergenic than carbohydrates, fats, or nucleic acids. Some allergens are haptens, nonproteinaceous substances with a molecular weight of less than 1,000 that can form complexes with carrier molecules in the body (see figure 15.11). Organic and inorganic chemicals found in industrial and household products, cosmetics, food, and drugs are commonly of this type. **Table 17.2** lists a number of common allergenic substances.

Allergens typically enter through epithelial portals in the respiratory tract, gastrointestinal tract, and skin. The mucosal surfaces of the gut and respiratory system present a thin, moist surface that is normally quite penetrable. The dry, tough keratin coating of skin is less permeable, but access still occurs through tiny breaks, glands, and hair follicles. It is worth noting that the organ of allergic expression may or may not be the same as the portal of entry.

Airborne environmental allergens such as pollen, house dust, dander (shed skin scales), or fungal spores are termed *inhalants*. Each geographic region harbors a particular combination of airborne substances that varies with the season and humidity (**figure 17.2a**). Pollen, the most common offender, is given off seasonally by the reproductive structures of pines and flowering plants (weeds, trees, and grasses). Unlike pollen, mold spores are released throughout the year and are especially profuse in moist areas of the home and garden. Airborne animal hair and dander (skin flakes), feathers, and the saliva of dogs and cats are common sources of

allergens. The component of house dust that appears to account for most dust allergies is not soil or other debris, but the decomposed bodies of tiny mites that commonly live in this dust (figure 17.2b). Some people are allergic to their work, in the sense that they are exposed to allergens on the job. Examples include florists, woodworkers, farmers, drug processors, welders, and plastics manufacturers whose work can aggravate inhalant and contact allergies.

Allergens that enter by mouth, called *ingestants,* often cause food allergies: *Injectant* allergies are an important adverse side effect of drugs or other substances used in diagnosing, treating, or preventing disease. A natural source of injectants is venom from stings by hymenopterans, a family of insects that includes honeybees and wasps. *Contactants* are allergens that enter through the skin. Many contact allergies are of the type IV, delayed variety discussed later in this chapter.

MECHANISMS OF TYPE I ALLERGY: SENSITIZATION AND PROVOCATION

What causes some people to sneeze and wheeze every time they step out into the spring air, while others suffer no ill effects? In order to answer this question, we must examine what occurs in the tissues of the allergic individual that does not occur in the normal person. In general, type I allergies develop in stages (**figure 17.3**). The initial encounter with an allergen provides a **sensitizing dose** that primes the immune system for a subsequent encounter with that allergen but generally elicits no signs or symptoms. The memory cells and immunoglobulin are then ready to react with a subsequent **provocative dose** of the same allergen. It is this dose that precipitates the signs and symptoms of allergy. Despite numerous anecdotal reports of people showing an allergy upon first contact with an allergen, it is generally believed that these individuals unknowingly had contact at some previous time. Fetal exposure to allergens from the mother's bloodstream is one possibility, and foods can be a prime source of "hidden" allergens such as penicillin.

The Physiology of IgE-Mediated Allergies

During primary contact and sensitization, the allergen penetrates the portal of entry (figure 17.3a). When large particles such as pollen grains, hair, and spores encounter a moist membrane, they release molecules of allergen that pass into the tissue fluids and lymphatics. The lymphatics then carry the allergen to the lymph nodes, where specific clones of B cells recognize it, are activated, and proliferate into plasma cells. These plasma cells produce immunoglobulin E (IgE), the antibody of allergy. IgE is different from other immunoglobulins in having an Fc receptor region with great affinity for mast cells and basophils. The binding of IgE to these cells in the tissues sets the scene for the reactions that occur upon repeated exposure to the same allergen (see figures 17.3b and 17.6).

The Role of Mast Cells and Basophils

The most important characteristics of mast cells and basophils relating to their roles in allergy are:

1. Their ubiquitous location in tissues. Mast cells are located in the connective tissue of virtually all organs, but particularly high concentrations exist in the lungs, skin, gastrointestinal

TABLE 17.2			
Common Allergens, Classified by Portal of Entry			
Inhalants	**Ingestants**	**Injectants**	**Contactants**
Pollen	Food	Hymenopteran	Drugs
Dust	(milk, peanuts,	venom (bee,	Cosmetics
Mold spores	wheat, shellfish,	wasp)	Heavy metals
Dander	soybeans,	Drugs	Detergents
Animal hair	nuts, eggs,	Vaccines	Formalin
Insect parts	fruits)	Serum	Rubber
Formalin	Food additives	Enzymes	Glue
Drugs	Drugs (aspirin,	Hormones	Solvents
Enzymes	penicillin)		Dyes

**National Allergy Bureau
Pollen and Mold Report**

Location: Sacramento, CA   Date: June 04, 2003
Counting Station: Allergy Medical Group of the North Area

Trees	Moderate severity	Total count: 41 / m³
Weeds	High severity	Total count: 64 / m³
Grass	High severity	Total count: 60 / m³
Mold	Low severity	Total count: 4,219 / m³

(a)

(b)

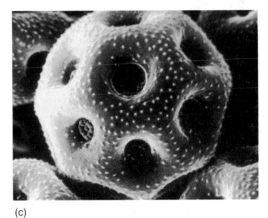

(c)

FIGURE 17.2

Monitoring airborne allergens. **(a)** The air in heavily vegetated places with a mild climate is especially laden with allergens such as pollen and mold spores. These counts vary seasonally. **(b)** Because the dust mite *Dermatophagoides* feeds primarily on human skin cells in house dust, these mites are found in abundance in bedding and carpets. Airborne mite feces and particles from their bodies are an important source of allergies. **(c)** Scanning electron micrograph of a single pollen grain from a rose (6,000×). Millions of these are released from a single flower.

tract, and genitourinary tract. Basophils circulate in the blood but migrate readily into tissues.

2. Their capacity to bind IgE during sensitization (figure 17.3). Each cell carries 30,000 to 100,000 cell receptors that attract 10,000 to 40,000 IgE antibodies.

3. Their cytoplasmic granules (secretory vesicles), which contain physiologically active cytokines (histamine, serotonin—introduced in chapter 14).

4. Their tendency to **degranulate** (figures 17.3*b* and 17.4), or release the contents of the granules into the tissues when properly stimulated by allergen.

Let us now see what occurs when sensitized cells are challenged with allergen a second time.

The Second Contact with Allergen

After sensitization, the IgE-primed mast cells can remain in the tissues for years. Even after long periods without contact, a person can retain the capacity to react immediately upon reexposure. The next time allergen molecules contact these sensitized cells, they bind across adjacent receptors and stimulate degranulation. As chemical mediators are released, they diffuse into the tissues and bloodstream. Cytokines give rise to numerous local and systemic reactions, many of which appear quite rapidly (figure 17.3*b*). The symptoms of allergy are not caused by the direct action of allergen on tissues but by the physiological effects of mast cell mediators on target organs.

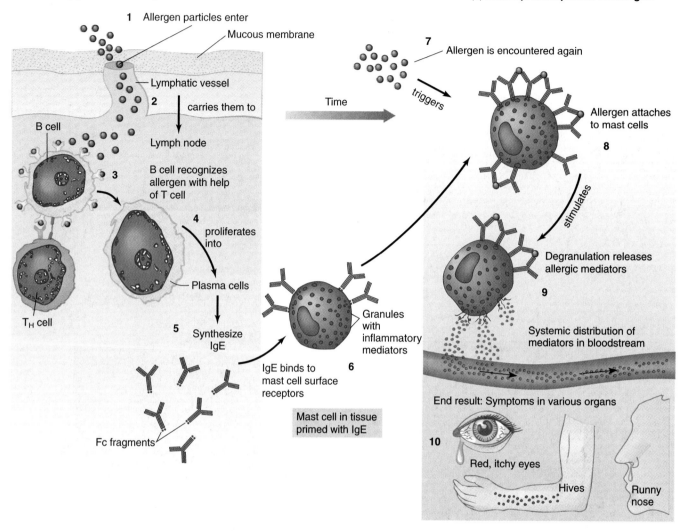

(a) Sensitization/IgE Production

1 Allergen particles enter

Mucous membrane

Lymphatic vessel

2 carries them to

B cell

Lymph node

3 B cell recognizes allergen with help of T cell

T_H cell

4 proliferates into

Plasma cells

5 Synthesize IgE

Fc fragments

Granules with inflammatory mediators

6 IgE binds to mast cell surface receptors

Mast cell in tissue primed with IgE

(b) Subsequent Exposure to Allergen

7 Allergen is encountered again

Time   triggers

Allergen attaches to mast cells

8

stimulates

9 Degranulation releases allergic mediators

Systemic distribution of mediators in bloodstream

End result: Symptoms in various organs

10 Red, itchy eyes

Hives

Runny nose

FIGURE 17.3

A schematic view of cellular reactions during the type I allergic response. **(a)** Sensitization (initial contact with sensitizing dose), 1–6. **(b)** Provocation (later contacts with provocative dose), 7–10.

CYTOKINES, TARGET ORGANS, AND ALLERGIC SYMPTOMS

Numerous substances involved in mediating allergy (and inflammation) have been identified. The principal chemical mediators produced by mast cells and basophils are histamine, serotonin, leukotriene, platelet-activating factor, prostaglandins, and bradykinin **(figure 17.4)**. These chemicals, acting alone or in combination, account for the tremendous scope of allergic symptoms. For some theories pertaining to this function of the allergic response, see **Medical Microfile 17.1**. Targets of these mediators include the skin, upper respiratory tract, gastrointestinal tract, and conjunctiva. The general responses of these organs include rashes, itching, redness, rhinitis, sneezing, diarrhea, and shedding of tears. Systemic targets include smooth muscle, mucous glands, and nervous tissue. Because smooth muscle is responsible for regulating the size of blood vessels and respiratory passageways, changes in its activity can profoundly alter blood flow,

blood pressure, and respiration. Pain, anxiety, agitation, and lethargy are also attributable to the effects of mediators on the nervous system.

Histamine* is the most profuse and fastest-acting allergic mediator. It is a potent stimulator of smooth muscle, glands, and eosinophils. Histamine's actions on smooth muscle vary with location. It *constricts* the smooth muscle layers of the small bronchi and intestine, thereby causing labored breathing and increased intestinal motility. In contrast, histamine *relaxes* vascular smooth muscle and dilates arterioles and venules. It is responsible for the **wheal*** *and flare* reaction in the skin (see figure 17.6a), pruritis (itching), and headache. More severe reactions (such as anaphylaxis) can be accompanied by edema and vascular dilation, which

---

*histamine (his′-tah-meen) Gr. *histio,* tissue, and amine.

*wheal (weel) A smooth, slightly elevated, temporary welt that is surrounded by a flushed patch of skin (flare).

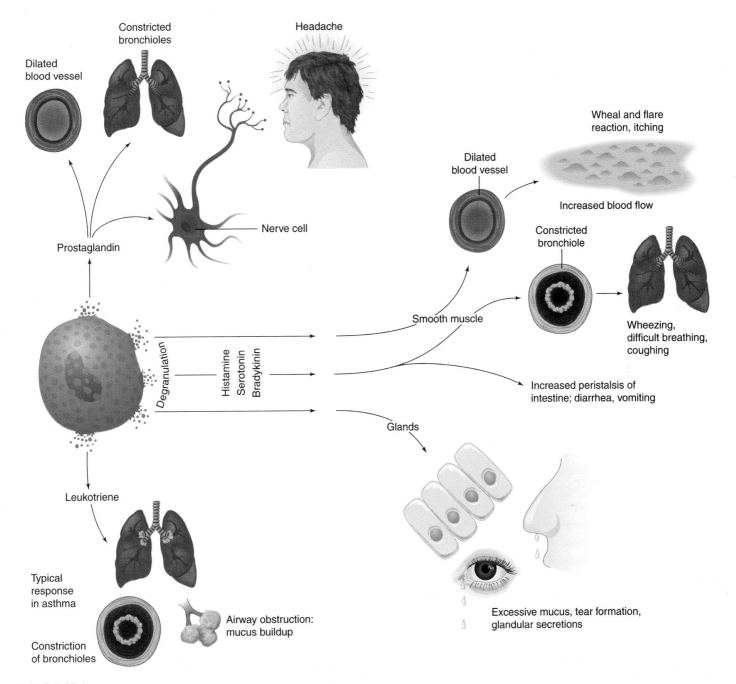

FIGURE 17.4

The spectrum of reactions to inflammatory cytokines released by mast cells and the common symptoms they elicit in target tissues and organs.
Note the extensive overlapping effects.

lead to hypotension, tachycardia, circulatory failure, and, frequently, shock. Salivary, lacrimal, mucous, and gastric glands are also histamine targets.

Although the role of **serotonin*** in human allergy is uncertain, its effects appear to complement those of histamine. In experimental animals, serotonin increases vascular permeability, capillary dilation, smooth muscle contraction, intestinal peristalsis, and respiratory rate, but it diminishes central nervous system activity.

Before the specific types were identified, **leukotriene*** was known as the "slow-reacting substance of anaphylaxis" for its property of inducing gradual contraction of smooth muscle. This type of leukotriene is responsible for the prolonged bronchospasm, vascular permeability, and mucous secretion of the asthmatic individual. Other leukotrienes stimulate the activities of polymorphonuclear leukocytes.

---

*leukotriene (loo″-koh-try′-een) Gr. *leukos,* white blood cell, and *triene,* a chemical suffix.

---

*serotonin (ser″-oh-toh′-nin) L. *serum,* whey, and *tonin,* tone.

Why would humans and other mammals evolve an allergic response that is capable of doing so much harm and even causing death? It is unlikely that this limb of immunity exists merely to make people miserable; it must have a role in protection and survival. What are the underlying biological functions of IgE, mast cells, and the array of potent cytokines? Analysis has revealed that, although allergic persons have high levels of IgE, trace quantities are present even in the sera of nonallergic individuals, just as mast cells and inflammatory chemicals are also part of normal human physiology. It is generally believed that one important function of

this system is to defend against helminth worms that are ubiquitous human parasites. In chapter 14, we learned that inflammatory mediators serve valuable functions, such as increasing blood flow and vascular permeability to summon essential immune components to an injured site. They are also responsible for increased mucous secretion, gastric motility, sneezing, and coughing, which help expel noxious agents. The difference is that, in allergic persons, the quantity and quality of these reactions are excessive and uncontrolled.

Platelet-activating factor is a lipid released by basophils, neutrophils, monocytes, and macrophages that causes platelet aggregation and lysis. The physiological response to stimulation by this factor is similar to that of histamine, including increased vascular permeability, pulmonary smooth muscle contraction, pulmonary edema, hypotension, and a wheal and flare response in the skin.

Prostaglandins* are a group of powerful inflammatory agents. Normally, these substances regulate smooth muscle contraction (for example, they stimulate uterine contractions during delivery). In allergic reactions, they are responsible for vasodilation, increased vascular permeability, increased sensitivity to pain, and bronchoconstriction. Certain anti-inflammatory drugs work by preventing the actions of prostaglandins.

Bradykinin* is related to a group of plasma and tissue peptides known as kinins that participate in blood clotting and chemotaxis. In allergy, it causes prolonged smooth muscle contraction of the bronchioles, dilatation of peripheral arterioles, increased capillary permeability, and increased mucous secretion.

SPECIFIC DISEASES ASSOCIATED WITH IgE- AND MAST CELL–MEDIATED ALLERGY

The mechanisms just described are basic to hay fever, allergic asthma, food allergy, drug allergy, eczema, and anaphylaxis. In this section, we cover the main characteristics of these conditions, followed by methods of detection and treatment.

Atopic Diseases

Hay fever is a generic term for **allergic rhinitis,*** a seasonal reaction to inhaled plant pollen or molds, or a chronic, year-round reaction to a wide spectrum of airborne allergens or inhalants (see table 17.2). The targets are typically respiratory membranes, and the symptoms include nasal congestion; sneezing; coughing; profuse mucous secretion; itchy, red, and teary eyes; and mild bronchoconstriction.

Asthma* is a respiratory disease characterized by episodes of impaired breathing due to severe bronchoconstriction. The airways of asthmatic people are exquisitely responsive to minute amounts of inhalant allergens, food, or other stimuli, such as infectious agents. The symptoms of asthma range from occasional, annoying bouts of difficult breathing to fatal suffocation. Labored breathing, shortness of breath, wheezing, cough, and ventilatory **rales*** are present to one degree or another. The respiratory tract of an asthmatic person is chronically inflamed and severely over-reactive to allergy chemicals, especially leukotrienes and serotonin from pulmonary mast cells. Other pathologic components are thick mucous plugs in the air sacs and lung damage that can result in long-term respiratory compromise. An imbalance in the nervous control of the respiratory smooth muscles is apparently involved in asthma, and the episodes are influenced by the psychological state of the person, which strongly supports a neurological connection.

The number of asthma sufferers in the United States is estimated at 10 million, with nearly one-third of them children. For reasons that are not completely understood, asthma is on the increase, and deaths from it have doubled since 1982, even though effective agents to control it are more available now than they have ever been before. A recent study of inner-city children has correlated high levels of asthma to contact with cockroach antigens in their living quarters. Nearly 40% of children age 10 or younger showed extreme sensitivity to the droppings and remains of these insects.

Atopic dermatitis is an intensely itchy inflammatory condition of the skin, sometimes also called **eczema.*** Sensitization occurs through ingestion, inhalation, and, occasionally, skin contact with allergens. It usually begins in infancy with reddened, vesicular, weeping, encrusted skin lesions. It then progresses in childhood and adulthood to a dry, scaly, thickened skin condition (**figure 17.5**). Lesions can occur on the face, scalp, neck, and inner surfaces of the limbs and trunk. The itchy, painful lesions cause considerable discomfort, and they are often predisposed to secondary bacterial

---

*prostaglandin (pross″-tah-glan′-din) From prostate gland. The substance was originally isolated from semen.

*bradykinin (brad″-ee-kye′-nin) Gr. *bradys,* slow, and *kinein,* to move.

*rhinitis (rye-nye′-tis) Gr. *rhis,* nose, and *itis,* inflammation.

---

*asthma (az′-muh) The Greek word for gasping.

*rales (rails) Abnormal breathing sounds.

*eczema (eks′-uh-mah; also ek-zeem′-uh) Gr. *ekzeo,* to boil over.

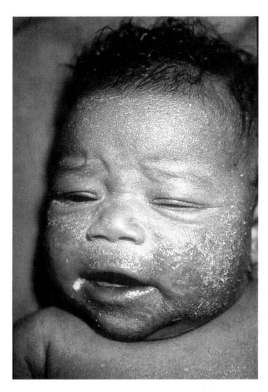

FIGURE 17.5

Atopic dermatitis, or eczema. Vesicular, encrusted lesions are typical in afflicted infants. This condition is prevalent enough to account for 1% of pediatric care.

infections. An anonymous writer once aptly described eczema as "the itch that rashes" or "one scratch is too many but one thousand is not enough."

Food Allergy

The ordinary diet contains a vast variety of compounds that are potentially allergenic. It is generally believed that food allergies are due to a digestive product of the food or to an additive (preservative or flavoring). Although the mode of entry is intestinal, food allergies can also affect the skin and respiratory tract. Gastrointestinal symptoms include vomiting, diarrhea, and abdominal pain. In severe cases, nutrients are poorly absorbed, leading to growth retardation and failure to thrive in young children. Other manifestations of food allergies include eczema, hives, rhinitis, asthma, and occasionally, anaphylaxis. Classic food hypersensitivity involves IgE and degranulation of mast cells, but not all reactions involve this mechanism. The most common food allergens come from peanuts, fish, cow's milk, eggs, shellfish, and soybeans.

Drug Allergy

Modern chemotherapy has been responsible for many medical advances. Unfortunately, it has also been hampered by the fact that drugs are foreign compounds capable of stimulating allergic reactions. In fact, allergy to drugs is one of the most common side effects of treatment (present in 5–10% of hospitalized patients). Depending upon the allergen, route of entry, and individual sensitivities, virtually any tissue of the body can be affected, and reactions range from mild atopy to fatal anaphylaxis. Compounds implicated most often are antibiotics (penicillin is number one in prevalence), synthetic antimicrobics (sulfa drugs), aspirin, opiates, and contrast dye used in X rays. The actual allergen is not the intact drug itself but a hapten given off when the liver processes the drug. Some forms of penicillin sensitivity are due to the presence of small amounts of the drug in meat, milk, and other foods and to exposure to *Penicillium* mold in the environment.

ANAPHYLAXIS: AN OVERPOWERING SYSTEMIC REACTION

The term **anaphylaxis,** or **anaphylactic shock,** was first used to denote a reaction of animals injected with a foreign protein. Although the animals showed no response during the first contact, upon reinoculation with the same protein at a later time, they exhibited acute symptoms—itching, sneezing, difficult breathing, prostration, and convulsions—and many died in a few minutes. Two clinical types of anaphylaxis are distinguished in humans. *Cutaneous anaphylaxis* is the wheal and flare inflammatory reaction to the local injection of allergen. *Systemic anaphylaxis,* on the other hand, is characterized by sudden respiratory and circulatory disruption that can be fatal in a few minutes. In humans, the allergen and route of entry are variable, though bee stings and injections of antibiotics or serum are implicated most often. Bee venom is a complex material containing several allergens and enzymes that can create a sensitivity that can last for decades after exposure.

The underlying physiological events in systemic anaphylaxis parallel those of atopy, but the concentration of chemical mediators and the strength of the response are greatly amplified. The immune system of a sensitized person exposed to a provocative dose of allergen responds with a sudden, massive release of chemicals into the tissues and blood, which act rapidly on the target organs. Anaphylactic persons have been known to die in 15 minutes from complete airway blockage.

DIAGNOSIS OF ALLERGY

Because allergy mimics infection and other conditions, it is important to determine if a person is actually allergic. If possible or necessary, it is also helpful to identify the specific allergen or allergens. Allergy diagnosis involves several levels of tests, including nonspecific, specific, *in vitro,* and *in vivo* methods.

A new test that can distinguish whether a patient has experienced an allergic attack measures elevated blood levels of tryptase, an enzyme released by mast cells that increases during an allergic response. Several types of specific *in vitro* tests can determine the allergic potential of a patient's blood sample. A differential blood cell count can indicate the levels of basophils and eosinophils—a higher level of these indicates allergy. The leukocyte histamine-release test measures the amount of histamine released from the patient's basophils when exposed to a specific allergen. Serological tests that use radioimmune assays (see chapter 16) to reveal the quantity and type of IgE are also clinically helpful.

Skin Testing

A useful *in vivo* method to detect precise atopic or anaphylactic sensitivities is skin testing. With this technique, a patient's skin is injected, scratched, or pricked with a small amount of a pure allergen extract. Hundreds of these allergen extracts contain common airborne allergens (plant and mold pollen) and more unusual allergens (mule dander, theater dust, bird feathers). Unfortunately, skin tests for food allergies using food extracts are unreliable in most cases. In patients with numerous allergies, the allergist maps the skin on the inner aspect of the forearms or back and injects the allergens intradermally according to this predetermined pattern (**figure 17.6a**). Approximately 20 minutes after antigenic challenge, each site is appraised for a wheal response indicative of histamine release. The diameter of the wheal is measured and rated on a scale of 0 (no reaction) to 4+ (greater than 15 mm). Figure 17.6*b* shows skin test results for a person with extreme inhalant allergies.

TREATMENT AND PREVENTION OF ALLERGY

In general, the methods of treating and preventing type I allergy involve

1. avoiding the allergen, though this may be very difficult in many instances;

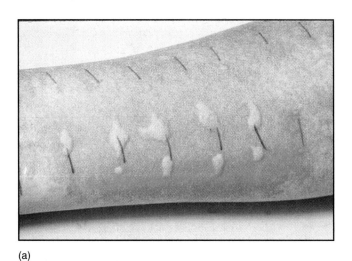

(a)

(b)

Environmental Allergens

No. 1 Standard Series				No. 2 Airborne Particles	
ID 8/85				ID 8/85	
+++	1. Acacia gum			+++	1. Ant
+++	2. Cat dander			++++	2. Aphis
++++	3. Chicken feathers			+++	3. Bee
++++	4. Cotton lint			+++	4. Housefly
++	5. Dog dander			✕	5. House mite
+	6. Duck feathers			+++	6. Mosquito
+	7. Glue, animal			++++	7. Moth
++	8. Horse dander			+++	8. Roach
✕	9. Horse serum			++	9. Wasp
+++	10. House dust #1			0	10. Yellow jacket
+	11. Kapok				Airborne mold spores
+	12. Mohair (goat)			++	11. *Alternaria*
+	13. Paper			+++	12. *Aspergillus*
++++	14. Pyrethrum			++	13. *Cladosporium*
+++	15. Rug pad, ozite			+++	14. *Hormodendrum*
+	16. Silk dust			0	15. *Penicillium*
+	17. Tobacco dust			+	16. *Phoma*
+	18. Tragacanth gum			+++	17. *Rhizopus*
+++++	19. Upholstery dust				18. ....
+++	20. Wool				

✕ - not done   ++ - mild reaction
0 - no reaction   +++ - moderate reaction
+ - slight reaction   ++++ - severe reaction

FIGURE 17.6

A method for conducting an allergy skin test. The forearm (or back) is mapped and then injected with a selection of allergen extracts. The allergist must be very aware of potential anaphylaxis attacks triggered by these injections. **(a)** Close-up of skin wheals showing a number of positive reactions (dark lines are measurer's marks). **(b)** An actual skin test record for some common environmental allergens with a legend for assessing them.

2. taking drugs that block the action of lymphocytes, mast cells, or chemical mediators; and

3. undergoing desensitization therapy.

It is not possible to completely prevent initial sensitization, since there is no way to tell in advance if a person will develop an allergy to a particular substance. The practice of delaying the introduction of solid foods apparently has some merit in preventing food allergies in children, though even breast milk can contain allergens ingested by the mother. Although rigorous cleaning and air conditioning can reduce contact with airborne allergens, it is not feasible to isolate a person from all allergens, which is the reason drugs are so important in control.

Therapy to Counteract Allergies

The aim of antiallergy medication is to block the progress of the allergic response somewhere along the route between IgE production and the appearance of symptoms **(figure 17.7).** Oral anti-inflammatory drugs such as corticosteroids inhibit the activity of lymphocytes and thereby reduce the production of IgE, but they also have dangerous side effects and should not be taken for prolonged periods. Some drugs block the degranulation of mast cells and reduce the levels of inflammatory cytokines. The most effective of these are diethylcarbamazine and cromolyn. Asthma and rhinitis sufferers can find relief with a new drug that blocks synthesis of leukotriene and a monoclonal antibody that inactivates IgE (Xolair).

Widely used medications for preventing symptoms of atopic allergy are **antihistamines,** the active ingredients in most over-the-counter allergy-control drugs. Antihistamines interfere with histamine activity by binding to histamine receptors on target organs. Most of them have major side effects, however, such as drowsiness. Newer antihistamines lack this side effect because they do not cross the blood-brain barrier. Other drugs that relieve inflammatory symptoms are aspirin and acetaminophen, which reduce pain by interfering with prostaglandin, and theophylline, a bronchodilator that reverses spasms in the respiratory smooth muscles. Persons who suffer from anaphylactic attacks are urged to carry at all times injectable epinephrine (adrenaline) and an identification tag indicating their sensitivity. An aerosol inhaler containing epinephrine can also provide rapid relief. Epinephrine reverses constriction of the airways and slows the release of allergic mediators.

Approximately 70% of allergic patients benefit from controlled injections of specific allergens as determined by skin tests. This technique, called **desensitization** or **hyposensitization,** is a therapeutic way to prevent reactions between allergen, IgE, and mast cells. The allergen preparations contain pure, preserved suspensions of plant antigens, venoms, dust mites, dander, and molds (but so far, hyposensitization for foods has not proved very effective). The immunologic basis of this treatment is open to differences in interpretation. One theory suggests that injected allergens stimulate the formation of high levels of allergen-specific IgG **(figure 17.8)** instead of IgE. It has been proposed that these IgG **blocking antibodies** remove allergen from the system before it can bind to IgE, thus preventing the degranulation of mast cells. It is also possible that allergen delivered in this fashion combines with the IgE itself and takes it from circulation before it can react with the mast cells.

CHAPTER CHECKPOINTS

Type I hypersensitivity reactions result from excessive IgE production in response to an exogenous antigen.

The two kinds of type I hypersensitivities are atopy, a chronic, local allergy, and anaphylaxis, a systemic, potentially fatal allergic response.

The predisposition to type I hypersensitivities is inherited, but age, geographic locale, and infection also influence allergic response.

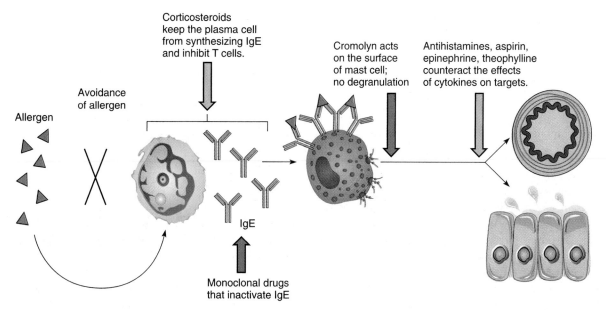

Corticosteroids keep the plasma cell from synthesizing IgE and inhibit T cells.

Cromolyn acts on the surface of mast cell; no degranulation

Antihistamines, aspirin, epinephrine, theophylline counteract the effects of cytokines on targets.

Avoidance of allergen

Allergen

IgE

Monoclonal drugs that inactivate IgE

FIGURE 17.7

Strategies for circumventing allergic attacks.

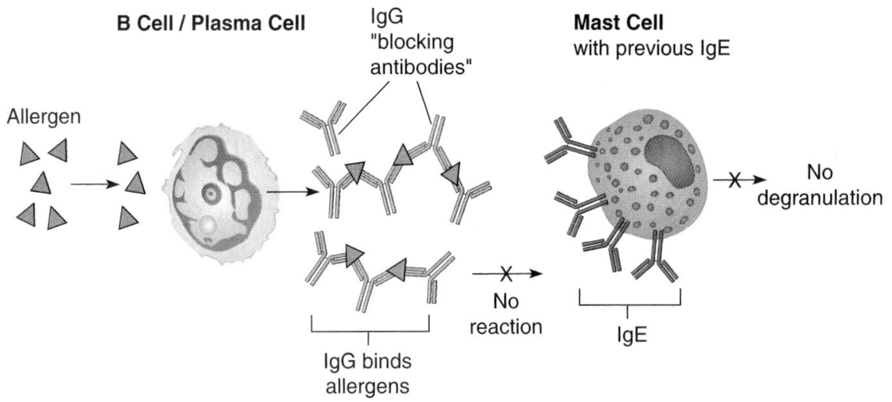

FIGURE 17.8

The blocking antibody theory for allergic desensitization. An injection of allergen causes IgG antibodies to be formed instead of IgE; these blocking antibodies cross-link and effectively remove the allergen before it can react with the IgE in the mast cell.

Type I allergens include inhalants, ingestants, injectants, and contactants.

The portals of entry for type I antigens are the skin, respiratory tract, gastrointestinal tract, and genitourinary tract.

Type I hypersensitivities are set up by a sensitizing dose of allergen and expressed when a second provocative dose triggers the allergic response. The time interval between the two can be many years.

The primary participants in type I hypersensitivities are IgE, basophils, mast cells, and agents of the inflammatory response.

Allergies are diagnosed by a variety of *in vitro* and *in vivo* tests that assay specific cells, IgE, and local reactions.

Allergies are treated by medications that interrupt the allergic response at certain points. Allergic reactions can often be prevented by desensitization therapy.

Type II Hypersensitivities: Reactions That Lyse Foreign Cells

The diseases termed type II hypersensitivities are a complex group of syndromes that involve complement-assisted destruction (lysis) of cells by antibodies (IgG and IgM) directed against those cells' surface antigens. This category includes transfusion reactions and some types of autoimmunities (discussed in a later section). The cells targeted for destruction are often red blood cells, but other cells can be involved.

HUMAN BLOOD TYPES

Chapters 14 and 15 described the functions of unique surface receptors or markers on cell membranes. Ordinarily, these receptors play essential roles in transport, recognition, and development, but they become medically important when the tissues of one person are placed into the body of another person. Blood transfusions and organ donations introduce alloantigens (molecules that differ in the same species) on donor cells that are recognized by the lymphocytes of the recipient. These reactions are not really immune dysfunctions as allergy and autoimmunity are. The immune system is in fact working normally, but it is not equipped to distinguish between the desirable foreign cells of a transplanted tissue and the undesirable ones of a microbe.

THE BASIS OF HUMAN ABO ANTIGENS AND BLOOD TYPES

The existence of human blood types was first demonstrated by an Austrian pathologist, Karl Landsteiner, in 1904. While studying incompatibilities in blood transfusions, he found that the serum of one person could clump the red blood cells of another. Landsteiner identified four distinct types, subsequently called the **ABO blood groups.**

Like the MHC antigens on white blood cells, the ABO antigen markers on red blood cells are genetically determined and composed of glycoproteins. These ABO antigens are inherited as two (one from each parent) of three alternative **alleles:*** A, B, or O. A and B alleles are dominant over O and codominant with one another. As **table 17.3** indicates, this mode of inheritance gives rise to four blood types (phenotypes), depending on the particular combination of genes. Thus, a person with an *AA* or *AO* genotype has type A blood; genotype *BB* or *BO* gives type B; genotype *AB* produces type AB; and genotype *OO* produces type O. Some important points about the blood types are:

1. They are named for the dominant antigen(s);
2. the RBCs of type O persons have antigens, but not A and B antigens; and
3. tissues other than RBCs carry A and B antigens.

The actual origin of the AB antigens and blood types are shown in **figure 17.9.** The A and B genes each code for an enzyme that adds a terminal carbohydrate to RBC receptors during maturation. RBCs of type A contain an enzyme that adds N-acetylgalactosamine to the receptor; RBCs of type B have an enzyme that adds D-galactose; RBCs of type AB contain both enzymes that add both carbohydrates; and RBCs of type O lack the genes and enzymes to add a terminal molecule.

ANTIBODIES AGAINST A AND B ANTIGENS

Although an individual does not normally produce antibodies in response to his or her own RBC antigens, the serum can contain antibodies that react with blood of another antigenic type, even though contact with this other blood type has *never* occurred. These preformed antibodies account for the immediate and intense quality of

---

*allele (ah-leel′) Gr. *allelon*, of one another. An alternate form of a gene for a given trait.

TABLE 17.3

Characteristics of ABO Blood Groups

Genotype	Blood Type	Antigen Present on Erythrocyte Membranes	Antibody in Plasma	Incidence of Type in United States		
				Among Whites (%)	Among Asians (%)	Among Blacks (%)
AA AO	A	A	Anti-b	41	28	27
BB, BO	B	B	Anti-a	10	27	20
AB	AB	A and B	Neither anti-a or anti-b	4	5	7
OO	O	Neither A nor B	Anti-a and anti-b	45	40	46

transfusion reactions. As a rule, type A blood contains antibodies (anti-b) that react against the B antigens on type B and AB red blood cells. Type B blood contains antibodies (anti-a) that react with A antigen on type A and AB red blood cells. Type O blood contains antibodies against both A and B antigens. Type AB blood does not contain antibodies against either A or B antigens[1] (table 17.3). What is the source of these anti-a and anti-b antibodies? It

---

1. Why would this be true? The answer lies in the first sentence of the paragraph.

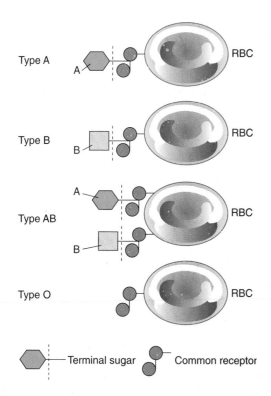

FIGURE 17.9

The genetic/molecular basis for the A and B antigens (receptors) on red blood cells. In general, persons with blood types A, B, and AB inherit a gene for the enzyme that adds a certain terminal sugar to the basic RBC receptor. Type O persons do not have such an enzyme and lack the terminal sugar.

appears that they develop in early infancy because of exposure to certain heterophile antigens that are widely distributed in nature. These antigens are surface molecules on bacteria and plant cells that mimic the structure of A and B antigens. Exposure to these sources stimulates the production of corresponding antibodies.[2]

Clinical Concerns in Transfusions

The presence of ABO antigens and a, b antibodies underlie several clinical concerns in giving blood transfusions. First, the individual blood types of donor and recipient must be determined. By use of a standard technique, drops of blood are mixed with antisera that contain antibodies against the A and B antigens and are then observed for the evidence of agglutination (**figure 17.10**).

Knowing the blood types involved makes it possible to determine which transfusions are safe to do. The general rule of compatibility is that the RBC antigens of the donor must not be agglutinated by antibodies in the recipient's blood (**figure 17.11**). The ideal practice is to transfuse blood that is a perfect match (A to A, B to B). But even in this event, blood samples must be crossmatched before the transfusion because other blood group incompatibilities can exist. This test involves mixing the blood of the donor with the serum of the recipient to check for agglutination.

Under certain circumstances (emergencies, the battlefield), the concept of universal transfusions can be used. To appreciate how this works, we must apply the rule stated in the previous paragraph. Type O blood lacks A and B antigens and will not be agglutinated by other blood types, so it could theoretically be used in any transfusion. Hence, a person with this blood type is called a **universal donor** Because type AB blood lacks agglutinating antibodies, an individual with this blood could conceivably receive any type of blood. Type AB persons are consequently called *universal recipients*. Although both types of transfusions involve antigen-antibody incompatibilities, these are of less concern because of the dilution of the donor's blood in the body of the recipient. Additional RBC markers that can be significant in transfusions are the Rh, MN, and Kell antigens (see next sections).

---

2. Evidence comes from germ-free chickens, which do not have antibodies against the antigens of blood types, whereas normal chickens possess these antibodies.

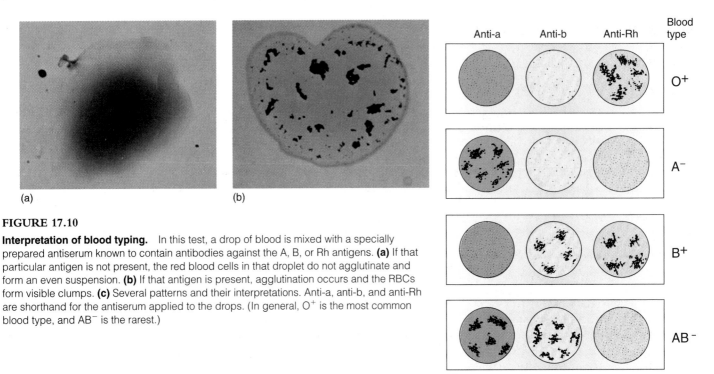

(a)   (b)

FIGURE 17.10

Interpretation of blood typing. In this test, a drop of blood is mixed with a specially prepared antiserum known to contain antibodies against the A, B, or Rh antigens. **(a)** If that particular antigen is not present, the red blood cells in that droplet do not agglutinate and form an even suspension. **(b)** If that antigen is present, agglutination occurs and the RBCs form visible clumps. **(c)** Several patterns and their interpretations. Anti-a, anti-b, and anti-Rh are shorthand for the antiserum applied to the drops. (In general, O^+ is the most common blood type, and AB^- is the rarest.)

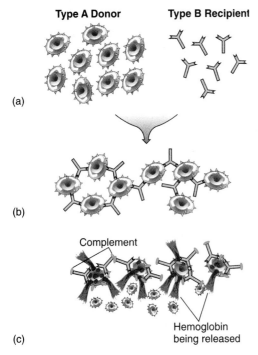

FIGURE 17.11

Microscopic view of a transfusion reaction. **(a)** Incompatible blood. The red blood cells of the type A donor contain antigen A, while the serum of the type B recipient contains anti-a antibodies that can agglutinate donor cells. **(b)** Agglutination particles can block the circulation in vital organs. **(c)** Activation of the complement by antibody on the RBCs can cause hemolysis and anemia. This sort of incorrect transfusion is very rare because of the great care taken by blood banks to ensure a correct match.

Transfusion of the wrong blood type causes various degrees of adverse reaction. The severest reaction is massive hemolysis when the donated red blood cells react with recipient antibody and trigger the complement cascade (figure 17.11). The resultant destruction of red cells leads to systemic shock and kidney failure brought on by the blockage of glomeruli (blood-filtering apparatus) by cell debris. Death is a common outcome. Other reactions caused by RBC destruction are fever, anemia, and jaundice. A transfusion reaction is managed by immediately halting the transfusion, administering drugs to remove hemoglobin from the blood, and beginning another transfusion with red blood cells of the correct type.

THE RH FACTOR AND ITS CLINICAL IMPORTANCE

Another RBC antigen of major clinical concern is the **Rh factor** (or D antigen). This factor was first discovered in experiments exploring the genetic relationships among animals. Rabbits inoculated with the RBCs of rhesus monkeys produced an antibody that also reacted with human RBCs. Further tests showed that this monkey antigen (termed Rh for rhesus) was present in about 85% of humans and absent in the other 15%. The details of Rh inheritance are more complicated than those of ABO, but in simplest terms, a person's Rh type results from a combination of two possible alleles—a dominant one that codes for the factor and a recessive one that does not. A person inheriting at least one Rh gene will be Rh^+; only those persons inheriting two recessive genes are Rh^-. This factor is denoted by a symbol above the blood type, as in O^+ or AB^- (see figure 17.10c). However, unlike the ABO antigens, exposure to normal flora does not sensitize Rh⁻ persons to the Rh factor. The only ways one can develop antibodies against this factor are through placental sensitization or transfusion.

Hemolytic Disease of the Newborn and Rh Incompatibility

The potential for placental sensitization occurs when a mother is Rh⁻ and her unborn child is Rh⁺. The obvious intimacy between mother and fetus makes it possible for fetal RBCs to leak into the mother's circulation during childbirth, when the detachment of the placenta creates avenues for fetal blood to enter the maternal

circulation. The mother's immune system detects the foreign Rh factors on the fetal RBCs and is sensitized to them by producing antibodies and memory B cells. The first Rh⁺ child is usually not affected because the process begins so late in pregnancy that the child is born before maternal sensitization is completed. However, the mother's immune system has been strongly primed for a second contact with this factor in a subsequent pregnancy (**figure 17.12a**).

In the next pregnancy with an Rh⁺ fetus, fetal blood cells escape into the maternal circulation late in pregnancy and elicit a memory response. The fetus is at risk when the maternal anti-Rh

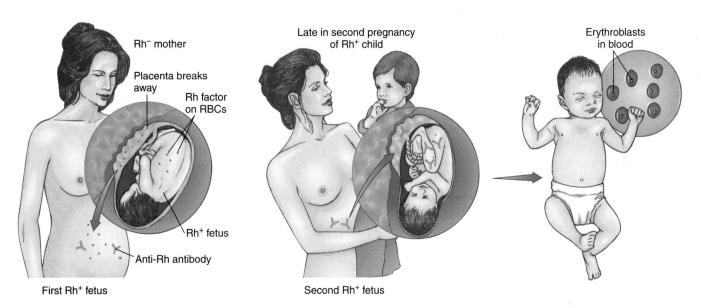

(a) **The development and aftermath of Rh sensitization**
Initial sensitization of the maternal immune system to fetal Rh⁺ factor occurs when fetal cells leak into the Rh⁻ mother's circulation late in pregnancy, or during delivery, when the placenta tears away. The child will escape hemolytic disease in most instances, but the mother, now sensitized, will be capable of an immediate reaction to a second Rh⁺ fetus and its Rh-factor antigen. At that time, the mother's anti-Rh antibodies pass into the fetal circulation and elicit severe hemolysis in the fetus and neonate.

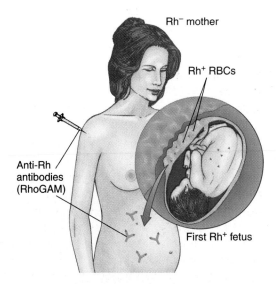

(b) **Prevention of erythroblastosis fetalis with anti-Rh immune globulin (RhoGAM)**
Injecting a mother who is at risk with RhoGAM during her first Rh⁺ pregnancy helps to inactivate and remove the fetal Rh-positive cells before her immune system can react and develop sensitivity.

FIGURE 17.12

Development and control of Rh incompatibility.

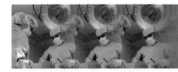

Think of it: Even though mother and child are genetically related, the father's genetic contribution guarantees that the fetus will contain molecules that are antigenic to the mother. In fact, with the recent practice of implanting one woman with the fertilized egg of another woman, the surrogate mother is carrying a fetus that has no genetic relationship to her. Yet, even with this essentially foreign body inside the mother, dangerous immunologic reactions such as Rh incompatibility are rather rare. In

what ways do fetuses avoid the surveillance of the mother's immune system? The answer appears to lie in the placenta and embryonic tissues. The fetal components that contribute to these tissues are not strongly antigenic, and they form a barrier that keeps the fetus isolated in its own antigen-free environment. The placenta is surrounded by a dense, many-layered envelope that prevents the passage of maternal cells, and it actively absorbs, removes, and inactivates circulating antigens.

antibodies cross the placenta into the fetal circulation, where they affix to fetal RBCs and cause complement-mediated lysis. The outcome is a potentially fatal **hemolytic disease of the newborn (HDN)** called *erythroblastosis fetalis* (eh-rith″-roh-blas-toh′-sis fee-tal′-is). This term is derived from the presence of immature nucleated RBCs called erythroblasts in the blood. They are released into the infant's circulation to compensate for the massive destruction of RBCs stimulated by maternal antibodies. Additional symptoms are severe anemia, jaundice, and enlarged spleen and liver.

Maternal-fetal incompatibilities are also possible in the ABO blood group, but adverse reactions occur less frequently than with Rh sensitization because the antibodies to these blood group antigens are IgM rather than IgG and are unable to cross the placenta in large numbers. In fact, the maternal-fetal relationship is a fascinating instance of foreign tissue not being rejected, despite the extensive potential for contact (**Medical Microfile 17.2**).

Preventing Hemolytic Disease of the Newborn

Once sensitization of the mother to Rh factor has occurred, all other Rh$^+$ fetuses will be at risk for hemolytic disease of the newborn. Prevention requires a careful family history of an Rh$^-$ pregnant woman. It can predict the likelihood that she is already sensitized or is carrying an Rh$^+$ fetus. It must take into account other children she has had, their Rh types, and the Rh status of the father. If the father is also Rh$^-$, the child will be Rh$^-$ and free of risk, but if the father is Rh$^+$, the probability that the child will be Rh$^+$ is 50% or 100%, depending on the exact genetic makeup of the father. If there is any possibility that the fetus is Rh$^+$, the mother must be passively immunized with antiserum containing antibodies against the Rh factor (*Rh$_0$ [D] immune globulin*, or **RhoGAM***). This antiserum, injected at 28 to 32 weeks and again immediately after delivery, reacts with any fetal RBCs that have escaped into the maternal circulation, thereby preventing the sensitization of the mother's immune system to Rh factor (figure 17.12*b*). Anti-Rh antibody must be given with each pregnancy that involves an Rh$^+$ fetus. It is ineffective if the mother has already been sensitized by a prior Rh$^+$ fetus or an incorrect blood transfusion, which can be determined by a

serological test. As in ABO blood types, the Rh factor should be matched for a transfusion, although it is acceptable to transfuse Rh$^-$ blood if the Rh type is not known.

OTHER RBC ANTIGENS

Although the ABO and Rh systems are of greatest medical significance, about 20 other red blood cell antigen groups have been discovered. Examples are the *MN, Ss, Kell,* and *P* blood groups. Because of incompatibilities that these blood groups present, transfused blood is screened to prevent possible cross-reactions. The study of these blood antigens (as well as ABO and Rh) has given rise to other useful applications. For example, they can be useful in forensic medicine (crime detection), studying ethnic ancestry, and tracing prehistoric migrations in anthropology. Many blood cell antigens are remarkably hardy and can be detected in dried blood stains, semen, and saliva. Even the 2,000-year-old mummy of King Tutankhamen has been typed A$_2$MN!

CHAPTER CHECKPOINTS

Type II hypersensitivity reactions occur when preformed antibodies react with foreign cell-bound antigens. The most common type II reactions occur when transfused blood is mismatched to the recipient's ABO type. IgG or IgM antibodies attach to the foreign cells, resulting in complement fixation. The resultant formation of membrane attack complexes lyses the donor cells.

Type II hypersensitivities are stimulated by antibodies formed against red blood cell (RBC) antigens or against other cell-bound antigens following prior exposure.

Complement, IgG, and IgM antibodies are the primary mediators of type II hypersensitivities.

The concepts of universal donor (type O) and universal recipient (type AB) apply only under emergency circumstances. Cross-matching donor and recipient blood is necessary to determine which transfusions are safe to perform.

Type II hypersensitivities can also occur when Rh$^-$ mothers are sensitized to Rh$^+$ RBCs of their unborn babies and the mother's anti-Rh antibodies cross the placenta, causing hemolysis of the newborn's RBCs. This is called hemolytic disease of the newborn, or erythroblastosis fetalis.

---

*RhoGAM Immunoglobulin fraction of human anti-Rh serum, prepared from pooled human sera.

Type III Hypersensitivities: Immune Complex Reactions

Type III hypersensitivity involves the reaction of soluble antigen with antibody and the deposition of the resulting complexes in basement membranes of epithelial tissue. It is similar to type II, because it involves the production of IgG and IgM antibodies after repeated exposure to antigens and the activation of complement. Type III differs from type II because its antigens are not attached to the surface of a cell. The interaction of these antigens with antibodies produces free-floating complexes that can be deposited in the tissues, causing an **immune complex reaction** or disease. This category includes therapy-related disorders (serum sickness and the Arthus reaction) and a number of autoimmune diseases (such as glomerulonephritis and lupus erythematosus).

MECHANISMS OF IMMUNE COMPLEX DISEASE

After initial exposure to a profuse amount of antigen, the immune system produces large quantities of antibodies that circulate in the fluid compartments. When this antigen enters the system a second

time, it reacts with the antibodies to form antigen-antibody complexes (**figure 17.13**). These complexes summon various inflammatory components such as complement and neutrophils, which would ordinarily eliminate Ag/Ab complexes as part of the normal immune response. In an immune complex disease, however, these complexes are so abundant that they deposit in the **basement membranes*** of epithelial tissues and become inaccessible. In response to these events, neutrophils release lysosomal granules that digest tissues and cause a destructive inflammatory condition. The symptoms of type III hypersensitivities are due in great measure to this pathologic state.

TYPES OF IMMUNE COMPLEX DISEASE

During the early tests of immunotherapy using animals, hypersensitivity reactions to serum and vaccines were common. In addition to anaphylaxis, two syndromes, the **Arthus reaction**[3] and **serum sickness,**

---

3. Named after Maurice Arthus, the physiologist who first identified this localized inflammatory response.

   Basement membranes are basal partitions of epithelia that normally filter out circulating antigen-antibody complexes.

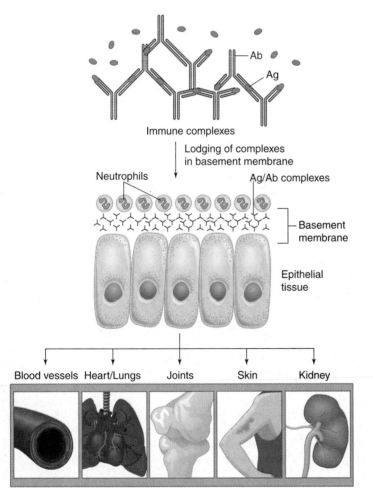

Steps:
1. Antibody combines with excess soluble antigen, forming large quantities of Ab/Ag complexes.

2. Circulating immune complexes become lodged in the basement membrane of epithelia in sites such as kidney, lungs, joints, skin.

3. Fragments of complement cause release of histamine and other mediator substances.
4. Neutrophils migrate to the site of immune complex deposition and release enzymes that cause severe damage in the tissues and organs involved.

Immune complexes

Lodging of complexes in basement membrane

Neutrophils   Ag/Ab complexes

Basement membrane

Epithelial tissue

Blood vessels Heart/Lungs Joints Skin Kidney

Major organs that can be targets of immune complex deposition

FIGURE 17.13

Pathogenesis of immune complex disease.

were identified. These syndromes are associated with certain types of passive immunization (especially with animal serum).

Serum sickness and the Arthus reaction are like anaphylaxis in requiring sensitization and preformed antibodies. Characteristics that set them apart are:

1. They depend upon IgG, IgM, or IgA (precipitating antibodies) rather than IgE;
2. they require large doses of antigen (not a minuscule dose as in anaphylaxis); and
3. they have delayed symptoms (a few hours to days).

The Arthus reaction and serum sickness differ from each other in some important ways. The Arthus reaction is a *localized* dermal injury due to inflamed blood vessels in the vicinity of any injected antigen. Serum sickness is a *systemic* injury initiated by antigen-antibody complexes that circulate in the blood and settle into membranes at various sites.

The Arthus Reaction

The Arthus reaction is usually an acute response to a second injection of vaccines (boosters) or drugs at the same site as the first injection. In a few hours, the area becomes red, hot to the touch, swollen, and very painful. These symptoms are mainly due to the destruction of tissues in and around the blood vessels and the release of histamine from mast cells and basophils. Although the reaction is usually self-limiting and rapidly cleared, intravascular blood clotting can occasionally cause necrosis and loss of tissue.

Serum Sickness

Serum sickness was named for a condition that appeared in soldiers after repeated injections of horse serum to treat tetanus. It can also be caused by injections of animal hormones and drugs. The immune complexes enter the circulation, are carried throughout the body, and are eventually deposited in blood vessels of the kidney, heart, skin, and joints (figure 17.13). The condition can become chronic, causing symptoms such as enlarged lymph nodes, rashes, painful joints, swelling, fever, and renal dysfunction.

AN INAPPROPRIATE RESPONSE AGAINST SELF, OR AUTOIMMUNITY

The immune diseases we have covered so far are all caused by foreign antigens. In the case of autoimmunity, an individual actually develops hypersensitivity to himself. This pathologic process accounts for **autoimmune diseases,** in which **autoantibodies** and, in certain cases, T cells mount an abnormal attack against self antigens. The scope of autoimmune diseases is extremely varied. In general, they can be differentiated as *systemic,* involving several major organs, or *organ-specific,* involving only one organ or tissue. They usually fall into the categories of type II or type III hypersensitivity, depending upon how the autoantibodies bring about injury. Some major autoimmune diseases, their targets, and basic pathology are presented in **table 17.4.** (For a reminder of hypersensitivity types, refer to table 17.1.)

Genetic and Gender Correlation in Autoimmune Disease

In most cases, the precipitating cause of autoimmune disease remains obscure, but we do know that susceptibility is determined by genetics and influenced by gender. Cases cluster in families, and even unaffected members tend to develop the autoantibodies for that disease. More direct evidence comes from studies of the major

TABLE 17.4

Selected Autoimmune Diseases

Disease	Target	Type of Hypersensitivity	Characteristics
Systemic lupus erythematosus (SLE)	Systemic	III	Inflammation of many organs; antibodies against red and white blood cells, platelets, clotting factors, nucleus
Rheumatoid arthritis and ankylosing spondylitis	Systemic	III	Vasculitis; frequent target is joint lining; antibodies against other antibodies (rheumatoid factor)
Scleroderma	Systemic	II	Excess collagen deposition in organs; antibodies formed against many intracellular organelles
Hashimoto's thyroiditis	Thyroid	II	Destruction of the thyroid follicles
Graves disease	Thyroid	II	Antibodies against thyroid-stimulating hormone receptors
Pernicious anemia	Stomach lining	II	Antibodies against receptors prevent transport of vitamin B_{12}
Myasthenia gravis	Muscle	II	Antibodies against the acetylcholine receptors on the nerve-muscle junction alter function
Type I diabetes	Pancreas	II	Antibodies stimulate destruction of insulin-secreting cells
Type II diabetes	Insulin receptor	II	Antibodies block attachment of insulin
Multiple sclerosis	Myelin	II	T cells and antibodies sensitized to myelin sheath destroy neurons
Goodpasture syndrome (glomerulonephritis)	Kidney	II	Antibodies to basement membrane of the glomerulus damage kidneys
Rheumatic fever	Heart	II	Antibodies to group A *Streptococcus* cross-react with heart tissue

histocompatibility gene complex. Particular genes in the class I and II major histocompatibility complex (see figure 15.3) coincide with certain autoimmune diseases. For example, autoimmune joint diseases such as rheumatoid arthritis and ankylosing spondylitis are more common in persons with the B-27 HLA type; systemic lupus erythematosus, Graves disease, and myasthenia gravis are associated with the B-8 HLA antigen. Why autoimmune diseases (except ankylosing spondylitis) afflict more females than males also remains a mystery. Females are more susceptible during childbearing years than before puberty or after menopause, suggesting a possible hormonal relationship.

The Origins of Autoimmune Disease

Very low titers of autoantibodies in otherwise healthy individuals suggest some normal function for them. A moderate, regulated amount of autoimmunity is probably required to dispose of old cells and cellular debris. Disease apparently arises when this regulatory or recognition apparatus goes awry. Attempts to explain the origin of autoimmunity include the following theories.

The *sequestered antigen theory* explains that during embryonic growth, some tissues are immunologically privileged; that is, they are sequestered behind anatomical barriers and cannot be scanned by the immune system **(figure 17.14a).** Examples of these

sites are regions of the central nervous system, which are shielded by the meninges and blood-brain barrier; the lens of the eye, which is enclosed by a thick sheath; and antigens in the thyroid and testes, which are sequestered behind an epithelial barrier. Eventually the antigen becomes exposed by means of infection, trauma, or deterioration, and is perceived by the immune system as a foreign substance.

According to the **clonal selection theory,** the immune system of a fetus develops tolerance by eradicating all self-reacting lymphocyte clones, called *forbidden clones,* while retaining only those clones that react to foreign antigens. Some of these clones may survive, and since they have not been subjected to this tolerance process, they can attack tissues with self antigens.

The *theory of immune deficiency* proposes that mutations in the receptor genes of some lymphocytes render them reactive to self or that a general breakdown in the normal T-suppressor function sets the scene for inappropriate immune responses.

Some autoimmune diseases appear to be caused by *molecular mimicry,* in which microbial antigens bear molecular determinants similar to normal human cells. An infection could cause formation of antibodies that can cross-react with tissues. This is one purported explanation for the pathology of rheumatic fever. Autoimmune disorders such as type I diabetes and multiple sclerosis

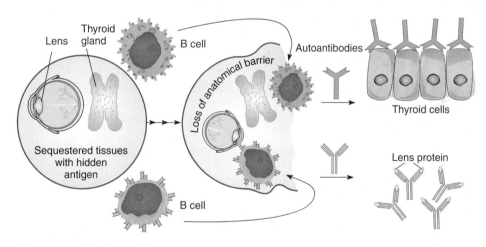

(a) **Sequestered Antigen Theory**

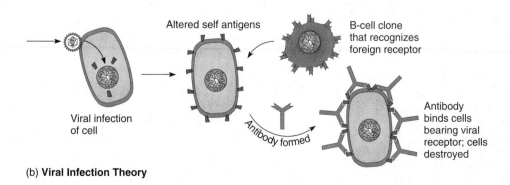

(b) **Viral Infection Theory**

FIGURE 17.14

Possible explanations for autoimmunity.   **(a)** Self antigens are sequestered and later incorrectly identified as a foreign antigen by B lymphocytes.
(b) Self antigens are altered by viral infection, which causes an immune response against the perceived foreign antigens.

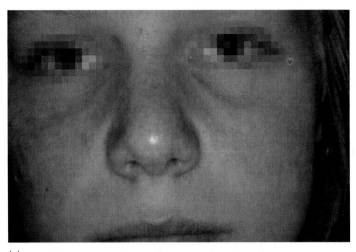

(a)

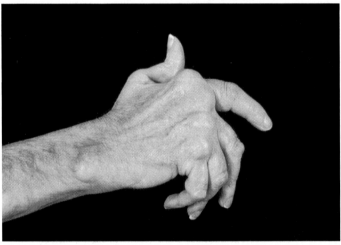

(b)

FIGURE 17.15

Common autoimmune diseases. **(a)** Systemic lupus erythematosus. One symptom is a prominent rash across the bridge of the nose and on the cheeks. These papules and blotches can also occur on the chest and limbs. **(b)** Rheumatoid arthritis commonly targets the synovial membrane of joints. Over time, chronic inflammation causes thickening of this membrane, erosion of the articular cartilage, and fusion of the joint. These effects severely limit motion and can eventually swell and distort the joints.

are likely triggered by *viral infection*. Viruses can noticeably alter cell receptors, thereby causing immune cells to attack the tissues bearing viral receptors (figure 17.14*b*).

Examples of Autoimmune Disease

Systemic Autoimmunities One of the most severe chronic autoimmune diseases is **systemic lupus erythematosus*** (SLE, or lupus). This name originated from the characteristic butterfly-shaped rash that drapes across the nose and cheeks (**figure 17.15*a*).

Although the manifestations of the disease vary considerably, all patients produce autoantibodies against a great variety of organs and tissues. The organs most involved are the kidneys, bone marrow, skin, nervous system, joints, muscles, heart, and GI tract. Antibodies to intracellular materials such as the nucleoprotein of the nucleus and mitochondria are also common.

In SLE, autoantibody-autoantigen complexes appear to be deposited in the basement membranes of various organs. Kidney failure, blood abnormalities, lung inflammation, myocarditis, and skin lesions are the predominant symptoms. One form of chronic lupus (called discoid) is influenced by exposure to the sun and primarily afflicts the skin. The etiology of lupus is still a puzzle. It is not known how such a generalized loss of self-tolerance arises, though viral infection or loss of T-cell suppressor function are suspected. The fact that women of childbearing years account for 90% of cases indicates that hormones may be involved. The diagnosis of SLE can usually be made with blood tests. Antibodies against the nucleus (ANA) and various tissues (detected by indirect fluorescent antibody or radioimmune assay techniques) are common, and a positive test for the lupus factor (an antinuclear factor) is also very indicative of the disease.

Rheumatoid arthritis,* another systemic autoimmune disease, incurs progressive, debilitating damage to the joints. In some patients, the lung, eye, skin, and nervous system are also involved. In the joint form of the disease, autoantibodies form immune complexes that bind to the synovial membrane of the joints and activate phagocytes and stimulate release of cytokines. Chronic inflammation leads to scar tissue and joint destruction. The joints in the hands and feet are affected first, followed by the knee and hip joints (figure 17.15*b*). The precipitating cause in rheumatoid arthritis is not known, though infectious agents such as Epstein-Barr virus have been suspected. The most common feature of the disease is the presence of an IgM antibody, called rheumatoid factor (RF), directed against other antibodies. This does not cause the disease but is used mainly in diagnosis. Some relief can be achieved with antiinflammatory agents, immunosuppressive drugs, and gold salt injections in some individuals.

Autoimmunities of the Endocrine Glands On occasion, the thyroid gland is the target of autoimmunity. The underlying cause of **Graves disease** is the attachment of autoantibodies to receptors on the follicle cells that secrete the hormone thyroxin. The abnormal stimulation of these cells causes the overproduction of this hormone and the symptoms of hyperthyroidism. In **Hashimoto thyroiditis,** both autoantibodies and T cells are reactive to the thyroid gland, but in this instance, they reduce the levels of thyroxin by destroying follicle cells and by inactivating the hormone. As a result of these reactions, the patient suffers from hypothyroidism.

The pancreas and its hormone, insulin, are other autoimmune targets. Insulin, secreted by the beta cells in the pancreas, regulates and is essential to the utilization of glucose by cells. **Diabetes mellitus** is caused by a dysfunction in insulin production or utilization (**figure 17.16).** Type I diabetes (also termed insulin-

---

*systemic lupus erythematosus (sis-tem′-ik loo′-pis air″-uh-theem-uh-toh′-sis) L. *lupus,* wolf, and *erythema,* redness.

*rheumatoid arthritis (roo′-muh-toyd ar-thry′-tis) Gr. *rheuma,* a moist discharge, and *arthron,* joint.

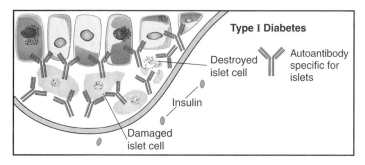

FIGURE 17.16

The autoimmune component in diabetes mellitus, type I.
Autoantibodies produced against the beta cells of the islets of Langerhans destroy the cells and greatly reduce insulin synthesis.

dependent diabetes) is associated with autoantibodies and sensitized T cells that damage the beta cells. A complex inflammatory reaction leading to lysis of these cells greatly reduces the amount of insulin secreted.

Neuromuscular Autoimmunities Myasthenia gravis* is named for the pronounced muscle weakness that is its principal symptom. Although the disease afflicts all skeletal muscle, the first effects are usually felt in the muscles of the eyes and throat. Eventually, it can progress to complete loss of muscle function and death. The classic syndrome is caused by autoantibodies binding to the receptors for acetylcholine, a chemical required to transmit a nerve impulse across the synaptic junction to a muscle (**figure 17.17**). The immune attack so severely damages the muscle cell membrane that transmission is blocked and paralysis ensues. Current treatment usually includes immunosuppressive drugs and therapy to remove the autoantibodies from the circulation. Experimental therapy using immunotoxins to destroy lymphocytes that produce autoantibodies shows some promise.

*myasthenia gravis (my″-us-thee′-nee-uh grah′-vis) Gr. *myo,* muscle, *astheneia,* weakness, and *gravida,* heavy.

Multiple sclerosis* (MS) is a paralyzing neuromuscular disease associated with lesions in the insulating myelin sheath that surrounds neurons in the white matter of the central nervous system. The underlying pathology involves damage to the sheath by both T cells and autoantibodies that severely compromises the capacity of neurons to send impulses. The principal motor and sensory symptoms are muscular weakness and tremors, difficulties in speech and vision, and some degree of paralysis. Most MS patients first experience symptoms as young adults, and they tend to experience remissions (periods of relief) alternating with recurrences of disease throughout their lives. Convincing evidence from studies of the brain tissue of MS patients points to a strong connection between the disease and infection with human herpesvirus 6 (see chapter 24). The disease can be treated passively with monoclonal antibodies that target T cells, and a vaccine containing the myelin protein has shown beneficial effects. Immunosuppressants such as cortisone and interferon B may also alleviate symptoms.

*sclerosis (skleh-roh′-sis) Gr. *sklerosis,* hardness.

CHAPTER CHECKPOINTS

Type III hypersensitivities are induced when a profuse amount of antigen enters the system and results in large quantities of antibody formation.

Type III hypersensitivity reactions occur when large quantities of antigen react with host antibody to form small, soluble immune complexes that settle in tissue cell membranes, causing chronic destructive inflammation. The reactions appear hours or days after the antigen challenge.

The mediators of type III hypersensitivity reactions include soluble IgA, IgG, or IgM, and agents of the inflammatory response.

Two kinds of type III hypersensitivities are localized (Arthus) reactions and systemic (serum sickness). Arthus reactions occur at the site of injected drugs or booster immunizations. Systemic reactions occur

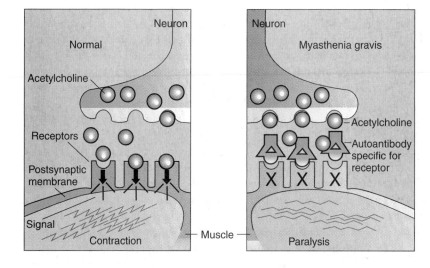

FIGURE 17.17

Mechanism for involvement of autoantibodies in myasthenia gravis. Antibodies developed against receptors on the postsynaptic membrane block them so that acetylcholine cannot bind and muscle contraction is inhibited.

when repeated antigen challenges cause systemic distribution of the immune complexes and subsequent inflammation of joints, lymph nodes, and kidney tubules.

Autoimmune hypersensitivity reactions occur when autoantibodies or host T cells mount an abnormal attack against self antigens. Autoimmune antibody responses can be either local or systemic type II or type III hypersensitivity reactions. Autoimmune T-cell responses are type IV hypersensitivity reactions.

Susceptibility to autoimmune disease appears to be influenced by gender and by genes in the MHC complex.

Autoimmune disease may be an excessive response of a normal immune function, the appearance of sequestered antigens, "forbidden" clones of lymphocytes that react to self antigens, or the result of alterations in the immune response caused by infectious agents, particularly viruses.

Examples of autoimmune diseases include systemic lupus erythematosus, rheumatoid arthritis, diabetes mellitus, myasthenia gravis, and multiple sclerosis.

FIGURE 17.18

Positive tuberculin test. Intradermal injection of tuberculin extract in a person sensitized to tuberculosis yields a slightly raised red bump greater than 10 mm in diameter.

Type IV Hypersensitivities: Cell-Mediated (Delayed) Reactions

The adverse immune responses we have covered so far are explained primarily by B-cell involvement and antibodies. A notable difference exists in type IV hypersensitivity, which involves primarily the T-cell branch of the immune system. Type IV immune dysfunction has traditionally been known as delayed hypersensitivity because the symptoms arise one to several days following the second contact with an antigen. In general, type IV diseases result when T cells respond to antigens displayed on self tissues or transplanted foreign cells. Examples of type IV hypersensitivity include delayed allergic reactions to infectious agents, contact dermatitis, and graft rejection.

DELAYED-TYPE HYPERSENSITIVITY

Infectious Allergy

A classic example of a delayed-type hypersensitivity occurs when a person sensitized by tuberculosis infection is injected with an extract (tuberculin) of the bacterium *Mycobacterium tuberculosis.* The so-called tuberculin reaction is an acute skin inflammation at the injection site appearing within 24 to 48 hours. So useful and diagnostic is this technique for detecting present or prior tuberculosis that it is the chosen screening device (see figure 19.17). Other infections that use similar skin testing are leprosy, syphilis, histoplasmosis, toxoplasmosis, and candidiasis. This form of hypersensitivity arises from time-consuming cellular events involving a specific class of T cells (T_{H1}) that receive the processed allergens from dendritic cells. Activated T_H cells release cytokines that recruit various inflammatory cells such as macrophages, neutrophils, and eosinophils. The build-up of fluid and cells at the site gives rise to a red papule (for example see **figure 17.18**). In a chronic infection (tertiary syphilis, for example), extensive damage to organs can occur through granuloma formation.

Contact Dermatitis

The most common delayed allergic reaction, contact dermatitis, is caused by exposure to resins in poison ivy or poison oak (**Spotlight on Microbiology 17.3**), to simple haptens in household and personal articles (jewelry, cosmetics, elasticized undergarments), and to certain drugs. Like immediate atopic dermatitis, the reaction to these allergens requires a sensitizing and a provocative dose. The allergen first penetrates the outer skin layers, is processed by dendritic cells (skin macrophages), and is presented to T cells. When subsequent exposures attract lymphocytes and macrophages to this area, these cells give off enzymes and inflammatory cytokines that severely damage the epidermis in the immediate vicinity (**figure 17.19a**). This response accounts for the intensely itchy papules and blisters that are the early symptoms (figure 17.19b). As healing progresses, the epidermis is replaced by a thick, horny layer. Depending upon the dose and the sensitivity of the individual, the time from initial contact to healing can be a week to 10 days.

T CELLS AND THEIR ROLE IN ORGAN TRANSPLANTATION

Transplantation or grafting of organs and tissues is a common medical procedure. Although it is life-giving, this technique is plagued by the natural tendency of lymphocytes to seek out foreign antigens and mount a campaign to destroy them. The bulk of the damage that occurs in graft rejections can be attributed to expression of cytotoxic T cells and other killer cells. This section will cover the mechanisms involved in graft rejection, tests for transplant compatibility, reactions against grafts, prevention of graft rejection, and types of grafts.

The Genetic and Biochemical Basis for Graft Rejection

In chapter 15, we discussed the role of major histocompatibility (MHC or HLA) genes and receptors in immune function. In general, the genes and receptors in MHC classes I and II are extremely important in recognizing self and in regulating the immune

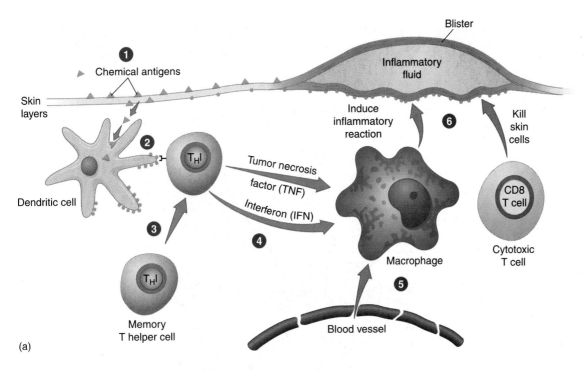

(a)

① Lipid-soluble chemicals are absorbed by the skin.

② Dendritic cells close to the epithelium pick up the allergen, process it, and display it on MHC receptors.

③ Previously sensitized T_H1 cells recognize the presented allergen.

④ Sensitized T_H1 cells are activated to secrete cytokines (IFN, TNF) that attract macrophages and cytotoxic T cells to the site. **⑤**

⑥ Macrophage releases mediators that stimulate a strong, local inflammatory reaction. Cytotoxic T cells directly kill cells and damage the skin. Fluid-filled blisters result.

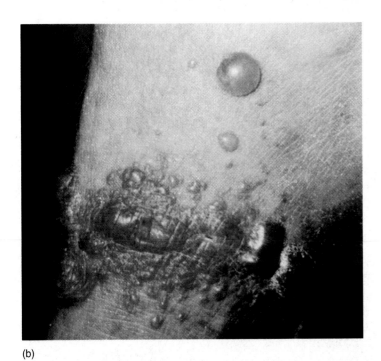

(b)

FIGURE 17.19

Contact dermatitis **(a)** Genesis of contact dermatitis **(b)** Contact dermatitis from poison oak, showing various stages of involvement: blisters, scales, and thickened patches.

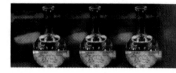

SPOTLIGHT ON MICROBIOLOGY 17.3
Pretty, Pesky, Poisonous Plants

As a cause of allergic contact dermatitis (affecting about 10 million people a year), nothing can compare with a single family of plants belonging to the genus *Toxicodendron*. At least one of these plants—either poison ivy, poison oak, or poison sumac—flourishes in the forests, woodlands, or along the trails of most regions of America. The allergen in these plants, an oil called urushiol, has such extreme potency that a pinhead-sized amount could spur symptoms in 500 people, and it is so long-lasting that botanists must be careful when handling 100-year-old plant specimens. Although degrees of sensitivity vary among individuals, it is estimated that 85% of all Americans are potentially hypersensitive to this compound. Some people are so acutely sensitive that even the most miniscule contact, such as handling pets or clothes that have touched the plant or breathing vaporized urushiol, can trigger an attack.

Humans first become sensitized by contact during childhood. Individuals at great risk (firefighters, hikers) are advised to determine their degree of sensitivity using a skin test, so that they can be adequately cautious and prepared. Some odd remedies include skin potions containing bleach, buttermilk, ammonia, hair spray, and meat tenderizer. Commercial products are available for blocking or washing away the urushiol. Allergy researchers are currently testing oral vaccines containing a form of urushiol, which seem to desensitize experimental animals. An effective method using poison ivy desensitization injection is currently available to people with extreme sensitivity.

Learning to identify these common plants can prevent exposure and sensitivity. One old saying that might help warns, "Leaves of three, let it be; berries white, run with fright." (See chapter opening photo of poison ivy.)

Poison oak

Poison sumac

response. These receptors also set the events of graft rejection in motion. The MHC genes of humans are inherited from among a large pool of genes, so the cells of each person can exhibit variability in the pattern of cell surface molecules (**figure 17.20**). The pattern is identical in different cells of the same person and can be similar in related siblings and parents, but the more distant the relationship, the less likely that the MHC genes and receptors will be similar. When donor tissue (a graft) displays surface receptors of a different MHC class, the T cells of the recipient (called the host) will recognize its foreignness and react against it.

T Cell–Mediated Recognition of Foreign MHC Receptors

Host Rejection of Graft   When the cytotoxic T cells of a host recognize foreign class I MHC receptors on the surface of grafted cells, they release interleukin-2 as part of a general immune mobilization. Receipt of this stimulus amplifies helper and cytotoxic T cells specific to the foreign antigens on the donated cells. The cytotoxic cells bind to the grafted tissue and secrete lymphokines that begin the rejection process within 2 weeks of transplantation (**figure 17.21a**). Late in this process, antibodies formed against the graft tis-

sue contribute to immune damage. A final blow is the destruction of the vascular supply, promoting death of the grafted tissue.

Graft Rejection of Host   In certain severe immunodeficiencies, the host cannot or does not reject a graft. But this failure may not protect the host from serious damage, because graft incompatibility is a two-way phenomenon. Some grafted tissues (especially bone marrow) contain an indigenous population called passenger lymphocytes. This makes it quite possible for the graft to reject the host, causing **graft versus host disease (GVHD)** (figure 17.21b). Since any host tissue bearing MHC receptors foreign to the graft can be attacked, the effects of GVHD are widely systemic and toxic. A papular, peeling skin rash is the most common symptom. Other organs affected are the liver, intestine, muscles, and mucous membranes. Previously GVHD occurred in approximately 30% of bone marrow transplants within 100 to 300 days of the graft. This percentage is declining as better screening and selection of tissues is developed.

Classes of Grafts
Grafts are generally classified according to the genetic relationship between the donor and the recipient. Tissue transplanted from one site on an individual's body to another site on his body is known as

an **autograft.** Typical examples are skin replacement in burn repair and the use of a vein to fashion a coronary artery bypass. In an **isograft,** tissue from an identical twin is used. Because isografts do not contain foreign antigens, they are not rejected, but this type of grafting has obvious limitations. **Allografts,** the most common type of grafts, are exchanges between genetically different individuals belonging to the same species (two humans). A close genetic correlation is sought for most allograft transplants (see next section). A **xenograft** is a tissue exchange between individuals of different species. Until rejection can be better controlled, most xenografts are experimental or for temporary therapy only.

Avoiding and Controlling Graft Incompatibility

Graft rejection can be averted or lessened by directly comparing the tissue of the recipient with that of potential donors. Several tissue matching procedures are used. In the *mixed lymphocyte reaction (MLR),* lymphocytes of the two individuals are mixed and incubated. If an incompatibility exists, some of the cells will become activated and proliferate. *Tissue typing* is similar to blood typing, except that specific antisera are used to disclose the HLA antigens on the surface of lymphocytes. In most grafts (one exception is bone marrow transplants), the ABO blood type must also be matched. Although a small amount of incompatibility is tolerable in certain grafts (liver, heart, kidney), a closer match is more likely to be successful, so the closest match possible is sought.

Drugs That Suppress Allograft Rejection Despite an international computerized hotline for matching recipients with donors and a greater availability of viable organs than in the past, an ideal match between donor and recipient is still the exception rather than the rule, and some sort of immunosuppressive therapy to overcome rejection is usually required. Rejection can be controlled with agents such as cyclosporin A, methotrexate, prednisone, and a monoclonal antibody OKT3. Except for cyclosporin A, intervention with drugs can be complicated by general suppression of the immune system (especially T cells) and frequent opportunistic infections.

Cyclosporin A is a polypeptide isolated from a fungus. It has dramatically improved the survival rate of allograft patients (kidney, heart, liver, and bone marrow) and has reduced the incidence of fatal infections. Although its action is not entirely understood, cyclosporin appears to block the activation of T helper cells and

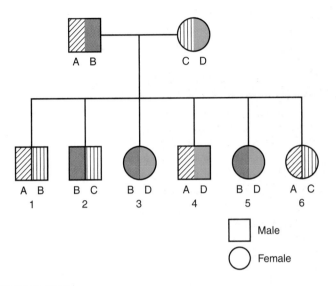

FIGURE 17.20

The pattern of inheritance of MHC (HLA) genes. A simplified version of the human leukocyte antigen (HLA) complex in a family. In this example, there are two genes in the complex, and each parent has a different set of genes (A/B and C/D). A child can inherit one of four different combinations. Out of six children, two sets (1 and 6, 3 and 5) have identical HLA genes and are good candidates for exchange grafts. Children sharing one gene (for example, 1, 4, and 6 share antigen A) are close matches, but two pairs of children (2 and 4, 3 and 6) do not match at all.

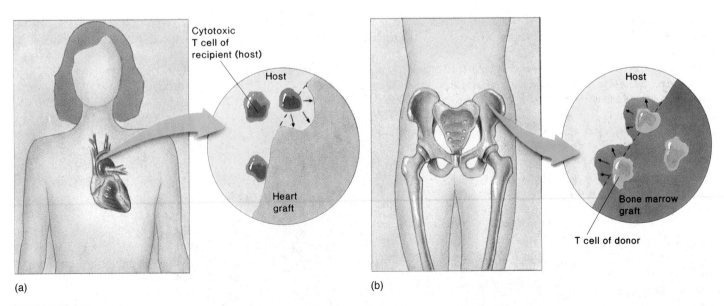

FIGURE 17.21

Potential reactions in transplantation. **(a)** The host's immune system (primarily cytotoxic T cells) encounters the cells of the donated organ (heart) and rejects the organ by secreting cytokines. **(b)** Grafted tissue (bone marrow) contains endogenous T cells that recognize the host's tissues as foreign and mount a cytokine attack. The recipient will develop symptoms of graft versus host disease.

interfere with the release of interleukin-2. What makes this drug so valuable is that it does not inhibit important lymphoid cells and phagocytes, and the body is better able to ward off infections. Its adverse effects of kidney toxicity and increased blood pressure can be reduced by adjusting the dose and monitoring blood levels of the drug. Because of its ability to inhibit undesirable T-cell activity, cyclosporin is also being used to treat autoimmune diseases such as type I diabetes and rheumatoid arthritis. Newer drugs aim to block the binding of IL-2 on T cells.

Types of Transplants

Today, transplantation is a recognized medical procedure whose benefit is reflected in several thousand transplants each year. It has been performed on every major organ, including parts of the brain. The most frequent transplant operations involve skin, heart, kidney, coronary artery, cornea, and bone marrow. The sources of organs and tissues are live donors (kidney, skin, bone marrow, liver), cadavers (heart, kidney, cornea), and fetal tissues. In the past decade, we have witnessed some unusual types of grafts. For instance, the fetal pancreas has been implanted as a potential treatment for diabetes, and fetal brain tissues for Parkinson disease. Part of a liver has been transplanted from a live parent to a child, and parents have donated a lobe from their lungs to help restore function in their children with severe cystic fibrosis.

Recent advances in stem cell technology have made it possible to isolate stem cells directly from the blood of donors without bone marrow sampling. Another potential source is the umbilical cord blood from a newborn infant. These have expanded the possibilities for treatment and survival.

Bone marrow transplantation is a rapidly growing medical procedure for patients with immune deficiencies, aplastic anemia,

leukemia and other cancers, and radiation damage. This procedure is extremely expensive, costing up to $200,000 per patient. Before bone marrow from a closely matched donor can be infused (**Medical Microfile 17.4**), the patient is pretreated with chemotherapy and whole-body irradiation, a procedure designed to destroy his own blood stem cells and thus prevent rejection of the new marrow cells. Within 2 weeks to a month after infusion, the grafted cells are established in the host. Because donor lymphoid cells can still cause GVHD, anti-rejection drugs may be necessary. An amazing consequence of bone marrow transplantation is that a recipient's blood type may change to the blood type of the donor.

CHAPTER CHECKPOINTS

Type IV hypersensitivity reactions occur when cytotoxic T cells attack either self tissue or transplanted foreign cells. Type IV reactions are also termed delayed hypersensitivity reactions because they occur hours to days after the antigenic challenge.

Type IV hypersensitivity reactions are mediated by T lymphocytes and are carried out against foreign cells that show both a foreign MHC and a nonself receptor site.

Examples of type IV reactions include the tuberculin reaction, contact dermatitis, and mismatched organ transplants (host rejection and GVHD reactions).

The four classes of transplants or grafts are determined by the degree of MHC similarity between graft and host. From most to least similar, these are: autografts, isografts, allografts, and xenografts.

Graft rejection can be minimized by tissue matching procedures, immunosuppressive drugs, and use of tissues that do not provoke a type IV response.

MEDICAL MICROFILE 17.4
The Mechanics of Bone Marrow Transplantation

In some ways, bone marrow is the most exceptional form of transplantation. It does not involve invasive surgery in either the donor or recipient, and it permits the removal of tissue from a living donor that is fully replaceable. While the donor is sedated, a bone marrow/blood sample is aspirated by inserting a special needle into an accessible marrow cavity. The most favorable sites are the crest and spine of the ilium (major bone of the pelvis). During this procedure, which lasts 1 to 2 hours, 3% to 5% of the donor's marrow is withdrawn in 20 to 30 separate extractions. Between 500 and 800 ml of marrow is removed. The donor may experience some pain and soreness, but there are rarely any serious complications. In a few weeks, the depleted marrow will naturally replace itself. Implanting the harvested bone marrow is rather convenient, because it is not necessary to place it directly into the marrow cavities of the recipient. Instead, it is dripped intravenously into the circulation, and the new marrow cells automatically settle in the appropriate bone marrow regions. The survival and permanent establishment of the marrow cells are increased by administering various growth factors and stem cell stimulants to the patient.

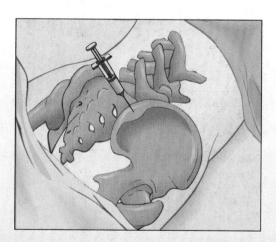

Removal of a bone marrow sample for transplantation. Samples are removed by inserting a needle into the spine or crest of the ilium. (The ilium is a prolific source of bone marrow.)

Immunodeficiency Diseases: Hyposensitivity of the Immune System

It is a marvel that development and function of the immune system proceed as normally as they do. On occasion, however, an error occurs and a person is born with or develops weakened immune responses. In many cases, these very "experiments" of nature have provided penetrating insights into the exact functions of certain cells, tissues, and organs because of the specific signs and symptoms shown by the immunodeficient individuals. The predominant consequences of immunodeficiencies are recurrent, overwhelming infections, often with opportunistic microbes. Immunodeficiencies fall into two general categories: *primary diseases,* present at birth (congenital) and usually stemming from genetic errors, and *secondary diseases,* acquired after birth and caused by natural or artificial agents **(table 17.5).**

PRIMARY IMMUNODEFICIENCY DISEASES

Deficiencies affect both specific immunities such as antibody production and less-specific ones such as phagocytosis. Consult **figure 17.22** to survey the places in the normal sequential development of lymphocytes where defects can occur and the possible consequences. In many cases, the deficiency is due to an inherited abnormality, though the exact nature of the abnormality is not known for a number of diseases. Because the development of B cells and T cells departs at some point, an individual can lack one or both cell lines. It must be emphasized, however, that some deficiencies affect other cell functions. For example a T-cell deficiency can affect B-cell function because of the role of T helper cells. In some deficiencies, the lymphocyte in question is completely absent or is present at very low levels, whereas in others, lymphocytes are present but do not function normally.

Clinical Deficiencies in B-Cell Development or Expression

Genetic deficiencies in B cells usually appear as an abnormality in immunoglobulin expression. In some instances, only certain immunoglobulin classes are absent; in others, the levels of all types of immunoglobulins (Ig) are reduced. A significant number of B-cell deficiencies are X-linked (also called sex-linked) recessive traits, meaning that the gene occurs on the X chromosome and the disease appears primarily in male children.

The term **agammaglobulinemia** literally means the absence of gamma globulin, the primary fraction of serum that contains immunoglobulins. Because it is very rare for Ig to be completely absent, some physicians prefer the term **hypogammaglobulinemia.***

---

*hypogammaglobulinemia (hy'-poh-gem-ah-glob-yoo-lin-ee'-mee-ah) Gr. *Hypo,* denoting a lowered level of Ig.

TABLE 17.5

General Categories of Immunodeficiency Diseases with Selected Examples

Primary Immune Deficiencies (Genetic)	Secondary Immune Deficiencies (Acquired)
B-Cell Defects (Low Levels of B Cells and Antibodies)	**From Natural Causes**
Agammaglobulinemia (X-linked, non-sex-linked)	Infection: AIDS, leprosy, tuberculosis, measles
Hypogammaglobulinemia	Other disease: cancer, diabetes
Selective immunoglobulin deficiencies	Nutrition deficiencies
	Stress
T-Cell Defects (Lack of All Classes of T Cells)	Pregnancy
Thymic aplasia (DiGeorge syndrome)	Aging
Chronic mucocutaneous candidiasis	
	From Immunosuppressive Agents
Combined B-Cell and T-Cell Defects (Usually Caused by Lack or Abnormality of Lymphoid Stem Cell)	Irradiation
Severe combined immunodeficiency disease (SCID)	Severe burns
X-SCIDI due to an interleukin defect	Steroids (cortisones)
Adenosine deaminase (ADA) deficiency	Drugs to treat graft rejection and cancer
Wiskott-Aldrich syndrome	Removal of spleen
Ataxia-telangiectasia	
Phagocyte Defects	
Chédiak-Higashi syndrome	
Chronic granulomatous disease of children	
Lack of surface adhesion molecules	
Complement Defects	
Lacking one of C components	
Hereditary angioedema	
Associated with rheumatoid diseases	

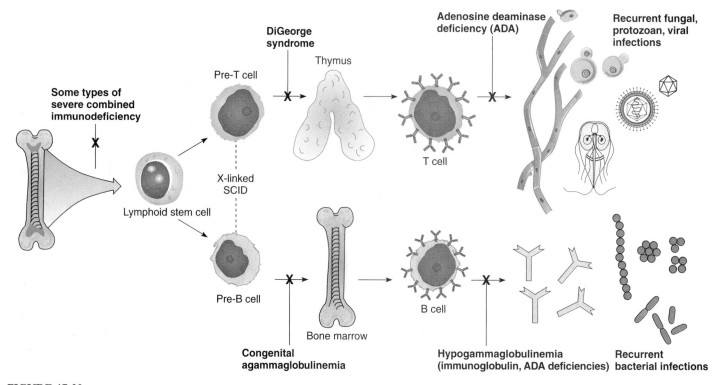

FIGURE 17.22

The stages of development and the functions of B cells and T cells, whose failure causes immunodeficiencies. Dotted lines represent the phases in development where breakdown can occur.

Both sex-linked and *autosomal*[4] recessive types of this rare syndrome occur. In X-linked agammaglobulinemia (XLA), mature B cells are absent, the level of antibodies is absent or greatly reduced, and lymphoid organs are incompletely developed. The actual genetic background for this disease involves some type of mutation in the gene for an essential enzyme (tyrosine kinase) required by the B cells to survive and mature (**figure 17.23**). T-cell function in these patients is usually normal. The symptoms of recurrent, serious bacterial infections usually appear about 6 months after birth. The bacteria most often implicated are pyogenic cocci, *Pseudomonas,* and *Haemophilus influenzae,* and the most common infection sites are the lungs, sinuses, meninges, and blood. Many Ig-deficient patients can have recurrent infections with viruses and protozoa, as well. Patients often manifest a wasting syndrome and have a reduced lifespan, but modern therapy has improved their prognosis. The current treatment for this condition is passive immunotherapy with immune serum globulin and continuous antibiotic therapy.

The lack of a particular class of immunoglobulin is a relatively common condition. Although genetically controlled, its underlying mechanisms are not yet clear. IgA deficiency is the most prevalent, occurring in about one person in 600. Such persons have normal quantities of B cells and other immunoglobulins, but they are unable to synthesize IgA. Consequently, they lack protection against local microbial invasion of the mucous membranes and suffer recurrent respiratory and gastrointestinal infections. The usual treatment using Ig replacement does not work, because conventional preparations are high in IgG, not IgA.

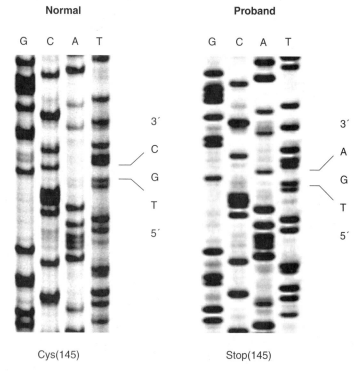

FIGURE 17.23

DNA sequencing detects immunodeficiency disease. The B cell defect called X-linked agammaglobulinemia can be detected by genetic analysis. A mutation of the tyrosine kinase gene on the X chromosome causes a C to be substituted by A, leading to a stop codon at that point (proband). The lack of a functioning enzyme results in death of the immature B cells. About 1 in 10,000 male children are born with this condition.

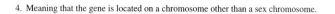

4. Meaning that the gene is located on a chromosome other than a sex chromosome.

Clinical Deficiencies in T-Cell Development or Expression

Due to their critical role in immune defenses, a genetic defect in T cells results in a broad spectrum of disease, including severe opportunistic infections, wasting, and cancer. In fact, a dysfunctional T-cell line is usually more devastating than a defective B-cell line because T helper cells are required to assist in most specific immune reactions. The deficiency can occur anywhere along the developmental spectrum, from thymus to mature, circulating T cells.

Abnormal Development of the Thymus The most severe of the T-cell deficiencies involve the congenital absence or immaturity of the thymus gland. Thymic aplasia, or **DiGeorge syndrome,** results when the embryonic third and fourth pharyngeal pouches fail to develop. Some cases are associated with a deletion in chromosome 22 **(figure 17.24).** The accompanying lack of cell-mediated immunity makes children highly susceptible to persistent infections by fungi, protozoa, and viruses. Common, usually benign, childhood infections such as chickenpox, measles, or mumps can be overwhelming and fatal in these children. Even vaccinations using attenuated microbes pose a danger. Other symptoms of thymic failure are reduced growth, wasting of the body, unusual facial characteristics, and an increased incidence of lymphatic cancer. These children can have reduced antibody levels, and they are unable to reject transplants. The major therapy for them is a transplant of thymus tissue.

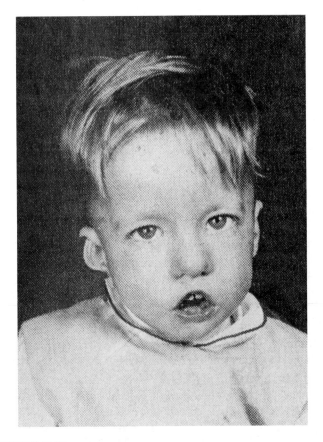

FIGURE 17.24

Facial characteristics of a child with DiGeorge syndrome. Typical defects include low-set, deformed earlobes; wide-set, slanted eyes; a small, bowlike mouth; and the absence of a philtrum (the vertical furrow between the nose and upper lip).

Severe Combined Immunodeficiencies: Dysfunction in B and T Cells

Severe combined immunodeficiencies (SCIDs) are the most dire and potentially lethal of the immunodeficiency diseases because they involve dysfunction in both lymphocyte systems. Some SCIDs are due to the complete absence of the lymphocyte stem cell in the marrow; others are attributable to the dysfunction of B cells and T cells later in development. Infants with SCID usually manifest the T-cell deficiencies within days after birth by developing candidiasis, sepsis, pneumonia, or systemic viral infections. This debilitating condition appears to have several forms. In the two most common forms, Swiss-type agammaglobulinemia and thymic alymphoplasia, the numbers of all types of lymphocytes are extremely low, the blood antibody content is greatly diminished, and the thymus and cell-mediated immunity are poorly developed. Both diseases are due to a genetic defect in the development of the lymphoid cell line.

A rarer form of SCID is **adenosine deaminase (ADA) deficiency,** which is caused by an autosomal recessive defect in the metabolism of adenosine. In this case, lymphocytes develop, but a metabolic product builds up abnormally and selectively destroys them. Infants with ADA deficiency are subject to recurrent infections and severe wasting typical of severe deficiencies. A small number of SCID cases are due to a developmental defect in receptors for B and T cells. An X-linked deficiency in interleukin receptors was responsible for the disease of David, the child in the "plastic bubble" **(Spotlight on Microbiology 17.5).** Another newly identified condition, *bare lymphocyte syndrome,* is caused by the lack of genes that code for class II MHC receptors.

Because of their profound lack of specific adaptive immunities, SCID children require the most rigorous kinds of aseptic techniques to protect them from opportunistic infections. Aside from life in a sterile plastic bubble, the only serious option for their long-time survival is total replacement or correction of dysfunctional lymphoid cells. Some infants can benefit from fetal liver or stem cell grafts. Although transplanting compatible bone marrow has been about 50% successful in curing the disease, it is complicated by graft versus host disease. The condition of some ADA-deficient patients has been partly corrected by periodic transfusions of blood containing large amounts of the normal enzyme. A more lasting treatment for both X-linked and ADA types of SCID would be gene therapy—insertion of normal genes to replace the defective genes (see figure 10.13). Although several children have benefitted by transfecting their bone marrow stem cells with the normal gene for ADA, this technique is still being worked out.

SECONDARY IMMUNODEFICIENCY DISEASES

Secondary acquired deficiencies in B cells and T cells are caused by one of four general agents:

1. infection,
2. organic disease,
3. chemotherapy, or
4. radiation.

The most recognized infection-induced immunodeficiency is **AIDS.** This syndrome is caused when several types of immune cells, including T helper cells, monocytes, macrophages, and antigen

SPOTLIGHT ON MICROBIOLOGY   17.5
An Answer to the Bubble Boy Mystery

David Vetter, the most famous SCID child, lived all but the last 2 weeks of his life in a sterile environment to isolate him from the microorganisms that could have quickly ended his life. When medical tests performed before birth had indicated that David might inherit this disease, he was delivered by cesarian section and immediately placed in a sterile isolette. From that time, he lived in various plastic chambers—ranging from room-size to a special suit that allowed him to walk outside. Remarkably, he developed into a well-adjusted child, even though his only physical contact with others was through special rubber gloves. When he was 12, his doctors decided to attempt a bone marrow transplant that might allow him to live free of his bubble prison. David was transplanted with his sister's bone marrow, but the marrow harbored a common herpesvirus called Epstein-Barr virus. Because he lacked any form of protective immunities against this oncogenic virus, a cancer spread rapidly through his body. Despite the finest medical care available, David died a short time later from the metastatic cancer.

In 1993, after several years of study, researchers discovered the basis for David's immunodeficiency. He had inherited an X-linked form (called XSCID) that arises from a defective genetic code in the receptors for interleukin-2, interleukin-4, and interleukin-7. The defect prevents the receptors on T cells and B cells from receiving the appropriate interleukin signals for growth, development, and reactivity. The end result is that both cytotoxic immunities and antibody-producing systems are shut down, leaving the body defenseless against infections and cancer.

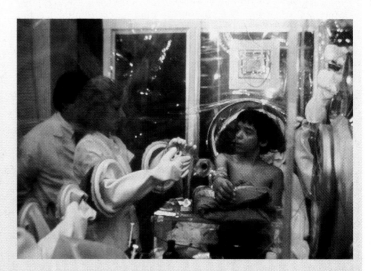

David Vetter, the boy in the plastic bubble.

presenting cells, are infected by the human immunodeficiency virus (HIV). It is generally thought that the depletion of T helper cells and functional impairment of immune responses ultimately account for the cancers and opportunistic protozoan, fungal, and viral infections associated with this disease. See chapter 25 for an extensive discussion of AIDS. Other infections that can deplete immunities are measles, leprosy, and malaria.

Cancers that target the bone marrow or lymphoid organs can be responsible for extreme malfunction of both humoral and cellular immunity (see next section). In leukemia, the massive number of cancer cells compete for space and literally displace the normal cells of the bone marrow and blood. Plasma cell tumors produce large amounts of nonfunctional antibodies, and thymus gland tumors cause severe T-cell deficiencies.

An ironic outcome of life-saving medical procedures is the possible suppression of a patient's immune system. For instance, some immunosuppressive drugs that prevent graft rejection by T cells can likewise suppress beneficial immune responses. Although radiation and anticancer drugs are the first line of therapy for many types of cancer, both agents are extremely damaging to the bone marrow and other body cells.

CHAPTER CHECKPOINTS

Immunodeficiency diseases occur when the immune response is reduced or absent.

Primary immune diseases are genetically induced deficiencies of B cells, T cells, the thymus gland, or combinations of these.

Secondary immune diseases are caused by infection, organic disease, chemotherapy, or radiation.

The best-known infection-induced immunodeficiency is AIDS.

Cancer: Cells Out of Control

The term **cancer** comes from the Latin word for crab and presumably refers to the appendage-like projections that a spreading tumor develops. A cancer is defined as the new growth of abnormal cells. The disease is also known by the synonym **neoplasm,*** or by more specific terms that usually end in the suffix -*oma*. **Oncology,** the field of medicine that specializes in cancer, is growing with such great rapidity that we can only survey major concepts as they relate to immunology, including the basic characteristics, classification, and origins of cancer.

CHARACTERISTICS AND CLASSIFICATION OF TUMORS AND CANCERS

An abnormal growth or tumor is generally characterized as benign or malignant. A **benign tumor** is a self-contained mass within an organ that does not spread into adjacent tissues. The mass is usually slow-growing, rounded, and not greatly different from its tissue of origin. Benign tumors do not ordinarily cause death unless they grow into a

---

*neoplasm (nee′-oh-plazm) Gr. *neo*, new, and *plasm*, formation.

critical space such as the heart valves or brain ventricles. The feature that most distinguishes a **malignant tumor** (cancer) is uncontrolled growth of abnormal cells within normal tissue. As a general rule, cancer originates in cells such as skin and bone marrow that have retained the capacity to divide, while mature cells that have lost this power (neurons, for instance) do not commonly become cancerous.

Most cancerous growths show other characteristics, including

1. disorganized behavior and independence from surrounding normal tissues,
2. permanent loss of cell differentiation, and
3. expression of special markers on their surface.

The cells of a malignant tumor also do not remain encapsulated but tend to spread both locally and distantly to other organs. As the initial (primary) tumor grows, it invades nearby tissues, enters lymphatic and blood vessels, and establishes secondary tumors in remote sites. This property of spreading is called **metastasis,** and the cancer is said to *metastasize.* Malignant tumors range in effects from those seen in pancreatic cancer, which spread rapidly and are almost always fatal, to others as occur in basal cell carcinoma of the skin, which are much less aggressive, more treatable, and seldom fatal.

Another tumor-classifying scheme relies on the tissue or cell of primary origin. Although cancers are commonly referred to simply as bladder cancer, breast cancer, or liver cancer, their technical names more clearly define the tumor origin. In general, cancers originating from epithelial tissues are called **carcinomas,*** and those originating from mesenchyme (embryonic connective tissue) are **sarcomas.*** Combining these terms with the tissue of origin supplies the full descriptive name. For example, adenocarcinoma develops in glandular tissue such as the pancreas or thyroid; squamous cell carcinoma arises in the skin epidermis; retinoblastoma occurs in the retina; lymphosarcoma in the lymph nodes; hepatoma (hepatic sarcoma) in the liver; and melanoma in the skin melanocytes. A special name, leukemia, denotes cancer of the blood-forming tissues. Considering the large number of cell types in a given organ, the great diversity of tumors (more than 100 clinical types) is to be expected.

Epidemiology of Cancer

One-third of all U.S. citizens will develop cancer some time during their lifetime. Cancer is the second leading cause of death in humans (following heart and circulatory disease). Annually, approximately 850,000 new cases are diagnosed, and 500,000 people die from various cancers. Cancer rates also vary according to gender, race, ethnic group, age, geographic region, occupational exposure to carcinogens, diet, and life-style. Cancers currently on the increase are melanoma (presumably due to increased levels of UV radiation entering the atmosphere), Kaposi's sarcoma (in AIDS patients), thyroid cancer, esophageal cancer, and prostate cancer (because of an improved method of detection). Lung cancer accounts for the largest number of deaths in both sexes. The second most common cause of cancer deaths is prostate cancer in men and breast cancer in women.

It is increasingly evident that susceptibility to certain cancers is inherited. At least 30 neoplastic syndromes that run in families have been described. A particularly stunning discovery is that some cancers in animals are actually due to endogenous viral genomes passed from parent to offspring in the gametes. Many cancers reflect an interplay of both hereditary and environmental factors. An example is lung cancer, which is associated mainly with tobacco smoking. But not all smokers develop lung cancer, and not all lung cancer victims smoke tobacco.

Proposed Mechanisms of Cancer

Cancer is clearly a complex disease associated with numerous physical, chemical, and biological agents **(table 17.6).** For years, there was no clear unifying explanation for how so many different

TABLE 17.6		
Selected Cancer-Triggering Agents and Their Diseases		
Agent	**Type of Exposure**	**Cancer Type**
Physical Agents		
Ionizing radiation	Occupational exposure, therapy	Leukemia, many others
UV radiation	Sunlight, suntanning lamps	Skin
Asbestos	Industrial, building insulation	Lung
Chemical Agents		
Tobacco	Smoking, chewing	Oral, lung
Alcohol	Ingestion	Oral, esophageal, liver
Arsenic	Manufacturing, mining	Lung, skin, liver, breast
Benzene	Various industrial processes	Leukemia
Polycyclic hydrocarbons	Medicines, industrial use (coal tar derivatives)	Lung, skin
Vinyl chloride	Manufacture of plastic	Liver
Aflatoxin	Fungus-contaminated foods	Liver
Aromatic amines	Manufacturing processes	Bladder
Heavy metal dust	Industry	Lung, nasal
Anabolic steroids	Medication	Liver
Estrogens	Medication	Vaginal, liver
Radon gas	Mines, homes	Lung
Biological Agents (Acquired Through Infection)		
Retrovirus	T-cell leukemia	
Papillomaviruses	Cervical cancer	
Polyoma viruses	Various tumors in mice and hamsters	
Epstein-Barr virus	Burkitt lymphoma; nasopharyngeal carcinoma	
Hepatitis B virus	Liver cancer	

---

*carcinoma (kar″-sih-noh′-mah) Gr. *karkinos,* crab, and *oma,* tumor.

*sarcoma (sar-koh′-mah) Gr. *sarcos,* flesh, and *oma,* tumor. Sometimes referred to as soft-tissue tumors.

factors could cause normal cells to become cancerous. But recent research findings link all cancers to genetic damage or inherited genetic predispositions that alter the function of certain genes found normally in all cells. The evidence for the interrelationship between genes and cancer is derived from the following observations:

1. Cancer cells often have damaged chromosomes;
2. a specific alteration in a gene can lead to cancer;
3. the predisposition for some cancers is inherited;
4. rates of cancer are highest in individuals who cannot repair damaged DNA;
5. mutagenic agents cause cancer;
6. cells contain genes that can be transformed to cancer-causing oncogenes; and
7. tumor-suppressor genes (anti-oncogenes) exist in the normal genome.

These discoveries and other advances in cancer research have provided the seeds for a general unifying theory for cancer.

Oncogene: A Normal Gene Gone Haywire? Cancer is another experiment of nature that continues to inform us about normal cell function. In the section on gene regulation in chapter 9, we learned that normal cell operations are strictly controlled by regulatory genes. Extracellular chemical signals are received by receptors, which transmit them to a regulatory complex in the nucleus that switches the appropriate genes on or off. During the rapid growth of a young individual, cells are extremely active, and the switch that initiates cell division is operating much of the time. By contrast, most adult cells exist in a nondividing state, except for divisions involved in repair and maintenance. This state of balance ensures that cells divide at a rate compatible with normal development and function.

Cancer results when the system that organizes cell division is short-circuited. Current theories explain that the gene complex controlling cell division contains a gene termed a **proto-oncogene** that regulates the onset of mitosis. The proto-oncogene is, in turn, regulated by another gene called a tumor-suppressor gene, or **anti-oncogene,** that prevents the proto-oncogene from acting continuously (**figure 17.25a**) and thereby keeps the cell division cycle operating normally. It is now clear that genetic disruptions (mutation, chromosomal alterations) can affect the normal actions of the proto-oncogenes and tumor-suppressor genes. If a genetic error keeps the proto-oncogene switched on, it converts to an **oncogene.** The oncogene is then programmed to override the normal mitotic controls and cause the cell to divide continuously (figure 17.25b). This process of cell immortalization is termed **transformation.** A number of happenings might explain how an oncogene goes permanently out of control and cannot be turned off. The anti-oncogene system may be missing or may have been inactivated by muta-

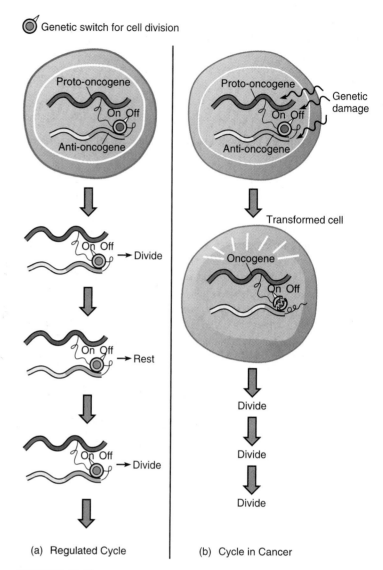

FIGURE 17.25

Common pathway for neoplasias. **(a)** Normal cell cycle and genetic controls. **(b)** Event that stimulates transformation into a cancer cell. Any agent that can alter the genetic control mechanisms can cause a cell to divide at incorrect times and in incorrect places.

tions, or new genetic material may have been added to it. Another possibility is that the oncogene products have been altered by mutation and no longer respond to regulation (remember that genes code for protein products). Some genes express their oncogenic potential after jumping from one site on a chromosome to another. New oncogenes are being discovered at a rapid pace, and their modes of action are gradually being deciphered (**Spotlight on Microbiology 17.6**).

SPOTLIGHT ON MICROBIOLOGY   17.6
Hot on the Trail of Cancer Clues

The outpouring of genetic discovery originating from the Human Genome Project has had a monumental impact on identification of genes involved in cancers. Remarkable progress is being made in identifying cancer genes, mapping them to specific areas of chromosomes, and developing special markers for these genes.

What has followed naturally from this new knowledge on exact locations of genes has been a tremendous activity in disclosing the ways these genes cause cancer. It is becoming clear that cancer development involves many complex interactions between genes, their products, and external signals received by the cell. There is no single mechanism. Consider the fact that at least eight different genetic defects lead to colon cancer and nearly that many are implicated in breast cancers. The one unifying pathway for all of these cancers, however, is a genetic alteration that disrupts the normal cell division cycle and transforms a normal cell into a cancer cell. In a large number of cases, cancer cells appear when the cell cycle clock in the nucleus can no longer respond appropriately to growth signals.

The Role of Oncogenes in the Cell Cycle

Cell chemicals called cyclins that control the phases in mitosis and cell growth can explain how some types of cancer arise. They act as chemical signals to promote cell division in a timely and orderly fashion. One of them, cyclin D1, is important for the early growth phase of cells. This substance turns on a set of enzymes called kinases that stimulate the replication of DNA. It turns out that the gene that codes for cyclin D1 (also called bc11) is a type of oncogene. When a defect in the gene causes the cyclin to be overproduced or produced at the wrong time, the cell goes into a nonstop division mode. Several other oncogenes apparently act at the level of the cyclins.

The Role of Tumor-Suppressor Genes

Another important model for cancer formation involves defective tumor-suppressor genes (TSGs), of which there are presently about a dozen known. One example is the human retinoblastoma (Rb) gene on chromosome 13, which prevents retinal cancer and other cancers. Individuals who have inherited defective genes, or whose normal genes have been mutated by carcinogens, can develop retinal tumors, osteosarcoma (a form of bone cancer), breast cancer, and a form of lung cancer. Molecular biologists determined that when the normal protein product of the Rb gene binds to a specific site on DNA, it blocks cellular growth and prevents the cells from growing out of control. However, when these genes are mutated or damaged, there is no suppression. Another fascinating tumor-suppressor gene, p53, becomes altered by simple missense mutations and has been implicated in 51 different tumors. p53 is a master regulator switch for the cell cycle that blocks the expression of the cyclin genes. When it is defective and the production of cyclin is altered, the cycle becomes unregulated and out of control. A second master regulator gene, P-TEN, codes for a protein to slow cell division. Disruption of this gene through mutation creates an aggressive cancer cell that is responsible for several rapidly growing tumors.

Programmed Cell Death

One normal developmental activity of cells is to undergo natural, programmed death, or **apoptosis.** This is one way that the body gets rid of abnormal, diseased, or old cells, and it accounts for various aspects of aging. Apoptosis is genetically coded by special "suicide genes" that are part of the genome of all cells. Now it appears that a disruption in this process may figure in some types of cancer. Because a cancer cell is an abnormal cell, it would ordinarily be programmed to die by apoptosis. But if genetic defects keep the suicide gene from being switched on, the cells never receive a death signal and thus become immortalized. Recent research on the role of viruses and cell death has shown that some viruses, such as the Epstein-Barr virus, contain a gene called C-MYC that interferes with the onset of apoptosis and thereby allows infected cells to survive. The EB virus is a known cause of Burkitt lymphoma. Other viruses that block cell death are human adenoviruses and papillomaviruses (warts).

Inability to Repair Mutations

Nearly one in every 200 people in the Western Hemisphere carries a defective gene that can lead to colon cancer, making this one of the most common genetic defects. If expressed, the defective gene leads to a condition called hereditary nonpolyposis colon cancer (HNPCC). The cause of this cancer is yet another variation on the theme of genetic defects. In this case, the defect occurs in the repair mechanisms for DNA. The genes that ordinarily regulate the repair of mismatched nucleotides are defective and can no longer detect and repair mutations. This condition leads to accumulated mutations and increases the likelihood of cancer.

Many discoveries in the basics of cancer genetics are already finding applications. In the future, it may be possible to use gene therapy to cure some types of cancer. Biotechnology and drug companies are already in the process of developing specialized therapies and cancer drugs based on molecular genetics (see chapter 10). A drug based on antisense technology has been successfully used to inactivate abnormal genes in melanoma cancers. Now that the exact locations and DNA sequences of oncogenes and tumor-suppressor genes are known, genetic probes and other markers are being developed at a rapid rate. The time is rapidly approaching when routine screening for susceptibility to many cancers will be as commonplace as routine medical tests. Other important developments are rapidly occurring in immunotherapy using monoclonal antibodies that inactivate receptors on cancer cells. This type of therapy has been used successfully on breast cancer.

The Role of Viruses in Cancer One documented oncogenic change occurs when a cell is infected by a retrovirus (**figure 17.26;** see chapter 25). Once inside a host cell, these RNA viruses utilize an enzyme called reverse transcriptase to synthesize viral DNA, which is subsequently inserted at a particular site on a host chromosome. Some retroviruses, such as the Rous sarcoma virus of chickens, carry viral oncogenes whose products cause transformation of host cells into cancer cells. Other retroviral genomes insert on host chromosomes at regulatory sites; their insertion at those sites can lead to activation of the cell's own proto-oncogenes. One type of cancer in humans, T-cell leukemia, is attributable to insertional mutagenesis by a retrovirus. DNA viruses are also involved in cancer. Human papillomavirus (HPV), which causes genital warts and is strongly implicated in cervical carcinoma, inserts its

genome directly into a chromosome in a way that overrides the usual cellular growth controls. Epstein-Barr virus (EBV), a cause of infectious mononucleosis and Burkitt lymphoma, induces a chromosomal alteration that can interfere with cell death.

Chromosomal Damage and Cancer A number of cancers have been associated with gross chromosomal changes or other aberrations that occur during mitotic divisions (**figure 17.27**). In the form of chromosomal disruption called *translocation* (figure 17.27*a*), a segment of one chromosome is transferred to another chromosome that is not its homologue. The abnormal placement of the translocated genes is believed to act as a strong oncogenic stimulus. Burkitt lymphoma (see figure 24.13), a malignant tumor of lymphocytes, is associated with the translocation of part of chromosome 8 to chromosome 14. Most cases of myelogenous leukemia are due to the translocation of a proto-oncogene on chromosome 9 to a different site on chromosome 22.

Certain cancers appear to result from a mistake in mitosis that produces cells bearing three rather than the usual pair of a particular chromosome. Such *trisomies* (figure 17.27*b*) occurring with chromosome 8 and 12 can lead to leukemia, and an extra copy of chromosome 7 is associated with melanoma. Several types of leukemia can be traced to the *loss* of a whole chromosome (figure 17.27*b*). The loss of part of a chromosome during cell division (chromosomal deletion) is also an important factor in some cancers. A large number of colorectal carcinomas manifest a deletion of a small part of chromosome 17 (figure 17.27*c*), and retinoblastoma occurs through deletions on chromosome 13 (see Spotlight on Microbiology 17.6).

Sufficient evidence indicates that some cells become cancerous because of *gene amplification* (figure 17.27*d*), a process whereby numerous extra copies of a gene are produced during DNA replication. When the copied gene is a proto-oncogene, the likelihood of oncogenic transformation is greatly increased. Certain tumors arising in squamous cell tissue or tissue of the lung, brain, and breast are linked to gene amplification.

The Role of Carcinogens What is the true etiologic role of chemical and physical carcinogenic agents (see table 17.6)? It appears that carcinogens genetically alter or damage DNA in a way that activates or deregulates a proto-oncogene and transforms it into an oncogene. Although physical agents such as X rays and UV radiation have well-developed powers to mutate or damage DNA (see figure 11.7), most chemical carcinogens operate in a stepwise fashion that requires at least two different stimuli. The first chemical, the *initiating stimulus,* is metabolized by the liver into a more reactive chemical that permanently alters the DNA of a particular target cell. This is the carcinogenic chemical targeted by the Ames test (see chapter 9). Before the altered cell becomes neoplastic, it must be acted upon by a second chemical stimulus, called a *promoter* or *co-carcinogen,* that completes the transformation to cancer.

Progression of Cancer

Transformation is marked by the appearance of aberrant histology and physiology in the malignant cell and its progeny. Cancer cells are frequently pleomorphic. They may appear as giant cells with bizarre shapes; show an abnormal, vacuolated cytoplasm; or pos-

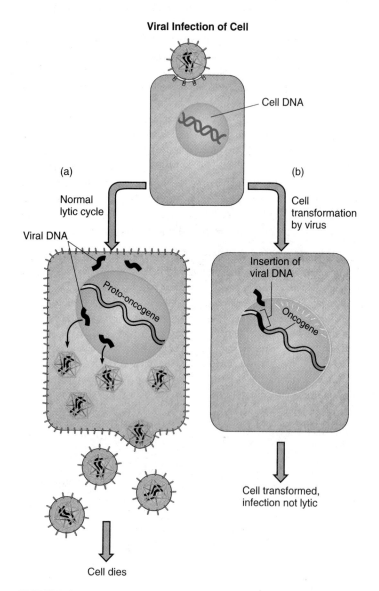

Viral Infection of Cell

Cell DNA

(a)

Normal lytic cycle

Viral DNA

Proto-oncogene

(b)

Cell transformation by virus

Insertion of viral DNA

Oncogene

Cell transformed, infection not lytic

Cell dies

FIGURE 17.26

Outcomes of viral infection. **(a)** A virus can go through the normal lytic cycle and destroy its host cell. **(b)** A possible mechanism for viral induction of cancer. DNA viruses and retroviruses can insert DNA into the host's genome. If it inserts into a sensitive site (near a proto-oncogene), the cell can be converted into a tumor cell.

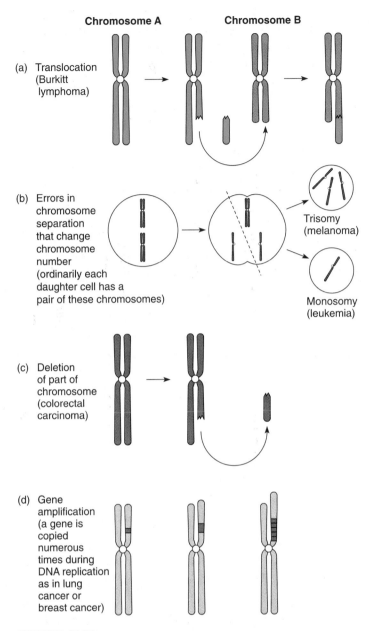

FIGURE 17.27

Alterations in chromosomes that can be implicated in some cancers (a-d).

(a) Translocation (Burkitt lymphoma)

(b) Errors in chromosome separation that change chromosome number (ordinarily each daughter cell has a pair of these chromosomes)

Trisomy (melanoma)

Monosomy (leukemia)

(c) Deletion of part of chromosome (colorectal carcinoma)

(d) Gene amplification (a gene is copied numerous times during DNA replication as in lung cancer or breast cancer)

sess enlarged, multiple nuclei. They express surface markers that may or may not be normal. Some markers are surface receptors for receiving stimuli from **growth factors,**[5] small polypeptides involved in cellular development that can favor cancerous growth. Some tumors secrete chemical factors that stimulate growth of additional circulatory vessels, an effect not unlike "self-feeding." An important feature of malignant cells is the loss of cohesiveness that keeps normal cells aggregated, so that they readily become dislodged from a tumor, metastasize into the circulation, and are carried to distant sites. Such invasive tumor cells also have the property of binding to and disrupting connective tissue barriers surrounding tissues and organs.

5. It is felt that certain oncogenes are the source of these growth factors.

THE FUNCTION OF THE IMMUNE SYSTEM IN CANCER

What role does the immune system play in controlling cancer? A concept that readily explains the detection and elimination of cancer cells is **immune surveillance.** Many experts postulate that cells with cancer-causing potential arise constantly in the body but that the immune system ordinarily discovers and destroys these cells, thus keeping cancer in check. Experiments with mammals have amply demonstrated that components of cell-mediated immunity interact with tumors and their antigens. The primary types of cells that operate in surveillance and destruction of tumor cells are cytotoxic T cells, natural killer (NK) cells, and macrophages. It appears that these cells recognize abnormal or foreign surface markers on the tumor cells and destroy them by mechanisms shown in figure 15.21. Antibodies help destroy tumors by interacting with macrophages and natural killer cells. This involvement of the immune system in destroying cancerous cells has inspired specific therapies (see Spotlight on Microbiology 17.6).

How do we account for the commonness of cancer in light of this powerful range of defenses against tumors? To an extent, the answer is simply that, as with infection, the immune system can and does fail. In some cases, the cancer may not be immunogenic enough; it may retain self markers and not be targeted by the surveillance system. In other cases, the tumor antigens may have mutated to escape detection. As we saw earlier, patients with immunodeficiencies such as AIDS and SCID are more susceptible to various cancers because they lack essential T-cell or cytotoxic functions. It may turn out that most cancers are associated with some sort of immunodeficiency, even a slight or transient one.

CHAPTER CHECKPOINTS

Cancer is caused by genetic transformation of normal host cells into malignant cells. These transformed cells perform no useful function but, instead, grow unchecked and interfere with normal tissue function.

Benign tumors are self-contained and slow-growing and do not differ greatly from their tissues of origin.

Malignant tumors are invasive, fast-growing, and very different from their tissues of origin. They metastasize into organs remote from the site of origin.

Possible causes of cancerous transformation include irregularities in mitosis, genetic damage, activation of oncogenes, and infection by retroviruses.

Cytotoxic T cells, NK cells, and macrophages identify and destroy transformed cancerous cells by recognizing and attaching to foreign surface markers on the transformed cells. Cancerous cells survive when these mechanisms fail.

Immunotherapy offers the most promise in treating cancers, compared with surgery, radiation, or conventional chemotherapy.

CHAPTER CAPSULE WITH KEY TERMS

I. **Immunopathology**

The study of disease states involving the malfunction of the immune system is called **immunopathology.**

   A. **Allergy,** or hypersensitivity, is an exaggerated, misdirected expression of certain immune responses.

   B. Abnormal responses to foreign antigens are characteristic of **immune complex reaction** or disease; undesirable reactions to foreign tissues are graft rejections.

   C. **Autoimmunity** involves abnormal responses to self antigens. A deficiency or loss in immune function is called **immunodeficiency.**

II. **Type I Hypersensitivities**

   A. Immediate-onset allergies involve contact with **allergens,** antigens that affect certain people; susceptibility is inherited; allergens enter through four portals: Inhalants are breathed in (pollen, dust); ingestants are swallowed (food, drugs); injectants are inoculated (drugs, bee stings); contactants react on skin surface (cosmetics, glue).

   B. *Mechanism:* On first contact with allergen, specific B cells react with allergen and form a special antibody class called IgE, which affixes by its Fc receptor to mast cells and basophils.

      1. This **sensitizing dose** primes the allergic response system.

      2. Upon subsequent exposure with a provocative dose, the same allergen binds to the IgE-mast cell complex.

      3. This causes **degranulation,** release of intracellular granules containing mediators (**histamine, serotonin, leukotriene, prostaglandin),** with physiological effects such as vasodilation and bronchoconstriction.

      4. Symptoms are rash, itching, redness, increased mucous discharge, pain, swelling, and difficulty in breathing.

   C. *Diagnosis of allergy* can be made by a histamine release test on basophils; serological assays for IgE; and skin testing, which injects allergen into the skin and mirrors the degree of reaction.

   D. *Control of allergy* involves drugs to interfere with the action of histamine, inflammation, and release of cytokines from mast cells. **Desensitization** therapy involves the administration of purified allergens.

III. **Type II Hypersensitivities**

   A. Type II reactions involve the interaction of antibodies, foreign cells, and complement, leading to lysis of the foreign cells. In transfusion reactions, humans may become sensitized to special antigens on the surface of the red blood cells of other humans.

   B. The **ABO blood groups** are genetically controlled: Type A blood has A antigens on the RBCs; type B has B antigens; type AB has both A and B antigens; and type O has neither antigen. People produce antibodies against A or B antigens if they lack these antigens. Antibodies can react with antigens if the wrong blood type is transfused.

   C. **Rh factor** is another RBC antigen that becomes a problem if an Rh$^-$ mother is sensitized by an Rh$^+$ fetus. A second fetus can receive antibodies she has made against the factor and develop **hemolytic disease of the newborn.** Prevention involves therapy with Rh immune globulin.

IV. **Type III Immune Complex Reactions**

   A. Exposure to a large quantity of soluble foreign antigens (serum, drugs) stimulates antibodies that produce small, soluble Ag/Ab complexes.

      1. These **immune complexes** are trapped in various organs and tissues, which incites a damaging inflammatory response.

      2. **Arthus reaction** is a local reaction to a series of injected antigens in the same body site; it may lead to tissue destruction.

      3. **Serum sickness** is a systemic disease resulting from repeated injections of foreign proteins; high levels of circulating immune complexes are deposited in various organs and cause tissue damage.

V. **Autoimmunity**

   A. In certain type II and III hypersensitivities, the immune system has lost tolerance to self molecules (autoantigens) and forms **autoantibodies** and sensitized T cells against them. Disruption of function can be systemic or organ specific.

   B. **Autoimmune diseases** are genetically determined and more common in females.

VI. **Type IV Cell-Mediated Hypersensitivity**

A delayed response to antigen involving the activation of and damage by T cells.

   A. *Delayed allergic response:* Skin response to allergens, including infectious agents. Example is tuberculin reaction; contact dermatitis is caused by exposure to plants (ivy, oak) and simple environmental molecules (metals, cosmetics); cytotoxic T cells acting on allergen elicit a skin reaction.

   B. *Graft rejection:* Reaction of cytotoxic T cells directed against foreign cells of a grafted tissue; involves recognition of foreign HLA by T cells and rejection of tissue.

      1. Host may reject graft; graft may reject host.

      2. Types of grafts include: **autograft,** from one part of body to another; **isograft,** grafting between identical twins; **allograft,** between two members of same species; **xenograft,** between two different species.

      3. All major organs may be successfully transplanted.

      4. Allografts require tissue match (HLA antigens must correspond); rejection is controlled with drugs.

VII. **Immunodeficiency Diseases**

Components of the immune response system are absent. Deficiencies involve B and T cells, phagocytes, and complement.

   A. Primary immunodeficiency is genetically based, congenital; defect in inheritance leads to lack of B-cell activity, T-cell activity, or both.

   B. B-cell defect is called **agammaglobulinemia;** patient lacks antibodies; serious recurrent bacterial infections result. In Ig deficiency, one of the classes of antibodies is missing or deficient.

   C. In T-cell defects, the thymus is missing or abnormal. In **DiGeorge syndrome,** the thymus fails to develop; afflicted children experience recurrent infections with eucaryotic pathogens and viruses; immune response is generally underdeveloped.

   D. In **severe combined immunodeficiency (SCID),** both limbs of the lymphocyte system are missing or defective; no adaptive immune response exists; fatal without replacement of bone marrow or other therapies.

   E. Secondary (acquired) immunodeficiency is due to damage after birth (infections, drugs, radiation). **AIDS** is the most common of these; T helper cells are main target; deficiency manifests in numerous opportunistic infections and cancers.

VIII. **Cancer and the Immune System**

   A. **Cancer** is characterized by overgrowth of abnormal tissue, also known as a **neoplasm.** It appears to arise from malfunction of **immune surveillance.**

      1. **Tumors** can be **benign** (a nonspreading local mass of tissue) or **malignant** (a cancer) that spreads (metastasizes) from the

tissue of origin to other particular sites by means of the circulation.
2. Malignant tumors may be **carcinomas,** originating from epithelial tissue, or **sarcomas,** originating from embryonic connective tissue.
3. Cancers occur in nearly every cell type (except mature, nondividing cells).

B. Cancer cells appear to share a common basic mechanism involving some type of gene alteration that turns a normal gene **(proto-oncogene)** into an **oncogene.**

1. The oncogene **transforms** the cell and triggers uncontrolled growth and abnormal structure and function.
2. Gene alterations may be due to chromosome disruptions, viral infection, or the actions of chemical and physical carcinogens.
3. Malignant cancer cells display special markers, respond to growth factors, and lose their cohesiveness.
4. Treatment is with surgery, chemicals, radiation, and immunotherapy.

MULTIPLE-CHOICE QUESTIONS

1. Pollen is which type of allergen?
   a. contactant
   b. ingestant
   c. injectant
   d. inhalant

2. B cells are responsible for which allergies?
   a. asthma
   b. anaphylaxis
   c. tuberculin reactions
   d. both a and b

3. Which allergies are T cell–mediated?
   a. type I
   b. type II
   c. type III
   d. type IV

4. The contact with allergen that results in symptoms is called the
   a. sensitizing dose
   b. degranulation dose
   c. provocative dose
   d. desensitizing dose

5. Production of IgE and degranulation of mast cells are involved in
   a. contact dermatitis
   b. anaphylaxis
   c. Arthus reaction
   d. both a and b

6. The direct, immediate cause of allergic symptoms is the action of
   a. the allergen directly on smooth muscle
   b. the allergen on B lymphocytes
   c. allergic mediators released from mast cells and basophils
   d. IgE on smooth muscle

7. Theoretically, type _____ blood can be donated to all persons because it lacks _____.
   a. AB, antibodies
   b. O, antigens
   c. AB, antigens
   d. O, antibodies

8. An example of a type III immune complex disease is
   a. serum sickness
   b. contact dermatitis
   c. graft rejection
   d. atopy

9. Type II hypersensitivities are due to
   a. IgE reacting with mast cells
   b. activation of cytotoxic T cells
   c. IgG-allergen complexes that clog epithelial tissues
   d. complement-induced lysis of cells in the presence of antibodies

10. Production of autoantibodies may be due to
    a. emergence of forbidden clones of B cells
    b. production of antibodies against sequestered tissues
    c. infection-induced change in receptors
    d. all of these are possible

11. Rheumatoid arthritis is an _____ that affects the _____.
    a. immunodeficiency disease, muscles
    b. autoimmune disease, nerves
    c. allergy, cartilage
    d. autoimmune disease, joints

12. A positive tuberculin skin test is an example of
    a. a delayed-type allergy
    b. acute contact dermatitis
    c. autoimmunity
    d. eczema

13. Contact dermatitis is caused by
    a. pollen grains
    b. chemicals absorbed by the skin
    c. microbes
    d. proteins found in foods

14. Which disease would be most similar to AIDS in its pathology?
    a. X-linked agammaglobulinemia
    b. SCID
    c. ADA deficiency
    d. DiGeorge syndrome

15. A general feature common to all cancers is a _____ that leads to _____.
    a. carcinogenic agent, carcinoma
    b. mutation, gene amplification
    c. genetic defect, uncontrolled cell growth
    d. proto-oncogene, anti-oncogene

16. The property whereby cancer cells display abnormal anatomy and growth patterns is called
    a. metastasis
    b. oncogenic
    c. carcinogenic
    d. malignant

17. A cancer associated with viral infections includes
    a. cervical carcinoma
    b. Burkitt lymphoma
    c. T-cell leukemia
    d. all of these

CONCEPT QUESTIONS

1. a. Define allergy and hypersensitivity.
   b. What accounts for the reactions that occur in these conditions?
   c. What does it mean when a reaction is immediate or delayed?
   d. Give examples of each type.

2. Describe several factors that influence types and severity of allergic responses.

3. a. How are atopic allergies similar to anaphylaxis?
   b. How are they different?

4. a. How do allergens gain access to the body?
   b. What are some examples of allergens that enter by these portals?

5. a. Trace the course of a pollen grain through sensitization and provocation in type I allergies.
   b. Include in the discussion the role of mast cells, basophils, IgE, and allergic mediators.
   c. Outline the target organs and symptoms of the principal atopic diseases and their diagnosis and treatment.

6. a. Describe the allergic response that leads to anaphylaxis. Include its usual causes, how it is diagnosed and treated, and two effective physiological targets for treatment.
   b. Explain how hyposensitization is achieved and suggest two mechanisms by which it might work.

7. a. What is the mechanism of type II hypersensitivity?
   b. Why are the tissues of some people antigenic to others?
   c. Would we be concerned about this problem if it were not for transfusions?
   d. What is the actual basis of the four ABO and Rh blood groups?
   e. Where do we derive our natural hypersensitivities to the A or B antigens that we do not possess?
   f. How does a person become sensitized to Rh factor? List consequences.

8. Explain the rules of transfusion. Illustrate what will happen if type A blood is accidently transfused into a type B person.

9. a. Contrast type II and type III hypersensitivities with respect to type of antigen, antibody, and manifestations of disease.
   b. What is immune complex disease?
   c. Differentiate between the Arthus reaction and serum sickness.

10. a. Explain the pathologic process in autoimmunity.
    b. Draw diagrams that explain five possible mechanisms for the development of autoimmunity.
    c. Describe four major types of autoimmunity, comparing target organs and symptoms.

11. Compare and contrast type I (atopic) and type IV (delayed) hypersensitivity as to mechanism, symptoms, eliciting factors, and allergens.

12. a. What is the molecular/cellular basis for a host rejecting the graft tissue?
    b. Account for a graft rejecting the host.
    c. Compare the four types of grafts.
    d. What does it mean to say that two tissues constitute a close match?
    e. Describe the procedure involved in a bone marrow transplant.

13. a. In general, what causes primary immunodeficiencies?
    b. Acquired immunodeficiencies?
    c. Why can T-cell deficiencies have greater impact than B-cell deficiencies?
    d. What kinds of symptoms accompany a B-cell defect?
    e. A T-cell defect?
    f. Combined defects?
    g. Give examples of specific diseases that involve each type of defect.

14. a. Name several medical conditions that require immunosuppressive drugs. Why is it necessary to inhibit immune responses?
    b. Name some immunosuppressive drugs, and explain what they do.

15. a. Define cancer, and give some synonyms.
    b. Differentiate between a benign tumor and a malignant tumor, and give examples.
    c. How are cancers technically differentiated?

16. a. Describe a possible mechanism to account for transformation of a normal cell to a cancer cell.
    b. Describe the possible genetic and environmental factors that increase oncogenesis.
    c. How do infectious agents such as viruses cause cancer?

17. Relate how the immune system is involved in cancer.

18. **Multiple Matching.** Choose the description that best fits the cytokines.

    _____ histamine   _____ leukotriene
    _____ prostaglandin   _____ platelet factor
    _____ bradykinin

    a. a peptide involved in blood clotting and chemotaxis
    b. a lipid that aggregates and lyses thrombocytes
    c. causes prolonged bronchospasm and mucous secretion in the lungs
    d. increases inflammation and sensitivity to pain
    e. a potent stimulus for smooth muscle and glandular secretion

CRITICAL-THINKING QUESTIONS

1. a. Discuss the reasons that the immune system is sometimes called a double-edged sword.
   b. Suggest a possible function of allergy.

2. A 3-week-old neonate develops severe eczema after being given penicillin therapy for the first time. Can you explain what has happened?

3. Can you explain why a person would be allergic to strawberries when he eats them, but shows a negative skin test to them?

4. a. Where in the course of type I allergies do antihistamine drugs work? Exactly what do they do?
   b. Cortisone?
   c. Desensitization?

5. a. Although we call persons with type O blood universal donors and those with type AB blood universal recipients, what problems might accompany transfusions involving O donors or AB recipients?
   b. What is the truly universal donor and recipient blood type? (Include Rh type.)
   c. Can you explain how a person could have a genotype for type A or B blood but a phenotype for type O?

6. Why would it be necessary for an Rh⁻ woman who has had an abortion, miscarriage, or an ectopic pregnancy to be immunized against the Rh factor?

7. a. Describe three circumstances that might cause antibodies to develop against self tissues.
   b. Can you explain how people with autoimmunity could develop antibodies against intracellular components (nucleus, mitochondria, and DNA)?

8. Would a person show allergy to poison oak upon first contact? Explain why or why not.

9. a. Looking at figure 17.20, predict which family members would be good allograft pairs and which would be incompatible.
   b. Why might a graft be rejected even between closely matched siblings?
   c. Transplanting a baboon heart into a baby is what kind of graft?

10. Why are primary immunodeficiencies considered experiments of nature?

11. Why are defective genes found on the X chromosome expressed most often in males?

12. a. Explain why babies with agammaglobulinemia do not develop opportunistic infections until about 6 months after birth.

b. Explain why people with B-cell deficiencies can benefit from artificial passive immunotherapy. Explain whether vaccination would work for them.
   c. Explain why the lack of a specific enzyme (ADA, tyrosine kinase) would lead to the death of lymphocytes.

13. a. Why are SCIDS children unable to reject grafts?
   b. What would be the major problem in most bone marrow transplantations for these children?
   c. Draw a general outline of the events that cause SCIDS in the child called David.
   d. Why did David acquire cancer?

14. In what ways can cancer be both a cause and a symptom of immunodeficiency?

15. What features of cancer cells account for metastasis?

16. In what ways could cancer be inherited?

17. If a chemical is found to be a mutagen by the Ames test, what other factors will be required for it to induce cancer?

18. Most cancer therapy removes or destroys the metastatic cells. Describe an alternate technology to prevent cancer. (Hint: You may have to refer to chapter 10.)

INTERNET SEARCH TOPICS

1. Locate information on registering to donate marrow and the process surrounding testing and transplantation.

2. Visit the student Online Learning Center at www.mhhe.com/talaro5. Go to chapter 17, Internet Search Topics, and log on to the available websites to:
   a. Review information on pollen count. Look up the data for the allergen count in your geographic area. Determine how the counts change between fall, winter, spring, and summer.

b. Determine which foods are most implicated in allergies. What measures must be taken by people with food allergies?
   c. Research allergies, autoimmunity, and immunodeficiencies.

Exponents

Dealing with concepts such as microbial growth often requires working with numbers in the billions, trillions, and even greater. A mathematical shorthand for expressing such numbers is with exponents. The exponent of a number indicates how many times (designated by a superscript) that number is multiplied by itself. These exponents are also called common *logarithms*, or logs. The following chart, based on multiples of 10, summarizes this system.

Exponential Notation for Base 10

Number	Quantity	Exponential Notation*	Number Arrived at By:	One Followed By:
1	One	10^0	Numbers raised to zero power are equal to one	No zeros
10	Ten	10^1**	10×1	One zero
100	Hundred	10^2	10×10	Two zeros
1,000	Thousand	10^3	$10 \times 10 \times 10$	Three zeros
10,000	Ten thousand	10^4	$10 \times 10 \times 10 \times 10$	Four zeros
100,000	Hundred thousand	10^5	$10 \times 10 \times 10 \times 10 \times 10$	Five zeros
1,000,000	Million	10^6	10 times itself 6 times	Six zeros
1,000,000,000	Billion	10^9	10 times itself 9 times	Nine zeros
1,000,000,000,000	Trillion	10^{12}	10 times itself 12 times	Twelve zeros
1,000,000,000,000,000	Quadrillion	10^{15}	10 times itself 15 times	Fifteen zeros
1,000,000,000,000,000,000	Quintillion	10^{18}	10 times itself 18 times	Eighteen zeros

Other large numbers are sextillion (10^{21}), septillion (10^{24}), and octillion (10^{27}).

*The proper way to say the numbers in this column is 10 raised to the nth *power, where* n *is the exponent. The numbers in this column can also be represented as* 1×10^n, *but for brevity, the* $1 \times$ *can be omitted.*

***The exponent 1 is usually omitted in numbers at this level.*

Converting Numbers to Exponent Form

As the chart shows, using exponents to express numbers can be very economical. When simple multiples of 10 are used, the exponent is always equal to the number of zeros that follow the 1, but this rule will not work with numbers that are more varied. Other large whole numbers can be converted to exponent form by the following operation: First, move the decimal (which we assume to be at the end of the number) to the left until it sits just behind the first number in the series (example: $3568. = 3.568$). Then count the number of spaces (digits) the decimal has moved; that number will be the exponent. (The decimal has moved from 8. to 3., or 3 spaces.) In final notation, the converted number is multiplied by 10 with its appropriate exponent: 3568 is now 3.568×10^3.

Rounding Off Numbers

The notation in the previous example has not actually been shortened, but it can be reduced further by rounding off the decimal fraction to the nearest thousandth (three digits), hundredth (two digits), or tenth (one digit). To round off a number, drop its last digit and either increase the one next to it or leave it as it is. If the number dropped is 5, 6, 7, 8, or 9, the subsequent digit is increased by one (rounded up); if it is 0, 1, 2, 3, or 4, the subsequent digit remains as is. Using the example of 3.528, removing the 8 rounds off the 2 to a 3 and produces 3.53 (two digits). If further rounding is desired, the same rule of thumb applies, and the number becomes 3.5 (one digit). Other examples of exponential conversions are shown on the following page:

Number	Is the Same As	Rounded Off, Placed in Exponent Form
16,825.	$1.6825 \times 10 \times 10 \times 10 \times 10$	1.7×10^4
957,654.	$9.57654 \times 10 \times 10 \times 10 \times 10 \times 10$	9.58×10^5
2,855,000.	$2.855000 \times 10 \times 10 \times 10 \times 10$ $\times 10 \times 10$	2.86×10^6

Negative Exponents

The numbers we have been using so far are greater than 1 and are represented by positive exponents. But the correct notation for numbers less than 1 involves negative exponents (10 raised to a negative power, or 10^{-n}). A negative exponent says that the number has been divided by a certain power of 10 (10, 100, 1,000). This usage is handy when working with concepts such as pH that are based on very small numbers otherwise needing to be represented by large decimal fractions—for example, 0.003528. Converting this and other such numbers to exponential notation is basically similar to converting positive numbers, except that you work from left to right and the exponent is negative. Using the example of 0.003528, first convert the number to a whole integer followed by a decimal fraction and keep track of the number of spaces the decimal point moves (example: $0.003528 = 3.528$). The decimal has moved three spaces from its original position, so the finished product is 3.528×10^{-3}. Other examples are:

Number	Is the Same As	Rounded Off, Express with Exponents
0.0005923	$\dfrac{5.923}{10 \times 10 \times 10 \times 10}$	5.92×10^{-4}
0.00007295	$\dfrac{7.295}{10 \times 10 \times 10 \times 10 \times 10}$	7.3×10^{-5}

Logarithmic Tables for the Base 2

Use this to calculate population growth as described on page 209.

(A) Powers of 2 from 0 through 20		(B) Logarithms for the base 2 and corresponding powers	
		Log	Number
2^0 =	1	0	1
2^1 =	2	1	2
2^2 =	4	2	4
2^3 =	8	3	8
2^4 =	16	4	16
2^5 =	32	5	32
2^6 =	64	6	64
2^7 =	128	7	128
2^8 =	256	8	256
2^9 =	512	9	512
2^{10} =	1024	10	1024
2^{11} =	2048	11	2048
2^{12} =	4096	12	4096
2^{13} =	8192	13	8192
2^{14} =	16384	14	16384
2^{15} =	32768	15	32768
2^{16} =	65536	16	65536
2^{17} =	131072	17	131072
2^{18} =	262144	18	262144
2^{19} =	524288	19	524288
2^{20} =	1048576	20	1048576

Significant Events in Microbiology

Date	Discovery/People Involved
1546	Italian physician Girolamo Fracastoro suggests that invisible organisms may be involved in disease.
1590	Zaccharias Janssen, a Dutch spectacle maker, invents the first compound microscope.
1660	Englishman Robert Hooke explores various living and nonliving matter with a compound microscope that uses reflected light.
1668	Francesco Redi, an Italian naturalist, conducts experiments that demonstrate the fallacies in the spontaneous generation theory.
1676	Antonie van Leeuwenhoek, a Dutch linen merchant, uses a simple microscope of his own design to observe bacteria and protozoa.
1776	An Italian anatomist, Lazzaro Spallanzani, conducts further convincing experiments that dispute spontaneous generation.
1796	English surgeon Edward Jenner introduces a vaccination for smallpox.
1838	Phillipe Ricord, a French physician, inoculates 2,500 human subjects to demonstrate that syphilis and gonorrhea are two separate diseases.
1839	Theodor Schwann, a German zoologist, and Matthias Schleiden, a botanist, formalize the theory that all living things are composed of cells.
1847–1850	The Hungarian physician Ignaz Semmelweis substantiates his theory that childbed fever is a contagious disease transmitted to women by their physicians during childbirth.
1853–1854	John Snow, a London physician, demonstrates the epidemic spread of cholera through a water supply contaminated with human sewage.
1857	French bacteriologist Louis Pasteur shows that fermentations are due to microorganisms and originates the process now known as pasteurization.
1858	Rudolf Virchow, a German pathologist, introduces the concept that all cells originate from preexisting cells.
1861	Louis Pasteur completes the definitive experiments that finally lay to rest the theory of spontaneous generation.
1867	The English surgeon Joseph Lister publishes the first work on antiseptic surgery, beginning the trend toward modern aseptic techniques in medicine.
1869	Johann Miescher, a Swiss pathologist, discovers in the cell nucleus the presence of complex acids, which he terms nuclein (DNA, RNA).

Date	Discovery/People Involved
1876–1877	German bacteriologist Robert Koch★ studies anthrax in cattle and implicates the bacterium *Bacillus anthracis* as its causative agent.
1881	Pasteur develops a vaccine for anthrax in animals.
	Koch introduces the use of pure culture techniques for handling bacteria in the laboratory.
	Walther and Fanny Hesse introduce agar-agar as a solidifying gel for culture media.
1882	Koch identifies the causative agent of tuberculosis.
1884	Koch outlines his postulates.
	Elie Metchnikoff,★ a Russian zoologist, lays groundwork for the science of immunology by discovering phagocytic cells.
	The Danish physician Hans Christian Gram devises the Gram stain technique for differentiating bacteria.
1885	Pasteur develops a special vaccine for rabies.
1887	Julius Petri, a German bacteriologist, adapts two plates to form a container for holding media and culturing microbes.
1890	A German, Emil von Behring,★ and a Japanese, Shibasaburo Kitasato, demonstrate the presence of antibodies in serum that neutralize the toxins of diphtheria and tetanus.
1892	A Russian, D. Ivanovski, is the first to isolate a virus (the tobacco mosaic virus) and show that it could be transmitted in a cell-free filtrate.
1895	Jules Bordet,★ a Belgian bacteriologist, discovers the antimicrobial powers of complement.
1898	R. Ross★ and G. Grassi demonstrate that malaria is transmitted by the bite of female mosquitoes.
	Germans Friedrich Loeffler and P. Frosch discover that "filterable viruses" cause foot-and-mouth disease in animals.
1899	Dutch microbiologist Martinus Beijerinck further elucidates the viral agent of tobacco mosaic disease and postulates that viruses have many of the properties of living cells and that they reproduce within cells.
1900	The American physician Walter Reed and his colleagues clarify the role of mosquitoes in transmitting yellow fever.
	An Austrian pathologist, Karl Landsteiner,★ discovers the ABO blood groups.

continued

★*These scientists were awarded Nobel prizes for their contributions to the field.*

Date	Discovery/People Involved	Date	Discovery/People Involved
1903	American pathologist James Wright and others demonstrate the presence of antibodies in the blood of immunized animals.	1959–1960	Gerald Edelman★ and Rodney Porter★ determine the structure of antibodies.
1905	Syphilis is shown to be caused by *Treponema pallidum,* through the work of German bacteriologists Fritz Schaudinn and E. Hoffman.	1972	Paul Berg★ develops the first recombinant DNA in a test tube.
1906	August Wasserman, a German bacteriologist, develops the first serologic test for syphilis.	1973	Herb Boyer and Stanley Cohen clone the first DNA using plasmids.
	Howard Ricketts, an American pathologist, links the transmission of Rocky Mountain spotted fever to ticks.	1975	A technique for making monoclonal antibodies is developed by Cesar Milstein, Georges Kohler, and Niels Kai Jerne.
1908	The German Paul Ehrlich★ becomes the pioneer of modern chemotherapy by developing salvarsan to treat syphilis.	1979	Genetically engineered insulin is first synthesized by bacteria.
1910	An American pathologist, Francis Rous,★ discovers viruses that can induce cancer.	1982	Development of first hepatitis B vaccine using virus isolated from human blood.
1915–1917	British scientist F. Twort and French scientist F. D'Herelle independently discover bacterial viruses.	1983	Isolation and characterization of human immunodeficiency virus (HIV) by Luc Montagnier of France and Robert Gallo of the United States.
1928	Frederick Griffith lays the foundation for modern molecular genetics by his discovery of transformation in bacteria.		The polymerase chain reaction is invented by Kerry Mullis.★
1929	A Scottish bacteriologist, Alexander Fleming,★ discovers and describes the properties of the first antibiotic, penicillin.	1987	The molecular genetics of antibody genes is worked out by Susumu Tonegawa.★
1933–1938	Germans Ernst Ruska★ and B. von Borries develop the first electron microscope.		First release of recombinant strain of *Pseudomonas* to prevent frost formation on strawberry plants.
1935	Gerhard Domagk,★ a German physician, discovers the first sulfa drug and paves the way for the era of antimicrobic chemotherapy.	1989	Cancer-causing genes called oncogenes are characterized by J. Michael Bishop, Robert Huber, Hartmut Michel, and Harold Varmus.
	Wendell Stanley★ is successful in inducing tobacco mosaic viruses to form crystals that still retain their infectiousness.	1990	First clinical trials in gene therapy testing.
			Vaccine for *Haemophilus influenzae,* a cause of meningitis, is introduced.
1941	Australian Howard Florey★ and Englishman Ernst Chain★ develop commercial methods for producing penicillin; this first antibiotic is tested and put into widespread use.	1991	Development of transgenic animals to synthesize human hemoglobin.
1944	Oswald Avery, Colin MacLeod, and Maclyn McCarty show that DNA is the genetic material.	1994	Human breast cancer gene isolated.
	Joshua Lederberg★ and E. L. Tatum★ discover conjugation in bacteria.	1995	First bacterial genome fully sequenced, for *Haemophilus influenzae.*
	The Russian Selman Waksman★ and his colleagues discover the antibiotic streptomycin.	1996	Dr. David Ho develops a "cocktail" of drugs to treat AIDS.
1953	James Watson,★ Francis Crick,★ Rosalind Franklin, and Maurice Wilkins★ determine the structure of DNA.	1997	Medical researchers at Case Western University construct human artificial chromosomes (HACs).
			Scottish researchers clone first mammal (a sheep) from adult nuclei.
1954	Jonas Salk develops the first polio vaccine.	1999	Heat-loving bacteria discovered 2 miles beneath earth in African gold mine.
1957	Alick Isaacs and Jean Lindenmann discover the natural antiviral substance interferon.	2000	A rough version of the human genome is mapped.
			250-million-year-old bacterium unearthed by Pennsylvania team.
	D. Carleton Gajdusek★ discovers the underlying cause of slow virus diseases.	2001	Mailed anthrax spores cause major bioterrorism event.
		2002	New York virologists create polio virus in a test tube.
		2003	New roles for small nuclear RNAs discovered.

Methods for Testing Sterilization and Germicidal Processes

Most procedures used for microbial control in the clinical setting are either physical methods (heat, radiation) or chemical methods (disinfectants, antiseptics). These procedures are so crucial to the well-being of patients and staff that their effectiveness must be monitored in a consistent and standardized manner. Particular concerns include the time required for the process, the concentration or intensity of the antimicrobic agent being used, and the nature of the materials being treated. The effectiveness of an agent is frequently established through controlled microbiological analysis. In these tests, a biological indicator (a highly resistant microbe) is exposed to the agent, and its viability is checked. If this known test microbe is destroyed by the treatment, it is assumed that the less resistant microbes have also been destroyed. Growth of the test organism indicates that the sterilization protocol has failed. The general categories of testing include:

Heat The usual form of heat used in microbial control is steam under pressure in an autoclave. Autoclaving materials at a high temperature (121°C) for a sufficient time (15–40 minutes) destroys most bacterial endospores. To monitor the quality of any given industrial or clinical autoclave run, technicians insert a special ampule of *Bacillus stearothermophilus,* a spore-forming bacterium with extreme heat resistance **(figure A.1).** After autoclaving, the ampule is incubated at 56°C (the optimal temperature for this species) and checked for growth.

Radiation The effectiveness of ionizing radiation in sterilization is determined by placing special strips of dried spores of *Bacillus sphaericus* (a common soil bacterium with extreme resistance to radiation) into a packet of irradiated material.

Filtration The performance of membrane filters used to sterilize liquids may be monitored by adding *Pseudomonas diminuta* to the liquid. Because this species is a tiny bacterium that can escape through the filter if the pore size is not small enough, it is a good indicator of filtrate sterility.

Gas sterilization Ethylene oxide gas, one of the few chemical sterilizing agents, is used on a variety of heat-sensitive medical and laboratory supplies. The most reliable indicator of a successful sterilizing cycle is the sporeformer *Bacillus subtilis,* variety *niger.*

Germicidal assays United States hospitals and clinics commonly employ more than 250 products to control microbes in the environment, on inanimate objects, and on patients. Controlled testing of these products' effectiveness is not usually performed in the clinic itself, but by the chemical or pharmaceutical manufacturer. Twenty-five standardized *in vitro* tests are currently available to assess the effects of germicides. Most of them use a non-spore-forming pathogen as the biological indicator.

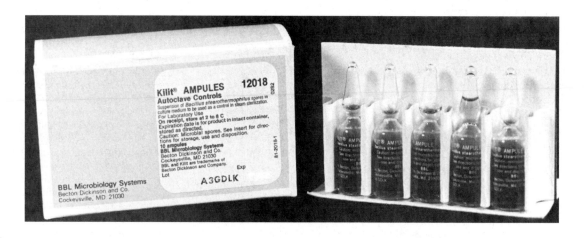

FIGURE A.1

The sterilizing conditions of autoclave runs may be tested by indicator bacteria in small broth ampules. One ampule is placed in the center of the autoclave chamber along with the regular load. After the process, the ampule is incubated to test for survival.

The Phenol Coefficient (PC)

The older disinfectant phenol has been the traditional standard by which other disinfectants are measured. In the PC test, a water-soluble phenol-based (phenolic) disinfectant (amphyl, lysol) is tested for its bactericidal effectiveness as compared with phenol. These disinfectants are serially diluted in a series of *Salmonella choleraesuis, Staphylococcus aureus,* or *Pseudomonas aeruginosa* cultures. The tubes are left for 5, 10, or 15 minutes and then subcultured to test for viability. The PC is a ratio derived by comparing the following data:

$$PC = \frac{\text{Greatest dilution of the phenolic that kills test bacteria in 10 min but not in 5 min}}{\text{Greatest dilution of phenol giving same result}}$$

The general interpretation of this ratio is that chemicals with lower phenol coefficients have greater effectiveness. The primary disadvantage of the PC test is that, because it is restricted to phenolics, it is inappropriate for the vast majority of clinical disinfectants.

Use Dilution Test

An alternative test with broader applications is the use dilution test. This test is performed by drying the test culture (one of those species used in the PC test) onto the surface of small stainless steel cylindrical carriers. These carriers are then exposed to varying concentrations of disinfectant for 10 minutes, removed, rinsed, and placed in tubes of broth. After incubation, the tubes are inspected for growth. The smallest concentration of disinfectant that kills the test organism on 10 carrier pieces is the correct dilution for use.

Filter Paper Disc Method

A quick measure of the inhibitory effects of various disinfectants and antiseptics can be achieved by the filter paper disc method. First, a small (1/2-inch) sterile piece of filter paper is dipped into a disinfectant of known concentration. This is placed on an agar medium seeded with a test organism (*S. aureus* or *P. aeruginosa*),

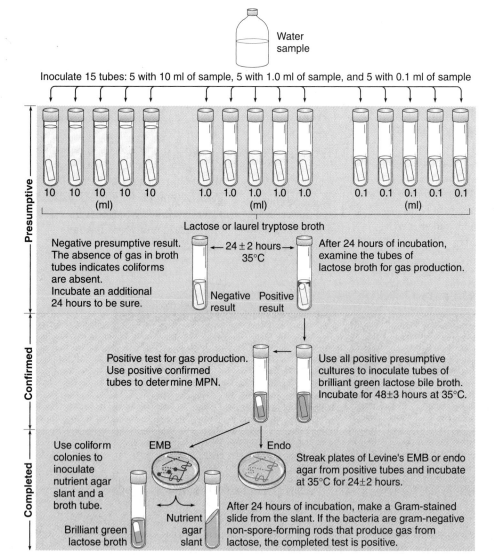

FIGURE A.2

The most probable number (MPN) procedure for determining the coliform content of a water sample. In the presumptive test, each set of five tubes of broth is inoculated with a water sample reduced by a factor of 10. After incubation, the sets of tubules are examined and rated for gas production (for instance, 0 means no tubes with gas, 1 means one tube with gas, 2 means two tubes with gas). Applying this result to the MPN table in appendix C will indicate the probable number of cells present in 100 ml of the water sample. Confirmation of coliforms can be achieved by confirmatory tests on additional media, and complete identification can be made through selective and differential media and Gram staining (see page 808).

and the plate is incubated. A zone devoid of growth (zone of inhibition) around the disc indicates the capacity of the agent to inhibit growth **(figure A.3)**. Like the antimicrobic sensitivity test (fig. 12.19, p. 376), this test measures the minimum inhibitory concentration of the chemical. In general, chemicals with wide zones of inhibition are effective even in high dilutions.

FIGURE A.3

A filter paper disc containing hydrogen peroxide produces a broad clear zone of inhibition in a culture of *Staphylococcus aureus*.

TABLE A.1

Most Probable Number Chart for Evaluating Coliform Content of Water

The Number of Tubes in Series, with Positive Growth at Each Dilution							
10 ml	1 ml	0.1 ml	MPN* per 100 ml of Water	10 ml	1 ml	0.1 ml	MPN* per 100 ml of Water
0	1	0	0.18	5	0	1	3.1
1	0	0	0.20	5	1	0	3.3
1	0	0	0.40	5	1	1	4.6
2	0	0	0.45	5	2	0	4.9
2	0	1	0.68	5	2	1	7.0
2	2	0	0.93	5	2	2	9.5
3	0	0	0.78	5	3	0	7.9
3	0	1	1.1	5	3	1	11.0
3	1	0	1.1	5	3	2	14.0
3	2	0	1.4	5	4	0	13.0
4	0	0	1.3	5	4	1	17.0
4	0	1	1.7	5	4	2	22.0
4	1	0	1.7	5	4	3	28.0
4	1	1	2.1	5	5	0	24.0
4	2	0	2.2	5	5	1	35.0
4	2	1	2.6	5	5	2	54.0
4	3	0	2.7	5	5	3	92.0
5	0	0	2.3	5	5	4	160.0

*Most probable number of cells per sample of 100 ml.

Classification of Major Microbial Disease Agents by System Affected, Site of Infection, and Routes of Transmission

Tables A.2 through A.8 provide a cross reference to the organismic approach of presenting pathogens used in this text. The tables list the primary organs and systems that serve as targets for infectious agents and the sources and routes of infection. Many infectious agents affect both the system they first enter and others. (For example, polio infects the gastrointestinal tract and also inflicts major damage on the central nervous system.) This reference highlights agents that have major effects on more than one system with a star (★). The chapter that contains detailed information on the listings is shown in parentheses in the far right column of each table.

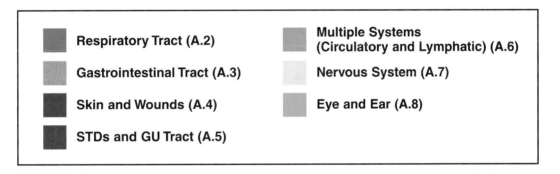

■ **Respiratory Tract (A.2)**	■	**Multiple Systems (Circulatory and Lymphatic) (A.6)**
■ **Gastrointestinal Tract (A.3)**	■	**Nervous System (A.7)**
■ **Skin and Wounds (A.4)**	■	**Eye and Ear (A.8)**
■ **STDs and GU Tract (A.5)**		

TABLE A.2

Infections of the Respiratory Tract (RT)

Name of Microbe	Name of Disease	Specific Targets of Infection	Source/Mode of Acquisition	(Ch)
Bacterial pathogens				
Streptococcus pyogenes★	Streptococcal pharyngitis	Upper RT membranes	Droplets	(18)
Viridans streptococci	Sinusitis	Upper RT and saliva	Normal flora	(18)
Bacillus anthracis★	Pulmonary anthrax	Upper and lower RT	Spores inhaled from animals/products	(19)
Corynebacterium diphtheriae★	Diphtheria	Upper RT membranes; also skin	Droplets	(19)
Haemophilus influenzae★	Pharyngitis	Upper RT membranes	Droplets	(20)
Streptococcus pneumoniae★	Pneumococcal pneumonia; otitis media, sinusitis meningitis	Lower RT, lungs; meninges	Droplets, mucus	(18)
Mycobacterium tuberculosis★	Tuberculosis	Lower RT, lungs	Droplets; droplet nuclei	(19)
Legionella pneumophila	Legionnaire disease Pontiac fever	Lower RT, lungs	Aerosols from aquatic habitats	(20)
Brucella spp.★	Brucellosis (undulant fever)	Respiratory tract	Aerosols from infected animals	(20)
Bordetella pertussis	Whooping cough	Upper, lower RT	Droplets	(20)
Klebsiella pneumoniae Pseudomonas aeruginosa	Opportunistic pneumonias	Lower RT, lungs	Droplets, normal flora, aerosols, fomites	(20)
Yersinia pestis★	Pneumonic plague	Lower RT, lungs	Droplets, animal aerosols	(20)
Mycoplasma pneumoniae	Primary atypical pneumonia	Lower RT, lungs	Droplets, mucus	(21)

continued

TABLE A.2 *continued*				
Infections of the Respiratory Tract (RT)				
Name of Microbe	**Name of Disease**	**Specific Targets of Infection**	**Source/Mode of Acquisition**	**(Ch)**
Bacterial pathogens *continued*				
Chlamydia pneumoniae	Chlamydial pneumonitis	Lower RT, lungs	Droplets, mucus	(21)
Chlamydia psittaci	Ornithosis (parrot fever)	Lower RT, lungs	Aerosols, droplets from birds	(21)
Coxiella burnetii	Q (for query) fever	Lower RT, lungs	Ticks, aerosols from animals, dust	(21)
Nocardia brasiliensis	Nocardiosis (a form of pneumonia)	Lower RT, lungs	Spores inhaled from environment	(19)
Fungal pathogens (chapter 22)				
Histoplasma capsulatum★	Histoplasmosis	Lower RT, lungs	Spores inhaled in dust	
Blastomyces dermatitidis★	Blastomycosis	Lower RT, lungs	Spores inhaled	
Coccidioides immitis★	Coccidioidomycosis (Valley fever)	Lower RT, lungs	Spores inhaled	
Cryptococcus neoformans★	Cryptococcosis	Lower RT, lungs	Yeasts inhaled, can be normal flora	
Sporothrix schenckii★	Sporothrix pneumonia	Lower RT, lungs	Spores on vegetation	
Pneumocystis (carinii) jiroveci	Pneumocystis pneumonia (PCP)	Lower RT, lungs	Spores widespread, normal flora	
Aspergillus fumigatus★	Aspergillosis	Lower RT, lungs	Spores widespread in dust, air	
Parasitic (protozoan, helminth) pathogens (chapter 23)				
Paragonimus westermani	Adult lung fluke	Lung tissue; intestine	Larvae ingested in raw or under-cooked crustacea	
Viral pathogens				
Influenza viruses	Influenza A, B, C	Mucosal cells of trachea, bronchi	Inhalation of droplets, aerosols from active cases	(25)
Hantavirus	Hantavirus pulmonary syndrome	Lung epithelia	Inhalation of airborne rodent excrement	(25)
Respiratory syncytial pneumovirus	RSV pneumonitis, bronchiolitis	Nasopharynx; upper and lower RT	Droplets and secretions from active cases	(25)
Parainfluenza virus Paramyxovirus	Parainfluenza	Upper, lower RT	Direct hand contact with secretions; inhalation of droplets	(25)
Adenovirus	Common cold	Nasopharynx; sometimes lower RT	Close contact with secretions from person with active infection; hand to nose	(24)
Rhinovirus	Common cold	Same as above		(25)
Coronavirus	Common cold	Same as above		(25)
Coxsackievirus	Common cold	Same as above		(25)
Mumps virus★	See table A.3			
Rubella virus★	See table A.4			
Measles virus★	See table A.4			
Chickenpox virus (VZV)★	See table A.4			

TABLE A.3

Diseases of the Gastrointestinal Tract (Food Poisoning and Intestinal, Gastric, Oral Infections)

Name of Microbe	Name of Disease	Specific Targets of Disease	Source/Mode of Acquisition	(Ch)
Bacterial pathogens				
Staphylococcus aureus	Staph food intoxication	Small intestine	Foods contaminated by human carriers	(18)
Clostridium perfringens	Perfringens food poisoning	Small intestine	Rare meats; toxin produced in the body	(19)
Clostridium difficile	Antibiotic-associated colitis	Large intestine	Spores ingested; can be normal flora; caused by antibiotic therapy	(19)
Listeria monocytogenes	Listeriosis	Small intestine	Water, soil, plants, animals, milk, food infection	(19)
Escherichia coli	Traveler's diarrhea O157: H7 enteritis	Small intestine	Food, water, rare or raw beef	(20)
Salmonella typhi★	Typhoid fever	Small intestine	Human carriers; food, drink	(20)
Salmonella species	Food infection, enteric fevers	Small intestine	Cattle, poultry, rodents; fecally contaminated foods	(20)
Shigella spp.	Bacillary dysentery	Large intestine	Human carriers; food, flies	(20)
Yersinia enterocolitica *Y. paratuberculosis*	Yersiniosis (gastroenteritis)	Small intestine, lymphatics	A zoonosis; food and drinking water	(20)
Vibrio cholerae	Epidemic cholera	Small intestine	Natural waters; food contaminated by human carriers	(21)
Helicobacter pylori	Gastritis, gastric intestinal ulcers	Esophagus, stomach, duodenum	Humans, cats (?)	(21)
Campylobacter jejuni	Enteritis	Jejunum segment of small intestine	Water, milk, meat, food infection	(21)
Streptococcus mutans	Dental caries	Tooth surface	Normal flora; saliva	(21)
Actinomyces israelii★	Actinomycosis	Oral membranes, cervicofacial area	Normal flora; enter damaged tissue	(19)
Fusobacterium, Treponema, Bacteroides	Various forms of gingivitis, periodontitis	Gingival pockets	Normal flora; close contact	(21)
Fungal pathogens (chapter 22)				
Aspergillus flavus	Aflatoxin poisoning	Liver	Toxin ingested in food	
Candida albicans★	Esophagitis	Esophagus	Normal flora	
Parasitic (protozoan, helminth) pathogens (chapter 23)				
Entamoeba histolytica★	Amebic dysentery	Caecum, appendix, colon, rectum	Food, water-contaminated fecal cysts	
Giardia lamblia (intestinalis)	Giardiasis	Duodenum, jejunum	Ingested cysts from animal, human carriers	
Balantidium coli	Balantidiosis	Large intestine	Swine, contaminated food	
Cryptosporidium spp.	Cryptosporoidosis	Small and large intestine	Animals, natural waters, contaminated food	
Toxoplasma gondii★	Toxoplasmosis	Lymph nodes	Cats, rodents, domestic animals; raw meats	
Enterobius vermicularis	Pinworm	Rectum, anus	Eggs ingested from fingers, food	
Ascaris lumbricoides	Ascariasis	Worms burrow into intestinal mucosa	Fecally contaminated food and water containing eggs	
Trichuris trichiura	Trichiuriasis, whipworm	Worms invade small intestine	Food containing eggs	
Trichinella spiralis★	Trichinosis	Larvae penetrate intestine	Pork and bear meat containing encysted larvae	
Necator americanus★ *Ancylostoma duodenale*★	Hookworms	Large and small intestine	Larvae deposited by feces into soil; burrow into feet	
Fasciola hepatica	Sheep liver fluke	Liver and intestine	Watercress	
Opisthorchis sinensis	Chinese liver fluke	Liver and intestine	Raw or undercooked fish or shellfish	
Taenia saginata	Beef tapeworm	Large intestine	Rare or raw beef or pork containing larval forms	
T. solium	Pork tapeworm	Pork larva can migrate to brain		

continued

TABLE A.3 *continued*

Diseases of the Gastrointestinal Tract (Food Poisoning and Intestinal, Gastric, Oral Infections)

Name of Microbe	Name of Disease	Specific Targets of Disease	Source/Mode of Acquisition	(Ch)
Viral pathogens				
Herpes simplex virus, type 1	Cold sores, fever blisters, gingivostomatitis	Skin, mucous membranes of oral cavity	Close contact with active lesions of human	(24)
Mumps virus★ (*Paramyxovirus*)	Parotitis	Parotid salivary gland	Close contact with salivary droplets	(25)
Hepatitis A virus	Infectious hepatitis	Liver	Food, water contaminated by human feces; shellfish	(25)
Rotavirus	Rotavirus diarrhea	Small intestinal mucosa	Fecally contaminated food, water, fomites	(25)
Norwalk agent (enterovirus)	Gastroenteritis	Small intestine	Close contact; oral-fecal, water, shellfish	(25)
Poliovirus★	Poliomyelitis	Intestinal mucosa, systemic	Food, water, mechanical vectors	(25)

TABLE A.4

Infections of Skin and Skin Wounds

Name of Microbe	Name of Disease	Specific Targets of Infection	Source/Mode of Transmission	(Ch)
Bacterial pathogens				
Staphylococcus aureus★	Folliculitis Furuncles (boils) Carbuncles Bullous impetigo Scalded skin syndrome	Glands, follicles, glabrous skin	Close contact, nasal and oral droplets; associated with poor hygiene, tissue injury	(18)
Streptococcus pyogenes★	Pyoderma (impetigo) Erysipelas Necrotizing fasciitis	Skin and underlying tissues	Contact with human carrier; invasion of tiny wounds in skin	(18)
Bacillus anthracis★	Cutaneous anthrax	Necrotic lesion in skin (eschar)	Spores from animal secretions, products, soil	(19)
Clostridium perfringens★	Gas gangrene	Skin, muscle, connective tissue	Introduction of spores into damaged tissue	(19)
Propionibacterium acnes	Acne	Sebaceous glands	Normal flora; bacterium is lipophilic	(19)
Erysipelothrix rhusiopathiae	Erysipeloid	Epidermal, dermal regions of skin	Normal flora of pigs, other animals; enters in injured tissue	(19)
Mycobacterium leprae★	Leprosy (Hansen disease)	Skin, underlying nerves	Long-term contact with infected person; skin is inoculated	(19)
Mycobacterium marinum	Swimming pool granuloma	Superficial skin	Skin scraped against contaminated surface	(19)
Pasteurella multocida	Pasteurellosis	Localized skin abscess	Animal bite or scratch	(20)
Bartonella henselae	Cat-scratch disease	Subcutaneous tissues, lymphatic drainage	Cat's claws carry bacteria, inoculate skin	(21)
Fungal pathogens (chapter 22)				
Blastomyces dermatitidis	Blastomycosis	Subcutaneous	Accidental inoculation of skin with conidia	
Trichophyton Microsporum Epidermophyton	Dermatomycoses: tineas, ringworm, athlete's foot	Cutaneous tissues (epidermis, nails, scalp, hair)	Contact with infected humans, animals, spores in environment	
Sporothrix schenckii★	Sporotrichosis	Epidermal, dermal, subcutaneous	Puncture of skin with sharp plant materials	
Madurella	Madura foot, mycetoma	Subcutaneous tissues, bones, muscles	Traumatic injury with spore-infested objects	

continued

TABLE A.4 *continued*

Infections of Skin and Skin Wounds

Name of Microbe	Name of Disease	Specific Targets of Infection	Source/Mode of Transmission	(Ch)
Fungal pathogens (chapter 22) *continued*				
Cladosporium *Fonsecaea*	Chromoblastomycosis	Deep subcutaneous tissues of legs, feet	Penetration of skin with contaminated objects	
Candida albicans★	Onychomycosis	Skin, nails	Chronic contact with moisture, warmth	
Malassezia furfur	Tinea versicolor	Superficial epidermis	Normal flora of skin	
Parasitic (protozoan, helminth) pathogens (chapter 23)				
Leishmania★ *tropica* *L. mexicana*	Cutaneous leishmaniasis	Local infection of skin capillaries	Sand fly salivary gland harbors infectious stage; transmitted through bite	(23)
Dracunculus medinensis	Dracontiasis	Starts in intestine, migrates to skin	Microscopic aquatic arthropod harbors larval form of guinea worm; is ingested	
Viral pathogens				
Varicella-zoster virus★	Chickenpox Shingles	Skin of face and trunk; becomes systemic and latent in nerves	Spread by respiratory droplets; recurrences from latent infection	(24)
Herpesvirus-6	Roseola infantum	Generalized skin rash	Close contact with droplets	
Parvovirus B19	Erythema infectiosum	Skin of cheeks, trunk	Close contact with droplets	
Measles virus★ *(Morbillivirus)*	Measles, rubeola	Skin on head, trunk, extremities	Contact with respiratory aerosols from active case	(25)
Rubella virus★ *(Rubivirus)*	Rubella	Skin on face, trunk, limbs	Respiratory droplets, urine of active cases	(25)
Human papillomavirus (See also table A.5)	Warts, verruca	Skin of hands, feet, body	Contact with warts, fomites	(24)
Molluscum contagiosum	See table A.5			(24)
Herpes simplex, type 2	See table A.5			(24)

TABLE A.5

Sexually Transmitted Diseases (STDs) and Infections of the Genitourinary (GU) Tract

Name of Microbe	Name of Disease	Specific Targets of Infection	Source/Mode of Acquisition	(Ch)
Bacterial pathogens				
Neisseria gonorrhoeae★	Gonorrhea	Vagina, urethra	Mucous secretions; STD	(18)
Haemophilus ducreyi	Chancroid	Genitalia, lymph nodes	Skin lesions; STD	(20)
Treponema pallidum★	Syphilis	Genitalia; mucous membranes	Chancres, skin lesions; STD	(21)
Chlamydia trachomatis★	Chlamydiosis (non-gonococcal urethritis), lymphogranuloma venereum	Vagina, urethra, lymph nodes	Mucous discharges; STD	(21)
Escherichia coli★	Urinary tract infection (UTI)	Urethra, bladder	Normal flora; intestinal tract	(20)
Leptospira interrogans★	Leptospirosis	Kidney, liver	Animal urine; contact with contaminated water, soil	(21)
Gardnerella vaginalis	Bacterial vaginosis	Vagina, occasionally penis	Mixed infection; normal flora; intimate contact	(20)
Fungal pathogens (chapter 22)				
Candida albicans	Yeast infections, candidiasis	Vagina, vulva, penis, urethra	Discharge from infected membranes; STD	
Parasitic (protozoan, helminth) pathogens (chapter 23)				
Trichomonas vaginalis	Trichomoniasis	Vagina, vulva, urethra	Genital discharges, direct contact; STD	

continued

TABLE A.5 *continued*

Sexually Transmitted Diseases (STDs) and Infections of the Genitourinary (GU) Tract

Name of Microbe	Name of Disease	Specific Targets of Infection	Source/Mode of Acquisition	(Ch)
Viral pathogens				
Molluscum contagiosum poxvirus	Molluscum contagiosum	Skin, mucous membranes	Pox lesions; STD	(24)
Herpes simplex, type 2	Genital herpes	Genitalia, perineum	Contact with vesicles, shed skin cells; STD	(24)
Human papillomavirus	Genital warts, condylomata	Membranes of the vagina, penis	Direct contact with warts; some fomites; STD	(24)
Hepatitis B virus★	Hepatitis B	Liver	Semen, vaginal fluids, blood; STD	(24)
Human immunodeficiency virus (HIV)★	AIDS	White blood cells, brain cells	Blood, semen, vaginal fluids; STD	(25)

TABLE A.6

Infections Affecting Multiple Systems (Circulatory and Lymphatic)

Name of Microbe	Name of Disease	Specific Targets of Disease	Source/Mode of Acquisition	(Ch)
Bacterial pathogens				
Staphylococcus aureus★	Toxic shock syndrome	Vagina, uterus, kidney, liver, blood	Normal flora; grow in environment created by tampons	(18)
	Osteomyelitis	Interior of long bones	Focal from skin infection	
Streptococcus pyogenes★	Scarlet fever	Skin, brain	Toxemia; complication of strep pharyngitis	(18)
	Rheumatic fever	Heart valves	Autoimmune reaction to strep toxins	
	Glomerulonephritis	Kidney	Immune reaction blocks filtration apparatus	
Streptococcus spp. (viridans)	Subacute endocarditis	Lining of heart and valves	Normal flora invades during dental work; colonizes heart	(18)
Corynebacterium diphtheriae★	Diphtheria	Heart muscle, nerves, occasionally skin	Droplets from carriers; local infection leads to toxemia	(19)
Mycobacterium tuberculosis★	Disseminated tuberculosis	Lymph nodes, kidneys, bones, brain, sex organs	Reactivation of latent lung infection	(19)
Yersinia pestis★	Plague	Lymph nodes, blood	Bite of infected flea	(20)
Francisella tularensis	Tularemia Rabbit fever	Skin, lymph nodes, eyes, lungs	Blood-sucking arthropods (ticks, mosquitoes); direct contact with animals and meat; aerosols	(20)
Borrelia recurrentis	Borreliosis, relapsing fever	Liver, heart, spleen, kidneys, nerves	Human lice; influenced by poor hygiene, crowding	(21)
B. hermsii	Same as above	Same as above	Soft tick bite; reservoir is wild rodents	(21)
B. burgdorferi	Lyme disease	Skin, joints, nerves, heart	Larval *Ixodes* ticks; reservoir is deer and rodents	(21)
Rickettsia prowazekii	Epidemic typhus	Blood vessels in many organs	Human lice, bites, feces, crowding, lack of sanitation	(21)
R. rickettsii	Rocky Mountain spotted fever	Cardiovascular, central nervous, skin	Wood and dog ticks; picked up in mountain or rural habitats	(21)
R. typhi	Endemic typhus	Similar to epidemic typhus	Rat flea vector; rodent reservoir; occupational contact	(21)
Coxiella burnetii★	See table A.2			

Fungal pathogens (chapter 22)

Many pathogenic fungi (*Histoplasma, Coccidioides, Blastomyces,* and *Cryptococcus*) have a systemic component to their infection, especially in immunocompromised patients. Because their primary infection involves the lungs, information on those fungi is listed in table A.2.

continued

TABLE A.6 *continued*

Infections Affecting Multiple Systems (Circulatory and Lymphatic)

Name of Microbe	Name of Disease	Specific Targets of Disease	Source/Mode of Acquisition	(Ch)
Parasitic (protozoan, helminth) pathogens (chapter 23)				
Leishmania spp.	Systemic leishmaniasis (kala-azar)	Spleen, liver, lymph nodes	Bite by sand (phlebotamine) fly; reservoir is dogs, rodents, wild carnivores	
Plasmodium spp.	Malaria	Liver, blood, kidney	Female *Anopheles* mosquito; human reservoir, mainly tropical	
Toxoplasma gondii ★	Toxoplasmosis	Pharynx, lymph nodes, nervous system	Cysts from cats and other animals are ingested; many hosts	
Strongyloides stercoralis	Strongyloidiasis Threadworm	Intestine, trachea, lungs, pharynx, subcutaneous; migrate through systems	Larvae deposited into soil or water by feces; burrow into skin	
Schistosoma spp.	Schistosomiasis, blood fluke	Liver, lungs, blood vessels in intestine, bladder	Larval forms develop in aquatic snails; penetrate skin of bathers	
Wuchereria bancrofti	Filariasis Elephantiasis	Lymphatic circulation, blood vessels	Various mosquito species carry larval stage; human reservoir	
Viral pathogens				
Cytomegalovirus (CMV)	CMV mononucleosis, congenital CMV	Lymph glands, salivary glands	Human saliva, milk, urine, semen, vaginal fluids; intimate contact	(24)
Epstein-Barr virus	Infectious mononucleosis	Lymphoid tissue, salivary glands	Oral contact, exposure to saliva, fomites	(24)
Hepatitis B virus ★	See table A.5			(24)
Flavivirus	Yellow fever	Liver, blood vessels, kidney, mucous membranes	Various mosquito species; human and monkey reservoirs	(25)
	Dengue fever (breakbone fever)	Skin, muscles, and joints	Several mosquito vectors	(25)
Filovirus	Ebola fever	Every organ and tissue except muscle and bone	Close contact with blood and secretions of victim; monkey reservoir?	(25)
Arenavirus	Lassa fever	Blood vessels, kidney, heart	Inhalation of airborne virus from rodent excreta	(25)
HIV ★	See table A.5			
HTLV I	Adult T-cell leukemia Sézary-cell leukemia	T cells, lymphoid tissue	Blood, blood products, sexual intercourse, IV drugs	(25)
HTLV II	Hairy-cell leukemia	B cells	Blood, shared needles	(25)

TABLE A.7

Diseases of the Nervous System

Name of Microbe	Name of Disease	Specific Targets of Disease	Source/Mode of Acquisition	(Ch)
Bacterial pathogens				
Neisseria meningitidis★	Meningococcal meningitis	Membranes covering brain, spinal cord	Contact with respiratory secretions or droplets of human carrier	(18)
Haemophilus influenzae★	Acute meningitis	Meninges, brain	Contact with carrier; normal flora of throat	(20)
Clostridium tetani	Tetanus	Inhibitory neurons in spinal column	Spores from soil introduced into wound; toxin formed	(19)
Clostridium botulinum★	Botulism	Neuromuscular junction	Canned vegetables, meats; an intoxication	(19)
Mycobacterium leprae★	Leprosy (Hansen disease)	Peripheral nerves of skin	Close contact with infected person; enters skin	(19)
Fungal pathogens (chapter 22)				
Cryptococcus neoformans★	Cryptococcal meningitis	Brain, meninges	Pigeon droppings; inhalation of yeasts in air and dust	
Parasitic (protozoan, helminth) pathogens (chapter 23)				
Naegleria fowleri	Primary acute meningoencephalitis	Brain, spinal cord	Acquired while swimming in fresh, brackish water; ameba invades nasal mucosa	
Acanthamoeba	Meningoencephalitis	Eye, brain	Ameba lives in fresh water; can invade cuts and abrasions	
Trypanosoma brucei	African trypanosomiasis, sleeping sickness	Brain, heart, many organs	Bite of tsetse fly, vector	
Trypanosoma cruzi	South American trypanosomiasis; Chagas disease	Nerve ganglia, muscle	Feces of reduviid bug rubbed into bite	
Viral pathogens (chapters 24 and 25)				
Poliovirus★	Poliomyelitis	Anterior horn cells; motor neurons	Contaminated food, water objects	
Rabies virus	Rabies	Brain, spinal nerves, ganglia	Saliva of infected animal enters bite wound	
Encephalitis viruses	St. Louis encephalitis Western equine encephalitis Eastern equine encephalitis	Brain, meninges, spinal cord	Bite of infected mosquito	
Bunyavirus	California encephalitis	Brain, meninges	Mosquito bite; reservoir is wild birds	
HIV★ (see table A.5)	AIDS, dementia	Destroys brain cells, peripheral nerves	Blood, semen, vaginal fluids; STD	
JC polyomavirus	Progressive multifocal leukoencephalopathy	Oligodendrocytes in cerebrum	Not well defined	
Prions	Spongiform encephalopathies Creutzfeldt-Jakob disease	Brain, neurons	Intimate contact with infected tissues	

TABLE A.8

Infections of the Eye and Ear

Name of Microbe	Name of Disease	Specific Targets of Infection	Source/Mode of Transmission	(Ch)
Bacterial pathogens				
Neisseria gonorrhoeae★	Ophthalmia neonatorum	Conjunctiva, cornea, eyelid	Vaginal mucus enters eyes during birth	(18)
Streptococcus pneumoniae★	Otitis media	Middle ear	Nasopharyngeal secretions forced into eustachian tube	(18)
Haemophilus ducreyi	Pinkeye	Conjunctiva, sclera	Contaminated fingers, fomites	(20)
Chlamydia trachomatis★	Inclusion conjunctivitis, trachoma	Conjunctiva, inner eyelid, cornea	Infected birth canal, contaminated fingers	(21)
Fungal pathogens (chapter 22)				
Aspergillus fumigatus	Aspergillosis	Eyelids, conjunctiva	Contact lens contaminated with spores	
Fusarium solani	Corneal ulcer, keratitis	Cornea	Eye is accidentally scratched, contact lens is contaminated	
Parasitic (protozoan, helminth) pathogens (chapter 23)				
Onchocerca volvulus	Onchocerciasis, river blindness	Eye, skin, blood	Black flies inject larvae into bite	
Loa loa	African eye worm	Subcutaneous tissues, conjunctiva, cornea	Bites by dipteran flies	
Viral pathogens				
Coxsackievirus	Hemorrhagic conjunctivitis	Conjunctiva, sclera, eyelid	Contact with fluids, secretions of carrier	(25)

Answers to Multiple-Choice Questions and Selected Matching Questions

Chapter 1
1. c
2. c
3. d
4. c
5. d
6. a
7. c
8. b
9. d
10. a
11. c
12. b
13. d
14. 1st col:
3, 7, 4, 2
2nd col:
8, 5, 6, 1

Chapter 2
1. c
2. c
3. b
4. c
5. a
6. e
7. a
8. d
9. a
10. c
11. c
12. a
13. b
14. c
15. c
16. b
17. b
18. d
19. b
20. c
21. c
22. c
23. a
24. d

Chapter 3
1. c
2. b
3. c
4. d
5. b
6. d

7. b
8. b
9. c
10. c
11. a
12. b
13. c
14. abf, df,
abf, ef, af, bef,
ac, bef

Chapter 4
1. d
2. a
3. c
4. a
5. c
6. b
7. c
8. d
9. b
10. d
11. c
12. d
13. c
14. b
15. a

Chapter 5
1. b
2. d
3. b
4. d
5. a
6. b
7. c
8. c
9. d
10. b
11. d
12. a
13. d
14. b
15. c
16. Matching:
b, e, c, h, g,
j, i, d, a, f

Chapter 6
1. c
2. d
3. d
4. b

5. d
6. a
7. a
8. d
9. b
10. b
11. c
12. d
13. d
14. a

Chapter 7
1. c
2. a
3. a
4. c
5. c
6. b
7. a
8. a
9. b
10. b
11. c
12. c
13. c
14. c
15. b
16. a

Chapter 8
1. b
2. a
3. c
4. d
5. d
6. b
7. b
8. c
9. b
10. b
11. a
12. c
13. a
14. d
15. b
16. c
17. c
18. c
19. c
20. Matching:
c, a, b, c,
b, a, c, c

Chapter 9
1. b
2. e
3. b
4. b
5. c
6. b
7. c
8. a
9. b
10. a
11. b
12. a
13. d
14. b
15. d
16. b
17. Matching:
e/h, f, b, g,
e, c, a, i,
c/e/h

Chapter 10
1. c
2. c
3. d
4. c
5. a
6. b
7. c
8. c
9. c
10. d
11. d
12. Matching:
h, c, f, a,
g, b, e, d

Chapter 11
1. d
2. c
3. b
4. a
5. c
6. b
7. b
8. d
9. c
10. b
11. d
12. c
13. d

14. a
15. b
16. c

Chapter 12
1. b
2. c
3. b
4. a
5. d
6. b
7. c
8. c
9. a
10. d
11. a
12. c
13. c
14. b

Chapter 13
1. a
2. d
3. b
4. d
5. c
6. d
7. b
8. c
9. c
10. c
11. d
12. c
13. a
14. a
15. d

Chapter 14
1. b
2. b
3. d
4. b
5. b
6. c
7. b
8. a
9. c
10. d
11. c
12. d
13. d
14. d
15. d

Chapter 15
1. a
2. d
3. c
4. a
5. c
6. c
7. c
8. a
9. c
10. a
11. c
12. a
13. c
14. b
15. e
16. e
17. Matching:
IgG   bghi
IgA   ac
IgD   be
IgE   bj
IgM   dfi

Chapter 16
1. b
2. c
3. d
4. b
5. c
6. c
7. d
8. b
9. c
10. a
11. a
12. d
13. c
14. b
15. Matching:
aei, adh,
cdh, bdh,
adh, bdh,
cj, ch, agh,
adh, ce

Chapter 17
1. d
2. d
3. d
4. c
5. b

6. c
7. b
8. a
9. d
10. c
11. d
12. a
13. b
14. d
15. c
16. d
17. d

Chapter 18
1. d
2. a
3. c
4. b
5. c
6. b
7. c
8. c
9. c
10. a
11. a
12. d
13. a
16. Matching:
h, j, d, l, m,
a, f, i, e, k,
b, c

Chapter 19
1. b
2. c
3. b
4. d
5. c
6. c
7. c
8. c
9. d
10. b
11. c
12. d
13. c
14. d
15. a
16. a

16. Matching:
1. a
2. b
3. c
4. b
5. c
6. d
7. bf
8. d
9. a

Chapter 20
1. b
2. b
3. d
4. b
5. c
6. d
7. d
8. b
9. d
10. d
11. c
12. d
13. a, c, d, e, h
14. all but c
15. Matching:
l, i, h, a, d,
g, k, e, j,
m, f, c, b

Chapter 21
1. b
2. b
3. c
4. d
5. d
6. e
7. d
8. c
9. b
10. d
11. a
12. d
13. c
14. d
15. d
16. a
17. b
21. Matching:
b, c, f, f, c,
d, a, d, g, a,
e, g, g

Chapter 22
1. c
2. a
3. a
4. d
5. d
6. a
7. a
8. d
9. c
10. c
11. c
12. c
13. d
14. b
15. Matching:
a/e, g, i, c,
h, l, b/k,
f/k, i, d, j

Chapter 23
1. c
2. b
3. d
4. b
5. b
6. d
7. c
8. d
9. c
10. a
11. b
12. b
13. b
14. a
15. a
16. c
17. Matching:
f, j, k, g, h,
c, i, l, a, d,
m, e, b

Chapter 24
1. d
2. c
3. b
4. a
5. d
6. d
7. c
8. b
9. a
10. b

Chapter 25
1. a
2. b
3. c
4. b
5. d
6. d
7. b
8. d
9. d
10. c
11. d
12. d
13. b
14. d
15. c
17. Matching:
c, adf, bei,
ac, hg, c,
af, h, hf,
a, ij

Chapter 26
1. c
2. b
3. b
4. c
5. d
6. c
7. c
8. d
9. d
10. b
11. d
12. c
13. a
14. b
15. d
16. d
17. d
18. b
19. b

Glossary

A

abiogenesis The belief in spontaneous generation as a source of life.

abiotic Nonliving factors such as soil, water, temperature, and light that are studied when looking at an ecosystem.

ABO blood group system Developed by Karl Landsteiner in 1904; the identification of different blood groups based on differing isoantigen markers characteristic of each blood type.

abscess An inflamed, fibrous lesion enclosing a core of pus.

abyssal zone The deepest region of the ocean; a sunless, high-pressure, cold, anaerobic habitat.

acellular vaccine A vaccine preparation that contains specific antigens such as the capsule or toxin from a pathogen and not the whole microbe. Acellular (without a cell)

acid-fast A term referring to the property of mycobacteria to retain carbol fuchsin even in the presence of acid alcohol. The staining procedure is used to diagnose tuberculosis.

acidic A solution with a pH value above 7 on the pH scale.

acidic fermentation An anaerobic degradation of pyruvic acid that results in organic acid production.

Acquired Immune Deficiency Syndrome See *AIDS*

actinomycetes A group of filamentous, funguslike bacteria.

active immunity Immunity acquired through direct stimulation of the immune system by antigen.

active site The specific region on an apoenzyme that binds substrate. The site for reaction catalysis.

activated lymphocyte A T or B cell that has received an immune stimulus, such as antigens or cytokines, and is undergoing rapid synthesis and proliferation.

active transport Nutrient transport method that requires carrier proteins in the membranes of the living cells and the expenditure of energy.

acute Characterized by rapid onset and short duration.

acyclovir A synthetic purine analog that blocks DNA synthesis in certain viruses, particularly the herpes simplex viruses.

adapt/adaptation The act of successfully adjusting to a new environment, often made possible by random genetic changes that provide an advantage under changed environmental conditions.

adenine (A) One of the nitrogen bases found in DNA and RNA, with a purine form.

adenosine deaminase (ADA) deficiency An immunodeficiency disorder and one type of SCIDS that is caused by an inborn error in the metabolism of adenine. The accumulation of adenine destroys both B and T lymphocytes.

adenosine triphosphate (ATP) A nucleotide that is the primary source of energy to cells.

adenovirus Nonenveloped DNA virus; means of transmission is human-to-human via respiratory and ocular secretions.

adhesion The process by which microbes gain a more stable foothold at the portal of entry; often involves a specific interaction between the molecules on the microbial surface and the receptors on the host cell.

adjuvant In immunology, a chemical vehicle that enhances antigenicity, presumably by prolonging antigen retention at the injection site.

adsorption A process of adhering one molecule onto the surface of another molecule.

aerobe A microorganism that lives and grows in the presence of free gaseous oxygen (O_2).

aerobic respiration Respiration in which the final electron acceptor in the electron transport chain is oxygen (O_2).

aerosols Suspensions of fine dust or moisture particles in the air that contain live pathogens.

aflatoxin From *Aspergillus flavus* toxin, a mycotoxin that typically poisons moldy animal feed and can cause liver cancer in humans and other animals.

agammaglobulinemia Also called hypogammaglobulinemia. The absence of or severely reduced levels of antibodies in serum.

agar A polysaccharide found in seaweed and commonly used to prepare solid culture media.

agglutination The aggregation by antibodies of suspended cells or similar-sized particles (agglutinogens) into clumps that settle.

agglutinin A specific antibody that cross-links agglutinogen, causing it to aggregate.

agglutinogen An antigenic substance on cell surfaces that evokes agglutinin formation against it.

agranulocyte One form of leukocyte (white blood cells), having globular, non-lobed nuclei and lacking prominent cytoplasmic granules.

AIDS Acquired immunodeficiency syndrome. The complex of signs and symptoms characteristic of the late phase of human immunodeficiency virus (HIV) infection.

alcoholic fermentation An anaerobic degradation of pyruvic acid that results in alcohol production.

algae Photosynthetic, plant-like organisms which generally lack the complex structure of plants; they may be single-celled or multicellular, and inhabit diverse habitats such as marine and freshwater environments, glaciers, and hot springs.

allele A gene that occupies the same location as other alternative (allelic) genes on paired chromosomes.

allergen A substance that provokes an allergic response.

allergy The altered, usually exaggerated, immune response to an allergen. Also called hypersensitivity.

alloantigen An antigen that is present in some but not all members of the same species.

allograft Relatively compatible tissue exchange between nonidentical members of the same species. Also called homograft.

allosteric Pertaining to the altered activity of an enzyme due to the binding of a molecule to a region other than the enzyme's active site.

alternative pathway Pathway of complement activation initiated by non specific components of tissue or bacterial substances such as endotoxin and polysaccharides.

amastigote The rounded or ovoid nonflagellated form of the *Leishmania* parasite.

Ames test A method for detecting mutagenic and potentially carcinogenic agents based upon the genetic alteration of nutritionally defective bacteria.

amination The addition of an amine (—NH_2) group to a molecule.

amino acids The building blocks of protein. Amino acids exist in 20 naturally occurring forms that impart different characteristics to the various proteins they compose.

aminoglycoside A complex group of drugs derived from soil actinomycetes that impairs ribosome function and has antibiotic potential. Example: streptomycin.

ammonification Phase of the nitrogen cycle in which ammonia is released from decomposing organic material.

amphibolism Pertaining to the metabolic pathways that serve multiple functions in the breakdown, synthesis, and conversion of metabolites.

amphipathic Relating to a compound that has contrasting characteristics, such as hydrophilic-hydrophobic or acid-base.

amphitrichous Having a single flagellum or a tuft of flagella at opposite poles of a microbial cell.

amplicon DNA strand that has been primed for replication during polymerase chain reaction.

anabolism The energy-consuming process of incorporating nutrients into protoplasm through biosynthesis.

anaerobe A microorganism that grows best, or exclusively, in the absence of oxygen.

anaerobic digesters Closed chambers used in a microbial process that converts organic sludge from waste treatment plants into useful fuels such as methane and hydrogen gases. Also called bioreactors.

anaerobic respiration Respiration in which the final electron acceptor in the electron transport chain is an inorganic molecule containing sulfate, nitrate, nitrite, carbonate, etc.

analog In chemistry, a compound that closely resembles another in structure.

anamnestic In immunology, an augmented response or memory related to a prior stimulation of the immune system by antigen. It boosts the levels of immune substances.

anaphylaxis The unusual or exaggerated allergic reaction to antigen that leads to severe respiratory and cardiac complications.

anion A negatively charged ion.

anneal Characteristic of changing binding properties in response to heating and cooling.

anoxygenic Any reaction that does not produce oxygen; usually in reference to the type of photosynthesis occurring in anaerobic photosynthetic bacteria.

antagonism Relationship in which microorganisms compete for survival in a common environment by taking actions that inhibit or destroy another organism.

anthrax A zoonotic disease of herbivorous livestock. The anthrax bacillus is a facultative parasite and can infect humans in a number of ways. In its most virulent form, it can be fatal.

antibiotic A chemical substance from one microorganism that can inhibit or kill another microbe even in minute amounts.

antibody A large protein molecule evoked in response to an antigen that interacts specifically with that antigen.

anticodon The trinucleotide sequence of transfer RNA that is complementary to the trinucleotide sequence of messenger RNA (the codon).

antigen Any cell, particle, or chemical that induces a specific immune response by B cells or T cells and can stimulate resistance to an infection or a toxin. See *immunogen.*

antigen binding site Specific region at the ends of the antibody molecule that recognize specific antigens. These sites have numerous shapes to fit a wide variety of antigens.

Antigen Presenting Cell (APC) A macrophage or dendritic cell that ingests and degrades an antigen and subsequently places the antigenic determinant molecules on its surface for recognition by CD4 T-lymphocytes.

antigenic determinant The precise molecular group of an antigen that defines its specificity and triggers the immune response.

antigenic drift Minor antigenic changes in the influenza A virus due to mutations in the spikes' genes.

antigenic shift Major changes in the influenza A virus due to recombination of viral strains from two different host species.

antigenicity The property of a substance to stimulate a specific immune response such as antibody formation.

antihistamine A drug that counters the action of histamine and is useful in allergy treatment.

anti-idiotype An anti-antibody that reacts specifically with the idiotype (variable region or antigen-binding site) of another antibody.

antimetabolite A substance such as a drug that competes with, substitutes for, or interferes with a normal metabolite.

antimicrobic A special class of compounds capable of destroying or inhibiting microorganisms.

anti-oncogene A gene that is responsible for regulating the function of a proto-oncogene. The interaction of these two genes keeps the cell division cycle operating normally.

antisense DNA A type of gene therapy which utilizes an oligonucleotide to bind to the sense strand of a specific piece of DNA, thereby inhibiting transcription.

antiseptic A growth-inhibiting agent used on tissues to prevent infection.

antiserum Antibody-rich serum derived from the blood of animals (deliberately immunized against infectious or toxic antigen) or from people who have recovered from specific infections.

antitoxin Globulin fraction of serum that neutralizes a specific toxin. Also refers to the specific antitoxin antibody itself.

apoenzyme The protein part of an enzyme, as opposed to the nonprotein or inorganic cofactors.

apoptosis The genetically programmed death of cells that is both a natural process of development and the body's means of destroying abnormal or infected cells.

appendages Accessory structures that sprout from the surface of bacteria. They can be divided into two major groups: those that provide motility and those that enable adhesion.

aquifer A subterranean water-bearing stratum of permeable rock, sand, or gravel.

arbovirus Arthropod-borne virus, including togaviruses, reoviruses, flaviviruses, and bunyaviruses. These viruses generally cause mild, undifferentiated fevers and occasionally cause severe encephalitides and hemorrhagic fever.

archaea Prokaryotic single-celled organisms of primitive origin that have unusual anatomy, physiology and genetics, and live in harsh habitats; when capitalized (**Archaea**) the term refers to one of the three domains of living organisms as proposed by Woese.

arthrospore A fungal spore formed by the septation and fragmentation of hyphae.

Arthus reaction An immune complex phenomenon that develops after repeat injection. This localized inflammation results from aggregates of antigen and antibody that bind, complement, and attract neutrophils.

artificial chromosome A large packet of DNA from yeasts (YAC), bacteria (BAC), or humans (HAC) that can be used to carry, transfer, or analyze isolated foreign DNA.

ascospore A spore formed within a saclike cell (ascus) of Ascomycota following nuclear fusion and meiosis.

ascus Special fungal sac in which haploid spores are created.

asepsis A condition free of viable pathogenic microorganisms.

aseptic technique Methods of handling microbial cultures, patient specimens, and other sources of microbes in a way that prevents infection of the handler and others who may be exposed.

asymptomatic An infection that produces no noticeable symptoms even though the microbe is active in the host tissue.

asymptomatic carrier A person with an inapparent infection who shows no symptoms of being infected yet is able to pass the disease agent on to others.

atmosphere That part of the biosphere that includes the gaseous envelope up to 14 miles above the earth's surface. It contains gases such as carbon dioxide, nitrogen, and oxygen.

atomic number (AN) A measurement that reflects the number of protons in an atom of a particular element.

atomic weight The average of the mass numbers of all the isotopic forms for a particular element.

atom The smallest particle of an element to retain all the properties of that element.

atopy Allergic reaction classified as type I, with a strong familial relationship; caused by allergens such as pollen, insect venom, food, and dander; involves IgE antibody; includes symptoms of hay fever, asthma, and skin rash.

ATP synthase A unique enzyme located in the mitochondrial cristae and chloroplast grana that harnesses the flux of hydrogen ions to the synthesis of ATP.

attenuate To reduce the virulence of a pathogenic bacterium or virus by passing it through a non-native host or by long-term subculture.

AUG (start codon) The codon that signals the point at which translation of a messenger RNA molecule is to begin.

autoantibody An "anti-self" antibody having an affinity for tissue antigens of the subject in which it is formed.

autoantigen Molecules that are inherently part of self but are perceived by the immune system as foreign.

autoclave A sterilization chamber which allows the use of steam under pressure to sterilize materials. The most common temperature/pressure combination for an autoclave is 121°C and 15 psi.

autograft Tissue or organ surgically transplanted to another site on the same subject.

autoimmune disease The pathologic condition arising from the production of antibodies against autoantigens. Example: rheumatoid arthritis. Also called autoimmunity.

autosome A chromosome of somatic cells as opposed to a sex chromosome of gametes.

autotroph A microorganism that requires only inorganic nutrients and whose sole source of carbon is carbon dioxide.

axenic A sterile state such as a pure culture. An axenic animal is born and raised in a germ-free environment. See *gnotobiotic.*

axial filament A type of flagellum (called an endoflagellum) that lies in the periplasmic space of spirochetes and is responsible for locomotion. Also called periplasmic flagellum.

Azidothymidine (AZT) A thymine analog used in the treatment of HIV infection. AZT inhibits the action of the reverse transcriptase enzyme, making it specific for HIV.

azole Five-membered heterocyclic compounds typical of histidine, which are used in antifungal therapy.

B

bacillus Bacterial cell shape that is cylindrical (longer than it is wide).

back-mutation A mutation which counteracts an earlier mutation, resulting in the restoration of the original DNA sequence.

bacteremia The presence of viable bacteria in circulating blood.

bacteria Category of prokaryotes with peptidoglycan in their cell walls and a single, circular chromosome. This group of small cells is widely distributed in the earth's habitats.

Bacteria (plural of bacterium) When capitalized can refer to one of the three domains of living organisms proposed by Woese, containing all non-archaea prokaryotes.

bacterial chromosome A circular body in bacteria that contains the primary genetic material. Also called nucleoid.

bacterial meningitis Inflammation of the meninges as a result of bacterial infection.

bactericide An agent that kills bacteria.

bacteriocin Proteins produced by certain bacteria that are lethal against closely related bacteria and are narrow spectrum compared with antibiotics; these proteins are coded and transferred in plasmids.

bacteriophage A virus that specifically infects bacteria.

bacteriostatic Any process or agent that inhibits bacterial growth.

bacterium A tiny unicellular prokaryotic organism that usually reproduces by binary fission and usually has a peptidoglycan cell wall, has various shapes, and can be found in virtually any environment.

barophile A microorganism that thrives under high (usually hydrostatic) pressure.

basement membrane A thin layer (1–6 μm) of protein and polysaccharide found at the base of epithelial tissues.

basic A solution with a pH value below 7 on the pH scale.

basidiospore A sexual spore that arises from a basidium. Found in basidiomycota fungi.

basidium A reproductive cell created when the swollen terminal cell of a hypha develops filaments (sterigmata) that form spores.

basophil A motile polymorphonuclear leukocyte that binds IgE. The basophilic cytoplasmic granules contain mediators of anaphylaxis and atopy.

bdellovibrio A bacterium that preys on certain other bacteria. It bores a hole into a specific host and inserts itself between the protoplast and the cell wall. There it elongates before subdividing into several cells and devouring the host cell.

benign tumor A self-contained mass within an organ that does not spread into adjacent tissue.

benthic zone The sedimentary bottom region of a pond, lake, or ocean.

beta-lactamase An enzyme secreted by certain bacteria that cleaves the beta-lactam ring of penicillin and cephalosporin and thus provides for resistance against the antibiotic. *See* penicillinase.

beta oxidation The degradation of long-chain fatty acids. Two-carbon fragments are formed as a result of enzymatic attack directed against the second or beta carbon of the hydrocarbon chain. Aided by coenzyme A, the fragments enter the tricarboxylic acid (TCA) cycle and are processed for ATP synthesis.

binary fission The formation of two new cells of approximately equal size as the result of parent cell division.

binomial system Scientific method of assigning names to organisms that employs two names to identify every organism—genus name plus species name.

bioamplification The concentration or accumulation of a pollutant in living tissue through the natural flow of an ecosystem.

biochemistry The study of organic compounds produced by (or components of) living things. The four main categories of biochemicals are carbohydrates, lipids, proteins, and nucleic acid.

bioenergetics The study of the production and use of energy by cells.

bioethics The study of biological issues and how they relate to human conduct and moral judgment.

biofilm A complex association that arises from a mixture of microorganisms growing together on the surface of a habitat.

biogenesis Belief that living things can only arise from others of the same kind.

biogeochemical cycle A process by which matter is converted from organic to inorganic form and returned to various non-living reservoirs on earth (air, rocks, and water) where it becomes available for reuse by living things. Elements such as carbon, nitrogen, and phosphorus are constantly cycled in this manner.

biological vector An animal which not only transports an infectious agent but plays a role in the life cycle of the pathogen, serving as a site in which it can multiply or complete its life cycle. It is usually an alternate host to the pathogen.

biomes Particular climate regions in a terrestrial realm.

bioremediation The use of microbes to reduce or degrade pollutants, industrial wastes, and household garbage.

biosphere Habitable regions comprising the aquatic (hydrospheric), soil-rock (lithospheric), and air (atmospheric) environments.

biotechnology The use of microbes or their products in the commercial or industrial realm.

biotic Living factors such as parasites, food substrates, or other living or once-living organisms that are studied when looking at an ecosystem.

blast cell An immature precursor cell of B and T lymphocytes. Also called a lymphoblast.

blocking antibody The IgG class of immunoglobulins that competes with IgE antibody for allergens, thus blocking the degranulation of basophils and mast cells.

blood cells Cellular components of the blood consisting of red blood cells, primarily responsible for the transport of oxygen and carbon dioxide, and white blood cells, primarily responsible for host defense and immune reactions.

B lymphocyte (B cell) A white blood cell that gives rise to plasma cells and antibodies.

booster The additional doses of vaccine antigen administered to increase an immune response and extend protection.

botulin *Clostridium botulinum* toxin. Ingestion of this potent exotoxin leads to flaccid paralysis.

bradykinin An active polypeptide that is a potent vasodilator released from IgE-coated mast cells during anaphylaxis.

broad spectrum A word to denote drugs that affect many different types of bacteria, both gram-positive and gram-negative.

Brownian movement The passive, erratic, nondirectional motion exhibited by microscopic particles. The jostling comes from being randomly bumped by submicroscopic particles, usually water molecules, in which the visible particles are suspended.

brucellosis A zoonosis transmitted to humans from infected animals or animal products; causes a fluctuating pattern of severe fever in humans as well as muscle pain, weakness, headache, weight loss, and profuse sweating. Also called undulant fever.

bubo The swelling of one or more lymph nodes due to inflammation.

Bubonic plague The form of plague in which bacterial growth is primarily restricted to the lymph and is characterized by the appearance of a swollen lymph node referred to as a bubo.

budding See *exocytosis*.

bulbar poliomyelitis Complication of polio infection in which the brain stem, medulla, or cranial nerves are affected. Leads to loss of respiratory control and paralysis of the trunk and limbs.

bulla A large, bubblelike vesicle in a region of separation between the epidermis and the subepidermal layer. The space is usually filled with serum and sometimes with blood.

C

calculus Dental deposit formed when plaque becomes mineralized with calcium and phosphate crystals. Also called tartar.

Calvin cycle The recurrent photosynthetic pathway characterized by CO_2 fixation and glucose synthesis. Also called the dark reactions.

cancer Any malignant neoplasm that invades surrounding tissue and can metastasize to other locations. A carcinoma is derived from epithelial tissue, and a sarcoma arises from proliferating mesodermal cells of connective tissue.

capsid The protein covering of a virus's nucleic acid core. Capsids exhibit symmetry due to the regular arrangement of subunits called capsomers. *See* icosahedron.

capsomer A subunit of the virus capsid shaped as a triangle or disc.

capsule In bacteria, the loose, gel-like covering or slime made chiefly of simple polysaccharides. This layer is protective and can be associated with virulence.

capsule antigen (K or fimbrial antigen) The capsular cell surface antigen that can be used to type or identify Gram-negative enteric bacteria.

carbohydrate A compound containing primarily carbon, hydrogen, and oxygen in a 1:2:1 ratio.

carbon cycle That pathway taken by carbon from its abiotic source to its use by producers to form organic compounds (biotic), followed by the breakdown of biotic compounds and their release to a non living reservoir in the environment (mostly carbon dioxide in the atmosphere).

carbon fixation Reactions in photosynthesis that incorporate inorganic carbon dioxide into organic compounds such as sugars. This occurs during the Calvin cycle and uses energy generated by the light reactions. This process is the source of all production on earth.

carbuncle A deep staphylococcal abscess joining several neighboring hair follicles.

carcinoma Cancers originating in epithelial tissue.

carotenoid Yellow, orange, or red photosynthetic pigments.

carrier A person who harbors infections and inconspicuously spreads them to others. Also, a chemical agent that can accept an atom, chemical radical, or subatomic particle from one compound and pass it on to another.

caseous lesion Necrotic area of lung tubercle superficially resembling cheese. Typical of tuberculosis.

catabolism The chemical breakdown of complex compounds into simpler units to be used in cell metabolism.

catalyst A substance that alters the rate of a reaction without being consumed or permanently changed by it. In cells, enzymes are catalysts.

catalytic site The niche in an enzyme where the substrate is converted to the product (also active site).

cation A positively charged ion.

CD4 – "Cluster of Differentiation, 4" An antigen associated with helper T-lymphocytes.

CD8 – "Cluster of Differentiation, 8" An antigen associated with cytotoxic T-lymphocytes.

cecum The intestinal pocket that forms the first segment of the large intestine. Also called the appendix.

cell An individual membrane-bound living entity; the smallest unit capable of an independent existence.

cell-mediated The type of immune responses brought about by T cells, such as cytotoxic, suppressor, and helper effects.

cellulitis The spread of bacteria within necrotic tissue.

cellulose A long, fibrous polymer composed of β-glucose; one of the most common substances on earth.

cephalosporins A group of broad-spectrum antibiotics isolated from the fungus *Cephalosporium*.

cercaria The free-swimming larva of the schistosome trematode that emerges from the snail host and can penetrate human skin, causing schistosomiasis.

cestode The common name for tapeworms that parasitize humans and domestic animals.

chancre The primary sore of syphilis that forms at the site of penetration by *Treponema pallidum*. It begins as a hard, dull red, painless papule that erodes from the center.

chancroid A lesion that resembles a chancre but is soft and is caused by *Haemophilus ducreyi*.

chemical bond A link formed between molecules when two or more atoms share, donate, or accept electrons.

chemical mediators Small molecules that are released during inflammation and specific immune reactions that allow communication between the cells of the immune system and facilitate surveillance, recognition and attack.

chemiosmotic hypothesis An explanation for ATP formation that is based on the formation of a proton (H^+)gradient across a membrane during electron transport. Movement of the protons back across an ATP synthase causes the formation of ATP.

chemoautotroph An organism that relies upon inorganic chemicals for its energy and carbon dioxide for its carbon. Also called a chemolithotroph.

chemoheterotroph Microorganisms that derive their nutritional needs from organic compounds.

chemokine Chemical mediators (cytokines) that stimulate the movement and migration of white blood cells.

chemostat A growth chamber with an outflow that is equal to the continuous inflow of nutrient media. This steady-state growth device is used to study such events as cell division, mutation rates, and enzyme regulation.

chemotactic factors Chemical mediators that stimulate the movement of white blood cells. See *chemokines.*

chemotaxis The tendency of organisms to move in response to a chemical gradient (toward an attractant or to avoid adverse stimuli).

chemotherapy The use of chemical substances or drugs to treat or prevent disease.

chemotroph Organism that oxidizes compounds to feed on nutrients.

chickenpox A papulo-pustular rash caused by infection by varicella-zoster virus.

chimera A product formed by the fusion of two different organisms.

chitin A polysaccharide similar to cellulose in chemical structure. This polymer makes up the horny substance of the exoskeletons of arthropods and certain fungi.

chloramphenicol A potent broad spectrum antibiotic which is no longer widely used because of its potential for causing lethal damage to the blood forming tissues.

chlorophyll A group of mostly green pigments that are used by photosynthetic eucaryotic organisms and cyanobacteria to trap light energy to use in making chemical bonds.

chloroplast An organelle containing chlorophyll that is found in photosynthetic eucaryotes.

cholesterol Best-known member of a group of lipids called steroids. Cholesterol is commonly found in cell membranes and animal hormones.

chromatic aberration Deviant focus of magnification due to refraction of the colored wavelengths that make up white light.

chromatin The genetic material of the nucleus. Chromatin is made up of nucleic acid and stains readily with certain dyes.

chromophore The chemical radical of a dye that is responsible for its color and reactivity.

chromosome The tightly coiled bodies in cells that are the primary sites of genes.

chronic Any process or disease that persists over a long duration.

chronic fatigue syndrome A collection of persistent, non-specific symptoms possibly linked to immune deficiency and/or viral infection.

cilium (plural: *cilia*) Eucaryotic structure similar to flagella that propels a protozoan through the environment.

class In the levels of classification, the division of organisms that follows phylum.

classical pathway Pathway of complement activation initiated by a specific antigen-antibody interaction.

clonal selection theory A conceptual explanation for the development of lymphocyte specificity and variety during immune maturation.

clone A colony of cells (or group of organisms) derived from a single cell (or single organism) by asexual reproduction. All units share identical characteristics. Also used as a verb to refer to the process of producing a genetically identical population of cells or genes.

cloning host An organism such as a bacterium or a yeast that receives and replicates a foreign piece of DNA inserted during a genetic engineering experiment.

coagulase A plasma-clotting enzyme secreted by *Staphylococcus aureus*. It contributes to virulence and is involved in forming a fibrin wall that surrounds staphylococcal lesions.

coagulase-negative staphylococci Opportunistic species that are usually part of the normal flora of the skin and mucous membranes. They cause infection when host defenses are low and lack many virulence factors such as coagulase.

coccobacillus An elongated coccus; a short, thick, oval-shaped bacterial rod.

coccus A spherical-shaped bacterial cell.

codon A specific sequence of three nucleotides in mRNA (or the sense strand of DNA) that constitutes the genetic code for a particular amino acid.

coenzyme A complex organic molecule, several of which are derived from vitamins (e.g., nicotinamide, riboflavin). A coenzyme operates in conjunction with an enzyme. Coenzymes serve as transient carriers of specific atoms or functional groups during metabolic reactions.

cofactor An enzyme accessory. It can be organic, such as coenzymes, or inorganic, such as Fe^{+2}, Mn^{+2}, or Zn^{+2} ions.

cold sterilization The use of nonheating methods such as radiation or filtration to sterilize materials.

coliform A collective term that includes normal enteric bacteria that are gram-negative and lactose-fermenting.

colinear Having corresponding parts of a molecule arranged in the same linear order as another molecule, as in DNA and mRNA.

colitis Inflammation and necrosis of the colon that may be caused by infection. Mild cases lead to diarrhea, while more severe cases can result in perforation of the caecum.

colony A macroscopic cluster of cells appearing on a solid medium, each arising from the multiplication of a single cell.

colostrum The clear yellow early product of breast milk that is very high in secretory antibodies. Provides passive intestinal protection.

commensalism An unequal relationship in which one species derives benefit without harming the other.

common (seed) warts Painless, elevated rough growths on the fingers or occasionally other parts of the body.

communicable infection Capable of being transmitted from one individual to another.

community The interacting mixture of populations in a given habitat.

competitive inhibition Control process that relies on the ability of metabolic analogs to control microbial growth by successfully competing with a necessary enzyme to halt the growth of bacterial cells.

complement In immunology, serum protein components that act in a definite sequence when set in motion either by an antigen-antibody complex or by factors of the alternative (properdin) pathway.

complementary DNA (cDNA) DNA created by using reverse transcriptase to synthesize DNA from RNA templates.

compounds Molecules that are a combination of two or more different elements.

concentration The expression of the amount of a solute dissolved in a certain amount of solvent. It may be defined by weight, volume, or percentage.

condylomata acuminata Extensive, branched masses of genital warts caused by infection with human papillomavirus.

congenital Transmission of an infection from mother to a fetus in utero.

congenital rubella Transmission of the rubella virus to a fetus in utero. Injury to the fetus is generally much more serious than it is to the mother.

conidia Asexual fungal spores shed as free units from the tips of fertile hyphae.

conjugation In bacteria, the contact between donor and recipient cells associated with the transfer of genetic material such as plasmids. Can involve special (sex) pili. Also a form of sexual recombination in ciliated protozoans.

conjunctivitis Sometimes called "pinkeye." A *Haemophilus* infection of the subconjunctiva that is common among children, is easily transmitted, and is treated with antibiotic eyedrops.

constitutive enzyme An enzyme present in bacterial cells in constant amounts, regardless of the presence of substrate. Enzymes of the central catabolic pathways are typical examples.

consumer An organism that feeds on producers or other consumers. It gets all nutrients and energy from other organisms (also called heterotroph). May exist at several levels, such as primary (feeds on producers), secondary (feeds on primary consumers).

contagious Communicable; transmissible by direct contact with infected people and their fresh secretions or excretions.

contaminant An impurity; any undesirable material or organism.

control locus The region of an operon which regulates the transcription of the structural elements of the operon.

convalescence Recovery; the period between the end of a disease and the complete restoration of health in a patient.

corepressor A molecule that combines with inactive repressor to form active repressor, which attaches to the operator gene site and inhibits the activity of structural genes subordinate to the operator.

covalent bond A chemical bond formed by the sharing of electrons between two atoms.

Creutzfeld-Jakob disease A spongiform encephalopathy caused by infection with a prion. The disease is marked by dementia, impaired senses and uncontrollable muscle contractions.

crista The infolded inner membrane of a mitochondrion that is the site of the respiratory chain and oxidative phosphorylation.

culture The visible accumulation of microorganisms in or on a nutrient medium. Also, the propagation of microorganisms with various media.

curd The coagulated milk protein used in cheese making.

cutaneous Second level of skin, including the stratum corneum and occasionally the upper dermis.

cutaneous anthrax The mildest form of anthrax, caused by the entrance of bacterial spores into small nicks or openings in the skin, their germination, and the formation of a dark necrotic lesion called an eschar.

cutaneous candidiasis A mycosis occurring in the skin and membranes of compromised patients and neonates as a result of infection by *Candida albicans.*

cyst The resistant, dormant, but infectious form of protozoans. Can be important in spread of infectious agents such as *Entamoeba histolytica* and *Giardia lamblia.*

cysteine Sulfur-containing amino acid that is essential to the structure and specificity of proteins.

cysticercus The larval form of certain *Taenia* species, which typically infest muscles of mammalian intermediate hosts. Also called bladderworm.

cystine An amino acid, HOOC—CH(NH$_2$)—CH$_2$—S—S—CH$_2$—CH (NH$_2$)COOH. An oxidation product of two cysteine molecules in which the —SH (sulfhydryl) groups form a disulfide union. Also called dicysteine.

cytochrome A group of heme protein compounds whose chief role is in electron and/or hydrogen transport occurring in the last phase of aerobic respiration.

cytokine A chemical substance produced by white blood cells and tissue cells that regulates development, inflammation, and immunity.

cytopathic effect The degenerative changes in cells associated with virus infection. Examples: the formation of multinucleate giant cells (Negri bodies), the prominent cytoplasmic inclusions of nerve cells infected by rabies virus.

cytoplasm Dense fluid encased by the cell membrane; the site of many of the cell's biochemical and synthetic activities.

cytosine (C) One of the nitrogen bases found in DNA and RNA, with a pyrimidine form.

cytotoxic Having the capacity to destroy specific cells. One class of T cells attacks cancer cells, virus-infected cells, and eucaryotic pathogens. See *killer T cells.*

D

death phase End of the cell growth due to lack of nutrition, depletion of environment, and accumulation of wastes. Population of cells begins to die.

debridement Trimming away devitalized tissue and foreign matter from a wound.

decomposer A consumer that feeds on organic matter from the bodies of dead organisms. These microorganisms feed from all levels of the food pyramid and are responsible for recycling elements (also called saprobes).

decomposition The breakdown of dead matter and wastes into simple compounds, that can be directed back into the natural cycle of living things.

decontamination The removal or neutralization of an infectious, poisonous, or injurious agent from a site.

deduction Problem-solving process in which an individual constructs a hypothesis, tests its validity by outlining particular events that are predicted by the hypothesis, and then performs experiments to test for those events.

definitive host The organism in which a parasite develops into its adult or sexually mature stage. Also called the final host.

degeneracy The property of the genetic code which allows an amino acid to be specified by several different codons.

degerm To physically remove surface oils, debris, and soil from skin to reduce the microbial load.

degranulation The release of cytoplasmic granules, as when cytokines are secreted from mast cell granules.

dehydration synthesis During the formation of a carbohydrate bond, the step in which one carbon molecule gives up its OH group and the other loses the H from its OH group, thereby producing a water molecule. This process is common to all polymerization reactions.

denaturation The loss of normal characteristics resulting from some molecular alteration. Usually in reference to the action of heat or chemicals on proteins whose function depends upon an unaltered tertiary structure.

dendritic cell A large, antigen-processing cell characterized by long, branchlike extensions of the cell membrane.

denitrification The end of the nitrogen cycle when nitrogen compounds are returned to the reservoir in the air.

dental caries A mixed infection of the tooth surface that gradually destroys the enamel and may lead to destruction of the deeper tissue.

deoxyribose A 5-carbon sugar that is an important component of DNA.

deoxyribonucleic acid (DNA) The nucleic acid often referred to as the "double helix." DNA carries the master plan for an organism's heredity.

dermatophytoses Superficial mycoses such as athlete's foot and ringworm associated with certain fungi with an affinity for the skin, hair, and nails.

desensitization See *hyposensitization.*

desiccation To dry thoroughly. To preserve by drying.

desquamate To shed the cuticle in scales; to peel off the outer layer of a surface.

diabetes mellitus A disease involving compromise in insulin function. In one form, the pancreatic cells that produce insulin are destroyed by autoantibodies, and in another, the pancreas does not produce sufficient insulin.

diapedesis The migration of intact blood cells between endothelial cells of a blood vessel such as a venule.

differential medium A single substrate that discriminates between groups of microorganisms on the basis of differences in their appearance due to different chemical reactions.

differential stain A technique that utilizes two dyes to distinguish between different microbial groups or cell parts by color reaction.

diffusion The dispersal of molecules, ions, or microscopic particles propelled down a concentration gradient by spontaneous random motion to achieve a uniform distribution.

DiGeorge Syndrome A birth defect usually caused by a missing or incomplete thymus gland that results in abnormally low or absent T-cells and other developmental abnormalities.

dimorphic In mycology, the tendency of some pathogens to alter their growth form from mold to yeast in response to rising temperature.

dinoflagellate A marine algae whose toxin can cause food poisoning. Overgrowth of this algae is responsible for the phenomenon known as "red tide."

diphtheria Infection by *Corynebacterium diphtheriae*. It is transmitted by human carriers or contaminated milk, and the primary infection is in the upper respiratory tract. Several forms of this infection may be fatal if untreated. Vaccination on the recommended schedule can prevent infection.

Diphtherotoxin Exotoxin responsible for the effects seen in diphtheria. The toxin affects the upper respiratory system, peripheral nervous system and the heart.

diplococcus Spherical or oval-shaped bacteria, typically found in pairs.

diploid Somatic cells having twice the basic chromosome number. One set in the pair is derived from the father, and the other from the mother.

direct, or total cell count 1. Counting total numbers of individual cells being viewed with magnification. 2. Counting isolated colonies of organisms growing on a plate of media as a way to determine population size.

disaccharide A sugar containing two monosaccharides. Examples: sucrose (fructose + glucose).

disease Any deviation from health, as when the effects of microbial infection damage or disrupt tissues and organs.

disinfection The destruction of pathogenic nonsporulating microbes or their toxins, usually on inanimate surfaces.

disseminated cytomegalovirus Systemic disease brought on by infection with cytomegalovirus. Often seen in AIDS patients and others susceptible to opportunistic infection.

division In the levels of classification, an alternate term for phylum.

DNA See *deoxyribonucleic acid*.

DNA fingerprint A pattern of restriction enzyme fragments which is unique for an individual organism.

DNA polymerase Enzyme responsible for the replication of DNA. Several versions of the enzyme exist, each completing a unique portion of the replication process.

DNA sequencing Determining the exact order of nucleotides in a fragment of DNA. Most commonly done using the Sanger dideoxy sequencing method.

DNA vaccine A newer vaccine preparation based on inserting DNA from pathogens into host cells to encourage them to express the foreign protein and stimulate immunity.

domain In the levels of classification, the broadest general category to which an organism is assigned. Members of a domain share only one or a few general characteristics.

droplet nuclei The dried residue of fine droplets produced by mucus and saliva sprayed while sneezing and coughing. Droplet nuclei are less than 5 μm in diameter (large enough to bear a single bacterium and small enough to remain airborne for a long time) and can be carried by air currents. Droplet nuclei are drawn deep into the air passages.

drug resistance An adaptive response in which microorganisms begin to tolerate an amount of drug that would ordinarily be inhibitory.

dry heat Air with a low moisture content which has been heated from 160 to several thousand degrees Celsius.

dysentery Diarrheal illness caused by exotoxins.

dyspnea Difficulty in breathing.

E

ecosystem A collection of organisms together with its surrounding physical and chemical factors.

ectoplasm The outer, more viscous region of the cytoplasm of a phagocytic cell such as an ameba. It contains microtubules, but not granules or organelles.

eczema An acute or chronic allergy of the skin associated with itching and burning sensations. Typically, red, edematous, vesicular lesions erupt, leaving the skin scaly and sometimes hyperpigmented.

edema The accumulation of excess fluid in cells, tissues, or serous cavities. Also called swelling.

electrolyte Any compound that ionizes in solution and conducts current in an electrical field.

electromagnetic radiation A form of energy that is emitted as waves and is propagated through space and matter. The spectrum extends from short gamma rays to long radio waves.

electron A negatively charged subatomic particle that is distributed around the nucleus in an atom.

electrophoresis The separation of molecules by size and charge through exposure to an electrical current.

electrostatic Relating to the attraction of opposite charges and the repulsion of like charges. Electrical charge remains stationary as opposed to electrical flow or current.

element A substance comprising only one kind of atom that cannot be degraded into two or more substances without losing its chemical characteristics.

ELISA Abbreviation for **e**nzyme-**l**inked **i**mmuno-**s**orbent **a**ssay, a very sensitive serological test used to detect antibodies in diseases such as AIDS.

emerging disease Newly identified diseases that are becoming more prominent.

encystment The process of becoming encapsulated by a membranous sac.

endemic disease A native disease that prevails continuously in a geographic region.

endergonic reaction A chemical reaction that occurs with the absorption and storage of surrounding energy. Antonym: exergonic.

endocarditis An inflammation of the lining and valves of the heart. Often caused by infection with pyogenic cocci.

endocytosis The process whereby solid and liquid materials are taken into the cell through membrane invagination and engulfment into a vesicle.

endoenzyme An intracellular enzyme, as opposed to enzymes that are secreted.

endogenous Originating or produced within an organism or one of its parts.

endoplasmic reticulum An intracellular network of flattened sacs or tubules with or without ribosomes on their surfaces.

endospore A small, dormant, resistant derivative of a bacterial cell that germinates under favorable growth conditions into a vegetative cell. The bacterial genera *Bacillus* and *Clostridium* are typical sporeformers.

endosymbiosis Relationship in which a microorganism resides within a host cell and provides a benefit to the host cell.

endotoxin A bacterial intracellular toxin that is not ordinarily released (as is exotoxin). Endotoxin is composed of a phospholipid-polysaccharide complex that is an integral part of gram-negative bacterial cell walls. Endotoxins can cause severe shock and fever.

energy of activation The minimum energy input necessary for reactants to form products in a chemical reaction.

energy pyramid An ecological model that shows the energy flow among the organisms in a community. It is structured like the food pyramid, but shows how energy is reduced from one trophic level to another.

enriched medium A nutrient medium supplemented with blood, serum, or some growth factor to promote the multiplication of fastidious microorganisms.

enteric Pertaining to the intestine.

enteroinvasive Predisposed to invade the intestinal tissues.

enteropathogenic Pathogenic to the alimentary canal.

enterotoxin A bacterial toxin that specifically targets intestinal mucous membrane cells. Enterotoxigenic strains of *Escherichia coli* and *Staphylococcus aureus* are typical sources.

enveloped virus A virus whose nucleocapsid is enclosed by a membrane derived in part from the host cell. It usually contains exposed glycoprotein spikes specific for the virus.

enzyme A protein biocatalyst that facilitates metabolic reactions.

enzyme induction One of the controls on enzyme synthesis. This occurs when enzymes appear only when suitable substrates are present.

enzyme repression The inhibition of enzyme synthesis by the end product of a catabolic pathway.

eosinophil A leukocyte whose cytoplasmic granules readily stain with red eosin dye.

epidemic A sudden and simultaneous outbreak or increase in the number of cases of disease in a community.

epidemic parotitis (mumps) Disease caused by infection with paramyxovirus and marked by swelling in and around the parotid salivary glands.

epidemiology The study of the factors affecting the prevalence and spread of disease within a community.

epimastigote The trypanosomal form found in the tsetse fly or reduviid bug vector. Its flagellum originates near the nucleus, extends along an undulating membrane, and emerges from the anterior end.

Epstein-Barr virus (EBV) Herpesvirus linked to infectious mononucleosis, Burkitts lymphoma and nasopharyngeal carcinoma.

erysipelas An acute, sharply defined inflammatory disease specifically caused by hemolytic *Streptococcus*. The eruption is limited to the skin but can be complicated by serious systemic symptoms.

erysipeloid An inflammation resembling erysipelas but caused by *Erysipelothrix,* a gram-positive rod. The self-limited cellulitis that appears at the site of an infected wound, usually the hand, comes from handling contaminated fish or meat.

erythema An inflammatory redness of the skin.

erythroblastosis fetalis Hemolytic anemia of the newborn. The anemia comes from hemolysis of Rh-positive fetal erythrocytes by anti-Rh maternal antibodies. Erythroblasts are immature red blood cells prematurely released from the bone marrow.

erythrocytes (red blood cells) Blood cells involved in the transport of oxygen and carbon dioxide.

erythrogenic toxin An exotoxin produced by lysogenized group A strains of β-hemolytic streptococci that is responsible for the severe fever and rash of scarlet fever in the nonimmune individual. Also called a pyrogenic toxin.

eschar A dark, sloughing scab that is the lesion of anthrax and certain rickettsioses.

essential nutrient Any ingredient such as a certain amino acid, fatty acid, vitamin, or mineral that cannot be formed by an organism and must be supplied in the diet. A growth factor.

ester bond A covalent bond formed by reacting carboxylic acid with an OH group:

$$\begin{matrix} & & O \\ & & \| \\ (R & \!\!-\!\!C\!\!-\!\!O\!\!-\!\! & R') \end{matrix}$$

Olive and corn oils, lard, and butter fat are examples of triacylglycerols—esters formed between glycerol and three fatty acids.

estuary The intertidal zone where a river empties into the sea.

ethylene oxide A potent, highly water-soluble gas invaluable for gaseous sterilization of heat-sensitive objects such as plastics, surgical and diagnostic appliances, and spices. Potential hazards are related to its carcinogenic, metagenic, residual, and explosive nature. Ethylene oxide is rendered nonexplosive by mixing with 90% CO_2 or fluorocarbon.

etiologic agent The microbial cause of disease; the pathogen.

eubacteria Term used for non-archaea prokaryotes, stands for "true bacteria."

eucaryotic cell A cell that differs from a procaryotic cell chiefly by having a nuclear membrane (a well-defined nucleus), membrane-bounded subcellular organelles, and mitotic cell division.

Eukarya One of the three domains (sometimes called superkingdoms) of living organisms, as proposed by Woese; contains all eukaryotic organisms.

eutrophication The process whereby dissolved nutrients resulting from natural seasonal enrichment or industrial pollution of water cause overgrowth of algae and cyanobacteria to the detriment of fish and other large aquatic inhabitants.

evolution Scientific principle that states that living things change gradually through hundreds of millions of years, and these changes are expressed in structural and functional adaptations in each organism. Evolution presumes that those traits which favor survival are preserved and passed on to following generations, and those traits which do not favor survival are lost.

exanthem An eruption or rash of the skin.

exergonic A chemical reaction associated with the release of energy to the surroundings. Antonym: endergonic.

exfoliative toxin A poisonous substance that causes superficial cells of an epithelium to detach and be shed. Example: staphylococcal exfoliatin. Also called an epidermolytic toxin.

exocytosis The process that releases enveloped viruses from the membrane of the host's cytoplasm.

exoenzyme An extracellular enzyme chiefly for hydrolysis of nutrient macromolecules that are otherwise impervious to the cell membrane. It functions in saprobic decomposition of organic debris and can be a factor in invasiveness of pathogens.

exogenous Originating outside the body.

exon A stretch of eucaryotic DNA coding for a corresponding portion of mRNA that is translated into peptides. Intervening stretches of DNA that are not expressed are called introns. During transcription, exons are separated from introns and are spliced together into a continuous mRNA transcript.

exotoxin A toxin (usually protein) that is secreted and acts upon a specific cellular target. Examples: botulin, tetanospasmin, diphtheria toxin, and erythrogenic toxin.

exponential Pertaining to the use of exponents, numbers that are typically written as a superscript to indicate how many times a factor is to be multiplied. Exponents are used in scientific notation to render large, cumbersome numbers into small workable quantities.

extrapulmonary tuberculosis A condition in which tuberculosis bacilli have spread to organs other than the lungs.

extremophiles Organisms capable of living in harsh environments, such as extreme heat or cold.

F

facilitated diffusion The passive movement of a substance across a plasma membrane from an area of higher concentration to an area of lower concentration utilizing specialized carrier proteins.

facultative Pertaining to the capacity of microbes to adapt or adjust to variations; not obligate.

Example: The presence of oxygen is not obligatory for a facultative anaerobe to grow. *See* obligate.

family In the levels of classification, a mid-level division of organisms that groups more closely related organisms than previous levels. An order is divided into families.

fastidious Requiring special nutritional or environmental conditions for growth. Said of bacteria.

fecal coliforms Any species of gram negative lactose positive bacteria (primarily *Escherichia coli*) that live primarily in the intestinal tract and not the environment. Finding evidence of these bacteria in a water or food sample is substantial evidence of fecal contamination and potential for infection (see *coliform*).

feedback inhibition Temporary end to enzyme action caused by an end product molecule binding to the regulatory site and preventing the enzyme's active site from binding to its substrate.

fermentation The extraction of energy through anaerobic degradation of substrates into simpler, reduced metabolites. In large industrial processes, fermentation can mean any use of microbial metabolism to manufacture organic chemicals or other products.

fermentor A large tank used in industrial microbiology to grow mass quantities of microbes that can synthesize desired products. These devices are equipped with means to stir, monitor and harvest products such as drugs, enzymes, and proteins in very large quantities.

fertility (F′) factor Donor plasmid that allows synthesis of a pilus in bacterial conjugation. Presence of the factor is indicated by F^+, and lack of the factor is indicated by F^-.

filament A helical structure composed of proteins that is part of bacterial flagella.

filariasis Illnesses caused by infection by nematodes. These illnesses include river blindness, elephantiasis, and eye worm.

fimbria A short, numerous surface appendage on some bacteria that provides adhesion but not locomotion.

first line of defense Inborn non-specific immunologic barriers that block the invasion of pathogens at the portal of entry, thereby keeping them out of sterile body compartments.

fixation In microscopic slide preparation of tissue sections or bacterial smears, fixation pertains to rapid killing, hardening, and adhesion to the slide, while retaining as many natural characteristics as possible. Also refers to the assimilation of inorganic molecules into organic ones, as in carbon or nitrogen fixation.

flagellum A structure that is used to propel the organism through a fluid environment.

flora Beneficial or harmless resident bacteria commonly found on and/or in the human body.

fluid mosaic model A conceptualization of the molecular architecture of cellular membranes as a bilipid layer containing proteins. Membrane proteins are embedded to some degree in this bilayer, where they float freely about.

fluorescence The property possessed by certain minerals and dyes to emit visible light when excited by ultraviolet radiation. A fluorescent dye combined with specific antibody provides a sensitive test for the presence of antigen.

focal infection Occurs when an infectious agent breaks loose from a localized infection and is carried by the circulation to other tissues.

folliculitis An inflammatory reaction involving the formation of papules or pustules in clusters of hair follicles.

fomite Virtually any inanimate object an infected individual has contact with that can serve as a vehicle for the spread of disease.

food chain A simple straight-line feeding sequence among organisms in a community.

food fermentations Addition to and growth of known cultures of microorganisms in foods to produce desirable flavors, smells, or textures. Includes cheeses, breads, alcoholic beverages, and pickles.

food infection A form of food-borne illness associated with ingesting living pathogenic microbes that invade the intestine. The damage caused by microbes growing in the body cause the main symptoms, which are usually gastroenteritis (example is salmonellosis).

food intoxication A form of food-borne illness that is the result of ingesting microbial toxins given off by bacteria growing in the food. The toxins cause the main symptoms (example is botulism).

food poisoning Any form of illness acquired from ingesting foods; sources include microbes, chemicals, plants, and animals (also food-borne disease).

food pyramid A triangular summation of the consumers, producers, and decomposers in a community and how they are related by trophic, number, and energy parameters.

food web A complex network that traces all feeding interactions among organisms in a community (see *food chain*). This is considered to be a more accurate picture of food relationships in a community than a food chain.

formalin A 37% aqueous solution of formaldehyde gas; a potent chemical fixative and microbicide.

frameshift mutation An insertion or deletion mutation which changes the codon reading frame from the point of the mutation to the final codon. Almost always leads to a non-functional protein.

fructose One of the carbohydrates commonly referred to as sugars. Fructose is commonly fruit sugars.

functional group In chemistry, a particular molecular combination that reacts in predictable ways and confers particular properties on a compound. Examples: —COOH, —OH, —CHO.

fungicide A chemical that can kill fungal spores, hyphae and yeast.

fungus Heterotrophic unicellular or multicellular eucaryotic organism which may take the form of a larger macroscopic organism, as in the case of mushrooms, or a smaller microscopic organism, as in the case of yeasts and molds.

furuncle A boil; a localized pyogenic infection arising from a hair follicle.

G

Gaia Theory The concept that biotic and abiotic factors sustain suitable conditions for one another simply by their interactions. Named after the mythical Greek goddess of earth.

GALT Abbreviation for gut-associated lymphoid tissue. Includes Peyer's patches.

gamma globulin The fraction of plasma proteins high in immunoglobulins (antibodies). Preparations from pooled human plasma containing normal antibodies make useful passive immunizing agents against pertussis, polio, measles, and several other diseases.

gamma interferon A protein produced by a virally infected cell that induces production of antiviral substances in neighboring cells. This defense prevents the production and maturation of viruses and thus terminates the viral infection.

gas gangrene Disease caused by a clostridial infection of soft tissue or wound. The name refers to the gas produced by the bacteria growing in the tissue. Unless treated early, it is fatal. Also called myonecrosis.

gastroenteritis Inflammation of the lining of the stomach and intestine. May be caused by infection, intestinal disorders, or food poisoning.

gel electrophoresis A laboratory technique for separating DNA fragments according to length by employing electricity to force the DNA through a gel-like matrix typically made of agarose. Smaller DNA fragments move more quickly through the gel, thereby moving further than larger fragments during the same period of time.

gene A site on a chromosome that provides information for a certain cell function. A specific segment of DNA that contains the necessary code to make a protein or RNA molecule.

gene probe Short strands of single-stranded nucleic acid that hybridize specifically with complementary stretches of nucleotides on test samples and thereby serve as a tagging and identification device.

gene therapy The introduction of normal functional genes into people with genetic diseases such as sickle-cell anemia and cystic fibrosis. This is usually accomplished by a virus vector.

generation time Time required for a complete fission cycle—from parent cell to two new daughter cells. Also called doubling time.

genetic engineering A field involving deliberate alterations (recombinations) of the genomes of microbes, plants, and animals through special technological processes.

genetics The science of heredity.

genital warts A prevalent STD linked to some forms of cancer of the reproductive organs. Caused by infection with human papillomavirus.

genome The complete set of chromosomes and genes in an organism.

genotype The genetic makeup of an organism. The genotype is ultimately responsible for an organism's phenotype, or expressed characteristics.

genus In the levels of classification, the second most specific level. A family is divided into several genera.

germ free See *axenic*.

germicide An agent lethal to non-endospore-forming pathogens.

germ theory of disease A theory first originating in the 1800s which proposed that microorganisms can be the cause of diseases. The concept is actually so well established in the present time that it is considered a fact.

giardiasis Infection by the *Giardia* flagellate. Most common mode of transmission is contaminated food and water. Symptoms include diarrhea, abdominal pain, and flatulence.

gingivitis Inflammation of the gum tissue in contact with the roots of the teeth.

gluconeogenesis The formation of glucose (or glycogen) from noncarbohydrate sources such as protein or fat. Also called glyconeogenesis.

glucose One of the carbohydrates commonly referred to as sugars. Glucose is characterized by its 6-carbon structure.

glutaraldehyde A yellow acidic liquid used in antimicrobial control. Because glutaraldehyde kills spores it is considered a sterilant.

glycan A polysaccharide.

glycerol A 3-carbon alcohol, with three OH groups that serve as binding sites.

glycocalyx A filamentous network of carbohydrate-rich molecules that coats cells.

glycogen A glucose polymer stored by cells.

glycolysis The energy-yielding breakdown (fermentation) of glucose to pyruvic or lactic acid. It is often called anaerobic glycolysis because no molecular oxygen is consumed in the degradation.

glycosidic bond A bond that joins monosaccharides to form disaccharides and polymers.

gnotobiotic Referring to experiments performed on germ-free animals.

Golgi apparatus An organelle of eucaryotes that participates in packaging and secretion of molecules.

gonococcus Common name for *Neisseria gonorrhoeae*, the agent of gonorrhea.

graft Live tissue taken from a donor and transplanted into a recipient to replace damaged or missing tissues such as skin, bone, blood vessels.

graft vs. host disease (GVHD) A condition associated with a bone marrow transplant in which T cells in the transplanted tissue mount an immune response against the recipient's (host) normal tissues.

Gram stain A differential stain for bacteria useful in identification and taxonomy. Gram-positive organisms appear purple from crystal violet-mordant retention, whereas gram-negative organisms appear red after loss of crystal violet and absorbance of the safranin counterstain.

grana Discrete stacks of chlorophyll-containing thylakoids within chloroplasts.

granulocyte A mature leukocyte that contains noticeable granules in a Wright stain. Examples: neutrophils, eosinophils, and basophils.

granuloma A solid mass or nodule of inflammatory tissue containing modified macrophages and lymphocytes. Usually a chronic pathologic process of diseases such as tuberculosis or syphilis.

Grave disease A malfunction of the thyroid gland in which autoantibodies directed at thyroid cells stimulate an overproduction of thyroid hormone (hyperthyroidism).

greenhouse effect The capacity to retain solar energy by a blanket of atmospheric gases that redirects heat waves back toward the earth.

group translocation A form of active transport in which the substance being transported is altered during transfer across a plasma membrane.

growth curve A graphical representation of the change in population size over time. This graph has four periods known as lag phase, exponential or log phase, stationary phase, and death phase.

growth factor An organic compound such as a vitamin or amino acid that must be provided in the diet to facilitate growth. An essential nutrient.

guanine (G) One of the nitrogen bases found in DNA and RNA in the purine form.

Guillain-Barré syndrome A neurological complication of influenza vaccination. Approximately one in 100,000 recipients of the vaccine will develop this autoimmune disorder which is marked by varying degrees of weakness and sensory loss.

gumma A nodular, infectious granuloma characteristic of tertiary syphilis.

gut-associated lymphoid tissue (GALT) A collection of lymphoid tissue in the gastrointestinal tract which includes the appendix, the lacteals, and Peyer's patches.

H

habitat The environment to which an organism is adapted.

halogens A group of related chemicals with antimicrobial applications. The halogens most often used in disinfectants and antiseptics are chlorine and iodine.

halophile A microbe whose growth is either stimulated by salt or requires a high concentration of salt for growth.

Hansen disease A chronic, progressive disease of the skin and nerves caused by infection by a mycobacterium that is a slow-growing, strict parasite. Hansen disease is the preferred name for leprosy.

H antigen The flagellar antigen of motile bacteria. *H* comes from the German word *hauch,* which denotes the appearance of spreading growth on solid medium.

haploid Having a single set of unpaired chromosomes, such as occurs in gametes and certain microbes.

hapten An incomplete or partial antigen. Although it constitutes the determinative group and can bind antigen, hapten cannot stimulate a full immune response without being carried by a larger protein molecule.

hashimoto thyroiditis An autoimmune disease of the thyroid gland that damages the thyroid follicle cells and results in decreased production of thyroid hormone (hypothyroidism).

hay fever A form of atopic allergy marked by seasonal acute inflammation of the conjunctiva and mucous membranes of the respiratory passages. Symptoms are irritative itching and rhinitis.

helical Having a spiral or coiled shape. Said of certain virus capsids and bacteria.

helminth A term that designates all parasitic worms.

helper T cell A class of thymus-stimulated lymphocytes that facilitate various immune activities such as assisting B cells and macrophages. Also called a T helper cell.

hemagglutinin A molecule that causes red blood cells to clump or agglutinate. Often found on the surfaces of viruses.

hemolysin Any biological agent that is capable of destroying red blood cells and causing the release of hemoglobin. Many bacterial pathogens produce exotoxins that act as hemolysins.

hemolytic disease Incompatible Rh factor between mother and fetus causes maternal antibodies to attack the fetus and trigger complement-mediated lysis in the fetus.

hemolyze When red blood cells burst and release hemoglobin pigment.

hemopoiesis The process by which the various types of blood cells are formed, such as in the bone marrow.

hepadnavirus Enveloped DNA viruses with a predisposition to affect the liver. Hepatitis B is the most serious form.

hepatitis Inflammation and necrosis of the liver, often the result of viral infection.

hepatitis A virus Enterovirus spread by contaminated food responsible for short-term (infectious) hepatitis.

hepatitis B virus (HBV) Hepadnavirus that is the causative agent of serum hepatitis.

hepatocellular carcinoma A liver cancer associated with infection with hepatitis B virus.

hepatocyte A liver cell.

heredity Genetic inheritance.

herd immunity The status of collective acquired immunity in a population that reduces the likelihood that nonimmune individuals will contract and spread infection. One aim of vaccination is to induce herd immunity.

hermaphroditic Containing the sex organs for both male and female in one individual.

herpes zoster A recurrent infection caused by latent chickenpox virus. Its manifestation on the skin tends to correspond to dermatomes and to occur in patches that "girdle" the trunk. Also called shingles.

herpetic keratitis Corneal or conjunctival inflammation due to herpesvirus type 1.

heterophile antigen An antigen present in a variety of phylogenetically unrelated species. Example: red blood cell antigens and the glycocalyx of bacteria.

heterotroph An organism that relies upon organic compounds for its carbon and energy needs.

hexose A 6-carbon sugar such as glucose and fructose.

hierarchies Levels of power. Arrangement in order of rank.

histamine A cytokine released when mast cells and basophils release their granules. An important mediator of allergy, its effects include smooth muscle contraction, increased vascular permeability, and increased mucus secretion.

histiocyte Another term for macrophage.

histone Proteins associated with eucaryotic DNA. These simple proteins serve as winding spools to compact and condense the chromosomes.

histoplasmin The antigenic extract of *Histoplasma capsulatum,* the causative agent of histoplasmosis. The preparation is used in skin tests for diagnosis and for conducting surveys to determine the geographic distribution of the fungus.

HLA An abbreviation for **h**uman **l**eukocyte **a**ntigens. This closely linked cluster of genes programs for cell surface glycoproteins that control immune interactions between cells and is involved in rejection of allografts. Also called the major histocompatibility complex (MHC).

holoenzyme An enzyme complete with its apoenzyme and cofactors.

hops The ripe, dried fruits of the hop vine (*Humulus lupulus*) that is added to beer wort for flavoring.

host Organism in which smaller organisms or viruses live, feed, and reproduce.

host range The limitation imposed by the characteristics of the host cell on the type of virus that can successfully invade it.

human diploid cell vaccine A vaccine made suing cell culture that is currently the vaccine of choice for preventing infection by rabies virus.

human herpesvirus-6 (HHV-6) The herpesvirus which causes roseola and may possibly be linked to chronic neurological diseases.

human immunodeficiency virus (HIV) A retro virus that causes acquired immunodeficiency syndrome (AIDS).

human papillomavirus (HPV) A group of DNA viruses whose members are responsible for common, plantar and genital warts.

humoral immunity Protective molecules (mostly B lymphocytes) carried in the fluids of the body.

hybridization A process that matches complementary strands of nucleic acid (DNA-DNA, RNA-DNA, RNA-RNA). Used for locating specific sites or types of nucleic acids.

hybridoma An artificial cell line that produces monoclonal antibodies. It is formed by fusing (hybridizing) a normal antibody-producing cell with a cancer cell, and it can produce pure antibody indefinitely.

hydatid cyst A sac, usually in liver tissue, containing fluid and larval stages of the echinococcus tapeworm.

hydration The addition of water as in the coating of ions with water molecules as ions enter into aqueous solution.

hydrogen bond A weak chemical bond formed by the attraction of forces between molecules or atoms—in this case, hydrogen and either oxygen or nitrogen. In this type of bond, electrons are not shared, lost, or gained.

hydrologic cycle The continual circulation of water between hydrosphere, atmosphere, and lithosphere.

hydrolase An enzyme that catalyzes the cleavage of a bond with the additions of —H and —OH (parts of a water molecule) at the separation site.

hydrolysis A process in which water is used to break bonds in molecules. Usually occurs in conjunction with an enzyme.

hydrophilic The property of attracting water. Molecules that attract water to their surface are called hydrophilic.

hydrophobic The property of repelling water. Molecules that repel water are called hydrophobic.

hydrosphere That part of the biosphere which encompasses water-containing environments such as oceans, lakes, rivers.

hypertonic Having a greater osmotic pressure than a reference solution.

hyphae The tubular threads that make up filamentous fungi (molds). This web of branched and intertwining fibers is called a mycelium.

hypogammaglobulinemia An inborn disease in which the gamma globulin (antibody) fraction of serum is greatly reduced. The condition is associated with a high susceptibility to pyogenic infections.

hyposensitization A therapeutic exposure to known allergens designed to build tolerance and eventually prevent allergic reaction.

hypothesis A tentative explanation of what has been observed or measured.

hypotonic Having a lower osmotic pressure than a reference solution.

I

icosahedron A regular geometric figure having 20 surfaces that meet to form 12 corners. Some virions have capsids that resemble icosahedral crystals.

immune complex reaction Type III hypersensitivity of the immune system. It is characterized by the reaction of soluble antigen with antibody, and the deposition of the resulting complexes in basement membranes of epithelial tissue.

immune surveillance The continual function of macrophages, cytotoxic T cells, and natural killer cells in identifying and destroying cancer cells within the body.

immunity An acquired resistance to an infectious agent due to prior contact with that agent.

immunoassays Extremely sensitive tests that permit rapid and accurate measurement of trace antigen or antibody.

immunocompetence The ability of the body to recognize and react with multiple foreign substances.

immunodeficiency Immune function is incompletely developed, suppressed, or destroyed.

immunodeficiency disease A form of immunopathology in which white blood cells are unable to mount a complete, effective immune response, which results in recurrent infections. Examples would be AIDS and agammaglobulinemia.

immunogen Any substance that induces a state of sensitivity or resistance after processing by the immune system of the body.

immunoglobulin The chemical class of proteins to which antibodies belong.

immunology The study of the system of body defenses that protect against infection.

immunopathology The study of disease states associated with overreactivity or underreactivity of the immune response.

immunotherapy Preventing or treating infectious diseases by administering substances that produce artificial immunity. May be active or passive.

immunotoxin An artificially prepared toxin attached to a monoclonal antibody specific to a tumor cell. Immunotoxins are used in cancer therapy to destroy tumors.

IMViC Abbreviation for four identification tests: **i**ndole production, **m**ethyl red test, **V**oges-Proskauer test (**i** inserted to concoct a wordlike sound), and **c**itrate as a sole source of carbon. This test was originally developed to distinguish between *Enterobacter aerogenes* (associated with soil) and *Escherichia coli* (a fecal coliform).

incidence In epidemiology, the number of new cases of a disease occurring during a period.

incineration Destruction of microbes by subjecting them to extremes of dry heat. Microbes are reduced to ashes and gas by this process.

inclusion A relatively inert body in the cytoplasm such as storage granules, glycogen, fat, or some other aggregated metabolic product.

incubate To isolate a sample culture in a temperature-controlled environment to encourage growth.

incubation period The period from the initial contact with an infectious agent to the appearance of the first symptoms.

indicator bacteria In water analysis, any easily cultured bacteria that may be found in the intestine and can be used as an index of fecal contamination. The category includes coliforms and enterococci. Discovery of these bacteria in a sample means that pathogens may also be present.

induced mutation Any alteration in DNA that occurs as a consequence of exposure to chemical or physical mutagens.

inducible enzyme An enzyme that increases in amount in direct proportion to the amount of substrate present.

inducible operon An operon that under normal circumstances is not transcribed. The presence of a specific inducer molecule can cause transcription of the operon to begin.

induction Process by which an individual accumulates data or facts and then formulates a general hypothesis that accounts for those facts.

infantile diarrhea An infection in newborns frequently caused by *E.coli* infection and often having a high mortality rate.

infection The entry, establishment, and multiplication of pathogenic organisms within a host.

infectious disease The state of damage or toxicity in the body caused by an infectious agent.

inflammation A natural, nonspecific response to tissue injury that protects the host from further damage. It stimulates immune reactivity and blocks the spread of an infectious agent.

inoculation The implantation of microorganisms into or upon culture media.

inorganic chemicals Molecules that lack the basic framework of the elements of carbon and hydrogen.

interferon Naturally occurring polypeptides produced by fibroblasts and lymphocytes that can block viral replication and regulate a variety of immune reactions.

interleukin A macrophage agent (interleukin-1, or IL-1) that stimulates lymphocyte function. Stimulated T cells release yet another interleukin (IL-2), which amplifies T-cell response by stimulating additional T cells. T helper cells stimulated by IL-2 stimulate B-cell proliferation and promote antibody production.

intermittent sterilization (tyndallization) A sterilization technique for items that cannot withstand the temperature/pressure extremes of an autoclave. Materials are subjected to free flowing steam (which kills vegetative cells) followed by an incubation period (which allows endospores to germinate). The cycle is repeated three times with the hope that, at completion, all spores will have germinated and all vegetative cells will have died.

intoxication Poisoning that results from the introduction of a toxin into body tissues through ingestion or injection.

intron The segments on split genes of eucaryotes that do not code for polypeptide. They can have regulatory functions. See *exon*.

in utero Literally means "in the uterus"; pertains to events or developments occurring before birth.

in vitro Literally means "in glass," signifying a process or reaction occurring in an artificial environment, as in a test tube or culture medium.

in vivo Literally means "in a living being," signifying a process or reaction occurring in a living thing.

iodophor A combination of iodine and an organic carrier that is a moderate-level disinfectant and antiseptic.

ion An unattached, charged particle.

ionic bond A chemical bond in which electrons are transferred and not shared between atoms.

ionization The aqueous dissociation of an electrolyte into ions.

ionizing radiation Radiant energy consisting of short-wave electromagnetic rays (X ray) or high-speed electrons that cause dislodgment of electrons on target molecules and create ions.

irradiation The application of radiant energy for diagnosis, therapy, disinfection, or sterilization.

irregular (in shape) Refers to bacteria that stain unevenly or display cell to cell variation in size and/or shape (pleomorphic) within a single species.

irritability Capacity of cells to respond to chemical, mechanical, or light stimuli. This property helps cells adapt to the environment and obtain nutrients.

isograft Transplanted tissue from one monozygotic twin to the other; transplants between highly inbred animals that are genetically identical.

isolation The separation of microbial cells by serial dilution or mechanical dispersion on solid media to create discrete colonies.

isotonic Two solutions having the same osmotic pressure such that, when separated by a semipermeable membrane, there is no net movement of solvent in either direction.

isotope A version of an element that is virtually identical in all chemical properties to another version except that their atoms have slightly different atomic masses.

J

jaundice The yellowish pigmentation of skin, mucous membranes, sclera, deeper tissues, and excretions due to abnormal deposition of bile pigments. Jaundice is associated with liver infection, as with hepatitis B virus and leptospirosis.

J chain A small molecule with high sulfhydryl content that secures the heavy chains of IgM to form a pentamer and the heavy chains of IgA to form a dimer.

K

Kaposi sarcoma A malignant or benign neoplasm that appears as multiple hemorrhagic sites on the skin, lymph nodes, and viscera and apparently involves the metastasis of abnormal blood vessel cells. It is a clinical feature of AIDS.

keratitis Inflammation of the cornea.

keratoconjunctivitis Inflammation of the conjunctiva and cornea.

mutagen Any agent that induces genetic mutation. Examples: certain chemical substances, ultraviolet light, radioactivity.

mutant strain A subspecies of microorganism which has undergone a mutation, causing expression of a trait that differs from other members of that species.

mutation A permanent inheritable alteration in the DNA sequence or content of a cell.

mutualism Organisms living in an obligatory, but mutually beneficial, relationship.

mycelium The filamentous mass that makes up a mold. Composed of hyphae.

mycetoma A chronic fungal infection usually afflicting the feet, typified by swelling and multiple draining lesions. Example: maduromycosis or Madura foot.

mycoplasma The smallest, self-replicating microorganisms. Mycoplasma naturally lack a cell wall. Most species are parasites of animals and plants.

mycorrhizae Various species of fungi adapted in an intimate, mutualistic relationship to plant roots.

mycosis Any disease caused by a fungus.

mycotoxicosis Illness resulting from eating poisonous fungi.

myonecrosis Another name for gas gangrene.

N

NAD/NADH Abbreviations for the oxidized/reduced forms of nicotinamide adenine dinucleotide, an electron carrier. Also known as the vitamin niacin.

nanobes Cell-like particles, found in sediments and other geologic deposits, that some scientists speculate are the smallest bacteria. Short for nanobacteria.

narrow-spectrum Denotes drugs that are selective and limited in their effects. For example, they inhibit either gram-negative or gram-positive bacteria, but not both.

nasopharyngeal carcinoma A malignancy of epithelial cells that occurs in older Chinese and African men and is associated with exposure to Epstein-Barr virus.

natural selection A process in which the environment places pressure on organisms to adapt and survive changing conditions. Only the survivors will be around to continue the life cycle and contribute their genes to future generations. This is considered a major factor in evolution of species.

necrosis A pathologic process in which cells and tissues die and disintegrate.

negative feedback Enzyme regulation of metabolism by the end product of a multienzyme system that blocks the action of a "pacemaker" enzyme at or near the beginning of the pathway.

negative stain A staining technique that renders the background opaque or colored and leaves the object unstained so that it is outlined as a colorless area.

nematode A common name for helminths called roundworms.

neoplasm A synonym for tumor.

neuramidase A glycoprotein found in the envelope of orthomyxoviruses which facilitates release of new viruses from the host cell.

neurotropic Having an affinity for the nervous system. Most likely to affect the spinal cord.

neutron An electrically neutral particle in the nuclei of all atoms except hydrogen.

neutralization The process of combining an acid and a base until they reach a balanced proportion, with a pH value close to 7.

neutrophil A mature granulocyte present in peripheral circulation, exhibiting a multilobular nucleus and numerous cytoplasmic granules that retain a neutral stain. The neutrophil is an active phagocytic cell in bacterial infection.

niche In ecology, an organism's biological role in or contribution to its community.

night soil An archaic euphemism for human excrement collected for fertilizing crops.

nitrification Phase of the nitrogen cycle in which ammonium is oxidized.

nitrogen base A ringed compound of which pyrimidines and purines are types.

nitrogen cycle The pathway followed by the element nitrogen as it circulates from inorganic sources in the non-living environment to living things and back to the non-living environment. The longtime reservoir is nitrogen gas in the atmosphere.

nitrogen fixation A process occurring in certain bacteria in which atmospheric N_2 gas is converted to a form (NH_4) usable by plants.

nitrogenous base A nitrogen-containing molecule found in DNA and RNA that provides the basis for the genetic code. Adenine, guanine and cytosine are found in both DNA and RNA while thymine is found exclusively in DNA and uracil is found exclusively in RNA.

nomenclature A set system for scientifically naming organisms, enzymes, anatomical structures, etc.

noncoliforms Lactose-negative enteric bacteria in normal flora.

noncommunicable An infectious disease that does not arrive through transmission of an infectious agent from host to host.

nonionizing radiation Method of microbial control, best exemplified by ultraviolet light, that causes the formation of abnormal bonds within the DNA of microbes, increasing the rate of mutation. The primary limitation of nonionizing radiation is its inability to penetrate beyond the surface of an object.

nonpolar A term used to describe an electrically neutral molecule formed by covalent bonds between atoms that have the same or similar electronegativity.

non-self Molecules recognized by the immune system as containing foreign markers, indicating a need for immune response.

nonsense codon A triplet of mRNA bases that does not specify an amino acid but signals the end of a polypeptide chain.

nonsense mutation A mutation that changes an amino acid-producing codon into a stop codon, leading to premature termination of a protein.

normal flora The native microbial forms that an individual harbors.

Norwalk agent One of a group of Calciviruses, Norwalk agent causes gastrointestinal distress and is commonly transmitted in schools, camps, cruise-ships and nursing homes.

nosocomial infection An infection not present upon admission to a hospital but incurred while being treated there.

nucleocapsid In viruses, the close physical combination of the nucleic acid with its protective covering.

nucleoid The basophilic nuclear region or nuclear body that contains the bacterial chromosome.

nucleolus A granular mass containing RNA that is contained within the nucleus of a eucaryotic cell.

nucleosome Structure in the packaging of DNA. Formed by the DNA strands wrapping around the histone protein to form nucleus bodies arranged like beads on a chain.

nucleotide The basic structural unit of DNA and RNA; each nucleotide consists of a phosphate, a sugar (ribose in RNA, deoxyribose in DNA), and a nitrogenous base such as adenine, guanine, cytosine, thymine (DNA only) or uracil (RNA only).

nucleus The central core of an atom, composed of protons and neutrons.

nucleotide A composite of a nitrogen base (purine or pyrimidine), a 5-carbon sugar (ribose or deoxyribose), and a phosphate group; a unit of nucleic acids.

numerical aperture In microscopy, the amount of light passing from the object and into the object in order to maximize optical clarity and resolution.

nutrient Any chemical substance that must be provided to a cell for normal metabolism and growth. Macronutrients are required in large amounts, and micronutrients in small amounts.

nutrition The acquisition of chemical substances by a cell or organism for use as an energy source or as building blocks of cellular structures.

O

obligate Without alternative; restricted to a particular characteristic. Example: An obligate parasite survives and grows only in a host; an obligate aerobe must have oxygen to grow; an obligate anaerobe is destroyed by oxygen.

Okazaki fragment In replication of DNA, a segment formed on the lagging strand in which biosynthesis is conducted in a discontinuous manner dictated by the $5' \rightarrow 3'$ DNA polymerase orientation.

oligodynamic action A chemical having antimicrobial activity in minuscule amounts. Example: Certain heavy metals are effective in a few parts per billion.

oligonucleotides Short pieces of DNA or RNA that are easier to handle than long segments.

oligotrophic Nutrient-deficient ecosystem.

oncogene A naturally occurring type of gene that when activated can transform a normal cell into a cancer cell.

oncology The study of neoplasms, their cause, disease characteristics, and treatment.

oocyst The encysted form of a fertilized macrogamete or zygote; typical in the life cycles of apicomplexan parasites.

operator In an operon sequence, the DNA segment where transcription of structural genes is initiated.

operon A genetic operational unit that regulates metabolism by controlling mRNA production. In sequence, the unit consists of a regulatory gene, inducer or repressor control sites, and structural genes.

opportunistic In infection, ordinarily nonpathogenic or weakly pathogenic microbes that cause disease primarily in an immunologically compromised host.

opportunistic pathogen See *opportunistic infection.*

opsonization The process of stimulating phagocytosis by affixing molecules (opsonins such as antibodies and complement) to the surfaces of foreign cells or particles.

optimum temperature The temperature at which a species shows the most rapid growth rate.

orbitals The pathways of electrons as they rotate around the nucleus of an atom.

order In the levels of classification, the division of organisms that follows class. Increasing similarity may be noticed among organisms assigned to the same order.

organelle A small component of eucaryotic cells that is bounded by a membrane and specialized in function.

organic chemicals Molecules that contain the basic framework of the elements carbon and hydrogen.

ornithosis Worldwide zoonosis carried in the latent state in wild and domesticated birds; caused by *Chlamydia psittacosis.*

osmophile A microorganism that thrives in a medium having high osmotic pressure.

osmosis The diffusion of water across a selectively permeable membrane in the direction of lower water concentration.

osteomyelitis A focal infection of the internal structures of long bones, leading to pain and inflammation. Often caused by *Staphylococcus aureus.*

oxidation In chemical reactions, the loss of electrons by one reactant.

oxidation-reduction Redox reactions, in which paired sets of molecules participate in electron transfers.

oxidative phosphorylation The synthesis of ATP using energy given off during the electron transport phase of respiration.

oxygenic Any reaction that gives off oxygen; usually in reference to the result of photosynthesis in eucaryotes and cyanobacteria.

P

palindrome A word, verse, number, or sentence that reads the same forward or backward. Palindromes of nitrogen bases in DNA have genetic significance as transposable elements, as regulatory protein targets, and in DNA splicing.

palisades The characteristic arrangement of *Corynebacterium* cells resembling a row of fence posts and created by snapping.

pandemic A disease afflicting an increased proportion of the population over a wide geographic area (often worldwide).

pannus The granular membrane occurring on the cornea in trachoma.

papilloma Benign, squamous epithelial growth commonly referred to as a wart.

papule An elevation of skin that is small, demarcated, firm, and usually conical.

parainfluenza A respiratory disease (croup) caused by infection with Paramyxovirus.

parasite An organism that lives on or within another organism (the host), from which it obtains nutrients and enjoys protection. The parasite produces some degree of harm in the host.

parenteral Administering a substance into a body compartment other than through the gastrointestinal tract, such as via intravenous, subcutaneous, intramuscular, or intramedullary injection.

paroxysmal Events characterized by sharp spasms or convulsions; sudden onset of a symptom such as fever and chills.

particle radiation Radiation in which high energy particles such as protons, neutrons and electrons provide energy which can be used for antimicrobial purposes.

passive carrier Persons who mechanically transfer a pathogen without ever being infected by it. For example, a health care worker who doesn't wash his/her hands adequately between patients.

passive immunity Specific resistance that is acquired indirectly by donation of preformed immune substances (antibodies) produced in the body of another individual.

passive transport Nutrient transport method that follows basic physical laws and does not require direct energy input from the cell.

pasteurization Heat treatment of perishable fluids such as milk, fruit juices, or wine to destroy heat-sensitive vegetative cells, followed by rapid chilling to inhibit growth of survivors and germination of spores. It prevents infection and spoilage.

pathogen Any agent, usually a virus, bacterium, fungus, protozoan, or helminth, that causes disease.

pathogenicity The capacity of microbes to cause disease.

pathology The structural and physiological effects of disease on the body.

pellicle A membranous cover; a thin skin, film, or scum on a liquid surface; a thin film of salivary glycoproteins that forms over newly cleaned tooth enamel when exposed to saliva.

pelvic inflammatory disease (PID) An infection of the uterus and fallopian tubes that has ascended from the lower reproductive tract. Caused by gonococci and chlamydias.

penicillinase An enzyme that hydrolyzes penicillin; found in penicillin-resistant strains of bacteria.

penicillins A large group of naturally occurring and synthetic antibiotics produced by *Penicillium* mold and active against the cell wall of bacteria.

pentose A monosaccharide with five carbon atoms per molecule. Examples: arabinose, ribose, xylose.

peptidase An enzyme that can hydrolyze the peptide bonds of a peptide chain.

peptide Molecule composed of short chains of amino acids, such as a dipeptide (two amino acids), a tripeptide (three), and a tetrapeptide (four).

peptide bond The covalent union between two amino acids that forms between the amine group of one and the carboxyl group of the other. The basic bond of proteins.

peptidoglycan A network of polysaccharide chains cross-linked by short peptides that forms the rigid part of bacterial cell walls. Gram-negative bacteria have a smaller amount of this rigid structure than do gram-positive bacteria.

perinatal In childbirth, occurring before, during, or after delivery.

period of invasion The period during a clinical infection when the infectious agent multiplies at high levels, exhibits its greatest toxicity and becomes well established in the target tissues.

periodontal Involving the structures that surround the tooth.

periplasmic space The region between the cell wall and cell membrane of the cell envelopes of gram-negative bacteria.

peritrichous In bacterial morphology, having flagella distributed over the entire cell.

petechiae Minute hemorrhagic spots in the skin that range from pinpoint- to pinhead-sized.

pertussis Infection by *Bordetella pertussis.* A highly communicable disease that causes acute respiratory syndrome. Pertussis can be life-threatening in infants, but vaccination on the recommended schedule can prevent infection. Also called whooping cough.

Peyer's patches Oblong lymphoid aggregates of the gut located chiefly in the wall of the terminal and small intestine. Along with the tonsils and appendix, Peyer's patches make up the gut-associated lymphoid tissue that responds to local invasion by infectious agents.

pH The symbol for the negative logarithm of the H ion concentration; p (power) or $[H^+]_{10}$. A system for rating acidity and alkalinity.

phage A bacteriophage; a virus that specifically parasitizes bacteria.

phagocytosis A type of endocytosis in which the cell membrane actively engulfs large particles or cells into vesicles.

phagolysosome A body formed in a phagocyte, consisting of a union between a vesicle containing the ingested particle (the phagosome) and a vacuole of hydrolytic enzymes (the lysosome).

phenetic Based on phenotype, or expression of traits.

phenotype The observable characteristics of an organism produced by the interaction between its genetic potential (genotype) and the environment.

phlebotomine Pertains to a genus of very small midges or blood-sucking (phlebotomous) sand flies and to diseases associated with those vectors such as kala-azar, Oroya fever, and cutaneous leishmaniasis.

phosphate An acidic salt containing phosphorus and oxygen that is an essential inorganic component of DNA, RNA, and ATP.

phospholipid A class of lipids that compose a major structural component of cell membranes.

phosphorylation Process in which inorganic phosphate is added to a compound.

photic zone The aquatic stratum from the surface to the limits of solar light penetration.

photoactivation (light repair) A mechanism for repairing DNA with ultraviolet light-induced mutations using an enzyme (photolyase) that is activated by visible light.

photoautotroph An organism that utilizes light for its energy and carbon dioxide chiefly for its carbon needs.

photolysis Literally, splitting water with light. In photosynthesis, this step frees electrons and gives off O_2.

photon A subatomic particle released by electromagnetic sources such as radiant energy (sunlight). Photons are the ultimate source of energy for photosynthesis.

photophosphorylation The process of electron transport during photosynthesis that results in the synthesis of ATP from ADP.

photosynthesis A process occurring in plants, algae, and some bacteria that traps the sun's energy and converts it to ATP in the cell. This energy is used to fix CO_2 into organic compounds.

phototrophs Microbes that use photosynthesis to feed.

phylum In the levels of classification, the third level of classification from general to more specific. Each kingdom is divided into numerous phyla. Sometimes referred to a division.

physiology The study of the function of an organism.

phytoplankton The collection of photosynthetic microorganisms (mainly algae and cyanobacteria) that float in the upper layers of aquatic habitats where sun penetrates. These microbes are the basis of aquatic food pyramids and together with zooplankton, make up the plankton.

pili Small, stiff filamentous appendages in gram-negative bacteria that function in DNA exchange during bacterial conjugation.

pinocytosis The engulfment, or endocytosis, of liquids by extensions of the cell membrane.

plague Zoonotic disease caused by infection with *Yersinia pestis*. The pathogen is spread by flea vectors and harbored by various rodents.

plankton Minute animals (zooplankton) or plants (phytoplankton) that float and drift in the limnetic zone of bodies of water.

plantar warts Deep, painful warts on the soles of the feet as a result of infection by human papillomavirus.

plaque In virus propagation methods, the clear zone of lysed cells in tissue culture or chick embryo membrane that corresponds to the area containing viruses. In dental application, the filamentous mass of microbes that adheres tenaciously to the tooth and predisposes to caries, calculus, or inflammation.

plasma The carrier fluid element of blood.

plasma cell A progeny of an activated B cell that actively produces and secretes antibodies.

plasmids Extrachromosomal genetic units characterized by several features. A plasmid is a double-stranded DNA that is smaller than and replicates independently of the cell chromosome; it bears genes that are not essential for cell growth; it can bear genes that code for adaptive traits; and it is transmissible to other bacteria.

platelets Formed elements in the blood which develop when megakaryocytes disintegrate. Platelets are involved in hemostasis and blood clotting.

pleomorphism Normal variability of cell shapes in a single species.

pluripotential Stem cells having the developmental plasticity to give rise to more than one type. Example: undifferentiated blood cells in the bone marrow.

***pneumocystis (carinii) jiroveci* pneumonia** A severe, acute lung infection caused by a fungus that is the leading cause of morbidity and mortality of AIDS patients.

pneumonia An inflammation of the lung leading to accumulation of fluid and respiratory compromise.

pneumococcus Common name for *Streptococcus pneumoniae*, the major cause of bacterial pneumonia.

pneumonic plague The acute, frequently fatal form of pneumonia caused by *Yersinia pestis*.

point mutation A change that involves the loss, substitution, or addition of one or a few nucleotides.

polar Term to describe a molecule with an asymmetrical distribution of charges. Such a molecule has a negative pole and a positive pole.

poliomyelitis An acute enteroviral infection of the spinal cord that can cause neuromuscular paralysis.

polyclonal In reference to a collection of antibodies with mixed specificities that arose from more than one clone of B cells.

polymer A macromolecule made up of a chain of repeating units. Examples: starch, protein, DNA.

polymerase An enzyme that produces polymers through catalyzing bond formation between building blocks (polymerization).

polymerase chain reaction (PCR) A technique that amplifies segments of DNA for testing. Using denaturation, primers, and heat-resistant DNA polymerase, the number can be increased several million-fold.

polymorphonuclear leukocytes (PMNLs) White blood cells with variously shaped nuclei. Although this term commonly denotes all granulocytes, it is used especially for the neutrophils.

polymyxin A mixture of antibiotic polypeptides from *Bacillus polymyxa* that are particularly effective against gram-negative bacteria.

polypeptide A relatively large chain of amino acids linked by peptide bonds.

polyribosomal complex An assembly line for mass production of proteins composed of a chain of ribosomes involved in mRNA transcription.

polysaccharide A carbohydrate that can be hydrolyzed into a number of monosaccharides. Examples: cellulose, starch, glycogen.

population A group of organisms of the same species living simultaneously in the same habitat. A group of different populations living together constitutes the community level.

porin Transmembrane proteins of the outer membrane of gram-negative cells that permit transport of small molecules into the periplasmic space but bar the penetration of larger molecules.

potable Describing water that is relatively clear, odor-free, and safe to drink.

portal of entry Characteristic route of entry for an infectious agent; typically a cutaneous or membranous route.

portal of exit Characteristic route through which a pathogen departs from the host organism.

positive stain Technique in which dye affixes to a specimen and imparts color to it. It takes advantage of the ready binding of bacterial cells to dyes.

pox The thick, elevated pustular eruptions of various viral infections. Also called pocks.

PPNG Penicillinase producing *Neisseria gonorrhoeae*.

prevalence The total cumulative number of cases of a disease in a certain area and time period.

primary infection An initial infection in a previously healthy individual that is later complicated by an additional (secondary) infection.

primary pulmonary infection (PPI) Disease that results from the inhalation of fungal spores and their germination in the lungs. This may serve as a focus of infection that spreads throughout the body.

primary response The first response of the immune system when exposed to an antigen.

primary structure Initial protein organization described by type, number, and order of amino acids in the chain. The primary structure varies extensively from protein to protein.

primers Synthetic oligonucleotides of known sequence that serve as landmarks to indicate where DNA amplification will begin.

prion A concocted word to denote "proteinaceous infectious agent"; a cytopathic protein associated with the slow-virus spongiform encephalopathies of humans and animals.

probes Small fragments of single-stranded DNA (RNA) that are known to be complementary to the specific sequence of DNA being studied.

procaryotic cell Small cells, lacking special structures such as a nucleus and organelles. All procaryotes are microorganisms.

prodromium A short period of mild symptoms occurring at the end of the period of incubation. It indicates the onset of an infection.

producer An organism that synthesizes complex organic compounds from simple inorganic molecules. Examples would be photosynthetic microbes and plants. These organisms are solely responsible for originating food pyramids and are the basis for life on earth (also called autotroph).

proglottid The egg-generating segment of a tapeworm that contains both male and female organs.

progressive multifocal leukoencephalopathy An uncommon, fatal complication of infection with JC virus (polyoma virus).

promastigote A morphological variation of the trypanosome parasite responsible for leishmaniasis.

promoter Part of an operon sequence. The DNA segment that is recognized by RNA polymerase as the starting site for transcription.

promoter region The site composed of a short signaling DNA sequence that RNA polymerase recognizes and binds to commence transcription.

properdin pathway A normal serum protein involved in the alternate complement pathway that leads to nonspecific lysis of bacterial cells and viruses.

prophage A lysogenized bacteriophage; a phage that is latently incorporated into the host chromosome instead of undergoing viral replication and lysis.

prophylactic Any device, method, or substance used to prevent disease.

prostaglandin A hormonelike substance that regulates many body functions. Prostaglandin comes from a family of organic acids containing 5-carbon rings that are essential to the human diet.

protease inhibitors Drugs that act to prevent the assembly of functioning viral particles.

protein Predominant organic molecule in cells, formed by long chains of amino acids.

proton An elementary particle that carries a positive charge. It is identical to the nucleus of the hydrogen atom.

proto-oncogene A gene that regulates the onset of mitosis.

protoplast A bacterial cell whose cell wall is completely lacking and that is vulnerable to osmotic lysis.

protozoa A group of single-celled, eucaryotic organisms.

pseudohypha A chain of easily separated, spherical to sausage-shaped yeast cells partitioned by constrictions rather than by septa.

pseudopods Protozoan appendage responsible for motility. Also called "false feet."

pseudomembrane A tenacious, noncellular mucous exudate containing cellular debris that tightly blankets the mucosal surface in infections such as diphtheria and pseudomembranous enterocolitis.

pseudopodium A temporary extension of the protoplasm of an ameboid cell. It serves both in ameboid motion and for food gathering (phagocytosis).

psychrophile A microorganism that thrives at low temperature (0°–20°C), with a temperature optimum of 0°–15°C.

pulmonary Occurring in the lungs. Examples include pulmonary anthrax and pulmonary nocardiosis.

pulmonary anthrax The more severe form of anthrax, resulting from inhalation of spores and resulting in a wide range of pathological effects, including death.

pure culture A container growing a single species of microbe whose identity is known.

purine A nitrogen base that is an important encoding component of DNA and RNA. The two most common purines are adenine and guanine.

purpura A condition characterized by bleeding into skin or mucous membrane, giving rise to patches of red that darken and turn purple.

pus The viscous, opaque, usually yellowish matter formed by an inflammatory infection. It consists of serum exudate, tissue debris, leukocytes, and microorganisms.

pyogenic Pertains to pus formers, especially the pyogenic cocci: pneumococci, streptococci, staphylococci, and neisseriae.

pyrimidine Nitrogen bases that help form the genetic code on DNA and RNA. Uracil, thymine, and cytosine are the most important pyrimidines.

pyrimidine dimer The union of two adjacent pyrimidines on the same DNA strand, brought about by exposure to ultraviolet light. It is a form of mutation.

pyrogen A substance that causes a rise in body temperature. It can come from pyrogenic microorganisms or from polymorphonuclear leukocytes (endogenous pyrogens).

Q

Q fever A disease first described in Queensland, Australia, initially dubbed Q for "query" to denote a fever of unknown origin. Q fever is now known to be caused by a rickettsial infection.

quaternary structure Most complex protein structure characterized by the formation of large, multiunit proteins by more than one of the polypeptides. This structure is typical of antibodies and some enzymes that act in cell synthesis.

quats A byword that pertains to a family of surfactants called quaternary ammonium compounds. These detergents are only weakly microbicidal and are used as sanitizers and preservatives.

Quellung test The capsular swelling phenomenon; a serological test for the presence of pneumococci whose capsules enlarge and become opaque and visible when exposed to specific anticapsular antibodies.

quinolone A class of synthetic antimicrobic drugs with broad-spectrum effects.

R

rabies The only rhabdovirus that infects humans. Zoonotic disease characterized by fatal meningoencephalitis.

rad A unit of measure for absorbed dose of ionizing radiation.

radiation Electromagnetic waves or rays, such as those of light given off from an energy source.

radioactive isotopes Unstable isotopes whose nuclei emit particles of radiation. This emission is called radioactivity or radioactive decay. Three naturally occurring emissions are alpha, beta, and gamma radiation.

radioimmunoassay (RIA) A highly sensitive laboratory procedure that employs radioisotope-labeled substances to measure the levels of antibodies or antigens in the serum.

reactants Molecules entering or starting a chemical reaction.

real image An image formed at the focal plane of a convex lens. In the compound light microscope, it is the image created by the objective lens.

receptor In intercellular communication, cell surface molecules involved in recognition, binding, and intracellular signaling.

recombinant DNA A technology, also known as genetic engineering, that deliberately modifies the genetic structure of an organism to create novel products, microbes, animals, plants, and viruses.

recombination A type of genetic transfer in which DNA from one organism is donated to another.

recycling A process which converts unusable organic matter from dead organisms back into their essential inorganic elements and returns them to their non-living reservoirs to make them available again for living organisms. This is a common term that means the same as mineralization and decomposition.

redox Denoting an oxidation-reduction reaction.

reduction In chemistry, the gain of electrons.

reemerging disease Previously identified disease that is increasing in occurrence.

refraction In optics, the bending of light as it passes from one medium to another with a different index of refraction.

regular (in shape) Refers to bacteria with uniform staining properties and which are consistent in shape from cell to cell within a particular species.

regulator DNA segment that codes for a protein capable of repressing an operon.

regulatory site The location on an enzyme where a certain substance can bind and block the enzyme's activity.

rennin The enzyme casein coagulase, which is used to produce curd in the processing of milk and cheese.

reovirus Respiratory enteric orphan virus. Virus with a double-stranded RNA genome and both an inner capsid and an outer capsid. Not a significant human pathogen.

replication In DNA synthesis, the semiconservative mechanisms that ensure precise duplication of the parent DNA strands.

replication fork The Y-shaped point on a replicating DNA molecule where the DNA polymerase is synthesizing new strands of DNA.

replicon A piece of DNA capable of replicating. Contains an origin of replication.

reportable disease Those diseases that must be reported to health authorities by law.

repressible operon An operon that under normal circumstances is transcribed. The buildup of the operon's amino acid product causes transcription of the operon to stop.

repressor The protein product of a repressor gene that combines with the operator and arrests the transcription and translation of structural genes.

reservoir In disease communication, the natural host or habitat of a pathogen.

resident flora The deeper, more stable microflora that inhabit the skin and exposed mucous membranes, as opposed to the superficial, variable, transient population.

resistance (R) factor Plasmids, typically shared among bacteria by conjugation, that provide resistance to the effects of antibiotics.

resolving power The capacity of a microscope lens system to accurately distinguish between two separate entities that lie close to each other. Also called resolution.

respiratory chain In cellular respiration, a series of electron-carrying molecules that transfers energy-rich electrons and protons to molecular oxygen. In transit, energy is extracted and conserved in the form of ATP.

respiratory syncytial virus (RSV) An RNA virus that infects the respiratory tract. RSV is the most prevalent cause of respiratory infection in newborns.

restriction endonuclease An enzyme present naturally in cells that cleaves specific locations on DNA. It is an important means of inactivating viral genomes, and it is also used to splice genes in genetic engineering.

reticuloendothelial system Also known as the mononuclear phagocyte system, it pertains to a network of fibers and phagocytic cells (macrophages) that permeates the tissues of all organs. Examples: Kupffer cells in liver sinusoids, alveolar phagocytes in the lung, microglia in nervous tissue.

retrovirus A group of RNA viruses (including HIV) that have the mechanisms for converting their genome into a double strand of DNA that can be inserted on a host's chromosome.

reverse transcriptase The enzyme possessed by retroviruses that carries out the reversion of RNA to DNA—a form of reverse transcription.

Reye syndrome A sudden, usually fatal neurological condition that occurs in children after a viral infection. Autopsy shows cerebral edema and marked fatty change in the liver and renal tubules.

Rh factor An isoantigen that can trigger hemolytic disease in newborns due to incompatibility between maternal and infant blood factors.

rhabdovirus Family of bullet-shaped viruses that includes rabies.

rhinovirus A picornavirus associated with the common cold. Transmission is through human-to-human contact, and symptoms typically are short-lived. Most effective control is effective hand washing and care in handling nasal secretions.

rhizobia Bacteria that live in plant roots and supply supplemental nitrogen that boosts plant growth.

rhizosphere The zone of soil, complete with microbial inhabitants, in the immediate vicinity of plant roots.

ribose A 5-carbon monosaccharide found in RNA.

ribonucleic acid (RNA) The nucleic acid responsible for carrying out the hereditary program transmitted by an organism's DNA.

ribosome A bilobed macromolecular complex of ribonucleoprotein that coordinates the codons of mRNA with tRNA anticodons and, in so doing, constitutes the peptide assembly site.

ribozyme A part of an RNA-containing enzyme in eucaryotes that removes intervening sequences of RNA called introns and splices together the true coding sequences (exons) to form a mature messenger RNA.

rickettsias Medically important family of bacteria, commonly carried by ticks, lice, and fleas. Significant cause of important emerging diseases.

Rifampin An antibiotic used primarily in the treatment of mycobacterial infection.

ringworm A superficial mycosis caused by various dermatophytic fungi. This common name is actually a misnomer.

RNA polymerase Enzyme process that translates the code of DNA to RNA.

rolling circle An intermediate stage in viral replication of circular DNA into linear DNA.

root nodules Small growths on the roots of legume plants that arise from a symbiotic association between the plant tissues and bacteria (Rhizobia). This association allows fixation of nitrogen gas from the air into a usable nitrogen source for the plant.

roseola Disease of infancy, usually self-limiting, caused by infection with human herpesvirus-6.

rosette formation A technique for distinguishing surface receptors on T cells by reacting them with sensitized indicator sheep red blood cells. The cluster of red cells around the central white blood cell resembles a little rose blossom and is indicative of the type of receptor.

rotavirus Virus with a double-stranded RNA genome and both an inner capsid and an outer capsid. Transmitted by fecal contamination, it is common in areas with poor sanitation. Rotavirus typically causes diarrheal disease and can be fatal.

rough endoplasmic reticulum (RER) Microscopic series of tunnels that originates in the outer membrane of the nuclear envelope and is used in transport and storage. Large numbers of ribosomes, partly attached to the membrane, give the rough appearance.

rubella Commonly known as German measles. Rubella is caused by *Rubivirus*, a member of the togavirus family. Postnatal rubella is generally a mild condition; congenital rubella poses a risk of birth defects and results when virus passes from infected mother to fetus.

rubeola (red measles) Acute disease caused by infection with Morbillivirus.

S

saccharide Scientific term for sugar. Refers to a simple carbohydrate with a sweet taste.

salmonelloses Illnesses caused by infection of the noncoliform *Salmonella* pathogens. The cause of typhoid fever, salmonella food poisoning, and gastroenteritis.

salpingitis Inflammation of the fallopian tubes.

sanitize To clean inanimate objects using soap and degerming agents so that they are safe and free of high levels of microorganisms.

saprobe A microbe that decomposes organic remains from dead organisms. Also known as a saprophyte or saprotroph.

sarcina A cubical packet of 8, 16, or more cells; the cellular arrangement of the genus *Sarcina* in the family Micrococcaceae.

sarcoma A fleshy neoplasm of connective tissue. Growth, usually highly malignant, is of mesodermal origin.

satellite phenomenon A type of commensal relationship in which one microbe produces growth factors that favor the growth of its dependent partner. When plated on solid media, the dependent organism appears as surrounding colonies. Example: *Staphylococcus* surrounded by its *Haemophilus* satellite.

saturation The complete occupation of the active site of a carrier protein or enzyme by the substrate.

schistosomiasis Infection by blood fluke, often as a result of contact with contaminated water in rivers and streams. Symptoms include fever, chills, diarrhea, and cough. Infection may be chronic.

schizogony A process of multiple fission whereby first the nucleus divides several times, and subsequently the cytoplasm is subdivided for each new nucleus during cell division.

scientific method Principles and procedures for the systematic pursuit of knowledge, involving the recognition and formulation of a problem, the collection of data through observation and experimentation, and the formulation and testing of a hypothesis.

scolex The anterior end of a tapeworm characterized by hooks and/or suckers for attachment to the host.

SCP Abbreviation for single-cell protein, a euphemistic expression for microbial protein intended for human and animal consumption.

sebaceous glands The sebum- (oily, fatty) secreting glands of the skin.

second line of defense An inborn, non-specific system of cells (phagocytes) and fluids (inflammation) that acts to detect and destroy foreign substances rapidly once they have entered the body.

secondary immune response The more rapid and heightened response to antigen in a sensitized subject due to memory lymphocytes.

secondary infection An infection that compounds a preexisting one.

secondary response The rapid rise in antibody titer following a repeat exposure to an antigen that has been recognized from a previous exposure. This response is brought about by memory cells produced as a result of the primary exposure.

secondary structure Protein structure that occurs when the functional groups on the outer surface of the molecule interact by forming hydrogen bonds. These bonds cause the amino acid chain to either twist, forming a helix, or to pleat into an accordion pattern called a β-pleated sheet.

secretion The process of actively releasing cellular substances into the extracellular environment.

secretory antibody The immunoglobulin (IgA) that is found in secretions of mucous membranes and serves as a local immediate protection against infection.

selective media Nutrient media designed to favor the growth of certain microbes and to inhibit undesirable competitors.

selectively toxic Property of an antimicrobic agent to be highly toxic against its target microbe while being far less toxic to other cells, particularly those of the host organism.

self Natural markers of the body that are recognized by the immune system.

self-limited Applies to an infection that runs its course without disease or residual effects.

semiconservative replication In DNA replication, the synthesis of paired daughter strands, each retaining a parent strand template.

semisolid media Nutrient media with a firmness midway between that of a broth (a liquid medium) and an ordinary solid medium; motility media.

semisynthetic Drugs which, after being naturally produced by bacteria, fungi, or other living sources, are chemically modified in the laboratory.

sensitizing dose The initial effective exposure to an antigen or an allergen that stimulates an immune response. Often applies to allergies.

sepsis The state of putrefaction; the presence of pathogenic organisms or their toxins in tissue or blood.

septic shock Blood infection resulting in a pathological state of low blood pressure accompanied by a reduced amount of blood circulating to vital organs. Endotoxins of all gram-negative bacteria can cause shock, but most clinical cases are due to gram-negative enteric rods.

septicemia Systemic infection associated with microorganisms multiplying in circulating blood.

septum A partition or cellular cross wall, as in certain fungal hyphae.

sequela A morbid complication that follows a disease.

sequencing Determining the actual order and types of bases in a segment of DNA.

serology The branch of immunology that deals with *in vitro* diagnostic testing of serum.

seropositive Showing the presence of specific antibody in a serological test. Indicates ongoing infection.

serotonin A vasoconstrictor that inhibits gastric secretion and stimulates smooth muscle.

serotyping The subdivision of a species or subspecies into an immunologic type, based upon antigenic characteristics.

serum The clear fluid expressed from clotted blood that contains dissolved nutrients, antibodies, and hormones but not cells or clotting factors.

serum sickness A type of immune complex disease in which immune complexes enter circulation, are carried throughout the body, and are deposited in the blood vessels of the kidney, heart, skin, and joints. The condition may become chronic.

severe acute respiratory syndrome (SARS) A severe respiratory disease caused by infection with a newly described Coronavirus.

severe combined immunodeficiencies A collection of syndromes occurring in newborns caused by a genetic defect that knocks out both B and T cell types of immunity. There are several versions of this disease, termed SCIDS for short.

sex pilus A conjugative pilus.

sexually transmitted disease (STD) Infections resulting from pathogens that enter the body via sexual intercourse or intimate, direct contact.

shigellosis An incapacitating dysentery caused by infection with *Shigella* bacteria.

side effects Unintended, usually harmful consequences of chemotherapy including damage to organs, allergy, and disruption of normal flora.

sign Any abnormality uncovered upon physical diagnosis that indicates the presence of disease. A sign is an objective assessment of disease, as opposed to a symptom, which is the subjective assessment perceived by the patient.

silent mutation A mutation that, because of the degeneracy of the genetic code, results in a nucleotide change in both the DNA and mRNA but not the resultant amino acid and thus, not the protein.

simple stain Type of positive staining technique that uses a single dye to add color to cells so that they are easier to see. This technique tends to color all cells the same color.

smallpox Disease caused by infection with variola virus. Smallpox has been eradicated worldwide, and variola virus exists today only in a few government laboratories.

smooth endoplasmic reticulum (SER) A microscopic series of tunnels lacking ribosomes that functions in the nutrient processing function of a cell.

sodoku A Japanese term pertaining to rat bite fever.

solution A mixture of one or more substances (solutes) that cannot be separated by filtration or ordinary settling.

solvent A dissolving medium.

somatic (O or cell wall antigen) One of the three major antigens commonly used to differentiate Gram-negative enteric bacteria.

source The person or item from which an infection is immediately acquired. See *reservoir.*

Southern blot A technique that separates fragments of DNA using electrophoresis and identifies them by hybridization.

species In the levels of classification, the most specific level of organization.

specificity Limited to a single, precise characteristic or action.

spheroplast A gram-negative cell whose peptidoglycan, when digested by lysozyme, remains intact but is osmotically vulnerable.

spike A receptor on the surface of certain enveloped viruses that facilitates specific attachment to the host cell.

spirillum A type of bacterial cell with a rigid spiral shape and external flagella.

spirochete A coiled, spiral-shaped bacterium that has endoflagella and flexes as it moves.

split gene Any gene in which the coding sequence is interrupted by intervening sequences which will not be translated (introns).

spongiform encephalopathies Transmissible, fatal, chronic infections of the nervous system caused by unconventional viruses now identified as prions (infectious proteins).

spontaneous generation Early belief that living things arose from vital forces present in nonliving, or decomposing, matter.

spontaneous mutation A mutation in DNA caused by random mistakes in replication and not known to be influenced by any mutagenic agent. These mutations give rise to an organism's natural, or background, rate of mutation.

sporadic Description of a disease which exhibits new cases at irregular intervals in unpredictable geographic locales.

sporangium A fungal cell in which asexual spores are formed by multiple cell cleavage.

spore A differentiated, specialized cell form that can be used for dissemination, for survival in times of adverse conditions, and/or for reproduction. Spores are usually unicellular and may develop into gametes or vegetative organisms.

sporicide A chemical agent capable of destroying bacterial endospores.

sporotrichosis A subcutaneous mycosis caused by *Sporothrix schenckii;* a common cause is the prick of a rose thorn.

sporozoite One of many minute elongated bodies generated by multiple division of the oocyst. It is the infectious form of the malarial parasite that is harbored in the salivary gland of the mosquito and inoculated into the victim during feeding.

sporulation The process of spore formation.

start codon The nucleotide triplet AUG that codes for the first amino acid in protein sequences.

starter culture The sizeable inoculation of pure bacterial, mold, or yeast sample for bulk processing, as in the preparation of fermented foods, beverages, and pharmaceuticals.

stasis A state of rest or inactivity; applied to nongrowing microbial cultures. Also called microbistasis.

stationary growth phase Survival mode in which cells either stop growing or grow very slowly.

stem cells Pluripotent, undifferentiated cells.

sterile Completely free of all life forms, including spores and viruses.

sterilization Any process that completely removes or destroys all viable microorganisms, including viruses, from an object or habitat. Material so treated is sterile.

storch Acronym for common infections of the fetus and neonate. Storch stands for **S**yphilis, **T**oxoplasmosis, **O**ther diseases (hepatitis B, AIDS and Chlamydiosis), **R**ubella, **C**ytomegalovirus and **H**erpes simplex virus.

strain In microbiology, a set of descendants cloned from a common ancestor that retain the original characteristics. Any deviation from the original is a different strain.

streptolysin A hemolysin produced by streptococci.

strict, or obligate anaerobe An organism which does not use oxygen gas in metabolism and cannot survive in oxygen's presence.

stroma The matrix of the chloroplast that is the site of the dark reactions.

structural gene A gene that codes for the amino acid sequence (peptide structure) of a protein.

subacute Indicates an intermediate status between acute and chronic disease.

subacute sclerosing panencephalitis (SSPE) A complication of measles infection in which progressive neurological degeneration of the cerebral cortex invariably leads to coma and death.

subcellular vaccine A vaccine against isolated microbial antigens rather than against the entire organism.

subclinical A period of inapparent manifestations that occurs before symptoms and signs of disease appear.

subculture To make a second-generation culture from a well-established colony of organisms.

subcutaneous The deepest level of the skin structure.

substrate The specific molecule upon which an enzyme acts.

subunit vaccine A vaccine preparation that contains only antigenic fragments such as surface receptors from the microbe. Usually in reference to virus vaccines.

sucrose One of the carbohydrates commonly referred to as sugars. Common table or cane sugar.

sulfonamide Antimicrobial drugs that interfere with the essential metabolic process of bacteria and some fungi.

superantigens Bacterial toxins that are potent stimuli for T cells and can be a factor in diseases such as toxic shock.

superficial mycosis A fungal infection located in hair, nails, and the epidermis of the skin.

superinfection An infection occurring during antimicrobic therapy that is caused by an overgrowth of drug-resistant microorganisms.

superoxide A toxic derivative of oxygen; ($O_2{}^-$).

suppressor T cell A class of T cells that inhibits the actions of B cells and other T cells.

surfactant A surface-active agent that forms a water-soluble interface. Examples: detergents, wetting agents, dispersing agents, and surface tension depressants.

sylvatic Denotes the natural presence of disease among wild animal populations. Examples: sylvatic (sylvan) plague, rabies.

symbiosis An intimate association between individuals from two species; used as a synonym for mutualism.

symptom The subjective evidence of infection and disease as perceived by the patient.

syncytium A multinucleated protoplasmic mass formed by consolidation of individual cells.

syndrome The collection of signs and symptoms that, taken together, paint a portrait of the disease.

synergism The coordinated or correlated action by two or more drugs or microbes that results in a heightened response or greater activity.

syngamy Conjugation of the gametes in fertilization.

syphilis A sexually transmitted bacterial disease caused by the spirochete *Treponema pallidum.*

systemic Occurring throughout the body; said of infections that invade many compartments and organs via the circulation.

T

tartar See *calculus.*

taxa Taxonomic categories.

taxonomy The formal system for organizing, classifying, and naming living things.

temperate phage A bacteriophage that enters into a less virulent state by becoming incorporated into the host genome as a prophage instead of in the vegetative or lytic form that eventually destroys the cell.

template The strand in a double stranded DNA molecule which is used as a model to synthesize a complementary strand of DNA or RNA during replication or transcription.

teratogenic Causing abnormal fetal development.

tetanospasmin The neurotoxin of *Clostridium tetani,* the agent of tetanus. Its chief action is directed upon the inhibitory synapses of the anterior horn motor neurons.

tetracyclines A group of broad-spectrum antibiotics with a complex 4-ring structure.

tertiary structure Protein structure that results from additional bonds forming between functional groups in a secondary structure, creating a three-dimensional mass.

tetanus A neuromuscular disease caused by infection with *Clostridium tetani.* Usual portals of entry include puncture wounds, burns, umbilical stumps, frostbite sites, and crushed body parts. Vaccination repeated at the recommended times can prevent infection. Also called lockjaw.

tetrads Groups of four.

theory A collection of statements, propositions, or concepts that explains or accounts for a natural event.

therapeutic index The ratio of the toxic dose to the effective therapeutic dose that is used to assess the safety and reliability of the drug.

thermal Related to temperature.

thermal death point The lowest temperature that achieves sterilization in a given quantity of broth culture upon a 10-minute exposure. Examples: 55°C for *Escherichia coli,* 60°C for *Mycobacterium tuberculosis,* and 120°C for spores.

thermal death time The least time required to kill all cells of a culture at a specified temperature.

thermocline A temperature buffer zone in a large body of water that separates the warmer water (the epilimnion) from the colder water (the hypolimnion).

thermoduric Resistant to the harmful effects of high temperature.

thermophile A microorganism that thrives at a temperature of 50°C or higher.

third line of defense An acquired, very focused immunologic defense that depends on the recognition of specific foreign materials and microbes by B and T lymphocytes. It is the primary source of immunities.

thrush *Candida albicans* infection of the oral cavity.

thylakoid Vesicles of a chloroplast formed by elaborate folding of the inner membrane to form "discs." Solar energy trapped in the thylakoids is used in photosynthesis.

thymine (T) One of the nitrogen bases found in DNA, but not in RNA. Thymine is in a pyrimidine form.

thymus Butterfly-shaped organ near the tip of the sternum that is the site of T-cell maturation.

tincture A medicinal substance dissolved in an alcoholic solvent.

tinea versicolor A condition of the skin appearing as mottled and discolored skin pigmentation as a result of infection by the yeast *Malassezia furfur.*

tinea Ringworm; a fungal infection of the hair, skin, or nails.

titer In immunochemistry, a measure of antibody level in a patient, determined by agglutination methods.

T lymphocyte (T cell) A white blood cell that is processed in the thymus gland and is involved in cell-mediated immunity.

tonsils A ring of lymphoid tissue in the pharynx which acts a repository for lymphocytes.

topoisomerases Enzymes that can add or remove DNA twists and thus regulate the degree of supercoiling.

toxemia An abnormality associated with certain infectious diseases. Toxemia is caused by toxins or other noxious substances released by microorganisms circulating in the blood.

toxigenicity The tendency for a pathogen to produce toxins. It is an important factor in bacterial virulence.

toxin A specific chemical product of microbes, plants, and some animals that is poisonous to other organisms.

toxinosis Disease whose adverse effects are primarily due to the production and release of toxins.

toxoid A toxin that has been rendered nontoxic but is still capable of eliciting the formation of protective antitoxin antibodies; used in vaccines.

trace elements Micronutrients (zinc, nickel, and manganese) that occur in small amounts, and are involved in enzyme function and maintenance of protein structure.

tracheostomy A surgically created emergency airway opening into the trachea.

trachoma Strain of *Chlamydia trachomatis* that attacks mucous membranes of the eye, genitourinary tract, and lungs in humans.

transcript A newly transcribed RNA molecule.

transcription mRNA synthesis; the process by which a strand of RNA is produced against a DNA template.

transduction The transfer of genetic material from one bacterium to another by means of a bacteriophage vector.

transfer RNA (tRNA) A transcript of DNA that specializes in converting RNA language into protein language.

transformation In microbial genetics, the transfer of genetic material contained in "naked" DNA fragments from a donor cell to a competent recipient cell.

transfusion Infusion of whole blood, red blood cells, or platelets directly into a patient's circulation.

transgenic technology Introduction of foreign DNA into cells or organisms. Used in genetic engineering to create recombinant plants, animals, and microbes.

transients In normal flora, the assortment of superficial microbes whose numbers and types vary depending upon recent exposure. The deeper-lying residents constitute a more stable population.

translation Protein synthesis; the process of decoding the messenger RNA code into a polypeptide.

transposon A DNA segment with an insertion sequence at each end, enabling it to migrate to another plasmid, to the bacterial chromosome, or to a bacteriophage.

traveler's diarrhea A type of gastroenteritis typically caused by infection with enterotoxigenic strains of *E. coli* that are ingested through contaminated food and water.

trematode A fluke or flatworm parasite of vertebrates.

tricarboxylic acid cycle (TCA or Krebs cycle) The second pathway of the three pathways that complete the process of primary catabolism. Also called the citric acid cycle.

trichinosis Infection by the *Trichinella spiralis* parasite, usually caused by eating the meat of an infected animal. Early symptoms include fever, diarrhea, nausea, and abdominal pain that progress to intense muscle and joint pain and shortness of breath. In the final stages, heart and brain function are at risk, and death is possible.

trichomoniasis Sexually transmitted disease caused by infection by the trichomonads, a group of protozoa. Symptoms include urinary pain and frequency, and foul-smelling vaginal discharge in females or recurring urethritis, with a thin milky discharge, in males.

triglyceride A type of lipid composed of a glycerol molecule bound to three fatty acids.

Trimethoprim A sulfa drug often used in the treatment of *Pneumocystis (carinii) jiroveci* pneumonia in AIDS patients. (often combined with sulfa drugs).

triplet See *codon.*

trophozoite A vegetative protozoan (feeding form) as opposed to a resting (cyst) form.

true pathogen A microbe capable of causing infection and disease in healthy persons with normal immune defenses.

trypomastigote The infective morphological stage transmitted by the tsetse fly or the reduviid bug in African trypanosomiasis and Chagas disease.

tubercle In tuberculosis, the granulomatous well-defined lung lesion that can serve as a focus for latent infection.

tuberculin A glycerinated broth culture of *Mycobacterium tuberculosis* that is evaporated and filtered. Formerly used to treat tuberculosis, tuberculin is now used chiefly for diagnostic tests.

tuberculin reaction A diagnostic test in which an intradermal injection of a purified protein (tuberculin) extract from *M. tuberculosis* elicits an immune response, seen as small red bump, in those persons previously exposed to tuberculosis.

tuberculoid leprosy A superficial form of leprosy characterized by asymmetrical, shallow skin lesions containing few bacterial cells.

tularemia Infection by *Francisella tularensis.* A zoonotic disease of mammals common to the northern hemisphere. Occasionally called rabbit fever. Portal of entry and symptoms are varied.

turbid Cloudy appearance of nutrient solution in a test tube due to growth of microbe population.

tyndallization Fractional (discontinuous, intermittent) sterilization designed to destroy spores indirectly. A preparation is exposed to flowing steam for an hour, and then the mineral is allowed to incubate to permit spore germination. The resultant vegetative cells are destroyed by repeated steaming and incubation.

typhoid fever Form of salmonelloses. It is highly contagious. Primary symptoms include fever, diarrhea, and abdominal pain. Typhoid fever can be fatal if untreated.

typhus Rickettsia infection characterized by high fever, chills, frontal headache, muscular pain, and a generalized rash within seven days of infection. In more severe cases, a personality change, low urine output, hypotension, and gangrene can cause complications. Mortality is high in older adults.

U

ultraviolet radiation Radiation with an effective wavelength from 240 nm to 260 nm. UV radiation induces mutations readily but has very poor penetrating power.

uncoating The process of removal of the viral coat and release of the viral genome by its newly invaded host cell.

undulant fever See *brucellosis*.

universal donor In blood grouping and transfusion, a group O individual whose erythrocytes bear neither agglutinogen A nor B.

Universal Precautions (UP) Center for Disease Control guidelines for health care workers regarding the prevention of disease transmission when handling patients and body substances.

uracil (U) One of the nitrogen bases in RNA, but not in DNA. Uracil is in a pyrimidine form.

urban plague Plague passed to humans through contact with domestic animals or other humans.

urinary tract infection Invasion and infection of the urethra and bladder by bacterial residents, most often *E. coli*.

V

vaccine Originally used in reference to inoculation with the cowpox or vaccinia virus to protect against smallpox. In general, the term now pertains to injection of whole microbes (killed or attenuated), toxoids, or parts of microbes as a prevention or cure for disease.

vacuoles In the cell, membrane-bounded sacs containing fluids or solid particles to be digested, excreted, or stored.

valence The combining power of an atom based upon the number of electrons it can either take on or give up.

variable region The antigen binding fragment of an immunoglobulin molecule, consisting of a combination of heavy and light chains whose molecular conformation is specific for the antigen.

varicella See *chickenpox*

varicella-zoster virus Herpesvirus responsible for the diseases chickenpox and shingles.

variolation A hazardous, outmoded process of deliberately introducing smallpox material scraped from a victim into the nonimmune subject in the hope of inducing resistance.

VDRL A flocculation test that detects syphilis antibodies. An important screening test. The abbreviation stands for **V**enereal **D**isease **R**esearch **L**aboratories.

vector An animal that transmits infectious agents from one host to another, usually a biting or piercing arthropod like the tick, mosquito, or fly. Infectious agents can be conveyed mechanically by simple contact or biologically whereby the parasite develops in the vector.

vector₂ A genetic element such as a plasmid or a bacteriophage used to introduce genetic material into a cloning host during recombinant DNA experiments.

vegetative In describing microbial developmental stages, a metabolically active feeding and dividing form, as opposed to a dormant, seemingly inert, nondividing form. Examples: a bacterial cell versus its spore; a protozoan trophozoite versus its cyst.

vehicle An inanimate material (solid object, liquid, or air) that serves as a transmission agent for pathogens.

verruca A flesh-colored wart. This self-limited tumor arises from accumulations of growing epithelial cells. Example: papilloma warts.

vesicle A blister characterized by a thin-skinned, elevated, superficial pocket inflated with serum.

vibrio A curved, rod-shaped bacterial cell.

viremia The presence of a virus in the bloodstream.

virion An elementary virus particle in its complete morphological and thus infectious form. A virion consists of the nucleic acid core surrounded by a capsid, which can be enclosed in an envelope.

viroid An infectious agent that, unlike a virion, lacks a capsid and consists of a closed circular RNA molecule. Although known viroids are all plant pathogens, it is conceivable that animal versions exist.

virtual image In optics, an image formed by diverging light rays; in the compound light microscope, the second, magnified visual impression formed by the ocular from the real image formed by the objective.

virucide A chemical agent which inactivates viruses, especially on living tissue.

virulence In infection, the relative capacity of a pathogen to invade and harm host cells.

virus Microscopic, acellular agent composed of nucleic acid surrounded by a protein coat.

vitamins A component of coenzymes critical to nutrition and the metabolic function of coenzyme complexes.

vulvovaginal candidiasis (VC) Vaginal STD caused by *Candida albicans* and often associated with disruption of the normal vaginal flora.

W

wart An epidermal tumor caused by papillomaviruses. Also called a verruca.

Western blot test A procedure for separating and identifying antigen or antibody mixtures by two-dimensional electrophoresis in polyacrylamide gel, followed by immune labeling.

wheal A welt; a marked, slightly red, usually itchy area of the skin that changes in size and shape as it extends to adjacent area. The reaction is triggered by cutaneous contact or intradermal injection of allergens in sensitive individuals.

whey The residual fluid from milk coagulation that separates from the solidified curd.

white piedra A fungus disease of hair, especially of the scalp, face, and genitals, caused by *Trichosporon beigelii*. The infection is associated with soft, mucilaginous, white-to-light-brown nodules that form within and on the hair shafts.

whitlow A deep inflammation of the finger or toe, especially near the tip or around the nail. Whitlow is a painful herpes simplex virus infection that can last several weeks and is most common among health care personnel who come in contact with the virus in patients.

whole blood A liquid connective tissue consisting of blood cells suspended in plasma.

whooping cough See *pertussis*.

Widal test An agglutination test for diagnosing typhoid.

wild type The natural, nonmutated form of a genetic trait.

wort The clear fluid derived from soaked mash that is fermented for beer.

X

xenograft The transfer of a tissue or an organ from an animal of one species to a recipient of another species.

Y

yaws A tropical disease caused by *Treponema pertenue* that produces granulomatous ulcers on the extremities and occasionally on bone, but does not produce central nervous system or cardiovascular complications.

yellow fever Best-known arbovirus. Yellow fever is transmitted by mosquitoes. Its symptoms include fever, headache, and muscle pain that can proceed to oral hemorrhage, nosebleeds, vomiting, jaundice, and liver and kidney damage.

Z

zoonosis An infectious disease indigenous to animals that humans can acquire through direct or indirect contact with infected animals.

zooplankton The collection of non-photosynthetic microoganisms (protozoa, tiny animals) that float in the upper regions of aquatic habitat, and together with phytoplankton comprise the plankton.

zygospore A thick-walled sexual spore produced by the zygomycete fungi. It develops from the union of two hyphae, each bearing nuclei of opposite mating types.

Credits

Photographs

Chapter 1

Opener: Associated Press/AP; **1.2a**: © Doug Sokell/Tom Stack & Associates; **1.2b**: Courtesy Tom Volk; **1.3a**: © Corale L. Brierley/Visuals Unlimited; **1.3b**: © Science VU/SIM, NBS/Visuals Unlimited; **1.3c**: Courtesy of General Electric Research and Development Center; **1.6a1**: © A.M. Siegelman/Visuals Unlimited; **1.6a2**: © Sinclair Stammers/SPL/Photo Researchers, Inc.; **1.6b1**: © Carolina Biological Supply/Phototake; **1.6b2**: © T.E. Adams/Visuals Unlimited; **1.6b3**: Courtesy Tom Volk; **1.8**: © Bettmann/Corbis; **1.9a**: © Kathy Park Talaro/Visuals Unlimited; **1.9b1**: © Kathy Park Talaro/Visuals Unlimited; **1.9b2**: © Science/VU/Visuals Unlimited; **1.12**: AKG/Photo Researchers, Inc.; **Microfile 1.2a,b,c**: Brian Smale/© 1998. Reprinted with permission of Discover Magazine; **1.13**: © Bettmann/Corbis.

Chapter 2

Opener: William R. Elliott and Rhonda Rimer; **2.6d**: Kathy Park Talaro; **2.10a,b,c**: © John W. Hole, Jr.; **Microbit 2.3**: © Don Fawcett/Visuals Unlimited; **2.22d**: From A.S. Moffat "Nitrogenase Structure Revealed," *Science*, 250:1513, 12/14/90. © 1990 by the AAAS. Photo by M.M. Georgiadis and D.C. Rees, Caltech.

Chapter 3

Opener: Courtesy Anne Fluery; **3.3b,d**: Kathy Park Talaro; **Microbits 3.1**: Courtesy Charles River Lab; **3.3f, 3.4b, 3.5b,3.6b,3.7a,b, 3.9a,b, 3.10,a,b**: Kathy Park Talaro; **3.11**: Courtesy of Harold J. Benson; **3.12a,b,c**: Kathy Park Talaro; **3.13**: © Kathy Park Talaro/Visuals Unlimited; **3.14**: Courtesy Lecia, Inc.; **3.19a**: © Carolina Biological Supply/Phototake, NYC; **3.19b,c**: © Abbey/Visuals Unlimited; **3.20**: © E.C.S. Chan/Visuals Unlimited; **3.21a**: © George J. Wilder/Visuals Unlimited; **3.21b**: © Abbey/Visuals Unlimited; **3.22**: © Molecular Probes, Inc.; **3.23c**: William Ormerod/Visuals Unlimited; **3.24a**: Department of Veterinary Science, Queen's University Belfast; **3.24b**: J.P. Dubey et al, *Clinical Microbiology Reviews*, © ASM, April 1998, Vol. II, #2, 281. Image courtesy of Dr. Jitender P. Dubey; **3.25**: © Dennis Kunkel/CNRI/Phototake, NYC; **Microbits 3.2**: Courtesy IBM Corporation. Unauthorized use not permitted; **3.26a1**: Kathy Park Talaro; **3.26a2**: Courtesy Harold Benson; **3.26b**: Kathy Park Talaro; **3.26c1**: © Jack Bostrack/Visuals Unlimited; **3.26c2**: © Jack Bostrack/Visuals Unlimited; **3.26c3**: © Manfred Kage/Peter Arnold, Inc.; **3.26d1**: © A.M. Siegelman/Visuals Unlimited; **3.26d2**: Kathy Park Talaro.

Chapter 4

Opener: Courtesy Kit Pogliano and Marc Sharp/UCSD; **4.2a**: Courtesy Julius Adler **4.3a**: Courtesy of Dr. Jeffrey C. Burnham; **4.3b**: From Reichelt and Baumann, *Arch. Microbiology*, 94:283-330 © Springer-Verlag, 1973; **4.3c**: From Noel R. Krieg in *Bacteriological Reviews*, March 1976, Vol. 40(1):87 fig 7; **4.3d**: From Preer et al., *Bacteriological Reviews*, June 1974, 38(2): 121, fig. 7. © ASM; **4.7c**: Courtesy Stanley F. Hayes, Rocky Mountain Laboratories, NIAID, NIH; **4.8a**: © Eye of Science/Photo Researchers, Inc; **4.8b**: Courtesy of Dr. S. Knutton from D.R. Lloyd and S. Knutton, *Infection and Immunity*, January 1987, p 86-92. © ASM; **4.9**: © L. Caro/SPL/Photo Researchers, Inc.; **4.12a**: Courtesy Harriet & Ephrussi-Taylor, Rockefeller University Press; **4.12b**: © John D. Cunningham/Visuals Unlimited; **4.13**: © Science VU-Charles W. Stratton/Visuals Unlimited; **4.15a2**: © S.C. Holt/Biological Photo Service; **4.15b2**: © T.J. Beveridge/Biological Photo Service; **4.18**: © E.S. Anderson/Photo Researchers, Inc.; **4.20**: © Paul W. Johnson/Biological Photo Service; **4.21b**: © Courtesy Kit Pogliano and Marc Sharp/UCSD; **4.21c**: © Lee D. Simon/Photo Researchers, Inc.; **4.23a,b**: © David M. Phillips/Visuals Unlimited; **4.23c**: From *Microbiological Reviews*, 55 (1): 25, fig. 2b, March 1991. Courtesy of Jorge Benach; **4.23d**: © R.G. Kessel-G. Shih/Visuals Unlimited; **4.24a**: © A.M. Siegelman/Visuals Unlimited; **4.24b**: From Braude, *Infectious Diseases and Microbiology*, 2E, Fig. 3, Page 257 © Saunders College Publishing Company; **4.27a,b**: Courtesy of Analytab Products, a division of Sherwood Medical; **4.29**: © Science VU-M.D. Maser/Visuals Unlimited; **Spotlight 4.3**: Courtesy Heide N. Schulz/Max Planck Institute for Marine Microbiology; **4.31**: From Baca and Paretsky, *Microbiological Reviews*, 47(2):133, fig. 16, June 1983 © ASM; **4.32**: © David M. Phillips/Visuals Unlimited; **4.33a**: Courtesy John Waterbury, WHOI; **4.33b,c**: © T.E. Adams/Visuals Unlimited; **4.34a**: From *ASM News*, 53(2), Feb. 1987, © American Society for Microbiology. Photo by H. Kaltwasser; **4.34b**: © Paul W. Johnson/Biological Photo Service; **4.35b**: GBF-German Research Centre for Biotechnology, Braunschweig, Germany; **4.37a**: © Wayne P. Armstrong, Palomar College; **4.37b**: Courtesy of Dr. Mike Dyall-Smith, University of Melbourne.

Chapter 5

Opener: From Wood and Bousfield: "Common Objects of the Microscope", 1900 George Routledge IV, London; **5.1a,b**: Eric Knoll; **5.3a,b**: RMF/FDF/Visuals Unlimited, Inc.; **5.4**: Courtesy of R.G. Garrison Ph.D. From *Fungal Dimorphism with Emphasis on Fungi Pathogenic for Human*, P.J. Szaniszlo (ed). Reprinted with permission of Plenum Publishing Corp.; **5.5**: © Don Fawcett/Visuals Unlimited; **5.6b**: © Science VU/Visuals Unlimited; **5.12b**: © Don Fawcett/Visuals Unlimited; **5.15a:,b**: Courtesy Dr. Judy A. Murphy, San Joaquin Delta College, Dept. of Microscopy, Stockton, CA; **5.16a**: © David M. Phillips/Visuals Unlimited; **5.17a**: Kathy Park Talaro; **5.17b**: © Everett S. Beneke/Visuals Unlimited; **Spotlight 5.2**: Courtesy Gregory M. Filip, Oregon State University; **5.23**: Kathy Park Talaro; **5.24a**: © A.M. Siegelman/Visuals Unlimited; **5.24b**: © John D. Cunningham/Visuals Unlimited; **5.25**: Neil Carlson, Department of Environmental Health and Safety; **5.26b**: © T.E. Adams/Visuals Unlimited; **5.26c**: © Jan Hinsch/Photo Researchers, Inc.; **5.26d**: Courtesy Howard Glasgow, Dept. of Botany, North Carolina State University; **5.28b**: © David M. Phillips/Visuals Unlimited; **5.31a**: BioMEDIA ASSOCIATES; **5.31b**: © Yuuji Tsukii, Protist Information Server, http://protist.i.hosei.ac.jp/Protist_menuE.html; **5.32b**: Michael Riggs et al, *Infection and Immunity*, Vol. 62, #5, May 1994, p. 1931 © ASM.

Chapter 6

Opener: Photo by Paula Bronstein, Getty Images; **6.2a**: © K.G. Murti/Visuals Unlimited; **6.2b**: © CDC/Phototake, NYC; **6.2c**: © A.B. Dowsette/SPL/Photo Researchers, Inc.; **6.3a**: Reprinted from Schaffer et al, *Proceedings of the National Academy of Science*, 41:1020, 1955; **6.3b**: © Omikron/Photo Researchers, Inc.; **6.6b**: © Dennis Kunkel/Phototake, NYC; **6.6d**: © K.G. Murti/Visuals Unlimited; **6.7c**: © Science VU-NIH, R. Feldman/Visuals Unlimited; **6.8a**: © Boehringer Ingelheim International GMBH; **6.8b**: Kathy Park Talaro; **6.9b**: Courtesy of Harold Fisher, University of Rhode Island; **6.12b**: © Lee D. Simon/Photo Researchers, Inc.; **6.14**: © Lee D. Simon/Photo Researchers, Inc.; **6.19**: © K.G. Murti/Visuals Unlimited; **6.20b**: © Chris Bjornberg/Photo Researchers, Inc.; **6.21a**: © Patricia Barber/Custom Medical Stock; **6.21b**: Massimo Battaglia, INeMM CNR, Rome, Italy; **6.22a**: Photo by Ted Heald, State of Iowa Hygenic Laboratory; **6.23a**: © E.S. Chan/Visuals Unlimited; **6.23b,c**: Courtesy Jack W. Frankel; **Microfile 6.1**: Science VU/Wayside/Visuals Unlimited; **6.24a**: Carroll M. Weiss/Camera M.D. Studios; **6.24b1**: © A.M. Siegelman/Visuals Unlimited; **6.24b2**: © Science VU/CDC/Visuals Unlimited; **6.24c1**: © K.G. Murti/Visuals Unlimited; **6.24c2**: Courtesy Fred Williams, U.S. Environmental Protection Agency; **6.24d**: Courtesy of Reference Laboratory, A PCC Laboratory.

Chapter 7

Opener: Yuuji Tsukii, Dr. Sc., Hosei University; **Spotlight 7.1a**: © Corale Brierley/Visuals Unlimited; **Spotlight 7.1b**: Courtesy E. E. Adams, Montana State University/National Science Foundation, Office of Polar Programs; **7.1a**: © Ralph Robinson/Visuals Unlimited; **7.1b**: Courtesy Jack Jones, US EPA; **7.10a**: © Pat Armstrong/Visuals Unlimited; **7.10b**: © Philip Sze/Visuals Unlimited; **Spotlight 7.4**: © Michael Milstein; **7.11a**: Courtesy of Sheldon Manufacturing, Inc.; **7.13**: © Science VU-Fred Marsik/Visuals Unlimited; **Spotlight 7.5b**: © Mike Abbey/Visuals Unlimited; **7.17a**: © Kathy Park Talaro/Visuals Unlimited; **7.19b**: Image courtesy Center for Applied Aquatic Ecology.

Chapter 8

Opener: Courtesy Nicholas C. DeMello/UCLA; **Microbits 8.2**: Courtesy Heinz W. Pley, Kevin M. Flaherty & David B. McKay.

Chapter 9

Opener: © A. Barrington Brown/Photo Researchers, Inc; **Historical Highlights 9.1**: © Lawrence Livermore Laboratory/SPL/Custom Medical Stock Photo; **9.1**: Royalty free image from Photodisc CD VO6, Nature, Wildlife and the Environment; **9.3**: © K.G. Murti/Visuals Unlimited; **9.17c**:

Courtesy Steven McKnight and O.L. Miller, Department of Biology, University of Virginia.

Chapter 10

Opener: Courtesy Peter Beyer, UNI-Freiburg; **10.3b:** Kathy Park Talaro; **Microbits 10.1:** © AFP/CORBIS; **10.7(left):** © Koester Axel/CORBIS SYGMA; **10.7(center):** Milton P. Gordon, Dept. of Biochemistry, University of Washington ; **10.7(right):** © Andrew Brookes/CORBIS; **Table 10.3a:** Courtesy David M. Stalker; **Table 10.3b:** Courtesy Richard Shade, Purdue University; **10.12:** Courtesy Brigid Hogan, Howard Hughes Medical Institute, Vanderbilt University; **Table 10:4(left):** Courtesy R.L. Brinster; **Table 10.4(right):** © Karen Kasmauski/Matrix; **Microbits 10.2:** © Marilyn Humphries; **10.16c:** Courtesy of Dr. Michael Baird, Lifecodes Corporation **10.18:** Courtesy of Tyson Clark, University of California, Santa Cruz.

Chapter 11

Opener: Kathy Park Talaro; **Historical Highlights 11.1:** © Bettmann/CORBIS; **11.5a:** © Science VU/Visuals Unlimited; **11.6:** © Raymond B. Otero/Visuals Unlimited; **11.8a:** © Science VU/Nordion International/Visuals Unlimited; **11.10:** Courtesy Trojan Technologies; **11.11b:** © Fred Hossler/Visuals Unlimited; **11.13a:** Image supplied courtesy of STERIS Corporation. SYSTEM 1 ® is a registered trademark of STERIS Corporation; **11.13b:** Photo by Robert Durell, © 2003, Los Angeles Times. Reprinted with permission; **11.16:** © Kathy Park Talaro/Visuals Unlimited; **Historical Highlights 11.3a:** © David M. Phillips/Visuals Unlimited; **Historical Highlights 11.3b:** Kathy Park Talaro; **11.18a:** Anderson Products: www.anpro.com; **Spotlight on Microbiology 11.4:** AFP/CORBIS.

Chapter 12

Opener: Associated Press/AP; **Historical Highlights 12.1:** © Bettmann/CORBIS; **Microbits 12.2:** © Kathy Park Talaro/Visuals Unlimited; **12.2e,f:** Reprinted with permission of Robert L. Perkins from A.S. Kleiner & R.L. Perkins, *Journal of Infectious Diseases*, 12:4, 1970, © University of Chicago Press; **12.10:** © Cabisco/Visuals Unlimited; **Spotlight on Microbiology 12.4:** Controller of Her Majesty's Stationary Office; **12.17:** © Kenneth E. Greer/ Visuals Unlimited; **12.19b,d:**Courtesy Alain Philippon; **12.20:** Image courtesy of AB Biodisk (Etest® is a registered

trademark of AB Biodisk and patented in all major markets); **12.21b:** Kathy Park Talaro; **12.21c:** Courtesy Alain Philippon.

Chapter 13

Opener: © AP Photo/Robyn Beck, Pool; **13.5:** © David M. Phillips/Visuals Unlimited; **Microbits 13.1:** © Eurelios/ Phototake; **13.20:** © Marshall W. Jennison.

Chapter 14

Opener: © Manfred Kage/Peter Arnold, Inc.; **14.3b:** © Ellen R. Dirksen/Visuals Unlimited; **14.11b:** Courtesy Steve Kunkel; **14.19a:** © David M. Phillips/Visuals Unlimited; **14.22e:** Reprinted from Wendell F. Rosse et al, "Immune Lysis of Normal Human and Paroxysmal Nocturnal Hemogloginuria (PNH) Red Blood Cells", *Journal of Experimental Medicine*, 123:969, 1966. Rockefeller University Press.

Chapter 15

Opener: Kurt E.J. Dittmar and Dr. Manfred Rohde; **15.14b:** © R. Feldmann/Rainbow; **15.22:** © Lennart Nilsson, "The Body Victorious" Bonnier Fakta; **15.23(1):** Royalty free image from Photodisc New Family CD; **15.23(2):** © Medical Perspectives/PhotoDisc; **15.23(3):** © Parenting Today/PhotoDisc; **15.23(4):** Creatas/PictureQuest.

Chapter 16

Opener: James Gillroy, British, 1757-1815, "The Cow-Pock", engraving, 1802, William McCallin McKee Memorial Collection, 1928. 1407. © 1991, The Art Institute of Chicago. All Rights Reserved; **16.7a:** From Gillies and Dodds, *Bacteriology Illustrated*, 5/e fig.13, page 24. Reprinted by permission of the publisher Churchill Livingstone; **16.7b:** Immuno-Mycologics, Inc.; **16.9:** Genelabs Diagnostics Pte Ltd; **16.11c:** Photo courtesy of CHEMICON® International, Inc.; **16.12b:** © Hank Morgan/Science Source/Photo Researchers, Inc.; **16.13a:** From Miguel Pedraza et al, *Laboratory Medicine*, Vol. 14, #1, January 1983; **16.13b:** From F. Latel et al., *Journal of Clinical Microbiology*, 23(6):1018, © American Society of Microbiology.

Chapter 17

Opener: © Runk/Schoenberger/Grant Heilman Photography; **17.2b:** © SPL/Photo Researchers, Inc.; **17.2c:** © David M. Phillips/Visuals Unlimited; **17.5:**© Kenneth E. Greer/Visuals

Unlimited; **17.6a:** © STU/Custom Medical Stock Photo; **17.10a,b:** © Stuart I. Fox; **17.15a:** © Diepgen TL, Yihune G et al. Dermatology Online Atlas published online www.dermis.net . Reprinted with permission; **17.15b:** © SIU/Visuals Unlimited; **17.18:** Kathy Park Talaro; **17.19b:** © Kenneth E. Greer/Visuals Unlimited; **Spotlight on Microbiology 17.3(top):** © Renee Lynn/Photo Researchers, Inc.; **Spotlight on Microbiology 17.3(bottom):** © Walter H. Hodge/Peter Arnold, Inc.; **17.23:** Robert Haire, Gary Litman, Scott Strong; **17.24:** Reprinted from R. Kretschmer, *New England Journal of Medicine*, 279:1295, 1968. © 1968 Massachusetts Medical Society. All rights reserved; **Spotlight on Microbiology 17.5:** Courtesy of Baylor College of Medicine, Public Affairs; **I.D(b):** © Fred Marsik/Visuals Unlimited; **I.F:** Courtesy Wadsworth Center, NYS Dept. of Health.

Line Art

Chapter 3

3.23: Courtesy William A. Jensen.

Chapter 4

4.36: Reprinted with permission from *Nature*, Vol. 196, pp. 1189-1192. Copyright © 1962 Macmillan Magazines Ltd.

Chapter 6

6.9a: From Westwood, et al., *Journal of Microbiology*, 34:67, 1964. Reprinted by permission of The Society for General Microbiology, United Kingdom.

Chapter 11

11.5b: From John J. Perkins, Principles and Methods of Sterilization in Health Science, 2e, 1969. Courtesy of Charles C. Thomas, Springfield, Illinois; **11.15b:** From Nolte, et al., Oral Microbiology, 4e. © 1982 Mosby. Reprinted by permission of Gloria-Mae Nolte.

Chapter 15

15.18: From *Immunology III*, Copyright © 1985 W.B. Saunders and Co. Philadelphia, PA. Reprinted by permission of Joseph A. Bellanti.

Index

Note: In this index, page numbers followed by a t designate tables; page numbers followed by an f refer to figures; page numbers set in **boldface** refer to definitions of terms or introductory discussions.